IRS 2004: CURRENT PROBLEMS IN ATMOSPHERIC RADIATION

Studies in
Geophysical Optics and Remote Sensing
Series Editor: Adarsh Deepak

A Lunar-Based Analytical Laboratory
  Robert W. Zumwalt and Charles W. Gehrke (Eds.)

Advances in Remote Sensing Retrieval Methods
  Adarsh Deepak, Henry E. Fleming, and
  Moustafa T. Chahine (Eds.)

Aerosols and Climate
  Peter V. Hobbs and M. Patrick McCormick (Eds.)

Aerosols and Their Climatic Effects
  Hermann E. Gerber and Adarsh Deepak (Eds.)

Applications of Remote Sensing for Rice Production
  Adarsh Deepak and K. R. Rao (Eds.)

*Atmospheric Aerosols: Global Climatology and
  Radiative Characteristics*
  Guillaume A. d'Almeida, Peter Koepke, and
  Eric P. Shettle

*Atmospheric Aerosols: Their Formation, Optical
  Properties, and Effects*
  Adarsh Deepak (Ed.)

Atmospheric Radiative Transfer
  Jacqueline Lenoble

Defense Conversion: A Critical East-West Experiment
  A. E. S. Green (Ed.)

Geographic Information Systems in Government
  Bruce K. Opitz (Ed.)

Hygroscopic Aerosols
  Lothar H. Ruhnke and Adarsh Deepak (Eds.)

IRS '84: Current Problems in Atmospheric Radiation
  Giorgio Fiocco (Ed.)

IRS '88: Current Problems in Atmospheric Radiation
  J. Lenoble and J. P. Geleyn (Eds.)

IRS '92: Current Problems in Atmospheric Radiation
  S. Keevallik and O. Krner (Eds.)

IRS '96: Current Problems in Atmospheric Radiation
  William L. Smith and Knut Stamnes (Eds.)

*IRS 2000: Current Problems in Atmospheric
  Radiation*
  William L. Smith and Yuriy M. Timofeyev (Eds.)

Light Absorption by Aerosol Particles
  Hermann E. Gerber and Edward E. Hindman
  (Eds.)

Microwave Remote Sensing of the Earth System
  Alain Chedin (Ed.)

Nucleation and Atmospheric Aerosols
  N. Fukuta and P. E. Wagner (Eds.)

Optical Properties of Clouds
  Peter V. Hobbs and Adarsh Deepak (Eds.)

Ozone in the Atmosphere
  Rumen D. Bojkov and Peter Fabian (Eds.)

Polar and Arctic Lows
  Paul F. Twitchell, Erik A. Rasmussen, and
  Kenneth L. Davidson (Eds.)

*Polarization and Intensity of Light in the
  Atmosphere*
  Kinsell L. Coulson

Proceedings in Atmospheric Electricity
  Lothar H. Ruhnke and John Latham (Eds.)

*Radiative Transfer in Scattering and Absorbing
  Atmospheres: Standard Computational Procedures*
  Jacqueline Lenoble (Ed.)

Remote Sensing Calibration Systems
  H. S. Chen

*RSRM '87: Advances in Remote Sensing Retrieval
  Methods*
  Adarsh Deepak, Henry E. Fleming, and John S.
  Theon (Eds.)

Spectral Line Shapes, Volume 4
  R. J. Exton (Ed.)

The Global Role of Tropical Rainfall
  J. S. Theon, T. Matsuno, T. Sakata, and
  N. Fugono (Eds.)

Tropical Rainfall Measurements
  John S. Theon and Nobuyoshi Fugono
  (Eds.)

*Understanding Climate: Selected Works of
  Yale Mintz*
  R. Bates, et al. (Eds.)

Volcanic Activity and Climate
  Kirill Ya. Kondratyev and Ignacio Galindo

IRS 2004: CURRENT PROBLEMS IN ATMOSPHERIC RADIATION

Proceedings of the
INTERNATIONAL RADIATION SYMPOSIUM
Busan, Korea, 23–28 August 2004

Edited by

Herbert Fischer
Forschungszentrum Karlsruhe
Institut für Meteorologie und Klimaforschung
Karlsruhe, Germany

and

Byung-Ju Sohn
School of Earth and Environmental Sciences
Seoul National University
Seoul, Korea

A. DEEPAK Publishing 2006
A Division of Science and Technology Corporation
Hampton, Virginia USA

Copyright © 2006 by *A. DEEPAK Publishing*
All rights reserved.
No part of this publication may be reproduced or
transmitted in any form or by any means, electronic
or mechanical, including photocopy, recording, or any
information storage and retrieval system, without
permission in writing from the publisher.

Authors may use original figures and tables from their
own contribution in their own future publications without
requesting permission from the publisher, so long as the
publisher is acknowledged as the copyright holder.

A. DEEPAK Publishing
A Division of Science and Technology Corporation
10 Basil Sawyer Drive
Hampton, Virginia 23666-1393 USA

Library of Congress Cataloging-in-Publication Data

   IRS 2004 (2004 : Pusan, Korea)
      IRS 2004 : current problems in atmospheric radiation : proceedings of
   the International Radiation Symposium, Busan, Korea, 23–28 August 2004 / edited by
   Herbert Fischer and Byung-Ju Sohn.
       p. cm. — (Studies in geophysical optics and remote sensing)
   Includes bibliographical references and indexes.
   ISBN-13: 978-0-937194-48-5
   ISBN-10: 0-937194-48-4
   1. Atmospheric radiation—Congresses. I. Title: Current problems in atmospheric radiation.
   II. Fischer, Herbert, 1942-   . III. Sohn, Byung-Ju. IV. Title. V. Series.
   QC912.3.I77 2004
     551.5'273—dc22   2006006016

CONTENTS

Preface xiv

List of Participants xvi

SESSION A: TOPICAL UNION SESSION

Convenors: **Teruyuki Nakajima, Herbert Fischer, Robert G. Ellingson, and Byung-Ju Sohn**

Exploring aerosol-cloud-climate interaction mechanisms using the new generation of earth observation system data
    Z. Li and T. Yuan 1

Three-dimensional radiative transfer
    B. Mayer and T. Zinner 5

Atmospheric parameter retrievals from high resolution infrared spectra in a cloudy atmosphere
    M. Höpfner 9

SESSION B: RADIATIVE TRANSFER THEORY AND MODELING

Convenors: **Robert F. Cahalan, Roger Davies, Peter Koepke, and Manuel Lopez-Puertas**

McSCIA: A Monte Carlo model for analysis of 3D features in a spherical atmosphere
    F. Spada and M. C. Krol 13

The development of new radiation code for CCSR/NIES AGCM
    M. Sekiguchi and T. Nakajima 17

Super fast radiative transfer forward model based on principal components
    X. Liu, W. L. Smith, D. K. Zhou, and A. M. Larar 19

Modeling UV-exposure inside a plant canopy during a vegetation period
    J. H. Schween and P. Koepke 23

Multiple Raman scattering in the Earth's atmosphere
    R. van Deelen, J. Landgraf, and I. Aben 27

Stochastic radiative transfer in multilayer broken clouds. Approach, validation and application
    E. I. Kassianov 31

Inferring domain-averaged cloud properties from the ARM observations and testing the PCLOS models
    Y. Ma and R. G. Ellingson 35

The value of limb measurements of the Earth's atmosphere
    J. I. van Gent 39

The use of the ARM WSI to estimate the atmospheric optical depth at night
    I. C. Musat and R. G. Ellingson 43

Convergence acceleration of a spherical harmonics method for strong anisotropic scattering
    V. P. Boudak 47

Simulations of spectral actinic flux density fields in scattered and overcast cloud conditions: Comparison with INSPECTRO aircraft measurements
    A. Kniffka, S. G. García, E. Jäkel, S. Schmidt, M. Wendisch, R. Scheirer, and T. Trautmann 51

Validation of a radiative transfer model for simulation of polarized radiances in a coupled ocean-atmosphere system
    L. Duforet, P. Dubuisson, B. Bonnel, F. Parol, and M. Vesperini 55

A quasi-analytic solution of the radiative transfer equation for three-dimensional, heterogeneous atmospheres
    H. Ishida and S. Asano 59

User-friendly software for satellite signal simulation based on line-by-line calculations of radiative fluxes in the atmosphere
    J. C. Ceballos, B. A. Fomin, and M. P. Corrêa 63

A new technique for developing k-distributions
    B. A. Fomin and M. P. Corrêa 67

Actual state of the UV radiation transfer model package STAR
    P. Koepke, D. Anwender, M. Mech, A. Oppenrieder, J. Reuder, A. Ruggaber, M. Schreier, H. Schwander, and J. Schween 71

Altitude effect on UV index deduced from the VELETA-2002 experimental campaign (Spain)
    J. Lorente, X. de Cabo, E. Campmany, Y. Sola, J. Gonzalez, J. Calbó, J. Badosa, L. Alados-Arboledas, A. Martinez-Lozano, V. Cachorro, A. Labajo, B. de la Morena, A. M. Diaz, M. Pujadas, H. Horvath, A. M. Silva, and G. Pavese 75

About simulation of radiation transfer in smoky atmosphere and environmental monitoring
    T. A. Sushkevich, S. A. Strelkov, E. V. Vladimirova, S. V. Maksakova, A. K. Kulikov, and E. I. Ignatijeva 79

Development of the Monte-Carlo radiative transfer model Promt
    U. Hamann, G. Seckmeyer, S. Raasch, and B. Mayer 83

SESSION C: MOLECULAR RADIATIVE PROPERTIES

Convenors: Shepard Clough, L. Rothman, and Jean-Marie Flaud

Theoretical calculation and simple parameterization for the water vapor millimeter wave foreign continuum
    Q. Ma and R. H. Tipping 87

Theoretical calculation of the collision-induced absorption in dry air in the far-infrared region
    R. H. Tipping and Q. Ma 91

Line strengths and half-widths of the N$_2$O bands in the 2.0- to 2.3-μm region at room temperature
    M. Fukabori, T. Aoki, T. Fujieda, and T. Watanabe 95

SESSION D: PARTICLE RADIATIVE PROPERTIES

Convenors: J. Vanderlei Martins, Michel Legrand, and Jhoon Kim

Retrieval of aerosol optical properties from dual-wavelength polarization lidar measurements
    T. Nishizawa, H. Okamoto, T. Takemura, N. Sugimoto, I. Matsui, and A. Shimizu 99

Thermal infrared radiometry, microphysical properties and geochemical nature of mineral dust
    M. Legrand and O. Pancrati 103

Aerosols and molecules within cloudy atmosphere
    I. N. Melnikova 107

Scaling properties of aerosol optical thickness retrieved from a long-term sun-photometer dataset
    M. Alexandrov, A. Marshak, B. Cairns, A. Lacis, and B. Carlson 111

Radiative transfer modelling for evaluation of the parameterization of cloud optical properties in the Meso-NH mesoscale model
    P. Dubuisson, J.-P. Chaboureau, J.-P. Pinty, and V. Giraud 115

The effects of nonsphericity and variation in ice crystal bulk density on 95 GHz cloud radar signals
    K. Sato and H. Okamoto 119

Optical properties of particles from laboratory fires: Comparison between measurements and Mie calculations
    K. Hungershöfer, O. Schmid, R. S. Parmar, G. Helas, M. O. Andreae, K. Zeromskiene, Y. Iinuma,
    A. Wiedensohler, H. Herrmann, J. Trentmann, and T. Trautmann 123

A code to compute the direct solar radiative forcing: Application to anthropogenic aerosols during the ESCOMPTE experiment
    P. Dubuisson, J.-C. Roger, M. Mallet, and O. Dubovik 127

Numerical calculations of gravito-photophoretic movement for aerosol aggregates
    A. A. Cheremisin and Y. V. Vassilyev 131

SESSION E: GENERAL REMOTE SENSING

Convenors: Yuriy M. Timofeyev, William L. Smith, and Kwang-Mog Lee

Multisensor ground-based remote sensing of radiatively important cloud parameters
    S. Y. Matrosov 135

Observations of aerosols and clouds with Mie-lidar network
    A. Shimizu, N. Sugimoto, and I. Matsui 139

Infrared spectral radiance validation using aircraft- and satellite-based sensor systems
  A. M. Larar, W. L. Smith, D. K. Zhou, X. Liu, H. Revercomb, R. Knuteson, and S. A. Mango 143

Deriving a global surface albedo from geostationary observations
  Y. M. Govaerts and A. Lattanzio 147

The atmospheric sensitivity of the airborne imaging spectrometer APEX
  J. W. Kaiser, J. Nieke, D. Schläpfer, J. Brazile, and K. I. Itten 151

A near global climatology of cirrus overlapping water clouds and retrieval of cirrus cloud properties from MODIS
  F. L. Chang and Z. Li 155

Quality, compatibility and synergy analyses of global aerosol products
  M.-J. Jeong, Z. Li, D.A. Chu, and S.-C. Tsay 159

Intercomparison of atmospheric parameters in the MLT region derived from the CRISTA-1 data by independent methods
  V. S. Kostsov, Yu. M. Timofeyev, A. V. Rakitin, M. Kaufmann, O. Gusev, and K. Grossmann 163

Simultaneous retrieval of aerosol and surface properties over bright targets including snow and ice using multi- and hyperspectral data
  W. Li, H. Eide, K. Stamnes, R. Spurr, T. Aoki, and M. Hori 167

Characterizing the atmosphere above turbid coastal waters using multi- and hyperspectral data
  K. Stamnes, H. Eide, W. Li, R. Spurr, and J. Stamnes 171

Simultaneous retrieval of aerosols and in-water constituents in turbid coastal waters using multi- and hyperspectral data
  H. Eide, W. Li, K. Stamnes, R. Spurr, and J. Stamnes 175

Comparison of satellite (TOMS, GOME) total ozone data with the measurements of Russian ground-based network
  D. V. Ionov, Yu. M. Timofeyev, A. M. Shalamiansky, and J. C. Lambert 179

Development of a polarised radiative transfer model in the oxygen A-band for satellite retrieval of cloud top pressure
  N. Fournier, P. Stammes, and M. Eisinger 183

Atmospheric and cloud parameters retrieval using IR spectral data
  D. K. Zhou, W. L. Smith, H. L. Huang, J. Li, X. Liu, A. M. Larar, L. Chiou, and S. A. Mango 187

Fourier transform spectroradiometers for the characterization of the composition and the radiative properties of the upper atmosphere
  L. Palchetti, G. Bianchini, B. Carli, F. Castagnoli, U. Cortesi, M. Pellegrini, and M. Trambusti ... 191

Retrieval of microphysical properties of mixed-phase clouds using active and passive sensors: A feasibility study for triple-band cloud radars at 9.4 GHz, 35 GHz, and 95 GHz
  Y. Yoshida and S. Asano 195

Satellite-derived surface reflectance using the image-based atmospheric correction
  K. Lee and Y. Kim 199

Characteristics of spectral emissivity for various types of surfaces derived from IMG spectrum data
   Y. Ota and R. Imasu 203

SESSION F, AND F-P: SATELLITE MEASUREMENTS, AND PRECIPITATION MEASUREMENTS FROM SPACE

Convenors: Jae H. Park, Johannes Schmetz, Michael D. King, W. Paul Menzel, Eric A. Smith, and Myoung-Hwan Ahn

Intra-seasonal variation of PSC compositions retrieved from ILAS-II data
   Y. Kim, J. H. Park, W. Choi, K.-M. Lee, S. T. Massie, T. Yokota, H. Nakajima, and Y. Sasano 207

Atmospheric infrared sounder on AIRS with emphasis on Level 2 products
   S.-Y. Lee, E. Fetzer, S. Granger, T. Hearty, B. Lambrigtsen, E. M. Manning, E. Olsen, and T. Pagano 211

Earth reflectance spectra from 300-1750 nm measured by SCIAMACHY
   P. Stammes, J. R. Acarreta, W. H. Knap, and L. G. Tilstra 215

Global comparison of MODIS and AVHRR-type aerosol retrievals over the ocean in the Terra/CERES-MODIS Single Scanner Footprint (SSF) data
   T. X.-P. Zhao and I. Laszlo 219

Aerosol and CO loading in the atmosphere observed by the MODIS and MOPITT-Russian forest fire case
   J. Kim, S. H. Choi, H. C. Lee, H. K. Cho, D. Edwards, S. H. Lee, H. S. Lim, and G. H. Choi 223

Characteristics of satellite-observed MODIS data during foggy days near the Inchon International Airport, Korea
   Y.-M. Kim and J.-M. Yoo 227

First results from Meteosat-8: Atmospheric radiative properties from geostationary orbit
   J. Schmetz, R. Borde, Y. Govaerts, K. Holmlund, M. König, and H. J. Lutz 231

A combined method of the TRMM precipitation radar and visible and infrared scanner for retrieval of cloud-precipitation interaction
   T. Kobayashi and A. Adachi 235

Satellite-based tropical warm pool surface heat budgets: Contrasts between 1997/98 El Nino and 1998/99 La Nina
   S.-H. Chou, M.-D. Chou, P.-H. Lin, P.-K. Chan, and K.-H. Wang 239

A simple approach to estimate land surface albedo from satellite measurements
   Y. Cui and T. Takamura 243

Application of σ-ASI to NAST-I
   A. Carissimo, G. Grieco, C. Serio, V. Cuomo, G. Masiello, and W. L. Smith 247

A comparison of satellite derived precipitation fields and Meteosat-5 observations of mesoscale convective systems over Indian Ocean
   T. F. Yang, I. Jobard, M. Capderou, and M. Desbois 251

Snow products derived from ADEOS-II/GLI data: Scientific implication
T. Aoki, M. Hori, A. Hachikubo, T. Tanikawa, H. Motoyoshi, Y. Iizuka, Y. Nakajima, F. Takahashi,
K. Stamnes, W. Li, H. Eide, R. Storvold, and J. Nieke 255

Aerosol optical thickness in fair-weather cumulus in the Amazon region estimated from ASTER data
G. Wen and R. F. Cahalan 259

Remote sensing of stratospheric chlorine species with the MIPAS/ENVISAT-experiment
N. Glatthor, T. von Clarmann, H. Fischer, M. Höpfner, and G. P. Stiller 263

Radiative characteristics of cirrus clouds as retrieved from AVHRR
S. Katagiri and T. Nakajima 267

Determination of broadband emissivity for arid lands
T. Schmugge, K. Ogawa, and S. Rokugawa 271

Bright-band height statistics observed by the TRMM precipitation radar
K. Okamoto, H. Sasaki, E. Deguchi, M. Thurai, and K. Matsukawa 275

Effects of multiple scattering for millimeter-wavelength weather radars
S. Kobayashi, S. Tanelli, and E. Im 279

A combined TMI-SSM/I rainfall estimation technique for land surfaces over western Africa
J. Schulz, C. Simmer, and M. Diederich 283

Precipitation retrieval from dual-view spaceborne passive microwave radiometers
F. J. Turk, S. Di Michele, P. Bauer, F. S. Marzano, A. Mugnai, L. Roberti, and A. Tassa 287

SESSION G: SURFACE MEASUREMENTS AND FIELD EXPERIMENTS

Convenors: Atsumu Ohmura, Thomas Ackerman, Brent N. Holben, and Guangyu Shi

Analysis of radiation measurements at the COVE sea platform
Z. Jin, T. P. Charlock, and K. Rutledge 291

Analysis of cloud variability and sampling errors in surface and satellite measurements
Z. Li, M. C. Cribb, F.-L. Chang, A. Trishchenko, and Y. Luo 295

UV-radiation and clouds: first results from the INSPECTRO project
R. Scheirer, B. Mayer, and S. Schmidt 299

REFIR measurements in the water vapour rotational band and comparison with a BOMEM AERI-type Fourier transform spectrometer
F. Esposito, R. Restieri, C. Serio, V. Cuomo, G. Masiello, G. Pavese, G. Bianchini, L. Palchetti,
M. Pellegrini, T. Maestri, and R. Rizzi 303

Automated retrieval of atmospheric aerosol and trace gases properties from large MFRSR datasets
M. Alexandrov, B. Carlson, A. Lacis, B. Cairns, and A. Marshak 307

Aerosol optical characteristics in Asia from measurements of SKYNET sky radiometers
K. Aoki, T. Nakajima, and T. Takamura 311

Optical properties of Asian dust measured at several sites in Japan
H. Kobayashi, K. Arao, T. Murayama, K. Iokibe, R. Koga, M. Yabuki, and M. Shiobara 315

Automatic image recording network of sand storms and dusty airs in Northeast Asia
K. Kinoshita, N. Iino, S. Hamada, H. Kikukawa, T. Batmunkh, J. Dulam, W. Ning, Z. Gang,
and A. Tupper 319

Ground-based measurements of UV doses and ozone amounts using filter instruments
H. Eide, A. Dahlback, K. Stamnes, B.-A. Høyskar, R. O. Olsen, F. J. Schmidlin, S. C. Tsay,
and J. Stamnes 323

New radiation and energy balance of the world and its variability
A. Ohmura 327

Continuous measurement of aerosol characteristics by ADEC sky-radiometer network
A. Uchiyama, A. Yamazaki, H. Togawa, and J. Asano 331

Aerosol optical depth measurement with MFRSR, UV-MFRSR, and sun photometer at Gwangju, Korea
J. E. Kim and Y. J. Kim 335

Relationship between aerosol properties and air pollutants
S. Mukai, I. Sano, and M. Yasumoto 339

Retrieval of aerosol microphysical properties from MFRSR observations
E. I. Kassianov, J. C. Barnard, and T. P. Ackerman 343

Measurements of aerosol optical properties at Pallas Gaw station in northern Finland
V. Aaltonen, H. Lihavainen, J. Hatakka, and Y. Viisanen 347

Uncertainty analysis of data from the Polar Atmospheric Emitted Radiance Interferometer (PAERI) during the South
Pole Atmospheric Radiation and Cloud Lidar Experiment (SPARCLE)
M. S. Town and V. P. Walden 351

Airborne measurements of upwelling and downwelling spectral actinic flux densities during the INSPECTRO campaign
E. Jäkel, M. Wendisch, S. Schmidt, T. Trautmann, and A. Kniffka 355

The French network for spectral measurement of solar UV irradiance
J. Lenoble, C. Brogniez, M. Houët, M. Legrand, J. Lenoble, A. de La Casinière, T. Cabot,
and F. Guirado 359

UV filter radiometers: Calibration, field measurements and applications
A. Los 363

Comparison between observed spectral albedos and theoretical ones for artificial snowpack
T. Tanikawa, M. Aniya, T. Aoki, A. Hachikubo, M. Hori, and O. Abe 367

Correction of the diffuse influence in spectrodirectional measurements
J. T. Schopfer, S. Dangel, J. W. Kaiser, M. Kneubühler, J. Nieke, G. Schaepman-Strub, M. E. Schaepman, and K. I. Itten 371

The radiate and light characteristics of the climate change during the second half of 20th century in Moscow
G. M. Abakumova, E. V. Gorbarenko, E. I. Nezval, and O. A. Shilovtseva 375

Water vapor estimated with GPS on Tibetan Plateau and its influence on radiation budget
Z. Sun, J. Liu, and H. Liang 379

Validation experiment for satellite remote sensing and numerical models of 'Yamase' clouds
M. Kojima, S. Asano, H. Iwabuchi, and S. Otake 383

Comparison of longwave effective cloud fraction with ARM cloudiness measurements
E. E. Takara and R. G. Ellingson 387

SESSION H: RADIATIVE BUDGET AND FORCING

Convenors: Thomas P. Charlock, Natalia Chubarova, Robert Kandel, and Kazuaki Kawamoto

All-sky aerosol direct forcing to SW and LW at TOA and surface using CERES Terra and the MATCH assimilation
T. P. Charlock, F. G. Rose, D. A. Rutan, D. W. Fillmore, and W. D. Collins 391

Long-term trend of surface shortwave radiation over China
T. Hayasaka, K. Kawamoto, J.-Q. Xu, and G.-Y. Shi 395

Aerosol optical/radiative forcing properties over East Asia determined from SKYNET radiation measurements
B. J. Sohn, D.-H Kim, T. Nakajima, and T. Takamura 399

AVHRR observations of the aerosol indirect effect for summertime stratiform clouds in the Northeastern Atlantic
M. A. Matheson, J. A. Coakley, and W. R. Tahnk 403

Estimates of the direct and indirect radiative forcing of climate by anthropogenic aerosols
J. A. Coakley, N. G. Loeb, T. P. Charlock, A. S. Ackerman, Y. J. Kaufman, and U. Lohmann 407

Satellite observations of cloud radiative forcing for the African tropical convective region
J. M. Futyan, J. E. Russell, and J. E. Harries 411

The visible and near infrared components of the shortwave radiation budget
I. Laszlo and R. T. Pinker 415

The NASA/GEWEX Surface Radiation Budget dataset: Results and analysis
S. J. Cox, S. K. Gupta, J. C. Mikovitz, M. Chiacchio, T. Zhang, and P. W. Stackhouse 419

CM-SAF surface radiation budget: First results from the initial operations phase
R. Hollmann, R. W. Müller, and A. Gratzki 423

Global spherical model of Earth radiation in planet scale
    T. A. Sushkevich, S. A. Strelkov, E. V. Vladimirova, A. N. Volkovich, V. V. Kozoderov, A. K. Kulikov, and S. V. Maksakova 427

UV radiation in past and future modelling with all atmospheric parameters including cloudiness
    J. Reuder, P. Koepke, and J. Schween 431

Comparative study of UVB and total radiation attenuation observed in a South American region
    M. P. Corrêa, J. C. Ceballos, M. J. Bottino, and G. Coronel 435

The influence of forest and peatbog fires on the optical and radiative regimes of the atmosphere and radiative forcing over Central Russia
    N. Y. Chubarova, G. M. Abakumova, E. V. Gorbarenko, E. I. Nezval, O. A. Shilovtseva, and A. N. Rublev 439

Correspondence of the low cloud microphysics to the aerosol amount over China
    K. Kawamoto, T. Hayasaka, and I. Uno 443

SESSION I: WEATHER AND CLIMATE APPLICATIONS

Convenors: V. Ramaswamy, Jean-Jacques Morcrette, and Chang-Hoi Ho

Study of cloud microphysical structure with cloud profiling radar and lidar: *Mirai* cruise
    H. Okamoto, T. Nishizawa, H. Kumagai, N. Sugimoto, T. Takemura, and T. Nakajima 447

Using satellite observations and reanalyses to evaluate climate and weather forecast models
    R. P. Allan 451

Modes of tropical water cycle variability
    J. J. Bates 455

Regional climate response induced by aerosol radiative forcing over Eurasia during boreal spring
    M.-K. Kim, W.-S. Lee, K. M. Lau, K.-M. Kim, Y. C. Sud, G. K. Walker, and M. Chin 461

Effective radius of cloud droplets by ground-based remote sensing: Implication to aerosol indirect effect
    B.-G. Kim 465

Simulation of climate change by aerosol direct and indirect effects with aerosol transport-radiation model
    T. Takemura, T. Nakajima, and T. Nozawa 469

Signals of climate variations in the WCRP/GEWEX SRB datasets and their connections with other climate indices
    T. Zhang, P. W. Stackhouse, S. K. Gupta, S. J. Cox, and C. Mikovitz 473

AUTHOR INDEX 478

SUBJECT INDEX 481

PREFACE

The International Radiation Symposium (IRS 2004) on Current Problems in Atmospheric Radiation was held on 23–28 August 2004, at Busan Exhibition and Convention Centre (BEXCO), Busan, Korea. It was the first in Asia since IRS 1972 in Sendai/Japan.

The Symposium was organized by the International Radiation Commission and hosted jointly by the Korean Meteorological Society (President: Dr. Hyo-Sang Chung) and the Centre of Atmospheric Environment Research of Seoul National University (Director: Prof. Dong-Kyou Lee). The conference focused on providing opportunities for exchanging advanced ideas and experiences on current problems in atmospheric radiation. It presented a platform for discussing and reviewing knowledge of spectroscopy, radiative transfer theory and modelling, remote sensing, and weather and climate applications. The Symposium attracted about 340 scientists from 25 countries who presented approximately 370 papers in plenary and poster sessions.

The Symposium afforded an ideal opportunity to honour the first recipients of the Commission's new awards, the Gold Medal and the Young Scientist Award, which were presented to Professor Richard Goody of Harvard University and Dr. Toshihiko Takemura of Kyushu University, respectively. The IRC Gold Medal is designed to honour a senior scientist who has made contributions of lasting significance to the field of radiation research. The IRC Young Scientist Award consists of a $500 cash award to a young scientist who has made recent noteworthy contributions to radiation studies and is regarded as becoming a leading radiation scientist in the future.

The proceedings of this conference are organized into the nine sessions of IRS 2004:

1. Topical Union Session
2. Radiative Transfer Theory and Modelling
3. Molecular Radiative Properties
4. Particle Radiative Properties
5. General Remote Sensing
6. Satellite Measurements (including Precipitation Measurements from Space)
7. Surface Measurements and Field Experiments
8. Radiative Budget and Forcing
9. Weather and Climate Applications

As at previous meetings, not all the authors provided manuscripts for these proceedings. Also, some of the submitted papers have not been accepted after the comprehensive reviewing process. Readers wishing copies of those not published here are encouraged to contact the corresponding author directly, using the address in the list of participants provided in this book.

We look forward to seeing you at the next International Radiation Symposium (IRS 2008) and to learning about your new research studies conducted during the coming years.

Herbert Fischer and Byung-Ju Sohn, Editors

ACKNOWLEDGMENTS

We would like to thank all those who contributed to the success of IRS 2004. Special thanks are expressed to the Korean Meteorological Society (KMS, President: Dr. Hyo-Sang Chung) and the Atmospheric Environmental Research Institute of Seoul National University (Director: Prof. Dong-Kyou Lee) for serving as conference hosts, and Meci International, Inc. for supporting the organization of the symposium.

Appreciation is extended to the Local Organizing Committee (LOC), including the Chair Prof. Byung-Ju Sohn, the Co-Chair Prof. Young-Seup Kim, the Secretary Dr. Jae-Cheol Nam as well as the members of the Steering Committee [Prof. Jae H. Park (Seoul Natl. Univ.), Dr. Sung-Nam Oh (METRI/KMA), Prof. Jhoon Kim (Yonsei Univ.), Dr. Myoung-Hwan Ahn (METRI/KMA), Prof. Myoung-Seok Suh (Kongju Natl. Univ.), Prof. Jung-Moon Yoo (Ewha Womans Univ.), Prof. Chang-Hoi Ho (Seoul Natl. Univ.), Prof. Jae-Hwan Kim (Busan Natl. Univ.), Prof. Tae-Yong Kwon (Kangnung Natl. Univ.), and Prof. Kwang-Mog Lee (Kyungpook Natl. Univ.)] and the members of the Advisory Committee [Prof. Hi-Ku Cho (Yonsei Univ.), Prof. Chong-Heum Kwak (Kongju Natl. Univ.), Prof. Dong-Kyou Lee (Seoul Natl. Univ.), Prof. Jong-Ghap Jhun (Seoul Natl. Univ.), and Dr. Hong-Yul Paik (Korea Aerospace Research Institute)]. The LOC took care of the planning of the symposium, making arrangements in Busan, and running the symposium smoothly.

The Scientific Program Committee consists of the IRC officers: President Prof. Herbert Fischer (Germany), Vice-President Prof. Teruyuki Nakajima (Japan), Secretary Prof. Robert G. Ellingson (USA), the chair of LOC (Prof. Byung-Ju Sohn) and the main conveners of the symposium sessions [Prof. Robert F. Cahalan (USA), Dr. Shepard A. Clough (USA), Dr. Jose Vanderlei Martins (USA), Prof. Yuriy M. Timofeyev (Russia), Prof. Jae H. Park (Korea), Prof. Atsumu Ohmura (Switzerland), Dr. Thomas Charlock (USA), and Dr. V. Ramaswamy (USA)]. The Scientific Program Committee organized the nine sessions, arranged for the co-conveners for the sessions and solved the delicate task of arranging the larger number of papers in a reasonable schedule of talks and poster presentations.

We thank also the responsible organizers of the accompanying program which gave some insight to the Korean culture and impressed particularly the colleagues from western countries.

The following national and international institutes, organizations and companies have sponsored IRS 2004: Seoul National University (SNU, Korea), BK21 School of Earth and Environmental Science of SNU (SNU/SEES, Korea), Climate Environment System Research Center (SNU, Korea), Korean Meteorological Research Institute (METRI/KMA, Korea), Korea Aerospace Research Institute (KARI, Korea), Forschungszentrum Karlsruhe (Germany), International Association for Meteorology and Atmospheric Sciences (IAMAS), Department of Energy (DoE, USA), EUMETSAT (Europe), Eko/Kipp & Zonen Alliance. Their generous support of this symposium provided the financial resources to support the travel, registration fees, and local subsistence expenses for a considerable number of scientists who were not able to get adequate funding in their countries. Also, the publication of these proceedings became possible. Further supporting organisations were Busan Metropolitan City, Korea Meteorological Administration (KMA), Korea Science and Engineering Foundation (KOSEF) and Korea National Tourism Organization.

The organization of the proceedings was performed mainly by the Scientific Program Committee, of course with the help of many co-conveners of the sessions and further acknowledged scientists. We are grateful to Ms. Won-Ju Na, Mrs. Seong-Chan Park, Hyun-Jong Oh, Eui-Seok Chung, Keun-Hyuk Ryu, Hyoung-Wook Chun, Ho-Sun Park, Ms. Hye-Suk Park, Hyo-Jin Han, and Seung-Hee Ham, who provided editorial assistance of these proceedings, and to Ms. Diana McQuestion of A. Deepak Publishing, who supported us in preparing the manuscripts in a suitable form. Certainly, we thank all the contributors to these proceedings.

Herbert Fischer and Byung-Ju Sohn, Editors

LIST OF PARTICIPANTS

Aaltonen, Veijo, Finnish Meteorological Institute, Sahaajankatu 20 E, FI-00880 Helsinki, Finland (E-mail: Veijo.Aaltonen@fmi.fi)

Ackerman, Thomas, DOE Atmospheric Radiation Measurement Program, Pacific Northwest National Laboratory, 902 Battelle Boulevard, P.O. Box 999, Richland, WA, 99352, USA (E-mail: ackerman@pnl.gov)

Ahn, Gi-Joon, Department of Atmospheric and Environmental Sciences, Kangnung National University, 123 Jibyeon-Dong, Kangnung-city, Kangwon, Korea (E-mail: yurojun@hotmail.com)

Ahn, Hyun-Jeong, School of Earth and Environmental Sciences, Seoul National University, San 56-1, Sillim-dong, Gwanak-gu, Seoul, Korea (E-mail: ahj1112@metri.re.kr)

Ahn, M.H, Remote Sensing Research Lab. Korea Meteorological Administration, Korea (E-mail: mhahn@kma.go.kr)

Ahn, Yu-Hwan, Ocean optics and Ocean color Remote sensing Laboratory, Korea Ocean Research and Development Institute, Seoul, 425-600, Korea (E-mail: yhahn@kordi.re.kr)

Alexandrov, Mikhail, Columbia University and NASA Goddard Institute for Space Studies, 2880 Broadway, New York, NY, 10025, USA (E-mail: malexandrov@giss.nasa.gov)

Allan, Richard P., Environmental Systems Science Centre, University of Reading, 3 Earley Gate, Whiteknights, Reading RG6 6AL, UK (E-mail: rpa@mail.nerc-essc.ac.uk)

Aoki, Kazuma, Toyama University, 3190 Gofuku, Toyama, 930-8555, Japan (E-mail: kazuma@edu.toyama-u.ac.jp)

Aoki, Teruo, Meteorological Research Institute, 1-1 Nagamine, Tsukuba, Ibaraki, 305-0052, Japan (E-mail: teaoki@mri-jma.go.jp)

Arai, Yutaka, Center for Climate System Research, University of Tokyo 4-6-1 Komaba, Meguro-ku, Tokyo, 153-8904, Japan (E-mail: yarai@ccsr.u-tokyo.ac.jp)

Asano, Shoji, Tohoku University, Aramaki, Aoba, Sendai, Miyagi, 980-8578, Japan (E-mail: asano@caos-a.geophys.tohoku.ac.jp)

Baek, Seon-Kyun, Meteorological Research Institute/KMA, 460-18, Shindaebang-dong, Tongjak-gu, Seoul, Korea (E-mail: sybang@kma.go.kr)

Bak, Joong-Hyun, Dept. of Atmospheric Science, Kyungpook National University, 1370 Sankyuk-dong, buk-gu, Daegu, Korea (E-mail: uarshaloe@hotmail.com)

Bakan, Stephan, Max-Planck-Institut für Meteorologie, Bundesstr. 55, D-20146 Hamburg, Germany (E-mail: bakan@dkrz.de)

Bang, So-Young, Meteorological Research Institute/KMA, 460-18, Shindaebang-dong, Tongjak-gu, Seoul, Korea (E-mail: sybang@kma.go.kr)

Bates, John, NOAA National Climatic Data Center, 191 Patton Ave., 28801, USA (E-mail: John.J.Bates@noaa.gov)

Behrens, Klaus, Deutscher Wetterdienst, Research and Development, Meteorological Observatory Lindenberg, Am Observatorium 12, OT Lindenberg, D-15848 Tauche, Germany (E-mail: klaus.behrens@dwd.de)

Boudak, Vladimir P., Moscow Power-Engineering Institute (Technical University), Light Engineering Department, Moscow, Protopopovsky, 14-14, 129090, Russia (E-mail: BudakVP@mpei.ru)

Cahalan, Robert F., NASA/GSFC, Code 913, Greenbelt, MD, 20771, USA (E-mail: Robert.F.Cahalan@nasa.gov)

Carleer, Michel, Chimie Quantique et Photophysique CP160/09 - Universite Libre de Bruxelles, Ave. F. D. Roosevelt, 50, B-1050, Belgium (E-mail: mcarleer@ulb.ac.be)

Carli, Bruno, IFAC-CNR, via Panciatichi, 64 – Firenze, 50127, Italy (E-mail: B.Carli@ifac.cnr.it)

Cha, Joo-Wan, Korea Global Atmosphere Watch Observatory, METRI, KMA, 1764-6, Seungeon-ri, Anmyeon-eup, Taean-gun, ChungNam, 357-961, Korea (E-mail: jwcha@kma.go.kr)

Chandrasekar, V., Dept. of Electrical Engineering, Colorado State University, Fort Collins, CO, 80523-1373, USA (E-mail: chandra@engr.colostate.edu)

Chang, Lim-Seok, School of Earth and Environment Science, Seoul National University, San 56-1, Sillim-dong, Gwanak-gu, Seoul, 151-742, Korea (E-mail: jls1@snupbl1.snu.ac.kr)

Charlock, Thomas P., NASA Langley Research Center, Hampton, VA, 23681, USA (E-mail: Thomas.P.Charlock@nasa.gov)

Chen, Hong Bin, Institute of Atmospheric Physics, Chinese Academy of Sciences, Institute of Atmospheric Physics, Chinese Academy of Sciences, Beijing, 100029, China (E-mail: chb@mail.iap.ac.cn)

Cheremisin, Aleksander Alekseevich, Krasnoyarsk State Technical University, Kirenskii str., 26, Krasnoyarsk, 660074, Russia (E-mail: cher@akadem.ru)

Chiu, Jui-Yuan, Joint Center for Earth Systems Technology, University of Maryland, Baltimore County, JCET/UMBC, 1000 Hilltop Circle, Baltimore, MD, 21250, USA (E-mail: chiuj@umbc.edu)

Cho, Chang-bum, Meteorological Research Institute, 460-18, Shindaebang-dong, Tongjak-gu, Seoul, Korea (E-mail: cbcho@metri.re.kr)

Cho, Hi-Ku, Dept. of Atmospheric Science, Yonsei University, 134 Sinchon-dong, Seodaemoon-gu, Seoul, Korea (E-mail: chk@atmos.yonsei.ac.kr)

Cho, Nam-Seo, METRI/KMA, 460-18, Shindaebang-dong, Tongjak-gu, Seoul, Korea (E-mail: pcanonjr@metri.re.kr)

Cho, Young-Jun, Department of Atmospheric and Environmental Sciences, Kangnung National University, 123 Jibyeon-Dong, Kangnung-city, Kangwon, Korea (E-mail: genesis99@naver.com)

Cho, Young Wook, METRI/KMA, 460-18, Shindaebang-dong, Tongjak-gu, Seoul, Korea (E-mail: kingcho76@metri.re.kr)

Choi, Byoung-Cheol, Korea Global Atmosphere Watch Observatory, METRI, KMA, 1764-6, Seungeon-ri, Anmyeon-eup, Taean-gun, Chung Nam, 357-961, Korea (E-mail: cbc@kma.go.kr)

Choi, Hye-Sun, School of Earth and Environmental Sciences, Seoul National University, San 56-1, Sillim-dong, Gwanak-gu, Seoul, Korea (E-mail: kenvie@strat.snu.ac.kr)

Choi, Hyun-Young, School of Earth and Environmental Sciences, Seoul National University, San 56-1, Sillim-dong, Gwanak-gu, Seoul, Korea (E-mail: chy0917@hanmail.net)

Choi, Jae-Cheon, METRI/KMA, 460-18, Shindaebang-dong, Tongjak-gu, Seoul, Korea (E-mail: jcchoi@kma.go.kr)

Choi, Sung-Hwa, Dept. of Atmospheric Science, Yonsei University, 134 Sinchon-dong, Seodaemoon-gu, Seoul, Korea (E-mail: shchoi78@yonsei.ac.kr)

Choi, Yong-Sang, School of Earth and Environmental Sciences, Seoul National University, San 56-1, Shillim-dong, Kwanak-gu, Seoul, 151-742, Korea (E-mail: yschoi@cpl.snu.ac.kr)

Chou, Ming-Dah, Department of Atmospheric Sciences, National Taiwan University, No. 1, Sec. 4, Roosevelt Road, Taipei, 106, Taiwan (E-mail: mdchou@atmos1.as.ntu.edu.tw)

Chou, Shu-Hsien, Department of Atmospheric Sciences, National Taiwan University, Taipei, 106, Taiwan (E-mail: shchou@atmos1.as.ntu.edu.tw)

Chubarova, Natalia, Moscow State University Geographical Department, Meteorological Observatory, Vorobyovy Gory, Moscow, 119992, Russia (E-mail: chubarova@imp.kiae.ru)

Chun, Hyoung-Wook, School of Earth and Environmental Sciences, Seoul National University, San 56-1, Shillim-dong, Kwanak-gu, Seoul, 151-742, Korea (E-mail: chunhw@eosat.snu.ac.kr)

Chun, Young-sin, Meteorological Research Institute, 460-18, Shindaebang-dong, Tongjak-gu, Seoul, Korea (E-mail: yschun@metri.re.kr)

Chung, Chu-Yong, METRI/KMA, 460-18, Shindaebang-dong, Tongjak-gu, Seoul, Korea (E-mail: cychung@metri.re.kr)

Chung, Chul, Scripps Institution of Oceanography, Mail code 0221, La Jolla, CA, 92093, USA (E-mail: cchung@fiji.ucsd.edu)

Chung, Eui-Seok, School of Earth and Environmental Sciences, Seoul National University, San 56-1, Sillim-dong, Gwanak-gu, Seoul, Korea (E-mail: chunges@eosat.snu.ac.kr)

Chung, Hyo-Sang, METRI/KMA, 460-18, Shindaebang-dong, Tongjak-gu, Seoul, Korea (E-mail: hschung0@metri.re.kr)

Clough, Shepard, Atmospheric & Environmental Research, Inc., 131 Hartwell Avenue, Lexington, MA, 02421, USA (E-mail: sclough@aer.com)

Coakley, James, College of Oceanic and Atmospheric Sciences, Oregon State University, 104 COAS Admin Building, Corvallis, OR, 97331-5503, USA (E-mail: coakley@coas.oregonstate.edu)

Collins, William, National Center for Atmospheric Research, P.O. Box 3000, Boulder, CO, 80307-3000, USA (E-mail: wcollins@ucar.edu)

Correa, Marcelo de Paula, Satellite and Environmental Systems Division – Centro de Previsão de Tempo e Estudos Climáticos, Instituto Nacional de Pesquisas Espaciais – Rod. Dutra, km 40 – Cachoeira Paulista – SP – 12.630-000, Brazil (E-mail: mpcorrea@cptec.inpe.br)

Cox, Stephen, Analytical Services & Materials, Inc., 1 Enterprise Parkway, Suite 300, Hampton, VA, 23666, USA (E-mail: s.j.cox@larc.nasa.gov)

LIST OF PARTICIPANTS

Cui, Yu, Graduate School of Science and Technology, Chiba University, Center for Environmental Remote Sensing, 1-33 Yayoi, Inage, Chiba, 263-8522, Japan (E-mail: cuiyu@ceres.cr.chiba-u.ac.jp)

DeSlover, Daniel, University of Wisconsin, 1225 W. Dayton St., Madison, WI, 53706, USA (E-mail: deslover@ssec.wisc.edu)

Dong, Xiquan, University of North Dakota, 4149 Campus Road, Box 9006, Grand Forks, ND, 58202-9006, USA (E-mail: dong@aero.und.edu)

Dubuisson, Philippe, ELICO/Universite du Littoral Cote d'Opale, Maison de la Recherche en Environnement Naturel, 32 Avenue Foch, Wimereux, 62930, France (E-mail: dub@mren2.univ-littoral.fr)

Duforet, Lucile, ELICO/Universite du Littoral Cote d'Opale, Maison de la Recherche en Environnement Naturel, 32 Avenue Foch, Wimereux, 62930, France (E-mail: lucile@mren2.univ-littoral.fr)

Eide, Hans, Department of Physics and Engineering Physics, Stevens Institute of Technology, Hoboken, NJ, 07030, USA (E-mail: heide@stevens.edu)

Ellingson, Robert G., Department of Meteorology, Florida State University, Tallahassee, FL 32306, USA (E-mail: bobe@met.fsu.edu)

Fischer, Herbert, IMK-ASF, Forschungszentrum Karlsruhe GmbH, Postfach 3640, Karlsruhe, 76021, Germany (E-mail: herbert.fischer@imk.fzk.de)

Flaud, Jean-Marie, LISA, University Paris 12&7 and CNRS, 61 Av. Général de Gaulle, Créteil, 94010, France (E-mail: flaud@lisa.univ-paris12.fr)

Fournier, Nicolas, KNMI, Royal Netherlands Meteorological Institute, 3730 AE, The Netherlands (E-mail: fournier@knmi.nl)

Friess, Udo, Space Research Centre, University of Leicester, EOS Group, Departments of Physics and Chemistry, Leicester LE1 7RH, UK (E-mail: uf5@le.ac.uk)

Fu, Qiang, Department of Atmospheric Sciences, University of Washington, Box 351640, Seattle, WA, 98195-1640, USA (E-mail: qfu@atmos.washington.edu)

Fukabori, Masashi, Meteorological Research Institute, 1-1 Nagamine, Tsukuba, Ibaraki, 305-0052, Japan (E-mail: fukabori@mri-jma.go.jp)

Fukushima, Hajime, School of High-technology for Human Welfare, Tokai University, 317 Nishino, Numazu, 410-0395, Japan (E-mail: hajime@wing.ncc.u-tokai.ac.jp)

Funke, Bernd, Instituto de Astrofisica de Andalucia (CSIC), Camino Bajo de Huetor, 24, Granada, 18008, Spain (E-mail: bernd@iaa.es)

Futyan, Joanna, Space and Atmospheric Physics Group, Blackett Laboratory, Imperial College, London, SW7 2BP, UK (E-mail: joanna.futyan@imperial.ac.uk)

Gatebe, Charles K., University of Maryland Baltimore County, NASA GSFC, Mail code 913, MD, 20903, USA (E-mail: gatebe@climate.gsfc.nasa.gov)

Govaerts, Yves, EUMETSAT, Am Kavalleriesand 31, D-64295, Germany (E-mail: govaerts@eumetsat.de)

Grieco, Giuseppe, DIFA Dip., Università della Basilicata, 85100 Potenza, Italy (E-mail: ggrieco@unibas.it)

Ha, Jong-Sung, Dept. of Atmospheric Science, Pusan National University 30 Jangjeon-dong, Geumjeong-gu, Busan 609-735, Korea (E-mail: jongsung@pusan.ac.kr)

Ha, Jung-Yup, Dept. of Atmospheric Science, Kyungpook National University, 1370 Sankyuk-dong, buk-gu, Daegu, Korea (E-mail: hazime98@hotmail.net)

Ha, Kyung-Ja, Department of Atmospheric Science, Pusan National University, 30 Jangjeon-dong, Geumjeong-gu, Busan, 609-735, Korea (E-mail: kjha@pusan.ac.kr)

Hagihara, Youichiro, School of High-technology for Human Welfare, Tokai University, 317 Nishino, Numazu, 410-0395, Japan (E-mail: you160@fksh.fc.u-tokai.ac.jp)

Ham, Seung-hee, School of Earth and Environmental Sciences, Seoul National University, San 56-1, Sillim-dong, Gwanak-gu, Seoul, Korea (E-mail: hamsh@eosat.snu.ac.kr)

Hamann, Ulrich, Institute of Meteorologie, University Hannover, Rosenbuschweg 10, Hannover, 30453, Germany (E-mail: super-uli@web.de)

Han, Hyo-jin, School of Earth and Environmental Sciences, Seoul National University, San 56-1, Sillim-dong, Gwanak-gu, Seoul, Korea (E-mail: hanhj@eosat.snu.ac.kr)

Haruma, Ishida, Center for Atmospheric and Oceanic Studies, Tohoku University, Aramaki-Aza-Aoba, Sendai, 980-8578, Japan (E-mail: ishida@caos.geophys.tohoku.ac.jp)

Hayasaka, Tadahiro, Research Institute for Humanity and Nature, 335 Takashima-cho, Kamigyo-ku, Kyoto, 602-0878, Japan (E-mail: hayasaka@chikyu.ac.jp)

Higurashi, Akiko, National Institute for Environmental Studies, 16-2 Onogawa, Tsukuba, Ibaraki, 305-8506, Japan (E-mail: hakiko@nies.go.jp)

Ho, Chang-Hoi, Seoul National University, Room 222, Bldg. 56, Seoul National University, Seoul, 151-747, Korea (E-mail: hoch@cpl.snu.ac.kr)

Hoepfner, Michael, Institut fur Meteorologie und Klimaforschung, Forschungszentrum Karlsruhe, Postfach 3640, 76021 Karlsruhe, Germany (E-mail: michael.hoepfner@imk.fzk.de)

Hong, Gi-Man, Dept. of Environmental Atmospheric Sciences, Pukyong National University, 599-1 Daeyeon 3-Dong, Nam-Gu, Busan, Korea (E-mail: hongkm@mail1.pknu.ac.kr)

Hong, Hyunkee, Dept. of Atmospheric Science, Yonsei University, 134 Sinchon-dong, Seodaemoon-gu, Seoul, Korea (E-mail: brunhilt77@hotmail.com)

Hong, Seong-Wook, Dept. of Atmospheric Science, Texas A&M University, College Station, TX 77843-3150, USA (E-mail: sesttiya@tamu.edu)

Hori, Masahiro, Japan Aerospace Exploration Agency, 1-8-10 Harumi, Chuo-ku, Tokyo, 104-6023, Japan (E-mail: hori@eorc.jaxa.jp)

Huang, Bormin, University of Wisconsin-Madison, 1225 W. Dayton St., WI, 53706, USA (E-mail: bormin@ssec.wisc.edu)

Huang, Hsuan-Chun, Center for Climate System Research, The University of Tokyo, 4-6-1 Komaba, Meguro-ku, Tokyo, 153-8904, Japan (E-mail: hsuanjun@ccsr.u-tokyo.ac.jp)

Huang, HungLung Allen, CIMSS, University of Wisconsin-Madison, 1225 West Dayton Street, Madison, WI, 53706, USA (E-mail: allenh@ssec.wisc.edu)

Hungershoefer, Katja, Institute for Meteorology, University of Leipzig, Stephanstr. 3, Leipzig, 04103, Germany (E-mail: hunger@uni-leipzig.de)

Hyun, Myung-Suk, Meteorological Research Institute, 460-18, Shindaebang-dong, Tongjak-gu, Seoul, Korea (E-mail: mshyun@metri.re.kr)

Iguchi, Takamichi, Center for Climate System Research, University of Tokyo, 4-6-1 Komaba, Meguro-ku, Tokyo, 153-8904, Japan (E-mail: iguchi@ccsr.u-tokyo.ac.jp)

Iguchi, Toshio, National Institute of Information & Communications Technology (NICT), Applied Research & Standards Department, 4-2-1 Nukui-Kita-machi, Koganei, Tokyo, 184-8795, Japan (E-mail: iguchi@nict.go.jp)

Im, Eun-Soon, Meteorological Research Institute, 460-18, Shindaebang-dong, Tongjak-gu, Seoul, Korea (E-mail: esim@metri.re.kr)

Imasu, Ryochi, Center for Climate System Research, University of Tokyo, 4-6-1 Komaba, Meguro-ku, Tokyo, 153-8904, Japan(E-mail: imasu@ccsr.u-tokyo.ac.jp)

Ishihara, Hironari, Fujitsu FIP corporation, Time 24 Building, 2-45 Aomi, Koto-ku, Tokyo 135-8686, Japan (E-mail: hironari@ilas2.nies.go.jp)

Iwabuchi, Hironobu, Japan Agency for Marine-Earth Science and Technology, 3173-25, Showa-machi, Kanazawa-ku, Tokohama-city, Kanagawa 236-0001, Japan (E-mail: hiro-iwabuhi@jamstec.go.jp)

Iwafune, Koji, Japan Aerospace Exploration Agency, 1-8-10, Harumi, Chuo-ku, 104-6023, Japan (E-mail: kadoba@eorc.jaxa.jp)

Jaekel, Evelyn, Leibniz-Institute for Tropospheric Research, Permoserstr. 15, 04318, Leipzig, Germany (E-mail: jaekel@tropos.de)

Jang, Yu-Lee, Dept. of Atmospheric Science, Yonsei University, 134 Sinchon-dong, Seodaemoon-gu, Seoul, Korea (E-mail: glass_822@naver.com)

Jhun, Jong-Ghap, School of Earth and Environmental Sciences, Seoul National University, San 56-1, Sillim-dong, Gwanak-gu, Seoul, Korea (E-mail: jgjhun@plaza.snu.ac.kr)

Je, Chang-Un, Dept. of Atmospheric Science, Kongju National University, 182 Shinkwan-dong, Kongju-city, ChungNam, Korea (E-mail: jecu7@yahoo.co.kr)

LIST OF PARTICIPANTS

Jee, Joon-Bum, Department of Atmospheric and Environmental Sciences, Kangnung National University, 123 Jibyeon-Dong, Kangnung-city, Kangwon, Korea (E-mail: chk@atmos.yonsei.ac.kr)

Jefferson, Anne, NOAA/CMDL, 325 Broadway, R/CMDL-1, Boulder, CO, 80305, USA (E-mail: anne.jefferson@noaa.gov)

Jeong, Eun-Joo, Department of Science Education, Ewha Womans University, 11-1 Daehyeon-dong, Seodaemun-ku Seoul, Korea (E-mail: jullyjeong79@hanmail.net)

Jeong, Myeong-Jae, Dept. of Meteorology, University of Maryland, 2207 CSS Bldg, College Park, MD, 20742, USA (E-mail: zli@atmos.umd.edu)

Jeong, Su-Jong, Seoul National University, 56 room 222, San 56-1, Sillim-dong, Gwanak-gu, Seoul, 151-742, Korea (E-mail: waterbell@cpl.snu.ac.kr)

Jo, Kwang-il, Dept. of Atmospheric Science, Kyungpook National University, 1370 Sankyuk-dong, buk-gu, Daegu, Korea (E-mail: kaienjo77@hanmail.net)

Kadobayushi, Shiori, Japan Aerospace Exploration Agency, 1-8-10, Harumi, Chuo-ku, 104-6023, Japan (E-mail: kadoba@eorc.jaxa.jp)

Kaiser, Johannes W., RSL, University of Zurich, Winterthurerstrasse 190, 8057 Zurich, Switzerland (E-mail: johannes@uni-bremen.de)

Kang, Jeon-Ho, Dept. of Atmospheric Science, Kongju National University, 182 Shinkwan-dong, Kongju-city, ChungNam, Korea (E-mail: jtiger02@kongju.ac.kr)

Kang, Sung-Dae, Meteorological Research Institute/KMA, 460-18, Shindaebang-dong, Tongjak-gu, Seoul, Korea (E-mail: sdkang@metri.re.kr)

Kassianov, Evgueni, Pacific Northwest National Laboratory, 902 Battele Boulevard, P.O. Box 999, Richland, WA, 99352, USA (E-mail: Evgueni.Kassianov@pnl.gov)

Katagiri, Shuichiro, Research Institute for Humanity and Nature, 335 Takashima-cho, Marutamachi-dori Kawaramachi nishi-iru, Kamigyo-ku, Kyoto, 602-0878, Japan (E-mail: katagiri@seg3rd.gr.jp)

Kato, Seiji, Hampton University, NASA Langley Research Center, Mail Stop 420, Hampton, VA, 23681-2199, USA (E-mail: s.kato@larc.nasa.gov)

Kawamoto, Kazuaki, Research Institute for Humanity and Nature, 335 Takashima-cho, Kamigyo-ku, Kyoto, 602-0878, Japan (E-mail: kawamoto@chikyu.ac.jp)

Kawata, Yoshiyuki, Kanazawa Institute of Technology, Ohgigaoka 7-1, Nonoichi, Ishikawa, 921-8501, Japan (E-mail: kawata@infor.kanazawa-it.ac.jp)

Kim, Bong-Geun, School of Earth and Environmental Sciences, Seoul National University, San 56-1, Sillim-dong, Gwanak-gu, Seoul, Korea (E-mail: kbgiskbg@hanmail.net)

Kim, Byong-Gon, Princeton University, NOAA/GFDL, Princeton University Forrestal Campus, Princeton, NJ, 08542, USA (E-mail: bgk@gfdl.noaa.gov)

Kim, Deok-Rae, Department of Atmospheric and Environmental Sciences, Kangnung National University, 123 Jibyeon-Dong, Kangnung-city, Kangwon, Korea (E-mail: kimdr78@hanmail.net)

Kim, Dongjoon, School of Earth and Environmental Sciences, Seoul National University, San 56-1, Sillim-dong, Gwanak-gu, Seoul, Korea (E-mail: djkim96@snu.ac.kr)

Kim, Eun-hee, Dept. of Atmospheric Science, Kongju National University, 182 Shinkwan-dong, Kongju-city, ChungNam, Korea (E-mail: cheaven@kongju.ac.kr)

Kim, Eun-Yun, METRI/KMA, 460-18, Shindaebang-dong, Tongjak-gu, Seoul, Korea (E-mail: eynhoklm@metri.re.kr)

Kim, Han-dol, KARI, COMS Program Office, 45 Eoeun-Dong, Yuseong-Gu, Daejeon, 305-333, Korea (E-mail: hkim@kari.re.kr)

Kim, Hyeong Seog, School of Earth and Environmental Sciences, Seoul National University, San 56-1, Sillim-dong, Gwanak-gu, Seoul, 151-742, Korea (E-mail: dol153@cpl.snu.ac.kr)

Kim, Hyung-Jin, Institute of Natural Sciences, Department of Atmospheric and Environmental Sciences, Kangnung National University, 123 Jibyeon-Dong, 210-702, Korea (E-mail: hjkim@kangnung.ac.kr)

Kim, Jae Hwan, Atmospheric Science Department, Pusan National University, 30 Jangjeon-dong, Geumjeong-gu, Busan, 609-735, Korea (E-mail: jaekim@pusan.ac.kr)

Kim, Jeong Eun, ADvanced Environmental Monitoring Research Center (ADEMRC) Kwangju Institute of Science and Technology (K-JIST), 1 Oryong-dong, Buk-gu, Korea (E-mail: jekim@kjist.ac.kr)

Kim, Jhoon, Yonsei University, Rm. 545, Dept. of Atmospheric Science, 134 Sinchon-dong, Seodaemoon-gu, 120-749, Korea (E-mail: jkim2@yonsei.ac.kr)
Kim, Jiyoung, School of Earth and Environmental Sciences, Seoul National University, San 56-1, Sillim-dong, Gwanak-gu, Seoul, Korea (E-mail: jykim2k2@snu.ac.kr)
Kim, Joo-Hong, School of Earth and Environmental Sciences, Seoul National University, San 56-1, Sillim-dong, Gwanak-gu, Seoul, Korea (E-mail: jhkim@cpl.snu.ac.kr)
Kim, Maeng-Ki, Kongju National University, Dept. of Atmospheric Science, Kongju National University, 182 Shinkwan-dong, Chungnam, 314-701, Korea (E-mail: mkkim@kongju.ac.kr)
Kim, Mee-Ja, Meteorological Research Institute/KMA, 460-18, Shindaebang-dong, Tongjak-gu, Seoul, Korea (E-mail: stat_mj@metri.re.kr)
Kim, Min-Jeong, University of Washington, Depart. of Atmospheric Sciences, University of Washington, Seattle, WA 98105, USA (E-mail: mjkim@atmos.washington.edu)
Kim, Sang-Woo, School of Earth and Environmental Sciences, Seoul National University, San 56-1, Sillim-dong, Gwanak-gu, Seoul, 151-742, Korea (E-mail: kimsw@air.snu.ac.kr)
Kim, So-Hee, Dept. of Atmospheric Science, Kongju National University, 182 Shinkwan-dong, Kongju-city, ChungNam, Korea (E-mail: green78@nate.com)
Kim, Won-sik, School of Earth and Environmental Sciences, Seoul National University, San 56-1, Sillim-dong, Gwanak-gu, Seoul, Korea (E-mail: wonsik@snu.ac.kr)
Kim, Yong-Seung, Korea Aerospace Research Institute (KARI), 45 Eoeun-Dong, Yuseong-Gu, Daejeon, Korea (E-mail: yskim@kari.re.kr)
Kim, Yoon-jae, Seoul National University, San 56-1, Sillim-dong, Gwanak-gu, Seoul, 151-742, Korea (E-mail: yoonjae@strat.snu.ac.kr)
Kim, Young-Mi, Department of Science Education, Ewha Womans University, 11-1 Daehyeon-dong, Seodaemun-ku Seoul, 120-750, Korea (E-mail: edicara@naver.com)
Kim, Young-Seup, Dept. of Environmental Atmospheric Sciences, Pukyong National University, 599-1 Daeyeon 3-Dong, Nam-Gu, Busan, Korea (E-mail: kimys@pknu.ac.kr)
King, Michael D., NASA Goddard Space Flight Center, Code 900, Greenbelt, MD, 20771, USA (E-mail: michael.d.king@nasa.gov)
Kinoshita, Kisei, Faculty of Education, Kagoshima University, Kagoshima, 90-0065, Japan (E-mail: kisei@edu.kagoshima-u.ac.jp)
Knapp, Kenneth, NOAA/NCDC, NCDC/RSAD, 151 Patton Ave., Asheville, NC, 28806, USA (E-mail: Ken.Knapp@noaa.gov)
Kniffka, Anke, Institute of Meteorology, Leipzig University, Stephanstr. 3, 04103, Leipzig, Germany (E-mail: kniffka@uni-leipzig.de)
Knuteson, Robert, University of Wisconsin-Madison, 1225 W. Dayton St., WI, 53706, USA (E-mail: robert.knuteson@ssec.wisc.edu)
Ko, Jung Woong, Meteorological Research Institute/KMA, 460-18, Shindaebang-dong, Tongjak-gu, Seoul, Korea (E-mail: kolee@metri.re.kr)
Kobayashi, Hiroshi, University of Yamanashi, 4-3-11, Takeda, Kofu, Yamanashi, 400-8511, Japan (E-mail: koba@js.yamanashi.ac.jp)
Kobayashi, Satoru, Atmospheric Radar Science & Engineering, NASA/Jet Propulsion Laboratory, California Institute of Technology, Pasadena, CA, 91109, USA (E-mail: satoru@radar-sci.jpl.nasa.gov)
Kobayashi, Takahisa, Meteorological Research Institute, 1-1, Nagamine, Tsukuba, Ibaraki, 305-0052, Japan (E-mail: kobay@mri-jma.go.jp)
Koepke, Peter, Meteorolog. Inst. Univ. Munich, Theresienstr. 37, 80333, Muenchen, Germany (E-mail: peter.koepke@lrz.uni-muenchen.de)
Kojima, Masaya, Center for Atmospheric and Oceanic Studies, Tohoku University, Aramaki Aza-Aoba Sendai, 980-8578, Japan (E-mail: kojima@caos-a.geophys.tohoku.ac.jp)
Koo, Gyo-Sook, Meteorological Research Institute/KMA, 460-18, Shindaebang-dong, Tongjak-gu, Seoul, Korea (E-mail: geogen@metri.re.kr)

LIST OF PARTICIPANTS

Kostsov, Vladimir, Saint-Petersburg State University, 1 Ulyanovskaya, 198504, Russia (E-mail: vlad@troll.phys.spbu.ru)

Kubota, Isao, Weathernews Inc., Japan (E-mail: ikubota@wni.com)

Kuji, Makoto, Nara Women's University, Japan, Kita-uoya Nishimachi, Nara, Japan (E-mail: makato@ics.nara-wu.ac.jp)

Kumagai, Hiroshi, National Institute of Information and Communications Technology, 4-2-1 NukuiKita, Koganei, Tokyo, 184-8795, Japan (E-mail: kumagai@crl.go.jp)

Kuo, Kwo-Sen, Caelum Research Corporation, NASA/GSFC-GPM Project Science (Code 912.1), Greenbelt, MD, 20771, USA (E-mail: kwo-sen.kuo@gsfc.nasa.gov)

Kwak, Chong-Heum, Dept. of Atmospheric Science, Kongju National University, 182 Shinkwan-dong, Kongju-city, ChungNam, 314-701, Korea (E-mail: chkwak@kongju.ac.kr)

Kwon, Byung-Hyuk, Dept. of Environmental Atmospheric Sciences, Pukyong National University, 599-1 Daeyeon 3-Dong, Nam-Gu, Busan, 608-737, Korea (E-mail: bhkwon@mail1.pknu.ac.kr)

Kwon, Eun-Han, School of Earth and Environmental Sciences, Seoul National University, San 56-1, Sillim-dong, Gwanak-gu, Seoul, Korea (E-mail: kwoneh@eosat.snu.ac.kr)

Kwon, Tae-Yong, Department of Atmospheric and Environmental Sciences, Kangnung National University, 123 Jibyeon-Dong, Kangnung-city, Kangwon, Korea (E-mail: tykwon@kangnung.ac.kr)

Kwon, Won-Tae, Meteorological Research Institute/KMA, 460-18, Shindaebang-dong, Tongjak-gu, Seoul, Korea (E-mail: wontk@metri.re.kr)

Landgraf, Jochen, Space Research Organization Netherlands, Sorbonnelaan 2, 3584 CA Utrecht, The Netherlands (E-mail: j.landgraf@sron.nl)

Larar, Allen, NASA Langley Research Center, 21 Langley Blvd., Mail Stop 401A, Hampton, VA, 23681, USA (E-mail: Allen.M.Larar@nasa.gov)

Laszlo, Istvan, National Oceanic and Atmospheric Administration, E/RA1, RM 711, WWBG, 5200, Auth Road, Camp Springs, MD, 20746-4304, USA (E-mail: Istvan.Laszlo@noaa.gov)

Lee, Byung-Il, School of Earth and Environmental Sciences, Seoul National University, San 56-1, Sillim-dong, Gwanak-gu, Seoul, Korea (E-mail: bilee@air.snu.ac.kr)

Lee, Dong-Kyou, School of Earth and Environmental Sciences, Seoul National University, San 56-1, Sillim-dong, Gwanak-gu, Seoul, Korea (E-mail: dklee@snu.ac.kr)

Lee, Eun-Joo, Kongju National University, 182 Shinkwan-dong, Kongju-city, ChungNam, Korea (E-mail: bigbangej1@hotmail.com)

Lee, Hui-kyo, School of Earth and Environmental Sciences, Seoul National University, San 56-1, Sillim-dong, Gwanak-gu, Seoul, Korea (E-mail: ns123@snu.ac.kr)

Lee, Hyun jin, Pusan National University, 30 Jangjeon-dong, Geumjeong-gu, Busan, 609-735, Korea (E-mail: hyunjin@pusan.ac.kr)

Lee, Jeongsoon, Satellite Technology Research Center, Korea Advanced Institute of Science and Technology, 373-1, Guseongdong, Yuseonggu, Daejun, 305-701, Korea (E-mail: jslee@satrec.kaist.ac.kr)

Lee, Jong-Eun, Dept. of Atmospheric Science, Kyungpook National University, 1370 Sankyuk-dong, buk-gu, Daegu, Korea (E-mail: jonga521@hanmail.net)

Lee, Joo-Hee, Korea Aerospace Research Institute (KARI), 45 Eoeun-Dong, Yuseong-Gu, Daejeon, 305-333, Korea (E-mail: jhl@kari.re.kr)

Lee, Jung Rim, Kongju National University, 182 Shinkwan-dong, Kongju-city, ChungNam, Korea (E-mail: jrlee@kongju.ac.kr)

Lee, Kwang-Mog, Dept. of Atmospheric Science, Kyungpook National University, 1370 Sankyuk-dong, buk-gu, Daegu, Korea (E-mail: kmlee@bh.knu.ac.kr)

Lee, Kwangjae, Korea Aerospace Research Institute, 45 Eoeun-dong, Yuseong-gu, Daejeon, 305-333, Korea (E-mail: yskim@kari.re.kr)

Lee, Kwon Ho, Advanced Environmental Monitoring Research Center, Kwangju Institute of Science & Technology, Oryong-dong, Buk-gu, Kwangju, 500-712, Korea (E-mail: envtech@kjist.ac.kr)

Lee, Kyu-Tae, Department of Atmospheric and Environmental Sciences, Kangnung National University, 123 Jibyeon-Dong, Kangnung-city, Kangwon, Korea (E-mail: ktlee@kangnung.ac.kr)

Lee, Myoung-Joo, Meteorological Research Institute/KMA, 460-18, Shindaebang-dong, Tongjak-gu, Seoul, Korea (E-mail: mjlee@metri.re.kr)
Lee, Sanghee, Korea Aerospace Research Institute, Remote Sensing Dep. 45 Eoeun-Dong, Youseong-Gu, Daejeon, 305-333, Korea (E-mail: sanghee@kari.re.kr)
Lee, Sung-Yung, Jet Propulsion Laboratory, 4800 Oak Grove Drive, Pasadena, CA, 91109, USA (E-mail: Sung-Yung.Lee@jpl.nasa.gov)
Lee, Tae Hee, Dept. of Atmospheric Science, Kongju National University, 182 Shinkwan-dong, Kongju-city, ChungNam, Korea (E-mail: pilot8157@empal.com)
Lee, Won-Hak, Department of Atmospheric and Environmental Sciences, Kangnung National University, 123 Jibyeon-Dong, Kangnung-city, Kangwon, Korea (E-mail: whlee@kangnung.ac.kr)
Lee, Woo Seop, Kongju National University, 182 Shinkwan-dong, Kongju-city, ChungNam, Korea (E-mail: wooseobi@kongju.ac.kr)
Lee, Yong-Keun, Texas A&M University, Department of Atmospheric Sciences TAMU 3150 Texas A&M University College Station, TX, 77843, USA (E-mail: yklee@ariel.met.tamu.edu)
Lee, Yun-Bok, School of Earth and Environmental Sciences, Seoul National University, San 56-1, Sillim-dong, Gwanak-gu, Seoul, Korea (E-mail: yblee95@cpl.snu.ac.kr)
Lee, Yun Gon, Global Environment Laboratory/Department of Atmospheric Sciences, Yonsei University, 134 Sinchon-dong, Seodaemoon-gu, Seoul, 120-749, Korea (E-mail: milogon@yonsei.ac.kr)
Legrand, Michel, LOA (UMR 8518), USTL, Villeneuve d'Ascq, 59655, France (E-mail: legrand@loa.univ-lille1.fr)
Leon, Jean-Francois, CNRS-Laboratoire d'Optique Atmospherique, Universite Lille 1, Cite scientifique BAT P5, Villeneuve d'Ascq cedex, 59655, France (E-mail: leon@loa.univ-lille1.fr)
Li, Jiangnan, Canadian Center for Climate and Analysis (CCCMA), PO Box 1700, University of Victoria, V8W 2Y2, Canada (E-mail: jiangnan.li@ec.gc.ca)
Li, Jun, University of Wisconsin-Madison, 1225 West Dayton Street, Madison, WI, 53706, USA (E-mail: Jun.Li@ssec.wisc.edu)
Li, Zhanqing, Dept. of Meteorology and ESSIC, University of Maryland, 2207 CSS Bldg, College Park, MD, 20742, USA (E-mail: zli@atmos.umd.edu)
Lim, Ju-Yeon, Meteorological Research Institute, 460-18, Shindaebang-dong, Tongjak-gu, Seoul, Korea (E-mail: rima@metri.re.kr)
Lim, Sanghun, Dept. of Electrical Engineering, Colorado State University, Fort Collins, CO, 80523-1373, USA (E-mail: dawnsky@engr.colostate.edu)
Liou, Kuo-Nan, University of California, Los Angeles, Department of Atmospheric Sciences and Institute of Radiation and Remote Sensing, Los Angeles, California, 90095, USA (E-mail: knliou@atmos.ucla.edu)
Liu, Guosheng, Dept. of Meteorology, Florida State University, Tallahassee, FL, 32306-4520, USA (E-mail: liug@met.fsu.edu)
Liu, Xu, NASA Langley Research Center, MS401, NASA Langley Research Center, Hampton, VA, 23681, USA (E-mail: Xu.Liu-1@nasa.gov)
Long, Chuck, Pacific Northwest National Laboratory, PO Box 999, MSIN: K9-24, Richland, WA, 99352, USA (E-mail: chuck.long@pnl.gov)
Lorente, Jeronimo, University of Barcelona, Dept. Astronomia i Meteorologia, Facultat de FISICA. C/ Martii Franques, 1, 08028 Barcelona, Spain (E-mail: jeroni@am.ub.es)
Lorenz, Anne, Institute of Hydrology and Meteorology, Technische University Dresden, Chemnitzerstr. 46b, D-01187 Dresden, Germany (E-mail: anne.lorenz@forst.tu-dresden.de)
Los, Alexander, Kipp & Zonen BV, Roentgenweg 1, 2624 BD Delft, The Netherlands (E-mail: alexander.los@kippzonen.com)
Luo, Yali, National Institute of Aerospace, 144 Research Drive, Hampton, VA, 23666, USA (E-mail: yali@nianet.org)
Luo, Yi, Canada Centre for Remote Sensing, 588 Booth Street, Ottawa, Ontario, K1A 0Y7, Canada (E-mail: yluo@nrcan.gc.ca)
Lutsko, Lydmila, Voejkov Main Geophysical Observatory, St. Petersburg, 194021, Russia (E-mail: lutsko@main.mgo.rssi.ru)

LIST OF PARTICIPANTS

Ma, Qiancheng, NASA/GISS, 2880 Broadway, New York, NY, 10025, USA (E-mail: crqxm@giss.nasa.gov)
Mace, Gerald, University of Utah, Salt Lake City, Utah, 84112-0110, USA (E-mail: mace@met.utah.edu)
Marshak, Alexander, NASA/GSFC, code 913, Greenbelt, MD, 20771, USA (E-mail: Alexander.Marshak@nasa.gov)
Martins, J. Vanderlei, University of Maryland Baltimore County/NASA GSFC, code 913 NASA Goddard Space Flight Center, Greenbelt, MD, 20770, USA (E-mail: martins@climate.gsfc.nasa.gov)
Masuda, Kooiti, Frontier Research System for Global Change, 3173-25 Showa, Kanazawa-ku, Yokohama, 236-0001, Japan (E-mail: masuda@jamstec.go.jp)
Masunaga, Hirohiko, Dept. of Atmospheric Science, Colorado State Univ., Fort Collins, CO, 80523, USA (E-mail: masunaga@atmos.colostate.edu)
Matheson, Mark A., Oregon State University, 104 COAS Admin Building, OSU, Corvallis, OR, 97331, USA (Email: mmatheso@coas.oregonstate.edu)
Matrosov, Sergey Y., CIRES, University of Colorado and NOAA Environmental Technology Lab., R/ET7, 325 Broadway, Boulder, CO, 80305, USA (E-mail: Sergey.Matrosov@noaa.gov)
Mattoo, Shana, NASA Goddard Space Flight Facility, Greenbelt, MD, 20771, USA (E-mail: mattoo@climate.gsfc.nasa.gov)
Mayer, Bernhard, Deutsches Zentrum fuer Luft- und Raumfahrt (DLR), Oberpfaffenhofen, Wessling, 82234, Germany (E-mail: bernhard.mayer@dlr.de)
McFarquhar, Greg, Department of Atmospheric Sciences, University of Illinois, 105 S. Gregory Street, Urbana, IL, 61801-3070, USA (E-mail: mcfarq@atmos.uiuc.edu)
Moon, Kyung-Jung, Dept. of Atmospheric Science, Yonsei University, 134 Sinchon-dong, Seodaemoon-gu, Seoul, Korea (E-mail: kjmoon3@freechal.com)
Mugnai, Alberto, Istituto di Scienze dell'Atmosfera e del Clima, Consiglio Nazionale delle Ricerche, Via del Fosso del Cavaliere, 100, 00133 Roma, Italy (E-mail: a.mugnai@isac.cnr.it)
Mukai, Makiko, Center for Climate System Research, University of Tokyo 4-6-1 Komaba, Meguro-ku, Tokyo, 153-8904, Japan (E-mail: mukai@ccsr.u-tokyo.ac.jp)
Na, Won-Ju, School of Earth and Environmental Sciences, Seoul National University, San 56-1, Sillim-dong, Gwanak-gu, Seoul, Korea (E-mail: nawj@eosat.snu.ac.kr)
Nakajima, Hideaki, National Institute for Environmental Studies, 16-2 Onogawa, Tsukuba, 305-8506, Japan (E-mail: hide@nies.go.jp)
Nakajima, Takashi, Japan Aerospace Exploration Agency, 1-8-10, Harumi, Chuo-ku, 104-6032, Japan (E-mail: nakajima@eorc.jaxa.jp)
Nakajima, Teruyuki, Center for Climate System Research, University of Tokyo, 4-6-1 Komaba, Meguro-ku, Tokyo, 153-8904, Japan (E-mail: teruyuki@ccsr.u-tokyo.ac.jp)
Nakamura, Kenji, Hydrospheric Atmospheric Research Center, Nagoya University, Furocho, Chikusaku, Naogya, 464-8601, Japan (E-mail: nakamura@hyarc.nagoya-u.ac.jp)
Nam, Jae-Cheol, Meteorological Research Institute/KMA, 460-18, Shindaebang-dong, Tongjak-gu, Seoul, Korea (E-mail: jcnam@metri.re.kr)
Neshyba, Steven, University of Puget Sound, CMB 1015, 1500 N. Warner, Tacoma, WA, 98416, USA (E-mail: nesh@ups.edu)
Nishizawa, Tomoaki, Center for Atmospheric and Oceanic Studies, Tohoku University, Aoba, Aramakiaza, Aoba, Sendai, 980-8578, Japan (E-mail: nisizawa@caos-a.geophys.tohoku.ac.jp)
Nobuta, Koji, Fujitsu FIP corporation, Time 24 Building, 2-45 Aomi, Koto-ku, Tokyo, 135-8686, Japan (E-mail: nobuta@ilas2.nies.go.jp)
Noda, Akira, Graduate School of Science, Tohoku University, Aramaki, Aoba, Sendai, Miyagi, 980-8578, Japan (E-mail: noda@wind.geophys.tohoku.ac.jp)
Noh, Yoojeong, Department of Meteorology, Florida State University, Tallahassee, FL, 32306-4520, USA (E-mail: yjnoh@met.fsu.edu)
Oh, Hyun-Jong, Meteorological Satellite Center, KMA, 460-18, Shindaebang-dong, Tongjak-gu, Seoul, 150-720, Korea (E-mail: ohj@kma.go.kr)
Oh, Sung-Nam, National Research Laboratory, Meteorological Research Institute, Korea Meteorological Administration, 460-18, Shindaebang-dong, Tongjak-gu, Seoul, 150-720, Korea (E-mail: snoh@metri.re.kr)

Oh, Tae-Hyeong, Meteorological Research Institute, 460-18, Shindaebang-dong, Tongjak-gu, Seoul, Korea
(E-mail: hyoung0203@metri.re.kr)
Ohmura, Atsumu, Institute for Atmospheric and Climate Science, ETH, Winterthurerstrasse 190, Zurich, CH-8057, Switzerland (E-mail: ohmura@env.ethz.ch)
Otake, Shinichi, Tohoku University, Aramaki, Aoba, Sendai, Miyagi, 980-8578, Japan
(E-mail: ethtake@caos-a.geophys.tohoku.ac.jp)
Okamoto, Hajime, Center for Atmospheric and Oceanic Studies, Tohoku University, Aramakiaza, Aoba, Sendai, 980-8578, Japan (E-mail: okamoto@caos-a.geophys.tohoku.ac.jp)
Okamoto, Ken'ichi, Dept. of Aerospace Engineering, Osaka Prefecture University, 1-1, Gakuen-cho, Sakai, Osaka, 599-8531, Japan (E-mail: okamoto@aero.osakafu-u.ac.jp)
Olson, William S., Joint Center for Earth Systems Technology/UMBC, NASA/Goddard Space Flight Center, Code 912, Greenbelt, MD, 20771, USA (E-mail: olson@agnes.gsfc.nasa.gov)
Omar, Ali, NASA Langley Research Center, Atmospheric Sciences, MS 401A, Hampton, VA, 23681, USA
(E-mail: ali.h.omar@nasa.gov)
Ota, Yoshifumi, Center for Climate System Research, The University of Tokyo, 4-6-1 Komaba, Meguro-ku, Tokyo, 153-8904, Japan (E-mail: ota@ccsr.u-tokyo.ac.jp)
Ou, Mi-Lim, Remote Sensing Research Laboratory, Meteorological Research Institute, Korean Meteorological Administration, 460-18, Shindaebang-dong, Dongjak-gu, Seoul, 156-720, Korea (E-mail: milim@kma.go.kr)
Park, Chang-Geun, Forecast Research Lab., 460-18, Shindaebang-dong, Tongjak-gu, Seoul, Korea
(E-mail: sphere95@metri.re.kr)
Park, E-Hyung, METRI/KMA, 460-18, Shindaebang-dong, Tongjak-gu, Seoul, Korea (E-mail: ehpark@metri.re.kr)
Park, Ho-sun, School of Earth and Environmental Sciences, Seoul National University, San 56-1, Sillim-dong, Gwanak-gu, Seoul, Korea (E-mail: parkhsm@eosat.snu.ac.kr)
Park, Hye-Sook, School of Earth and Environmental Sciences, Seoul National University, San 56-1, Sillim-dong, Gwanak-gu, Seoul, Korea (E-mail: parkhsf@eosat.snu.ac.kr)
Park, Jae Hyung, School of Earth and Environmental Sciences, Seoul National University, San 56-1, Sillim-dong, Gwanak-gu, Seoul, Korea (E-mail: j.h.park@strat.snu.ac.kr)
Park, Myung-Sook, School of Earth and Environmental Sciences, Seoul National University, San 56-1, Sillim-dong, Gwanak-gu, Seoul, Korea (E-mail: jina@cpl.snu.ac.kr)
Park, Sang-Min, Dept. of Atmospheric Science, Kyungpook National University, 1370 Sankyuk-dong, buk-gu, Daegu, Korea (E-mail: psm6420@lycos.co.kr)
Park, Seong-Chan, School of Earth and Environmental Sciences, Seoul National University, San 56-1, Sillim-dong, Gwanak-gu, Seoul, Korea (E-mail: parksc@eosat.snu.ac.kr)
Park, Seung-Hwan, Dept. of Atmospheric Science, Yonsei University, 134 Sinchon-dong, Seodaemoon-gu, Seoul, Korea (E-mail: parze@hanmail.net)
Park, Su-hee, Climate Research Laboratory/METRI, 460-18, Shindaebang-dong, Tongjak-gu, Seoul, Korea
(E-mail: suhee@metri.re.kr)
Park, Sung-Hee, School of Earth and Environmental Sciences, Seoul National University, San 56-1, Sillim-dong, Gwanak-gu, Seoul, Korea (E-mail: parksh@eosat.snu.ac.kr)
Park, Tae-Won, School of Earth and Environmental Sciences, Seoul National University, San 56-1, Sillim-dong, Gwanak-gu, Seoul, Korea (E-mail: park2760@cpl.snu.ac.kr)
Philipona, Rolf, Observatory Davos, Dorfstrasse 33, Davos Dorf, 7260, Switzerland (E-mail: r.philipona@pmodwrc.ch)
Pincus, Robert, NOAA-CIRES Climate Diagnostics Center, 325 Broadway, R/CDC1; Boulder, CO, 80305, USA
(E-mail: Robert.Pincus@colorado.edu)
Postylyakov, Oleg, Institute of Atmospheric Physics, RAS, Pyzhevsky per.3, Moscow, 109017, Russia
(E-mail: oleg-962@yandex.ru)
Ramaswamy, V., NOAA/GFDL, P.O. Box 308 Princeton, NJ, 08542, USA (E-mail: V.Ramaswamy@noaa.gov)
Revercomb, Henry, University of Wisconsin, Space Science and Engineering Center, 1225 West Dayton Street, Madison, WI, 53706, USA (E-mail: hankr@ssec.wisc.edu)
Rho, Chae-Shik, National Academy of Science, Korea (csrho@yahoo.co.kr)

LIST OF PARTICIPANTS

Rosenfeld, Daniel, The Hebrew University of Jerusalem, Institute of Earth Sciences, The Hebrew University, Jerusalem, 91904, Israel (E-mail: daniel.rosenfeld@huji.ac.il)

Rowe, Penny, Department of Chemistry, Box 351700, University of Washington, Seattle, WA, 98195-1700, USA (Email: prowe@u.washington.edu)

Russell, J. E., Imperial College Space and Atmospheric Physics, Blackett Lab., Prince Consort Road, London SW7 2BW, UK (E-mail: j.e.russell@imperial.ac.uk)

Ryu, Geun-Hyeok, School of Earth and Environmental Sciences, Seoul National University, San 56-1, Sillim-dong, Gwanak-gu, Seoul, Korea (E-mail: ryukh@eosat.snu.ac.kr)

Sano, Itaru, Kinki University, Kowakae 3-4-1, Higashi-Osaka, 577-8502, Japan (E-mail: sano@info.kindai.ac.jp)

Sato, Kaori, Center for Atmospheric and Oceanic Studies, Tohoku University, Aramaki Aza-Aoba Sendai, 980-8578, Japan (E-mail: ksato@caos-a.geophys.tohoku.ac.jp)

Scheirer, Ronald, Institut fuer Physik der Atmosphaere, German Aerospace Center, Oberpfaffenhofen, 82234, Germany (E-mail: ronald.scheirer@dlr.de)

Schmetz, Johannes, EUMETSAT, Am Kavalleriesand 31, 64295, Germany (E-mail: Schmetz@eumetsat.de)

Schulz, Jörg, Deutscher Wetterdienst (German Weather Service), Satellite Application Facility on Climate Monitoring, Dept. of Climate and Environment, P.O. Box 10 04 65, 63004 Offenbach, Germany (E-mail: joerg.schulz@dwd.de)

Schwarzkopf, M. Daniel, NOAA/GFDL, P.O. Box 308 Princeton, NJ, 08542, USA (E-mail: Dan.Schwarzkopf@noaa.gov)

Schween, Jan H., Meteorological Institute, Unversity of Munich, Theresienstrasse 37, 80333, Muenchen, Germany (E-mail: jan.schween@lrz-uni.muenchen.de)

Seo, Eun-Jin, Meteorological Research Institute/KMA, 460-18, Shindaebang-dong, Tongjak-gu, Seoul, Korea (E-mail: sej1004@kma.go.kr)

Seo, Won-Ick, Meteorological Research Institute/KMA, 460-18, Shindaebang-dong, Tongjak-gu, Seoul, Korea (E-mail: wiseo@metri.re.kr)

Serio, Carmine, University of Basilicata, DIFA, Via dell'Ateneo Lucano 10, 85100, Italy (E-mail: serio@unibas.it)

Shimizu, Atsushi, National Institute for Environmental Studies, 16-2 Onogawa, Tsukuba, 305-8506, Japan (E-mail: shimizua@nies.go.jp)

Shin, In-Chul, Meteorological Research Institute/KMA, 460-18, Shindaebang-dong, Tongjak-gu, Seoul, Korea (E-mail: icshin98@metri.re.kr)

Smith, Eric A., GPM Project Science/Laboratory for Atmospheres, NASA/Goddard Space Flight Center, Code 912.1, Greenbelt, MD, 20771, USA (E-mail: eric.a.smith@nasa.gov)

Smith, William Leo, NASA Langley Research Center (retired), 117 Creek Circle, Seaford, VA, 23696, USA (E-mail: bill.l.smith@cox.net)

So, Chul-Whan, Dept. of Atmospheric Science, Yonsei University, 134 Sinchon-dong, Seodaemoon-gu, Seoul, Korea (E-mail: cwso@hanmir.com)

Sohn, Byung-Ju, School of Earth and Environmental Sciences, Seoul National University, San 56-1, Sillim-dong, Gwanak-gu, Seoul, 151-747, Korea (E-mail: sohn@snu.ac.kr)

Sohn, Eun-Ha, Meteorological Research Institute/KMA, 460-18, Shindaebang-dong, Tongjak-gu, Seoul, Korea (E-mail: icshin98@metri.re.kr)

Song, Byung-hyun, METRI/KMA, 460-18, Shindaebang-dong, Tongjak-gu, Seoul, Korea (E-mail: song@kma.go.kr)

Song, Won-Young, School of Earth and Environmental Sciences, Seoul National University, San 56-1, Sillim-dong, Gwanak-gu, Seoul, Korea (E-mail: wysong@strat.snu.ac.kr)

Spada, Francesco, IMAU - Utrecht University, Princetonplein 5 – Utrecht, 3584CC, The Netherlands (E-mail: f.spada@phys.uu.nl)

Spinhirne, James D., NASA, Goddard SFC/912, 20769, USA (E-mail: f.spada@phys.uu.nl)

Stammes, Piet, KNMI - Royal Netherlands Meteorological Institute, P.O. Box 201, De Bilt, 3730 AE, The Netherlands (E-mail: stammes@knmi.nl)

Stamnes, Knut, Department of Physics and Engineering Physics, Stevens Institute of Technology, Castle Point on Hudson, Hoboken, NJ, 07030, USA (E-mail: kstamnes@stevens.edu)

Sugita, Takafumi, National Institute for Environmental Studies, 16-2 Onogawa, Tsukuba, 305-8506, Japan (E-mail: tsugita@nies.go.jp)

Suh, Myoung-Seok, Kongju National University, 182 Shinkwan-dong, Kongju-city, ChungNam, Korea
 (E-mail: sms416@kongju.ac.kr)
Suzuki, Kentaroh, Center for Climate System Research, University of Tokyo 4-6-1 Komaba, Meguro-ku, Tokyo, 153-8904, Japan (E-mail: kenta@ccsr.u-tokyo.ac.jp)
Takara, Ezra E., Department of Meteorology, Florida State University, Tallahassee, Florida, 32306-4520, USA
 (E-mail: etakara@met.fsu.edu)
Takemura, Toshihiko, Research Institute for Applied Mechanics, Kyushu University, 6-1 Kasuga-koen, Kasuga, Fukuoka, 816-8580, Japan (E-mail: toshi@riam.kyushu-u.ac.jp)
Tanikawa, Tomonori, Graduate School of Life and Environmental Sciences, University of Tsukuba, 1-1-1, Tennoudai, Tsukuba, Ibaraki, 305-8572, Japan (E-mail: tanikawa@atm.geo.tsukuba.ac.jp)
Tipping, Richard, Department of Physics and Astronomy, University of Alabama, Tuscaloosa, AL, 35487-0324, USA
 (E-mail: rtipping@bama.ua.edu)
Town, Michael S., Depart. Atmos. Sci., University of Washington, ATG Bldg. Rm. 408, Box 351640, Seattle, WA, 98195, USA (E-mail: mstown@u.washington.edu)
Trishchenko, Alexander, Canada Centre for Remote Sensing, Natural Resources Canada, 588 Booth Str, Ottawa, Ontario, K1A 0Y7, Canada (E-mail: trichtch@ccrs.nrcan.gc.ca)
Tsay, Si-Chee, NASA Goddard Space Flight Center, Code 913, Greenbelt, Maryland, 20771, USA
 (E-mail: si-chee.tsay-1@nasa.gov)
Tsushima, Yoko, Frontier Research System for Global Change, 3173-25 Showamachi, Kanazawa-ku, Yokohama City, Kanagawa, 236-0001, Japan (E-mail: tsussi@jamstec.go.jp)
Turner, David, Pacific Northwest National Laboratory, Climate Physics Group, MS-K9-24, PO Box 999, Richland, WA, 99352, USA (E-mail: dave.turner@pnl.gov)
Uchida, Kiyotaka, EKO Instruments Co., Ltd., Sasazuka Center Bldg. 2-1-6, Sasazuka Shibuya-ku, Tokyo 151-0073, Japan
 (E-mail: info@eko.co.jp)
Uchida, Shinichi, Japan Aerospace Exploration Agency, 1-8-10, Harumi, Chuo-ku, 104-6023, Japan
 (E-mail: uchidas@eorc.jaxa.jp)
Uchiyama, Akihiro, JMA Meteorological Research Institute, 1-1 Nagamine, Tsukuba, Ibaraki, 305-0052, Japan
 (E-mail: uchiyama@mri-jma.go.jp)
Ueno, Naho, Center for Climate System Research, University of Tokyo, 4-6-1 Komaba, Meguro-ku, Tokyo, 153-8904, Japan
 (E-mail: ueno@ccsr.u-tokyo.ac.jp)
Um, Junshik, University of Illinois at Urbana-Champaign, 105 S. Gregory Street, Urbana, IL, 61801, USA
 (E-mail: junum@atmos.uiuc.edu)
van Deelen, Rutger, National Institute for Space Research (SRON), Sorbonnelaan 2, 3584 CA Utrecht, The Netherlands
 (E-mail: R.van.Deelen@sron.nl)
van Diedenhoven, Bastiaan, National Institute for Space Research (SRON), Sorbonnelaan 2, 3584 CA Utrecht, The Netherlands (E-mail: b.van.diedenhoven@sron.nl)
van Gent, Jeroen, KNMI Royal Netherlands Meteorological Institute, P.O. Box 201 De Bilt, NL-3730 AE, The Netherlands
 (E-mail: Jeroen.van.Gent@knmi.nl)
Varghese, Saji, Tata Institute of Fundamental Research, TIFR Centre, Indian Institute of Science Campus, Bangalore, 560 012, India (E-mail: saji@math.tifrbng.res.in)
Viollier, Michel, LMD/IPSL/CNRS, ECOLE POLYTECHNIQUE, 91128 Palaiseau Cedex, France
 (E-mail: viollier@lmd.polytechnique.fr)
Walden, Von P., University of Idaho Department of Geography, Moscow, Idaho, 83844-3021, USA
 (E-mail: vonw@uidaho.edu)
Walter, Holger, National Institute of Space Research (SRON), Sorbonnelaan 2, 3584 CA Utrecht, The Netherlands
 (E-mail: H.Walter@sron.nl)
Weber, Sebastian, University of Leipzig, Stephanstr. 3, Leipzig, 04103, Germany (E-mail: mai99jty@studserv.uni-leipzig.de)
Wielicki, Bruce Anthony, NASA Langley Research Center, Mail Stop 420, NASA Langley Research Center, Hampton, VA, 23681, USA (E-mail: b.q.wielicki@nasa.gov)

LIST OF PARTICIPANTS

Won, Jae-Gwang, Seoul National University, Shilim 9-dong, Kwanak-gu, Seoul, 151-747, Korea
(E-mail: wonjg@air.snu.ac.kr)
Xu, Kuan-Man, NASA Langley Research Center, Mail Stop 420, Hampton, VA, 23681, USA
(E-mail: Kuan-Man.Xu@nasa.gov)
Yamazaki, Akihiro, JMA Meteorological Research Institute, 1-1 Nagamine, Tsukuba, Ibaraki 305-0052, Japan
(E-mail: akyamaza@mri-jma.go.jp)
Yang, Koon-Ho, Korea Aerospace Research Institute, 45 Eoeun-Dong, Youseong-Gu, Daejeon, 305-333, Korea
(E-mail: khyanf@kari.re.kr)
Yang, Ping, Texas A&M University, 3150 TAMU, College Station, Texas, TX, 77843, USA
(E-mail: pyang@ariel.met.tamu.edu)
Yang, Song, George Mason University, NASA/GSFC, Code 912.1, Greenbelt, MD, 20771, USA
(E-mail: ysong@agnes.gsfc.nasa.gov)
Yang, Taifeng, Laboratoire de Météorologie Dynamique, Ecole Polytechnique, Palaiseau, 91128, France
(E-mail: yang@lmd.polytechnique.fr)
Yokota, Tatsuya, National Institute for Environmental Studies, 16-2 Onogawa, Tsukuba, Ibaraki, 305-8506, Japan
(E-mail: yoko@nies.go.jp)
Yong, Sang-Soon, Korea Aerospace Research Institute, 45 Eoeun-Dong, Youseong-Gu, Daejeon, 305-333, Korea
(E-mail: ssyong@kari.re.kr)
Yoo, Joo-Young, Department of Science Education, Ewha Womans University, 11-1 Daehyeon-dong, Seodaemun-ku Seoul, Korea (E-mail: wndud11@hotmail.com)
Yoo, Jung-Moon, Ewha Womans University, Department of Science Education, Ewha Womans University, 11-1 Daehyeon-dong, Seodaemun-ku Seoul, 120-750, Korea (E-mail: yjm@mm.ewha.ac.kr)
Yoon, Jong-Min, Dept. of Atmospheric Science, Yonsei University, 134 Sinchon-dong, Seodaemoon-gu, Seoul, Korea
(E-mail: cromx2@yonsei.ac.kr)
Yoon, Mi-Young, Department of Science Education, Ewha Womans University, 11-1 Daehyeon-dong, Seodaemun-ku Seoul, Korea (E-mail: arrow46@daum.net)
Yoon, Soon-Chang, School of Earth and Environmental Sciences, Seoul National University, San 56-1, Sillim-dong, Gwanak-gu, Seoul, Korea (E-mail: yoon@snu.ac.kr)
Yoshida, Yukio, Center for Atmospheric and Oceanic Studies, Tohoku University, Aramaki Aza-Aoba Sendai, 980-8578, Japan (E-mail: yyoshida@caos-a.geophys.tohoku.ac.jp)
Youn, Dae-ok, School of Earth and Environmental Sciences, Seoul National University, San 56-1, Sillim-dong, Gwanak-gu, Seoul, Korea (E-mail: ydo@strat.snu.ac.kr)
Yun, Ja-young, Kongju National University, 182 Shinkwan-dong, Kongju-city, Chung Nam, Korea
(E-mail: jyoungs@kongju.ac.kr)
Zavyalov, Vladimir, Space Dynamic Laboratory, Department of Physics, Utah State University, Logan, UT, 84322, USA
(E-mail: zavyal@cc.usu.edu)
Zhang, Taiping, AS&M/NASA Langley Research Center, Mail Stop 936, Hampton, VA, 23681-2199, USA
(E-mail: t.zhang@larc.nasa.gov)
Zhang, Yuying, University of Utah, Rm819 WBB, 135S 1460E, University of Utah, Salt Lake City, Utah, 84112-0110, USA
(E-mail: zyuying@met.utah.edu)
Zhao, Xuepeng, QSS Group Inc. & NOAA/NESDIS/ORA, E/RA1, RM 7121, WWBG, 5200 Auth Rd., MD, 20746-4304, USA (E-mail: Xuepeng.Zhao@noaa.gov)
Zhou, Daniel, NASA Langley Research Center, MS 401A, Hampton, VA, 23681, USA (E-mail: daniel.k.zhou@nasa.gov)
Zhuravleva, Tatiana Borisovna, Institute of Atmospheric Optics, SB RAS, 1, Akademicheskii ave, Tomsk, 634055, Russia
(E-mail: ztb@iao.ru)

EXPLORING AEROSOL-CLOUD-CLIMATE INTERACTION MECHANISMS USING THE NEW GENERATION OF EARTH OBSERVATION SYSTEM DATA

Zhanqing Li and Tianle Yuan
Department of Meteorology and ESSIC
University of Maryland, College Park, 20742, MD, USA
Email: zli@atmos.umd.edu

ABSTRACT

Since the Twomey effect was proposed in 1977, it has been a common belief that cloud particle size only decreases with aerosol loading. Using NASA's MODIS products, an opposite trend is found, together with a general finding that cloud particle size may increase or decrease with aerosol loading depending on water vapor supply, cloud regime and atmospheric circulation. Convective clouds formed in moist regions show an overwhelmingly positive dependence, as opposed to the negative dependence for stratiform clouds in water-limited regions. The slope of the dependence is driven primarily by water vapor amount that explains 70% of the variance. A new mechanism is proposed to explain this new effect. Consideration of this effect in climate models could lower the modeled estimation of aerosol indirect forcing to agree better with the estimations constrained by the global temperature records. The effect may also help reconcile controversial findings concerning if pollutants suppress or enhance precipitation: suppression in dry regions while enhancement in humid regions. This might offer a new explanation for the "south flood and north drought" pattern in Asia.

1. INTRODUCTION

Aerosol indirect effects (AIE) refer to any aerosol-induced alternations of cloud microphysics, cloud duration, precipitation, etc. The largest of all the uncertainties about global climate forcing is probably the indirect effect of aerosols on clouds which can alter both the energy and water cycles (Ramaswamy et al., 2001; Ramanathan et al., 2001). While different types of AIE have been proposed (Twomey, 1977; Albrecht, 1989; Hansen, et al. 1997; Kaufman and Fraser, 1997; Rosenfeld, 2000; Andreae et al., 2004; Koren et al., 2004), almost are rooted to a fundamental change in cloud droplet size by aerosol. Twomey hypothesized (Twomey, 1977) that cloud particle size is reduced by adding more CCN for fixed liquid water amount and it has been referred as the theoretical foundation of the observed AIEs, either using satellite, ground or in-situ measurements (Leaitch et al., 1996; Nakajima et al., 2001; Breon et al., 2002; Schwartz et al., 2002; Feingold et al., 2003; Kim et al., 2003; Penner et al., 2004). However, the rate of change in cloud droplet size with aerosol varies considerably from one study to another (Nakajima et al., 2001; Breon et al., 2002) and from the same study but for different cases (Feingold et al., 2003). Despite the differences in methods and/or data used, natural variation is anticipated. Because of the uncertainties in AIE, aerosol indirect forcing (AIF) estimations, the difference between radiation budgets with and without AIE, show large disparities. (Ramaswamy et al., 2001; Anderson, et al, 2003; Charlson et al., 1992; Hansen et al., 2001; Menon et al., 2002; Ghan et al., 2001; Harvey and Kaufmann, 2002; Knutti et al., 2002; Forest et al., 2002). Numbers from -0.7 to -4.4Wm^{-2} have been reported using either forward (Charlson et al., 1992; Hansen et al., 2001; Menon et al., 2002; Ghan et al., 2001, Takemura et al. 2005) or inversion (Harvey and Kaufmann, 2002; Knutti et al., 2002; Forest et al., 2002) methods. Several explanations (Lohmann and Lesins, 2002; Liu and Daum, 2002; Rotstayn and Liu, 2003) for the discrepancies have been proposed, yet the gap is to be closed.

In this investigation, we will focus on a new target of locally developed cumulus clouds over humid and warm regions of Southwest US. Implications and analysis of our new discovered results will also be discussed.

2. A NEW DEPENDENCE OF CLOUD DROPLET SIZE ON AEROSOLS LOADING

It's well known that the Bermuda high system dominates the Gulf of Mexico in the summer. This high pressure system brings a large amount of moisture (Fig. 1) from the sea together with aerosols to the land and causes many atmospheric activities. Locally generated cumuli are visible everyday from satellite imagery. This feature gives us an opportunity to study cloud-aerosol interaction repeatedly and definitely because of the rate of the occurrence and the localness of the cloud generation.

We use MODIS 10-km resolution aerosol optical thickness (AOT) data together with MODIS cloud products to correlate cloud droplet effective radius (DER) with AOT nearby, no further than 10-km, given the unique capability of MODIS to retrieval both quantities over land. Other MODIS parameters, liquid water path (LWP), precipitable water (PW) and cloud optical depth (COD), are also employed.

A target region is selected using a MODIS reflectance image to include a patch of broken cumulus clouds with any shape. It is chosen such that its area coverage is large enough to include a statistically meaningful set of pixels, but can assure a relatively homogenous dynamic and thermodynamic condition at the same time. A target region usually encompasses an area of order of 1000 km^2.

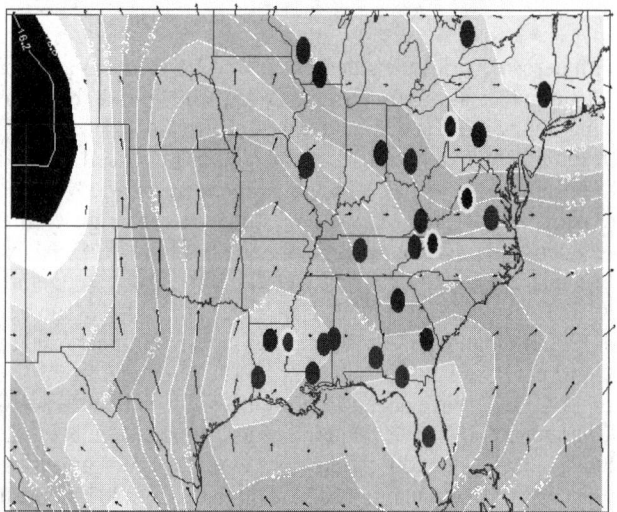

Fig. 1. Shaded contours are for precipitable water; every dot represents a case selected; arrows are wind directions from NCEP reanalysis; the color of a dot represents the slope of DER-AOT regression line of a case, red for large positive slope, blue for negative slope, intermediate values are given in other colors.

For all the cloudy pixels selected, a set of screening criteria is applied to throw out pixels not suited for our purpose. PW, which is column-integrated water vapor amount, is retrieved for both cloudy and clear pixels using 0.93 μm channel. To represent the environmental conditions of water vapor, only clear pixel values are used. MODIS cloud, PW and aerosol products are of different resolutions, 1-km for cloud and PW products and 10-km for aerosol product. PW and cloud products are then averaged into 10-km resolution data for clear and cloudy pixels respectively. It is worthy of noting that channels used to retrieve each of these products are different, which makes them independent data sets.

Given the similar environmental conditions for each case selected, we plot (Fig. 2) DER against AOT as a means of expressing AIE, which is widely used. In a plot, each dot represents a 10km by 10km pixel inside which we have one retrieval for DER and one for AOT. Each panel in Fig. 2 represents one of more than 40 cases we select in US. The slope of DER-AOT plots, which denotes how much change of DER can be brought about with a unit of change in AOT, varies, and most cases have positive values contrary to previous findings where slopes are negative. Our results show a large dynamic range of slope values from negative, neutral to positive, and majority of them are of positive values. This sharp difference is to be explained.

The slope varies greatly even for the same area. We decide there must be some factor(s) that drives the changes of slopes, which is a result of complex aerosol-cloud interaction process. Precipitable water derived from MODIS turns out to be one dominant factor among all available variables we examined. As plotted in Fig. 3, the slopes of DER-AOT for all the cases correlate strongly with corresponding PW values. The correlation coefficient is 0.83, which means about 70% of the variance in the slopes is explained by PW. Cloud droplet sizes can be increased with aerosol loading where moisture is abundant!

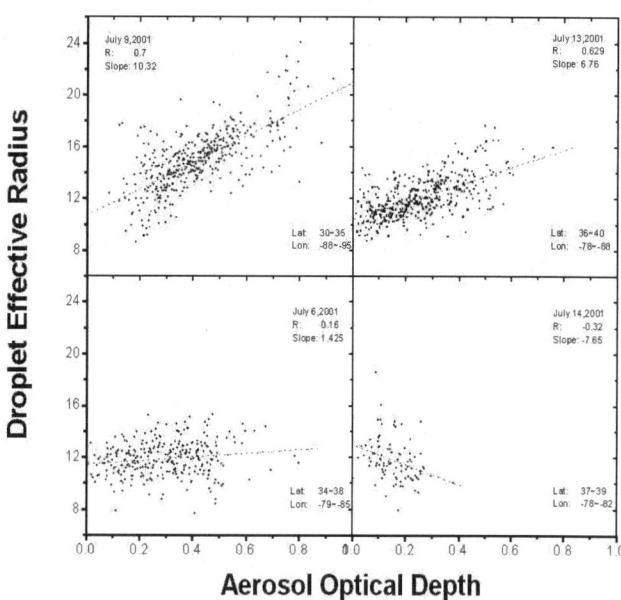

Fig. 2. Four representative cases; slopes vary from negative, neutral to positive.

3. ANALYSIS OF POSSIBLE SATELLITE RETRIEVAL PITFALLS

Quantities derived from satellite measurements are influenced by fractional cloud, viewing geometry effects and so on. Errors along however, do not warrant a concern. The real concern is the possibility that the two quantities used to correlate each other are connected to a third factor by certain satellite retrieval problem. We consider four potential 'third factors': cloud contamination, aerosol swelling effect, viewing geometry effect and cloud dynamics effect. Fractional clouds could contaminate both aerosol and cloud retrievals so that AOT and DER appear high or low simultaneously everywhere; Aerosols could swell more at a more moist region where cloud may develop deeper, which could form another false correlation; Due to 3-D effect, pixels with larger viewing zenith angles or relative azimuth angles may make AOT and DER retrievals both larger; lastly, the correlation could be interpreted as a result of cloud producing aerosols where deeper, therefore larger DER, clouds 'produce' more aerosols because of either entrainment at the sides or convergence of aerosols.

Our analyses wipe out all four possibilities. We also use another technique (Rosenfeld, 2000) to study the AIE and the positive relationship holds. We are convinced that the correlation discovered here is most likely a manifestation of real physical mechanisms. On the other hand, in-situ measurements are needed to definitely wipe out other possibilities and find out the physical processes.

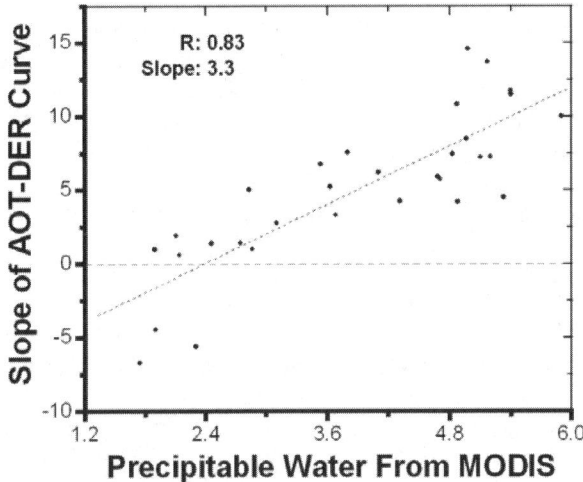

Fig. 3. Variation of the slope of DER-AOT with precipitable water. 70% of the variation in the slope is accounted by PW. Positive slopes are expected for moist regions in southwest US.

4. HYPOTHESIS

There has been a model study for smoke aerosols in Brazil that predicted the possibility of increasing DER with AOT (Feingold, 2001). Their hypothesis is, however, not applicable for our results because of aerosol types and moisture conditions.

We hypothesize here that heavier loading of hygroscopic aerosols may enhance the development of cumulous clouds. The enhancement makes collision process more likely because of three reasons. First, a shorter free moving distance between droplets due to larger number of drops makes collision more likely; second, anthropogenic aerosols tend to increase the relative dispersion of cloud droplet spectrum (Liu and Daum, 2002), which is favorable for droplet collision and coalescence; third, if large CCNs are present, polluted clouds are much more likely to have giant droplets (Han et al., 2002). The circulation pattern we provide clearly indicates that giant sea salt particles may be introduced to inland. Some indirect evidence is found to support the hypothesis. Droplet number concentration (DNC) is found to be negatively correlated with AOT, which can be an indicator of coalescence. The slopes of DNC-AOT correlate with the slopes of DER-AOT well at a correlation coefficient of –0.8 (Fig. 4).

5. IMPLICATIONS

Findings here may have significant implications towards resolving the AIE puzzle. First, the effect discovered here can offset part of cooling AIF estimated based on past parameterization of AIE if this effect exists over large part of world. Similar analyses over Asia, Europe and even over ocean show the same results. The widespread existence of this effect is also visible from Fig. 1.

Second, the findings clearly have profound implication for aerosol's effect on precipitation. Precipitation may be enhanced over moist regions (Kaufman et al., 2002) together with urban heat island effect while suppressed over dry regions (Rosenfeld, 2000). For this reason, the 'south flooding and north drought' pattern of precipitation in China can be explained differently than before (Menon et al., 2002)). High loading of aerosols in dry northern China may suppress precipitation while enhance it in moist southern China.

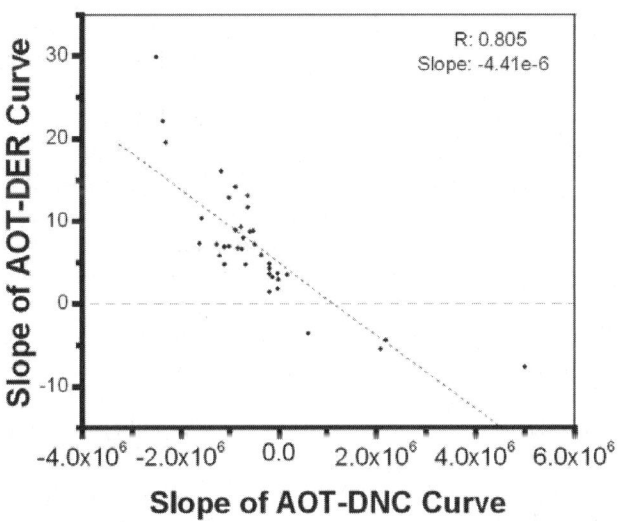

Fig. 4. The slopes of AOT-DER are plotted against the slopes of AOT-DNC.

6. REFERENCES

Albrecht, B.A., 1989: Aerosols, cloud microphysics, and fractional cloudiness, *Science,* 245, 1227-1230.

Anderson, T. L., et al., 2003: Climate Forcing by Aerosols—a Hazy Picture *Science*, 300, 1103-1104.

Andreae, M. O., et al., 2004: Smoking rain clouds over the amazon, *Science*, 303, 1337-1342.

Bréon, F.-M., D. Tanré, and S. Generoso, 2002: Aerosol effect on cloud droplet size monitored from satellite, *Science,* 295, 834-838.

Charlson, R. J., et al., 1992: Climate forcing by anthropogenic aerosols, *Science*, 255, 423-430.

Feingold, G., 2003: Modeling of the first indirect effect: Analysis of measurement requirements, *J. Geophy. Res. Lett.*, 30, 1997-2000

Feingold, G., et al., 2001: Analysis of smoke impact on clouds in Brazilian biomass burning regions: An extension of Twomey's approach, *J. Geophy. Res.* 106, 22907-22922.

Forest, C. E., Peter H. Stone, Andrei P. Sokolov, Myles R. Allen, and Mort D. Webster, 2002: Quantifying Uncertainties in Climate System Properties with the Use of Recent Climate Observations, *Science,* 295, 113-117.

Ghan, S. J., R. C. Easter, J. Hudson, and F.-M. Breon, 2001: Evaluation of aerosol indirect radiative forcing in MIRAGE, *J. Geophys. Res.*, 106, 5317-5334.

Han, Q., W. B. Rossow, J. Zeng, and R. Welch, 2002: Three Different Behaviors of Liquid Water Path of Water Clouds in Aerosol–Cloud Interactions, *J. Atmos. Sci.*, 59, 726-735

Hansen, J., M. Sato, and R. Ruedy, 1997: Radiative forcing and climate response, *J. Geophys. Res.*, 102, 6831-6864.

Hansen, J. E., and M. Sato, 2001: Trends of measured climate forcing agents, *Proc. Natl. Acad. Sci. U.S.A.,* 98, 14778-14783.

Harvey, L. D. D., and R. K. Kaufmann, 2002: Simultaneously Constraining Climate Sensitivity and Aerosol Radiative Forcing, *J. Climate,* 15, 2837–2861.

Hindman, E., P. V. Hobbs, and L. F. Radke, 1977: Cloud condensation nucleus size distributions and their effects on cloud droplet size distributions, *J. Atmos. Sci.*, 34, 951-956.

Kaufman, Y. J. and R. S. Fraser, 1997: The effect of smoke particles on clouds and climate forcing, *Science*, 277, 1636-1639.

Kim, B.-G., S. E. Schwartz, and M. Miller, 2003: Effective Radius of Cloud Droplets by Ground-Based Remote Sensing: Relationship to Aerosol, *J. Geophy. Res.*, 108, 4740-4762.

Knutti, R., T. F. Stocker, F. Joos, and G.-K. Plattner, 2001: Constraints on radiative forcing and future climate change from observations and climate model ensembles, *Nature,* 416, 719-723.

Koren, I., et al., 2004: Measurement of the effect of Amazon smoke on inhibition of cloud formation, *Science*, 303, 1342-1345.

Leaitch, W. R., et al., 1996: Physical and chemical observations in marine stratus during 1993 NARE: Factors controlling cloud droplet number concentrations, *J. Geophy. Res.*, 101, 29123-29135.

Liu, Y. G., and P. Daum, 2002: Anthropogenic aerosols: Indirect warming effect from dispersion forcing, *Nature*, 419, 580-581.

Lohmann U., and G. Lesins, 2002: Stronger Constraints on the Anthropogenic Indirect Aerosol Effect, *Science,* 298, 1012-1015.

Menon, S., A. D. Del Genio, D. Koch, and G. Tselioudis, 2002: GCM simulations of the aerosol indirect effect: Sensitivity to cloud parameterization and aerosol burden, *J. Atmos. Sci.*, 59, 692-713.

Menon, S., J. Hansen, Larissa Nazarenko, and Yunfeng Luo 2002: Climate Effects of Black Carbon Aerosols in China and India, *Science*, 297, 2250-2253.

Nakajima, T., et al., 2001: A possible correlation between satellite-derived cloud and aerosol microphysical parameters, *Geophy. Res. Lett.*, 28, 1171-1174.

Penner, J. E., et al., 2004: Observational evidence of a change in radiative forcing due to the indirect aerosol effect, *Nature*, 427, 231-234.

Ramanathan, V., et al., 2001: Aerosol, climate, and the hydrologic cycle, *Science*, 294, 2119-2124

Ramaswamy, V., et al., 2001: Climate Change 2001: The Scientific Basis. Contribution of Working Group I to the Third Assessment Report of the Intergovernmental Panel on Climate Change, *(Cambridge Univ. Press,* New York, 2001), pp. 349-416.

Rosenfeld, D., 2000: Suppression of rain and snow by urban and industrial air pollution, *Science*, 287, 1793-1796.

Rosenfeld, D., and G. Feingold, 2003: Explanation of the discrepancies among satellite observations of the aerosol indirect effects, *Geophy. Res. Lett.*, 30, 1776-1779.

Rotstayn, L. D., and Y. G. Liu, 2003: Sensitivity of the First Indirect Aerosol Effect to an Increase of Cloud Droplet Spectral Dispersion with Droplet Number Concentration, *J. Climate*, 16, 3476-3481.

Schwartz, S. E., Harshvardhan, C. Benkovitz, *Proc. Natl. Acad. Sci. U.S.A.,* 99, 1784

Shepherd, J. M. and S. J. Burian, 2003: Detection of Urban-Induced Rainfall Anomalies in a Major Coastal City, *Earth Interactions*, 7, paper 4

Takemura, T., T. Nozawa, S. Emori, T. Y. Nakajima, and T. Nakajima, 2005: Simulation of climate response to aerosol direct and indirect effects with aerosol transport-radiation model. *J. Geophys. Res.*, doi: 10.1029/2004JD005029.

Twomey, S., 1977: The influence of pollution on the shortwave albedo of clouds, *J. Atmos. Sci.*, 34:1149-1152.

THREE-DIMENSIONAL RADIATIVE TRANSFER

Bernhard Mayer and Tobias Zinner

Deutsches Zentrum für Luft- und Raumfahrt (DLR), Institute of Atmospheric Physics (IPA)
D-82234 Wessling, Germany; eMail bernhard.mayer@dlr.de

ABSTRACT

An attempt is made to give an overview of the current status of three-dimensional (3D) radiative transfer, with some emphasis on cloud remote sensing, although it is close to impossible to cover this wide field on only four pages. Some examples of experimental evidence for 3D effects in radiative transfer are presented. It is outlined that 3D models have reached a mature status where they can be used for accurate forward simulations of radiances, irradiances, and actinic fluxes. Considerable efforts are made to provide realistic cloud structures as model input. Finally, some real applications of 3D radiative transfer are shown, in particular retrievals of inhomogeneous clouds from radiance observations.

1. INTRODUCTION

Radiation is the driving force of atmospheric circulation. It controls weather and climate, triggers the most important reactions in atmospheric chemistry, and is exploited by remote sensing methods to study atmospheric state and composition. The Earth's atmosphere and surface are complex and highly variable in all three spatial dimensions and in time. On the contrary, in most radiative transfer applications the world is considered simple, one-dimensional, and plane-parallel which is often justified by computational constraints e.g. of general circulation models. Clouds are one example where the plane-parallel homogeneous approximation is known to cause considerable bias and horizontal inhomogeneity needs to be considered, at least with an approximation. 3D radiative transfer includes all problems where an atmospheric or surface property varies horizontally, including clouds, surface albedo or BRDF (Bidirectional Reflectance Distribution Function), topography, and even small scale features like vegetation. 3D radiative transfer codes are well established. It is often argued that 3D codes are too slow for practical applications. However, the computational cost is linked to a gain in information, and e.g. for domain-averaged solar irradiance, the computational cost of Monte Carlo methods is not necessarily higher than that of their one-dimensional counterparts. Although considerable progress has been made on approximations during recent years, more work is required to solve actual problems with those.

The basic and crucial question in 3D radiative transfer is the availability of accurate and realistic input data. While high-resolution topography, surface albedo, or BRDF usually do not pose problems, 3D cloud structures are a real challenge: Clouds have been shown to vary on all spatial scales, from the synoptic scale of several hundreds of km down to the centimeter scale. Favourably, variability below the mean free photon path (usually 20m or more) seems to be of little concern for radiation (Marshak et al., 1998). But still, memory requirements can be enormous for large model domains. Even more important, the experimental observation of cloud structure in sufficient detail is only possible by a combination of various instruments and is limited to case studies or campaigns. Considerable effort has therefore been devoted to the generation of realistic cloud input, using statistical (e.g. power spectra, bounded cascade) and physics-based cloud models (e.g. Large Eddy Simulation, LES) as well as observations (e.g. satellite and surface-based remote sensing), and combinations of those.

At the time of the last IRS in 2000 the relevance of 3D radiative transfer was well recognized. During recent years research has made considerable progress on systematically quantifying 3D effects, on taking them into account in general circulation and cloud-resolving models, and on developing remote sensing methods that directly determine 3D cloud structure rather than considering it a nasty disturbance of the measurement. These methods are an important further step from the mere recognizing of such effects towards an appropriate, physically more correct treatment of clouds and their characteristic variability.

2. THREE-DIMENSIONAL EFFECTS

Evidence for 3D effects is found in a variety of experimental data. A well-known example is shown in Fig. 1. Observations by two pyranometers, one for global horizontal irradiance and one mounted on a sun-tracking device to measure direct irradiance, are presented. The left panel shows the direct-horizontal irradiance. The fluctuations indicate highly inhomogeneous clouds changing rapidly with time. The direct horizontal irradiance is basically fluctuating between the cloudless sky value and zero, except for relatively thin cloud edges where some radiation is transmitted. The thick solid line is a simulation of the direct horizontal irradiance with the libRadtran model package by Mayer and Kylling (2005). For the simulation, the midlatitude summer atmosphere was used. Precipitable water was scaled to 15mm. The aerosol optical thickness was defined using the Angstrøm formula $\tau = \beta \cdot \lambda^{-\alpha}$ where λ is the wavelength in μm, $\alpha = 1.3$, and $\beta = 0.02$. β was tuned to this value to match the observations of the direct irradiance. The right panel shows the observed global irradiance (thin solid) and the simulation of the cloudless sky irradiance (thick solid), based on the same input as the direct irradiance. The observed global irradiance exceeds the cloudless model simulation by more than 50%. The highest values occur when the direct sun is not shaded and the diffuse component is enhanced due to cloud scattering and reflections off cloud sides. The enhancement is dramatic: not only the extraterrestrial irradiance $E_0 \cos \theta_0$ (thick dashed) is exceeded but even the solar constant $E_0 = 1340 W/m^2$ (thick dotted; $1368 W/m^2$, corrected for the Sun-Earth distance for this particular day). Once the irradiance is averaged over longer time periods, neither of these curves is exceeded. This is to be expected from energy conservation unless the clouds are stationary in space, as in the case of orographic clouds forming e.g. around mountain tops.

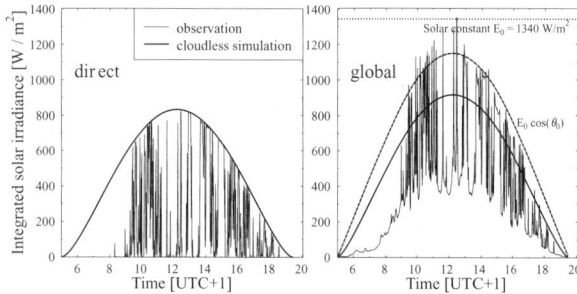

Fig. 1: Evidence for three-dimensional effects in pyranometer observations at Garmisch-Partenkirchen, May 6, 1994; (left) direct horizontal irradiance; (right) global irradiance.

The observed enhancement is a clear indication of 3D effects which cannot be reproduced by one-dimensional approximations. To simulate this observation, photons are required to propagate from a cloudy area to a cloudless area where the direct sun may reach the ground. In other words, the net horizontal photon transport plays an important role. This is not considered by the usual one-dimensional approximations, e.g. the plane-parallel or independent column approximations (Cahalan et al., 1994). Only the approach by Nack and Green (1974) – a modified independent column approximation – reproduces the enhancement due to broken clouds in first approximation.

3D effects are not always as obvious as in the above example. The main reason is that often the atmospheric and in particular the cloud properties are not known accurately enough to constrain the measurements. In many cases, observations can be explained by "effective" one-dimensional clouds. Closure experiments where the cloud properties are observed accurately enough to constrain the radiation measurements are scarce, due to the high demands to the observation systems. 3D effects, however, become obvious when the radiation is clearly outside the bounds set by the one-dimensional model, as in the example above. There is a variety of other examples proving 3D effects. Considerable effort has been spent on remote sensing. Loeb and Coakley Jr. (1998); Buriez et al. (2001) and references therein studied the variation of reflectivity and the retrieved cloud optical thickness with solar zenith angle or satellite viewing angles. For a plane-parallel cloud, no dependence is expected. The correlations found by the authors clearly prove the influence of cloud inhomogeneity. Using multi-angle MISR data Zhao and Di Girolamo (2004) studied the dependence of cloud fraction on viewing angle and used the observed variability as a measure for cloud inhomogeneity. Varnai and Marshak (2002) directly observed the influence of cloud top structure on the retrieval of optical thickness with MODIS. 3D effects and their influence on radiative transfer and remote sensing have therefore been identified. Their consideration in radiation algorithms of general circulation models (GCMs) or retrieval schemes, however, is a completely different issue and only the very first steps have been made into this direction.

3. 3D RADIATIVE TRANSFER

3D radiative transfer codes are well established. The most common technique for 3D radiative transfer is the Monte Carlo method. Only few other methods are in use. The most popular of those is the freely available SHDOM (Evans, 1998). An extensive Intercomparison of 3D Radiation Codes (I3RC) where more than 20 models were compared for a variety of atmospheric conditions demonstrated that properly coded models agree on the 1% level (Cahalan et al., 2005). A validation with observations, however, is complicated due to the above mentioned reasons.

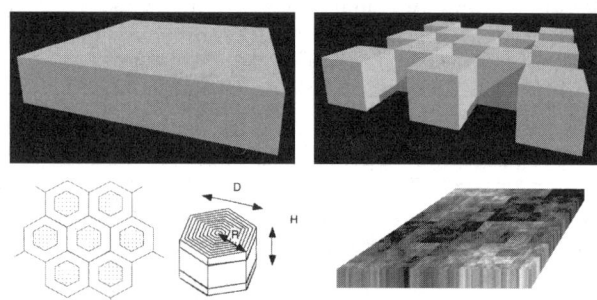

Fig. 2: Examples for cloud structures used in the early years of 3D radiative transfer, see text.

One of the key problems in 3D radiative transfer is to feed the model with realistic input. Figure 2 shows some examples of what has been used over the years, including plane-parallel clouds (upper left), simple geometric models like regular arrays of cubes or cuboids (upper right, (McKee and Cox, 1974)), arrays of hexagonal prisms which are probably more representative for real cumulus clouds (lower left, adapted from Los et al. (1997)), and statistical models like the often-used bounded cascade (lower right, (Cahalan et al., 1994)). Today, cloud models include much more detail, like cloud top structure, vertical variability, etc. An example is given in the next section.

4. NEW TECHNIQUES

Cloud inhomogeneity may cause considerable uncertainty in remote sensing. For an efficient retrieval of cloud properties from measured radiances, two assumptions are required: (1) the cloud properties (and the radiance) are constant over the field of view of the satellite instrument; no information below this scale is available from the observation itself; the lack of sub-pixel information about cloud inhomogeneity causes the so-called plane-parallel bias which is usually an under-estimation of the cloud optical thickness; the plane-parallel bias obviously increases with increasing pixel size as larger pixels usually contain more inhomogeneity; and (2) pixels are assumed independent of each other; net photon flux from a pixel to its neighbours is neglected to allow a retrieval on a pixel-by-pixel basis; the corresponding uncertainty increases with decreasing pixel size; hence a higher resolution of a satellite instrument does not necessarily imply a better accuracy of the derived cloud products. Figure 3 qualitatively illustrates the uncertainty of the retrieved cloud properties as a function of the resolution of the satellite instrument. Lower resolution instruments are subject to a large plane-parallel bias, high-resolution observations are subject to larger independent-pixel errors when evaluated on a pixel-by-pixel basis.

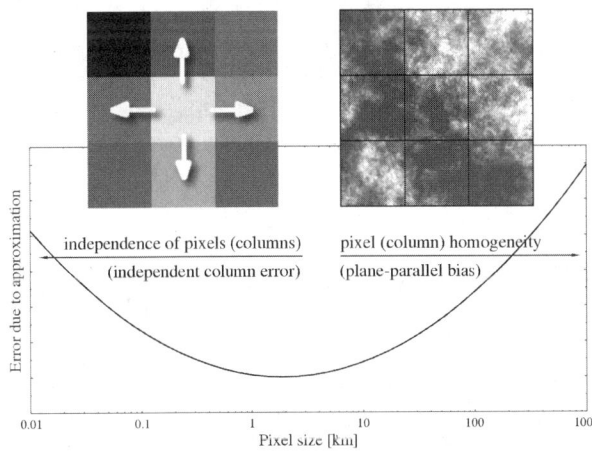

Fig. 3: Schematic diagram of the uncertainties in remote sensing, see text.

Several studies have addressed uncertainties due to the neglect of cloud inhomogeneity in radiation calculations or in remote sensing applications, by comparing 3D calculations and 1D approximations for given clouds. To include 3D effects in a retrieval, however, is a completely different story and only few attempts have been made so far. One example is shown in Fig. 4. Basis of the retrieval are high-resolution observations of the Compact Airborne Spectrographic Imager (CASI), an imaging spectrometer with a spatial resolution of 15m on a low-flying aircraft. An iterative retrieval was developed, to derive cloud structure at this high resolution. The retrieval starts from an initial guess of the cloud optical thickness, complemented by the assumption of adiabatically increasing liquid water content (LWC) with height, to obtain profiles of extinction coefficient and effective radius, assuming a constant number concentration of cloud droplets. In a next step, the observed radiance is sim-

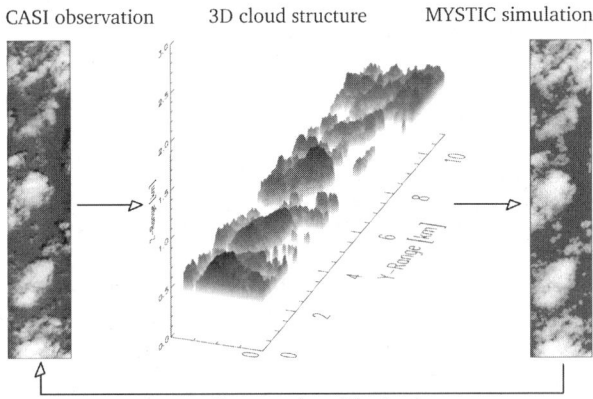

Fig. 4: Iterative retrieval of cloud properties from high-resolution aircraft observations.

ulated using our 3D radiative transfer model MYSTIC (Mayer, 1999). A comparison between observed and simulated radiances (by means of standard deviation or power spectra) shows if the cloud properties have been derived correctly which is probably not the case for the first guess. In a next step the cloud structure is modified, the radiance is simulated and again compared to the observations. This process is repeated until satisfactory agreement is obtained. For the variation of the cloud properties, a variety of methods can be imagined. We decided to use a step-wise deconvolution of the observation with an approximate point spread function, applying the Richardson-Lucy algorithm (Richardson, 1972). Certainly, the derived cloud field is not unambiguous because the observed radiances do not include the complete information about the cloud structure. However, in contrast to traditional remote sensing algorithms we are at least sure that the derived cloud has the same radiative properties (concerning reflection into the direction of the satellite) as the actual clouds; reflectance and transmittance of the cloud structure are probably also much more realistic than the result derived from standard retrievals. We use the thus derived cloud properties e.g. as basis for studies on the remote sensing of inhomogeneous clouds. Details of the retrieval are provided by Zinner (2004). A publication has been submitted to JGR.

In recent years, several approaches have been made to directly retrieve inhomogeneous cloud structure. To mention but a few, Davis et al. (1999) proposed an off-beam lidar to study cloud geometrical thickness and inhomogeneous cloud structure. In a very innovative approach Davis (2002) proposed the concept of spherical, rather than plane-parallel clouds, as a model for broken cumulus. Evans et al. (2003) suggested an in-situ lidar to directly measure volume-averaged properties of inhomogeneous clouds. Marshak et al. (2000) defined the normalized difference cloud index to measure the optical thickness of inhomogeneous clouds.

SUMMARY

Cloud inhomogeneity and the corresponding effects of 3D radiative transfer have been recognized as an important issue which needs to be considered in cloud physics, climate, atmospheric chemistry, and remote sensing applications. A variety of interesting and novel approaches have emerged since the last IRS in 2000, promising a bright future for 3D radiative transfer. For most practical purposes, however, 1D approximations are still required and will be required for a while, until computers become fast enough or fast yet accurate 3D approximations like diffusion theory will become available.

ACKNOWLEDGMENTS

We thank Marc Schröder and the group of Jürgen Fischer from FU Berlin for the CASI data. Tobias Zinner was supported by the EC funded CLOUDMAP project, EVK2-2000-00547.

References

Buriez, J.-C., M. Doutriaux-Boucher, F. Parol, and N. Loeb, 2001: Angular variability of the liquid water cloud optical thickness retrieved from ADEOS-POLDER, J. Atmos. Sci., 58, 3007–3018.

Cahalan, R., W. Ridgway, W. Wiscombe, and T. Bell, 1994: The albedo of fractal stratocumulus clouds, J. Atmos. Sci., 51, 2434–2455.

Cahalan, R., et al., 2005: The International Intercomparison of 3D Radiation Codes (I3RC): Bringing together the most advanced radiative transfer tools for cloudy atmospheres, Bull. Amer. Meteor. Soc., accepted.

Davis, A., R. Cahalan, D. Spinhirne, M. McGill, and S. Love, 1999: Off-beam Lidar: an emerging technique in cloud remote sensing based on radiative green-function theory in the diffusion domain, Phys. Chem. Earth (B), 24, 177–185.

Davis, A., 2002: Cloud remote sensing with sideways-looks: theory and first results using Multispectral Thermal Imager (MTI) data, in Algorithms for Multispectral, Hyperspectral, and Ultraspectral Imagery VIII, vol. 4725 of *S.P.I.E. Proceedings*, pp. 397–405.

Evans, K., 1998: The spherical harmonics discrete ordinate method for three-dimensional atmospheric radiative transfer, J. Atmos. Sci., 55, 429–446.

Evans, K., P. Lawson, P. Zmarzly, D. O'Connor, and W. Wiscombe, 2003: In situ cloud sensing with multiple scattering lidar: Simulations and demonstration, J. Atmos. Ocea. Technol., 20, 1505–1522.

Loeb, N. and J. Coakley Jr., 1998: Inference of marine stratus cloud optical depths from satellite measurements: Does 1D theory apply?, J. Climate, 11, 215–233.

Los, A., M. van Weele, and P. Duynkerke, 1997: Actinic fluxes in broken cloud fields, J. Geophys. Res., 102, 4257–4266.

Marshak, A., A. Davis, W. Wiscombe, and R. Cahalan, 1998: Radiative effects of sub-mean free path liquid water variability observed in stratiform clouds, J. Geophys. Res., 103, 19 557–19 567.

Marshak, A., Y. Knyazikhin, A. Davis, W. Wiscombe, and P. Pilewskie, 2000: Cloud - vegetation interaction: Use of normalized difference cloud index for estimation of cloud optical properties, Geophys. Res. Lett., 27, 1695–1698.

Mayer, B., 1999: I3RC phase 1 results from the MYSTIC Monte Carlo model, in Intercomparison of three-dimensional radiation codes: Abstracts of the first and second international workshops, University of Arizona Press, ISBN 0-9709609-0-5.

Mayer, B. and A. Kylling, 2005: Technical Note: The libRadtran software package for radiative transfer calculations: Description and examples of use, Atmos. Chem. Phys. Discuss., 5, 1319–1381.

McKee, T. and S. Cox, 1974: Scattering of visible radiation by finite clouds, J. Atmos. Sci., 31, 1885–1892.

Nack, M. and A. Green, 1974: Influence of clouds, haze, and smog on the middle ultraviolet reaching the ground, Appl. Opt., 13, 2405–2415.

Richardson, W. H., 1972: Bayesian-based iterative method of image restoration, J. Opt. Soc. Am., 62, 55–59.

Varnai, T. and A. Marshak, 2002: Observations of the three-dimensional radiative effects that influence MODIS cloud optical thickness retrievals, J. Atmos. Sci., 59, 1607–1618.

Zhao, G. and L. Di Girolamo, 2004: A cloud fraction versus view angle technique for automatic in-scene evaluation of the MISR cloud mask, J. Appl. Meteor., 43, 860–869.

Zinner, T., 2004: Fernerkundung inhomogener Bewölkung und deren Einfluss auf die solare Strahlungsbilanz, Ph.D. thesis, Ludwig-Maximilians-Universität München.

ATMOSPHERIC PARAMETER RETRIEVALS FROM HIGH RESOLUTION INFRARED SPECTRA IN A CLOUDY ATMOSPHERE

M. Höpfner

Institut für Meteorologie und Klimaforschung, Forschungszentrum Karlsruhe
Postfach 3640, 76021 Karlsruhe, Germany (michael.hoepfner@imk.fzk.de)

ABSTRACT

In the mid-infrared, not only gases, but also liquid and solid materials show distinct spectral features. Thus, when observing the atmosphere with high-resolution FTIR spectrometers signatures of trace gas molecules are mixed with those of aerosols and cloud particles. We report on the development of a scheme to retrieve information of trace gases and clouds from such measurements. Baseline for this algorithm is the radiative transfer model KOPRA (Karlsruhe Optimised and Precise Radiative transfer Algorithm) which includes absorption/emission and scattering processes for various observational geometries. Comparison with a multiple scattering code is used to assess the range of the model's applicability with respect to various kinds of clouds. As example, the retrieval of particle composition and size parameters from measurements of polar stratospheric clouds (PSCs) by ground based solar-absorption FTIR-spectroscopy is presented. Further, from PSC observations with the balloon-borne limb-emission spectrometer MIPAS-B (Michelson Interferometer for Passive Atmospheric Sounding), the strong influence from scattered radiation of tropospheric origin on the spectra was proven. With MIPAS on Envisat in orbit there exists the possibility to detect and characterise PSCs on a global scale even during polar night. In addition to the altitude region of cloud occurrence, it is possible to identify various types of PSCs.

1. INTRODUCTION

At the Institut für Meteorologie und Klimaforschung, Karlsruhe, a variety of ground-based (from Kiruna and Izana), balloon-(MIPAS-B), aircraft-(MIPAS-STR) and space-borne (MIPAS-Envisat) high-resolution Fourier transform spectrometers working in the mid-infrared spectral region are deployed to analyse the atmospheric composition primarily with respect to trace gas concentrations. However, also aerosols and clouds exhibit a strong effect on infrared spectra by scattering and absorption of photons from background sources out of and by scattering and emission of light into the line-of-sight of the observing instrument. On the one hand these effects have to be taken into consideration for the correct retrieval of trace gas abundances. On the other hand the signatures of the particles can be used to gain information about their chemical and microphysical properties. In the following we first describe the tools which have been developed for cloud analysis from high-resolution mid-IR spectroscopy: the radiative transfer model and the retrieval implementation. Then we show examples from analysis of measurements: ground-based Fourier transform infrared (FTIR) solar absorption measurements of polar stratospheric clouds (PSCs) and balloon-born limb-emission observation of PSCs. MIPAS/ENVISAT with its unique capability to observe thin clouds continuously either in polar regions or over the tropics is discussed at the end.

2. THE FORWARD MODEL

The Karsruhe Optimized and Precise Radiative transfer Algorithm (KOPRA) which has initially been developed for the retrieval of trace gas profiles from MIPAS/Envisat observations (Höpfner et al., 1998), was upgraded to be used also in a retrieval environment for aerosol and cloud properties (Höpfner, 2004). First, KOPRA includes a Mie model which provides aerosol optical properties as well as the partial derivative of these properties with respect to the input quantities, i.e., size distribution parameters and refractive indices. It supports altitude dependent mono- and bi- modal lognormal particle distributions. The coupling of a Mie model to the forward code offers the possibility to retrieve aerosol parameters directly from the observed radiances without having to determine extinction coefficient profiles first. Second, single scattering in a non-plane parallel atmosphere is implemented in KOPRA to account for the important term of scattering of thermal radiation from the troposphere in case of limb-sounding observations. This approach has been validated by comparing KOPRA results with those from a multiple scattering code (Höpfner and Emde, 2004). The main result was that for high clouds, which are in limb direction optically thin (e.g., PSCs and subvisual cirrus), the maximum model error of the single scattering approach is within a few percent. For in limb direction optically thick clouds the quality of single scattering calculations strongly depends on the single scattering albedo (ω_0). The differences range from only 2-3% for $\omega_0 = 0.24$ up to 10-30% for $\omega_0 =$

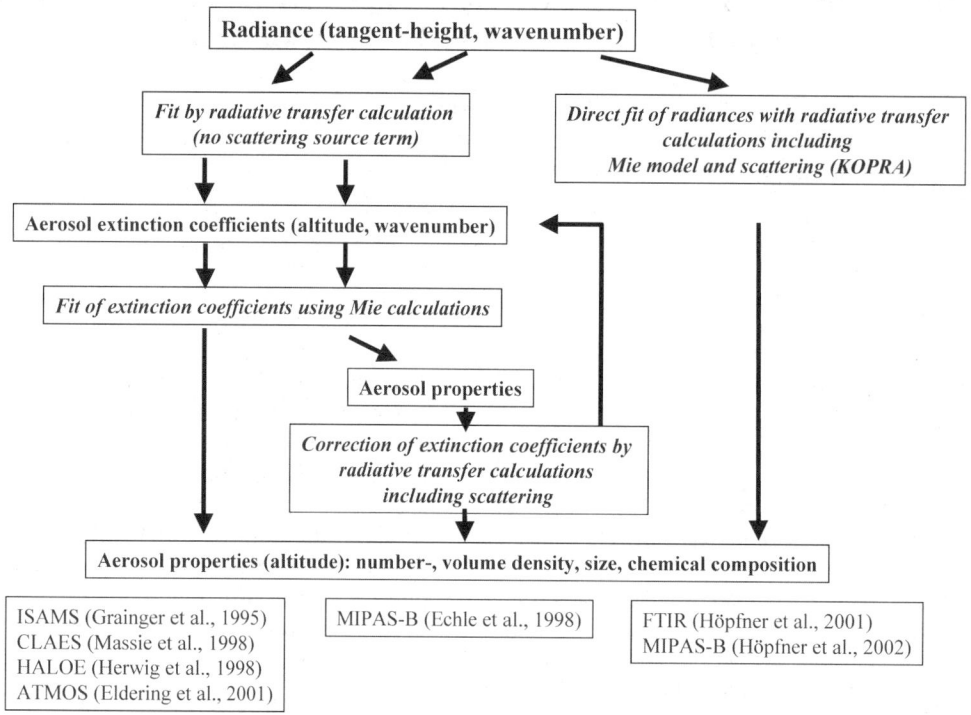

Fig. 1. Overview of schemes used to derive aerosol/cloud properties from IR limb sounding observations.

0.84. For the latter case larger errors appear for clouds which are optically thick in limb, but not in nadir direction and which, thus, still scatter a large amount of lower tropospheric radiation into the instrumental line-of-sight. A further outcome of the study was that in case of large single scattering albedo and clouds which are optically thick in limb, but thin in nadir direction the continuum radiance due to the cloud can be by a factor of up to 1.7 enhanced with respect to the blackbody radiation at cloud top altitude. This is a consequence of the scattering of radiation from the warm troposphere and the earth's surface into the line-of-sight of a limb viewing instrument.

3. THE INVERSION MODEL

Fig. 1 shows different schemes how aerosol or cloud information has been derived from infrared limb sounding observations: in a widely-used approach (Fig. 1, left pathway) extinction coefficient profiles are derived from the radiance measurements. Then, in a second step, from these (wavelength and altitude dependent) extinctions aerosol parameters are derived. However, this approach is problematic in case of limb emission sounding when scattering effects cannot be neglected. In this case the derived aerosol extinction depends largely on the correct assumption on the scattering coefficient and the phase function. This was first noticed by Echle et al., 1998 who tried to correct this effect by an iterative approach (Fig. 1, middle). By using a Mie model internally in the forward code including scattering we overcome these inherent problems of the two step approach by directly retrieving aerosol parameters from the measured radiances (Fig. 1, right).

4. GROUNDBASED FTIR PSC OBSERVATION

An example for the application of the discussed scheme are ground based solar absorption FTIR measurements of PSCs in January 2000 (Höpfner et al., 2001). Fig. 2 shows a comparison between measurements without and with PSCs in the line-of-sight of the instrument. The additional extinction due to the particles is clearly visible. We performed broadband spectral fits to the PSC observations probing various kinds of PSC compositions. As shown in Fig. 2 the assumption of ice leads to the best agreement between measurement and calculation. We could derive volume densities of around 60 $\mu m^3 cm^{-3}$ and mean particle radii of 1–2 μm. This supports the spectral finding that the PSC where composed of ice since the volume density is compatible with 2.5 ppmv of water vapour at 20 hPa. HNO_3 concentrations are much low (about 10 ppbv) to produce such a strong signal in the spectrum. These measurements are compared with air-borne lidar observations. These showed mountain wave water ice PSCs where the line-of-sight of the FTIR crossed the PSC location, thus, confirming our analysis.

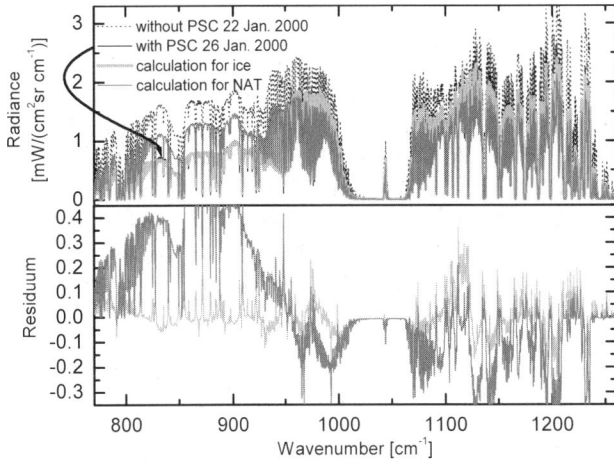

Fig. 2. Groundbased solar absorption FTIR observations from Kiruna without and with PSCs compared to best fit calculated spectra for ice and NAT. The residual spectra are: light grey for the calculation of ice and dark grey for the calculation for NAT minus the PSC observation.

5. MIPAS-B PSC OBSERVATION

The balloon-borne limb sounder MIPAS-B observed synoptic-scale PSCs on 11 Jan. 2001 from Kiruna/Sweden (Höpfner et al., 2002). Fig. 3 shows two measurements with an tangent altitude of about 20 km. The curve with the lower radiances was measured when pointing in southern direction with no PSCs in the line-of-sight. The spectrum with the high radiance background was gathered by pointing northwards where PSCs where located. The effect of PSCs on the measurement is threefold: (1) a strong background continuum, (2) a spectral structure (kink) in the background signal at 820 cm^{-1}, and (3) spectrally high resolved absorption-like features. These features (3) are downward pointing water vapour lines and inverted wings of CO_2 emission lines which can only be explained by a significant contribution of tropospheric mid-IR radiance from below scattered by the PSC into the line-of-sight of the instrument—a phenomenon which has never been observed before. Fig. 4 illustrates this effect. Despite the influence on the spectrally high resolved features, there is also a significant contribution from the tropospheric radiation on the background continuum part. In our simulation of the balloon measurements the scattered radiation makes up more than half of the radiance background. Without considering this contribution in the retrieval, we got too high PSC volume density since in this case all the background radiance has to be generate by the aerosol emission. The same problems would arise when 2-step retrieval approaches, like those described in paragraph 3 would be applied. When allowing for scattering the volume density decreases to values which are comparable to in-situ observations. This is probably an explanation for too large extinction coefficients derived from spaceborne limb emission measurements compared with solar absorption or in-situ observations (Hervig and Deshler, 1998, Massie et al., 1998, Lambert et al., 1996). Mind, however, that the strong scattering effects appear only for particles with radii larger than 0.7–1 µm. For smaller particles the single scattering albedo is too small.

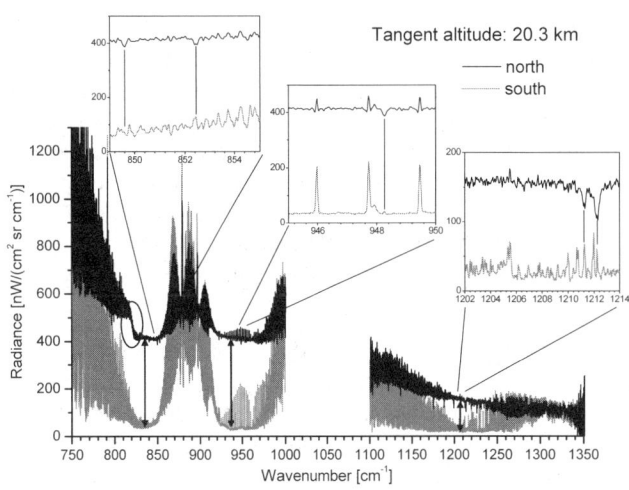

Fig. 3. MIPAS-B limb emission spectrum taken on 11 Jan. 2001. The top curve (black) is an observations with and the bottom curve (grey) without PSCs in the line-of-sight.

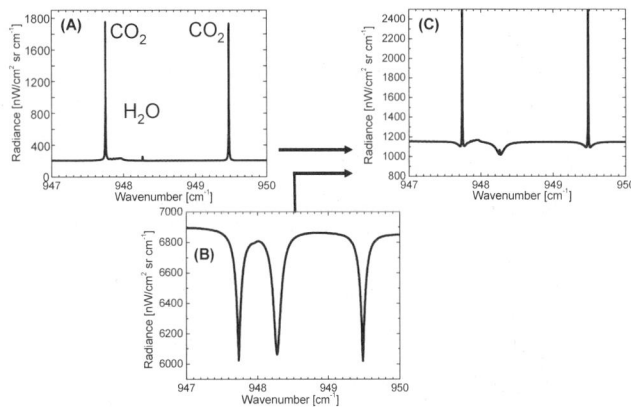

Fig. 4. Superposition of radiation emitted by the stratospheric trace gases and particles (A) and the scattered tropospheric spectrum (B) explain the measured radiance in limb direction (C).

The kink in the observed PSC spectrum at around 820 cm^{-1} could not be modelled by Höpfner et al., 2002. It is, however, likely connected to the ν_2-band of the NO_3^- ion of NAT.

6. MIPAS-ENVISAT

MIPAS on Envisat is a FTIR limb-emission spectrometer which has been launched into a sun-synchronous polar orbit on 1 March 2002. Continuous measurements exist for the period from September 2002 until March 2004 when operation was suspended due to problems with the drive velocity of the interferometer slides. In January 2005 measurements have resumed on a campaign-based mode.

Due to its mode of operation MIPAS is especially suited for the observation of PSCs. It has produced the first continuous measurements of PSCs even during polar night: in the Arctic winter 2002/2003 (Spang et al., 2004) and in the Antarctic winter 2004 (Höpfner et al. 2004). Despite daily maps of PSC top height, MIPAS allows to distinguish spectroscopically between different types of PSC (Höpfner et al. 2004) and, thus, to follow the development of the different PSC types during the winter on a vortex-wide basis (Höpfner et al., 2005).

Despite PSCs, MIPAS observations contain information on cirrus, and especially on tropical subvisual cirrus clouds. Work is in progress to determine a climatology of cirrus appearance and to derive size information from the spectra.

7. REFERENCES

Echle, G., T. von Clarmann, and H. Oelhaf, 1998: Optical and microphysical parameters of the Mt. Pinatubo aerosol as determined from MIPAS-B mid-IR limb emission spectra, *J. Geophys. Res.*, 103, D15, 19193-19211.

Eldering, A., F. W. Irion, A. Y. Chang, M. R. Gunson, F. P. Mills, and H. M. Steele, 2001: Vertical profiles of aerosol volume from high-spectral resolution infrared transmission measurements. I. Methodology, *Appl. Opt.*, 40, 18, 3082-3091.

Grainger, R. G., A. Lambert, C. D. Rodgers, and F. W. Tayler, 1995: Stratospheric aerosol effective radius, surface area and volume estimated from infrared measurements, *J. Geophys. Res.*, 100, D8, 16507-16518.

Hervig, M. E., and T. Deshler, 1998: Stratospheric aerosol surface area and volume inferred from HALOE, CLAES, and ILAS measurements, *J. Geo-phys. Res.*, 103, D19, 25345-25352.

Höpfner, M., G. P. Stiller, M. Kuntz, T. v. Clarmann, G. Echle, B. Funke, N. Glatthor, F. Hase, H. Kemnitzer, and S. Zorn, 1998: The Karlsruhe optimized and precise radiative transfer algorithm. Part II: Interface to retrieval applications. *SPIE*, 3501, 186-195.

Höpfner, M., T. Blumenstock, F. Hase, A. Zimmermann, H. Flentje, and S. Fueglistaler, 2001: Mountain polar stratospheric cloud measurements by ground based FTIR solar absorption spectroscopy, *J. Geophys. Res. Lett.*, 28, 11, 2189-2192.

Höpfner, M., H. Oelhaf, G. Wetzel, F. Friedl-Vallon, A. Kleinert, A. Lengel, G. Maucher, H. Nordmeyer, N. Glatthor, G. Stiller, T. v. Clarmann, H. Fischer, C. Kröger, and T. Deshler, 2002: Evidence of scattering of tropospheric radiation by PSCs in mid-IR limb emission spectra: MIPAS-B observations and KOPRA simulations, *J. Geophys. Res. Lett.*, 29, 8, 10.1029/2001GL014443.

Höpfner, M., 2004: Study on the impact of polar stratospheric clouds on high resolution mid-IR limb emission spectra, *J. Quant. Spectrosc. Radiat. Transfer*, Vol. 83, No. 1, 93-107, doi: 10.1016/S0022-4073(02)00299-6.

Höpfner, M., and C. Emde, 2004: Comparison of single and multiple scattering approaches for the simulation of limb-emission observations in the mid-IR, *J. Quant. Spectrosc. Radiat. Transfer*, 91, 3, 275-285, doi: 10.1016/j.jqsrt.2004.05.066.

Höpfner, M., T. von Clarmann, H. Fischer, N. Glatthor, U. Grabowski, S. Kellmann, M. Kiefer, A. Linden, G. Mengistu Tsidu, M. Milz, T. Steck, G. P. Stiller, D.-Y. Wang, P. Massoli, F. Cairo, and A. Adriani, 2004: Determination of PSC properties from MIPAS/ENVISAT limb emission measurements during the Antarctic winter 2003, *Proc. XX Quadrennial Ozone Symposium*, 1–8 June 2004, Kos, Greece, 974-975.

Höpfner, et al., 2005: manuscript in preparation.

Lambert, A., et al., 1996: Validation of aerosol measurements from the improved stratospheric and mesospheric sounder, *J. Geophys. Res.*, 101, D6, 9811-9830.

Massie, S. T., D. Baumgardner, and J. E. Dye, 1998: Estimation of polar stratospheric cloud volume and area densities from UARS, stratospheric aero-sol measurement II, and polar ozone and aerosol measurement II extinction data, *J. Geophys. Res.*, 103, D5, 5773-5783.

Spang, R., and J. J. Remedios, 2003: Observations of a distinctive infra-red spectral feature in the atmospheric spectra of polar stratospheric clouds measured by the CRISTA instrument, *J. Geophys. Res. Lett.*, 30, 16, 1875, doi:10.1029/2003GL017231.

Spang, R., J. J. Remedios, L. J. Kramer, L. R. Poole, M. D. Fromm, M. Müller, G. Baumgarten, and P. Konopka, 2004: Polar stratospheric cloud observations by MIPAS on ENVISAT: detection method, validation and analysis of the northern hemisphere winter 2002/2003, *Atmos. Chem. Phys. Discuss.* 4, 6283.

MCSCIA: A MONTE CARLO MODEL FOR ANALYSIS OF 3D FEATURES IN A SPHERICAL ATMOSPHERE

F. Spada and M. C. Krol
f.spada@phys.uu.nl
Institute for Marine and Atmospheric Research Utrecht
Princetonplein 5, 3584 CC, Utrecht, the Netherlands

ABSTRACT

The trend for new satellites is to increase the horizontal spatial resolution of their instruments. However the information about trace gas concentrations in UV-vis retrievals, is carried by solar radiation. So a study of the volume of air sampled by the radiation received by the satellite is needed to asses the limitations of the spatial resolution for UV-vis retrievals. A Monte Carlo radiative transfer model, McSCIA, is used to make a first study on this problem.

1. INTRODUCTION

In order to study the composition of the atmosphere, satellites can be used to sample the solar radiation as it is reflected by the Earth and the atmosphere. Due to interaction of these gases with atmospheric constituents this UV-vis-NIR radiation contains information about the composition of the atmosphere.

The GOME instrument (Burrows et al., 1999), launched in April 1995, measures the entire UV-vis spectrum at moderately high resolution. The vertically integrated amount of ozone is obtained by differential optical absorption spectroscopy (DOAS) by fitting the ozone absorption cross-section with the spectral features that are measured in the reflected solar radiation. The SCIAMACHY instrument (Bovensmann et al., 1999), launched in March 2002, uses the same DOAS principle but samples the atmosphere also in limb mode. In this way, stratospheric profiles can be sampled and more information about the tropospheric ozone can be obtained by subtracting the stratospheric ozone columns from the total columns measured in nadir. GOME and SCIAMACHY don't focus only on ozone: spectral signatures of NO_2, CH_2O, SO_2, and BrO have been identified in the reflected solar radiances and have been exploited to retrieve column amounts of these gases. SCIAMACHY extends the wavelength range of GOME to the near-infrared which enables the retrieval of column amounts of the greenhouse gases CO_2, CH_4, N_2O and H_2O. Additionally, CO columns can be obtained from this wavelength region. SCIAMACHY is the first space borne platform that exploits this wavelength region and efforts are ongoing in the fields of calibration and validation of these pioneering satellite data.

Meanwhile, new CCD sensors to measure the UV-vis radiance from space have been developed. The OMI instrument (Stammes et al., 1999, Laan et al., 2000, Smorenburg et al., 1999) onboard of the AURA platform, launched in July 2004, measures the reflected solar radiation at high spectral and horizontal resolution. The horizontal resolution is about 13x24 km, but can zoom to 13x13 km. Such high resolution allows for a detailed mapping of the trace gas distributions in our atmosphere. Moreover, due to the high spatial resolution, more pixels will be free of clouds and thus allow useful retrievals of vertical amounts of trace gases.

However, since the column amounts of trace gases are not determined directly, but indirectly via their interaction with solar radiation, one might ask the question whether the information measured from space represents the gas column in the pixel located underneath the spacecraft. Since radiation does not necessarily travel from the sun to the earth and then directly to the satellite instrument, the information carried by the radiation reaching the satellite may be more diffuse in nature. In addition, most nadir satellite measurements are conducted by looking at the earth under a certain angle (often up to 30 degrees off-nadir) and the majority of the measurements are taken when the sun is not overhead the sampled air-volume but at considerable solar zenith angles. These factors, which are the subject of the current study, place a lower limit to the pixel size of nadir looking instruments. Reducing the pixel sizes even further would not increase the actual resolution, since the radiation that carries the information is not restricted to the sub-satellite pixel but integrates information about the trace gas amount while it is scattered by the earth surface, aerosols, and air molecules. Absorption, and subsequent use of a DOAS-like technique, occurs over the entire photon trajectories.

Although this smoothing of trace gas absorption is well known from theory, it is largely ignored in discussions that aim at nadir-looking instruments with higher horizontal resolution. This can in part be explained by the fact that retrieval algorithms that translate measured radiances into vertical trace gas columns are normally developed using plane parallel radiative transfer models. These models assume a horizontal homogeneous atmosphere and are not able to treat three-dimensional variations in trace gas amounts. When these effects are important, a three-dimensional radiative transfer model should be employed.

These models are far more computer-time demanding than a plane parallel model and not practical for an operational retrieval algorithm.

Therefore, a 3D spherical radiative transfer model has been developed to study nadir, but also limb, measurements at the UV/Vis. wavelengths. The model (McSCIA) uses a backward (adjoint) Monte Carlo approach (Cashwell and Everett, 1959, Collins and Wells, 1970, Adams and Kattawar, 1978, Lenoble and Chen, 1992, Oikarinen et al., 1999, Evans and Marshak personal communication, 2004) applying the so-called Equivalence Theorem (ET) (Partain et al., 2000, Feigelson, 1984, van de Hulst, 1980, Irvine, 1964) to separate scattering and absorption. Photon trajectories, calculated in a non-absorbing atmosphere, are convolved with the trace gas profiles, which enables a fast calculation for different absorber distributions.

McSCIA has been validated against MCC++ (Postylyakov, 2004), using the result of the inter-comparison paper by Loughman et al. (2004).

We will use this model to study the real volume that is scanned by the photons, focusing especially on the multiple scattering. In fact while the single scattering smoothing can be easily calculated (is purely geometric) the influence of multiple scattering needs radiative transfer calculations.

2. ATMOSPHERE

As a first step we choose a setting that wouldn't show a very accentuated multiple scattering. To do so we selected a cloud free, purely Rayleigh scattering atmosphere that extends from the surface to a 100 km altitude.

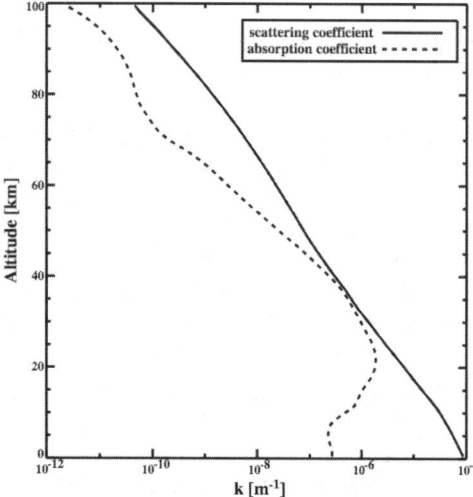

Fig. 1. Atmospheric profile at 330 nm for a purely Rayleigh scattering atmosphere with Ozone absorption. The full line represents the resolved scattering coefficient [m^{-1}], while the dashed line shows the absorption scattering coefficient [m^{-1}] (one point for each km).

A wavelength of 330 nm (normally used for ozone retrieval) was chosen and the vertical structure was resolved with 100 homogeneous layers of 1 km depth each (Fig. 1). The ground reflection was approximated by using a Lambertian surface, with an albedo of 0.3.

The satellite is in a nadir looking position, with no azimuth angle between its viewing line and the solar rays, that are coming with a zenith angle of 40° (Fig. 2).

3. EXPERIMENT DESCRIPTION

The Equivalence Theorem of Irvine (1964) is very powerful and can help radiative transfer models to treat absorption. As discussed by van de Hulst (1980), it states that it does not matter whether the constituents doing the scattering and those doing absorption are identical. This means that if, in the atmosphere, we distinguish two constituents

- haze, that provides conservative scattering
- gas, that provides absorption along the path between scattering events

we can decide to treat absorption, as it would occur

1. all at the scattering points, using single scattering albedo values less than one, or
2. along the path between scattering events, using an exponential decrease of the intensity (I) along the trajectory.

McSCIA is able to use the Equivalence Theorem approach in the manner described by the second option above. Thus we run the model in a purely scattering atmosphere and store the scattering events of each photon that travels between the sun and the satellite.

Next, absorption is introduced, by a convolution of the photon trajectories with the absorption profile. On top of a background ozone profile, we introduced perturbations of ozone (10%) at a height of 30 km. We varied the perturbation by changing the size of a circular spot at nadir (see Fig. 2) from 0 (no perturbation) to infinity (a 10% increase in ozone in the entire layer).

In order to quantify the effect of radiative smoothing on the retrieval, we introduce the cumulative perturbation contribution (PC):

$$PC = \frac{I_0 - I(r)}{I_0 - I_\infty} \bullet 100$$

In this equation I_0 is the intensity when the spot has radius of 0 km (i.e., no perturbation), I_∞ is the intensity when the spot is covering all the layer (maximum perturbation), and I(r) is the intensity when the spot has radius r. PC represents the cumulative contribution, in percent, to the maximum perturbation, as a function of the spot radius.

Fig. 2 depicts some quantities that are important for the further discussion. The observational layer is at 30 km. The satellite line of sight is **VO**, with O being the sub-satellite point, and intersects the observational layer at point A. The sunray reach O following **SO** with a solar zenith angle VOS = 40°, and intersecting in B the layer of observation.

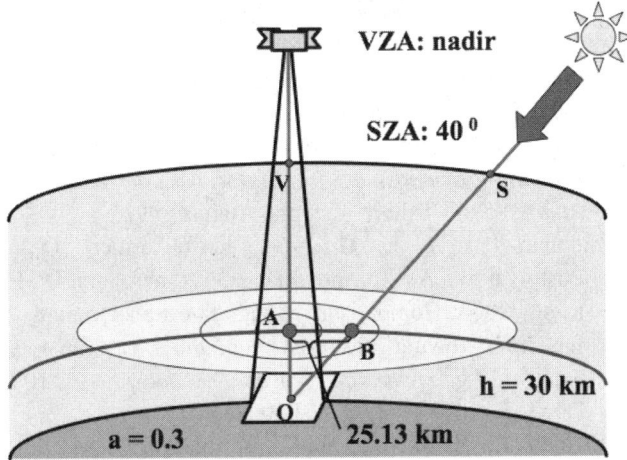

Fig. 2. Geometry of the experiment. The observational layer is at 30 km. The satellite is looking in nadir VO and see the observational layer in A and the ground in O (sub-satellite point). The solar zenith angle VOS is 40°, and OS intersect the observational layer in B. AB = 25 km. The ground albedo is a=0.3.

One important hypothesis that is used is that the field of view of the instrument is very small, so that can be approximated by a single direction. Of course this is not true for a real satellite, but we approximate the real view of the satellite considering that it is seeing a spot of a certain range (e.g., 13x24 or 13x13 km for OMI).

4. RESULTS

We analyze the ozone perturbation at an altitude of 30 km, i.e., around the maximum of the ozone layer. Thus, we expect to observe a transition at **AB** = 25.1 km, i.e., a spot size that is just seen by photons that travel along trajectories SOV.

Fig. 3 depicts the PC as function of the spot size. If only single scattering (SS) is considered (the full line in Fig. 3), it is observed that the region just under the satellite (A) is very important. Even a very small spot at location A, makes the PR value to jump form 0 to over 40%. As expected the value of PR for SS goes to 100% after the geometrical value of 25.13 km. In fact, the interpretation of the majority of the SS PR value can be made by considering just the geometrical path of the SS rays that travel from the sun to the satellite.

The calculated single scattering contribution to the total observed radiation at 330 nm is about the 40%. Thus the behavior of the total radiation mimics that of SS with two important differences.

First, the importance of the layer sub-satellite point A, for total scattering, is about 50%. This means that, of all the total photons that contribute to multiple and single scattering, almost all of them seem to cross point A, i.e., the last scattering event usually occur under the 30 km layer along **OA**, while for the single scattering the contribution of the radiation coming from **AV** is bigger.

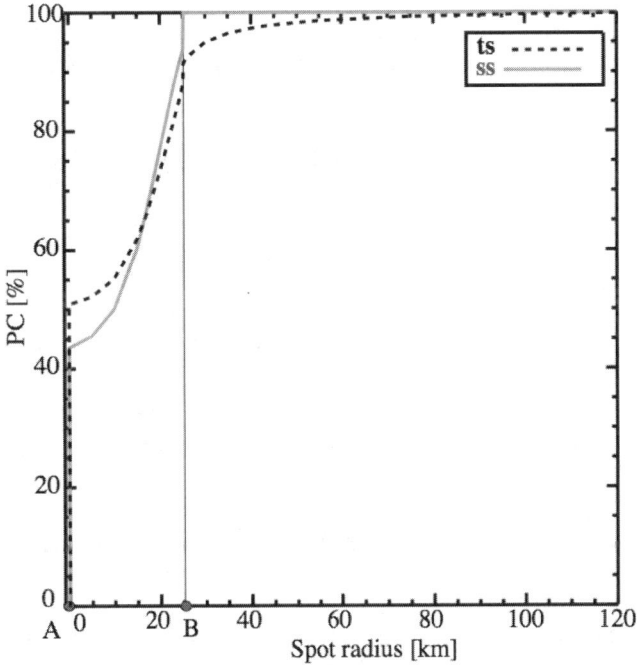

Fig. 3. The cumulative perturbation contribution [%] as a function of the radius of the Ozone perturbation spot [km] for single scattering (full line) and total (multiple+single) scattering (dashed line). The points A and B (Fig. 2) are also noted.

Second, and more important, at a spot size of 25.13 km only 90% of the maximum absorption is observed. This implies that 10% of the absorption at the 30 km altitude layer, is caused by multiple scattered photons that cross the 30 km layer at a distance from A that is larger than 25.13 km.

5. CONCLUSIONS

The quick evolution of satellite technologies allows the development high spatial resolution mapping of trace gases that are needed by the community. However, the concentrations of these gases are not calculated directly, but are the result of the measurement of solar radiation that interacted with them. However, a large part of the radiation does not follow the geometric path sun—scattering point—instrument, but rather is diffused by multiple scattering before reaching the satellite. This simple case study shows that even in a very favorable situation as this (no clouds, no aerosol) 10% of the absorption occurs outside the volume that is geometrically seen by the satellite instrument. So we point out the necessity to increase our knowledge on these "smoothing" effects in order to be able to interpret current and future remote sensing measurements based on backscattered UV-vis radiation.

Future studies with our MC-ET approach will focus on the 3D absorption features at different altitudes, wavelengths and in atmospheres with different scattering characteristics.

6. ACKNOWLEDGMENTS

We are grateful to Jochen Landgraf for the discussions we had about this topic.

7. REFERENCES

Adams, C. N., and G. W. Kattawar, 1978: Radiative transfer in spherical shell atmospheres 1. rayleigh scattering. *Icarus*, 35.

Bovensmann, H., J. P. Burrows, M. Buchwitz, J. Frerick, S. Nol, V. V. Rozanov, K. V. Chance, and A. P. H. Goede, 1999: Sciamachy: Mission objectives and measurement modes. *J. Atmos. Sci.*, 56, 127–150.

Burrows, J. P., M. Weber, M. Buchwitz, V. Rozanov, A Ladstatter-Weissenmayer, A. Richter, R. De Beek, R. Hoogen, K. Bramstedt, K. W. Eichmann, M. Eisinger, and D. Perner, 1999: The global ozone monitoring experiment (gome): Mission concept and first scientific results. *J. Atmos Sci.*, 56, 151–175.

Cashwell, E. D., and C. J. Everett. Monte Carlo Method for random walk problems. *Pergamon Press*, 1959.

Collins, D. G., and M. B. Wells, 1970: Flash, a monte carlo procedure for use in calculating light scattering in a spherical-shell atmosphere. *Report rra-t704*.

Feigelson, E. M, 1984: Radiation in a cloudy atmosphere. Kluwer.

Irvine, W. M., 1964: The formation of absorption bands and the distribution of photon optical paths in a scattering atmosphere. *Bulletin of the Astronomical Institutes of The Netherlands*, 17, 266–279.

Laan, E., J. de Vries, B. Kruizinga, H. Visser, P. Levelt, G. H. J. van den Oord, A. Maelkki, G. Leppelmeier, and E. Hilsenrath, 2000: "Ozone monitoring with the OMI instrument". *Proceedings of the SPIE's 45th Annual Meeting – The International Symposium on Optical Science and Technology*, SPIE Vol. 4132-41, 334-343.

Lenoble, J., and H. B. Chen, 1992: Monte Carlo studies of the effects of stratospheric aerosols and clouds on zenith sky absorption measurements. *The International Radiation Symposium, Int. Assoc. of Meteorol. and Atmos. Phys.*, Tallinn, Estonia, August 1992.

Loughman, R. P., E. Griffioen, L. Oikarinen, O. V. Postylyakov, A. Rozanov, D. E. Flittner, and D. F. Rault, 2004: Comparison of radiative transfer models for limb-viewing scattered sunlight measurements. *Journal of geophysical research, D, Atmospheres*, 109, D06303, doi:10.1029/2003JD003854.

Oikarinen, L., E. Sihvola, and E. Kyrola, 1999: Multiple scattering radiance in limb-viewing geometry. *Journal of Geophysical Research*, 104, 31261–31274.

Partain, P. T., A. K. Heidinger, and G. L. Stephens, 2000. High spectral resolution atmospheric radiative transfer: Application of the equivalence theorem. *Journal of Geophysical Research*, 105, 2163–2177.

Postylyakov, O. V., 2004: Linearized vector radiative transfer model mcc++ for a spherical atmosphere. *Journal of Quantitative Spectroscopy and Radiative Transfer*, 88, 297–317.

Smorenburg, C., H. Visser, and K. Moddemeier, 1999: "OMI-EOS: Wide field imaging spectrometer for ozone monitoring." *Proceedings of Europto/SPIE conference*, SPIE vol. 3737.

Stammes, P., P. Levelt, J. de Vries, H. Visser, B. Kruizinga, C. Smorenburg, G. Leppelmeier, and E. Hilsenrath, 1999: "Scientific requirements and optical design of the Ozone Monitoring Instrument on EOS-CHEM." *Proceedings of SPIE Conference on Earth Observing Systems IV*, SPIE Vol. 3750, p. 221-232.

van de Hulst, H. C., 1980: Multiple light scattering. *Academic press*.

THE DEVELOPMENT OF NEW RADIATION CODE FOR CCSR/NIES AGCM

M. Sekiguchi
Tokyo University of Marine Science and Technology
Tokyo, 135-8533, Japan

T. Nakajima
Center for Climate System Research, University of Tokyo
Tokyo, 153-8841, Japan

ABSTRACT

In this study, the gaseous absorption process of the broadband radiative transfer code "*mstrn8*", which was developed by Center for Climate System Research (CCSR), is upgraded. At first, the absorption line database is replaced by HITRAN2001 from HITRAN 1992, and the absorption continuum program is also replaced by MT_CKD_1.0 from CKD_0 which is adopted by LOWTRAN 7. This upgrade decreases the error in the lower troposphere about 0.3 K/day in heating rate. Next the treated gas absorption bands in *mstrn8* are not enough and the error of heating rate in longwave region is 5.2%. So, the selection rule of the treated gas absorption bands is developed and several bands are newly considered. Furthermore, an optimization method is adopted in *mstrn8* to decrease the number of quadrature points for wavenumber integration using the correlated k-distribution method and to increase the computational efficiency in each band. We improve this method to select an optimal initial condition to get a suitable result. With these improvements, the new radiation code computes the flux with errors less than 2.0 W/m^2 and heating rate with errors less than 1.0 K/day in the upper stratosphere in longwave region and less than 0.2 K/day in shortwave region. The cold bias in the simulation by the CCSR/NIES AGCM is largely decreased by the improved radiation code developed in the present study.

1. INTRODUCTION

The radiative forcing can be estimated within an accuracy of about 0.1 W/m^2 by the progress of optical modeling of the earth's atmosphere and the progress of computers. Under this situation there is a large demand for an accurate yet rapid radiation transfer scheme accurate for atmospheric dynamics modeling. The broadband radiative transfer code "*mstrn8*", which was developed by Center for Climate System Research (CCSR) about eight years ago and was implemented in the CCSR/NIES atmospheric general circulation model (AGCM) (Nakajima et al., 2000), has a error of 10% in the radiative heating rate calculation and start having a difficulty to meet this demand. Most difficult part of accurate and efficient radaiative transfer calculation is to evaluate the light absorption by atmospheric molecules. It is known that CCSR/NIES AGCM has a cold bias about 10 K around the tropopause that seems to be caused by the error of the *mstrn8* code. In this study, we improve the gaseous absorption process of the radiation code to solve these difficulties met by the *mstrn8*.

2. MODEL DESCRIPTION

For *mstrn8*, the line parameter database is adapted to HITRAN92 (Rothman et al., 1992), the continuum absorption program is implemented that of LOWTRAN 7. In the case of the line absorption, H_2O, CO_2, O_3, N_2O and CH_4 are considered and self and foreign broadenings, ozone and oxygen in ultraviolet-visible spectrum are treated as the continuum absorption. 16 halocarbons and heavy molecules are also considered (Shi et al., 1992). The spectrum, which is considered from 0.2 to 200 μm in this code, divided into 18 bands and the number of integration points is selectable, 37 (low-resolution version) or 55 (high-resolution version). The correlated-k distribution method is adapted and the number of quadrature points is decreased and the position and weights are determined by the optimization, the sequential quadratic programming (SQP) method, which an objective function is decreased. The objective function is determined the sum of root mean square error of radiative flux at TOA and surface and heating rate at each level in standard atmospheres. In the case of overlapping, the number of integration points, position and weights are also determined by the optimization method. An initial condition is set at the points of correlated overlapping. This code can also treat Rayleigh and Mie scattering and absorption/emission for particulate matter. The absorption, scattering and emission are calculated and tabulated beforehand. The radiative transfer solver is applied the discrete-ordinate method/adding method.

The maximum error of heating rate of mstrn8 is about 0.8 K/day in longwave region at tropopause. The cold bias in troposphere at tropics is known by the results of CCSR/NIES AGCM, which is implemented *mstrn8*. It is considered the error of mstrn8 makes this bias.

3. MODIFICATIONS AND THEIR EFFECTS

In this chapter, model modifications for *mstrn8* are introduced. The code, which is upgraded all modification below from *mstrn8* is called *mstrnX*.

3.1. Treated Gas Species and Their Coefficients

The database of line parameters is replaced by HITRAN 2000 (Rothman et al., 2003) from HITRAN 1992, and the absorption continuum program is also replaced by MT_CKD_1.0 (see http://www.rtweb.aer.com/) from CKD_0, which is adopted by LOWTRAN 7. By these updates, the light absorption by water vapor is changed in longwave region and absorption by the Chappuis band of ozone is substantially revised in shortwave region. The effects are up to 0.3 K/day in the troposphere in longwave region and up to 0.1 K/day in the upper stratosphere in shortwave region.

The heating rate calculated by the *mstrn8* has errors about 5.2% in longwave region and 3.6% in shortwave region compared to the result evaluated with all the absorption lines of major seven gas species, so that a new criterion for selecting absorption lines is needed. We determine the evaluation function, which is the sum of root mean square error of radiative flux at TOA and surface and heating rate at each level in standard atmospheres, evaluate the effect of each gas and select the gas species in each band. By this criterion, *mstrnX* is newly considered about CO_2 around 10 μm, ozone of rotational band and N_2O around 17 μm, etc., and the errors are estimated under 0.2%.

3.2. Optimization Method

An optimization method is adopted in *mstrn8* to decrease the number of quadrature points for wavenumber integration using the correlated *k*-distribution method and to increase the computational efficiency in each band. And the quadrature points of *k*-distribution is optimized for accurate calculation of the heating rate up to altitude of 30 km. We extend the maximum altitude up to 70 km for accurate calculation of the heating rate. For this purpose we adopted a new non-linear optimization method of the *k*-distribution and studied an optimal initial condition and the cost function for non-linear optimization. The cost function is determined same as the evaluation function discussed in chapter 3.1. We cloud select the number of integration points according to the degree of error, 40 points for 0.1% and 59 points for 0.05%.

4. RESULTS AND DISCUSSIONS

With these improvements, the new radiation code computes the flux with errors less than 2.0 W/m^2 and heating rate with errors less than 1.0 K/day in the upper stratosphere in longwave region and less than 0.2 K/day in shortwave region. Under doubling CO_2 condition, the results using the cost function in standard atmospheres are not represented precisely. In this case, an increase of several points can be decreased the error to same extent in standard atmospheres using the cost function, which is also considered the doubling condition. The cold bias in the simulation by the CCSR/NIES AGCM is largely decreased (~ 3K) by the improved radiation code developed in the present study.

The effects of the increasing treated gases and the updating the absorption coefficients are relatively small to the effects of the improvement optimization. Halocarbons are important especially for global warming simulation, so their effects should be carefully treated. In HITRAN 2000, 27 species of halocarbons and heavy molecules are described and their spectra are highly dependent to wavelength. In future work, we will consider their effects.

5. REFERENCES

Clough, S. A., M. J. Iacono, and J. L.Moncet, 1992: Line-by-line calculations of atmospheric fluxes and cooling rates: Application to water vapour, *J. Geophys. Res.,* 97, 15761-15785.

Nakajima, T., M. Tsukamoto, Y. Tsushima, A. Numaguti, and T. Kimura, 2000: Modeling of the radiative process in an atmospheric general circulation model. *Appl. Opt.,* 39, 4869-4878.

Rothman, L. S., R. R. Gamache, R. H. Tipping, C. P. Rinsland, M. A. H. Smith, D. C. Benner, V. M. Devi, J.-M. Flaud, C. C. Peyret, A. Perrin, A. Goldman, S. T. Massie, L. R. Brown, and R. A. Toth, 1992: The HITRAN database: Editions of 1991 and 1992, *J. Quant. Spectrosc. Radiat. Transfer,* 48, 469-507.

Rothman, L. S., A. Barbe, D. Chris Benner, L. R. Brown, C. Camy-Peyret, M. R. Carleer, K. Chance, C. Clerbaux, V. Dana, V. M. Devi, A. Fayt, J.-M. Flaud, R. R. Gamache, A. Goldman, D. Jacquemart, K. W. Jucks, W. J. Lafferty, J.-Y. Mandin, S. T. Massie, V. Nemtchinov, D. A. Newnham, A. Perrin, C.P. Rinsland, J. Schroeder, K. M. Smith, M. A. H. Smith, K. Tang, R. A. Toth, J. Vander Auwera, P. Varanasi, and K. Yoshino, 2003: The HITRAN molecular spectroscopic database: edition of 2000 including updates through 2001, *J. Quant. Spectrosc. Radiat. Transfer,* 82, 5-44.

Shi, G., 1992: Radiative forcing and greenhouse effect due to atmospheric trace gases, *Sci, Sin. Ser. B,* 35, 217-229.

SUPER FAST RADIATIVE TRANSFER FORWARD MODEL BASED ON PRINCIPAL COMPONENTS

Xu Liu[1a]*, William L. Smith[b], Daniel K. Zhou[a], and Allen Larar[a]
[a]NASA Langley Research Center, MS401A, Hampton, VA 23681, USA
[b]Hampton University, VA 23668, USA

ABSTRACT

This paper presents a novel Principal Component-based Radiative Transfer Model (PCRTM). Only a few hundred radiative transfer calculations are needed for hyperspectral sensors with thousands of channels. The model is very accurate and flexible. Its execution speed is a factor of 3–30 times faster than channel-based fast models. Due to its high speed and compressed spectral information format, it has great potential for super fast one-dimensional physical retrievals and for Numerical Weather Prediction (NWP) large volume radiance data assimilation applications. The model has been successfully developed for the NAST-I and AIRS instruments.

1. INTRODUCTION

Modern sensors such as AIRS, TES, CrIS, IASI, and GIFTS have thousands of channels and provide a wealth of information on the atmospheric and surface properties. One of the key components of analyzing these data is the radiative transfer forward model, which relates the atmospheric states (such as temperature and trace gas profiles) to the radiances observed from the advanced sensors. Due to the high spectral resolution, a large number of radiative transfer calculations through inhomogeneous atmosphere are needed. Often times, only a subset of channels is used with existing RT models in order to handle the corresponding computational constraints.

There are several ways to minimize the computational time needed to perform radiative transfer calculations. One of the approaches, which the TES science team uses, is to store absorption coefficients as a function of atmospheric pressure and temperature at a monochromatic frequency grid in a large lookup table. As a result, the optical depths in a particular atmospheric layer can be calculated simply by interpolations and multiplications. In this approach, time-consuming calculations of spectral line shapes and intensities are avoided, but the forward model still has to perform numerous monochromatic radiative transfer calculations through an inhomogeneous atmosphere in order to obtain Top Of Atmosphere (TOA) radiances. To minimize the calculations needed, TES science algorithm selects narrow micro-windows to perform retrievals for a specific trace gas (TES ATBD 2002).

Another forward model approach is to predict effective layer optical depths by using an efficient fast parameterization. The effective layer optical depth is a channel-averaged quantity, which contains Instrument Line Shape (ILS) function or Sensor Response Function (SRF), therefore only one radiative transfer calculation per channel is needed. The effective optical depth is derived in such a way that the additive property of optical depth (or the multiplicative properties of transmittances) between different atmospheric layers and different atmospheric gases holds. The effective channel layer optical depth is predicted using average layer temperatures, pressure-weighted layer temperatures, satellite zenith angle and column densities of trace gases. There are several fast model parameterizations based on effective optical depth for satellite remote sensing applications (Smith 1969, McMillin et al., 1976, Saunders et al., 1999, Strow et al., 2003, Eyre and Woolf 1988). Optical Path Transmittance (Optran), Stand-Alone Radiative Transfer Algorithm (SARTA) and RTTOV are three well-known fast models of this kind.

Another kind of fast parameterization is to predict channel transmittances or radiances by using a few representative monochromatic transmittances or radiances (Edwards and Francis 2000, Goody et al., 1989, West et al., 1990, Mlawer et al., 1997, Wiscombe and Evans, 1977, Gerste 1993, Chou et al., 1995, Lacis and Oinas, 1991). Correlated K Distribution (CKD), Exponential Sum Fitting Transmittance (ESFT), Radiance Sampling Method (RSM) and Optimal Spectral Sampling (OSS) are examples this type of fast RT model approaches. CKD and ESFT have advantages of being monochromatic (can extend to include multiple scattering) and very fast (only few monochromatic calculations per channels are needed). They are also very accurate for a single atmospheric layer and a single absorption gas. However, these models are usually trained on one atmospheric layer and the dependency of channel transmittance on pressure, temperature and gas amounts are introduced later on by assuming a good correlation between vertical layers and no correlations between overlapping gases. Extending them to vertically inhomogeneous atmosphere leads to limited accuracy. Extensive efforts

---

*Xu.Liu-1@nasa.gov, Tel: 757-864-1398

were made to extend these methods to remote sensing applications (Edwards and Francis, 2000, Gerste, 1993, Chou et al., 1995, Armbruster and Fischer, 1996, Tjemkes and Schmetz, 1997). One way to overcome this deficiency is to include both layer transmittance and space-to-layer transmittance in the training (Armbruster and Fischer, 1996). Another way is to predict the TOA channel radiance using statistically selected TOA monochromatic radiances (Tjemkes and Schmetz, 1997). This approach has the advantage that it treats overlapping gases and inhomogeneous layers in a simple manner. The drawback of the RSM is that it needs relative large number of monochromatic radiances to approximate TOA channel radiances due to the method used to determine the monochromatic frequency locations. The OSS method developed by AER (Moncet et al., 2001, Liu et al., 2003, Moncet and Uymin, 2003) overcomes this deficiency by fitting TOA channel radiance using a robust ESFT minimization (Wiscombe and Evans, 1977). Typically, one to twenty monochromatic radiative calculations are needed to predict channel radiance. OSS is also capable of predicting layer or space-to-layer transmittances. Moncet, et al. (2003) have extended the search method to a more robust Monte Carlo approach.

All the fast models described above are channel-based forward models. The PCRTM on the other hand does not predict channel radiance directly. Instead, it predicts Principal Component (PC) scores, which have much smaller dimension as compare to the number of channels.

2. DESCRIPTION AND THEORETICAL BASIS OF PCRTM FORWARD MODEL

The PCRTM compresses channel transmittances or radiances into a few orthogonal eigenvectors or PCs. These radiative transfer properties can be easily regenerated via a linear combination of a few significant PCs. The linear coefficients are called Principal Component Scores. These PC scores can be obtained by projecting the channel radiances onto each of the PCs. Usually 100-250 PCs are enough to regenerate channel radiances to 0.01 K accuracy. For a given atmospheric state, PC scores are generated using approaches similar to other fast forward model, i.e., a non-linear function of atmospheric state or a linear function of a few monochromatic transmittances or radiances. The later approach has the advantage that the radiative transfer equation transforms the atmospheric temperature, moisture and trace gas profiles into radiances and these radiances have a linear relationship with the PC scores:

$$\vec{R}^{ch} = \sum_{i=1}^{N_{PC}} Y_i \vec{U}_i + \vec{\varepsilon} = \sum_{i=1}^{N_{PC}} \left(\sum_{j=1}^{N_{mono}} a_j R_j^{mono} \right) \vec{U}_i + \vec{\varepsilon} \quad (1)$$

Where $\vec{R}^{ch}$ is the channel spectrum vector, $\vec{U}_i$ is the ith PC vector, N_{pc} is the number of significant PCs. Y_i is the PC score, it is generated by linearly combining monochromatic radiances R^{mono}. $\vec{\varepsilon}$ is a vector containing forward model errors. Usually, the values of $\vec{\varepsilon}$ are smaller than 0.04 K. The PC score is a dot product of PC vectors with the channel radiance vector (see equation 2).

$$Y_i = U^T_{N_{ch} \times 1} R^{ch}_{N_{ch} \times 1} = \sum_{j=1}^{N_{ch}} U(j,i) \times R^{ch}(j) \quad (2)$$

The channel radiance is a linear combination of monochromatic radiances within the frequency span of that channel. The weights are simply the normalized ILS or SRF at the monochromatic frequency grid. Both the pre-calculated PC vectors and the ILS (or SRF) do not vary from one atmospheric state to another. Therefore, the PC score is a linear function of monochromatic radiances (equation 3).

$$Y_i = \sum_{j=1}^{nch} U(j,i) \times \left[\sum_{k=1}^{N} \phi_k R^{mono}(k) \right] = \sum_{l=1}^{N_{mono}} a_l R^{mono}(l) \quad (3)$$

As mentioned before, there is a lot of redundant information among thousands to millions of monochromatic radiances or transmittances. There is no need to use all of them to predict Y_i in equation 3. Unlike methods described above, the PCRTM method selects the location of monochromic frequencies in a very straightforward way by using a correlation function (Liu, 2004).

3. RESULTS

We have implemented the PCRTM forward model for both AIRS and the NPOESS Airborne Sounder Testbed Interferometer (NAST-I) instruments. The PCRTM fast model has very high accuracy as compared to the line-by-line calculations. The RMS errors between the LBLRTM[22] and PCRTM for both AIRS and NAST-I are usually less than 0.04K.

To test the accuracy of the PCRTM model, we have compared the observed AIRS radiance with radiances calculated using the PCRTM model. The AIRS spectrum was taken from a clear sky overpass of Aqua over Atmospheric Radiation Measurement Tropical Western Pacific (ARM-TWP) site on 12/08/2002. The surface pressure and ozone profile are from ECMWF model. Surface temperature is from UMBC retrieval (Strow et al., 2003). Water and temperature profiles are mainly from collocated radiosondes over the ARMS-TWP site. The water profile above 200 mb and the temperature profile above 60 mb are taken from the ECMWF model analyses. Fig. 1 shows the observed AIRS radiance (top panel), the PCRTM calculated radiance (middle panel) and the difference between the two (bottom panel). Large residuals exist near 1050 cm^{-1}. It is clear that ozone profile from ECMWF analysis does not represent the true atmospheric ozone state. The spikes on the top and bottom panels are due to the undetected "popping" noise of the AIRS instrument. Overall, the agreement is good between the PCRTM calculation and the AIRS observation.

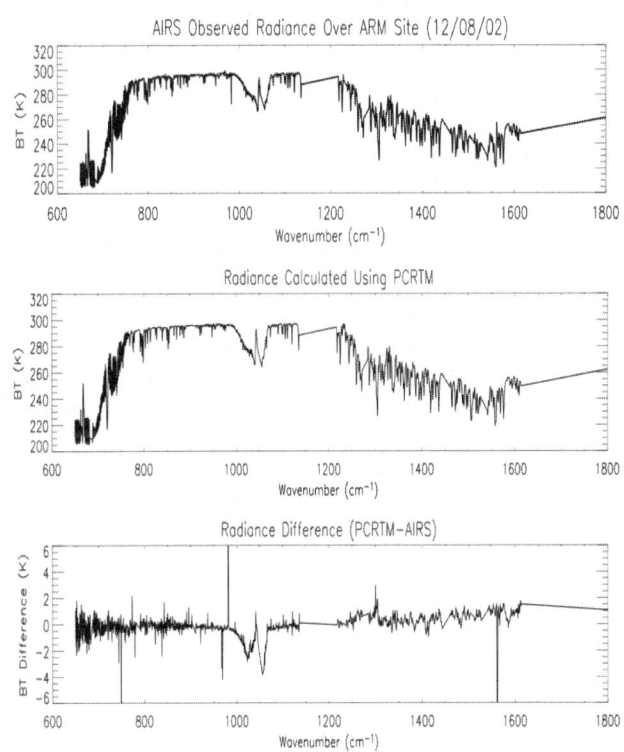

Fig. 1. Comparison of the observed AIRS radiance over ARM site on 12/08/2002 with radiances calculated using PCRTM model.

4. CONCLUSIONS

PCRTM is a much faster than channel-based RTM since it predicts PC scores (or "Super Channel" magnitudes) directly instead of channel radiance or transmittance individually. The parameterization of PCRTM is physical and can be trained to any desired accuracy relative to LBL codes. The choice of using radiances as predictors is derived theoretically from the properties of PC scores. The PCRTM provides both PCs scores and associated jacobian, therefore it is recommended that the physical inversion of state vector be done in the PC domain directly. Information from all channels is transformed into PC scores and the retrieval process can take advantage of maximum information content and retain the best signal to noise ratio of the observed spectrum. This way, there is no need to select sub-set of channels for the sake of computational speed limitation. For quality control purpose, channel radiances can be generated with a simple PC transformation. Due to its fast speed and high accuracy, PCRTM has great potential for the assimilation of the complete spectrum of observed radiances in NWP. Furthermore, the use of the PCRTM may enable cloud parameters in the NWP operation.

5. ACKNOWLEDGEMENTS

This project is supported by funds from NASA and NPOESS Integrated Program Office (IPO). We would like to thank AER for providing the line-by-line radiative transfer mode (LBLRTM) and United Kingdom Meteorology Office for providing ECMWF profiles used in this study.

6. REFERENCES

Armbruster, W., and J. Fischer, 1996: Improved method of exponential sum fitting of transmissions to describe the absorption of atmospheric gases, *Applied Optics.*, 12, 1931-1941.

Chou, M., W. Ridgway, and M. Yan, 1995: Parameterizations for water vapor IR radiative transfer in both the middle and lower atmospheres," *J. Atmos. Sci.*, 52, 1159-1167.

Clough, S. A., and M. J. Iacono, 1995: Line-by-line calculation of atmospheric fluxes and cooling rates: 2. Application to carbon dioxide, ozone, methane, nitrous oxide and the halocarbons," *J. Geophys. Res.*, 100, 16579-16593.

Edwards, D. P., and G. L. Francis, 2000: Improvements to the correlated-k radiative transfer method: Application to satellite infrared sounding, *J. Geophys. Res.*, 105, D14, 18135-18156.

Eyre, J. R., and H. M. Woolf, 1988: Transmittance of atmospheric gases in the microwave region: a fast model, *Appl. Opt.*, 27, 3244-3249.

Gerste. M., 1993: Tropical Obtaining the cumulative k-distribution of a gas mixture from those of its components, *J. Quant. Spectrosc. Radiat. Transfer*, 49, 15-38.

Goody, R., R. West, L. Chen, and D. Crisp, 1989: The correlated-k method for radiation calculations in nonhomogeneous atmospheres, *J. Quant. Spectrosc. Radiat. Transfer*, 42, 539-550.

Lacis, A., and V. Oinas, 1991: A description of the correlated k-distribution method for modeling non-grey gaseous absorption, thermal emission and multiple scattering in vertically inhomogeneous atmospheres, *J. Geophys. Res.*, 96, 9027-9063.

Liu, X., J.-L. Moncet, D. K. Zhou, and W. L. Smith, 2003: A Fast and Accurate Forward Model for NAST-I Instrument, *Fourier Transform Spectroscopy and Optical Remote Sensing of Atmosphere, OSA Topical Meetings*, Quebec, Canada.

Liu, X., W. L. Smith, D. K. Zhou, and Allen Larar, 2004: Super Fast PC-based Radiative Transfer Model, *Advanced High Spectral Resolution Infrared Observations Workshop*, May 24–26, 2004 in Ravello, Italy.

McMillin, L. M., and H. E. Fleming, 1976: Atmospheric transmittance of an absorbing gas, a computationally fast and accurate transmittance model for absorbing gases with constant mixing ratios in inhomogeneous atmospheres, *Appl. Opt.*, 15, 358-367.

Mlawer, E. K., S. J. Taubman, P. D. Brown, M. J. Iacono, and S. A. Clough, 1997: Radiative transfer for inhomogeneous atmospheres: RRTM, a validated correlated-k model for the longwave, *J. Geophys. Res.*, 102, 16663-16682.

Moncet, J.-L., X. Liu, R. Helene, H. Snell, S. Zaccheo, R. Lynch, J. Eluszkiewicz, Y. He, G. Uymin, C. Lietzke, J. Hegarty, S. Boukabara, A. Lipton, and J. Pickle, 2001: Algorithm Theoretical Basis Document (ATBD) for the Cross Track Infrared Sounder (CrIS) Environmental Data Records (EDR). V1.2.3, *AER Document Released to ITT and Integrated Program Office*.

Moncet, J.-L., and G. Uymin, 2003: High spectral resolution Infrared radiance modeling using Optimal Spectral Sampling (OSS) method," *ITSC XIII Proceedings, Sainte Adele, Canada 29 October 2003 – 4 November 2003*.

Saunders, R., M. Matricardi, and P. Brunel, 1999: An improved fast radiative transfer model for assimilation of satellite radiance observations, *Q. J. R. Meteorol. Soc.*, 125, 1407-1425.

Smith, W. L., 1969: A polynomial representation of carbon dioxide and water vapor transmission. U.S. Department of Commerce, Environmental Science Services Administration, National Environmental Satellite Center, Washington, DC, ESSA Technical Report NESC 47, 20 pp.

Strow, L., S. E. Hannon, S. D. Souza-Machado, H. E. Motteler, and D. Tobin, 2003: An overview of t he AIRS radiative transfer model, *IEEE Trans Geosci. Remote Sensing*, 41, 379-389.

TES Level 2 Algorithm Theoretical Basis Document, *Version 1.16*, JPL D-16474, June 27, 2002.

Tjemkes, S. A., and J. Schmetz, 1997: Synthetic satellite radiances using the radiance sampling method," *J. Geophys. Res.*, 102, 1807-1818.

West, R., D. Crisp, and L. Chen, 1990: Mapping transformations for broadband atmospheric radiation calculations," *J. Quant. Spectrosc. Radiat. Transfer*, 43, 191-199.

Wiscombe, W. J., and J. W. Evans, 1977: Exponential-Sum Fitting of Radiative Transmission Functions," *J. of Computational Phys.*, 24, 416-444.

MODELLING UV-EXPOSURE INSIDE A PLANT CANOPY DURING A VEGETATION PERIOD

J. H. Schween and Peter Koepke
Meteorologisches Institut der Universität München
Theresienstrasse 37, 80333 München, Germany

ABSTRACT

Plants are constantly exposed to solar radiation. Thus, the observed raised levels of UV-radiation lead also to a raised exposure for plants. In the recent years, so-called 'non-parasitic leaf spots' (NPLS) have led to yield losses for barley of up to 40%. It is assumed that this disease is caused by raised UV-radiation levels. For this reason a model has been developed to calculate UV-exposure within plant stands for longer periods. By use of plant-specific parameters this model can be applied for different plant types, for forests as well as for crops like barley.

The model calculates the variation of irradiance due to different sun positions, and due to atmospheric parameters such as total ozone content, aerosol amount and properties, as well as clouds. The model takes into account the plant growth, which is important for annual plants like barley. As an example, the UV-exposure has been modelled for a real barley stand in southern Germany for the growing season of the year 2002. The plant-weighted UV-irradiances have been calculated on an hourly basis for 10 levels within the canopy, and from this the UV-exposure as the cumulative UV-radiation of all preceding days. Since plants have repair mechanisms for damage due to UV-radiation, additionally UV-impact was modelled by accounting mainly for UV-irradiance during the most recent days under the assumption that its impact on the plant decays exponentially.

1. INTRODUCTION

Plants are constantly exposed to solar UV-Radiation. In recent years, the so-called 'non-parasitic leaf spots' (NPLS) led in Bavarian barley fields to yield losses of up to 40%. It is assumed that this disease is caused by raised UV levels. Since much effort is necessary to measure radiation in plant stands, and even impossible without changing the shadow conditions in narrow canopies, a model has been developed to calculate the UV-radiation with high temporal resolution during a growing season. There exist a large number of models to describe the visible or PAR spectral range of radiation within plant canopies. Most of them focus on energy exchange with the atmosphere (e.g., Norman 1982) or have the aim of making predictions about the biomass production or agricultural yield of crops. For the UV-spectral range there exist only few attempts to model the irradiance on the plants (e.g., Grant 1999). When modelling UV-radiation within a plant stand three aspects have to be considered:

- Models that describe the energy exchange of a whole plant stand only need to describe the radiation available on average over the area. By contrast a model for the description of UV-exposure should consider the maximum UV-radiation exposure of single leaves.
- In the UV-spectral range the amount of diffuse radiation lies in the range of 50% to 90% of the total irradiance due to the strong Rayleigh and aerosol scattering at short wavelengths. By contrast in the visible range most of the energy is coming directly from the sun, in general about 80%. The strong scattering in the atmosphere in the UV-spectral range leads to a large amount of multi scattering processes, which makes the simulation of UV-radiation in the atmosphere a difficult task, so that a complex model is necessary.
- The reflectivity of the leaves in the UV is below 3%. Accordingly a model for the UV-radiation in a plant canopy does not necessarily need to describe multiple scattering between leaves.

2. MODEL AND INPUT PARAMETERS

2.1 Radiation in the Atmosphere

The radiation transfer within the atmosphere is calculated with the matrix operator model STAR (Schwander et al., 2001). It describes all multi-scattering processes in the atmosphere. Input parameters are position of the sun (given via geographical position, date and time), total ozone content, optical thickness for molecules and aerosols, and spectral behaviour of aerosol scattering and absorption. The result is the radiance field at given wavelengths for the whole sky as well as direct radiance from the sun for cloud-free conditions.

To model the radiation under a partly cloud covered sky the probabilities for cloud-free lines of sight from Lund and Shanklin (1973) are used. They are given for different zenith angles, cloud types and cloud coverages. For the three different cloud types low-, mid- and high level clouds typical optical thicknesses are assumed giving typical transmittances of the clouds for every zenith angle. From this an expectation value of the transmittance of the cloud covered sky can be calculated. This technique is valid if the source of radiation lies above the cloud level and if there is no multi scattering between clouds. This is true for direct solar radiation. It is roughly valid for diffuse radiation originating mainly from rayleigh scattering in the upper atmosphere. Diffuse radiation originating from aerosol scattering is not described precisely with this technique. But since Rayleigh scattering forms the major part of the diffuse radiation in the UV this can be regarded as a good approximation to reality.

The expectation values of the transmittances are calculated separately from observed cloud coverage of the three classes high-, mid- and low level clouds. The radiance fields modelled by STAR are then multiplied with these transmittances giving the probable radiance field.

2.2 Radiation Within the Plant Stand

Inside the plant canopy the radiation is modelled according to the statistical approach of Norman (see Ross 1981). Input parameters are stem and leaf area distribution and the leaf angle distribution. From these parameters the probability to see the sky in a certain direction is calculated. Since the leaves absorb practically all radiation in the UV spectral range, this probability multiplied by the apparent radiance from that direction gives the expectation value for the UV-radiance in the canopy from that direction. For every level within the canopy the radiances from the upper hemisphere are integrated under consideration of a leaf orientation giving the irradiance. This integration is done separately for every wavelength. The resulting spectra are multiplied by an action spectrum describing the effect of the UV-radiation on the plants (Caldwell, 1971).

As explained before, not only the average radiation exposure but also its maximum and the possible range are of interest. The described model is a statistical model and gives only average values, but the range of radiation inside the canopy is mainly produced by the occurrence of sunflecks. Since irradiance from diffuse sky radiation is the sum of radiances from many directions, the expectation value describes the real situation very well. The range of radiation in the canopy can be described by including or omitting the direct solar radiation. The average irradiance Eavg at level z thus lies between Esun for a leaf in the sun and Esdw for a leaf in the shadow:

$$\quad (0.1)$$

With $E_{dir}(h)$ the direct solar irradiance at canopy top, $E_{diff}(z)$ the diffuse solar irradiance at level z inside the canopy and $P_S(z,\theta)$ the probability to see the sky from level z in a direction with zenith angle θ.

Integrating the UV-irradiances E in time gives the UV-exposure H of the leaves. This exposure is a cumulative quantity and accordingly grows at all times. By contrast, it must be assumed that plants recover from the radiative stress of a certain day. There are hints that NPLS appear especially when after a long period with overcast conditions the sun shines again (see Baumer et al., 2001). This means that not the cumulative exposure of the whole vegetation period but the exposure of the most recent days triggers the appearance of NPLS. To describe this, a fictive recovery is modelled by assuming that the impact of the UV-radiation decays exponentially with a time constant Tx. The resulting UV-impact Hx(t) at time t is:

$$H_x(t) = \int_{-\infty}^{t} E(s) \cdot e^{-(t-s)/T_x} \, ds \quad (0.2)$$

2.3 Input Parameters

The model was used to calculate the UV-exposure during the vegetation period of the year 2002 in one of the barley fields of the Bavarian State Research Center for Agriculture (LFL) at Frankendorf near Munich, Germany. Total ozone content was taken from the TOMS database, and information about cloud types and coverage from the German Meteorological Service (DWD), which maintains a station at Munich airport 14 km to the west. The standard SYNOP cloud information from the DWD was transformed into cloud coverages for low-, mid- and high level clouds.

There were no plant parameters, such as spatial leaf area distribution, available for the whole vegetation period in Frankendorf due to the high effort necessary. For this reason, data of leaf area density and stem area density published by Ross (1981) were used. These profiles were adapted to the situation in Frankendorf by taking into account real plant development stages and assigning the Ross-data to the respective day. The original Ross-data are available for 8 development stages. Between these dates a linear interpolation was performed to obtain the necessary information for every day of the vegetation period. For the leaf angles an elliptical leaf angle distribution was assumed. The parameter x was assumed to be 1.7 following the estimate of Campbell and van Evert (1994).

3. RESULTS

With the described methods and parameters the irradiance for horizontal leaves were calculated for the period April 25 to June 10 of the year 2002 for every daytime hour. In Fig. 1 the height and time dependence of the calculated UV-irradiances for May 15, 2002 are shown as an example between 7:00 and 12:00 local time. In accordance with the spatial distribution of the leaves the radiation decays inside the canopy. Above 60 cm this decay is due to ears and stems, below 60 cm it is due to the leaves. The shift of Eavg from the curves of Esun at canopy top and Esdw deep inside the canopy reflects the decay of the probability for a place lit by the sun.

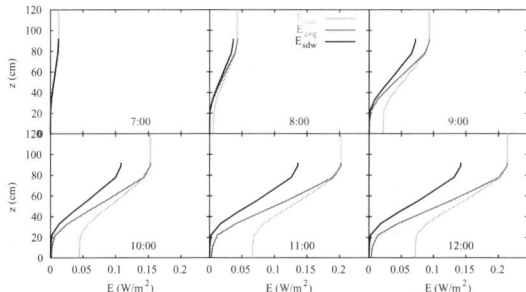

Fig. 1. Profiles of the 'plant'-weighted UV-irradiances for leaves in the sun (E_{sun}), in the shadow (E_{sdw}) and the average leaf (E_{avg}) for May 17, 2002 without clouds. The different panels show the situation between 7:00 and 12:00 local time.

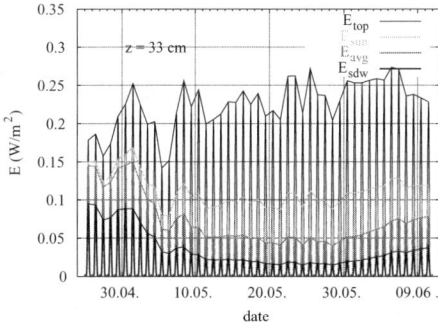

Fig. 2. Course of the 'plant'-weighted UV-irradiances on horizontal leaves at 33 cm inside the canopy without clouds. E_{top} is irradiance at the top of the canopy. Thick lines show the daily course, and thin lines interconnect the daily maximum values at local noon.

Fig. 2 shows the course of the UV-irradiance under the assumption of cloud-free conditions on horizontal leaves as an example for the height of 33 cm in the canopy during the vegetation period. Since this height was reached on April 21, the figure starts a few days later with April 25 when the effects of the growing leaves become visible. The thick lines show the course of irradiance during one day with zero values during night and a daily maximum at local noon. The thin lines interconnect these maximum values to accentuate the course due to varying meteorological conditions and the growing canopy. The variation of Etop is mainly determined by the total ozone content of the atmosphere and the sun position in the sky. The relatively small values on May 5 and May 6 are related to a relatively high total ozone content of about 420 DU. The absolute ozone minimum during the period was on May 1, leading to a relative maximum in the UV-irradiance. Due to the rising elevation of the sun during the following weeks, however even days with higher ozone amount led to higher UV-irradiances. Beside this modulation by total ozone content and solar position, the irradiances inside the canopy at 33 cm decay strongly till May 10 and weaken the following days until May 27. The reason is the strong increase of leaf area above 33 cm until May 10; the following merely weak increase until May 27 is a consequence of the decrease of leaf area due to ripening. With the increase of the leaf area the average UV-irradiance (Eavg) is shifted from the value for sunlit (Esun) leaves towards the value for the shadow (Esdw). As before, this shift reflects the decreasing probability for a sunlit place with increasing leaf area above. In the same manner Eavg moves back towards Esun when the leaf area decreases again after May 27. Analogous patterns can be found for the other levels above and below 33 cm, but are not shown in a figure.

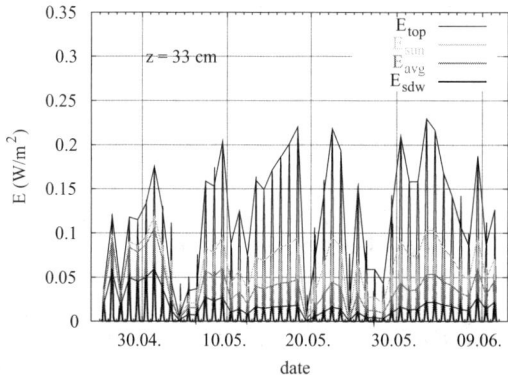

Fig. 3. Course of the Irradiance on horizontal leaves at 33 cm inside the canopy under consideration of observed clouds.

The picture changes considerably if clouds are included, as shown in Fig. 3 for the same conditions as in Fig. 2. Some of the maxima (e.g., May 1, May 8, or June 5) remain visible, because during these days the radiation was influenced only weakly by a few clouds. Very noticeable are the minima in the irradiance around April 26 and May 5, and on May 19 and May 28. These days are characterized by high daytime cloud coverage of low-level clouds. The modulation of the UV-radiation at canopy top is so large that the influence of the growing plants on the irradiance at 33 cm becomes relative weak. As for the case without clouds the course at other levels within the canopy shows in principle the same behavior.

Fig. 4 shows the course of the total time-integrated exposure, again as an example for 33 cm within the plant stand. The integration starts with April 25. The stair structure reflects the integral of the diurnal course of the radiation: every day leads to a rise of the exposure while during the night no change occurs. Without clouds (left panel of Fig. 4) the exposure at canopy top increases nearly linearly besides the stair structure and a slight concave tendency. The slight upward bend is due to the rising solar noon elevation and accordingly increasing irradiance. Inside the canopy this increase is in the beginning slightly reduced, and after May 5 strongly reduced by the increasing leaf area above the 33 cm level. The visible upward bend later on is due to the decreasing leaf area.

If clouds are included in the model (right panel of Fig. 4) the exposure is obviously smaller. Especially the periods with high cloudiness during daytime around April 26, May 5, May 19, and May 28 lead to a smaller or even negligible increase of the exposure with time. The result is 40–45% smaller exposure at the end of the growing season. During the period the difference from the case without clouds lies in the range 14–55% and depends strongly on time and position within the canopy.

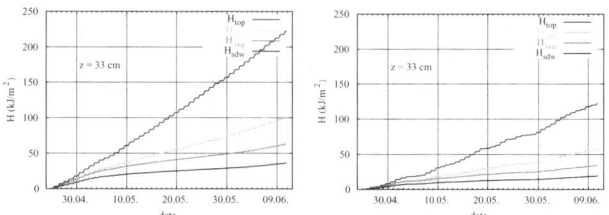

Fig. 4. Course of the UV-exposure of horizontal leaves at 33 cm inside the canopy without (left) and with consideration of observed clouds (right).

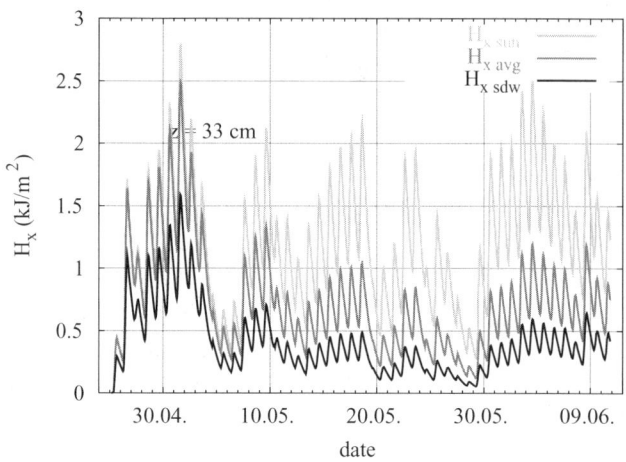

Fig. 5. Course of the exposure impact H_x with a virtual recovery of the plants for horizontal leaves at 33 cm inside the canopy by use of a regeneration time constant T_x of one day.

As described before, to consider the possible effects of repair mechanisms, an UV-impact Hx was modelled with a virtual recovery where the impact of irradiances further in the past are weighted smaller. The regeneration time constant Tx for this model was assumed to be 24 hours because the plants react very sensitively to variations on the order of one day. The result is shown, again for 33 cm, in Fig. 5. The peaks reflect the daily course with its maximum at noon. Due to the assumed regeneration time of one day the effect of the noon maximum of the irradiation decays rapidly, and only a small increase in Hx remains from day to day as long as there are no clouds (e.g., May 14–19). During the strongly clouded periods Hx falls to low values in a short time (e.g., May 4). But it increases again rapidly if clouds vanish and solar radiation increases (e.g., May 7). In total the UV-impact Hx does not increase above a certain level and reaches large values only at times with intense solar radiation. This result provides an explanation of the findings of Baumer, et al. (2001) who observed that the NPLS appear in particular when the irradiance increases strongly after a period with high cloudiness and low radiation and when the plants are at a certain stage of development.

4. ACKNOWLEDGEMENTS

The presented investigations have been performed in the framework of the Bavarian Joint Research Programme on Elevated UV-Radiation (BayFORUV) funded by Bayerisches Staatsministerium für Landesentwicklung und Umweltfragen (StMLU). We thank Sebastian Trepte from the DWD observatory Hohenpeissenberg for providing the cloud data.

5. REFERENCES

Baumer, M., A. Behn, P. Doleschel, K. Fink, and J. Wybranietz: 2001, `Notreife durch nicht-parasitäre Blattverbräunung'. Getreide 2(7), 92-97.

Caldwell, M. M.: 1971, "Solar Ultraviolet radiation and the growth and development of higher plants." Photophysiology 6, 131-177.

Campbell, G. S., and F. van Evert: 1994, `Resource Capture by Crops', Chapt. "Light interception by plant canopies: efficiency and architecture." *Nottingham University Press.*

Grant, R. H.: 1999, "Ultraviolet-B and photo-synthetically active radiation environment of inclined leaf surfaces in a maize canopy and implications for modeling." *Agricultural and Forest Meteorology,* 95, 187-201.

Lund, I. A., and M. D. Shanklin: 1973, "Universal Methods for Estimating Probabilities of Cloud-Free-Lines-Of-Sight through the Atmosphere." *Journal of Applied Meteorology,* 12, 23-35.

Norman, J.: 1982, "Biometeorology" in "Integrated pest management," Chapt. "Simulation of Microclimates," pp. 65-99. New York: Academic Press.

Ross, J.: 1981, "The Radiation Regime and Architecture of Plant Stands." *Junk Publishers,* The Hague-Boston-London.

Schwander, H., A. Kaifel, A. Ruggaber, and P. Koepke: 2001, "Spectral radiative transfer modeling with minimized computation time by use of neural-network technique." *Appl. Opt.,* 40(3), 331-335.

MULTIPLE RAMAN SCATTERING IN THE EARTH'S ATMOSPHERE

R. van Deelen, J. Landgraf, and I. Aben
SRON – National Institute for Space Research, the Netherlands
Utrecht, 3584 CA, the Netherlands

ABSTRACT

Inelastic rotational Raman scattering complicates the interpretation of satellite measurements of backscattered solar radiation. This is because rotational Raman scattering fills-in Fraunhofer lines and absorption features in an atmospheric spectrum of ultraviolet and visible radiation. An accurate simulation of this filling-in is essential in the analysis of measurements by space-borne instruments, such as the Global Ozone Monitoring Experiment (GOME), which has a spectral resolution of 0.2 nm.

Atmospheric radiative transfer models which take Raman scattering into account have been developed over the past few years. However, none of the existing models includes multiple Raman scattering. To study the effect of multiple orders of Raman scattering, the doubling-adding method was extended to simulate the exact filling-in structures. For the numerical implementation, the scalar approximation of radiative transfer was used.

With this extended doubling-adding model the contribution of multiple Raman scattering to the total filling-in could be determined. Results are presented for the spectral range 390–400 nm, which contains two Ca II Fraunhofer lines. The multiple inelastically scattered light in this range was found to contribute 0.2% to the total signal and reached 0.6% at the center of the Fraunhofer lines for GOME-type instruments. For remote sensing applications which make use of the exact amplitude and spectral shape of the filling-in, the neglect of multiple Raman scattering may cause a significant bias of the retrieval product.

1. INTRODUCTION

Raman scattering features are clearly present in ultraviolet and visible earthshine measurements by space-borne spectrometers with a spectral resolution better than 1 nm. For instance, in measurements by the Global Ozone Monitoring Experiment (GOME), which has an instrument spectral resolution of approximately 0.2 nm, these Raman features can be as large as 20% of the measured signal (see Fig. 1). GOME, which is mounted on ERS-2, measures a daily solar spectrum and earthshine spectra in nadir viewing geometry. The measurements are performed in the spectral range 240–790 nm. Similar operational satellite instruments are SCIAMACHY on ENVISAT, with a spectral resolution comparable to GOME, and the Ozone Monitoring Experiment (OMI) on EOS-AURA, with a spectral resolution of approximately 0.5 nm. The interpretation of the earthshine measurements by these space-borne spectrometers requires an accurate radiative transfer model which takes inelastic Raman scattering into account.

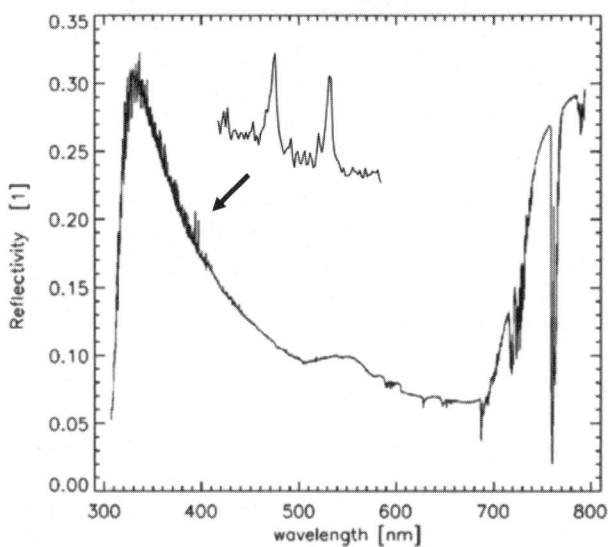

Fig. 1. Reflectivity spectrum (ratio of an earthshine spectrum to the solar spectrum) measured by GOME. The measurements were taken in 1995 above the Netherlands under clear sky condition. The spectral range 390–400 nm is enlarged to show the two pronounced Raman scattering features caused by the filling-in of the Ca II K and H lines in the solar spectrum.

Light-scattering by molecules such as N_2 and O_2 is frequently described by elastic Rayleigh scattering. This only provides an effective description (Young, 1982): in each scattering event, approximately 4% of the radiation is scattered inelastically due to rotational Raman scattering (see Fig. 2).

Rotational Raman scattering has the tendency to smooth spectral structures in an earthshine spectrum. For example, at the location of the Fraunhofer lines, more radiation is scattered from the continuum to the Fraunhofer line-center than the other way around. This makes

Fraunhofer lines shallower in an earthshine spectrum than expected from a purely elastic scattering atmosphere (i.e., a Rayleigh scattering atmosphere). This effect is known as filling-in or as the Ring effect (Grainger and Ring, 1962). Not only Fraunhofer lines, but also strong absorption features of, e.g., ozone are filled-in.

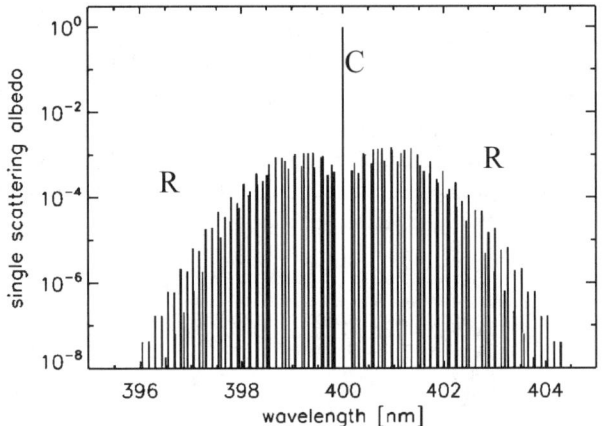

Fig. 2. Single scattering albedo of air for incident light at 400 nm, in the absence of absorption. Approximately 96 % of the light is scattered elastically (C, Cabannes line) and 4 % of the light is scattered inelastically (R, rotational Raman lines).

Over the past years, accurate radiative transfer models have been developed which take rotational Raman scattering into account (e.g., Joiner et al., 1995; Vountas et al., 1998; Landgraf et al., 2004). All these models use different approaches, but they have in common that a Rayleigh scattering atmosphere is considered which is corrected for one order of Raman scattering. The first model which accounts for two orders of Raman scattering was presented by Aben, et al. (2000), but due the second order of scattering approximation the continuum height was not properly accounted for in this model (Aben et al., 2001; Stam et al., 2002).

To study the effect of multiple Raman scattering, a reference model was developed by Van Deelen et al. (2005) which takes multiple scattering processes, both elastic and inelastic, into account. The doubling-adding method (e.g., Hansen and Travis, 1974; De Haan et al., 1987) was extended to include inelastic scattering. With this model it was possible to determine the exact spectral shape of the filling-in and to quantify the contribution of multiple Raman scattering.

2. DOUBLING-ADDING MODEL

The doubling-adding model offers a different approach than solving the radiative transfer equation directly. Based on the reflection and transmission properties of individual atmospheric model layers, the corresponding properties of the model atmosphere can be calculated using the doubling and adding equations (e.g., Lacis and Hansen, 1974). To study the main effect of multiple Raman scattering, the radiative transfer equation is solved in its scalar approximation. This means that any effect of polarization is neglected.

To extend the doubling-adding method to include inelastic scattering, the reflection and transmission functions had to be redefined. These functions relate the incoming radiation to the reflected and transmitted radiation of a given model layer. In the elastic scattering case, only scattering from the angle of incoming radiation to the angle of the emerging radiation, at each individual wavelength, had to be considered; now, also scattering from other wavelengths to each wavelength has to be taken into account. For example, the reflection function can be written as $R(\lambda,\Omega;\lambda',\Omega')$, which relates λ', Ω', the wavelength and direction of the incoming radiation at the top of the model layer, to λ, Ω, the wavelength and direction of the emerging radiation at the top of the model atmosphere layer.

As a consequence, each product of reflection and transmission functions in the doubling-adding scheme, C = AB, needs to be modified to

$$C(\lambda,\Omega;\lambda',\Omega') = \frac{1}{\pi}\int_0^\infty d\lambda'' \int_0^{2\pi} d\Omega'' \mu'' \quad (1)$$
$$A(\lambda,\Omega;\lambda'',\Omega'') B(\lambda'',\Omega'';\lambda',\Omega'),$$

where not only integration over all internal angles Ω'' takes place, but also integration over all possible wavelengths λ''. Here, μ'' is the cosine of the zenith angle. With this modification, the general form of the doubling and adding equations, which yield the reflection and transmission properties of combined model layers, remains exactly the same as in the elastic scattering case.

For the numerical implementation of the extended doubling-adding model, it was critical to adopt an optimized wavelength grid to reduce the computational effort. This internal wavelength grid contains only the relevant wavelengths of all possible λ'' in Eq.(1). More information on the trade-off between accuracy and computing effort can be found in Van Deelen, et al. (2005).

The benefits of the doubling-adding approach are clear. Instead of applying increasingly complex corrections for each order of Raman scattering, the doubling-adding approach only requires to be initiated with the exact single scattering solution of an optically thin model layer including inelastic scattering. Then, with the doubling-adding scheme, the multiple scattered radiation, both elastic and inelastic, comes out in a straightforward way.

3. MODEL RESULTS

In this section, model results are presented for the spectral range 390–400 nm. This range contains the strong Ca II K and H Fraunhofer lines at 393 nm and 397 nm respectively, which introduce the most prominent Raman scattering features in the GOME spectrum. In this wavelength range, ozone absorption is negligible. This means that the observed filling-in features are caused by the Fraunhofer lines only.

A model atmosphere of two layers was used, composed of 78% NB2B, 21% OB2B, 1% Ar and ozone (360 Dobson units in total). The air density profile, ozone density profile and temperature profile were adapted from the US standard atmosphere (NOAA, 1976). Furthermore, the surface was considered to be black. The doubling-adding calculations were performed on a 1 cmP-1P grid and later convoluted with a Gaussian with a full width half maximum of 0.2 nm to simulate GOME measurements. The input solar spectrum was taken from Chance and Spurr (1997).

For the first time, the effect of multiple Raman scattering on the spectral shape of the filling-in structures can be presented with the doubling-adding model. In Fig. 3A the reflectivity spectrum versus wavelength is shown. The solid line shows the result of the doubling-adding model including all orders of Raman scattering. The dotted line shows the result for a model calculation where only the elastic scattering component, i.e., the Cabannes line in Fig. 2, is taken into account. Obviously, filling-in does not occur and a considerable fraction of the light is missed when inelastic scattering processes are neglected. The difference between the solid and the dotted line, which is shown in Fig. 3B, can be interpreted as the inelastic fraction, which is the fraction of scattered radiation due to one or more orders of Raman scattering. Percentages ranging from about 5% to more than 20% show that rotational Raman scattering cannot be ignored in the interpretation of GOME-type measurements.

With a slight modification to the doubling-adding model, it was possible to determine the contribution of one order of Raman scattering (Van Deelen et al., 2005). Subtracting this from the model result where all scattering processes were taken into account (Fig. 3A), the contribution of multiple Raman scattering could be isolated. The result is shown in Fig. 3C. Multiple Raman scattering contributes 0.2% to the total signal. At the location of the Ca II K and H lines, the multiple Raman scattering fraction reaches approximately 0.6%. Important to note is that the exact percentages depend on wavelength, instrument spectral resolution, viewing geometry, solar zenith angle, choice of model atmosphere and surface albedo.

Calculations for other spectral ranges were also performed. In the range 280-290 nm ozone absorption is so strong that the single scattering approximation suffices. For longer wavelengths, multiple scattering becomes more important. In the range 320–330 nm, ozone absorption and scattering are of comparable strength. The lack of strong Fraunhofer lines in this range though prevents strong filling-in features to appear. Still, the multiple Raman scattering fraction of the continuum here is approximately 0.5%. At longer wavelengths than 350 nm ozone absorption becomes weaker and, as mentioned before, can even be neglected around 390–400 nm. The strength of scattering by air molecules decreases with wavelength to the fourth power and, therefore, reduces the importance of (multiple) Raman scattering toward longer wavelengths.

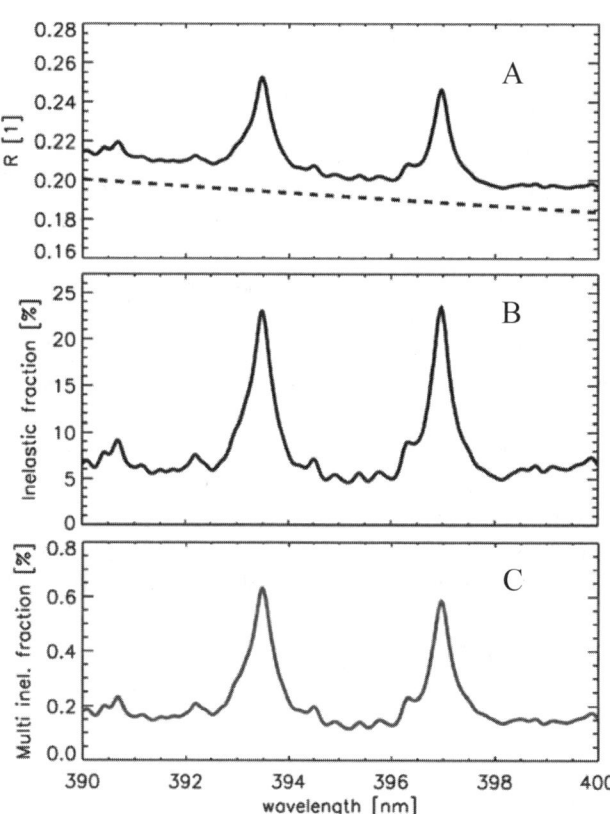

Fig. 3. (A) Reflectivity for the spectral range 390–400 nm, (B) inelastic scattering contribution, and (C) multiple inelastic scattering contribution. Results are shown for a solar zenith angle of 60 degrees, a viewing angle of 30 degrees and an azimuth angle difference of 10 degrees.

4. CONCLUSION

For the first time, a radiative transfer model is presented which takes multiple rotational Raman scattering into account. With this model we were able to quantify the contribution of multiple Raman scattering to the continuum, which was approximately 0.5% in the spectral range 320–330 nm and 0.2% in the spectral range 390–400 nm. At wavelengths where filling-in occurs, this contribution is generally larger. For example at the Ca II Fraunhofer lines, the multiple Raman scattering contribution was found to be 0.6% for instruments with a spectral resolution similar to GOME.

In applications which make use of the exact amplitude and spectral shape of the filling-in features, the neglect of multiple Raman scattering may cause a significant bias of the retrieval product. Also, when future instruments obtain better spectral resolution, the need for accurate radiative transfer models which include (multiple) inelastic Raman scattering will increase.

5. REFERENCES

Aben, I., R. Tamboer, and D. M. Stam, 2000: The Ring effect in skylight polarization spectra. In *IRS 2000: Current Problems in Atmospheric Radiation,* A. Deepak Publishing, Hampton, Virginia, 273-275.

Aben, I., D. M. Stam, and F. Helderman, 2001: The Ring effect in skylight polarisation. *Geophys. Res. Lett.,* 28, 519-522.

Chance, K. V, and R. J. D. Spurr, 1997: Ring effect studies: Rayleigh scattering, including molecular parameters for rotational Raman scattering, and the Fraunhofer spectrum. *Appl. Opt.,* 36, 5224-5230.

De Haan, J. F., P. B. Bosma, and J. W. Hovenier, 1987: The adding method for multiple scattering calculations of polarized light, *Astron. Astrophys.,* 183, 371-391.

Grainger, J. F., and J. Ring, 1962: Anomalous Fraunhofer line profiles. *Nature,* 193, 762-762.

Hansen, J. E and L. D. Travis, 1974: Light scattering in planetary atmospheres, *Space Sci. Rev.,* 1974, 16, 527-610.

Joiner, J., P. K. Bhartia, R. P. Cebula, E. Hilsenrath, R. D. McPeters, and H. Park, 1995: Rotational Raman scattering (Ring effect) in satellite backscatter ultraviolet measurements. *Appl. Opt.,* 34, 4513-4524.

Lacis, A. A. and J. E. Hansen, 1974: A parameterization for the absorption of solar radiation in the Earth's atmosphere. *J. Atmos. Sci.,* 31, 118-133.

Landgraf, J., O. P. Hasekamp, R van Deelen, and I. Aben, 2004: Rotational Raman scattering of polarized light in the Earth's atmosphere: a vector radiative transfer model using the radiative transfer perturbation theory approach. *J. Quant. Spectrosc. Radiat. Transfer,* 87, 399-433.

NOAA, 1976: U.S. standard atmosphere, 1976, Report NOAA-S/T76-1562, National Oceanic and Atmospheric Administration, Washington, DC, U.S. *Government Printing Office.*

Stam, D. M., I. Aben, and F. Helderman, 2002: Skylight polarization spectra: Numerical simulation of the Ring effect. *J. Geophys. Res.,* 107, 4419-4434.

Van Deelen, R., J. Landgraf, and I. Aben, 2005: Multiple elastic and inelastic scattering in the Earth's atmosphere: a doubling-adding method to include rotational Raman scattering by air. *J. Quant. Spectrosc. Radiat. Transfer,* in press.

Vountas, M., V. V. Rozanov, and J. P. Burrows, 1998: Ring effect: impact of rotational Raman scattering on radiative transfer in Earth's atmosphere. *J. Quant. Spectrosc. Radiat. Transfer,* 60, 943-961.

Young, A. T., 1982: Rayleigh scattering. *Phys. Today,* 35, 42-48.

STOCHASTIC RADIATIVE TRANSFER IN MULTILAYER BROKEN CLOUDS: APPROACH, VALIDATION AND APPLICATION

E. I. Kassianov
Pacific Northwest National Laboratory
Richland, WA 99354, USA

ABSTRACT

We describe a new statistical treatment of solar radiative transfer in multilayer broken clouds. The proposed approach is a logical development of the statistical approaches originally suggested for single-layer broken clouds. A new statistically inhomogeneous Markovian model and the stochastic radiative transfer equation have been used to derive equations for the mean radiance of solar radiation. The obtained equations for the mean direct radiance agree with corresponding equations previously derived for (i) broken clouds with maximum and random overlap, and (ii) overcast clouds with vertical inhomogeneity. It is demonstrated that the obtained equations for the mean direct and diffuse radiances could provide reasonable accuracy for the ensemble- and domain-averaged radiative properties.

1. INTRODUCTION

Cloud fields are stochastic scattering media due to their inhomogeneous internal structure and stochastic geometry. A radiation field transformed by such a cloud field also becomes random. This fact dictates the necessity to use statistical methods to study the cloud-radiation interaction. The ultimate goal of a statistical approach is to suggest a relatively simple and practical treatment of the stochastic radiative transfer problem and to establish the relationship between the statistical parameters of clouds and radiation. Although the necessity of such statistical treatment was acknowledged long ago (e.g., Stephens et al., 1991) and several approaches were suggested (e.g., Cahalan et al., 2005 and bibliography therein), the solution of the complex problem as a whole is still far from completed. Here we outline the statistical treatment of solar radiative transfer in multilayer broken clouds. The proposed approach is a logical development of the statistical ones originally suggested for single-layer broken clouds (e.g., Titov, 1990; Malvagi et al., 1993). The term "broken clouds" means that the cloud field has stochastic geometry and deterministic optical parameters inside an individual cloud.

2. APPROACH

Real clouds have strong horizontal and vertical variability. To extend the Markovian approach (e.g., Titov, 1990; Malvagi et al., 1993) for *multi-layer* broken clouds, we have to specify a statistical relationship between cloud layers. It was demonstrated (e.g., Hogan and Illingworth, 2000) that (i) the overlap of clouds at two levels tends to fall rapidly as their vertical separation is increased, and (ii) the degree of overlap as a function of level separation can be described by a simple inverse-exponential expression. In our approach, we assume that the statistical relationship between two adjusted layers is described by inverse-exponential expression. We represent broken clouds as a set of correlated cloud layers: each layer is homogeneous in vertical but inhomogeneous in horizontal dimensions. Such a representation allows one to use methods developed for a single layer of broken clouds.

The suggested approach is based on the stochastic radiative transfer equation and a new statistically inhomogeneous Markovian model of broken clouds (Kassianov, 2003). The term "statistical inhomogeneity" is understood to mean that cloud statistics depend on the vertical coordinate. The vertical profile of the cloud fraction $p(z)$ is one of the input parameters of this statistically inhomogeneous model. There are three additional input parameters. The first additional parameter $A(z)$ characterizes the mean horizontal cloud size in each layer. The second parameter $A^{up}(z)$ describes the statistical relationship between two adjacent layers in an upward direction. The third parameter $A^{dw}(z)$ specifies the statistical relationship between two adjacent layers in a downward direction.

This statistically inhomogeneous model has three appealing features. First, all input parameters can be derived from observations. This allows one to correctly compare theory predictions with field data. Further, the model flexibility makes it possible to describe the different combinations of random and maximum overlap, which are typical for the majority of multilayer broken clouds. For example, if clouds are perfectly independent for any two adjacent layers (the random cloud overlap) then $A^{up}(z) \gg 1$ and $A^{dw}(z) \gg 1$. If $p(z)$=const and clouds are perfectly dependent for any two adjacent layers (the maximum cloud overlap), then $A^{up}(z)=0$ and $A^{dw}(z)=0$. Hereafter we will use subscripts, *up*, and, *dw*, for upward and downward radiation, respectively. Finally, the statistically inhomogeneous model is a generalization of the statistically homogeneous ones (e.g., Titov, 1990; Malvagi et al., 1993) that have been originally suggested for one-layer broken clouds. This circumstance makes it possible to apply a beautiful theory and an abundance of elegant numerical methods that have been developed for the statistically homogeneous models.

The statistically inhomogeneous model of broken clouds and the stochastic transfer equation were used to derive approximated equations for the mean solar radiance (Kassianov, 2003). It was assumed that, for each kth layer, the domain-averaged optical properties are constant (piecewise constant approximation), e.g., the extinction coefficient $\sigma(\mathbf{r}) = \sigma(z) = \sigma_k$, the single scattering albedo $\omega_0(\mathbf{r}) = \omega_0(z) = \omega_{0,k}$ and the scattering phase function $g(\mathbf{r},\omega,\omega') = g(z,\omega,\omega') = g_k(\omega,\omega')$, where $\mathbf{r} = (x,y,z)$ and $\omega = (a,b,c)$. Also it was assumed that a parallel unit flux of solar radiation is incident on the upper boundary of a given cloud field in direction $\omega_\oplus$.

The derived equation for the mean solar radiance $\langle I(z,\omega) \rangle$ has the form

$$\langle I(z,\omega) \rangle = \frac{1}{|c|} \int_{E_z} \omega_0(\xi)\phi(z,\xi)d\xi \int_{4\pi} g(\xi,\omega,\omega')f(\xi,\omega')d\omega' + \langle j(z,\omega) \rangle \delta(\omega - \omega_\oplus) \quad (1)$$

where $E_z = (hb, z)$ if $c>0$, and $E_z = (z, ht)$ if $c<0$, ht and hb are the top height and the base height of cloud field, respectively; $\langle j(z,\omega) \rangle$ is the mean direct (unscattered) radiance, $f(z,\omega) = \sigma(\mathbf{r})\langle \kappa(\mathbf{r})I(\mathbf{r},\omega) \rangle$ is the mean collision density, $\kappa(\mathbf{r})$ is random indicator function ($\kappa(\mathbf{r}) = 1$ inside clouds and $\kappa(\mathbf{r}) = 0$ outside the clouds), and $\delta(\cdot)$ is Dirac's delta function.

The derived integral equation for the mean collision density (Kassianov, 2003) is given by

$$f(\mathbf{x}) = \int_X k(\mathbf{x},\mathbf{x}') f(\mathbf{x}') d\mathbf{x}' + \Psi(\mathbf{x}) \quad (2)$$

where

$$k(\mathbf{x},\mathbf{x}') = \frac{\omega_0(z) g(z,\omega,\omega') \eta(\mathbf{r},\mathbf{r}')}{2\pi |\mathbf{r}-\mathbf{r}'|^2} \delta\left(\frac{\mathbf{r}-\mathbf{r}'}{|\mathbf{r}-\mathbf{r}'|} - \omega\right)$$

$$\Psi(\mathbf{x}) = \sigma(z) p(z) v(z,\omega) \delta(\omega - \omega_\oplus),$$

X is the phase space of coordinates and directions, $\mathbf{x} = (\mathbf{r}, \omega)$. All functions $\phi(\mathbf{r},\mathbf{r}')$, $\eta(\mathbf{r},\mathbf{r}')$ and $v(z,\omega) = \langle \kappa(\mathbf{r}) j(\mathbf{r},\omega) \rangle / p(z)$ are defined by the recurrent expressions. Note that the closed system of equations (1)–(2) can be solved by using any appropriate numerical method or analytic technique (e.g., the spherical harmonic method). We have developed Monte Carlo algorithms for solving this system.

3. VALIDATION

Mean Direct Radiance. We have demonstrated that the general analytical equations for the direct radiance (derived by us) are equivalent to those obtained by others for the three limiting cases: broken field with *maximum* overlap, broken field with *random* overlap and overcast *vertically inhomogeneous* field.

Mean Diffuse Radiance. We have estimated the accuracy of the approximated equations for diffuse radiance by comparing the ensemble-averaged radiative properties obtained by two independent methods (*numerical averaging* method and *analytical averaging* method). In our analysis we used the three-dimensional (3D) cloud fields provided by (i) the stochastic Boolean model, (ii) large-eddy simulation model, and (iii) satellite cloud retrieval (Kassianov et al., 2003a).

Numerical averaging method. In each of these 3D realizations we calculated radiative properties by using a Monte Carlo method and periodic boundary conditions. The mean radiative properties were obtained after appropriate processing. Since the *full* 3D cloud geometry is used in the radiative calculations, the calculated mean radiative properties are considered the reference values.

Analytical averaging method. Approximated equations for the mean radiance, which have been derived by analytically averaging the stochastic radiative transfer equation, are also used for estimating mean radiative properties. Contrary to the numerical averaging method, another Monte Carlo technique was applied for solving these equations. Since only the *bulk* cloud statistics (the vertical profiles of cloud fraction p_k and parameters A_k, A_k^{up} and A_k^{dw}) are used in the radiative calculations, the mean radiative properties obtained by this method are considered as approximations of the true radiative properties. Note, that using the approximated equations allows one to significantly speed up (by a factor of 10 to 100) calculations of the ensemble-averaged radiative properties.

4. VALIDATION RESULTS

Here we show some of validation results (Kassianov et al., 2003a) obtained for marine low-level cumulus clouds (Fig. 1). The 3D cloud field was obtained from collocated and coincident satellite and ground-based radar observations in the tropical Pacific region at the island of Nauru (Kassianov et al., 2003b). The reconstructed 3D geometry of broken clouds corresponds to the domain ~30x30x2 km with 0.275–km horizontal resolution (total number of pixels is 110x110).

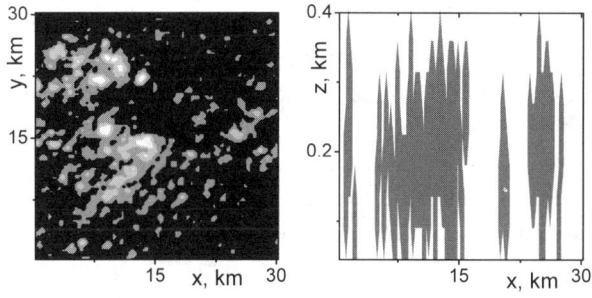

Fig. 1. The horizontal (left) and the vertical (right) distributions of broken clouds. Brightness in the horizontal distributions (a) is proportional to the geometrical thickness of clouds.

For this field, the bulk geometrical statistics (*domain-averaged*) were derived (Fig. 2) and then used as input data to calculate the mean radiative properties. It is easily seen that cloud statistics (cloud geometry) show substantial vertical variability.

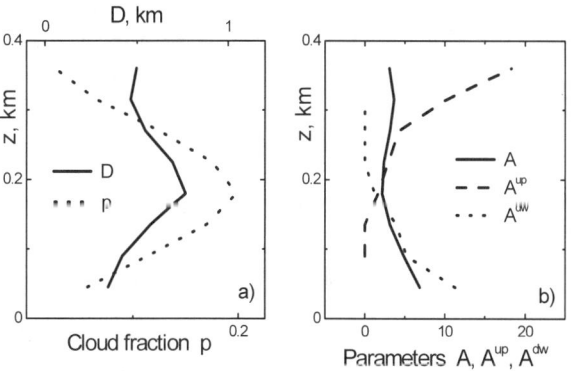

Fig. 2. The vertical profiles of the cloud fraction, p, and the mean cloud horizontal chord, D (a). The vertical distribution of parameters A, A^{up} (upward direction) and A^{dw} (downward direction) (b).

Figs. 3 and 4 show the mean radiative properties calculated by two independent methods. The full 3D cloud geometry was used in the first method (reference), and only the bulk cloud statistics (Fig. 2) were applied for the second method (approximation). While the radiative calculations were performed for a set of solar zenith angles (SZAs), below we demonstrate results for SZA=0 (high Sun) and SZA=70 (low Sun).

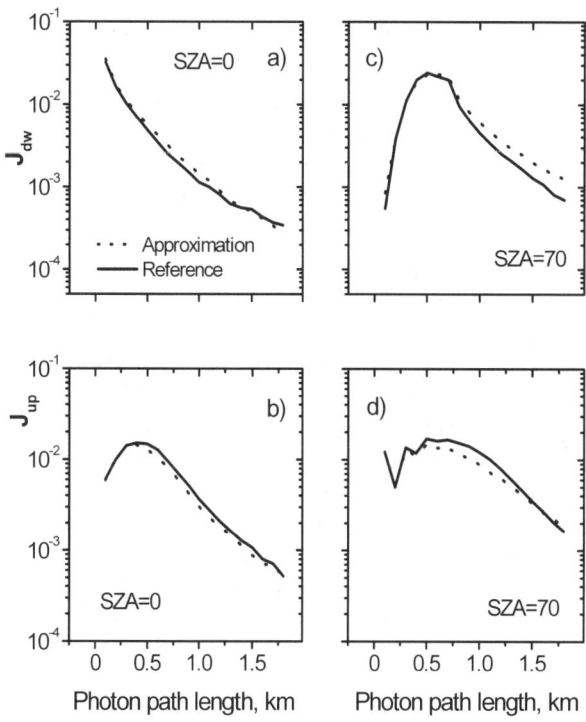

Fig. 3. The mean photon path length distributions in the transmitted (a,c) and reflected (b,d) radiation.

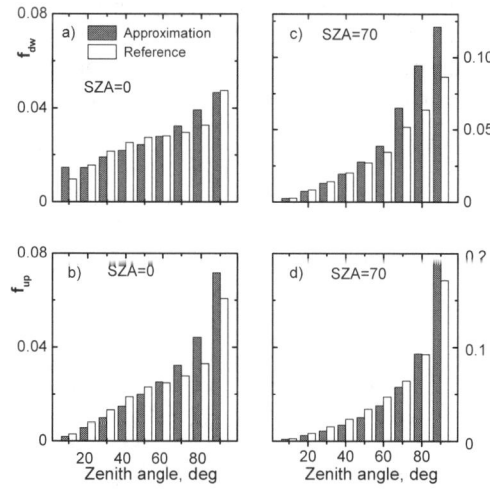

Fig. 4. The mean angular distribution histograms (azimuth-averaged) of the transmitted (a,c) and reflected (b,d) radiation.

As can be seen (Figs. 3 and 4), the mean radiative properties, which have been obtained by exact and approximated methods, agree qualitatively and quantitatively. The similar agreement was obtained for other radiative properties (e.g., fluxes) and different cloud fields (Kassianov et al., 2003a; Kassianov et al., 2005).

5. CONCLUDING REMARKS

The results above are, in our view, a significant step forward in the development of new parameterizations for the representation of 3D broken clouds (multi-layer) and their radiative properties. Using simple probabilistic assumptions (Markovian approximation), we have introduced a simplified but realistic representation of the 3D structure of broken clouds. The model parameters represent observable bulk cloud features/statistics. Validation tests show good predictive performance for the mean radiative properties; moreover, the model is reasonably efficient in its parameterization thanks to the simplifying assumptions. These equations, that *analytically* link statistics of 3D multi-layer broken clouds to their mean radiative properties, can form the basis of remote sensing. The new stochastic treatment of solar radiative transfer in *multi-layer* broken clouds needs more evaluation over additional cloud fields, but the obtained results suggest it has considerable promise.

6. ACKNOWLEDGMENTS

This work was supported by the U.S. Department of Energy's Office of Biological and Environmental Research as part of the Atmospheric Radiation Measurement (ARM) Program. We express our gratitude to Professor G. Titov for initiating the development of this approach and Dr. Cahalan for the invitation to give a talk at the IRS'2004.

7. REFERENCES

Cahalan R. F., et al., 2005: The International Intercomparison of 3D Radiation Codes (I3RC): Bringing together the most advanced radiative transfer tools for cloudy atmospheres, *Bull. Amer. Meteor. Soc.*, in press.

Hogan, R. J., and A. J. Illingworth, 2000: Deriving cloud overlap statistics from radar, *Q.J.R. Meteorol. Soc.*, 126, 2903-2909.

Kassianov, E. I., 2003: Stochastic radiative transfer in multilayer broken clouds. Part I: Markovian approach, *J. Quant. Spectrosc. Radiat. Transfer*, 77, 373-393.

Kassianov, E. I., T. P. Ackerman, R. T. Marchand, and M. Ovtchinnikov, 2003a: Stochastic radiative transfer in multilayer broken clouds. Part II: Validation tests, *J. Quant. Spectrosc. Radiat. Transfer*, 77, 395-416.

Kassianov, E. I., T. P. Ackerman, R. T. Marchand, and M. Ovtchinnikov, 2003b: Satellite multi-angle cumulus geometry retrieval: Case Study, *J. Geophys. Res.*, 108, doi:10.1029/2002JD002350.

Kassianov, E. I., T. P. Ackerman, and P. Kollias, 2005: The role of cloud-scale resolution on radiative properties of oceanic cumulus clouds, *J. Quant. Spectrosc. Radiat. Transfer*, 91, 211-226.

Malvagi, F., R. N. Byrne, G. C. Pomraning, and R. C. J. Somerville, 1993: Stochastic radiative transfer in a partially cloudy atmosphere. *J. Atmos. Sci.*, 50, 2146-2158.

Stephens, G. L., P. M. Gabriel, and S. C. Tsay, 1991: Statistical radiative transport in one-dimensional media and its application to the terrestrial atmosphere. *Transport Th. Statist. Phys.*, 20, 139-175.

Titov, G. A., 1990: Statistical description of radiation transfer in clouds. *J. Atmos. Sci.*, 47, 24-38.

INFERRING DOMAIN-AVERAGED CLOUD PROPERTIES FROM THE ARM OBSERVATIONS AND TESTING THE PCLOS MODELS

Yingtao Ma[a] and Robert G. Ellingson[b]

[a]Department of Meteorology, University of Maryland, College Park, MD 20742
Email: ytma@atmos.umd.edu

[b]Department of Meteorology, Florida State University, Tallahassee, FL 32306
Email: bobe@met.fsu.edu; Tel: (850) 644-8583

1. INTRODUCTION

In climate studies, longwave radiation fluxes and heating rate are usually calculated as the weighted average of clear and overcast values, i.e.,

$$F = (1 - N)F_{clear} + NF_{overcast} \quad (1)$$

where, F is downwelling flux of radiant energy at the surface; Subscripts clear and overcast denote fluxes for those conditions. N is the absolute cloud fraction, usually assumed to be the fractional coverage of the vertical projections of plane-parallel clouds. However, real cloud fields are non-uniform in both morphological and microphysical senses. The error due to neglecting the 3D effects can be climatically significant. (Harshvardhan and Weinman, 1982; Ellingson, 1982; Heidinger and Cox, 1996; Han and Ellingson, 1999; Takara and Ellingson, 2000).

Three characteristics of 3D clouds have been found to be important for longwave radiative transfer. They are: (1) the 3D geometrical structure of the cloud fields (Geometrical effect), (2) the horizontal variation of cloud optical depth (Variable optical depth effect), and (3) the vertical variation of cloud temperature (Non-isothermal cloud effect). At zenith angles $\theta > 0$, vertically extended clouds project greater areas than the PPH clouds. This is denoted as the geometrical effect. By neglecting the geometrical effect the PPH approximation will underestimate the downwelling longwave flux.

One way to incorporate the 3D geometrical effect in climate studies is through the use of an effective cloud fraction (Ellingson 1982; Han and Ellingson 1999). That is,

$$F = (1 - N_e) F_{clear} + N_e F_{overcast} \quad (2)$$

The effective cloud fraction, N_e, is the plane-parallel cloud fraction that generates the same flux as the detailed models for a given broken cloud field after taking into account the effects of geometric shapes, size, spatial distribution and absolute amount (N) of clouds. These effects may be integrated into N_e through a single cloud field property—the Probability of Clear Line Of Sight (PCLOS). The PCLOS also plays an important role in accounting for longwave 3D effects caused by variable cloud optical depth and vertical change of cloud temperature. In this study we use a variety of data from the Atmospheric Radiation Measurement (ARM; Stokes and Schwartz, 1994) program to test several different PCLOS models.

2. THE PROBABILITY OF A CLEAR LINE OF SIGHT (PCLOS)

For a given absolute cloud fraction, N, the PCLOS is a function of cloud shape, aspect ratio and distribution of the clouds in the cloud field. Assuming the clouds have identical base heights, are randomly distributed in an infinitely large horizontal plane, and are right cylinders or semi-ellipsoids, the PCLOS at zenith angle θ is given as (Ellingson, 1982; Ma, 2004)

$$p(\theta) = (1-N)^{f(\theta)}$$

$$\text{where } f(\theta) = \begin{cases} 1 + \frac{4}{\pi}\beta \tan\theta, & \text{for right-cylindrical clouds.} \\ \frac{1}{2}\left(\sqrt{1 + 4\beta^2 \tan^2\theta} + 1\right), & \text{for semi-ellipsoidal clouds} \end{cases} \quad (3)$$

β is the cloud aspect ratio defined as the ratio of the cloud height to the horizontal size. Theoretically, the PCLOS decreases monotonically from (1-N) at zenith to zero at the horizon.

3. ARM CLOUD OBSERVATIONS: FROM TIME SERIES TO DOMAIN AVERAGE

The ARM cloud sensing instruments are fixed at the surface at the ARM Southern Great Plains (SGP) site, and most can only sample the cloud field in the zenith direction. Assuming the statistical properties of the cloud field do not change as the clouds move with the mean wind allows us to infer the domain-averaged properties from a time series of observations. This raises two questions: (1) For a given wind condition, over how long a period or over how many individual clouds does one need to average to obtain representative statistics? and (2) What sampling rate will give the most accurate estimate?

Assuming that the cloud field is a homogeneous and generated by an isotropic random process, the target domain is a rectangular area and the sampling transect is located at the center of the target area, the sampling error can be estimated from the correlation function of the cloud field, the number of observations and the interval between observations. Using the cloud fraction as an example, the sampling error can be expressed as:

$$\sigma_{\hat{N}}^2 = \sigma_N^2 \left[\frac{1}{n} + \frac{2}{n^2} \sum_{i=1}^{n-1} (n-i)\, r(i\Delta l) \right] \quad (4)$$

where $\hat{N}$ is the estimate of the cloud fraction along the single transect; $\sigma_{\hat{N}}^2$ denotes the variance of $\hat{N}$, which is a function of the number of sample points, n, and the correlation function of the cloud field, r; Δl is the interval between two consecutive observations; $\sigma_{\hat{N}}^2$ is the variance of the cloud field. (Kagan, 1997)

The correlation function in equation (4) describes the statistical spatial structure of the cloud field. Fig. 1 shows the estimated correlation functions for 45 days of single layer cumulus cloud fields over the ARM CART site from July 2000 through October 2001. Also shown in the figure is the fitted correlation model, which is assumed to be a negative exponential function. The e-folding parameter for the correlation model, ρ_0 is estimated to be about 1.3 km. The value is obtained by fitting the model to the mean observed correlation function. When fitted to each of the 45 cases, the standard deviation of ρ_0 is 0.7 km.

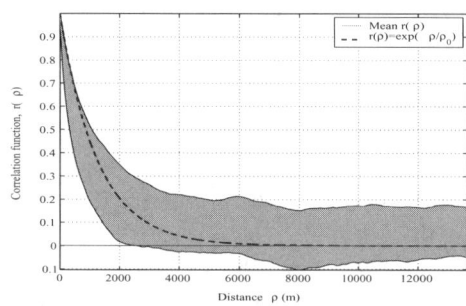

Fig. 1. Observed and modeled correlation functions for single layer cumulous cloud field of the ARM CART site.

Based on an analysis of four cloud resolving model generated cloud fields with (4) and the measured correlation function, we conclude that: (1) For a domain size = $100\rho_0 \times 100\rho_0$, a transect of length $50\rho_0$ can achieve a relative root-mean-square (RRMS) error of about 30% when inferring N (assuming a 10 m/s wind speed, $50\rho_0$ corresponds to about 100min); (2) When the sampling interval $\Delta l < \rho_0$, further increasing sampling rate does not significantly improve the accuracy; and (3) If $\Delta l > 3\rho_0$, the observations can be seen as independent of each other.

4. DETERMINING THE CLOUD FIELD PARAMETERS FROM THE ARM CLOUD OBSERVATIONS

4.1 The PCLOS

The PCLOS is estimated from Total Sky Imager (TSI) data by determining the fraction of clear pixels within each 1° annular ring from zenith to instrument horizon for each image on days when only cumulus clouds were present. PCLOS(θ) for a given period is determined by averaging over several images. Fig. 2 shows $p(\theta)/(1-N)$ for 86 cumulus cloud fields selected from July 2000 to October 2001. This normalized PCLOS is the conditional probability of a clear line of sight given that the line of sight reaches the cloud base level in the (1-N) portion of the cloud field. Alternatively, $1 - p(\theta)/(1-N)$ is the probability of seeing cloud sides at an angle θ given that the line of sight reaches the cloud base level in the (1-N) portion of the cloud field. Generally, the curve decreases from 1 at zenith to 0 at horizon. Note that some cases have a conditional probability larger than 1 at some angles. This is likely caused by the presence of a cloud streak or an inhomogeneity in the cloud field.

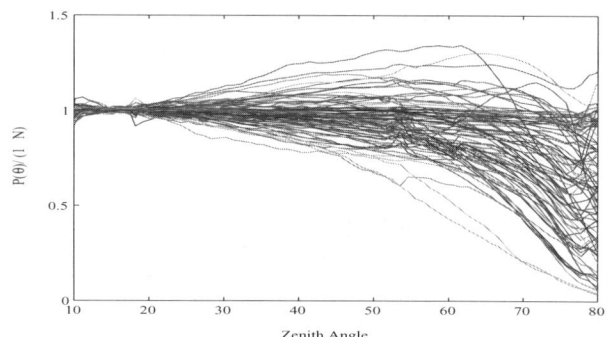

Fig. 2. The PCLOS inferred from the TSI.

4.2 Cloud Thickness

Several laser instruments at the ARM-CART site directly measure cloud base height. The Active Remotely-Sensed Clouds Locations (ARSCL) products give a good estimate of the cloud base height by merging these laser observations. Due to the clutter problem, sometimes, it is difficult to determine the top height of the fair weather cumulus from the cloud radar (MMCR) observations. In this study, we combined the information from the MMCR and the relative humidity (RH) profiles to infer the needed cloud top height. First, an initial cloud top height is obtained from the ARSCL data. This top height is checked with the available RH profiles. If the RH profiles clearly indicate a cloud top layer and this layer is much different from the MMCR measurement, then we take the RH height as the final cloud top height. Fig. 3 shows histograms of the obtained cloud thicknesses. The correction based on the RH profiles mainly eliminates some larger cloud thicknesses reported by the MMCR. Fig. 4 shows the aspect ratios (thickness/horizontal size) of the single layer cumuli over the ARM SGP site. For most of the fair weather cumulus clouds, the aspect ratio has value less than 1.

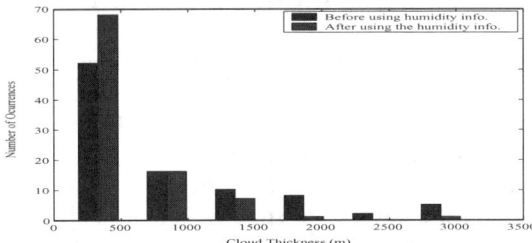

Fig. 3. Histograms of the cloud thickness determined before and after taking into account the relative humidity information, for 93 cases of single layer cumulus fields over the ARM CART site.

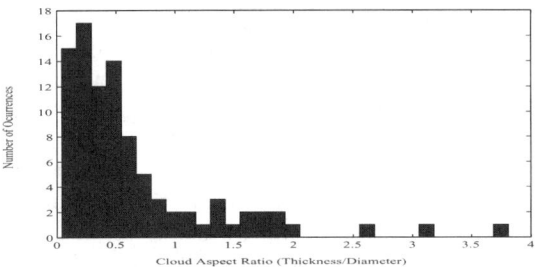

Fig. 4. Histogram of the cloud aspect ratio for the 93 cases.

4.3 Cloud Spacing and Horizontal Size

Assuming the cloud field properties do not change significantly as they move past at the mean wind speed, the spacing and horizontal sizes are estimated as the products of the wind speed and the lengths of the time intervals from a time series of observations of the cloud field. The wind speed at cloud height is determined from the 915 MHz Radar Wind Profiler (RWP915) data. The Narrow Field of View Sensor (NFOV) data is used to infer the spacing and cloud time interval lengths. Fig. 5 shows the histograms of the cloud sizes and spacing. For most selected cases, the median cloud size and spacing are less than 1000 m and 2000 m, respectively.

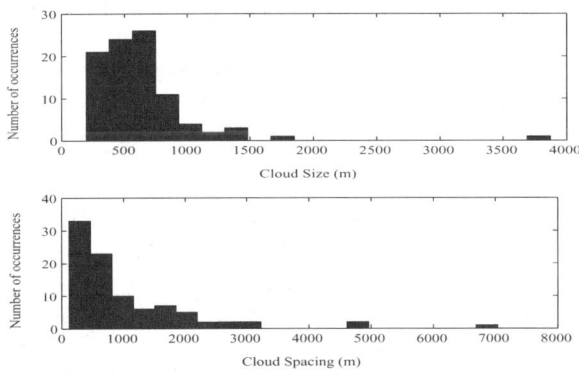

Fig. 5. Histograms of the cloud sizes and spacing for the data used in this study.

4.4 Absolute Cloud Fraction

The absolute cloud amount is defined as the fractional coverage of the vertical projections of the clouds on the surface. We inferred the absolute amount by using the same assumptions as for cloud spacing, and estimated $N=l/L$, where L is the total length of a time series of observation and l is the length of the cloud segments.

The ARM NFOV, TSI, Whole Sky Imager (WSI) and ARSCL data have been used to obtain N. When using the TSI and the WSI, the image sequence of the central circle of a field-of-view of 20° is used to calculate N. Fig. 6 gives a comparison of the four N's. The N_{WSI} agree well with the N_{TSI} but with a little larger variance, while the N_{NFOV} and the N_{ARSCL} tend to overestimate the cloud fraction relative to N_{TSI} or N_{WSI}. The cause of these biases may be due to the sensitivity of the instruments to the various clouds and the cloud decision algorithms used to infer cloudiness. The TSI and the WSI detect only visible and relatively thick clouds, while the NFOV and the laser instruments are sensitive to thin and sub-visible high clouds.

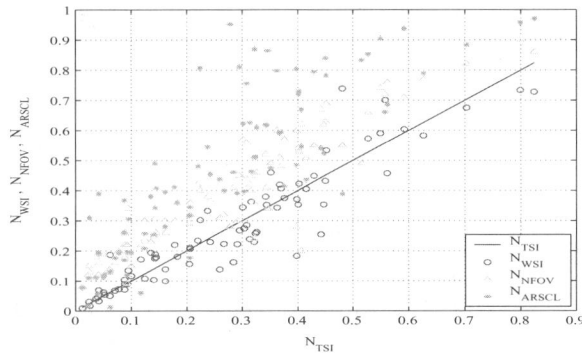

Fig. 6. A comparison of the absolute cloud fraction estimated from the TSI, WSI, NFOV and the ARSCL cloud base data.

5. COMPARING THE VARIOUS PCLOS MODELS WITH THE TSI OBSERVATIONS

Fig. 7 shows the differences between model (PCLOS$_{model}$) and observed PCLOS (PCLOS$_{TSI}$.) All curves in the figures are averages over 38 non-streak cases selected from July 2000 through October 2001 whose cloud thicknesses were confirmed with the relative humidity data. For most models, the average difference between the model PCLOS and the TSI observations is within ±0.1; the standard deviation of the differences is less than 0.2. Most models tend to underestimate the PCLOS in $30° < \theta < 70°$, and those models that assume a Poisson cloud distribution give better results. Cloud inclination angel has large impact on the modeling of the PCLOS, but we do not account for that in this study.

Fig. 8 shows the statistics of various model predictions of the cloud side effect and that inferred from the TSI observations. The cloud side effect (*CSE*) is defined as

$$CSE = 2 \int_0^1 [1 - N - P_{clr}(\mu)] u du \quad (5)$$

In the plot, the 16th column is the TSI *CSE*, which indicates that for those fair weather cumulus clouds over the ARM CART site, the mean flux departure at the surface due to the cloud side effects is about 3.7±2.5 W/m^2 (assuming the cloud base height is 1.5 km and midlatitude summer conditions). Among the model predictions, we see that the randomly distributed hemispheres generated better results.

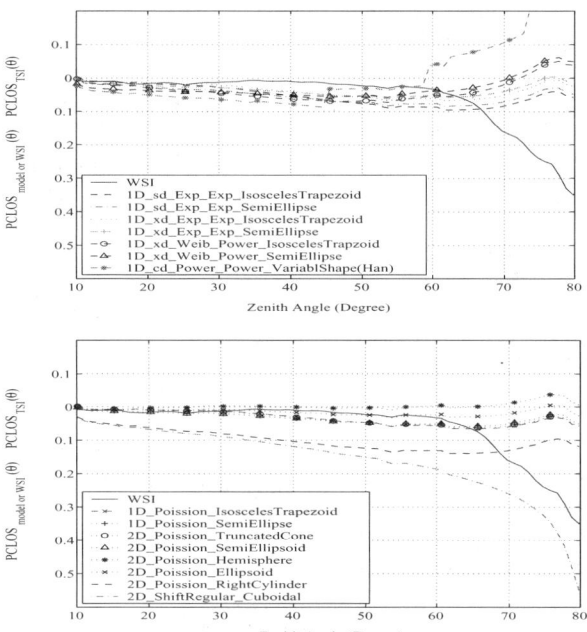

Fig. 7. Average (PCLOS$_{model}$ - PCLOS$_{TSI}$) for 38 cases.

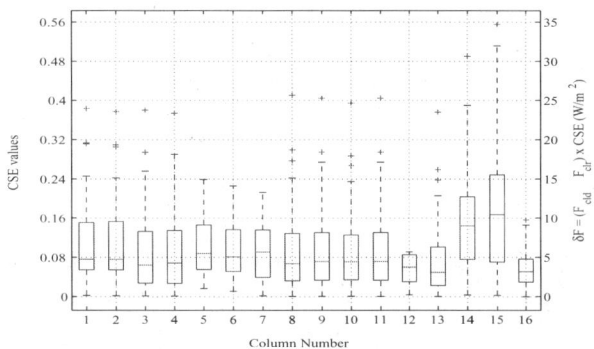

Fig. 8. Statistics of the model predictions of the *CSE* values and those inferred from the TSI observations (column No. 16). Column No. 12 is the model that assumes the clouds are randomly distributed hemispheres.

6. SUMMARY AND CONCLUSIONS

To evaluate a cloud sampling strategy using observations from a fixed point, an evaluation technique was developed and tested with the CRM/LES model data. For a homogeneous random field, the accuracy of the domain-averaged value inferred from a single transect depends on the covariance structure of the field. A field with larger correlation scale needs a longer sampling transect to achieve the desired accuracy. Observations made within the distance of correlation scale contain redundant information and are not independent of each other. For the clouds sampled in this study, the e-folding parameter for the correlation model is about 1.3 km.

Compared with the PCLOS inferred from the ARM total sky images, the PCLOS models can roughly capture the shape of PCLOS. Among the models, those assuming a Poisson cloud distribution and round-top cloud shape tend to give better results. However, since the clouds sampled in this study are relatively small, these results should not be expected to hold for all cumulus clouds. We plan to extend the results by looking at more mature, tropical fair-weather cumulus clouds.

7. REFERENCES

Ellingson, R. G., 1982: On the effects of cumulus dimensions on longwave irradiance and heating rate calculations. *J. Atmos. Sci.*, 39, 886-896.

Han, D., and R. G. Ellingson, 1999: Cumulus cloud formulations for longwave radiation calculations. *J. Atmos. Sci.*, 56, 837-851.

Harshvandhan, 1982, and J. A. Weinman, 1982: Infrared radiative transfer through a regular array of cuboidal clouds. *J. Atmos. Sci.*, 39, 431-439

Heidinger, A. K., and S. K. Cox, 1996: Finite-cloud effects in longwave radiative transfer. *J. Atmos. Sci.*, 53, 953-963.

Kagan, R. L, 1997: *Averaging of Meteorological Fields*. Translated from Russian. Ed. by L. S. Gandin and T. M. Smith. *Kluwer Academic,* 279 pp.

Ma, Y., 2004: Longwave radiation transfer through 3D cloud fields: Testing of the probability of clear line of sight models with the ARM cloud observations. Ph.D. dissertation, University of Maryland at College Park, 174 pp.

Stokes, G. M., and S. E. Schwartz, 1994: The Atmosphere Radiation Measurement (ARM) program: Programmatic background and design of the cloud and radiation test bed. *Bull. Amer. Meteor. Soc.*, 75, 1201-1221.

Takara, E. E., and R. G. Ellingson, 2000: Broken cloud field longwave-scattering effects. *J. Atmos. Sci.*, 57, 1298-1310.

THE VALUE OF LIMB MEASUREMENTS OF THE EARTH'S ATMOSPHERE

J. I. van Gent
Royal Netherlands Meteorological Institute
NL-3772GK Utrecht, Netherlands
Jeroen.van.Gent@knmi.nl

ABSTRACT

In this presentation, limb spectra from the SCIAMACHY satellite instrument are presented. Examples of major spectral features are shown and their significance for the investigation of the structure and composition of the Earth's atmosphere is discussed.

1. INTRODUCTION

Many satellite-borne instruments have observed the Earth's atmosphere in nadir-view (e.g., TOMS and GOME). Although this viewing geometry allows for a high horizontal resolution and global coverage, the vertical resolution that can be obtained is relatively low (about 10 km). Better vertical resolution is retrieved with solar occultation measurements (SAGE, POAM), but coverage of the atmosphere is limited to sunset and sunrise. Observations with a limb line of sight combine the advantages of global coverage and high vertical resolution.

With the launch of the SCIAMACHY instrument onboard ESA's ENVISAT satellite, the advantages of nadir and limb measurements are combined into a single instrument (Bovensmann *et al.*, 1999). By alternating limb and nadir measurements, SCIAMACHY observes the same volume of air in two different geometries, increasing the information

In this paper, the value of limb measurements is illustrated by presenting limb spectra from SCIAMACHY in the ultraviolet, visual and near infrared. Some prominent spectral features are shown along with their significance for the understanding of the Earth's atmosphere.

2. SCIAMACHY

ENVISAT was launched on 1 March 2001. The satellite travels a sun-synchronous polar orbit with mean local time of 10:00 a.m. at the descending node. For the larger part of each 100 minute orbit, SCIAMACHY alternates nadir and limb measurements, allowing the same volume of air to be observed from these to geometries within a time interval of approximately 7 minutes. At the largest possible swath with of 960 km, global coverage is reached within 3 days. The SCIAMACHY detectors cover a wide spectral range of 240 nm to 2380 nm. Spectral resolution ranges from 0.24 nm in the UV to 1.5 nm in the near-infrared. Therewith the instrument is optimized for derivation and monitoring of the distribution of a large number of atmospheric trace gasses, among which O_3, NO_2, H_2O, and CH_4. Other atmospheric properties that may be determined are cloud characteristics and surface reflection.

2.1 Limb Observations

When in limb observation mode, SCIAMACHY observes the horizon, with the purpose to gain information on the atmosphere around the line-of-sight tangent point. During a typical limb observation session, SCIAMACHY's line-of-sight scans the horizon in the flight direction for approximately 53 seconds. During this time, the line-of-sight tangent altitude covers a –3 km to 93 km range at 3 km intervals, in a pattern of alternating forward and backward scans (Fig. 1).

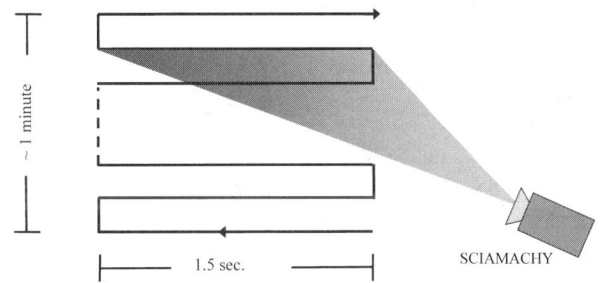

Fig, 1. Scan geometry of a typical SCIAMACHY limb measurement session.

Measurement integration times vary between 3/8 and 1.5 seconds, depending on the wavelength interval and the observed scene.

3. SCIAMACHY LIMB SPECTRA

Examples of SCIAMACHY limb spectra are shown in Fig. 2. A large number of absorption features from a host of atmospheric constituents can be seen (Kaiser *et al.*, 2004a). At higher line-of-sight altitudes, the overall shape of the spectra is largely determined by the wavelength dependence

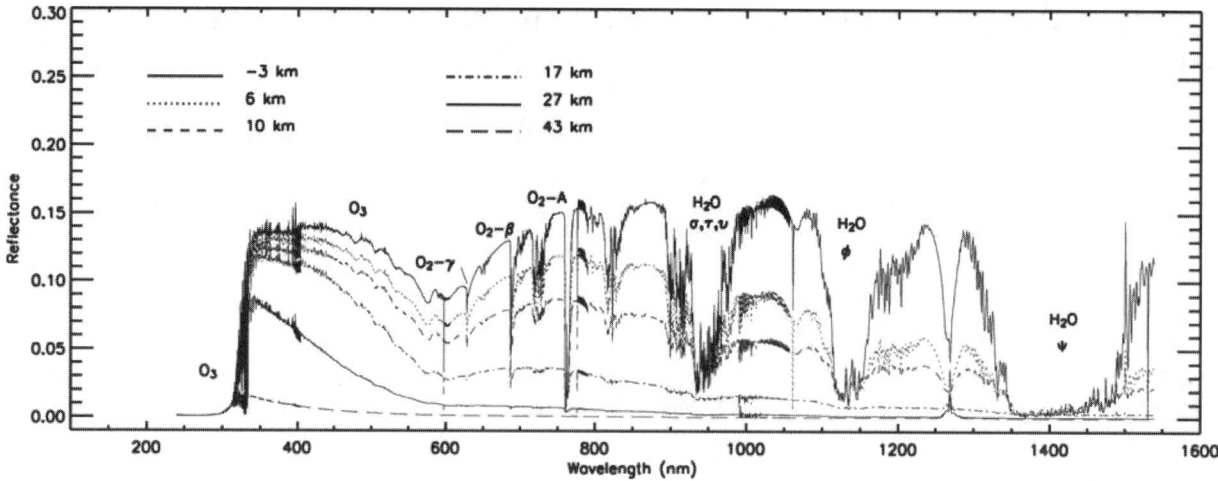

Fig. 2. Examples of SCIAMACHY limb spectra.

of the Rayleigh-scattering cross-section. Lower altitudes show additional contributions from multiply scattered light, as well as from cloud and surface reflection. Among the numerous features visible in the spectra, the most prominent are those due to tropospheric water vapor in the near infrared. While dominating the spectra at tropospheric tangent altitudes, water is almost absent in the stratospheric spectra. The visible spectrum is mainly characterized by oxygen molecular bands (Par. 3.1). The shape of the ultraviolet and short wavelength end of the visible range is determined by the broad bands of ozone, described next.

3.1 Ozone

The UV and visual spectral range show broad absorption bands of ozone. Between 200 and 300 nm, the majority of solar light entering the Earth's atmosphere is absorbed by this molecule in the Hartley molecular band. More spectral structure can be seen in the ozone Huggins (300–360 nm) and Chappuis band, the latter extending from 440 to approximately 740 nm.

Because of the large variation of optical depth with altitude and wavelength, the Huggins and Chappuis bands prove excellent candidates for the retrieval of the vertical ozone distribution in the stratosphere. Examples of the vertical profile of ozone, derived from SCIAMACHY limb measurements, are shown in Fig. 3. Cloud cover and large scattering optical depth hinder an accurate derivation of the tropospheric profile of ozone from limb and nadir measurements. However, it has been shown (van der A, 2001) that the combined ozone retrieval from limb and nadir spectra of SCIAMACHY leads to a significant improvement of the accuracy of the stratospheric profile. The larger amount of information on the stratospheric profile in the limb measurements constrains the tropospheric concentration, much more effectively than a nadir measurement alone. Error levels of a few percent in ozone concentrations retrieved from nadir observations only, can be reduced to as little as 0.1 percent when nadir and limb observations are combined.

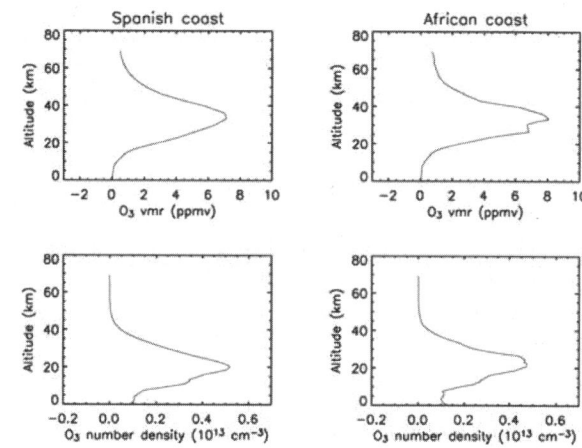

Fig. 3. Two examples of vertical ozone distribution, retrieved from SCIAMACHY limb observations (orbit 11514) with the Sciarali algorithm at KNMI. The profiles are expressed in volume mixing ratio (top panels) and number density.

3.2 Satellite Pointing

Accurate knowledge of the pointing direction of the limb line-of-sight is crucial for reliable retrieval of vertical trace gas distributions. The pointing information is provided by the satellite's orientation and scan mirror positions. Independent methods exist to validate the provided pointing directions, since pointing information is available in the limb spectra themselves (Janz *et al.*, 1996; Kaiser *et al.*, 2004b). This can be seen in Fig. 4. When one considers limb spectra in the UV-B spectral range, the observed reflectance does not monotonically decreases with tangent altitude, but rather shows a maximum (Fig. 4). The altitude of this maximum depends on the sun position and observed wavelength. By considering several of this "knee-shaped" reflectance profiles at different wavelengths, the tangent altitudes of the lines-of-sight can be derived.

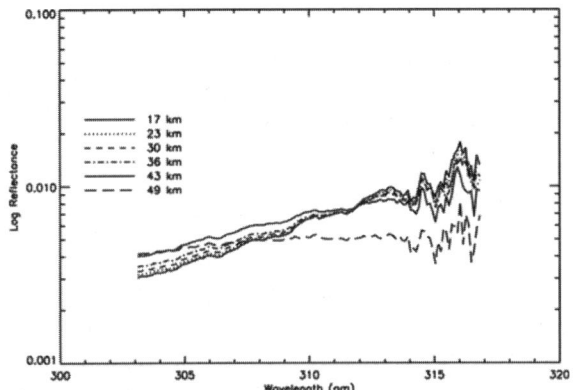

Fig. 4. Ultraviolet limb spectra for several tangent altitudes. Below approx. 310 nm, reflectance does not vary monotonically with tangent altitude, but shows a maximum: the knee-effect.

3.3 Oxygen

The coupling of electronic transitions from the ground state of molecular oxygen with rotational-vibrational states results in absorption bands in the near infrared and visible spectral domain (Fig. 2). The depth of these bands is sensitive to cloud cover fraction, cloud top pressure and surface reflection (Fig. 5).

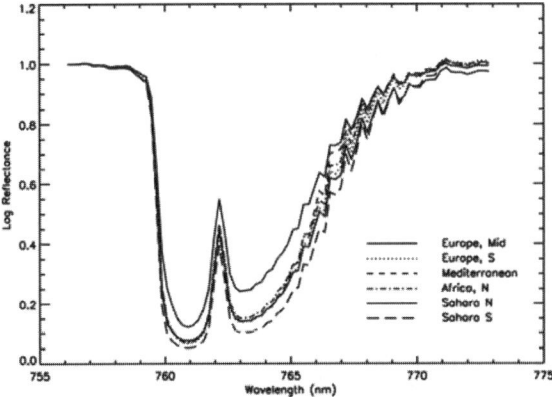

Fig. 5. O_2 A-band absorption for several SCIAMACHY ground pixels. Cloud cover and variation surface influence the depth of the feature.

In particular the O_2 A-band, centered at 762 nm, is used to derive cloud fraction and top pressure from nadir measurements (Fig. 6; Koelemeijer *et al.*, 2001). In limb observations, the O_2-A feature may be used to derive vertical pressure and temperature profiles.

Another prominent spectral feature of oxygen is at 1270 nm (1.27 µm). Ground state absorption at this wavelength leads to the production of excited oxygen, O_2 ($a^1\Delta_g$), at tropospheric altitudes. In the stratosphere and mesosphere, O_2 ($a^1\Delta_g$) is produced by photolysis of ozone. Subsequent electronic recombination to the ground state results in 1.27 µm emission in the limb spectra at higher tangent altitudes (Fig. 7). Since this airglow is a result of the photolysis of ozone, its strength can be used to derive the vertical distribution of ozone in the mesosphere (e.g., Llewellyn *et al.*, 2003).

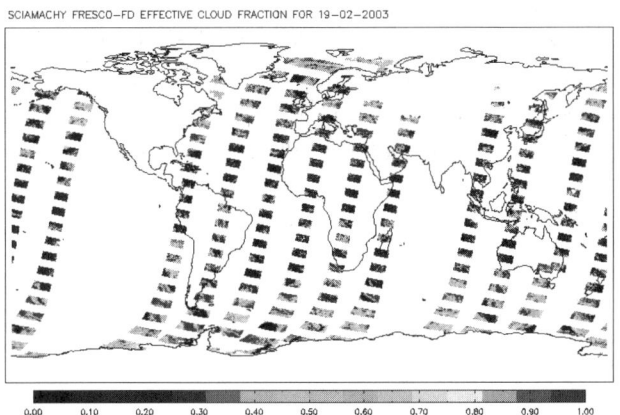

Fig 6. Cloud cover fraction, derived from SCIAMACHY oxygen A-band observations with the FRESCO cloud algorithm, operated at KNMI.

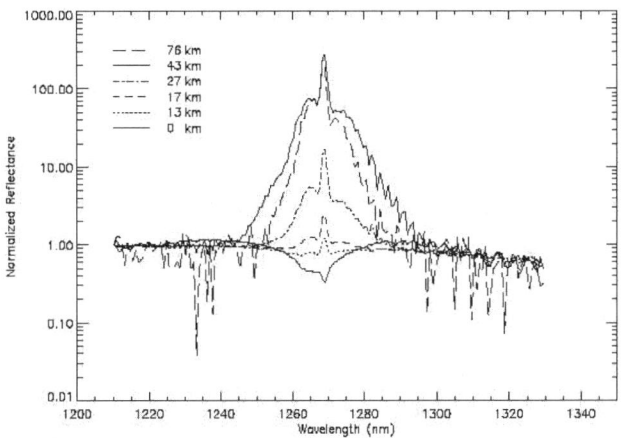

Fig. 7. The O_2 ($a^1\Delta_g$) 1.27 µm. feature for several line-of-sight tangent altitudes. The spectra are normalized at the short wavelength edge.

4. CONCLUSION

SCIAMACHY limb spectra provide excellent possibilities to improve knowledge on atmospheric structure and composition. The data already has, among other applications, contributed to the quality enhancement of stratospheric ozone profiles, and will do so for other atmospheric constituents. Many other applications are foreseen or in progress.

5. REFERENCES

van der A. R. J., 2001: Improved ozone profile retrieval from combined nadir/limb observations of SCIAMACHY, *J. Geophys. Res.*, 106(D13), 14583 14594.

Bovensmann, H. J. P., J. P. Burrows, and M. Buchwitz et al., 1999: SCIAMACHY: Mission objectives and measurement modes, *J. Atmos. Sci.*, 56(2), 127 150.

Janz, S. J, E. Hilsenrath, D. Flittner, and D. Heath, 1996: Rayleigh Scattering Attitude Sensor, *SPIE Proc.*, 2831, 146 153.

Kaiser, J. W, K.-U Eichmann, S. Nöel, M. W. Wuttke, J. Skupin, C. von Savigny, A. V. Rozanov, V. V. Rozanov, H. Bovensmann, and J. P. Burrows, 2004a: SCIAMACHY Limb Spectra. *Adv. Space Res.*, 34, no. 4, 715 720.

Kaiser, J. W, C. von Savigny, K.-U. Eichmann, S. Noël, H. Bovensmann, J. Frerick, and J. P. Burrows, 2004b: Satellite Pointing Retrieval from Atmospheric Limb Scattering of Solar UV-B Radiation, *Can. J. Phys./Rev. Can. Phys.*, 82(12): 1041 1052.

Koelemeijer, R. B. A. Stammes, P. Hovenier, J. W., and J. F. de Haan, 2001: A fast method for retrieval of cloud parameters using oxygen A band measurements from the Global Ozone Monitoring Experiment, *J. Geophys. Res.*, 106, 3475 3490.

Llewellyn, E. J., D. A, Degenstein, N. D, Lloyd, R. L. Gattinger, S. Petelina, I. C. McDade C. B. Haley, H. Solheim, C. von Savigny, C. Sioris, W. F. Evans, J. K. Strong, D. P. Murtagh, and J. Stegman, 2003: First Results from the OSIRIS Instrument on-board Odin, Proceedings of the 28AM on Optical Studies of the Upper Atmosphere, Oulu, Finland. *Sodankylä Geophysical Observatory Publications*, 92, 41 47.

THE USE OF THE ARM WSI TO ESTIMATE THE ATMOSPHERIC OPTICAL DEPTH AT NIGHT

I. C. Musat (imusat@met.fsu.edu) and R. G. Ellingson
Florida State University
Tallahassee, FL, 32306, USA

ABSTRACT

Broadband observations of starlight with a whole sky imager are used for determining the nighttime aerosol optical depth. The main difficulty in such measurements consists of accurately separating the star flux value from the non-stellar diffuse light. The monochromatic extinction at the ground due to aerosols is extracted from heterochromatic measurements. The total error is between 2.6 and 3% rms. Comparison with aerosol optical depth measured by other methods shows good agreement.

1. MEASUREMENT TECHNIQUE

The whole sky imager (WSI) observes the night sky vault with exposure times of 1 minute, and does not compensate for Earth's rotation. Digital images with stars visible simultaneously are used to infer star irradiance at the ground. Sequences of images taken as the stars pass through different altitudes above the horizon are used to infer aerosol optical depth variation during the night.

The star irradiance is inferred by an "On/off" technique, as the amount of radiation above the sky background, geometrically associable with the star center position (known from catalog) and with the spread of the star profile over a few pixels, which is due to the compounded effects of turbulence, diurnal rotation during the exposure time and multi-wavelength nature of the observation. The sky background value for a particular star, defined as value of the sky without the star present in it, is observable "as is" only when the star is close to the zenith direction, otherwise it is "enhanced" by atmospheric scattering of night-sky extended sources and by observational geometry.

The difficulties inherent for observations with WSI are the high sky background, the low resolution (poor separation between stars) and coarse pixelization of star profiles (undersampling). Approximately 120 stars, not all present simultaneously on the sky, were measured on 300 clear nights from 1998 to 2003.

2. HETEROCHROMATICITY

When applied to WSI measurements, which are made at wavelengths between 400 and 900 nm, simple monochromatic Langley plots, as those derived with monochromatic observations with photometers, do not yield an accurate value of the aerosol optical depth. An inter-play exists between the different spectral output of stars, known from their Morgan-Keenan spectral type, and the aerosol spectral opacity. The wavelength-dependent aerosol optical depth (AOD) is modeled following a well-known empirical law, (AOD=$\beta \lambda^{-\alpha}$), where "alpha" is the Angstrom exponent and "beta" is the turbidity parameter. The AOD obtained at the center of the observing bandwidth, between 500 and 600 nm, is the most reliable monochromatic parameter, constructed from a heterochromatic measurement.

3. TEMPORAL/SPATIAL VARIABILITY

Rather than trying to obtain the optical thickness value towards each star direction, the method used in this study assumes that all irradiances from stars visible at the same time can be used to obtain the AOD, that is, the optical thickness in the vertical direction. This AOD should be comparable with the same amount determined with more precise methods, as the nighttime LIDAR. The comparison holds during the majority of nights measured because the aerosol spatial coherence distance, of a little under 200 km (Anderson et al., 2003), is also the maximum WSI observable distance, measured on an aerosol layer at 5 km height above the ground, between directions of stars situated as low as 3 degrees elevation above the horizon, on the opposite ends of a big meridian circle.

4. NON-STELLAR DIFFUSE BACKGROUND

At low elevations, stellar irradiance value is more difficult to assess, because of the diffuse scattered light, which has non-stellar origin (airglow emission, integrated starlight of non-resolved stars and zodiacal light), but also because of the molecular and aerosol scattering of the airglow, and of the other extended sources. A correction for irradiances of stars observed at low elevations is used. Principally, the sky background value is computed as the median pixel in an elliptical aperture placed on the sky such that more pixels covering the same zenith angle to be included, that is, an elliptical aperture with the larger axis perpendicular on the diameter of the lens; this step allows the most probable value per pixel of the air scattered airglow at that zenith distance to be included. Secondly, an atlas of sky brightness (Leinert et al., 1998) is used to account for the part, in the sky median pixel value, that represents the airglow multiple scattering on air molecules.

5. WSI EQUATION

For a given star, the wavelength averaged irradiance F_i is given by equation (1),

$$F_i(\theta) = F_{0555} \int_{\Delta\lambda} f_0(\lambda, class) R(\lambda) T_{\lambda gases}(\theta) T_{\lambda aerosols}(\theta) d\lambda$$

where F_{0555} is the 555nm top of atmosphere monochromatic flux; T_λ is the monochromatic transmissivity; θ is the zenith angle; R_λ is normalized instrument response function and $f(\lambda, class)$ is the normalized spectrum for the Morgan-Keenan spectral class and luminosity classification of the star (i.e., $F_{0\lambda}/F_{0555}$); $T_{aerosols} = \exp(-X\tau)$ and $\tau = AOD = \beta \lambda^{-\alpha}$.

Using an iterative method (Kythe and Puri, 2002) with an a priori knowledge of atmospheric gases opacity, and a first guess for aerosol parameters ($\beta = 0.001$ and $\alpha = 0.1$), all the possible pairs of different colored star irradiances measured at the same moment are reduced to find the best transmission correction factor and best effective wavelength, in the atmospheric filter, for their color.

6. AEROSOL OPTICAL DEPTH COMPUTATION UNCERTAINTY

Uncertainties in the instrument calibration, background corrections, absolute flux of the individual stars, and the gaseous transmittance uncertainties govern the absolute accuracy of the aerosol estimation. The compounded effect of different errors introduced in the AOD computation (Musat, 2004) by the method used is given in the Table 1.

Table 1. Errors in AOD Computation from WSI Star Irradiances

Uncertainties:	Random	Systematic
Absolute Flux calibration constant		0.5%
Star flux accuracy (includes error from dark current subtraction, readnoise, digitization noise, flatfield division and bias subtraction for the CCD frame)	< 0.03 rms	
Solid angle per pixel	0.71%	
Pixel surface projection on constant height layer	0.22%	
Background sky correction (for large zenith angle only)		0.26%
Vega monochromatic flux	$0.002\ 10^{-7}$ mW/m2/nm rms	
Catalog Magnitude	0.01mag	
Numerical integration		10^{-12}
Monochromatic absolute flux (other stars)		1%
Star calibrated measurement (CM)	1.11%(Vega) 1.37% (other stars)	
Relative airmass (<84 deg.zenith)	< 0.065%	
LOWTRAN model sensitivity (without aerosols)		0.72%
Kernel	0.72%	
Per step iteration (from CM & Kernel)	(1.32-1.55)%	
Convergence error		10^{-5}
AOD at centre of bandpass	(2.6-3.1)%	

Fig. 1 shows a sequence of AOD from the Sun photometer CIMEL (500 nm, daytime) and WSI (night) for 17–24 Aug. 1998, while Fig. 2 presents Raman Lidar AOD (355 nm, line), in comparison with AOD at 355 nm (diamond symbol) simulated from the WSI measurement, for 14–16 Dec. 1998.

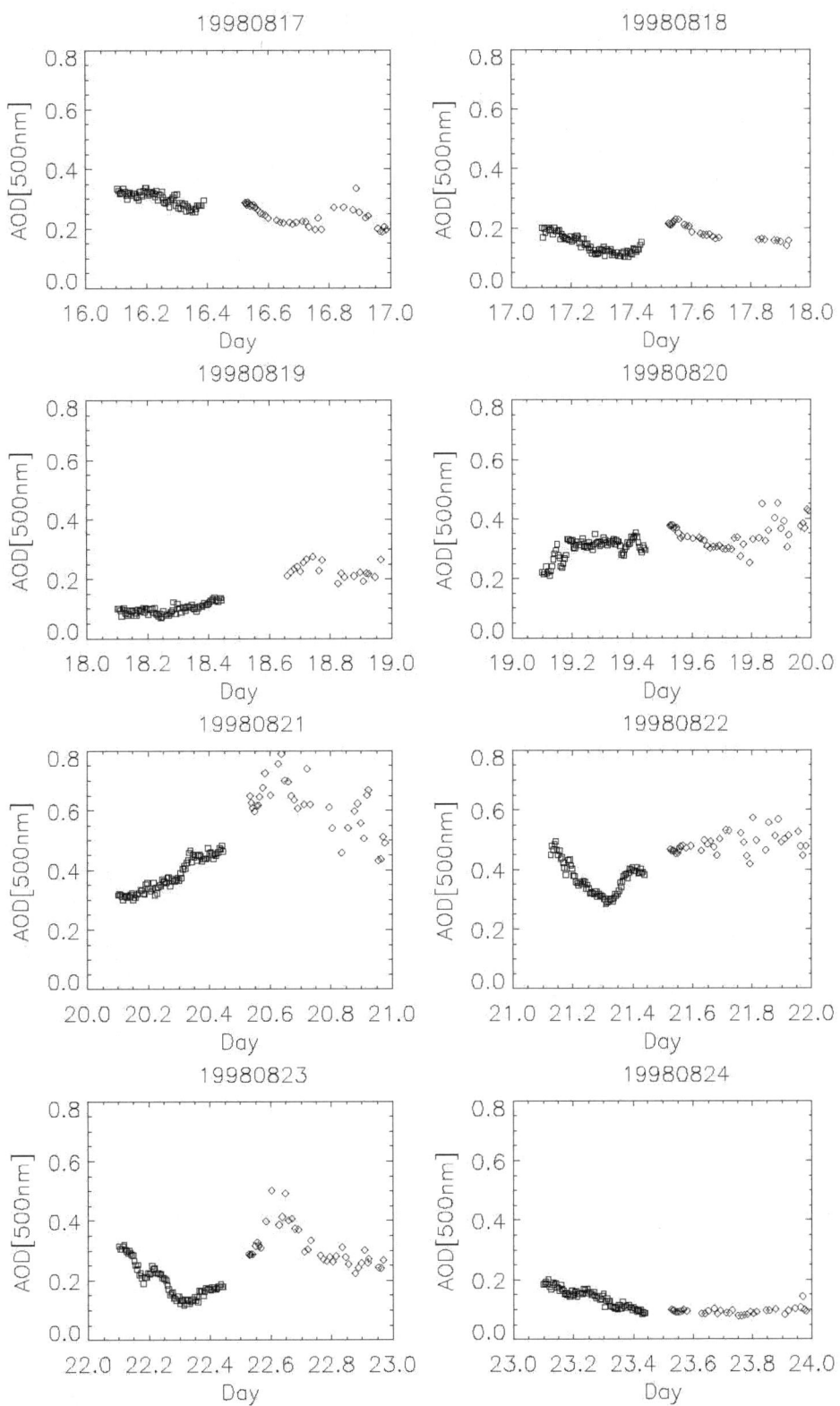

Fig. 1. From left to right, from top to bottom, a sequence of AOD from the Sun photometer CIMEL (500 nm, square symbol, daytime) and WSI (500 nm, diamond symbol, nighttime) for 17–24 Aug. 1998.

7. ACKNOWLEDGEMENTS

Data used in this paper were obtained from the Atmospheric Radiation Measurement (ARM) Program sponsored by the U.S. Department of Energy, Office of Science, Office of Biological and Environmental Research, Environmental Sciences Division.

8. REFERENCES

Anderson, T. L., R. Charlson, D. M. Winker, J. A. Ogren, and K. Holmen, 2003: Mesoscale variations of tropospheric aerosols, *J. Atmos. Sci.*, 60, 119-136.

Kythe, P. K., and P. Puri, 2002: Computational methods for linear integral equations, *Birkhauser,* 508 pp.

Leinert, Ch., S. Bowyer, L. K. Haikala, M. S. Hanner, M. G. Hauser, A.-Ch. Levasseur-Regourd, I. Mann, K. Mattila, W. T. Reach, W. Schlosser, H. J. Staude, G. N. Toller, J. L. Weiland, J. L. Weinberg, and A. N. Witt, 1998: The 1997 reference of diffuse night sky brightness, *Astron. Astrophys. Suppl. Ser.*, 127, 1-99.

Musat, I. C., 2004: Short-term variability of atmospheric extinction during the night, under clear sky conditions, investigated by broadband stellar photometry, *Ph.D. Dissertation*, Univ. of Maryland, College Park, 229 pp.

CONVERGENCE ACCELERATION OF A SPHERICAL HARMONICS METHOD FOR STRONG ANISOTROPIC SCATTERING

V. P. Boudak

Moscow Power-Engineering Institute (Technical University), Light Engineering Department
Moscow, 111250, Krasnokazarmennaya, 14, Russia

ABSTRACT

The singularity in the radiance angular distribution is internally proper to the description of the radiation transfer processes in the ray approximation, which essentially reduces convergence of the solution of the radiation transfer equation (RTE) by any numerical method. The method of the elimination of this solution singularity is offered in the paper. It is based on the representation of the solution as the sum of the small angle approximation (SAA) and the smooth part. The generalization of the SAA for the case of arbitrary geometry is developed.

1. INTRODUCTION

The methods of the RTE solution are variations of a finite element method. However nowadays there are no accessible application packages of the RTE solution for the arbitrary three-dimensional area similar to the solution of partial equations. It is connected with the features of a physical model of the radiation transfer—ray approximation. In particular, owing to the physically selected direction of the radiation propagation in space the radiance angular distribution contains a singularity.

Such singularities of the radiance angular distribution are of the key character that requires for accommodating it the development of special methods of the RTE solution. Chandrasekhar (1950) offered to subtract the direct nonscattered component from the solution and to state the equation for the smooth remainder that eliminates σ-singularity of the radiance angular distribution.

Let's consider the singularities of the boundary-value problems of the transport theory on the example of a light field calculation of the flat unidirectional source by a spherical harmonics (SH) method

$$\begin{cases} \mu \dfrac{\partial L(\tau,\hat{\mathbf{l}})}{\partial \tau} + L(\tau,\hat{\mathbf{l}}) = \dfrac{\Lambda}{4\pi} \oint x(\hat{\mathbf{l}},\hat{\mathbf{l}}')L(\tau,\hat{\mathbf{l}}')d\hat{\mathbf{l}}', \\ L(\tau,\hat{\mathbf{l}})\big|_{\tau=0,\mu\geq 0} = \delta(\hat{\mathbf{l}}-\hat{\mathbf{l}}_0),\ L(\tau,\hat{\mathbf{l}})\big|_{\tau=\tau_0,\mu\leq 0} = 0; \end{cases} \quad (1)$$

where $L(\tau,\hat{\mathbf{l}}) \equiv L(\tau,\mu,\varphi)$ - radiance of a light field in the direction $\hat{\mathbf{l}} = \{\sqrt{1-\mu^2}\cos\varphi, \sqrt{1-\mu^2}\sin\varphi, \mu\}$, $\mu = (\hat{\mathbf{l}},\hat{\mathbf{z}})$, on the optical depth $\tau = \int_0^z \varepsilon(\xi)d\xi$; $\hat{\mathbf{l}}_0$ - the incident direction of the external radiation on the upper bound of the slab, $\mu_0 = (\hat{\mathbf{l}}_0,\hat{\mathbf{z}})$; ε – attenuation coefficient, Λ – single scattering albedo of the medium; τ_0 – slab optical thickness. The axis OZ of a Cartesian frame is located perpendicularly downwards to the slab border. Hereinafter the unit vectors are marked by the symbol "^".

The main point of the SH method is the presentation of the angular dependences of all the functions in RTE as a series on the spherical harmonics:

$$L(\tau,\mu,\varphi) = \sum_{k=0}^{\infty}\sum_{m=0}^{k}\frac{2l+1}{4\pi}C_k^m(\tau)Q_k^m(\mu)e^{-im\varphi}, \quad (2)$$

$$x(\hat{\mathbf{l}},\hat{\mathbf{l}}') = \sum_{k=0}^{\infty}(2l+1)x_k P_k(\hat{\mathbf{l}}\cdot\hat{\mathbf{l}}'), \quad (3)$$

where $Q_l^n(\mu) = \sqrt{\dfrac{(l-n)!}{(l+n)!}}\,P_l^n(\mu)$ – renormalized Legendre polynomials.

On the basis of the addition theorem, recursion relations and orthogonality of the spherical harmonics RTE are reduced to the infinite set of the ordinary differential equations:

$$\frac{1}{2k+1}\frac{d}{d\tau}\left[\sqrt{(k-m+1)(k+m+1)}\,C_{k+1}^m \right.$$
$$\left. +\sqrt{(k-m)(k+m)}\,C_{k-1}^m\right] + (1-\Lambda x_k)C_{k+1}^m(\tau) = 0. \quad (4)$$

For the solution it is necessary to make the set finite that strongly smoothes the radiance angular distribution. Chandrasekhar (1950) offered to subtract the direct nonscattered radiation $\delta(\mu-\mu_0)\delta(\varphi)$. The rest part of the solution is a smooth function that allows making the set finite. In this case the set of the equations is presented in the matrix form convenient for further transformations

$$\overset{\leftrightarrow}{\mathrm{A}}\frac{d}{d\tau}\vec{\mathrm{C}}(\tau) + \overset{\leftrightarrow}{\mathrm{D}}\vec{\mathrm{C}}(\tau) = \vec{\mathrm{f}}(\tau), \quad (5)$$

where $\vec{\mathrm{C}}(\tau)$ is a column vector of the required coefficients $C_{k+1}^m(\tau)$, $\overset{\leftrightarrow}{\mathrm{A}}$ and $\overset{\leftrightarrow}{\mathrm{D}}$ - relevant set matrices (4), $\left[\vec{\mathrm{f}}^m(\tau)\right]_k = e^{-\tau/\mu_0} x_k Q_k^m(\mu_0)$. The double-headed arrow marks matrices, the right arrow – the column vector. The obvious coefficient m will be further omitted.

However all the natural formations, whether it were atmosphere or ocean, have suspended particles with the size much greater than the wave length, that according to the Mie theory gives a strong anisotropic light scattering on them. In the conditions of a strong scattering anisotropy in the small angles the radiation is indistinguishable from the direct radiation, and method of Chandrasekhar (1950) becomes ineffective.

2. MODIFICATION OF SH METHOD

Let's present the solution as the sum of the SAA and smoothly varying addition

$$L(\tau,\mu,\varphi) = L_{SAA}(\tau,\hat{\mathbf{l}}) + \tilde{L}(\tau,\mu,\varphi), \quad (6)$$

where the solution in SAA $L_{SAA}(\tau,\hat{\mathbf{l}})$ will be taken in the form of Goudsmit and Saunderson (1940).

Since SAA (Boudak and Sarmin (1990)) takes into account all the singularities of the exact solution, therefore the addition $\tilde{L}(\tau,\mu,\varphi)$ is a smooth angular function at any degree of a scattering anisotropy. This form of the SAA is analytically represented as series on Legendre polynomials

$$L_{SAA}(\tau,\hat{\mathbf{l}}) = \sum_{k=0}^{\infty} \frac{2k+1}{4\pi} Z_k(\tau) P_k(\hat{\mathbf{l}} \cdot \hat{\mathbf{l}}_0), \quad (7)$$

where $Z_k(\tau) = \exp\left\{-\frac{(1-\Lambda x_k)\tau}{\mu_0}\right\}$, that saves the equations structure of the SH method (5), only adding other free terms in a right-hand part:

$$\vec{\mathbf{A}}^m \frac{d}{d\tau} \vec{\mathbf{C}}^m(\tau) + \vec{\mathbf{D}} \vec{\mathbf{C}}^m(\tau)$$
$$= (\vec{\mathbf{A}}^m/\mu_0 - \vec{\mathbf{1}}) \vec{\mathbf{D}} \vec{\mathbf{Q}}^m \vec{Z}(\tau) + a_{N+1}^m \vec{Z}_{N+1}(\tau)$$

where $\vec{Z}_{N+1} = \left\{\underbrace{0...0}_{N}, Z_{N+1}(\tau)\right\}$, $\vec{\mathbf{Q}}^m = \{Q_k^m(\mu_0)\}$,

$$a_{N+1}^m = \frac{\sqrt{(N+1-m)(N+1+m)}}{2N+1} \frac{1-\Lambda x_{N+1}}{\mu_0} Q_{N+1}^m(\mu_0).$$

The obtained set permits the analytical solution, which can be written in the matrix form as follows

$$e^{-\vec{\mathbf{B}}\tau_0} \vec{\mathbf{C}}(\tau_0) = \vec{\mathbf{C}}(0) + \frac{1}{\mu_0} \int_0^{\tau_0} e^{-\vec{\mathbf{B}}t} (\vec{\mathbf{1}} - \mu_0 \vec{\mathbf{A}}^{-1}) \vec{\mathbf{D}} \vec{\mathbf{Q}} \vec{Z}(t) dt$$
$$+ a_{N+1}^m \int_0^{\tau_0} e^{-\vec{\mathbf{B}}t} \vec{\mathbf{A}}^{-1} \vec{Z}_{N+1}(t) dt \equiv \vec{\mathbf{C}}(0) + \vec{J}(\tau_0), \quad \vec{\mathbf{B}} = \vec{\mathbf{A}}^{-1}\vec{\mathbf{D}}. \quad (8)$$

Matrix exponential curve is presented as:
$$e^{-\vec{\mathbf{B}}t} = \vec{\mathbf{U}} e^{-\vec{\Gamma}t} \vec{\mathbf{U}}^{-1}$$

that allows transforming the solution (8) to the simple analytic form

$$\vec{J}(\tau_0) = \left\{\sum_{j=1}^{N+1} (\vec{\mathbf{T}})_{ij} \frac{[1-\exp(-(\Gamma_{ii}+d_j)\tau_0)]}{(\Gamma_{ii}+d_j)\mu_0}\right\} + a_{N+1}^m$$
$$\times (\vec{\Gamma} + d_{N+1}\vec{\mathbf{1}})^{-1} (\vec{\mathbf{1}} - \exp(-(\vec{\Gamma}+d_{N+1}\vec{\mathbf{1}})\tau_0)) \vec{\mathbf{U}}^{-1} \vec{\mathbf{A}}_{N+1}^{-1},$$

where $d_j = (1-\Lambda x_{j-1})$.

Taking into account (6) the boundary conditions will take a form:

$$\begin{cases} \tilde{L}(\tau,\mu_0,\mu,\varphi)\big|_{\tau=0,\mu\geq 0} = 0, \\ \tilde{L}(\tau,\mu_0,\mu,\varphi)\big|_{\tau=\tau_0,\mu<0} = -L_{SAA}(\tau_0,\mu,\varphi). \end{cases} \quad (9)$$

Representing (9) in the matrix Marshak's form it will be received a complete set of the algebraic equations

$$\begin{cases} -\vec{\mathbf{S}}\vec{\mathbf{U}}^{-1}\vec{\mathbf{C}}(0) + \vec{\mathbf{H}}\vec{\mathbf{U}}^{-1}\vec{\mathbf{C}}(\tau_0) = \vec{\mathbf{S}}\vec{J}(\tau_0), \\ [\vec{\mathbf{1}} \ \ \vec{\mathbf{G}}]\vec{\mathbf{P}}\vec{\mathbf{C}}(0) = \vec{0}, \ [\vec{\mathbf{1}} \ -\vec{\mathbf{G}}]\vec{\mathbf{P}}\vec{\mathbf{C}}(\tau_0) = -\vec{Y}(\tau_0), \end{cases} \quad (10)$$

where the matrices $\vec{\mathbf{S}}, \vec{\mathbf{U}}, \vec{\mathbf{H}}, \vec{\mathbf{G}}, \vec{\mathbf{P}}$ are completely identical to the relevant matrices from Karp et al. (1980) the elements of a column vector $\vec{Y}(\tau_0)$ are expressed by the formula

$$Y_j^m(\tau) = Z_{2j-1}(\tau) Q_{2j-1}^m(\mu_0) - \sum_{l=0}^{\infty} G_{jl}^m Z_{2l-2}(\tau) Q_{2l-2}^m(\mu_0),$$

G_{jl}^m are elements of the matrix $\vec{\mathbf{G}}$. It is used in (10) scale transformation from Karp et al. (1980).

The analytical solution and scale transformation strongly improve the condition of the system matrix (10), and the set is solved without problems. In Fig. 1 there are the diagrams of the calculations comparison of the radiance angular distributions on the slab upper bound: g is the parameter of the phase function in the Henyey-Greenstein form, N is the number of the spherical harmonics, M is the number of the zenith terms. The diagram shows the essential acceleration of the convergence of the presented solution (solid line) in comparison with the method Karp et al. (1980) (dotted). At the increase of the harmonics number both solutions converge to each other. It is easy to notice, that the offered solution does not require any smoothing algorithms that are so prevalent in the conventional SH method. The convergence of the offered method is higher with increase of a scattering anisotropy, an incidence angle and an optical depth.

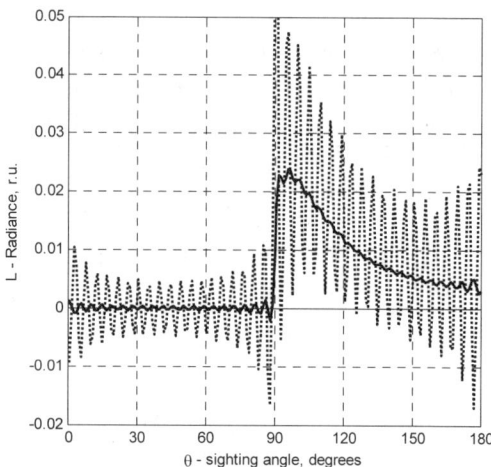

Fig. 1. Radiance angular distribution: $\tau_0=20$, $\theta_0=40°$, $\Lambda=0.9$, $g=0.97$, $N=79$, $M=8$.

3. THE SMALL ANGLE APPROXIMATION

The main problem of the generalization of the present method on other geometries is that SAA in the form Goudsmit and Saunderson (1940) is possible only for the flat geometry at the almost normal incidence of the radiation on the medium. Therefore further development of a method is possible only through the generalization of SAA.

For the first time SAA was formulated by Wentzel (1922), where the term SAA (*kleine Streuwinkel*) was introduced. In this paper for the expression of the radiance of the arbitrary scattering multiplicity

$$L_n(\tau,\hat{\mathbf{l}}) = \left(\frac{\Lambda}{4\pi}\right)^n \underbrace{\int_0^{\xi_n}\cdots\int_0^{\xi_1}}_{n} \underbrace{\oint\cdots\oint}_{n} \exp\left(-\xi-\sum_{k=1}^{n}\zeta_k\right) \quad (11)$$
$$\cdot x(\hat{\mathbf{l}},\hat{\mathbf{l}}_n)\cdot\ldots\cdot x(\hat{\mathbf{l}}_2,\hat{\mathbf{l}}_1)L_0(\hat{\mathbf{l}}_1)d\hat{\mathbf{a}}_1\cdots d\hat{\mathbf{a}}_n d\zeta_1\cdots d\zeta_n$$

where $L_0(\mathbf{l}_1)$ is the distribution of the incident radiation, the assumption is made, that at the strong anisotropic scattering the trajectory length of scattered and nonscattered rays is enough indistinguishable from each other. In such approach it is possible to take out the Bouguer exponential curve in (11) from an integral, that allows to execute the integration on the space variables. As the result the problem is reduced to the calculation of the multiple convolution on a spatial angle

$$L(\tau,\hat{\mathbf{l}}) = \sum_{n=0}^{\infty} L_n(\tau,\hat{\mathbf{l}}) = e^{-\tau/\mu_0}\sum_{n=0}^{\infty}\frac{(\Lambda\tau/\mu_0)^n}{n!}\Phi_n(\hat{\mathbf{l}}), \quad (12)$$

$$\Phi_n(\hat{\mathbf{l}}) = \left(\frac{1}{4\pi}\right)^n \underbrace{\oint\cdots\oint}_{n} x(\hat{\mathbf{l}},\hat{\mathbf{l}}_n)\cdot\ldots\cdot x(\hat{\mathbf{l}}_2,\hat{\mathbf{l}}_1)d\hat{\mathbf{a}}_1\cdots d\hat{\mathbf{a}}_n.$$

The problem of the evaluation of the multiple convolution on a spatial angle $\Phi_n(\hat{\mathbf{l}})$ turned out to be enough complicated, that generated three SAA analytic forms.

In the first form offered by Bothe (1929) the central limit theorem was used for the convolution evaluation, according to which the multiple convolutions lead to a gaussoid. At a scattering phase function is a sharper function of an angle, than the radiance distribution, RTE is reduced to the diffusion equation, the solution of which is a gaussoid. The similar approach features well the integrated performances of a light field, but is inapplicable to the description of the radiance angular distribution.

Goudsmit and Saunderson (1940) used the addition theorem for surface harmonic, that in representations (2) and (3) gives the expression

$$\Phi_n(\hat{\mathbf{l}}) = \sum_{k=0}^{\infty}\frac{2k+1}{4\pi}x_k P_k(\hat{\mathbf{l}}\cdot\hat{\mathbf{l}}_0),$$

which allows writing the exact expression of the radiance distribution (2), but only for the flat geometry.

The practical scantiness of a flat geometry case stimulated the attempts to search other SAA forms. Kompaneets (1947), Moliere (1948), Snyder and Scott (1949), offered the third form. It is based on the replacement of the convolution on a sphere (spatial angle) by the convolution on a plane tangential to this sphere.

$$\Phi_n(\hat{\mathbf{l}}) \approx \int_{-\infty}^{+\infty} x(\mathbf{l}_\perp - \mathbf{l}_{\perp n})\cdot\ldots\cdot x(\mathbf{l}_{\perp 2}-\mathbf{l}_{\perp 1})L_0(\mathbf{l}_{\perp 1})d\mathbf{l}_{\perp 1}\cdots d\mathbf{l}_{\perp n}$$

where the index $\perp$ means a projection of a vector to a plane, perpendicular to the incidence direction.

Where from the expression (12) gains a form

$$L(z,\mathbf{l}_\perp) = \frac{e^{-\varepsilon z/\mu_0}}{\pi}\int_{-\infty}^{+\infty} L_0(k)\exp\left(\frac{\Lambda\varepsilon z x(k)}{\mu_0}\right)J_0(kl_\perp)kdk, \quad (13)$$

where $L_0(k)$ is Fourier spectrum of $L_0(z,\mathbf{l}_{\perp 1})$.

The analytical view of this SAA form has allowed solution of the greatest number of problems of the ocean and atmosphere optics.

Wang and Guth (1951) surveyed the connection of all the SAA forms and showed, that the Goudsmit-Saunderson form is the most general, from which the others follow at the strong anisotropic scattering and the restriction by a small sighting angles

$$P_k(\cos\theta) \approx J_0(k\theta), \quad \sum_{k=0}^{\infty}\frac{2k+1}{4\pi} \to \frac{1}{2\pi}\int_{-\infty}^{+\infty}kdk. \quad (14)$$

4. SMALL ANGLE MODIFICATION OF THE SH METHOD

Let's consider the generalization of Goudsmit-Saunderson SAA form in the case of arbitrary sources on the basis of the slab irradiating by a flat unidirectional source under an arbitrary incidence angle. Let's transfer from RTE to the infinite set of the ordinary differential equations of the SH method. However in this case we will not truncate the numbers of the series terms and assume

- a continuous dependence of the series coefficients $C^m(k,\tau)$ by the index k, which in integer points coincides with values of the coefficients $C_k^m(\tau)$ in (2);

- at a strong anisotropy of the angular distribution its spectrum $C_k^m(\tau)$ is a slowly monotonically decreasing function of the indices k, that allows to expand it in a Taylor series preserving two or three first terms

$$C^m(\tau,k\pm 1) \approx C^m(\tau,k)\pm\frac{\partial C^m(\tau,k)}{\partial k}+\frac{1}{2}\frac{\partial^2 C^m(\tau,k)}{\partial k^2}; \quad (15)$$

- owing to the anisotropy of the radiance angular distribution the basic contribution to the solution is given by the terms with $k\gg 1$ and its anisotropy is much greater then its asymmetry $k\gg m$.

These assumption reduce the infinite set of the SH method (4) to one partial equation

$$\mu_0\frac{\partial C^m}{\partial\tau}+\frac{\sqrt{1-\mu_0^2}}{2}\frac{\partial}{\partial\tau}\left[\frac{\partial C^{m+1}}{\partial\kappa}+\frac{\partial C^{m-1}}{\partial\kappa}\right.$$

$$+\frac{1}{\kappa}\left((m+1)C^m - (m-1)C^m\right)\right] = -(1-\Lambda x_k)C^m(\tau,\kappa),$$

permitting the analytical solution

$$C_k^m(\tau) = \frac{e^{-\tau/\mu_o}}{2\pi} \int_0^{2\pi} \cos m\varphi \exp\left[\frac{\Lambda\tau}{\mu_o}\int_0^\infty x(\rho)e^{-\zeta}d\zeta\right]d\varphi, \quad (16)$$

where $\rho = \sqrt{\kappa^2 + a^2\zeta^2 - 2\kappa a\zeta\cos\varphi}$, $\kappa = \sqrt{k(k+1)}$, $a = \tan\theta_0$.

The obtained solution will be termed as a small angle modification of SH method (MSH). On the diagram of a Fig. 2 there is a comparison of the exact solution RTE (solid line), Goudsmit-Saunderson approximation (dotted) and MSH (dashdot). It is easy to notice, that MSH as against the Goudsmit-Saunderson approximation describes the rotation of the maximum of the radiance angular distribution from an incidence direction on the upper slab border to a vertical in the medium depth.

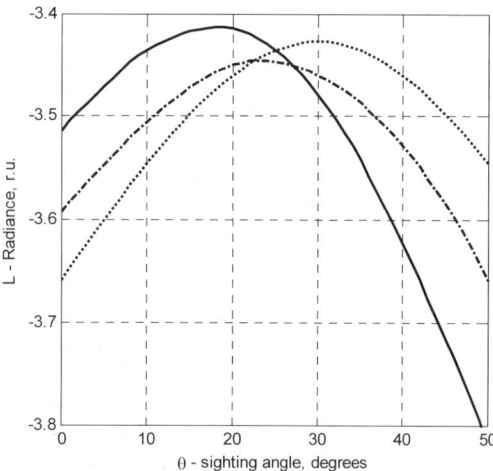

Fig. 2. Radiance angle distribution: $\tau=30$, $\theta_0=30°$, $\Lambda=0.8$, $\gamma=0.97$.

It is possible to show, that MSH is the most general form of SAA and all the forms follow from it under the requirements of a normal incidence, small sighting angle and strong scattering anisotropy. MSH neglects only the variance of the trajectory length of the scattered rays and their backscattering. MSH contains all the singularities of exact solution of RTE (Boudak and Sarmin (1990)).

Applying the formulated approach we succeed in the solutions of RTE for all the fundamental sources:

- point isotropic (Boudak and Sarmin (1990))

$$\ddot{C}_k(r) = \frac{e^{-\varepsilon r}}{r^2}\exp\left\{\varepsilon\Lambda r\int_0^1 \vec{x}(\xi\kappa)d\xi\right\},$$

- polarized flat unidirectional (Astakhov et al. (1994))

$$\ddot{C}_k^m(\tau) = e^{-\tau/\mu_0}\exp\left(\Lambda\tau\vec{x}_k/\mu_0\right),$$

- polarized point unidirectional (Boudak and Veklenko (2002))

$$\ddot{C}_{kl}^m(r) = \frac{e^{-\varepsilon r}}{2\pi r^2}\int_0^{2\pi}\exp\left\{-im\psi + \Lambda\varepsilon r\int_0^1 \vec{x}(\rho)d\zeta\right\}d\psi,$$

where $\rho = \sqrt{\lambda^2(1-\zeta^2) + \kappa^2\zeta^2 - 2\zeta(1-\zeta)\lambda\kappa\cos\psi}$,
$\lambda = \sqrt{l(l+1)}$, that allows to extend the present method of the convergence acceleration to the arbitrary medium geometry.

5. CONCLUSIONS

The analytic form of MSH as a series on the surface harmonic allows extending this method to the combination with any numerical method: discrete ordinates, characteristics and Monte-Carlo. The polarization problems are solved completely similarly, but the boundary conditions in the SH method are better used in Mark's form.

6. REFERENCES

Astakhov I. E., V. P. Boudak, D. V. Lisitsin, and V. A. Selivanov, 1994. Solution of the vector radiative transfer equation in small angle modification of a spherical harmonics method, *Atm. and Ocean. Opt.*, 7, pp.753-761 (in Russian).

Bothe W., 1929. Die Streuabsorption der Elektronenstrahlen, *Zeit.f. Physik*, 54, pp.161-178.

Boudak V. P., and B. B. Veklenko, 2002. Polarization of light field generated by a point unidirectional source in a turbid anisotropically scattering medium, *Atm. and Ocean. Opt.*, 15, pp.790-794.

Boudak V. P., and S. E. Sarmin, 1990. The radiative transfer equation solution by a method of small angle modification of spherical harmonics, *Atm. Opt.*, 3, pp. 981-987 (in Russian).

Chandrasekhar S., 1950. Radiative Transfer, Clarendon Press, Oxford.

Goudsmit S., and J. L. Saunderson, 1940. Multiple scattering of electrons, *Phys. Rev.*, 57, pp. 24-29.

Karp A. H., J. Greenstadt, and J. A. Fillmore, 1980. Radiative transfer through an arbitrary thick, scattering atmosphere, *J. Quant. Spectr. Rad. Trans.*, 24, pp. 391-406.

Kompaneets A. S., 1947. A multiple scattering of thin bundles of the fast electrons, *Zhurnal Eksperimental'noi i Teoreticheskoi Fiziki*, 17, pp. 1059-1062 (in Russian).

Moliere G., 1948. Theorie der Streuung schneller geladener Teilchen. Mehrfach - und Vielfachstreuung, *Zeit.f. Natur.*, 3a, pp. 78-97.

Snyder H. S., and W. T. Scott, 1949. Multiple scattering of fast charged particles, *Phys. Rev.*, 76, pp. 220-225.

Wang M. C., and E. Guth, 1951. On the theory of multiple scattering, particularly of charged particles, *Phys. Rev.*, 84, pp. 1092-1111.

Wentzel G., 1922. Zur theorie der Streuung von β-Strahlen, *Ann. d. Phys.*, 69, pp. 335-368.

SIMULATIONS OF SPECTRAL ACTINIC FLUX DENSITY FIELDS IN SCATTERED AND OVERCAST CLOUD CONDITIONS: COMPARISON WITH INSPECTRO AIRCRAFT MEASUREMENTS

Anke Kniffka and Sebastián Gimeno García
Institute for Meteorology, University of Leipzig, Stephanstr. 3, 04103 Leipzig, Germany
kniffka@uni-leipzig.de

Evelyn Jäkel, Sebastian Schmidt, and Manfred Wendisch
Institute for Tropospheric Research, 04318 Leipzig, Germany

Ronald Scheirer and Thomas Trautmann
German Aerospace Centre, Remote Sensing Technology Institute, 82234 Wessling, Germany

ABSTRACT

The influence of spatial atmospheric inhomogeneities on the actinic flux density field is analyzed by means of three-dimensional radiative transfer simulations. For this purpose aircraft in-situ measurements of spectral actinic flux densities and microphysical cloud parameters obtained during INSPECTRO are used. Based on these measurements a set of model atmospheres is created that represent specific atmospheric situations including all the necessary properties of the scattering objects in a spatially resolved form. The presented study is focused on the two different atmospheric situations: total overcast case and broken cloud field case. For the overcast cloud case the measurements and simulation results show a reasonably good agreement within the measurement un-certainty. The optical properties of the cloud cover were calculated by either accounting for measured droplet size spectra or with a parameterization based on the measured effective droplet radius and liquid water content of the clouds. The model results reveal the well-known effect of enhancement of the downwelling component of the actinic flux density right below the cloud top. This enhancement is confirmed by the measurements. Implementation of a spatially variable spectral surface albedo instead of a grey value leads to a slightly improved agreement between the measurements and the three-dimensional simulations. For the analysis of a broken cloud field airborne measurements below and above the cloud layer were used. As expected, significant local discrepancies between the measurements and simulation results appear in this case. However the statistical properties of the measured and calculated spectral actinic flux density fields resemble each other closely.

1. INTRODUCTION

The intention of the EU-project INSPECTRO (INfluence of clouds on the SPECtral actinic flux in the lower TROposphere) is to measure the influence of inhomogeneities of atmospheric aerosol and clouds on the radiative transport in the atmosphere. In the following, mainly the influence on the actinic flux density field is studied using airborne measurements. The three-dimensional radiative transfer simulations are carried out with the models SHDOM (Spherical Harmonics Discrete Ordinate Method, see Evans, 1998) and LMCM (Leipzig Monte Carlo Model, see Gimeno García and Trautmann, 2003), whereas the one-dimensional transfer simulations are carried out with DISORT (Discrete Ordinate Method, see Stamnes et al., 1988). Prior to the simulations, the optical properties of the atmospheric components are determined in a spatially resolved form. The results for the single components are combined to form the optical properties of the atmosphere.

2. DATA RETRIEVAL

The optically relevant constituents of the atmosphere and several variables of radiation were determined by means of airborne and ground based measurements. The results presented here are based on the data that were obtained during the first of two measurement campaigns. This campaign took place in Norwich, East Anglia, GB, in September 2002. Four ground stations and 3 aircraft, 1 balloon and 1 ultralight aircraft formed the campaign setup. The flight patterns covered an area of approximately 800 km². The following measurements were carried out continuously during the flights: the actinic flux densities and surface albedo were measured with the Actinic Flux Density Meter (AFDM) and the Albedometer, see Jäkel, et al. (2005). Simultaneously the cloud parameters were obtained using an Nevzorov hot wire probe, a Fast Forward Scattering Spectrometer Probe (Fast-FSSP) and a Particle Volume Monitor (PVM). Measurements of the aerosol parameters were carried out with a Passive Cavity Aerosol Spectrometer Probe (PCASP-X). All devices were mounted on the Partenavia aircraft operated by enviscope GmbH (Frankfurt, Germany).

3. DATA RECONSTRUCTION

For the analysis below mostly the measurements of the Partenavia were used. This includes droplet size distribution, liquid water content (LWC) and effective radius (r_{eff}) of the cloud droplets, aerosol size distributions and the albedo as well as actinic flux density measurements. Prior to the calculation of radiative transfer, the optical properties of the atmosphere are to be defined. The optical properties of the input atmosphere in a certain volume element can be specified by few variables, which are the extinction coefficient β_{ext}, single scattering albedo ω and the scattering phase function or the asymmetry parameter g, respectively. In order to construct such an atmosphere, the properties of the particular scattering components such as aerosol particles and cloud droplets are to be calculated separately. The scattering coefficient σ_{sca} and the asymmetry parameter g_p of a single aerosol particle are derived from Mie theory for this analysis unlike the air molecules which are treated with Rayleigh theory. The necessary refractive indices are taken from Shettle and Fenn (1979). The chemical composition of the aerosol particles is defined for different types of aerosol (rural, urban, oceanic, maritime), which influences in our studies only the refractive index of the particles. In this study, the maritime type was used since the measurement site was located close to the sea. In order to obtain the parameters of the complete ensemble of particles in a grid cell, the scattering coefficient β_{sca} and g at certain radii and weighted by the number size distribution of the aerosols, are integrated. β_{ext} is calculated alike. Within each grid cell the number size distribution can be chosen according to measurements, or in case of few data, can be taken from a standard profile (Lenoble and Brogniez, 1984). The optical properties of cloud droplet size distributions are also calculated using Mie theory. For each grid cell either mono- or bimodal distributions are chosen, depending on the measured values. In case of missing droplet spectra information, the parameterization of Slingo (1989) is a very good approximation, for which the LWC and r_{eff} of the grid volume have to be known. Ozone is a very important absorbing gas in the atmosphere. The absorption cross section of an ozone molecule is assumed to depend on pressure and temperature after Daumont, et al. (1992). The number concentration profile of the molecules is gathered from McClatchey, et al. (1971) in standardized form, but is scaled to the particular ozone column content retrieved from measurements of the Global Ozone Monitoring Experiment spectrometer on the European Remote Sensing Satellite (GOME, ERS-2). Additionally, the appropriate background is composed of a standardized atmosphere from McClatchey, et al. (1971). As a last step, the optical properties of the particular species are combined to form the necessary parameters for the whole atmosphere. A more detailed description of the data reconstruction can be found at Kniffka and Trautmann (2004).

4. CLOUD SCENARIOS

4.1 Stratiform Cloud

Two special cloud scenarios shall be considered. The first scenario will be concerned with an overcast cloud measured during the INSPECTRO campaign on 09/14/02. The layer reached from approximately 600 m to 1060 m in vertical direction and contained an appreciable amount of liquid water (0.5 g/m³ on average). For the analysis, the model cloud was considered to consist of several vertical layers. Within one layer, the cloud parameters were kept homogeneous. The optical properties of the cloud's layers were calculated in two different ways, first with a parameterization following Slingo (1989) which is a very efficient calculation method and second through using the more detailed Mie theory. The aerosol particles were spatially subdivided in the same manner as the cloud droplets, and their optical properties were calculated using Mie theory. The atmosphere's aerosol content was very low during the measurement (maximal optical thickness 0.027 at $\lambda = 550$ nm) and the ozone content was 283 DU. So two model atmospheres were created which differ only in the treatment of the cloud droplets. As can be seen in Fig. 1, both simulations resemble the measurement very closely. In this profile of the actinic flux density, the measurement with the surrounding error interval is displayed in orange color, the dashed lines show the SHDOM simulations, whereas the blue line corresponds to the simulation based on droplet spectra. The red line indicates the simulation based on the Slingo parameterization. Near the upper edge of the cloud (indicated by the black horizontal line at 1060 m), an enhancement of the actinic flux density is observed in the measurement which is caused by the very thin uppermost cloud layer. When carrying out the same calculations but with more dense optical properties for this layer, this enhancement will not occur (calculations not shown). This enhancement is due to the conversion of direct light into diffuse light caused by the droplet multiple scattering process taking place within the upper few decameters near the cloud top. It can be shown, that under suitable conditions a very strong enhancement of actinic flux density can occur in the uppermost layer of a cloud, see Madronich (1987).

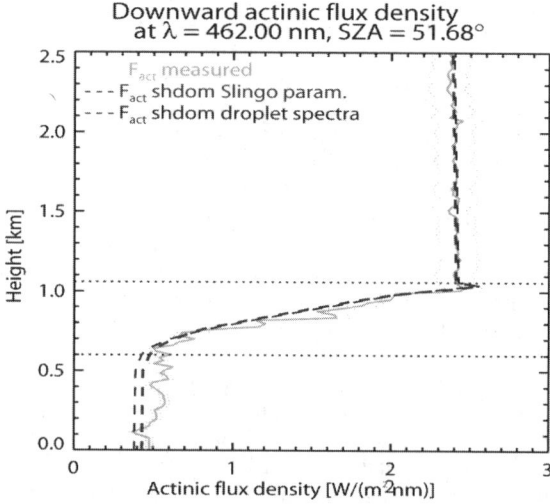

Fig. 1. Profiles of the actinic flux density at 462.0 nm, stratiform cloud cover, 09/14/2002.

Below the cloud layer the actinic flux density field depends also distinctly on the surface albedo. In order to retrieve a most realistic ground field, the measurements of the albedometer on the aircraft were used. The albedo-meter measures the areal surface albedo. This albedo is the result of a superposition of backscattering effects caused by different ground structures. We have reduced the perturbations of the clear atmosphere in between according to Wendisch, et al. (2004). With this method a spectral areal surface albedo could be derived from clear sky measurements for the region below the flight tracks. The effects of the ground albedo are not very pronounced in the studied wavelength region (see Fig. 2) but nevertheless they should not be neglected. Particularly in the UV and the middle wavelength region it is important to have realistic albedo values and the used method prevents the modeler from guessing an inaccurate value.

RADIATIVE TRANSFER THEORY AND MODELING

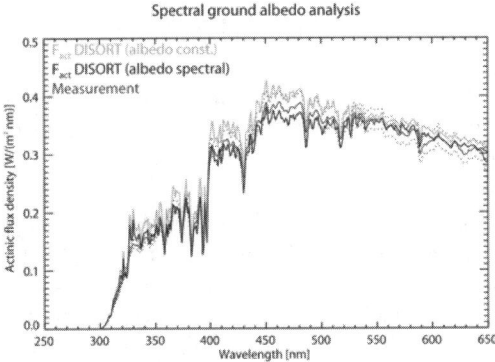

Fig. 2. Spectra of the actinic flux density for constant and spectral ground albedo, below cloud cover.

4.2 Inhomogeneous Cloud Cover

Here the scattered cloud cover as it was observed on September the 28th, 2002 is analyzed. The cloud field was constructed with the model CLABAUTAIR (Scheirer and Schmidt, 2004), by applying an autocorrelation function technique to the measured data (LWC, r_{eff}) along the flight tracks (lines of dense data) and taking characteristic turbulence length scales into account. The generated cloud field possessed a small extension in height of approx. 150–200 m. Its optical thickness however reached values up to 33.75. The cloud cover was rather inhomogeneous both in horizontal and in vertical direction. Like the day before, the aerosol content was very low near the ground but remained constant up to 1400 m altitude, so that the aerosol optical thickness was 0.019 at 550 nm. The ozone column content was 260 DU. In this cloud case Slingo's parameterization was used since the information about the cloud droplets consisted of r_{eff} and LWC. The information on the aerosol particles was applied like in section 4.1. In combination with the cloud information a three-dimensional simulation grid results, in which the optical properties of the atmosphere are provided at each grid point. The simulation field measured 10 x 20 x 60 km³ which corresponded to 50 x 100 x 60 cells in x-, y- and z-direction.

For the analysis, the structure of the simulated and the measured actinic flux density fields was investigated in order to detect possible similarities. The flight pattern of this measurement contained two horizontal flight legs, above and below the cloud field. In Table 1 an overview about the moments of the resulting actinic flux density fields is given. The mean values and the variances of the measured and the simulated actinic flux density fields are quite similar, both above and below the cloud field. Remarkably the deviations of the moments do not worsen below the cloud field. Here the variances of the simulations are bigger. This is probably caused by local differences in the reconstructed and the real cloud field. The values of the two higher moments skewness and kurtosis (not shown in the table) do not correspond which is obvious, since the skewness measures the asymmetry of a distribution and the kurtosis indicates its dilation or compression. The higher moments are very sensitive to deviations in the analyzed distributions, especially if the distributions are not monomodal (see Fig. 3).

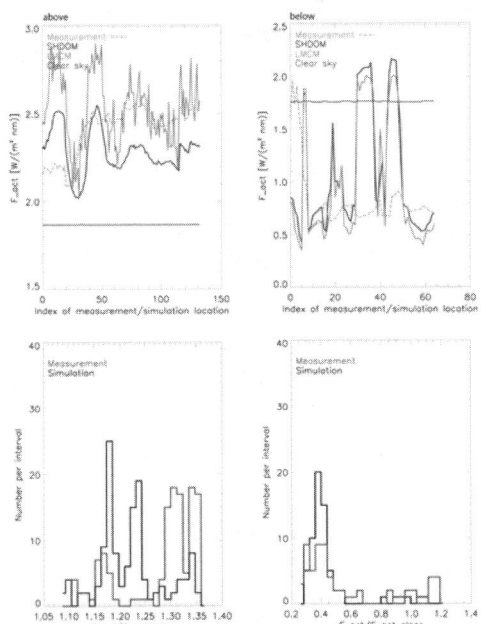

Fig. 3. Actinic flux densities along flight path and the corresponding frequency distributions, above and below cloud cover, 09/28/2002.

The two upper graphs in Fig. 3 show the actinic flux densities as they were measured or simulated along the flight path for a wavelength of 550 nm. The dashed green line denotes the measured points, the corresponding SHDOM simulations are displayed in blue, the turquoise line shows the LMCM simulations. The simulated clear-sky value is displayed in red color. The course of simulated and measured values along the flight path is much smoother above the cloud top when compared to the values below the cloud. This is due to the dominance of direct light. Below the cloud field, parts of the direct light have already been scattered and have partly been absorbed or they have penetrated through the cloud gaps. At the rims of the gaps an amplification of the actinic flux density occurs frequently because diffuse radiation can enter the cloud gaps unhampered. In addition to the light that directly enters the gaps from above, the actinic flux density can reach values above the clear sky value. Due to these effects, the actinic flux density values below the cloud field fluctuate very much along the flight path. This is also true for the measured values although their course seems to be much smoother. This is due to an averaging procedure which was necessary for fitting the highly resolved measured data into the simulation grid for this illustration. The LMCM calculations are in good agreement with the SHDOM results below the cloud layer. The absolute values of measurement and simulation show relatively large differences, but nevertheless the curves above the cloud field proceed very similar. Also below the cloud field an enhancement of radiation above the clear sky value is visible. The two lower figures depict the related frequency distributions of the actinic flux densities of SHDOM simulations and measurements. At first sight, the distributions of the data above the cloud top seem to be very different. But still the structure of the distributions is quite similar. The distributions seem to be divided into three major parts with gaps in between. It should be pointed out that a local point-by-point agreement between the 3D measured and observed actinic flux density fields cannot be expected, since the mapping of the reconstructed 3D cloud structure onto the simulation grid may cause spatial shifts and averagings as compared to the real cloud. Below the cloud field, the two distributions bear much more resemblance. In this case also a division into several parts is noticeable.

Table 1. Statistical Moments

	above		below	
	Mean	Variance	Mean	Variance
SHDOM	2.28255	0.01541	1.05371	0.32761
Measure.	2.40759	0.01746	0.82282	0.09562
LMCM	2.52295	0.03627	0.98169	0.34953

5. DISCUSSION

The analysis of the INSPECTRO data from the first measurement campaign shows a good agreement of measured and simulated actinic flux densities in case of an atmosphere with stratiform cloud cover. By means of an atmospheric model and three-dimensional radiative transport simulations realistic actinic flux density fields can thus be simulated. Two methods concerning cloud droplet treatment were tested, including either the efficient parameterization following Slingo or the more precise Mie theory. The radiative transfer calculations, however, show only slightly different results. In the measured as well as in the simulated profiles a small enhancement of actinic flux density directly below the cloud top could be observed. The enhancement was not detected in the irradiances, neither in the measured nor in the simulated profiles. The introduction of a realistic spectral ground albedo leads to an improvement of the simulation results below the cloud layer. Above the cloud top, the actinic flux density field showed for this case no significant differences when taking the spectral ground albedo into account.

The analysis of the inhomogeneous cloud field was carried out mostly by means of statistical methods. The frequency distributions of the actinic flux densities below and above the cloud cover prove the similar structures of the fields. The displayed time series of the flight leg a-bove the cloud top indicates a strong similarity of the real and the reconstructed cloud field concerning their spatial composition. Expectedly, the local errors are quite big, since even a small spatial displacement of a cloud segment leads to significant changes of radiation fields. Also the spatial resolution of the simulation grids as well as the integration time interval of the measurements showed very strong influences.

6. ACKNOWLEDGEMENTS

The work of A. Kniffka and E. Jäkel was funded by the Deutsche Forschungsgemeinschaft under grants TR 315/3-1, JA 836/13-3 and WE 1900/6-1 and -2.

7. REFERENCES

Daumont D., J. Brion, J. Charbonnier, and J. Malicet (1992): Ozone UV spectroscopy 1: Absorption cross-sections at room temperature. *J. Atmos. Chem.*, 15, 145-155.

Evans, K. F. (1998): The spherical harmonic discrete ordinate method for three-dimensional atmospheric radiative transfer. *J. Atmos. Sci.*, 55, 429-446.

Gimeno García, S., and T. Trautmann (2003): Radiative transfer modeling in inhomogeneous clouds by means of the Monte Carlo method. *Wissenschaftliche Mitteilungen aus dem Institut für Meteorologie der Universität Leipzig*, 30, 29-43.

Jäkel, E., M. Wendisch, A. Kniffka, and T. Trautmann (2004): A new airborne system for fast measurements of up and downwelling spectral actinic flux densities. *Applied Optics*, 44/3, 434-444.

Kniffka, A., and T. Trautmann (2004): Verwendung von mikrophysikalischen Messungen zur Charakterisie-rung von Aerosol und Wolken für Strahlungsübertragungsrechnungen. W*issenschaftli-che Mitteilungen aus dem Institut für Meteorologie der Universität Leipzig*, 34, 13-29.

Lenoble, J., and C. Brogniez (1984): A comparative review of radiation aerosol models. *Beitr. Phys. Atmosph.*, 57, 1-20.

Madronich, S. (1987): Photodissociation in the atmosphere 1. Actinic flux and the effects of ground reflections and clouds. *J. Geophys. Res.*, 92 (D8), 9740-9752.

McClatchey, R. A., R. W. Fenn, J. E. A. Selby, F. E. Volz, and J. S. Garing (1971): Optical properties of the atmosphere. AFCRL-71-0279, *Environmental Research Papers*, 354.

Scheirer R., and S. Schmidt (2004): CLABAUTAIR: a new algorithm for retrieving three-dimensional cloud structure from airborne microphysical measurements. *Atmospheric Chemistry and Physics Discussions*, 4, 8609-8625.

Shettle, E. P., and R. W. Fenn (1979): Models for the aerosols of the lower atmosphere and the effect of humidity variations on their optical properties. AFGL-TR-79-0214, *Environmental Research Papers*, 676.

Slingo, A. (1989): A GCM parameterization for the shortwave radiative properties of water clouds. *J. Atmos. Sci.*, 46, 1419-1427.

Stamnes, K., S. Tsay, W. Wiscombe, and K. Jayaweera (1988): Numerically stable algorithm for discrete-ordinate-method radiative transfer in multiple scattering and emitting layered media. *Appl. Opt.*, 27, 2502-2509.

Wendisch M., P. Pilewskie, E. Jäkel, S. Schmidt, J. Pommier, S. Howard, H. H. Jonsson, H. Guan, M. Schröder, and B. Mayer (2004): Airborne measurements of areal spectral surface albedo over different sea and land surfaces. *J. Geophys. Res. – Atmospheres*, 109 (D8), D08203.

VALIDATION OF A RADIATIVE TRANSFER MODEL FOR SIMULATION OF POLARIZED RADIANCES IN A COUPLED OCEAN-ATMOSPHERE SYSTEM

L. Duforet and P. Dubuisson
ELICO, Universite du Littoral Cote d'Opale, 32 av Foch, 62930 Wimereux, France

B. Bonnel, F. Parol, and M. Vesperini
LOA, Université des Sciences et Technologies de Lille, 59655, Villeneuve d'Ascq, France

ABSTRACT

A high resolution radiative transfer model, from visible to microwave wavelengths, is presented for the ocean-atmosphere system. It allows computations of polarized radiances for an absorbing and scattering atmosphere including aerosols. Any anisotropic and polarized surfaces can be included so a rough sea-surface description was developed to take into account interactions with the atmospheric radiance field. Inter-comparisons are realized with a reference model, the Successive Orders Code. Discrepancies are smaller than the instrumental noise, so computational efficiency is demonstrated. Then, simulations are compared with POLDER airborne measurements above sea surface. Agreements between observations and numerical simulations are quite good in respect with the uncertainties on aerosol properties.

1. INTRODUCTION

In ocean color studies, information about chlorophyll concentration or particles suspended in water are retrieved from the water-leaving radiance. To obtain this quantity from satellite observations, the diffuse atmospheric radiance and the reflected signal onto the sea-surface have to be removed from the total leaving radiance. Accurate radiative transfer models, in the ocean-atmosphere system, are very useful because they allow computations of these different contributions. This study consists on developing such model on a wide spectral range from visible to microwave wavelengths. This model takes also into account polarization whose important key role, in atmospheric studies, was shown by POLDER. Finally, a validation is carried out through inter-comparisons with a reference model and experimental data.

2. RADIATIVE TRANSFER MODELING

The equation describing the transfer of monochromatic polarized radiation at wavelength λ through a plane-parallel layer is given by:

$$\mu \frac{d\vec{L}(\delta,\mu,\varphi)}{d\delta} = \vec{L}(\delta,\mu,\varphi) - \frac{\omega_0}{4\pi} \int_0^{2\pi} \int_{-1}^{1} Z(\delta,\mu,\mu',\varphi-\varphi')$$
$$\times \vec{L}(\delta,\mu',\varphi')d\mu'd\varphi' - \frac{\omega_0}{4\pi} Z(\mu,\mu_s,\varphi-\varphi_s)\vec{E}_s e^{\frac{-\delta}{\mu_s}} \quad (1)$$

The intensity vector $\vec{L}$ stands for the radiance at an optical depth δ along the $\vec{s}$ direction related to μ the cosine of the polar angle and φ, the azimuthal angle. ω_0 is the single scattering albedo, Es the solar constant outside the Earth's atmosphere, the subscript s refers to solar quantities. The real components of this vector (I, Q, U, V) are called the Stokes parameters. I represents the intensity of light; Q and U describe the linear polarization state and V the circular polarization state. The degree of polarization P, ratio of the polarized and total radiance, is expressed by:

$$P = \frac{\sqrt{Q^2 + U^2 + V^2}}{I} \quad (2)$$

The scattering of a polarized radiation by a volume-element, over an angle Θ related to μ, μ' and $\varphi - \varphi'$ is expressed by the scattering matrix $F(\Theta)$. $F(\Theta)$ is defined in the scattering plane (plane through the incident and scattered beam), whereas the Stokes parameters are described in the local meridian plane (plane through the direction of propagation and the zenith). For this reason, scattering processes are expressed by the phase matrix $Z(\Theta)$ instead of $F(\Theta)$ (Hovenier et al., 1983). It is written as:

$$Z(\Theta) = L(\pi - \sigma_2)F(\Theta)L(-\sigma_1), \quad (3)$$

with $L(\pi-\sigma_2)$ used for turning the incident meridian plane to the scattering plane and $L(\sigma_1)$ used for turning the scattering plane to the meridian plane of the scattered ray.

In the mathematical developments, the Stokes parameters and the phase matrix, azimuth-dependent quantities, are expanded in a Fourier cosine and sine series. It allows to reduce the number of variables treated at one time and simplify the radiative transfer equation computations. The azimuthally independent phase matrix component can be calculated with the following expression:

$$Z^m(\pm u, u') = (-1)^m \sum_{l=m}^{M} P_m^l(\pm u) S^l P_m^l(u') \quad (4)$$

with

$$S^l = \begin{bmatrix} \alpha_1^l & \beta_1^l & 0 & 0 \\ \beta_1^l & \alpha_2^l & 0 & 0 \\ 0 & 0 & \alpha_3^l & \beta_2^l \\ 0 & 0 & -\beta_2^l & \alpha_4^l \end{bmatrix}. \quad (5)$$

The matrix $P^l_m(\pm u)$ is defined through the generalized spherical functions and the elements of S^l are the expansion coefficients. M is the maximum number of harmonics used in the expansion. u is equal to μ absolute value. The advantage of such mathematical development is important: in the computations of our model, the phase matrix can be expressed in expansion coefficients without considering the geometrical conditions.

For molecular scattering, the required order M is 2, but phase functions associated with aerosols are rather or less asymmetric and the number of harmonics required is important. In practical, the summation (4) is truncated at M terms, it results that the phase function is not restituted with accuracy. The delta-M approximation is used to remove the sharp, forward-scattering peak and so reduce the number of terms taken into account in the expansion (Wiscombe et al., 1977). Some preliminary computations have shown that, with the aerosol phase functions used in our model, 24 terms are needed to restitute aerosol phase functions correctly with the delta-M approximation. Based on such mathematical developments, the radiative transfer equation is solved by means of the Adding Doubling method (De Haan et al., 1987).

3. ANISOTROPIC POLARIZED ROUGH SEA-SURFACE FROM VISIBLE TO MICROWAVE WAVELENGTHS

3.1 Fresnel's Reflection

If the sea is absolutely calm, its flat surface is like a mirror and the sun beam is reflected towards the specular point according to Fresnel's formulas. These formulas describe the intensity reflection coefficients for the parallel and perpendicular polarization; these coefficients are respectively r_l and r_r:

$$r_l = \frac{m(\lambda)\cos\theta_i - \cos\theta_t}{m(\lambda)\cos\theta_i + \cos\theta_t} \quad (6)$$

and

$$r_r = \frac{\cos\theta_i - m(\lambda)\cos\theta_t}{\cos\theta_i + m(\lambda)\cos\theta_t}, \quad (7)$$

with θ_i the angle between the incident radiation and the normal to the surface; θ_t is the refraction angle and m (λ) the water complex refractive index with λ the wavelength of the incident ray. Using the Snell-Descartes' law in the expressions (6) and (7), Fresnel's formulas depend only θ_i and m (λ). From r_r and r_l, we deduce the reflection matrix R(Θ).

3.2 Interaction wave Slope Probability Density

In case of a smooth surface, only a single bright spot can be seen at the horizontal specular point. However, in usual cases, the pattern called glitter is observed. It is due to the roughness of the sea caused by wind. Waves can be considered like a set of water facets, each facet which is so inclined as to reflect a sun ray towards the receiver, becomes a highlight. The thousands highlights constitute the glitter pattern. From aerial photographs of the glitter, Cox and Munk expressed the wave-slope probability density (Cox and Munk, 1954). The probability density of having a wave with a slope θ_n with respect to the zenith is:

$$p(\theta_n) = \frac{1}{2\pi\sigma^2}\exp\left[-\frac{\tan^2(\theta_n)}{2\sigma^2}\right], \quad (8)$$

with

$$2\sigma^2 = 0.003 + 0.00512W \pm 0.004, \quad (9)$$

where σ^2 is the wave-slope variance and W is the isotropic wind speed. For infrared and microwave wavelengths, the wave-slope probability density becomes the interaction probability density. p (θ_n) is just multiply by cos ω/cos θ_n with ω the angle between the incident ray to the facet and the zenith (Freund et al., 1997). To compute the anisotropic polarized reflectance, each elements of the reflection matrix are multiplied by the probability density corresponding to the water-facet slope on which the Fresnel's reflection occurs. Then the rotation of the planes is applied like explained Eq. (3)

3.3 Polarized Sea-Surface Emissivity

To compute the polarized emissivity for one single facet, it requires only the knowledge of the reflection matrix. In fact, in the infrared spectral range, a quite good approximation consists in not considering the energy transmitted. So the energy absorbed is,

$$A = 1 - R, \quad (10)$$

with 1 the identity matrix. From Kirchhoff's law, we obtain that the emissivity E is equal to the absorbed energy A. To obtain the emissivity of the entire rough sea-surface E have to be integrated over all of the viewed direction θ_r between $\left[0, \frac{\pi}{2}\right]$.

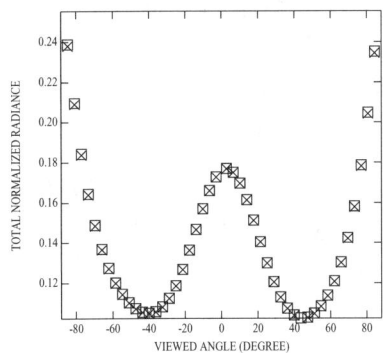

Fig. 1. Total normalized radiance (= radiance (in W. m^{-2}. sr^{-1}) * π / Es). Squares and crosses are normalized radiances computed with the Successive Orders model and our model, respectively. w = 7 m/s and θ_s = 3°.

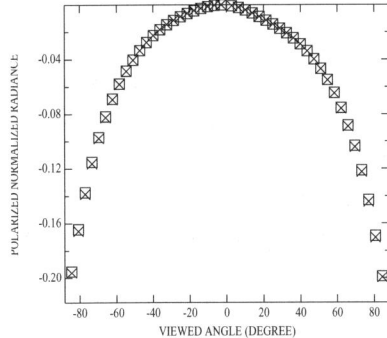

Fig. 2. Same as Fig. 1 for polarized normalized radiance.

4. INTER-COMPARISONS WITH A REFERENCE MODEL

Inter-comparisons with a reference model, the Successive Orders Code (SOC) (Deuze et al., 1989) are realized to test the validity of the rough sea-surface model and its coupling with the atmosphere. Computations of the atmospheric path radiance by the Adding Doubling method have already been validated (De Haan et al., 1987), so numerical inter-comparisons are only carried out on a simple atmospheric case. Thus, upward radiances are computed at the top of the atmosphere composed of one layer without aerosols and the molecular absorption is not considered (ω_0 = 1). Note that further simulations will even be performed for a more realistic atmosphere with gas absorption but directly compared with instrumental measurements (see section 5) to present an application of our model.

Fig. 1 and Fig. 2 compare radiances for different viewed angles in the solar principal plane, the wind speed w equal to 7 m/s and the solar zenith angle θ_s is 3°. Note that in the SOC, the solar constant Es is equal to π so simulated quantities called normalized radiances are in sr^{-1}. To obtain radiances in W.m^{-2}.sr^{-1}, simulation results have to be multiple by Es/π. Absolute discrepancies are evaluated at each viewed angles for various wind speeds and zenith angles θ_s. Maximum absolute discrepancies in W.m^{-2}.sr^{-1} are presented in Table 1. As an example, the Noise Equivalent Radiance (NEΔL) of Meris, one of the most accurate sensor, is equal to 0.025 Wm^{-2}.sr^{-1} at 440 nm (IOCCG, 1997). It is the same order of magnitude than the absolute discrepancies between the two models. The accuracy of our ocean-surface model is demonstrated.

Table 1. Maximum Absolute Discrepancies, in Wm^{-2}. sr^{-1}, Between the Successive Orders Code and Our Model, at 440 nm

w (m/s)	θ_s = 60°	θ_s = 40°	θ_s = 20°	θ_s = 3°
2	0.0170	0.0050	0.0020	0.0008
7	0.0250	0.0060	0.0030	0.0015
14	0.0270	0.0125	0.0040	0.0020

5. INTER-COMPARISONS WITH MEASUREMENTS

Simulations are compared with measurements, in clear-sky, from POLDER airborne radiometer during FRENCH campaign (Field Radiation Experiment Natural Cirrus and High-level clouds, October 2001, South-Western France). Molecule and aerosol scattering is accounted in the model as well as gaseous absorption through the correlated k-distribution (Dubuisson et al., 2004).

The aerosol optical parameters and phase matrix are from the World Meteorological Organization models (WMO, 1986). The aerosol vertical extinction profile is supposed mainly composed of maritime aerosols concentrated just above the sea and their optical thickness δ_a is fixed from the PHOTONS/AERONET photometers. Pressure, temperature profiles and the surface wind speed are obtained from radiosonde observations.

Inter-comparisons are realized in three POLDER spectral bands (443, 765, and 864 nm). Results are similar for 765 nm spectral band so only 864 nm and 443 band are shown in Fig. 3 and Fig. 4 respectively which show a general quite good agreement between experimental and simulated data. Nevertheless, small discrepancies are observed. A first explanation concerns the aerosol optical thickness and wind speed uncertainties because photometers and radiosonde observations are not located near the observation area. New simulations are obtained for w = 10 m/s and δ_a^{440} = 0.14 but discrepancies do not decrease. A second explanation concerns the aerosol model. In next studies, simulations could be lead for another aerosol type, with a phase function less sharp in the straight-forward direction and/or with an aerosol more absorbing. Third, differences may be due to the anisotropic wind which is not accounted in because the isotropic probability density is used in our surface model. Previous studies show that the shape of the glitter pattern varies with wind direction and so influence the intensity of reflected radiance in a given viewed direction.

6. CONCLUSIONS

The numerical accuracy of our model is demonstrated through inter-comparisons with reference and experimental data. One of the first improvements will consist on extent the atmospheric gaseous absorption from 4 microns to microwave wavelengths to dispose of a suitable tool on a very wide spectral range. Thus, this model could be used in data processing of various new satellites like PARASOL and CALIPSO.

7. ACKNOWLEDGEMENTS

This work is supported by the Centre National des Etudes Spatiales (CNES) and the Conseil Regional du Nord Pas-de-Calais.

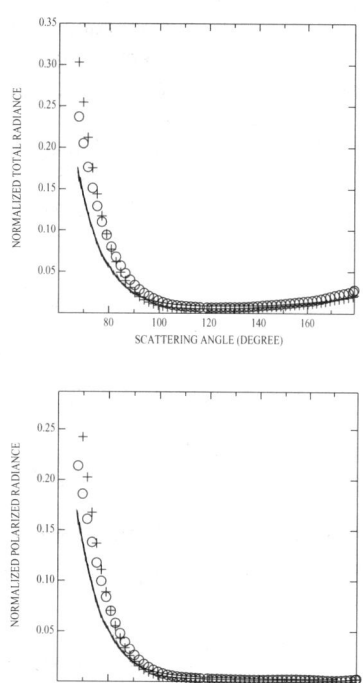

Fig. 3. Inter-comparisons between POLDER (solid curve) and simulated radiances (crosses for $\delta_a = 0.035$, w = 7 m/s and circle dots for $\delta_a = 0.07$, w = 10 m/s) at 864 nm.

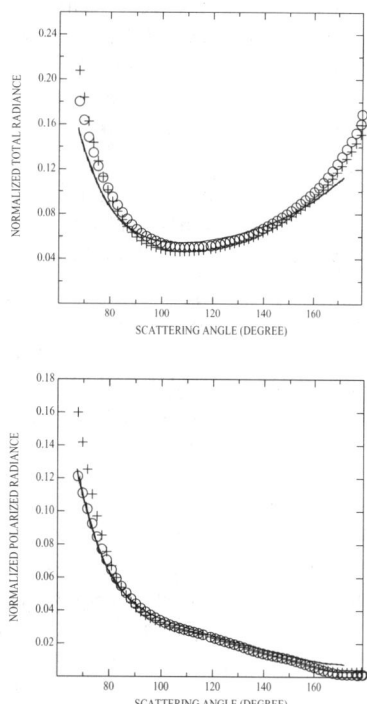

Fig. 4. Same as Fig. 3, at 443 nm (crosses for $\delta_a = 0.07$, w = 7 m/s and circle dots for $\delta_a = 0.14$, w = 10 m/s.

8. REFERENCES

Cox, C. and, W. Munk, 1954: Measurement of the Roughness of the Sea Surface from Photographs of the Sun's Glitter. *J. Opt. Soc. Am.*, 44, 838-850.

De Haan, J. F., P. B. Bosma, and J. W. Hovenier, 1987: The adding method for multiple scattering calculations of polarized light. *Astron. Astrophys.*, 183, 371-391.

Deuze, J. L., M. Herman, and R. Santer, 1989: Fourier series expansion of the transfer equation in the atmosphere-ocean system. *J. Quant. Spectrosc. Radiat. Transfer,* 41, 483-494.

Dubuisson, P., D. Desailly, M. Vesperini, and R. Frouin, 2004: Water vapor retrieval over ocean using near-infrared radiometry. *J. Geophys. Res.*, 109, D19106.

Freund, D. E, R. I. Joseph, D. J. Donohue, and K. T. Constantikes, 1997: Numerical computations of rough sea surface emissivity using the interaction probability density. *J. Opt. Soc. Am.*, 14, 1836-1849.

Hovenier, J. W., and C. V. M. van der Mee, 1983: Fundamental relationships relevant to the transfer of polarized light in a scattering atmosphere. *Astron. Astrophys.*, 128, 1-16.

International Ocean-Colour Coordinating Group (IOCCG), 1997: Minimum Requirements for an Operational Ocean-Colour Sensor for the Open Ocean. *Rep. 1*, 45 pp.

Wiscombe, W. J., 1977: The Delta-M Method: Rapid Yet Accurate Radiative Flux Calculations for Strongly Asymmetric Phase Functions. *J. Atmos. Sci.*, 34, 1408-1422.

World Meteorological Organization (WMO), 1986: A preliminary cloudness standard atmosphere for radiation computation. Rep. 24, 53 pp., Geneva.

A QUASI-ANALYTIC SOLUTION OF THE RADIATIVE TRANSFER EQUATION FOR THREE-DIMENSIONAL, HETEROGENEOUS ATMOSPHERES

H. Ishida and S. Asano
Center for Atmospheric and Oceanic Studies, Tohoku University
Sendai, 980-8578, Japan

ABSTRACT

A new solution of the three-dimensional (3-D) radiation transfer equation (RTE) is developed. The remarkable feature of the solution is an application of scaling-function expansion, which is suitable to represent irregular and inhomogeneous variations in space, for discretization of terms that depend on the horizontal coordinates in the 3-D RTE. Using the present solution, the effects of horizontal inhomogeneity of cloud on the corresponding radiation field are estimated.

1. INTRODUCTION

For estimating the Earth's radiation budget and for improving the atmospheric remote sensing, it is important to accurately calculate the radiative transfer in the atmosphere. Clouds in nature often have inhomogeneous structures, which raise difficulty in exact calculations of radiative transfer in clouds. It is necessary to estimate quantitatively the effects of cloud inhomogeneity on radiation field. For the purpose, we have developed a new method to solve the three-dimensional (3-D) radiative transfer equation (RTE). The solution is explicit and quasi-analytical, and therefore it is usable for quantitative analysis of the relation between cloud inhomogeneity and radiation field. A characteristic feature of the present solution is the use of a scaling function for decomposition of variables; the scaling function is appropriate to represent fine and irregular spatial variation of such parameters as extinction coefficients of clouds. Further, the calculation efforts can be reduced, compared with the conventional Fourier expansion technique. Here, we briefly describe the scheme of the solution, and we present some computed examples for the radiation field in a hypothetical inhomogeneous cloud layer.

2. DESCRIPTION OF THE SOLUTION

2.1 Discretization of the Radiative Transfer Equation

In general, the RTE for a 3-D inhomogeneous atmosphere can be formulated into a five-dimensional (3 for the position and 2 for the radiance direction) integro-differential equation, which is usually impossible to be solved analytically. To solve the equation, discretization technique is useful in order to separate the variables and to reduce the number of independent variables. We apply Fourier expansion for the terms that depend on azimuth angle, and we use Gaussian quadrature discretization for the terms that depend on zenith angle. Both of the discretization techniques are widely used for radiative transfer in one-dimensional (plane-parallel) atmospheres. In the present algorithm, the terms that depend on the horizontal coordinates are also discretized by applying a scaling function decomposition. The decomposition enables an arbitrary function (e.g., a spatial distribution function of cloud extinction coefficients) to be represented in terms of the expansion coefficients. Details of the scaling function are explained in the next subsection. After discretization of all the variables, except the vertical coordinates, the 3-D RTE can be finally formulated into a one-dimensional first-order differential equation in a vector-matrix form, which is similar to the familiar one-dimensional (1-D) RTE. Therefore, we can apply the computational techniques that have been developed for 1-D RTE to the solution of the 3-D RTE.

2.2 Scaling Function Decomposition

The significant characteristics of a scaling function for the decomposition are as follows: first, value of a scaling function is not zero only within a finite and small domain, whereas the value is zero in the other domain. This means that the shape of a scaling function is like a wave packet. Second, an arbitrary continuous function $\alpha(x)$, which corresponds radiance or the extinction coefficient in the RTE, can be expressed by a series of the scaling function set $\{\phi(x - kb_0)\}$ in the form;

$$\alpha(x) = \sum_{k \in \mathbb{Z}} c_k \phi(x - kb_0), \quad (1)$$

where $\phi(x)$ is a scaling function, and k is an integer number. Then, $\phi(x - kb_0)$ means a translation of the scaling function $\phi(x)$ by kb_0 along the x coordinate; here b_0 is referred to as "the sampling rate". In other words, the set $\{\phi(x - kb_0)\}$ can compose the basis of a function subspace. The function set is not always orthogonal. The expansion coefficients $\{c_k\}$ of $\alpha(x)$ in Eq.(1) are uniquely determined by taking the inner product with the scaling function as follows,

$$c_k = \frac{1}{b_0} \int \alpha(x) \tilde{\phi}(x - kb_0) dx, \quad (2)$$

where $\tilde{\phi}(x)$ is the dual function of $\phi(x)$ that satisfies the relation,

$$\frac{1}{b_0} \int \phi(x - kb_0) \tilde{\phi}(x - k'b_0) dx = \begin{cases} 1 & (k = k') \\ 0 & (k \neq k') \end{cases}. \quad (3)$$

The dual function also has the same characteristics as the scaling function.

Equation (2) indicates that c_k can be considered to represent the magnitude of fluctuation of $\alpha(x)$ within a small domain around kb_0, because the domain where $\tilde{\phi}(x)$ is not zero is finite and small. This also means that c_k has no correlation to values of $\alpha(x)$ at sufficiently separated positions and a local fluctuation of the function $\alpha(x)$ does not affect all of the coefficients c_k of the scaling function series. These characteristics suggest that the scaling function decomposition is more appropriate for the discretization of fine and irregular fluctuations of spatial cloud distributions in the atmosphere than the other methods. For example, the Fourier expansion (Stephens, 1988) needs much larger number of expansion coefficients to represent a fine spatial fluctuation of cloud distributions.

As the scaling function, we employ a 4th-order B-spline function. One of the advantages to use this function is that not only the function itself but also its derivatives are continuous. This makes the discretization accurate and easy. Another advantage is that the expansion coefficients by the B-spline function take real number only. The profiles of the 4th-order B-spline function and its dual are shown in Fig. 1.

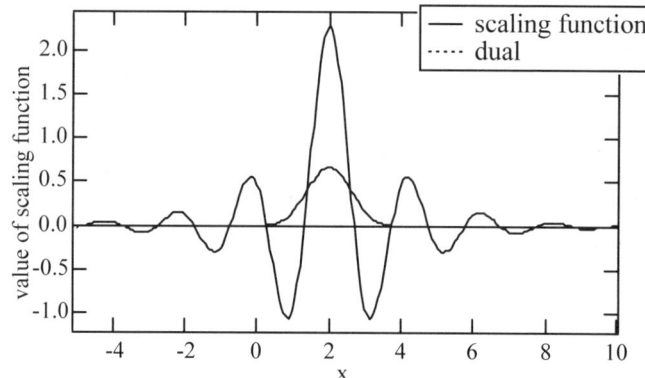

Fig. 1. Shape of the 4th-order B-spline function and its dual used in the present study.

2.3 Doubling-Adding Method

To calculate the discretized 3-D RTE formulated in a vector-matrix form, we apply the doubling-adding scheme by using the transmission and reflection matrices for inhomogeneous atmospheric layers, because this scheme is stable in computation with large-sized matrices. Further, the transmission and reflection matrices contain information of the correlation between cloud inhomogeneity and radiation field. Therefore, the scheme enables us to investigate quantitatively the effects of cloud inhomogeneity by analyzing the matrices.

3. AN EXAMPLE OF CALCULATION

In this section, we present an example of calculation by the present method for a hypothetical simple cloud model.

It is assumed that the geometrical shape of the cloud is plane-parallel and the cloud exists only in one layer with thickness of 100 m. Furthermore, the cloud layer along the y (horizontal) and z (vertical) coordinate is assumed to be uniform. The cloud is assumed to contain only liquid water particles whose number concentration varies only along the x-coordinate. The horizontal distribution of the cloud optical thickness along the x-coordinate is illustrated in the upper panel of Fig. 2. Here we treat the visible solar radiation at wavelength 500 nm. The single scattering properties of cloud particles were calculated by Mie theory. For further simplification, absorption by gaseous constituents and cloud particles is not considered.

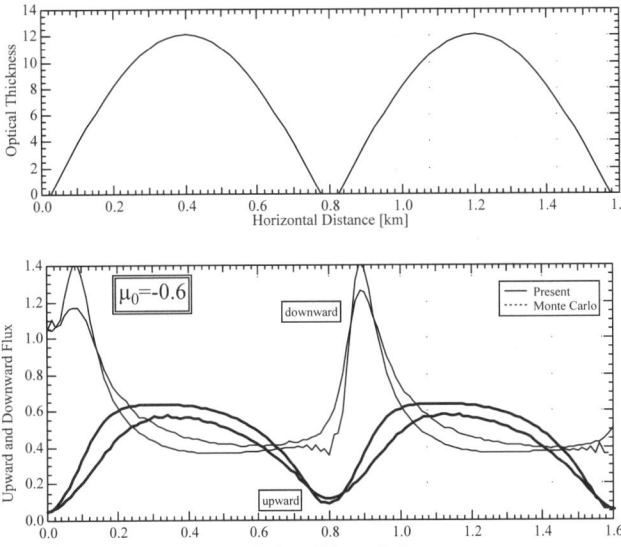

Fig. 2. *Upper panel*: The horizontal distribution of the optical thickness of the assumed cloud layer used in the radiative transfer calculation. *Lower panel*: The corresponding upward flux at the cloud top and downward flux at the bottom.

The calculated upward flux at the top of the cloud layer and the downward flux at the bottom are shown in the lower panel of Fig. 2 for the cosine of solar zenith angle of 0.6; the sun light is incoming from the upper left of the figure. In the figure, the fluxes calculated by Monte Carlo method are also illustrated for comparison: the grid size for the Monte Carlo calculation is set to 25 m. The flux values are normalized by the value of the incident flux. The general feature of the radiation field calculated by the present method is the same as that by the Monte Carlo method. It means that the present method is reasonable and efficient to calculate the radiation field in inhomogeneous clouds. However, some discrepancies are seen, especially, in the downward fluxes around the optically thin positions. It is seemed that the present solution exaggerates the effect of cloud inhomogeneity more evidently than the Monte Carlo calculation does, otherwise the Monte Carlo calculation tends to smooth out the effects of cloud inhomogeneity.

Fig. 3 shows an example of the radiant intensity distribution calculated for a complicated cloud model that has the optical thickness distribution along the x-coordinates as illustrated in the upper panel. Other assumptions are the same as the case of Fig. 2. The cosine of zenith angle of the upward and downward intensities is set to be 0.95 and -0.95, respectively, with the azimuth angle of 0.0, i.e., in the principal plane. The horizontal distribution of the upward intensity almost corresponds with that of the optical thickness, but the distribution of the downward intensity is more complicated. Fig. 4 shows more clearly the correlation between the optical thickness and the upward and downward intensities along the x-coordinate. The horizontal variability of the downward intensity around optically thin positions tends to be rapid with finer scales than that of the cloud optical thickness, whereas the variability of the intensity around optically thick positions is smoothed out by the cloud inhomogeneity effects. This feature implies a difficulty in retrieval of cloud properties from the measurement of transmitted solar radiation.

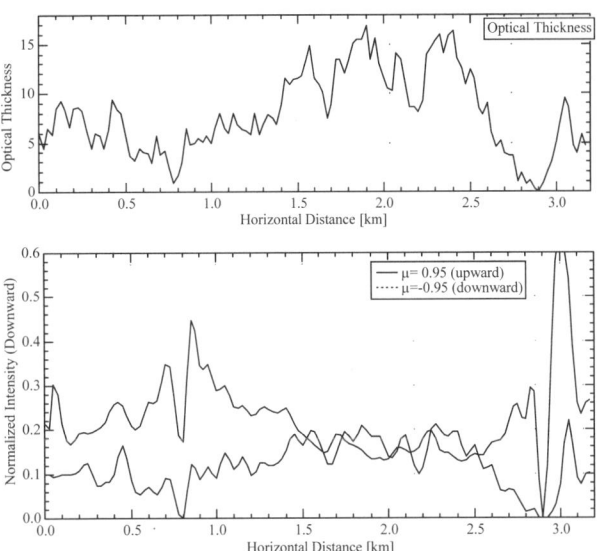

Fig. 3. Upper panel: The horizontal distribution of cloud optical thickness in the assumed cloud layer. Lower panel: The calculated upward and downward intensities in the principal plane of azimuth angle 0. The incident direction of the solar radiation is set to be μ_0=-0.6 and ϕ_0=0.0.

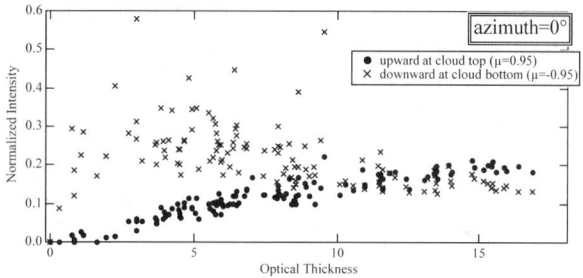

Fig. 4. Correlation between the cloud optical thickness and the upward and downward intensities along the *x*-coordinates..

4. SUMMARY

We have developed the new method for solving the radiative transfer equation in a three-dimensionally inhomogeneous atmosphere. The remarkable feature of the present method is the adoption of the scaling function decomposition for horizontal variables of the RTE. The scaling function is appropriate for calculation of radiation field in clouds whose horizontal variability is of small scale and irregular. Comparison with the Monte Carlo calculation suggests that the present method is reasonable and efficient to calculate the radiative transfer in inhomogeneous clouds.

The calculated results indicate that cloud inhomogeneity raises difficulty in retrieval of cloud properties by remote sensing, especially, from measurements of the transmitted solar radiation. The effects of cloud inhomogeneity should be quantitatively estimated. The present method can be helpful to investigate the relationship between cloud inhomogeneity and the corresponding variability of radiation field.

5. REFERENCE

Stephens, G. L., 1988, Radiative transfer through arbitrarily shaped optical media. Part I: A general method of solution. *J. Atmos. Sci.*, 45, 1818-1836.,

USER-FRIENDLY SOFTWARE FOR SATELLITE SIGNAL SIMULATION BASED ON LINE-BY-LINE CALCULATIONS OF RADIATIVE FLUXES IN THE ATMOSPHERE

J. C. Ceballos, B. A. Fomin, and M. P. Corrêa

Satellite and Environmental Systems Division – Centro de Previsão de Tempo e Estudos Climáticos
Instituto Nacional de Pesquisas Espaciais – Rod. Dutra, km 40 – Cachoeira Paulista – SP – 12.630-000 – Brazil

ABSTRACT

A user-friendly software for rigorous simulation of satellite signal is presented. It is a FORTRAN-90 package that runs under UNIX and MS-Windows platforms, designed for satellite data validations, assessment of atmospheric models being used in these systems and also as a teaching program for faculty and post-graduate students. The present version simulates satellite signal for stratified plane-parallel clear and cloudy/aerosol atmospheres in the microwave and thermal infrared as well as in the solar spectrum (near infrared, visible and ultraviolet: radiances, without and/or with scattering). User can use any response function and realistic optical properties of atmosphere and surface. The program is based on line-by-line and Monte-Carlo methods.

1. INTRODUCTION

At the present time there are many spectroradiometers flying onboard satellites for global monitoring of atmospheric/surface properties from space. These spectroradiometers (hereafter sensors), which are designed for solution of a wide variety of practical tasks, have very different spectral bands in the MW, IR, VIS and UV spectral regions. We describe in this work user-friendly and universal software for sensors signals simulation (FLISS – **F**ast **Li**ne-by-line satellite **S**ignal **S**imulator), which is useful for the development and validation of sensors data processing system as well as for satellite experiments planning. Moreover, it is useful as a teaching program for students and post-graduate students. It is based on the fast line-by-line and Monte-Carlo codes for treatment of both the gaseous absorption and the particulate scattering in the slab atmosphere. In this version all the fundamental scattering/absorption atmospheric properties are taken into account except the line mixing, NONLTE and polarization effects, which will be included in further improvements. It is worthwhile to stress that at present only this software, based on a recently developed algorithm [Fomin, 2005], can rigorously treat cloud/aerosol scattering in narrow and wide spectral bands.

2. ALGORITHM DESCRIPTION

The spectral database HITRAN-11 v [*Rothman et al.,* 2003] and continuum model for N_2, O_2, O_3, CO_2 and H_2O (CKD-2.4) have been used in FLISS, based on the same physical assumptions used in the well known and carefully validated Line-by-Line Radiation Transfer Model (LBLRTM) [*Mlawer et al.,* 1997]. However, FLISS makes use of other line-by-line algorithms for gas absorption calculations [Fomin, 1994] and vertical integration [*Fomin et al.,* 2004]. FLISS has two versions: FLISS-L, for longwave spectral region, which considers only absorption; and FLISS-S that treats shortwave processes (in a plane-parallel atmosphere) by means of the Monte-Carlo method [Fomin and Mazin, 1998]. Both versions provide spectral radiances, while L version provides brightness temperature and S version atmospheric albedo. FLISS-L has an accurate treatment of the gaseous absorption (supported by HITRAN database) and FLISS-S is used for proper calculations in scattering media. The line-by-line method is used here not only to increase the calculation accuracy but also to allow for application of FLISS in different tasks without previous development of the correspondent gas absorption parameterization. The FLISS principal feature is that it does not need detailed constraint information except that concerning physical composition of the atmosphere and satellite-view geometry. For instance, once defined the lower and upper limits of a working spectral interval, the spectral resolution in FLISS is varied automatically with frequency so that the narrowest spectral line could be adequately included.

FLISS was designed for being user-friendly software, allowing easy introduction of an ASCII database defining cloud/aerosol optical as well as surface reflecting properties. For user it is necessary only to point out the path to the ASCII file(s) where the scattering/absorption coefficients, the Stokes parameters and/or the spectral surface albedo are written. The wavenumber and angular grids in those files are freely defined by the user. If the cloud/aerosol optical model is complex and consists of several components with different profiles, FLISS assesses the proper mixing of these fractions during the calculations.

Fig. 1 shows the structure of the software package. Surface albedo, atmospheric profiles and cloud/aerosols data are included in database. Spectral response function of satellites sensors as MODIS, GOES, AIRS and HIRS are also included in this library.

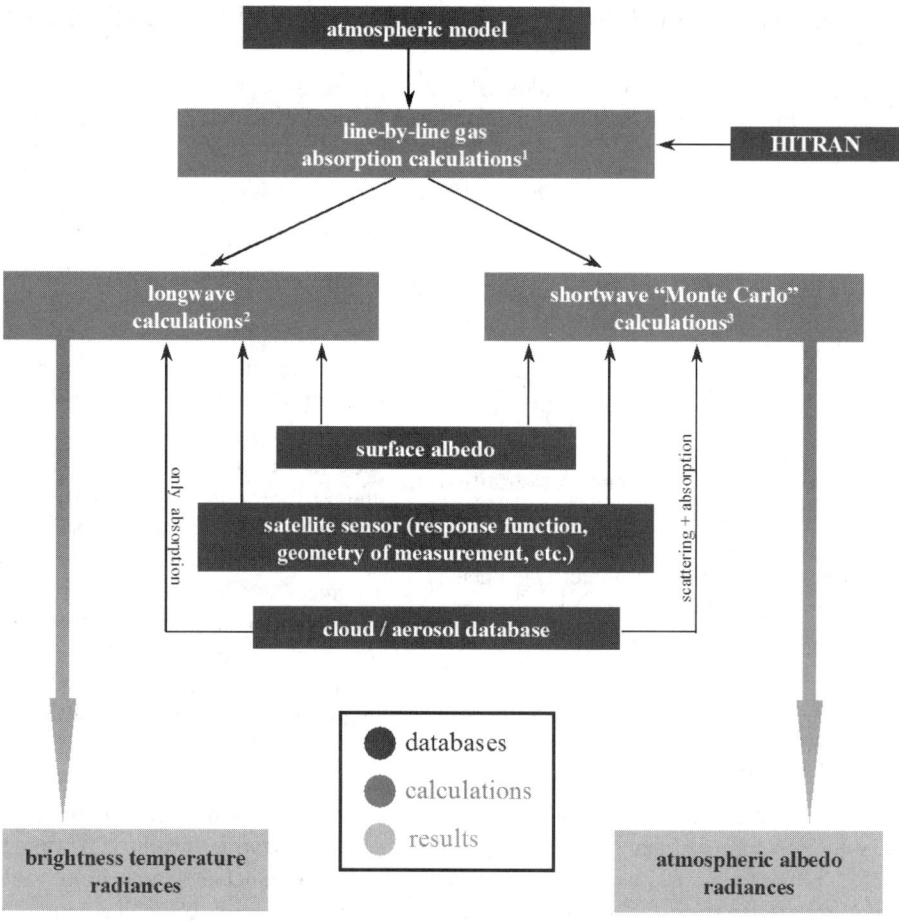

Fig. 1. Structure of the software package.

FLISS package shows user-friendly structure. Figs. 2 illustrate some screenshots. Top pictures (2a and 2b) display FLISS running under different operational systems: UNIX (left figure) and Microsoft XP Windows (right figure) platforms. User should only change parameters according his/her needs. Bottom figures illustrate "help files" with the available database (2c – FLISS-S and 2d – FLISS-L).

3. APPLICATION EXAMPLE

Fig. 3 illustrates simulation for thermal infrared channels of MODIS (Terra) sensor, nadir line-of-sight, allowing comparison with assessments performed by the software RTTOV [*Saunders et al., 1999*]. Channel response functions were taken in account. Atmospheric profiles for O_2, O_3, H_2O, CO_2 are those reported in RTTOV. Fig. 3 includes several cases of satellite signal assessment: 1) RTTOV version 7 calculations (crosses), declaring four active gases (O_2, O_3, H_2O, CO_2); 2) FLISS-L (circles), using the same atmospheric profile; 3) FLISS-L (a: triangles) including three additional gases (CH_4, N_2O, CO), and 4) FLISS-L with these seven gases, reducing CH_4, N_2O, CO concentrations to half the initial value. FLISS calculation for the 16 infrared MODIS channels takes about 8 minutes in a Pentium 4, 2.8 GHz, 512 Mb RAM.

It can be seen an excellent agreement between RTTOV and FLISS in atmospheric channel windows as well as in the 6.7 μm water vapour band. Differences between RTTOV-7 and the four-gas FLISS calculations (cases 1 and 2) suggest the influence of other atmospheric gases (not explicitly included in RTTOV user screen). As expected, the inclusion of three additional gases (cases 3 and 4) does not affect the 14 μm band (CO_2+H_2O) but the impact of CH_4 and N_2O bands on specific MODIS channels is clearly seen. Fig. 3 makes evident that varying concentrations allow for proper tuning with RTTOV brightness temperatures (case 4).

RADIATIVE TRANSFER THEORY AND MODELING

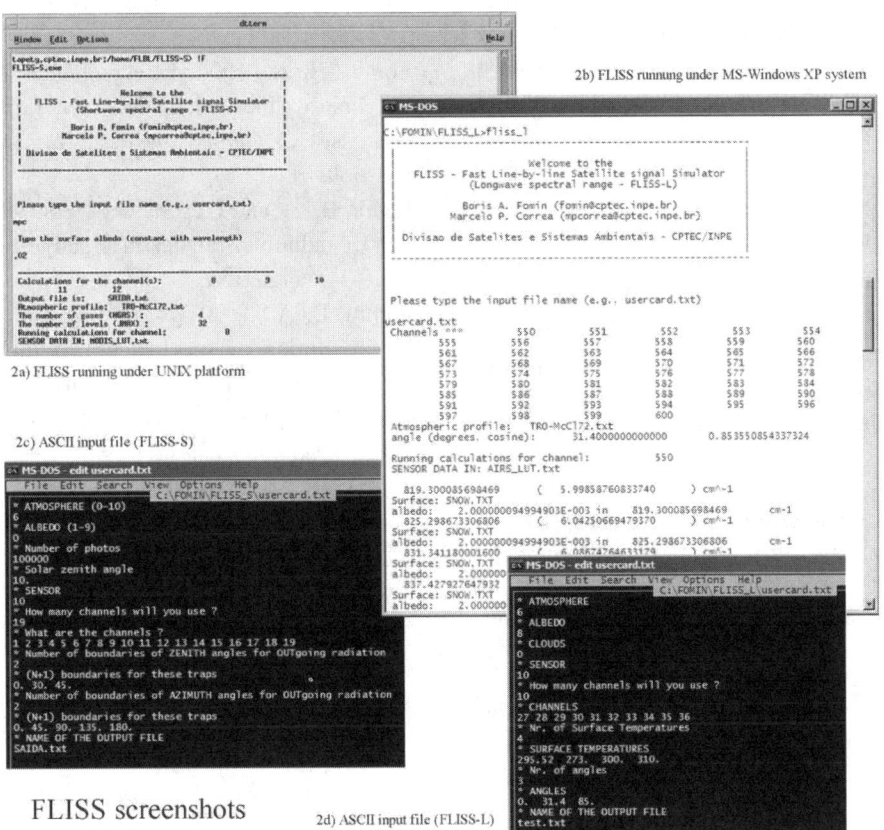

Fig. 2. FLISS screenshots.

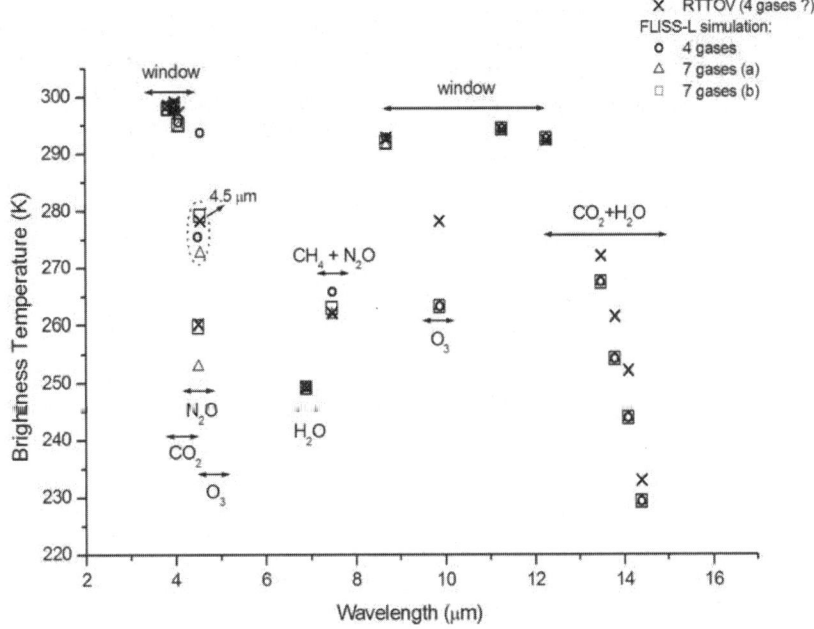

Fig. 3. FLISS and RTTOV comparison.

Differences observed in 9.6 μm channel (affected by O_3) as well as in 14 μm CO_2+H_2O band have not a clear origin, being a matter of further analysis. In particular, comparison with "satellite truth" (MODIS multispectral imagery) is being carried on.

4. CONCLUSIONS

The FLISS' line-by-line procedure provides accurate calculations of radiative transfer [Fomin, 1994]; at the same time, it has a distinct user-friendly structure, simplifying its use in research and teaching. In despite of its complexity and spectral completeness, it has a fast performance compatible even with personal computers [*Fomin et al., 2004*].

A preliminary comparison with other satellite-simulator softwares shows fairly good agreement in clear-sky infrared spectrum. Further tests of simulation for FLISS-L and FLISS-S are being performed, including comparison with: a) other codes, and b) actual information provided by satellites and campaign measurements. A preliminary version of FLISS can be obtained contacting the authors (fomin@cptec.inpe.br; mpcorrea@cptec.inpe.br or ceballos@cptec.inpe.br).

5. ACKNOWLEDGEMENTS

To Wagner F. A. Lima (DSA/CPTEC), for kind calculations of RTTOV-7 MODIS spectra. M. P. Corrêa thanks to Brazilian Research Foundation FUNCATE and Korea Meteorological Society to the support for IRS'04. B. A. Fomin and J. C. Ceballos had partial support by Brazilian CNPq.

6. REFERENCES

Fomin, B. A., 1994: Effective interpolation technique for line-by-line calculations of radiation absorption in gases, *J. Quant. Spectrosc. Radiat. Transfer, 53*, 663-669.

Fomin B. A., and I. P. Mazin, 1998: Model for investigation of radiative transfer in cloudy atmosphere, *Atm. Res., 47-48*, 127-153.

Fomin, B. A., T. A. Udalova, and E. A. Zhitnitskii, 2004: Evolution of spectroscopic information over the last decade and its effect on line-by-line calculations for validation of radiation codes for climate models, *J. Quant. Spectrosc. Radiat. Transfer, 86*, 73-85.

Fomin, B. A., Monte-Carlo algorithm for line-by-line calculations of thermal radiation in multiple scattering layered atmospheres, submitted for *J. Quant. Spectrosc. Radiat. Transfer.*

Mlawer, E. J., J. Taubman, P .D. Brown, M. J. Iacono, and S. A. Clough, 1997: Radiative transfer for inhomogeneous atmospheres: RRTM, a validated correlated-k model for the longwave, *J. Geophys. Res., 102*, 16663-16682.

Rothman, L. S., et al., 2003: The HITRAN molecular spectroscopic database: edition of 2000 including updates through 2001, *J. Quant. Spectrosc. Radiat. Transfer, 82*, 5-44.

Saunders R. W., M. Matricardi, and P. Brunel, 1999: A fast radiative transfer model for assimilation of satellite radiance observations – RTTOV-5. ECMWF *Research Dept. Tech. Memo.*, 282 (available from the librarian at ECMWF).

A NEW TECHNIQUE FOR DEVELOPING K-DISTRIBUTIONS

B. A. Fomin and M. P. Corrêa

Satellite and Environmental Systems Division – Centro de Previsão de Tempo e Estudos Climáticos
Instituto Nacional de Pesquisas Espaciais – Rod. Dutra, km 40 – Cachoeira Paulista – SP – 12.630-000 – Brazil

ABSTRACT

This work is devoted to a new technique, which finds the shortest k-distribution series achievable in practice. Its novelty consists in the use of real atmospheric flux calculations to guide both the position of spectral bands and the k terms within each band. Wavenumber subintervals which have similar atmospheric absorption behavior are chosen, then a representative absorption coefficient is set to the value which best fit the results of line-by-line calculations of fluxes. This method of choosing one absorption coefficient to represent a large number of wavenumber subinterval contrasts with other published methods, and is responsible in large part for the improved computational efficiency. Moreover this technique works as well in the stratosphere as the troposphere and it permits a more accurate treatment of cloud scattering and absorption properties than other methods.

1. INTRODUCTION

Despite on essential efforts, up to now, huge computational resources are required for the calculation of radiative transfer in the atmosphere, for weather and climate prediction, and the processing of radiance data retrieved by satellites. So a number of radiation codes have been developed in the past decade. But all these codes are based on the same correlated-k distribution method [e.g., *Lacis and Oinas*, 1991], which has been developed more than a dozen years ago. Consequently, all these codes have the same well-known difficulties in a treatment of overlapping absorption by different species and in cooling/heating rate simulation in upper atmosphere. For example, in the longwave codes by *Mlawer et al.* [1997] and *Cusack et al.* [1999] 256 and 33 k-distribution terms are used, respectively. But the both codes have comparable accuracy: cooling rate errors in stratosphere are 0.75 and 1.5 K day^{-1}.

It should be mentioned that in all these codes cloud absorption is treated separately from the gas absorption. For this a few broad spectral bands are used with effective scattering and absorption coefficients of cloud media in each band. Unfortunately, the absorption coefficient and, consequently, the single-scattering co-albedo of cloudy media, $1-\omega$, strongly depend on wavenumber and vary by several orders of magnitude within these bands. So, as it should be stressed, "no optimal method has been found for deriving ω over a broad band" [*Chou and Suarez*, 2002].

A new k-distribution technique, which is free of these difficulties, has been developed recently. Moreover it finds the shortest k-distribution series achievable in practice. By means of this technique a fast k-distributions model (FKDM) suitable for use in weather and climate prediction has been created using 23 and 15 k-distribution terms for the longwave and shortwave regions, respectively. The molecule species represented in the model are H_2O, CO_2, O_2, O_3, N_2O, CH_4, and CFC-11, 12, 113.

2. TECHNIQUE APPLICATIONS

The technique and FKDM are described in detail in [*Fomin*, 2004; *Fomin and Correa*, 2005]. Here we consider some applications. Fig. 1, where CO_2 cooling rate in the 15 µm band for the midlatitude summer atmosphere up to 120 km is calculated, illustrates its possibility to create vertically unrestricted fast and accurate parameterization of the radiation. Fig. 2 shows each k-term weight separated by number in previous figure. It should be stressed that none other method can obtain such kind of weights from ~10^{-5} up to ~10^{-1}.

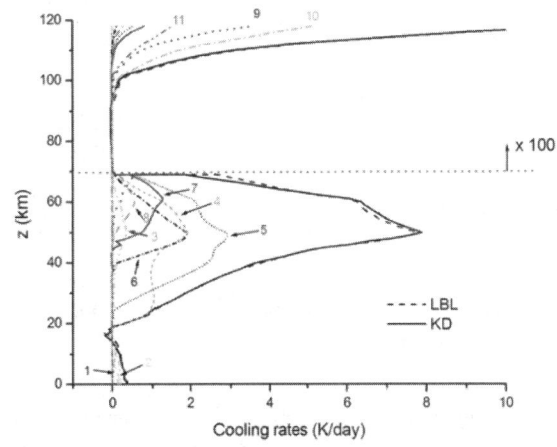

Fig. 1. CO_2 cooling rate in the 15 µm band. Calculations considering line-by-line (LBL – dashed line) and k-distribution (KD – solid line) techniques.

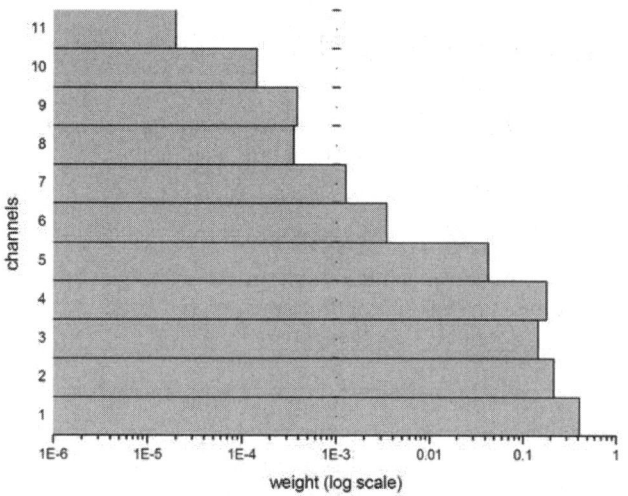

Fig. 2. Band weights detailed in previous figure.

Fig. 3, where erythemal UV fluxes (0.2–0.4 μm) are considered, illustrates the technique efficiency. Ozone absorption and molecular scattering are considered in these calculations. The fluxes were calculated using line-by-line and one-term fit approximation for tropical and subarctic-summer clear-sky conditions and three different ozone concentrations (150, 300, and 600 DU). Representative ozone absorption cross-sections for this fit as a function of

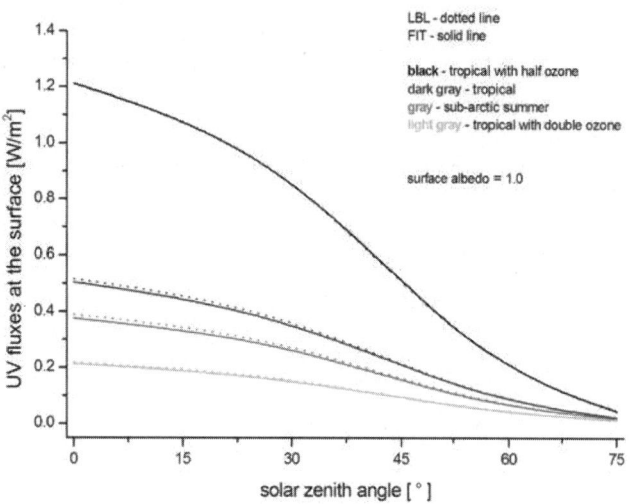

Fig. 3. Erythemal UV fluxes (0.2–0.4 μm) for several solar zenital angles.

ozone amount along the direct solar radiation is shown in Fig. 4 The fit accuracy is better than 10% for different conditions (surface albedo between 0.0–1.0, ozone amounts between 150–600 DU, tropical, midlatitude and subarctic atmospheres). This simple one-term fit may be useful for medical applications.

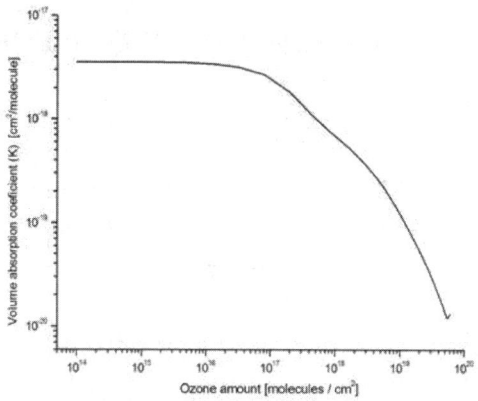

Fig. 4. Representative ozone absorption cross-section.

As mentioned above the technique is effective for taking into account overlapping absorption by different species. It also gives a possibility to consider radiation absorption by cloud droplets and ice crystals similarly to absorption by molecular species. This feature provides more accurate treatment of cloud optical properties by taking into account correlation between water vapor and liquid water (or ice) absorption (Fig. 5). It has been found that "the neglect this correlation in a radiation model can essentially distort simulated fluxes and heating rates" [*Fomin and Correa*, 2005].

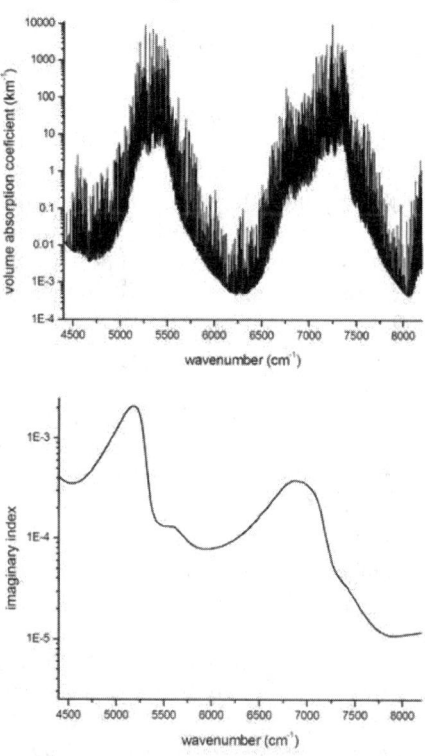

Fig. 5. Water vapor absorption at bottom of tropical atmosphere (upper plot) and imaginary index of refraction for liquid water (lower plot).

3. CONCLUSIONS

An assessment of this technique accuracy and speed has shown it to be a more efficient technique than the correlated-k method. Moreover this technique works as well in the stratosphere as the troposphere and it permits a more accurate treatment of cloud scattering and absorption properties than other methods. So it is recommended for radiation flux and heating/cooling rate calculations in GCMs and for simulations radiations in the on-line data processing systems for sattelite experiments.

4. ACKNOWLEDGEMENTS

The authors are grateful to Luiz Augusto Toledo Machado and Juan Ceballos for their help. This work has been supported by the CNPq foundation (grant 301263019), Brazil.

5. REFERENCES

Chou, M.-D., and M. J. Suares, 2002: A solar radiation parameterization for atmospheric studies, Technical report series on global modeling and data assimilation, *NASA/TM-1999-10460*, 15, 42-42.

Cusack, S., J. M. Edwards, and J. M. Crowther, 1999: Investigating k distributing method for parametrizing gaseos absorption in the Hadley Centre Climate Model, *Journal of Geophysical Research*, 104, 2051-2057.

Fomin, B. A., 2004: A k-distribution technique for radiative transfer simulation in inhomogeneous atmosphere I: FKDM, fast k-distribution model for the longwave, *Journal of Geophysical Research*, 109, D02110, doi: 10.1029/2003 JD003802.

Fomin, B. A., and M. P. Correa, 2005: A k-distribution technique for radiative transfer simulation in inhomogeneous atmosphere: 2. FKDM, fast k-distribution model for the shortwave. *Journal of Geophysical Research*, 110, D02106, doi:10.1029/ 2004JD005613.

Lacis, A. A., and V. Oinas, 1991: A description of the correlated k-distribution method for modeling nongray gaseous absorption, thermal emission, and multiple scattering in vertically inhomogeneous atmospheres, *Journal of Geophysical Research*, 96, 9027-9074.

Mlawer, E. J., J. Taubman, P. D. Brown, M. J. Iacono, and S. A. Clough, 1997: Radiative transfer for inhomogeneous atmospheres: RRTM, a validated correlated-k model for the longwave, *Journal of Geophysical Research*, 102, 16663-16682.

ACTUAL STATE OF THE
UV RADIATION TRANSFER MODEL PACKAGE STAR

P. Koepke, D. Anwender, M. Mech, A. Oppenrieder, J. Reuder, A. Ruggaber, M. Schreier, H. Schwander, and J. Schween

Meteorologisches Institut der Universität München
Theresienstrasse 37, 80333 München, Germany

ABSTRACT

The radiative transfer model STAR that is freely available in the versions STARsci and STARneuro has been improved and completed. The software package has been extended by the module RadOnInc that models UV irradiances on arbitrarily tilted surfaces. It uses radiances, modelled with STARsci and separates the effects of local and regional albedo on the radiation field. A module PHOLY has been added to determine user-defined photolysis frequencies from the spectral actinic flux output of STARsci. Moreover, the appended data sets for surface albedo have been extended by an ocean albedo considering the effects of waves and solar zenith angle, the existing user friendly input mask has been adapted and bugs have been removed.

1. INTRODUCTION

The radiative transfer model STAR (Ruggaber et al., 1994), for the determination of the radiation field in the UV spectral range up to 400 nm and with lower spectral resolution up to 700 nm, combines the matrix operator model of Nakajima and Tanaka (1986) with an initialisation model, which calculates the input data for the radiation transfer model on realistic data bases for all atmospheric constituents. In a later version (MIM, 2001) UV modelling capabilities have been extended with respect to selectable spectral resolution, the possibility to take into account specific sensor characteristics and to calculate UV radiation simultaneously at different altitudes. A user friendly and self explaining mask to combine sets of input data has been developed that uses standard proxy data in cases of information not provided by the user (see Fig. 1).

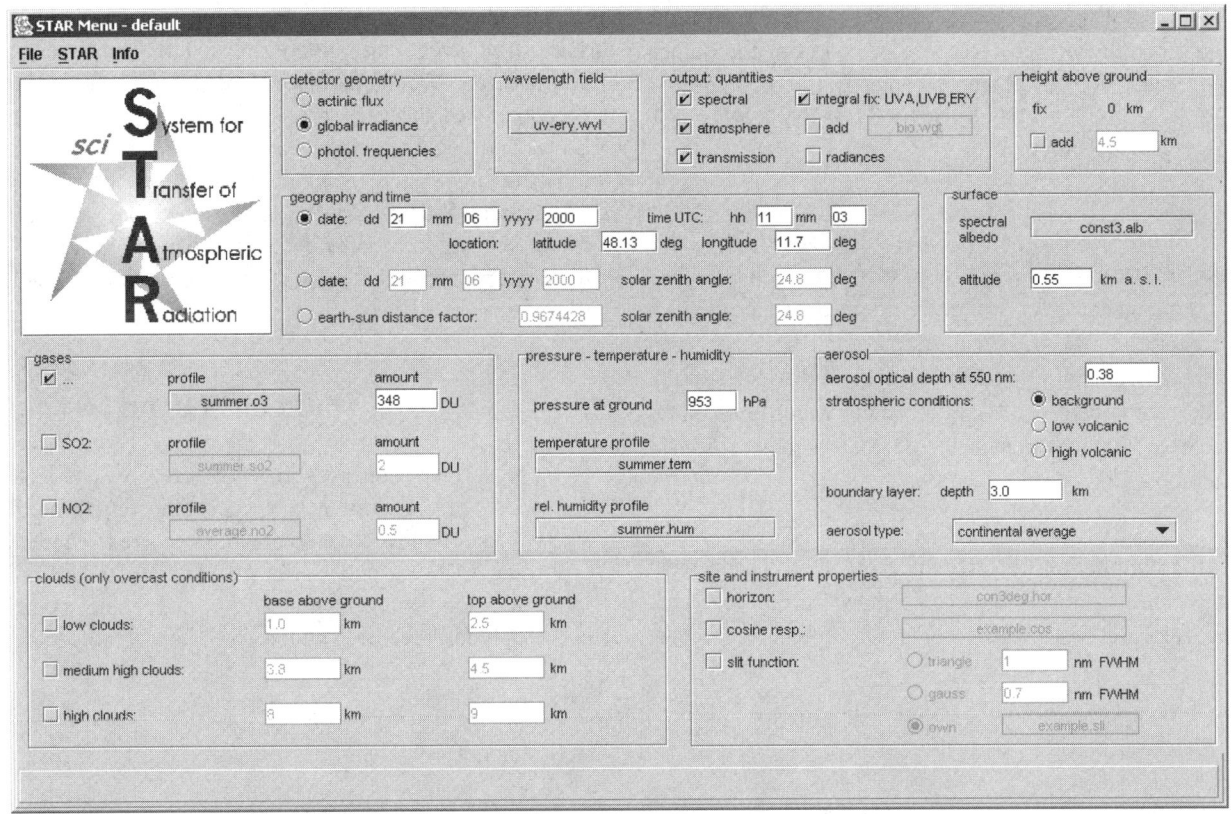

Fig. 1. Input mask for STARsci.

This program exists as a science version, STARsci, (MIM, 2001) which models all radiation quantities in detail, but takes clouds only as homogeneous layers into account. A second version, STARneuro (Schwander et al., 2001), models the radiation quantities using a neural network algorithm. Thus this version is very fast and in addition able to consider the effect of broken clouds (Schwander et al., 2002) on irradiances with respect to horizontal oriented surfaces. Both versions of STAR are able to determine integral radiation quantities for variable spectral weighting functions.

The radiation transfer model STAR has been compared to other algorithms, with very good agreement (Koepke et al., 1998; DeBacker et al., 2001; van Weele et al., 2000).

The software package described above has been amended and completed by the algorithms presented below.

2. IRRADIANCES ON TILTED SURFACES

In general UV irradiances are measured and modelled for horizontally oriented surfaces. However, most biologically relevant receivers (e.g., the skin of the human body or the leaves of plants) are mostly tilted. Thus it is essential to model UV irradiances on flat but arbitrarily oriented, inclined surfaces.

To do this, an algorithm called RadOnInc, has been developed that integrates radiances modelled with STARsci over the hemisphere for arbitrarily oriented surfaces (Mech and Koepke, 2004). This algorithm, which has been used for detailed sensitivity studies (Koepke and Mech, 2005), can be combined with any model which calculates radiances, but is included in the actual software package STAR.

An example for the irradiance on tilted surfaces, in combination with measurements, is shown in Fig. 2. (Mech and Koepke, 2003)

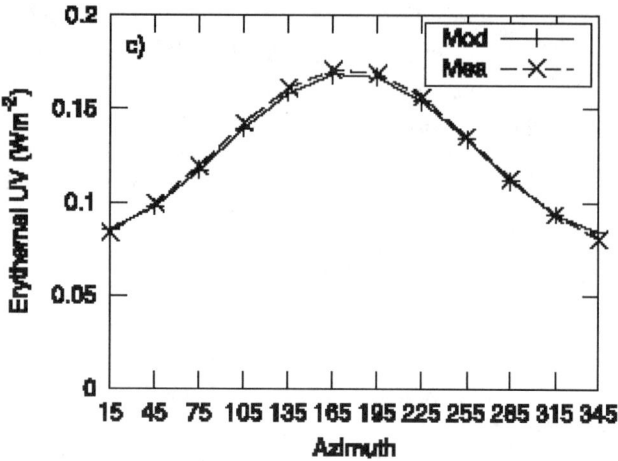

Fig. 2. Modelled and measured erythemal weighted UV irradiances for a flat receiver tilted by 45° as function of azimuth, with 0° in north. Solar elevation is 60°, ozone amount 337 DU, albedo 9%.

The irradiance on horizontally oriented surfaces is influenced by the surface albedo only via multiple scattering. Thus the reflection properties of the surfaces in the surrounding area of about 30 km have to be considered (Deguenther et al., 1998), determined as regional albedo. The irradiance on tilted receivers, however, gets contribution of photons directly reflected at the surface in the vicinity of the receiver, determined by the local albedo. Local and regional albedo in RadOnInc can be assumed independently since they may have different values, e.g., for a cross-country ski run in a snow free surrounding. Moreover, the local albedo can be separated into parts with different properties, as it is the case for a position at a beach near the water.

3. SURFACE ALBEDO

The surface properties in STAR are given as isotropic reflection with spectral values depending on surface type. The values for snow covered conditions have been modified according to Doda and Green (1981), also the values for sand (Feister and Grewe, 1995), and some additional surfaces types have been added.

The reflection at an ocean surface with waves has been taken into account for the local albedo with the angle dependent bidirectional reflection (Koepke, 1985). Thus the illumination of a tilted receiver considers specular reflection.

For the regional albedo, however, values for an isotropic albedo are necessary. They have been determined for the ocean surface with waves by modelling radiation fields for sun and sky, using average atmospheric conditions, and calculating the upward emerging radiation at the ocean surface by anisotropic reflection. The ratio of these two fluxes, the exitance from the surface and the global irradiance from the sky represents the value for isotropic ocean albedo. These albedo values have been modelled as function of solar elevation, height of the waves given via wind speed (Cox and Munk, 1954) and the wavelength of the UV radiation. An example for the isotropic albedo in the UV for an ocean with waves is shown in Fig. 3.

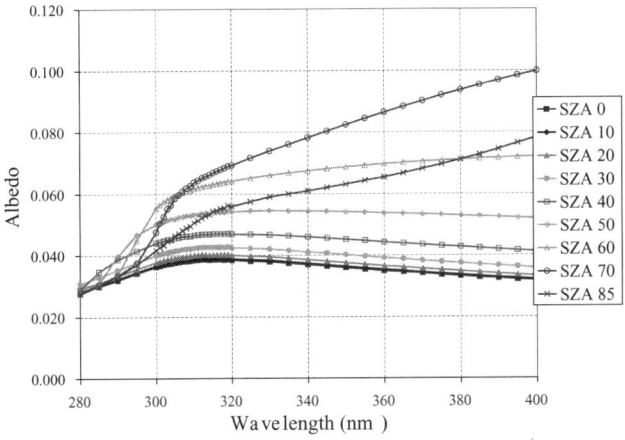

Fig. 3. Spectral albedo of an ocean surface with waves related to a wind speed of 7 m/s for different solar zenith angles.

The albedo increases with increasing solar zenith angle, due to the increase of specular reflected direct sun. But the albedo decreases again for larger solar zenith angles, since the contribution of the direct sun decreases and is even negligible for very low sun. This is also the case for the short wavelength range with strong ozone absorption, with the consequence of reduced variability of the albedo with solar zenith angle for short wavelengths. Even for conditions with strong reflection of the direct sun, the ocean albedo in the UV spectral range is lower than in the visible due to the reduced contribution of the direct sun.

4. PHOTOLYSIS FREQUENCIES

Modelling of 21 predetermined photolysis frequencies was already possible with the previous version of STAR. Advances in laboratory measurements of molecular absorption cross sections and quantum yields have indicated partially distinct deviations from earlier data sets used in STAR (e.g., Matsumi et al., 2002). To correct for this shortcoming a new photolysis module (PHOLY) has been developed. Temperature and wavelength dependent data sets of absorption cross section and quantum yields can be supplied as text files by the user. PHOLY uses these data sets for the determination of photolysis frequencies on the basis of spectral actinic flux calculated by STAR.

5. CLOSING

The radiation transfer package STAR has been improved. It is freely available. If you are interested visit: http://www.meteo.physik.uni-muenchen.de/strahlung/uvrad/Star/STARinfo.htm.

6. ACKNOWLEDGEMENTS

Parts of the modifications of the STAR software package result from research projects. Thus the authors appreciate the support by the Bavarian State Ministry of the Environment, Public Health and Consumer Protection and by the German Federal Office for Radiation Protection.

7. REFERENCES

Cox, C., and W. Munk, 1954: measurement of the roughness of the sea surface from photographs of the Sun´s glitter, *J. Opt. Soc. America*, 44, 838-850.

De Backer, H., P. Koepke, A. Bais, X. de Cabo, T. Frei, D. Gillotay, C. Haite, A. Heikkilä, A. Katzanzidis, T. Koskela, R. E. Kyrö, B. Lapeta, J. Lorente, K. Masson, B. Mayer, H. Plets, A. Redondas, A. Renaud, G. Schauberger, A. Schmalwieser, H. Schwander, and K. Vanicek, 2001, Comparison of measured and modelled

UV indices for the assessment of health risks, *Meteorolog. Applications*, 8 (3), 267-277.

Deguenther M., R. Meerkoetter, A. Albold, and G. Seckmeyer, 1998: Case study on the influence of inhomogeneous surface albedo on UV irradiance, *Geophys. Res. Lett.*, 25, 4287-4296.

Doda, D. D., and A. E. S. Green, 1981: Surface reflection measurements in the ultraviolet from an airborne platform. Part 2, *Appl. Optics*, 20, 636-642.

Feister, U., and R. Grewe, 1995: Spectral albedo measurements in the UV and visible region over different types of surfaces, *Photochem. Photobiol.*, 62, 736-744.

Koepke, P., 1985: The reflectance factors of a rough ocean with foam, *Int. J. Remote Sens.*, 6, 787-799.

Koepke, P., A. Bais, D. Balis; M. Buchwitz, H. De Backer, X. De Cabo, P. Eckert, P. Eriksen, D. Gillotay, A. Heikkilä, T. Koskela, B. Lapeta, Z. Litynska, J. Lorente, B. Mayer, A. Renaud, A. Ruggaber, G. Schauberger, G. Seckmeyer, P. Seifert, A. Schmalwieser, H. Schwander, K. Vanicek, and M. Weber, 1998, Comparison of Models Used for UV Index Calculations, *Photochem. Photobiol.*, 67(6), 657-662.

Koepke, P., and M. Mech, 2005, UV radiation on arbitrarily oriented surfaces: Variation with atmospheric and ground properties, *Theor. Appl. Climatol.*, in print.

Matsumi, Y., F.-J. Comes, G. Hancock, A. Hofzumahaus, A. J. Hynes, M. Kawasaki, and A. R. Ravishankara, 2002: Quantum yields for production of O(1D) in the ultraviolet photolysis of ozone. *J. Geophys. Res.*, 107 (D3), 10.129/2001JD000510.

Mech M., and P. Koepke, 2004, Model for UV irradiance on arbitrarily oriented surfaces, *Theor. Appl. Climatology*, 77, 151-158.

MIM 2001: Information to the software package STAR: http://www.meteo.physik.uni-muenchen.de/strahlung/uvrad/Star/starprog.html

Nakajima, T., and M. Tanaka, 1986: Matrix formulations for the transfer of solar radiation in a plane-parallel scattering atmosphere, *J. Spectrosc. Radiat. Transfer*, 40, 51-69.

Ruggaber A., R. Dlugi, and T. Nakajima, 1994: Modelling radiation quantities and photolysis frequencies in the troposphere, *J. Atmos. Chem.*, 18, 171-210.

Schwander, H., A. Kaifel, A. Ruggaber, and P. Koepke, 2001, Spectral radiative transfer modelling with minimized computation time by use of neural-network technique, *Appl. Opt.*, 40, 331-335

ALTITUDE EFFECT ON UV INDEX DEDUCED FROM THE VELETA-2002 EXPERIMENTAL CAMPAIGN (SPAIN)

J. Lorente, X. de Cabo, E. Campmany, and Y. Sola, Dept. de Astronomía y Meteorología,
Universitat de Barcelona, Spain (jeroni@am.ub.es)
J. A. González, J. Calbó, and J. Badosa, Dept. de Física. Universitat de Girona, Spain
L. Alados-Arboledas, Grupo de Física la Atmósfera, Universidad de Granada, Spain
A. Martínez-Lozano, Grupo de Radiación Solar, Universitat de Valencia, Spain
V. Cachorro, Grupo de Óptica Atmosférica, Universidad de Valladolid, Spain
A. Labajo, Instituto Nacional de Meteorología, Spain
B. de la Morena, Estación de Sondeos Atmosféricos de El Arenosillo, INTA, Huelva, Spain
A. M. Díaz, Dept. de Física Básica, Universidad de La Laguna, Spain
M. Pujadas, Dept. de Impacto Ambiental de la Energía. CIEMAT, Spain
H. Horvath, Experimental Physics Institute, University of Wien, Austria
A. M. Silva, Dept. de Física. Universidade de Évora, Portugal
G. Pavese, CNR-IMAA, Italy

ABSTRACT

During the 2002 summer the Veleta-2002 experimental campaign was undertaken at the Sierra Nevada Massif, close to Granada in the South-Eastern of Spain. Measurements of UV solar spectral irradiance, UVB broad-band irradiance, ozone, aerosol optical depth (AOD) and other optical properties of the atmosphere were carried out simultaneously at both slopes of the Sierra Nevada Massif, at different altitudes between 10 and 3398 m above sea level (a.s.l.). During the campaign the UV altitude effect was determined. Results for cloudless conditions show that UV altitude effect decreases with increasing wavelength. For small zenith angles, the altitude effect is around 9 % km^{-1} in the UVB solar irradiance and 6% km^{-1} in the UVA range while this effect on UV Index (UVI), based on biologically effective irradiance, ranges 11–14 %km^{-1} depending on the SZA. There is a strong relation between atmospheric turbidity and altitude effect, so those intervals are higher when AOD increases.

UVI values show an important diurnal evolution, mainly in hours with high values of AOD. At noon, the maximum UVI observed at different altitudes ranges from around 8 at the lower measurement point to 12 at Veleta Peak (3398 m a.s.l.).

1. INTRODUCTION

The knowledge of solar UV radiation reaching the Earth's surface has a great interest because of its significant role in atmospheric and biological processes. This radiation is influenced by diverse factors like the atmospheric ozone content, aerosol particles, cloudiness, surface albedo and altitude. This last factor is the goal of our study. Solar radiation increases with the altitude because of the decrease of its attenuation due to absorption and scattering by air molecules, aerosols and clouds. The *UV altitude effect* (UVAE) is defined as the increase of UV solar irradiance with altitude due to the shorter path length of solar beam through the atmosphere and it is generally expressed as the relative irradiance increase (in % per km) between the higher to the lower site (Blumthaler et al., 1997). Several UVAE investigations have been carried out in different places during recent years showing a wide range of values (2–25% km^{-1}) depending on UV wavelength, solar zenith angle, air mass type, surface cover and season (Blumthaler et al., 1997; McKenzie et al., 2001; Pfeifer et al., 2003). Atmospheric turbidity is one of the main important factors contributing to UVAE variation. When the air in the surface of lower sites is much polluted the UVAE increases considerably. McKenzie et al. (1997) report significant smaller UVAE values from comparisons between two clear sites, Lauder (New Zealand) and Mauna Loa Observatory (Hawaii), showing the effect of low altitude pollution on UV differences.

At the beginning of summer 2002, the VELETA-2002 experimental campaign (Alados-Arboledas et al., 2002) was undertaken at the Sierra Nevada Massif, close to Granada city in the South-Eastern of Spain (Fig. 1).

One of the main goals of the campaign was the evaluation of aerosol impact on UV solar radiation. Measurements of UV solar spectral global irradiance, UVB broad-band irradiance, aerosol optical depth (AOD) and other optical properties of the atmosphere were carried out simultaneously at three altitudes: Armilla (680 m a.s.l.), Sabinas (2200 m a.s.l.) and Veleta (3398 m a.s.l.).

In order to assess the quality of the measurements from the whole campaign, a common calibration with laboratory lamps and exhaustive intercomparisons between different kind of spectroradiometers (Brewer MK-IV, Bentham DM150, Oriel and Optronic 754) and broadband UVB radiometers (Yankee UVB-1) were performed during the first stage of this field campaign.

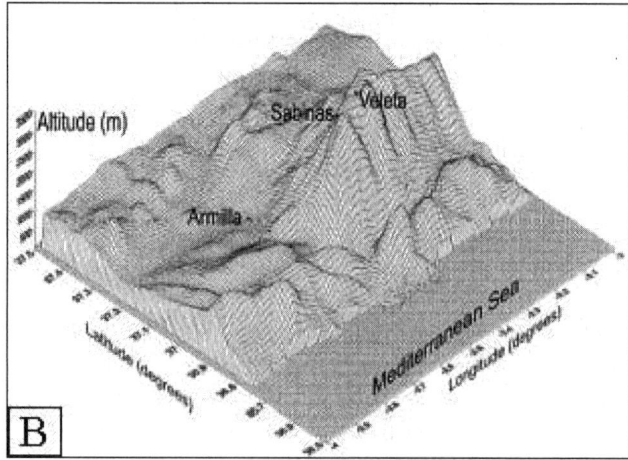

Fig. 1. Details of the Veleta 2002 campaign. A) Geographical location. B) Topography of the zone showing different points of measurements.

Fig. 2 shows one of these intercomparisons and spectral irradiances measured simultaneously at different altitudes during the second stage. Due to the specific optical characteristics of each instrument, the spectra of solar irradiance have been standardised to a 1nm FWHM triangular slit using the SHICrivm program (Slaper, 1997). This software tests and calibrates the outputs of different kinds of sensors.

In addition daily sounding profiles of temperature, humidity, wind and ozone as well as LIDAR measurements were achieved.

Here we present some of the results of UVAE calculations for UV spectral, UVB integrated and biologically effective irradiances at the Northern slope of the Sierra Nevada Massif. This last was both obtained from spectral measurements weighting by the erythema action spectrum (McKinlay and Diffey, 1987), also known as CIE action spectrum. Frequently this biologically effective irradiance is expressed as UV index (UVI), multiplying its values in W m^{-2} by 40/(W m^{-2}) (Vanicek et al., 2000).

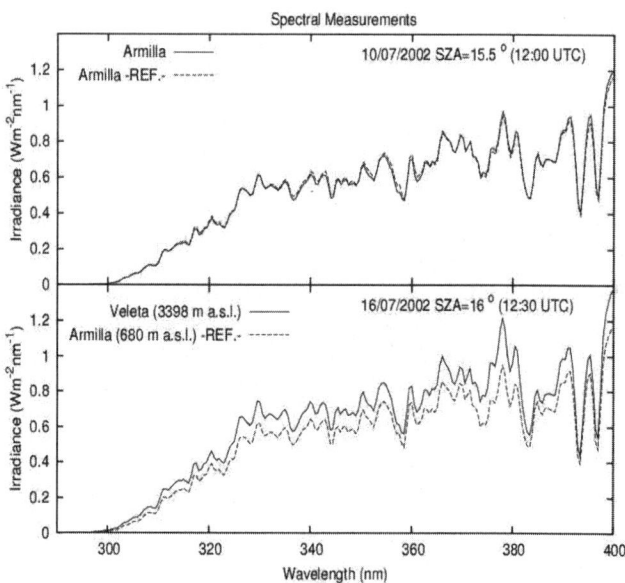

Fig. 2. Spectral measurements during the campaign. On the top, it is shown the spectral distribution of UV solar irradiance corresponding to two different spectro-radiometers measuring at the same site (Armilla) during the intercomparison stage. On the bottom, the spectral distributions corresponding to simultaneous measurements at two different altitudes: Armilla and Veleta Peak.

2. RESULTS OF UV ALTITUDE EFFECT CALCULATIONS

Simultaneous measurements of UV spectral irradiance at different altitudes for cloudless conditions showed a considerable variation of UVAE.

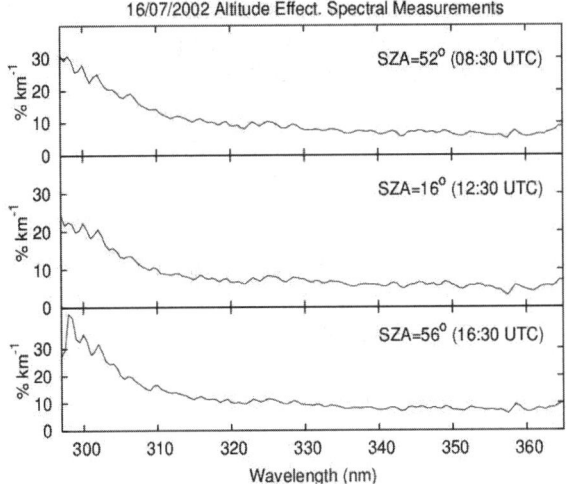

Fig. 3. Spectral altitude effect obtained from relative differences (% km^{-1}) between spectral measurements at Veleta Peak and Armilla for three different SZA during the same day.

Fig. 3 shows the decrease of the UVAE with the increase of wavelength. It shows a different behaviour depending on the spectral band. In the UVB wavelength interval there is a high dependence of UVAE with wavelength since it ranges from 30% km^{-1} to 5% km^{-1}. Differences are highest when SZA increases. In the UVA wavelength range (320–400 nm), the altitude effect variations with SZA or wavelength are less and values are between 5 and 10% km^{-1}.

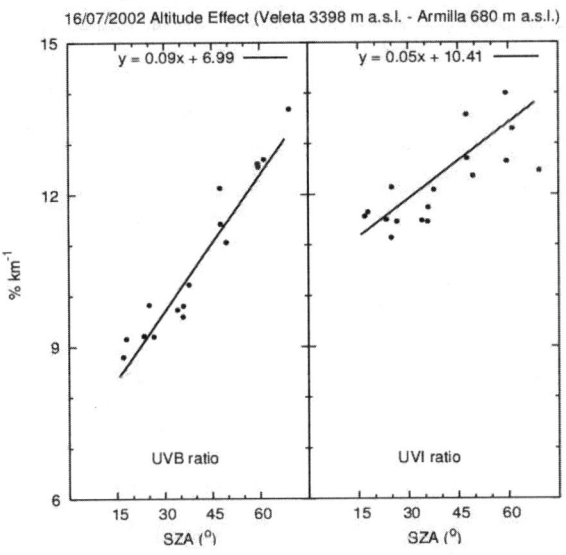

Fig. 4. Altitude effect versus SZA obtained from relative difference (% km^{-1}) between integrated magnitudes (UVB and UVI) of spectral irradiances measured by spectroradiometers.

UVAE shows values from 8 %km^{-1} to 13% km^{-1} for integrated UVB (Fig. 4). In the case of UVI based on biologically effective irradiance, the slope of the lineal tendency is less due to the influence of wavelengths in the UVA wavelength interval. The UVAE ranges 11–14 % km^{-1} but the measurements do not show such a good correlation.

These last values are higher to the 8% km^{-1} recommended by COST Action 713 (Vanicek et al., 2000). One of the reasons of this fact is that the UVAE has a strong dependence with atmospheric turbidity. Little variations of the amount or the optical properties of the aerosols in one of the points of reference can introduce differences in UVAE since there are changes in the solar spectral irradiance measured.

3. ALTITUDE EFFECT AND AEROSOL OPTICAL DEPTH

In the literature there are several works about the UVAE based on the study of data measured at different place of the world. The result is a wide range of values depending on the specific atmospheric turbidity during the period of measurement.

The Veleta-2002 campaign lasted some days, so different atmospheric situations were observed. From the spectral irradiance and photometric measurements, AOD has been determined to characterize the turbidity.

The highest variations were recorded at the lowest point due to the pollution of the city. On the other hand at highest altitudes, differences are less significant (Fig. 5).

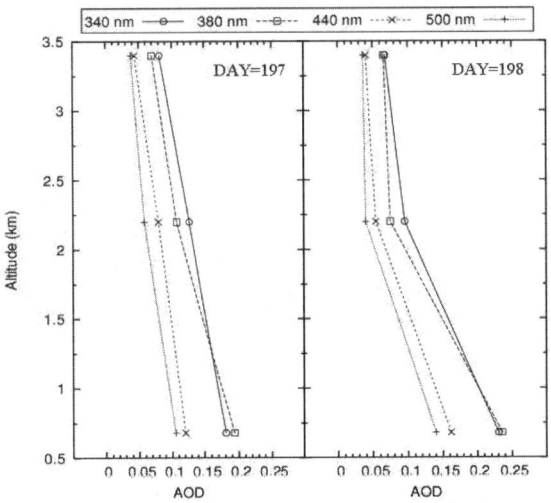

Fig. 5. AOD profiles from simultaneous measurements at the points of observation during two days of the campaign at 12:30 UTC (SZA=16°).

These variations in AOD are reflected in the spectral altitude effect (Fig. 6). When the turbidity increases at the lowest point, the UVAE is higher. Although there are differences in the entire UV spectrum, these are high in UVB range and less appreciable in UVA one.

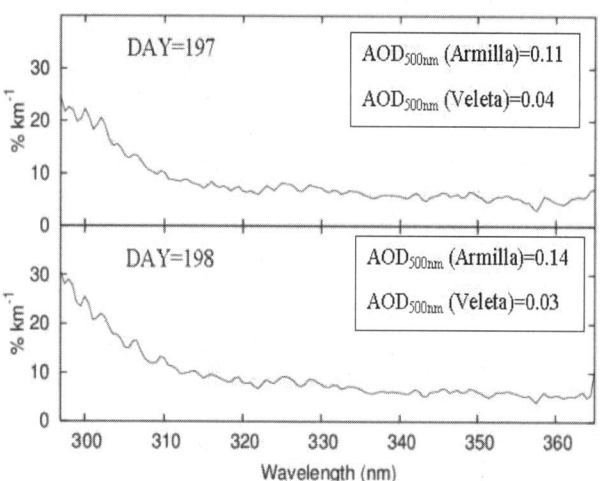

Fig. 6. Spectral altitude effect for two days of the campaign.

For measurements with the same SZA, 16°, UVAE in UVI is 11.6% km^{-1} in day 197 and 14.0% km^{-1} in day 198 when the turbidity was high.

After studying the UVAE calculated from different altitudes, that it can not be described by a single number because of the close relation with aerosols (type and amount).

4. CONCLUSIONS

UV altitude effect observed during the VELETA 2002 campaign shows a significant and complex variability with wavelength and SZA, increasing strongly for high turbidity conditions in the lower places. The main features of altitude effect are that it increases at shorter wavelengths and with high SZA, respectively. SZA dependence produces an important diurnal evolution of UVAE, mainly in hours with high values of AOD.

Therefore, in agreement with other studies, this effect can not be described by a single percentage of increase applied to the altitude variation. Therefore, in order to apply a realistic correction of UV index for different altitude sites, an appropriate characterisation of the air mass turbidity will be necessary.

5. AKNOWLEDGMENTS

This work was supported by "La Dirección General de Ciencia y Tecnología" from the Education and Research Spanish Ministry through coordinated project No: CLI200-0903-C08.

6. REFERENCES

Alados-Arboledas, L. F. J. Olmo, J. Lorente, J. A. Martínez, V. Cachorro, C. González Frías, B. De la Morena, J. P. Díaz, and M. Pujadas, 2002: VELETA 2002 Field Campaign. *Proc. 3ªAsamblea Hispano Portuguesa de Geodesia y Geofísica*, Valencia 1189-1193.

Blumthaler, M., W. Ambach, and R. Ellinger, 1997: Increase in solar UV radiation with altitude. *J. Photochem. Photobiol. B*, 39, 130-134.

McKenzie, R. L., P. V. Johnston, D. Smale, B. A. Bodhaine, and S. Madronich, 2001: Altitude effects on UV spectral irradiance deduced from measurements at Lauder, New Zealand, and at Mauna Loa Observatory, Hawai. *J. Geophys. Res.*, 106(D19), 22845-22860.

McKinlay, A. F., and B. L. Diffey, 1987: A reference action spectrum for ultraviolet induced erythema in human skin. *CIE J.*, 6, 17-22.

Pfeifer, M., P. Koepke, J. Reuder, and F. Wagner, 2003: Dependence of UV radiation on altitude and aerosol optical depth: example Bolivia. *Geophysical Research Abstracts*, Vol. 5 08862, EGS.

Slaper, H., 1997: Methods for intercomparing instruments. Advances in solar ultraviolet spectroradiometry, European Commission. *Air pollution research report 63*, Luxembourg, 155-164

Vanicek, K., T. Frei, Z. Litynska, and A. Schmalwieser, 2000: UV-Index for the Public. *COST-713 Action*. European Communities, 27 pp.

ABOUT SIMULATION OF RADIATION TRANSFER IN SMOKY ATMOSPHERE AND ENVIRONMENTAL MONITORING

Tamara A. Sushkevich, Sergey A. Strelkov, Ekaterina V. Vladimirova, Svetlana V. Maksakova,
Alexey K. Kulikov, and Ekaterina I. Ignatijeva
Keldysh Institute of Applied Mathematics, Russian Academy of Sciences
4, Miusskaya Ploschadj, Moscow, 125047 Russia, Email: tamaras@keldysh.ru

ABSTRACT

The different models and approximations of the optical transfer operator are employed in multidimentional problems of radiative correction for remote sensing of ocean, in theories of light and image propagation in turbin media, and in theoretical and computational fundamentals of optical-electronic remote-sensing systems. In this paper we present new results—the sets of basic models to compute the optical radiation transfer in atmosphere-smoke system (SAS). The basic sets of the influence functions (IF) of the atmosphere and smoke are determined. These sets of IF are invariant with respect to the reflection and refraction properties of the air-smoke boundary. The mathematical modeling method is proposed to compute the solar radiation field in the atmosphere-smoke of the separated solutions for the atmosphere and the smoke.

Our achievements in this direction are in elaboration of one-dimensional models of the solar radiation transfer in the SAS on the basis of the following two approaches. The iteration method of characteristics (Sushkevich et al., 1990) is used by us for the two-media SAS. The radiation field in the SAS is calculated using the optical transfer operator through the influence functions of the atmosphere (IFA) and smoke (IFS) in our second model (Sushkevich, 1994). This new original approach is formulated in the context of a classic linear-system approach. In this model the problems to determine the IFA and IFS are identical to the common-used one-dimensional problem of the radiative transfer theory in a planar layer illuminated by an external parallel flux. The proposed method of solution of these problems depends on the optical-physical characteristics of media. The atmosphere can be cloudy or cloudless. The smoke is considered as a finite or semi-infinite layer.

The representation of the solution to boundary-value problem as a functional is the optical transfer operator of the radiation transfer system which establishes the explicit relationship between the recording radiation and the "scenarios" (the optical image) at the dividing boundary of two media. In turn, by the use of the influence functions, the "scenarios" is described clearly through the characteristics of the reflection and transmission of the dividing boundary at the given its illumination. The influence functions are invariant about the conditions of the illumination and the properties of the dividing boundary.

The influence functions method has such its general purpose as to find tools to integrate the radiation fields in homogeneous or inhomogeneous finite and semi-infinite layers in a combined model of the SAS. Through the IFA and IFS it is possible to estimate the backscattering term of the radiation transfer in the atmosphere and smoke. The techniques elaborated also enable to calculate the reflection coefficients of a smoke surface layer in view of a multiscattering radiation in this layer. Having the IFA and IFS sets calculated, an opportunity appears to recognize structures of any radiation field in the SAS and to study in details mechanisms of forming the atmospheric radiation fields under the smoke influence.

1. INTRODUCTION

Information and mathematical techniques as well as software tools of computer applications are initial to realize the proposed research designed for mathematical modeling of natural processes and calculation procedures on the parallel computer systems. These techniques and applications are linked with the environmental and technological safety, industrial pollutions, natural disasters and anthropogenic influences, emergency situations.

Additionally to the common-used hypothesis of global warming due to as assumed by the "green-house gases" effect, the problem of studying the optical and meteorological as well as climatic behavior of the complicated "atmosphere-ocean-land-ice-biosphere" system with many its feedbacks is seen to be urgent. Thorough analysis of data sets obtained from global observing systems is important for diagnostic and predictability purposes. These data sets are typically analyzed along with modeling results to use remote sensing and monitoring techniques while employing the related scenarios of emergency situations and natural disasters evolution in the atmosphere and on the Earth surface. The relevant understanding, regular monitoring and predictability of weather and climate phenomena and their extreme manifestation under natural and anthropogenic influences are of great importance to sustainable development and assessment of consequences of evolution of the Earth as a planet and to investigate the life-supporting systems for the human community.

All the listed factors open up an opportunity to evolve the subsequent models of an interaction between radiation and natural media based on particular approximations of the optical transfer operator (OTO) as a comprehensive mathematical tool for the simulation techniques. These models are employed in multi-dimensional problems of radiation correction of the ocean from space, in theories of light and image propagation in turbid media, and in theoretical and computational fundamentals of optical-electronic remote sensing systems. New results are presented here given by sets of basic models to compute the optical radiation transfer in the atmosphere-ocean system (SAO). The basic sets of the influence functions (IF) of the atmosphere and ocean are determined. These sets of the IFs are invariant with respect to the reflection and refraction properties of the air-water boundary. The newly defined techniques are proposed to compute the solar radiation field in the atmosphere-ocean system as separated solutions for the atmosphere and the ocean.

Multidimensional spherical and plane-parallel models of the radiation transfer in the atmosphere - surface system (SAS) are under way together with the relevant computer applications. These models are developed to find numerical solutions of the boundary value-problem (BVP) for the stationary equation of the monochromatic or quasi-monochromatic radiation transfer in the atmosphere with scattering, absorbing, emitting and refracting properties. The spatial structure of the atmospheric media (both in spherical and plane-parallel approaches) is underlied by any inhomogeneous reflecting surface (land, ocean, etc.) with the specified boundary conditions for cloud or hydrometeor layers.

If to consider the atmosphere as a channel or a unit of an optical system in the radiation transfer theory, the optical transfer operator (OTO) can be defined in a linear approach assuming the applicability of the superposition law for the integral expression in the equation of transfer. This approach is called by us as the method of the influence functions (Sushkevich et al., 1990; Sushkevich, 1994; Sushkevich, 1999; Sushkevich, 2000; Sushkevich, 1996; Sushkevich and Strelokov, 1999; Sushkevich and Vladimirova, 2003).

Linear-system approximation with models for the optical transfer function and the point-spread function formulated in physical terms is commonly employed in multidimensional problems of radiation correction for remote sensing of objects and the natural environment, in processing of optical information, in theories of sight and image propagation in turbid media, and in theoretical and computational fundamentals of optical-electronic remote-sensing systems. Problems of radiation propagation in three-dimensional plane layers with horizontally non-uniform reflecting boundaries are more complicated, because certain theoretical principles implicit in the theory of linear systems, such as the invariance principle, optical reciprocity theorem, and invariance with respect to planar translations, are not valid.

The constructed base of a variety of mathematical models using the vector IF and the vector OTO values can allow to apply the proposed new algorithms of numerical modeling of the polarized radiation transfer in the systems "atmosphere – land", "atmosphere – ocean", "atmosphere – cloud", "atmosphere – hydrometeors", "atmosphere – vegetative community", "atmosphere – smoke". Besides that, the radiation correction procedures are outlined in the related methods of remote sensing applications. Additional applications are concerned the theory of vision and the transfer theory of images through any opaque polarizing media.

The radiation field in the SAO is calculated using the optical transfer operator characterized by the influence functions of the atmosphere (IFA) and ocean (IFO) in our models. This new original approach is formulated in the context of a classic linear-system approach. The problems set up in these models to determine the IFA and IFO are identical to a routine one-dimensional problem of the radiation transfer theory in a planar layer illuminated by an external parallel flux of incoming radiation. The proposed techniques to solve the formulated problems depend on the optical and physical characteristics of natural media. The atmosphere can be cloudy or cloudless. The ocean is considered as a finite or semi-infinite layer.

The ocean is a rather conservative medium as compared to much more changeable optical and meteorological parameters of the atmosphere. The proposed methods of IFA and IFO are effective just for such situations of different optical properties of these natural media. In the final run, the relevant research is dedicated to finding the sunlight distribution in the oceanic waters by using remote sensing data for the bioproductivity assessment.

2. MATHEMATICAL STATEMENT OF THE PROBLEM

Consider a plane layer, horizontally unbounded $(-\infty < x, y < \infty)$ and of finite height $(0 \leq z \leq H)$, $r_\perp = (x, y)$. The set of all directions $s = (\mu, \varphi)$, $\mu = \cos\vartheta \in [-1,1]$, and $\varphi \in [0, 2\pi]$ is a unit sphere $\Omega = \Omega^+ \cup \Omega^-$, where Ω^+, Ω^- are the hemispheres of propagation directions for descending, transmitted radiation $(\mu \in [0,1])$ and ascending, reflected radiation $(\mu \in [-1,0])$, respectively. For the convenience of representation of the boundary conditions, we introduce sets
$t = \{(z, r_\perp, s): z = 0, s \in \Omega^+\}$,
$b = \{(z, r_\perp, s): z = H, s \in \Omega^-\}$.

The first boundary-value problem for the three-dimensional transfer equation is a problem with a non-reflecting boundary:

$$K\Phi = 0, \quad \Phi\big|_t = 0, \quad \Phi\big|_b = E(r_\perp, s) \qquad (1)$$

and linear operators. The transfer operator is

$$D \equiv (s, grad) + \sigma_{tot}(z) =$$

$$= D_z + \sin\vartheta\cos\varphi\frac{\partial}{\partial x} + \sin\vartheta\sin\varphi\frac{\partial}{\partial y},$$

$$D_z \equiv \mu\frac{\partial}{\partial z} + \sigma_{tot}(z)$$

and the collision integral is

$$S\Phi \equiv \sigma_{scat}(z)\int_\Omega \gamma(z,s,s')\Phi(z,r_\perp,s')\,ds',$$

$$K \equiv D - S,$$

where $\sigma_{tot}(z)$ and $\sigma_{scat}(z)$ are the extinction and scattering coefficients; $\gamma(z,s,s')$ is the single scattering phase function.

For function $f(s^-;H,r_\perp,s)$ with parameter $s^- \in \Omega^-$ we define a linear functional – "a superposition integral"

$$(\Theta,f)(s^-;z,r_\perp,s) \equiv$$

$$\equiv \frac{1}{2\pi}\int_{\Omega^-}ds''\int_{-\infty}^{\infty}f(s^-;H,r'_\perp,s'') \times$$

$$\times \Theta(s'';z,r_\perp - r'_\perp,s)\,dr'_\perp,$$

where the kernel is IF $\Theta(s^-;z,r_\perp,s)$ which solves the boundary-value problem

$$K\Theta = 0, \quad \Theta|_t = 0, \quad \Theta|_b = f_\delta(s^-;r_\perp,s);$$

$$f_\delta(s^-;r_\perp,s) = \delta(r_\perp)\delta(s-s^-).$$

We can obtain a solution to problem (1) in the form of a linear functional (Numerical solution of atmospherical optics problems, 1984; Sushkevich et al., 1990, 1994; Sushkevich, 1994):

$$\Phi(z,r_\perp,s) = (\Theta,E).$$

To describe a single event of interaction between the radiation and reflecting boundaries, we use the operator

$$R\Phi \equiv \int_{\Omega^+}\Phi(H,r_\perp,s^+)q(r_\perp,s,s^+)\,ds^+.$$

3. RESULTS

With a radiation source conditioned by the back-ground radiation of the layer or other factors being prescribed at the boundary $z = H$, we describe the formation process of the illumination as resulting from multiple reflections of radiation at the boundary and including multiple scattering in the medium described in terms of the general boundary-value problem

$$K\Phi = 0, \quad \Phi|_t = 0, \quad \Phi|_b = \varepsilon R\Phi + \varepsilon E(r_\perp,s). \qquad (2)$$

We can seek a solution to problem (2) in the form of linear functionals, their kernels being IFs including the effects of multiple scattering and repeated reflection, or series with respect to the multiplicity of reflection from the underlying surface:

$$\Phi = (\Theta, VE), \quad VE \equiv \sum_{n=0}^{\infty}G^nE, \quad Gf \equiv R(\Theta,f).$$

The asymptotically exact solution to the boundary-value problem in the transfer theory

$$K\Phi = 0, \quad \Phi|_t = 0, \quad \Phi|_b = 0, \qquad (3)$$

$$\Phi|_{d1} = \varepsilon(R_1\Phi + T_{21}\Phi + E_1),$$

$$\Phi|_{d2} = \varepsilon(R_2\Phi + T_{12}\Phi + E_2)$$

with the predetermined illumination of dividing border $E = \{E_1, E_2\}$ obtained by the influence functions method. The horizontally inhomogeneous dividing border of two media, transmitting and reflecting of a radiation, passes at a level $z = h$ within the layer. For the convenience of recording of the boundary conditions, we introduce sets

$$d2 = \{(z,r_\perp,s):\ z = h,\ s \in \Omega^+\};$$

$$d1 = \{(z,r_\perp,s):\ z = h,\ s \in \Omega^-\}.$$

To describe a single event of the interaction between the radiation and dividing boundaries, we use the operators of reflection R_1, R_2 and transmission T_{12}, T_{21}, where the indices 1 and 2 to be related to the layers with $z \in [0,h]$ and $z \in [h,H]$ respectively.

To solve the problem (3), we introduce the perturbation theory series

$$\Phi(z,r_\perp,s) = \sum_{n=1}^{\infty}\varepsilon^n\Phi_n(z,r_\perp,s)$$

with the parameter $0 < \varepsilon \leq 1$, fixing a single event of the passage through dividing boundary. The components of the vectors $\Phi_n = \{\Phi_n^1, \Phi_n^2\}$ satisfy a set of recurrent boundary-value problems $(n \geq 1)$ for the first medium with $z \in [0,h]$:

$$K\Phi_n^1 = 0, \quad \Phi_n^1|_t = 0, \quad \Phi_n^1|_{d1} = F_n^1,$$

$$F_n^1 = \begin{cases} E_1 & \text{for } n = 1, \\ R_1\Phi_{n-1}^1 + T_{21}\Phi_{n-1}^2 & \text{for } n \geq 2, \end{cases}$$

and for the second medium with $z \in [h,H]$:

$$K\Phi_n^2 = 0, \quad \Phi_n^2|_b = 0, \quad \Phi_n^2|_{d2} = F_n^2,$$

$$F_n^2 = \begin{cases} E_2 & \text{for } n = 1, \\ R_2\Phi_{n-1}^2 + T_{12}\Phi_{n-1}^1 & \text{for } n \geq 2. \end{cases}$$

Using the vector of the influence functions $\Theta = \{\Theta_1, \Theta_2\}$ we introduce the vectorial operation, describing a single event of the interaction between the radiation and the dividing boundary and including multiple scattering in either media $(\mathbf{f} = \{f_1, f_2\})$

$$[\Pi \mathbf{f}](h, r_\perp, s) \equiv P(\mathbf{\Theta}, \mathbf{f}) =$$
$$= \begin{bmatrix} R_1(\Theta_1, f_1) + T_{21}(\Theta_2, f_2) \\ T_{12}(\Theta_1, f_1) + R_2(\Theta_2, f_2) \end{bmatrix},$$

where $(\mathbf{\Theta}, \mathbf{f}) = \{(\Theta_1, f_1), (\Theta_2, f_2)\}$ is a vectorial functional; the matrix of the operators is

$$P = \begin{bmatrix} R_1 & T_{21} \\ T_{12} & R_2 \end{bmatrix}.$$

One can prove that two successive n-approximations are connected by the recurrent relations $\mathbf{\Phi}_n = (\mathbf{\Theta}, P\mathbf{\Phi}_{n-1})$ and for $n \geq 1$ the following representation exists $\mathbf{\Phi}_n = (\mathbf{\Theta}, \Pi^{n-1} \mathbf{E})$.

As a result we obtain the asymptotically exact solution

$$\mathbf{\Phi} = \sum_{n=1}^{\infty} \mathbf{\Phi}_n = (\mathbf{\Theta}, Z\mathbf{E}); \qquad (4)$$

$$Z\mathbf{E} \equiv \sum_{n=0}^{\infty} \Pi^n \mathbf{E} \qquad (5)$$

is the sum of a von Neumann series with respect to the multiplicity of radiation passage through dividing boundary including the contribution of multiple scattering in either media.

The representation of the solution to boundary-value problem (3) as a functional (4) is the optical transfer operator of the radiation transfer system which establishes the explicit relationship between the recording radiation and the "scenarios" (the optical image) at the dividing boundary of two media. In turn, by the use of the influence functions, the "scenarios" (5) is described clearly through the characteristics of the reflection and transmission of the dividing boundary at the given its illumination. The influence functions are invariant about the conditions of the illumination and the properties of the dividing boundary.

4. CONCLUSION

The influence functions method has been shown to be one of the most effective techniques of solution of the boundary value problems of the radiation transfer. The presented constructive approach is effective for mathematical modeling of radiation transfer in natural media to solve the multidimensional problems using multiprocessor computers with parallel structure.

Three main conclusions can be extracted from many previous results concerning the state-of-the-art of the problem of measurement data registration and remote sensing. The first is related to finding implicit ways of the registration of any refraction surface from spacecrafts. The second is concerned the opportunity to make these procedures by the explicit way of calculating through the application of the IF method. The third is given by the method of value function and adjoint equations. The IF term enables to integrate any type of singularity and diffusive characteristics of the relevant sources and boundaries.

5. ACKNOWLEDGMENTS

The work has been supported by the Russian Foundation for Basic Research (project 03-01-00132).

6. REFERENCES

Sushkevich T. A., S. A. Strelkov, and A. A. Ioltukhovsky, 1990: Characteristic Method to the Atmospheric Optics Problems. (in Russian) Moscow, Nauka, 296 p. (Book review in *Transport theory and statistical physics*, Vol. 22, 4, 587-591 (1993)).

Sushkevich, T. A., 1994: Solution of the general boundary-value problem in the transfer theory for a plane layer with a horizontal nonuniformity. In *Physics-Doklady*, Vol. 39, No. 11, 740-744.

Sushkevich, T. A., 1999: About solution of the problems of the spacecraft data atmospheric correction. In *Earth observation and remote sensing*, No. 6, 49-66.

Sushkevich, T. A., 2000: Linear-system approach and the theory of the optical transfer operator. In *Atmospheric and oceanic optics*. Vol. 13, No. 4, 421-440.

Sushkevich T. A., 1996: A solution to the boundary-value problem of transport theory for a planar layer with a horizontally inhomogeneous boundary of an interface between two media. In *Doklady Mathematics*, Vol. 54, No. 2, 775-779.

Sushkevich T. A., and S. A. Strelkov, 1999: The influence function of the general boundary value vectorial problem in transfer theory. In *Doklady Mathematics*, Vol. 59, No. 1, 132-136.

Sushkevich T. A., and E. V. Vladimirova, 2003: About a model of the reflecting boundary consideration in the radiative transfer problems for the spherical shell. In *Siberian Journal of Computational Mathematics*, Vol. 6, No. 1, 73-88.

DEVELOPMENT OF THE MONTE-CARLO RADIATIVE TRANSFER MODEL PROMT

U. Hamann, G. Seckmeyer, and S. Raasch

Institute for Meteorology and Climatology, University of Hanover

30419 Hanover, Germany

B. Mayer

Institute of Atmospheric Physics, German Aerospace Center (DLR)

82230 Wessling, Germany

ABSTRACT

The Program for Monte-Carlo radiative Transfer simulation *Promt* simulates the atmospheric radiative transfer regarding the effects of Rayleigh scattering, gaseous absorption and Mie scattering and absorption by clouds. For one-dimensional (1D) cases *Promt* is compared to the sophisticated library for radiative transfer *libRadtran* (Mayer, 2005). Also a first comparison with the Monte-Carlo model MYSTIC (Mayer, 1999) is presented, where the cloud field is taken from the parallelised LES model PALM (Raasch and Schröter, 2001). In future *Promt* shall be coupled with the LES model PALM to investigate the inter-actions between radiative transfer and convection.

1. INTRODUCTION

Solar radiation makes life on earth possible, it drives the atmospheric circulation and determines the climate of the earth. Detailed knowledge of radiative transfer is therefore important for climate simulation and weather prediction as well as for remote sensing.

For many operational applications 1D models are used as their demand for calculation time is quite small. But in order to take horizontal inhomogeneities of clouds into account three-dimensional (3D) radiative transfer models are necessary. The Monte-Carlo method enables to regard arbitrary cloud shapes and to simulate radiative transfer with high spectral and special resolution (O´Hirok and Gautier, 1998). Apart from the cloud shape, also the extinction of the cloud is highly variable.

3D cloud fields including their microphysical properties have to be provided for 3D radiative transfer calculation. Usually these cloud fields are derived by remote sensing from satellite, cloud radar, airborne measurements, Large Eddy Simulation (LES) or statistical cloud generation models.

Promt uses the Monte-Carlo approach to simulate the radiative transfer through inhomogeneous cloud field. In particular for the comparison with MYSTIC a marine stratocumulus field generated by PALM is used, where the simulated meteorological situation is a cold air outbreak (Schröter et al., 2005).

2. MODEL DESCRIPTION

The program *Promt* mainly consists of two parts, the calculation of the optical properties of the atmosphere in dependence of the wavelength and the Monte-Carlo simulation of the photon flight. The optical properties are calculated as follows:

The Rayleigh scattering cross-section used in *Promt* (Bucholtz, 1995) takes a wavelength dependent aniso-tropy factor into account. *Promt* uses two routines of the program Grimaldi (Scheirer and Macke, 2003), which calculate monochromatic or spectrally averaged absorption coefficients as a function of temperature and pressure. Furthermore the program MIEV0.f (Wiscombe, 1980) is used to calculate the optical properties of cloud droplets.

The Monte-Carlo simulation of the photon flight uses the maximum cross-section method for simulating the way to the next scattering event. Then the new photon direction is determined according to the cumulative phase function in the natural coordinate system of the photon, which is then rotated to the fixed coordinate system of the atmosphere. Absorption is simulated by reducing the photon weight according to the Lambert-Beers law, every time a photon moves.

3. COMPARISON TO LIBRADTRAN

To validate the model, results of *Promt* are compared to those of *libRadtran* for 1D cases. All following simulations, including the 3D simulations, were done for a solar zenith angle of 0° and a surface albedo of 0.

3.1 Rayleigh Scattering

As a first test the transmittance is simulated by *Promt* and *libRadtran* using only Rayleigh scattering, see Fig. 1. The deviation between *Promt* and *libRadtran* is less than 0.6% for wavelength smaller than 300 nm, for all other regions it is even less than 0.2%. These small deviations may be caused by slightly different Rayleigh scattering cross-sections or the dependence of the Rayleigh phase function on the anisotropy factor, which is not yet included in *Promt*.

3.2 Absorption and Rayleigh Scattering

There is a general non trivial problem to account for absorption over larger wavelength regions, which is illustrated in Fig. 2.

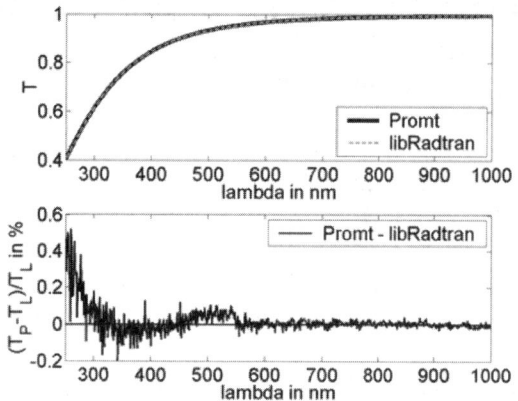

Fig. 1. Comparison of Rayleigh scattering simulation of *Promt* and *libRadtran*.

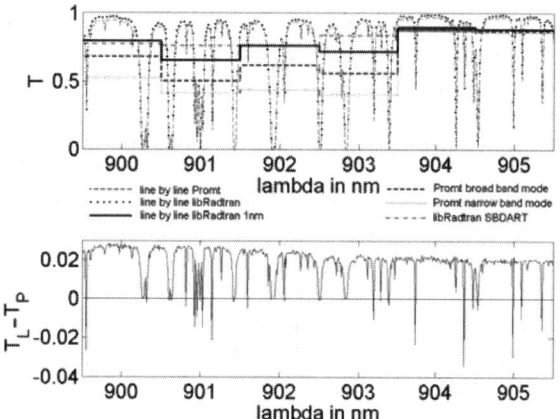

Fig. 2. Comparison of different methods to account for absorption. The gray lines are line-by-line simulations of *Promt* and *libRadtran*. The thick lines are various methods, which are using averaged absorption coefficients. The lower diagram shows the difference of the line-by-line simulations.

The absorption coefficient varies rapidly with wavelength. As result the transmittance may range from 0 to nearly 1 within 1 nm, but in order to model the solar radiative transfer wavelength intervals of thousands of nanometres have to be simulated.

Line-by-line simulations possess such a high spectral resolution that every single absorption line is resolved. The thin lines in the upper diagram of Fig. 2 show the results of line-by-line simulations of *libRadtran* and *Promt*, the lower diagram shows the difference. (The difference is stated here, as percentage information is highly variable for transmittances near 0.) In the plotted wavelength region *Promt* simulates a transmittance, which is in average 0.017 higher than it is simulated by *libRadtran*.

The use of spectral averaged absorption coefficients causes an uncertainty, which must be accepted in order to reduce CPU time to a reasonable amount, if a band parameterisation is required. Fig. 2 also shows some results of simulations, which use averaged absorption coefficients:

libRadtran can be run using absorption coefficients, which are calculated by LOWTRAN. In the following this simulation is denoted as *libRadtran*-LOWTRAN.

Promt contains a routine *lilibroad*, which originates from Grimaldi (Scheirer, 2002). It averages the transmittance through the layers of the model atmosphere and regains the absorption cross-section back from the averaged transmittance.

The most accurate result these simulations can be compared with is a line-by-line calculation, which is aggregated to the 1nm grid. Such a result is plotted as thick solid line in Fig. 2.

That way *Promt-lilibroad* and a *libRadtran*-LOWTRAN are compared to a line-by-line calculation for 500 to 2500 nm, see Fig. 3.

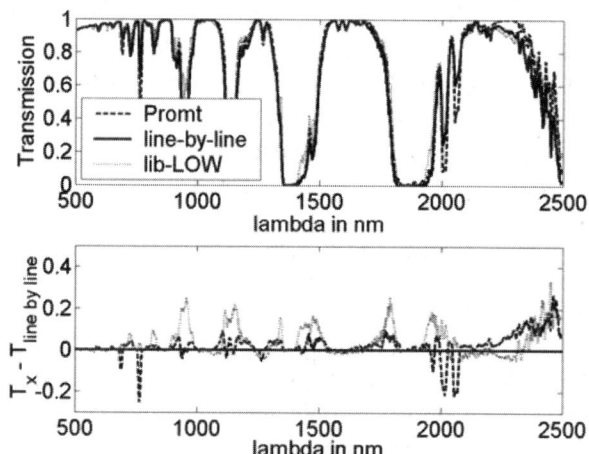

Fig. 3. The transmittance is simulated by *libRadtran*-LOWTRAN and *Promt-lilibroad* using averaged absorption coefficients. Also shown is a line-by-line simulation of *libRadtran*. The lower diagram shows the differences to the line-by-line simulation. The data is plotted as 10 nm gliding mean.

In this comparison the absorption of O_3 is not accounted for, as the absorption line database HITRAN 1996, which is used by *Promt-lilibroad* and also by the program GENLN2 for the *line-by-line* calculation of *libRadtran*, does not contain information of the Chappuisband.

The line-by-line method simulates an irradiance of 830.9 W/m^2 from 500 to 2500 nm. *Promt-lilibroad* underestimates this value by 3.7%, *libRadtran*-LOWTRAN overestimates this value by 3.4%.

In a second comparison the irradiance from 250 nm to 3000 nm is calculated including also the absorption of O_3. *Promt* simulated an irradiance of 1062.7 W/m^2 and *libRadtran*-LOWTRAN simulated 1084.0 W/m^2.

3.3 Clouds, Absorption and Rayleigh Scattering

The transmittance through an atmosphere including a 1 km thick horizontal homogeneous cloud layers is simulated by *Promt-lilibroad* and *libRadtran*-LOWTRAN, see Fig. 4. In *libRadtran* the cloud optical properties are calculated according to a parameterization (Hu and Stamnes, 1993). The liquid water content within the cloud layer is 1 g/m^3, and the droplet radius is 10 μm.

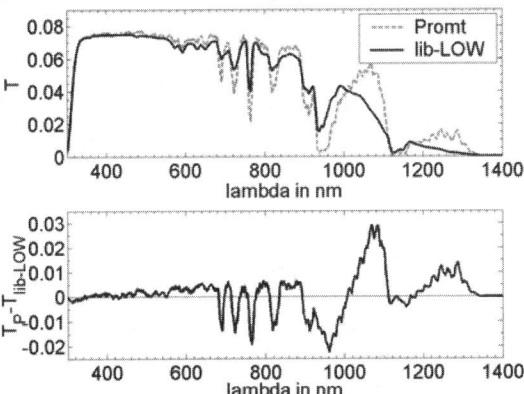

Fig. 4. The transmittance through a 1 km thick cloud layer simulated by *Promt-lilibroad* and *libRadtran*-LOWTRAN. The data is plotted as 10 nm gliding mean.

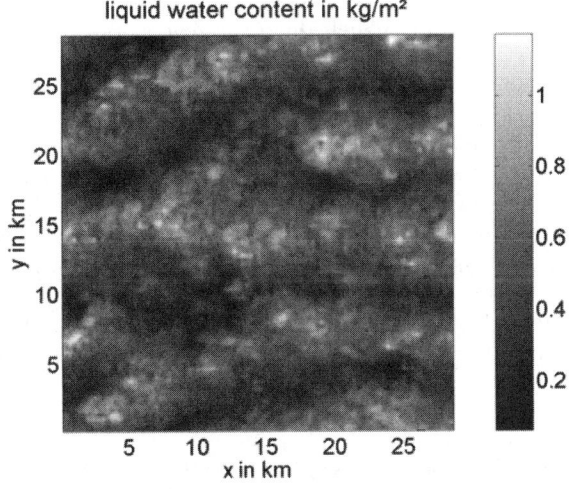

Fig. 5. Vertically integrated liquid water content of a cloud field generated by PALM.

Promt simulates a stronger gaseous absorption than *libRadtran*-LOWTRAN, e.g., around 950 nm, but a weaker absorption by liquid water of the clouds, e.g., around 1070 nm. These effects partly compensate, so that the simulations of the integrated solar irradiance from 290 to 3000 nm deviate only by 1.7%, where *Promt-lilibroad* simulates 64.3 W/m^2 and *libRadtran*-LOWTRAN 63.2 W/m^2.

3. COMPARISON TO MYSTIC

In this comparison a 3D cloud field generated by PALM is used, see Fig. 5. The transmittance for 550 nm simulated by *Promt* is shown in Fig. 6. Regions with high transmittance correspond well to regions with low liquid water content in Fig. 5. The transmittance field simulated by MYSTIC (not shown) is very similar. The horizontally averaged transmittance is 0.1427 for *Promt* and 0.1445 for *libRadtran*.

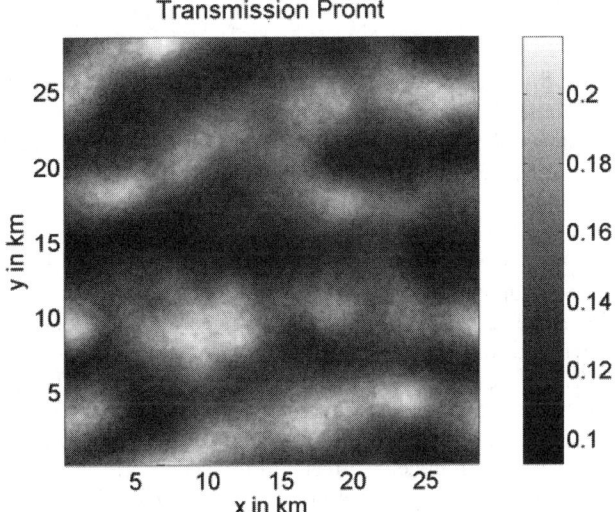

Fig. 6. Transmittance for 550 nm simulated by *Promt*.

Fig. 7 shows the difference between the simulation of *Promt* and MYSTIC. The standard deviation is 0.0025, which is mainly due to the statistical simulation uncertainty of the Monte-Carlo method.

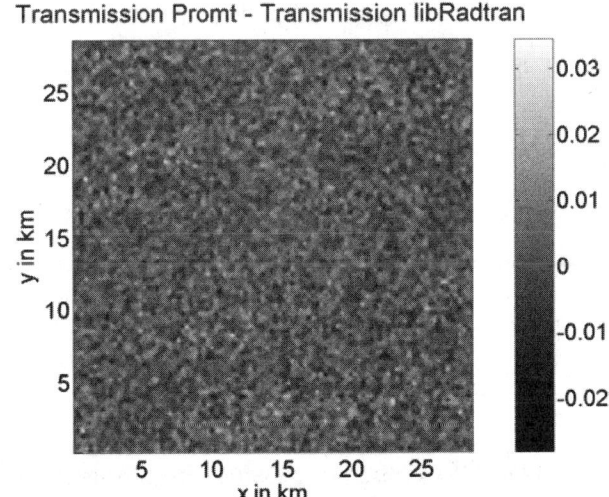

Fig. 7. Difference between the transmittance fields simulated by *Promt* and MYSTIC for 550 nm.

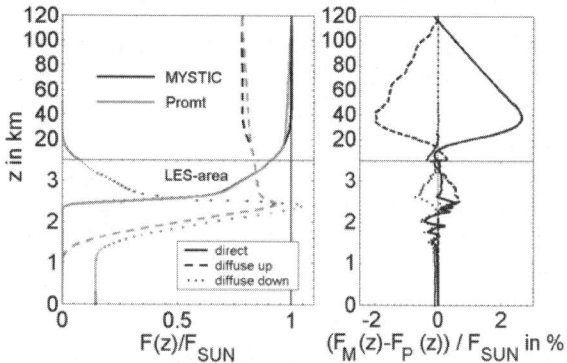

Fig. 8. Horizontal averaged normalised spectral irradiance for 550 nm in dependence of height, simulated by *Promt* and MYSTIC for the LES cloud field in Fig. 5.

The horizontal mean of spectral irradiance is also simulated as a function of height. It is divided into its direct, diffusive upward and diffusive downward com-ponents, see Fig. 8. The right diagram shows the difference between *Promt* and MYSTIC. The largest deviations are observed at 40 km, where *Promt* simulates a 2.7% higher diffusive upward and a 1.9% lower direct irradiance than MYSTIC. It is concluded, that this is a problem of *Promt*, as *Promt* simulates a reduced direct irradiance and therefore scattering, but no corresponding diffusive downward irradiance in these heights. The agreement in the LES-area is good with deviations less than 1%.

A comparison of the solar irradiance at the ground and as function of height is performed, too. The fields of irradiance look similar to Fig. 6 for both models, the values ranging from 90 to 215 W/m^2. The horizontal averaged irradiance is 137.4 W/m^2 for MYSTIC and 142.1 W/m^2 for *Promt*. The differences are probably caused by the different methods of calculating the optical properties of the clouds. The deviations of the vertical profiles are basically the same as for 550 nm.

4. SUMMARY AND OUTLOOK

The Monte-Carlo model *Promt* was compared to *libRadtran* and MYSTIC.

For 1D cases the results of *Promt* are in a good agreement with those of the well established 1D radiative transfer model *libRadtran*. Regarding Rayleigh-scattering the transmittance simulated by *Promt* deviates less than 0.6% from the result of *libRadtran*. *Promt* simulates a solar irradiance with respect to gaseous absorption, which is about 2% smaller than those simulated by *libRadtran*, when it uses LOWTRAN absorption coefficients. Simulating the solar irradiance for a 1 km thick horizontal homogeneous cloud layer, the results of *Promt* deviates 1.8% from those of *libRadtran*.

Results of 3D simulations of *Promt* are in good agreement with those of the Monte Carlo model MYSTIC. The transmittances simulated for 550nm are nearly identical for *Promt* and MYSTIC apart from the sta-tistical noise of the Monte Carlo method. The structures of the simulated solar irradiance fields are consistent, however the horizontal average mean simulated by *Promt* is 3.4% higher than it is simulated by MYSTIC, which is probably due to the different methods of calculating the cloud optical properties. The profiles of the horizontal averaged irradiance differ in the upper atmosphere up to 2.7%, but correspond well in the lower atmosphere.

Promt and PALM are well suitable for investigation of the interactions of cloud formation and radiative transfer, because both programs are parallelised and therefore the computational effort becomes realizable.

5. REFERENCES

Buchholtz, A. 1995: Rayleigh-scattering calculation for the terrestrial atmosphere. In *Applied Optics 34 (15)*, Washington, DC, 2765-2773.

Hu, Y. X., and K. Stamnes, 1993: An accurate parameterization of the radiative properties of water clouds suitable for use in climate models. In *Journal of Climate 6 (4)*, American Meteorological Society, Boston, MA, 728-742.

Mayer, B., 1999: I3RC phase 1 results from the MYSTIC Monte Carlo model. In *Intercomparison of three-dimensional radiation codes: Abstracts of the first and second international workshop*, University of Arizona Press, Tucson, AZ.

Mayer, B., and A. Kylling, accepted 2005: Technical Note: The libRadtran software package for radiative transfer calculations: Description and example of use. In *Atmospheric Chemistry and Physics Discussions*, European Geosciences Union, Katlenburg-Lindau, Germany.

O'Hirok, W., and C. Gautier, 1998: A three dimensional radiative transfer model to investigate the solar radiation within a cloudy atmosphere. In *Journal of the Atmospheric Sciences 55*, Boston, MA, 2162-2179 and 3065-3076.

Raasch, S., and M. Schröter, 2001: A Large Eddy Simulation model performing on Massively Parallel Computers. In *Meteorologische Zeitschrift 10(5)*, Stuttgart, Germany, 363-372.

Scheirer R., and A. Macke, 2003: Cloud inhomogeneity and Broadband solar fluxes. In *Journal of Geophysical Research 108 (D19)*, Washington, DC, Art. No. 4599.

Schröter M., S. Raasch, and H. Jansen, accepted 2005: Cell broadening revised: Results from high resolution large eddy simulation of cold air outbreaks. In *Journal of the Atmospheric Sciences*, Boston, MA.

Wiscombe, W. J., 1980, Improved Mie scattering algorithm. In *Applied Optics 19 (9)*, Washington, DC, 1505-1509.

THEORETICAL CALCULATION AND SIMPLE PARAMETERIZATION FOR THE WATER VAPOR MILLIMETER WAVE FOREIGN CONTINUUM

Q. Ma

NASA/GISS & Dept. Applied Physics and Applied Mathematics, Columbia University
2880 Broadway, New York, NY 10025, USA
qma@giss.nasa.gov

R. H. Tipping

Dept. of Physics & Astronomy, University of Alabama
Tuscaloosa, AL 35487, USA

ABSTRACT

We present theoretical calculations of the millimeter wave foreign continuum due to colliding pairs of H_2O-N_2 and H_2O-O_2 molecules. The calculations are based on an idea that the millimeter water vapor continuum is caused by all lines of the pure rotational band of H_2O, but not individual ones. By treating this band as a whole, one is able to calculate the continuum directly. In practice, this idea is realized by using three powerful tools: the Lanczos algorithm, the coordinate representation, and the Monte Carlo method. Based on this procedure, we have calculated the foreign continuum from the H_2O-N_2 and H_2O-O_2 pairs for a range of temperatures relevant to the atmosphere. The calculated results of H_2O-N_2 are compared to laboratory measurements and to widely used empirical models. For easy use, we fit our results for the absorption coefficient to a simple analytic function of frequency applicable up to 450 GHz and temperature ranging from 200 K to 300 K.

1. INTRODUCTION

It is well known that besides contributions from local lines of H_2O and O_2, continuum absorptions are important at millimeter wavelengths. There are two continua: water vapor continuum and dry air continuum. The former consists of the self- and foreign–continuum. A good knowledge of the water vapor millimeter wave foreign-continuum absorption is essential to the accuracy of satellite retrievals, especially in dry air environments. At present, our understanding of the problem is not satisfactory. There are only a few laboratory measurements. Most of them were carried out at room temperature and above. There are no data available for lower temperatures which are more common in the atmosphere. In addition, the self-continuum data provided by different groups differ by large amounts and the situation is even worse for the foreign-continuum because these laboratory data contain even larger uncertainties (Rosenkranz, 1998, 1999). On the other hand, there are several empirical models available, such as the Liebe's MPM89 and MPM93 models (Liebe, 1989, 1993), and a more recent one proposed by Rosenkranz which is a combination of the MPM93 self-continuum plus the MPM89 foreign-continuum (Rosenkranz, 1998). Unfortunately, none of them meets the desired accuracy, or works satisfactorily for all atmospheric conditions. With respect to theory, there is a lack of theoretical work heretofore from which one is able to predict the foreign-continuum quantitatively well.

In the present study, we present theoretical progress with which one is able to calculate the millimeter wave foreign continuum due to colliding pairs of H_2O-N_2 and H_2O-O_2 molecules. The progress is based on one simple idea. The idea is that because the millimeter water vapor continuum is caused by all absorption lines of the pure rotational band of H_2O, one does not need to distinguish contributions from each of individual lines of the band, rather one can treat this band as a whole. This idea is realized in three steps. First of all, with the Lanczos algorithm calculations are reduced to evaluating ensemble averages over the band. Then, by introducing the coordinate representation, these ensemble averages become multi-dimensional integrations. Finally, the Monte Carlo method is used to evaluate these integrations.

2. THEORY

2.1 Absorption Coefficient and the Spectral Density

With the binary collision approximation valid in atmospheric environments, the absorption coefficient at the frequency ω (in cm^{-1}) can be expressed as

$$\alpha(\omega) = n_{pair} \frac{4\pi^2}{3\hbar c} \omega \tanh(\frac{\beta\hbar\omega}{2})\{F(\omega) + F(-\omega)\}, \quad (1)$$

where n_{pair} is the pair number density. The spectral density $F(\omega)$ is given by

$$F(\omega) = -\frac{1}{\pi} \operatorname{Im} Tr\{\vec{\mu}^\dagger \sqrt{\rho} \frac{1}{\omega-\mathcal{L}} \sqrt{\rho}\, \vec{\mu}]\}. \quad (2)$$

By dividing all degrees of freedom of the two interacting molecules into internal and translational degrees, we are able to write $F(\omega)$ as

$$F(\omega) = -\frac{1}{\pi} \operatorname{Im} Tr_r\{\rho_{iso} Tr_{ab}[\vec{\mu}^\dagger \sqrt{\rho_a \rho_b} \frac{1}{\omega-\mathcal{L}} \sqrt{\rho_a \rho_b}\, \vec{\mu}]\}, \quad (3)$$

where the whole trace is divided into a trace over internal degrees and a trace over the translational degree. It turns out that to carry out the Tr_{ab} is a difficult job.

2.2 The Lanczos Algorithm and Continued Fractions

With the Lanczos algorithm, one can define the starting vector $|v> \equiv \sqrt{\rho_a \rho_b}\, \vec{\mu}$ and write the inner trace of $F(\omega)$ in terms of a continued fraction,

$$Tr_{ab}[\vec{\mu}^\dagger \sqrt{\rho_a \rho_b}\, \frac{1}{\omega - \mathcal{L}} \sqrt{\rho_a \rho_b}\, \vec{\mu}]$$
$$= <v|v> \cfrac{1}{\omega - \alpha_1 - \cfrac{\beta_2^2}{\omega - \alpha_2 - \cfrac{\beta_3^2}{\omega - \alpha_3 - \cdots}}}, \quad (4)$$

where L is a Liouville operator associated with the total Hamiltonian and α_1, β_2, α_2, β_3, and α_3 are parameters that can be derived from the matrix elements $<v|L|v>$, $<v|L^2|v>$, $<v|L^5|v>$.

In order to guarantee convergence, the line space is divided into two sub-spaces (+ and –) constructed by the positive and negative resonance lines. Accordingly, the original starting vector $|v>$ is divided into two vectors: $|v>_+$ and $|v>_-$. Then, the inner trace of $F(\omega)$ consists of two terms

$$Tr_{ab}[\vec{\mu}^\dagger \sqrt{\rho_a \rho_b}\, \frac{1}{\omega - \mathcal{L}} \sqrt{\rho_a \rho_b}\, \vec{\mu}]$$
$$= {}_+<v|\frac{1}{\omega - \mathcal{L}}|v>_+ + {}_-<v|\frac{1}{\omega - \mathcal{L}}|v>_-. \quad (5)$$

Each term can be expressed as a continued fraction, so there are two sets of α_n and β_n^2 to be evaluated. For simplicity, in the following we don't distinguish these two terms unless this becomes necessary.

2.3 The Coordinate Representation

At this stage, the introduction of the coordinate representation plays a crucial role because with the standard method, to evaluate $<v|L^n|v>$ with $n \geq 2$ is intractable. With the standard method, the basis set in Hilbert space is constructed by $|j\tau m> \otimes |\ell n>$, the states of two interacting molecules. On the other hand, instead of choosing the internal states, one can select the orientations of the pair of molecules as the basis set in Hilbert space $|\delta(\Omega_a - \Omega_{a\varsigma})> \otimes |\delta(\Omega_b - \Omega_{b\varsigma})>$ where $\Omega_{a\varsigma}$ and $\Omega_{a\varsigma}$ represent orientations of H_2O and N_2 (or O_2), respectively. By introducing this coordinate representation, the potential becomes a diagonal operator and the matrix elements become 9-dimensional integrations. No matter how complicated the potential is, the integrands are always ordinary functions.

2.4 The Interaction Potential Models

In the present study we use a realistic potential model consisting of the isotropic Lennard-Jones, the dipole-quadrupole, the quadrupole-quadrupole, the induction and the atom-atom interactions

$$V(r, \Omega_a, \Omega_b) = V_{iso}(r) + V_{dq}(r, \Omega_a, \Omega_b) + V_{qq}(r, \Omega_a, \Omega_b) \\ + V_{induction}(r, \Omega_a, \Omega_b) + V_{atom-atom}(r, \Omega_a, \Omega_b), \quad (6)$$

where the expression for the atom-atom interaction adopted here is given by

$$V_{atom-atom}(r, \Omega_a, \Omega_b) = V_0 \sum_{i \in a} \sum_{j \in b} 4\varepsilon_{ij} \{(\frac{\sigma_{ij}}{r_{ij}})^{12} - (\frac{\sigma_{ij}}{r_{ij}})^6\}. \quad (7)$$

In the above expression, ε_{ij} and σ_{ij} are parameters. The expressions for the other terms are well known. Values of the dipole and quadrupole of H_2O used here are $\mu = 1.8546$ D, $\Theta_{bb} = -0.13$ DÅ, $\Theta_{cc} = -2.50$ DÅ, and $\Theta_{aa} = 2.63$ DÅ. For N_2 and O_2, $\Theta = 1.466$ DÅ and 0.39 DÅ, respectively. Values of the isotropic and anisotropic parts of the polarizability of N_2 and O_2 are $\alpha_{(N_2)} = 11.74$ a.u., $\gamma_{(N_2)} = 4.75$ a.u. and $\alpha_{(O_2)} = 10.87$ a.u., $\gamma_{(O_2)} = 7.30$ a.u., respectively. We note that the atom-atom potential contains an isotropic part which can be characterized by two L-J effective parameters ε_{eff} and σ_{eff} whose values are 3.64 Å, 131 K for $H_2O - N_2$ and 3.66 Å, 159 K for $H_2O - O_2$.

2.5 The Monte Carlo Calculations

A Monte Carlo algorithm (VEGAS) is used to carry out the 9-dimensional integrations. By taking approximately 10^7 random selections of a set of the 9 angular variables, the Monte Carlo method yields converged results (Ma, 2002). The matrix elements of interest depend on r, the separation between two molecules. For each of the matrix elements, we calculate 80 values corresponding to 80 separations r simultaneously. These values of r are chosen to cover the whole interaction range well. For a specified temperature T and a specified set of the potential parameters, the CPU times required to derive the set of matrix elements are reasonable. After all of these matrix elements are calculated, the explicit expression for the continued fraction is known.

2.6 Poles and Residues of the Continued Fractions

With the current cut-off, the continued fractions are explicitly given by Eq. (4). Then, their poles are the roots of a cubic equation

$$(\omega - \alpha_1)(\omega - \alpha_2)(\omega - \alpha_3) - (\omega - \alpha_1)\beta_3^2 - (\omega - \alpha_2)\beta_2^2 = 0, \quad (8)$$

and the residues are given by

$$R_i = \frac{[(Z_i - \alpha_2)(Z_i - \alpha_2) - \beta_3^2](Z_j - Z_k)}{Z_1^2(Z_2 - Z_3) + Z_2^2(Z_3 - Z_1) + Z_3^2(Z_1 - Z_2)}, \quad (9)$$

where i, j, and k are in cyclic order. It is worth mentioning that poles whose values are real and within $[-30\text{cm}^{-1}, 30\text{cm}^{-1}]$ are the ones of interest in calculations because only these

MOLECULAR RADIATIVE PROPERTIES

poles make contributions to the continuum absorption in the millimeter region.

2.7 Contributions to α(ω) from Poles

After all poles and corresponding residues are available, the difficult job is finished. Then, with the quasistatic approximation, the remaining trace Tr_r of $F(\omega)$, a classical ensemble average over the translational motion, can be expressed as an integration over r,

$$F(\omega) = -4\mu^2 \text{Im} \int_0^\infty \{M_+ \sum_{i=1}^{3} \frac{R_i^{(+)}(r)}{\omega - z_i^{(+)}(r)} + M_- \sum_{i=1}^{3} \frac{R_i^{(-)}(r)}{\omega - z_i^{(-)}(r)}\} e^{-\beta V_{iso}(r)} r^2 dr, \quad (10)$$

where $M_+ = {}_+\langle v|v\rangle_+$ and $M_- = {}_-\langle v|v\rangle_-$ and the superscripts (+) and (−) are used to distinguish quantities accociated with $|v\rangle_+$ and $|v\rangle_-$, respectively. The above is a typical Cauchy integral and can be easily evaluated. $F(-\omega)$ which is given by a similar expression can also be easily evaluated. Through $F(\omega)$ and $F(-\omega)$ respectively, both the positive and negative poles of interest make contributions to $\alpha(\omega)$.

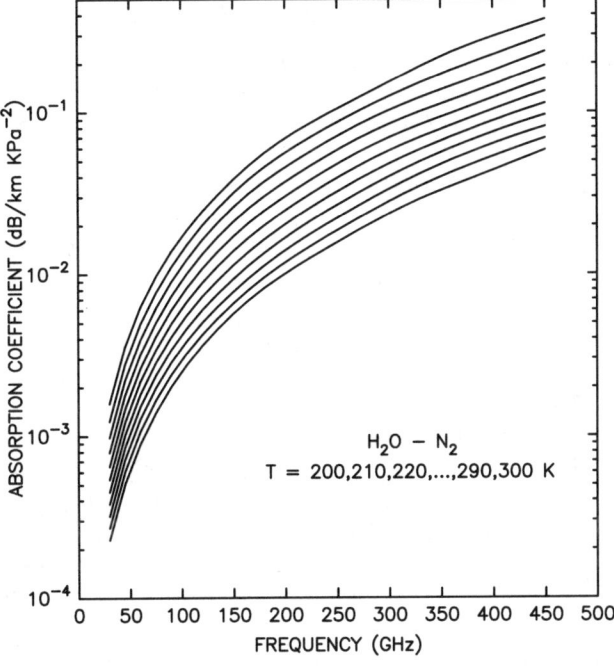

Fig. 1. The calculated H_2O-N_2 millimeter wave continuum (in units of dB/km kPa^{-2} for T = 200, 210, 300 K; these are represented by 11 lines from the top to the bottom, respectively.

3. RESULTS AND DISCUSSIONS

3.1 Results for H_2O-N_2

We calculate $\alpha(f,T)$ of H_2O-N_2 for f = 0 – 450 (in GHz) at eleven T ranging from 200 K to 300 K and plot the results in Fig. 1. We plot the results at T = 296 and 330 K in Figs. 2 and 3. Measurements and the results obtained from MPM89 and MPM93 are also plotted. The theoretical results are in good agreement with experiment. In general, the theoretical values are between those of MPM89 and MPM93, while the T dependence is closer to MPM93.

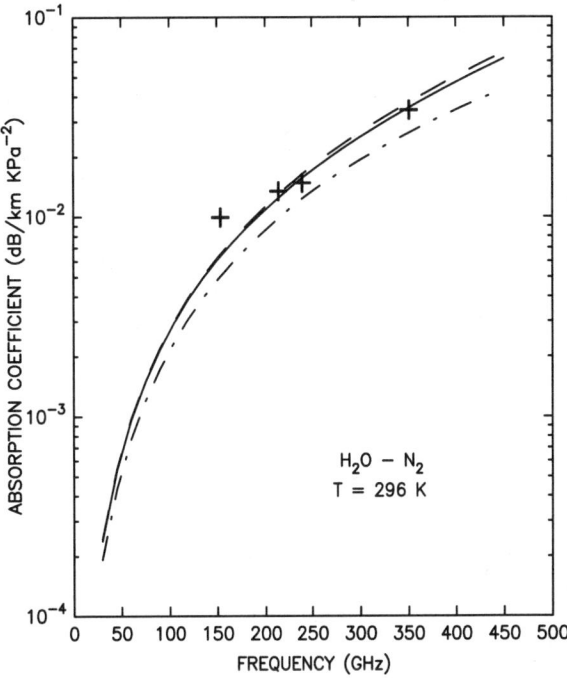

Fig. 2. The calculated H_2O-N_2 continuum for T = 296 K is represented by the solid line. The values derived from MPM89 and MPM93 are the dot-dashed and dashed lines, respectively. The values from measurements at 153.0, 213.525, 239.37, and 350.3 GHz (Godon, 1992; Bauer, 1993, 1995; Kuhn, 2002) are represented by a symbol +.

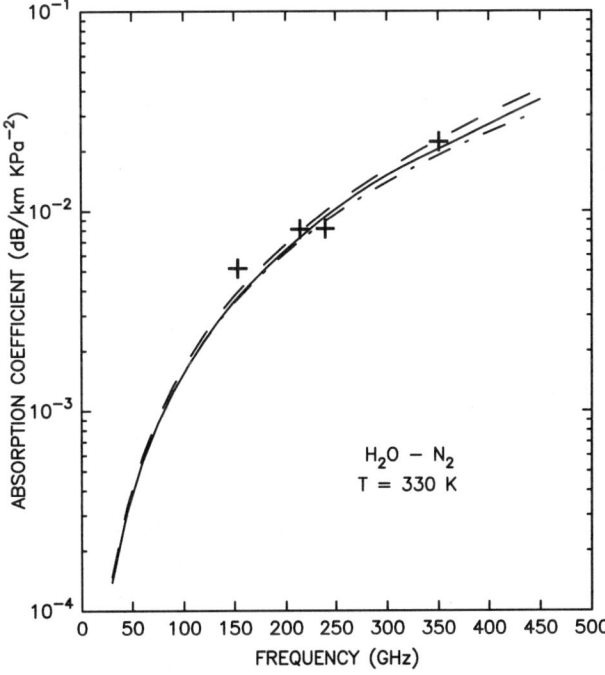

Fig. 3. The same as Fig. 2, except for T = 330 K.

3.2 Results for H_2O-O_2 and H_2O-Air

In a similar way, we calculate the continuum absorption resulting from pairs of H_2O-O_2. Then, by assuming that the air consists of 79% H_2O-N_2 and 21% H_2O-O_2, one is able to derive results for H_2O-air. The magnitudes of H_2O-air are less than those of H_2O-N_2.

3.3 Simple Parameterization

By fitting the results calculated for eleven T ranging from 200 K to 300 K and for a certain range of f, a simple analytical formula for the continuum (in dB/km) can be derived as

$$\alpha(f,T) = A P_{H_2O} P_{N_2} (300/T)^B f^C, \qquad (9)$$

where P_{H_2O} and P_{N_2} (or P_{O_2} and P_{air}) are in kPa. Values of the parameters A, B, and C applicable for 0 – 450 GHz are listed in the following table.

Pair	0 – 450 GHz		
	A	B	C
$H_2O - N_2$	2.225×10^{-7}	4.677	2.033
$H_2O - O_2$	6.077×10^{-8}	4.570	2.047
H_2O – Air	1.885×10^{-7}	4.670	2.034

3.4 Discussions

In comparison between the present formula and MPM89 which is given in a similar way

$$\alpha(f,T) = 2.057 \times 10^{-7} P_{H_2O} P_{N_2} (300/T)^3 f^2, \qquad (10)$$

the calculated frequency dependence parameter C is close to, but not exactly quadratic as in MPM89. However, the main difference is the temperature index. As listed in the table, the calculated value of the temperature index B for H_2O-N_2 is around 4.7 which differs significantly from 3 of MPM89. By analyzing the expression for $\alpha(\omega)$ given by Eq. (1), one can conclude the temperature index B = 3 of MPM89 comes from the product of two factors $n_{pair} \times \tanh(\beta\hbar\omega/2)$ which varies as $(1/T)^3$ in the millimeter region. This implies that in MPM89 one assumes that $F(\omega) + F(-\omega)$ does not depend on T. Such assumption lacks a physical basis.

In summary, we have shown that the calculation of the water vapor millimeter wave foreign continuum becomes possible using a combination of the new idea and three powerful tools: the Lanczos algorithm, the coordinate representation, and the Monte Carlo method. We also have provided simple formulas for the millimeter community based on our recent calculations. However, there is room for improvement of the present results by using more accurate interaction potentials for H_2O-N_2 and H_2O-O_2. Therefore, feedback from the community is very welcome.

4. ACKNOWLEDGMENTS

We acknowledge financial support from NASA under grants NAG5-13337 and FCCS-547. Q. Ma wishes to acknowledge financial support from the Biological and Environmental Research Program (BER), U.S. Department Energy, Interagency Agreement No. DE-AI02-93ER61744. We also would like to thank the NERSC (Livermore, CA) for computer time and facilities provided.

5. REFERENCES

Bauer, A., M. Godon, J. Carlier, Q. Ma, and R. H. Tipping, 1993: Absorption by H_2O and H_2O-N_2 mixtures at 153 GHz, *J. Quant. Spectrosc. Radiat. Transfer,* 50, 463-475.

Bauer, A., M. Godon, J. Carlier, and Q. Ma, 1995: Water vapor absorption in the atmospheric window at 239 GHz, *J. Quant. Spectrosc. Radiat. Transfer,* 53, 411-423.

Godon, M., J. Carlier, and A. Bauer, 1992: Laboratory study of water vapor absorption in the atmospheric window at 213 GHz, *J. Quant. Spectrosc. Radiat. Transfer,* 47, 275-285.

Kuhn, T., A. Bauer, M. Godon, S. Bühler, and K. Künzi, 2002: Water vapor continuum: absorption measurements at 350 GHz and model calculations, *J. Quant. Spectrosc. Radiat. Transfer,* 74, 545-562.

Liebe, H. J., 1989: MPM – an atmospheric millimeter-wave propagation model, Int. J. Infrared & Millimeter Waves 10, 631-650.

Liebe, H. J., G. A. Hufford, and M. G. Cotton, 1993: Propagation modeling of moist air and suspended water/ice particles at frequencies below 1000 GHz, *AGARD Conf. Proc.* 542, 3.1-3.10.

Ma, Q., and R. H. Tipping, 2002: Water vapor millimeter wave foreign continuum: A Lanczos calculation in the coordinate representation. *J. Chem. Phys.* 117, 10581-10596.

Rosenkranz, P. W., 1998: Water vapor microwave continuum absorption: A comparison of measurements and models. *Radio Sci.* 33, 919-928.

Rosenkranz, P. W., 1998: Correction to "Water vapor microwave continuum absorption: A comparison of measurements and models." *Radio Sci.* 34, 1025.

THEORETICAL CALCULATION OF THE COLLISION-INDUCED ABSORPTION OF DRY AIR IN THE FAR-INFRARED REGION

R. H. Tipping

Department of Physics and Astronomy, University of Alabama
Tuscaloosa, AL 35487, USA

Q. Ma

NASA/Goddard Institute for Space Studies and Columbia University
2880 Broadway, New York, NY 10025, USA

ABSTRACT

Using the molecular parameters we obtained from the theoretical calculation of the collision-induced fundamental bands of N_2–N_2 and O_2–O_2 pairs, we have calculated the corresponding absorption in the far-infrared spectral region. By slightly increasing the isotropic polarizability matrix element of N_2 and the hexadecapole moment matrix element of O_2, we can obtain results that are in good agreement with experimental measurements. These parameters can then be used to calculate the collision-induced absorption for N_2–O_2 pairs, and combined to obtain results for dry air. For convenience, a correction factor for scaling the absorption coefficient for N_2–N_2 to that of dry air is presented.

INTRODUCTION

In the Earth's atmosphere, both in the far-infrared and infrared regions, there are two main sources for the continuous absorption of radiation over a range of frequencies: the self and foreign water continua and collision-induced absorption (CIA), both scaling as ρ^2 for a pure gas or as the product of the ρ's for a binary mixture. In most cases for zenith measurements, the former continua dominate and are included in most atmospheric radiative transfer programs. However, for low humidity conditions or long-path limb measurements through the stratosphere, the collision-induced absorptions of O_2 and N_2 can both be important and must also be included.

In a series of papers we have previously presented new experimental data and theoretical analyses of the collision-induced fundamental absorption spectra of N_2–N_2, O_2–O_2, and N_2–O_2 pairs (Boissoles et al., 1994; Moreau et al., 2000; Moreau et al., 2001a, b). In a recent paper (Boissoles et al., 2003) we have extended this analysis to the translation-rotation spectral region from (0 to 400 cm^{-1}). While a number of published experimental results and previous theoretical studies are available for N_2–N_2 for a range of T, for O_2–O_2 there exists only one experimental result at T = 300 K. No experimental data are available for O_2–N_2 mixtures. Because of the availability of more extensive experimental results for N_2–N_2, most radiation transfer codes include only this contribution. The neglect of the contribution from O_2–O_2 collision pairs in the atmosphere is not very important, both because of its smaller magnitude and the fact that the number density of O_2 is smaller than that of N_2 by a factor of 4. However, the absorption by N_2–O_2 pairs is not negligible and should be included.

In this paper, we give a brief discussion of the theoretical expressions used to model the collision-induced absorption for N_2–N_2 and O_2–O_2 in the mm and far-IR regions. Then, using the parameters that give the best fits, we can generate theoretically the CIA spectra for N_2–O_2 pairs. We then present results for air, assuming that N_2 (79%) and O_2 (21%) are uniformly mixed. In order to facilitate application, we calculate a correction factor $R(\omega,T)$ as a function of frequency and temperature, which along with accurate results for N_2–N_2, enables one to model accurately the CIA for dry air.

THEORY

In (Boissoles et al., 1994; Moreau et al., 2000; Moreau et al., 2001a, b; Boissoles et al., 2003), we have given the details of how one can calculate the CIA of pure or binary mixtures of diatomic gases. We write the absorption coefficient $\alpha(\omega)$ as a function of wavenumber ω (cm^{-1}) as

$$\alpha(\omega) = \omega(1 - e^{-\beta\hbar\omega}) A(\omega) \tag{1}$$

where $\beta = 1/kT$. We note that in the far-IR, the stimulated emission factor must be included, whereas in the fundamental region, it has been neglected. By considering only binary collisions valid for atmospheric conditions, the function $A(\omega)$ can be written for a pure gas or mixture, respectively as

$$A(\omega) = \rho^2 G(\omega)/2 \tag{2a}$$

$$A(\omega) = \rho_1 \rho_2 G(\omega). \tag{2b}$$

The function $G(\omega)$ consists of a linear superposition of individual components, $G_n(\omega)$, where in the present case, because all the vibrational quantum numbers are zero, n is specified by the set of integers

$$n \equiv \{J_1, J_1', J_2, J_2', \lambda_1, \lambda_2, \Lambda, L\} \quad (3)$$

where J_i and J_i' are the initial and final rotational quantum numbers, and the other four integers characterize the induced dipole moment mechanism. Explicitly, we include the long-range isotropic quadrupolar induction described by the coefficients (2,0,2,3) and (0,2,2,3); the anisotropic quadrupolar mechanism (2,2,2,3), (2,2,3,3), and (2,2,4,3); the isotropic hexadecapolar mechanism (4,0,4,5) and (0,4,4,5); and the anisotropic hexadecapolar mechanism (4,2,4,5), (4,2,5,5), (4,2,6,5), (2,4,4,5), (2,4,5,5), and (2,4,6,5).

Because of other theoretical approximations, we do not include the short-range dipole moment components because this would involve additional, unknown parameters.

The zeroth moment of the spectrum, Γ_0, in units of cm^{-1} amagat^{-2} for a pure gas is defined by

$$\Gamma_0 \equiv \frac{1}{\rho^2} \int_{-\infty}^{\infty} \frac{\alpha(\omega)}{\omega(1 - e^{-\beta\hbar\omega})} d\omega = \sum_n \Gamma_0(n). \quad (4)$$

Explicit expressions for the individual contributions to the moment have been presented previously and are not repeated here (Boissoles et al., 1994; Moreau et al., 2000; Moreau et al., 2001a, b). We note that this theoretical definition of the zeroth moment is equivalent to the experimental moment (Stone et al., 1984)

$$\Gamma_0 = \int_0^{\infty} \frac{\alpha(\omega)\coth(\beta\hbar\omega/2)}{\omega} d\omega \quad (5)$$

by using the principle of detailed balance. For later comparisons we define the integrated intensity S in units of cm^{-2} in the usual way

$$S = \int_0^{\infty} \alpha(\omega) d\omega. \quad (6)$$

Unlike the zeroth moment, the integrated intensity does not have an explicit theoretical expression although it is widely used.

2.1 N_2-N_2 Spectra

As the first step in our analysis of the translation-rotation band of N_2-N_2, we use the molecular parameters obtained previously from the global fit to the fundamental spectra of N_2-N_2, (Boissoles et al., 1994) to calculate the zeroth moment. The theoretical values are consistently lower than the experimental results (Stone et al., 1984) by approximately 9–14% over the range of T from 228 K to 343 K. As in our previous work, by knowing the strengths and positions of all the components, we can generate the spectrum by multiplying each component by a normalized line shape and summing. Because in this case the isotropic quadrupolar induction mechanism is dominant, we can easily improve the agreement by increasing slightly the isotropic polarizability matrix element, α_{00}. Given the experimental uncertainties as well as other simplifications in the theory, we conclude that with the larger value of α_{00} we can accurately model the translation-rotational spectra of N_2-N_2.

2.2 O_2–O_2 Spectrum

The only experimental data available for the translation-rotational spectrum of O_2–O_2 are those taken at T = 300 K (Bosomworth and Gush, 1965). We calculated separately the contributions from the isotropic quadrupole and hexadecapole mechanisms, together with their sum. As in the fundamental band, the anisotropic contributions are small, but the total is considerably smaller than the experimental results. We can increase the value of the previous hexadecapole moment (ϕ_{00} = 4.8 ea$_0^4$) by a factor of 1.7 (ϕ_{00} = 7.48 ea$_0^4$) without affecting the corresponding agreement with the fundamental band significantly. We find that the theoretical integrated intensity, S = 8.25 x 10^{-5} cm^{-2} amagat^{-2}, is lower than the experimental value, S = 8.64 x 10^{-5} cm^{-2} amagat^{-2}, by only 4.5%. Finally, we note that with this increased value for the hexadecapole moment matrix element, this mechanism contributes 150% of that due to the quadrupole mechanism and, as a result, gives rise to a much broader spectrum as observed.

2.3 O_2–N_2 Mixtures

Because of the absence of experimental data, the collision-induced spectra for mixtures must be generated by theory, using the parameters obtained as discussed above by fitting the N_2–N_2 and O_2–O_2 spectra separately. However, these spectra are precisely what are needed to model the collision-induced absorption in air. In general, if one includes the anisotropic contributions, one cannot separate the system into an "active" molecule (one making a rotational transition) and an "inactive" molecule (one not making a rotational transition). This separation was possible in the fundamental region because the vibrational bands of N_2 and O_2 occur at different, non-overlapping frequencies. However, because the anisotropic contributions are small (approximately 15%), it is still possible to treat the contributions from $\alpha(O_2)Q(N_2)$ plus $\alpha(O_2)\phi(N_2)$, for which N_2 is active, and $\alpha(N_2)Q(O_2)$ plus $\alpha(N_2)\phi(O_2)$ for which O_2 is active separately. In this case, the O_2 active contribution is very much smaller than that of the N_2 active contribution, as expected from the much smaller magnitude of the quadrupole moment matrix element of O_2.

By comparing the sum of the two active contributions with the total result, one can see that the approximation for the absorption coefficient used previously (Pardo et al., 2001)

$$\alpha(O_2 + N_2)_{exact} \approx \alpha(N_2 \text{ active - } O_2)$$
$$+ \alpha(O_2 \text{ active - } N_2) \quad (7)$$

is not too bad.

3. CIA IN DRY AIR

For air we write the absorption coefficient, $\alpha_{air}(\omega)$, as the sum of three terms due to N_2–N_2, O_2–O_2, and N_2–O_2, respectively, assuming $\rho(N_2) = 0.79\,\rho_t$, and $\rho(O_2) = 0.21\,\rho_t$, where ρ_t is the total number density of air. Factoring out the absorption of N_2–N_2, this can be rewritten as

$$\alpha_{air}(\omega) = \alpha_{N2-N2}(\omega)\,[1 + \varepsilon(\omega,T)]$$
$$\equiv \alpha_{N2-N2}(\omega)\,R(\omega,T). \quad (8)$$

The ratio $R(\omega,T)$ shown in Fig. 1 is relatively constant out to approximately 120 cm^{-1}, then increases with frequency and this increase is greater at lower T. The constant value $R(\omega,T) \approx 1.35$ is very close to the value (1.29) (Pardo et al., 2001) necessary to give improved agreement with their atmospheric observations. One can understand qualitatively this agreement for the low-frequency constant value by assuming that the "normalized" profiles for O_2–O_2, N_2–N_2, and N_2–O_2 are not very different

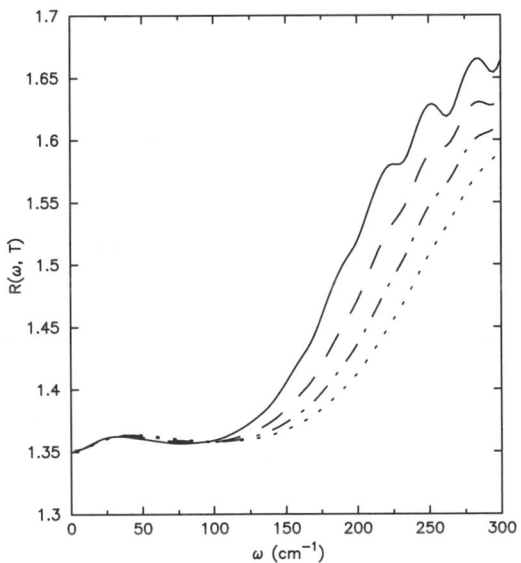

Fig. 1. The correction factor $R(\omega,T)$ as a function of ω (in cm^{-1}) for T = 200 K, 250 K, 300 k, and 350 K, from top to bottom.

4. DISCUSSION AND CONCLUSIONS

We have analyzed the collision-induced translation-rotation spectra for N_2–N_2 and O_2–O_2 and showed that one can get good agreement between theory and experiment by increasing slightly the isotropic polarizability matrix element for N_2 and increasing the hecadecapole matrix element for O_2 by a factor 1.7. Using the parameters obtained allows us to calculate theoretically the corresponding absorption spectra for N_2–O_2 mixtures at different temperatures, and thus we can calculate the corresponding absorption coefficient for air. By factoring out the absorption for N_2–N_2 we can obtain a correction function, $R(\omega,T)$, which can easily be incorporated into atmospheric models in order to go from pure N_2–N_2 to dry air.

5. ACKNOWLEDGMENT

We acknowledge financial support from NASA under grant NAG5-13337.

6. REFERENCES

Boissoles, J., R. H. Tipping, and C. Boulet, 1994: Theoretical Study of the Collision- Induced Fundamental Absorption Spectra of N_2-N_2 Pairs for Temperatures Between 77 K and 297 K, *J. Quant. Spectrosc. Radiat. Transfer,* 51, 615-627.

Boissoles, J., C. Boulet, R. H. Tipping, A. Brown, and Q. Ma, 2003: Theoretical Calculation of the Translation-Rotation Collision-Induced Absorption in N_2-N_2, O_2-O_2, and N_2-O_2 Pairs, *J. Quant. Spectrosc. Radiat. Transfer,* 82, 505-516.

Bosomworth D. R., and H. P. Gush, 1965: Collision-Induced Absorption of Compressed Gases in the Far Infrared, *Can. J. Phys.* 43, 751-757.

Moreau, G., J. Boissoles, C. Boulet, R. H. Tipping, and Q. Ma, 2000: Theoretical Study of the Collision-induced Fundamental Absorption Spectra of O_2-O_2 Pairs for Temperatures between 193 K and 273 K, *J. Quant. Spectrosc. Radiat. Transfer,* 64, 87-107.

Moreau, G., J. Boissoles, R. Le Doucen, C. Boulet, R. H. Tipping, and Q. Ma, 2001a. Experimental and Theoretical Study of the Collision-Induced Fundamental Absorption Spectra of N_2-O_2 and O_2-N_2 Pairs, *J. Quant. Spectrosc. Radiat. Transfer,* 69, 245-256.

Moreau, G., J. Boissoles, R. Le Doucen, C. Boulet, R. H. Tipping, and Q. Ma, 2001b: Metastable Dimer Contributions to the Collision-Induced Fundamental Absorption Spectra of N_2 and O_2 Pairs, *J. Quant. Spectrosc. Radiat. Transfer,* 70, 99-113.

Pardo, J. R., E. Serabyn, and J. Cernicharo, 2001: Submillimeter Atmospheric Transmission Measurements on Mauna Kea during Extremely Dry El Niño Conditions Implication for Broadband Opacity Contributions, *J. Quant. Spectrosc. Radiat. Transfer,* 68, 419-433.

Stone, N. W. B., L. A. A. Read, A. Anderson, I. R. Dagg, and W. Smith, 1984: Temperature-Dependent Collision-Induced Absorption in Nitrogen, *Can. J. Phys.* 62, 338-347.

LINE STRENGTHS AND HALF-WIDTHS OF THE N_2O BANDS IN THE 2.0- TO 2.3-μm REGION AT ROOM TEMPERATURE

M. Fukabori[1], T. Aoki[2], and T. Fujieda
Meteorological Research Institute
Tsukuba, Ibaraki, 305-0052, Japan

T. Watanabe
Toray Research Center, Inc.
Otsu, Shiga, 520-8567, Japan

ABSTRACT

Line strengths and half-widths in the $4\nu_1$, $3\nu_1+2\nu_2$, $2\nu_1+\nu_3$, $\nu_1+2\nu_2+\nu_3$, and $2\nu_3$ bands of N_2O were determined from spectra obtained by a high-resolution Fourier transform spectrometer at room temperature. The squared vibrational transition dipole moments and the coefficients of the Herman-Wallis factors were also determined for these bands. Data were used to validate the two recent versions of the HITRAN database. The measured line strengths agree with the values of HITRAN04 rather than with those of HITRAN2K. The self-, N_2-, and O_2-broadened half-widths were in good agreement with the results of recent high-resolution experiments within the experimental errors.

1. INTRODUCTION

Nitrous oxide (N_2O) is an important greenhouse gas in the Earth's atmosphere. The very strong infrared absorption bands of the ν_3 and ν_1 fundamental bands at 4.5 and 7.78 μm are important for studies of thermal balance in the terrestrial atmosphere. Infrared spectra of N_2O were used to observe the total amount and the vertical profiles of N_2O. In order to calculate atmospheric transmittance, accurate knowledge of the absorption line parameters, such as the line strengths and half-widths of the absorption bands of N_2O, is required. Several spectroscopic databases are available for the above purposes. HITRAN databases (e.g., Rothman et al., 2003; Rothman et al., 2005) are widely used to calculate the infrared and near infrared spectra. However, the accuracies of the compiled values in the HITRAN databases could be insufficient for accurate remote sensing, especially for the near infrared region.

In the near infrared region, N_2O gives rise to five main absorption bands ($4\nu_1$, $3\nu_1+2\nu_2$, $2\nu_1+\nu_3$, $\nu_1+2\nu_2+\nu_3$, and $2\nu_3$) in the 2.0- to 2.3-μm region. The line strengths of these bands have not been investigated until the late 1990's, because of their weakness. Recently, two high-resolution experiments of the near-infrared absorption bands of N_2O have been carried out to determine the absorption line parameters (Toth, 1999; Daumont et al., 2001). Although the spectroscopic databases have been revised several times based on experimental studies, the line strengths of the $4\nu_1$, $3\nu_1+2\nu_2$, $2\nu_1+\nu_3$, $\nu_1+2\nu_2+\nu_3$, and $2\nu_3$ bands of N_2O were not revised from the initial version of the AFCRL (Air Force Cambridge Research Laboratories) line compilation (McClatchey et al., 1973) until the HITRAN04 database (Rothman et al., 2005). The line strengths of the $4\nu_1$, $3\nu_1+2\nu_2$, $2\nu_1+\nu_3$, $\nu_1+2\nu_2+\nu_3$, and $2\nu_3$ bands of N_2O were updated in the latest HITRAN04 database, based on the high-resolution experiment (Toth, 1999). The line strengths of HITRAN04 for the $4\nu_1$, $3\nu_1+2\nu_2$, $2\nu_1+\nu_3$, $\nu_1+2\nu_2+\nu_3$, and $2\nu_3$ bands of N_2O are largely different from those of HITRAN2K (Rothman et al., 2003). The Herman-Wallis factors were taken into account for these bands of HITRAN04 for the first time. In order to validate the HITRAN databases, we determined the line strengths, self-, N_2-, and O_2-broadened half-widths in the $4\nu_1$, $3\nu_1+2\nu_2$, $2\nu_1+\nu_3$, $\nu_1+2\nu_2+\nu_3$, and $2\nu_3$ bands of N_2O at room temperature. The squared vibrational transition dipole moments and the coefficients of the Herman-Wallis factors were also determined using the measured line strengths. The self-, N_2-, and O_2-broadened half-widths were compared with the experimental results previously reported in the literature.

2. EXPERIMENT AND ANALYSIS

All the spectra were measured using a high-resolution Fourier transform spectrometer (Bruker IFS 120HR) with a resolution of 0.01 cm^{-1}. A CaF_2 beam splitter was used in conjunction with an InSb detector and a tungsten lamp as the light source. Ten spectra were obtained with pure N_2O samples and mixtures of N_2O in N_2 and O_2. The gas sample was 99.99% pure N_2O purchased from Sumitomo Seika Chemicals Co., Ltd. Ultra-high-purity samples of N_2 and O_2 were used for the buffer gases. Nippon Sanso Co. supplied the N_2 samples and Sumitomo Seika Chemicals Co., Ltd. supplied the O_2 samples. We used the absorption

---

[1] fukabori@mri-jma.go.jp
[2] Present affiliation: National Institute for Environmental Studies, Tsukuba, Ibaraki 305-8506, Japan

cells with path lengths of 20.0, 67.1, and 100.0 cm. The room temperature was stabilized and the wall temperature of the cell (297.5K) was measured with a thermocouple. The temperature variation was kept within 1K during the measurements. Sample pressures were measured with an MKS Baratron pressure gauge with a 1000-torr head. The estimated uncertainty of the sample pressure readings was 0.75%. The signal-to-noise ratios of the spectra were better than 500.

A nonlinear least-squares spectrum fitting technique was employed to determine the line strengths, self-, N_2-, and O_2-broadened half-widths. The Voigt profile was assumed as the line shape. An example of the spectral fit to the R(29) to R(34) lines of the $2\nu_3$ band of N_2O is illustrated in Fig. 1. The experimental spectrum was reproduced well by the Voigt profile.

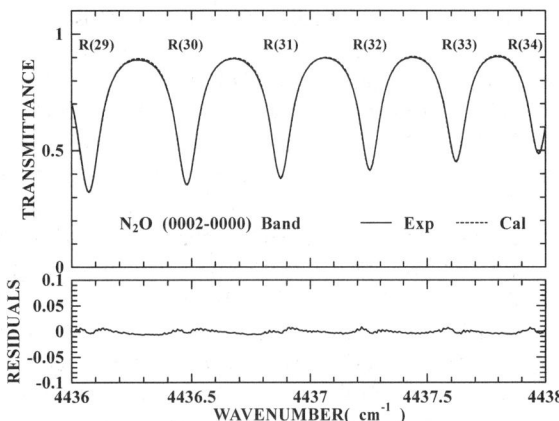

Fig. 1. Comparison of measured and calculated spectra of the R(29)-R(34) lines in the $2\nu_3$ band of N_2O. The solid and broken lines in the upper panel indicate the measured and calculated spectra. The gas pressure of N_2O was 380 torr and the path length was 20 cm. The lower panel presents the residuals between the measured and calculated spectra.

The individual line strengths S_i are expressed by the following formula:

$$S_i = \{8\pi^3 10^{-36}/[3hcg_V Q_V Q_R]\}\{\nu_i f \exp[-(E_V+E_R)hc/kT]\}$$
$$\times L_i |R|^2 [1-\exp(-hc\nu_i/kT)], \quad (1)$$

where $|R|^2$ is the square of the transition dipole moment, L_i is the Hönl-London factor, ν_i is the line position, f=0.9903 is the isotopic abundance for $^{14}N_2^{16}O$, and g_V is the degeneracy factor. Q_V and Q_R are the vibrational and rotational partition functions. E_V and E_R are the lower state vibrational and rotational energy levels. T is the temperature, h is the Planck constant, k is the Boltzmann constant, and c is the speed of light. The line strength is given in units of cm^{-1}/(molecule/cm^2) and the square of the transition dipole moment is in Debye^2. $|R|^2$ can be expressed as

$$|R|^2 = |R_V|^2 F(m), \quad (2)$$

where $|R_V|^2$ is the squared vibrational transition dipole moment and F(m) is the Herman-Wallis factor. The Herman-Wallis factor, which allows for vibration-rotation interaction on the line strengths, has been expressed in several forms for application in analyses of N_2O. The following expression employed in previous papers (Toth, 1993; Toth, 1999) was adopted for the R and P branches:

$$F(m) = [1+a_1 m+a_2 J'(J'+1)]^2, \quad (3)$$

where the running index m=J''+1 for the R-branch and m=-J'' for the P-branch. J' and J'' are the upper and lower state rotational quantum numbers, respectively.

3. RESULTS AND DISCUSSION

3.1 Line Strengths and Squared Transition Dipole Moments

Measured line strengths in the $4\nu_1$, $3\nu_1+2\nu_2$, $2\nu_1+\nu_3$, $\nu_1+2\nu_2+\nu_3$, and $2\nu_3$ bands of N_2O were different from the compiled values in HITRAN2K. On the basis of the high-resolution experiment (Toth, 1999), the line strengths compiled in the HITRAN04 database were revised for the above bands. Measured line strengths in the $4\nu_1$, $3\nu_1+2\nu_2$, $2\nu_1+\nu_3$, and $\nu_1+2\nu_2+\nu_3$ bands were in agreement with the values of HITRAN04 within the experimental errors. Our results in the R-branch lines of the $2\nu_3$ band were clearly 2% smaller than the values of HITRAN04.

Fig. 2 illustrates the variation of the squared transition dipole moments with the rotational quantum number for the $2\nu_3$ band of N_2O. The squared vibrational transition dipole moment in this study was close to the calculated value using the line strengths of HITRAN04, however the coefficients of the Herman-Wallis factor in HITRAN04 were slightly different from the experimental results. The squared transition dipole moments are independent of m in the HITRAN2K database. This implies that the Herman-Wallis factor was not taken into account for this band in HITRAN2K.

Almost similar results were obtained for the $4\nu_1$, $3\nu_1+2\nu_2$, $2\nu_1+\nu_3$, and $\nu_1+2\nu_2+\nu_3$ bands. Table I compares the squared vibrational transition dipole moments in this study together with the experimental results of Toth for the $4\nu_1$, $3\nu_1+2\nu_2$, $2\nu_1+\nu_3$, $\nu_1+2\nu_2+\nu_3$, and $2\nu_3$ bands of N_2O. Our values agreed with the high-resolution results of Toth within 4%.

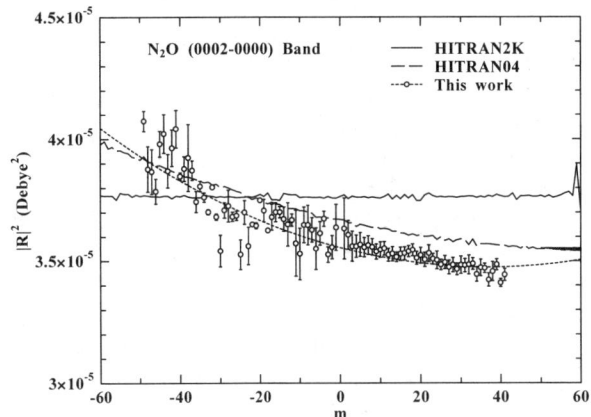

Fig. 2. Variation of the squared transition dipole moment with the rotational quantum number for the $2\nu_3$ band of N_2O. The solid and dash-dotted lines represent the HITRAN2K values and the HITRAN04 values. Open circles indicate the squared transition dipole moments measured in this study. The dotted line represents the fitted values to our data.

Table 1. Squared Vibrational Transition Dipole Moments for the $4\nu_1$, $3\nu_1+2\nu_2$, $2\nu_1+\nu_3$, $\nu_1+2\nu_2+\nu_3$, and $2\nu_3$ Bands of N_2O

Band Center (cm^{-1})	(v'$_1$v'$_2$l'$_2$v'$_3$)	|Rv(N$_2$O)|^2 Toth (1999) (x10^{-6})	|Rv(N$_2$O)|^2 This work (x10^{-6})
5105.65	4000	1.369	1.344±0.011
5026.34	3200	1.323	1.355±0.009
4730.828	2001	21.07	20.37±0.07
4630.164	1201	2.993	3.081±0.025
4417.379	0002	36.60	35.21±0.04

3.2 Coefficients of the Herman-Wallis Factor

The linear and quadratic coefficients of the Herman-Wallis factor were derived from the transition dipole moment using the least-squares method. Columns 3 and 4 in Table II list the linear and quadratic coefficients of the Herman-Wallis factors for the $4\nu_1$, $3\nu_1+\nu_2$, $2\nu_1+\nu_3$, $\nu_1+2\nu_2+\nu_3$, and $2\nu_3$ bands of N_2O. The absolute values and signs of quadratic coefficients (a_2) of the Herman-Wallis factor in this study were very close to the results obtained by Toth, whereas those of linear coefficients (a_1) of the $2\nu_1+\nu_3$ and $\nu_1+2\nu_2+\nu_3$ bands were slightly different from the results of Toth.

3.3 Self-Broadened Half-Widths

Figure 3 depicts the comparison of the self-broadened half-widths. The self-broadened half-widths reported by Toth (1971) and Margolis (1972) in the early 1970's are larger than the high-resolution experimental results of Lacome et al. (1984), Margottin-Maclou et al. (1985), and Toth (1993). The results of Margottin-Maclou et al., which are not shown in Fig. 2, were in good agreement with those of Lacome et al. The self-broadened half-widths based on experimental results from Toth (1993) are compiled in the HITRAN database. Although our values are 5–10% less than those of Lacome et al., Margottin-Maclou et al. and Toth (1993), our values are in agreement with the high-resolution experiments within the experimental errors.

Table 2. Coefficients of the Herman-Wallis Factors for the $4\nu_1$, $3\nu_1+2\nu_2$, $2\nu_1+\nu_3$, $\nu_1+2\nu_2+\nu_3$, and $2\nu_3$ Bands of N_2O

(v'$_1$v'$_2$l'$_2$v'$_3$)		a_1 (10^{-4})	a_2 (10^{-5})
4000	Toth (1999)	3.06	-3.04
	This work	5.94	-3.23
3200	Toth (1999)	3.00	2.21
	This work	7.20	2.20
2001	Toth (1999)	1.98	-0.91
	This work	-0.658	-0.609
1201	Toth (1999)	5.57	2.49
	This work	1.08	4.11
0002	Toth (1999)	-4.98	0.36
	This work	-5.15	0.827

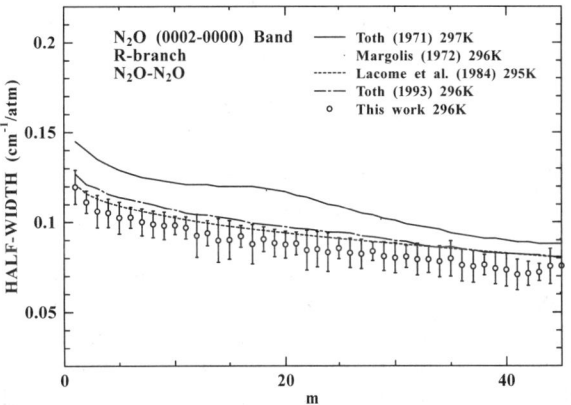

Fig. 3. Self-broadened half-widths of the R-branch lines in the $2\nu_3$ band of N_2O. Open circles represent the measured self-broadened half-widths of the R-branch lines of this study. The solid line indicates the results of Toth (1971) and the dotted line indicates the values of Margolis (1972). The broken line indicates the results of Lacome et al. (1984) and the dash-dotted line indicates the values of Toth (1993).

3.4 N_2- and O_2-Broadened Half-Widths

Although several experiments have been conducted to determine the N_2- and O_2-broadened half-widths in the infrared region, experimental results of the O_2-broadened half-widths are sparse. The air-broadened half-widths in the latest HITRAN database are based on the experimental results of the N_2- and O_2-broadened half-widths reported by Lacome et al., Toth (2000), and Nemtchinov et al. (2004). Fig. 4 depicts the N_2- and O_2-broadened half-widths of the R-branch lines in the $2\nu_3$ band of N_2O. The N_2-broadened half-widths obtained in this study agree well with the recent high-resolution experiments. The previous values (Toth, 1971; Margolis, 1972), which are not shown in Fig. 4, reported in the early 1970's were about 10% larger than the recent high-resolution results for the N_2-broadened half-widths. The O_2-broadened half-widths obtained in this study are in agreement well with the high-resolution experimental results within the experimental errors. The air-broadened half-widths derived from the N_2- and O_2-broadened half-widths in this study are in good agreement with the values of the HITRAN databases.

These results will contribute to the improvement of the accuracies of spectroscopic databases.

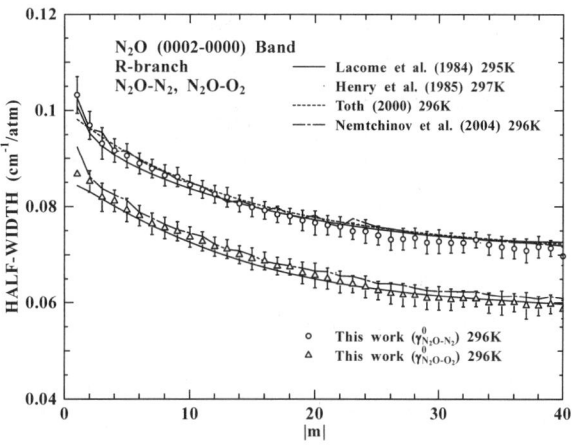

Fig. 4. N_2- and O_2-broadened half-widths of the R-branch lines in the $2\nu_3$ band of N_2O. Open circles and triangles represent the measured N_2- and O_2-broadened half-widths of the R-branch lines of this study. The solid line indicates the results of Lacome, et al. (1984) and the dotted line indicates the values of Henry (1985). The broken line indicates the results of Toth (2000) and the dash-dotted line indicates the values of Nemtchinov, et al. (2004).

4. REFERENCES

Daumont, L., J. Vander Auwera, J.-L. Teffo, V. I. Perevalov, and S. A. Tashkun, 2001: Line intensity measurements in $^{14}N_2^{16}O$ and their treatment using the effective dipole moment approach. *J. Mol. Spectrosc.* 208, 281-291.

Henry, A., M. Margottin-Maclou, and N. Lacome, 1985: N_2- and O_2-broadening parameters in the ν_3 band of $^{14}N_2^{16}O$. *J. Mol. Spectrosc.* 111, 291-300.

Lacome, N., A. Levy, and G. Guelachvili, 1984: Fourier transform measurement of self-, N_2-, and O_2- broadening of N_2O lines: temperature dependence of linewidths. *Appl. Opt.* 23, 425-435.

Margolis, J. S., 1972: Intensity and half width measurements of the (00^02-00^00) band of N_2O. *J.Quant. Spectrosc. Radiat. Transfer,* 12, 751-757.

Margottin-Maclou, M., P. Dahoo, A. Henry, and L. Henry, 1985: Self-broadening parameters in the ν_3 band of $^{14}N_2^{16}O$. *J. Mol. Spectrosc,* 111, 275-290.

McClatchey, R. A., W. S. Benedict, S. A. Clough, D. E. Burch, R. F. Calfee, K. Fox, L. S. Rothman, and J. S. Garing, 1973: *AFCRL Atmospheric Absorption Line Parameters Compilation,* AFCRL–TR–0096, AFCRL, Bedford, MA.

Nemtchinov, V., C. Sun, and P. Varanasi, 2004: *J. Quant. Spectrosc. Radiat. Transfer,* 83, 267-284.

Rothman, L. S., A. Barbe, D. Chris Benner, L. R. Brown, C. Camy-Peyret, M. R. Carleer, K. Chance, C. Clerbaux, V. Dana, V. M. Devi, A. Fayt, J.-M. Flaud, R. R. Gamache, A. Goldman, D. Jacquemart, K. W. Jucks, W. J. Lafferty, J.-Y. Mandin, S. T. Massie, V. Nemtchinov, D. A. Newnham, A. Perrin, C. P. Rinsland, J. Schroeder, K. M. Smith, M. A. H. Smith, K. Tang, R. A. Toth, J. Vander Auwera, P. Varanasi, and K. Yoshino, 2003: The HITRAN molecular spectroscopic database: edition of 2000 including updates through 2001. *J. Quant. Spectrosc. Radiat. Transfer,* 82, 5-44.

Rothman, L. S., D. Jacquemart, A. Barbe, D. Chris Benner, M. Birk, L. R. Brown, M. R. Carleer, C. Chackerian, Jr, K. Chance, V. Dana, V. M. Devi, J.-M. Flaud, R. R. Gamache, A. Goldman, J.-M. Hartmann, K. W. Jucks, A. G. Maki, J.-Y. Mandin, S. T. Massie, J. Orphal, A. Perrin, C. P. Rinsland, M. A. H. Smith, J. Tennyson, R. N. Tolchenov, R. A. Toth, J. Vander Auwera, P. Varanasi, and G. Wagner, 2005: The HITRAN 2004 molecular spectroscopic database. *J. Quant. Spectrosc. Radiat. Transfer* (in print).

Toth, R. A., 1971: Self-broadened and N_2 broadened linewidths of N_2O. *J. Mol. Spectrosc.* 40, 605-615.

Toth, R. A., 1993: Line strengths (900-3600 cm^{-1}), self-broadened linewidths, and frequency shifts (1800-2360 cm^{-1}) of N_2O. *Appl. Opt.,* 32, 7326-7365.

Toth, R. A., 1999: Line positions and strengths of N_2O between 3515 and 7800 cm^{-1}. *J. Mol. Spectrosc,* 197, 158-187.

Toth, R. A., 2000: N_2- and air-broadened linewidths and frequency-shifts of N_2O. *J. Quant. Spectrosc. Radiat. Transfer,* 66, 285-304.

RETRIEVAL OF AEROSOL OPTICAL PROPERTIES FROM DUAL-WAVELENGTH POLARIZATION LIDAR MEASUREMENTS

Tomoaki Nishizawa and Hajime Okamoto,
Center for Atmospheric and Oceanic Studies, Tohoku University, Sendai, 980-8578, Japan

Toshihiko Takemura,
Research Institute for applied mechanics, Kyushu University, Fukuoka, 816-8580, Japan

Nobuo Sugimoto, Ichiro Matsui, and Atsushi Shimizu
National Institute for Environmental Studies, Tsukuba, 305- 0052, Japan

ABSTRACT

We developed a forward algorithm that can distinguish three types of aerosols, i.e., water-soluble, sea-salt and dust, and that can retrieve the vertical profiles of extinction coefficient of each aerosol type from dual-wavelength polarization lidar. We applied the algorithm to the lidar data measured in the Pacific Ocean near Japan during two weeks in May 2001. The results show that most of water-soluble and sea-salt aerosols were concentrated below the altitude of 1 km. A few air-masses dominated by water-soluble and dust particles were sometimes found above the altitude of 1 km. We also compared the retrieved time-height distributions with those simulated by three dimensional aerosol transport model. The simulated distributions of each aerosol type are roughly consistent with those retrieved in this study.

1. INTRODUCTION

The ship-borne measurement was conducted with the research vessel MIRAI of JAMSTEC (Japanese Maritime Science and Technology Center) in the Pacific Ocean near Japan from May 14 to May 27, 2001. In the cruise, a dual wavelength lidar with polarization function of NIES (National Institute for Environmental Studies) and a 95-GHz cloud profiling radar of NICT (National Institute of Information and Communications Technology) were installed on the vessel. It is the first time that the ship-borne measurement with both lidar and cloud profiling radar was conducted. The analysis of the lidar and radar data will provide information about the temporal and spatial distribution of aerosols and clouds over the sea, and further will be helpful for the validation with products from such numerical models as aerosol transport and chemical models and cloud-resolving models.

We developed a forward algorithm to retrieve vertical profiles of extinction coefficient at the wavelength (λ) of 532 nm of water-soluble (σ_{WS}), sea-salt (σ_{SS}) and dust aerosols (σ_{DS}) from the dual-wavelength polarization lidar data. The lidar has three channels that are attenuated backscattering coefficients at $\lambda = 532$ nm ($\beta_{obs,532}$) and 1064 nm ($\beta_{obs,1064}$) and depolarization ratio at $\lambda = 532$ nm. The algorithm uses the three-channel data and estimates type of aerosols and σ_{WS}, σ_{SS} and σ_{DS}. And also we applied the algorithm to the lidar data measured in the MIRAI cruise.

2. ALGORITHM

One of advantages of the algorithm is that extinction-to-backscattering ratio (S), number concentration and optical thickness of aerosols can be obtained from σ_{WS}, σ_{SS} and σ_{DS}. The inversion algorithm developed by *Fernald* (1984) has been widely used to retrieve the vertical profile of aerosol extinction coefficient from lidar data. In their algorithm, S-value should be prescribed and assumed to be vertically invariable. Our algorithm does not need them. Another advantage is that the algorithm estimates the profiles of aerosol microphysics upward from the surface. This feature enables to obtain aerosol microphysics below cloud bottom. Thus, it is expected that this algorithm can provide knowledge related to aerosol-cloud interaction around cloud bottom layers.

In the algorithm, we assume the followings; the volume size distribution of aerosols is bimodal-shape of lognormal distribution. There are two models of aerosols, i.e., sea-salt model and dust model. The sea-salt model is consisted of two aerosol components, i.e., water-soluble with a mode radius in accumulation-mode and sea-salt with that in coarse-mode. The dust model is also consisted of two aerosol components, i.e., water-soluble with a mode radius in accumulation-mode and dust with that in coarse-mode. The microphysical and optical properties for water-soluble, sea-salt and dust aerosols are assumed by using the results of the other studies (*Hess et al.*, 1998; *Smirnov et al.*, 2002).

We retrieve the vertical profiles of σ_{WS}, σ_{SS} and σ_{DS} sequentially from the lowest layer to the highest layer, by repeating the following steps; Step1) We retrieve the optimal values of extinction coefficients of the two components for each model, i.e., sea-salt model and dust model, that can simulate the β_{obs} measured at the two wavelengths. Step 2) The aerosol model is determined by using aerosol depolarization ratio (δ_a). Step 3) The optical thickness of aerosols is estimated from the retrieved σ_{WS}, σ_{SS} and σ_{DS} to correct the attenuation. In the step 2, we can estimate the aerosol depolarization ratio by removing the polarization effect due to molecule scattering from the depolarization measurement with the lidar. Then, the retrieved backscattering coefficient of aerosols is needed, and hence the δ_a is estimated for each model. We adopt the sea-salt model when the δ_a-values computed for the sea-salt model and the dust model are smaller than 0.1. In contrary, we adopt the dust model when the δ_a-values computed for both of the models are greater than 0.1. There might be cases that the δ_a-values computed for the dust model and the sea-salt model do not match the conditions, due to the assumption needed in this algorithm and/or measurement uncertainty. Then, we do not specify the model and treat it as 'unknown' model. The values of σ_{WS} and σ_{DS} retrieved using the dust model are adopted as the estimation.

3. APPLICATION

First of all, we removed the lidar data contaminated by clouds by using the lidar and radar data. We set a threshold of 5.0×10^{-3} km^{-1}ster^{-1} in $\beta_{obs,1064}$ for lidar data and noise level plus 0.5 dB for the radar data empirically, and find the cloud layer where $\beta_{obs,1064}$-value is larger than the threshold or the value of radar signal is larger. In the retrieval, the data above cloud bottom layer are removed.

Fig. 1 shows the results of the application of the algorithm to the lidar data measured in the whole period of the Mirai cruise. The most of sea-salt aerosols exist in the planetary boundary layer, which is below the altitude of 1km. Most of water-soluble aerosols are also concentrated in the layer. The figure further shows that there are the layers of dust and water-soluble above 1 km. Fig. 2 shows the vertical profiles of σ_{WS}, σ_{SS}, σ_{DS} and σ_{all} averaged over the whole observation period, where σ_{all} is total extinction coefficient of aerosols at $\lambda = 532$ nm. The extinction coefficient of each aerosol component is almost in the range of $0.03 < \sigma_{WS} < 0.05$, $0.02 < \sigma_{SS} < 0.04$ and $0.005 < \sigma_{DS} < 0.015$ km^{-1}. As the sum, σ_{all} is in the range from 0.03 km^{-1} to 0.07 km^{-1}. The values of σ_{all} are almost consistent with those measured over sea (e.g., *Sugimoto et al.*, 2001).

We investigated the differences in the aerosol concentration between in clear sky and under clouds. The Fig. 3 provides the vertical profile of sea-salt aerosols under and without cloud layers. The figure indicates that the extinction coefficients of sea-salt aerosols under cloud layers are about twice in maximum larger than those without cloud layers. The similar feature was also found for water-soluble aerosols. The larger values of σ_{SS} under cloud layers might be due to the larger relative humidity under cloud layers than without cloud layers. This suggests the possible link how aerosol particles relate to the generation of clouds.

Further, we compared the time-height distributions of each aerosol component retrieved in this study with those simulated from three-dimensional aerosols transport-radiation model SPRINTARS (Spectral Radiation-Transport Model for Aerosols Species) (*Takemura et al.*, 2003). Fig. 4 shows the distribution of σ_{all} simulated by SPRINTARS for the whole period of the MIRAI cruise. It is found that the simulated distributions of σ_{all} are roughly similar to that retrieved in this study. The simulated distributions of each component also roughly matched those retrieved in this study. The simulation by the SPRINTARS indicates that the dust aerosols might be mainly transported from the Gobi desert, and further indicates that the water-soluble aerosols found above 1 km might be mainly transported from the seaboard of China. The comparison study further indicates that the σ_{WS} and σ_{DS} simulated by the SPRINTARS are overestimated and σ_{SS} is underestimated.

4. SUMMARY

We developed an algorithm that can determine the aerosol types and can estimate the vertical profiles of each aerosol component. The algorithm was applied to the lidar data measured in the MIRAI cruise. The major findings are as follows: (1) Most of water-soluble and sea-salt aerosols are concentrated in the planetary boundary layer below the altitude of 1 km. (2) A few aerosol-rich air-masses dominated by water-soluble and dust particles were sometimes found above the altitude of 1 km. (3) The extinction coefficients of water-soluble and sea-salt aerosols under cloud layers were larger than those under clear-sky condition. This suggests the possible link how aerosol particles relate to the generation of clouds. (4) The distributions of each aerosol component were roughly consistent with those simulated by the SPRINTARS.

PARTICLE RADIATIVE PROPERTIES

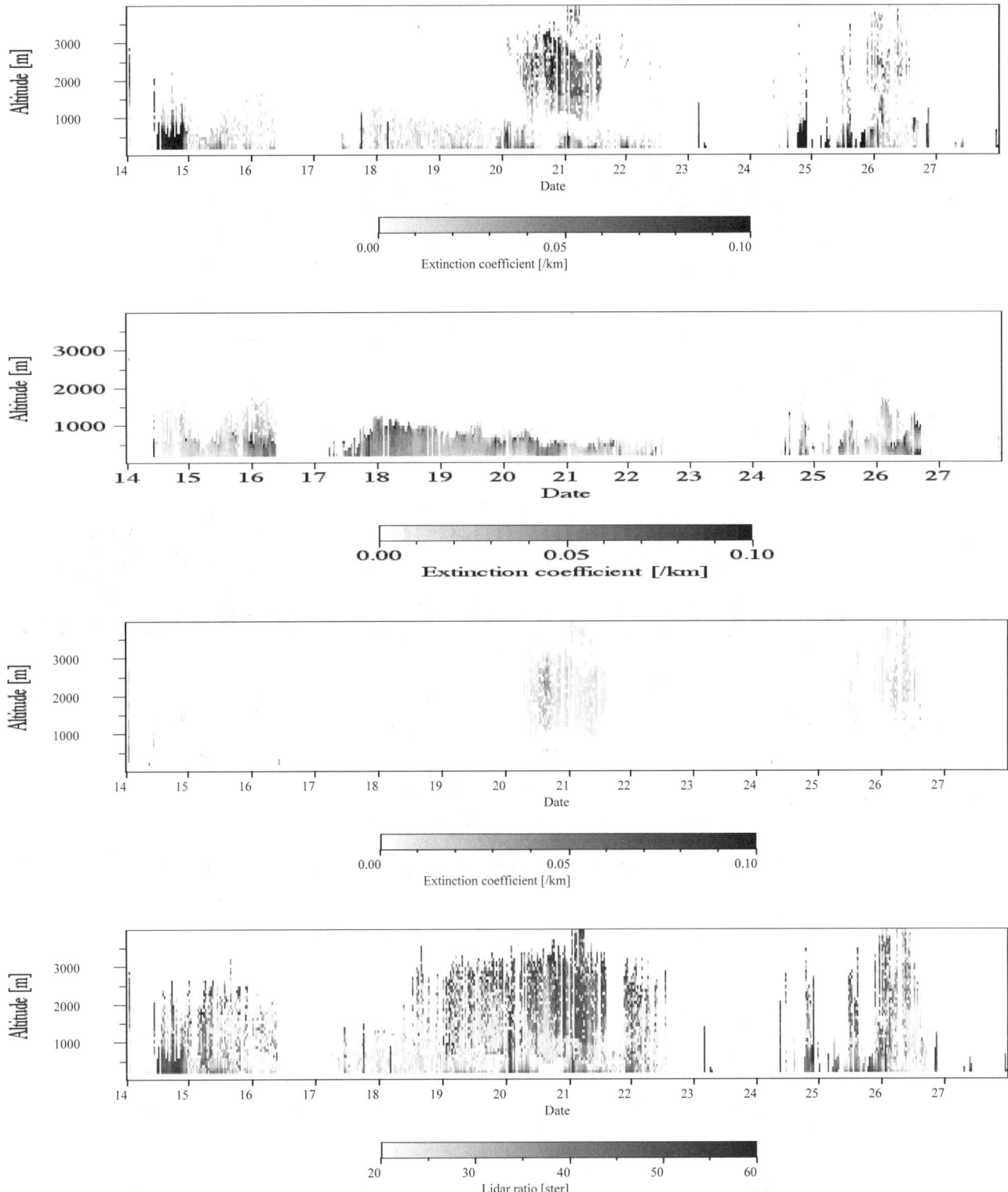

Fig. 1. The time-height cross-sections of σ_{WS}, σ_{SS}, σ_{DS} and S at $\lambda = 532$ nm retrieved from the lidar measurements in the whole observation period of the Mirai cruise.

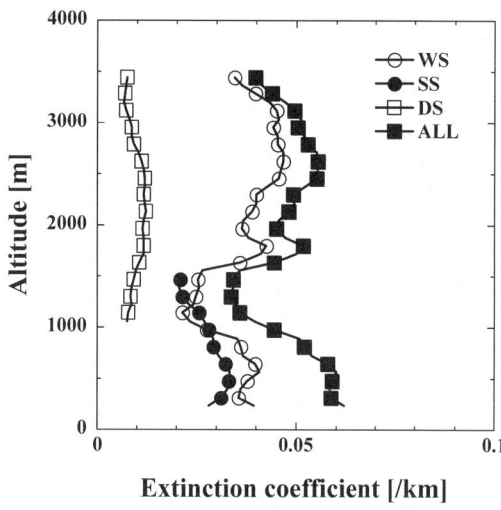
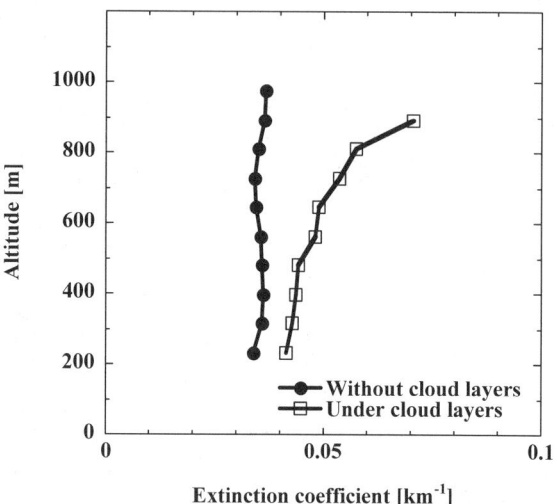

Fig. 2. The vertical profiles of σ_{WS} (WS), σ_{SS} (SS), σ_{DS} (DS) and σ_{all} (ALL) averaged over the whole observation period of the MIRAI cruise.

Fig. 3. The vertical profiles of σ_{SS} under and without cloud layers below 1 km.

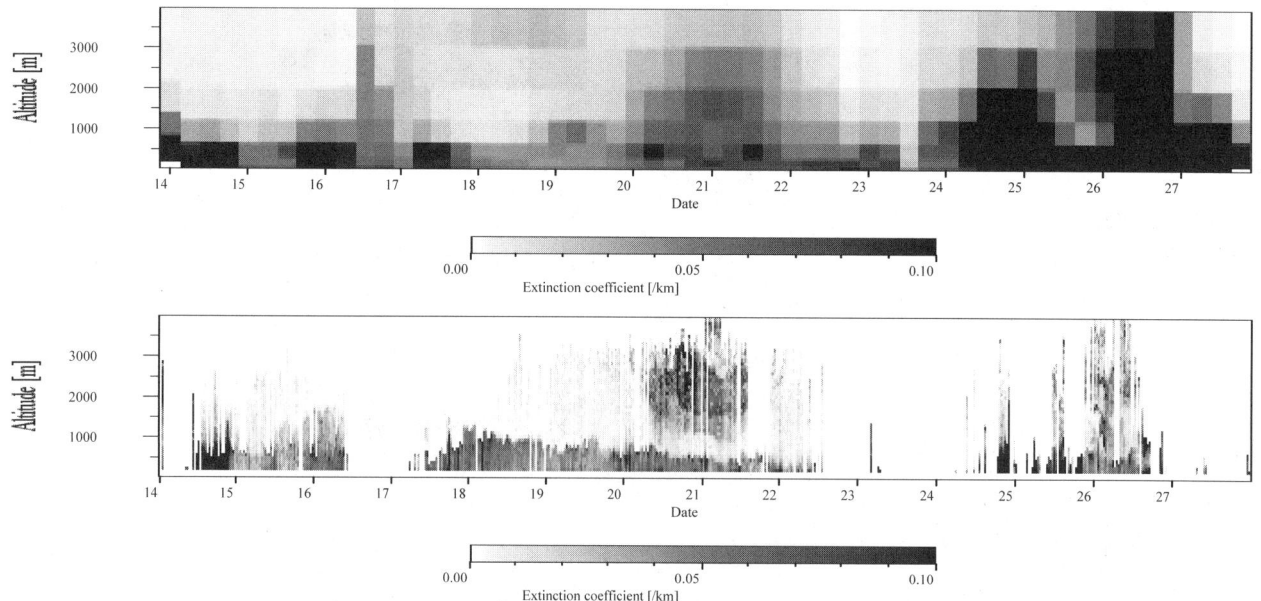

Fig. 4. The time-height cross-sections of total extinction coefficient of aerosols at $\lambda = 532$ nm (σ_{ALL}) simulated by the SPRINTARS (Upper) and retrieved in this study (Lower) in the whole period of the Mirai cruise.

5. REFERENCES

Fernald, F. G., 1984: Analysis of atmospheric lidar observation: some comments, *Appl. Opt.*, 23, 652, 1984

Hess, M., P. Koepke, and I. Schult, Optical properties of aerosols and clouds: The software package OPAC, *Bull. Amer. Meteor. Soc.*, 79, 831-844, 1998.

Smirnov, A., B. N. Holben, Y. J. Kaufman, O. Dubovik, T. F. Eck, I. Slutsker, C. Pietras, and R. N. Halthore, Optical properties of atmospheric aerosol in maritime environments, *J. Atmos. Sci.*, 59, 501-523, 2002.

Sugimoto, et al., 2001: Latitudinal distribution of aerosols and clouds in the western Pacific observed with a lidar on board the research vessel Mirai, *Geophys. Res. Lett.*, 28, 4187-4190.

Takemura, et al., 2000: Global three-dimensional simulation of aerosol optical thickness distribution of various origins, *J. Geophys. Res.*, 105(D14), 17,853-17,873.

THERMAL INFRARED RADIOMETRY, MICROPHYSICAL PROPERTIES AND GEOCHEMICAL NATURE OF MINERAL DUST

Michel Legrand
Laboratoire d'Optique Atmosphérique, Université de Lille-1
59655 Villeneuve d'Ascq cedex, France, legrand@loa.univ-lille1.fr

Ovidiu Pancrati
CARTEL, Université de Sherbrooke, 2500 Boulevard de l'Université, Local A6
J1K 2R1 Sherbrooke, Quebec, Canada, Ovidiu.Pancrati@USherbrooke.ca

ABSTRACT

In this paper we present results derived from sky measurements in the thermal infrared (TIR) with the 4-channel ground-based radiometer CLIMAT. The measurements were performed during the Sahelian campaign NIGER-98, in the dry season of 1998, in the dusty region of Niamey (Niger). Cloud contamination is identified with the help of satellite data. Effects due to water vapor and mineral dust can be separated by means of their different spectral signatures with the available radiometric channels. The resulting TIR radiance of dust corrected from the water vapor changes is closely correlated to the photometric dust optical depth. The spectral signature of mineral dust derived from the campaign is compared with calculated signatures. It is shown to be strongly dependent on dust composition. The measured signature is in good agreement with simulations based on independent determinations of dust properties, including particle size distribution derived from AERONET photometric network and Sahelian dust composition from the literature.

1. INTRODUCTION

The best known method of aerosol remote sensing uses ground-based photometers pointing at the Sun. The spectral range is then limited to the shortwave. The specific optical and microphysical properties of mineral dust result in the possibility to expand the useful range to the thermal infrared (TIR). The size distribution of dust comprises a significant fraction of large supermicronic particles and its mineral components are characterized by peaks of absorption around 10 μm. Few radiometric measurements in the TIR have been applied to mineral dust so far (Fouquart et al., 1987; Highwood et al., 2003). We describe results of remote sensing of mineral dust using the radiometer CLIMAT (Conveyable Low noise Infrared radiometer for Measurements of Atmosphere and Target surfaces) (Sicard et al., 1999; Legrand et al., 2000) measuring sky radiation from the ground surface in four channels in the TIR (Fig. 1).

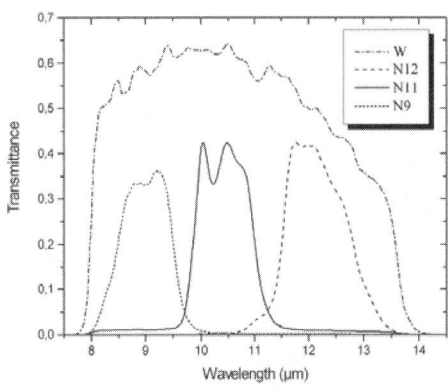

Fig. 1. Spectral transmittance of the channels of the radiometer CLIMAT.

2. THE EXPERIMENT AND THE DATA

2.1 The Campaign NIGER-98

The radiometer CLIMAT was operated during the Sahelian field campaign NIGER-98 carried out during the dry season, from 9 February to 28 May, 1998. The site of measurements was Banizoumbou (13°32'N, 2°39'E), 50 kilometers east of Niamey, in a region frequently affected by dust outbreaks in this season. The campaign was jointly realized by the LISA (Univ. of Paris-12), the LOA (Univ. of Lille-1), the University of Niamey and the IRD.

Dust optical depths at 440, 670, 870 and 1020 nm were derived from measurements realized at this site using a hand-held Sun-photometer CIMEL.

Ancillary data obtained independently are used in this study, including:

(i) aerosol optical depth from the AERONET station of Banizoumbou (PI D. Tanré) (limited to the period 9–17 February) as well as particle size distribution retrieved using methods described by Nakajima et al. (1996) and Dubovik et al. (2000);

(ii) atmospheric temperature and water vapor profiles from the daily 12-UTC balloon soundings of the station of Niamey Airport;

(iii) IDDI (Infrared Difference Dust Index) cloud mask archived at the LOA (Legrand et al., 2001).

2.2 The Measured TIR Radiances

The optical head of the radiometer CLIMAT was set on an horizontal platform and aimed by a robot successively at a target and at a blackbody intended to control the measurements in situ (Brogniez et al., 2003). The blackbody temperature measured by a platinum probe is compared to the brightness temperature (BT) derived from the radiometric measurements. The difference ΔT has a mean value of 0.1 K and a standard deviation σ of 0.1 K, for all channels. The ΔT values have been used for a slight correction of the radiometric calibration coefficients measured in laboratory. The measuring protocol consisted in hourly measurements in the four channels, aiming at the blackbody then at the sky zenith. The delivered data were converted into radiance by means of the calibration coefficients and into BT using the radiance-BT formulae obtained for each channel (Legrand et al., 2000).

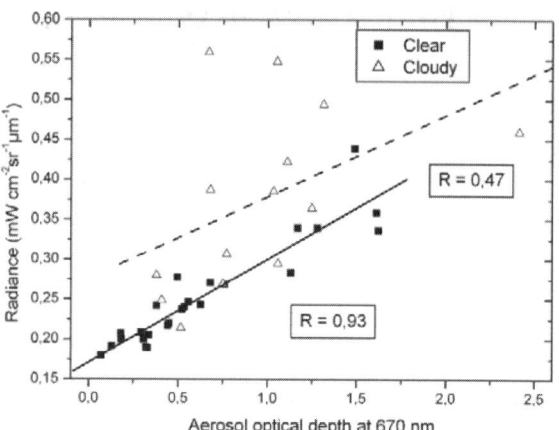

Fig. 2. Correlation between measured radiance (channel W) and dust optical depth at 670 nm, for clear and cloudy cases, on the criterion of cloud presence in IDDI images at 12 UTC.

3. DATA PROCESSING

3.1 Identification of Cloud Contamination

We have eliminated the cloud-contaminated measurements by using the cloud mask from an IDDI archive created using the Meteosat IR images, available for the period 1983–2000. The state (cloudy or clear) of the pixel of Banizoumbou is used to reject cloudy cases. Fig. 2 compares the coincident measurements of the radiometer (channel W) and of photometer, for clear and cloudy days. The comparison of the correlation coefficients confirms the validity of the method.

Fig. 3 displays time series from 13 February to 31 March, 1998, for cloud-free days at 12 UTC, of normalized CLIMAT radiance (channel W), aerosol optical depth at 670 nm and water vapor amount. The peaks of TIR radiance and of dust optical depth, in close coincidence, reveal the succession of dust outbreaks over the site of measurements.

3.2 Correction of Water Vapor Variations

The sky radiation in the 8–13-µm atmospheric window is due to emission of water vapor, aerosol and clouds. Cloud contamination being eliminated from the dataset, it is still necessary to correct the TIR radiance from changes of water vapor amount.

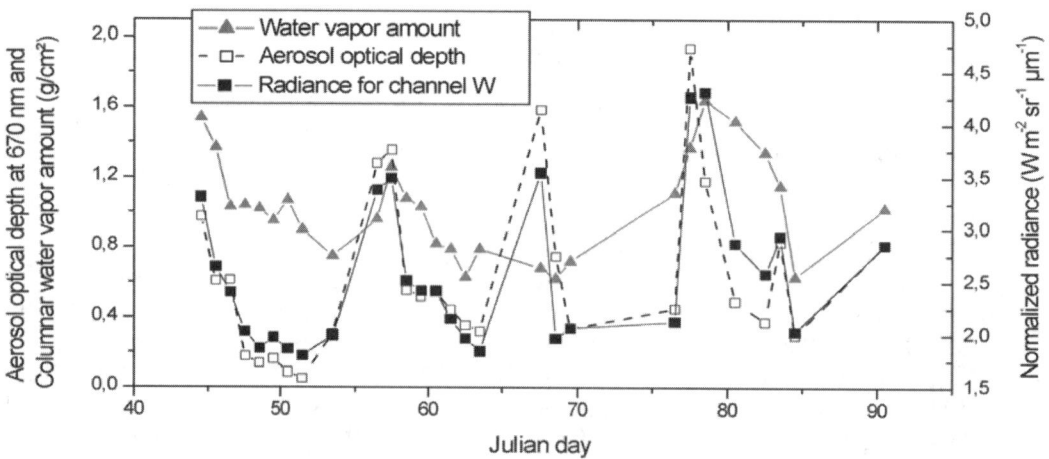

Fig. 3. For cloud-free days at 12 UTC: (i) radiometric measurements (TIR normalized radiance of channel W), (ii) photometric dust optical depth at 670 nm, (iii) water vapor amount from radiosoundings (Niamey Airport).

Calculating the TIR radiance from zenith at ground surface using the code of radiative transfer MODTRAN 4.1 (Anderson et al., 1995) with a dry tropical atmosphere and mineral dust following a model included in this code (Longtin et al., 1988); we can see that the radiance varies quasi linearly with the water vapor amount w and the dust optical depth δ_a. So the normalised TIR radiance R_n follows a linear relation

$$R_n = R_0 + S_a \delta_a + S_w w \qquad (1)$$

where S_a and S_w are sensitivity coefficients to aerosol and to water vapor, respectively, which can be fitted by applying a double linear regression between R_n and the measured values of δ_a and w. Thus, after the computation of S_w, it is possible to correct the effect of the varying water vapor amount on the radiance. The corrected normalized radiance at w_0 is then

$$R_n^{cor} = R_n + S_w(w - w_0) \qquad (2)$$

This correction is applied using the value w_0 = 1 g.cm^{-2}. The results, presented in Fig. 4 (channel W), show that the changes of water vapor amount affect the radiometric signal. The correction enhances the correlation of radiance upon aerosol optical depth from ρ = 0.91 to ρ = 0.95.

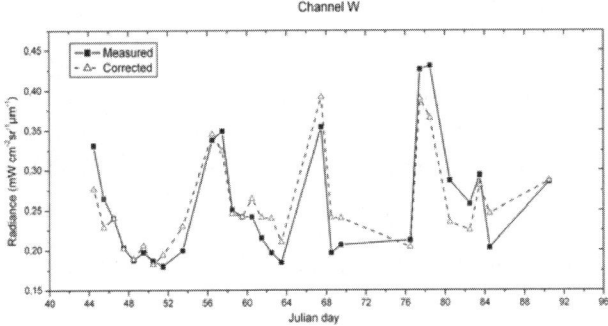

Fig. 4. Comparison between normalized radiances (for channel W), before and after correction of water vapor effect.

3.3 Spectral Signatures

Spectral signatures for water vapor and for mineral dust arise from the determination of these sensitivities in each channel of the radiometer. In Fig. 5, the measured sensitivities for these components are reported for the narrow channels N12, N11 and N9. The water vapor signature shows a minimum sensitivity in channel N11. Dust signature is characterized by a sensitivity in channel N9 larger by a ratio of 3 than in channel N12.

4. SPECTRAL SIGNATURE AND MINERAL DUST PROPERTIES

The spectral signature displayed in Fig. 5 results from the radiative characteristics of the mineral dust which depend themselves on particle size, shape, structure and composition and on vertical profile of dust. They can be compared with signatures calculated using the code of radiative transfer MODTRAN 4.1 in which different dust models can be used. Such simulations are carried using the measured meteorological profiles and normalizing the water vapor content to 1 g.cm^{-2}. Dust is assumed homogeneously distributed from the surface to 2000 m. Dust particles are assumed spherical and made up of pure minerals (quartz and clays) externally mixed in varying ratios.

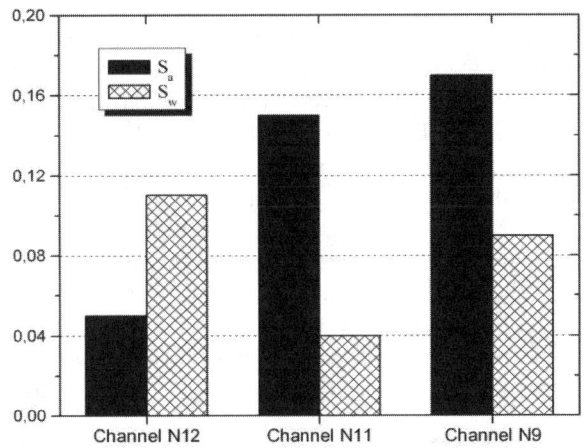

Fig. 5. Measured spectral signatures of water vapor (in mW.cm^{-2}.sr^{-1}.µm^{-1}) and mineral dust (in W.kg^{-1}.sr^{-1}.µm^{-1}) in the narrow channels of CLIMAT.

Simulations using a dust model with a fixed mineral composition, varying only the particle size distribution, show sensitivities increasing with the fraction of supermicronic particles. The sensitivity increments are almost the same for each channel so that this effect could be described as well using a single channel.

Simulations using a dust model with a fixed particle size distribution, varying only the mineral composition, show significant changes of signature shape, according to the absorption spectra of the mineral species. Fig. 6 describes the spectra of quartz, kaolinite and illite: the prevailing minerals in Sahelian dust (Caquineau et al. 2002). The peaks at 8.7 µm for quartz and at 9.3 µm for kaolinite and illite are consistent with the signature of Fig. 5.

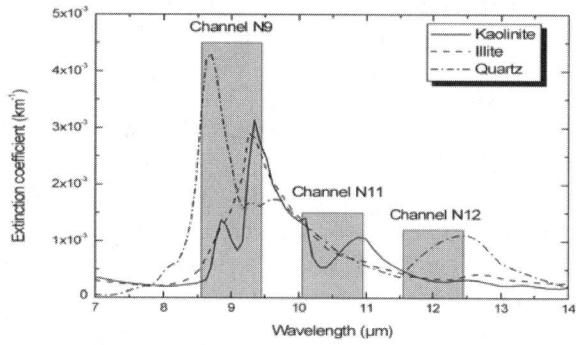

Fig. 6. Spectral variations of the extinction coefficient (with a particle per unit air volume) for quartz, illite and kaolinite, along with the location of the narrow channels of the radiometer CLIMAT.

In order to obtain simulations consistent with the signature of Fig. 5, we use the mean particle size distribution for February-March at Banizoumbou, derived from AERONET database obtained from the code SKYRAD.pack (Nakajima et al. 1996). We use the mineral composition reported by Caquineau (1997), for dust collected at Sal island, originating from the Sahel, south of 20°N, with weight fractions of 71.6% kaolinite, 9.1% illite and 19.3% quartz. Fig. 7 shows the agreement, between the simulated signature (normalized to channel W) and the experimental one. The best agreement is obtained with experiment using 85% kaolinite, 10% illite and 5% quartz.

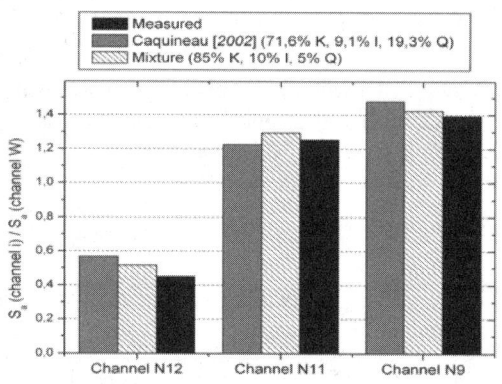

Fig. 7. Simulated spectral signatures of Sahelian mineral dust compared to the CLIMAT-measured signature.

5. REFERENCES

Anderson, G. P., and CoAuthors: FASCODE /MODTRAN/ LOWTRAN: Past/Present/Future, *18th Annual review conference on atmospheric transmission models,* 6–8 June, 1995.

Brogniez, G., C. Piétras, M. Legrand, P. Dubuisson, and M. Haeffelin: A high-accuracy multiwavelength radiometer for in situ measurements in the thermal infrared. Part 2: qualification in field experiments, *J. Atmos. Oceanic Technol.,* 20, 1023-1033, 2003.

Caquineau, S.: Les sources des aérosols sahariens transportés au-dessus de l'Atlantique tropical nord: localisation et caractéristiques minéralogiques, *Thesis,* Univ. Paris-7, 181 pp., 1997.

Caquineau, S., A. Gaudichet, L. Gomes, and M. Legrand: Mineralogy of Saharan dust over northwestern tropical Atlantic ocean in relation with sources regions, *J. Geophys. Res.,* 107, 10.1029/2000JD000247, 2002.

Dubovik, O., A. Smirnov, B. N. Holben, M. D. King, Y. F. Kaufman, T. F. Eck, and I. Slutsker: Accuracy assessment of aerosol optical properties retrieval from AERONET sun and sky radiance measurements, *J. Geophys. Res.,* 105, 9791-9806, 2000.

Fouquart, Y., B. Bonnel, G. Brogniez, J. C. Buriez, L. Smith, J.-J. Morcrette, and A. Cerf: Observations of Saharan aerosols: results of ECLATS field experiment. Part 2: broadband radiative characteristics of the aerosols and vertical radiative flux divergence, *J. Climate Appl. Meteor.,* 26, 38-52, 1987.

Highwood, E. J., J. M. Haywood, M. D. Silverstone, S. M. Newman, and J. P. Taylor: Radiative properties and direct effect of Saharan dust measured by the C-130 aircraft during Saharan Dust Experiment (SHADE). 2: terrestrial spectrum, *J. Geophys. Res.,* 108, 10.1029/ 2002JD002552, 2003.

Legrand, M., C. Piétras, G. Brogniez, M. Haeffelin, N. K. Abuhassan, and M. Sicard: A high-accuracy multiwavelength radiometer for in situ measurements in the thermal infrared. Part 1: Characterization of the instrument, *J. Atmos. Oceanic Technol.,* 17, 1203-1214, 2000.

Legrand, M., A. Plana-Fattori, and C. N'doumé: Satellite detection of dust using the IR imagery of Meteosat 1. Infrared difference dust index, *J. Geophys. Res.,* 106, 18,251-18,274, 2001.

Longtin, D. R., E. P. Shettle, J. R. Hummel, and J. D. Pryce: A wind dependent desert aerosol model: radiative properties, *Scientific report No 6,* Hanscom Air Force Base, Massachusetts, USA, 1988.

Nakajima, T., G. Tonna, R. Rao, P. Boi, Y. Kaufman, and B. Holben: Use of sky brightness measurements from ground for remote sensing of particulate polydispersions, *Appl.Opt.,* 35, 2672-2686, 1996.

Sicard, M., P. R. Spyak, G. Brogniez, M. Legrand, N. K. Abuhassan, C. Pietras, and J.-P. Buis: Thermal-infrared field radiometer for vicarious cross-calibration: characterization and comparisons with other field instruments, *Opt. Eng.,* 38, 345-356, 1999.

AEROSOLS AND MOLECULES WITHIN CLOUDY ATMOSPHERE

Irina N. Melnikova
Head of Laboratory, Research Centre for Ecological Safety, Russian Academy of Sciences
Korpusnaya Str., 18, St. Petersburg, 197110, Russia, Email: Irina.Melnikova@pobox.spbu.ru

ABSTRACT

Optical thickness and single scattering albedo retrieved from the measurements of solar radiation demonstrate disagreement with theoretically calculated values according to the scattering theory for a cloudy atmosphere. Namely, the optical thickness shows an evident spectral dependence and the single scattering albedo points to the true absorption exceeding model values. The influence of the multiple scattering of solar radiation together with the break of the axiomatical correlation between characteristic dimensions within cloud media is proposed as an explanation of the disagreement.

1. INTRODUCTION

The results of the single scattering albedo retrieval from the radiation observations in the cloudy atmosphere revealed a strong absorption within clouds even in the short wavelength region outside of atmospheric gaseous absorption bands (Melnikova, 1989, 1992; Melnikova and Mikhailov, 1994; Kondratyev et al., 1998). Namely, a single scattering co-albedo is: $1-\omega_0 \sim 0.0005$ for clean cloud; $1-\omega_0 \sim 0.005$—for intermediate case and even $1-\omega_0 \sim 0.05$ for polluted clouds with sand, dust or soot. Similar values of the single scattering co-albedo were obtained from the satellite radiative observation by Melnikova and Nakajima (2000) and from the ground observations (Melnikova et al. 2000). All retrieved values of the single scattering co-albedo are much greater than it is expected from scattering Mie theory for the elementary volume of cloud.

Optical thickness τ and scattering coefficient α obtained from the airborne radiative observation indicate a distinct decreasing with wavelength. The optical thickness retrieved from the satellite measurements by POLDER gives the same result for seven satellite images and for all considered pixels (about 3000). Values τ at wavelengths longer than 0.8 μm are twice less than at wavelength 0.4 μm. Asano (1994) has found the similar feature from the airborne observation of radiation in cloudy atmosphere: the optical thickness of stratus clouds appeared twice higher in visual range than in near-IR range. The interpretation of the UV radiation observations in the cloudy sky by Mayer et al. (1998) also demonstrates an unexpected strong extinction: the cloud optical thickness in the UV region has been retrieved equal to several hundreds. However, it is to be pointed out, that the calculation for the cloud droplets using scattering Mie theory does not show any spectral dependence of the scattering coefficient within cloud. The additive account of the molecular scattering gives no significant increasing in the short-wavelength region (less than 0.001%).

The contradiction may be provoked by the break of the correspondence of the elementary volume dimensions for different components of cloud media (droplets, molecules and aerosol particles) with the axiomatic conditions accepted in Mie theory. Special attention is to be directed toward correctly accounting for the multiple scattering by molecules and aerosols within the elementary volume which dimensions are defined comparing with droplets.

2. LINEAR CHARACTERISTICS

Remind the axiomatics concerned dimensions of the linear characteristics of the turbid media following Shifrin (1951). The wavelength is a real size defining relations between dimensions of the problem. Let it equal $\lambda=0.5$ μm. Introduce the following dimensions: distances between the molecules within the density fluctuation (aerosol particles or droplets) satisfy the condition $d<<\lambda$; size of the density fluctuation (aerosol particles or droplets) $a \geq \lambda$; size of the macro-volume of media (in our case—cloud size) $Z>>\lambda$; distances between fluctuations of densities (also between aerosol and droplet particles) $l>>\lambda$. According to Shifrin (1951) four dimensionless parameters are defined as following:

$x_1 = d/\lambda$ – describes the particle (density fluctuations, aerosol and droplet particles) structure;

$x_2 = a/\lambda$ – describes the scattering properties of the particle;

$x_3 = l/\lambda$ – describes the turbid media structure;

$x_4 = Z/\lambda$ – describes the scattering properties of the whole media volume.

The following inequalities are evident for the media consisting of one kind of particles:

$$x_1 < x_2 < x_3 < x_4 \quad \text{and also} \quad x_1 << 1; x_4 >> 1. \quad (1)$$

Density fluctuations, aerosol and droplet particles could be found in cloudy media. Their sizes are: 0.01 μm, 0.1–1.0 μm and 10–100 μm correspondingly, that is to say $x_2 \sim 0.02$, 0.2–2.0 and 20–200 correspondingly.

The distances between the particles could be estimated as follows: 0.05–0.5μm for density fluctuations, 300 μm for aerosol particles and 1000 μm for droplets that yields $x_3 \sim 0.1$–1; 600 and 2000 correspondingly.

Thus, it is clear, that inequalities (1) are not valid within cloud media if $x_3\sim0.01$ is considered for density fluctuations and aerosols and $x_2\sim20-200$ for droplets. As it mentioned by Shifrin (1951) in case $a \sim l$ the separation of the problem to the light scattering by one particle and the following solving the problem of multiple scattering is impossible. In clouds it is necessary to solve the problem of multiple scattering by droplets, molecules and aerosols simultaneously.

Following Van de Hulst (1957) the initial condition of the scattering theory is an approximation of the independent particles, i.e., a distance between the scatters should be less than $3a$, for the electromagnetic waves, which scattered by different particles, do not interfere. Another condition is approximation of a single scattering for the same initial light beam to illuminate all the scatters. So, the size of the elementary volume should satisfy a single occurrence of interaction with the electromagnetic wave. It is clear that the elementary volume is different for particles of different size: for cloud droplets ($a\sim10$ μm) it is about 1000 μm^3, for density fluctuation of gas molecules ($a\sim10^{-2}$ μm) – 10^{-6} μm^3, and for aerosol particles ($a\sim0,1-1,0$ μm) – $10^{-3}-1$ μm^3. Thus, the elementary volume sizes differ by 6–10 orders of the magnitude for different particles composing the cloud media.

3. EMPIRICAL FORMULAS

Strictly speaking, the equation of the radiative transfer for the complex multi-component medium is to be inferred from Maxwell equations accounting all components of cloudy atmosphere. However, we do not aim here to consider the mathematical aspect of the problem, thus we propose the empirical approach, presented in previous studies (Melnikova 1989, 1992; Melnikova and Vasilyev, 2004).

To account for the mutual influence of the scattering and absorption by different components, the empirical relations for correction of Mie values of scattering and absorption coefficients are proposed by Melnikova (1989, 1992):

$$\alpha = (\alpha'_M + \alpha'_A)\tau_D^p \omega_0^q + \alpha_D$$
$$\kappa = (\kappa'_M + \kappa'_A)\tau_D^p \omega_0^q \qquad (2)$$

where ω_0 is the single scattering albedo, τ_D and α_D are the optical thickness and volume scattering coefficient caused only by droplets scattering (values at wavelengths $\lambda>0.8$ μm), α'_M, α'_A, κ'_M, κ'_A are the scattering and absorption coefficients of molecules and aerosol particles in the clear sky (α'_M is a coefficient of Rayleigh scattering at corresponding wavelength and altitude of the atmosphere),

they coincide with model magnitudes; p and q are empiric coefficients equal to: $p=2$ and $q=\tau_D^2$, as per the estimations in previous studies (Melnikova, 1989, 1992; Melnikova and Vasilyev, 2004). The item $\kappa_M'\tau_D^p\omega_0^q$ in right hand of Eqs. (2) differs from zero only within the molecular absorption bands. Remind that the problem is considered only for $\tau>>1$.

These magnitudes are due to the fact that the mean number of scattering events in the cloud of optical thickness τ is proportional to τ^2 (Minin, 1981; Yanovitskij, 1997).

The idea of including value τ characterizing dimensions of the whole media (cloud) in Eq. (2) seems surprising. Nevertheless, according to Shifrin (1951) the size of the media in whole should be accounted in the case of breaking the axiomatic relation between dimensions of the elementary volume and distances between molecules (fluctuations of density).

Eq. (2) could be used for the transformation of the values obtained from measurements to model values. It is more convenient in the following form:

$$\alpha_R + \alpha_A = \frac{\widetilde{\alpha} - \widetilde{\alpha}_D}{\tau_D^2 \widetilde{\omega}_0^{\tau_D^2}}$$
$$k_M + k_A = \frac{\widetilde{k}}{\tau_D^2 \widetilde{\omega}_0^{\tau_D^2}} \qquad (3)$$

Fig. 1 illustrates the result of applying Eqs. (3) to values of scattering and absorption coefficients retrieved from spectral radiative observations in (Melnikova, 1989, 1992; Melnikova and Mikhailov, 1994). Digits marking curves in Fig. 1 correspond to different radiative experiments, which are described in details by Melnikova and Vasilyev (2004).

Difference [$\alpha(\lambda)-\alpha(0.8)$] converted with Eqs. (3) is evidently close to the spectral dependence of Rayleigh scattering coefficient for the clear sky at the corresponding altitude. Hence, it should be emphasized that the scattering coefficient and optical thickness of cloud layer are characterized by the transparent spectral dependence.

It is seen from Fig. 1b that the magnitudes of the volume absorption coefficient practically coincide with the ones customary calculated with Mie theory for cloud droplets. The molecular absorption bands become sharper and more recognizable. The values of the single scattering albedo corresponding to the absorption coefficients presented in Fig. 1b are about 0.99998 that is typical of the magnitudes for the cloud layer simulation using scattering theory (Grassl, 1975).

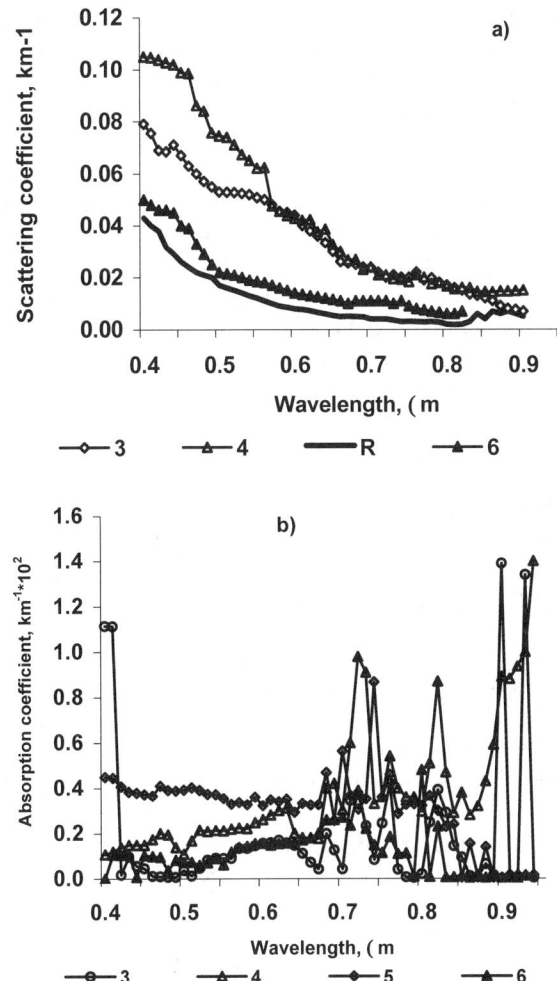

Fig. 1. Volume coefficients (a – scattering and b – absorption), transformed using Eq. (3). The curve marked with letter *R* characterizes the molecular scattering at altitude 1 km.

4. MULTIPLE SCATTERING OF RADIATION AS A REASON OF CLOUD ANOMALOUS ABSORPTION

The presented consideration concerns the *external mixture* i.e. the case, when aerosol particles are between the cloud droplets. When aerosol particles are within the droplets (the *internal mixture*) the aerosol absorption is correctly accounted in the calculation with the formulas for one-component medium. Basing on the obtained results it could be concluded that the *anomalous absorption* by clouds points to the external mixture of the atmospheric aerosols and cloud droplets because in the opposite case the radiation absorption by clouds coincides with the theoretical values.

The aerosols consisting from non-hygroscopic particles such as sand, soot etc. could exist in cloud between droplets with a higher probability than the hygroscopic aerosols (salt, sulphates); hence, they increase the shortwave absorption of radiation by cloud. Hygroscopic particles, being the nuclei of condensation increase the droplet number, which in turn increases the cloud optical thickness and causes the cloud cooling.

Let the spectral range be 0.4–1.0 μm and assume the mean value of the aerosol volume absorption coefficient equal to 0.08 km^{-1}, the volume scattering coefficient equal to 30 km^{-1}, the geometrical thickness equal to $\Delta z = 1$ km. Thus on the basis of Eqs (2), aerosol absorption is estimated to increase up to 15%, in comparison with the layer of clear atmosphere.

The molecule absorption in ozone Chappuis band increases up to 6–10% and in the oxygen band 0.76 μm up to 10% that coincides with the results of Dianov-Klokov (1973). This effect turns out stronger for thicker clouds, and it quantitatively explains the anomalous absorption by clouds.

Fig. 2 demonstrates the spectral dependence of the radiative flux divergence for clear (a) and cloudy (b) atmosphere obtained from the airborne spectral observations (Melnikova and Vasilyev, 2004). Dates, sites of observation and type of the ground surface are indicated in this figure. The results evidently expose a significant radiation absorption within clouds, which exceeds the radiation absorption of dusty atmosphere in Kara-Kum Desert after sand storm.

Some observational data corresponds to the conservative scattering in clouds at the separate wavelengths, but it is a rare case.

Indeed, the high content of the carbonaceous and mineral compounds of the atmospheric aerosols and their significant yield to the radiative regime of the atmosphere were obtained in the experimental studies (Boers et al. 1996; Bott et al. 1996).

The non-hygroscopic particles could be injected to the atmosphere as a result of the industrial escapes, sand storms, volcanic eruptions, and fires. The escapes of aerosol flue are extending up to 3000 km and they are keeping their radiation activity in the optical ranges (Mazin and Khrgian, 1989), thus these sources seem enough for the cloud anomalous absorption displaying in the global scale.

In conclusion, the careful accounting for the optical properties of all atmospheric components is necessary for constructing an adequate optical model (Vasilyev and Ivlev, 2002) of the cloudy atmosphere.

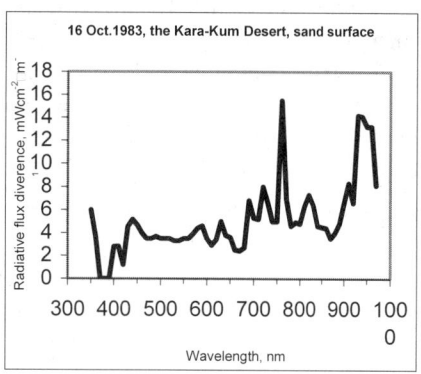
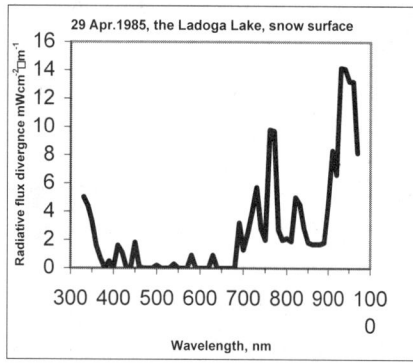
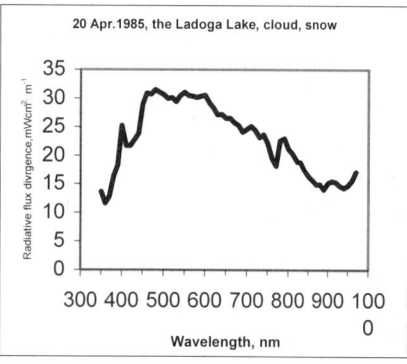

Fig. 2. Spectral values of the radiative flux divergence according to airborne observations

5. REFERENCES

Asano S., Cloud and radiation studies in Japan. Cloud radiation interactions and their parameterization in climate models. *WCRP-86* (WMO/TD No. 648). Geneva. WMO, 72-73, 1994.

Boers R, J. B. Jensen, P. B. Krummel, and H. Gerber, Microphysical and short-wave radiative structure of wintertime stratocumulus clouds over the Southern Ocean. *Q. J. R. Meteorol. Soc.*, 122, 1307–1339, 1996.

Bott A., T. Trautmann, and W. Zdunkowski, A numerical model of the cloud-topped planetary boundary layer: Radiation, turbulence and spectral microphysics in marine stratus. *Q. J. R. Meteorol. Soc.*, 122, 635–667, 1996.

Dianov-Klokov B. G., E. I. Grechko, and G. P. Malkov, Airborne measurements of the effective photon free pass from radiation reflected and transmitted by clouds in the oxygen band 0.76 μm Izv. *Acad. Sci. USSR. Atmosphere and Ocean Physics*, 9, 524–537, 1973 (in Russian).

Grassl H., Albedo reduction and radiative heating of clouds by absorption aerosol particles. *Beitr. Phys. Atmos.*, 48, 199–209, 1975.

Hulst HC Van de, *Light scattering by a small particles.* John Willey, New York, 1957.

Kondratyev, K., V. Binenko, and I. Melnikova, Absorption of solar radiation by clouds and aerosols in the visible wavelength region. *Meteorology and Atmospheric Physics*, 65, 1-10, 1998.

Mayer B., A. Kylling, S. Madronich, and G. Seckmeyer, Enchanced absorption of UV radiation due to multiple scatering in clouds: Experimental evidence and theoretical explanation. *J. Geophys. Res.*, 103 (D23), 31241-31254, 1998.

Mazin I. P., and A. Kh. Khrgian (eds.), *Clouds and cloudy atmosphere.* Reference book. Gidrometeoizdat, Leningrad, 1989 (in Russian).

Melnikova I. N., Light absorption in cloud layers. In: *Atmospheric Physics Problems.* St. Petersburg State University Press, 20: pp. 18–25, 1989, (in Russian).

Melnikova I., Spectral optical parameters of cloud layers. Application to experimental data. Part II. *Atmospheric Optics*, 5, 178-185 1992 (in Russian).

Melnikova I., and V. Mikhailov, Spectral scattering and absorption coefficients in stratus derived from aircraft measurements. *J. Atmos. Sci.* 51, 925-931, 1994.

Melnikova, I., P. Domnin, V. Mikhailov, and V. Radionov, Optical characteristics of clouds derived from measurements of reflected or transmitted solar radiation. *J. Atmos. Sci.*, 57, 2145-2143, 2000.

Melnikova I., and A. Vasilyev, *Short-wave solar radiation in the Earth atmosphere. Calculation Observation Interpretation.* Springer-Verlag GmbH&Co.KG, Heidelberg, 2004, 310 pp.

Melnikova I. N., and T. Nakajima, Single scattering albedo and optical thickness of stratus clouds obtained from "POLDER" measurements of reflected radiation. Earth Observations and Remote Sensing. 1–16, 2000. (Bilingual).

Minin I. N., Leningrad School of the radiative transfer theory. *Astrophysics. Academy of Sciences of Armenian Republic*, 17, 585–618 1981 (in Russian).

Shifrin K. S., *Light scattering in turbid media.* Moscow. 1951 288 p. (in Russian).

Vasilyev A. V., and L. S. Ivlev, On optical properties of polluted clouds. *Atmosphere and Ocean Optics*, 15, 157–159, 2002.

Yanovitskij E. G., *Light scattering in inhomogeneous atmospheres.* Springer-Verlag, New-York, 1997.

SCALING PROPERTIES OF AEROSOL OPTICAL THICKNESS RETRIEVED FROM A LONG-TERM SUN-PHOTOMETER DATASET

M. Alexandrov[1,2], A. Marshak[3], B. Cairns[1,2], A. Lacis[2], and B. Carlson[2]

[1]Columbia University, 2880 Broadway, New York, NY 10025, USA
[2]NASA Goddard Institute for Space Studies, 2880 Broadway, New York, NY 10025, USA
[3]NASA Goddard Space Flight Center, Code 613.2, Greenbelt, MD 20771, USA

ABSTRACT

The statistical scale-by-scale analysis was applied to the aerosol optical thickness (AOT) retrieved from the Multi-Filter Rotating Shadowband Radiometer (MFRSR) measurements made during 1993–2003. The MFRSR data were collected in 1993–2003 from the instrument operated by DOE Atmospheric Radiation Measurement program and located at the Central Facility of the Southern Great Plains (SGP) site in Oklahoma. These data have 20 sec time resolution. The scaling exponents obtained by computing second-order structure functions appear to exhibit a well-established seasonal cycle with maximum in winter-spring. In addition to this, the small-scale (corresponding to lags shorter than 40–50 min) exponents have a decreasing trend in 1993–1994, which we attribute to dissipation of stratospheric aerosols induced by Mt. Pinatubo eruption in 1991.

1. INTRODUCTION

In our earlier paper (Alexandrov et al., 2004a) the two-point statistical analysis, for the first time, has been applied to the aerosol optical thickness (AOT) retrieved from MFRSR network data. The data used in that study were collected in September 2000 from the dense local network operated by DOE Atmospheric Radiation Measurement (ARM) program and located in Oklahoma and Kansas. These data have 20 sec temporal resolution. The instrument sites form an irregular grid with the mean distance between neighboring sites about 80 km. We found that temporal variability of AOT can be separated into two well-established scale-invariant regimes: (1) microscale (0.5–15 km) where fluctuations are governed by 3D turbulence; and (2) intermediate scale (15–100 km) characterized by a transition towards large-scale 2D turbulence. The spatial scaling of AOT was determined by comparison of retrievals between different instrument sites (distance range 30–400 km). Analyzing scaling properties of AOT time series we discovered a good correlation between AOT spectral exponents (at both local and transition scales) and the aerosol scaling heights as well as between spectral exponents and concavity/convexity of the site topography. We investigate how simultaneous determination of AOT scaling in space and time can provide means to examine validity of Taylor's frozen turbulence hypothesis.

For the study presented in this paper we selected a longest single-instrument MFRSR dataset from the SGP local network. These data have been collected by the instrument C1 at the SGP Central Facility since 1993 (with some gaps). As we will show below this dataset allows us to observe seasonal variability in AOT scaling exponents, as well as the effects of stratospheric volcanic aerosol (caused by Mt. Pinatubo eruption) on small-scale variability of AOT.

The MFRSR makes precise simultaneous measurements of the direct solar beam extinction and horizontal diffuse flux, at six wavelengths (nominally 415, 500, 615, 670, 870, and 940 nm) at short intervals (20 sec in our case) throughout the day (cf., Harrison et al., 1994) for a description of the operational details). Besides water vapor at 940 nm, the other gaseous absorbers within the MFRSR channels are NO_2 (at 415, 500, and 615 nm) and ozone (at 500, 615, and 670 nm). Aerosols and Rayleigh scattering contribute atmospheric extinction in all MFRSR channels. The single-instrument retrievals used in this study have been made using our new retrieval method (Alexandrov et al., this volume). Automated cloud screening technique (Alexandrov et al., 2004b) allowed us to make effective retrievals from multi-year MFRSR datasets. For this study we selected the AOT in 870 nm channel, the product the least affected by retrieval assumptions due to absence of gaseous absorption at this wavelength (calibration is the only remaining uncertainty).

We describe the AOT datasets in the framework of scale-invariant (fractal) statistics. The particular structure of aerosol datasets suggests that the commonly used 1-point Gaussian statistics can be complimented by 2-point statistics that characterize correlations between values of AOT at different scales. This means that in addition to the mean and the standard deviation of the AOT distribution, the aerosol variability can be evaluated by determination of power-law scaling exponents over a range of scales.

2. METHOD

The mean and standard deviation are essentially one-point statistics, i.e., the order of data values within a time interval is being ignored and replaced by a purely statistical set of parameters characteristic of the interval as a whole. Thus, the interval effectively collapses into a point with some average properties assigned to it. This approach leads to a scale truncation of the data variability, e.g., the sequence of daily means and standard deviations does not contain information about hourly or minute variability. In this study we add another parameter to the traditional two which will relate AOT variabilities at different scales. It comes from the scale analysis approach that has become popular in recent years for studies of various geophysical phenomena (cf. Alexandrov et al., 2004a; Davis et al., 1994 for examples in the literature).

The technique used in this study is based on the notion of structure function. A structure function of order $q>0$ is defined as

$$S_q(r) = <|\tau(t+r) - \tau(t)|^q>.$$

Here, $\tau(t)$ is the physical field (AOT in our case), the angular brackets stand for averaging in t, implying that an ergodicity hypothesis is valid and an ensemble average over realizations is equivalent to an average over t (time, or space). Computation of structure functions is straightforward, and (unlike with Fourier spectral methods) data gaps can be easily handled. S_2 is often referred to as "the" structure function in turbulence literature, or as a variogram in geostatistics. For scale invariant fields S_q are power-law functions of the lag r:

$$S_q(r) \sim r^{\zeta(q)},$$

where $\zeta(q)$ is a monotonically non-decreasing concave function of q. The values of $\zeta(q)$ can be determined from the slope of the $S_q(r)$ in log-log scales. Using $\zeta(q)$, a non-increasing hierarchy of exponents H_q is defined:

$$H_q = \zeta(q)/q.$$

If exponents H_q do not depend on q, the field is classified as monofractal (monoaffine), and is otherwise classified as multifractal (multiaffine). $H_1=1$ correspond to almost everywhere differentiable functions while $H_1=0$ to graphs filling a 2D area. White noise is the classic example of a time series with $H=0$. The Wiener-Khinchine theorem (Monin and Yaglom, 1975) states that the second-order structure function S_2 is in Fourier duality with the Fourier energy spectrum $E(k)$ (see Davis et al. (1994) for the energy spectrum formalism). This results in the relationship between the corresponding scaling exponents:

$$\beta = \zeta(2)+1 = 2H_2+1,$$

where β is the energy spectrum exponent.

Ensemble averaging of structure functions can reduce noise and provide a better estimate of their typical behavior. In this study dealing with a single instrument dataset we are not able to average structure functions over the whole network for a given day as we did in (Alexandrov et al., 2004a). Thus, we have to sacrifice some temporal resolution for data quality and take averages over days within a month. This results in time series of monthly values of scaling exponents.

3. RESULTS

Fig. 1 shows individual structure functions (in grey) and the results of averaging (in black) for four different months (May and October, 1993, and March and August, 1997) selected in order to show some larger and smaller values of the scaling exponents encountered during 1993–2003 period. The scale break at time scales of 2500–3000 sec is clearly seen in some of the plots. Taking a typical wind speed of 5 m/sec as a conversion factor, we obtain an estimate of the scale break's spatial scale of 12–15 km. In our earlier paper we attributed this effect to the beginning of the transition from small-scale 3D turbulence to the large-scale 2D turbulence regime. The scale break prompts us to derive two sets of scaling exponents: small- and large-scale denoted as $H_{[2,3]}$ and $H_{[3,4]}$ respectively (numbers in brackets correspond to decimal logarithms of the scale

PARTICLE RADIATIVE PROPERTIES

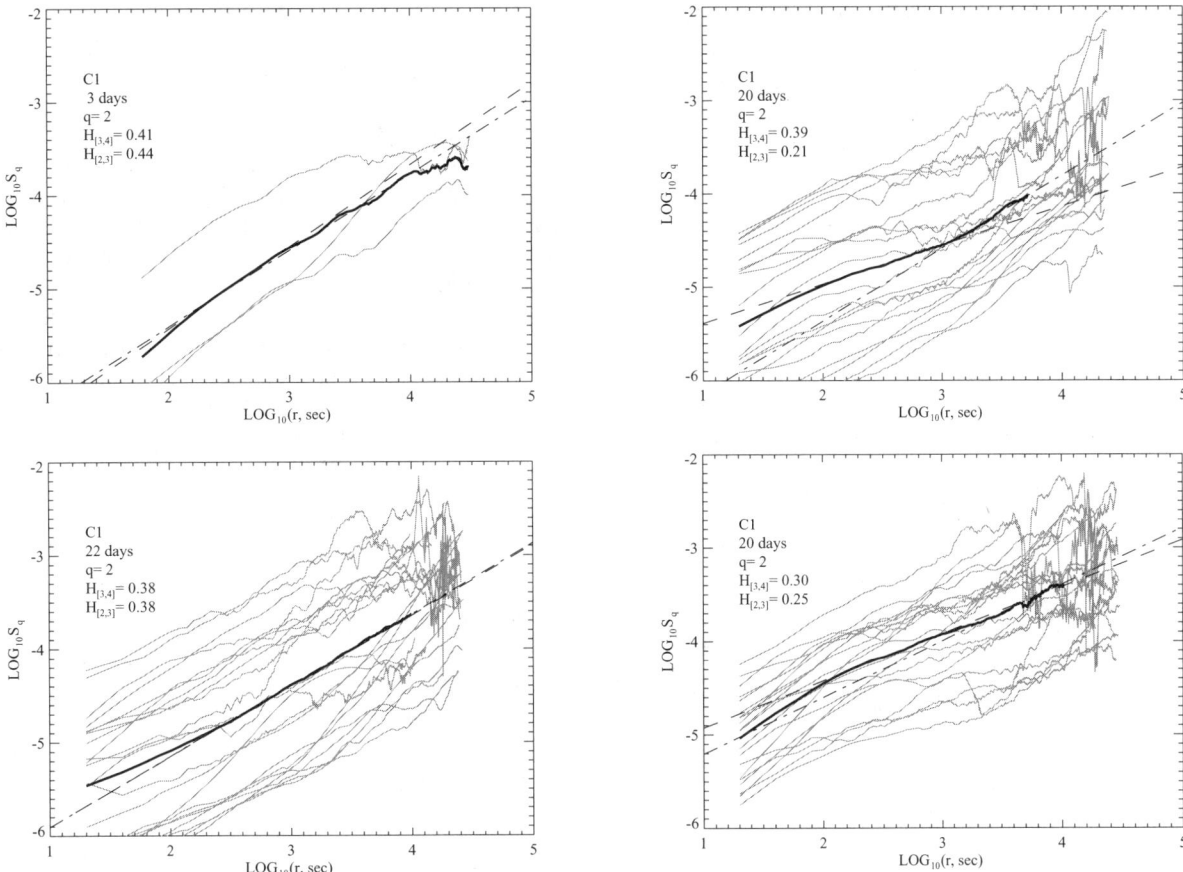

Fig. 1. Structure functions (second order) for individual days (grey) and averaged over the month (black) for May and October, 1993 (top), and March and August, 1997 (bottom). Here $H_{[2,3]}$ denotes small scale exponent (for $10^2 < r < 10^3$ sec), and $H_{[3,4]}$ stands for the large scale exponent (for $10^3 < r < 10^4$ sec).

ranges limits). Fig. 2 presents time series of both these exponents for the whole 11-year period. We should note that not all the data presented in this plot is of equal quality. The early 1993, as it is seen in the top left plot in Fig. 1, suffered from lack of clear or partially clear days suitable for determination of scaling exponents (we require a day to have at least 500 clear data points, i.e., total of 2.8 hours of clear sky, to be considered). During almost the whole year of 1994 the C1 MFRSR had persistent alignment problems that resulted in artificial oscillations in measured AOT, and therefore in the structure functions. We do not know to what extent these oscillations affected the scaling exponents, however, the extremely low values of H in 1994 may be an artifact caused by this instrument problem. We believe that despite the above issues, the dataset as a whole reflects the actual variability of AOT scaling properties. The small-scale exponent variability in particular exhibits the most remarkable features: well-established seasonality with maximum in winter-spring and the decreasing inter annual trend in 1993–1995. The latter can be attributed to presence of thick stratospheric aerosol layer during the period following Mt. Pinatubo eruption in 1991 (cf. Goodman et al., 1994). While the most of aerosols contributing to the measured AOT are normally located in the convective boundary layer with 1–3 km height, the stratospheric particles are spread over a much thicker layer. Thus, as we demonstrated in our earlier work that the layer thickness is related to smoothness of AOT time series (the thicker the layer the higher the scaling exponent), it is not unexpected to see higher values of H when a considerable portion of aerosol is located in stratosphere rather than in the boundary layer. The seasonal cycle in both scaling exponents cannot be explained just using the mixed layer height argument, since the boundary layer is not thicker in winter or spring than in summer. We may suggest larger contribution to AOT of aerosols from free troposphere and/or stratosphere during the cold season, or seasonal differences in aerosol hygroscopic growth, however this matter needs more serious investigation. We should also note that the data presented in Fig. 2 shows a decreasing trend in the small-scale scaling exponent during July–October 2000, which, while presenting a deviation from the statistical seasonality, is in agreement with the September 2000 trend in the daily exponents averaged over all Extended Facilities reported by Alexandrov et al. (2004a).

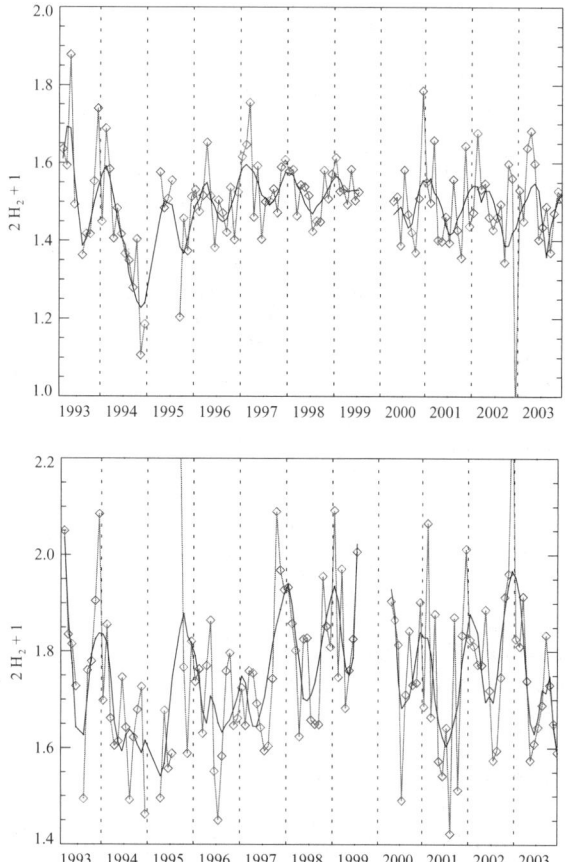

Fig. 2. Time series of monthly mean small- and large-scale (top and bottom respectively) scaling exponents $2H_2 + 1$. The black curves are obtained by smoothing the data using 5-degree polynomial fit in 9-month moving window.

4. CONCLUSIONS

The time series are presented of monthly scaling exponents of 870 nm AOT derived from MFRSR measurements made at the Central Facility of DOE ARM Program SGP site in northern Oklahoma. The values of the scaling exponents reflect atmospheric dynamics and aerosol processes however establishment of a definitive relationship between them is not yet achieved. Both short- and large-scale scaling exponents (corresponding to time intervals respectively shorter or longer than 40–50 minutes) exhibit the same seasonal cycle with maximum in winter-spring. In addition to this, the small-scale exponents have a decreasing trend in 1993–1994, which we believe reflects the dissipation of stratospheric aerosols induced by Mt. Pinatubo eruption in 1991.

5. REFERENCES

Alexandrov, M., A. Marshak, B. Cairns, A. Lacis, and B. Carlson, 2004a: Scaling properties of aerosol optical thickness fields retrieved from ground-based and satellite measurements, *J. Atmos. Sci.,* 61, 1024-1039.

Alexandrov, M. D., A. Marshak, B. Cairns, A.A. Lacis, and B. E. Carlson, 2004b: Automated cloud screen-ing algorithm for MFRSR data, *Geophys. Res. Lett.,* 31, L04118, doi:10.1029/2003GL019105.

Davis, A., A. Marshak, W. Wiscombe, and R. Cahalan, 1994: Multifractal characterizations of nonstationarity and intermittency of geophysical fields: Observed, retrieved or simulated, *J. Geophys. Res.,* 99, 8055-8072.

Goodman, J., K. G. Snetsinger, R. F. Pueschel, G. V. Ferry, and S. Verma, 1994: Evolution of Pinatubo aerosol near 19 km altitude over western North America, *Geophys. Res. Lett.,* 21, 1129-1132.

Harrison, L., J. Michalsky, and J. Berndt, 1994: Automated multifilter shadow-band radiometer: instrument for optical depth and radiation measurement, *Appl. Opt.,* 33, 5118–5125.

Monin, A. S., and A. M. Yaglom, 1975: *Statistical Fluid Mechanics: mechanics of turbulence, Vol. 2,* MIT Press, 874 pp.

RADIATIVE TRANSFER MODELLING FOR EVALUATION OF THE PARAMETERIZATION OF CLOUD OPTICAL PROPERTIES IN THE MESO-NH MESOSCALE MODEL

P. Dubuisson
ELICO, Université du Littoral Côte d'Opale, 32 av. Foch, 62930 Wimereux, France

J.-P Chaboureau and J.-P Pinty
Laboratoire d'Aérologie, Observatoire Midi-Pyrénées, Toulouse, France

V. Giraud
Laboratoire de Météorologie Physique, Université Blaise Pascal, Clermont-Ferrand, France

ABSTRACT

A fast yet accurate radiative transfer code has been developed to evaluate the quality of the Meso-NH cloud scheme. This code has been applied to a study case using a situation observed during the Special Observing Period of the Mesoscale Alpine Programme with gravity waves over the eastern Alps. A comparison between visible and infrared radiances computed from predicted Meso-NH model fields and AVHRR measurements is presented. A global agreement is observed between simulated and observed cloud cover, especially for the spatial distribution of low clouds. However, the Méso-NH model overpredicts the cloud water content. In addition, cirrus clouds are not well represented in the model.

1. INTRODUCTION

Cloud systems can be simulated using mesoscale models assuming accurate predictions of the vertical motions in the atmosphere and a correct representation of the condensed phase amount. Satellite observation allows a validation of the model by comparing brightness temperatures or reflectances measured by the satellite with synthetic ones computed using predicted mesoscale model fields. A model-to-satellite approach needs to simulate radiative quantities at the top of the atmosphere using an accurate radiative transfer code. The first aim of this work was to develop a radiative transfer code to evaluate the cloud scheme used in the Meso-NH mesoscale model. The code needs to take accurately into account both scattering and absorption/emission. The second step of this study is to compare the simulated radiances, obtained from predicted Meso-NH model fields, with visible and infrared imagery of the Advanced Very High Resolution Radiometer (AVHRR / NOAA-14). Due to the large number of simulations needed to generate a mesoscale situation, the code requires fast computation times. Consequently, a fast yet accurate radiative transfer code has been developed to simulate the top of atmosphere reflectance and brightness temperature in the 5 channels of the AVHRR (0.63, 0.84, 3.8, 10.8 and 12 μm). This spectral range allows testing assumptions about cloud properties through a spectral analysis. The code is based on the correlated k-distribution. This method allows an accurate treatment of interactions between absorption and scattering with manageable computer times. From atmospheric fields predicted by the Meso-NH model top of the atmosphere radiances can be simulated in each AVHRR channel. As an application, a study case has been analysed using a situation observed during the Special Observing Period of the Mesoscale Alpine Programme (MAP). A comparison between visible and infrared radiances computed from predicted Meso-NH model fields and AVHRR measurements is presented. An evaluation of the microphysical and optical properties of the cloud scheme used in the Meso-NH model is discussed.

2. RADIATIVE TRANSFER MODELING

The radiative transfer code, referred to as FASDOM (Dubuisson et al., 2005), has been developed using the correlated k-distribution (Lacis et Oinas, 1991) to approximate the atmospheric absorption. The primary advantage of the correlated k-distribution is that this technique can be used for multiple scattering, especially for cloudy atmospheres, with reasonable computational times. For a layer at pressure P and temperature T, the transmittance $T_{\Delta\nu}$ in a spectral interval $\Delta\nu$ is approximated by exponential series over a limited number of terms (N=10) as:

$$T_{\Delta\nu}(P,T) \approx \sum_{i=1}^{N} a_i \exp[-k_i(P,T)u(P,T)], \qquad (1)$$

with a_i and k_i the weight and the absorption coefficients, and u the absorber amount. For a given spectral interval $\Delta\nu$, coefficients of the exponential series are estimated for a reference pressure P_0 and temperature T_0, from a line-by-line model (LBL) (Dubuisson, et al, 1996) and a spectroscopic database (HITRAN). Coefficients are computed from the three following steps: (1) absorption coefficients k_ν are computed with the LBL code; (2) the absorption coefficient range is subdivided into N sub-interval, defined as $k_i \pm \Delta k_i$, where k_i is the mean absorption coefficients for each class i with i=1,N; (3) coefficients a_i are calculated from absorption coefficients k_ν, using the spectral response of the sensor f_ν as:

$$a_i = \frac{\int_{\Delta v} \delta_{iv} f_v dv}{\int_{\Delta v} f_v dv}, \quad (2)$$

with $\delta_{iv} = 1$ if $k_i - \Delta k_i < k_v < k_i + \Delta k_i$, and $\delta_{iv} = 0$ otherwise. Coefficients a_i define the fraction of the interval for which the absorption coefficient is between $k_i - \Delta k_i$ and $k_i + \Delta k_i$. Moreover, following the concept of the correlated k-distribution, the exponential sum is assumed to be correlated at any pressure level with that at the reference pressure P_0 and temperature T_0, for a given spectral interval. A scaling approximation can be then applied to account for the vertical dependence of the absorption:

$$k_i(P,T) = k_i(P_0, T_0) \left(\frac{P}{P_0}\right)^n \left(\frac{T_0}{T}\right)^m \quad (3)$$

The coefficients a_i and k_i have been calculated for water vapor which is the main absorber for the AVHRR channels. Absorptions by O_3 and Uniformly Mixed Gases UMG (CO_2, O_2, CH_4 and N_2O) have non-negligible effects and these gases are treated from a simple correction by calculating an average optical thickness with the LBL model.

Radiances are estimated by solving the Radiative Transfer Equation (RTE) for each term of the summation in Eq. (1), in the assumption of an atmosphere stratified into plane and parallel layers with the Meso-NH vertical resolution (50 layers). The RTE is solved using the Discrete Ordinate Method (DOM) (Stamnes et al., 1988) allowing accounting for scattering and absorption/emission, as well as their interactions. Clouds are represented from their optical properties: optical thickness, single scattering albedo and moments of the phase function development. Optical properties for liquid water clouds are computed for each AVHRR channel using Mie theory with an effective radius ranging from 4 to 40 µm. Ice clouds are defined using optical properties for ice crystals from Baum et al., 2000, with a mean effective size ranging from 8.9 to 78.5 µm. Rayleigh scattering is also accounted for. Note that cloud fractional coverage is not considered in this work.

The FASDOM code has been validated from the LBLDOM reference code (Dubuisson et al., 1996) for which the RTE is resolved at each step of the line-by-line model using the DOM. Comparisons have been performed for FASDOM using realistic clouds, with liquid water or ice droplets, and atmospheric profiles (Dubuisson et al., 2005). Relative deviations between FASDOM and LBLDOM are less than 1% on the top of the atmosphere radiances, for each AVHRR channel. In addition, deviations in brightness temperatures between FASDOM and LBLDOM are generally less than 0.2 K. This accuracy is adequate for the purpose of this study.

3. A MODEL-TO-SATELLITE APPROACH

Méso-NH model:

The MESO-NH model is a non-hydrostatic mesoscale model jointly developed by Météo-France and the Centre National de la Recherche Scientifique (CNRS) (Lafore et al., 1998). The explicit cloud scheme used is a bulk mixed-phase microphysical scheme (Pinty and Jabouille, 1998), which predicts the mixing ratio of six atmospheric water categories: water vapour, cloud water, rain water, non-precipitating ice, snow and graupel. The main outputs needed for the FASDOM code are vertical profiles of temperature, pressure, water vapour, ice and liquid water content and cloud effective radii, as well as surface reflectance and emissivity.

Case Study:

Predicted fields by the Méso-NH model (Fig. 1b) have been used as inputs for the FASDOM code for a gravity waves situation observed during the Special Observing Period (SOP) of the Mesoscale Alpine Programme (MAP) (Bougeault et al., 2001), over the eastern Alps – 25 October 1999 (Volkert et al., 2003). AVHRR data are available for this case study (Fig. 1a).

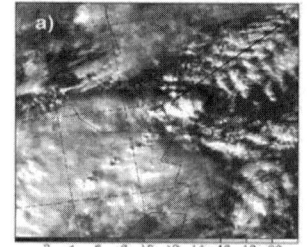
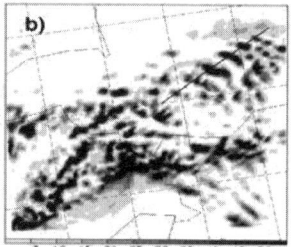

Fig. 1. a) AVHRR observations at 1422 UTC (visible channel); b) Simulated vertically integrated liquid cloud water content at 1430 UTC (Volkert et al., 2003)

4. COMPARISON BETWEEN AVHRR OBSERVATIONS AND MESO-NH SIMULATIONS

For the previous case study, synthetic radiances and brightness temperatures (BT) have been simulated for the five channels of the AVHRR, from Méso-NH fields. Comparisons between solar radiances (at 0.63 µm) and brightness temperatures (at 11 µm) are presented in Fig. 2. There is a global agreement (spatial and vertical distribution) between simulated and observed cloud-cover, especially for liquid clouds. However, the Méso-NH model overpredicts the radiance at 0.63 µm suggesting too large optical thicknesses for liquid water clouds. Spatial distribution of cirrus clouds in the south-west of the image is not well represented. Moreover, the Méso-NH model overpredicts ice contents over liquid water clouds in the north-east of the image. Discrepancies are more noticeable with BTs at 11 µm.

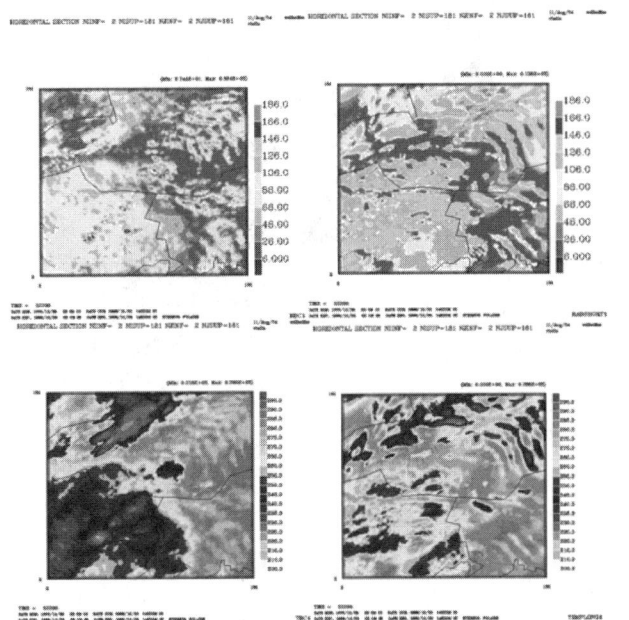

Fig. 2. Comparison between AVHRR observation (left column) and FASDOM simulation (right column). First line presents the radiances at 0.63 μm and the second line the brightness temperature at 11 μm.

Parameterizations of cloud optical properties have been evaluated from spectral variations of the radiance and brightness temperature. Comparisons on radiances, brightness temperature and Brightness Temperature Difference (BTD) from the two thermal channels at 11 and 12 μm are presented in Fig. 3. This spectral analysis mainly shows that the Méso-NH overpredicts radiances at 0.63 and 3.8 μm, suggesting too large liquid water contents. There is an agreement for liquid clouds on the magnitude of BTDs which depend on cloud effective radius. However, Fig. 3 also shows many negative points for AVHRR BTDs, especially for low brightness temperatures at 11 μm (220K < BT < 260K). These differences are mainly due to the cirrus cloud parameterization. Moreover, aerosols can have an influence on results. Unfortunately, aerosol optical thicknesses were not predicted with the méso-NH model for this case study.

5. CONCLUSIONS

A fast yet accurate radiative transfer code (FASDOM) has been developed with applications to the Méso-NH model. Comparisons with AVHRR observations have shown a global agreement between observed and simulated radiances, especially for low clouds. The spectral analysis suggests that (1) the cloud liquid water content is generally overestimated, (2) the cirrus cloud scheme and optical properties have to be improved, and (3) effects of cloud fractional coverage have to be analyzed. The same analysis has been performed for a case study over the South Atlantic Convergence Zone off the Brazilian coast, leading to equivalent results (Chaboureau et al., 2004).

6. REFERENCES

Baum, B. A., D. P. Kratz, P. Yang, S. C. Ou, Y. Hu, P. F. Soulen, and S. C. Tsay, 2000, Remote sensing of cloud properties using MODIS Airborne Simulator imagery during SUCCESS. I. Data and models, *J. Geophys. Res.*, Vol. 105, No. D9, 11767-11780.

Bougeault P., P. Binder, A. Buzzi, R. Dirks, R. Houze, J. Kuettner, R. B. Smith, R. Steinacker, and H. Volkaert, 2001, The MAP special observing period, *Bull. Am. Meteorol. Soc.*, 82, 433-462.

Chaboureau J.-P., P. Dubuisson, J. Pardo, C. Prigent, J.-P. Pinty, P. J. Mascart, and E. Richard, 2004, Cloud scheme assessment of the MESO-NH mesoscale model using satellite observations from the visible to the microwaves, *14th International Conference on Cloud and Precipitation,* Bologne, Italy, 18-23 July.

Dubuisson P., J. C. Buriez, and Fouquart Y., 1996, High spectral resolution solar radiative transfer in absorbing and scattering media, application to the satellite simulation, *J. Quant. Spectrosc. Radiat. Transfer,* 55, 103-126.

Dubuisson P., V. Giraud, O. Chomette, H. Chepfer, and J. Pelon, 2005, Fast radiative transfer modeling for infrared imaging radiometry, *J. Quant. Spectrosc. Radiat. Transfer,* in press.

Lacis, A. A., and V. Oinas, 1991, A description of the correlated k-distribution method, *J. Geophys. Res.,* 96, 9027-9064.

Lafore J. P., et al., 1998, The Meso-NH atmospheric simulation system. Part. I: adiabatic formulation and control simulations, *Annales Geophysicae,* 16, 90-109.

Stamnes, K., S. Tsay, W. Wiscombe, and K. Jayaweera, 1988, Numerically stable algorithm for discrete-ordinate-method radiative transfer in multiple scattering and emitting layered media, *Appl. Opt.,* 27, 2502-2509.

Pinty J.-P., and P. Jabouille, Proceedings of the AMS conference on cloud physics, Everett, Washington, 1998.

Volkert H., C. Keil, C. Kiemle, G. Poberaj, J.-P. Chaboureau, and E. Richard, 2003, Gravity waves over the eastern Alps: A synopsis of the 25 October 1999 event (IOP 10) combining in situ and remote-sensing measurements with a high-resolution simulation, *Q. J. R. Meteorol. Soc.,* 129, 777-797.

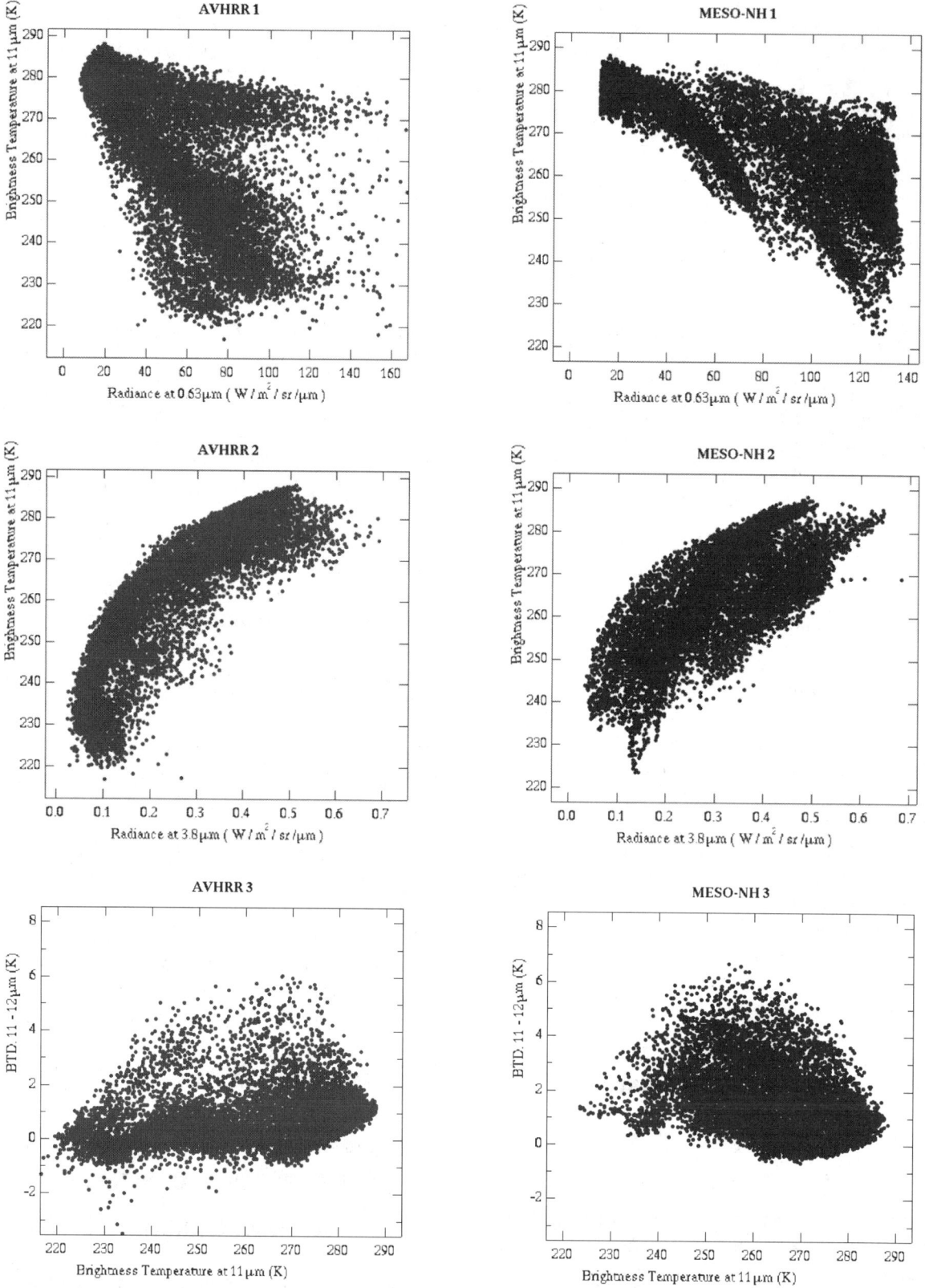

Fig. 3. Comparison between AVHRR observations (left column in red) and FASDOM simulations from predicted fields of Méso-NH (right column in blue); radiances, brightness temperatures or Brightness Temperatures Differences (BTD) are presented.

THE EFFECTS OF NONSPHERICITY AND VARIATION IN ICE CRYSTAL BULK DENSITY ON 95 GHZ CLOUD RADAR SIGNALS

Kaori Sato and Hajime Okamoto
Center for Atmospheric and Oceanic Studies, Graduate School of Science, Tohoku University
Aoba, Aramaki aza, Aoba-ku, Sendai 980-8578, Japan
ksato@caos-a.geophys.tohoku.ac.jp

ABSTRACT

We examine the effects of nonsphericity and change in bulk density (ρ_b) of several ice particle types on the reflectivity factor (Z_e) and the linear depolarization ratio (*LDR*) for vertically pointing 95GHz cloud radar based on the discrete dipole approximation (DDA). The variation in ρ_b does not have a significant impact on Z_e and *LDR*, while nonsphericity plays an important role especially for effective radius (r_{eff}) larger than 100μm. It is shown that the additional use of *LDR* with Z_e and Doppler velocity is effective to determine particle type and reduce the uncertainties in the retrievals of microphysical properties of ice clouds, e.g., r_{eff} and *IWC*. In addition, the Maxwell-Garnett formula together with Mie theory for a porous sphere circumscribing nonspherical ice particles is evaluated, and shown to deviate significantly from the reference solution by the DDA, e.g., more than 10 dB at r_{eff}=100 μm.

1. INTRODUCTION

The representation of radiative properties for ensemble of nonspherical ice particles have captured much interest [Macke, 1993; Yang and Liou, 1996; Mishchenko et al., 2000], and is one of the major cause of uncertainty in making precise estimate of climatic impact of ice clouds. In order to reduce such uncertainty, and with the general recognition on the importance of the vertical information, intensive efforts have been made to develop algorithms using active sensors such as lidars and cloud radars to retrieve microphysical properties of ice clouds. Although the synergy use of radar and lidar is quite attractive for such purpose [Okamoto, 2003], recent studies show that ground based lidar cannot penetrate low level clouds when its optical thickness exceeds 3, while cloud radar has much wider applicability and can detect more than 90% of the clouds. Thus, applicability of the synergy method is rather limited to less multi-layered scenes [e.g., Okamoto et al., in preparation]. Radar backscattering is not directly related to the estimate of radiative forcing of ice clouds. Nevertheless, cloud radar is a relevant tool for such estimation through its promising ability to retrieve ice microphysics. Therefore, accurate estimation of backscattering properties at this wavelength is essential. Our main objective is to examine the backscattering properties of ensemble of nonspherical ice particles at 95 GHz to retrieve microphysical properties of ice clouds with the single use of ground based 95 GHz cloud radar. In section 2, the ice particle geometries are described briefly. In section 3, the results for the DDA calculation and the MG-Mie method are shown. In section 4, the combined use of Z_e, *LDR*, and Doppler velocity for the retrieval of r_{eff} and *IWC* are discussed. Finally, summary of our results is given in section 5.

2. THEORETICAL PROCEDURE

Here, the ice particle geometries for the DDA [Draine, 1988] and the MG-Mie method are described. The DDA is an ideal method that can estimate backscattering by nonspherical particles with sufficient accuracy as long as the validity criteria are fulfilled [Okamoto, 2002]. The MG-Mie method is used for its convenience of treating nonspherical particles with air inclusion. For the DDA calculation, we considered three types of shape, i.e., plate, column, and bullet rosette, ranging from 1 to 1000 μm in melted equivalent radius (r_{eq}). The information on the size dependence of aspect ratio is from Mitchell [1996] for plates with ρ_b=0.92g/cm^3 (solid ice), and from Heymsfield and Miloshevich [2003] for column and bullet rosette with ρ_b=0.81g/cm^3 where the relations between area and size are converted to that of aspect ratio to size. In order to take into account the decrease in ρ_b from that of solid ice, an air pocket is created within the ice particles. The ice particles are randomly oriented in horizontal space to the normal incident wave (hereafter 2-D), and in three-dimensional space (hereafter 3-D) for plate and column and in 3-D for bullet rosette. The MG-Mie method is applied to two types of imaginary sphere for different purposes. The first model (hereafter MG-Mie 1) is porous sphere of ice-air mixture that circumscribes the ice particle considered for the DDA study. This is to validate the applicability of the MG-Mie method itself where the Maxwell-Garnett Formula is applied to sizes out of the Rayleigh region at 95 GHz. The density of such porous sphere is expressed as effective density (ρ_e) in this study and it becomes much smaller than ρ_b at large particle size. Another model (hereafter MG-Mie 2) is a porous sphere with ρ_e of Brown and Francis [1995], which is widely used for the MG-Mie method in retrieving microphysical properties of ice clouds.

3. RESULTS

3.1 Effect of Variation in Ice Crystal Bulk Density

We first investigate the effect of variation in ρ_b on Z_e and LDR for 2-D and 3-D column, and bullet rosette in this sub-section. Z_e and LDR for our models are compared to that of the same particle geometry but with $\rho_b=0.92$g/cm^3. In addition, lower ρ_b of 0.6g/cm^3 is also considered, which is reported for large column and bullet rosette in cold region [Heymsfield, 1972]. Z_e of 2-D column are estimated for three cases of ρ_b mentioned above (Fig. 1). It turns out that ρ_b does not play a role in Z_e, i.e., the differences between the three are less that 1dB for all the particle types considered here. Such effect on LDR (figures not shown) is smaller than 1dB when our model is compared to the solid ice case and the maximum difference among the three are 2dB.

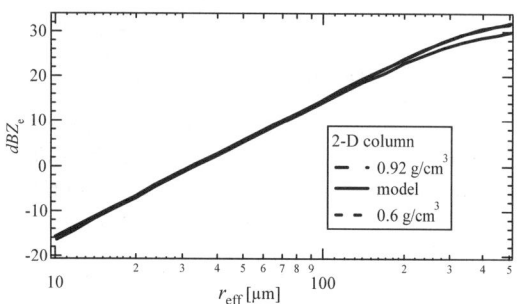

Fig. 1. The radar reflectivity factor Z_e of 2-D column with several ρ_b.

3.2 The Applicability of the MG-Mie Method

The MG-Mie method has been widely used to treat the nonspherical ice crystals with porous structure, since this can be easily applied to arbitrary shaped particles and requires less computing memory and time compared with the DDA. Here, we test the applicability of the MG-Mie method in obtaining Z_e against the DDA solutions (Fig. 2).

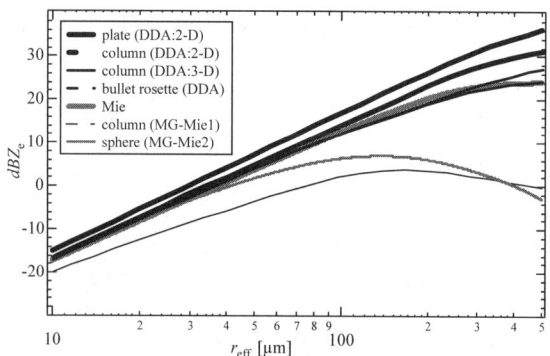

Fig. 2. Z_e for the 2-D (thick black lines) and 3-D (thin black lines) particles obtained by the DDA. Results for the MG-Mie 1 and 2 (thin gray lines), and Mie (thick gray line) are also shown.

Z_e for both the MG-Mie 1 and 2 deviates from that of the DDA to a great extent from very small sizes. It can be explained by the fact that the Maxwell-Garnett formula requires Rayleigh approximation in its derivation. This result suggests the need for accurate treatment of scattering characteristic of ice particles by reliable theory such as the DDA with its criteria fulfilled.

3.3 The Role of Nonsphericity on dBZ_e and LDR

Z_e for the nonspherical ice particles based on the DDA are compared to that of sphere with the same mass (Fig. 2). For $r_{eff}<100$ μm, the difference is smaller than 1dB for most particle types except that for column and plate of 2-D, which are 2dB and 4dB at maximum, respectively. This is due to the larger projected geometrical cross sections of 2-D particles compared to that of 3-D particles and sphere. For $r_{eff}>100$ μm, the effect of nonsphericity become more important especially for particles of 2-D, e.g., the differences between 2-D plate and sphere becomes 5 dB and 15 dB at 100 μm$<r_{eff}<300$ μm and 300 μm$<r_{eff}$, respectively.

LDR is known as an indicator of nonsphericity. LDR for the same nonspherical particles are estimated except that for 2-D plate, which is under the detection limit of -25dB (Fig. 3). Although all the particles considered are often specified as pristine particles, LDR shows strong dependence on its shape and orientation. This suggests the possibility of using LDR in the retrieval algorithm to infer particle type. Further detailed discussion about the effects of nonsphericity and orientation on Z_e and LDR will be provided from us [in preparation].

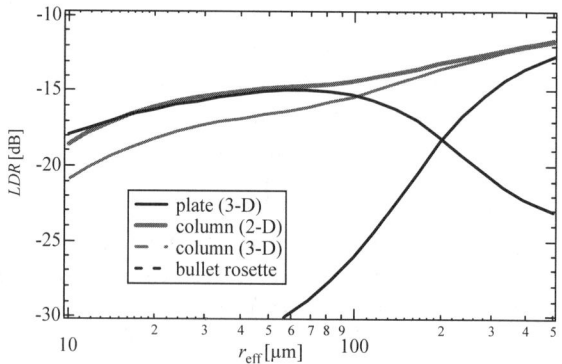

Fig. 3. LDR for the nonspherical ice particles.

4. IMPLEMENTATION OF LDR WITH DOPPLER VELOCITY AND Z_e IN THE RETRIEVAL ALGORITHM

Many retrieval algorithms using 95 GHz cloud radar have been developed to investigate cirrus microphysical properties. Yet, the retrieved quantities are much affected by the assumptions made, especially by the geometry of ice particles. In this section, the advantage of the algorithm that combines LDR with Z_e and Doppler velocity is discussed. First we give the definition of the reflectivity-weighted particle fall velocity V_{tz}.

$$V_{tz} = \frac{\int_{r_{eq,min}}^{r_{eq,max}} V_t(r_{eq}) Z_e(r_{eq}) \frac{dn(r_{eq})}{dr_{eq}} dr_{eq}}{\int_{r_{eq,min}}^{r_{eq,max}} Z_{eq}(r_{eq}) \frac{dn(r_{eq})}{dr_{eq}} dr_{eq}}, \quad (1)$$

where V_t is the ice particle fall velocity in quiet air calculated according to Heymsfield et al., [2002] and Z_e is based on the DDA for the same ice particles described in section 2. dn/dr_{eq} is the modified Gamma size distribution with the dispersion of the distribution to be 2, and the positive sign of the velocity denotes downward motion. The expected r_{eff} and IWC are estimated from the measurements of Z_e and V_{tz} where dBZ_e is set to -20dB in the following calculations. It is found that without the knowledge of LDR, the uncertainty in the retrieved r_{eff} and IWC for a given V_{tz} increases with V_{tz}. The difference in r_{eff} among the particle types are especially pronounced when plates (2-D and 3-D) and other particle types are compared, e.g., in the case of plate and column of 2-D, the differences in the expected r_{eff} are 50 µm and 200 µm at $V_{tz} \cong 0.5$m/s and $V_{tz} \cong 1.0$m/s, respectively (Fig. 4). The uncertainty in the retrieval of IWC results from the errors in the retrieved r_{eff}. Similarly

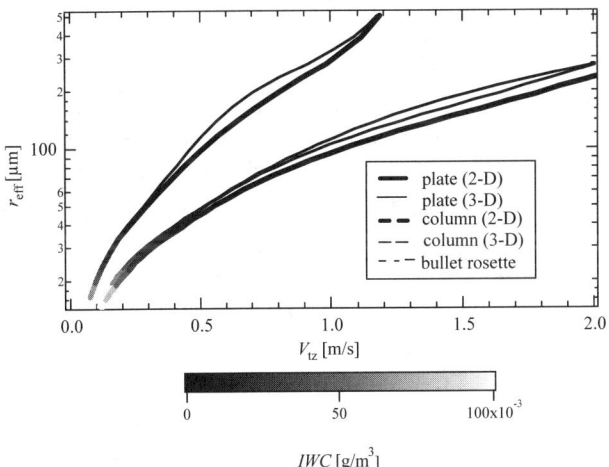

Fig. 4. r_{eff} and IWC as a function of V_{tz} for nonspherical particles. IWC is the case for observed dBZ_e = -20dB.

to the case of r_{eff}, such uncertainty is large when plate of 2-D and others are compared, which are different for more than one order of magnitude for $V_{tz} \geq 0.5$m/s.

In order to see the potential of the combinational use of LDR to Z_e and V_{tz} for reducing the above-mentioned uncertainties in the retrieved r_{eff} and IWC, we examine the dependence of LDR on V_{tz} and r_{eff} (Fig. 5). As expected from Fig. 3, LDR is effective in discriminating the particle types and indeed can help to reduce the uncertainties in the retrieval of r_{eff} and IWC, e.g., the difference in LDR for 2-D column and bullet rosette is more than 10dB for a wide range of V_{tz}. There are some limitations where LDR for some particles are similar, for instance, the discrimination of 2-D column and 3-D plate for small V_{tz}.

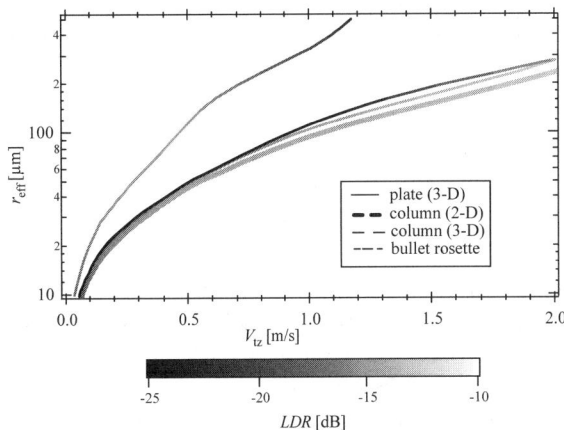

Fig. 5. The same as Fig. 4 but for r_{eff} and LDR.

In practice, it may be necessary to consider mixtures of different ice particle types. The differences in r_{eff} for a given V_{tz} for several mixing ratios of column and plate of 2-D are examined as an example (Fig. 6). The accuracy of the retrieval of r_{eff} shows strong dependence on the mixing ratio and a weak dependence on V_{tz}. Similar tendency is found for the accuracy in the retrieved IWC (Fig. 7). These results show that LDR removes the uncertainty in r_{eff} and IWC, which is otherwise more than 100% for both r_{eff} and IWC at V_{tz}>0.4m/s, when the mixture of the particle types are specified in order to avoid the possibility of multi-solutions for a given V_{tz}, LDR, and Z_e.

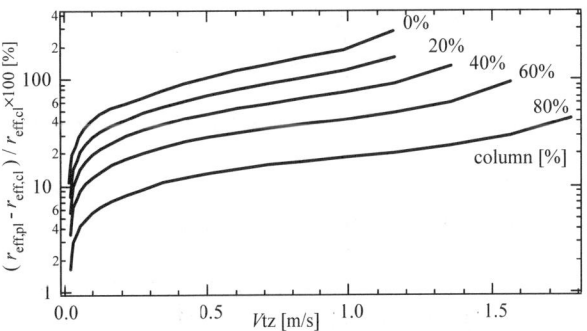

Fig. 6. Δr_{eff} as a function of V_{tz} for mixtures of column and plate of 2-D. Δr_{eff} is based on the case of 100% of 2-D column.

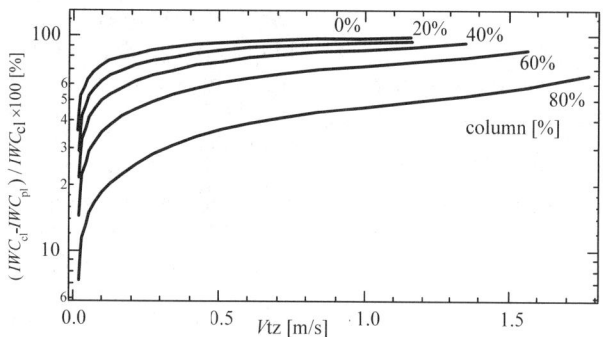

Fig. 7. The same as Fig. 6 but for ΔIWC.

5. SUMMARY

We apply the DDA to examine the effects of nonsphericity and variation in ρ_b on Z_e and LDR at 95 GHz. We consider plate and column of 2-D and 3-D, and bullet rosette of 3-D. Further, the use of LDR in addition to Z_e and Doppler velocity for the retrieval of r_{eff} and IWC of ice clouds are investigated. In addition, we examine the performance of treating Z_e for nonspherical particles by that of their circumscribed spheres (the MG-Mie method). Our findings are briefly summarized as follows.

1. Columns and bullet rosette with ρ_b of 0.81g/cm³ are compared to that of solid ice and low case of 0.6g/cm³ is also considered based on the DDA. We found that ρ_b does not play a role in Z_e and LDR, i.e., within 1dB for dBZ_e, and 2dB for LDR.

2. The MG-Mie method underestimates Z_e almost throughout the whole size range considered. Its deviation from the reference solution of the DDA is significant for r_{eff}>50 μm, and is more than 10dB at r_{eff}=100 μm. The limitation of the applicability of the MG-Mie to small sizes originates from the requirement of Rayleigh approximation in the Maxwell-Garnett mixing rule. This result suggests that the look-up tables in the retrieval algorithms for ice clouds should be based on rigorous scattering theories such as the DDA.

3. The uncertainties in retrieving r_{eff} and IWC by measurements of Z_e and Doppler velocity increase with velocity. Analyses for single usual type and for mixtures of different particle types show that the uncertainties in r_{eff} and IWC are more than 100% in both cases. The additional information of LDR is effective in reducing such uncertainty.

This study is limited to few particles types and in nature the situations are more complex. Further studies are necessary to extend our method to anvils and other convectively generated tropical ice clouds.

6. ACKNOWLEDGEMENT

We thank B. T. Draine and P. J. Flatau for providing the DDA code DDSCAT 6.1.

7. REFERENCES

Brown, P., and P. Francis, 1995: Improved measurements of the ice water content in cirrus using a total-water probe. *J. Atmos. Oceanic. Technol.*, 12, 410-414.

Draine, B. T., and P. J. Flatau, 1994: Discrete-dipole approximation for scattering calculations. *J. Opt. Soc. Am. A.*, 11, 1491–1499.

Heymsfield, A. J., 1972: Ice crystal terminal velocities. *J. Atmos. Sci.*, 29, 1348-1357.

Heymsfield, A. J., et al., 2002: A general approach for deriving the properties of cirrus and stratiform ice cloud particles. *J. Atmos. Sci.*, 59, 3-29.

Heymsfield, A. J., and L. M. Miloshevich, 2003: Parameterizations for the cross-sectional area and extinction of cirrus and stratiform ice particles. *J. Atmos. Sci.*, 60, 936-956.

Macke, A., 1993: Scattering of light by polyhedral ice crystals. *Appl. Opt.*, Vol. 32(15), 2780-2788, 1993.

Mischenko, M. L., J. W. Hovenier, and L. D. Travis (Eds), 2000: Light Scattering by Nonspherical particles; Theory, Measurements, and Applications, 690 pp., *Academic,* San Diego, Calif.

Mitchell D. L., 1996: Use of mass-and area-dimensional power laws for determining precipitation particle terminal velocities. *J. Atmos. Sci.*, 53, 1710-1723.

Okamoto, H., 2002: Information content of the 95GHz cloud radar signals; Theoretical assessment of effects of non-sphericity and error evaluations of the discrete dipole approximation, *J. Geophys. Res.*, 107(D22), 4628, doi:10.1029/2001JD001386.

Okamoto, H., et al., 2003: An algorithm for retrieval of cloud microphysics using 95-GHz cloud radar and lidar. *J. Geophys. Res.*, 108(D7), 4226, doi:10.1029/2001JD001225.

Yang, P., and K. N. Liou, 1996: Geometric-Optics-integral-equation method for light scattering by nonspherical ice crystals. *Appl. Opt.*, 35, 6568-6584.

OPTICAL PROPERTIES OF PARTICLES FROM LABORATORY FIRES: COMPARISON BETWEEN MEASUREMENTS AND MIE CALCULATIONS

Katja Hungershöfer
University of Leipzig, Institute for Meteorology, Stephanstr. 3, 04103 Leipzig, Germany
hunger@uni-leipzig.de

Otmar Schmid, Ravindra Singh Parmar, Günter Helas, and Meinrat O. Andreae
Max Planck Institute for Chemistry, Department of Biogeochemistry, 55128 Mainz, Germany

Kristina Zeromskiene, Yoshiteru Iinuma, Alfred Wiedensohler, and Hartmut Herrmann
Institute for Tropospheric Research, 04318 Leipzig, Germany

Jörg Trentmann
University of Mainz, Institute for Atmospheric Physics, 55099 Mainz, Germany

Thomas Trautmann
German Aerospace Centre, Remote Sensing Technology Institute, 82234 Wessling, Germany

ABSTRACT

Despite a great deal of scientific research, information on optical properties of biomass burning aerosol is still difficult to acquire, although this is necessary with regard to a quantification of the direct influence on the radiation budget. Within the framework of the project EFEU a series of controlled biomass combustion experiments were carried out and a multitude of physical and chemical properties of the gaseous and particulate emissions were determined. Here a comparison between the measured optical properties and Mie simulations are discussed for combustion of musasa, i.e., an African hardwood.

1. INTRODUCTION

Vegetation fires are a significant source for atmospheric trace gases and aerosol particles. The direct and indirect effect of these emitted particles on the atmospheric radiation is known, but not finally clarified (Houghton et al., 2001). For that reason a series of controlled burn experiments with typical fuel types (spruce, oak, musasa (African hardwood), savanna grass and peat) have been carried out at the combustion facility of the Max Planck Institute for Chemistry in Mainz, Germany, within the framework of the 'Impact of Vegetation Fires on the Composition and Circulation of the Atmosphere' (EFEU) project (Wurzler et al., 2001).

Here we discuss a comparison between measured and calculated aerosol optical properties for one selected EFEU fire in order to improve the modeling of the aerosol optical properties and thus to better quantify the direct effect of biomass burning aerosol.

2. MEASUREMENTS

The combustion of the different biofuels during the EFEU campaigns was carried out at the combustion laboratory at the Max Planck Institute for Chemistry in Mainz, Germany (Lobert et al., 1990). Through the exhaust stack of the laboratory oven the emitted gases and particles were drawn into a 32.6 m^3 metal container, operated as continuous flow mixing chamber, from which the sampling was performed. An internal fan guaranteed homogeneous mixing inside of this container and a 1:20 dilution stage was employed to avoid instrument saturation. A typical burn for a given biofuel lasted about one hour and the average residence time of aerosol particles in the buffer container was approximately 8 minutes (Schmid et al., 2004).

Gas phase measurements of CO and CO_2 with a BINOS analyzer were used to monitor the burning conditions. The number size distribution was measured with a Scanning Mobility Particle Sizer (SMPS) and an Aerosol Particle Sizer (APS) for particle radii between 0.006 – 0.4 µm and 0.25 – 5 µm, respectively and fitted with a log-normal distribution with four modes. In case of the musasa fire an effective radius of 0.202 µm was found. Typical effective radii of between 0.101 and 0.162 µm were derived for fresh ambient biomass burning aerosol from the count median diameters given in Reid et al. (2005a) assuming an average geometric standard deviation of 1.7 (Reid et al., 2005a). Thus the large effective radius of 0.202 µm for the musasa burn indicates the presence of relatively large particles possibly due to enhanced coagulation during the 8 minutes in the mixing container at elevated concentration levels. For the total particle concentration, measured with a condensation particle counter (CPC, TSI 3255A), we observed very high concentrations with a temporal average of around $1.4 \cdot 10^6$ particles per cm^3 in the mixing chamber resulting in a concentration of $7 \cdot 10^4$ cm^{-3} after the 1:20 dilution. Information about the size-resolved elemental carbon (EC) fraction was provided from samples with a five-stage Berner type impactor that were analyzed with a thermographic method which is based on the VDI 2465, Part 2 (1999) with EC temperature cut at 650°C rather than 850°C. The total volume scattering and backscattering coefficients at three different wavelengths (λ=0.45, 0.55 and 0.7 µm) were measured with an integrating nephelometer (TSI, model 3563) and corrected for angular non-idealities. Volume absorption coefficients were determined with a photoacoustic spectrometer (PAS) at 0.532 µm and a particle soot absorption photometer (PSAP) at 0.565 µm. Measurements with both instruments agreed within experimental uncertainties, but due to the higher degree of accuracy (5%) (Schnaiter et al., 2005), the PAS data are used for the comparison with model results in Section 4. Values of the single scattering albedo at λ=0.55 µm were calculated from the measured scattering coefficient at λ=0.55 µm and the PAS results, neglecting the small wavelength difference. Mass specific scattering and absorption coefficients were derived by dividing the volume specific measurements with the mass concentrations calculated from the size distribution, using the volume averaged density of the model particles described below. A humidified tandem differential mobility analyzer (H-TDMA) was employed to determine the hygroscopic properties of the particles at initial dry diameter of 50, 100, 150, 250, and 325 nm and a relative humidity of 85%. In case of the musasa combustion the hygroscopic growth factors, defined as the ratio of the dry and wet particle diameters at 85% relative humidity, were below 1.07, showing that the particles were almost hydrophobic (Maßling et al., 2003).

While the gas phase measurements, the size distribution, the total particle concentration and the optical properties used in Section 3 and 4 were available as two minutes averages, the size-resolved EC fractions from the chemical analysis were only available as averages over the whole burning time.

3. THEORETICAL MODELING

To calculate the optical properties of the biomass burning aerosol, the size, shape and composition of the particles have to be known. In addition to the measured particle size we assumed internally mixed spheres composed of non-absorbing organic carbon (OC) and strong absorbing black carbon (BC). The mean mass fraction of BC over the whole experiment is known from the chemical analysis assuming that elemental and black carbon fractions are equal. To estimate a time-dependent BC fraction the measured mean EC mass fraction from different EFEU combustion experiments were plotted against the corresponding average burning conditions, described by

$\Delta CO/\Delta CO_2$, where ΔCO and ΔCO_2 are the CO and CO_2 concentrations above the background level. Based on this data, a curve fit of the form $y=0.83 \cdot x^{-0.96}$ with $x=\Delta CO/\Delta CO_2$ and y as BC mass fraction was calculated that reproduced the measurements with a coefficient of determination of r^2=0.98. This power expression was used to derive a time dependent BC fraction from the gas phase measurements for every two minutes. The result as well as the $\Delta CO/\Delta CO_2$ trace in case of the musasa fire, are given in Fig. 1.

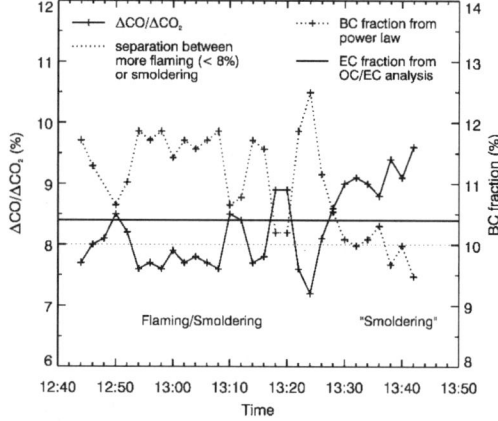

Fig. 1. Burning conditions during the musasa combustion described by the emission ratio of CO relative to CO_2 ($\Delta CO/\Delta CO_2$) (left axis). The time-dependent BC mass fraction derived from the power law is also shown (right axis).

As can be seen in Fig. 1 BC mass fractions between 9.5 and 12.5% are resulting with the highest values in the flaming dominated periods ($\Delta CO/\Delta CO_2$<8%) and the lowest values during the smoldering dominated phases ($\Delta CO/\Delta CO_2$>8%). In this way we got a mass-averaged BC fraction of 11.1%, which is 0.7% above the average of 10.4% resulting from the thermograph analysis. For the refractive index of OC and BC we set the imaginary part of OC to zero and modified the real part of OC until we obtained good agreement with the measured scattering coefficient. A literature value for soot of 1.75 (Fenn et al., 1985) was taken as real part of BC whereas the imaginary part was changed in order to find consistency with the absorption coefficient. The resulting effective refractive index was derived with the rule of Maxwell-Garnett (Maxwell-Garnett, 1905), assuming particle densities of 1.2 g/cm^3 and 1.8 g/cm^3 for OC and BC, respectively (Ross et al., 1998). The volume averaged particle density was determined from the densities of OC and BC and the BC mass fraction. In case of the musasa combustion the refractive index of OC and BC was determined as 1.50–0i and 1.75–0.178i, respectively. For a BC fraction of 10.4%, this resulted in an effective index of 1.51–0.012i at λ=0.55 µm and a density of 1.25 g/cm^3.

In the following section the results of Mie calculations for two model cases are shown. For case A, a variable BC fraction derived from the power law was used. For case B, the mean BC mass fraction from the chemical analysis was assumed and kept constant in time. In both model cases the chemical composition was assumed to be independent of particle size.

4. RESULTS FOR COMBUSTION OF MUSASA

The burning conditions during the musasa experiment (Fig. 1) can be summarized as follows: The emission ratio of CO relative to CO_2 ($\Delta CO/\Delta CO_2$) varied between 7 and 10%. Since an emission ratio of 8% is often used as limit between a flaming ($\Delta CO/\Delta CO_2 < 8\%$) and a smoldering ($\Delta CO/\Delta CO_2 > 8\%$) dominated fire, the emission ratio in Fig. 1 indicates that we had a mixture of flaming and smoldering combustion with alternating periods slightly dominated by one of the two phases. Only towards the end of the experiment the smoldering became more important.

The temporal development of the measured scattering and absorption coefficients and the single scattering albedo at $\lambda=0.55$ µm as well as the accompanying model results are shown in Figs. 2–4.

The mass scattering coefficient (Fig. 2) varies between 2.4 and 5.1 m^2/g with a mean of 4.4 m^2/g. This average is higher than the mean for ambient young biomass burning aerosol, which is around 3.6 m^2/g (Houghton et al., 2001) and is more suitable for aged biomass burning aerosol (Reid et al., 2005b). Both models reproduce the mean scattering coefficient well, but the observed temporal evolution is only modeled satisfyingly from the start of the measurements up to 13:00. Furthermore, the use of the time-dependent BC fraction has no obvious effect on the scattering coefficient and does therefore not lead to an improvement of the model in case of scattering.

The observed mass absorption coefficient (Fig. 3) exhibits two maxima around 13:14 and 13:24, which can be explained by a higher contribution of flaming combustion as indicated by the minima in $\Delta CO/\Delta CO_2$ (Fig. 1). For the mean absorption coefficient we calculate a value of 0.36 m^2/g which lies in the range between 0.2 and 0.7 m^2/g that is typically found in smoldering dominated fires whereas reported values for a mixed phase combustion (as encountered here) are typically slightly larger between 0.6 and 1.0 m^2/g (Reid et al., 2005b). While the average absorption was modeled quite well with a constant BC fraction (case B), the temporal structure is much better described in model case A.

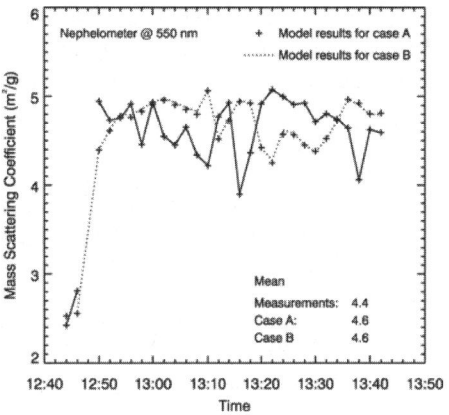

Fig. 2. Comparison of the mass scattering coefficient at $\lambda=0.55$ µm between measured and modeled values for a musasa fire. Fire average values are also given.

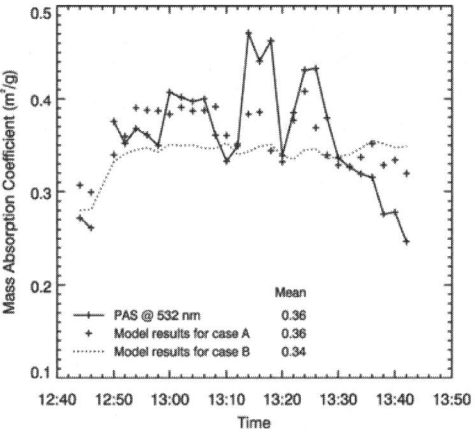

Fig. 3. Comparison of the mass absorption coefficient at $\lambda=0.55$ µm between the measurements and the two model cases. Fire average values are also given.

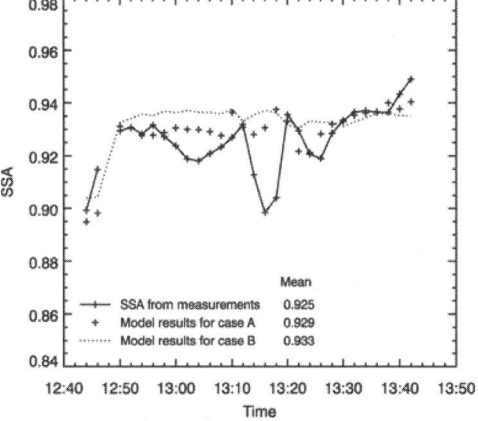

Fig. 4. Comparison of the single scattering albedo at $\lambda=0.55$ µm for measured and modeled values. Temporal averages are given as well.

Hence, much of the observed temporal variability in aerosol absorption can be attributed to changes in BC content, which are due to varying burning conditions. The observed single scattering albedo at $\lambda=0.55$ µm (Fig. 4) varies between 0.89 and 0.95 with a mean value of 0.93, which is in the typical range of a smoldering combustion (Reid et al., 2005b). Again, including particle chemistry (case A) improves the correlation between data and model, but discrepancies remain.

5. DISCUSSION AND CONCLUSIONS

We have shown a comparison between measurements and Mie calculations of the mass scattering coefficient, the mass absorption coefficient and the single scattering albedo for biomass burning particles at $\lambda=0.55$ µm for controlled combustion of musasa, an African hardwood. Although the age of the particles was only a few minutes and the burning conditions were a mixture of flaming and smoldering combustion, the aerosol optical properties were more typical for smoldering or aged particles. We attribute this at least in part to unnaturally enhanced particle growth due to coagulation in the mixing chamber (particle concentration ~ 10^6 particles per cm^3). The measured average scattering Ångström exponent (1.4 and 2.0 between 0.45–0.55 µm and 0.55–0.7 µm, respectively) and the mean backscatter ratio at 0.55 µm (b=0.12) are also indicative of relatively large particles. Values of fresh smoke are typically between 2 and 2.5 in case of the Ångström exponent and around 0.18 for the backscatter ratio, but particle growth during aging reduces the Ångström exponent as much as one and results in lower backscatter ratios with values around 0.12 (Reid et al., 2005b). To simulate the optical properties with Mie calculations, the BC mass content was estimated based on the measured burning conditions ($\Delta CO/\Delta CO_2$) and the measured fire-integrated EC mass fraction. The use of this BC parameterisation improves the modelling of the absorption coefficient and single scattering albedo. This stresses the impact of burning conditions on light-absorbing properties of smoke aerosol even for relatively constant burning conditions as encountered here. The aerosol scattering can be modelled quite well using average BC contents. The derived refractive index of 1.53-0.012i at $\lambda=0.55$ µm for a BC fraction of 10.4% is in agreement with values from other fires (Reid et al., 2005b). Additional closure studies and the application to the other EFEU experiments are needed to further constrain the complex refractive indices for fresh biomass burning aerosol for different biofuels and burning conditions.

6. ACKNOWLEDGEMENTS

This work was funded by the Federal Ministry of Education and Research (BMBF), Germany within the AFO 2000 Program under Grant 07 ATF 46 (EFEU).

7. REFERENCES

Fenn, R. W., S. A. Clough, W. O. Gallery, R. E. Good, F. X. Kneizys, J. D. Mill, L. S. Rothman, E. P. Shettle, and F. E. Volz, 1985: *Handbook of Geophysics and Space Environment*.

Houghton, J. T., Y. Ding, D. J. Griggs, M. Noguer, P. J. van der Linden, and D. Xiaosu, 2001: *Climate Change 2001: The Scientific Basis*, Cambridge University Press.

Lobert, J. M., D. H. Scharffe, W.-M. Hao, and P. J. Crutzen, 1990: Importance of biomass burning in the atmospheric budgets of nitrogen-containing gases, *Nature*, 346, 552-554.

Maßling, A., A. Wiedensohler, B. Busch, C. Neusüß, P. Quinn, T. Bates, and D. Covert, 2003: Hygroscopic properties of different aerosol types over the Atlantic and Indian oceans, *Atmos. Chem. Phys*, 3, 1377-1397.

Maxwell-Garnett, J.C., 1905: Colours in metal glasses and in metallic films, *Philos. Trans. R. Soc. London*, 203, 385-420.

Reid, J. S., R. Koppmann, T. F. Eck, and D. P. Eleuterio, 2005a: A review of biomass burning emissions, part II: Intensive physical properties of biomass burning particles, *Atmos. Chem. Phys.*, 5, 799-825.

Reid, J. S., T. F. Eck, S. A. Christopher, R. Koppmann, O. Dubovik, D. P. Eleuterio, B. N. Holben, E. A. Reid, and J. Zhang, 2005b: A review of biomass burning emissions, Part III: Intensive optical properties of biomass burning particles, *Atmos. Chem. Phys.*, 5, 827-849.

Ross, J. L., P. V. Hobbs, and B. Holben, 1998: Radiative characteristics of regional haze dominated by smoke from biomass burning in Brazil: Closure tests and direct radiative forcing, *J. Geophys. Res.*, 103 (D24), 31925-31941.

Schmid, O., et al., 2004: Physical Properties of Biomass Burning Aerosol from Various Fuel Types, *8th Int. Conf. on Carbonaceous Particles in the Atmosphere*, Sept 14-16, Vienna, 2004.

Schnaiter, M., et al., Measurement of Wavelength-Resolved Light Absorption by Aerosols Utilizing a UV-VIS Extinction Cell, *Aerosol Sci. Technol.*, accepted 2005.

VDI 2465 Part 2, 1999: Measurement of soot (Ambient Air) – Thermographical determination of elemental carbon after thermal desorption of organic carbon, VDI Handbuch Reinhaltung der Luft 4.

Wurzler, S., et al., 2001: Impact of vegetation fires on the composition and circulation of the atmosphere: Introduction of the research project EFEU, *J. Aerosol Sci.*, 32(1), 199-200.

A CODE TO COMPUTE THE DIRECT SOLAR RADIATIVE FORCING: APPLICATION TO ANTHROPOGENIC AEROSOLS DURING THE ESCOMPTE EXPERIMENT

P. Dubuisson and J.-C. Roger
ELICO, Université du Littoral Côte d'Opale, 32 av. Foch, 62930 Wimereux, France

M. Mallet
LEPI, Université de Toulon et du Var, Av. Georges Pompidou, B.P.56, 83162 La Valette du Var, France

O. Dubovik
University of Maryland and NASA/GSFC Code 923, Greenbelt, MD 20771, USA

ABSTRACT

A fast yet accurate radiative transfer code (GAME) has been developed to calculate solar fluxes with absorption, scattering as well as their interactions. Using this code, a methodology has been defined to characterize the aerosol properties and to estimate the local direct aerosol forcing. Data from the ESCOMPTE experiment, which took place in the vicinity of Marseille (France) during summer 2001, provide the possibility to realize a local study during pollution events. Major results are: (1) a retrieval of the microphysical and optical properties of POM for one of the first time; (2) a retrieval of microphysical and optical properties for soot particles different of those usually used; (3) a large impact of anthropogenic aerosols (BC, Sulphate and POM) on the daily direct forcing.

1. INTRODUCTION

The direct radiative effect of aerosols represents one of the largest sources of uncertainty in estimating regional climate changes. Each aerosol type presents its own chemical and microphysical properties, which control its interaction with the solar radiation, leading to different direct radiative forcings. As reported by Haywood and Boucher [2000] and *IPCC* [2001], certain aerosols (such as sulphate, ammonium, organic carbon and nitrate) exert a cooling direct effect at the top of the atmosphere (TOA), while other, as black carbon aerosol exerts on the contrary a heating direct effect at the TOA.

The aim of this work is to present a solar radiative transfer scheme (GAME: Global Atmospheric ModEl) to calculate the aerosol forcing at a local scale (0D and 1D). This code also allows a validation of the 3D models through inter-comparisons with satellite observations. In the GAME code, the Radiative Transfer Equation (RTE) is accurately solved using the Discrete Ordinate Method to account for multiple scattering, and the correlated k-distribution is applied to account for gaseous absorption as well as interactions with scattering. Radiative fluxes, radiances and heating rates can be calculated at each level of the atmosphere, in the 0.2–3 μm spectral range.

This code has been applied to urban aerosols, using in situ measurements collected during the ESCOMPTE experiment (July 2001), in a south-eastern part of France near the city of Marseille (Cros et al., 2003). Due to urban and industrial emissions, and due to a strong solar radiation, high and frequent pollution events are regularly observed in this area. The radiative fluxes at each altitude are computed with GAME and the daily direct aerosol forcing deduced for the aerosol and its components. Contribution of different aerosol species to the direct forcing is evaluated (black carbon, ammonium sulphate and particulate organic matter for the accumulation and coarse modes). Especially, individual direct radiative forcing at surface, at the top of the atmosphere and into the atmosphere is presented.

2. RADIATIVE TRANSFER MODELING

Solar fluxes (0.3 to 5 μm) are calculated using a fast yet accurate radiative transfer code, referred to as the GAME code (Dubuisson et al., 2004). This code accounts for absorption and scattering, as well as their interactions, from the Discrete Ordinates Method (DOM) (Stamnes et al., 1988). This accurate method allows calculating fluxes at any level with the assumption of a vertically inhomogeneous atmosphere, stratified into plane and homogeneous layers (20 layers).

Gaseous absorption (H_2O, CO_2, O_2 and O_3) is treated from the correlated k-distribution (Lacis et Oinas, 1991). For an absorber amount u, the transmission function is approximated by a transmission integral over the density distribution of absorption coefficient strengths:

$$T(u) = \int_0^\infty f(k) \exp(-ku) dk . \quad (1)$$

For each spectral interval Δν, the function f(k) represents the probability of absorption coefficient k to occur in the spectral interval between k and k+dk. For a layer at pressure P and temperature T, the transmittance is then calculated by an exponential summation over a limited number N of absorption classes as:

$$T_{\Delta v}(P,T) = \sum_{i=1}^{N} a_i \exp[-k_i(P,T)u(P,T)] . \quad (2)$$

The weight a_i represents the probability associated to the mean absorption coefficient k_i for each class i. The coefficients a_i and k_i are determined from the density distribution f(k) which can be obtained from analytic expressions. Indeed, the transmission formula in (1) has the standard form of a Laplace transform and the probability density function f(k) can be formally identified as the inverse Laplace transform of T(u). The Malkmus band model possesses an inverse Laplace transform and it provides convenient analytical expressions for the transmission function T(u) (see Lacis et Oinas, 1991). For the GAME code, band model parameters are directly estimated by least-square fitting the band model T(u) with reference calculations $T_{ref}(u)$. This reference code LBLDOM (Dubuisson et al, 1996) includes the DOM at each step of a line-by-line (LBL) model. The maximal deviations between T(u) and $T_{ref}(u)$ are then on the order of 1% for a spectral resolution of 100 cm^{-1}. Simulations showed that these deviations can reach 10-20% if band model parameters are obtained directly from statistical line spectrum from a spectroscopic database.

Coefficients of the exponential series (a_i and k_i) are then estimated for a reference pressure P_0 and temperature T_0 with the LBLDOM code. Following the concept of the correlated k-distribution, the exponential sum is assumed to be correlated at any pressure level with that at P_0 and T_0 for a given spectral interval. A scaling approximation can then be applied to account for the vertical dependence of the absorption as:

$$k_i(P,T) = k_i(P_0,T_0)(\frac{P}{P_0})^n (\frac{T_0}{T})^m . \quad (3)$$

The primary advantage of the correlated k-distribution technique is that it can be used for multiple scattering, with manageable computational time (about 4 seconds per case). The GAME code has a fixed spectral resolution Δv of 100 cm^{-1} (0.6 to 5 µm) or 400 cm^{-1} (0.2 to 0.6 µm) with an exponential number N equal to 7 per interval with gaseous absorption. The Radiative transfer Equation (RTE) is then solved for each term of the exponential series. As a comparison, the LBL spectral resolution is about 0.01 cm^{-1} in the solar spectrum and calculations need to solve about 10000 RTE over a spectral interval of 100 cm^{-1}.

The GAME code also includes Rayleigh and aerosol scattering. For each atmospheric layer, the DOM needs to know the gaseous absorption optical thickness (calculated with the correlated k-distribution) as well as scattering parameters such as moments of the phase function, single scattering albedo and extinction optical thickness for aerosols and molecular scattering. Note that radiative fluxes, radiances and heating rates can be simulated with GAME.

The accuracy has been evaluated during the ICRCCM program (Halthore et al., accepted, 2005): InterComparison of Radiation Codes in Climate Models.

During this program, 16 solar radiative transfer codes have been tested with varying spectral resolution ranging from line-by-line to broadband models. An example of results is presented in Table 1 for two aerosol optical thicknesses and a tropical atmosphere. Differences on fluxes between GAME and the mean of the 16 codes are always less than 1.5 W/m^2 and the GAME accuracy is similar to the other codes.

Table 1. Comparison Between ICRCCM (Halthore et al., 2005) and GAME Fluxes

Fluxes (W/m^2)	LOW Aerosols			HIGH Aerosols		
	Mean	SD	GAME	Mean	SD	GAME
Direct Down Surface	785.0	5.6	784.9	692.5	5.9	692.1
Diffuse Down Surface	115.7	1.7	114.6	188.2	1.7	186.8
Diffuse Up Surface	181.0	1.7	179.9	176.9	1.7	175.8
Diffuse Up TOA	207.6	2.4	208.3	210.6	2.4	211.2

Mean, SD: mean and Standard Deviation from radiative transfer codes used during the ICRCCM program
LOW aerosols: optical thickness = 0.08 at 550 nm.
HIGH aerosols: optical thickness = 0.24 at 550 nm.

3. MICROPHYSICAL PROPERTIES

A first application has been done during the ESCOMPTE experiment (Cros et al., 2003). The aerosol chemical properties were measured at Vallon d'Ol (near Marseille in the southern part in France) using ionic chromatography (ions) and thermal method (black and organic carbon). The aerosol size distribution was estimated from BERNER impactor at ambient relative humidity. For each chemical massic size distributions (Soot, P.O.M., Sulphate, Sea Salt, Nitrate, Dust), the mass mean radii and the standard deviations (σ) of each component have been defined (Mallet et al., 2003). Mass mean radii have been transformed into number mean radii (r_n). Results are presented in Table 2. A fine mode for Soot, P.O.M., Sulphate, Dust, and a coarse mode for P.O.M., Nitrate, Sea Salt and Dust have been retrieved. Soot presents differences with commonly used values. P.O.M. characteristics are obtained for one of the first time.

Table 2. Mean Microphysical Properties (Dry State) from ESCOMPTE Data (9 Days)

	Fine mode		Coarse mode	
	r_n (μm)	σ	r_n (μm)	σ
Soot	0,027	1,95		
POM	0,028	1,84	0,313	1,96
Sulphate	0,040	1,74		
Sea Salt			0,350	1,75
Nitrate			0,505	1,61
Dust	0,036	1,97	0,36	1,98

Table 3. Mean Optical Properties at 550 nm (Wet State)

	Kext	Kscat	ω0	g
Soot	0,0116	0,0039	0,336	0,434
POM nuc.	0,0094	0,0092	0,972	0,552
POM acc.	0,0022	0,0020	0,946	0,738
Sulphate	0,0260	0,0252	0,970	0,585
Sea Salt	0,0014	0,0014	0,990	0,734
Nitrate	0,0018	0,0017	0,930	0,742
Dust nuc.	0,0016	0,0016	0,960	0,563
Dust acc.	0,0031	0,0028	0,890	0,719

4. OPTICAL PROPERTIES

Using Mie theory, the main optical properties of aerosols (the extinction and scattering coefficients K_{ext} and K_{scat}, the single-scattering albedo ω_o, and the asymmetry parameter g) have been computed (Table 3) assuming an external mixture on this zone (Mallet et al., 2004). Almost 90 % (mean value) of the aerosol extinction at 550nm is due to anthropogenic aerosol (during pollution events):

- Sulfate: 45%
- Soot: 20%
- POM: 20%
- Nitrate: 5% and 10% from natural aerosols (Sea Salt and Dust).

We also computed the optical properties assuming an internal mixture (Fig. 1). In that case, and compared to the external case, g is slightly higher, and ω_o lower. With our values of Soot and regarding to usual ones, the weight of absorption decreases compared to the scattering. Moreover, the solar energy is more scattered by Soot particles in the forward direction.

5. LOCAL DIRECT AEROSOL FORCING

The direct aerosol forcing is computed at the surface level (BOA) and at the top of the atmosphere (TOA). Fluxes F have been computed using the GAME code. The direct forcing BOA and TOA are defined by:

$\Delta F\downarrow = F\downarrow$ (BOA, with aerosol) $- F\downarrow$ (BOA, molecular)
$\Delta F\uparrow = F\uparrow$ (TOA, with aerosol) $- F\uparrow$ (TOA, molecular)

For the GAME code, vertical profiles of temperature, pressure, ozone and relative humidity are available from in-situ measurements. Aerosol optical properties profiles have been calculated. Surface albedo has been estimated from MODIS data. At last, aerosol optical thicknesses from the AERONET/PHOTON network (Holben et al., 1998; Dubovik et al., 2002) have been used (Fig. 2).

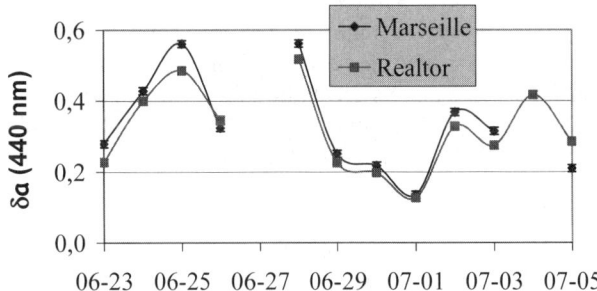

Fig. 2. Aerosol optical thickness δ_a at 440 nm from AERONET/PHOTON network. Note that all AERONET wavelengths have been used for aerosol forcing calculations.

BOA and TOA daily direct aerosol forcing presented in Fig. 3 show a large aerosol forcing in this peri-urban zone. Table 4 presents a large predominance of the anthropogenic aerosols (Soot and Sulphate are the main contributors; POM has a non negligible impact). At the surface, aerosols lead to a mean daily forcing of -32 W/m², in correlation with the aerosol optical thicknesses.

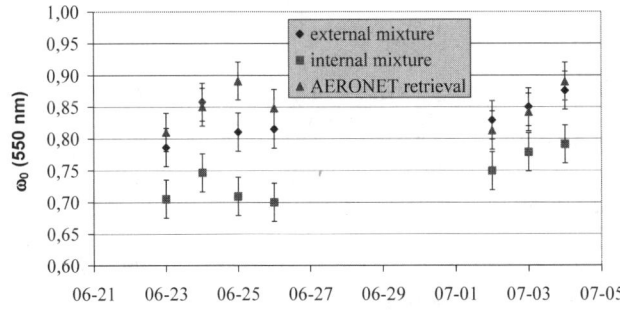

Fig. 1. Single scattering albedo ω_o at 550 nm (internal and external mixture); AERONET/PHOTON retrievals are also reported (Dubovik et al., 2002).

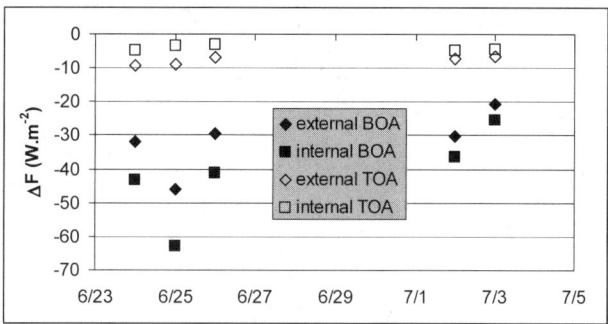

Fig. 3. BOA and TOA daily direct aerosol forcing ΔF in case of external and internal mixtures. Note that the direct radiative forcing is directly depending of the value of the total flux. Thus, taking into account the upward and downward transmission, values of the impact at BOA is higher than those at TOA.

6. CONCLUSIONS

A radiative transfer code has been developed allowing simultaneously (1) the retrieval of aerosol optical properties and (2) the computation of solar aerosol forcing. The next step will be to take into account the thermal spectral range. From the aerosol chemical characteristics, acquired during the ESCOMPTE experiment, the microphysical and optical properties of each aerosol components during pollution events, in a first step, and the daily direct aerosol forcing for each component (Roger et al., submitted JGR), in a second step, have been determined. Major results are: (1) a retrieval of the microphysical and optical properties of POM for one of the first time; (2) microphysical and optical properties for soot particles different of those usually used; (3) a large impact of anthropogenic aerosols (BC, Sulphate and POM) on the daily direct forcing. This radiative transfer scheme allows accounting for a vertical structure of the polluted low troposphere and will be a useful tool in the frame of urban pollution and air quality studies.

Table 4. Example of Daily Direct Aerosol Forcing ΔF (24th of June) for an External Mixing State

ΔF (W.m^{-2}) total	BOA	TOA
Soot	**-15.5**	**2.3**
POM nuc.	**-2.6**	**-1.8**
POM acc.	-0.6	-0.3
Sulfate	**-9.9**	**-6.3**
Nitrate	-0.9	-0.3
Sea Salt	-0.4	-0.3
Dust nuc.	-0.1	-0.1
Dust acc.	-0.2	-0.1

Note: total row values are -32.4 (BOA) and -9.0 (TOA).

7. ACKNOWLEDGEMENTS

The authors are undebted to Dr Bernard Cros and Dr Pierre Durand for the organization of ESCOMPTE experiment. They gratefully thank the different Institutes in France (CNRS, INSU, ADEME, and Ministry of Environment) for their support. Logistic help from local agencies (AIRMAREX and AIRFOBEP) is gratefully acknowledged.

8. REFERENCES

Dubovik O., B. N. Holben, T. F. Eck, A. Smirnov, Y. J. Kaufman, M. D. King, D. Tanré, and I. Slutsker, 2002, Variability of absorption and optical properties of key aerosol types observed in worldwide locations, *J. Atmos. Sci.*, 59, 590-608.

Dubuisson P., J. C. Buriez, and Y. Fouquart, 1996, High spectral resolution solar radiative transfer in absorbing and scattering media, application to the satellite simulation, *J. Quant. Spectrosc. Radiat. Transfer*, 55, 103-126.

Dubuisson P., D. Dessailly, M. Vesperini, and R. Frouin, 2004, Water Vapor Retrieval Over Ocean Using Near-Infrared Radiometry, *J. Geophys. Res.*, Vol. 109, D19106, doi:10.1029/2004/JD004516.

Cros, B., P. Durand, H. Cachier, Ph. Drobinski, E. Frejafon, C. Kottmeïer, P. E. Perros, V.-H. Peuch, J. L. Ponche, D. Robin, F. Saïd, G. Toupance, and H. Wortham, 2003, The ESCOMPTE program: an overview, *Atmos. Res.*, 69, 3-4, 241-279

Halthore R., et al., 2005, Intercomparison of shortwave radiative transfer codes and measurements, *J. Geophys. Res.*, accepted.

Haywood, J. M., and O. Boucher, 2000, Estimates of the direct and indirect radiative forcing due to tropospheric aerosols: a review, *Rev. Geophys.*, 38, 513-543.

Holben, B. N., et al., 1998, AERONET-A federated instrument network and data archive for aerosol characterization, *Remote Sens. Environ.*, 66, 1-16.

Intergovernmental Panel on Climate Change, Climate Change, 2001, edited by J. T. Houghton et al., *Cambridge Univ. Press*, New York.

Lacis, A. A., and V. Oinas, 1991, A description of the correlated k-distribution method, *J. Geophys. Res.*, 96, 9027-9064.

Mallet M., J.-C. Roger, S. Despiau, O. Dubovik, and J. P. Putaud, Microphysical and optical properties of aerosol particles in urban zone during ESCOMPTE, 2003, *Atmospheric Research*, 69, 73-97.

Mallet M., J.-C. Roger, S. Despiau, J. P. Putaud, and O. Dubovik, A study of the mixing state of black carbon in urban zone, *J. Geophys. Res.*, 109, D04202, doi:10.1029/2003JD003940, 2004.

Stamnes K., S. Tsay, W. Wiscombe, and K. Jayaweera, 1988, Numerically stable algorithm for discrete-ordinate-method radiative transfer in multiple scattering and emitting layered media, *Appl. Opt.*, 27, 2502-2509.

NUMERICAL CALCULATIONS OF GRAVITO-PHOTOPHORETIC MOVEMENT FOR AEROSOL AGGREGATES

Alexander A. Cheremisin
Krasnoyarsk State Technical University
Kirensky str., 26, Krasnoyarsk, 660074, Russia. Email: cher@akadem.ru

Yuri V. Vassilyev
Research Institute of Physics & Engineering at the Krasnoyarsk State University
P.O. Box 8678, Akademgorodok, 13a, Krasnoyarsk, 660036, Russia. Email: ygrab@online.ru

ABSTRACT

In this paper we discuss formation of aerosol layers under the influence of gravito-photophoretic forces. We examine particles of various sizes, absorption power of solar and terrestrial IR radiation and thermal accommodation coefficients over surface of particles. We show how the effect of gravito-photophoretic forces can explain structure of the stratospheric and mesospheric aerosol layers.

1. INTRODUCTION

Stable aerosol layers are observed at different altitudes in the stratosphere and the mesosphere of the Earth: the Junge layer at the altitude about 20 km, the noctilucent clouds at altitudes 80–83 km in the cold polar summer mesosphere, the polar stratospheric clouds.

According to the data of our space observations in ultraviolet spectral band by the method of tangent sensing, the stable aerosol layers exist under unperturbed atmospheric conditions at altitudes 50, 70, 93 km in the equatorial zone and at the middle latitudes. After large space vehicles launches, such as Space Shuttle, an anthropogenic aerosol layer is formed along the active trajectory of launch at altitudes near 100 km (Cheremisin et al., 2000).

These and other known facts of aerosol stratification in the atmosphere can be interpreted within the framework of existing models which are based on the idea of sedimentation-diffusion equilibrium in the gravity field (Turco et al., 1982). But these models leave some questions unresolved. Aerosol particles can be formed, and exiting ones can grow, in an environment that is saturated with sulfuric acid or water vapor, and hence when atmospheric temperature is sufficiently low. Therefore, it is difficult to explain existences of layers at 50 km near stratopause and above 70 km.

An alternative concept used to explain the existence of such aerosol layers and to correctly describe the vertical transport of aerosol particles is based on photophoretic forces. This concept is based first of all on the results of laboratory observations of levitating particles sustained by absorbed light. The theoretical study of photophoretic forces has a long history, but its application to atmospheric processes such as the vertical transport of aerosol particles (Rohatschek, 1996; Pueschel et al., 2000) or aerosol layers (Cheremisin et al., 2002; Beresnev et al., 2003) was initiated quite recently.

2. THEORETICAL MODELING

Presently, there are three known mechanisms of lifting force inducing under influence of photophoresis: (1) is a classical photophoretic effect (ΔT-forces) for ideal spherical particles. According to the new estimates, the photophoretic force is low as compared to the gravity (Tehranian et al., 2001, Beresnev et al., 2003); (2) is gravito-photophoresis of $\Delta\alpha$-type when there is sufficient variation of the thermal accommodation coefficient α over the particle surface (Rohatschek, 1996), and (3) is gravito-photophoresis of ΔT-type for aerosol aggregates consisting of individual particles differing in temperature due to a difference in their physical properties and in radiation absorption cross-section (Cheremisin et al., 2002). The latter two mechanisms generate much greater lifting force than the former one. For monospherical particles strongly absorbing sunlight, estimates show that gravito-photophoretic forces of $\Delta\alpha$-type are capable of lifting particles of micrometer radius from the troposphere into the mesosphere (Rohatschek, 1996). Therefore it was concluded that these forces can provide vertical transport of fractal soot (Pueschel et al., 2000).

For nonspherical or cluster particles the description of gravito-photophoresis requires appropriate computing methods of mathematical simulation. We have developed the technique for calculation of particle motion in free-molecule regime by using Monte Carlo method. Our technique differs from that based on the hydrodynamic equations with so-called slip-flow boundary conditions when Knudsen number is small (Filippov, 2001). For the middle and the upper atmosphere where the Knudsen number is large enough, the Monte Carlo method is more applicable.

The generalized form of motion equation for the spherical, nonspherical and aggregated particles is

$$\begin{bmatrix} m_p \\ \hat{I}_p \end{bmatrix} \cdot \frac{d}{dt} \begin{bmatrix} v \\ \omega \end{bmatrix} = \begin{bmatrix} F_f \\ M_f \end{bmatrix} + \begin{bmatrix} F_{vis} \\ M_{vis} \end{bmatrix} + \begin{bmatrix} G \\ 0 \end{bmatrix} = \begin{bmatrix} F_f \\ M_f \end{bmatrix} + \begin{bmatrix} \hat{v} & \hat{\sigma} \\ \hat{\eta} & \hat{\xi} \end{bmatrix} \cdot \begin{bmatrix} v \\ \omega \end{bmatrix} + \begin{bmatrix} G \\ 0 \end{bmatrix}$$

where v and ω are the linear and the angular velocities of the particle. m_p and $\hat{I}_p$ are its mass and tensor of inertia. G is the gravity. There are a photophoretic force F_f and its torque M_f to describe the photophoretic and the gravito-photophoretic effects. Dependences of a viscous force F_{vis} and its torque M_{vis} on v and ω, within a linear approximation, can be represented by the four 3×3 matrixes $\hat{v}, \hat{\eta}, \hat{\sigma}, \hat{\xi}$. The tensors $\hat{\eta}, \hat{\sigma}$ are to describe the connections between rotation and translation. Here the torques produced by the forces are shown relative to a particle's center of mass.

Photophoretic force causes the movement of particle absorbing solar and terrestrial IR radiation when its temperature differs from temperature of the surrounding gas. The coefficients of molecular heat transfer are calculated by using the technique based on Monte Carlo method.

3. RESULTS

A simple type of clusters for demonstration of gravito-photophoretic effect is the bispherical particle. The velocities of sedimentation for the bispherical particles consisting of spheres of 1, 2 and 3 µm in diameter, calculated with and without taking into account the photophoretic forces of $\Delta\alpha$-type, are shown in Fig. 1.

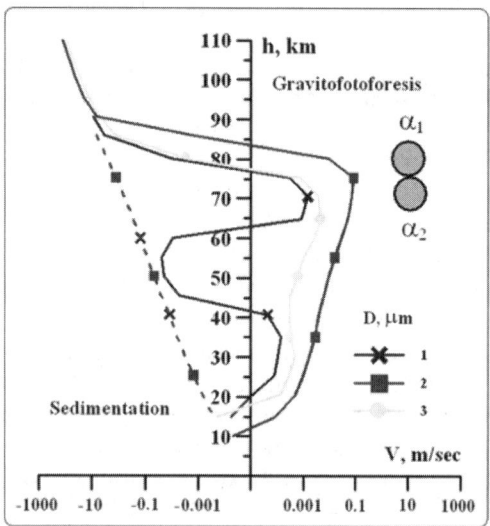

Fig. 1. The velocities of sedimentation of bispherical particles with different thermal accommodation coefficients $\alpha_1 = 0.74, \alpha_2 = 0.86$ for upper and down spheres.

Both spheres strongly absorb visible (solar) and terrestrial IR radiation. Dashed line show the case when the gravito-photophoretic force equals zero. Solid lines present the particles movement under influence of gravito-photophoretic force. If $V > 0$ the particles move up, when $V < 0$ they go down.

For the calculation of total mass of bispherical particle the average density of spheres of 1000 kg/m^3 is used. But it is supposed that the spheres are not identical and the centre of gravity of bispherical particle is displaced down by one fourth of radius from the geometrical centre along the symmetry axis. There is a stable orientations of particles in space when particles are moving in accordance with the motion equation, and a gravito-photophoretic effect arises. Generally, clusters particle's movement is accompanied by its rotation.

The gravito-photophoretic forces of $\Delta\alpha$-type for bispherical particles are shown in Fig. 2. Both of spheres represent the particles (black particles) which, similarly to particles of soot, strongly absorb both visible (solar) and terrestrial IR radiation. The force is induced owing to the difference in the value of accommodation coefficient, α_1 for first sphere and α_2 for another. The temperatures of the spheres are equal, $T_1 = T_2$, but differ from the temperature of atmospheric gas.

We have divided the force F by gravity G and parameter $\beta = (m_0/m)[(\alpha_2 - \alpha_1)/(\alpha_2 + \alpha_1)]$, where m is the total mass of bispherical particle, m_0 is reference mass when density of the both spheres is equal to $\rho_0 = 1000$ kg/m^3. If averaged density of spheres is equal to ρ_0 we have $0 \leq \beta < 1$. At altitudes less than 60-70 km the force is proportional to β.

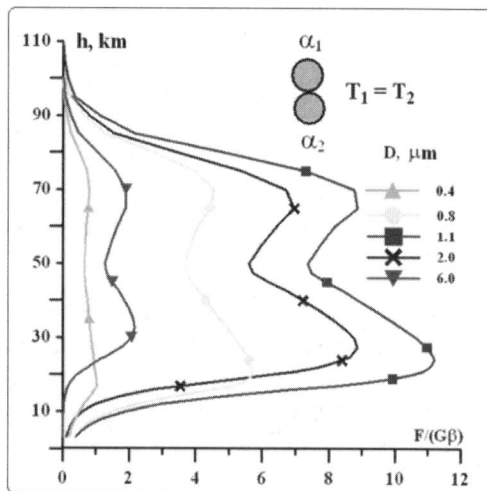

Fig. 2. The normalized gravito-photophoretic force of $\Delta\alpha$-type for the bispherical particles of the same type as in Fig. 1.

Random molecular perturbations of the particle orientation and reduction of the force at high pressure for Knudsen number $Kn < 1$ where taken into account; therefore, the results of these calculations are more accurate than those in the work (Cheremisin et al., 2002).

The large lifting forces can be induced for bispherical particle, which similar to sulphate containing aerosols strongly absorb terrestrial IR radiation and weakly absorb sunlight (transparent particles) (Fig. 3).

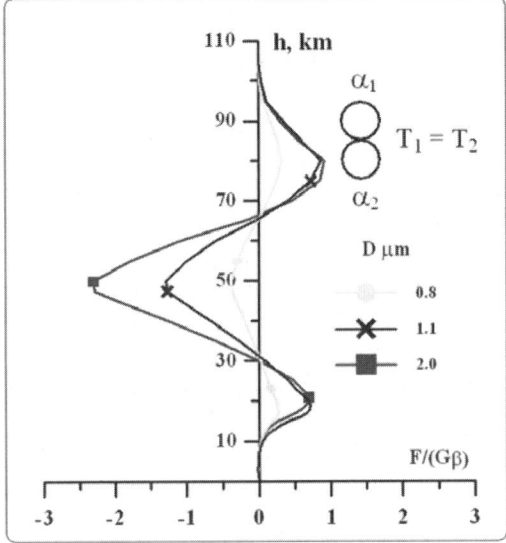

Fig. 3. The normalized gravito-photoretic force of $\Delta\alpha$-type for bispherical particles. Both of the spheres similar to sulphate containing aerosols strongly absorb only terrestrial IR radiation. $\alpha_1 = 0.74$, $\alpha_2 = 0.86$.

It was assumed that the absorption cross section in the visible band is equal to zero, and in an infra-red range it has the same value as that of soot. The temperatures of the particles are equal. The force is induced owing to the difference in the value of accommodation coefficient.

The forces for bispherical particles are close to those for monospherical particles when corresponding values of the force parameter β are equal (Cheremisin A. A., Vassilyev Yu.V., Horvath H., 2005, person. comm).

The mechanism of ΔT gravito-photophoresis can induce large lifting forces when spheres differ in their physical properties and in radiation absorption power for solar and IR radiation, even if the values of accommodation coefficient are the same for both spheres (Fig. 4). The force arises because the temperatures of particles differ from each other. The heat transfer through contacts of adjoining particles of cluster was neglected.

Altitudes for which the value of velocity is equal to zero correspond to the level at which the gravito-photoretic forces are balanced by the gravity. A stable layer is formed if $F/G = 1$ and the value F/G decreases with increasing altitude (Cheremisin et al., 2002). This is the case at several altitudes, as can be seen from Fig. 5.

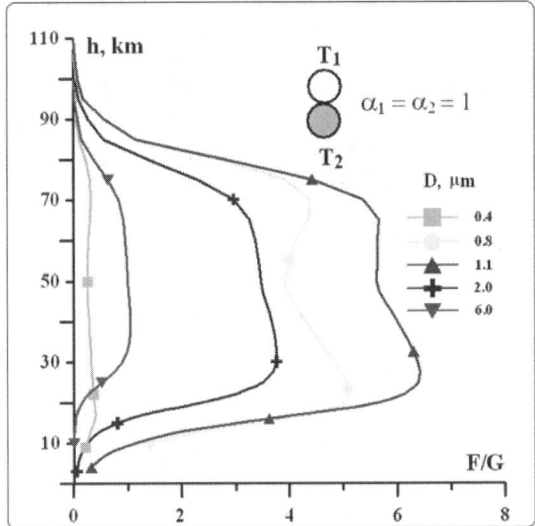

Fig. 4. Gravito-photoretic force of ΔT-type for bispherical particles ($\alpha_2 = \alpha_1 = 1$, $T_1 < T_2$). Down sphere is black in visible and in IR, upper is transparent in visible and black in IR.

Every point of black area in Fig. 5 corresponds to altitude of suspended particle of certain size for given value of β. There are the threshold values of β for stratospheric and mesospheric layers formation. With increase of value of β the thickness (altitude range) of layers is going up. Gravito-photophoresis explains the existence of narrow aerosol layers. The thickness of the layers can be less than 1 km.

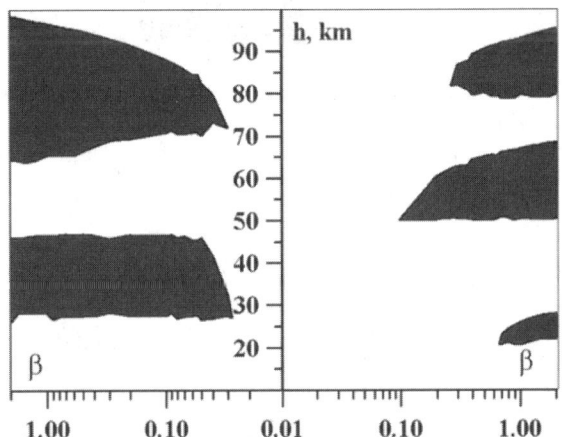

Fig. 5. The thickness of the layers in the equator on dependence of β for bispherical particles. Black particles strongly absorbing solar and IR radiation (left). Transparent particles weakly absorbing solar radiation and black in IR (right).

It takes a small value of β (a certain differences in accommodation coefficient over surface of particle) for the black particles to form the layers in stratosphere and at altitudes above 70 km (observations, e.g., see Cheremisin et al., 2000). Therefore they easily expand over the globe. The transparent particles are also capable of forming global layers above 50 km (observations, e.g., see Rozenberg, et. al., 1982; Cheremisin et al., 2000) But it requires a rather large difference in accommodation coefficient over surface for transparent particles to be suspended at altitudes about 20 km and about 80 km

The layers at altitudes 80–83 suspended by gravito-photophoresis occur in summer polar mesosphere of both hemispheres (observations, e.g., see Thomas et al., 2003).

The layer at the altitude about 20 km is formed in the equatorial zone (Fig. 6). Layers can also be formed in the polar areas in winter, but their altitudes differ from 20 km (observations, e.g., see Hitchman et al., 1994).

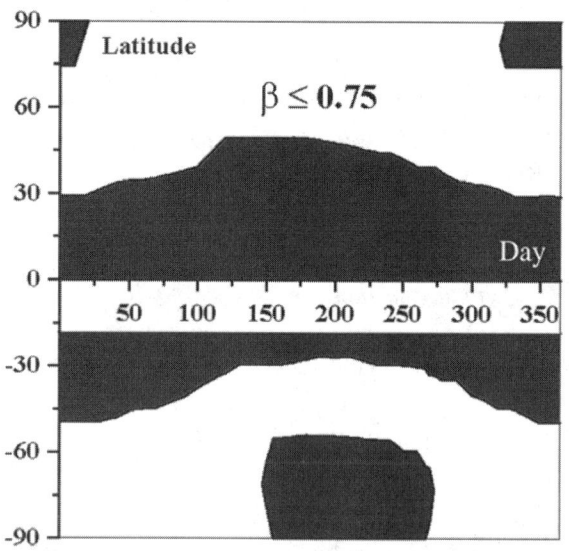

Fig. 6. The layer distribution at altitudes about 20 km depending on latitude and day of year. Transparent particles weakly absorbing solar radiation and black in IR.

4. SUMMARY

The procedure based on the Monte Carlo method for calculation of the non-spherical and aggregated particles movement is developed. Photophoretic forces, drag forces, connections between rotation and translation are taken into account

The obtained aerosol stratification under gravito-photophoresis qualitatively corresponds to the vertical distribution of aerosol scattering in the stratosphere and the mesosphere observed in the visible and ultraviolet wavelength range. It is true for layers at altitudes about 20 km (the Junge layer), 50 km, 70 km and 80–83 km (noctilucent clouds).

Both mechanisms of gravito-photophoresis of $\Delta\alpha$-type and of ΔT-type can induce large lifting forces for aerosol aggregates.

5. ACKNOWLEDGEMENTS

This research is supported by the grant 04-05-64390a by the Russian Fund of Fundamental Research. We also thank Local Organizing Committee of IRS-2004 for Travel Support.

6. REFERENCES

Beresnev, S. A., F. D. Kovalev, L. B. Kochneva, V. A. Runkov, P. E. Suetin, and A. A. Cheremisin, 2003: On the possibility of particle's photophoretic levitation in stratosphere. *Atmos. Oceanic Opt. (Engl. transl.)*, 16, 44-48.

Cheremisin, A., L. Granitskii, V. Myasnikov, and N. Vetchinkin, 2000: Improved aerosol scattering in the upper atmosphere, according to data of ultraviolet observations from space, with instrumental smoothing taken into account. *Proc. SPIE*, 4341, 383-389.

Cheremisin, A. A., Yu.V. Vassilyev, and A. V. Kushnarenko, 2002: Photophoretic forces for bispherical aerosol particles. *Proc. SPIE*, 5027, 23-34.

Filippov, A. V., 2001: Phoretic Motion of Arbitrary Clusters of N Spheres. *J. Colloid Interface Sci.*, 241 (2), 479-491.

Hitchman, M. H., M. McKay, and C. R. Trepte, 1994: A climatology of stratospheric aerosol. *J. Geophys. Res.*, 99 (D10), 20, 689-20, 700.

Pueshel, R. F., S. Verma, H. Rohatschek, G. V. Ferry, N. Boiadjieva, S. D. Howard, and A. W. Strawa, 2000: Vertical transport of anthropogenic soot aerosol into the middle atmosphere. *J. Geophys. Res.*, 105, 3727-3736.

Rohatschek, H., 1996: Levitation of mesospheric and stratospheric aerosols by gravito-photophoresis. *J. Aerosol Sci.*, 27, 467-475.

Rozenberg, G. V., I. G. Melnikova, and T. G. Megrelishvili, 1982: Aerosol stratification and its variability. *Izv. Akad. Nauk SSSR, Fiz. Atmos. Okeana*, 18 (4), 363-372.

Tehranian, S., F. Giovane, J. Blum, Y.-L. Xu, and B. A. S. Gustafson, 2001: Photophoresis of micrometer-sized particles in the free-molecular regime. *Int. J. Heat Mass Transfer*, 44, 1649-1657.

Thayer, J. P., G. E. Thomas, and F.-J. Lübken, 2003: Foreword: Layered phenomena in the mesopause region. *J. Geophys. Res.*, 108 (D8), 8434-8438.

Turco, R. P., R. C. Whitten, and O. B. Toon, 1982: Stratospheric Aerosols: Observation and Theory. *Rev. Geophys. Space Phys.*, 20, 233-279.

MULTISENSOR GROUND-BASED REMOTE SENSING OF RADIATIVELY IMPORTANT CLOUD PARAMETERS

S. Y. Matrosov

Cooperative Institute for Research in Environmental Sciences, University of Colorado at Boulder and National Oceanic and Atmospheric Administration's Environmental Technology Laboratory, Boulder, Colorado, 80305, USA

ABSTRACT

Multi-sensor remote sensing methods are potentially more accurate than those using single measurements. This invited paper presents a brief overview of remote sensing approaches that use different types of measurements from vertically pointing ground-based instruments to retrieve cloud microphysical properties that are radiatively important. The applicability of different approaches to various cloud types including liquid water clouds and ice clouds is discussed and the potential retrieval uncertainties are estimated.

1. INTRODUCTION

Clouds are among the least certain components of climatic models. The main cloud properties that are responsible for their radiative impact include extinction coefficient (and its vertical integral—optical thickness), the single scattering albedo, phase function (or its moments such as the asymmetry factor) and temperature. These parameters are used to calculate heating/cooling rates that reflect cloud contribution in the models. Cloud radiative properties are determined by their macrophysical characteristics such as cloud height and geometrical thickness, and, more importantly, by their microphysical parameters. The most important parameters are cloud content (liquid or ice water content, depending on cloud phase) and the cloud particle size distribution. Typically the whole distribution is not available from remote sensing measurements, and the size information is usually given in terms of some parameters describing this distribution. The single most important size distribution parameter is the cloud particle characteristic size, which is often given in terms of median or mean diameter of the equal-volume sphere.

Cloud microphysical retrievals are also important for various cloud physics applications since they can provide important information on cloud evolution and precipitation formation. In the last 10–15 years a number of different multi-sensor/multi-parameter methods have been suggested to retrieve cloud microphysical properties. These methods take advantage of new powerful active and passive instrument tools that have became available for the cloud studies over the last few decades.

2. LIQUID WATER CLOUD RETRIEVALS

For several decades, microwave radiometer (MWR) measurements conducted at or in the vicinity of frequencies of 30 and 22 GHz have been used to obtain an estimate of cloud liquid water path (LWP). A typical value of uncertainty of the LWP retrievals is about 15–25 gm^2, though the use of an additional measurement at around 183 GHz can potentially reduce retrieval uncertainties for lower values of LWP.

The introduction in the mid 1980s – early 1990s of sensitive K_a and W band radars for cloud studies allowed retrievals of high-resolution vertical profiles of liquid water content (LWC). Typically, a radar-radiometer approach for liquid water clouds uses measurements of vertical profiles of radar reflectivity, Z_e, and the normalizing value of LWP derived from MWR to retrieve vertical profiles of LWC and the effective radius of cloud drops, r_e. Sometimes, cloud model information is used to refine retrievals, and typical assumptions include those about shape of the drop size distribution and the character of vertical profiles of drop concentration. Several examples of liquid cloud retrieval radar-radiometer retrievals methods are given by Frisch et al. (1995), Dong et al. (1997), Loehnert et al. (2001).

Retrievals of LWC vertical profiles using radar data have a serious limitation. When drizzle-size drops are present in clouds, the retrievals can be significantly biased. Fig. 1 shows an example of liquid cloud retrievals for a marine stratiform cloud. During the first part of the

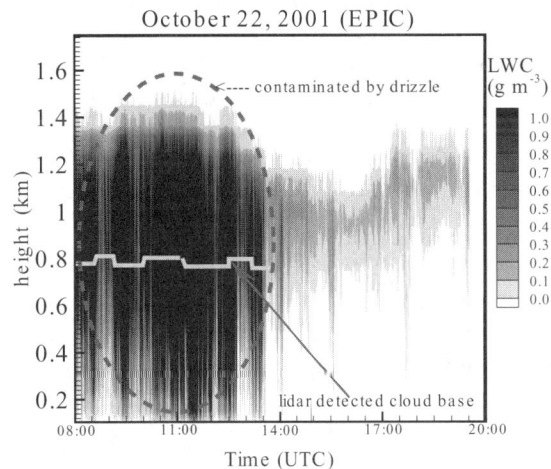

Fig. 1. An example of LWC retrievals in a marine stratus.

observation period, drizzle was observed in this cloud and the retrievals produced unrealistically high values of LWC, and radar echoes were reaching the lowest radar resolution volume. Drizzle was absent during the second half of this observation, and the retrievals provided realistic values of

LWC and quasi-adiabatic profile shapes. Drizzle areas with significant influence on retrieval results in clouds can be identified and filtered out by several means including reflectivity thresholding at about −17 dBZ (Matrosov et al., 2004) and analysis of the shape of the radar reflectivity profile. Marine liquid water clouds usually contain drizzle-size drops (100 μm < r < 250 μm) more often than clouds observed over land. Vertical profiles of r_e and the extinction coefficient can be calculated after retrievals of LWC if some further assumptions are made.

3. ICE CLOUD RETRIEVALS

Ice cloud microphysical retrievals are generally more complicated compared to liquid water cloud retrievals. One important factor contributing to this relative complexity is that for ice clouds there is no convenient measurement as there is in the case of liquid water clouds when MWR data provide a straightforward estimate of integrated water content. Microwave radiometers operating at millimeter and centimeter wavelengths are not sensitive to ice phase, so the use of infrared (IR) measurements is required for active-passive remote sensing approaches in ice cloud retrievals.

Though some earlier optical approaches for retrievals of ice cloud parameters were suggested in late the1970s (e.g., Platt and Dilley, 1979), the active development of multi-sensor methods for this cloud type began in early 1990s when sensitive cloud radars became readily available. One such multi-sensor method uses vertically pointing radar reflectivities (Z) and IR spectral measurements of downwelling radiation in the vicinity of a 10–12 μm transparency "window" to retrieve ice cloud layer-mean values of particle characteristic size and ice water path (IWP) (Matrosov et al. 1992; Mace et al., 1998). This ZIR method (as it is often called) also requires estimates of the atmospheric IR transmittance and background radiation, which are calculated using the water vapor information available from MWR data.

The layer-mean ZIR method was used for a number of years by the US Department of Energy's Atmospheric Radiation Measurements (ARM) program for ice cloud retrievals. This method was later generalized for retrievals of vertical profiles of ice water content (IWC) and particle characteristic size (Matrosov, 1999). Fig. 2 shows an example of ice cloud retrievals using the ZIR-profile method.

The field of particle concentrations (Fig. 2c), was calculated from retrieved IWC and median size fields assuming the gamma-function particle size distribution (PSD). As can be seen from Fig. 2, smaller particles at higher concentrations are usually observed near the cloud top. The highest IWC values are often seen in fall streaks, and they are associated with large particles.

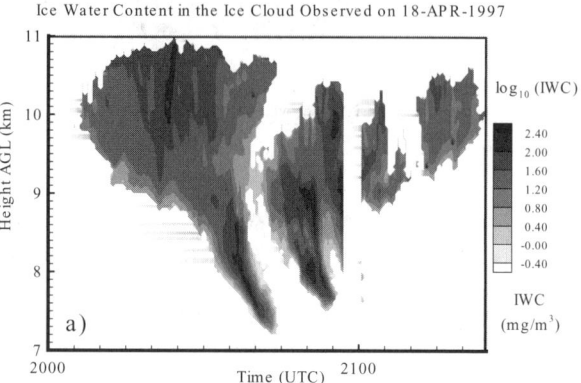

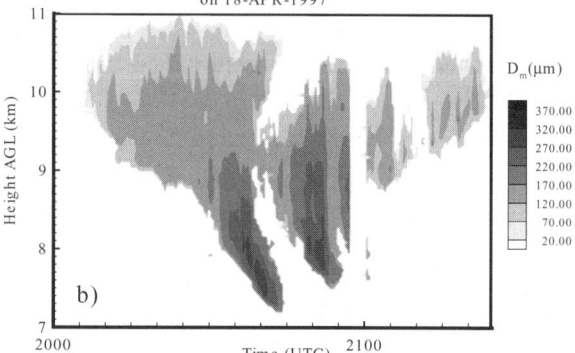

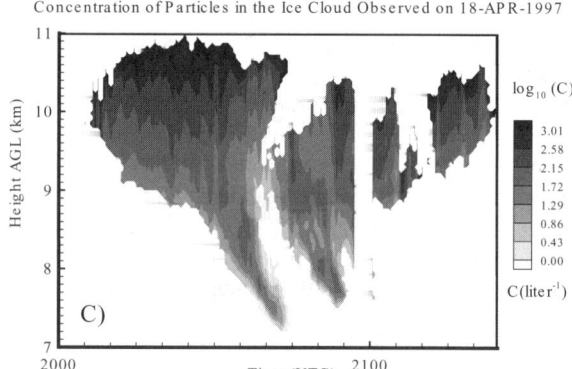

Fig. 2. Examples of retrievals of IWC (a), median particle size (b) and concentration (c) using the ZIR-profile method.

Another popular approach used to retrieve vertical profiles of ice cloud microphysical parameters involves measurements from vertically pointing collocated radar and lidar (e.g., Intrieri et al., 1993; Wang and Sassen, 2002; Donovan et al., 2001). With this approach, the particle size information is retrieved using the radar/lidar backscatter ratio, and then, the IWC values are calculated based on the absolute magnitude of backscatter coefficients. Special procedures are, however required to account for attenuation of lidar signals in clouds.

Both ZIR and lidar-radar methods provide a powerful tool to retrieve ice cloud microphysics. However, the use of optical instruments (i.e., visible or IR lidars and radiometers) results in a substantial limitation of the applicability range for these methods. They can only be used for optically thin ice clouds that are unobstructed by liquid clouds when viewed from the ground. This limitation eliminates a substantial fraction of ice clouds that are either a part of multi-level cloud systems or are optically thick, so that optical signals become completely attenuated. Only millimeter and longer wavelength radars can penetrate liquid clouds without significant attenuation and observe ice clouds aloft.

A new Doppler radar approach was recently developed (Matrosov et al., 2002; Mace et al., 2002) for retrievals of vertical profiles. Although technically this is not a multi-sensor approach, it still uses different types of input information. The Doppler radar methods use the vertical Doppler velocity measurements to infer particle size information and then the IWC information is retrieved based on reflectivity measurements. To minimize the influence of vertical air motion on size retrievals, Doppler velocity measurements are averaged over intervals of 20–30 minutes. This averaging degrades the time resolution of Doppler radar retrievals compared to ZIR or radar-lidar retrievals. However, the Doppler-based retrievals are applicable to virtually all ice clouds excluding ones that have strong up- or downdraft vertical motions.

Fig. 3 shows comparisons of IWC retrievals in a single layer ice cloud to which both the ZIR-profile and the Doppler radar methods could be applied. This cloud was observed on 24 July 2002 in southern Florida. There is a general agreement between the results from both methods, though Doppler radar retrievals are smoother because of time averaging of the input information. The relative deviation of IWC retrievals from these two methods is 69%. This is of the order of the uncertainty of IWC retrievals for which theoretical estimates yielded a value of about 60–70% (Matrosov et al., 2002). The relative standard deviation between the two methods for the median particle size is about 35%, which is also close to the retrieval uncertainty of this parameter.

Since a number of different versions of ZIR, radar-lidar and Doppler radar algorithms are now in use by different research groups, retrieval result intercomparisons must be performed on a much larger scale than that shown above. The extensive intercomparisons could better determine the consistency (or lack thereof) for certain algorithms and they would help to assess the applicability ranges for different approaches. Such an intercomparison effort is now being undertaken by the cloud properties working group of the ARM science team. About 15 different ice cloud retrieval algorithms are being tested using the common observational data sets.

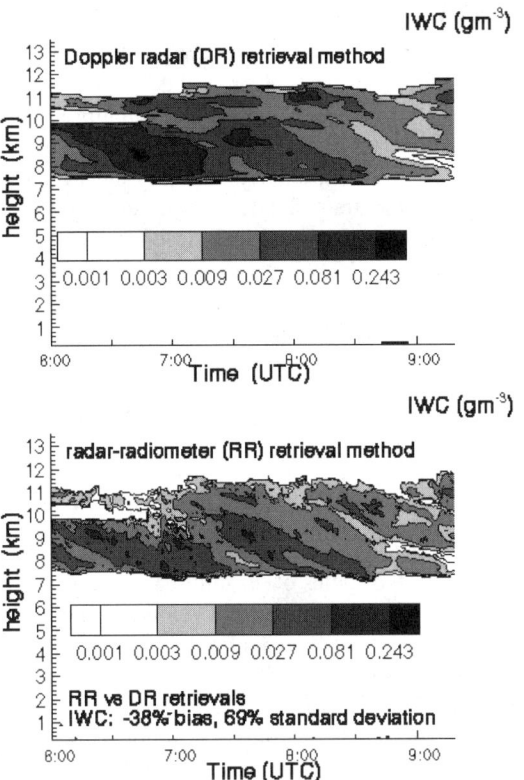

Fig. 3. Comparisons of IWC retrievals using the ZIR profile radar-radiometer (RR) and Doppler radar (DR) methods.

4. MIXED-PHASE CLOUDS

Although there have been recently significant advances in the field of multi-sensor ground-based microphysical retrievals of single phase (liquid or ice) clouds, developments of retrievals for mixed phase clouds are considerably lagging behind. In part, this can be explained by complex issues of separating ice and liquid phases in mixed-phase clouds. Some retrieval methods for special kinds of mixed phase clouds, however, have been suggested. One such special mixed-phase cloud type is a predominantly ice cloud with a thin layer of super-cooled liquid that is often observed near the cloud top. Such thin liquid layer is usually identifiable with lidar measurements and, if it is thin enough, separate retrievals are possible (Wang et al., 2004). When super-cooled liquid drops and ice particles are mixed in the same cloud volume, radar parameters are usually dominated by an ice phase, which is due to typical differences in characteristic sizes (i.e., ice particles are usually much larger than liquid water drops). In such clouds, the Doppler radar retrieval methods discussed above provide information about the ice phase component of mixed-phase clouds. Sometimes liquid and ice phases can be differentiated with the use of Doppler spectra measurements and separate liquid and ice phase retrievals can be performed (Shupe et al., 2004).

5. STATISTICS OF CLOUD PARAMETERS

Long-term operations of multi-sensor suites of instrumentation on a given site and subsequent microphysical retrievals provide an opportunity to generate statistical information on cloud parameters and their variability. Such information is unique and valuable for adequately representing clouds in different models. Well suited for generating statistical information on cloud properties are ARM Cloud and Radiation Testbed sites that are located in the different climatic zones (e.g., Mace et al., 2001) and other long-time remote sensor deployments such as the one during the Surface Heat Budget of the Arctic (SHEBA) experiment. The SHEBA retrievals provided the data for estimating the annual cycle of Arctic cloud parameters (Shupe et al., 2001).

6. CONCLUSIONS

During the last 10–15 years, significant progress has been made in the field of ground-based multi-sensor / multi-parameter retrievals of cloud microphysical parameters as new advanced remote sensing instruments have become available. Most of the remote sensing methods developed during this period are applicable to single-phase (liquid or ice) clouds, although several new approaches have been recently offered for mixed-phase clouds. Further validations, including comparisons of retrievals with in situ cloud measurements, are needed to better understand retrieval uncertainties and applicability ranges of different remote sensing approaches. Other outstanding issues include a need for thorough intercomparisons of different methods that are applicable to the same cloud types to ensure mutual consistency and the lack of significant biases between the retrieval results from these methods. This becomes especially important as different methods are being used to generate statistical information on cloud parameters.

7. ACKNOWLEDGEMENTS

This study was supported by the office of Science, U.S. Department of Energy, Grant DE-FG02-05ER63954.

8. REFERENCES

Dong, X., T. P. Ackerman, E. E.Clothiaux, P. Pilewskie, and Y. Han, 1997: Microphysical and radiatiative and radiatiative properties of boundary layer stratus from ground-based measurements. *J. Geophys. Res.*, 102, 23869-23843.

Donovan, D. P., and A. C. A. P. van Lammeren, 2001: Cloud effective particle size and water content retrievals using combined lidar and radar observations. *J. Geophys. Res.*, 106, 27425-27464.

Frisch, A. S., C. W. Fairall, and J. B. Snyder, 1995: measurement of stratus cloud and drizzle parameters in ASTEX with a Ka-band Doppler radar and microwave radiometer. *J. Atmos. Sci.*, 52, 2788-2799.

Intrieri, J. M., G. L Stephens, W. Eberhard, and T. Uttal, 1993: A method for determining cirrus cloud particle sizes using a lidar and radar backscatter technique. *J. Appl. Meteor.*, 32, 1074-1082.

Loehnert, U., S. Crewell, A. Macke, and Simmer, 2001: Profiling of cloud liquid water by combining active and passive microwave measurements with cloud model statistics. *J. Atmos. Oceanic Technol.*, 18, 1354-1366.

Mace, G. G., T. P. Ackerman, P. Minnis, and D. F. Young, 1998: Cirrus layer microphysical properties derived from surface-based millimeter radar and infrared interferometer data. *J. Geophys. Res.*, 103, 23207-23216.

Mace, G. G., E. E. Clothiaux, and T. P. Ackerman, 2001: The composite characteristics of cirrus clouds: Bulk properties revealed by one year of continuous cloud radar data. *J. Climate*, 14, 2185-2203.

Mace, G. G., A. J. Heymsfield, and M. Poellot, 2002: On retrieving the microphysical properties of cirrus clouds using moments of millimeter-wavelength Doppler spectrum. *J. Geophys. Res.*, 107, 4815, doi:10.1029/2001JD001308.

Matrosov, S. Y., T. Uttal, J. B. Snider, and R. A. Kropfli, 1992: Estimation of ice cloud parameters from ground-based infrared radiometer and radar measurements. *J. Geophys. Res.*, 97, 11567-11574.

Matrosov, S. Y., 1999: Retrievals of vertical profiles of ice cloud microphysics from radar and IR measurements using tuned regressions between reflectivity and cloud parameters. *J. Geophys. Res.*, 104, 16741-16753.

Matrosov, S. Y., A. V. Korolev, and A. J. Heymsfield, 2002: Profiling cloud mass and particle characteristic size from Doppler radar measurements. *J. Atmos. Oceanic Technol.*, 19, 1003-1018.

Matrosov, S. Y., T. Uttal, and D. A. Hazen, 2004: Evaluation of reflectivity-based estimates of water content in stratiform marine clouds. *J. Appl. Meteor.*, 43, 405-419.

Platt, C. M. R., and A. C. Dilley, 1979: Remote sounding of high clouds: Emissivity of cirrostratus. *J. Appl. Meteor.* 18, 144-1150.

Shupe, M. D., T. Uttal, S. Y. Matrosov, and A. S. Frisch, 2001: Cloud water contents and hydrometeor sizes during the FIRE Arctic clouds experiment. *J. Geophys. Res.*, 106, 15015-15028.

Shupe, M. D., P. Kollias, S. Y. Matrosov, and T. L. Schneider, 2004: Deriving mixed-phase cloud properties from Doppler radar spectra. *J. Atmos. Oceanic Technol.*, 21, 660-670.

Wang, Z., and K. Sassen, 2002: Cirrus cloud microphysical property retrieval using lidar and radar measurements. *J. Appl. Meteor.*, 41, 218-229.

Wang, Z., K. Sassen, D. N. Whiteman, and B. B. Demoz, 2004: Studying altocumulus with ice virga using ground-based active and passive remote sensors. *J. Appl. Meteor.*, 43, 449-460.

OBSERVATIONS OF AEROSOLS AND CLOUDS WITH MIE-LIDAR NETWORK

Atsushi Shimizu, Nobuo Sugimoto, and Ichiro Matsui
National Institute for Environmental Studies
16-2 Onogawa, Tsukuba, 305-8506 Japan

ABSTRACT

National Institute for Environmental Studies (NIES) and collaborating universities/institutes are operating 12 dual wavelength polarization Mie-scattering LIDARs in Asian region. Automated continuous LIDARs are located at Tsukuba, Nagasaki, Amami-Ohshima, Miyakojima, Fukuejima, Sapporo and Toyama in Japan, at Suwon in Korea, at Beijing, Hefei and Hohhot in China, and at Sri Samrong in Thailand. They observe vertical distributions of clouds in whole troposphere and aerosols in lower troposphere 96 times everyday regardless of weather conditions. The LIDAR can distinguish water clouds and ice clouds using depolarization ratio. It estimates vertical distribution of extinction coefficient of aerosols in clear sky conditions. Ratios of extinction caused by dust particles and by spherical aerosols are also calculated. As the LIDAR system is continuously operated, aerosol events and evolutions in various time scales can be detected. In the spring season in northern hemisphere, northeast Asian region is repeatedly covered by Asian dust. LIDAR can determine the time of beginning and finishing of the dust event, the altitude and thickness of dust layer, status of mixing of dust particles and spherical particles. Monthly averaged profiles of extinction coefficient by dust in March showed year-to-year variations corresponding to the variations of meteorological conditions in this region. Time-height sections of aerosol distribution are also useful for validation of 3 dimensional numerical models such as CFORS (Chemical weather FORecasting Systems).

1. INTRODUCTION

In the Asian region, there are several major sources of aerosols, which affect local atmospheric environment or regional climate. A large source in the spring season in the northern hemisphere is Asian dust (Kosa), which is mainly generated in inner Mongolian region or west part of China. There is also a large emission of anthropogenic aerosols (sulfate, nitrate) near the coastal region of East Asia. Sometimes smoke plumes produced by boreal forest fires cover this area. Therefore understanding of distributions and variations of each component of these aerosols is an important issue for atmospheric science in Asian region. LIDAR is a powerful instrument to measure mixed or vertically layered aerosol structure. NIES and collaborators developed a network of Mie-scattering LIDARs and described spatial and temporal distributions of 2 components of aerosols, dust and spherical particles. In this paper, LIDAR observation and data analysis method are described. Then year-to-year variations of aerosol distribution are shown and comparison of numerical model and observation are presented.

2. LIDAR OBSERVATION AND ANALYSIS METHOD

Specifications of NIES-type are listed in Table 1. Although some specs are different among LIDARs, all systems are operated in the same way.

Table 1. Specifications of LIDARs

Energy of laser (Nd:YAG)	20–50mJ (532nm), 20–100mJ (1064nm)
Pulse repetition	10Hz
Beam direction	Zenith (fixed)
Diameter of telescope	20–35cm
Detection unit	PMT (532nm, $\parallel$ and $\perp$), APD (1064nm)
A-D conversion Vertical resolution	Digital oscilloscope 6m

A Linux based PC controls laser and digital oscilloscope. The PC starts emission of laser beam at 00, 15, 30, and 45 min in every hour and laser emission is continued for 5 minutes. Then accumulated backscattering signal during 3000 shots (10Hz * 5 minutes) is stored in the PC. As the system is installed in a room or a container and a transparent plate glass covers above telescope, it can be operated even in rainy or cloudy times. Observed results are transferred to NIES through the Internet or telephone lines everyday, and quicklook figures of backscattering intensity for 2 wavelengths and depolarization ratio are displayed at WWW pages in NIES. Locations of NIES-type LIDARs are indicated in Fig. 1.

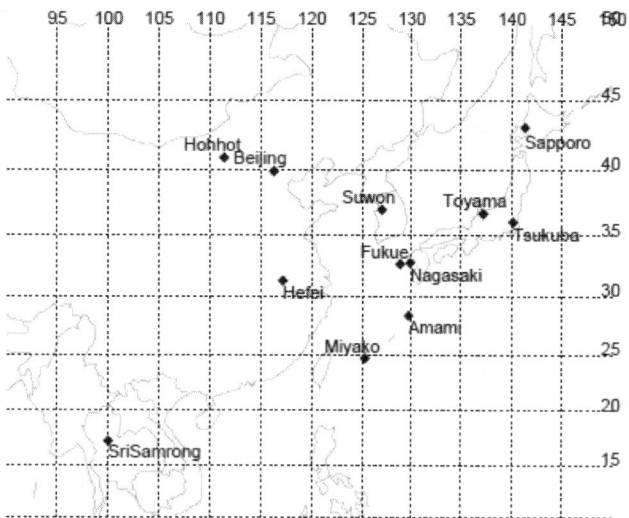

Fig. 1. Locations of NIES-Type LIDARs as of February 2004.

Ninety-six profiles per day are independently processed. At first cloud base height is detected using vertical profile of scattering intensity. If a profile contains scattering from a cloud, it is not used for further analysis. A profile during rainy conditions is also eliminated using a signal caused by strong scattering at wet window glass. Finally, Fernald's method (Fernald, 1984) is applied on profiles obtained in clear sky conditions with an assumption of constant LIDAR ratio (S1). In this study we employed S1 = 50 sr. Once extinction coefficient (α) is obtained, particle depolarization ratio (δ) is calculated (Browell et al., 1990). If we accept following assumptions, mixing ratio of dust (R) can be estimated using observed δ (Shimizu et al., 2004).

1. Aerosols consist of dust and spherical aerosols only, and they are always externally mixed.

2. Both aerosol components have fixed values of δ.

Dust extinction coefficient (α_d) and spherical aerosols extinction coefficient (α_s) are expressed using R as follows:

$$\alpha_d = R \times \alpha, \quad \alpha_s = (1-R) \times \alpha.$$

3. VARIATION OF AEROSOL DISTRIBUTION

Climatological vertical distributions of aerosols are independently determined for dust and spherical aerosols with the method mentioned above. As observations had been conducted since March 2001 in Beijing, Nagasaki and Tsukuba, we can derive long-term variations of aerosol distribution. Fig. 2 shows vertical profiles of dust extinction coefficient at 3 observatories in March 2001, 2002 and 2003.

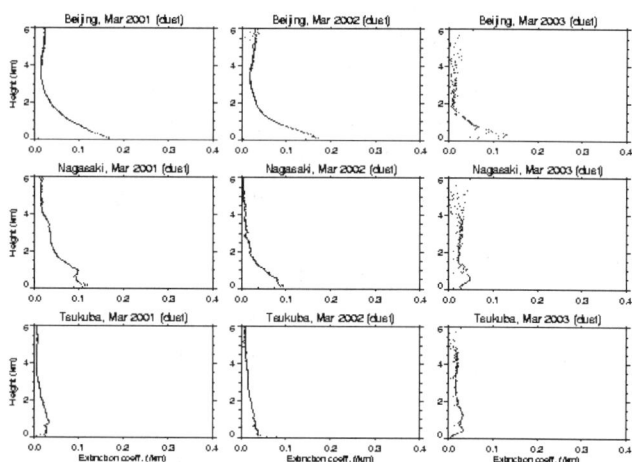

Fig. 2. Monthly mean profiles of dust extinction coefficient. Results in March 2001, 2002 and 2003 are shown from left to right, at Beijing (top), Nagasaki (middle) and Tsukuba (Bottom).

In Beijing and Nagasaki, dust extinction coefficient was large in 2001 and 2002, but it was small in 2003. There is a little difference among profiles in 3 years at Tsukuba. Vertical distributions of spherical aerosols shown in Fig. 3 also indicate that there are year-to-year variations of extinction caused by spherical aerosols.

In the spring season, extinction by dust is larger than that by spherical aerosols at Beijing. To the contrary, both components are comparable or extinction coefficient by spherical aerosols is dominant at Nagasaki and Tsukuba.

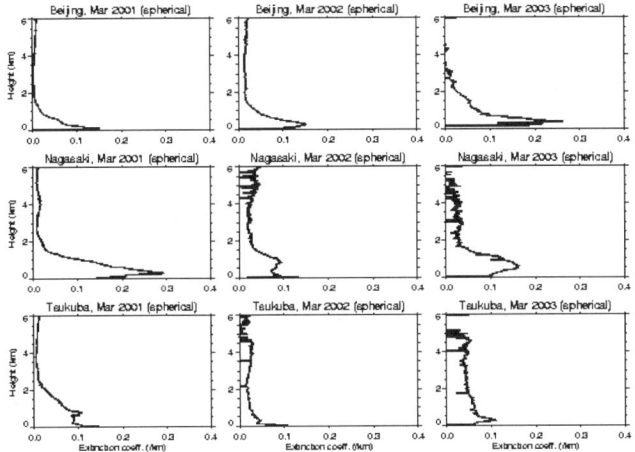

Fig. 3. Same as Fig. 2, but for profiles of spherical aerosols extinction coefficient.

4. COMPARISON WITH NUMERICAL MODELS

An advantage of separation method is that obtained profiles can be compared effectively with results of numerical models, which produce distributions of independent components. In this study, Chemical Forecasting system (CFORS) (Uno et al., 2003) developed at Kyushu University is employed for comparison with LIDAR observation results. CFORS is a regional transport model, which treats emission, transport and deposition of natural/anthropogenic aerosols and minor constituents in East Asia. Mass concentrations of dust and sulfate in CFORS are compared with LIDAR extinction coefficient caused by dust and spherical aerosols, respectively. In November 2002, a major Asian dust event occurred and LIDARs detected dust layers. Fig. 4 indicates time-height sections of observed extinction coefficients by dust and by spherical aerosols.

A dust layer near the surface appeared on November 11 at Beijing, on 12 at Nagasaki and on 13 at Tsukuba. The depth of dust layer was almost 2 km. Simultaneously, spherical aerosols were concentrated near the surface at Beijing. In Nagasaki, spherical aerosols were distributed widely in vertical direction. Fig. 5 shows results of CFORS. Dust distributed below 2 km at 3 locations, but peak concentration occurred slightly earlier in the model than in the LIDAR results. Enhancement of spherical aerosols occurred on November 13 and 15 in Beijing which corresponds with the LIDAR observation. In Nagasaki,

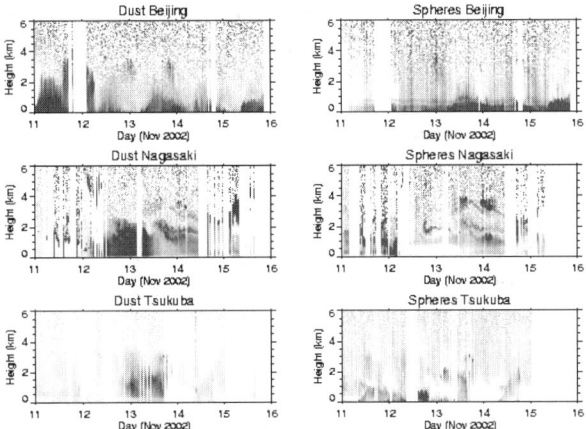

Fig. 4. Time-height sections of extinction coefficient by dust (left) and spherical aerosols (right) observed by LIDAR during Asian dust event in November 2002 at Beijing (top), Nagasaki (middle) and Tsukuba (Bottom). Values for white and black tone are 0/m and 2×10^{-4}/m.

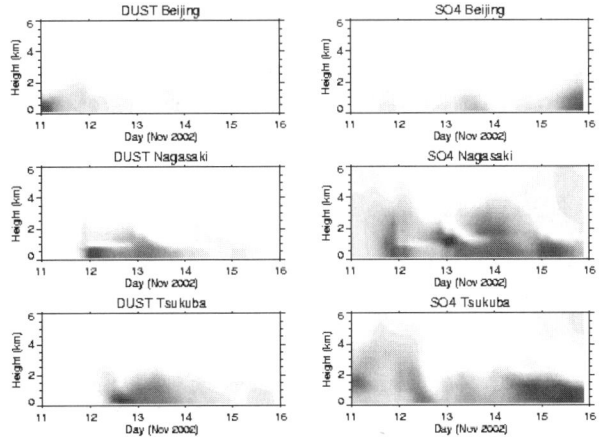

Fig. 5. Same as Fig. 4, but for mass concentrations of dust (left) and sulfate (right) calculated by CFORS. Values for white and black tone are 0μg/m^3 and 5000 μg/m^3 for dust or 20μg/m^3 for sulfate.

model also calculated floating layer of sulfate. Although there were little similarity between LIDAR and CFORS in Tsukuba, it should be noted that during cloud cover period extinction coefficients were not presented in Fig. 4. One major problem in this comparison is that units are different in extinction coefficient (/m) from LIDAR results and mass concentration ($\mu g/m^3$) from the model. Conversion factors between them for each component should be determined by other kinds of observation to understand clearly the characteristics of the model.

5. CONCLUDING REMARKS

In Asian region 12 multi-wavelength polarization LIDARs are continuously operated regardless of whether conditions based on an unified strategy. Observed results are displayed on WWW at NIES everyday. Fernald's method and separation algorithm are employed to determine vertical distributions of dust and spherical aerosols. Monthly mean profiles of extinction coefficients showed year-to-year variations. Separated profiles are useful for comparison between LIDAR observations and calculated distribution by a numerical model.

6. ACKNOWLEDGEMENT

At present CFORS is operated at NIES with the courtesy of Itsushi Uno, Kyushu University.

7. REFERENCES

Browell, E. V., Butler, C. F., Ismail, S., Robinette, P. A., Carter, A. F., Higdon, N. S., Toon, O. B., Schoeberl, M. R., and Tuck, A. F., 1990: Airborne lidar observation in the wintertime arctic stratosphere: Polar stratospheric clouds. *Geophys. Res. Lett.*, 17(4), 385–388.

Fernald, F. G., 1984: Analysis of atmospheric lidar observations: Some comments. *Appl. Opt.*, 23(5), 652–653.

Shimizu, A., N. Sugimoto, I. Matsui., K. Arao, I. Uno, T. Murayama, N. Kagawa, K. Aoki, A. Uchiyama, and A. Yamazaki, 2004: Continuous observations of Asian dust and other aerosols by polarization lidars in China and Japan during ACE-Asia. *J. Geophys. Res.*, 109, D19S17, doi:10.1029/2002JD003253.

Uno, I., G. R. Carmichael, D. G. Streets, Y. Tang, J. J. Yienger, S. Satake, Z. Wang, J.-H. Woo, S. Guttikunda, M. Uematsu, K. Matsumoto, H. Tanimoto, and K. Yoshioka, 2003: Regional chemical weather forecasting system CFORS: Model descriptions and analysis of surface observations at Japanese island stations during the ACE-Asia experiment. *J. Geophys. Res.*, 108(D23), 8668, doi:10.1029/2002JD002845.

INFRARED SPECTRAL RADIANCE VALIDATION USING AIRCRAFT- AND SATELLITE-BASED SENSOR SYSTEMS

Allen M. Larar[1], William L. Smith[2], Daniel K. Zhou[1], Xu Liu[1], Henry Revercomb[3], Robert Knuteson[3], and Stephen A. Mango[4]

[1]NASA Langley Research Center, Hampton, VA 23681, USA; Allen.M.Larar@nasa.gov
[2]Hampton University, Hampton, VA 23668, USA
[3]University of Wisconsin, Madison, WI 53706, USA
[4]NPOESS Integrated Program Office, Silver Spring, MD 20910, USA

ABSTRACT

High-altitude aircraft flights of the National Polar-orbiting Operational Environmental Satellite System (NPOESS) Airborne Sounding Testbed (NAST) and Scanning High-resolution Interferometer Sounder (S-HIS) systems are supported by the Integrated Program Office (IPO) as part of risk mitigation activities for future NPOESS sensors. The NAST-Interferometer (NAST-I) is a high spectral and spatial resolution (0.25 cm^{-1} and 0.13 km nadir footprint per km of aircraft altitude, respectively) cross-track scanning (2.3 km swath width per km of altitude) Fourier Transform Spectrometer (FTS) observing within the 3.7–15.5 micron spectral range. The S-HIS is also a cross-track scanning FTS which measures emitted thermal radiation at high spectral and spatial resolution (~0.5 cm^{-1} and 0.1 km nadir footprint per km of aircraft altitude) between 3.3 and 18 microns, across a 2 km swath width per km of aircraft altitude. NAST-I and S-HIS infrared spectral radiances are used to characterize the atmospheric thermodynamic state and provide information on radiatively active trace gases (e.g., O_3 & CO), clouds, and the terrestrial surface during experimental campaigns. In this study we address some of the challenges associated with validating high resolution remote sensing systems, through comparison of NAST-I and S-HIS radiances measured during recent field experiment campaigns with other aircraft- and satellite-based radiance measurements and corresponding calculations performed using Line-by-Line (LBL) and "Fast" forward radiative transfer models.

1. INTRODUCTION

Field validation measurements from high-altitude airborne interferometric sensors are critical for successful space-based instrument validation, since only observations from such platforms can provide the proper spatial & temporal context needed as well as be used to emulate expected satellite measurements for the instrument being validated. The higher spectral and spatial resolution aircraft sensor data can be spectrally and spatially convolved, respectively, to simulate what should be measured by the concurrent satellite observations during overpass events. The much higher spatial resolution of the aircraft sensor data can play an important role in validating satellite-derived data products under the conditions of variable surface and atmospheric radiance (e.g., due to clouds) within the satellite sensor footprint. Some of the challenges associated with validating high resolution remote sensing systems are addressed in this study. Data obtained with the National Polar-orbiting Operational Environmental Satellite System (NPOESS) Airborne Sounder Testbed-Interferometer (NAST-I) during recent field campaigns will be compared with that obtained from several other independent sensors; most importantly, using data from the airborne Scanning High-resolution Interferometer Sounder (S-HIS), MODIS Airborne Simulator (MAS), Cloud Physics Lidar (CPL), along with the Aqua-satellite-based Atmospheric Infrared Sounder (AIRS) and Moderate-resolution Imaging Spectroradiometer (MODIS).

2. FIELD EXPERIMENT CASE STUDY

The *Pacific THORPEX Observing System Test (PTOST)* campaign was based out of Hickam AFB, Hawaii during 18 February to 13 March, 2003. Objectives of the field experiment phase focused on evaluating the potential of various in situ and remote sensing observation systems providing observations needed to accelerate improvements in operational weather predictions. The NASA ER-2 and the NOAA G-IV served as the aircraft platforms for this experiment carrying the NAST-I, NAST-M, S-HIS, MAS, and CPL remote sensors along with dropsondes and ozone in-situ measurements, respectively. Flights targeted frontal boundaries and storm systems, as well as satellite sensor validation underflights of the AQUA, TERRA, and ICESat satellites.

3. INSTRUMENT SYSTEMS

Data from several different remote sensors were incorporated into this analysis: two high-resolution, airborne Michelson interferometers (NAST-I and S-HIS); one high-resolution, satellite-based discrete-channel grating spectrometer (AIRS); and several broadband radiometer systems, both satellite-based (i.e., MODIS and GOES) and aircraft-based (i.e., MAS) systems. The NAST-Interferometer, NAST-I (Cousins and Smith, 1997; Gazarik et al., 1998; Prutzer et al., 1998), high spectral resolution (0.25 cm^{-1}) data are collected over the 3.7–15.5 micron spectral range, using a step and stare scanning mirror to obtain +/- 48.4° cross-range coverage with thirteen atmospheric scene views. The instrument's instantaneous field of view (IFOV) translates into a 0.13 km ground footprint at nadir for each 1.0 km of aircraft altitude (i.e., 2.6 km footprint from a 20 km ER-2 altitude). The Scanning High-resolution Interferometer Sounder, S-HIS (Revercomb et al., 1998), measures emitted thermal radiation at high spectral resolution between 3.3 and 18 microns, with 1.5 kilometer resolution (at nadir) across a 30 kilometer ground swath from a nominal altitude of 15 kilometers on-board the Proteus aircraft. The MODIS Airborne Simulator (MAS) is a scanning broadband spectrometer which flies on a NASA ER-2 high altitude research aircraft and acquires high spatial resolution imagery employing 50 spectral bands within the 0.55 to 14.3 micron range. The MAS spectrometer is mated to a scanner sub-assembly which collects image data with an IFOV of 2.5 mrad, giving a ground resolution of 50 meters from the 20 kilometer ER-2 altitude, and a cross track scan width of 85.92 degrees. AIRS (Aumann et al., 2003) is a high spectral resolution grating spectrometer with 2378 bands in the thermal infrared between 3.7–15.4 µ that is operational aboard the NASA EOS AQUA satellite. In the cross-track direction, a ±49.5 degree swath centered on the nadir is scanned. Each scan line contains 90 IR footprints, with a resolution of 13.5 km at nadir and 41 km x 21.4 km at the scan extremes from the nominal 705.3 km orbit. The Moderate Resolution Imaging Spectroradiometer (MODIS) is a broadband sensor, also aboard Aqua, that provides high radiometric sensitivity in 36 spectral bands ranging in wavelength from 0.4 µ to 14.4 µ, with a nominal band-dependent nadir resolution </= 1 km.

4. VALIDATION METHODOLOGY AND RESULTS

The objective of the analysis herein is to infuse multiple spatially- and temporally-coincident data sources from several independent sensors. Spatial co-registration of the datasets is verified by comparing geo-referenced 2-d horizontal images at select wavelengths, e.g., surface observations in the 11 micron window region. The focus herein is restricted to only the coincident portion of the flight track, wherein co-registration is quite obvious by noting spatial characteristics of prominent features (e.g., clouds) consistent in all observations. Simulated observations are based upon line-by-line (LBL)–based calculations using the best estimate of surface and atmospheric state. Along-track variations are compared by extracting spatially-coincident data (i.e. nadir fields-of-view) from the remote sensors and applying the MODIS (or GOES) broadband spectral response functions (SRFs) to the high-resolution measurements (i.e., NAST-I, S-HIS, and AIRS) so that all data to be compared represent broadband (or equivalent) observations. The figures shown in this manuscript illustrate some example results from one flight day (i.e., 03/10/03) of the PTOST campaign. Fig. 1 shows on the left a GOES 10 visible image of the Hawaii

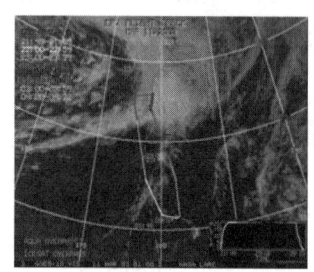
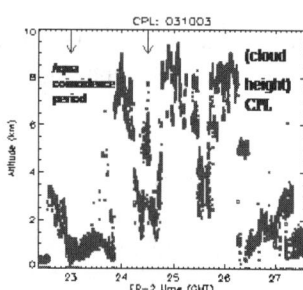

Fig. 1. GOES visible image (left) and Cloud Physics Lidar, CPL, (right) cloud height data as a function of aircraft flight time, with Aqua coincidence period noted, for the 03/10/03 PTOST flight.

flight region with the ground tracks of the Aqua satellite and ER-2 aircraft over-layed, and on the right a time series of cloud height from the Cloud Physics Lidar (CPL) is depicted along with the coincidence period with Aqua denoted. This figure shows both clear and cloudy scenes to exist along the ground-track of the sensors. Fig. 2 shows the corresponding along-track comparisons for the 7.2 μ water vapor (MODIS band 28) on the left and the 8.55 μ window region (MODIS band 29) on the right. These figures clearly delineate clear and cloudy scenes (similar to Fig. 1) and demonstrate the need for space and time coincidence for satellite sensor validation, as the water vapor comparison shows inter-platform agreement worsening as a function of time away from the satellite/aircraft coincidence period (i.e., left side of plot).

space-coincident points are included. A histogram-filtering technique was then implemented to remove "outliers" in such scatter plot comparisons; this enables comparing "same scene" views, i.e., both clear or both cloudy, to remove differences introduced from spatial and/or temporal mismatches between the sensors. Fig. 4 shows the corresponding scatter plots after implementation of this histogram-filtering approach. The inter-instrument agreement shown in these figures (>/~ tenths of degrees) demonstrates the level of radiometric calibration consistency across the sensors being compared.

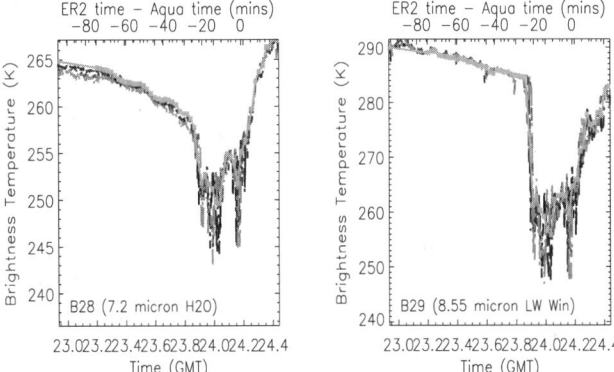

Fig. 2. Broadband radiance time series along the Aqua satellite nadir track for all sensors having Modis band 28 (left) and Modis band 29 (right) spectral response functions applied.

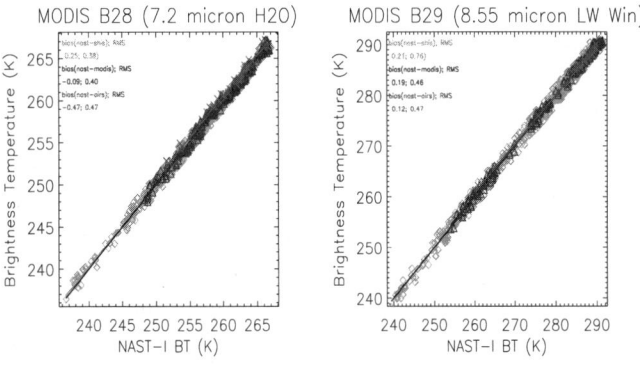

Fig. 4. Scatter plot relationships, as shown in Fig. 3, after implementation of a histogram-filtering technique (see text).

Fig. 3 shows the corresponding scatter plots for broadband radiance signals from the various sensors, for water vapor (left) and the window region (right) when all

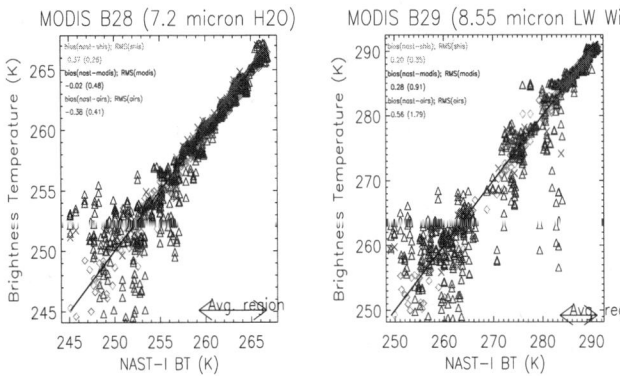

Fig. 3. Scatter plots for broadband radiance signals for water vapor (left) and the window region (right) when all space coincident points are included.

Spectra from the high-resolution remote sensors (i.e., NAST-I, S-HIS, and AIRS) are also compared for select portions of the aircraft/satellite coincident tracks. Principal Component Analysis (PCA) filtering ("trained" on each sensor's entire dataset) is applied to the measured radiances prior to averaging over the clear FOVs to achieve maximum noise reduction over the coincident FOVs to be examined. The resulting spectra are then compared having their original SRFs and on a common spectral grid. Accounting for SRF- and view-induced differences is essential for such validation inter-comparisons. The SRF-induced differences are not only due to, for example, different effective spectral resolutions and wing transfer function shapes of the different sensors, but also due to the FTS producing continuous spectra and the grating yielding only discrete channel locations across their respective bandpasses.

5. CONCLUSION AND FUTURE WORK

This analysis has shown instrument systems and simulation agreement to be, in general, within acceptable levels considering measurement and forward model uncertainties. Modis offsets observed within this analysis are of similar magnitude and direction to those reported in earlier studies (e.g. Tobin et al., 2003; Moeller et al., 2003) and are likely due to calibration offsets or spectral response function (SRF) spectral shifts. Spatial and temporal coincidence between observing systems is crucial to differentiate between measurement uncertainty and geophysical variability. Aside from collocated sensor(s) on the same platform, space-based sensor validation is best achieved using high-altitude aircraft sensors in a multi-sensor data fusion approach; this can eliminate errors from spatial/temporal/view-angle mismatches and forward radiative transfer model uncertainties. High resolution FTS systems enable (very-well-calibrated) emulation of other high-resolution or broadband infrared instrument systems. NAST-I and S-HIS are proven high spatial/spectral/temporal resolution sensors well-suited for infrared radiance and atmospheric state measurement validation.

The case study examined herein can be analyzed in further detail, bringing in more independent measurements from the other in-situ and passive and active remote sensors that also participated in this experimental campaign. Further examination of data from other flight days or different portions of the flight presented herein will also be part of this continued analysis.

6. ACKNOWLEDGMENT

The authors wish to acknowledge the NPOESS Integrated Program Office, NASA Langley Research Center, and the various team members and their respective institutions for the continued, enabling support of the NAST program. We also wish to express our sincere appreciation to the many people, too many to name herein, that have contributed to make these field campaigns a success.

7. REFERENCES

Cousins, D., and W. L. Smith, "National Polar-Orbiting Operational Environmental Satellite System (NPOESS) Airborne Sounder Testbed-Interferometer (NAST-I)," *Proceedings from the SPIE Application of Lidar to Current Atmospheric Topics II Conference,* July 27–August 1, 1997, San Diego, CA, Vol. 3127, pp. 323-331, 1997.

Gazarik, M., L. Candell, E. Leonard, and S. Prutzer, "NPOESS airborne sounder testbed interferometer (NAST-I) signal processing and initial flight test results," *Proceedings from the SPIE Earth Observing Systems III Conference,* Vol. 3439, pp. 503-514, 1998.

Prutzer, S., D. P. Ryan-Howard, J. G. Garcia, and D. R. Bold, "NPOESS Airborne Sounder Testbed (NAST) IR Spectrometer Design," *Proceedings of the IEEE Aerospace Conference,* Snowmass, CA, 1998.

Revercomb, H. E., V. P. Walden, D. C. Tobin, J. Anderson, F. A. Best, N. C. Ciganovich, R. G. Dedecker, T. Dirkx, S. C. Ellington, R. K. Garcia, R. Herbsleb, R. O. Knuteson, D. LaPorte, D. McRae, and M. Werner, "Recent Results From Two New Aircraft-Based Fourier-Transform Interferometers: The Scanning High-resolution Interferometer Sounder and the NPOESS Atmospheric Sounder Testbed Interferometer," *ASSFTS Conference,* Toulouse, France, October 1998.

Aumann, Hartmut H., Moustafa T. Chahine, Catherine Gautier, Mitchell D. Goldberg, Eugenia Kalnay, Larry M. McMillin, Hank Revercomb, Philip W. Rosenkranz, William L. Smith, David H. Staelin, L. Larrabee Strow, and Joel Susskind. AIRS/AMSU/HSB on the Aqua Mission: Design, science objectives, data products, and processing systems. *IEEE Transactions on Geoscience and Remote Sensing,* V. 41, No. 2, 2003, pp 253-264.

Tobin et al., OSA FTS mtg., February, 2003.

Moeller et al., SPIE SD03, August, 2003.

Smith, W. L., H. E. Revercomb, H. B. Howell, and H. M. Wolf, "HIS-A Satellite Instrument to Observe Temperature and Moisture Profiles with High Vertical Resolution," in *Fifth Conference on Atmospheric Radiation, American Meteorological Society,* Boston, 1983.

DERIVING A GLOBAL SURFACE ALBEDO FROM GEOSTATIONARY OBSERVATIONS

Y. M. Govaerts
EUMETSAT
D64295 Darmstadt, Germany (yves.govaerts@eumetsat.de)

A. Lattanzio
Makalumedia
D64295 Darmstadt, Germany (alessio.lattanzio@eumetsat.de)

ABSTRACT

This paper investigates the possibility to derive a spatially consistent surface albedo product from different geostationary spacecrafts. The consistency analysis relies on the comparison of albedo derived over the common area observed by two adjacent satellites. So far, Meteosat-7, -5 and GMS-5 data have been processed. There is a good consistency between products retrieved from Meteosat-5 and -7 observations. Improvements in the data processing of GMS-5 VIS band are still required to secure a consistent surface albedo product between Meteosat-5 and GMS-5.

1. INTRODUCTION

Although the potential of space-based observations to derive globally surface albedo maps has long been recognised, it is routinely retrieved only since 2001 from the Moderate-Resolution Imaging Spectroradiometer (MODIS) and the Multiangle Imaging SpectroRadiometer (MISR) onboard the Terra platform (Jin et al., 2003; Martonchik et al., 1998). Consequently, inter-annual surface albedo variability and its impact on the atmosphere as predicted by Charney (1975) are still poorly quantified. This situation results in part from the scarceness of space instruments dedicated to land surface observations before the late 1990s, when systematic space-borne observations of the land surface were essentially limited to data acquired by geo-stationary meteorological satellites and a few polar platforms. Nevertheless, Pinty et al. (2000b) demonstrated the potential of geostationary satellites for the generation of reliable surface albedo maps. The high temporal sampling of geostationary satellites allows to account for both the atmospheric scattering effects and the anisotropy of the surface reflectance when data are accumulated during the course of the day (Pinty et al., 2000a). This novel approach opens new avenues for the exploitation of geostationary satellite observations for climate studies since their corresponding archives often cover two decades or more.

The systematic exploitation of these archived data for the retrieval of long time series of surface albedo data sets raises however several issues. First, the instruments onboard these satellites were not originally conceived to support climate monitoring activities and are not adequately calibrated. Second, a single geostationary satellite only observes a part of the globe, limited to an area of about 60° around the sub-satellite point. A global view of the Earth from the geostationary orbit, with the exception of the poles, is ensured by a suite of operational meteorological satellites located at regular interval along the Equator (Fig. 1). All radiometers onboard these geostationary platforms observe the Earth with a broad solar channel, referred to as the VIS band, ranging approximately from 0.5 up to 0.9μm. In principle, the method proposed by Pinty et

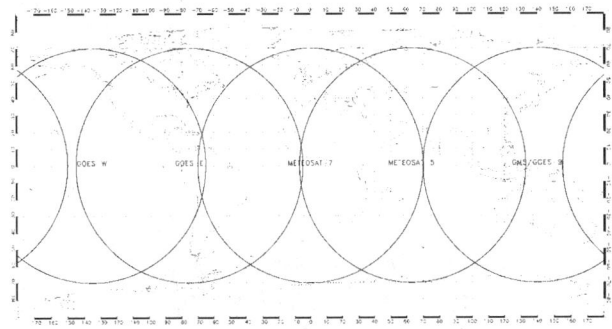

Fig. 1. Location of the current operational geostationary satellites with systematic archived data. Circles show the 70° viewing angle limit.

al. (2000a) could be applied to any of these geostationary satellites, irrespective of their position along the Equator. This paper investigates the possibility to derive a spatially consistent surface albedo product from Meteosat-7, -5 and GMS-5 observations. The consistency analysis relies on the comparison of albedo derived over the common area observed by two adjacent satellites.

2. SURFACE ALBEDO RETRIEVAL FROM GEOSTATIONARY SATELLITES

Surface albedo, or more precisely Directional Hemispherical Reflectance (DHR), is the integral of the Bidirectional Reflectance Factor (BRF) at the surface over all angles of the upward hemisphere. Retrieving DHR from space observations requires thus to account for scattering and absorption processes in the atmosphere, and to document the angular anisotropy of the surface. The accuracy achieved in the albedo estimation depends on the density of the angular sampling and the reliability of the atmospheric correction. It is also difficult to accurately characterise the surface without simultaneously quantifying the interactions of the radiation field with the atmosphere because the surface constitutes one of the boundary conditions of the radiation transfer problem that needs to be solved when analysing remote sensing data. Therefore, retrieval approaches should treat the surface and atmosphere as a coupled radiative system.

Numerous approaches have been proposed in the past to derive surface albedo from geostationary observations, most of them neglecting surface anisotropy and/or prescribing atmospheric conditions. Pinty et al. (2000a) proposed a new method for the simultaneous characterisation of surface anisotropy and atmospheric scattering properties, explicitly accounting for the radiative coupling between these two systems (Fig. 2). The approach relies on a daily accumulation of geostationary observations in the VIS band under different illumination

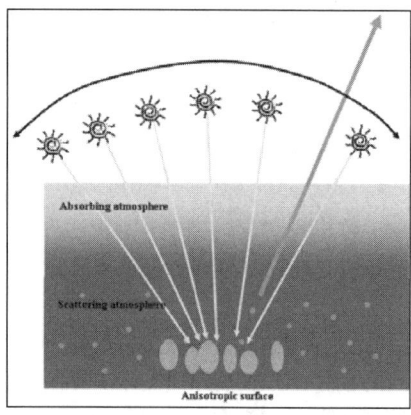

Fig. 2. Schematic representation of the Geostationary Surface Albedo (GSA) algorithm concept. The forward model is composed of an upper absorbing gaseous layer and a lower scattering one inverted against daily accumulation of observations acquired at different illumination angles.

conditions to assess the scattering properties of the surface and the atmosphere, assuming that, i) surface and atmospheric scattering properties are constant throughout the day, ii) continental aerosol type are applicable everywhere and all year long, iii) surface anisotropy can be represented with the simple BRF model proposed by Rahman et al. (1993) and finally, iv) the Helmholtz reciprocity principle is valid over terrestrial surfaces at a spatial resolution of a few kilometres. On a daily basis, the Geostationary Surface Albedo (GSA) algorithm estimates the DHR value, corrected for atmospheric effects, for a Sun zenith angle fixed at 30° and the Bi-Hemispherical Reflectance (BHR) corresponding to isotropic illumination. A simple composite procedure is applied over consecutive 10-day periods to produce geographically complete maps of surface albedo (Pinty et al., 2000b).

3. RESULTS

One 10-day period has been processed to demonstrate the possibility to derive a global surface albedo map from Meteosat-5/-7 and GMS-5 data collected during a period ranging from 1 to 10 May 2001. The characteristics of the Meteosat-5 and -7 radiometer VIS band should be, in principle, similar, as their radiometers have been produced in the same batch, according to identical specifications. Both instruments are routinely calibrated with the same vicarious method that relies on simulated radiance over bright desert targets (Govaerts et al., 2004). The intercalibration consistency between the two instruments needs also to be addressed as uncertainties in the characterisation of the VIS band have already been reported (Govaerts, 1999). To this end, Top of Atmosphere (TOA) BRFs derived for both instruments have been collected under identical viewing zenith angles for differences in sun zenith angles not exceeding 2°. There is a good agreement between both instruments as can be seen in Fig. 4 (LEFT), except over dark sea surface. Over terrestrial surfaces, observations are consistent, Meteosat-5 reflectances overestimating only by about 1–2% those observed by Meteosat-7. A similar comparison has been performed between GMS-5 and Meteosat-5 (Fig. 4 RIGHT). The GMS-5 VIS band data processing still requires some improvements.

The global product derived from the three geostationary satellites, remapped on a common grid, is shown in Fig. 5. As Meteosat and GMS instruments have slightly different spectral responses (Fig. 3), the respective albedo products has been converted into a

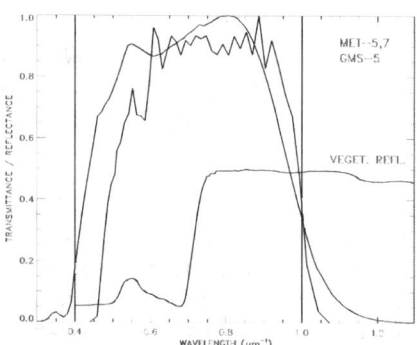

Fig. 3. Sensor Spectral Response (SSR) of the geostationary VIS band radiometer used in this study. The green solid line shows the typical reflectance of green vegetation. The vertical solid black line shows the common spectral interval.

GENERAL REMOTE SENSING

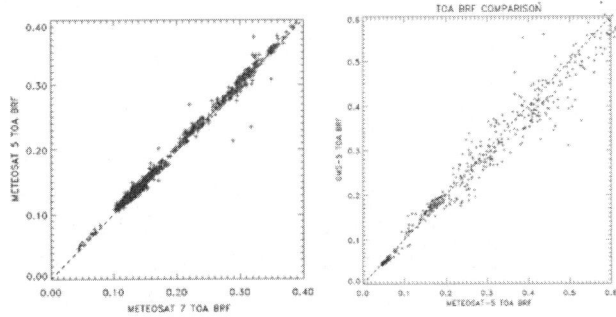

Fig. 4. LEFT: Scatter plot of TOA BRFs derived from Meteosat-7 and -5 along the 31.5±E latitudinal transect, i.e., with identical viewing zenith angles and differences in sun zenith angles not exceeding ±2°. RIGHT: Idem but between Meteosat-5 and GMS-5.

Table 1. Number of Days Processed During the 1–10 May 2001 Period for Each Satellite and Mean Number of Images Available per Day (<Img/Day>)

Satellite	Nbr Days	<Img/Day>
MET-7	10	17.6
GMS-5	10	9.6
MET-5	10	16.8

common spectral interval ranging from 0.4 up to 1.0 μm. Table 1 indicates the number of days available for each spacecraft and the mean number of images available per day. As can be seen on Fig. 1, pairs of adjacent spacecrafts observe common areas. Surface albedo comparison over these common regions offers a unique opportunity to evaluate the consistency of a same product retrieved from two radiometers located at two different places, i.e., pixels are observed with different viewing angle. On the average, surface albedo values retrieved from Meteosat-7 observations exceed by about 0.015, or 6% in terms of relative difference, those retrieved from Meteosat-5 (Govaerts et al., 2004) (Fig. 6, LEFT). Problems related to the accuracy of the instruments characterisation are not excluded at this stage to explain the mean difference (Govaerts, 1999), which is within the calibration error reported in Govaerts et al. (2004). The density plot between Meteosat-5 and GMS-5 exhibits bias as a result of a data processing problem of this latter instrument (Fig. 6, RIGHT).

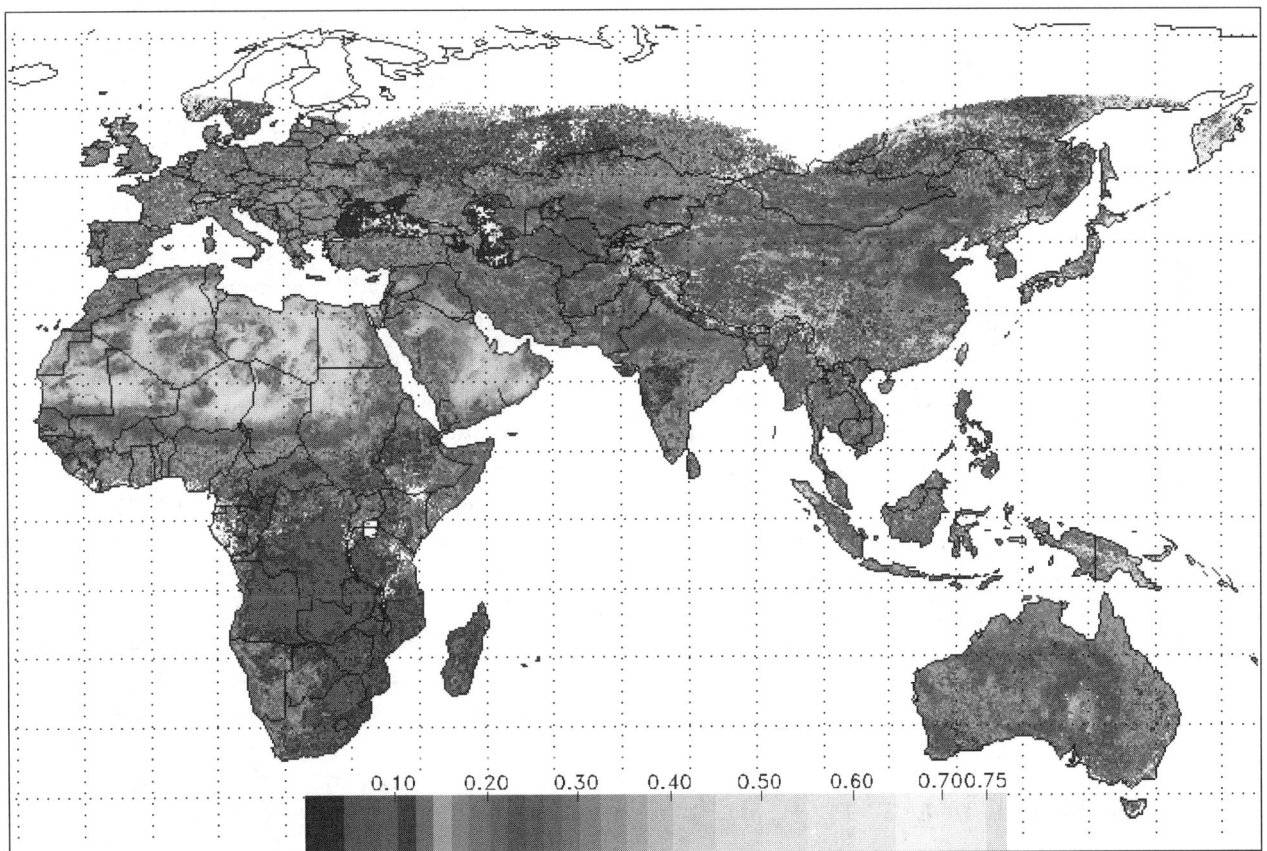

Fig. 5. Surface albedo map derived at EUMETSAT with the GSA algorithm from Meteosat-5/-7 and GMS-5 observations acquired May, 1–10, 2001.

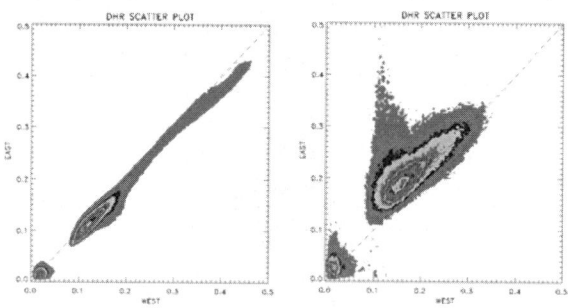

Fig. 6. LEFT: Density plot between the surface albedo retrieved from Meteosat-7 (WEST) and Meteosat-5 (EAST) in May 2001. RIGHT: Idem but between Meteosat-5 (WEST) and GMS-5 (EAST).

4. CONCLUSIONS

Recently, the Global Climate Observing System (GCOS) committee recognised the need for establishing a benchmark for assessing land surface albedo product and implementing a system for the retrieval of surface albedo from existing and archived geostationary satellites to form a global climatology of albedo for the entire period of available measurements (GCOS, 2003). The results presented in this paper constitute a first step in that direction. It has already been possible to demonstrate the consistency of this product between two Meteosat satellites. Calibration of GMS-5 still requires some improvements prior assessing the consistency between albedo derived from Meteosat and GMS. The processing of GOES data is already ongoing and the resulting surface albedo will be added to this product. The use of the GSA algorithm has permitted to identify several problems in the quality of the archived geostationary data. Some of them have been fixed and other will be fixed in the future to the benefit of a quantitative exploitation of the archive. The generation of the GSA product provides thus a quantitative mechanism to assess the quality of these archives, also demonstrating its contribution to the GCOS recommendations.

5. ACKNOWLEDGEMENTS

The authors are grateful to the Japan Meteorological Agency for providing the GMS-5 images and support their processing.

6. REFERENCES

Charney, J. G. (1975). Dynamics of deserts and droughts in the Sahel. *Quarterly Journal of the Royal Meteorological Society 101*, 193–202. GCOS (2003). Second report on the adequacy of the global observing systems for climate. Technical Report GCOS-82 (WMO/TD No. 1143), World Meteorological Organization.

Govaerts, Y., A. Lattanzio, B. Pinty, and J. Schmetz (2004). Consistent surface albedo retrieval from two adjacent geostationary satellites. *31*, doi:10.1029/2004GL020418.

Govaerts, Y. M. (1999). Correction of the Meteosat-5 and -6 VIS band relative spectral response with Meteosat-7 characteristics. *International Journal of Remote Sensing, 20*, 3677–3682.

Govaerts, Y. M., M. Clerici, and N. Clerbaux (2004). Operational Calibration of the Meteosat Radiometer VIS Band. *IEEE Transactions on Geoscience and Remote Sensing, 42*, 1900–1914.

Jin, Y., C. B. Schaaf, F. Gao, X. Li, A. H. Strahler, W. Lucht, and S. Liang (2003). Consistency of MODIS surface bidirectional reflectance distribution function and albedo retrievals: 1. Algorithm performance. *Journal of Geophysical Research, 108*, 4158, doi:10.1029/2002JD002803.

Martonchik, J. V., D. J. Diner, B. Pinty, M. M. Verstraete, R. B. Myneni, Y. Knyazikhin, and H. R. Gordon (1998). Determination of land and ocean reflective, radiative, and biophysical properties using multiangle imaging. *IEEE Transaction on Geosciences and Remote Sensing, 36*, 1266–1281.

Pinty, B., F. Roveda, M. M. Verstraete, N. Gobron, Y. Govaerts, J. V. Martonchik, D. J. Diner, and R. A. Kahn (2000a). Surface albedo retrieval from Meteosat: Part 1: Theory. *Journal of Geophysical Research, 105*, 18099–18112.

Pinty, B., F. Roveda, M. M. Verstraete, N. Gobron, Y. Govaerts, J. V. Martonchik, D. J. Diner, and R. A. Kahn (2000b). Surface albedo retrieval from Meteosat: Part 2: Applications. *Journal of Geophysical Research, 105*, 18113–18134.

Rahman, H., B. Pinty, and M. M. Verstraete (1993). Coupled surface atmosphere reflectance (CSAR) model. 2. Semiempirical surface model usable with NOAA Advanced Very High Resolution Radiometer Data. *Journal of Geophysical Research, 98*, 20,791–20,801.

THE ATMOSPHERIC SENSITIVITY OF THE AIRBORNE IMAGING SPECTROMETER APEX

J. W. Kaiser, J. Nieke, D. Schläpfer, J. Brazile, and K. I. Itten

Remote Sensing Laboratories, University of Zurich, Winterthurerstrasse 190, 8057 Zurich, Switzerland

Email: jkaiser@geo.unizh.ch (J.W. Kaiser)

ABSTRACT

The new airborne imaging spectrometer APEX will be available in 2006. It observes the visible and shortwave infrared spectral ranges. With its unique combination of spectral and spatial resolution, atmospheric molecular spectroscopy is feasible with unprecedented spatial resolution and coverage. Sensitivity calculations show, that atmospheric columns of NO_2 and CH_4 will be quantified with a spatial resolution of about 30 m. Such observations open new possibilities in the characterization of sub-satellite pixel distributions and air pollution monitoring. H_2O and CO_2 may also be observed. Additionally, a large set of atmospheric aerosol and cloud properties can be measured.

1. INTRODUCTION

The Airborne Prism EXperiment (APEX) is an airborne dispersive push broom imaging spectrometer for the hyperspectral observation of ground reflectances [e.g. Nieke et al., 2004]. APEX is currently being built in a joint Swiss/Belgian project funded through the ESA PRODEX program. It will be operated by VITO, Belgium, starting from 2006.

The spectrometer records up to 511 spectral points in the wavelength range 380–2500nm with a sampling interval of 0.4 to 10 nm. A CCD and a CMOS are used as detectors in the visible/near infrared (VNIR) and shortwave infrared (SWIR) spectral ranges, respectively. The ground pixel size ranges from 2 to 5 m corresponding to flight altitudes of 4 to 10 km. Since APEX is a push broom instrument all spectral points for all 1000 across-track pixels are observed simultaneously. A dedicated calibration laboratory is being built to facilitate the absolute calibration of the instrument. For more details visit the instrument homepage http://www.apex-esa.org.

APEX achieves an unprecedented combination of good spatial and spectral resolutions, coverage, and signal to noise ratio. Thus a large number of atmospheric parameters may also be retrieved from its measurements. In this paper, we give an overview of potential atmospheric applications of APEX and calculate theoretical precisions for the retrieval of columns of the trace gases NO_2, O_3, and CH_4.

2. OVERVIEW OF ATMOSPHERIC APPLICATIONS

Even though APEX primarily aims at the observation of the surface reflectance, atmospheric applications have already been taken into account during the definition of the instrument requirements [Schläpfer and Schaepman, 2002]. Figs. 1 and 2 give an overview of the optical depths of the total zenith atmosphere of various atmospheric constituents in the spectral range of APEX. The values are calculated with MODTRAN [e.g., Berk et al, 1998], modeling mid-latitudinal atmospheric profiles for summer and 360 p.p.m. CO_2. They are convolved to the spectral resolution of APEX. The center band wavelengths are indicated by vertical red lines.

2.1 Trace Gases

Owing to APEX' good spectral resolution the spectral signatures of several trace gases can be observed. The optical depths of water vapor, oxygen, and carbon dioxide obviously display strong signatures that can be analyzed in the observed spectra. They can be used for the atmospheric correction of the observation, i.e., the retrieval of the surface reflectance.

Nitrogen dioxide, ozone, and methane have optical depths that are hardly discernible in the total optical depth. They are, however, of particular importance for the monitoring of local and regional air pollution [e.g., Schaub et al., 2005, Heldstab et al., 2004]. Section 3 of this work quantitatively analyses their expected retrieval precisions.

2.2 Aerosols

The optical depth of typical aerosols displays a spectrally smooth signature. This characteristic is sometimes used to distinguish aerosol parameters from the underlying land surface reflectance [e.g., von Hoyningen-Huene et al., 2003, for SeasWiFS]. However, the optical depth varies considerably over the relatively large spectral range observed by APEX, which is indirectly used by the MODIS aerosol retrieval over land [Kaufman et al., 1997]. Additionally, the observations at small wavelengths, starting with 380 nm show a relatively (as compared to the surface reflectance) strong aerosol influence, which is used, e.g., by Höller et al. [2004] for analyzing observations by the Global Imager (GLI) aboard JAXA's ADEOS-II satellite. Since all these different algorithms can be applied to APEX observations, the aerosol product can be improved by combining information extracted with several algorithms.

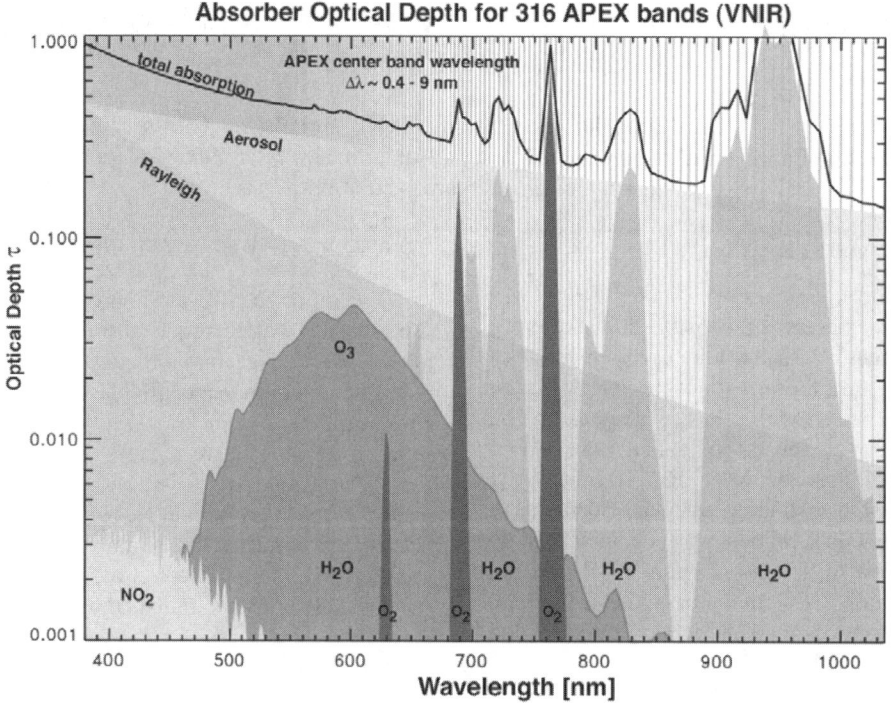

Fig. 1. Atmospheric Constituents' Optical Depths Seen by APEX' VNIR Channel.

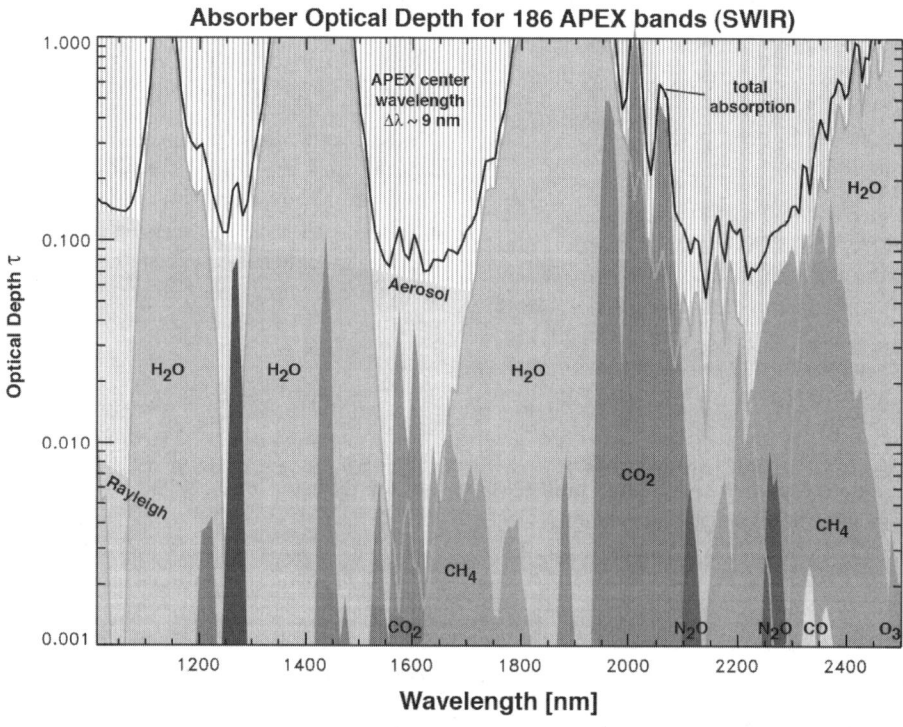

Fig. 2. Atmospheric Constituents' Optical Depths Seen by APEX' SWIR Channel.

2.3 Clouds

APEX will observe variations in the radiance field due to the scattering by structured clouds at spatial scales to the scattering by structured clouds at spatial scales ranging from about 1 m to several km simultaneously, e.g., von Savigny et al. [2002] have analyzed the radiative smoothing using a time series of zenith sky radiances under clouds. With APEX such analyses can be extended by observing a 2-dimensional field of up-welling radiances over a large spectral range.

3. THEORETICAL PRECISIONS FOR TRACE GAS OBSERVATIONS

3.1 Method

Theoretical retrieval precisions are calculated for the column densities of NO_2, O_3, and CH_4 assuming a least squares retrieval algorithm. The analysis is performed separately for each molecule, considering the fit windows listed in Table 2. The flow of the analysis is pictured in Fig. 3. The assumed parameters are detailed in Table 1.

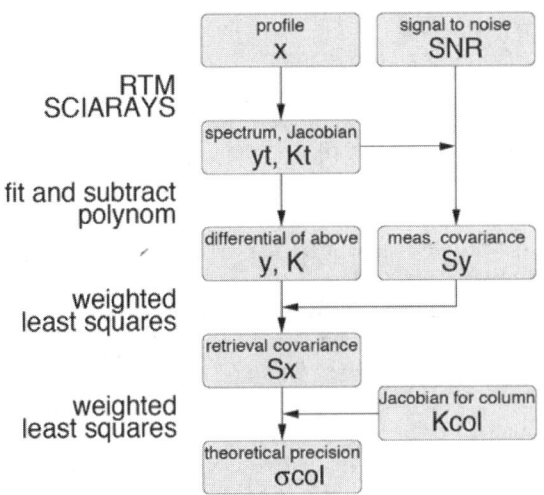

Fig. 3. Flow of Analysis.

Table 1. Measurement and Retrieval Scenario

atmospheric profiles	mid-latitude
solar zenith angle	30 deg
surface albedo	30%
flight height	6km
signal/noise ratio	150–470
subtracted polynomial	3rd order

First, the observed spectrum y_t and its Jacobian K_t, i.e. the matrix of derivatives w.r.t. the trace gas concentrations in the assumed trace gas profile x, are simulated with a radiative transfer model. Fitting a polynomial to the spectrum y_t and subtracting it from the spectrum separates the differential molecular absorption signature y from the other, broadband effects in a DOAS-type fashion. The Jacobian K of y is calculated analogously by fitting and subtracting polynomials.

Typical signal to noise ratios of APEX are taken from the critical design review documents of APEX. They are combined with the simulated spectrum y_t to obtain the measurement covariance matrix S_y, which is assumed to be diagonal. The least squares formalism provides the inverse of the retrieval error covariance matrix S_x:

$$S_x^{-1} = K^T S_y^{-1} K.$$

The matrix S_x^{-1} cannot be inverted directly as the observation does not provide enough information to resolve the layers of the atmospheric profile. Thus real retrievals will have to be regularized with additional information, e.g., by scaling an a priori profile in a DOAS-type algorithm [Platt, 1994] or with an optimal estimation scheme [Rodgers, 2000]. Then any retrieved profile can be interpreted as indirect measurement of the total trace gas column. For the purpose of estimating the total column retrieval precision inherent to the APEX measurements the least squares formalism can be applied to S_x^{-1}: The variance σ_{col}^2 of the total column retrieval error is expressed as:

$$\sigma_{col}^2 = (K_{col}^T S_x^{-1} K_{col})^{-1},$$

where K_{col} denotes the Jacobian of the profile-to-column conversion. Note, that it contains only one row as only one parameter is retrieved, i.e. the trace gas column. Finally, σ_{col} is interpreted as theoretical retrieval precision. It should be noted that it represents the best case in which the measurement noise dominates the total retrieval error.

All calculations are performed with an adapted version of the toolbox SCIARAYS, which contains a radiative transfer model, an instrument model, and inversion routines [Kaiser and Burrows, 2003].

3.2 Results

The calculated theoretical retrieval precisions in absolute units [cm^{-2}] are summarized in Table 2. Column 4 gives the precision for retrievals from individual pixels of the APEX observation. For the assumed flight altitude of 6 km. This corresponds to a ground pixel size of 3 x 3 m^2. Additionally, column 5 shows the precisions for an average of 100 pixels. This would correspond to a ground pixel size of, e.g., 30 x 30 m^2. It enhances the retrieval precision by a factor of 10.

Table 2. Theoretical Column Retrieval Precisions

	fit window [nm]	atmospheric column [cm^{-2}]	theoretical precision [cm^{-2}] 3m x 3m	theoretical precision [cm^{-2}] 30m x 30m
NO_2	420–480	0.5–1.0 x 10^{16}	6 x 10^{15}	6 x 10^{14}
O_3	550–620	1 x 10^{19}	2 x 10^{18}	2 x 10^{17}
CH_4	1640–1740	2 x 10^{19}	1 x 10^{19}	1 x 10^{18}

Column 3 provides a comparison to typical atmospheric values observed by SCIAMACHY with a ground pixel size of 30 x 60 km^2 [Bovensmann et al., 2003, and references therein].

The range of typical atmospheric NO_2 columns represents both polluted and background situations. Substantially larger values are expected once the sub-SCIAMACHY-pixel structure of the NO_2 pollution is resolved. The NO_2 retrieval from individual APEX pixels will have a precision of the background's order of magnitude. It is expected that such analyses are only useful in the case of strong local emission sources. However, retrievals with a spatial resolution of 30m will yield NO_2 column information with a precision of about 10% in most observational situations.

Considering that the tropospheric contribution to the total column of O_3 is relatively small, the retrieval precision of 20% for 3m spatial resolution is not considered to be useful. The precision of 2% for 30m spatial resolution might be sufficient to study extreme events of ozone smog.

The retrieval of CH_4 with 3m ground resolution shows a precision of 50%, which may facilitate the qualitative detection of strong CH_4 sources. Quantitative studies are possible with observations averaged to about 30 m spatial resolution. Additional information on CH_4 can be expected from the absorption band around 2350 nm wavelength.

All mentioned precision values correspond to the maximal potential accuracy achieved under idealized experimental conditions. The realization of such accuracy when analyzing real measurements will strongly depend and the actual conditions, e.g., underlying surface reflectivity. Anyway it will be a challenging task.

4. CONCLUSIONS

The airborne imaging spectrometer APEX will record atmospheric radiance spectra in the wavelength range 380–2500 nm with a spectra resolution varying between 0.5 and 10 nm. The spectra are recorded simultaneously for 1000 across-track pixels with a pixel size of 2–5 m depending on flight altitude. The instrument will become operationally available in 2006.

The observations offer a unique opportunity to quantitatively observe the tropospheric distributions of the atmospheric trace gases NO_2 and CH_4 with a resolution of a few meters while covering regions of several kilometers extent. This is a new possibility to monitor local and regional air pollution. Additionally, the distributions within entire satellite pixels may be characterized to validate the satellite instrument products and facilitate the down/up-scaling. Trace gas profile information may be obtained by observing the same scene with several flight altitudes.

5. ACKNOWLEDGMENTS

The work in this paper is being carried out under ESA/ESTEC contract no. 15440/01/NL/SFe. The support of the University of Zurich is acknowledged.

6. REFERENCES

Berk, A., et al., MODTRAN cloud and multiple scattering upgrades with applications to AVIRIS. *Remote Sensing of Environment*, 65, 367–375, 1998,

Bovensmann, H., et al., SCIAMACHY on ENVISAT: In-flight optical performance. *Proc. SPIE*, 5235:160–173, 2003.

Heldstab, J., et al., Modelling of NO_2 and benzene ambient concentrations in Switzerland 2000 and 2020. Swiss Agency for the Environment, Forests and Landscape SAEFL, CH-8003 Berne, 2004.

Höller, R., et al., The GLI 380-nm channel application for satellite remote sensing of tropospheric aerosol. In *The 2004 EUMETSAT Meteorological Satellite Conference*, 2004.

Kaiser, J. W., and J. P. Burrows. Fast weighting functions for retrievals from limb scattering measurements. *J. Quant. Spectrosc. Radiat. Transfer*, 77(3):273–283, 2003.

Kaufman, Y. J., et al. The MODIS 2.1 µm channel correlation with visible re- flectance for use in remote sensing of aerosol. *IEEE Trans. Geosci. Remote Sensing*, 35(5): 1286–1298, 1997.

Nieke, J., et al., APEX: Current status of the airborne dispersive pushbroom imaging spectrometer. *Proc. SPIE*, 5542, 2004.

Platt, U. Differential Optical Absorption Spectroscopy (DOAS). In: *Air Monitoring by Spectroscopic Techniques, Chem. Anal.*, 127:27–76, John Wiley, New York, 1994.

Rodgers, C. D. Inverse Methods for Atmospheric Sounding: Theory and Practice. *World Scientific*, Singapore, 2000.

Schaub, D., et al., A transboundary transport episode of nitrogen dioxide as observed from GOME and its impact in the Alpine region. *Atmos. Chem. Phys.*, 5:23–37, 2005.

Schläpfer, D., and M. Schaepman. Modeling the noise equivalent radiance requirements of imaging spectrometers based on scientific applications. *Appl. Opt.*, 41(27):5691–5701, 2002.

von Hoyningen-Huene, W., et al., Retrieval of aerosol optical thickness over land surfaces from top-of-atmosphere radiance. *J. Geophys. Res. A*, 108(D9): 4260, 2003.

von Savigny, C., et al., Time-series of zenith radiance and surface flux under cloudy skies: Radiative smoothing, optical thickness retrievals and large-scale stationarity. *Geophys. Res. Lett.*, 29(17):1825–1828, 2002.

A NEAR GLOBAL CLIMATOLOGY OF CIRRUS OVERLAPPING WATER CLOUDS AND RETRIEVAL OF CIRRUS CLOUD PROPERTIES FROM MODIS

Fu-Lung Chang and Zhanqing Li

Earth System Science Interdisciplinary Center, University of Maryland, College Park, Maryland 20742, USA

ABSTRACT

This study presents a new method for the determination of cirrus-overlapping-water cloud properties. Both single-layered and overlapped cirrus cloud amounts and their associated optical properties are derived from a near-global analysis of the Moderate Resolution Imaging Spectrometer (MODIS) data. The near-global analysis highlighted, 1) a bimodal distribution of cloud top heights, 2) an enhanced total low-cloud amount due to cirrus overlapping, and 3) the differentiated cirrus and low-cloud τ_{VIS} retrievals from the new method.

1. INTRODUCTION

The frequent occurrence of cirrus overlapping lower-water clouds poses a major challenge in retrieving their optical properties from space. While unprecedented MODIS cloud imager provides much better information on cirrus cloud-top heights (King et al., 2003; Platnick et al., 2003), no information is provided concerning if there is any lower cloud beneath the cirrus cloud and so no appropriation is made to separate the cirrus cloud optical depth from the total column-integrated cloud optical depth.

This paper presents a new retrieval method that takes full advantage of the MODIS data for the identification of overlapped cirrus and lower water clouds and determination of their individual optical depth, top height and emissivity. The cirrus cloud top height is determined from the CO_2-slicing retrieval and its underlying low cloud top height, if retrieved, is determined from its neighboring pixels that are identified as low clouds. To retrieve separated cirrus and lower-water cloud optical depths, the method adopts an ice-over-water dual-layer cloud radiative transfer model. An automated-iterative retrieval procedure follows by adjusting the cirrus and lower-water cloud optical depths until both computed 0.65-μm (VIS) and 11-μm (IR) radiances from the dual-layer model match with the MODIS-observed radiances. The method is valid for pixel-by-pixel retrievals when cirrus cloud optical depths are less than ~4 (emissivity ~0.85). For more than two-layer clouds, its validity depends on the thickness of the upper-layer cloud. A preliminary validation is conducted by comparing against ground-based active remote sensing data. Some error analyses are presented. It is also demonstrated that retrievals based on a single-layer algorithm can result in systematic biases in the retrieved cloud top and optical properties. Such biases can be removed or lessened considerably by applying the new algorithm.

2. METHODOLOGY

The algorithm developed by Chang and Li (2005a) is used to determine the existence of overlapped cirrus and water clouds and to retrieve their individual optical properties. It is based on the combined information from multispectral CO_2-slicing channels and conventional VIS and IR window channels. Initially, the algorithm utilizes both a cirrus cloud optical depth (τ'_{VIS}) estimated from the 11-μm emissivity and a total-column cloud optical depth estimated from the VIS reflectance (0.65-μm for land and 0.86 μm for ocean). The cirrus cloud top height/temperature (T_{hc}) is estimated from the CO_2-slicing retrieval; the low cloud top height/temperature (T_{lc}) is estimated from the average of nearby single-layer low-cloud pixels identified within a ±125 km domain. To determine τ'_{VIS}, an effective 11-μm emissivity (ε_{hc}) is first computed by (Menzel et al., 1992; Wylie et al., 1994):

$$\varepsilon_{hc} = \frac{R - R'}{R(T_{hc}) - R'}, \quad (1)$$

where R is the 11-μm observed radiance, $R(T_{hc})$ is the computed equivalent blackbody radiance at the cirrus T_{hc} and R' is the background clear-sky radiance for single-layer cirrus or the underlying low-cloud radiance. From ε_{hc}, an IR optical depth (τ_{IR}) is derived by (Minnis et al., 1993)

$$\tau_{IR} = -\mu \ln(1 - \varepsilon_{hc}), \quad (2)$$

where μ is the cosine of satellite zenith angle. A ratio factor ξ (Minnis et al., 1993; Rossow and Schiffer, 1999) is used to relate τ'_{VIS} at VIS wavelengths and τ_{IR}:

$$\tau'_{VIS} = \tau_{VIS}. \quad (3)$$

Mean value adopted for ξ is equal to 2.13 for ice cloud and 2.56 for water cloud (Rossow and Schiffer 1999).

The parameterized τ'_{VIS} is dictated by IR radiative transfer for the cirrus cloud and the VIS-retrieved τ_{VIS} is dictated by the total cloud column. The IR-based τ'_{VIS} is significantly smaller than τ_{VIS} if the cirrus overlaps lower clouds as shown in Fig. 1a for a typical cirrus-overlapping stratus system observed on April 2, 2001 over a 50-km region in north Oklahoma centered at the Atmospheric Radiation Measurement Program (ARM) SGP site (97.5°W, 36.6°N) of the U.S. Department of Energy. The differences between τ_{VIS} and τ'_{VIS} are used to determine if a dual-layer cloud model should be called upon to retrieve separate optical properties for the overlapped cirrus and lower-water cloud layers. In Fig. 1b, a single-layer cirrus cloud system observed on 6 March 2001 is illustrated, where $\tau'_{VIS} \cong \tau_{VIS}$.

For a cirrus-overlapping-low cloud case,

$$R' = \varepsilon_{lc} R(T_{lc}) + (1 - \varepsilon_{lc}) R_{clr}, \quad (4)$$

where $R(T_{lc})$ is the computed equivalent blackbody radiance at low-cloud T_{lc} and ε_{lc} is the low-cloud emissivity. When $\tau_{VIS} > \tau'_{VIS}$ and there exists neighboring low-cloud T_{lc}, a two-layer cirrus-over-water cloud radiative transfer model is applied to the VIS channel to retrieve a low-cloud optical depth (τ_{lc}). The best-fit value of τ_{lc} is decided until the model-computed VIS reflectance matches the observation, which determines ε_{lc}. Because retrievals of τ_{hc} and τ_{lc} are mutually dependent, an iterative process is needed.

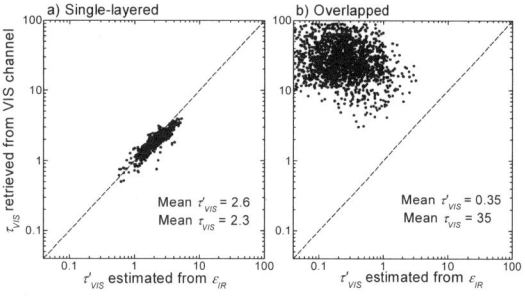

Fig. 1. Comparisons of MODIS-retrieved τ_{VIS} and τ'_{VIS} for, a) single-layer cirrus, and b) cirrus-overlapping-stratus clouds.

From the retrieved ε_{hc}, all high clouds with Pc < 500 hPa are classified into three categories, namely, 1) High1: single-layer cirrus (ε_{hc} < 0.85), 2) High2: cirrus overlapping low clouds, and 3) High3: thick high clouds ($\varepsilon_{hc} \geq 0.85$). Fig. 2 shows a flow chart illustrating our processing paths

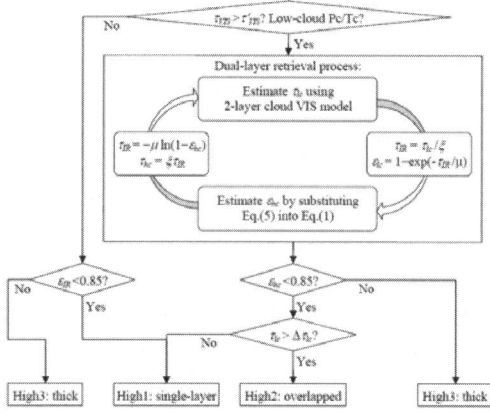

Fig. 2. Schematic illustration of the retrieval algorithm.

for High1, High2 and High3 clouds. The High3 category includes two processing paths with (rightmost path) and without (leftmost path) the identification of nearby low cloud. In either case, the High3 cloud is considered as a single thick layer cloud with a total-column τ_{VIS}. But some High3 clouds may also overlap lower clouds. For all lower clouds with Pc $\geq$ 500 hPa, no decision was made regarding overlapping. They are separated into a Low1 category for Pc > 600 hPa and an *ad hoc* Mid category for 500 hPa $\leq$ Pc $\leq$ 600 hPa. Selection of this interval for the Mid category was mainly to illustrate an extremely low occurrence of MODIS Pc falling between 500–600 hPa as revealed in this study. Detailed dual-layer radiative transfer calculations and the overlapped retrieval algorithm are described in Chang and Li (2005a), which also shows preliminary validation in comparisons with the ARM Active Remote Sensing Cloud Locations (ARSCL) product. For overcast cases, MODIS high and low cloud top temperatures, Tc (Fig. 3a), and pressures, Pc (Fig. 3b), are compared against the ARSCL means and standard deviations.

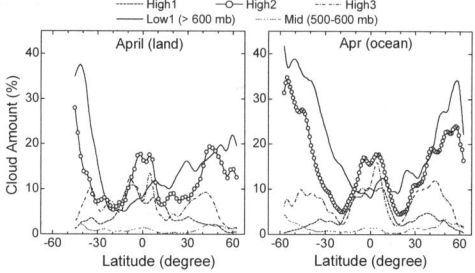

Fig. 3. Comparisons of, a) Tc, and b) Pc for overlapped high (open squares, T_{hc} and P_{hc}) and low (solid squares, T_{lc} and P_{lc}) cloud cases observed over ARM SGP site.

3. NEAR GLOBAL CLOUD PROPERTIES

The Terra/MODIS data are from the 5-km overcast scenes obtained in January, April, July and October 2001 (overpass time ~10:30 am). Global data were processed excluding polar winter regions and solar zenith angles greater than 75°. The data are sampled every fourth day (i.e., days 2, 6, 10, 14, 18, 22, 26, and 30) in each month. All 5-km overcast scenes are classified into five cloud categories, i.e., High1, High2, High3, Mid and Low1. The retrieved low clouds overlapped by the High2 cirrus are hereafter referred to as Low2. They are listed in Table 1.

Table 1. Cloud Categories

High1	Single-layer cirrus cloud
High2	Overlapped cirrus cloud
High3	High thick cloud
Mid	Middle cloud
Low1	Single-layer low cloud
Low2	Overlapped low cloud

The near-global monthly-mean total cloud amount is around 60% with the lowest in July (~57%) and the highest in January (~61%). The total 5-km overcast cloud amount is about 45% (~48% in January and ~42% in July). Fig. 4 shows the frequency occurrence of Pc for all overcast clouds (solid lines) obtained in April 2001. A distinct bimodal PC distribution is seen with a demarcation at about 500–600 hPa. Almost identical bimodal Pc distributions are found in the other three months (January, July and October). The Pc distributions in Fig. 4 are also plotted for: 1) single-layer Low1, Mid, and High1 clouds (connected with the same dashed curves), 2) overlapped High2 (dotted lines with circles) and Low2 (dotted lines with triangles), and 3) thick High3 (dash-dotted lines). Note that Low2 and High2 are from the same overlapped cloud categories; they have the same total frequency but different Pc. It should be pointed out that the bimodal Pc distribution is not resulted by our dual-layer cloud retrievals. The MODIS standard Pc product reveals the same bimodal distribution without the inclusion of our Low2 clouds. This bimodal Pc distribution forms the foundation of our overlapped cloud retrieval algorithm.

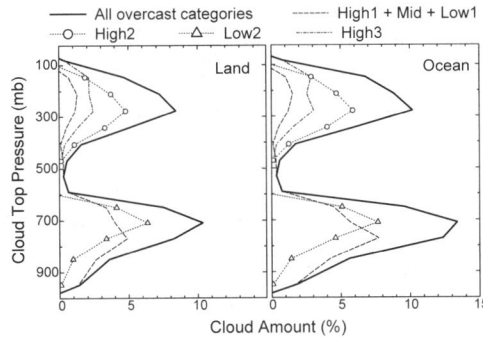

Fig. 4. Distributions of Pc observations for all overcast clouds, High1+Mid+Low1, High2, Low2, and High3 in April 2001.

Fig. 5 shows the monthly frequency distribution of τ_{VIS} for the different categories and their total. Both High1 (single-layer) and High2 (overlapped) cirrus clouds have the smallest τ_{VIS} among all cloud categories, whose means are ~1.5 and standard deviations ~1.0. The distributions of Low1 and Low2 τ_{VIS} are much broader with similar means (± standard deviations) of ~11 (± 10) for Low1 clouds and ~14 (± 13) for Low2 clouds. The High3 clouds have the largest mean τ_{VIS} ~22 (standard deviations ~20) and Mid clouds (Pc = 500-600 hPa) have the second largest mean τ_{VIS} ~16 (standard deviations ~20).

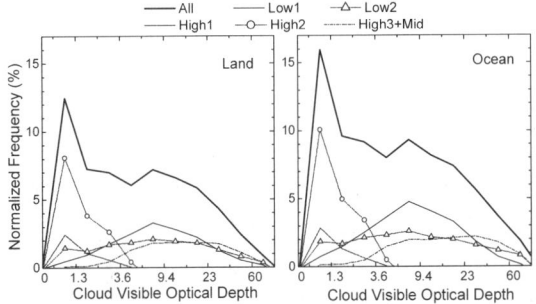

Fig. 5. Distributions of τ_{VIS} observations for all overcast clouds, Low1, Low2, High1, High2, and High3+Mid.

Fig. 6 shows the variations of 1°-latitudinal zonal-mean overcast cloud amounts for the five categories. The monthly means are given in Table 2. More clouds are found over oceans than over land. The two most dominant categories are Low1 and High2. The overlapped High2 accounts for about 50% of total high clouds and 30% of total overcast clouds. In general, high clouds occur most frequently over the ITCZ and midlatitudes and less frequently in subtropics. Tropical high clouds are often associated with extensive anvil cirrus clouds covering large spatial domain. Frequent High3 clouds are in low latitudes due to tropical convection and in midlatitudes due to mesoscale cyclones.

Table 2. Monthly-Mean Overcast Cloud Amounts (%) Over Ocean/Land for the Five Categories

	Jan 2001	Apr 2001	Jul 2001	Oct 2001
High1	4.3/ 5.5	4.6/ 4.8	3.4/ 3.7	4.0/ 5.2
High2	12.3/14.0	14.3/12.6	9.9/ 9.0	12.8/12.7
High3	7.6/ 9.3	7.9/ 7.6	7.2/ 7.6	7.9/ 9.1
Mid	2.3/ 1.4	1.2/ 1.0	1.8/ 2.2	1.5/ 1.1
Low1	22.4/13.6	18.8/14.0	20.4/16.2	20.6/14.5

Fig. 7 shows the zonal-mean Pc and τ_{VIS} for each cloud categories for April. Mean Pc for High1, High2, and High3 clouds vary between 200-350 mb; it is highest (~200 mb) in the tropics and lower toward higher latitudes. Mean Pc for Low1 and Low2 vary between 700-800 mb. Differences between high and low cloud Pc are largest in the tropics and smaller towards high latitudes. Standard deviations for the high and low clouds are similar ~100 mb. Mean τ_{VIS} for High1 and High2 are ~1.0–1.8. There is little difference in terms of zonal means. Thick High3 clouds have the largest τ_{VIS} (~22 ± 20); Mid clouds have the second largest τ_{VIS} (~16 ± 20); and Low1 (~11 ± 10) and Low2 (~14 ± 13) clouds have smaller τ_{VIS}. These differences are mainly due to the different geometrical vertical extents. The Low2 τ_{VIS} is slightly larger than the Low1 τ_{VIS}, which is probably due to the possibility that those Low2 clouds near high clouds are indeed thicker than those more uniform low clouds further away from high clouds or due to cirrus contamination.

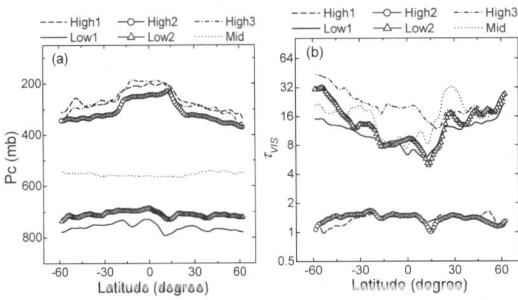

Fig. 7. Zonal-mean a) Pc and b) τ_{VIS} for six cloud categories.

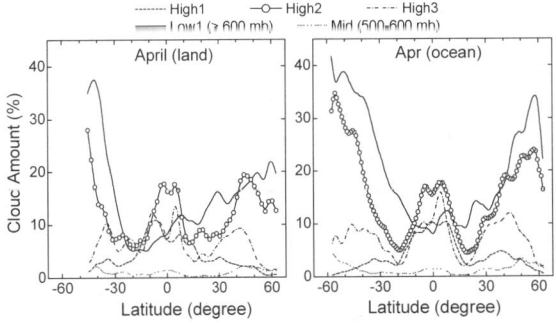

Fig. 6. Zonal-mean cloud amounts for High1, High2, High3, Mid, and Low1 categories.

Based on the Terra/MODIS 5-km overcast cloud data, Chang and Li (2005b) derived the new-developed global statistics of the frequency occurrence, Pc/Tc, τ_{VIS} and ε. Out of the 61% (52%) of high clouds identified by MODIS, our retrievals reveal that 41% (35%) are thin cirrus clouds (ε < 0.85 and Pc < 500 hPa) and the remaining 20% (17%) are thick high clouds ($\varepsilon \geq 0.85$). Out of the 41% (35%) of thin cirrus, 29% (27%) are found to overlap with lower water clouds and 12% (8%) are single-layer cirrus. The total low cloud amount is thus increased to 68% (39%+29%) over land and 75% (48%+27%) over ocean, which is greater than those reported by MODIS and ISCCP.

In comparisons with MODIS standard products, Fig. 8 shows the joint frequency distributions of Pc and τ_{VIS} from a) MODIS products, and b) our retrievals for the four-month total overcast amounts. The MODIS products show the single-layer retrievals; whereas ours include both single-layer and overlapped retrievals obtained from the same data. Note that the MODIS Pc products only provide the height of the topmost cloud layer as viewed from space. Through our retrievals, approximately half of the high-top clouds are overlapped with low clouds. Our total low-cloud amount is ~30% more than the MODIS products. If one uses the MODIS Pc data alone to determine high-cloud and low-cloud amounts, the fraction of low cloud amount would be substantially underestimated.

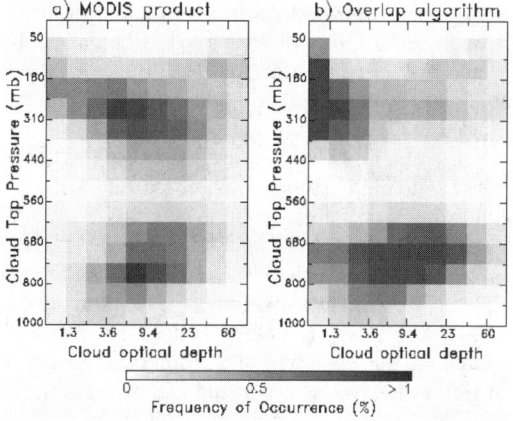

Fig. 8. Joint frequency distributions of Pc and τ_{VIS} from a) MODIS products, and b) our retrievals.

Since the MODIS τ_{VIS} are retrieved for the entire cloud column and do not differentiate high and low clouds, attributing τ_{VIS} to a single (highest) cloud top at Pc would substantially overestimate τ_{VIS} for the true cirrus cloud overlapping a low thick cloud. Another bias may also arise in choosing a cloud microphysical model for retrieving τ_{VIS} for the overlapped ice-over-water clouds. Regarding such overlapped clouds as single-layer ice or water clouds can lead to significant biases.

4. CONCLUDING REMARKS

The MODIS products only provide the highest cloud top for both single-layer and overlapped clouds. The optical depths of overlapped cirrus clouds would be overestimated by a factor of ~7 due to their thick water clouds underneath. Use of these cloud top data would underestimate low clouds by about 30%. The ISCCP uses 11-μm channel that is difficult in cirrus detection. Its bulk emission may be overwhelmed by much warmer low clouds when cirrus overlapping low clouds. This leads to an underestimation of ISCCP high cloud amounts, but more middle clouds.

In light of the substantial differences in cloud vertical structure, much caution is warranted in both validating general circulation models (GCMs) and improving their cloud parameterization schemes. At present, results of cloud simulations from many GCMs have been validated against ISCCP total cloud amounts, promoting many improvements in the models. More attention is now being paid to more detailed comparisons concerning the vertical distribution of clouds occurrence in different layers. For example, the current ARM Cloud Parameterization and Modeling (ARMCPM) working group has collected and analyzed 10 sets of GCM-simulated cloud layer data and compared them to the statistics of the ISCCP data (Zhang et al., 2005). It was found that most GCMs produce substantially less middle and low clouds than ISCCP. The GCMs' middle cloud amounts are closer to our new retrieval products, whereas the GCMs' low cloud amounts are quite different from our retrievals. Clearly, it is critical to sort out these differences in order to improve the performance of GCMs and other types of models of higher resolution.

5. ACKNOWLEDGMENTS

The authors thank the MODIS Atmosphere Team and NASA DAAC for providing the data. This study was supported by DOE grant DE-FG02-01ER63166 under the ARM program and NASA grant NNG04GE79G.

6. REFERENCES

Chang, F.-L., and Z. Li, 2005a: A new method for detection of cirrus-overlapping-water clouds and determination of their optical properties, *J. Atmos. Sci.*, in press.

Chang, F.-L., and Z. Li, 2005b: A near-global climatology of single-layer and overlapped clouds and their optical properties retrieved from Terra/MODIS data using a new algorithm, *J. Climate*, in press.

King, M. D., and Coauthors, 2003: Cloud and aerosol properties, precipitable water, and profiles of temperature and humidity from MODIS, *IEEE Trans. Geosci. Remote Sens.*, 41, 442-458.

Minnis, P., P. W. Heck, and D. F. Young, 1993: Inference of cirrus cloud properties using satellite-observed visible and infrared radiances. Part II: Verification of theoretical cirrus radiative properties, *J. Atmos. Sci.*, 50, 1305-1322.

Platnick, S., and Coauthors, 2003: The MODIS cloud products: Algorithms and examples from Terra, *IEEE Trans. Geosci. Remote Sens.*, 41, 459-473.

Rossow, W. B., and R. A. Schiffer, 1991: ISCCP cloud data products, *Bull. Amer. Meteor. Soc.*, **72**, 2-20.

Rossow, W. B., and R. A. Schiffer, 1999: Advances in understanding clouds from ISCCP, *Bull. Amer. Meteor. Soc.*, 80, 2261-2287.

Zhang, M. H., and coauthors, 2005: Comparing clouds and their seasonal variations in 10 atmospheric general circulation models with satellite measurements, Special Issue of ARMCPM, *J. Geophys. Res.*, in press.

QUALITY, COMPATIBILITY AND SYNERGY ANALYSES OF GLOBAL AEROSOL PRODUCTS

M.-J. Jeong and Z. Li
Department of Meteorology and Earth System Science Interdisciplinary Center, University of Maryland
College Park, Maryland 20742, USA

D. A. Chu and S.-C. Tsay
Laboratory for Atmospheres, NASA Goddard Space Flight Center
Greenbelt, Maryland 20771, USA

ABSTRACT

A number of global aerosol products of varying quality, strengths and weakness have been produced, but little insight has been gained about their compatibility and quality. This study presents a comparison of some prominent global aerosol products derived from the Advanced Very High Resolution Radiometer (AVHRR) (Mishchenko et al., 1999). Total Ozone Mapping Spectrometer (TOMS) (Herman et al., 1997; Torres et al., 2002), and Moderate Resolution Imaging Spectroradiometer (MODIS) (Tanré et al., 1997). While TOMS and AVHRR Aerosol Optical Thickness (AOT) reveal some common features, substantial differences exist. Taking the respective advantages of AVHRR and TOMS products, we developed a simple algorithm to classify aerosol type(s). For the same types of aerosols, AOTs from the two products are more compatible. AOT at 0.55 μm over land was estimated from TOMS AOT (0.38 μm) based on relationships established for each aerosol type. Comparisons against AERONET show reasonable agreements. Likewise, MODIS and AVHRR AOT also differ substantially. While some of the differences stem from cloud screening, use of different aerosol models was found to be a leading factor, which introduces uncertainties up to a factor of two. Values of the MODIS and AVHRR Angstrom Exponent (AE) have little correlation. For monthly mean data, AVHRR AE data seems to suffer from severe random-like errors. Difference in aerosol models seems to explain a significant portion of discrepancies, although various other factors such as errors in spectral AOT also contribute to the discrepancy.

1. INTRODUCTION

Satellite has played a vital role in understanding the effects of aerosols on the earth's climate (Kaufman et al., 2002). Extensive aerosol products have been generated from a suite of advanced instruments such as MODIS, MISR, POLDER. These products alone, however, cannot meet the needs of climate change studies due to their short observation periods. In this regard, aerosol products derived from AVHRR and TOMS offer a unique advantage for monitoring the planet for more than two decades. Inter-comparison of AVHRR and MODIS aerosol products helps us acquire longer term global aerosol climatology by connecting the aerosol products based on the past and current sensors as well as to understand uncertainties in the satellite estimated aerosol characteristics.

In this study, we investigate consistency and compatibility among TOMS, AVHRR and MODIS aerosol products, explore any synergy among them, and evaluate the inherent discrepancies between MODIS and AVHRR AOTs in terms of differences of the aerosol models employed by the two different algorithms. The study is fully described in Jeong and Li (2005) and Jeong et al., (2005).

2. COMPATIBILTY AND SYNERGY BETWEEN THE AVHRR AND TOMS AEROSOL PRODUCTS

The AVHRR and TOMS aerosol products exhibit a good synergy, which is exploited here. For example, TOMS data alone have difficulty in differentiating between dust and biomass burning aerosols, which can be compensated for by the AVHRR AE pertaining to aerosol size. Taking advantage of their respective strengths, we developed an algorithm to classify aerosol types into dust, biomass burning, a mixture of the two, sulfate/pollution, and sea-salt, (Jeong and Li, 2005). Using this algorithm, regions under the dominant influence of various types of aerosols are determined from the two satellite products alone (Fig. 1). Prior to MODIS and MISR, it was difficult to gain such information from satellite observations. The performance of this algorithm is influenced by the quality of each aerosol product (especially the AVHRR Angstrom Exponent [AE] and the TOMS Aerosol Index [AI]).

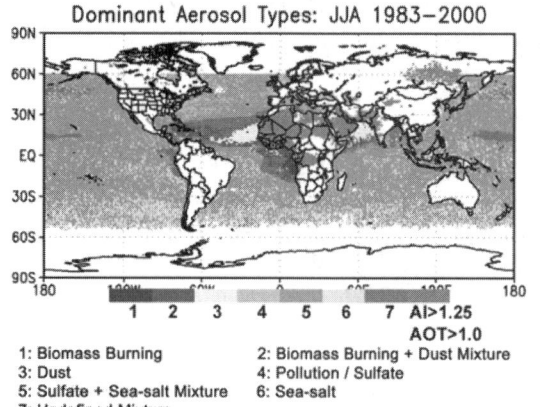

Fig. 1. Global distribution of dominant aerosol types classified from AVHRR and MODIS. Land areas with TOMS AOT greater than 1 and AI greater than 1.25 are shaded to indicate major aerosol sources.

As an application of the classification and exploitation of the synergy, the two AOT products are integrated to generate an AOT product at a common wavelength (0.55 µm) of truly global coverage over both ocean and land (Fig. 2). The original TOMS and AVHRR AOT data were stratified according to the aerosol types inferred by our aerosol type identification algorithm; then, we were able to acquire better relationships between the two AOTs compared to the one without the data stratification. After then, over land, we converted TOMS AOT at 0.38 µm into the AOT at 0.55 µm using the regression equations derived for the respective aerosol types from TOMS and AVHRR AOTs. Aerosol types over land had to be determined according to geographical location and TOMS AI.

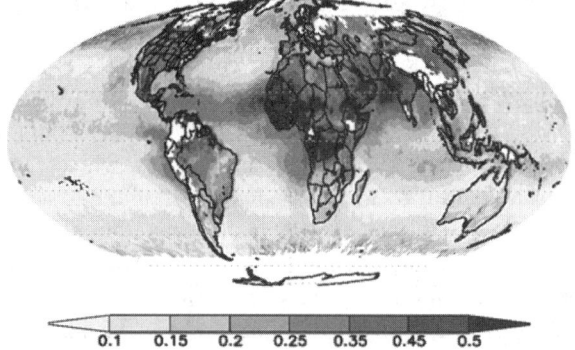

Fig. 2. Global map of seasonal mean AOT at 0.55 µm (JJA, 1983–2000). AOT over land was estimated from TOMS AOT, TOMS AI, and AVHRR AOT. AOT over the ocean is taken from the AVHRR AOT.

The overall range of uncertainty of the estimated AOT for each aerosol type is $\pm 0.08 \pm 0.20 \cdot AOT$. It was determined as the range within which 95% of developmental data points reside. The maximum error due to the assumption on aerosol type over land is 0.21*[TOMS AOT], which would be the case when biomass burning aerosol was selected instead of dust, and vice versa. These inferred AOTs are compared to AERONET measurements (Table 1), and most of the estimations (>70%) fall within the overall range of uncertainty ($\pm 0.08 \pm 0.20 \cdot AOT$).

Table 1. Comparison of the Monthly Mean Estimated AOT at 0.55 µm Over Land Against AERONET AOT

Region	Slope	Intercept	R*
South America	1.07	0.10	0.81
South Africa	0.59	0.16	0.71
North Africa	0.56	0.14	0.62
North Central America	0.27	0.17	0.34
Far East Asia	0.66	0.11	0.61
Arabia	0.70	0.10	0.64

[Est. AOT] = Slope * [AERONET AOT] + Intercept
*R: correlation coefficient

3. COMPATIBILITY BETWEEN THE AVHRR AND MODIS AEROSOL PRODUCTS

In light of large discrepancies among various satellite-based global aerosol products, two prominent monthly global aerosol products retrieved from AVHRR (Mishchenko et al., 1999) and MODIS (Tanré et al., 1997) measurements are compared and factors leading to their discrepancies are explored. Comparisons of the monthly AOT at 1x1 degree resolution showed substantial scattering and moderate systematic differences. However, their regional means are much better correlated with the general tendency that the AVHRR values are smaller than the MODIS values, especially for heavy aerosol loadings (Table 2). The difference in cloud screening is likely a factor (Myhre et al., 2004, Jeong et al., 2005), but other factors also come into play, for example, use of different aerosol models differentiated in size distribution function and refractive index.

Table 2. Relationship Between AVHRR and MODIS AOTs for Selected Regions Over Ocean

Region	Slope	Intercept	R
Africa / Arabia	0.55	0.06	0.92
East Asia / NW Pacific	0.44	0.06	0.85
Open Oceans	0.89	0.01	0.89
North Atlantic	0.89	0.01	0.81
Central America	0.89	0.01	0.80
South America	0.69	0.05	0.40

[AVHRR AOT] = Slope * [MODIS AOT] + Intercept

The MODIS retrieval algorithm employs 20 combinations of aerosol size distributions given by bi-lognormal (BL) functions with variable refractive index. The AVHRR algorithm used a modified power (MP) law size distribution with a fixed refractive index. Extensive model simulations were conducted to investigate the impact of the differences in the size distribution function and the refractive index on the AOT discrepancies (Fig. 3). It is found that differences in the size distribution function can bring about substantial AOT discrepancies of up to a factor of 2, while different refractive indices cause a moderate systematic difference. The discrepancies depend on the similarity in aerosol size modes selected by the two algorithms. More drastic underestimation of AOT by the AVHRR relative to the MODIS is more likely induced by differences in cloud screening including misclassification of heavy aerosols as clouds in the AVHRR product.

from random-like errors with low signal-to-noise ratio. In comparison, the MODIS AE product is of better quality in terms of spatial variation and its correlation with the AOT (Fig. 4). We attempted to understand the discrepancies between AE derived from MODIS and AVHRR by simulating the effects of aerosol size distribution function and refractive indices on AE retrieval. The results also point to a big contribution by different aerosol models used in the AVHRR and MODIS retrieval algorithms. The influence of aerosol size distribution on the estimation of aerosol effective radius from AE is also evaluated (Fig. 5). For a given AE, the corresponding aerosol effective radii may differ by more than 1μm among the various size distribution functions. This result clearly shows that AE is not an absolute measure of aerosol size; therefore, AE should be used in relative sense and may be used as proxy of aerosol size only when the size distributions can be assumed as not too different from each other.

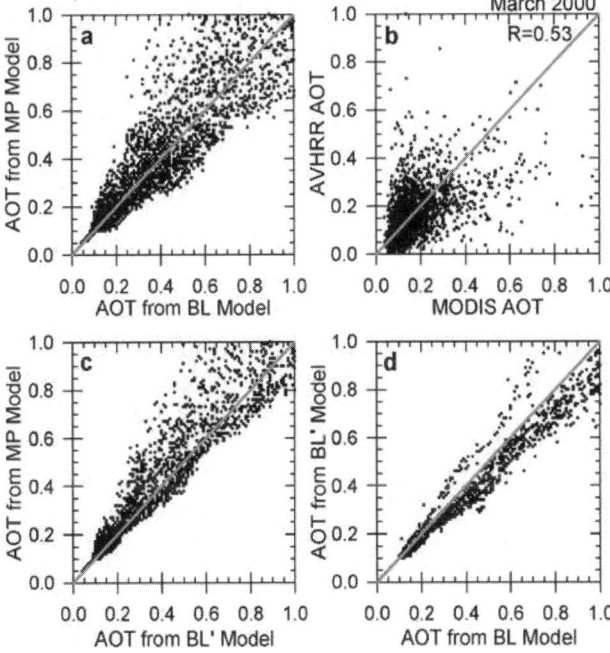

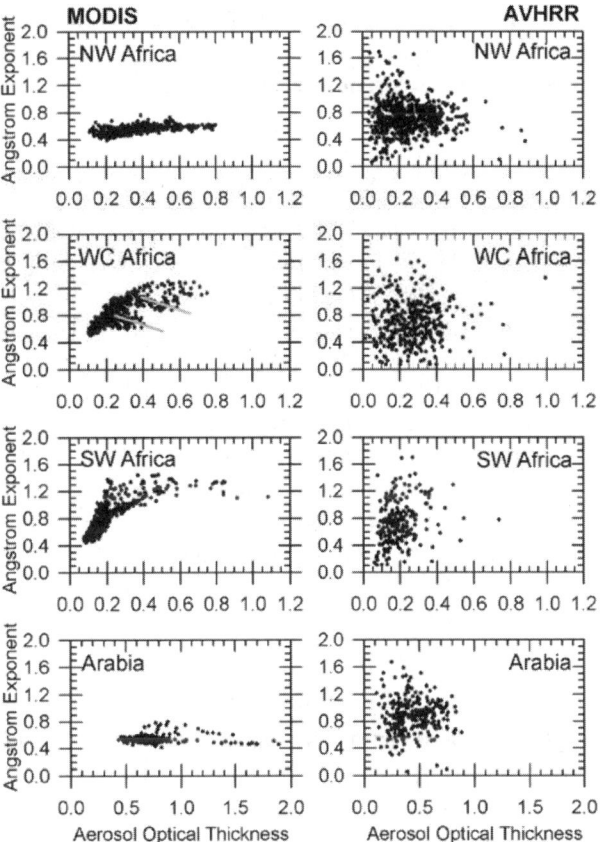

Fig. 3. (a) Scatter plot of AOT from MP models versus that from BL models. (b) Scatter plot of observed AOT from MODIS and AVHRR (global, March 2000). (c) The same as Fig. 3(a) but refractive index for BL models were replaced by a single fixed value (i.e., m=1.5-0.003i) as used in the MP models, which are referred to as BL' models. (d) Analogous to Fig. 3(a) and (c) except for BL' versus BL models. Gray solid line is the one-to-one line.

Larger discrepancies exist in the AE derived from the MODIS and the AVHRR. The AVHRR retrievals suffer

Fig. 4. Scatter plots of Angstrom exponent versus AOT. Left panels are based on MODIS data while the right panels are from AVHRR data for the same period (July, 2000). Gray lines provided in the WC Africa region for MODIS indicate possible signals from dusts co-existing with biomass burning aerosols in this region.

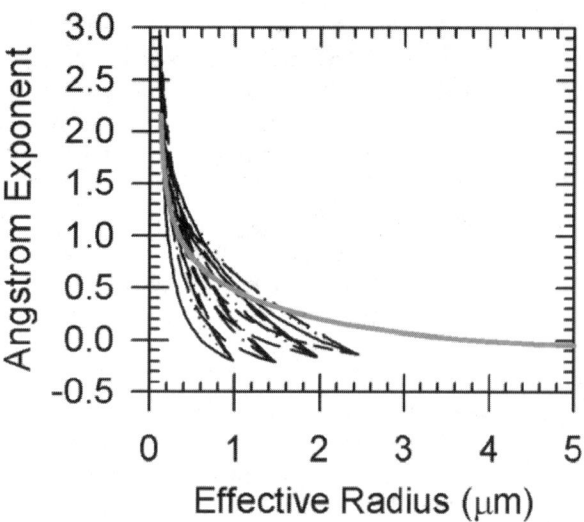

Fig. 5. Angstrom exponent versus effective radius for modified power size distributions (thick gray line) and for various combinations of bi-modal size distributions (thin lines with various types). Each line stands for different combinations of small and large modes that compose bi-modal log-normal size distributions.

4. ACKNOWLEDGEMENTS

The datasets under study include Aerosol Optical Thickness (AOT) and Angstrom Exponent (AE) derived from AVHRR under the Global Aerosol Climatology Project (GACP; http://gacp.giss.nasa.gov/; Mishchenko et al., 1999; Geogdzhayev et al., 2002), AOT (Torres et al., 1998, 2002), and Aerosol Index (AI; Herman et al., 1997) from TOMS (http://toms.gsfc.nasa.gov/). AERONET (http://aeronet.gsfc.nasa.gov/; Holben et al., 1998, 2001) Level 2.0 data and monthly MODIS aerosol products (Tanré et al., 1997) were also used for the comparison of AVHRR aerosol product. We thank them for providing AVHRR, TOMS, AERONET and MODIS data and helps in using the datasets.

5. REFERENCES

Geogdzhayev, I. V., M. I. Mishchenko, W. B. Rossow, B. Cairns, and A. A. Lacis, 2002: Global two-channel AVHRR retrievals of aerosol properties over the ocean for the period of NOAA-9 observations and preliminary retrievals using NOAA-7 and NOAA-11 data, *J. Atmos. Sci.*, 59(3), 262-278.

Herman, J. R., P. K. Bhartia, O. Torres, C. Hsu, C. Seftor, and E. Celarier, 1997: Global distribution of UV-absorbing aerosols from Nimbus 7/TOMS data, *J. Geophys. Res.*, 102(D14), 16911-16922.

Holben, B. N., T. F. Eck, I. Slutsker, D. Tanré, J. B. Buis, A. Setzer, E. Vermote, J. A. Reagan, Y. J. Kaufman, T. Nakajima, F. Lavenu, I. Jankowiak, and A. Smirnov, 1998: AERONET – a federated instrument network and data archive for aerosol characterization, *Remote Sens. Environ.*, 66, 1-16.

Holben, B. N., D. Tanré, A. Smirnov, T. F. Eck, I. Slutsker, N. Abuhassan, W. W. Newcomb, J. S. Schafer, B. Chatenet, F. Lavenu, Y. J. Kaufman, J. Vande Castle, A. Setzer, B. Markham, D. Clark, R. Frouin, R. Halthore, A. Kareli, N. T. O'Neill, C. Pietras, R. T. Pinker, K. Voss, and G. Zibordi, 2001: An emerging ground-based aerosol climatology: Aerosol optical depth from AERONET, *J. Geophys. Res.*, 106(D11), 12067-12097.

Jeong, M.-J., and Z. Li, 2005: Quality, compatibility and synergy analyses of global aerosol products derived from the advanced very high resolution radiometer and total ozone mapping spectrometer, *J. Geophys. Res.*, 110, D10S08, doi:10.1029/2004JD004647.

Jeong, M.-J., Z. Li, D. A. Chu, and S.-C. Tsay, 2005: Quality and compatibility analyses of global aerosol products derived from the advanced very high resolution radiometer and moderate resolution imaging Spectroradiometer, *J. Geophys. Res.*, 110, D10S09, doi:10.1029/2004JD004648.

Kaufman, Y. J., D. Tanré, and O. Boucher, 2002: A satellite view of aerosols in the climate system, *Nature*, 419, 215-223.

Mishchenko, M. I., I. V. Geogdzhayev, B. Cairns, W. B. Rossow, and A. A. Lacis, 1999: Aerosol retrievals over the ocean by use of channels 1 and 2 AVHRR data: sensitivity analysis and preliminary results. *Appl. Optics*, 38, 7325-7341.

Myhre, G. F. Stordal, M. Johnsrud, A. Ignatov, M. I. Mishchenko, I. V. Geogdzhayev, D. Tanré, J-L. Deuzé, P. Goloub, T. Nakajima, A. Higurashi, O. Torres, B. N. Holben, 2004: Intercomparison of satellite retrieved aerosol optical depth over ocean, *J. Atmos. Sci.*, 61, 499-513.

Tanré, D., Y. J. Kaufman, M. Herman, and S. Mattoo, 1997: Remote sensing of aerosol properties over oceans using the MODIS/EOS spectral radiances, *J. Geophys. Res.*, 102(D14), 16971-16988.

Torres, O., P. K. Bhartia, J. R. Herman, Z. Ahmad, and J. Gleason, 1998: Derivation of aerosol properties from satellite measurements of backscattered ultraviolet radiation: Theoretical basis, *J. Geophys. Res.*, 103(D14), 17099-17110.

Torres, O., P. K. Bhartia, J. R. Herman, A. Sinyuk, Paul Ginoux, and B. Holben, 2002: A long-term record of aerosol optical depth from TOMS observations and comparison to AERONET measurements, *J. Atmos. Sci.*, 59, 398-413.

INTERCOMPARISON OF ATMOSPHERIC PARAMETERS IN THE MLT REGION DERIVED FROM THE CRISTA-1 DATA BY INDEPENDENT METHODS

V. S. Kostsov, Yu.M. Timofeyev, and A. V. Rakitin
Research Institute of Physics, St. Petersburg State University, St. Petersburg-Petrodvorets, 198504, Russia

M. Kaufmann[*], O. Gusev, and K. Grossmann
Bergische Universitaet, Gesamthochschule Wuppertal, Wuppertal, 42097, Germany
[*]Now at Instituto de Astrofisica de Andalucia (CSIC), Granada, 18080, Spain

ABSTRACT

Limb infrared spectra recorded during the first mission of the Cryogenic Infrared Spectrometers and Telescopes for the Atmosphere (CRISTA-1) are used for the derivation of temperature, CO_2 and O_3 vertical profiles in the mesosphere and lower thermosphere (MLT) by two independent methods, developed at St. Petersburg State University (Russia) and at the University of Wuppertal (Germany). The methods are based on principally different approaches to account for the non-local thermodynamic equilibrium conditions relevant to the vibrational states of the CO_2 and O_3 molecules. Besides, the methods used measurements in different spectral regions (4.3 μm and 15 μm) for the derivation of CO_2 profiles. The agreement between CO_2 volume mixing ratio (vmr) and O_3 vmr mean profiles is very good. Both methods clearly demonstrate that CO_2 abundance begins to deviate from the lower atmosphere mixing ratio at significantly lower altitudes than predicted by models: at 70–75 km (St. Petersburg University results) and at 75–80 km (Wuppertal University results). The correlation coefficient for the two sets of independent CO_2 number density profiles is 0.7–0.87 in the whole altitude range considered. There are systematic differences between mean CO_2 vmr profiles and mean O_3 vmr profiles obtained by the two methods, but these differences are within the error limits. In all, the performed intercomparison is the successful validation of the atmospheric state parameters in the MLT region derived from CRISTA-1 spectra.

1. INTRODUCTION

The information on the CO_2 and O_3 distribution in the mesosphere and lower thermosphere is very important due to several reasons. Carbon dioxide plays a dominant role in the energy balance of the atmosphere in this altitude region. The altitude, up to which the CO_2 volume mixing ratio is nearly constant, is an indicator for turbulent mixing (*Chabrillat et al.*, 2002). The knowledge about the CO_2 vertical distribution is necessary for the traditional methods of temperature sounding of the atmosphere based on 15 μm and 4.3 μm band measurements. Absorption of ultraviolet radiation by ozone and molecular oxygen provides the energy which maintains the temperature structure and wind field of the middle atmosphere, thus the distribution of ozone plays a major role in determining the mean circulation of the upper atmosphere (*Prather*, 1981).

In 1994 and 1997 two experiments were carried out with the Cryogenic Infrared Spectrometers and Telescopes for the Atmosphere (CRISTA) instrument. The instrument measured limb infrared atmospheric radiance in a wide range of tangent altitudes. As a result of the interpretation of these measurements, a large amount of data on the thermal regime and composition of the atmosphere has been obtained (*Offermann et al.*, 1999; *Riese et al.*, 1999; *Grossmann et al.*, 2002; *Kaufmann et al.*, 2002, 2003; *Kostsov and Timofeyev*, 2001, 2003).

In the present paper we compare the results of the derivation of the CO_2 and O_3 profiles in the mesosphere from the CRISTA spectra by two independent methods (*Kaufmann et al.*, 2002, 2003; *Kostsov and Timofeyev*, 2003). The validation of the results of satellite remote sensing of the upper stratosphere, mesosphere and lower thermosphere is rather problematic, if compared to the lower atmosphere, first of all due to the relatively small number of independent direct measurements of atmospheric parameters, which are suitable for validation purposes. Therefore, the intercomparison of the results derived from the same experimental data but by different methods is very important. It should be stressed that the CO_2 vertical profiles were obtained not only by different methods but also from different data sets of the same experiment (4.3 μm and 15 μm measurements). For the derivation of O_3 profiles, both methods used the same data set (9.6 μm measurements).

It should be stressed that most of model CO_2 profiles in the mesosphere-lower thermosphere (MLT) region, which are used at present, are characterized by deviation from uniformly-mixed values starting at 90-100 km.

The CO_2 vertical profiles derived from the CRISTA-1 and CRISTA-2 experimental data (*Kaufmann et al.*, 2002; *Kostsov and Timofeyev*, 2003) are characterized by considerably lower altitudes, where the volume mixing ratio starts to decrease. This fact and the lack of data for traditional validation stimulate the study for indirect validation of the CO_2 retrieval results based on the intercomparison of CO_2 profiles derived by different methods from different radiance data of the same experiment.

2. DESCRIPTION OF METHODS

Hereafter we denote the method of the retrieval of the CO_2 and O_3 profiles developed at St. Petersburg State University as "the SPB method" and the one developed at the University of Wuppertal as "the WUP method".

The retrievals by the WUP method are performed iteratively in two steps. At the first step the vibrational excitation of CO_2 (O_3 as well) is calculated using a NLTE models. At the second step this excitation is frozen and the 4.3 μm measured radiances (9.6 μm radiances for O_3) are simulated with a forward radiance model. The CO_2 and O_3 number densities are derived using the onion peeling technique.

The SPB method considers the measured spectral limb radiances I at wavenumber ν and tangent altitude z_t as a functional of several parameters:

$$I(\nu, z_t) = A\left[T_k(z), p(z), n(z), T_\nu^s(z)\right]$$

where A denotes a nonlinear forward operator, z is the vertical coordinate, p is pressure, T_k is kinetic temperature, n is the CO_2 (O_3) number density, T_v is the vibrational temperature and "s" is the index identifying the vibrational states of the CO_2 (O_3) molecules which give origin to the transitions forming the absorption bands. The expression is linearized and solved by the optimal estimation method with respect to kinetic and vibrational temperatures, pressure and CO_2 (O_3) number density. The a priori information on the retrieved profiles and several additional constraints are applied according to (*Kostsov and Timofeyev,* 2001). This information is of general character and does not require modeling of the mechanisms driving the non-equilibrium populations. The algorithm of the derivation of O_3 profiles from 9.6 μm measurements uses the temperature and pressure data, which were obtained from 15 μm measurements in the same experiment.

3. RESULTS AND DISCUSSION

The major attention in our intercomparison of CO_2 is paid to the number density due to the following reasons: (1) both methods solved the inverse problem which was formulated with respect to this quantity and not volume mixing ratio; (2) the CO_2 volume mixing ratio was calculated using pressure and temperature profiles obtained by the different methods and the differences between these profiles (especially pressure) make additional contribution to error budget.

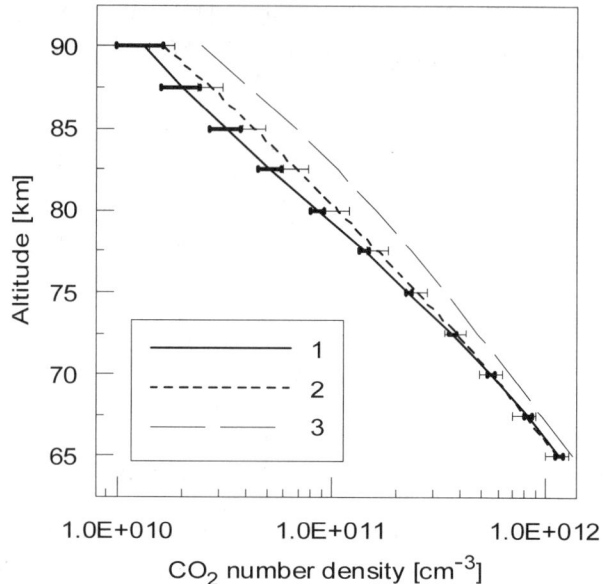

Fig. 1. Mean CO_2 number density profiles obtained by the different methods (1 – SPB, 2 – WUP, 3 – a priori profile).

Fig. 1 presents the mean CO_2 number density profiles obtained by the different methods and the a priori profile used in the SPB method. The mean profiles are obtained by averaging of all 341 individual profiles retrieved by each method. The figure clearly demonstrates a very good agreement of the results up to 75 km altitude and a systematic difference starting at 75 km and increasing with altitude. It should be stressed that the two mean profiles are significantly lower than the a priori profile used in the SBP retrieval algorithm. Fig. 2 shows the difference in percent (relative to the WUP results) and the combined error corridor. Since the random error component is strongly suppressed for the mean profiles, the combined error includes only the systematic error components of two methods. One can see that below 75 km altitude the difference is less than 10% and it is considerably lower than the combined error. In the vicinity of 80 km the difference is practically equal to the combined error value and constitutes 20%. The difference is growing up to 87.5 km reaching 27%. At 90 km a small decrease of the difference is observed (to 20%). So, in the entire considered altitude range the difference does not exceed the limits of combined errors.

A clear indication of the agreement or disagreement of the retrieval results delivered by the independent methods can be obtained by the calculation of the correlation coefficient. The value of the correlation coefficient is the estimate of the coincidence disregard of any systematic shift of the retrieved profiles. The obtained values are 0.7-0.87 at all altitudes, which is the demonstration of a good agreement of the results.

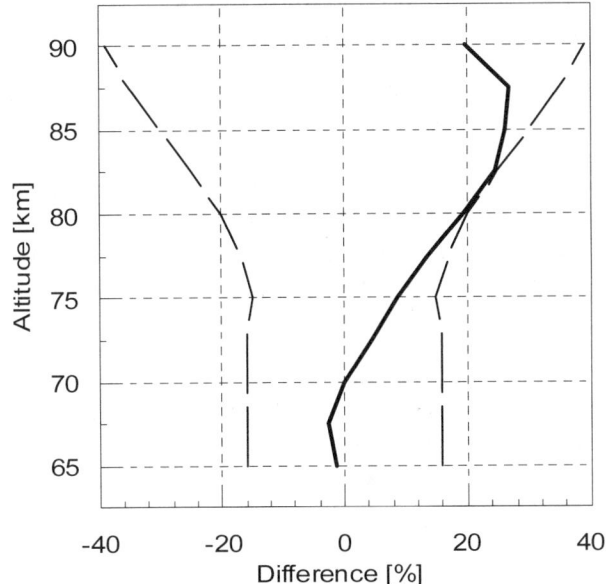

Fig. 2. The difference between CO_2 mean profiles (solid line) and the combined error corridor (dashed lines).

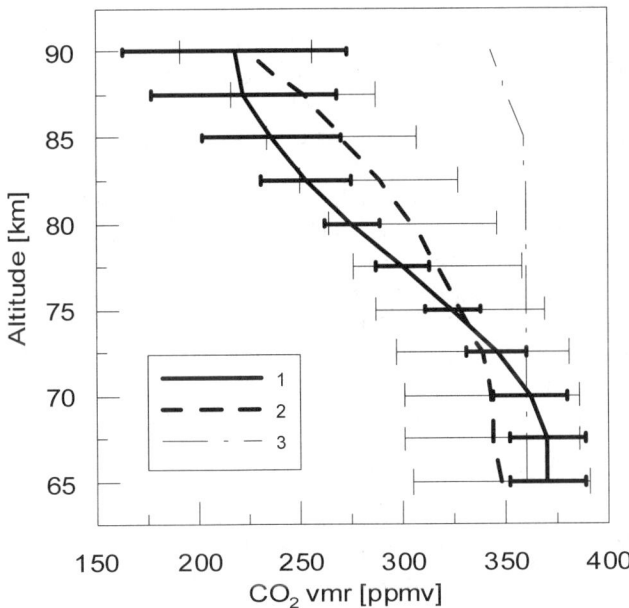

Fig. 3. The mean CO_2 volume mixing ratio profiles obtained by different methods (1 – SPB, 2 – WUP). 3 – profile by Fomichev et al. (1998) (used as a priori profile by the SPB method).

One of the steps of the intercomparison is the analysis of the latitudinal behavior of the CO_2 number density. We calculated the mean number density for every altitude over 5-degree latitudinal zones. Both methods show a nearly identical latitudinal behavior of the number density values (at 85 km and 90 km altitude—neglecting the systematic shift).

Since the CO_2 content expressed in terms of volume mixing ratio (vmr) is an important parameter and this very parameter is mainly used in different atmospheric models, now we compare the CO_2 vmr profiles obtained by the two methods. Fig. 3 presents the CO_2 vmr mean profiles obtained by both methods and one model profile. First, we stress that the two profiles obtained by the considered methods agree very well within error bars corresponding to systematic errors. The difference between the obtained profiles has a kind of systematic character: the SPB data are higher in the region 65-75 km, and lower in the upper layers. Excellent agreement is observed at 90 km altitude where both methods produce the CO_2 vmr value of about 220 ppmv. It is clearly visible that the two profiles obtained from CRISTA data start to deviate from the well-mixed value considerably lower than model profile. The model profile compiled by *Fomichev et al.* (1998) on the basis of different observations starts to deviate from well-mixed value above 85 km.

The intercomparison of ozone profiles was made separately for day- and nighttime (solar zenith angle values less than 80^0 and greater than 110^0 correspondingly). The altitude range 60–90 km was considered. The mean O_3 vmr profiles obtained by the two methods are shown in Fig. 4. Daytime mean profiles demonstrate very good agreement up to 70 km altitude. In the upper layers the O_3 values obtained by SPB method are systematically higher than obtained by WUP method. Nighttime mean profiles are in very good agreement higher than 82 km. Below, the systematic difference is observed. Despite systematic difference, the mean profiles (daytime and nighttime as well) agree within the combined error limits. The qualitative character of the difference of mean profiles obtained by the two methods (related to SPB results) and of the combined error corridor is similar to one presented in Fig. 2 for CO_2 intercomparison. However, the combined error in case of ozone retrieval is larger than in the CO_2 case and constitute: about 20% in the altitude range 60–70 km, increasing from 20% to 40% in the altitude range 70–80 km, and 40% above 80 km.

The values of correlation coefficient are not so high as in the case of CO_2 intercomparison, if daytime and nighttime retrievals are considered separately. The highest values are 0.7–0.8, which are observed at 60–65 km and 82–90 km (daytime) and 80–90 km (nighttime).

Zonal mean values (averaging over 5-degree latitude zones) of O_3 vmr obtained by the two methods are in excellent agreement at all altitudes for nighttime conditions. The daytime zonal mean values are in good agreement at all altitudes except the vicinity of 80 km, where the results of SPB method show local maximum of ozone in the

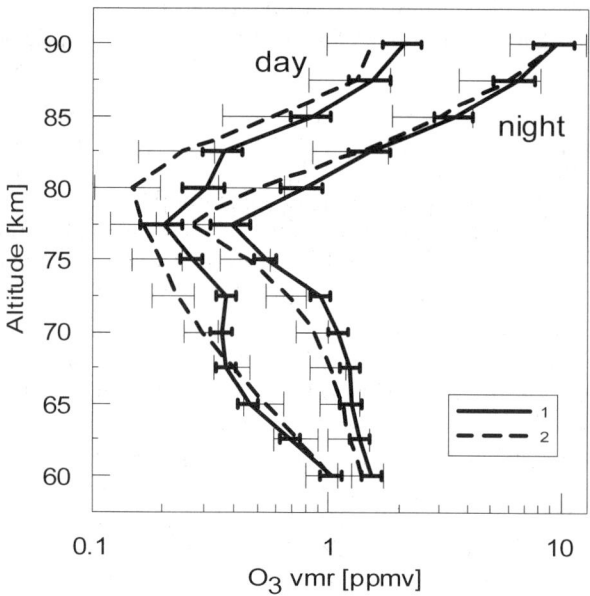

Fig. 4. The mean O_3 volume mixing ratio profiles obtained by different methods (1 – SPB, 2 – WUP).

equatorial region while WUP method gives practically constant ozone for all considered latitudes. However for nighttime, both methods show local maximum of ozone in the equatorial region at 80 km altitude.

4. CONCLUSION

The high altitude limb IR spectra registered by the CRISTA instrument during the experiment in 1994 have been used for the derivation of CO_2 and O_3 profiles in the mesosphere by two independent methods developed at the Universities of St. Petersburg (Russia) and Wuppertal (Germany).

The results of the intercomparison of CO_2 and O_3 number density and vmr demonstrated good agreement within the combined error limits.

The systematic differences between CO_2 number density and vmr profiles do not disprove the important conclusion made in papers (*Kaufmann et al.*, 2002; *Kostsov and Timofeyev*, 2003) about the deviation of the CO_2 abundance from the well mixed value significantly lower than predicted by atmospheric models and detected by several previous measurements.

5. ACKNOWLEDGMENTS

The work of the St. Petersburg team was supported by Russian Foundation for Basic Research (grant 03-05-64830) and by Russian Ministry of Education in the frame of the program "The Universities of Russia" (grant yp.01.01.044). The work at Wuppertal University was supported by BMBF in the frame of the AFO2000 project through grant 07ATF10.

6. REFERENCES

Chabrillat, S., G.Kockarts, and D.Fonteyn, 2002: Impact of molecular diffusion on the CO_2 distribution and the temperature in the mesosphere. *Geophys. Res. Lett.*, 29, 19-1 – 19-4.

Fomichev, V. I., J. P.Blanchet, and D. S.Turner, 1998: Matrix parameterization of the 15 μm CO_2 band cooling in the middle and upper atmosphere for variable CO_2 concentration. *J. Geophys. Res.*, 103, 11,505-11,528.

Grossmann, K., D. Offerman, O. Gusev, J. Oberheide, M. Riese, and R. Spang, 2002: The CRISTA 2 mission. *J. Geophys. Res.*, 107, doi:10.1029/2001JD000667.

Kaufmann, M., O. A. Gusev, K. U. Grossmann, R. G. Roble, M. Hagan, C. Hartsough, and A. A. Kutepov, 2002: The vertical and horizontal distribution of CO_2 densities in the upper mesosphere and lower thermosphere as measured by CRISTA. *J. Geophys. Res.*, 107, doi: 10.1029/2001JD000704.

Kaufmann, M., O. A. Gusev, K. U. Grossmann, F. J. Martín-Torres, D. R. Marsh, and A. A. Kutepov, 2003: Satellite observations of daytime and nighttime ozone in the mesosphere and lower thermosphere. *J. Geophys. Res.*, 108, doi: 0.1029/2002JD002800.

Kostsov, V. S., and Yu. M. Timofeyev, 2001: Interpretation of Satellite Measurements of the Outgoing Nonequilibrium IR Radiation in the CO_2 15-μm Band: 1. Description of the Method and Analysis of Its Accuracy. *Atmospheric and Oceanic Physics*, 37, 728-738, (Engl. transl.).

Kostsov, V. S., and Yu. M.Timofeyev, 2003: Mesospheric carbon dioxide content as determined from the CRISTA-1 experimental data. *Atmospheric and Oceanic Physics*, 39, 359-370, (Engl. transl.).

Offermann, D., K. U. Grossmann, P. Barthol, P. Knieling, M. Riese, and R. Trant, 1999: The Cryogenic Infrared Spectrometers and Telescopes for the Atmosphere (CRISTA) experiment and middle atmosphere variability. *J. Geophys. Res.*, 104, 16311-16325.

Prather, M., 1981: Ozone in the Upper Stratosphere and Mesosphere. *J. Geophys. Res.*, 86, 5325-5338.

Riese, M., R. Spang, P. Preusse, M. Ern, M. Jarisch, D. Offermann, and K. Grossmann, 1999: Cryogenic Infrared Spectrometers and Telescopes for the Atmosphere (CRISTA) data processing and atmospheric temperature and trace gas retrieval. *J. Geophys. Res.*, 104, 16349-16367.

SIMULTANEOUS RETRIEVAL OF AEROSOL AND SURFACE PROPERTIES OVER BRIGHT TARGETS INCLUDING SNOW AND ICE USING MULTI- AND HYPERSPECTRAL DATA

Wei Li, Hans Eide, and Knut Stamnes
Light and Life Laboratory
Department of Physics and Engineering Physics
Stevens Institute of Technology
Hoboken, NJ 07030, USA

Robert Spurr
Harvard-Smithsonian Center for Astrophysics
Cambridge, MA 02138, USA

Teruo Aoki
Meterological Research Institute
1-1, Nagamine, Tsukuba, Ibaraki 305-0052, Japan

Masahiro Hori
Japan Aerospace Exploration Agency
104-6023 Harumi 1-8-10,Chuo-Ku,Tokyo, Japan

ABSTRACT

Retrieval of surface properties of highly reflecting targets such as snow and ice is a challenging problem due to the influence of aerosols, which vary considerably in space and time. Also, accounting for the bidirectional properties of a bright surface such as snow is very important for reliable retrievals. The main purpose of the work described in this paper is to explore the opportunities and possibilities offered by multi- and hyperspectral data such as those provided by the MODIS, GLI, the Advanced Land Imager ALI), and Hyperion sensors to retrieve reliable aerosol and surface properties. Over snow and ice surfaces these properties include aerosol optical depth, single scattering albedo, the mean size of snow grains and ice "particles" (inclusions), and the spectral and broadband snow/ice albedo. In particular the following question will be addressed: To what extent can multi- and hyperspectral data help improve our knowledge of snow and ice parameters that are important for understanding global climate change?

concentration is possible because snow reflectance depends primarily on impurity concentration in the visible, but on snow grain size in the NIR. Snow grain size and impurity concentration are important parameters because they allow us to determine the *spectral albedo* of snow, which is a crucial climate parameter, required to determine climate system response to enhanced greenhouse gas emissions.

Aerosol products over bright surfaces such as snow and ice are currently not being retrieved by most satellite sensors. Thus, the quality of snow/ice products will be compromised by this lack of knowledge about the atmospheric contribution to the measured TOA radiances.

This implies that it is very difficult if not impossible to make reliable assessments of snow/ice parameters retrieved from the measured TOA radiances. Atmospheric correction relies on the presence of nearby dark pixels, and if there are no nearby dark pixels, climatology is used. Also, the snow surface is frequently assumed to be a Lambertian reflector, which is not true.

1. INTRODUCTION

It is well recognized that snow and ice has a strong impact on the surface energy balance in any part of the world. Fortunately, satellite remote sensing provides a very useful tool for estimating spatial and temporal changes in snow and ice cover, and for retrieving snow/iceproperties. In particular, retrieval of snow grain size and impurity

2. THEORETICAL ASPECTS: FORWARD MODEL

Our algorithm development is based on the discrete-ordinate-method (DISORT, Stamnes et al., 1988) to compute the top-of-the-atmosphere (TOA) radiances, because this method allows radiances to be computed at arbitrary user-specified polar and azimuthal angles. We treat the atmosphere and the snow as two adjacent slabs.

The optical properties of each atmospheric layer is specified in terms of scattering and absorption by molecules and aerosols, and the snow pack is included as an additional slab at the bottom of the atmosphere. Thus, the bidirectional reflectance of the snow is automatically accounted for—no Lambertian assumption needed, because we use a RT model for the Coupled Atmosphere-Snow (CAS) system based on DISORT (CAS-DISORT).

2.1 Retrieval Principles: Aerosols

"Atmospheric Correction" means *quantifying the aerosol contribution to TOA radiance*, because Rayleigh scattering can be computed accurately, and absorption by trace gases is unimportant in the GLI channels at 380 and 460 nm, that we use for aerosol retrieval. But, quantifying the aerosol contribution to the TOA radiance is difficult due to large spatial and temporal variation in aerosol properties, and the high albedo of the snow surface. Retrieving incorrect aerosol model implies TOA reflectance errors as large as 20% for weakly absorbing aerosols, and 50% or more for strongly absorbing aerosols. Such large errors will cause retrieval failure. Thus, proper retrieval of aerosol properties is critically important for accurate retrieval of snow grain size and impurity concentration.

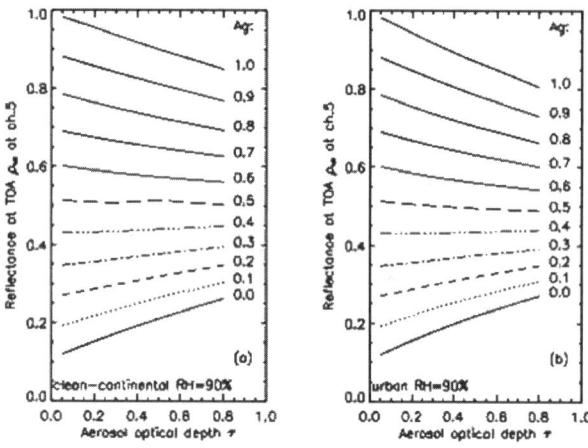

Fig. 1. TOA reflectance at 460 nm (GLI channel 5) versus aerosol optical depth and surface reflectance. Left: non-absorbing aerosols. Right: absorbing aerosols.

How can we do aerosol retrieval over bright surfaces like snow and ice? As illustrated in Fig. 1, retrieval is possible because the reflectance in GLI channel 5 *decreases* almost linearly with optical depth when the albedo is *higher* than about 0.5. But it *increases* almost linearly with optical depth when the albedo is *lower* than about 0.5. Thus, aerosol retrieval is feasible over bright as well as dark targets.

3. SIMULTANEOUS RETRIEVAL OF AEROSOL AND SNOW/ICE PROPERTIES

3.1 Vertically Homogeneous Snow Pack

An accurate cloud mask and surface classification scheme are important, but difficult to obtain over bright surfaces such as snow and ice. Below we show examples of retrievals based on a suite of algorithms developed for the GLI sensor, but modified for application also to MODIS (Hori et al., 2001; Stamnes, 2003). This suite of algorithms provides, (i) cloud mask, surface classification and temperature, (ii) aerosol model and optical depth, (iii) snow grain size and impurity concentration, and (iv) the spectral snow albedo.

Fig. 2 shows the monthly snow/ice extent retrieved from the ADEOS-II/GLI sensor in 2003. Note that the algorithms are designed to provide a complete surface classification, and thus discriminate between snow/ice, open water, snow-covered land and so on. Fig. 3 shows the corresponding retrieved grain size for the snow-covered areas. These figures illustrate that it is feasible to obtain the information required to assess the spectral albedo of snow.

3.2 Vertically Inhomogeneous Snow Pack

So far we have assumed that the snow pack is homogeneous. Typically the grain size *increases with depth* because new snow (small grains) accumulates on the top of older snow (larger grains). Because the photon penetration depth depends strongly on wavelength, light at say 1640 nm will penetrate less deeply than light at 865 nm. Thus, photons at 1640 nm will *"see" only the small grains* at the top of the snow whereas photons at 865 nm will penetrate deeper and *"see" larger grains* on average. Assuming that the snow pack is vertically homogeneous (when in reality it is *not*), we may erroneously conclude that light at 865 nm yields a much larger grain size than light at say 1640 nm. Retrieval of vertical information about the grain size will be done by extending CAS-LIDORT—our existing *linearized* discrete ordinate RT model (LIDORT) for the coupled atmosphere snow (CAS) system applicable to a *homogeneous* snow pack—to an *inhomogeneous* (multi-layered) snow pack. The advantage of this approach is that we may use state-of-the-art iterative inversion schemes (e.g., optimal estimation theory based on Bayes' theorem) to explore systematically the information content of the data collected. Such methods are *well-suited for use with multi- and hyperspectral data*, and they allow for rigorous error analysis yielding an optimum number of retrieved quantities with known error budgets.

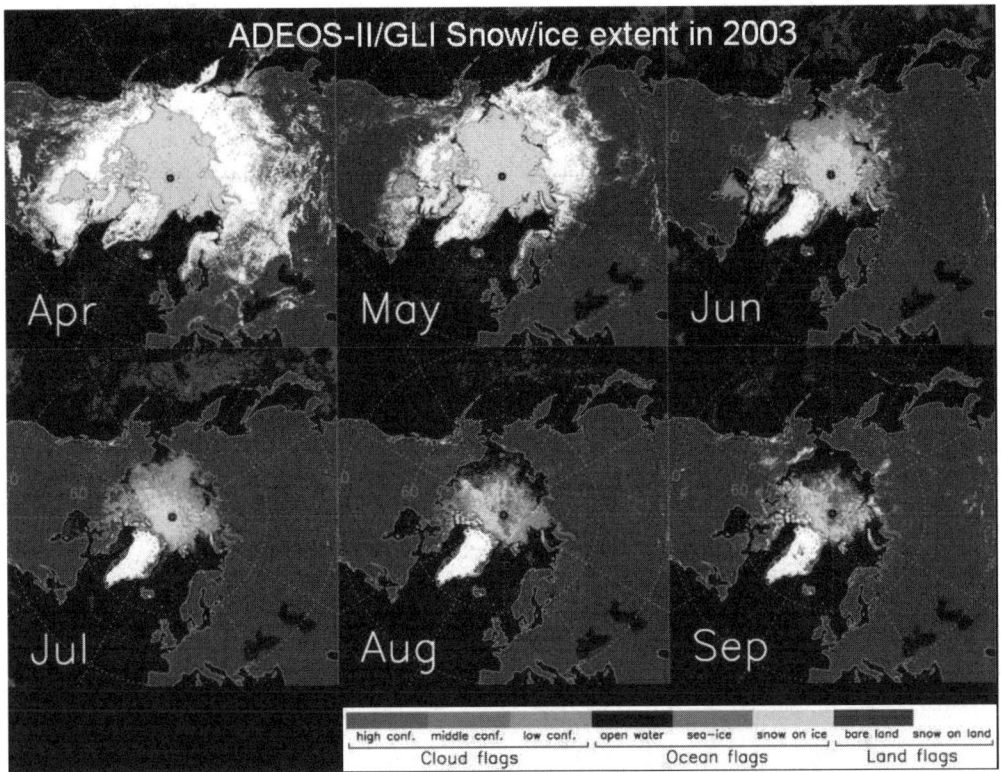

Fig. 2. Monthly snow/ice cover extent in northern hemisphere in 2003.

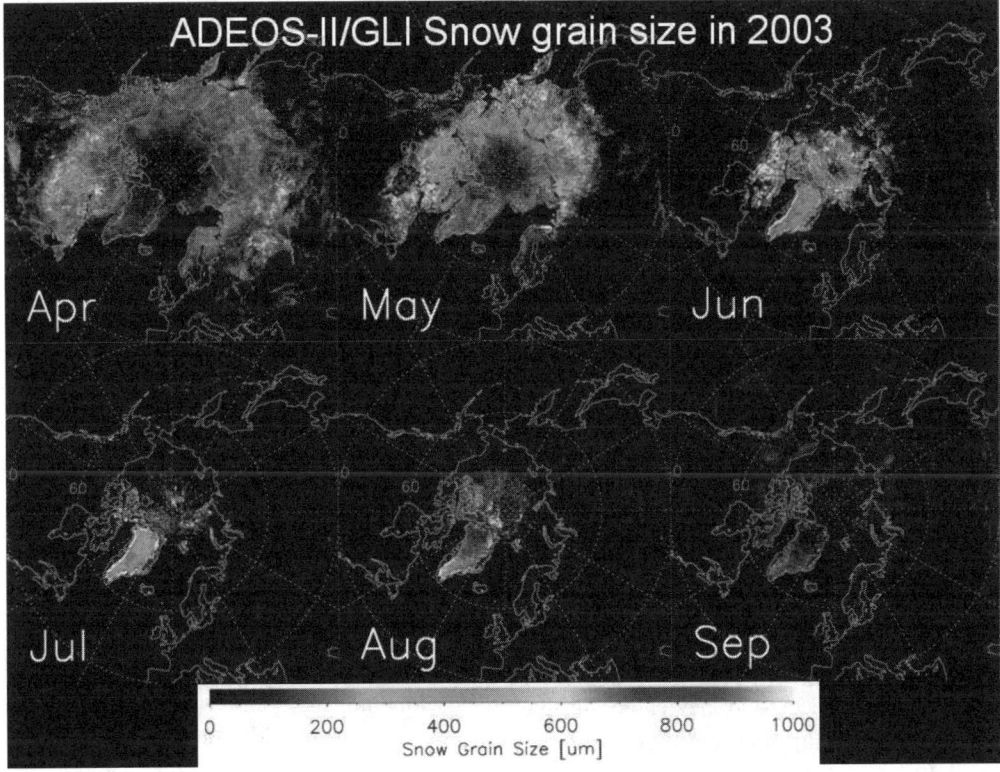

Fig. 3. Seasonal variation of snow grain size in northern hemisphere in 2003.

4. DISCUSSION

Field measurements and model simulations indicate that snow exhibits significant bi-directional reflectance, which is more pronounced at large viewing angles. To relate the reflectance of snow to physical properties such as grain size, we use a comprehensive radiative transfer model based on the DISORT multiple scattering algorithm (Stamnes et al., 1988; Stamnes et al., 2000) to compute the bi-directional reflectance of the snow surface instead of assuming a Lambertian reflector. We employ the optical properties of snow obtained by Wiscombe and Warren (1980) and Warren and Wiscombe (1980). Mie theory is used to get the extinction coefficient and the phase function, which means the snow grains are assumed to be spherical. The snow cover is considered as a homogeneous layer with a single grain size when we carry out the radiative transfer calculations. Of course, the real snow particles are not spherical. Thus, the particle radius used here is called the effective grain-size. It will generally be different from the measured grain size. Also these effective grain sizes represent average (depth-weighted) values. The depth of observation depends on the radiation penetration depth, which varies with wavelength.

5. SUMMARY

We have discussed retrieval of aerosol and snow/ice properties emphasizing the use of *linearized* forward radiative transfer models for the coupled atmosphere-snow/ice system, because this allows us to compute weighting functions (Jacobians) required in state-of-the-art inversion schemes such as optimal estimation theory based on Bayes' theorem. Such inversion methods are, (i) well-suited for use with multi- and hyperspectral data, and (ii) allow for rigorous error analysis yielding an optimum number of retrieved quantities with known error budgets.

We showed examples of retrievals with a suite of algorithms developed for the GLI sensor that can be used to provide (i) Cloud Mask and Surface Classification, (ii) Aerosol Optical Properties, (iii) Snow Grain Size and Impurity Concentrations. These algorithms have been modified for use with MODIS data, and they can easily be extended for use with other multi- and hyperspectral data. By using several NIR wavelengths with different photon penetration depths, we are extending this approach to provide depth information about snow properties, and thereby accurate spectral albedo of snow. In fact, preliminary results show that it is possible to obtain information about the vertical structure of the snow grain size (Li et al., 2001; Zhou et al., 2003) by using a combination of several wavelengths available on GLI and MODIS. Since the methodology described above is generic in nature it can also be applied to retrieve parameters from vegetated land surfaces, although such applications have not been discussed here.

6. REFERENCES

Hori, M., T. Aoki, K. Stamnes, B. Chen, and W. Li, Preliminary validation of the GLI Cryosphere Algorithms with MODIS daytime data, *Polar Meteorol. Glaciol.*, 15, 1-20, 2001.

Li, W., K. Stamnes, B. Chen, and X. Xiong, Retrieval of the depth dependence of snow grain size from near-infrared radiances at multiple wavelengths, *Geophys. Res. Lett.*, 28, 1699-1702, 2001.

Stamnes, K., Snow grain size retrieval in polar and mid-latitude regions, ATBD (CTSK2b1). *NASDA internal document,* 3.4.2-1-3.4.1-27, March, 2003.

Stamnes, K., S.-C. Tsay, W. J. Wiscombe, and K. Jayaweera, Numerically stable algorithm for discrete-ordinate-method radiative transfer in multiple scattering and emitting layered media, *Appl. Opt.*, 27, 2502-2509, 1988.

Stamnes, K., S.-C. Tsay, W. J. Wiscombe, and I. Laszlo, DISORT, A General-Purpose Fortran Program for Discrete-Ordinate-Method Radiative Transfer in Scattering and Emitting Layered Media: Documentation of Methodology, *Report,* available from ftp://climate1.gsfc.nasa.gov/wiscombe/, 2000.

Wiscombe, W. J., and S. G. Warren, A model for the spectral albedo of snow. I. Pure snow, *J. Atmos. Sci.,* 37, 2712-2733, 1980.

Warren, S. G., and W. J. Wiscombe, A model for the spectral albedo of snow. II. Snow containing atmospheric aerosols, *J. Atmos. Sci.,* {\bf 37}, 2734-2745, 1980.

Zhou, X., S. Li, and K. Stamnes, Effects of vertical inhomogeneity on snow spectral albedo and its implication for optical remote sensing of snow, *J. Geophys. Res.,* 108, D23, 4738, doi:10.1029/2003JD003859, 2003.

CHARACTERIZING THE ATMOSPHERE ABOVE TURBID COASTAL WATERS USING MULTI- AND HYPERSPECTRAL DATA

Knut Stamnes, Hans Eide, and Wei Li,
Light and Life Laboratory
Department of Physics and Engineering Physics
Stevens Institute of Technology
Hoboken, NJ 07030, USA

Robert Spurr
Harvard-Smithsonian Center for Astrophysics
Cambridge, MA 02138, USA.

Jakob Stamnes
Department of Physics
University of Bergen, Norway

ABSTRACT

Retrieval of aerosol properties and marine constituents in turbid coastal waters from space is a challenging problem, mainly because it is difficult to discriminate the atmospheric contribution to the radiance measured by an instrument deployed on a satellite from the signal coming from the water. In clear open ocean water it is reasonable to assume that the water is black in the near-infrared (NIR). This NIR black-pixel assumption (BPA) can be used to isolate the atmosphere from the water, but the BPA is not valid in turbid coastal waters. We review a method for simultaneous retrieval of aerosol and chlorophyll concentrations in clear open ocean waters. As a first step in extending this method to turbid coastal waters, we discuss an improved way of selecting aerosol models. This consists of using a combination of large and small particle models in such a way that it is possible to quantify the sensitivity of the top-of-the-atmosphere (TOA) radiance to the fraction of large versus small particles. By computing the Jacobians it becomes possible to retrieve this fraction as well as the aerosol optical depth and marine parameters. This approach allows us to employ standard iterative inversion schemes that are well suited for multi- and hyperspectral measurements. Our results indicate that it is important to use forward models that accurately treat the radiative transfer in the coupled atmosphere-ocean system, and to select carefully a set of suitable aerosol models. There are several advantages of using a continuum set of aerosol model rather than a discrete one as will be discussed.

These data may be used to remotely evaluate: (1) water quality, (2) transport of sediments and adhered pollutants, (3) primary production, upon which commercial shellfish and finfish populations depend for food, and (4) harmful algal blooms that pose a threat to public health and economies of affected areas. Government agencies are interested in monitoring coastal waters using optical remote sensing. This interest is broadly driven by: (i) resource management concerns over the impact of coastward shifts in population and land use on the ecosystems of estuaries, wetlands, near shore benthic environments and fisheries, (ii) recognition of the need to understand short time scale global change due to urbanization of sensitive land-margin ecosystems, and (iii) national security issues.

In order to use satellite ocean color data, the influence of the atmosphere on the total top of the atmosphere (TOA) radiance measured by the sensor must be removed in a process referred to as atmospheric correction. Atmospheric correction is crucial to the eventual accuracy of the derived ocean color parameters because the radiances emanating from the water's surface and used in bio-optical algorithms to estimate geophysical products comprise at most 10\% of the total TOA radiance. Several atmospheric correction algorithms exist, yet most employ assumptions that are only valid for open ocean conditions, e.g., negligible water-leaving radiances in the near infrared. These atmospheric correction algorithms do not perform routinely well in nearshore coastal waters where environmental conditions violate one or more of these assumptions.

1. INTRODUCTION

Improvements in ocean color data quality and availability have enhanced their potential use in the operational monitoring of estuarine and coastal waters.

2. ATMOSPHERIC CORRECTION ISSUES

Most ocean color algorithms consist of two steps. First, "atmospheric correction" is performed to obtain the water-leaving radiance. Second, the oceanic chlorophyll concentration is retrieved from this water-leaving radiance. In the visible more than 90% of the radiance measured by the satellite sensor typically comes from the atmosphere. This makes atmospheric correction a very challenging task. It is important because a small uncertainty in the atmospheric correction may lead to a big error in the inferred chlorophyll concentration, and difficult because aerosol optical properties vary considerably in space and time. Clearly, the two-step retrieval approach has several limitations. Use of an ``atmospheric correction'' to decouple the atmosphere from the water surface, based on the commonly adopted near-infrared black-pixel approximation (BPA), leads to many retrieval uncertainties that affect the accuracy of both the retrieved marine and atmospheric properties. In reality, surface reflection/transmission and atmospheric scattering/absorption are coupled. Ignoring this coupling severely limits the two-step approach.

2.1 Simultaneous Retrieval of Aerosol Properties and Marine Parameters

Our goal is to remove unnecessary assumptions inherent in current algorithms leading to uncertainties that are difficult to quantify. This is accomplished by using CAO-DISORT: a rigorous discrete-ordinates solution (DISORT) of the radiative transfer equation pertinent for the *coupled* atmosphere-ocean (CAO) system (Jin and Stamnes, 1994; Stamnes et al., 2003). The algorithm employs two parameters to select an aerosol model from a suite of candidate models, and a third parameter to retrieve the corresponding aerosol optical depth. We express the satellite-measured reflectance as:

$$\rho_{tot}^{meas}(\lambda) \approx \rho_{atm}(\lambda) + t(\lambda)\rho_w(\lambda) \quad (1)$$

where $t(\lambda)$ is the diffuse atmospheric transmittance, $\rho_w(\lambda)$\$ is the water-leaving reflectance, and $\rho_{atm}(\lambda)$ represents the contribution from the atmosphere. We use the parameter $\varepsilon_{ms}(\lambda, 865)$ defined as:

$$\varepsilon_{ms}(\lambda, 865) \equiv \frac{[\rho_{atm}(\lambda) - \rho_{ray}(\lambda)]}{[\rho_{atm}(865) - \rho_{ray}(865)]} \quad (2)$$

to obtain initial estimates of the aerosol model. Then we employ a second parameter defined as:

$$\rho_{diff}(\lambda_j) \equiv \frac{[\rho_{tot}^{meas}(\lambda_j) - \rho_{tot}^{LUT}(\lambda_j)]}{\rho_{tot}^{meas}(\lambda_j)} \times 100 \quad (3)$$

to finalize the aerosol model selection. Here $\rho_{tot}^{LUT}(\lambda_j)$ is the TOA reflectance for wavelength λ_j inferred from lookup tables (LUTs) based on the retrieved aerosol optical properties and chlorophyll concentration. The third parameter is defined as:

$$\gamma_{diff}(\lambda, 865) \equiv \gamma(\lambda) - \gamma(865) \quad (4)$$

where $\gamma(\lambda) = \rho_{atm}(\lambda) / \rho_{ray}(\lambda)$. Fig. 1 illustrates that $\varepsilon_{ms}(\lambda, 865)$ is a good parameter for selecting aerosol model, while $\gamma_{diff}(\lambda, 865)$ is a good one for retrieving aerosol optical depth. Fig. 2 shows an example of retrieval and Fig. 3 illustrates that we can quantify the error in all channels.

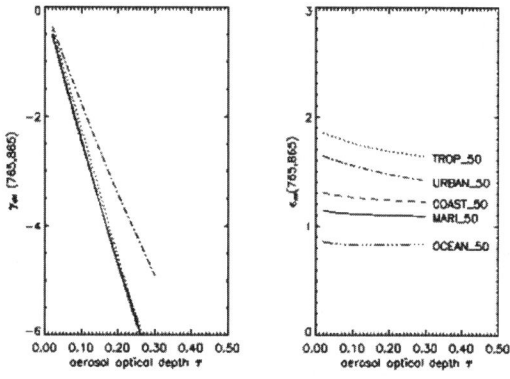

Fig. 1. Simulated values of $\gamma_{diff}(765, 865)$ and $\varepsilon_{ms}(765, 865)$ as a function of aerosol optical depth τ_{865} for several aerosol models including maritime, tropospheric, coastal, urban, and oceanic at RH = 50\%. The various aerosol models in the left panel are labeled in the same way as indicated in the right panel.

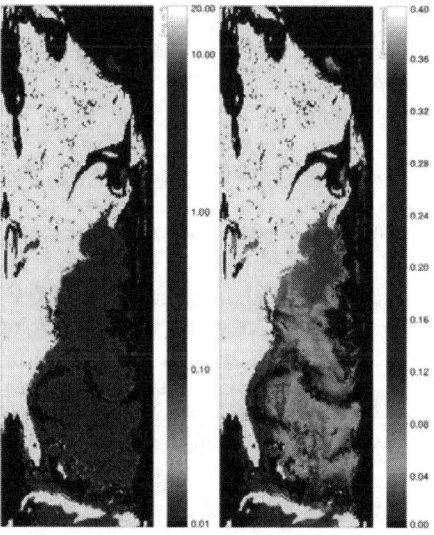

Fig. 2. Retrieved chlorophyll concentration and aerosol optical depth τ_{865} for a SeaWiFS image obtained off the East Coast of the United States. The yellow color indicates land areas. Left panel: chlorophyll concentration. Right panel: τ_{865}.

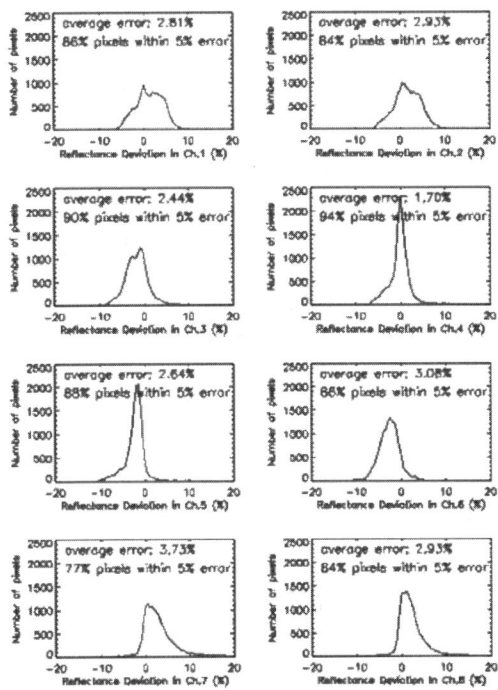

Fig. 3. Relative deviation between measured and inferred reflectances, $\rho_{diff}(\lambda_j)$ [see Eq.(3)], for the SeaWiFS image presented in Fig. 12. The eight panels correspond to the eight wavelengths available for the SeaWiFS instrument.

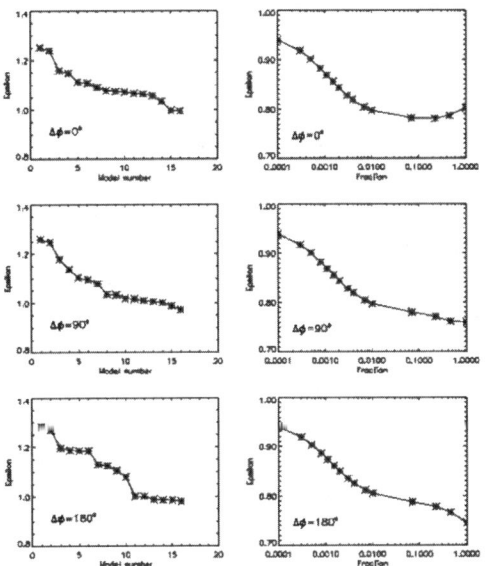

Fig. 4. Parameter $\varepsilon_{ms}(765, 865)$ used for model selection. Left panel: *discrete* set of aerosol models. Right panel: *continuous* set of aerosol models. SZA = 45°, viewing angle = 30°.

3. DISCRETE VERSUS CONTINUOUS AEROSOL MODELS

The MODIS ocean color group has adopted two basic aerosol models: (i) a small particle "Tropospheric" model consisting of 70% water-soluble and 30% dust-like particles, (ii) a large particle "Oceanic" model consisting of sea salt particles. A combination of these two small ("Tropospheric") and large ("Oceanic") particle models yields two additional models, and (iii) a "Coastal" aerosol model with 99.5% small and 0.5% large particles, and (iv) a "Maritime" aerosol model with 99% small and 1% large particles. This yields 4 aerosol models: 2 mono-modal and 2 bi-modal, but by allowing for 4 different relative humidities one arrives at a total of 16 "discrete" candidate aerosol models: half *mono-modal*, the other half *bi-modal*.

3.1 Continuum Set of Aerosol Models

We would like to be able to compute the sensitivity of the TOA radiance to the change in aerosol model, but with a discrete set of models we *cannot easily compute partial derivatives* required for use in iterative inversion algorithms that rely on the use of Jacobians (for example, optimal estimation theory based on Bayes' theorem). Since aerosol models are believed to be generally bi-modal in nature, we would like to construct a set of models that would allow us to be able to change the fraction of small versus large particle populations in a *continuous* fashion, because this would allow us to compute the Jacobians analytically in a fast and reliable manner using a *linearized* discrete ordinate radiative transfer (LIDORT) model for the coupled atmosphere ocean (CAO) system.

There are several advantages of using continuous aerosol models. It allows for all candidate models to be bi-modal and the fraction of large versus small, can be retrieved. Only half of the discrete aerosol models used by the current MODIS ocean color group are bi-modal, and the bi-modality is restricted to only two fractions of small versus large particles: 0.990 and 0.995. It allows for fast and accurate computation of Jacobians using a linearized radiative transfer code for the coupled atmosphere-ocean system (CAO-LIDORT), which in turn allows for use of *state-of-the-art iterative inversion schemes* such as optimal estimation theory based on Bayes' theorem. Such inversion methods allow for a systematic exploitation of the information content of the data collected, they are *well suited for use with multi- and hyperspectral data*, and they allow for *rigorous error analyses* yielding an optimum number of retrieved quantities with known error budgets.

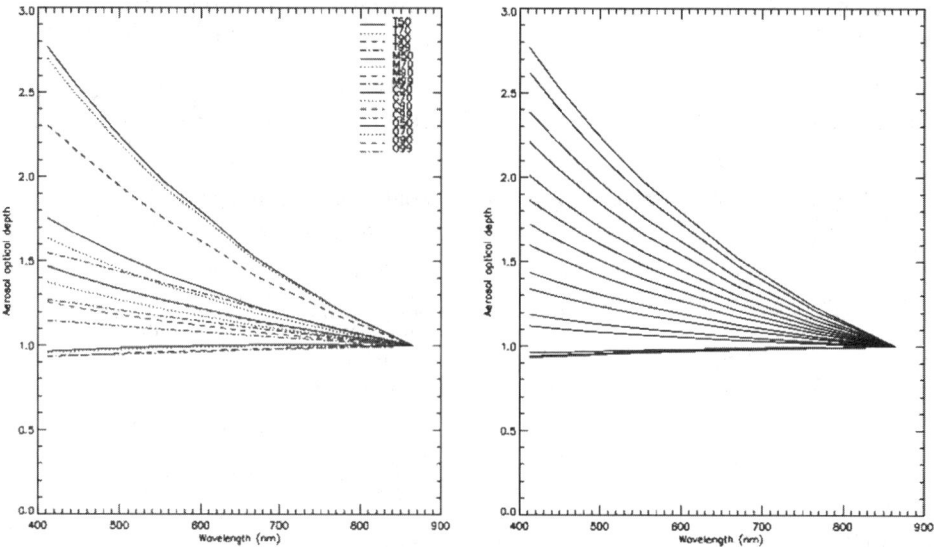

Fig. 5. Extrapolation of aerosol optical depth from the NIR into the visible. Left panel: *Discrete* MODIS aerosol models. Right panel: *Continuous* Models.

4. SUMMARY

We have discussed issues associated with characterization of the atmosphere over turbid waters and emphasized the advantage of selecting aerosol models in an optimum manner by adopting a "continuous" rather than a "discrete" set of models. To illustrate this advantage we discussed how an existing algorithm for simultaneous retrieval of aerosol parameters and chlorophyll concentrations in *clear waters* based on an accurate radiative transfer code for the coupled atmosphere-ocean system (CAO-DISORT) can be extended for use in *turbid coastal waters* by adopting a linearized radiative transfer code for the coupled atmosphere-ocean system (CAO-LIDORT). This new way of characterizing the atmosphere has the following advantages: All models are bimodal as suggested by available data, and they allow for a flexibility in terms of model selection not afforded by a limited set of discrete models. The aerosol optical properties can be computed by the more realistic external mixing rule rather than the internal mixing rule for any combination of small and large particles. In conjunction with the CAO-LIDORT forward model this approach allows for computation of Jacobians required for use state-of-the-art in iterative inversion schemes yielding simultaneous retrieval of: (i) aerosol optical properties that are accurate and reliable with known error budgets, and (ii) marine parameters and water-leaving radiances that are accurate and reliable with known error budgets.

5. REFERENCES

Jin, Z., and K. Stamnes, Radiative transfer in nonuniformly refracting media such as the atmosphere/ocean system, *Appl. Opt.*, 33, 431-442, 1994.

Stamnes, K., W. Li, B. Yan, H. Eide, A. Barnard, W. S. Pegau, and J. J. Stamnes 2003: Accurate and self-consistent ocean color algorithm: simultaneous retrieval of aerosol optical properties and chlorophyll concentrations, *Appl. Opt.*, 42, 939-951.

SIMULTANEOUS RETRIEVAL OF AEROSOLS AND IN-WATER CONSTITUENTS IN TURBID COASTAL WATERS USING MULTI- AND HYPERSPECTRAL DATA

Hans Eide, Wei Li, and Knut Stamnes
Light and Life Laboratory
Department of Physics and Engineering Physics
Stevens Institute of Technology
Hoboken, NJ 07030, USA

Robert Spurr
Harvard-Smithsonian Center for Astrophysics
Cambridge, MA 02138, USA.

Jakob Stamnes
Department of Physics
University of Bergen, Norway.

ABSTRACT

A new method for simultaneous retrieval of aerosol properties and marine constituents in turbid coastal waters is described. This method is an extension of an approach previously developed for simultaneous retrieval of aerosol properties and chlorophyll concentrations in Case 1 waters to turbid waters. This extension is accomplished by employing near-infrared (NIR) channels not available on the SeaWiFS and MERIS instruments to help retrieve aerosol parameters over turbid waters. Both forward and inverse modeling strategies of the method are discussed. Synthetic data, as well as multi- and hyperspectral images of data obtained over clear, as well as turbid waters, are used to test the validity of the new retrieval approach. Our results show that it is possible to extend the traditional atmospheric correction scheme to the longer wavelengths, but also that it is important to use forward models that accurately treat the radiative transfer in the coupled atmosphere-ocean system. With a carefully selected set of suitable aerosol models, the atmospheric contribution to TOA radiances above turbid waters can be accurately determined.

1. INTRODUCTION

Our ability to retrieve accurate information from ocean color in coastal areas is largely limited by how accurately we can determine the atmospheric contribution to the top-of-the-atmosphere (TOA) measured radiances. Improved aerosol retrieval algorithms for turbid waters are motivated by this need for accurate atmospheric correction that can enable highly reliable estimates for the water-leaving radiances. Currently the atmospheric correction routines employed in operational ocean color retrieval algorithms are in reality limited to scenes where the atmosphere is relatively clear and with non-absorbing aerosols. These algorithms also have a major drawback in that they rely on the so-called "black pixel approximation" (BPA), which assumes that the underlying surface can be considered to be optically black in the wavelengths used for atmospheric correction.

The improved spectral capabilities of new satellite instruments provide several additional channels that can be used for characterizing a complex atmosphere above the targets. A correctly determined atmospheric contribution facilitates accurate retrievals of multiple in-water constituents, which are typically present in coastal waters. Near-shore waters are of high importance for coastal nations in terms of environmental monitoring, national security, and resource management.

2. PROBLEMS WITH CURRENT AEROSOL RETRIEVALS

Current attempts to retrieve aerosol information over turbid waters face difficulties because: (1) the black pixel approximation (BPA) fails for the channels usually employed for this purpose because of significant contribution to the TOA radiances from scattering by sediments in the water (i.e., the BPA is not valid), and (2) only an algorithm accounting for radiative transfer in the *coupled* atmosphere-ocean system can yield an answer that is radiometrically correct in all cases. Further more, three major sources of uncertainty remains in current atmospheric correction algorithms for ocean color even for clear water: (1) the aerosols are usually assumed to be non-absorbing; (2) a simple semi-empirical model is used to quantify the ocean reflectance and the water-leaving radiance is assumed to be isotropic; and (3) the decoupling of the atmospheric and oceanic radiative transfer problems makes it difficult to quantify the retrieval error (Yan et al., 2002).

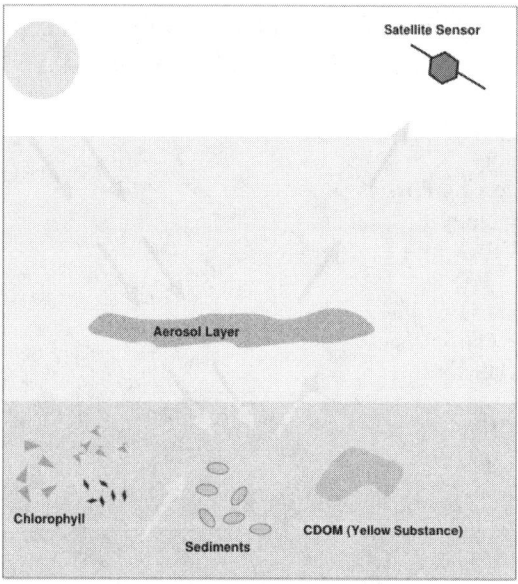

Fig. 1. A schematic illustration showing aspects of the retrieval problem. With an accurate forward model, the problem can be approached with traditional inverse modeling techniques.

3. RETRIEVAL BASED ON COUPLED MODELS

An accurate radiative transfer model (RTM) for the coupled atmosphere ocean (CAO) system becomes an essential tool when the BPA fails. The RTM-CAO approach provides a Forward Model that can be used with standard inversion methods. For example, the RTM-CAO approach can quantify the dependence of the TOA radiance on marine constituents because it enables calculation of weighting functions (Jacobians), and also quantification of the retrieval errors and their sources. Thus, the RTM-CAO approach allows for construction of robust inversion algorithms using state-of-the-art methods, including optimal estimation theory and simulated annealing (Stamnes et al., 2003, Frette et al., 1998).

3.1 Improved Approach for MODIS Channels

Compared to most ocean color instruments (e.g., SeaWiFS), MODIS has channels at longer wavelengths that are useful for atmospheric correction. This is because: (1) at the longer wavelengths even very turbid water has negligible reflectance due to the shallow penetration depth (the BPA holds); (2) yet these channels are sensitive to aerosol properties; and (3) one can easily discriminate land and cloud areas from water using these channels. The additional NIR channels on MODIS that can be used for this purpose are the channels at 1.24, 1.64 and 2.13 μm. Retrieval of aerosols using the longer NIR wavelengths are advantageous because even turbid waters will work as "black pixels", and we may use the information thus retrieved as a "first-guess" in an iterative approach to explore the validity of extrapolating aerosol information retrieved at longer NIR wavelengths into the visible. An accurate description of the atmospheric contribution will allow us to retrieve multiple in-water parameters, and to do so more accurately.

Fig. 2 illustrates the rationale behind the aerosol retrievals. In the left panel we have plotted the parameter γ_{diff} (1240 nm, 1640 nm) versus aerosol optical depth (τ_{865}) for the 16 different aerosol models used in our algorithm. This parameter is defined as:

$$\gamma_{diff}(\lambda_1,\lambda_2) = \gamma(\lambda_1) - \gamma(\lambda_2), \quad (1)$$

where

$$\gamma(\lambda) = \rho_{atm}(\lambda)/\rho_{ray}(\lambda), \quad (2)$$

and ρ_{atm} is the calculated TOA reflectance for an atmosphere containing aerosols, and ρ_{ray} is the TOA reflectance for a pure Rayleigh scattering atmosphere. In the right panel we have plotted the parameter ε_{ms}(1240 nm,1640 nm), which is defined as:

$$\varepsilon_{ms}(\lambda_1,\lambda_2) = [\rho_{atm}(\lambda_1) - \rho_{ray}(\lambda_1)]/[\rho_{atm}(\lambda_2) - \rho_{ray}(\lambda_2)] \quad (3)$$

versus aerosol optical depth (τ_{865}) for the same 16 aerosol models. The two parameters γ_{diff} and ε_{ms} form the basis for our aerosol retrieval procedure. Our aerosol model has 4 types of aerosols, maritime, tropospheric, coastal, and oceanic, each at RH = 50%, 70%, 90%, 99%, thus a total of 16 aerosol models. Retrieving the aerosol optical depth becomes a matter of finding the appropriate aerosol model based on the ε_m -value (which is insensitive to the aerosol optical depth), and then using the determined aerosol model and the calculated γ_{diff} –value to arrive at the aerosol optical depth.

4. RESULTS

We make extended use of synthetic (or simulated) data in addition to real data, because we have found that this approach is very useful for evaluation purposes since it makes it easy to run benchmark comparisons between different algorithms. Using synthetic data and from preliminary tests on Hyperion images, we have found that the longer wavelength channels discriminate water from land and clouds better, and that the BPA holds. Characterizing aerosols over shallow or highly sediment-loaded water is also possible. Thus, the problem of aerosol retrieval and atmospheric correction over turbid water seems to be tractable, but low signal to noise levels for the longer NIR channels might necessitate averaging over areas greater than the pixel size resolution. This scheme is

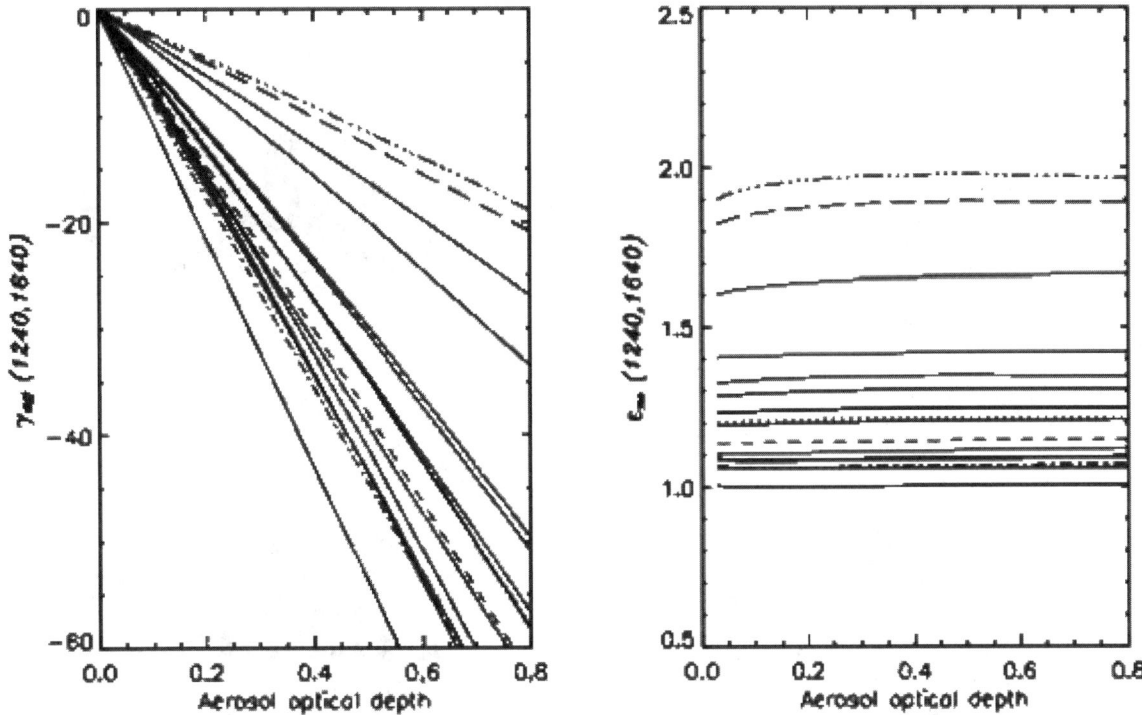

Fig. 2. Simulated values of γ_{dif} (1240/1640) and ε_{ms} (1240/1640) as a function of aerosol optical depth (τ_{865}). The different lines correspond to our 16 aerosol models.

expected to work because of the high temporal resolution (e.g., of the Hyperion and MODIS data), and the fact that atmospheric conditions tend to vary over a larger scale than what is the case for water.

Fig. 4 shows results using our algorithm. We have picked two different 21x21 pixel portions from the same Hyperion image scene shown in Fig. 3. One portion is from the upper part of the image and is believed to be clear water (Case 1), whereas the other portion is from the lower portion of the image where discharges from a nearby river is making the water turbid.

When doing retrievals by averaging over several adjacent pixels, the final retrieved value representing the whole area is simply based on picking the value that was retrieved the most inside the area. In our example, the retrieved aerosol optical depth is the same over each of our sample areas (left histogram), and so is the retrieved aerosol model (right histogram), but the retrieved aerosol model is by a less clear margin over the turbid water (which could just as well mean there *is* an actual difference in the aerosols over the two areas even though they are spatially close to one another).

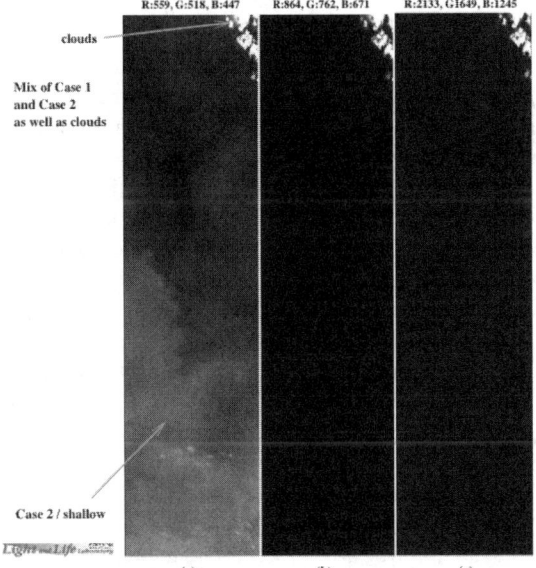

Fig. 3. Reference image made from visual channels, (a) image made from the SeaWIFS atmospheric correction channels, (b) image made using the longer wavelength channels available with MODIS and Hyperion, and (c) The actual wavelengths are indicated on top of each figure.

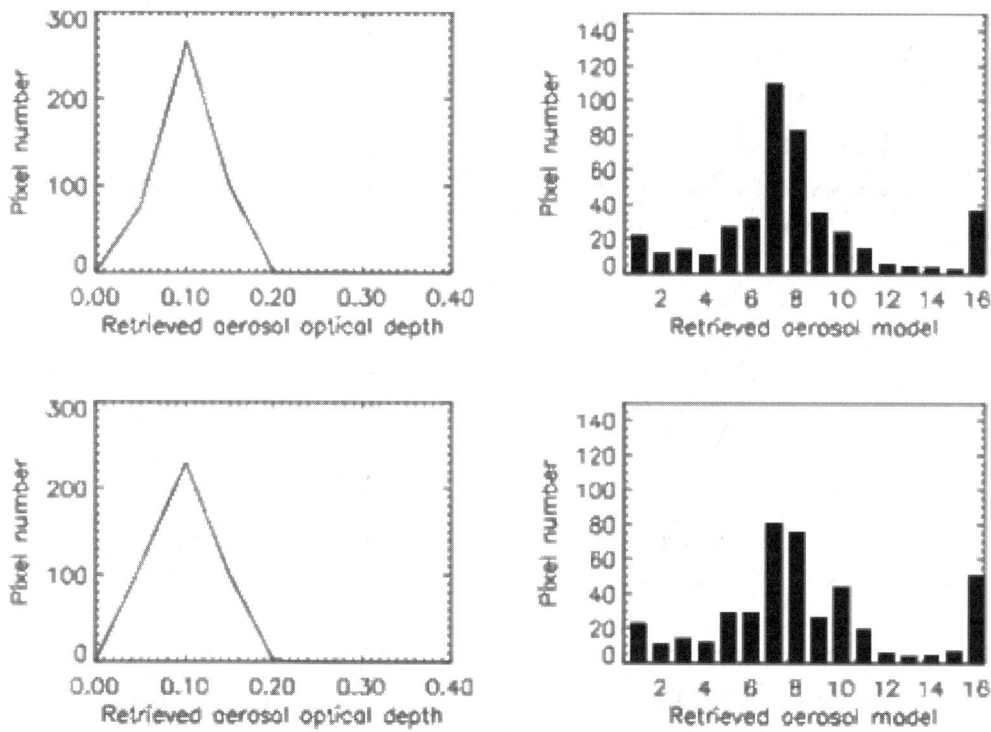

Fig. 4. Retrieval results from 21 x 21 pixel areas (roughly 600 × 600 meters). The upper row is for clear water; the lower row is for turbid water. Averaging over the 21$\times$21 pixel areas there are no problems selecting aerosol model and associated optical depth.

5. SUMMARY

We have illustrated how aerosol retrievals (i.e. atmospheric correction) can be performed adequately over clear as well as turbid waters using NIR channels (for which the BPA holds) available on satellite instruments such as MODIS and Hyperion. Retrieving aerosol properties over several neighboring pixels (averaging) can alleviate a potential problem arising from low signal-to-noise values in the NIR channels. Having determined the atmospheric contribution over turbid water, we are now poised to do retrievals of the in-water constituents. To this end we have developed an improved bio-optical model that is based on the empirical data available in the SeaBASS database (Paper submitted, 2004. See also http://odin.mat.stevens-tech.edu/remote_sensing/). Utilizing a linearized forward model for the coupled atmosphere-ocean system, we are proceeding to retrieve multiple components from turbid waters using rigorous inversion methods and analysis.

6. REFERENCES

Frette, O., J. J. Stamnes, and K. Stamnes, Optical remote sensing of marine constituents in coastal waters: A feasibility study, *Applied Optics*, 37, 8218-8326, 1998.

Stamnes, K., W. Li, B. Yan, H. Eide, A. Barnard, W. S. Pegau, and J. J. Stamnes 2003: Accurate and self-consistent ocean color algorithm: simultaneous retrieval of aerosol optical properties and chlorophyll concentrations, *Appl. Opt.*, 42, 939-951.

Yan, B., K. Stamnes, M. Toratani, W. Li, and J. J. Stamnes, 2002: Evaluation of a reflectance model used in the SeaWiFS ocean color algorithm: Implications for chlorophyll concentration retrievals, *Appl. Opt.*, 41, 6243-6259.

COMPARISON OF SATELLITE (TOMS, GOME) TOTAL OZONE DATA WITH THE MEASUREMENTS OF RUSSIAN GROUND-BASED NETWORK

D. V. Ionov and Yu. M. Timofeyev
Department of Atmospheric Physics, St.Petersburg State University
St.Petersburg, 198504, Russia

A. M. Shalamiansky
Main Geophysical Observatory
St.Petersburg, 194021, Russia

J.-C. Lambert
Belgian Institute for Space Aeronomy
Brussels, B-1180, Belgium

ABSTRACT

The paper presents the comparison of ERS-2 GOME and EarthProbe TOMS overpass total ozone measurements with daily averaged ground-based data by M-124 filter radiometer, collected from the 17 locations over Russia for the period of 1996–2001. Apart from the data of TOMS, which has demonstrated relatively good agreement with M-124, the satellite total ozone measurements by GOME display some shift, producing about 3% underestimation on the average. However, this feature was found to be reduced when looking at the same dataset, first processed by the GOME Data Processor version 2.7 and then by GDP 3.0.

1. INTRODUCTION

The regular satellite monitoring of global ozone distribution is carried out by NASA TOMS instrument since 1978 (Heath et al., 1975). In 1995 another instrument with a similar goal has started its observations—the GOME onboard ESA ERS-2 satellite (Burrows et al., 1999). The quality of resulting satellite products on total ozone is usually validated against independent well-controlled ground-based measurements, with corresponding improvements to the retrieval algorithms being introduced and data reprocessing being run, if necessary. Both TOMS and GOME ozone data products have passed through a number of corrections and updates; the GOME data processor (GDP) has switched from version 2.7 to 3.0 in 2002 (ESA, 2002), and then to 4.0 in 2004 (ESA, 2004); the release of TOMS data version 8 after 7 took place in 2004 (http://toms.gsfc.nasa.gov).

Russia and NIS (New Independent States) have a network of regular ground-based measurements of total ozone, which enumerates more than 40 stations. The network is equipped with filter ozonometers M-124 (Dudko et al., 1985) calibrated against Dobson spectrophotometer, which is regularly compared with WMO standard. The network covers a wide range of latitude—from 38 to 78 degrees north, and the developed processing algorithm allows performing nearly all-weather observations of ozone vertical column.

In this study the data of simultaneous TOMS (version 7) and GOME (version 2.7 and 3.0) overpass measurements were compared to the daily averaged ground-based data, collected over the 17 locations (14–141 E/43–78 N) for the period of 1996–2001. The overall number of comparisons exceeded 14000.

2. RUSSIAN GROUND-BASED NETWORK OF TOTAL OZONE MEASUREMENTS

The Russian and NIS countries network of total ozone (TO) observations is equipped with filter ozonometers M-124 measuring the direct sun or scattered zenith radiation. Two spectral intervals with 302 and 326 nm maxima and half-width of about 20 nm are used for the observations. TO is retrieved from direct sun measurements at zenith angles 20–70° and zenith scattered (both clear-sky and cloudy) radiation measurements at zenith angles 20–85°. The method provides measurements at high latitude stations and practically all-weather conditions. All the M-124 ozonometers are calibrated against the standard (for Russian network) measurement instrument—the Dobson spectrophotometer No. 108, which is regularly compared with the WMO standard. Intercomparisons in Boulder (1988), Hradec Králové (1993), Kalavrita (1997) and Hohenpaissenberg (2001) showed that the measurement-scale drift of the Dobson No. 108 did not exceed 1.0%. A permanent control of M-124 measurement scale is the obligatory part of instrument verification and provides TO measurements with errors less than 3–4%.

In general, the measurements of TO with M-124 ozonometer apply Dobson like technique, based on the registration of the ratio of solar radiation fluxes within and outside the ozone absorption band. Unlike the Dobson spectrophotometer, the optical filters of M-124 select rather wide spectral intervals—about 20 nm half-width. Thus, the Beer-Lambert-Bouguer's law for the incoming direct solar UV radiation in the filter absorption band n should be used in its integral form:

$$I_n = c \int_{\lambda_{n1}}^{\lambda_{n2}} \omega_{n\lambda} I_{0\lambda} e^{-\alpha_\lambda \mu X - \beta_\lambda m - \delta_\lambda m'} d\lambda \quad (1)$$

where I_n is the instrument reading with spectral sensitivity $\omega_{n\lambda}$ in the spectral interval λ_{n1}, λ_{n2}; $I_{0\lambda}$ is the extraterrestrial solar radiation; α_λ, β_λ, δ_λ are the ozone absorption, molecular scattering and aerosol extinction coefficients, respectively; μ and m are the optical ozone and air masses, respectively; m' is the aerosol air mass, X is the TO content, and c is an instrument constant.

TO is then determined from the ration of two instruments readings with filter 1 and 2:

$$K_H K_t \frac{I_1}{I_2} = K_s \frac{\int_{\lambda_1}^{\lambda_2} \omega_{1\lambda} I_{0\lambda} e^{-\alpha_\lambda \mu X - \beta_\lambda m - \delta_\lambda m'} d\lambda}{\int_{\lambda_1}^{\lambda_2} \omega_{2\lambda} I_{0\lambda} e^{-\alpha_\lambda \mu X - \beta_\lambda m - \delta_\lambda m'} d\lambda} \quad (2)$$

where K_s and K_t are the instrument calibration and temperature coefficients, respectively; K_H is the state of cloudiness coefficient, which is set to 1 for direct solar measurements.

The measured quantities are in the left part of (2), thus X is derived under assumption that δ_λ is constant in the spectral range of the measurements, and other quantities—I_0, α, β, μ, m and m'—are calculated in a way, similar to Dobson.

For zenith sky measurements, the K_s in the equation (2) is replaced by an empirical coefficient K_z, which is calculated based on the ratios of radiation intensities I_1 and I_2, simultaneously measured in direct Sun (s) and zenith sky (z) mode:

$$K_z = \frac{[I_1/I_2]_s}{[I_1/I_2]_z} \quad (3)$$

The numerous observations have shown, that K_z depends mostly on the ozone air mass (μ), but not on TO (X), and this function may be expressed by the polynomial of 2nd order.

This method was originally developed for clear sky zenith observations, but it was found to be suitable for cloudy conditions, as well. Long-term measurements carried all over the network of M-124 stations have revealed a following feature—within a fixed μ and X, the instrument readings ratio $[I_1/I_2]_z$ is changing negligibly and depends only on cloud optical density at the sky zenith. Moreover, this dependence persists at every value of μ and X. Thus, the K_H in (2), which characterize the degree of change in $[I_1/I_2]_z$ ratio due to the cloudiness, is determined by the visual estimation of cloud optical density through the color scale and homogeneity of clouds at the sky zenith.

3. SPACE-BORNE TOTAL OZONE MEASUREMENTS – TOMS AND GOME

The long-term daily mapping of global ozone have been started by NASA with Total Ozone Mapping Spectrometer (TOMS) on board Nimbus-7 (1978–1993), and continued with TOMS on board Meteor-3 (1993–1994), and then with TOMS on board EarthProbe, launched in 1996 and operational now (McPeters et al., 1998). TOMS observes backscattered ultraviolet radiance at six ~1 nm bands centered on 308.6, 313.5, 317.5, 322.3, 331.2 and 360.4 nm. A mirror scans ±51° across track in 35 steps of 3°. With its 500-km ascending polar orbit, TOMS provides daily global coverage beyond 60° north, with a ground pixel less then 40 × 40 km size.

The Global Ozone Monitoring Experiment (GOME) was launched on April 21, 1995 on board the second European Remote Sensing Satellite (ERS-2) (Burrows et al., 1999). ERS-2 flies in a sun-synchronous polar orbit at mean altitude of 780 km with an inclination of 98°, and a local time of 10:30 of the equator crossing at the descending node. GOME is a nadir-viewing grating spectrometer that measures the solar irradiance and the solar radiation backscattered from both the atmosphere and the earth's surface. The instrument is operating in UV-visible range of spectra (240–790 nm) with a moderate spectral resolution of 0.2 to 0.4 nm sensed by four individual linear detector arrays each with 1024 detector pixels. The field of view may be varied in size from 320 km × 40 km (forward scan) to 960 km × 40 km (back scan). With the large swath, global coverage is achieved every three days at the equator and earlier at higher latitudes.

4. COMPARISON OF SATELLITE DATA WITH RUSSIAN GROUND-BASED MEASUREMENTS

The data of ground-based TO observations over 17 Russian locations (14-141 E/43-78 N) for the period of 1996–2001 was compared to correlative satellite measurements by TOMS (EarthProbe) and GOME (ERS-2). The map of ground-based sites providing the data for this study is presented in Fig. 1. The total number of pairs of observations in comparison has exceeded 14000.

The data of TO by TOMS (V7) in comparison with ground-based M-124 observations over Russia in 1996-2001 is presented in Fig. 2. An individual results of comparison for each of the 17 stations with TOMS overpass data is given in Table 1. The RMS deviation between TOMS and M-124 (σ) vary from 4.8% at Arhangelsk (65°N) to 8.1% at Samara (53°N). Overall, the average and RMS deviations are: $\Delta_{TOMS-M124} = -0.3\%$, $\sigma_{TOMS-M124} = 5.6\%$.

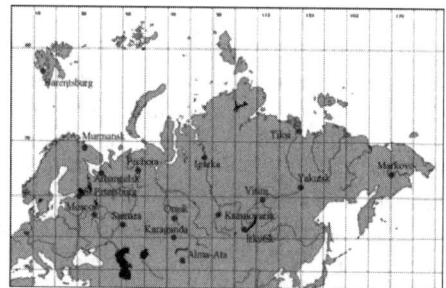

Fig. 1. Locations of 17 Russian sites that contributed to the comparison of TOMS and GOME TO data with ground-based measurements by M-124 in 1996–2001.

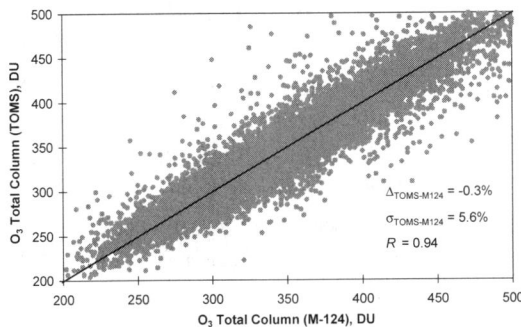

Fig. 2. Satellite TO by TOMS (V7) versus ground-based measurements over Russia in 1996–2001 by M-124 (Δ, σ and *R* correspond to the average, RMS deviation and correlation coefficient, respectively).

The data of TO by GOME (V2.7, 3.0) have been searched for correlative measurements within 500 km from the ground-based, for the same dataset used in comparison with TOMS. Overall, the average and RMS deviations between M-124 and GOME (V3.0) are: $\Delta_{GOME-M124}$ = –2.5%, $\sigma_{TOMS-M124}$ = 7.0%. This presents less agreement, than it was found before for TOMS (Δ= –0.3%, σ = 5.6%). However, the data of GOME V3.0 is in a definitely better agreement with ground-based observations, compared to the same dataset, processed by GOME Data Processor (GDP) 2.7 (see Table 2). The table presents results of comparisons between ground-based (M-124) and satellite (TOMS V7, GOME V2.7 and 3.0)—the average (Δ), RMS (σ) deviation—for the entire dataset and separate classes of data, as well (different Sun elevation, satellite pixel cloudiness and season). Thus the overall estimate of Δ and σ for the data of GOME V2.7 is –3.0% and 7.4%, respectively. All of the 3 satellite datasets demonstrate a clear dependence on pixel cloudiness, with an underestimation of TO increasing at high cloud fraction value. The latter is likely due to the effects of cloud screening, which makes a part of tropospheric ozone invisible for satellite sensor. Besides, the deviations of GOME V2.7 from ground-based observations depend much on the season – corresponding Δ varies from –6.0% in autumn to +0.6% in spring. However, this seasonal feature was found to be much reduced in the same dataset processed with GDP 3.0; instead, the data of TOMS V7 is almost free of such seasonal effect.

The two-step DOAS approach adopted in GOME consists of the spectral fitting of slant column amount followed by its conversion into vertical column amount using a calculated Air Mass Factor (AMF). Compared to GDP 2.7, GDP 3.0 included a new determination of effective absorption temperature derived by spectral analysis, better atmospheric databases, and AMFs determined iteratively using a neural network trained on column- and latitude-classified atmospheric profiles, taken from the TOMS V7 climatological database (ESA, 2002). Compared to GDP 3.0, the current version GDP 4.0 includes an improved correction for ozone absorption distortion due to inelastic rotational Raman scattering by air molecules, a new cloud treatment for the retrieval of three auxiliary pieces cloud information, and further improvements to the AMF calculation using on-the-fly radiative transfer modeling (ESA, 2004).

Table 1. Comparison of Satellite TO Data by TOMS (V7) with Measurements by M-124 at 17 Locations of Russian Ground-based Network in 1996-2001. N – Number of Pairs in Comparison, Δ and σ – Average and RMS Deviation, Accordingly (TOMS-M124)

Location	Lat, °N	N	Δ, %	σ, %
Vladivostok	43.1	646	+1.8	5.4
Voronezh	51.7	973	+0.3	5.3
Irkutsk	52.3	1360	–1.4	5.4
Nikolaevsk	53.1	726	–0.1	5.5
Samara	53.2	437	+0.3	8.1
Omsk	54.9	1071	–0.6	5.5
Moscow	55.7	702	–3.1	5.2
Krasnovarsk	56.0	411	+0.9	5.2
Ekaterinburg	56.8	1156	–0.6	5.2
Vitim	59.4	778	–0.9	6.8
St.Petersburg	60.0	1434	+1.3	5.1
Arhangelsk	64.6	1078	–0.4	4.8
Pechora	65.1	1183	+0.0	5.9
Igarka	67.5	608	–0.2	6.3
Olenek	68.5	340	+0.0	5.0
Murmansk	69.0	845	–2.1	6.0
Barentsburg	78.1	653	+0.3	4.9
OVERALL:	-	14401	-0.3	5.6

Table 2. The Average (Δ) and RMS (σ) Deviations of Satellite TO Data by TOMS (V7) and GOME (V2.7, 3.0) from the Ground-based Measurements (M-124) over Russia in 1996-2001 (SZA – Sun Zenith Angle, CF – Cloud Fraction)

Satellite Dataset	Δ, %			σ, %		
	TOMS V7	GOME V2.7	GOME V3.0	TOMS V7	GOME V2.7	GOME V3.0
OVERALL	-0.3	-3.0	-2.5	5.6	7.4	7.0
SZA ≤ 70°	-0.4	-2.6	-2.5	5.3	6.9	6.8
SZA > 70°	-0.2	-4.1	-2.3	6.2	8.6	7.6
CF ≤ 50%	-0.1	-2.6	-2.3	5.4	7.1	6.8
CF > 50%	-1.7	-5.3	-3.6	6.8	9.0	7.8
winter	+0.0	-2.7	-1.0	6.5	8.1	7.0
spring	-0.2	+0.6	-0.8	5.5	6.3	6.9
summer	-0.3	-4.6	-4.3	5.3	7.3	7.2
autumn	-0.7	-6.0	-3.6	5.2	8.3	6.8

Retrieval of TO by TOMS is based on a comparison between the measured normalized radiance (the ratio of the backscattered Earth radiance to the incident solar irradiance) and radiance derived by radiative transfer calculations for different ozone amounts and the conditions of the measurement (McPeters et al., 1998). It is implemented by using radiative transfer calculations to generate a table of backscattered radiance as a function of total ozone, viewing geometry, surface pressure, surface reflectivity, and latitude. Given the computed radiance for the particular observing conditions, the total ozone value can be derived by interpolation in radiance as a function of ozone. Compared to algorithm V7, the new one (V8) uses a set of fixed standard ozone profiles similar to the previous algorithm and then updates the solution based on a new climatology of ozone and temperature profiles that are dependent on month, latitude, and total ozone.

Within our study, we've recently got an opportunity to include into analysis the datasets of re-processed TO measurements by GOME (GDP 4.0) and TOMS V8. The general view of comparison between ground-based measurements and satellite data is shown in Fig. 3, which display an improvement in GOME data accuracy, but also significant drop in the agreement with TOMS (preliminary results).

5. SUMMARY

Here we compared the data of Russian ground-based total ozone network observations with correlative satellite measurements by ERS-2 GOME and EP TOMS. The comparison provides an independent estimate for the evolution of satellite data quality by continuous re-processing—GDP 2.7, 3.0 and 4.0, TOMS V7 and 8.

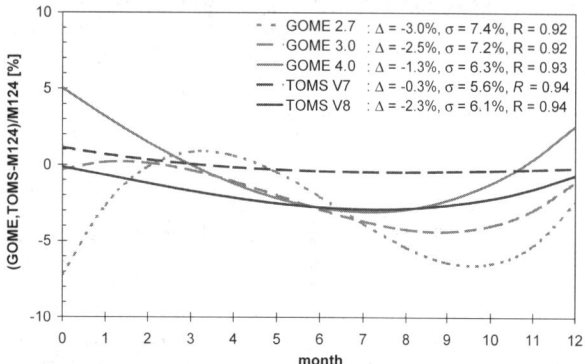

Fig. 3. Relative difference of satellite TO data (TOMS V7 and 8, GOME V2.7, 3.0 and 4.0) from the ground-based measurements (M-124) over Russia in 1996–2001 (Δ, σ and R correspond to the average, RMS deviation and correlation coefficient, respectively).

6. ACKNOWLEDGEMENTS

This work was partly supported by joint INTAS and ESA grant (INTAS-IESA-99-1511), the President of Russia grant (MK-2686.2003.05), and the grants of St. Petersburg Administration (PD02-1.5-96, PD03-1.5-43, PD05-1.5-46). TOMS overpass data have been downloaded from official TOMS homepage at http://toms.gsfc.nasa.gov. GOME level 2 data was provided by ESA within "Announcement of Opportunity" project (ERS AO3-174). GOME data search and extraction was carried out with a software tool, developed by K. Bramstedt (University of Bremen).

7. REFERENCES

Burrows, J. P., M. Weber, M. Buchwitz, V. V. Rozanov, A. Ladstaetter-Weissenmayer, A. Richter, R. de Beek, R. Hoogen, K. Bramstedt, K.-U. Eichmann, M. Eisinger, and D. Perner, 1999: The Global Ozone Monitoring Experiment (GOME): Mission concept and first scientific results, *J. Atmos. Sci.*, 56, 151-175.

Dudko, B. G., V. V. Lagina, and N. Z. Prosjankina, 1985: The new ozonometer M-124, *Sov. Meteorol. Hydrol.*, 12, 113-114.

ESA (European Space Agency), 2002: ERS-2 GOME GDP 3.0 implementation and delta validation, *Technical Note ERSE-DTEX-EOAD-TN-02-0006*, Lambert, J.-C. (Ed.), 138 pp.

ESA (European Space Agency), 2004: Delta validation report for ERS-2 GOME Data Processor upgrade to version 4.0, *Technical Note ERSE-CLVL-EOPG-TN-04-0001*, Lambert, J.-C. and D. S. Balis (Ed.), 82 pp.

Heath, D. F., A. J. Krueger, H. R. Roeder, and B. D. Henderson, 1975: The Solar Backscatter Ultraviolet and Total Ozone Mapping Spectrometer (SBUV/TOMS) for Nimbus 7, *Opt. Eng.*, 14(4), 323-331.

McPeters, D. F., P. K. Bhartia, A. J. Krueger, J. R. Herman, C. G. Wellemeyer, C. J. Seftor, G. Jaross, O. Torres, L. Moy, G. Labow, W. Byerly, S. L. Taylor, T. Swissler, and R. P. Cebula, 1998: Earth Probe Total Ozone Mapping Spectrometer (TOMS) data products user's guide, *NASA Reference Publication 1998-206895*, 60 pp.

DEVELOPMENT OF A POLARISED RADIATIVE TRANSFER MODEL IN THE OXYGEN A-BAND FOR SATELLITE RETRIEVAL OF CLOUD TOP PRESSURE

N. Fournier and P. Stammes
Royal Netherlands Meteorological Institute (KNMI)
De Bilt, 3730AE, The Netherlands (fournier@knmi.nl)

M. Eisinger
European Space Agency/ESTEC
Noordwijk, 2200 AG, The Netherlands

ABSTRACT

A monochromatic radiative transfer model that includes polarisation is extended with oxygen absorption in the O_2 A-band. Moreover, improvement in the model run-time is sought by developing a parallel implementation coded using Message Passing Interface. The model is then applied in the 755–775 nm window for several cloud types and geometries. The calculated radiances are used as input of a cloud retrieval algorithm to assess the accuracy of its retrievals from the O_2 A-band. Effective cloud fraction and cloud top pressure are retrieved with a respective accuracy of 0.05 and 50 hPa. As corroborated through a comparison with SCIAMACHY in-flight data, the cloud retrieval algorithm performs well but would be improved by taking into account Rayleigh scattering.

1. INTRODUCTION

As radiative scattering by clouds strongly influences retrievals of tropospheric gases and aerosols from satellite spectrometers (GOME, SCIAMACHY or GOME-2), accurate co-located cloud information is required. The O_2 A-band at 760 nm is the strongest band of O_2 in the Visible Near-Infrared and thus, is well suited to provide cloud information. Since oxygen is a well-mixed gas, the measured column amount of oxygen yields the cloud top pressure.

In this study, radiances from the DAK (Doubling-Adding KNMI) radiative transfer model are used as input to the FRESCO (Fast Retrieval Scheme for Clouds from the Oxygen A-band) cloud retrieval algorithm. The accuracy of FRESCO to retrieve cloud parameters is evaluated for different atmospheric scenarios. This required the extension of the DAK model with oxygen absorption in the O_2 A-band, its optimisation to allow faster simulations and its application for specific clouds and geometries for the case of the GOME-2 instrument on-board Metop (ESA/EUMETSAT).

Moreover, to investigate the influence of Rayleigh scattering on cloud retrieval in the O_2 A-band, SCIAMACHY in-flight data from a cloud-free scene are compared with DAK reflectances and the transmittances used by FRESCO.

2. FRESCO CLOUD ALGORITHM

The FRESCO method is a simple, fast, and robust algorithm to provide cloud information for cloud correction of ozone (Koelemeijer et al., 2001). FRESCO uses the reflectance in three 1-nm wide windows of the O_2 A-band: 758–759 nm, 760–761 nm, and 765–766 nm. The measured reflectance is compared to a modelled reflectance, as computed for a simple cloud model. In this model the cloud is assumed to be a Lambertian reflector with albedo $A_c=0.8$ below a clear atmosphere, in which only O_2 absorption is taken into account. To simulate the spectrum of a partly cloudy pixel, a simple atmospheric transmission model is used, in which the atmosphere above the ground surface (albedo A_s) or cloud (for the cloudy part of the pixel) is treated as a purely absorbing, non-scattering, medium. The retrieved parameters are the effective cloud fraction (between 0 and 1) and the cloud top pressure with an accuracy of 0.05 and 50 hPa, respectively. Besides the latter retrieval errors in cloud top pressure is most likely due to the fact that absorption by O_2 inside the cloud is neglected, both retrievals are particularly sensitive to errors in the assumed surface and cloud albedo (Koelemeijer et al., 2001). FRESCO is used in the fast-delivery processing of GOME ozone data and in the SCIAMACHY ozone processor TOSOMI, which provides ozone data from SCIAMACHY within the ESA TEMIS project (see http://www.temis.nl).

3. RADIATIVE TRANSFER MODELLING IN THE O_2 A-BAND

The DAK model is designed for line-by-line calculations of radiance, polarisation, and irradiance at top-of-atmosphere, and inside the atmosphere. It consists of an atmospheric shell around a monochromatic multiple scattering kernel, based on the polarised doubling-adding method (De Haan et al., 1987; Stammes, 2001). DAK is commonly used in the UV-visible spectrum. In this study,

DAK is extended with O_2 absorption in the O_2 A-band, using the latest HITRAN 2001 database for the O_2 cross-sections (HITRAN, 2003). Fig. 1 shows the calculated reflectance (R) at specific geometries for clear sky and a surface albedo of 0.05 for a Mid-Latitude Summer atmosphere. Fig. 2 gives the degree of linear polarisation (P) showing that the wavelength dependence of P and R is opposite.

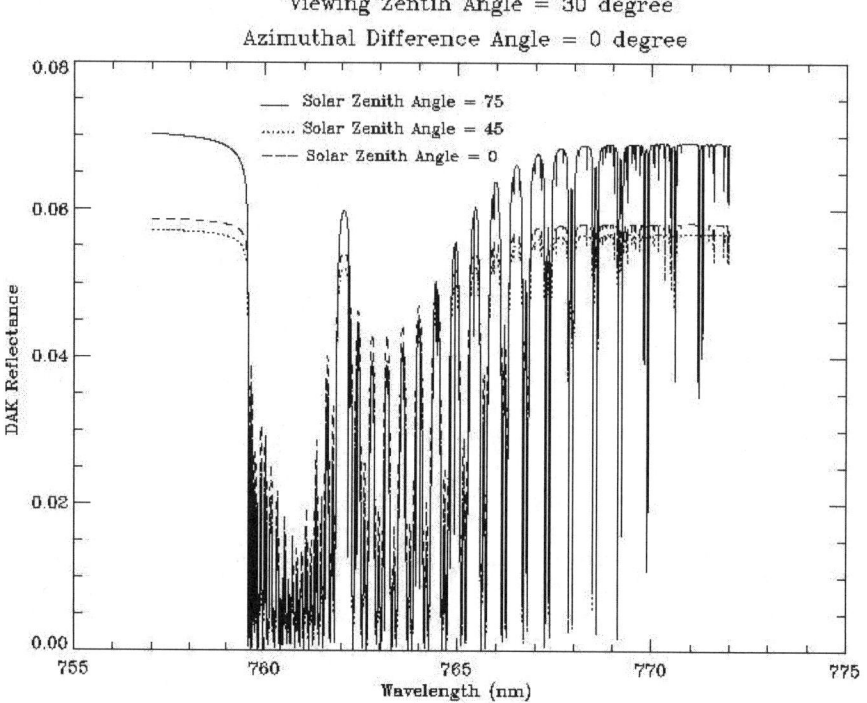

Fig. 1. Reflectance spectra at TOA (θ=30 and Φ- Φ_0=0) from the DAK model.

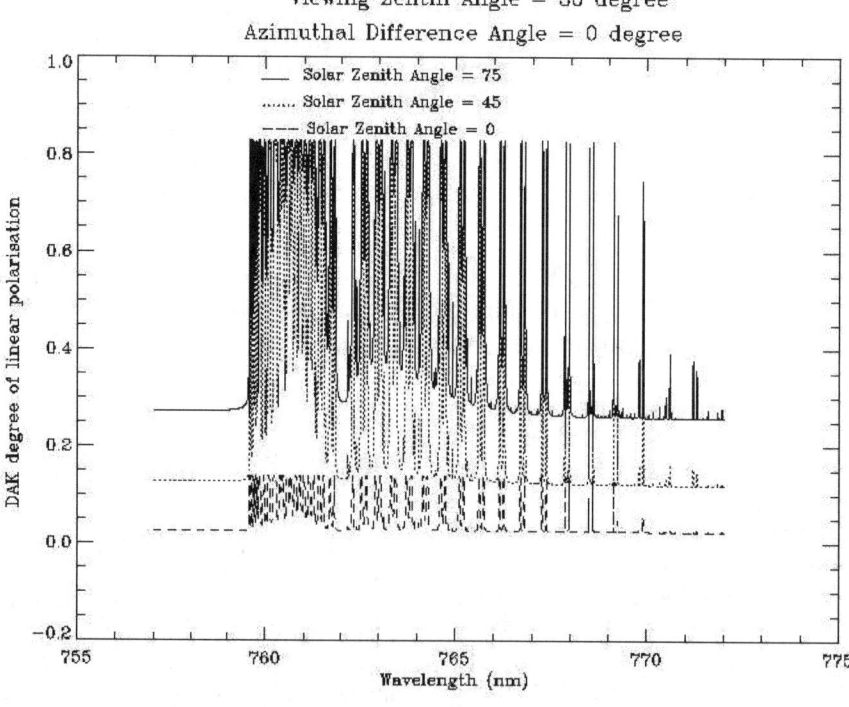

Fig. 2. Degree of linear polarisation at TOA (θ=30 and Φ- Φ_0=0) from the DAK model.

3.1 DAK Optimisation

However, such calculations, together with the description of a Mie scattering cloud and linear polarisation impose a major computational requirement. Indeed, such simulation, at 0.01 nm resolution in the 757–72 nm region, is executed in 10 days with a classic 3.2 GHz workstation (8 Gbytes memory). Therefore, improvement in the model run-time was sought by developing a parallel implementation of DAK coded in a data-parallel approach using Message Passing Interface (Pacheco, 1996) on the KNMI Sun Fire 15K Platform (900 MHz, 48 Gbytes memory). The implementation distributes the calculations at the different wavelengths over multiple processors. For example, running the code on this machine with 30 processors led to a speedup by a factor of 22. However, an optimal speed-up can be reached by using the 68 available processors of the parallel Platform while no other codes are running simultaneously. However, as the number of processors increases, the communications increase and outweigh the advantages of distributing the code on a larger number of processors. Although the parallelisation's performance is limited by these communications, the speedup obtained with the parallel code is an important foundation for simulating, in a reasonable CPU time, scenarios including polarisation or explicit clouds at a fine spectral resolution.

4. FRESCO RETRIEVALS

FRESCO information on cloud top pressure and effective cloud fraction is derived from the reflectivity in and around the Oxygen A-band. In this particular study, the reflectivity will not be deduced from the radiance measured by a satellite instrument, but it will be calculated from the DAK model for specific scenarios and geometry. A pixel representative of a Metop test orbit is used with a viewing and solar zenith angle of 45 degrees and an azimuthal difference angle of 0 degree.

Table 1. FRESCO Retrieval of Effective Cloud Fraction (c) and Cloud Top Altitude (z_c; km) for Different input Scenarios from DAK

Cases :		NoCld	LR-8k	Mie-8k
DAK	c	0.0	1.0	1.0
	z_c	0.0	8.0	8.0
FRESCO	c	0.05	1.0	1.0
	z_c	1.23	7.56	8.08
Cases :		LR-2k	Mie-2k	MixCld
DAK	c	1.0	1.0	0.4
	z_c	2.0	2.0	2.0
FRESCO	c	1.0	1.0	0.58
	z_c	1.59	2.03	1.98

Table 1 shows DAK prescribed and FRESCO retrieved values of effective cloud fraction (c) and cloud top altitude (z_c) for different scenarios. The scenarios considered are *NoCld* (no clouds and A_s=0.05); *LR-8k* (full cloud cover; cloud assumed to be a Lambertian Reflector at 8 km with A_c=0.8); *Mie-8k* (full cloud cover; Mie scattering cloud at 8 km with an optical thickness of 40.0); *LR-2k* (full cloud cover; cloud assumed to be a Lambertian Reflector at 2 km with A_c=0.8); *Mie-2k* (full cloud cover; Mie scattering cloud at 2 km with an optical thickness of 40.0) and *MixCld* (cloud cover of 40%; Mie scattering cloud at 2 km). The coefficients of the scattering matrix of the Mie cloud are calculated using the Meerhoff Mie code (De Rooij and Van der Stap, 1984) assuming a two-parameter gamma cloud droplet size distribution with an effective radius r_{eff}=10 µm and an effective variance v_{eff}=0.2 (Hovenier et al., 2004).

The retrievals from FRESCO are in good agreement with the cloud input conditions of the DAK scenarios. However, discrepancies arise between DAK and FRESCO in the case of low cloud fractions (cloud free or mixed cloud scenarios). As the major difference between both codes is the consideration of Rayleigh scattering, it would be worth taking this process into account in FRESCO. This would not affect significantly the fully cloudy pixels but would produce lower retrieved values of cloud top altitude in the case of low cloud fractions (inferior to 50%). Indeed, in the latter case, the Rayleigh effect is larger as most of the light is scattered before reaching the lower atmosphere where absorption is important and hence, the bands will be weaker.

5. RAYLEIGH SCATTERING INFLUENCE IN THE O_2 A-BAND

To demonstrate the usefulness of an accurate transmittance description in FRESCO, Fig. 3 compares a clear-sky Sahara spectrum measured by SCIAMACHY (August 23, 2002) and the spectrum from the FRESCO transmission database and DAK, respectively (normalized at 758 nm). Both approaches compare well with SCIAMACHY in-flight data. The differences between DAK and SCIAMACHY are partly due to instrument calibration errors but, also, to aerosols and surface albedo effects.

The main difference between DAK and the transmittance used by FRESCO is the consideration of

Rayleigh scattering. In DAK, Rayleigh scattering and O_2 absorption are included in 65 homogeneous layers between 0 and 100 km. However, this former process is not taken into account through the transmission database used by FRESCO to retrieve the cloud parameters. Thus, DAK produces bands slightly weaker (less absorption) which corroborates the fact that including Rayleigh scattering in the FRESCO Transmission database could significantly improve cloud top altitude retrievals for scenes with low values of cloud fraction.

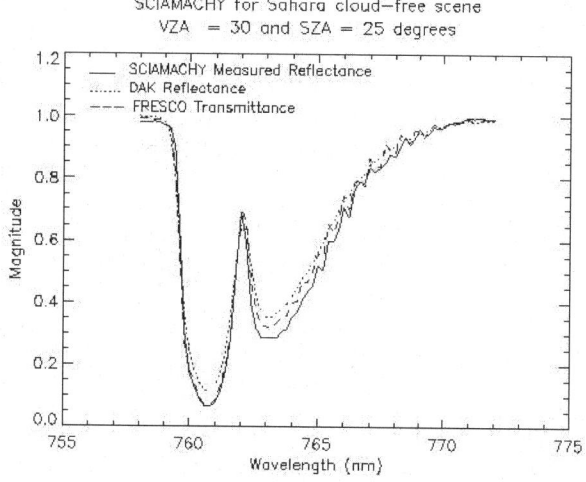

Fig. 3. Comparison between FRESCO transmittance and the in-flight measured reflectance of SCIAMACHY at a specific location of the Sahara in the O_2 A-band.

6. CONCLUSION

A radiative transfer model has been extended with the O_2 A-band using the latest HITRAN database for the O_2 cross-sections. The model has then been parallelised using MPI and applied in a line-by-line mode with multiple scattering in the 757–772 nm region. The obtained radiances have been combined with the SCIAMACHY slit function for different atmospheric scenarios and used as input of the FRESCO algorithm to retrieve the corresponding effective cloud fraction and cloud top altitude. This shows that FRESCO succeeds well to retrieve cloud information. However, for low cloud fraction, the cloud algorithm would benefit of taking into account Rayleigh scattering to retrieve cloud top pressure more accurately. This has been corroborated through the comparison with SCIAMACHY in-flight data.

7. ACKNOWLEDGMENTS

This work has been partly funded by ESA/ESTEC under the project *17332/03/NL/GS*.

8. REFERENCES

De Haan, J. F., P. B. Bosma, and J. W. Hovenier, 1987: The adding method for multiple scattering computations of polarized light, *Astronomy Astrophysic*, 183, p.371.

De Rooij, W. A., and C. C. A. H. Van der Stap, 1984: Expansion of Mie scattering matrices in generalized spherical functions, *Astronomy Astrophysic*, 131, 237-248.

HITRAN, 2003: Special Issue of the *Journal of Quantitative Spectroscopy and Radiative Transfer*, 82, number 1-4.

Hovenier, J. W., C. Van der Mee, and H. Domke, 2004: Transfer of polarized light in planetary atmospheres, *Kluwer Academic Publishers*, p. 222.

Koelemeijer, R. B. A., P. Stammes, J. W. Hovenier, and J. F. de Haan, 2001: A fast method for retrieval of cloud parameters using oxygen A band measurements from GOME, *Journal of Geophysical Research*, 106, 3475-3490.

Pacheco P., 1996: Parallel Programming with MPI, Morgan Kaufmann.

Stammes, P., 2001: Spectral radiance modeling in the UV-visible range. In *IRS 2000: International Radiation Symposium 2000: Current problems in Atmospheric Radiation*, A. Deepak Publishing, Virginia, 385-388.

ATMOSPHERIC AND CLOUD PARAMETERS RETRIEVAL USING IR SPECTRAL DATA

Daniel K. Zhou[*], William L. Smith[‡], Hung-Lung A. Huang[#], Jun Li[#], Xu Liu[*],
Allen M. Larar[*], Linda Chiou[*], and Stephen A. Mango[†]

[*]NASA Langley Research Center, Hampton, VA 23681, USA
[‡]Hampton University, Hampton, VA 23668, USA
[#]University of Wisconsin – Madison, WI 53706, USA
[†]NPOESS Integrated Program Office, Silver Spring, MD 20910, USA

ABSTRACT

High resolution infrared radiance spectra obtained from near nadir observations have provided atmospheric, surface, and cloud property information. A fast radiative transfer model, including a fast cloud model, is used for atmospheric profile and cloud parameter retrieval. The retrieval algorithm is presented along with its application to recent field experiment data from the NPOESS Airborne Sounder Testbed—Interferometer (NAST-I). The retrieval accuracy dependence on cloud properties is discussed. Preliminary simulations demonstrate that accurate temperature and moisture retrievals can be achieved below optically thin clouds. For optically thick clouds, accurate temperature and moisture profiles down to cloud top level are obtained. For both optically thin and thick cloud situations, the cloud top height can be retrieved with an accuracy of ~1.0 km. Preliminary NAST-I retrieval results from the recent Atlantic-THORPEX Regional Campaign (ATReC) are presented and compared with coincident observations obtained from dropsondes and the nadir-pointing Cloud Physics Lidar (CPL).

1. INTRODUCTION

The retrievals of atmospheric state (i.e., temperature and moisture profiles) obtained from infrared radiometric measurements will contain intolerable error near and below the cloud level if the attenuation of infrared radiation emitted from the Earth's surface and the atmosphere below the clouds are not properly accounted for in the retrieval process. Since the vast cloudy regions affect the global infrared measurements, a great deal of effort has gone into the cloud detection and cloud-clearing processes. As the fast molecular and cloud transmittance models have been developed recently, the infrared radiances under cloudy conditions can be simulated accurately and fast enough for retrieval processing. Retrieval schemes can be expanded to deal with the observations under both cloud-free and cloudy conditions. We report that the EOF statistical regression scheme (e.g., Smith and Woolf, 1976; Zhou et al., 2002) has been expanded to include realistic cloud parameters (e.g., cloud top height, optical depth) to deal with the cloudy and/or cloud-free observations; thus, not only atmospheric state is inferred, but also the cloud parameters. The NAST has been successfully operating on high altitude aircraft since 1998. NAST-I simulated and measured data are used in this study. The retrieval accuracy depending on cloud properties is discussed. Preliminary NAST-I retrieval results are compared with coincident observations obtained from the nadir-pointing Cloud Physics Lidar (CPL) and dropsondes.

2. RADIANCE SIMULATIONS, TRAINING, AND REGRESSION

The infrared radiances measured under cloudy conditions are simulated by the Optimal Spectral Sampling fast molecular radiative transfer model (Liu et al., 2003; Moncet et al., 2003), and the physically based cloud radiative transfer model using DISORT calculations performed for a wide variety of cloud microphysical properties (e.g., Yang et al., 2001).

Single-layer-cloud statistics for EOF regression are not sufficient because multiple layers of clouds are commonly present in the observations. In this work, a maximum of 2 cloud levels is used; a single cloud layer (either ice or water) and other (potentially) optically thick clouds can be assumed at the lower level when the radiosonde detects lower level clouds. These cloud layers, along with the radiosonde profile, are used to simulate NAST-I radiances. Cirrus clouds are assumed to be at the higher level and the cloud microphysical properties are simulated using a random number generator to specify cloud visible optical depth within a pre-specified range. Parameterization of balloon and aircraft cloud microphysical data base (Heymsfield et al., 2003) is used to specify cloud effective particle radius and cloud optical depth. A random error of 10% is added to parameterized effective radius to account for real data scatter. At the lower cloud level, the opaque cloud representation (i.e., isothermal/saturated) is assumed and the profile is treated as isothermal below the lower cloud level. This lower level cloud is represented as an equivalent cloud-free isothermal temperature condition in the radiative transfer calculation.

Detailed description of NAST-I retrieval methodology and optimal EOF number selection can be found elsewhere (e.g., Zhou et al., 2002). NAST-I EOF statistical regression methodology has been expanded to include cloud parameters. Regression relations are generated not only for predicting thermodynamic parameters but also cloud top height and cloud microphysical properties. Because the radiance is highly non-linear with respect to cloud height, statistics are formulated for one class of data which contains all cloud height conditions and seven other classes for which the cloud height has been stratified to within ~1.5 km of the mean for that class. The classes are also separated by the cloud phase (ice and/or water). The final cloud height class to be used for the retrievals is obtained by iteration beginning with the unclassified class to predict the initial cloud height stratification for the retrievals. Usually, the final cloud height class is defined within five iterations of the cloud height prediction process. The cloud phase is determined by the signatures of micro-window channels and initial retrieved cloud top temperature. However, sufficient numbers of radiosondes, ~800 soundings per each cloud height group per cloud phase, are used to ensure that the observations are well covered by the statistics representation. For semitransparent and/or scattered clouds, with an effective optical depth (defined here as the exponential of a negative argument which is the product of the fractional cloud amount times the cloud visible optical depth) is less than one, the correct profile below the cloud is attempted to be retrieved. If a lower level cloud underlies the semitransparent and/or scattered upper level cloud, the lower level cloud is treated as an equivalently clear isothermal condition as described for the opaque cloud condition retrieval. EOF regression enables both the cloud height and the cloud microphysical properties of the highest-level cloud to be estimated.

3. RETRIEVAL SIMULATION WITH CLOUDS

Retrieval simulations have been performed over a set of winter hemispheric data (Cloudy soundings from November 1st to January 10th, 1995–2003; latitude from 33°N to 54°N; longitude from 58°W to 85°W), which is also used for statistic training for the recent Atlantic-THORPEX Regional Campaign (ATReC) from November 18th to December 15, 2003. In order to evaluate the retrieval ability through the thin cirrus clouds, the radiances are simulated with only one level cloud having an optical depth within a certain range. The retrievals are performed (with dependent EOF regression coefficients) over simulated radiances with instrument noise. As shown in Fig. 1, the different EOF statistical regression coefficients are used for analyses and specified as (1) clear conditions only (denoted as CLE), (2) clear and an equivalently clear isothermal conditions (denoted as MIX), (3) cloudy without cloud height grouping (denoted as CLD), and (4) cloudy with cloud height grouping (denoted as GRP). The retrieval results are compared with the retrieval over clear radiances with clear regression coefficient (as a reference, denoted as REF). The statistical analyses are performed over a set of near 6000 soundings. The GRP retrieval accuracy above the clouds is somewhat independent of the cloud optical depth; and, as expected, the MIX retrievals are better than the CLE retrievals. However, the retrieval accuracy under the clouds is greatly improved over the optically thinner clouds, and the accuracy of GRP retrievals for the optical depth of less than one is close to that of the reference (clear sounding retrieval). EOF regression enables thermodynamic properties to be inferred through thin cirrus clouds.

The cloud parameters, such as the cloud top pressure (P_c in mb), visible optical depth (τ vis.), particle effective diameter (D_e in μm), and cloud phase (i.e., ice or liquid), are used in the cloud radiative transfer calculations. For the winter hemispheric data used here, most of the cirrus clouds are in the form of ice particles, so the statistical analysis over the cloud phase is excluded for this case. The statistical results from the cloud parameter retrievals are listed in Table 1. Since the effective cloud feature present in the spectral radiances at the cloud top level, the cloud parameter retrieval accuracy is somewhat independent of the cloud optical thickness. EOF regression enables both the cloud height and the cloud microphysical properties of the highest-level cloud to be inferred.

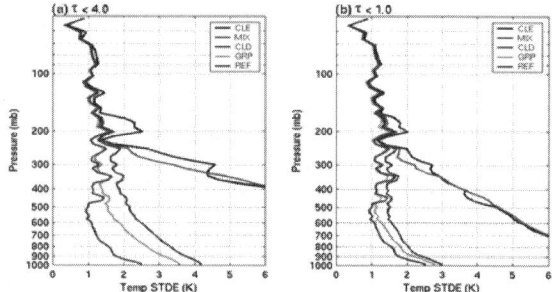

Fig. 1. Retrieval accuracy analyses through semi-transparent clouds: (a) for optical depth (visible) less than 4, and (b) for optical depth (visible) less than one.

Table 1. Cloud Parameters Retrieval Accuracy over Dependent Samples

	τ=0-1; D_e=25-58 μm		τ=0-4; D_e=25-90 μm	
	CLD	GRP	CLD	GRP
	Standard Deviation Error			
P_c (mb)	105	30	109	31
τ vis.	0.78	0.17	0.54	0.40
D_e (μm)	5.7	4.6	7.6	5.6
	Mean Deviation Error			
P_c (mb)	-27.0	-9.2	-24.8	-9.2
τ vis.	-0.03	-0.02	-0.09	-0.04
D_e (μm)	-0.4	-0.2	-0.5	-0.2

4. NAST-I RETRIEVALS AND INTER-COMPARISONS

NAST-I instrumentation, measurements, calibration, and radiance validation are documented elsewhere (e.g., Cousins and Smith, 1997; Zhou et al., 2002; Smith et al., 2005). NAST-I provides relatively high spectral resolution (0.25 cm^{-1}) measurements in the spectral region of 645–2700 cm^{-1}. Retrievals from the recent ATReC (Shapiro and Thorpe, 2004) are used to demonstrate this inversion methodology. These, together with the radiosonde and dropsonde released from the NOAA G-4 aircraft that flew below the NASA ER-2 aircraft provide a unique data set for detailed analysis of retrieval resolution and accuracy. During this field campaign, cloud properties were also provided by the nadir-pointing Cloud Physics Lidar (e.g., McGill et al., 2002) on board the NASA ER-2 aircraft. All coincident observations obtained during this experiment are used to understand the atmospheric state and cloud microphysical properties for validating NAST-I retrievals.

The experiment of December 5th 2003 is chosen to test and demonstrate this EOF regression with a realistic cloud radiative transfer model. As shown in Fig. 2a, the target scenes covered a variety of conditions desired by the experiment scientific objectives. These included a variety of cloud conditions, such as medium-level altocumulus under the Aqua satellite track (yellow dashed line) flown by the ER-2 and Citation, as well as low-level cumulus, thunderstorms, and extensive high cirrus in the ATReC region covered by the ER-2 and G-4. In addition, turbulence was reported by both the ER-2 and Citation. The ER-2 flight track is plotted over the GOES-8 satellite infrared image. NAST-I retrievals from a variety of cloudy conditions are performed by the methodology described above. Fig. 2b shows the image of NAST-I retrieved cloud top heights.

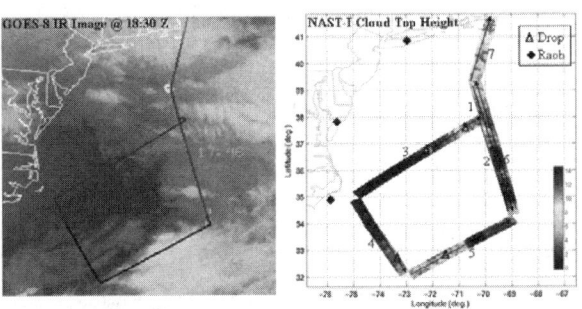

Fig. 2. GOES-8 infrared image shows a variety of cloud conditions in the region covered by the ER-2 and the G-4. The ER-2 flight track is plotted over the image to compare with the NAST-I horizontal distribution of retrieved cloud top heights. The arrow lines and numbers are the ER-2 flight legs; and the triangles are the dropsondes released from the G-4.

Fig. 3a plots NAST-I retrieved cloud top height from the nadir observations against CPL measured cloud top heights of the top 2 layers; and Fig. 3b shows the cloud optical depth inferred from NAST-I measurements against that of CPL 1064-nm channel measurements. It is noted that NAST-I horizontal resolution (at the cloud height) of a linear resolution (at nadir) is 13% of the distance between the aircraft altitude and the cloud height (i.e., 1.56 km when the cloud height is at 8 km and the ER-2 is at 20 km) while the CPL horizontal resolution is about 0.2 km; furthermore, the NAST-I vertical resolution is about 1 km while the CPL vertical resolution is 0.03 km. Despite the differences of the instruments and of their spatial resolutions, the cloud top heights inferred from NAST-I compare very well with CPL measurements for a variety of cloud conditions. The measurement sensitivity and accuracy of cloud optical depth inferred from the infrared measurement is expected to be much poorer than that measured by the CPL because of the nature of the instruments. In general, NAST-I data compare favorably to CPL data.

NAST-I retrieved temperature and relative humidity vertical cross sections are shown in Figs. 3c and 3d, respectively. The areas wiped off are under the clouds where the cloud optical depth is larger than one. The variation of atmospheric conditions is well captured by NAST-I retrievals indicated not only by the regions above optically thick clouds but extended regions below the optically thin clouds as well. These soundings are also validated by the dropsondes released from the G-4 aircraft. The dropsondes indicated by triangles in Fig. 1b are used to reveal the retrieval sounding accuracy under cloudy conditions. The cloud top height is obtained from NAST-I retrievals. Sounding comparisons show a good agreement above the clouds; the sounding comparison continues to show a good agreement under the (optically thin) cloud to the second layer cloud if it indicated by the CPL. As shown in Figs. 3 and 4, retrievals of temperature and moisture above the clouds are not disturbed by the clouds below, and retrievals are relative accurate under the optically thin clouds ($\tau<1.0$). The retrievals from cloud contaminated infrared radiances provide abundant atmospheric state information.

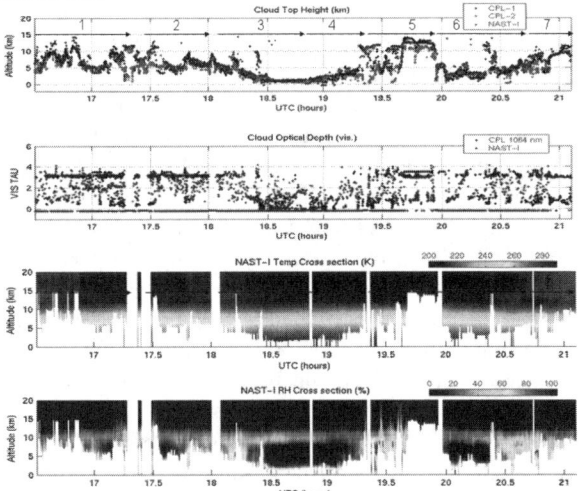

Fig. 3. Panel (a): NAST-I retrieved cloud top height compared with the CPL measured cloud top heights of top 2 layers. Panel (b) NAST-I retrieved cloud optical depth (effective visible) compared with the CPL measurement. Panels (c) and (d) plot NAST-I retrieved temperature and relative humidity vertical cross sections, respectively. The areas wiped off are under the top layer clouds where the cloud visible optical depth is larger than one and under the lower "opaque" cloud.

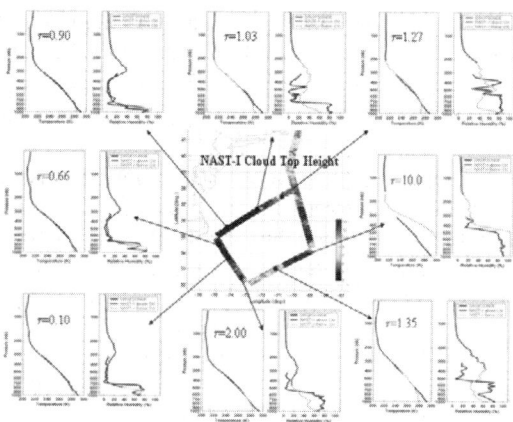

Fig. 4. Red profiles are NAST-I retrievals above the clouds while green profiles are under the cloud. Retrievals are compared with the co-located dropsounds (blue curves). The retrieval accuracy below the clouds depends on the cloud optical depth.

5. CONCLUSION AND FUTURE WORK

Clouds greatly complicate the interpretation of infrared sounding data. The new hyperspectral resolution infrared sounding systems alleviate much of the ambiguity between cloud, atmospheric temperature, and moisture contributions. However, in heavily clouded situations, the thermodynamic profile information to be retrieved is limited to the atmosphere above the clouds. The results of this study indicate some success in the ability to retrieve information below scattered and partially transparent cirrus clouds (i.e., clouds with effective optical depths of less than one). The thermodynamic profile information might be obtained by a combination of cloud clearing and by direct retrieval from the clouded radiances using a realistic cloud radiative transfer model. Results achieved with airborne NAST-I observations show that accuracies close to those achieved in totally cloud-free conditions can be achieved down to cloud top levels. The accuracy of the profile retrieved below cloud top level is dependent upon the optical depth and fractional coverage of the clouds. This EOF regression has laid an initial step dealing with infrared sounding data under cloudy conditions, which might be further improved by a physical iteration inversion. The correct implementation still requires a considerable research development effort. However, cloudy sky radiative transfer models now exist which should enable the extraction of profile information from cloud contaminated radiances suitable for numerical weather prediction (NWP) application. These cloudy observations for NWP analyses are under investigation.

6. ACKNOWLEDGMENT

The NAST-I program is supported by the NPOESS Integrated Program Office. The authors express sincere thanks to the NAST-I team members from various organizations. The authors also acknowledge support from NASA Langley Research Center. CPL data were kindly provided by Mr. Dennis Hlavka of NASA Goddard Space Flight Center.

7. REFERENCES

Cousins, D., and W. L. Smith, 1997: National Polar-Orbiting Operational Environmental Satellite System (NPOESS) Airborne Sounder Testbed-Interferometer (NAST-I), in *Proceedings, SPIE Application of Lidar to Current Atmospheric Topics II*, A. J. Sedlacek, and K. W. Fischer, Eds., 3127, 323–331.

Heymsfield, A. J., S. Matrosov, and B. Baum, 2003: Ice water path–optical depth relationships for cirrus and deep stratiform ice cloud layers, *J. of Appl. Meteorol.*, 42, 1369–1390.

Liu, X., J.-L. Moncet, D. K. Zhou, and W. L. Smith, 2003: A Fast and Accurate Forward Model for NAST-I Instrument, *Fourier Transform Spectroscopy and Optical Remote Sensing of Atmosphere, OSA Topical Meetings*, Quebec, Canada.

McGill, M. J., D. L. Hlavka, W. D. Hart, J. D. Spinhirne, V. S. Scott, and B. Schmid, 2002: The Cloud Physics Lidar: Instrument description and initial measurement results, *Applied Optics*, 41, 3725–3734.

Moncet, J. L., et al., 2003: Algorithm theoretical basis document (ATBD) for the Cross Track Infrared Sounder (CrIS) environmental data records (EDR), V1.2.3, *AER Document Number P882-TR-E-1.2.3-ATBD-03-01*.

Shapiro, M. A., and A. J. Thorpe, 2004: THORPEX: A global atmospheric research program for the beginning of the 21st century, *WMO Bulletin*, 53, 222–226.

Smith, W. L., and H. M. Woolf, 1976: The use of eigenvectors of statistical co-variance matrices for interpreting satellite sounding radiometer observations, *J. Atmos. Sci.*, 33, 1127–1140.

Smith, W. L., D. K. Zhou, A. M. Larar, S. A. Mango, H. B. Howell, R. O. Knuteson, H. E, Revercomb, and W. L. Smith Jr., 2004: The NPOESS Airborne Sounding Testbed Interferometer-Remotely Sensed Surface and Atmospheric Conditions during CLAMS, *J. Atmos. Sci.*, 61, in press.

Yang, P., B. C. Gao, B. A. Baum, Y. Hu, W. Wiscombe, S.-C. Tsay, D. M. Winker, and S. L. Nasiri, 2001: Radiative Properties of cirrus clouds in the infrared (8–13 um) spectral region, *J. Quant. Spectros. Radiat. Transfer*, 70, 473-504.

Zhou, D. K., W. L. Smith, J. Li, H. B. Howell, G. W. Cantwell, A. M. Larar, R. O. Knuteson, D. C. Tobin, H. E. Revercomb, and S. A. Mango, 2002: Thermodynamic product retrieval methodology for NAST-I and validation, *Applied Optics*, 41, 6957–6967.

FOURIER TRANSFORM SPECTRORADIOMETERS FOR THE CHARACTERIZATION OF THE COMPOSITION AND THE RADIATIVE PROPERTIES OF THE UPPER ATMOSPHERE

L. Palchetti, G. Bianchini, B. Carli, F. Castagnoli, U. Cortesi, M. Pellegrini, and M. Trambusti
Istituto di Fisica Applicata "Nello Carrara" – CNR
Via Panciatichi 64, 50127 Firenze, Italy

ABSTRACT

Fourier transform spectroscopy in the IR spectral region has been proven to be a valuable tool for the identification and quantification of the composition and the radiative properties of the atmosphere, with special attention to components involved in natural and anthropogenic processes.

Laboratory experiments, field campaigns and data interpretation have been carried out at IFAC-CNR to address questions related to global and regional environmental changes.

1. INTRODUCTION

In the study of the Earth's atmosphere, one of the most challenging research task is to improve our understanding of all the processes, which affects the Earth's radiation budget and therefore the climate. This requires a better knowledge of the atmospheric composition, chemistry and dynamics at regional and global scale. IR Fourier transform spectrometers (FTS) are a valuable tool for a direct measurement of the emitted radiance, spectrally resolved, which allows also a remote measurement, in the troposphere and lower stratosphere, of greenhouse gases concentrations, such as ozone, carbon dioxide, water vapour, etc (Sinha et al, 1995).

At IFAC-CNR, the study of atmospheric properties has been performed by using two FTS: an high resolution spectrometer, named SAFIRE-A (Spectroscopy of the atmosphere using far-infrared emission airborne) (Bianchini et al., 2004), and a low resolution spectrometer, named REFIR-PAD (Radiation explorer in the far infrared—prototype for applications and development) (Carli et al., 1999; Palchetti et al., 2004). Both instruments were built in house for operation on stratospheric platforms. SAFIRE-A is aimed at the determination of stratospheric constituents and operates from about 20 km flight altitude on-board the Russian aircraft M55—Geophysica. REFIR-PAD is aimed at the measurement of the outgoing Far IR radiation at the top of the atmosphere and the improvement of our knowledge of the principal drivers of this flux, e.g., temperature structure, water vapour, and clouds throughout the troposphere-surface system. It operates from about 35 km float altitude on-board a stratospheric balloon.

This paper describes the performances of these two instruments and gives an example of possible applications in which the two instruments could be simultaneously deployed for the characterization of the Earth's atmosphere and in particular for the chemical and radiative characterization of the upper troposphere/lower stratosphere (UTLS).

2. SAFIRE-A

The SAFIRE-A spectrometer provides an unique combination of different measuring schemes and parameters, including limb, nadir and space direction line of sight, variable spectral resolution up to 0.004 cm^{-1}, two spectral channels in the region from 10 to 250 cm^{-1} with bandwidth from 1 to several cm^{-1}, and the possibility of operating both atmospheric radiance and polarization measurements. Fig. 1 shows the instrument on-board the stratospheric aircraft M55-Geophysica.

Fig. 1. SAFIRE-A on-board the Russian stratospheric aircraft M55-Geophysica.

Currently, a filter set aimed at the study of ozone chemistry is used. The measured species, filter characteristics and corresponding noise equivalent spectral radiance (NESR) are shown in Table 1. The detection limit for each of the measured chemical specie is shown in Table 2.

In the high spectral resolution radiance operating mode, SAFIRE-A has been deployed in several campaigns, amongst which a mid-latitude and a polar campaign aimed to the validation of the Michelson interferometer for passive atmospheric sounding (MIPAS) aboard the environmental satellite (ENVISAT) data and the airborne polar experiment—Geophysica aircraft in Antarctica (APE-GAIA Antarctic) campaign with the objective of the study of the southern polar vortex region.

Table 1. Measurement Specifications of SAFIRE-A

Waven. (cm^{-1})	Bandw. (cm^{-1})	Chemical species	NESR (nW/cm^2 sr cm^{-1})
23	1	O_3, ClO, N_2O, HNO_3	0.22
118	2	H_2O, OH, HOCl, O_3	10
125	2	H_2O, HCl, O_3	13

Table 2. Specie Detection Limits

Chemical species	Detection limit
Ozone	0.1 ppmv
Nitrous oxide	5.0 ppbv
Nitric acid	0.5 ppbv
Chlorine monoxide	0.1 ppbv
Water vapour	1.0 ppmv
Hydrochloric acid	0.5 ppbv
Hydroxyl radical	0.5 ppbv

In these campaigns the instrument retrieved 2-D maps of volume mixing ratio (VMR) of trace atmospheric constituents with higher spatial resolution compared to spaceborne instruments, thus providing a valuable information for validation and for study of local phenomena involved in atmospheric chemistry. Table 3 summarizes information on the performed field campaigns and Fig. 2 shows an example of the measurement capability of this instrument.

Table 3. Field Campaigns Performed with SAFIRE-A

Field campaign	Location	Period
APE-GAIA Antarctic	Ushuaia (Argentina)	19 Sept.-12 Oct. 1999
ENVISAT Mid-Lat Validation	Forlì (Italy)	13-22 Jul. 2002 08-28 Oct. 2002
ENVISAT Arctic Validation	Kiruna (Sweden)	28 Feb. - 18 Mar. 2003

The comparison with in-situ measurements simultaneously acquired aboard the M-55 has allowed to verify the degree of quality obtained in the retrieved profiles, see Fig. 3 and, consequently, to use SAFIRE-A data for validation purposes. A substantially good agreement is found in most of the comparisons with level 2 data of MIPAS-ENVISAT, showing an overall consistency between the airborne limb measurements and the satellite data. An example of comparison is given in Figs. 4 and 5 for O_3 and NHO_3, respectively.

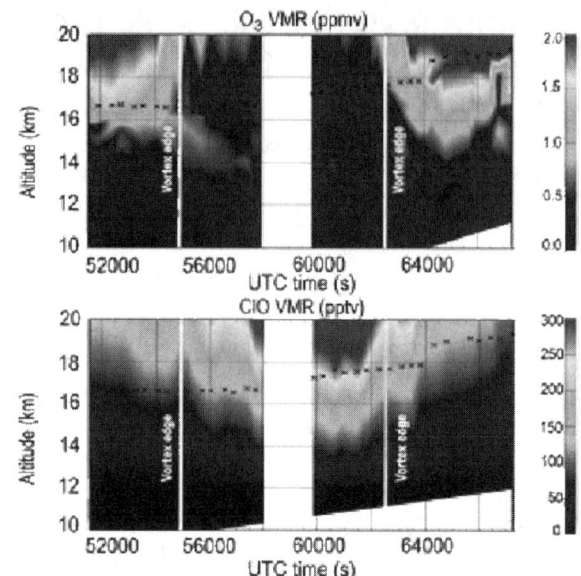

Fig. 2. 2-D VMR distributions of Ozone and Chlorine Monoxide along the flight, as measured by SAFIRE-A during the APE-GAIA flight on 23.09.1999.

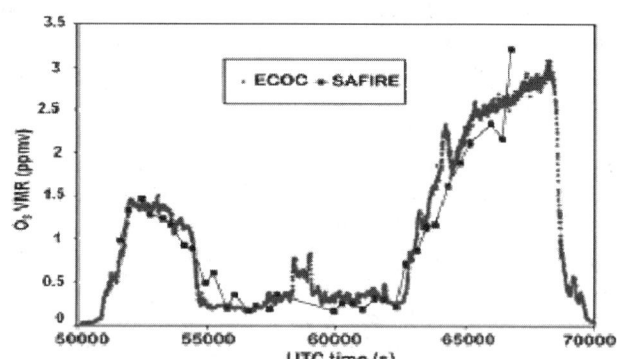

Fig. 3. Comparison between O_3 VMR values at flight altitude measured by SAFIRE-A(---) and by the in-situ sensor ECOC (ElectroChemical zone Cell) (-•-).

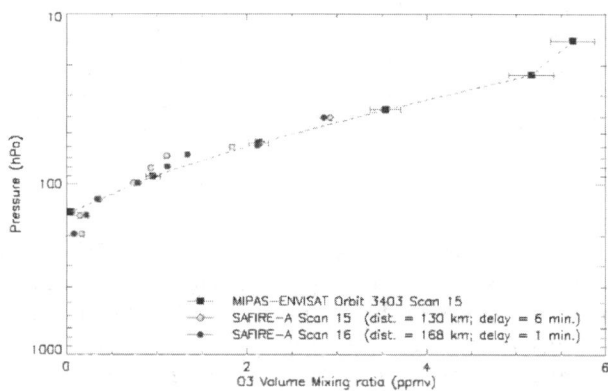

Fig. 4. ENVISAT validation flight on 02.03.2003: comparison of O_3 VMR vertical profiles retrieved by SAFIRE-A and MIPAS.

GENERAL REMOTE SENSING

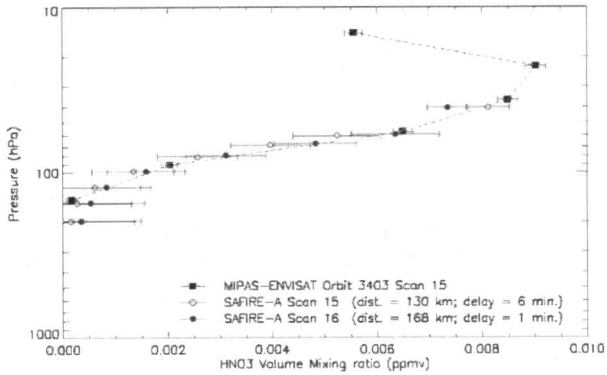

Fig. 5. ENVISAT validation flight on 02.03.2003: comparison of HNO_3 VMR vertical profiles retrieved by SAFIRE-A and MIPAS.

In the low resolution polarization mode, SAFIRE-A can perform the measurement of the degree of polarization of observed radiation calculated as the ratio between the difference of vertical (V) and horizontal (H) components, and the horizontal component itself. Fig. 6 shows an example for the polar stratospheric clouds (PSC) observed during the AGE-GAIA campaign.

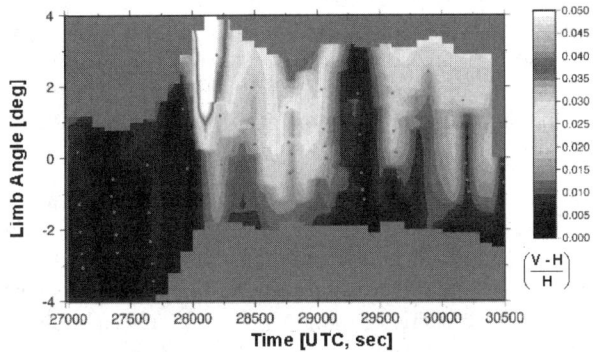

Fig. 6. Relative difference between vertical and horizontal polarization components of the atmospheric signal measured by SAFIRE-A at different limb angles during the 02.10.1999 flight.

3. REFIR-PAD

REFIR-PAD operates in the spectral range of 100-1100 cm^{-1} with a resolution of 0.5 cm^{-1}. It is a compact prototype designed both for laboratory applications and for field campaigns, in particular for operations on-board stratospheric balloon platforms.

Laboratory tests, performed in air and under vacuum conditions, have allowed to study the trade-off among all the instrument parameters and to test the new optical design of the interferometer with particular attention to the photolithographic beam splitters and the room-temperature pyroelectric detectors. Table 4 summarizes the instrument specification and Fig. 7 the instrument performance in term of NESR measured under vacuum conditions.

Spectroscopic performances were inferred by measuring the transmittance of air inside the interferometer optical path. It can be evaluated by measuring the signal with the instrument into air after the baseline measurement is taken under vacuum. The transmittance value is given by the ratio between the spectra acquired in the two measured conditions.

Table 4. REFIR-PAD Specifications

Parameter	Value
Spectral resolution	0.25 cm^{-1} (double-sided interferogram) 0.1 cm^{-1} (single-sided interferogram)
Optical throughput	0.0097 cm^2 sr
Line of sight	Zenith and 0° to +30° elevation angle
Beam splitters	4 wire-grid photolithographic polarisers or 2 multilayer films
Detector type	2 room-temperature DLATGS
Ref. interferometer	Single-mode laser diode, λ=780 nm
Dimensions	62 cm x 50 cm x 26 cm
Power supply	50 W
Mass	60 kg

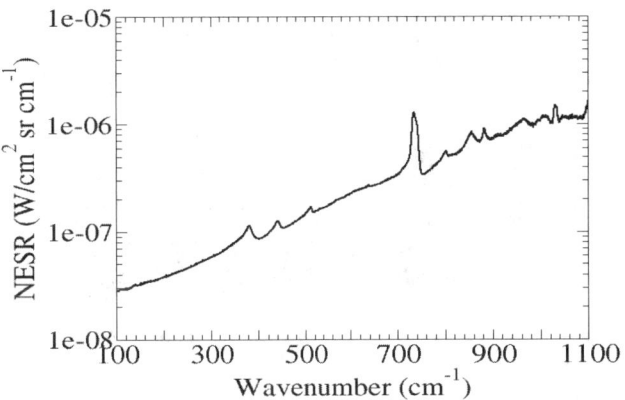

Fig. 7. REFIR-PAD NESR measured under vacuum conditions.

Fig. 8 shows the results in the water vapour rotational band and in the CO_2 vibration-rotational band. The results are averaged on 20 measurements taken under vacuum and 20 measurements taken in air in the following conditions: 1009 hPa pressure, 21 °C temperature and about 39% RH. The figure also shows the comparison with a simulated spectrum based on HITRAN'96 database. The comparison confirms the good accordance with the model apart some discrepancies that are to be related to the residual calibration effect due to phase errors, which cannot be compensated in this kind of measurements.

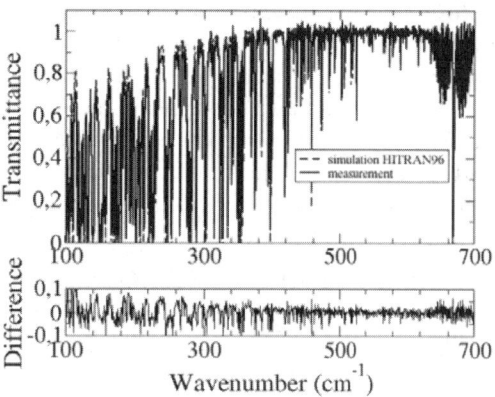

Fig. 8. Comparison with REFIR-PAD measurements and simulation (HITRAN'96).

Recently, REFIR-PAD was integrated on-board the stratospheric gondola, which hosts the infrared atmospheric sounding interferometer (IASI balloon) of the Laboratoire de Physique Moléculaire et Applications (LPMA, Paris). The first launch opportunity is planned for June 2005 from Teresina (Northern Brazil). A picture of the instrument on-board the gondola is shown in Fig. 9.

Fig. 9. REFIR-PAD on-board the IASI-LPMA stratospheric gondola.

4. FUTURE APPLICATIONS

As a future development, lower spectral resolution (0.1 cm^{-1}) and broadband measurement capabilities of a nadir looking configuration of SAFIRE-A could be exploited to retrieve information on structure and properties of high cirrus clouds and to make a contribution to the study of their role in both Earth's radiation budget and upper tropospheric chemistry. In this framework, an interesting possibility is the synergy that is provided by an instrumentation package including also REFIR-PAD for the nadir characterization of the emitted radiation with wide spectral range measurements.

The two instruments, operated simultaneously, will allow to sounding one of the less known region of the atmosphere, e.g., the UTLS, which is roughly a layer ±5 km above and below the tropopause. This region is a layer where the greenhouse gases are more effective for entrapping radiation and the feedback processes act with more intensity. In spite of the importance of UTLS, its observation has been limited and inaccurate, because it requires remote measurements from high-altitude platforms or from space with high vertical and horizontal resolution. These kind of measurements are difficult to be performed with observing techniques, which at nadir have a lower vertical resolution and at limb have a lower horizontal resolution. By combining two instruments such as SAFIRE-A and REFIR-PAD on-board the same high-altitude platform, it will be possible to collect measurements by looking simultaneously both at nadir and at limb, and therefore improving resolution in the characterization of the UTLS region.

5. ACKNOWLEDGMENTS

Financial support for the SAFIRE-A experiment was provided by the Italian Space Agency (ASI), which funded the instrument optimization and the field operation through the contracts for the APE-ENVISAT project, and by the Italian National Program for Antarctic Research, which financed the instrument participation to the APE-GAIA campaign. Financial support for the REFIR project was provided for the feasibility study by European Union. The development of the prototype was partially supported by ASI and the European Space Agency.

The authors wish to thank Prof. P. A. R. Ade and Dr. C. Lee (University of Wales, Cardiff, UK) for the polarizer beam splitters currently used in both prototypes. The authors also gratefully acknowledge the technical support of Mrs. M. G. Baldecchi and Mr. G. Valmori (IFAC-CNR, Firenze, Italy).

6. REFERENCES

Bianchini, G., U. Cortesi, L. Palchetti, and E. Pascale, 2004, SAFIRE-A: optimised instrument configuration and new assessment of spectroscopic performances, *Applied Optics*, 43, 2962-2977.

Carli, B., A. Barbis, J. E. Harries, and L. Palchetti, 1999, Design of an efficient broad band far infrared FT spectrometer, *Applied Optics*, 38, 3945-3950.

Palchetti, L., G. Bianchini, M. Pellegrini, F. Esposito, R. Restieri, and G. Pavese, 2004, Radiometric performances of the Fourier transform spectrometer for the Radiation Explorer in the Far-InfraRed (REFIR) space mission, in Sensors, Systems, and Next-Generation Satellites VIII, *Proc. SPIE*, 5570, 433-444.

Sinha, A., and J. E. Harries, 1995, Water vapor greenhouse trapping: The role of the far infrared absorption, *Geophys. Res. Lett.*, 22, 2147-2150.

RETRIEVAL OF MICROPHYSICAL PROPERTIES OF MIXED-PHASE CLOUDS USING ACTIVE AND PASSIVE SENSORS: A FEASIBILITY STUDY FOR TRIPLE-BAND CLOUD RADARS AT 9.4 GHZ, 35 GHZ, AND 95 GHZ

Yukio Yoshida
National Institute for Environmental Studies, Center for Global Environmental Research
Tsukuba, Ibaraki, 305-8506, Japan

Shoji Asano
Center for Atmospheric and Oceanic Studies, Tohoku University
Sendai, 980-8578, Japan

1. INTRODUCTION

Low-level boundary-layer clouds are important components of the Earth's climate system because of their effects on radiation budget. Many studies on the cloud-radiation interaction often assumed that low-level clouds are mostly in liquid phase. In reality, however, even the low-level clouds are often partially glaciated or of mixed-phase in the mid-latitude and high-latitude regions in cold season (e.g., Riley, 1998). In order to estimate their effect on the radiation budget, we have to know their microphysical properties such as liquid-water-content (LWC), ice-water-content (IWC), and characteristic sizes and shapes of cloud particles. The vertical profiles of these cloud microphysical properties and the cloud-boundary altitudes are also important parameters for estimating the radiative heating rate profile and cloud radiative forcing, which greatly affect the Earth's climate system. However, there have been only a few simultaneous observations of the radiative and microphysical properties of mixed-phase clouds.

In this study, we develop a new algorithm to retrieve the vertical profiles of microphysical parameters of mixed-phase clouds from the ground-based observation by using a "triple-band radar system", which consists of three radars of X-band (wavelength 31.98 mm; frequency 9.4 GHz), Ka-band (8.48 mm; 35 GHz), and W-band (3.15 mm; 95 GHz), together with a microwave-radiometer. For simplicity, the target cloud is restricted to non-precipitating clouds, but we consider the cloud in mixed-phase condition, for which water-droplets and ice-particles are coexisting in the same cloud volume.

2. THEORETICAL BACKGROUND

2.1 Radar Equation

The radar equation for a cloud sub-layer with a finite thickness can be expressed as follows,

$$P_\lambda(R_i) = \int_{R_{b,i}}^{R_{t,i}} \frac{C_\lambda}{R^2} Z_{e,\lambda}(R) \exp\left[-2\int_0^R \sigma_\lambda(R')dR'\right]dR, \quad (1)$$

where P is the received radar power, R is the range, and R_i denotes the altitude of the center of the i-th sub-layer. The sub-script λ indicates wavelength dependence. $R_{t,i}$ and $R_{b,i}$ are the top and bottom boundaries of the i-th sub-layer, respectively. The equivalent radar reflectivity factor Z_e, in units of [mm^6 m^{-3}], and the extinction coefficient σ, in units of [m^{-1}], are defined, respectively, as follows,

$$Z_{e,\lambda}(R) = \frac{\lambda^4}{\pi^5 |K_{ref,\lambda}|^2} \int N(R;r) C_{bk,\lambda}(R;r)dr, \quad (2)$$

$$\sigma_\lambda(R) = \int N(R;r) C_{ext,\lambda}(R;r)dr + \beta_{gas,\lambda}(R), \quad (3)$$

where C_{bk} and C_{ext} are the back-scattering and extinction cross section as a function of the cloud particle radius r, respectively, β_{gas} denotes gaseous absorption coefficient, $N(r)$ denotes the size distribution of cloud particles. To simplify Eq. (1), we expand the exponential term into a polynomial of σ, and neglect terms of the order of σ^2 and higher. The approximation may be reasonable for clouds measured at the considering radar wavelengths, for which the value of σ is small enough. Under the assumption of homogeneous cloud microphysical properties in the target sub-layer, Eq. (1) is expressed as follows,

$$P_\lambda(R_i) = \frac{C_\lambda}{R_i^2} Z_{obs,\lambda}(R_i) \exp\left[-2\int_0^{R_{b,i}} \sigma_\lambda(R')dR'\right], \quad (4)$$

$$Z_{obs,\lambda}(R_i) = Z_{e,\lambda}(R_i)[c_{0,i} + c_{1,i}\sigma_\lambda(R_i)], \quad (5)$$

where $c_{0,i}$ and $c_{1,i}$ are the function of $R_{t,i}$ and $R_{b,i}$, and they relate the expansion coefficients of the exponential term.

2.2 Scattering Properties of Ice-Particles

The single scattering properties, such as extinction cross section and back-scattering cross section, of water-droplets and ice-particles are calculated by Mie theory and the discrete dipole approximation (DDA; Draine and Flatau, 2003), respectively, under the assumption of 3D-random orientation. Based on in-situ observations, the width D of an ice-particle may be related to the length L, and several dimensional relationships have been proposed in previous studies (e.g., Auer and Veal, 1970). According to these relationships, the aspect ratio of ice-particle D/L deviates from unity with the increase of ice-particle maximum dimension. Since the triple-band radars have less sensitivity to smaller particles with maximum dimension less than about 200 μm (e.g., Hogan et al., 2000), we consider hexagonal ice-particles with several aspect ratios of D/L = 1/3, 1/4, 1/5, 3/1, 4/1, and 5/1; the ratios correspond to those of ice-particles with maximum dimension larger than 200 μm.

We assume that the particle size distributions of mixed-phase clouds can be expressed by the following gamma distribution functions as,

$$N(r) = \sum_{phase} N_{0,phase} r^{\mu_{phase}} \exp\left[-\frac{\mu_{phase}+3}{r_{eff,phase}} r\right] \quad (6)$$

where μ is the dispersion parameter of the size distribution, r is the cloud particle radius, r_{eff} is the effective radius. The summation over $phase$ indicates that the size distribution of mixed-phase clouds is expressed as the sum of those for water-droplets and ice-particles (suffixed by wat and ice, respectively, hereafter). According to the $in\text{-}situ$ measurements of cloud microphysical properties, the values of μ are reported to be about 2 for water clouds (Pruppacher and Klett, 1997) and between 0 to 2 for ice clouds (Kosarev and Mazin, 1991). We assume a fixed value of $\mu_{wat} = 2$ for water-droplets and $\mu_{ice} = 1$ for ice-particles, respectively. As shown in Eqs. (2) and (3), Z_e and σ are expressed by the sum of the back-scattering and extinction cross sections over the size distribution, respectively. As a result, the attenuated radar reflectivity factor Z_{obs} is a function of LWC, $r_{eff,wat}$, IWC, $r_{eff,ice}$, and D/L.

2.3 Sensitivity of Cloud Microphysical Parameters to Radar Reflectivity Factors

We first consider the sensitivity of water-droplets to radar reflectivity factor. Because we neglect precipitation as rain and drizzle, cloud droplets are small enough to be assumed as Rayleigh-scatterer at the radar wavelengths. As a result, Z_e depends on $r_{eff,wat}$ and LWC, and it is independent of the radar wavelengths. On the other hand, σ is independent of $r_{eff,wat}$, and it strongly depends on the radar wavelengths due to the difference of water absorptivity at the wavelengths. Fig. 1 shows the dual-wavelength ratio (DWR) of water-droplets as a function of $r_{eff,wat}$ and LWC. The ratio $DWR_{\lambda_1/\lambda_2}$ is defined as follows,

$$DWR_{\lambda_1/\lambda_2} = 10 \log\left(Z_{obs,\lambda_1}/Z_{obs,\lambda_2}\right), \quad (7)$$

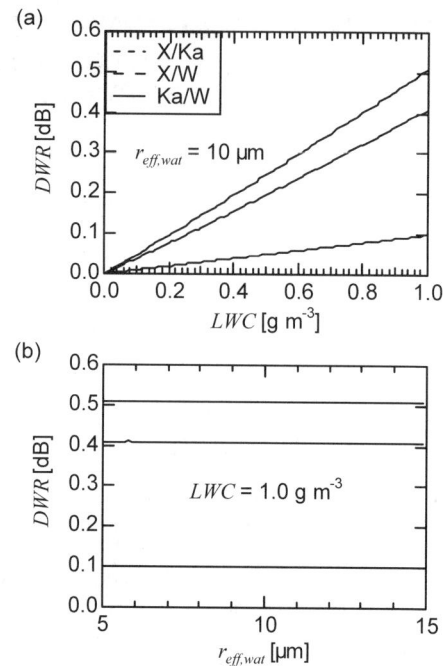

Fig. 1. The relationship of DWR for the different radar-band pairs with (a) LWC and (b) $r_{eff,wat}$ for water-droplets.

where the suffix λ_1/λ_2 represents a pair of the dual-wavelengths λ_1 and λ_2, for which $\lambda_1 > \lambda_2$. Because water-droplets can be treated as Rayleigh-scatterer, $DWR_{\lambda_1/\lambda_2}$ for water-droplets are related to the different degree of attenuation between the two wavelengths. Since the attenuation of radar wave by water-droplets is proportional to LWC, $DWR_{\lambda_1/\lambda_2}$ for water-droplets is also proportional to LWC.

Next, we consider the sensitivity of ice-particles to radar reflectivity factor. Unlike water-droplets, large ice-particles are non-Rayleigh-scatterer, and Z_e of ice clouds depends on $r_{eff,ice}$, IWC, D/L, and radar wavelength. σ also shows the size dependency caused by ice-particles scattering effects. Fig. 2 shows the $DWR_{\lambda_1/\lambda_2}$ of ice-particles with $D/L = 1/3$ as a function of $r_{eff,ice}$ and IWC. The results without radar wave attenuation are also shown in Fig. 2(b) (gray line). The attenuation effect becomes large as $r_{eff,ice}$ increased, and neglect of this effect brings $r_{eff,ice}$ overestimation. $DWR_{\lambda_1/\lambda_2}$ for ice-particles is related to the different degree of scattering rather than absorption between two wavelengths, because ice-particles hardly absorb the radar waves. The magnitude of $DWR_{\lambda_1/\lambda_2}$ for ice-particles depends on $r_{eff,ice}$ and D/L through the scattering properties.

If we consider a mixed-phase condition, Z_e and σ can be expressed, respectively, as the sums of the corresponding Z_e and σ of water-droplets and ice-particles. For mixed-phase clouds, Z_e is controlled by that of ice-particles, because ice-particles are relatively larger than water-droplets. On the other hand, σ is influenced by both water-droplets and ice-particles. Thus, the attenuated radar reflectivity factor Z_{obs} for mixed-phase clouds has no sensitivity to $r_{eff,wat}$, which affects $Z_{e,wat}$ but not σ_{wat}. Also, DWRs for mixed-phase clouds depend on both the attenuation by water-droplets and the scattering by ice-particles, i.e., LWC, $r_{eff,ice}$ and D/L.

We prepare look-up-tables (LUT) calculated for Z_e and σ of water-droplets and ice-particles with different values of D/L, and with $LWC = IWC = 1.0$ g m^{-3}, as a function of $r_{eff,wat}$ and $r_{eff,ice}$, respectively. The LUTs are used in the following retrieval procedure.

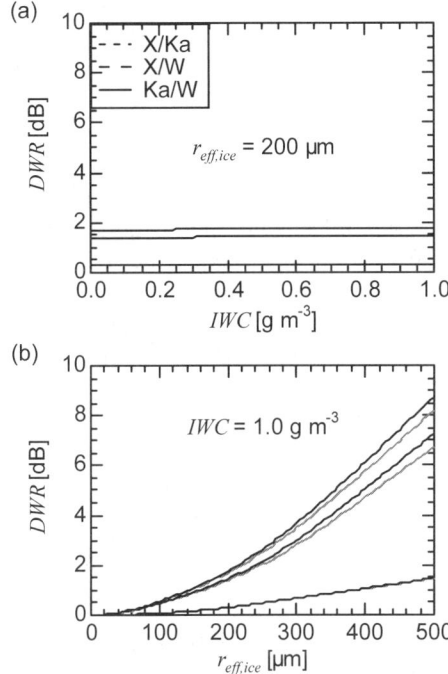

Fig. 2. The relationships of DWR for the different radar-band pairs with (a) IWC and (b) $r_{eff,ice}$ for ice-particles with D/L = 1/3. In Fig. 2(b), the results without radar wave attenuation are also shown as gray line.

3. DESCRIPTION OF THE ALGORITHM

As mentioned before, we use the microwave-radiometer as well as the triple-band radar for retrieving the vertical distribution of mixed-phase cloud microphysical properties. The microwave-radiometer can detect column-integrated LWC or liquid-water-path (LWP) only when the cloud contains water-droplets. However, we cannot a priori know the mixing ratio of water-droplets and ice-particles for each cloud sub-layers. On the other hand, the observed radar reflectivity factor Z_{obs} of mixed-phase cloud sub-layers depend on LWC, IWC, $r_{eff,ice}$, and D/L. To retrieve these four unknown parameters from the triple-band radars,

we have to make some assumption in the algorithm. Here, we try to estimate the vertical profiles of LWC, IWC, and $r_{eff,ice}$ which satisfies the observed radar reflectivity factors and the microwave brightness temperatures by assuming a value of D/L in each sub-layer. Without the radar wave attenuation due to scattering by ice-particles, $DWR_{\lambda 1/\lambda 2}$ for mixed-phase clouds can be written as follows,

$$DWR_{\lambda_1/\lambda_2} \approx 10 \log \left(\frac{Z_{e,ice,\lambda_1}(R_i)\{c_{0,i} + c_{1,i}\sigma_{\lambda_1,wat}(R_i)\}}{Z_{e,ice,\lambda_2}(R_i)\{c_{0,i} + c_{1,i}\sigma_{\lambda_2,wat}(R_i)\}} \right)$$
$$= DWR_{ice,\lambda_1/\lambda_2} + DWR_{wat,\lambda_1/\lambda_2} \qquad (8)$$
$$= DWR_{ice,\lambda_1/\lambda_2}(r_{eff,ice}; D/L) + A_{\lambda_1/\lambda_2} LWC,$$

where $A_{\lambda 1/\lambda 2}$ indicates the proportional constant of $DWR_{\lambda 1/\lambda 2}$ to LWC for water clouds (Fig. 1a). Firstly, we fix a value of D/L (D/L = 1/3 as an initial value) for the sub-layer, and estimate $r_{eff,ice}$ and LWC from $DWR_{X/Ka}$ and $DWR_{X/W}$ as initial values. Next, $r_{eff,ice}$ and LWC are determined from DWRs by an iteration procedure with considering attenuation by scattering of ice-particles. Finally, we estimate IWC from $Z_{obs,X}$. Then, we estimate the cloud microphysical parameters of the next cloud sub-layer in the same manner, after the correction for attenuation of the radar signals in sub-layers under the target sub-layer. After retrieving the cloud microphysical parameters throughout the cloud layer, we compare the microwave brightness temperature calculated from the retrieved microphysical parameters with the observed one. We repeat the retrieving procedure by changing the assumed value of D/L (in order of 1/4, 1/5, 3/1, 4/1, 5/1) until the difference between calculated and measured brightness temperatures should be less than the observed error of the microwave-radiometer.

4. SENSITIVITY STUDY OF THE ALGORITHM

We examine performance of the newly developed algorithm through sensitivity simulations for assumed clouds. The assumed microphysical profiles are vertically variable from a sub-layer to next sub-layer, but homogeneous in each sub-layer. For simplicity, gaseous absorption is neglected in the sensitivity study.

We first investigate retrieval errors resulting from biases in the radar signals. We assume that the triple-band radars are properly calibrated each other, and the absolute calibration error is supposed to be, at most, ±1 dB in radar signals. Further, the same amounts of errors are assigned to all of the radar signals. This brings no error in $DWR_{\lambda 1/\lambda 2}$. Fig. 3 shows an example of the retrieved microphysical profiles for the assumed mixed-phase cloud. Here, we set $\mu_{wat} = 2$, $\mu_{ice} = 1$, and D/L = 1/3 for the assumed (true) cloud. In the retrieval procedure, we added the biases of ±1 dB to the radar signals calculated from the assumed profiles. The biases do not disturb the retrieval of $r_{eff,ice}$, because $r_{eff,ice}$ is estimated from DWRs with no error, while the retrieved profiles of IWC are shifted from the assumed profiles due to the biases. Although LWC is estimated from DWR, the retrieved LWC profiles show large fluctuations. Since the back-scattering effect by ice-particles on the radar signals is larger than the attenuation effect due to water-droplets, small errors in IWC and $r_{eff,ice}$ can bring large errors in LWC. The retrieval errors for LWC, IWC, and $r_{eff,ice}$ due to the biases are estimated to ±10%, ±20%, and ±0.7%, respectively.

We also examined the retrieval errors caused from such assumed parameters as μ_{wat}, μ_{ice}, and D/L, for various assumed microphysical profiles (figures not shown). As a consequence, the total errors in the retrievals of LWC, IWC, and $r_{eff,ice}$ are estimated to be about ±27%, ±27%, and ±10%, respectively. While, the retrieval errors in LWP and IWP are estimated about ±10% and ±23%, respectively.

5. SUMMARY

We have developed the algorithm to retrieve microphysical profiles of mixed-phase clouds from the surface observation by using the triple-band radar system and the microwave-radiometer. A set of X-band (9.4 GHz), Ka-band (35 GHz), and W-band (95 GHz) radars is a suitable selection for observing cloud vertical structures. To precisely treat the scattering of radar waves, we considered non-sphericity effect and radar wave attenuation by non-spherical ice-particles, which were sometimes neglected in the previous studies. The algorithm can retrieve cloud microphysical profiles with the iteration procedures by changing the ice-particle shape parameter D/L in each sub-layer, until the retrieved profiles satisfy the observed radar signals and microwave brightness temperatures. From the sensitivity study for the assumed cloud microphysical profiles, the uncertainty of the present algorithm is estimated to be about ±27%, ±27%, and ±10% for LWC, IWC, and $r_{eff,ice}$, respectively. At present, unfortunately, no triple-band radar observation and in-situ microphysical observation are available to validate the present algorithm for mixed-phase clouds.

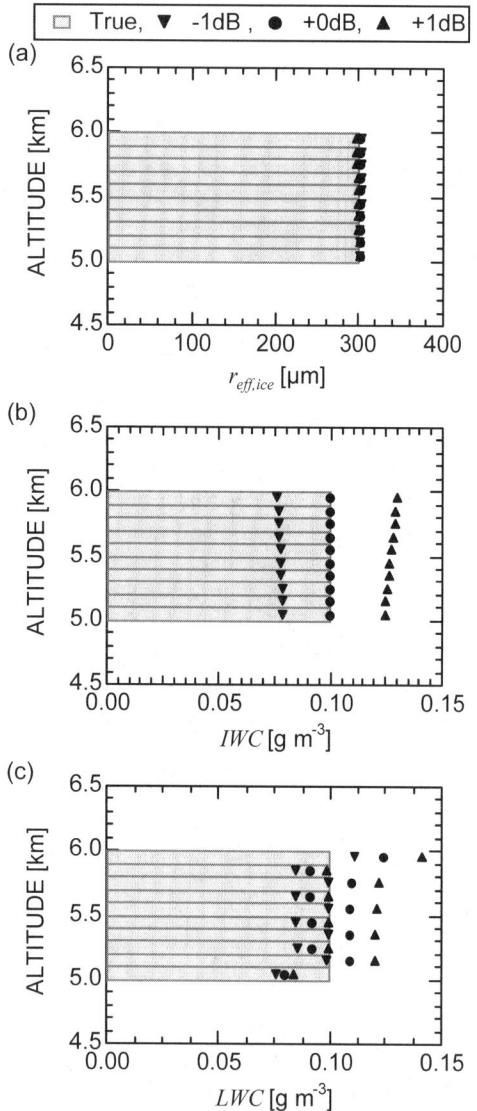

Fig. 3. Example of the retrieved microphysical profiles of (a) $r_{eff,ice}$, (b) IWC, and (c) LWC for the assumed mixed-phase cloud model. The shaded bars indicate the true (assumed) profile, and the markers indicate the profiles retrieved with and without biases of ±1 dB in the radar signals of Z_{obs}, respectively. In the retrieval procedure, $r_{eff,wat}$ is set to a fixed value of 10 μm.

6. REFERENCES

Auer, A. H., Jr., and D. L. Veal, 1970: The dimension of ice crystals in natural clouds. *J. Atmos. Sci.*, 27, 919-926.

Draine, B. T., and P. J. Flatau, 2003: User guide for the discrete dipole approximation code DDSCAT.6.0. http://arxiv.org/abs/astro-ph/0309069.

Hogan, R. J., A. J. Illingworth, and H. Sauvageot, 2000: Measuring crystal size in cirrus using 35- and 94-GHz radars. *J. Atmos. Ocean. Technol.*, 17, 27-37.

Kosarev, A. L., and I. P. Mazin, 1991: An empirical model of the physical structure of upper-layer clouds. *Atmos. Res.*, 26, 213-228.

Pruppacher, H. R., and J. D. Klett, 1997: Microphysics of clouds and precipitation. *Kluwer Academic Publishers*, 954 pp.

Riley, J. T., 1998: Mixed-phase icing conditions: A review. *U.S. Dept. of Transportation Rep.*, DOT/FAA/AR-98/76, 45 pp.

SATELLITE-DERIVED SURFACE REFLECTANCE USING THE IMAGE-BASED ATMOSPHERIC CORRECTION

Kwangjae Lee and Yongseung Kim

Korea Aerospace Research Institute, 45 Eoeun-dong Yuseong-gu, Daejeon 305-333, Korea
kjlee@kari.re.kr, yskim@kari.re.kr

ABSTRACT

The purpose of this study is to examine the image-based atmospheric correction method and to expand its application for the data from the high resolution satellite sensor, such as the forthcoming KOrea Multi-Purpose SATellite-2 (KOMPSAT-2) Multi-Spectral Camera (MSC). We have tested the several atmospheric correction models using the Korean image data of Landsat 7 and IKONOS since the spectral bands of MSC are quite similar to those of two instruments. Test results were compared to the *in-situ* measurements of surface reflectance. In the case of Landsat ETM+, substantial differences are present between Top-Of-the Atmosphere (TOA) reflectance and *in-situ* measurements, the results showed that Case 1 based on COST model gives most accurate results among three cases. The accuracy of Case 2_1 is very close to Case 1 and its values are smaller than *in-situ* data. No notable features appear between some bands in the Case 3_1 and *in-situ* data. Also Case 1 and Case 2_1 generated acceptable results with IKONOS image. It is expected from this study that we will be able to develop the suitable atmospheric correction methods for KOMPSAT-2 MSC data.

1. INTRODUCTION

Digital analysis of remote sensing data has become an important component of a wide variety of earth science studies. Also remote sensing data have been used for various earth science applications, such as geology, mapping, atmosphere, ocean, environment, etc. Moreover, a major benefit of multi-temporal remote sensing data is its applicability to change detection (Robinove et al., 1981; Jensen and Toll, 1982; Fung, 1990; Chavez and MacKinnon, 1994). However, in order to fully realize the potential of spectral data for such application, it is necessary to convert satellite-recorded digital counts to values independent of atmospheric conditions, that is, values of surface reflectance. Typical ways to correct for atmospheric effects is to use *in-situ* atmospheric measurements and radiative transfer code (RTC) (Chavez, 1996). However, the problem of this type of correction is that it requires *in-situ* field measurements during satellite over-flight. Ideally, a model that utilizes ground truth data is the most accurate in terms of correction for atmospheric effects. However, in the most case, data users have to utilize the remote sensing data that have already been collected and archived. In this case, the simple Dark Object Subtraction (DOS) technique may be useful because it requires only information contained in the image data. Most DOS techniques assume that there is a high probability that there are at least a few pixels within an image which should be black (0% reflectance), thus the digital image system should not detect any radiance at dark shadow areas, and a DN value of zero should be assigned to them. However, these shadow areas will not be completely dark, because of atmospheric effects. For that reason, minimum DN values are subtracted in all bands so that zero values appear in the data, the effects of atmospheric scattering will be somewhat minimized. This simple DOS technique has been utilized for the past few decades (Vincent, 1972; Rowan et al., 1974; etc). An improved DOS model proposed by Chavez (1988) is based on the assumption, that a completely black or zero reflectance surfaces usually does not exist and a minimum reflectance value of 1 or 2 percent is more realistic, was utilized for many applications. However, when this model was evaluated with low-altitude, aircraft-based measurements for seven dates over a 1 year period by Moran et al. (1992), it produced greater error in estimates of NIR reflectance than no correction at all. After her announcement, Chavez (1996) presented an entirely images-based procedure that expanded on the DOS model (COST model) by including a simple multiplicative correction for transmittance using published and unpublished data from Moran et al. (1992). In the results, the COST model which is atmospheric transmittance along the path from the sun to the ground surface generated results as good as the Herman-Browning Code (HBC) model.

Table 1. Characteristics of Landsat ETM+, IKONOS, and KOMPSAT-2 MSC System

Satellites \ Items	Landsat ETM+	IKONOS	KOMPSAT-2 MSC
Spatial resolution	Pan : 15m MS : 30m (band 1, 2, 3, 4, 5, 7) MS : 60m (band 6)	Pan : 1m MS : 4m	Pan : 1m MS : 4m
Spectral bands	Pan : 450-900 nm MS1 : 450-520 nm MS2 : 520-600 nm MS3 : 630-690 nm MS4 : 760-900 nm	Pan : 450-900 nm MS1 : 450-520 nm MS2 : 520-600 nm MS3 : 630-690 nm MS4 : 760-900 nm	Pan : 450-900 nm MS1 : 450-520 nm MS2 : 520-600 nm MS3 : 630-690 nm MS4 : 760-900 nm
Revisit time	16 day	3 day	3 day
Altitude	705 km	682 km	685 km
Swath width	185 km	11 km	15 km
Dynamic range	8 bits	11 bits	10 bits

In this study, we examined the image-based atmospheric correction models using the data from Landsat ETM+ and IKONOS that have quite similar spectral characteristics to the forthcoming KOMPSAT-2 MSC and the *in-situ* measured surface reflectance data during satellite overflight. KOMPSAT-2 MSC is scheduled to be launched in 2005 and expected to provide 1 m panchromatic image and 4 m multi-spectral image with four bands. Table 1 shows the characteristics of Landsat ETM+, IKONOS, and KOMPSAT-2 MSC system.

2. DATA AND METHODS

In order to examine the image-based atmospheric correction Landsat ETM+ data obtained on 22 November 2002 was utilized. Fig. 1 shows the study area. 22 ground control points (GCPs) were extracted from the digital map of 1:25,000 for geometric correction.

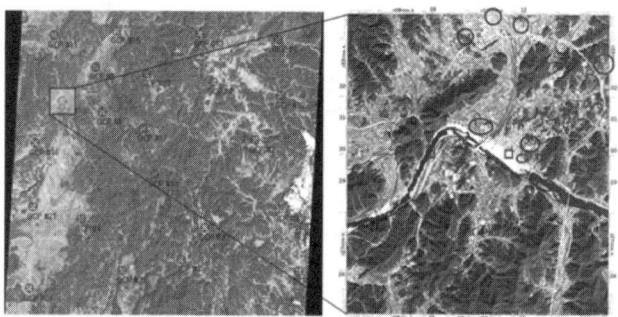

Fig. 1. Landsat ETM+ image (November 22, 2002) and 22 GCP extracted from the digital map of 1:25,000 (left) and location of *in-situ* measurement (right).

FR Portable Spectroradiometer (FieldSpec FR) was utilized to collect the field spectral data during Landsat ETM+ overflight. This instrument collects data from 350nm to 2500nm (Table 2), but we only used data from 450nm to 900nm for this study (Fig. 2). *In-situ* field measurements are used to evaluate the accuracy of each method that computes the surface reflectance from the ETM+ satellite image.

Table 2. FieldSpec FR Specification

Spectral Bandwidth	1.4 nm in the 350 nm to 1000 nm 2.0 nm in the 1000 nm to 2500 nm
Spectral Range	350 nm to 2500 nm
Spectral Resolution	2 nm @700 nm 10 nm@1500 nm, and 10 nm@2100 nm
Sensor Linearity	±1%
Wavelength Accuracy	±1%
Scan Time	A new spectrum is generated every 0.1 seconds for the entire spectral range
Operational Weight	7 kg+2.2 kg battery pack
Operational Size	35×29×13 cm

Three approaches have been experimented in this study; each of which calculates surface reflectance from Equation (1) using different assumptions for T_v^{dir}, T_v^{diff}, T_z^{dir}, and T_z^{diff} (Table 3).

Fig. 2. The measured surface reflectance from six different targets using FieldSpec FR.

The path radiance (L_{path}) was assumed to be the dark-object radiance minus the radiance contributed by 0.01 surface reflectance. These path radiance values were used to do a dark-object subtraction correction.

$$\rho_{surface} = \frac{\pi(L_{sat} - L_{path})}{T_v(E_{sun}\cos(\theta_z)T_z + E_{down})} = \rho_{surface} = \frac{\pi(L_{sat} - L_{path})}{T_v E_{sun}\cos(\theta_z)(T_z^{dir} + T_z^{diff})} \quad (1)$$

Where,
L_{sat} Spectral radiance at the satellite sensor.

L_{path} Upwelling atmospheric spectral radiance scattered in the direction of and at the sensor entrance pupil and within the sensor's field of view, i.e., the path radiance.

E_{sun} Solar spectral irradiance on a surface perpendicular to the sun's rays outside the atmosphere. E_{sun} contains the correction of the earth-sun distance.

T_v Atmospheric transmittance along the path from the ground surface to the sensor ($T_z^{dir} + T_v^{diff}$).

T_z Atmospheric transmittance along the path from the sun to the ground surface.

θ_z Angle of incidence of the direct solar flux onto the earth's surface. Downwelling spectral irradiance at the surface due to the scattered solar flux in the atmosphere.

Table 3. Each Transmittance Value for Three Cases used in Equation (1)

Method	T_v^{dir}	T_v^{diff}	T_z^{dir}	T_z^{diff}
Case 1	1	0	$\cos(\theta_z)$	0
Case 2	1	0	$\cos(\theta_z)$	$\frac{1}{4}T_z^{dir}\frac{1}{3}T_v^{dir}\frac{1}{2}T_z^{dir}$
Case 3	$e^{-\tau/\cos\theta_v}$	$\frac{1}{4}T_v^{dir}\frac{1}{3}T_v^{dir}\frac{1}{2}T_v^{dir}$	$e^{-\tau/\cos\theta_z}$	$\frac{1}{4}T_z^{dir}\frac{1}{3}T_z^{dir}\frac{1}{2}T_z^{dir}$

Three approaches in this study are based on the COST model, and Table 4 is estimated results from Cess et al. (1991) and Sohn et al. (1998).

Table 4. Estimates of Aerosol, Rayleigh and Total Optical Thickness used for this Study ($\tau_{0.5um} = 0.4$)

Landsat TM+	Aerosol (τ_{Aer})	Rayleigh (τ_{Ray})	Total
τ_{band1}	0.4×1.24=0.496	0.226	0.722
τ_{band1}	0.4×1=0.4	0.099	0.499
τ_{band1}	0.4×0.81=0.324	0.050	0.374
τ_{band1}	0.4×0.65=0.26	0.023	0.283

3. RESULTS AND DISCUSSION

The scatter plot (Fig. 3) and Table 5, while substantial difference are present between TOA reflectance and *in-situ* measurements, the results showed that Case 1 based on COST model gives most accurate results among three cases. The accuracy of Case 2_1 is very close to Case 1 and its values are smaller than *in-situ* data. No notable features appear between some bands in the Case 3_1 and *in-situ* data.

Table 5. Estimates of Surface Reflectance for Six Different Targets Using Landsat ETM+ Data

ETM+	Case	Paddy field	Field	Sand bar	Barren land	asphalt	Concrete
Band1	DN	71	67	90	78	68	78
	TOA	0.146	0.137	0.191	0.163	0.139	0.163
	Case 1	0.126	0.108	0.211	0.158	0.113	0.158
	Case 2_1	0.103	0.089	0.171	0.128	0.092	0.128
	Case 2_2	0.097	0.084	0.161	0.121	0.087	0.121
	Case 2_3	0.088	0.076	0.144	0.108	0.079	0.108
	Case 3_1	0.330	0.280	0.563	0.416	0.293	0.416
	Case 3_2	0.291	0.240	0.496	0.366	0.258	0.366
	Case 3_3	0.232	0.198	0.394	0.292	0.206	0.292
	Measured	0.117	0.072	0.260	0.168	0.163	0.245
Band2	DN	56	53	88	71	53	68
	TOA	0.122	0.115	0.204	0.161	0.115	0.153
	Case 1	0.152	0.138	0.310	0.226	0.138	0.211
	Case 2_1	0.124	0.112	0.250	0.183	0.112	0.171
	Case 2_2	0.117	0.106	0.235	0.172	0.106	0.161
	Case 2_3	0.105	0.095	0.210	0.154	0.095	0.144
	Case 3_1	0.214	0.193	0.439	0.319	0.193	0.298
	Case 3_2	0.189	0.171	0.387	0.282	0.171	0.263
	Case 3_3	0.151	0.137	0.308	0.225	0.137	0.210
	Measured	0.170	0.127	0.354	0.268	0.178	0.305
Band3	DN	66	57	109	89	53	84
	TOA	0.136	0.115	0.238	0.191	0.106	0.179
	Case 1	0.219	0.178	0.413	0.323	0.160	0.300
	Case 2_1	0.177	0.144	0.333	0.260	0.130	0.242
	Case 2_2	0.166	0.136	0.313	0.245	0.122	0.228
	Case 2_3	0.149	0.122	0.279	0.219	0.110	0.203
	Case 3_1	0.217	0.177	0.411	0.321	0.159	0.298
	Case 3_2	0.192	0.156	0.362	0.283	0.141	0.263
	Case 3_3	0.154	0.126	0.288	0.226	0.113	0.210
	Measured	0.222	0.074	0.420	0.337	0.192	0.330
Band4	DN	61	61	90	79	49	76
	TOA	0.191	0.191	0.295	0.256	0.148	0.245
	Case 1	0.336	0.336	0.537	0.461	0.253	0.440
	Case 2_1	0.271	0.271	0.431	0.370	0.204	0.354
	Case 2_2	0.254	0.254	0.405	0.348	0.192	0.332
	Case 2_3	0.227	0.227	0.361	0.310	0.172	0.297
	Case 3_1	0.258	0.258	0.411	0.353	0.195	0.337
	Case 3_2	0.228	0.228	0.362	0.311	0.172	0.297
	Case 3_3	0.182	0.182	0.288	0.248	0.138	0.237
	Measured	0.337	0.449	0.503	0.420	0.220	0.365

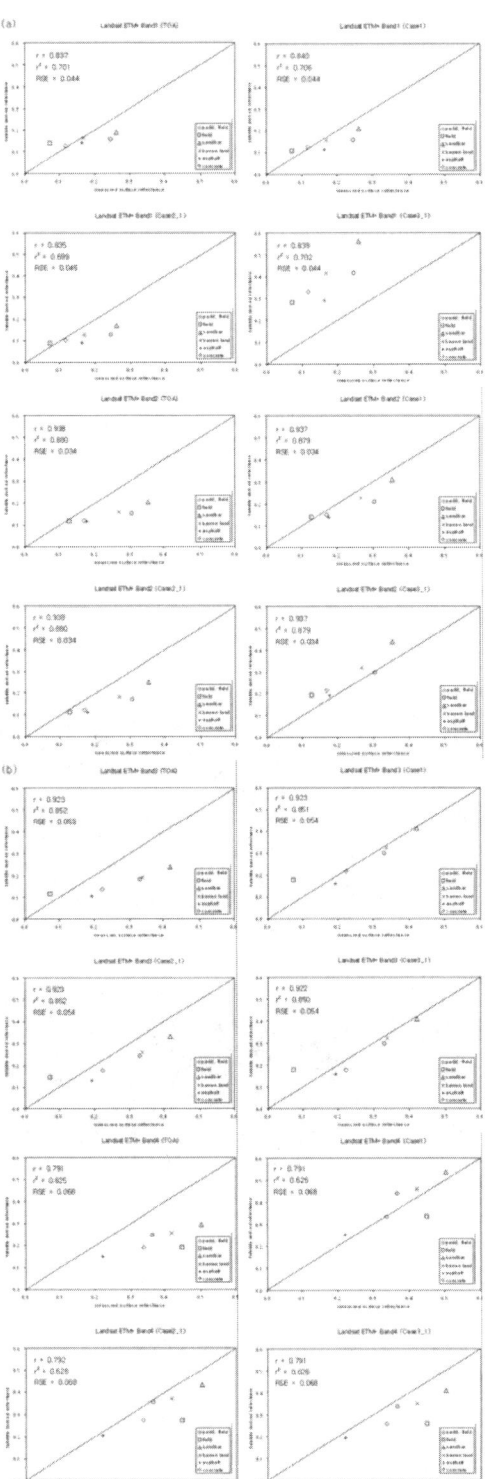

Fig. 3. Comparisons of satellite-derived reflectance and *in-situ* measurements for each band: (a) band 1 and band 2, (b) band 3 and band 4.

Fig. 4 shows the results of comparison between TOA and Case 1 histogram pattern.

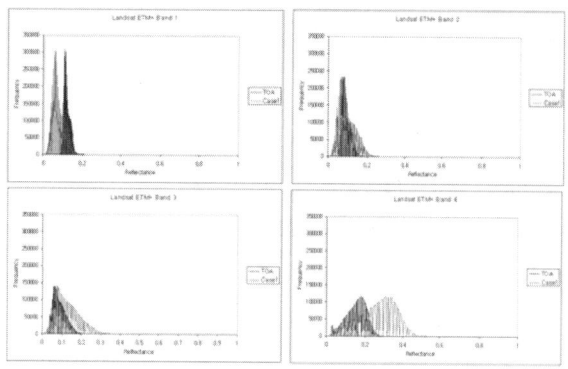

Fig. 4. A Comparison of TOA and Case 1 histogram distribution pattern.

Also Case 1 and Case 2_1 generated acceptable results with IKONOS image. It is apparent from Fig. 5 that two cases have substantially removed the atmospheric effects on the images.

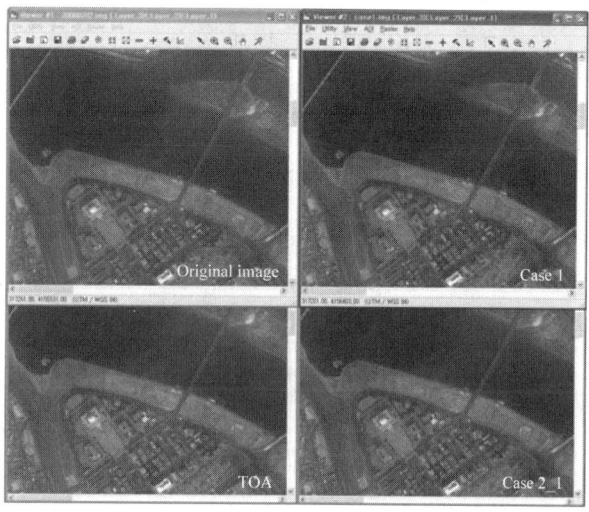

Fig. 5. Comparison of image-based atmospheric correction using IKONOS image.

4. CONCLUSIONS

In this study, three cases were experimented to examine the image-based atmospheric correction model using Landsat ETM+ that has quite similar spectral characteristics to the forthcoming KOMPSAT-2 MSC and *in-situ* measurements of surface reflectance. The results showed that Case 1 based on COST model was the most accurate in terms of correction for atmospheric effects. However, since Case 2 accounts for more physical components and produces promising results comparable to Case 1, we need further studies for better performance of such model.

5. ACKNOWLEDGEMENT

This research was supported by Ministry of Science and Technology (MOST). *In-situ* measurement data were obtained from Korea Institute of Geoscience & Mineral Resources (KIGAM).

6. REFERENCES

Cess, R. D., E. G. Dutton, J. J. Deluisi, and F. Jiang, 1991: Determining surface solar absorption from broadband satellite measurements for clear skies: Comparison with surface measurements. *Journal of Climate*, 236-247.

Chavez, P. S., Jr., 1988: An improved dark-object subtraction technique for atmospheric scattering correction of multispectral data. *Remote Sensing of Environment*, 24: 459-479.

Chavez, P. S., Jr., 1996: Image-based atmospheric corrections-Revisited and improved. *Photogrammetric Engineering & Remote Sensing*, 62(9): 1025-1036.

Chavez, P. S., Jr., and D. K. MacKinnon, 1994: Automatic detection of vegetation changes in the southwestern United States using remotely sensed images. *Photogrammetric Engineering & Remote Sensing*, 60(5): 571-583.

Fung, T., 1990: An assessment of TM imagery for land-cover change detection. *IEEE transactions on Geoscience and Remote Sensing*, 28(4): 681-684.

Jensen, J. R., and D. L. Toll, 1982: Detecting residential land-use development at the urban fringe. *Photogrammetric Engineering & Remote Sensing*, 48(4): 629-643.

Moran, M. S., R. D. Jackson, P. N. Slater, and P. M. Teillet, 1992: Evaluation of simplified procedures for retrieval of land surface reflectance factors from satellite sensor output. *Remote Sensing of Environment*, 41: 169-184.

Robinove, C. J., P. S. Chavez, Jr., D. Gehring, and R. Holmgrem, 1981: Arid land monitoring using Landsat albedo difference images. *Remote Sensing of Environment*, 11: 133-156.

Rowan, L. C., P. H. Wetlaufer, A. F. H. Goetz, F. C. Billingsley, and J. H. Stewart, 1974: Discrimination of rock types and detection of hydrothermally altered areas in south-central Nevada by the use of computer-enhanced ERTS images. *U.S. Geological Survey Professional Paper 883*, 35.

Sohn, B. J., D. S. Shin, and S. S. Lee, 1998: Optical characteristics of the Yellow Sand from ground-based solar radiation measurements near the Yellow Sea. *Proceedings of International Symposium on Remote Sensing*, Korea, 9-13.

Vincent, R. K., 1972: An ERTS multispectral scanner experiment for mapping iron compounds. *In proceedings of the Eighth International Symposium on Remote Sensing of Environment*, Ann Arbor, MI, 1239-1247.

CHARACTERISTICS OF SPECTRAL EMISSIVITY FOR VARIOUS TYPES OF SURFACES DERIVED FROM IMG SPECTRUM DATA

*Yoshifumi Ota and Ryoichi Imasu

*(ota@ccsr.u-tokyo.ac.jp), Center for Climate System Research, The University of Tokyo
5-1-5 Kashiwanoha, Kashiwa-shi, Chiba 277-8568, Japan

ABSTRACT

Spectral dependency of surface emissivity is characterized by surface materials and conditions. Infrared radiance spectra observed over surfaces contain information that is related to these surface properties. This paper presents an analytical procedure to estimate the spectral surface emissivity from infrared radiance spectra observed using a high-resolution spectrometer such as a Fourier Transform Spectrometer (FTS). It presents some results from analyses applying the procedure to data observed using the Interferometric Monitor for Greenhouse gases (IMG) sensor. Results show that emissivity values for various types of surfaces were consistent with those expected from satellite imagery data. Some interesting spectral emissivity characteristics were obtained for some types of surfaces.

1. INTRODUCTION

Infrared surface emissivity, particularly in the wavelength of the atmospheric window region, is an important parameter in the surface radiation budget and for precise remote sensing of the atmosphere and surface. Recently, global maps of broad band infrared emissivity have been obtained from the Moderate Resolution Imaging Spectroradiometer (MODIS) and the Advanced Spaceborne Thermal Emission and Reflection radiometer (ASTER) data analyses (Ogawa et al., 2003; Zhou et al., 2003). However, surface emissivity has spectral characteristics for each surface material and condition (Masuda et al., 1988; Salisbury and D'Aria, 1992); spectral emissivity data are still necessary to characterize the surface more precisely. Furthermore, surface emissivity data are important for improving the accuracy of temperature and gas profile retrievals from infrared radiance measured by recent satellite-based high spectral resolution spectrometers such as a Fourier Transform Spectrometer (FTS).

Thermal infrared radiance observed from space is a function of surface type and atmospheric parameters: surface temperature, surface emissivity, atmospheric temperature, and absorbing gas concentration profiles. Thermo-dynamic surface temperature data must be obtained independently from other parameters to derive information related to surface emissivity from observed radiance data. However, it is difficult to derive the surface temperature separately from emissivity without any assumptions.

This paper presents a data analysis procedure to estimate surface emissivity for each wavelength (wavenumber) under a simple assumption about the spectral feature of emissivity. Section 3 describes the principle of that procedure. Some results from analyses of actually observed satellite data are presented in section 4.

2. INSTRUMENTS AND DATA

The Interferometric Monitor for Greenhouse gases (IMG) was an FTS launched aboard the Advanced Earth Observing Satellite (ADEOS) satellite in August 1996 (Kobayashi et al., 1999). It measured upwelling infrared radiance from the Earth by nadir looking during its 10-month operation period up to June 1997. The horizontal size of the instantaneous field of view (IFOV) of IMG was 8 km × 8 km. The spectral range and resolution were 600–3000 cm^{-1} and 0.1 cm^{-1}, respectively. The range covered the entire thermal infrared atmospheric window. Although the main objective of the IMG project was to observe atmospheric greenhouse gases, we anticipate obtaining useful information related to surface properties from IMG data measured in the atmospheric window region. Fig. 1 shows an example of the IMG spectrum in the wavenumber range of the atmospheric window region observed over the

Sahara Desert. It shows the equivalent back body temperatures on both sides of an ozone absorption band located at 1000–1070 cm^{-1}, which differ by about 10°C, implying a strong spectral dependency of surface emissivity in this region.

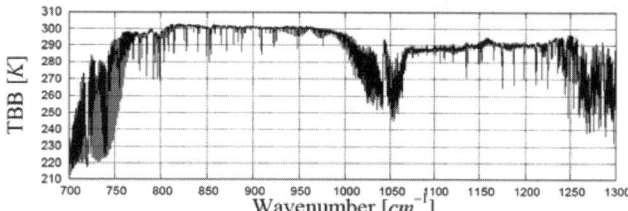

Fig. 1. IMG spectrum data (Band 3) measured over the Sahara Desert region (27.7°N, 4.6°W, 25 Dec., 1996).

3. METHOD

The upwelling radiance at wavenumber k in the infrared wavelength region that is measured at the top of the atmosphere under no cloud conditions can be written as

$$I_k^{obs} = \varepsilon_k \tau_k^{total} B_k(T^s) + \int_{\tau_k^{total}}^{1} B_k(T(p)) d\tau_k \\ + (1-\varepsilon_k)\tau_k^{total} \int_{\tau_k^{total}}^{1} B_k(T(p)) d\tau_k^*, \quad (1)$$

where $B_k(T)$ is the Planck function at atmospheric temperature $T(p)$ or at surface skin temperature T^s, τ_k is the atmospheric transmittance from a pressure level p to the top of the atmosphere $\left(\tau_k^* = \tau_k^{total}/\tau_k\right)$, and ε_k is the surface emissivity. First and second terms on the right-hand side represent respective upwelling radiation emissions by the surface and the atmosphere. The third term represents the contribution of the downwelling radiation emitted from the atmosphere. This radiation is reflected by the surface and can be merged to the upwelling radiation.

Solving (1) for ε_k, we can finally obtain a form as

$$\varepsilon_k = \frac{I_k^{obs} - \int_{\tau_k^{total}}^{1} B_k(T) d\tau_k - \tau_k^{total} \bar{I}_k^{\downarrow}}{\tau_k^{total}\left(B_k(T^s) - \bar{I}_k^{\downarrow}\right)}. \quad (2)$$

Spectral emissivity is obtainable from Eq. (2) if the surface temperature and the vertical profiles of atmospheric composition are known or estimated independently. For evaluation of surface emissivity from a thermal infrared radiance measured by a FTS, it is expected that atmospheric corrections can be made based on the radiance data itself because FTS offers a large number of spectral channels that are available for analysis of both emissivity and atmospheric parameters. This paper introduces an atmospheric correction method by which the atmospheric temperature and water vapor profiles were retrieved from the observed infrared spectrum itself. For this analysis, radiance spectral channels were selected so that the atmospheric total transmittances of the first guess profile are less than 1%. Fig. 2 shows the selected channels. The retrieval method was based on the approach described by Rodgers (2000). The statistical covariance matrix including both diagonal and off-diagonal elements was used for *a priori* information because the channels shown in Fig. 2 are sensitive only for analysis above the middle troposphere.

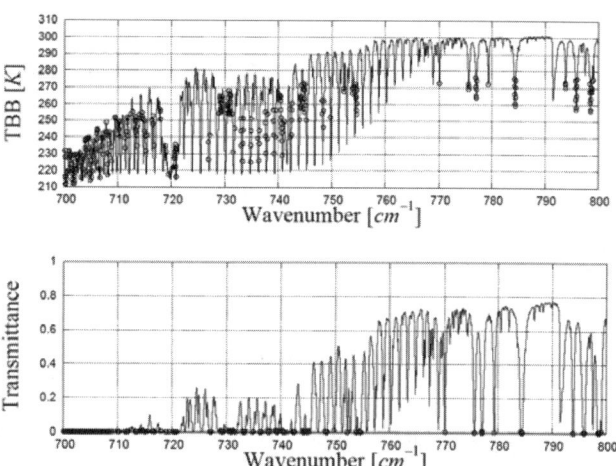

Fig. 2. First-guess radiance and transmittance spectra (solid lines) used for the IMG spectrum analysis in Fig. 1. Open circles show spectral channels used for atmospheric correction, where the atmospheric transmittance is less than 1%.

Fig. 3 shows the averaging kernel functions for temperature and water vapor, respectively. The surface skin temperature was estimated based on one channel data that showed the warmest brightness temperature in the atmospheric window region after the atmospheric correction. Therefore, the estimated surface emissivity is scaled relative to the channel that has been assumed as unit emissivity.

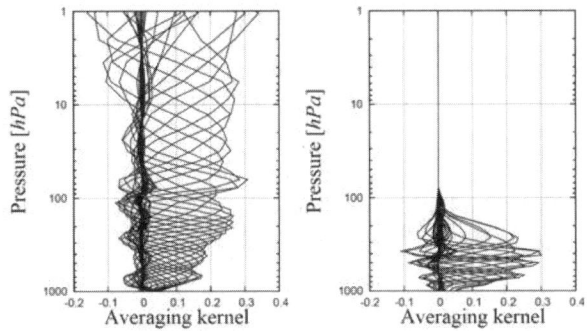

Fig. 3. Averaging kernel functions for temperature (left) and water vapor (right).

Visible imageries of the Ocean Color Temperature and Scanner (OCTS) that had been operated simultaneously with IMG sensor on the ADEOS satellite were used for verifying a clear sky condition and for identifying the surface type in the IFOV of IMG. Atmospheric transmittance calculations used LBLRTM Ver.8.4 (Clough and Iacono, 1995) with the HITRAN 2000 database (Rothman et al., 2003).

4. RESULTS AND DISCUSSION

Examples of spectral emissivity analyzed from IMG spectrum data observed over the Sahara Desert region are shown in Fig. 4. It is remarkable that the obtained emissivity spectra show the characteristic feature of quartz reststrahlen bands. As described by Salisbury and D'Aria (1992), this feature results from the existence of silicate stretching vibration bands in the atmospheric window region (8–14 μm). Especially, both the strong asymmetric stretching vibration band between 8 μm and 10 μm and the weaker symmetric stretching vibration band between 12.2 μm and 13 μm are visible in Fig. 4. However, these emissivity spectrum band shapes varied scene-by-scene even over the desert region, thereby indicating the variation of the surface property (e.g., the components, the structures of the mineral molecules and crystal lattices, and surface covering conditions) that are identifiable in the visible OCTS imageries.

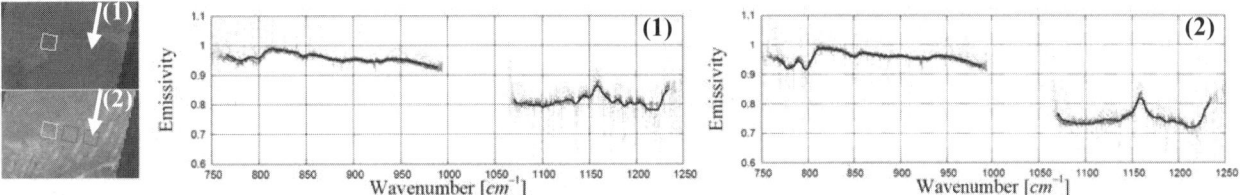

Fig. 4. Examples of surface emissivity analyses using IMG spectrum data observed over the Sahara Desert region. Panels on the left side are OCTS images; the IFOV of IMG are indicated by arrows and squares. Middle and right side figures show the analyzed spectral emissivity. Numbers correspond to those on the left side panels. In figures of emissivity, dots show the emissivity analyzed at the IMG channels and solid lines show those convolved at a lower resolution of 3.0 cm^{-1}.

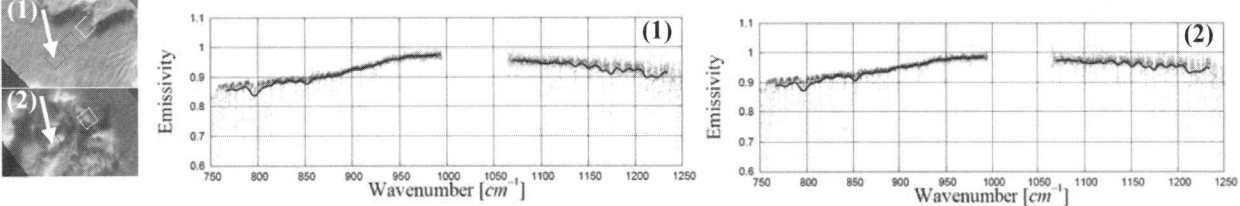

Fig. 5. Same as in Fig. 4, but for the snowy (or icy) data observed over the Antarctic region.

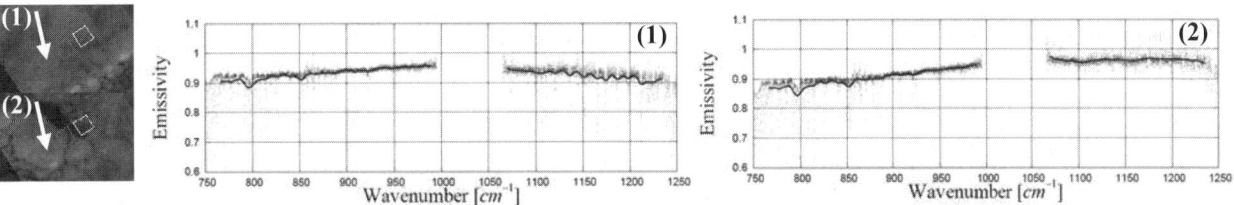

Fig. 6. Same as in Fig. 4, but for the icy data observed above the sea near the Antarctic region.

Emissivity in the spectral region of ozone absorption band (9.6 µm) was relatively noisy because the atmospheric transmittance was small as a result of ozone absorption and because emissivity analysis is sensitive to the retrieved ozone concentration. Consequently, the ozone concentration profile must be retrieved accurately and the spectral residual must be minimized in this spectral region for more precise analysis of emissivity.

Figs. 5 and 6 show some other examples of data for icy surfaces. The wavelength dependency of the emissivities corresponded to that of the refractive index of ice. However, absolute values of emissivity differ somewhat from those of MODIS and ASTER analyses. Although the absolute values of emissivity for ice and snow depend on the surface condition, some part of this difference may result from the assumption adopted in this study: emissivity is unity at the wavelength where the brightness temperature is warmest in the window region. Validation experiments are necessary in future studies to verify the reason for the difference.

5. CONCLUSION

A data analysis procedure was presented to evaluate the relative spectral emissivity of the surface from a high-resolution infrared spectrum measured by a satellite-based FTS. In that procedure, all calculations were made under the assumption that emissivity is unity at the wavelength where the brightness temperature is highest in the window region. Applying this procedure to IMG spectrum data, it has been shown that the spectral dependency of the estimated emissivity is consistent with those expected from surface properties identified with OCTS imagery data.

Consequently, we conclude that the data analysis procedure presented in this study is useful to obtain the spectral emissivity of the surface on a relative-unit basis. This method is inferred to be applicable to analysis of similar spectrometry data measured by IMG such as the Tropospheric Emission Spectrometer (TES) and the Infrared Atmospheric Sounding Interferometer (IASI).

6. REFERENCES

Clough, S. A., and M. J. Iacono, 1995: Line-by-Line calculations of atmospheric flux and cooling rate 2: Application to carbon dioxide, ozone, methane, nitrous oxide, and the halocarbons. *J. Geophys. Res.*, 100, 16519-16535.

Kobayashi H., A. Shimota, C. Yoshigahara, I. Yoshida, Y. Uehara, and K. Kondo, 1999: Satellite-borne high-resolution FTIR for lower atmosphere soundings and its evaluation. *IEEE Trans. Geosci. Remote Sens.*, 37, 1496-1507.

Masuda, K., T. Takashima, and T. Takayama, 1988: Emissivity of pure sea waters for the model sea surface in the infrared window region. *Remote Sens. Environ.*, 24, 313-329.

Ogawa, K., T. Schmugge, F. Jacob, and A. French, 2003: Estimation of land surface window (8–12 µm) emissivity from multi-spectral thermal infrared remote sensing—A case study in a part of Sahara Desert, *Geophys. Res. Lett.*, 30(2), 1067, doi:10.1029/2002GL016354.

Rodgers, C. D., 2000: Inverse Methods for Atmospheric Sounding: Theory and Practice. *World Sci. Singapore*, 238 pp.

Rothman, L. S., A. Barbe, D. C. Benner, L. R. Brown, C. Camy-Peyret, M. R. Carleer, K. Chance, C. Clerbaux, V. Dana, V. M. Devi, A. Fayt, J.-M. Flaud, R. R. Gamache, A. Goldman, D. Jacquemart, K. W. Jucks, W. J. Lafferty, J.-Y. Mandin, S. T. Massie, V. Nemtchinov, D. A. Newnham, A. Perrin, C. P. Rinsland, J. Schroeder, K. M. Smith, M. A. H. Smith, K. Tang, R. A. Toth, J. Vander Auwera, P. Varanasi, and K. Yoshino, 2003: The HITRAN Molecular Spectroscopic Database: Edition of 2000 Including Updates through 2001. *J. Quant. Spectrosc. Radiat.*, 82, 5-44.

Salisbury, J. W., and D. M. D'Aria, 1992: Emissivity of terrestrial materials in the 8–14 µm atmospheric windows. *Remote Sens. Environ.*, 42, 83-106.

Zhou, L., R. E. Dickinson, Y. Tian, M. Jin, K. Ogawa, H. Yu, and T. Schmugge, 2003: A sensitivity study of climate and energy balance simulations with use of satellite-derived emissivity data over Northern Africa and the Arabian Peninsula. *J. Geophys. Res.*, 108(D24), 4795, doi:10.1029/2003JD004083.

INTRA-SEASONAL VARIATION OF PSC COMPOSITIONS RETRIEVED FROM ILAS-II DATA

Y. Kim, J. H. Park, and W. Choi
Seoul National University
Seoul, 151-747, Korea

K. -M. Lee
Kyungpook National University
Daegu, 702-701, Korea

S. T. Massie
National Center for Atmospheric Research
Boulder, Colorado, 80305, USA

T. Yokota, H. Nakajima, and Y. Sasano
National Institute of Environmental Studies
Tsukuba, Ibaraki, 305-8506, Japan

ABSTRACT

Using a transmittance-ratio technique for Improved Limb Atmospheric Spectrometer II (ILAS-II) transmittance data, relative extinction coefficients of Antarctic PSCs (Polar Stratospheric Clouds) were determined for the period of June–August of 2003. Properties of PSCs were determined by simulating the relative extinction coefficients assuming a single mode log-normal distribution of PSCs with the Mie theory. The results indicate that there were various mixtures of PSC compositions in the Antarctic cases. Our analyses include more than 90 PSC observed cases, where retrieved compositions seemed to be dependent on the temperature of PSCs. β-NAT was the major component for most of cases in June when temperature changed in the range of 190 to 196K. In the middle of July, as temperatures became close to T_{ICE} (185K), Nitric Acid Water (NAW, HNO_3/H_2O) PSCs were most frequently observed PSC types. As temperature decreased to minimum temperature near the 180K in late July and August, ICE-type PSC became dominant PSC type.

1. INTRODUCTION

It has been known that PSC plays an important role in ozone depletion process through the heterogeneous chemical reaction and denitrification processes. Among observations of PSCs, Fahey et al. (1989) and Dye et al. (1992) supported the presence of nitric acid in PSCs, Carslaw et al. (1994), Tabazadeh et al. (1994) show that PSC particles could be also composed of supercooled ternary solution (STS) of $H_2SO_4/HNO_3/H_2O$ droplets. When the temperature of PSC drops below T_{NAT}, these ternary droplets change to binary droplets of nitric acid and water (Nitric Acid Water, HNO_3/H_2O) absorbing HNO_3 in the atmosphere. Recent infrared spectra analysis (Toon and Tolbert, 1995) and in situ measurements (e.g., Dye et al., 1996; Schreiner et al., 1999; and Larsen et al., 2000) strongly supported the presence of STS PSCs.

Lee et al. (2003) reported satellite-based remote sensing of PSCs in the Arctic region in 1997 using the ILAS occultation spectra. The composition of PSCs is derived by comparing the observed aerosol extinction spectra with those of simulations using the Mie theory. Lee et al. (2003) found three types of PSCs from the Arctic measurements, NAW (Nitric Acid Water, HNO_3/H_2O, or), α-NAT, and β-NAT (α and β form of NAT) particles, but for each limb measurement a single composition was identified because of weak spectral signal of PSCs.

In the following sections of this paper, analysis techniques, input data, and computational results are presented. Detailed error analysis of current results is not reported.

2. ANAYSIS METHOD

The observed relative extinction coefficients are derived from ILAS-II infrared and mid-infrared transmittance data using transmittance-ratio technique. The theoretical relative extinction coefficients can be calculated using the refractive indices of known PSC types and Mie-theory. Non-linear Least Square Fit (NLSF) method was used to determine the properties of PSCs by comparing simulated and observed relative extinction coefficients and by minimizing the difference between the two (see Lee et al., 2001; 2003).

2.1 Satellite Data

Solar occultation observations of ILAS-II occurred from June to August of 2003, for Antarctic region from 65.5°S to 87.1°S. The Infrared Channel (IR Ch.1) is consisted of 44 pixels in the range of 850~1,610 cm^{-1} (6.21~11.77 μm) with each pixel width of 0.129 μm. The Mid-infrared Channel (IR Ch.2) is consisted of 22 pixels in the range of 1,754~3,333 cm^{-1} (3.00~5.70 μm) having the same spectral interval as the infrared.

2.2 Transmittance-Ratio Technique

Two transmittance profiles, with and without PSC signal, are used for the transmittance-ratio calculation (hereafter called PSC profile and REF profile, respectively). Each profile is represented mathematically as,

$$\tau_{psc}(z,\lambda) = \tau_{psc}^g(z,\lambda) \times \exp[-\beta_{psc}(z,\lambda)L_{psc}] \quad (1)$$

$$\tau_{ref}(z,\lambda) = \tau_{ref}^g(z,\lambda) \times \exp[-\beta_{ref}(z,\lambda)L_{ref}] \quad (2)$$

where, β_{psc} and L_{psc} are extinction coefficient and optical path length of PSC. τ_{psc}^g and τ_{ref}^g are gaseous absorption.

Fig. 1 shows the case of July 1; (a) profiles of τ_{psc} and τ_{ref} (displayed as solid and dashed lines) of the 7 window pixels of the Infrared channel (pixel # 0,8,17,29,38,59,65), (b) the relative optical depths (Eq. 3), and (c) the mean of 7 optical depths normalized by the peak value of each profile (see Lee et al., 2003 for definitions).

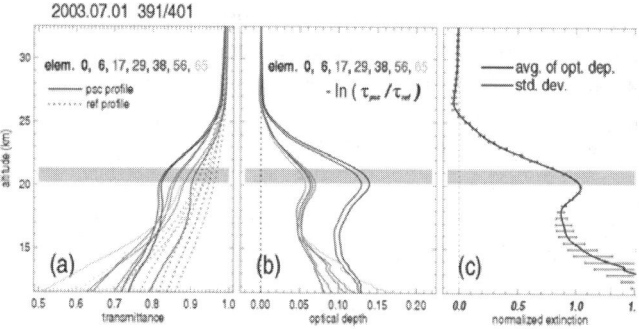

Fig. 1. The PSC case of July 1st, 2003; (a) two vertical transmittance profiles of τ_{psc} and τ_{ref}, (b) the relative optical depths profile for each pixel element, and (c) the mean of normalized optical depths.

With assumptions that (1) two background gaseous absorptions are nearly same and (2) the extinction by PSC is negligible in the REF profile, the equation of relative optical depth is derived as,

$$\beta_{psc}(z,\lambda)L_{psc} \cong -\ln[\tau_{psc}(z,\lambda)/\tau_{ref}(z,\lambda)] \quad (3)$$

When PSC and REF profiles are appropriately selected satisfying the assumptions, relative extinction coefficients β_{psc} are obtained from two transmittance profiles. Fig. 2 shows relative extinction coefficients in the case of 1st July (profile #391). Other cases are analyzed in the same manner.

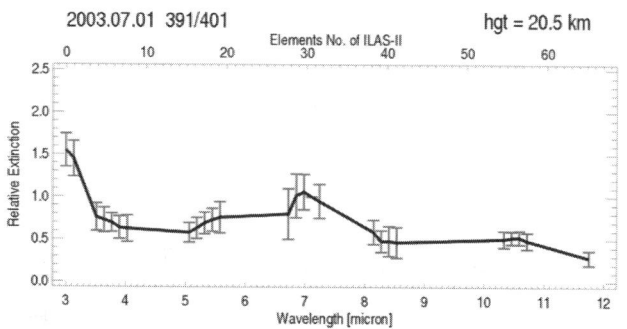

Fig. 2. The relative extinction coefficients are normalized by the value of the element 29 (7.11 μm) (vertical bars indicate estimated measurement errors).

2.3 Refractive Index

Table 1. Refractive Index of PSC Candidates

PSC type		phase	Optical Data
α-NAT	Nitric Acid Trihydrate PSC type Ia	solid	Richwine et al. (1995)
β-NAT	Nitric Acid Trihydrate PSC type Ia	solid	Toon et al. (1994)
NAD	Nitric Acid Dihydrate PSC type Ia	solid	Niedziela et al. (1998)
ICE	PSC type II	solid	Clapps et al. (1995)
SAW	Sulfuric Acid Water	Liquid (4 kinds)	Niedziela et al. (1998, 1999)
LTA	Liquid Ternary Aerosol (PSC type Ib)	Liquid (6 kinds)	Norman et al. (2002)
NAW	Nitric Acid Water PSC type Ib	Liquid (6 kinds)	Norman et al. (1999)

Refractive indices of several known PSC types are shown in Table 1. Total 20 PSC compositions are considered; α-Nitric Acid Trihydrate (NAT), β-NAT, Nitric Acid Dehydrate (NAD), ICE type PSC as solid types, Nitric Acid Water (NAW, HNO_3/H_2O) types of Nitric weight 35, 40, 45, 50, 54, 63%, Sulfuric Acid Water (SAW) types of 32, 43, 55, 66%, and Liquid Ternary Aerosol (LTA, $HNO_3/H_2SO_4/H_2O$) types of nitric/sulfuric weight 45/10, 35/22, 31/28, 26/29, 24/33, 15/40% as liquid-type PSCs. The data sources and names used for derivations of compositions for the PSCs observed by ILAS-II are summarized.

Fig. 3 shows simulated relative extinction coefficients for aerosols of 4 compositions (i.e., β-NAT, Ice Water at 180K, LTA N15 S40%, and NAW 45%). The solid dots on the element axis indicate the elements where relative extinctions are determined from ILAS-II data. The extinction coefficients are calculated for a set of selected conditions (i.e., the geometric mean standard deviation sig(=σ) =1.6 and five different values of the mean radius rad (= Ri). All extinctions are normalized by the element 29 value (wavelength = 7.113 μm).

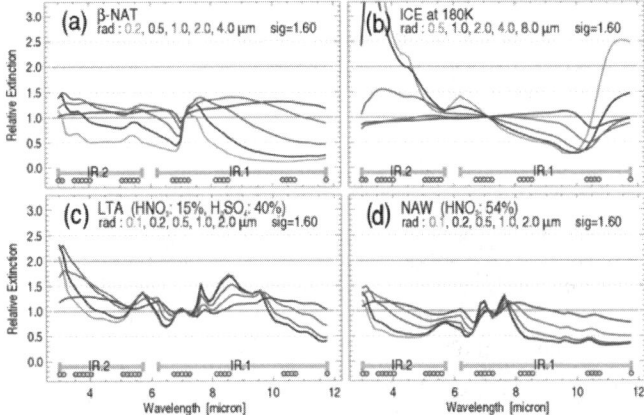

Fig. 3. Simulated relative extinction coefficients for aerosols of 4 compositions (i.e., β-NAT, Ice Water at 180K, LTA N15 S40%, and NAW 45%, to demonstrate that their shapes are distinct as a function of element number (or wavelength). (These are representative cases from all available optical data reported in the Table 1.).

3. RESULTS

3.1 PSC Properties Retrieval

Four cases of spectral simulations are shown in Fig. 4. Each case is shown for the date and the peak altitude. Shown in each figure are two compositions for each case; observed extinction is indicated as solid red line, the simulated total extinction is shown as crosses (the first composition contribution as blue dotted line and the second composition as green dotted line), and residual between the two is indicted as yellow circles. The overall good agreements between measured and simulated values using the transmittance-ratio technique imply that single or dual compositions can be identified correctly with the limited observed data and laboratory measurements of refractive indices.

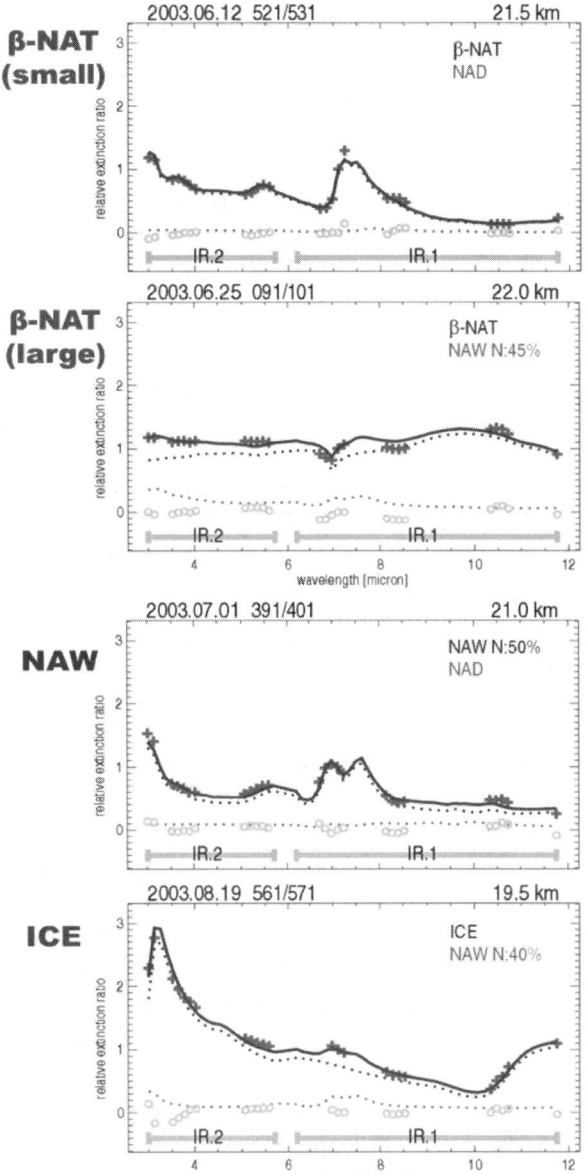

Fig. 4. NLSF results of representative cases. Comparisons are made for measured relative extinction coefficients (indicated by crosses) and simulated extinction coefficients (red solid lines). The residuals of the two are indicated by yellow circles.

3.2 Intra-Seasonal Variation of PSC Compositions

Total 90 Antarctic PSC cases are analyzed for a period of June to August 2003. Fig. 5 shows the major compositions (relatively large extinctions of the two PSCs). The composition of PSC seems to be separated according to temperature of PSC and change with time in the austral winter.

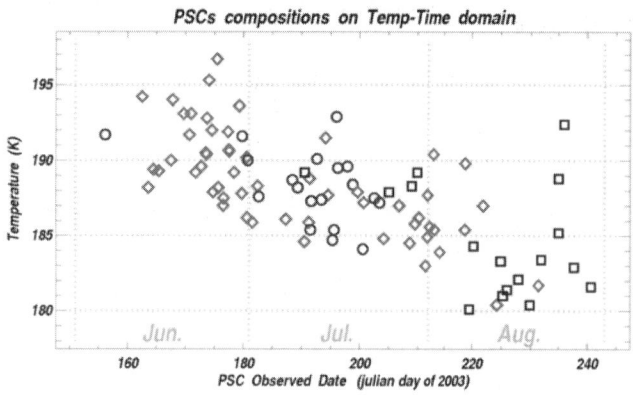

Fig. 5. 1st mode PSC composition as function of temperature and date. Compositions are shown as diamond, circle, and square shapes, respectively for β-NAT, NAW, and ICE type.

β-NAT is most frequently observed PSC type during the period from middle of June to early August of 2003 when temperatures were in the range of 185~195K. In the middle of July, NAW (circle) became representative PSC type. At this time, temperatures were a few degrees below T_{ICE} (185K) to 190K in most of cases. As temperature decreased to the near minimum (180K) in August 2003, PSCs seem to be frozen in almost cases and ICE-type PSC became dominant PSC type. The change of compositions for a period over the winter seems to be closely related to the variation of environmental temperatures. (Fig. 5)

3.3 Dual Compositions

In the cases of June 2003, β-NAT is the major component for most of cases and the other minor component is any one of NAW, LTA, α-NAT, NAD, SAW, and Ice-Water.

Table 2 shows RMS residual between the observed and simulated coefficients for a case of June 12. Error analysis results are not shown in this report.

Table 2. RMS Values of Residuals Between the Observed and Simulated Coefficients for June 12, #521

Single Composition		Dual Compositions and rms values	
β-NAT	3.60	β-NAT & NAD	2.28
NAW	6.12	β-NAT & SAW	2.55
α-NAT	6.19	β-NAT & ICE	2.73
NAD	8.03	β-NAT & LTA	2.74
ICE	11.80	β-NAT & NAW	3.25
LTA	13.08		

This mixture of two compositions is assumed to be two separate patches of PSCs along the optical path or mixed clusters of two different types in a same air mass.

4. REFERENCES

Dye, J. E., D. Baumgardner, B. W. Grandrud, K. Drdla, K. Barr, D. W. Fahey, L. A. Delnegro, A. Tabazadeh, H. H. Jonsson, J. C. Wilson, M. Loewenstein, J. R. Podolske, and K. R. Chan, 1996: In-situ observations of an Antarctic polar stratospheric cloud: Similarities with Arctic observations, *Geophys. Res. Lett.*, 23(15), 1913-1916.

Larsen, N., I. S. Mikkelsen, B. M. Knudsen, J. Schreiner, C. Voigt, K. Mauersberger, J. M. Rosen, and N. T. Kjome, 2000: Comparison of chemical and optical in situ measurements of polar stratospheric cloud particles, *J. Geophys. Res.*, 105(D1), 1491-1502

Lee, K.-M, J. H. Park, Y. Kim, W. Choi, H.-K. Cho, S. T. Massie, Y. Sasano, and T. Yokota, 2003: Properties of polar stratospheric clouds observed by ILAS in early 1997, *J. Geophys. Res.*, 108(D7), 4228, doi:10.1029/2002JD002854

Norman, M. L., J. Qian, R. E. Miller, and D. R. Worsnop, 1999: Infrared Complex Refractive Indices of Supercooled Liquid HNO_3/H_2O Aerosols, *J. Geophys. Res.*, 104(D23), 30,571-30,584.

Toon, O. B., M. A. Tolbert, B. G. Koehler, A. M. Middlebrook, and Joseph Jordan, 1994: Infrared optical constants of H_2O ice, amorphous ntric acid solutions, and nitric acid hydrates, *J. Geophys. Res.*, 99(D12), 25,631-25,654.

Toon, O. B., and M. A. Tolbert, 1995: Spectroscopic evidence against nitric acid trihydrate in polar stratospheric clouds, *Nature*, 375, 218-221.

ATMOSPHERIC INFRARED SOUNDER ON AIRS WITH EMPHASIS ON LEVEL 2 PRODUCTS

Sung-Yung Lee (Sung-Yung.Lee@jpl.nasa.gov), Eric Fetzer, Stephanie Granger, Thomas Hearty, Bjorn Lambrigtsen, Evan M. Manning, Edward Olsen, and Thomas Pagano

Jet Propulsion Laboratory, California Institute of Technology

Pasadena, California, USA

ABSTRACT

The Atmospheric InfraRed Sounder (AIRS) was launched aboard EOS Aqua in May of 2002. AIRS is a grating spectrometer with almost 2400 channels covering the 3.74 to 15.40 micron spectral region with a nominal spectral resolution ($v/\delta v$) of 1200, with some gaps. In addition, AIRS has 4 channels in the NIR/VIS region. The AIRS operates in conjunction with the microwave sounders Advanced Microwave Sounding Unit (AMSU-A) and Humidity Sounder of Brazil (HSB). The microwave sounders are mainly used for cloud clearing of IR radiances, or to remove the effect of cloud on the IR radiances.

AIRS has been very stable, radiometrically, as well as spectrally, meeting all stability requirements. All level 1 products (calibrated radiances) and level 2 products (retrieved geophysical parameters) have been released to the public. The primary level 2 products of the AIRS sounding suite include temperature, water vapor and ozone profiles, as well as cloud and surface parameters. AIRS science team members are also working on retrieving atmospheric aerosols, SO_2, CO, and CO_2.

Use of the AIRS data, with its high spatial resolution and accuracy, can be expected to lead to significant advances in atmospheric research and climate studies, as the data sets are beginning to see extensive use in the research community. We will describe the AIRS data products and discuss the significant effort under way to improve the accuracy and scope of the retrievals. And the quality control that has been developed to make the data useful for assimilation and research at this early stage of the mission.

1. INTRODUCTION

The Atmospheric Infrared Sounder (AIRS) is one of six instruments launched onboard (one word?) the Aqua spacecraft on May 4, 2002, from Vandenberg Air Force Base in California. The AIRS sounding system is made up of three instruments: AIRS, the Advanced Microwave Sounding Unit (AMSU-A) and the Humidity Sounder of Brazil (HSB).[1]

The AIRS suite of instruments will observe the global water and energy cycles, climate variations and trends, and the climate system response to increased greenhouse gases by making highly accurate measurements of air temperature and humidity, clouds, and surface temperature.

The EOS Aqua platform is in a polar, sun-synchronous orbit with a nominal altitude of 705 km, an inclination of 98.2° and an orbital period of 98.8 minutes. The repeat cycle is 233 orbits (16 days) with a ground track repeatability of +/- 20 km. The platform will have equatorial crossing times of 1:30 AM (descending) and 1:30 PM (ascending).

AIRS is a continuously operating cross-track scanning infrared sounder. AIRS has almost 2400 channels in the 3.4 to 15 μm infrared region with some spectral gaps, and a nominal spectral resolution of about 1200. The instantaneous field of view of the AIRS channels is 1.1° x 0.6°, which corresponds to about 15 km at the nadir. AIRS has scan rate of 8/3 seconds.

AMSU-A is a 15-channel microwave temperature sounder. AMSU-A has 2 channels, 23.8 GHz and 31.4 GHz, which provide surface and moisture information (total precipitable water and cloud liquid water), 12 channels in the 50–58 GHz oxygen absorption band which provide the primary temperature sounding capabilities, and one channel at 89 GHz which provides surface and moisture information. Like AIRS, AMSU-A is a cross-track scanner, but with scan rate of 8 seconds. AMSU-A footprints (45 km at nadir) are approximately three times as large as those of AIRS. The result is three AIRS scans per AMSU-A scan and nine AIRS footprints per AMSU-A footprint.

Special attention is paid to ensure good collocation of AIRS footprints to footprints of the MW instruments. Normally, instruments scan perpendicular to the satellite track as the satellite travels. This introduces an angle between the cross–track and the line connecting the footprints, called the helix angle, which depends on scan rate. Since AMSU scans three times slower than AIRS (and HSB), the AMSU helix angle is significantly larger than that of AIRS. Even HSB and AIRS have slightly different helix

---

[1] HSB scanner stopped in early Feb 2003 after 8 months of operation. Attempts to restart the scanner have not been successful yet.

angles due to slightly different scan timing. Although they have the same scan cycle of 2.67 seconds, their ground scan speed is slightly different. AMSU-A and HSB were rotated slightly on the platform to compensate for the difference in helix angles.

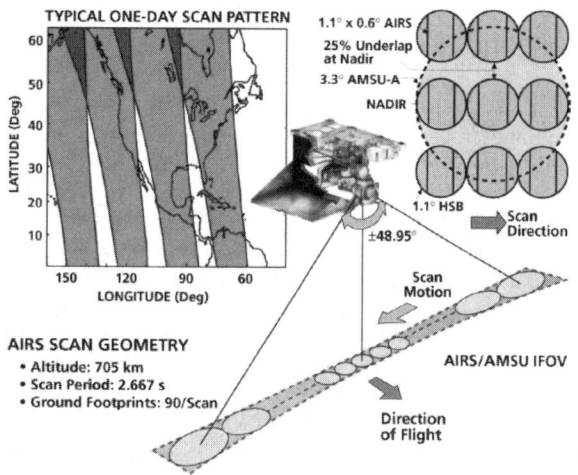

Fig. 1. Scan pattern of AIRS sounding suite.

2. AIRS LEVEL 1 DATA PRODUCTS

Fig. 2 is a plot of noise equivalent delta temperature (NeDT) at 250 K for the AIRS spectral channels. Many of the channels have NeDT of 0.1 to 0.2K, with the exception of the longer-wavelength IR channels. Many of the longwave channels are sensitive to stratospheric temperature where effects of clouds or the surface are negligible. AIRS performs retrievals at the AMSU footprint-scale, averaging those channels (not clear to me). Therefore, the relatively large values of NeDT have only a marginal effect.

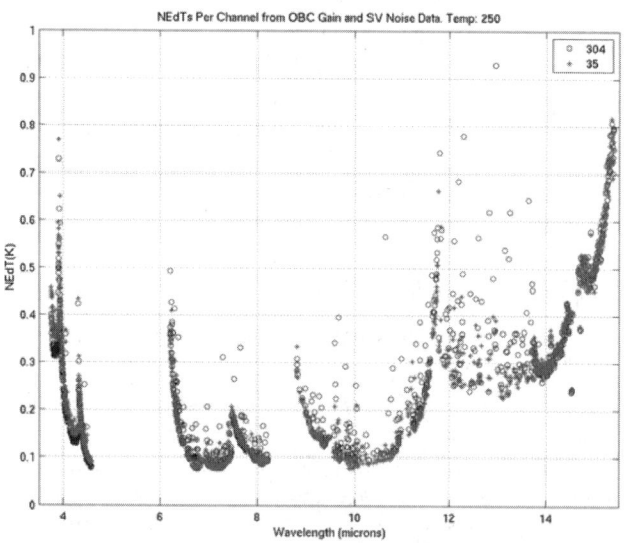

Fig. 2. Instrument noise characteristics of AIRS.

Fig. 3 (courtesy of George Aumann) illustrates the radiometric stability of the AIRS 2616 cm-1 channel over the 16–month period starting in September 2002. The figure has the daily mean and bias of skin temperature estimated from the 2616 cm-1 channel with respect to the NCEP RTG (define!) SST. The mean bias is -0.64K and is stable at the rate of 5 mK per year. The data is collected from night (descending) orbits over clear ocean footprints with satellite zenith angle less than 35 degrees.

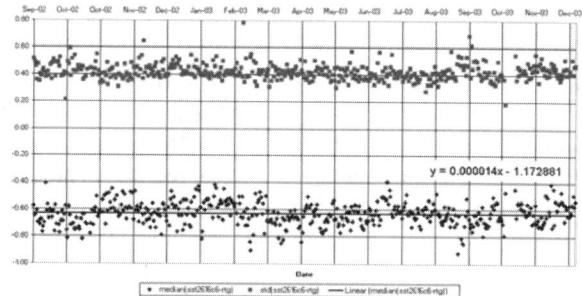

Fig. 3. Radiometric stability of AIRS.

Fig. 4 (courtesy of S. Gaiser) shows the spectral stability of AIRS. The AIRS spectral calibration software package estimates focal plane displacement from the shape of a selected set of absorption lines. The stability of this displacement is a direct measure of spectral stability. The AIRS was turned off when strong solar storm activity was observed at the end of October 2003. It took a couple of weeks to stabilize the spectral properties of AIRS. This shows up on the figure as a glitch starting at the end of October 2003, and lasting about three weeks. With the exception of this time period, the spectral properties are extremely stable. The figure seems to indicate a very small negative slope. The effect of this trend on spectral frequencies and radiances is estimated to be very small. We will need to observe the trend for a longer period of time before any analysis of the trend itself can be attempted.

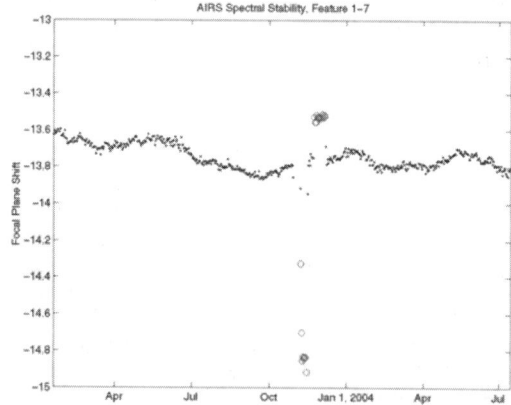

Fig. 4. Spectral stability of AIRS.

3. AIRS LEVEL 2 DATA PRODUCTS

The AIRS retrieval algorithm is described in Susskind et al [2]. This is a unified algorithm with many components from other AIRS science team members. First, MW only retrievals are performed and cloud clearing is attempted using the MW only retrieval as a first guess, then the regression retrieval is attempted to further refine the retrievals, and then the final physical retrieval is attempted.

It is estimated that only a few percent of AIRS observations are clear enough that the net effect of cloud on AIRS radiances is smaller than 1K. Through cloud clearing, that is, removing the effect of cloud on the radiances before attempting a geophysical retrieval, we can raise the yield to somewhere between 40 and 60% of AIRS observations at AMSU resolution, depending on the degree of tolerance for outliers. The cloud clearing is an extrapolation of cloudy radiances to clear radiances and exaggerates noise, especially when the contrast between AIRS footprints is small or when the cloud is optically thick. Also, the cloud clearing assumes uniformity (uniformity of what, temperature and moisture fields?) between AIRS footprints, which is not always met. When the uniformity assumption fails and is not detected, this algorithm produces outliers. Therefore, we work very hard to detect and understand outliers.

The standard AIRS level 2 parameters include temperature, water vapor, and ozone profiles, surface parameters such as surface skin temperature, and cloud parameters including effective cloud fraction and cloud top pressure. AIRS produces a combined IR/MW product almost 90% of all AMSU footprints in version 4.0. This does not imply that all AIRS products written to the output files are valid. There are many flags for version 4 that indicate the quality of particular products. It is important that data users are familiar with the quality flags. For example, the flag Qual_H2O should be examined when water vapor products are used. Also, surface skin temperature will have high quality only when the flag Qual_Surface indicates so.

All Level 2 products with the exception of cloud fraction are produced at the AMSU-A footprint resolution (45 km at nadir). The spatial resolution of AIRS cloud fraction is at the AIRS spatial resolution, or 15 km at nadir.

The AIRS Level 2 files range in size from 4.7 to 20 MB, with each file containing approximately 6 minutes of satellite coverage data, for a total of 240 files per day, per instrument, per processing level. The Level 2 files are broken down further into standard and support files; the size of a single 6 minute Level 2 standard file is 4.7 MB and for a single Level 2 support file is 18 MB. The total daily volume for Level 2 standard products is about 1.1 GB per day; for support products, the total is about 4.3 GB per day.

4. VALIDATION OF AIRS PRODUCTS

The version 3 AIRS level 2 data products became public in early 2003, available from the Goddard DAAC. This was the first release of level 2 data products. The validation of AIRS data products is proceeding in stages. The AIRS level 1b radiances, as well as the temperature profiles, are provisionally validated over non-polar ocean surfaces for version 3. The term "provisionally validated" means that the product in question can be used for scientific analysis with caution. The version 4 products are scheduled for public release in early 2005. More products will be validated as well as some products over non-frozen land surfaces.

The AIRS project set the goal of achieving "global sounding from satellite with the accuracy of radiosondes." With the upcoming version 4 release we will have achieved this goal over certain unfrozen land surfaces, as well as over unfrozen ocean. Fig. 5 shows the bias and the RMS difference of AIRS temperature profiles with respect to ECMWF analyses over the unfrozen ocean in a prerelease version of version 4 data products. We are achieving the goal of 1K RMS in 1 km layers in the troposphere.

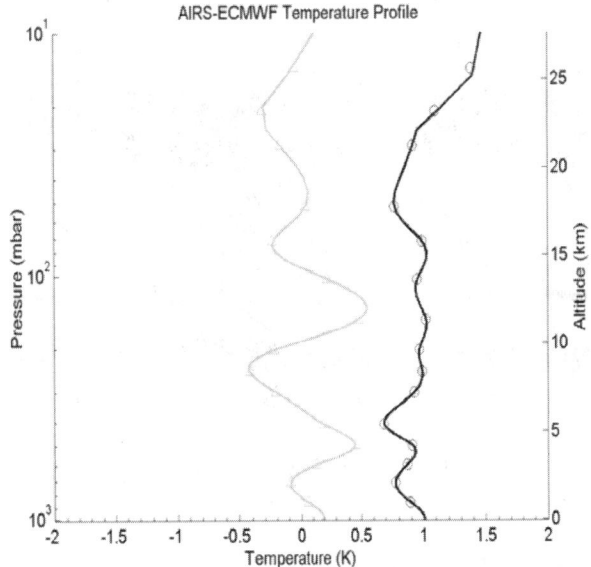

Fig. 5. Statistics w.r.t. ECMWF analysis.

Each release of AIRS data is accompanied by a validation report, a user guide, and other documents. The validation report for version 4 will include validation of water vapor, cloud products, ozone as well as temperature.

5. AIRS LEVEL 3 PRODUCTS

Level 3 products will be released with the version 4 AIRS data release. Many of the level 2 data products are average at a spatial resolution of 1x1 for each day. Ascending orbits (day side) will be averaged separately from the descending (night side) orbits. Since level 3 products include counts as well as means, multi-day level 3 products for any number of days can be generated from the daily level 3 products. The temporal resolution of the products generated at Goddard DAAC is daily, 8-day (half of the 16 day Aqua orbit repeat cycle) and monthly.

Fig. 6. Total precipitable water vapor for May 2003.

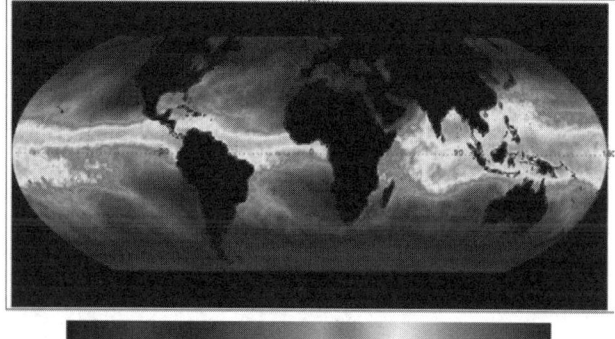

Fig. 7. MW only total precipitable water vapor for May 2003.

The above two figures are maps of monthly mean total precipitable water vapor from the descending orbits for May 2003. Fig. 6 is the AIRS/AMSU combined product, while Fig. 7 is MW only product. Notice that MW only products are used as a guess to the combined product. Since cloud clearing is known to fail for precipitating or very cloudy scenes, the MW product tends to be wetter than the combined product. This is why we have MW only products as well as combined products in the level 3. Although combined products are of higher quality, its sampling characteristics can become troublesome issues.

6. OTHER SOUNDER DEVELOPMENT AT JPL

JPL is developing two more sounders for possible future missions. Both are under Instrument Incubator Program (IIP) and applicable to MEO as well as GEO missions. T. Pagano is working with Ball Aerospace on Spaceborne InfraRed Atmospheric Sounder (SIRAS). This is a grating spectrometer like AIRS. B. Lambrigtsen is working on the Geostationary Synthetic Thinned Aperture Radiometer (GeoSTAR). This will have AMSU–like channels working from geostationary obit.

7. ACKNOWLEDGMENT AND CLOSING REMARKS

The authors wish to express sincere appreciation to our colleagues on the AIRS team for providing support and inputs for this work. This is only an introduction to AIRS related activities. Further information on AIRS can be found on AIRS project web page at http://www.jpl.nasa.gov/. AIRS data can be accessed through http://daac.gsfc.nasa.gov/atmodyn/airs/.

This work was performed at the Jet Propulsion Laboratory, California Institute of Technology, under contract with the National Aeronautics and Space Administration, through the office of the Earth Systems Enterprise.

8. REFERENCES

H. H. Aumann, et al., "AIRS/AMSU/HSB on the Aqua Mission: Design, Science Objectives, Data Products, and Processing Systems," *IEEE Special Issue on the EOS Aqua Mission,* Feb. 2003, pp 253–264.

J. Susskind, et al., "Retrieval of Atmospheric and Surface Parameters from AIRS/AMSU/HSB Data in the presence of Clouds," *IEEE Special Issue on the EOS Aqua Mission,* Feb. 2003, pp 390–409.

Lambrigtsen B., and S.-Y. Lee, "Coalignment and Synchronization of the AIRS Instrument Suite," *IEEE Special Issue on the EOS Aqua Mission,* Feb. 2003, pp 343–351.

EARTH REFLECTANCE SPECTRA FROM 300–1750 NM MEASURED BY SCIAMACHY

P. Stammes, J. R. Acarreta, W. H. Knap, and L. G. Tilstra
Royal Netherlands Meteorological Institute (KNMI)
P.O. Box 201, 3730 AE, De Bilt
The Netherlands

ABSTRACT

SCIAMACHY onboard ESA's Envisat satellite, launched in 2002, is a spectrometer covering the wavelength range 240–2380 nm with a spectral resolution between 0.2 and 1.5 nm. SCIAMACHY offers for the first time a contiguous spectral view of the Earth in the shortwave domain. Here we show some first top-of-atmosphere reflectance spectra of typical Earth scenes.

1. INTRODUCTION

Since the launch of GOME/ERS-2 in 1995, reflectance spectra of the Earth with moderately high spectral resolution of 0.2–0.4 nm have become available for the range 240–800 nm. SCIAMACHY is the successor of GOME, but with increased capabilities (Bovensmann et al., 1999). It has a much wider spectral range, including the near-infrared, and also has limb and occultation modes. SCIAMACHY is able to detect trace gases relevant to the ozone layer such as ozone and BrO, air pollution gases such as NO_2 and CO, and greenhouse gases such as H_2O, CO_2, and CH_4.

2. DESCRIPTION OF SCIAMACHY

SCIAMACHY has 8 spectral channels with diode-array detectors, having 1024 pixels per channel. Channels 1–5 cover the range 240–1000 nm, with a resolution of 0.2 to 0.5 nm, channel 6 covers the range 1000–1750 nm with a resoluton of 1.5 nm, whereas channel 7 and 8 cover the ranges 1940–2040 nm and 2265–2380 nm, with a respective resolution of 0.2 and 0.3 nm. SCIAMACHY measures both in nadir and limb, by alternatively switching its viewing mode. Here we only consider nadir data with a spatial resolution of 240×30 km^2.

SCIAMACHY observes the sun every day. Therefore, the Earth reflectance can be determined from the Earth radiance and solar irradiance measurements by SCIAMACHY. This accounts for some calibration errors, but not all. From comparison of the SCIAMACHY measured reflectances with UV radiative transfer model calculations (Tilstra et al., 2004) and with MERIS reflectances between 440 and 880 nm (Acarreta and Stammes, 2005) it appears that the SCIAMACHY reflectances are about 10–20 % too low. The reason for this radiometric calibration error probably lies in the pre-flight measurements. Ad-hoc correction for this radiometric error is possible by using a reflectance correction factor, but this has not been done in the results shown hereafter.

3. OBSERVED REFLECTANCE SPECTRA

The quantity used here is the Earth's reflectance:

$$R = \pi I / (\mu_0 E),$$

where I is the radiance reflected at top-of-atmosphere (TOA), in W/m^2/sr/nm, E is the solar irradiance incident at TOA perpendicular to the beam, in W/m^2/nm, and μ_0 is the cosine of the solar zenith angle. We focus on SCIAMACHY channels 2–6, i.e., the spectral range 300–1750 nm, because channels 1, 7, and 8 currently suffer from dark current problems.

We have selected five typical Earth scenes from Envisat orbit 2509, 23 August 2002: cloud-free ocean, cloud-free Sahara desert, cloud-free vegetated land, thick water clouds, and thick ice clouds. Scenes were selected using colocated images from MERIS, which is also onboard Envisat. See Figs. 1–3 for the spectra, which are discussed below. Note that the spectra are split into two parts: 300–800 nm and 800–1800 nm.

3.1 Ocean Scene

At UV and visible wavelengths the cloud-free ocean reflectance is dominated by atmospheric Rayleigh scattering. The combination of Rayleigh scattering and ozone absorption causes the characteristic peak in the reflectance around 330 nm, with a value of about 0.3. At 800 nm the reflectance in the continuum is only 0.02, and at 1750 nm only 0.01. Since water itself is almost black beyond 700 nm, this reflectance is due to aerosol and Rayleigh scattering, with possibly contributions from the ocean surface (glitter, white caps).

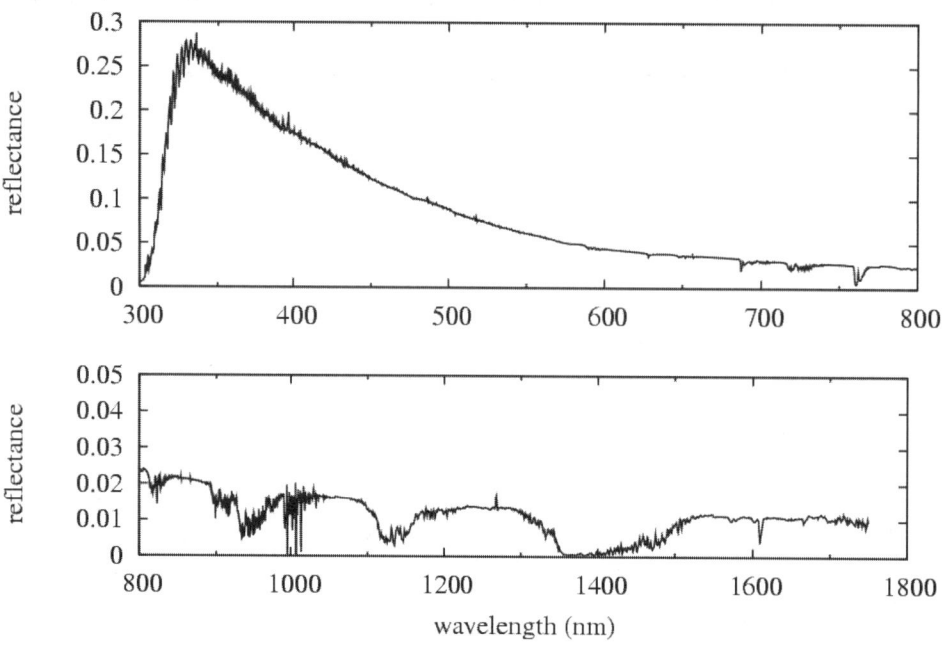

Fig. 1. SCIAMACHY nadir reflectance spectrum of a cloud-free ocean scene (45.2° N, 4.4° W), with $\mu_0 = 0.77$.

3.2 Land Scenes

Also the reflectances of the cloud-free land scenes contain the Rayleigh peak at 330 nm. The Sahara spectrum is dominated by the steadily rising continuum reflectance beyond 500 nm, which attains a value of almost 0.6 at 1700 nm. The vegetation spectrum shows a decreasing behaviour from 330 to 700 nm, except for a weak bump from 500–600 nm, which gives the green colour to vegetation. From 700–750 nm a steep rise of the reflectance occurs, which is known as the red edge. The vegetation reflectance remains about 0.2 from 800 to 1300 nm, and then decreases. Both land spectra show strong gaseous absorption bands beyond 700 nm (cf. Sect. 4).

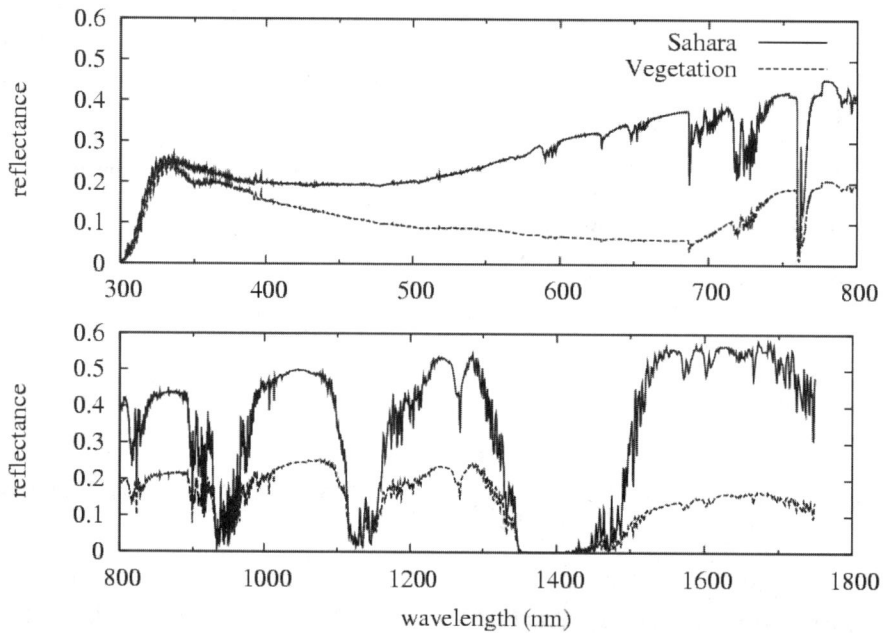

Fig. 2. SCIAMACHY nadir reflectance spectra of cloud-free land scenes: Sahara desert (31.1° N, 17.3° E) and vegetated land (53.9° N, 28.6° E). Here $\mu_0 = 0.87$ for desert and 0.74 for vegetation.

3.3 Clouds

The cloud spectra shown in Fig. 3 were chosen as being representative of optically thick clouds. The ice cloud spectrum and water cloud spectrum have a very similar behaviour from 300 to about 1200 nm. In that range the continuum reflectance is almost constant with wavelength. However, large differences between water and ice clouds occur from 1500 to 1750 nm. There ice clouds absorb more than water clouds, and have a different absorption spectrum.

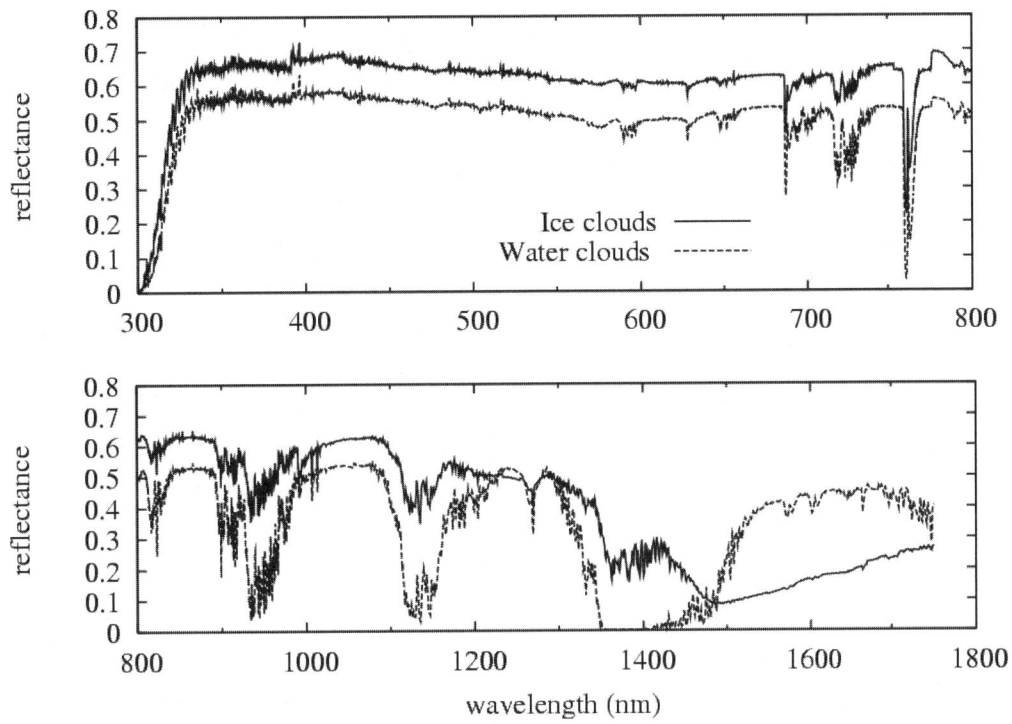

Fig. 3. SCIAMACHY nadir reflectance spectra observed over optically thick water and ice clouds: water clouds (60.1° N, 2.7° W) and ice clouds (14.8° N, 15.4° W). Here $\mu_0 = 0.64$ for the water clouds and 0.86 for the ice clouds.

This is due to the different refractive index of water and ice, and to the larger size of ice particles (Knap et al., 2002; Acarreta et al., 2004). The H_2O bands in the ice cloud spectrum are weaker than in the water cloud spectrum, due to the higher altitude of ice clouds.

4. COMPARISON WITH MODTRAN

To verify the atmospheric absorption bands in the SCIAMACHY spectra, we simulated a clear-sky reflectance spectrum with the MODTRAN4 radiative transfer model (version 1.1, Berk et al., 2000). We chose a triangular slit function chosen for the MODTRAN calculations with a FWHM of 10 cm^{-1}. This resolution is comparable to that of SCIAMACHY. The step size chosen was 5 cm^{-1}. In MODTRAN multiple scattering was included according to the 8-streams DISORT option. The atmospheric composition was that of the Midlatitude Summer standard atmosphere. No clouds or aerosols were included.

Fig. 4 shows the calculated nadir reflectance for a clear sky case with surface albedo 0.50. In this figure the major gaseous absorption bands have been indicated. As can be seen from comparison with the SCIAMACHY spectra, most easily with that of water clouds, the absorption bands of O_3, H_2O, O_2, CO_2, and CH_4 are indeed well reproduced by MODTRAN. However, some absorption bands, e.g., due to NO2 around 425 nm and O_2-O_2 at 477 nm and 570 nm, are missing in the MODTRAN simulation. Furthermore, the numerous small peaks in the UV and visible parts of the SCIAMACHY spectra, which are due to the Ring effect (rotational Raman scattering in the atmosphere), are not found in the MODTRAN simulations. From the comparison with the MODTRAN spectrum, it also appears that there are calibration errors in the SCIAMACHY spectra, which appear as jumps at the channel overlaps at 400 nm, 780 nm, and 1000 nm.

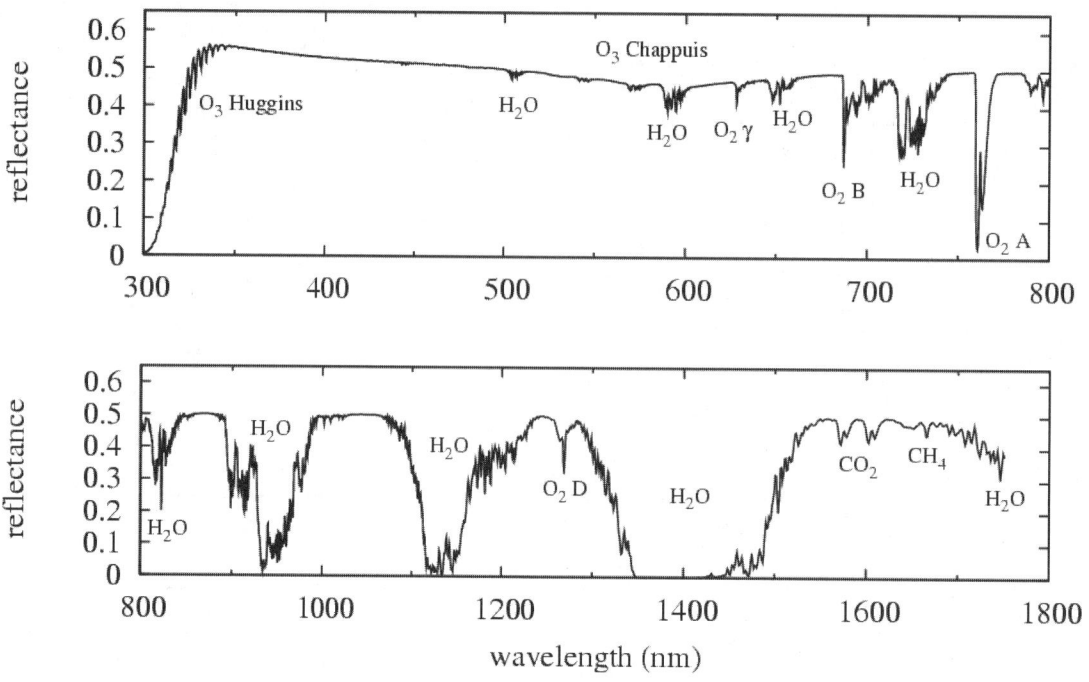

Fig. 4. MODTRAN4 simulation of nadir reflectance of a clear scene with surface albedo 0.50; $\mu_0 = 0.87$.

5. ACKNOWLEDGEMENTS

We are glad to acknowledge the assistance of Ankie Piters (KNMI) and the personnel of the Netherlands SCIAMACHY Data Center (http://neonet.knmi.nl/neoaf) in making the SCIAMACHY data accessible. SCIAMACHY data have been provided by ESA/DLR. The Modtran models have been kindly provided by G. Anderson and A. Berk (Air Force Research Lab, Hanscom, MA).

6. REFERENCES

Acarreta, J. R., P. Stammes, and W. H. Knap, "First retrieval of cloud phase from SCIAMACHY spectra around 1.6 micron," *Atmos. Res.*, Vol. 72, 89-105 (2004).

Acarreta, J. R., and P. Stammes, 2005, "Calibration comparison between SCIAMACHY and MERIS on board ENVISAT," *IEEE Geoscience and Remote Sensing Letters (GRSL)*, Vol. 2, 31-35, doi: 10.1109/LGRS.2004.838348.

Berk, A., et al., "MODTRAN4 user's manual," 1 June 1999 (Last revised 17 April 2000), Air Force Research Laboratory, Hanscom AFB, MA.

Bovensmann, H., J. P. Burrows, M. Buchwitz, J. Frerick, S. Noël, V. V. Rozanov, K. V. Chance, and A. P. H. Goede, 1999: "SCIAMACHY: mission objectives and measurement modes," *J. Atmos. Sci.* 56, 127-150.

Knap, W. H., P. Stammes, and R. B. A. Koelemeijer, "Cloud thermodynamic phase determination from near-infrared spectra of reflected sunlight," *J. Atmos. Sci.*, 59, 83-96 (2002).

Tilstra, L. G., G. van Soest, M. de Graaf, J. R. Acarreta, and P. Stammes, "Reflectance comparison between SCIAMACHY and a radiative transfer code in the UV," In: *Envisat Validation Workshop Proceedings (ACVE-2)*, ESA Special Publication SP-562, May 3-7, 2004, Frascati, Italy.

GLOBAL COMPARISON OF MODIS AND AVHRR-TYPE AEROSOL RETRIEVALS OVER THE OCEAN IN THE TERRA/CERES-MODIS SINGLE SCANNER FOOTPRINT (SSF) DATA

Tom X.-P. Zhao[1,2] and Istvan Laszlo[2]
[1]ESSIC/UMD, College Park, MD 20742, USA
[2]NOAA/NESDIS/ RA, Camp Springs, MD 20746, USA

ABSTRACT

The MODIS and AVHRR-Type aerosol retrievals over the ocean are compared and evaluated on a global scale by using the two aerosol products in the Terra/CERES-MODIS SSF data. The objective is to compare the consistency of the two aerosol products during their overlap period, and build a connection between them. Cloud and surface roughness effects on the two aerosol retrievals are identified, which need further investigation.

1. INTRODUCTION

Two aerosol products derived respectively from the MODIS and an AVHRR-like aerosol retrieval algorithm are available in the Terra/CERES-MODIS SSF data (Geier et al., 2003) for the evaluation of the improvement offered by the more advanced MODIS aerosol retrieval relative to the simple AVHRR aerosol retrieval. They also can be used for building connection and examining consistency between the 20-year historical AVHRR aerosol data and the new MODIS aerosol data.

2. THE SSF DATA

The entire year of 2001 Edition-1A SSF aerosol data were analyzed in this paper. There are two aerosol products (I and II) in the SSF dataset. Product I was obtained by averaging the standard MODIS aerosol products (see Remer et al., 2004) in a CERES footprint according to the CERES Point Spread Function (PSF). The MODIS Collection 3 aerosol product (MOD04) was used in the Edition-1A SSF data. Product II was derived from the independent two-channel (hereafter, AVHRR-type) aerosol retrieval algorithm of the NOAA/NESDIS (see Ignatov et al., 2004). The re-sampled MODIS clear-sky radiances at 1km resolution (MOD02 L1B Radiance) were averaged in a CERES footprint with a 20km resolution at nadir according to the CERES PSF and, then, input to the AVHRR-type aerosol algorithm to perform the retrieval for the footprint.

Table 1 of Ignatov et al. (2004) compared the two SSF aerosol retrievals in detail. The sampling approach, cloud screenings, surface reflection computation, and aerosol model assumptions are the four potential contributors to the differences in the two retrievals. The current analyses focus only on the last three contributors. To minimize the sampling issues in the analysis, only those SSF footprints that have both MODIS and AVHRR-type aerosol retrievals available were used. Moreover, the analyses used only monthly mean values of the two SSF aerosol products, which are subject to less sampling issues compared to daily and weekly mean values.

3. COMPARISON

Scatter plot, linear regression, and probability distribution function (PDF) techniques were used in the inter-comparison and analysis. The comparison focused on the retrieved aerosol optical thicknesses, τ_1 (at λ_1=0.66 µm channel) and τ_2 (at λ_2=1.60 µm channel), and the derived Ångström exponent α ($=-\ln[\tau_1/\tau_2]/\ln[\lambda_1/\lambda_2]$). Specifically, January, April, July, and October were used to check the seasonal variations.

Three CERES parameters, the Clear Strong Index (CSI), Cloud Fraction (CF), and Surface Wind Speed (SWS) were used to facilitate the comparison and analysis. A high CSI or low CF (both with values between 0% and 100%) means a CERES footprint is clearer and has less cloud effect (including cloud contamination, cloud edge effect, aerosol enhancement in the moist environment near cloud, etc.). The SWS was used to describe the surface roughness of a CERES footprint. The two SSF aerosol products were compared for four cases: 1) the original SSF data, 2) the strong clear case (with minimal cloud effect) defined by sampling the CERES footprints with CSI > 90%, 3) the smoothest surface condition determined by sampling the CERES footprints with SWS < 1m/s, 4) the clearest and smoothest case defined by sampling the CERES footprints with both CSI > 90% and SWS < 1 m/s. These cases were designed to identify the cloud and surface roughness effects on the two SSF aerosol retrievals.

3.1 Global Inter-Comparison

As an example, Fig. 1 displays the time series of global monthly mean τ_1 during 2001 for the four cases. The MODIS τ_1 is larger than the AVHRR-type τ_1 year-round in Case 1.

The difference (between the solid lines and the corresponding dashed lines) is reduced for Cases 2–4 with the AVHRR-type τ_1 becoming slightly larger than the MODIS τ_1. This result implies there is a difference in the response of the two aerosol retrievals to the cloud effect. The remaining difference in the ideal Case 4 is associated with the difference in the aerosol model and surface assumptions and the retrieval procedures in the two retrieval algorithms since the other effects have been minimized through the specified samplings. It should be pointed out the discrepancies between the two τ's for the ideal Case 4 do not represent the global mean difference anymore due to the resulting limited sample size. However, it does indicate that the assumptions used by the two algorithms can make a significant difference in the regions covered by these limited sample points.

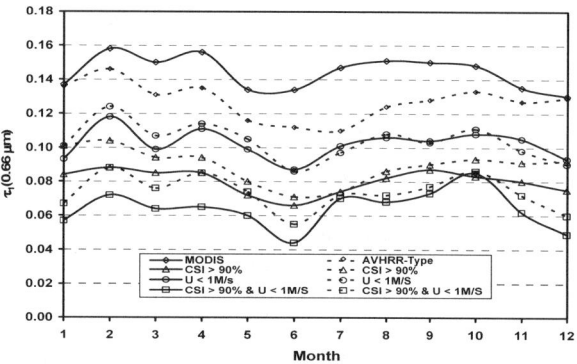

Fig. 1. Time series of global monthly mean τ_1 during 2001 for the four cases defined in the text. The solid line is for MODIS and the dashed line is for AVHRR-Type.

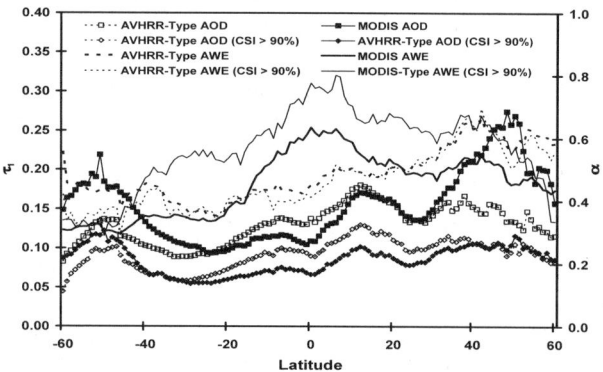

Fig. 2. Four-month averaged zonal mean τ_1 and α. AOD is aerosol optical depth and AWE is Ångström wavelength exponent.

The zonal mean τ_1 and α averaged for four months (January, April, July, and October) are shown in Fig. 2. The major difference between the MODIS and AVHRR-type τ_1 are at high latitudes of both hemispheres (the maximum difference of τ_1 can be up to ~0.1). These differences are reduced significantly for the strong clear case (the maximum is reduced to ~ 0.03) mainly due to the reduction of the MODIS values, which suggests the cloud effects associated with the MODIS retrieval are stronger than those associated with the AVHRR-type, especially over the high latitudes. The zonal mean differences in α are more noticeable than in τ_1, with the MODIS α being larger than (can be up to ~0.2) the AVHRR-type α in the tropics but smaller at highest latitudes. The differences at the high latitudes are reduced significantly in the strong clear case but enhanced over the tropics. The different behaviors of α at high latitudes and over the tropics suggest the cloud effects on the two retrievals are different in these two regions, which needs further studies.

The monthly mean 1 x 1 values of the two aerosol products were also compared in scatter plots for the four cases. A linear regression (e.g., $\tau_M = a + b\,\tau_A$) was performed. For the majority of grids in Case 1, the two τ agree reasonably well for each season. However, there are still some outliers and, for most of them, the MODIS values are significantly larger than the AVHRR-type values. The outliers are reduced in the other three cases and the RMS error and the correlation of the linear regression are also improved as shown in Table 1. Since errors in aerosol model and surface treatment mainly affect the slope (b) and the offset (a), respectively, in the regression equation (see Zhao et al., 2002; 2003), the non-unity slope b in the ideal Case 4 of Table 1 is mainly due to the different assumptions in the aerosol model rather than the different surface treatment in the two retrieval algorithms.

The scatter-plot comparison of α shows that the two aerosol products do not agree as well as τ in the case of original data since α is more sensitive than τ to the retrieval uncertainties, especially to those associated with aerosol model assumptions. Unlike τ, the two α's compare even worse for the less disturbed cases; for example, the RMS and correlation are gradually degraded from Case 1 to Case 4 (not shown here). This indicates that cloud and surface roughness effects may mask the difference in α caused by the difference in the aerosol model assumptions in the two aerosol retrievals. The difference in α caused by the aerosol model assumptions in the two retrieval algorithms becomes more evident after the other effects have been minimized as in Case 4. The MODIS α associated with a *dynamic aerosol model* is, in general, larger than the AVHRR-type α associated with a *fixed aerosol model*.

3.2 Ensemble AERONET Validation

The two SSF aerosol products were also compared to the AERONET observations (Holben et al., 1998) at thirteen AERONET sites selected as the baseline validation sites for the operational AVHRR aerosol retrieval (see Zhao et al., 2002). Quality assured Level 2 AERONET data were used as "ground truth". The spatial match-up window consists of an outer circle with a 150 km radial distance from the site, excluding an inner circle (with a fixed radius of 25 km) to reduce the effects of coastline or shallow water influences. The temporal match-up window is within ±1 hour of satellite overpass. The 110 match-up points found in 2001 over the 13 sites were put together to make an ensemble (or global) validation. The ensemble mean validations are shown in Fig. 3 for τ_1, τ_2 and α and two scenarios: 1) the original match-up points, 2) the strong clear scenario determined by selecting the match-up records with a criterion of CSI > 90% from the original match-up records at each match-up point.

Table 1. Linear Regression ($\tau_M = a + b\ \tau_A$) Statistics Derived for the 1°x1° Values of the SSF MODIS and AVHRR-type τ_1 in January, April, July, and October. RMS and r are the Root Mean Square Error and the Correlation Coefficient of the Regression, Respectively. Cases 1, 2, 3, and 4 are the Original Data, the Strong Clear, the Smoothest Surface, and the Clearest and Smoothest Cases

Month	Case	a	b	r	RMS
Jan.	1	0.02	0.90	0.749	0.049
	2	0.00	0.81	0.895	0.030
	3	0.00	0.84	0.891	0.043
	4	0.00	0.83	0.927	0.025
April	1	0.03	0.93	0.694	0.078
	2	0.02	0.70	0.879	0.034
	3	0.01	0.82	0.849	0.056
	4	0.00	0.81	0.948	0.030
July	1	0.04	0.88	0.726	0.071
	2	0.03	0.60	0.808	0.034
	3	0.02	0.78	0.816	0.058
	4	0.02	0.64	0.854	0.043
Oct.	1	0.05	0.74	0.538	0.074
	2	0.02	0.70	0.742	0.040
	3	0.01	0.77	0.820	0.051
	4	0.01	0.72	0.892	0.036

The MODIS and AVHRR-type τ of the strong clear scenario agree better with the AERONET than the original match-ups. The absolute difference of the MODIS and AVHRR-type τ is almost the same for the two scenarios. These results suggest that both MODIS and AVHRR-type retrievals are subject to some cloud effects as indicated in the above global inter-comparison. The AERONET τ stays the same in the two data populations, but both satellite values drop in cloud-free conditions. If the aerosols were really different in the two situations then the AERONET value should also drop. Thus, it is reasonable to conclude the cloud effects on the original match-up data of the two SSF aerosol products should be mainly due to the residual sub-pixel cloud contamination rather than the real aerosol signals in the vicinity of clouds.

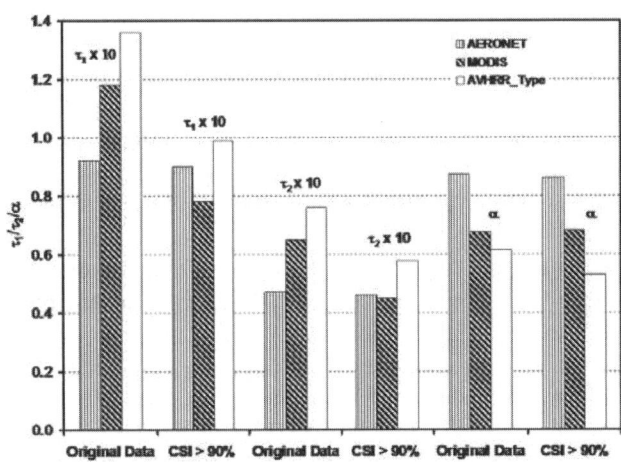

Fig. 3. Bar plots of the two SSF and the AERONET aerosol data for τ_1, τ_2, and α. The ensemble mean values of all match-ups are displayed for two scenarios described in the text.

The ensemble validation shows that the τ derived from the simple independent two-channel AVHRR-type retrieval is comparable to that derived from the multi-channel MODIS aerosol retrieval (e.g., $\Delta\tau_1 \sim 0.02\ \&\ \Delta\tau_2 < 0.02$) for this sample of stations that are dominated by clean, background marine aerosols. This is not unexpected since most of the additional information in a multi-channel inversion is applied to retrieve the size parameter, not to improve the basic τ retrieval.

The difference in α between the AERONET and MODIS for the strong clear scenario is almost the same as for the original match-ups, but the corresponding α difference between the AERONET and AVHRR-type increases in the strong clear scenario. Moreover, the difference between the MODIS and AVHRR-type α also increases in the strong clear scenario. This enhanced difference is due to the differences in the two retrieval algorithms (such as the aerosol model assumptions). The ensemble comparison with the AERONET "ground truth" suggests that cloud contamination masks the difference in α caused by the difference in aerosol model assumptions. The α (which is more sensitive to aerosol model assumptions) derived from the multi-channel MODIS retrieval (with more consistent treatment of aerosol model by using more channels) is better than that derived from the simple AVHRR-type retrieval (with less consistent treatment of aerosol model due to the limited number of channels). Thus, the improvement in the aerosol retrievals from the multi-channel algorithm relative to the simple two-channel algorithm is mainly realized in the aerosol size parameter rather than in the aerosol optical thickness, at least in clean background marine aerosol conditions.

4. SUMMARY

As a final summary of the global comparison and analysis, the annual means of the global MODIS τ_1, τ_2, and α in the SSF data of 2001 are given in Table 2 along with their absolute and relative differences with the corresponding AVHRR-type values. The values for the original SSF data and the strong clear condition (CSI > 90%) are listed together to demonstrate that the cloud effects cannot be neglected even for the global annual mean values. The difference in the optical thicknesses ($\Delta\tau$) and size parameter ($\Delta\alpha$) between the two SSF aerosol products are less than 0.02. The values of $\Delta\tau$ are reduced to less than 0.01 if the cloud effects are minimized but the value of $\Delta\alpha$ is increased up to 0.13. More detailed comparison and analyses of the two SSF aerosol data have been documented in two companion papers (Zhao et al., 2005a,b).

Table 2. Annual Means of τ_1, τ_2, and α of the Global SSF MODIS Data. Their Differences from the Corresponding AVHRR-type Values are also Given for the Original Data and the Strong Clear Conditions

	Original Data	Strong Clear
τ_1(MODIS)	0.144	0.080
$\Delta\tau_1$(AVHRR-MODIS)	-0.017	0.009
$\Delta\tau_1/\tau_1$(MODIS)	-11.8%	11.3%
τ_2(MODIS)	0.100	0.056
$\Delta\tau_2$(AVHRR-MODIS)	-0.018	0.006
$\Delta\tau_2/\tau_2$(MODIS)	-18.0%	10.7%
α(MODIS)	0.419	0.493
$\Delta\alpha$(AVHRR-MODIS)	0.019	-0.129
$\Delta\alpha/\alpha$(MODIS)	4.5%	-26.2%

5. ACKNOWLEDGMENTS

Drs. Patrick Minnis and Lorraine Remer at the NASA/Langley and NASA/GSFC, respectively, are acknowledged for their collaboration. The authors would like to thank Dr. Jerry Sullivan of NOAA/NESDIS/Office of Research and Applications for proofreading the manuscript. We also appreciate the effort of the CERES, MODIS, AERONET, and SIMBIOS scientists in producing the data used here. The funding support is from the NASA Radiation Program and the IPO/NPOESS project.

6. REFERENCES

Geier, E. B., R. Green, D. P. Kratz, P. Minnis, W. F. Miller, S. K. Nolan, and C. B. Franklin, 2003: CERES Data Management System: Single Satellite Footprint TOA/Surface Fluxes and Clouds (SSF) Collection Document. Release 2, Version 1, 212 pp., Radiation and Aerosol Branch, Atmospheric Sciences Research, NASA Langley Research Center, Hampton, Virginia.

Holben, B. N., T. F. Eck, I. Slutsker, D. Tanre, J. P. Buis, A. Setzer, E. Vermote, J. A. Reagan, Y. J. Kaufman, T. Nakajima, F. Lavenu, I. Jankowiak, and A. Smirnov, 1998: AERONET-A federated instrument network and data archive for aerosol characterization, *Remote Sens. Environ.*, 66, 1-16.

Ignatov, A., P. Minnis, N. Loeb, B. Wielicki, W. Miller, S. Sun-Mack, D. Tanre, L. Remer, I. Laszlo, and E. Geier, 2004: Two MODIS aerosol products over ocean on the Terra and Aqua CERES SSF datasets, *J. Atmos. Sci.*, in press.

Remer, L. A., Y. J. Kaufman, D. Tanre, S. Mattoo, D. A. Chu, J. V. Martins, R.-R. Li, C. Ichoku, R. C. Levy, R. G. Kleidman, T. F. Eck, E. Vermote, and B. N. Holben, 2004: The MODIS aerosol algorithm, products and validation, *J. Atmos. Sci.*, in press.

Zhao, X., L. Stowe, A. Smirnov, D. Crosby, J. Sapper, and C. R. McClain, 2002: Development of a global validation package for satellite oceanic aerosol optical thickness retrieval based on AERONET observations and its application to NOAA/NESDIS operational aerosol retrievals, *J. Atmos. Sci.*, 59, 294-312.

Zhao, X., I. Laszlo, P. Minnis, and R. Remer, 2005a: Comparison and analysis of two aerosol retrievals over the ocean in the Terra/CERES-MODIS Single Scanner Footprint (SSF) Data: Part I – global evaluation, *J. G. R.*, in revision.

Zhao, X., I. Laszlo, P. Minnis, and R. Remer, 2005b: Comparison and analysis of two aerosol retrievals over the ocean in the Terra/CERES-MODIS Single Scanner Footprint (SSF) Data: Part II – regional evaluation, *J. G. R.*, in revision.

AEROSOL AND CO LOADING IN THE ATMOSPHERE OBSERVED BY THE MODIS AND MOPITT – RUSSIAN FOREST FIRE CASE

J. Kim, S. H. Choi, H. C. Lee, and H. K. Cho
Yonsei University, Seoul 120-749, Korea (Rep. of)

D. Edwards
ACD, NCAR, Boulder, CO, U.S.A.

S. H. Lee, H. S. Lim, and G. H. Choi
Korea Aerospace Research Institute, Daejon 305-333, Korea (Rep. of)

ABSTRACT

Intense fires in the southeast part of Russia in May, 2003 are studied with the satellite data from MODIS and MOPITT. Reasonably good correlation between CO and AOD are found for the forest fire case near the source region. This multi-instrument approach to monitor the aerosol in the atmosphere is expected to contribute to the classification of the aerosol characteristics in the atmosphere.

1. INTRODUCTION

Aerosols have been studied extensively by using both the satellite and ground-based observation as aerosol is one of the major uncertainty in climate change (IPCC, 2001). Since 1999, the MODIS provides global observation of aerosol onboard the Terra and Aqua satellites. The satellite data on AOD have provided useful information on the estimation of aerosol loading and the tracking of the yellow sand events. Furthermore, the MODIS provides retrieved AOD along with the fine mode fraction over ocean and darker land surfaces. However, there still is lack of AOD information over the land surfaces with higher reflectance which usually corresponds to the source region of aerosol. The current MODIS product distinguishes the fine and coarse mode aerosol. In the meantime, there still exists need to distinguish the fine mode aerosol into carbonaceous and sulfate, for example. MOPITT onboard the Terra satellite provides quantitative information of carbon monoxide (CO). Measurements of CO whose principal sources arise from anthropogenic and natural emissions such as biomass burning and forest fires, is very useful for tracing fire emissions in the atmosphere. Wild fires and biomass burning emit CO into the atmosphere and results in high AOD. As the measurements of CO become available together with the fine mode AOD from satellite remote sensing, the possible correlation between these two quantities are expected to contribute to distinguish carbonaceous from fine mode aerosol which also includes sulfate. Thus, large scale forest fires would be an excellent case to see the correlation between the CO and AOD as a first stage. There have been persistent forest fires from spring to autumn in the southern areas of eastern Russia bordering China and Mongolia. They start with snow-melt and rapid drying of forest fuel at the onset of dry weather in March and April (Edwards et al., 2004). Statistics on the forest fires show two peaks in May and September in fire counts (International Forest Fire News (IFFN), 2003).

In May of 2003, this Russian forest fires have affected the air quality of Korea, located near the southeastern part of Russia. In this study, intense fires in the southeast part of Russia in May, 2003 are studied with the satellite data from MODIS and MOPITT. The correlation between CO and AOD are analyzed throughout the East Asia region. This multi-instrumental approach to monitor the aerosol together with the information on composition in the atmosphere is expected to contribute to the classification of the aerosol characteristics in the atmosphere, carbonaceous aerosol in particular.

2. DATA

AOD and CO column densities are directly proportional to total column amount of aerosol and CO over a unit area, respectively, thus have similar physical meaning despite different physical quantities and thus in different units. To compare AOD data with CO column densities, three cases are selected to see the difference among the clear sky, coarse mode (yellow sand) and fine mode aerosol(forest fire) cases. As these two datasets have different orbit characteristics, spatial and temporal coverage, datasets are reprocessed at 1x1 degree grid for comparison and simple mathematics between the two. To have full spatial coverage in this East Asian region and to avoid instantaneous biases of the CO column density measurements, 8 day average datasets are selected.

Considering the lifetime of typical aerosol in the atmosphere, this 8-day average has limitation in identifying instantaneous picture of aerosol source and sink, but it is the current limit to have meaningful spatial coverage and still provides general feature of the aerosol.

AOD and CO column densities are compared in Fig. 1 only for the forest fire cases among the cases stated above. The fine mode aerosol of forest fire case is selected which have affected the air quality badly over Korea throughout the whole week of May, 2003(Lee et al., 2004). The satellite image and trajectory analysis in this period showed the southward movement of aerosols clearly from the source region.

The AOD at 0.5 micron measured at Anmyeon from the AERONET showed values exceeding 3.0 during this period. Although the data plot is not shown here, hourly mean values of PM10 at Seoul were measured to be up to 250 g/m^3, and the visibility was reduced to about less than 2 km during the whole week of May, 2003, compared to normal 6 to 10 km values at Yonsei University, Seoul, Korea. The direct component irradiance was reduced down to about 20% of the diffuse component due to heavy aerosol loading in this period based on the measurements from the Rotating Shadowband Radiometer at the same station. The air quality over Korea during this Russian forest fire events can be seen in the satellite data of AOD and CO column densities as shown in Fig. 1(a) and (b).

The CO column densities are enhanced significantly over the region for the forest fire cases when compared with other cases, clear sky case in particular. The CO concentrations at 700 hPa from the MOPITT for May, 2003 also show enhanced values, which indicate the high rise of plumes into the atmosphere. The CO column density in clear sky condition may indicate the background CO level over Korea. In general, the concentrations of CO at surface in Korea show seasonal variation with its maximum in winter due to increase of fuel burning with concentrations ranging from 0.5 to 1.3 ppm in urban areas, and somewhat lower values and less seasonal variations in natural background areas.

Fig. 2 shows the ratios of CO concentration at 700 to 850 hPa taken from the MOPITT measurements which are expected to provide information on the vertical motion. The region with higher ratio indicates stronger upward motion compared to other regions with lower values. The ratio shows enhanced values as the emitted CO moves along the trajectory toward the south. However, the values of ratios are close to 1 which implies that the CO concentrations at two levels are somewhat correlated. Vertical cross section of CO concentration in Fig. 3 shows the enhanced emission of CO into the atmosphere near the source region, which also supports the analysis from Fig. 2. Spatial correlation between the AOD and CO will be discussed later.

3. ANALYSIS AND RESULTS

Both the fine mode aerosol and CO arises from the anthropogenic emission or burning. However, sulfates contribute only to the fine mode aerosol. The major sink for the aerosol is dry and wet deposition, while the sink for CO is the oxidation by OH producing CO_2 and HO_2 as a third body reaction with O_2. One of the important difference between the two quantities are lifetime: hours to week for aerosol and several months for CO. These similarities and differences in source and sink of the aerosol and CO can be utilized to analyze and deduce additional information from the satellite observation. For example, a region with higher correlation between the two can be regarded as a source region for carbonaceous aerosols. The difference in lifetime of the two suggests changes of the correlation with respect to time during transport, and can be also utilized to enhance the aerosol and CO signals and to locate the source and sink of plumes in the atmosphere. As the lifetime of CO is long compared to that of aerosol in the atmosphere, the ratio of AOD to CO can be used to trace the emission source and sink regions in the atmosphere. This ratio is expected to be larger near the source region and smaller near the sink, because the aerosol is deposited to the ground leaving the CO in the atmosphere.

Fig. 4(a) shows an example of the AOD to CO ratio for the forest fire case. Considering the persistent residence of CO in the atmosphere, higher ratio is expected near the source region, and lower values near the sink region. Thus the AOD and CO data are complimentary each other. Foreward trajectory analysis using the HYSPLIT model for the source region (Fig. 4(b)) shows actual movement of air parcels during the event. Backward trajectory for the sink region (not shown here) also showed similar pattern toward the source region as expected. North to south movement of the forest fire emissions toward the Korean Peninsula and the East Sea is evident in the analysis, which affected the air quality over Korea during the whole week.

In order to see the regional pattern similarity between the fine-mode AOD and CO column density, pattern correlation between the two images are obtained over 7x7 pixels as shown in Fig. 5. Although the similar features can be obtained from pixel by pixel calculation, this resolution is better in viewing clear structure of the pattern correlation. Most of the regions show relatively low values close to zero.

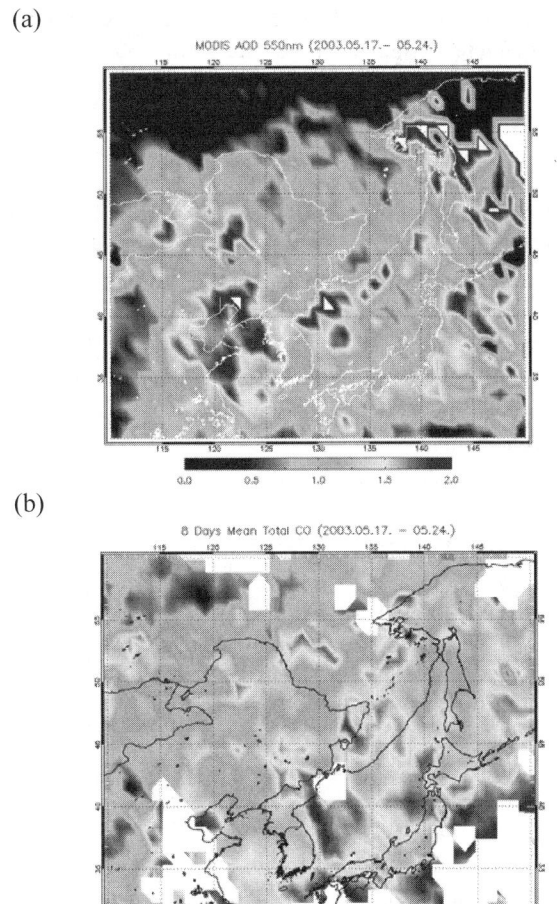

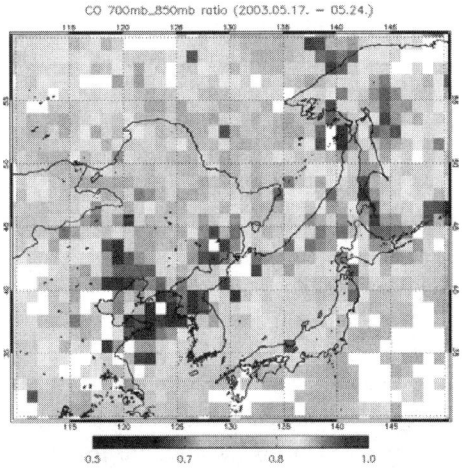

Fig. 1. (a) AOD from MODIS, and (b) CO column densities from MOPITT, respectively, for the Russian forest fire cases in May, 2003.

Fig. 2. Ratio of CO concentration at 700 to 850 hPa, measured from MOPITT during May, 2003.

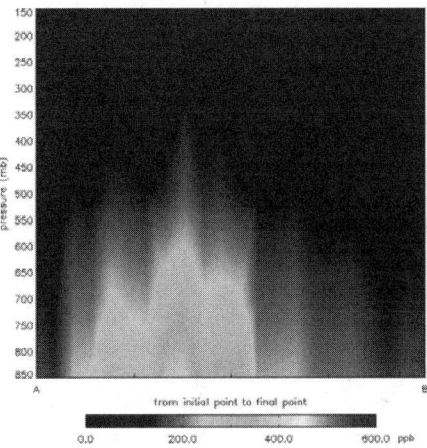

Fig. 3. Vertical cross section of CO concentrations along the trajectory from source to the Korean Peninsula.

However, in the source region near the border of southern Russia, China and Mongolia, highly positive correlation was obtained, and in the Korean Peninsula, highly negative correlation is obtained. These highly negative values can be regarded as different movement of aerosol and CO from the emission source due to different lifetime in the atmosphere. Fig. 1(b) show enhanced CO levels in the East Sea of Korea while enhanced aerosol in the Yellow sea of the Korean Peninsula. In the sense of tracing emissions from the biomass burning, it would be helpful to see the magnitudes of the correlation, which would provide information on the transport of highly-correlated air masses. Thus the region with highly correlated values indicates either source or sink region of forest fire emissions. Positive or negative correlation is an indication of the same or different movement of aerosol and CO. This correlation analysis together with the previous AOD/CO ratio help to trace the emission source and sink regions in the atmosphere. The pattern correlation indicates the horizontal movements of the emission while the AOD/CO ratio infers vertical movements.

4. CONCLUSION

To analyze the aerosol and CO loading in the atmosphere for the case of the Russian forest fire in May 2003, AOD from MODIS and total column CO density from MOPITT were analyzed. HYSPLIT trajectory model was used to aid the view of the transport during the event. Reasonable correlation between CO and AOD are obtained for the forest fire case especially near the source region. The analysis can be used to trace long-range transport of fire emissions. This type of correlation studies between CO and AOD would help distinguish carbonaceous from fine mode particles. AOD/CO ratio can help in analyzing the MOPITT CO data to locate the source and sink region.

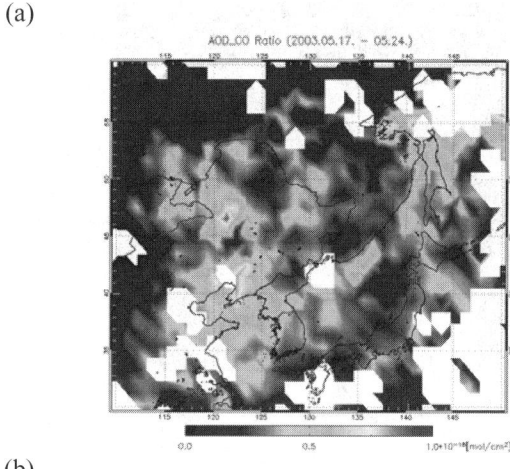

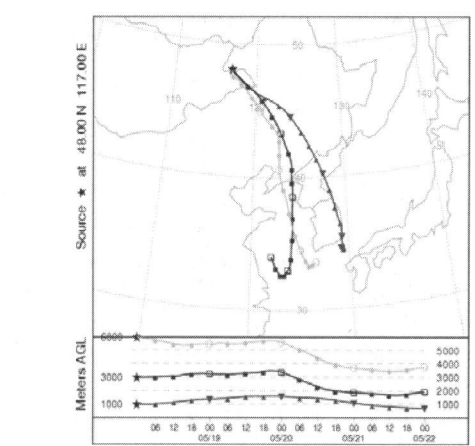

Fig. 4. (a) Ratio of fine-mode AOD and CO column density for the Russian forest fire case, (b) foreward trajectory analysis using the HYSPLIT model for the source region.

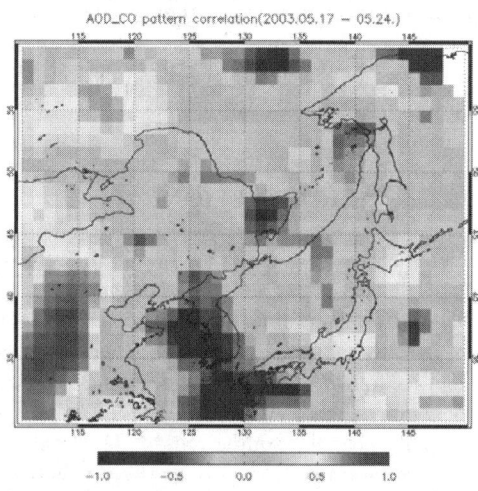

Fig. 5. Pattern correlation between the find-mode AOD and total CO column density for the forest fire case.

In the future study, time lagged correlation needs to be considered which requires more frequent observation. Combining MODIS and MOPITT with other satellite data such as SCIAMACHY would provide better coverage and help us to locate source and sink for aerosol and CO. However, when we deal with the same CO measurements with the two different instruments, the calibration between the MOPITT and SCIAMACHY needs to be considered carefully. Further analysis on pattern correlation between the AOD and the CO can provide information on classification for carbonaceous aerosols for the fine mode aerosol. Comparison of the current results with the chemical transport model will help to understand the detail physical and chemical mechanisms.

5. ACKNOWLEDGMENT

This work is supported by the COMeS project of KARI and METRI, Korea. Authors gratefully acknowledge the NOAA Air Resources Laboratory (ARL) for the provision of the HYSPLIT transport and dispersion model used in this publication.

6. REFERENCES

D. P. Edwards, L. K. Emmons, D. A. Hauglustaine, A. Chu, J. C. Gille, Y. J. Kaufman, G. P´etron, L. N. Yurganov, L. Giglio, M. N. Deeter, V., Yudin, D. C. Ziskin, J. Warner, J. -F. Lamarque, G. L. Francis, S. P. Ho, D. Mao, J. Chen, E. I. Grechko, and J. R. Drummond, Observations of carbon monoxide and aerosols from the Terra satellite: Northern Hemisphere variability, *J. Geophys. Res.*, 109, D24202, doi:10.1029/2004JD004727, 2004.

IFFN, Russian Federation 2002 Fire Special, *International Forest Fire News*, 28, Parts I–IV, 2–32, 2003.

IPCC, 2001: Climate Change 2001: The Scientific Basis. Contribution of Working Group I to the Third Assessment Report of the Intergovernmental Panel on Climate Change, Houghton, J. T., Y. Ding, D. J. Griggs, M. Noguer, P. J. van der Linden, X. Dai, K. Maskell, and C. J. Johnson (eds.), Cambridge University Press, Cambridge, United Kingdom and New York, NY, USA, 881 pp., 2001.

Lee, Kwon H., Jeong E. Kim, Young J. Kim, J. Kim, and Wolfgang von Hoyningen-Huene, Impact of the Smoke Aerosol from Russian Forest Fires on the Atmospheric Environment over Korea during May 2003, in press, *Atm. Environ.*, 2004.

CHARACTERISTICS OF SATELLITE-OBSERVED MODIS DATA DURING FOGGY DAYS NEAR THE INCHON INTERNATIONAL AIRPORT, KOREA

Y.-M. Kim and J.-M. Yoo
Ewha Womans University
Seoul, South Korea

ABSTRACT

Simultaneous observations of MODIS radiometer onboard the Aqua and Terra satellites and weather station at ground near the Inchon International Airport (37.2–37.7 N, 125.7–127.2 E) had been utilized in order to analyze the characteristics of satellite-observed data under fog and clear-sky conditions. The differences in brightness temperature between 3.75 μm and 11.0 μm were used as threshold values for remote sensing fog detection. The $Tb_{3.7\mu m}-Tb_{11\mu m}$ value during day-time was greater when it was fog condition than that when it was clear-sky, but $Tb_{3.7\mu m}-Tb_{11\mu m}$ value during night-time was opposite to day-time. Since the 3.75 μm-near IR was affected by solar and IR radiations in the day-time, both IR and visible have been used to detect fog. Reflectance at 0.65 μm during fog condition was higher than during clear-sky.

1. INTRODUCTION

A low visibility by fog causes traffic accidents over the land, sea, and air. Accurate forecasts about the fog occurrence and dissipation may bring economical effect because they can prevent a lot of accidents and disasters in advance. Fog observations at ground weather stations have limitation because of sparse stations and the difficulty of night-time observations (Ahn et al., 2003). The remote sensing of fog from satellite observations provides its information over vast areas in a short period.

The brightness temperature, emitted by fog (i.e., a cloud near surface), is similar to surface temperature. Emission of opaque water is 1.0 in the infrared channel at 11 μm, and 0.8–0.9 at 3.7 μm. So, the brightness temperature at 11 μm under fog is close to surface temperature, but the brightness temperature at 3.7 μm becomes lower than the surface temperature (Eyre et al., 1984).

The purpose of this study is to analyze the characteristics of the MODIS channels for foggy and clear days during the period of 2002–2003 near the Inchon International Airport. Here we derived the difference (i.e., $Tb_{3.7\mu m}-Tb_{11\mu m}$) in brightness temperature between 3.7 μm and 11 μm during foggy days and nights, and also other threshold value of solar reflectance in the day-time.

2. DATA AND METHOD

The MODIS that has 36 channels offers satellite data by passing over the equator four times a day (TERRA-10:30, 22:30 ECT AQUA-1:30, 13:30 ECT). This study used the MODIS data of channels 1, 2, 3, 4, 5, 6, 7, 20, 31 near the Inchon International Airport.

The brightness temperature of the MODIS channels during foggy days is different from that during clear days because of the infrared absorption and emission by water particle. Thus, fog in this study could be detected by comparing characteristics of each channel during fog with those during clear-sky condition. Fog is defined that horizontal visibility is below 1 km, while clear-sky is defined that the amount of clouds is zero. Based on the foggy and clear days classified by ground observation from the weather forecast center in Yongjong-do, foggy and clear-sky days were chosen separately with simultaneous satellite observations of the MODIS over the area of this study.

This study examined the threshold value of $Tb_{3.7\mu m}-Tb_{11\mu m}$ based on the above fog condition and clear-sky, together with the MODIS data of solar and infrared radiation.

Table 1. Values of Brightness Temperature (Tb) in the MODIS Channels of 3.75 μm, 11 μm, 12 μm, 13.3 μm and the Tb Difference (Tb$_{3.7\mu m}$–Tb$_{11\mu m}$) Between 3.75 μm and 11 μm During the Period from Dec. 15, 2002 to Dec. 14, 2003 at the Inchon International. The Number in Parentheses Means Standard Deviation

Weather		case	T (K)				T$_{3.7-11}$
			3.7 μm	11 μm	12 μm	13.3 μm	
day	fog	4	298.4 (10.0)	262.6 (3.9)	262.0 (3.9)	249.9 (5.6)	35.86 (10.43)
	clear	4	296.0 (5.8)	281.7 (6.9)	281.8 (6.7)	259.0 (2.9)	14.37 (2.27)
night	fog	12	275.4 (6.4)	272.1 (6.5)	271.8 (6.9)	255.8 (5.8)	3.29 (3.35)
	clear	12	277.9 (5.3)	273.2 (6.0)	273.3 (6.0)	256.8 (5.5)	4.67 (1.48)

3. RESULT

The cases of foggy and clear days under same time zones during each day and night were compared and analyzed (Fig. 1).

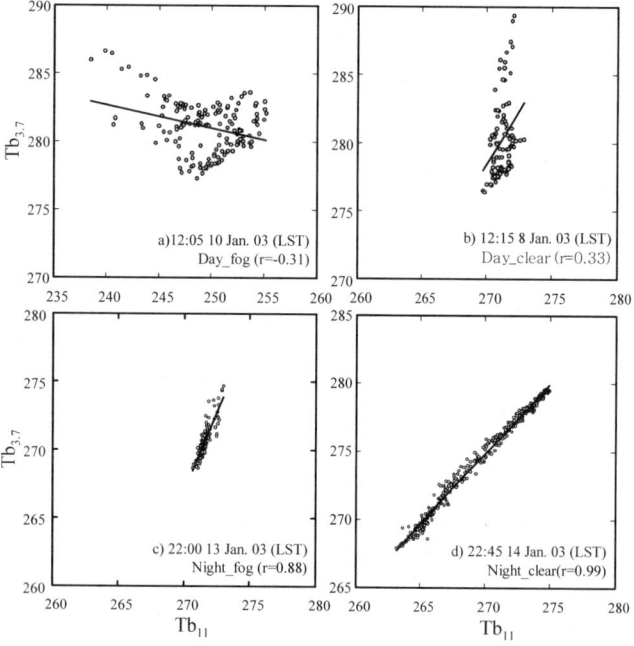

Fig. 1. Scatter diagrams in brightness temperature of 3.7 μm versus 11 μm temperature near the Inchon International Airport.

The cases of 12:05 LST January 10, 2003 and 12:15 LST January 8 were chosen for day-time fog and clear-sky conditions, respectively. Also night-time cases for fog and clear-sky were 22:00 LST January 13, 2003 and 22:45 LST January 14, 2003, respectively.

The values of brightness temperature at 3.7 μm (Tb$_{3.7\mu m}$) were higher by about 30 K than those of the Tb$_{11\mu m}$. The wavelength at 3.7 μm is located in the boundary between infrared (IR) and solar radiations. So, the Tb$_{3.7\mu m}$ value was enhanced due to the diffusion and emission effect of solar radiation by fog particles, and the correlation between Tb$_{3.7\mu m}$ and Tb$_{11\mu m}$ became low. The day-time case of clear-sky on January, 8, 2003 was different from that of fog on January 10, 2003. Similarly to the foggy case, there was a low correlation between surface (Tb$_{11\mu m}$) and the channels which reflected upper atmospheric temperatures. But, the correlation between 3.7 μm and 11 μm for day-time clear-sky was higher than that for day-time fog. The correlation (0.88) between Tb$_{3.7\mu m}$ and Tb$_{11\mu m}$ when fog occurred at night on January 13, 2003, compared with the case of day-time fog, was relatively high. Since there was only IR radiation without solar radiation at night, a high correlation between them was expected. The correlation (0.99) between 3.7 μm and 11.0 μm for night-time clear-sky on January 14, 2003 was remarkably higher than that (0.88) for night-time foggy case.

Using the data of ground observation at the Inchon International Airport and MODIS IR channels, the average values during fog and clear-sky were derived in this study (Table 1). The Tb$_{3.7\mu m}$–Tb$_{11\mu m}$ value during day-time fog was higher than that during day-time clear-sky, and during night-time fog.

The radiance and reflectance during fog at 11:05 LST on January 7, 2004 were compared with those during clear-sky at 12:10 LST on January 4, 2004 (Figs. 2a–b). Fig. 2c–d shows scatter diagrams of brightness temperatures for Tb$_{11\mu m}$ and Tb$_{3.7\mu m}$–Tb$_{11\mu m}$. The radiance and reflectance during fog were higher by 15 Wm^{-2}μm^{-1}sr^{-1} and 0.03 respectively, than those during clear-sky around 0.65 μm (Figs. 2a-b). The Tb$_{11\mu m}$ values during fog and clear-sky were about 276 K (Figs. 2c–d). As reflectance is a function of solar altitude, it is consider this specific for fog detection in day-time. The research of solar reflectance which varies throughout the day will work to using RTM (Radiative Transfer Model).

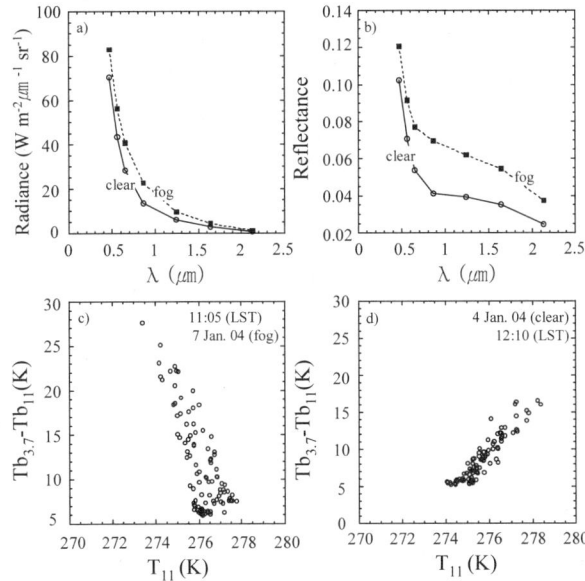

Fig. 2. The MODIS data of, a) radiance, b) reflectance, c) $Tb_{3.7\mu m}-Tb_{11\mu m}$ versus $Tb_{11\mu m}$ during foggy day, and d) $Tb_{3.7\mu m}-Tb_{11\mu m}$ versus $Tb_{11\mu m}$ during clear-sky near the Inchon airport.

4. CONCLUSION

In this study, we have investigated possible remote sensing of fog near the Inchon International Airport using the MODIS IR and solar. The threshold value of $Tb_{3.7\mu m}-Tb_{11\mu m}$ during day-time was greater by about 21 K when it was foggy than when it was clear. However, the value during night-time fog was less by 1.5 K than during night-time clear-sky. Based on these features of the satellite data, the threshold values of $Tb_{3.7\mu m}-Tb_{11\mu m}$ during day and night were derived. However, the values included errors when clouds existed above a layer of fog.

The near-IR channel at 3.7 μm was affected by solar and IR radiation in the day-time. Since fog particles resulted in high reflectance of solar radiation, the reflectance during fog was higher than that during clear-sky. The value of reflectance was sensitive to the change of solar zenith angle (Kokhanovsky, 2004).

In summary, the threshold values of $Tb_{3.7\mu m}-Tb_{11\mu m}$ have been derived in this study for detecting fog during day and night near the Inchon International Airport. However, the day-time value needs to be compensated by solar radiation a reflectance, together with radiative transfer simulation.

5. ACKNOWLEDGMENTS

This research is supported by the project 'Development of Meteorological Data Processing System for COMS' of the Korea Meteorological Administration.

6. REFERENCES

Ahn, M. H., Sohn E.-H., and Hwang B.-J.: A New Algorithm for Sea Fog/Stratus Detection Using GMS-5 IR Data. *Advances in atmospherics in atmospheric sciences*, 20(6), 899–913.

Eyre, J., Brownscombe, J., and R. Allam, 1984: Detection of fog at night using Advanced Very High Resolution Radiometer (AVHRR) imagery. *Meteorological Magazine*, 113, 266-271.

Kokhanovsky, A., 2004, Optical properties of terrestrial cloud. *Earth-science Reviews*, 64, 189-241.

FIRST RESULTS FROM METEOSAT-8: ATMOSPHERIC RADIATIVE PROPERTIES FROM GEOSTATIONARY ORBIT

J. Schmetz, R. Borde, Y. Govaerts, K. Holmlund, M. König, and H.-J. Lutz
EUMETSAT, 64295 Darmstadt, Germany

ABSTRACT

The paper provides a short overview and first results of the second generation of European geostationary meteorological satellites that started their operational phase on 29 January 2004 with Meteosat-8.

The new satellite series combines multispectral imaging in twelve spectral channels with an unprecedented imaging repeat cycle of 15 minutes for the Earth's 'full disk'. With four channels in the solar spectrum and eight channels in the thermal infrared it combines remote sensing capabilities known so far only from polar orbiting satellites in a geostationary orbit, thus enabling novel observations of rapidly developing atmospheric weather phenomena. Four MSG-type satellites will serve the community for the next decade and a half. The satellites have an improved radiometric and calibration performance with on-board calibration for all thermal IR channels and a vicarious calibration of the four solar channels using primarily stable and well-characterised desert targets. This paper provides a few examples of novel results from Meteosat-8: i) multispectral radiance measurements in the clear atmosphere depict unstable air in terms of instability indices, and thus provide lead-time before the onset of convection. ii) large areas of aerosol can be observed over land as well as over oceans. The data hold promise for new aerosol products; and iii) the upper level wind field divergence can be directly inferred from image data by using the temporal evolution of the active parts of high level deep convective clouds. Those are depicted by relatively small cloud particles in the convective updrafts. The inferred divergence fields are consistent with the results from simultaneously observed atmospheric motion vectors obtained from tracking features in the 6.2 µm channel.

1. INTRODUCTION

Meteosat-8, the first of a new series of European geostationary satellites (Meteosat Second Generation – MSG), started its routine operations in January 2004 after an extensive commissioning phase. Meteosat-8 provides significantly enhanced services to the operational user community and wealth of new observations for research. Meteosat-8 takes images with a twelve-channel imager, called SEVIRI (Spinning Enhanced Visible and Infrared Imager). SEVIRI observes the full disk of the Earth with an unprecedented imaging repeat cycle of 15 minutes. SEVIRI has eight channels in the thermal infrared (IR) at 3.9, 6.2, 7.3, 8.7, 9.7, 10.8, 12.0, and 13.4 µm, four channels in the solar spectrum at 0.6, 0.8, 1.6 µm, and a broad-band high resolution visible. The image sampling distance is 3 km at nadir for all channels except for the High Resolution Visible channel with 1 km. All channels have a 10 bits digitisation. The advent of Meteosat-8 is the most significant upgrade of the meteorological observing capabilities in geostationary orbit since the launch of the GOES-8 in 1994 (Menzel and Purdom, 1994). A detailed introduction to the new Meteosat Second Generation Series has been provided by Schmetz et al. (2002), including information on the on-board black-body calibration of thermal IR channels; inter-calibration with other satellites confirm that the accuracy goal of 1 K in equivalent brightness temperature has been reached. The solar calibration is described in Govaerts and Clerici, 2004.

As additional scientific payload the MSG satellites carry a Geostationary Earth Radiation Budget (GERB) instrument observing the broadband thermal infrared and solar radiances exiting the Earth-atmosphere system (Harries et al., 2000).

2. SCENES ANALYSIS AND PRODUCT DERIVATION

The first step in the derivation of products from imagery is a 'scenes analysis' which provides also the basis for detailed information on cloud. The operational algorithm at EUMETSAT follows the work of Saunders and Kriebel (1988), i.e. it is thresholding technique. For each pixel it provides information on cloud cover, cloud top temperature, cloud top pressure/height and cloud type and phase (Lutz, 2003, Lutz et al., 2003). The multispectral radiance measurements of clouds provide a clear distinction between ice and water. Specifically the images depict the presence of super-cooled water cloud droplets and the phase change in time, which is important to aviation forecasting.

The intermediate Cloud Analysis product constitutes the basis for the derivation of most other products, for instance for the generation of the Atmospheric Motion Vectors (AMV) and the calculation of Clear Sky Radiances (CSR), both of which are used in operational numerical weather prediction. Other operational products derived centrally at EUMETSAT include tropospheric humidity, total ozone, and atmospheric instability monitoring which is included in the section on examples below. The applications ground segment for the MSG satellites is de-centralised, i.e., there is a central processing and also an operational processing in dedicated Satellite Application Facilities which focus on thematic areas. For a summary of the SAFs the reader is referred to Schmetz et al. (2002) or the web site www.eumetsat.de.

3. EXAMPLES OF OBSERVATIONS

3.1 Diurnal Cycle of Marine Strato Cumulus

Fig. 1 shows observations of the diurnal cycle of marine stratocumulus over the South Atlantic. Cloud cover shows the well-known early morning maximum and a minimum in the late afternoon (Turton and Nicholls, 1985). The multispectral imagery of Meteosat-8 enables also the estimation of cloud droplet radii and with the frequent imagery one can also observe the diurnal cycle. In principle one would expect cloud cover and effective radii to be in phase. Future investigations will address this topic.

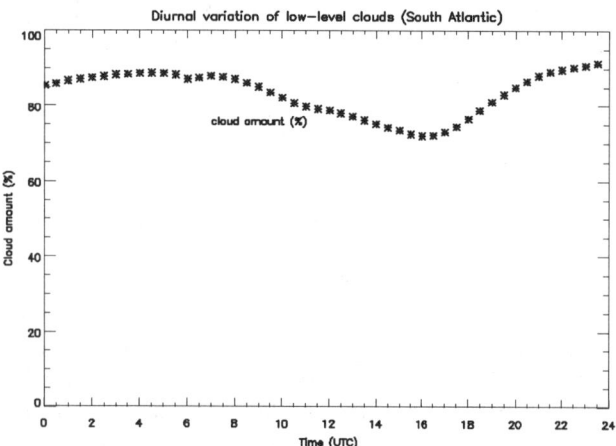

Fig. 1. Diurnal cycle of cloud cover for a marine stratocumulus over the South Atlantic. The location of the area is 5°S and 19°W and 10°W to 3°E with the center at 12°S and 3°W, i.e., UTC is close to local time.

3.2 Aerosol

Aerosol observations over land are difficult in the solar spectral channels because of the relatively low contrast between aerosol-free and aerosol-loaden atmosphere. Fig. 2 demonstrates the potential of Meteosat-8: The figure shows an optically thick aerosol layer over the Saharan desert which is well depicted by a simple difference between channel 8.7 μm and channel 10.8 μm. Considering the additional channels and the fact that the frequent images help the detection of moving dust clouds, one can suppose that Meteosat-8 holds promise for new aerosol detection methods over land. An important application of Saharan aerosol product would be the forecasting of Hurricane developments because Saharan air layer associated with dust outbreaks may influence the intensity of tropical cyclones (Dunion and Velden, 2004).

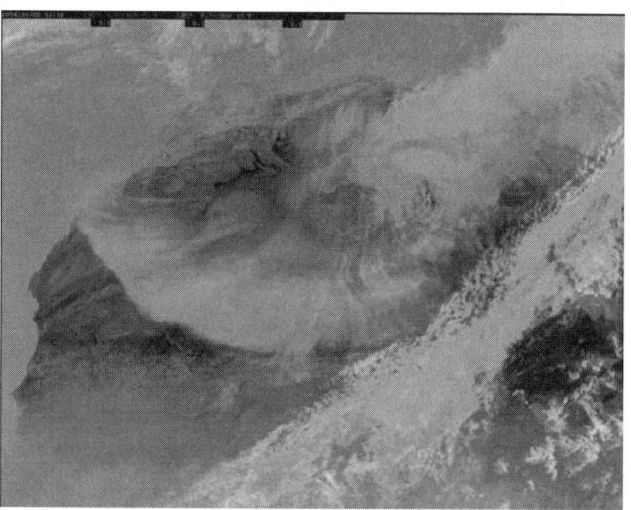

Fig. 2. A difference image between channel 7 (8.7 μm) and channel 9 (10.8 μm) depicting a large aerosol cloud cover over the Saharan desert on 3 March 2004.

3.3 Atmospheric Instability

The Global Instability Index (GII) product consists of a set of air mass parameters indicating the stability of the clear atmosphere. The retrieval is performed over a scale of about 30 km. The idea for the GII emerged from the successful applications and experience by NOAA/NESDIS with GOES lifted index products (Menzel et al., 1998). The GII product includes information on total precipitable water and several other empirical instability indices as, e.g., the Lifted Index (König et al., 2002). Six SEVIRI channels at 6.2, 7.3, 8.7, 10.8, 12.0 and 13.4 μm, are used in the retrieval. Fig. 3 shows an example for the lifted index, which is the difference between the observed temperature at 500 hPa and the temperature of an air parcel that has adiabatically risen to 500 hPa. Negative values indicate instability.

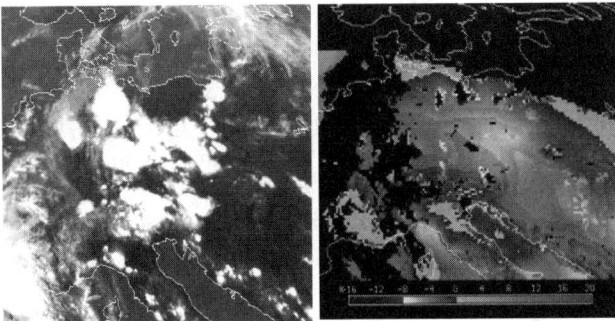

Fig. 3. Lifted Index observed with Meteosat-8 on 5 June 2003 at 0915 UTC. The left panel shows the cloud image 5 hours later, when a convective cloud systems had formed over the area indicated as unstable.

3.4 Upper Level Wind Divergence

Fig. 4 shows a time series of the upper tropospheric wind field divergence maxima for two tropical cloud systems of scale of 500 km (Schmetz et al., 2004). While one system has an active phase of only five hours between about 1300h and 2000h, the second system is active from the early morning into the night. The absolute divergence maxima reach in both cases $4.5 \cdot 10^{-4}$ s^{-1}. The divergence maxima have been obtained by tracking the migration of the convective systems manually, i.e. the advection of the systems has been taken out (Lagrangian approach). The tracking includes cloud and water vapour structures. The mean of the divergence level has been estimated 150 hPa with a standard deviation of about 50 hPa. It is noted that tracking of clear sky water vapour and thin cloud tracers implies a tracking of a deep layer mean; as such the pressure height has to be considered as 'representative height' and not as exact level of the inferred divergence.

It is also possible to derive a divergence field from the equation:

$$\nabla \cdot \vec{v} = \frac{dA}{A} \frac{1}{dt} \quad (1)$$

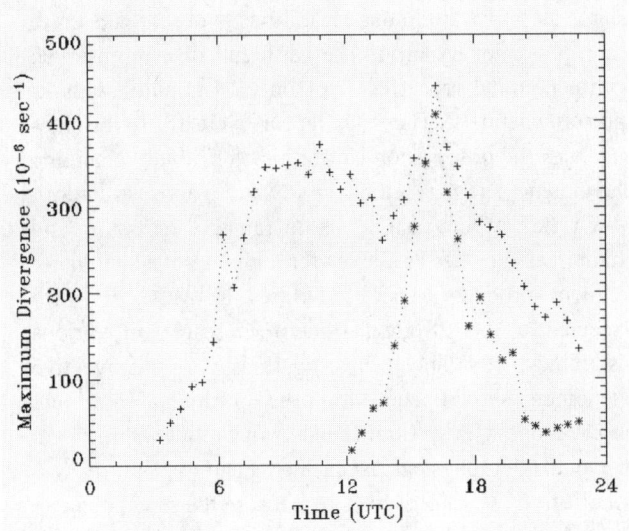

Fig. 4. Diurnal cycle of high level wind field maximum divergence in two tropical deep convective systems directly inferred from Meteosat-8 atmospheric motion vectors (AMVs). AMVs were derived by tracking features in the 6.2 µm channel.

where A is the cloud cover, t the time and v the wind vector. The method had been proposed originally by Fraedrich et al. (1976). Schmetz et al. (1997) then considered the application to Meteosat data.

In order to employ Equation 1 one needs to infer the cloud cover and its change from successive satellite images. It is essential to use only the relative changes of coverage with convectively active cloud at high level. Those can be defined by pixels where the brightness temperature in the 6.2 µm WV channel is higher than the brightness temperature at 10.8 µm (i.e., $T_{WV} > T_{IR}$) (Schmetz et al., 1997). Using those pixels and their evolution in time it is possible to infer the upper level wind field divergence in deep convective systems directly from multi-spectral image data without the need to derive atmospheric motion vectors first. This novel application has been described in detail by Schmetz et al. (2004) and could provide a key observable in future studies of the diurnal evolution of deep convective tropical cloud systems.

4. CONCLUDING REMARKS

The Meteosat Second Generation (MSG) satellites with Meteosat-8 being the first in a series of four satellites, provide a large step forward in our capabilities to observe the earth from geostationary orbit. The earth full disk is observed with a nominal repeat cycle of only 15 minutes with 12 spectral channels. The high repeat cycle of 15 minutes enhances the observation capabilities for rapidly changing phenomena such as cloud and water vapor structures, which helps Nowcasting, short range forecasting and Numerical Weather Prediction through improved and more frequent products. The capabilities of MSG are also expected to be of great value to research in various disciplines. Notably investigations of convective phenomena should benefit from the operational 15 minute observations with spectral channels that allow the retrieval of cloud microphysical parameters. Land application will benefit from MSG observations because the same area can be observed under varying sun illumination throughout a day. The frequent imaging also increases the probability to obtain a clear-sky radiance.

The meteorological products derived from MSG image data provide continuity for existing products as well as new products and scope for new applications. A novelty is the establishment of the network of Satellite Application Facilities (SAFs), which constitute a part of the Applications Ground Segment for MSG. This SAF network provides a broad and structured basis within the user community to utilize the new capabilities of MSG. Additional information on MSG is available on the EUMETSAT webpage under www.eumetsat.de (go to Meteosat Second Generation) and www.eumetsat.de/saf/.

5. REFERENCES

Dunion, J. P., and C. S. Velden, 2004: The Impact of the Saharan Air Layer on Atlantic Tropical Cyclone Activity. *Bull. American Meteor. Soc.*, Vol. 85, No. 3, 353 – 365.

Govaerts, Y. M., and M. Clerici, 2004: Evaluation of radiative transfer simulations over bright desert calibration sites, *IEEE Transactions on Geoscience and Remote Sensing*, 42, 176–187

Holmlund, K., 'The Atmospheric Motion Vector retrieval Scheme for Meteosat Second Generation.' *Proc. Fifth International Winds Workshop, Lorne, Australia*, EUMETSAT, EUM P 28, 73 – 80, 2000.

Harries, J. E., 2000: The geostationary Earth Radiation Budget Experiment: Status and science. *Proc. 2000 EUMETSAT Meteorological Satellite Data Users' Conference*, Bologna, Italy, EUMETSAT EUM P29, 62-71.

König, M., 2003: Atmospheric Instability Parameters Derived from MSG SEVIRI Observations. *Proceedings of 'The 2003 EUMETSAT Meteorological Satellite Conference,'* Weimar, Germany, EUMETSAT, EUM P39, 155 – 161.

Lutz, H. J., 2003: Scenes and Cloud Analysis from Meteosat Second Generation (MSG) Observations. *Proceedings of 'The 2003 EUMETSAT Meteorological Satellite Conference,'* Weimar, Germany, EUMETSAT, EUM P39, 265 – 7272.

Lutz, H. J., T. Inoue, and J. Schmetz, 'Comparison of a split-window and a multi-spectral cloud classifica-tion for MODIS observations,' *J. Meteorol. Society of Japan*, 81, 623 – 631, 2003.

Menzel, W. P., and J. F. W. Purdom, 'Introducing GOES-I: The first of a new generation of Geostationary Operational Environmental Satellites.' *Bull. Amer. Meteor. Soc.*, 75, 757 – 781, 1994.

Menzel, W. P., F. C. Holt, T. J. Schmit, R. M. Aune, A. J. Schreiner, G. S. Wade, and D. G. Gray, 'Application of GOES-8/9 soundings to weather forecasting and nowcasting.' *Bull. American Meteor. Soc.*, 79, 2059 – 2077, 1998.

Saunders R. W., and Kriebel K. T., 'An improved method for detecting clear sky and cloudy radiances from AVHRR data.' *Int. J. of Remote Sensing*, Vol. 9, pp. 123, 1988.

Schmetz, J., S. A. Tjemkes, M. Gube, and L. van de Berg, 1997: Monitoring deep convection and convective overshooting with Meteosat Advances in Space Research, Vol. 19, No. 3, 433–441.

Schmetz, J., P. Pili, S. A. Tjemkes, D. Just, J. Kerkmann, S. Rota, and A. Ratier. 'An Introduction to Meteosat Second Generation (MSG),' *Bull. American Meteor. Soc.*, 83(7), 977–992, 2002

Schmetz, J., R. Borde, M. König, and H.-J. Lutz, 2004: Upper tropospheric divergence fields in a tropical convective system observed with Meteosat-8, *Proceedings of the 7th Intern. Winds Workshop,* Helsinki, Finland, June 2004, EUM P42, in print.

Turton, J. D., and S. Nicholls, 1987: A study of the diurnal variation of stratocumulus using a multiple mixed layer model. *Q. J. R. Meteorol. Soc.*, 113, 969–1009.

A COMBINED METHOD OF THE TRMM PRECIPITATION RADAR AND VISIBLE AND INFRARED SCANNER FOR RETRIEVAL OF CLOUD-PRECIPITATION INTERACTION

Takahisa Kobayashi and Ahoro Adachi
Meteorological Research Institute
1-1, Nagamine, Tsukuba, Ibaraki 305-0052, Japan

ABSTRACT

This paper presents a method for determining drizzle and clouds properties by combined use of a 95 GHz cloud profiling radar and a visible/near-IR multi-spectral radiometer from space. When clouds and drizzle coexist, both the radar and the radiometer detect the sum of scattering signals from cloud drops and drizzle. However, the radar reflectivity factor measured with a radar is mostly determined from scattering by drizzle drops. The visible/near-IR radiometer, on the other hand, is more sensitive to cloud drops than drizzle. This different sensitivity of the sensors to drop size can be used to determine the properties of drizzle and clouds, discriminatively. To examine the potential of the combined method, we have made simulations for drizzling clouds in which clouds and drizzle coexist. Results show that the method obtains an accuracy of around 10~20 % in the drizzle reflectivity factor from a radar, cloud optical thickness and the effective radius from a radiometer when the liquid water content of drizzle is smaller than that of clouds.

1. INTRODUCTION

Precipitation is projected to increase during 21st century associated with the global warming. However, little is known about how climate change results in precipitation modification because of current luck of reliable techniques to measure the cloud-precipitation interaction.

When clouds and precipitation coexist (precipitation clouds), both the radar and radiometer detect the sum of signals scattered from cloud drops and raindrops. However, the radar reflectivity factor measured with the radar is mostly determined from scattering by raindrops. The visible/near-IR radiometer, on the other hand, is more sensitive to cloud drops than raindrops. The different sensitivity of the sensors to drop size can be used to determine the properties of precipitation and clouds, discriminatively.

The precipitation radar (PR) operated at a frequency of 14 GHz on the Tropical Rainfall Measuring Mission (TRMM) provides the 3-dimentional rain structures over the wide area of ocean and land. The visible infrared scanner (VIRS) on board the TRMM satellite observes cloud properties. A combined use of the PR and the VIRS can be furthermore used to derive cloud-precipitation interaction. In addition, a millimeter-wave radar will be equipped on the CloudS Mission [*Stephens et al.*, 2002]. The millimeter radar will be also equipped on the Global Precipitation Mission (GPM) and the Earth Cloud, Aerosol and Radiation Explorer (EarthCARE) mission which is jointly planned by the National Space Development Agency (NASDA) of Japan and the European Space Agency (ESA). Expected sensors on board are a cloud profiling radar (CPR), a lidar, a passive solar Multi Spectral Imager (MSI), etc. Measurements of clouds are of primary importance of the CPR in the mission. The CPR is furthermore useful to detect backscattering signals from drizzle. The multi spectral imager with 7 bands from visible to infrared wavelengths measures Mie scattering signals from cloud drops and is used to derive cloud properties. These sensors have different sensitivity to drop size. This different sensitivity can be used to measure both cloud and drizzle properties simultaneously when drizzle and clouds coexist. In this paper, we will examine a potential of a combined use of the CPR and the MSI for measuring drizzle and clouds.

2. SIMULATIONS

To examine the contribution of clouds and drizzle to the CPR signals, we made simulations of backscattering signals from drizzling clouds, based on the conventional radar equation assuming single scattering. Backscattering cross section of drops was calculated using Mie theory. Attenuation was neglected. We also made radiative transfer simulations for reflected radiances of the passive radiometer at wavelengths of 0.86 and 2.16 μm.

The cloud drop size distribution (N_c) is assumed to be a log-normal distribution given by

$$N_c(r) = \frac{N_0}{\sqrt{2\pi}\sigma r} \exp[-\ln^2(r/r_0)/2\sigma^2],$$

where r is the drop radius (μm), and r_0 is the mode radius. N_0 is the number density at $r = r_0$. σ is the logarithmic width of the drop size distribution. Here, σ is taken to be 0.35 and r_0 is assumed to be 5.89 μm which corresponds to the drop size distribution with the effective radius r_e of 8 μm. Drizzle is assumed to have M-P type drop size distribution (N_D) given by

$$N_D(r) = N_{D0} \exp(-\Lambda r).$$

The parameter Λ is taken to be 0.092 μm^{-1} (Model A) and 0.056 μm^{-1} (Model B). These values correspond to the rain rate of 0.00001 and 0.0001 mm/h [Marshall and Palmer, 1948]. Total liquid water content of drizzling cloud W_t is the sum of drizzle and cloud portion. Fractional amount of liquid water content (LWC) of drizzle is given by f.

$$W_D = fW_t$$
$$W_c = (1-f)W_t,$$

where W_D and W_c are LWC of drizzle and a cloud, respectively.

The backscattering cross sections of cloud and drizzle drops for the passive visible and near-IR Radiometer are in the Mie regime and are proportional to r^2. On the other hand, the CPR mostly detects Rayleigh scattering signals which is proportion to r^6. The different sensitivity of the CPR and radiometer can be used to discriminate signals scattered from clouds and drizzle.

Fig. 1 shows the reflectivity factor (Z) and the optical thickness τ at a wavelength of 2.16 μm as a function of drop radius for the drizzling cloud model A. The values are normalized so that the peak values are unity. Solid curves are the reflectivity (right curve) and the optical thickness (left curve) for drizzling clouds. Dotted curves are Z (right curve) for drizzle and τ (left curve) for a cloud. There appear considerable difference in the range of drop size that the CPR and the radiometer are most sensitive. The CPR is sensitive to drops of around 60 μm in radius while the radiometer is most sensitive to the drops of around 6 μm. Drops of $r > 19$ μm have significant collection efficiencies in interacting with smaller drops [Feingold et al., 1999]. Drizzle is often defined as drops of the radius greater than 20 μm [Gerber, 1996]. Fig. 1 shows that the Z at $r =15$ μm is about 20 times less than the peak values. Drops of less than 5 μm in size have negligible effects on Z. The shoulder at 7~15 μm in Z is a contribution of clouds and is small as 1/20 of the peak. The reflectivity factor of drizzling clouds almost coincides with that of drizzle and is therefore almost determined from drizzle drops. The CPR is sensitive to the scattering by drizzle and almost insensitive to cloud drops.

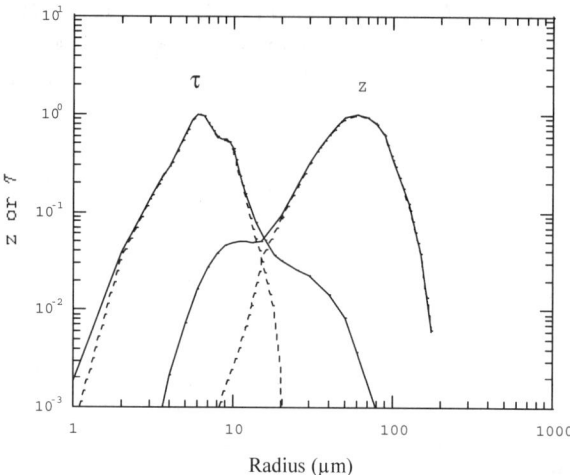

Fig. 1. Normalized values of the optical thickness at 2.16 μm (left, solid curves) and the reflectivity factor of drizzling clouds (right, solid curves). Dashed lines are the cloud optical thickness (left curve) and the drizzle reflectivity (right curve).

Small drops of radius from 2 to 15 μm are the major contributors to the optical thickness in contrast to Z. The optical thickness of the drizzling cloud ($τ_t$) almost coincides with that of clouds, and is therefore almost determined by the scattering from cloud drops. The reflection of cloud at visible and near-IR wavelengths is almost proportional to cloud optical thickness. The passive radiometer is, therefore, sensitive to cloud drops and is almost insensitive to drizzle.

Fig. 2 shows the cloud contribution to the total reflectivity factor of drizzling clouds (Z_t) and $τ_t$ at 0.86 μm as a function of f for Model B. A fractional contributions C_Z and $C_τ$ are plotted and are defined by

$$C_Z = \frac{Z_c}{Z_t},$$

for the reflectivity factor and

$$C_τ = \frac{τ_c}{τ_t},$$

for optical thickness, where Z_c and $τ_C$ is the reflectivity factor and the optical thickness of cloud, respectively. The liquid water content of drizzling cloud is fixed at 0.2 g/m^3. The contribution of the scattering from clouds to Z_t increases with a fractional amount of cloud in the drizzling clouds but the contribution is limited for very small values of f. The scattering by cloud drops contributes to Z_t by only 5% when the cloud contains 90% ($f=0.1$) of LWC of drizzling clouds. Even for a drizzling cloud with liquid water contents of clouds as high as 99%, the scattering by drizzle contributes to Z_t by 60%. Significant contributions of large drizzle drops on Z_t mask the scattering from clouds and make the CPR difficult to measure cloud properties. In other words, the CPR can be used to discriminate drizzle from cloud drops in the drizzling cloud. For optical thickness, on the other hand, the contribution of drizzle to the optical thickness is very limited. The scattering by drizzle drops contributes to τ_t by only a few percent if drizzle contains 10% (f = 0.1) of LWC of drizzling clouds. Even for a drizzling cloud with liquid water contents of drizzle as high as 50 %, the scattering by cloud drops contributes to τ_t by 80%. For $f \ll 0.5$, drizzle has negligible effects on τ and hardly affects τ_t. Significant contributions of cloud drops on τ_t and the negligible contributions of drizzle, make the radiometer difficult to measure drizzle properties. Thus, cloud properties in drizzling clouds can be estimated from measurements using radiometer.

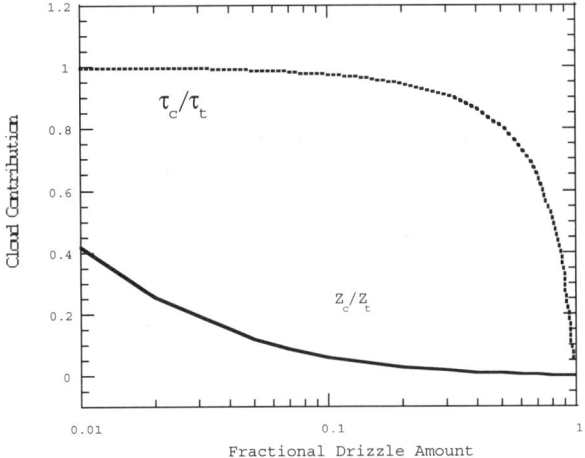

Fig. 2. Cloud contributions to Z (solid curve) and τ at 0.86 μm (dashed curve) for drizzling cloud as a function of fractional drizzle amount for Model B.

3. CLOUD PROPERTIES RETRIEVAL

Here, we discuss the accuracy of the cloud properties derived from the radiometer in drizzling clouds. Although the contribution of drizzle to the radiance detected with the radiometer is small, it may lead to a bias in the cloud radiative properties depending on f. The cloud radiative properties can be expressed in terms of optical thickness (τ_c) and the effective radius (r_e). The effective radius is defined by

$$r_e = \frac{\int_0^\infty r^3 N_c(r) dr}{\int_0^\infty r^2 N_c(r) dr}.$$

These two parameters can be determined from the simultaneous measurements of reflected solar radiances at visible and near-IR wavelengths from space [e.g., *Nakajima and King*, 1990]. This technique is based on the different absorption properties of water at visible and near-IR wavelength regions. We apply this technique to drizzling clouds.

Fig. 3 shows the reflection function R (τ, μ=0.866, ϕ=0, μ_0,=0.5) at wavelengths of 0.86 μm and 2.16 μm for various values of τ and r_e. Here, μ_0 is the cosine of the solar zenith angle, μ the cosine of viewing zenith, ϕ the azimuth angle relative to the solar azimuth. Fig. 3 suggests that the cloud optical thickness is mostly determined by the reflection function at 0.86 μm. While the effective radius is sensitive to the reflection function at 2.16 μm but is almost insensitive to that at 0.86 μm. The different sensitivity of the parameters can be used to derive τ_c and r_e simultaneously. Black circles are calculated values of the drizzling cloud model B for f = 0.1, 0.2, 0.3, 0.4, and 0.5. Corresponding cloud optical thickness at 0.86 μm is given by 20.18 x (1–f) (circles at right to left) at 0.86 μm. The value of 20.18 is the optical thickness of the drizzling cloud with f = 0. Total liquid water content is fixed to be 0.1 g/m^3. The effective radius of the cloud model is assumed to be r_e = 8 μm. The black circle at f = 0.1 (most right point) is very close to the curve of r_e = 8 μm and τ = 20 and the bias error in r_e and τ_c is negligible. As the fractional drizzle amount increases, the circles deviate from the curve of exact value of r_e = 8 μm associated with increases in the effects of the scattering by drizzle drops.

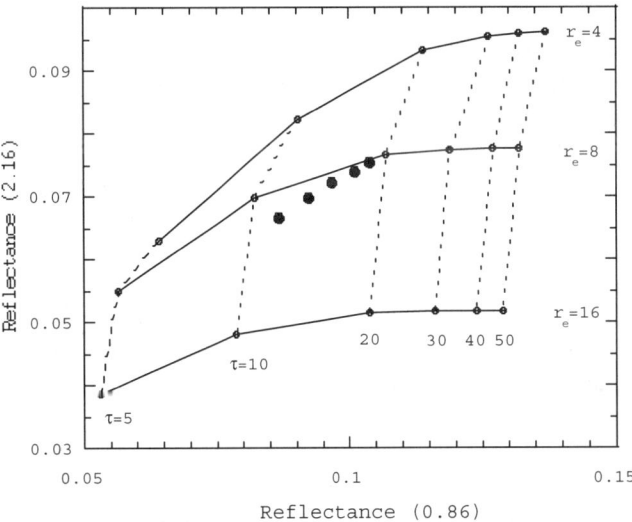

Fig. 3. The reflection function R (τ, μ=0.866, ϕ=0, μ_0,=0.5, ϕ_0=0) at wavelengths of 0.86 μm and 2.16 μm for various values of τ_c and r_e. Black circles are calculated values of the drizzling cloud Model B for f=0.1, 0.2, 0.3, 0.4, and 0.5. B.

The radiometer tends to overestimate the values with increase in the coexisting drizzle portion. The increased tendency to overestimate τ_c and r_e becomes predominant for $f \gg 0$. The bias error in the cloud optical thickness is 15~25% at $f = 0.5$. For $f < 0.3$, the errors in τ_c is less than 10%. The bias error in r_e is better and is about 15% at $f = 0.5$. For f less than 0.4, the error in r_e is less than 10%.

As mentioned previously, the cloud reflectivity factor can be identified from the CPR measurements. The estimate of the liquid water content of drizzle is also needed to determine the value of f. Once we specify the reflectivity factor of drizzle, we can calculate the LWC of drizzle assuming an appropriate Z-LWC relationship. This relationship, however, significantly depends on the drop size distribution. An accurate estimate of LWC needs information on the drop size. One of useful techniques to estimate a size is to use a Doppler spectrum [*Gossard et al.*, 1997]. The CPR is planned to have a Doppler capability with a velocity accuracy better than 1 m/s in the EarthCare mission. The terminal velocity of a drop is related to its size, and can be used to estimate drop size. This technique has been applied to a wind profiler for measuring raindrop size distributions [e.g., *Kobayashi and Adachi*, 2001, 2005].

4. CONCLUSIONS

We have demonstrated a technique of a combined use of a cloud profiling radar and a passive visible/near-IR multi-spectral radiometer for measuring drizzling clouds from space. The CPR is sensitive to small cloud drops. When clouds and drizzle coexist, however, the CPR is difficult to derive cloud properties because the CPR signals are approximately proportional to D^6 and therefore the CPR is more sensitive to drizzle than cloud drops. In general, a small amount of larger drizzle drops contribute to Z much more than a large amount of smaller cloud drops do and masks the returns from clouds. If the fractional amount of LWC of drizzle is more than 10%, the scattering by cloud drops hardly affect the total reflectivity factor of drizzling clouds. The signal measured with a cloud profiling radar is mostly determined from scattering by drizzle drops. On the other hand, a passive visible/near-IR radiometer is more sensitive to cloud drops than drizzle. For drizzling cloud in which 50% of the LWC is due to drizzle, the drizzle contribution to the total optical thickness is less than 20%. Consequently, the passive radiometer detects most signals reflected from clouds. The LWC of drizzle may be derived accurately from the CPR signals with the Doppler capability.

Applications of the present method to the millimeter-wave radar which will be equipped on the EarthCare, CloudSat and on the Global Precipitation Mission, will improve our understanding of the cloud-drizzle interaction.

5. ACKNOWLEDGEMENTS

This study was partly supported by the Japan Aerospace Exploration Agency (JAXA) TRMM program.

6. REFERENCES

Feingold, G., W. R. Cotton, S. M. Kreidenweis, and J. T. davis, The impact of giant cloud condensation nuclei on drizzle formation in stratocumulus: implications for cloud radiative properties, *J. Atmos. Sci.*, 56, 4100-4117, 1999.

Kobayashi, T., and A. Adachi, Measurements of raindrop breakup by using UHF wind profilers, *Geophys. Res. Lett.*, 28, 4071-4074, 2001.

Kobayashi, T., and A. Adachi, Retrieval of arbitrarily shaped raindrop size distributions from wind profiler measurement, *J. Atmos. Oceanic Technol.*, 22, 433-442, 2005.

Gerber, H., Microphysics of marine stratocumulus clouds with two drizzle modes, *J. Atmos. Sci.*, 53, 1649-1662, 1996.

Gossard, E. E., J. B. Snider, E. E. Clothiaux, B. Martner, J. S. Gibson, R. A. Kropfli, and A. S. Frisch, The potential of 8-mm radars for remotely sensing cloud drop size distributions, *J. Atmos. Oceanic Technol.*, 14, 76-87, 1997.

Marshall, J. S., and W. M. Palmer, The distribution of raindrops with size, *J. Meteor.*, 5, 165-166, 1948.

Nakajima, T., and M. D. King, Determination of the optical thickness and effective particle radius of clouds from reflected solar radiation measurements. Party I: Theory, *J. Atmos. Sci.*, 47, 1878-1893, 1990.

Stephens, G. L., D. G. Vane, R. J. Boain, G. G. Mace, K. Sassen, Z. Wang, A. J. Illingworth, E. J. O'Connor, W. B. Rossow, S. L. Durden, S. D. Miller, R. T. Austin, A. Benedetti, C. Mitrescu, and the CloudSat Science Team, The CloudSat mission and the A-train, *Bull. Amer. Meteor. Soc.*, 83, 1771-1790, 2002.

SATELLITE-BASED TROPICAL WARM POOL SURFACE HEAT BUDGETS: CONTRASTS BETWEEN 1997/98 EL NINO AND 1998/99 LA NINA

S.-H. Chou, M.-D. Chou, and P.-H. Lin
Department of Atmospheric Sciences, National Taiwan University, Taipei 106, Taiwan

P.-K. Chan
Science Systems and Applications, Inc., Lanham, MD 20706, USA

K.-H. Wang
Central Weather Bureau, Taipei 100, Taiwan

ABSTRACT

The change of net surface heating (F_{NET}) between 1997/98 El Nino and 1998/99 La Nina is found to be significantly larger in the tropical eastern Indian Ocean than that in the tropical western Pacific. For the former region, reduced evaporative cooling arising from weakened winds during the El Nino is generally associated with enhanced solar heating due to reduced cloudiness, leading to enhanced interannual variability of F_{NET}. For the latter region, reduced evaporative cooling due to weakened winds is generally associated with reduced solar heating arising from increased cloudiness, and vice versa. Consequently, the interannual variability of F_{NET} is reduced.

1. INTRODUCTION

The tropical western Pacific and eastern Indian Oceans are characterized by the highest sea surface temperature (SST), frequent heavy rainfall, strong atmospheric heating and weak mean winds with highly intermittent westerly wind bursts. The heating drives the global climate and plays the key role in the El Nino–Southern Oscillation (ENSO) and the Asian-Australian monsoon (Webster et al. 1998). To understand the interannual variations of SST, the information on the seasonal to interannual variability of the surface heat budgets and the correlation of surface heating with SST tendency in the tropical warm pool is particularly important. Chou et al. (2004) have investigated the seasonal to interannual variations of the surface heat budgets and their relationships to SST tendency over the tropical eastern Indian and western Pacific Oceans (30°S–30°N, 90°E–170°W) during a 3-yr period of October 1997–September 2000. This period included the strong 1997/98 El Nino warm event, and the moderate 1998/99 La Nina cold event. In this paper, we present the major results of Chou et al. (2004).

2. DATA SOURCES

The data used are 1°x1° lat-lon monthly mean SST, surface radiative and turbulent fluxes, 10-m wind speed and specific humidity, and outgoing longwave radiation (OLR) for October 1997–September 2000 cover the domain 30°S–30°N, 90°E–170°W. Deep convection is inferred from the OLR of NOAA's polar-orbiting satellites. The radiative fluxes are taken from the Goddard Satellite-retrieved Surface Radiation Budget (GSSRB), derived from Japan Geostationary Meteorological Satellite-5 radiances (Chou et al., 2001). Compared to the surface observations at the Atmospheric Radiation Measurement (ARM) site on Manus Island, daily downward shortwave (SW) radiation has a bias of 6.7 W m^{-2} with a standard deviation (SD) error of 28.4 W m^{-2}, while daily downward surface longwave (LW) radiation has a bias of +2.3 W m^{-2} with a SD error of 6.6 W m^{-2}.

The turbulent fluxes are taken from the Goddard Satellite-based Surface Turbulent Fluxes, version 2 (GSSTF2; Chou et al., 2003). The GSSTF2 contains daily, monthly, and climatological means of turbulent fluxes and the input parameters used in the derivation of fluxes over global oceans. The GSSTF2 has a spatial resolution of 1° x 1° lat-lon and covers the period July 1987–December 2000. Daily turbulent fluxes are derived from the Special Sensor Microwave/Imager (SSM/I) inferred surface air humidity (Chou et al., 1997), SSM/I surface winds (Wentz, 1997), SST and 2-m air temperature of NCEP/NCAR reanalysis, based on Monin-Obukhov similarity theory (Chou et al., 2003). Compared to those of research ship measurements over the tropical oceans, daily latent heat fluxes have a negative bias of -2.6 W m^{-2}, and a SD error of 29.7 W m^{-2}, daily sensible heat fluxes have a positive bias of 7.0 W m^{-2} and a SD error of 6.2 W m^{-2}, and daily wind stresses have a positive bias of 0.005 N m^{-2} and a SD error of 0.019 N m^{-2}.

3. ANNUAL MEANS

The spatial distributions of mean surface heat budgets over the warm pool for the 3-yr (October 1997 to September 2000) mean condition are shown in Fig. 1. The sensible heat flux is generally very small and is not shown. It can be seen from Fig. 1 that the magnitude of solar heating (F_{SW}) is larger than that of evaporative cooling (F_{LH}) but the spatial variation of the latter is significantly larger than the former. The range of the annual-mean is 180–240 W m^{-2} for F_{SW} and 80–190 W m^{-2} for F_{LH}. As a result, the net surface heating (F_{NET}) are dominated by the variability of F_{LH}. For the 3-yr mean condition within ~10–15° of the equator, the oceans gain heat up to ~ 50–75 W m^{-2}, due to the large solar heating and small evaporative cooling.

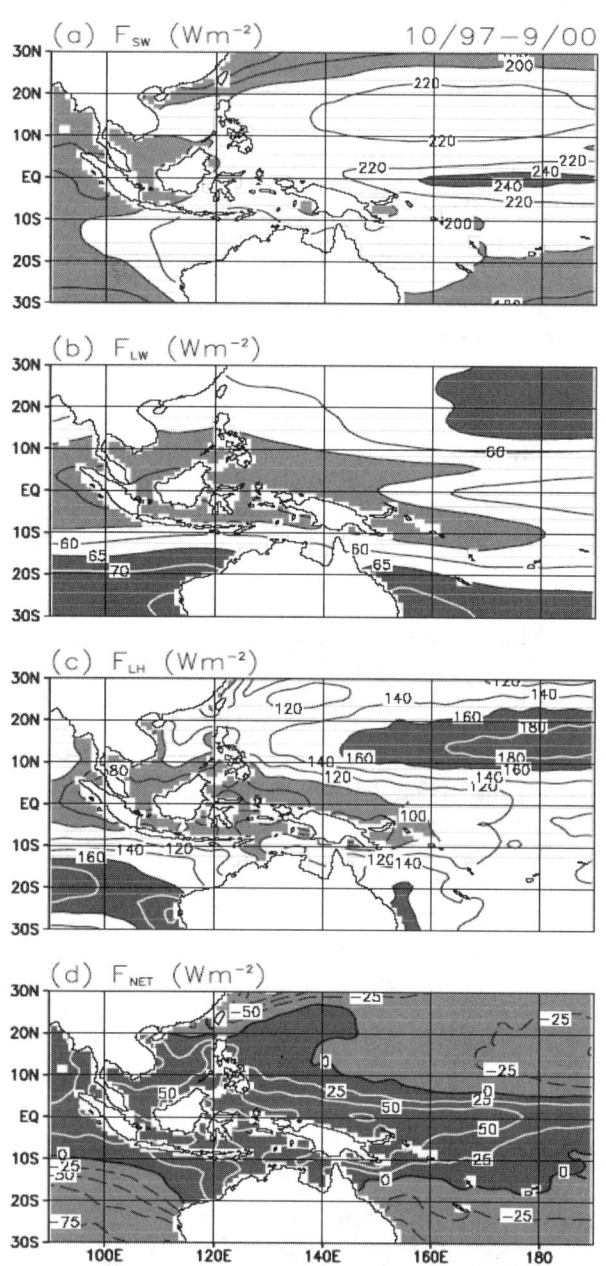

Fig. 1. Surface (a) solar heating, (b) IR cooling, (c) evaporative cooling, and (d) net heating, averaged over the 3-yr period Oct 1997–Sep 2000. Regions of large values are marked with dark shading, and regions of small values are marked with light shading.

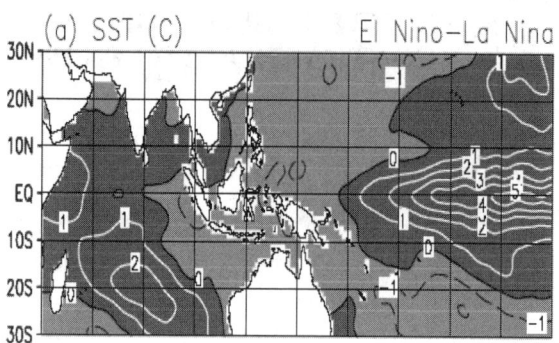

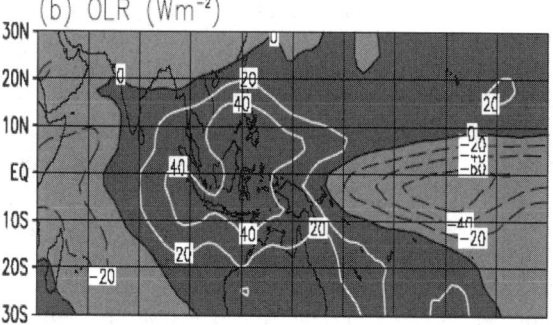

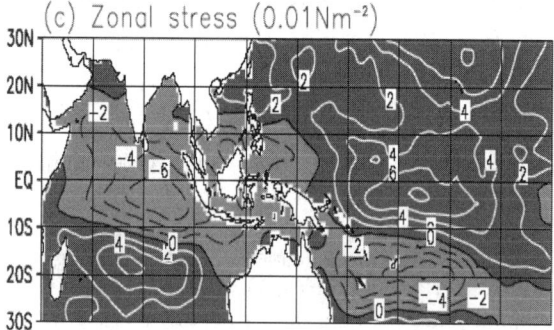

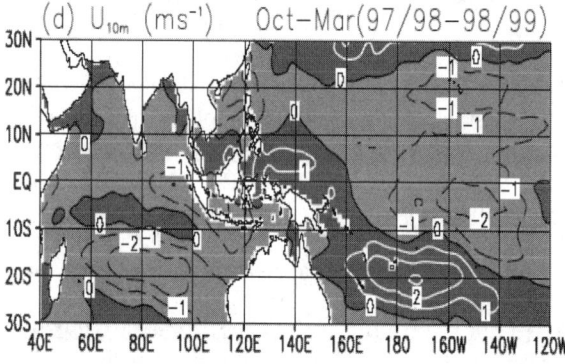

Fig. 2. 1997/98 El Nino – 1998/99 La Nina differences of (a) SST, (b) OLR, (c) zonal wind stress, and (d) 10-m wind speed for the boreal winter of Oct–Mar. Positive values are marked with dark shading, and negative values are marked with light shading.

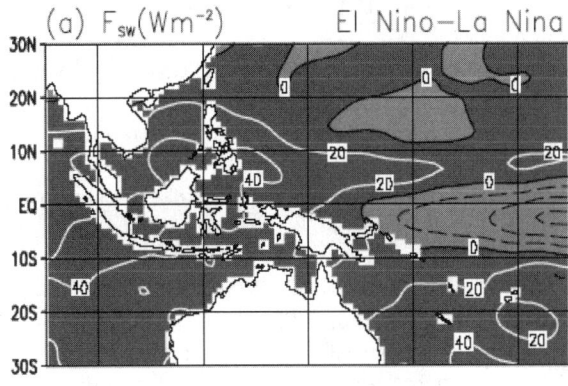
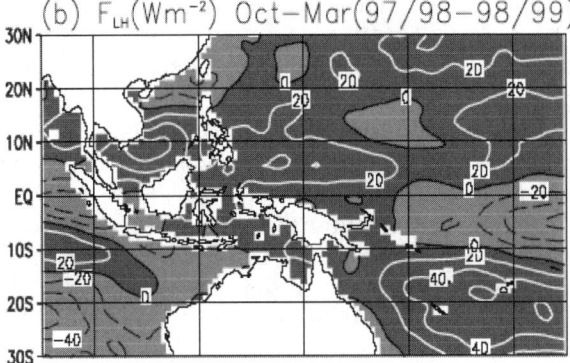
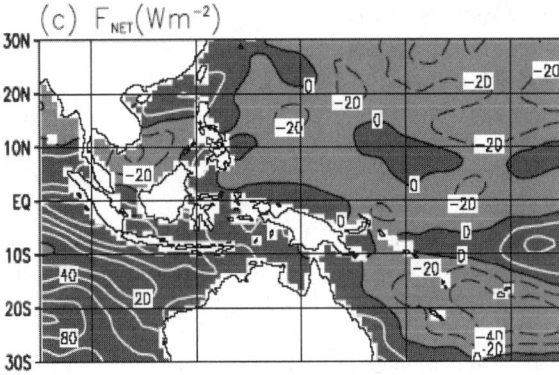
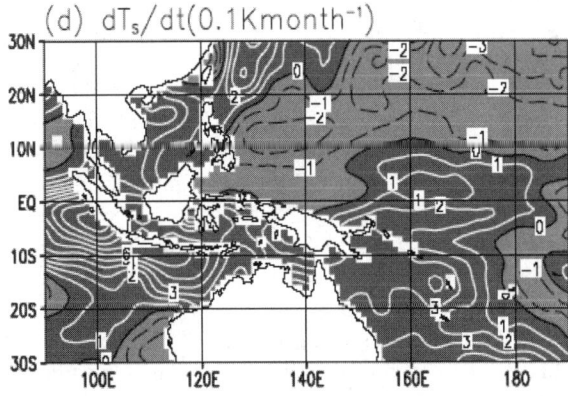

Fig. 3. As in Fig. 2 except for (a) solar heating, (b) evaporative cooling, (c) net surface heating, and (d) SST tendency.

Poleward of this region, the heat loss by the ocean increases with latitudes primarily due to an increase in evaporative cooling. The spatial distribution of annual-mean solar heat-ing (Fig. 1a) resembles only slightly that of clouds (or OLR, not shown). It reflects the fact that solar heating depends not only on clouds but also on the seasonal and latitudinal variations of insolation at the top of the atmosphere. The LW cooling (F_{LW}) depends on clouds, SST, and the temperature and humidity of the atmospheric boundary layer. Although the range of F_{LW} is small (Fig. 1b), its spatial distribution resembles closely that of clouds (or OLR). The spatial distribution of the evaporative cooling (Fig. 1c) is determined primarily by wind speed and secondarily by Q_s–Q_{10m} in the warm pool.

4. 1997/98 El NINO AND 1998/99 LA NINA

The central and eastern equatorial Pacific warms significantly during May 1997–May 1998 corresponding a strong El Nino, but cools slightly during July 1998–March 2001 corresponding to a moderate La Nina (Bell et al., 1999, 2000). To understand the changes in SST and atmospheric circulation over the tropical Pacific and Indian Oceans, the 1997/98 El Nino-minus-1998/99 La Nina differences of SST, OLR, zonal wind stress and 10-m wind speed over a domain of 30°S–30°N, 40°E–120°W for the boreal winter of October–March are analyzed in Fig. 2. Compared to the La Nina, the SST during the El Nino increases by ~2–5°C in the central and eastern equatorial Pacific and by ~1°C in the western equatorial Indian Ocean, but decreases by <1°C near the maritime continent (Fig. 2a). Correspondingly, the OLR decreases by 40–60 and 20 Wm^{-2}, respectively, in the central equatorial Pacific and the western equatorial Indian Ocean, but increases by 20–40 Wm^{-2} in the maritime continent (Fig. 2b). This indicates shifting of the convection center from the maritime continent eastward to the central equatorial Pacific and westward to the western equatorial Indian Ocean.

In the equatorial region during the El Nino, the zonal wind stress increases (more westerly as trade winds weaken) east of ~150°E but decreases (more easterly as trade winds strengthen) west of ~150°E (Fig. 2c). In the equatorial region, the changes in OLR and zonal wind stress indicate a weakened Walker circulation in both the Pacific and Indian Oceans during the El Nino. Compared to the La Nina, the surface wind speed generally decreases by ~1–2 m s^{-1} during the El Nino, except for the regions extending from the southern section of the South China Sea to the maritime continent and further to the SPCZ where cloudiness decreases (Fig. 2d).

Fig. 3 shows that the change of F_{NET} between 1997/98 El Nino and 1998/99 La Nina during the boreal winter is significantly larger in the tropical eastern Indian Ocean than in the tropical western Pacific. For the tropical eastern Indian Ocean, reduced evaporative cooling arising from weakened winds during the El Nino is generally associated with enhanced solar heating due to reduced cloudiness, leading to an enhanced interannual variability of F_{NET}. For the tropical western Pacific, reduced evaporative cooling due to weakened winds is generally associated with reduced solar heating arising from increased cloudiness, and vice versa. Consequently, the interannual variability of F_{NET} is reduced. The change in dT_S/dt is generally larger in the tropical eastern Indian Ocean than in the tropical western Pacific. The correlation between interannual variations of F_{NET} and dT_S/dt is very weak for both regions. This result for the equatorial Pacific is consistent with that of Wang and McPhaden (2001) who found that all terms in the heat balance of the oceanic mixed layer contributed to the SST variation in the equatorial Pacific during the 1997/98 El Nino and 1998/99 La Nina.

The change in the zonal wind stress suggests some important interannual changes in ocean dynamics. Compared to the La Nina, the change of ocean dynamics during the El Nino may include an increase of upwelling (downwelling) in the eastern (western) equatorial Indian Ocean and an increase in the interhemispheric heat transport from the south to the north Indian Ocean due to increased easterly wind forcing (Webster et al, 1999; Yu and Rienecker, 2000; Loschnigg and Webster, 2000). It may also include an increase in heat transport from the equatorial western pacific to the equatorial central Pacific and a decrease in the heat transport from the Pacific to Indian Ocean by Indonesian throughflow due to increased westerly wind forcing in the equatorial Pacific (Godfrey, 1996; Lukas et al., 1996; Meyers, 1996). In addition, the solar radiation penetration through the oceanic mixed layer in the weak wind regions of the equatorial warm pool may change due to the change in surface wind, as suggested by previous studies (e.g., Anderson et al., 1996; Sui et al., 1997; Godfrey et al., 1998; Chou et al., 2000). Thus, we suggest that interannual changes of the surface heating and ocean dynamics all play important roles in the interannual variation of SST in this climatically important region.

5. ACKNOWLEDGMENTS

SHC and MDC would like to acknowledge the support of the National Sciences Council of Taiwan. This study was originally support by the TRMM Program, Global Modeling and Analysis Program, Radiation Sciences Program, and NASA Office of Earth Science, USA. The GSSTF2 turbulent fluxes were obtained from http://daac.gsfc.nasa.gov/hydrology/hd_gsstf2.0.shtml, and GSSRB radiative fluxes were obtained from http://daac.gsfc.nasa.gov/hydrology/hd_gssrb.shtml.

6. REFERENCES

Anderson, S. P., R. A. Weller, and R. B. Lukas, 1996: Surface buoyancy forcing and the mixed layer of the western Pacific warm pool: observations and 1D model results. *J. Climate,* 9, 3056-3085.

Bell, G. D., and co-authors, 1999: Climate assessment for 1998. *Bull. Amer. Meteor. Soc.,* 80, S1–S48.

Bell, G. D., and co-authors, 2000: Climate assessment for 1999. *Bull. Amer. Meteor. Soc.,* 81, S1–S50.

Chou, M.-D., P.-K. Chan, and M. M.-H. Yan, 2001: A sea surface radiation dataset for climate applications in the tropical western Pacific and South China Sea. *J. Geophys. Res.,* 106, 7219-7228.

Chou, S.-H., C.-L. Shie, R. M. Atlas, and J. Ardizzone, 1997: Air-sea fluxes retrieved from Special Sensor Microwave Imager data. *J. Geophys. Res.,* 102, 12705-12726.

Chou, S.-H., W. Zhao, and M.-D. Chou, 2000: Surface heat budgets and sea surface temperature in the Pacific warm pool during TOGA COARE. *J. Climate,* 13, 634-649.

Chou, S.-H., E. Nelkin, J. Ardizzone, R. M. Atlas, and C.-L. Shie, 2003: Surface turbulent heat and momentum fluxes over global oceans based on the Goddard satellite retrievals, version 2 (GSSTF2). *J. Climate,* 16, 3256-3273.

Chou, S.-H., M.-D. Chou, P.-K. King, P.-H. Lin, and K.-H. Wang, 2004: Tropical warm pool surface heat budgets and temperature: Contrast between 1997/98 El Nino and 1998/99 La Nina. *J. Climate,* 17, 1845-1858.

Godfrey, J. S, 1996: The effect of the Indonesian throughflow on ocean circulation and heat exchange with the atmosphere: A review. *J. Geophy. Res.,* 101, 12217-12237.

Loschnigg, J., and P. Webster, 2000: A coupled ocean-atmosphere system of sst modulationfor the Indian Ocean. *J. Climate,* 13, 3342-3360.

Meyers, G., 1996: Variation of Indonesian throughflow and the El Nino-Southern Oscillation. *J. Geophy. Res.,* 101, 12255-12263.

Sui, C.-H., X. Li, K.-M. Lau, and D. Adamac, 1997: Mutiscale air-sea interactions during TOGA COARE. *Mon. Wea. Rev.,* 125, 448-462.

Wang, W., and M. J. McPhaden, 2001: Surface layer temperature balance in the equatorial Pacific during the 1997-98 El Nino and 1998-99 La Nina. *J. Climate,* 14, 3393-3407.

Webster, P. J., T. Palmer, M. Yanai, V. Magana, J. Shukla, R. A. Toma, and A. Yasunari, 1998: Monsoons: Processes, Predicability and the prospects for prediction. *J. Geophys. Res.,* 103, (C7), 14451-14510.

Wentz, F. J., 1997: A well calibrated ocean algorithm for SSM/I. *J. Geophys. Res.,* 102, 8703-8718.

Yu, L., and M. Rienecker, 2000: Indian Ocean warming of 1997-98. *J. Geophys. Res.,* 105, 16923-16939.

A SIMPLE APPROACH TO ESTIMATE LAND SURFACE ALBEDO FROM SATELLITE MEASUREMENTS

Yu Cui
Graduate School of Science and Technology, Chiba University
Chiba-shi, Chiba, 263-8522, Japan

Tamio Takamura
Center for Environmental Remote Sensing, Chiba University
Chiba-shi, Chiba, 263-8522, Japan

ABSTRACT

In this study, the POLarization and Directionality of the Earth's Reflectance (POLDER-1) bidirectional reflectance distribution function (BRDF) database, which is produced and distributed by Medias-France, are used to estimate surface spectral albedo. Except for the cover types of water bodies and snow-ice, fifteen out of a total seventeen International Geosphere Biosphere Programme (IGBP) land cover types are analyzed one by one, month-by-month, An optimal viewing angle is found at the back scattering direction, and its corresponding reflectance, the so-called hot spot is extracted from each pixel in the BRDF database. An empirical model is built for estimating the spectral albedo from the corresponding hot spot directional reflectance. The most interesting thing is that beyond the expectations, the hot spot-albedo relations appears almost in the same shape, no matter the type of variation of the surface cover and its seasonal change. This method can be used for instantaneous inference of surface albedo at global scale in the case of hot spot information exists. For more convenient applications for other viewing angles of the satellite measurements, further studies are needed.

1. INTRODUCTION

Surface albedo is an important parameter for satellite remote sensing, global radiative transfer, and climate modeling. Compared to its importance, it is not well understood because of its diversity and complicated anisotropic reflectance property. There have been many efforts on modeling the BRDF [e.g., Verhoef, 1984, Verstraete et al., 1990] to correct the anisotropic effect through great numbers of field and satellite measurements. With the operations of the multi-angle sensor such as ATSR-2 (on-board the ERS-2 satellite), MISR (on-board the EOS-Terra/Aqua satellite) and POLDER (on-board the ADEOS I and ADEOS II satellite), our understanding on the anisotropy reflectance property of the surface has been greatly improved. Some global surface albedo products derived from these satellites measurements with fine spatial resolutions have been produced [Wanner et al., 1996, Pinty et al., 2000].

For correcting the anisotropic effect of the surface from satellites data, BRDF models are necessary to be used. However, the present existing models such as empirical model [Walthall et al., 1985] and physical model [Chen et al., 1997] need at least three and six parameters to be determined respectively for each channel by model inversion. This is an ill posed inversion problem and a certain amount of observations should be collected in order to get comparatively accurate estimations. Therefore the albedo products usually have temporal resolutions of several days to one month. The long period needed for the data collections causes problems. Clouds make the data collection difficult; different atmosphere condition such as precipitation and snowing makes data changing sharply; all of these make the model inversion unstable. The most unavoidable problem is the conflict between long-term data collection and the necessity of unchanged surface objects. All these problems are possibly causing the failure of the model inversion, which is the main reason of the discontinuities in the global albedo products.

Observations obtained by most sensors are with single observation angle, also these data cover a much longer period than multi-angle sensor does, so it is important to make sufficient use of single observation angle data while multi-angle sensors give us chances to study the surface reflectance and its anisotropy. Also, if albedo from one single observation angle can be estimated, it is much easier to estimate the real time albedo, and also decrease the spatial discontinuities come along with long term data collections.

Our efforts here attempt to find out a relationship between one single view angle and hemispherical reflectance by using the multi-angle observations obtained from POLDER-1. In the following section, some basic concepts are described. Section 3 gives the descriptions of the datasets used, coming up with the data processing. In Section 4, we present and give some discussions on the results. Finally, we give out the conclusions of this work.

2. BRDF AND OPTIMAL ANGLE FOR OBSERVATION

The diffuse reflection of surface used to be treated as isotropy until Nicodemus et al. [1977] first defined BRDF, which is used to represent reflectance of the target as a function of illumination angle and observation angle. For most natural surface except water body and snow cover, the anisotropy reflectance properties observed by satellite sensors are thought caused by the shadow effect [Hapke, 1993; Hapke et al., 1996]. When solar and viewing directions coincide, shadows tend to be hidden, and a surge of reflectance is observed, which is the so-called hot spot. Hot spot of the earth surface has been observed and reported by many researches, via satellite [Grant et al., 2003] and aircraft observations [Coca et al., 2004]. Viewing from the hot spot direction, all the shadows are hidden by the illuminated elements, which cast the shadows. Then the sensor received signal at this direction, is dependent on the inherent reflectance of elements of the surface scene. As the phase angle becomes larger, when the observation direction diverges from the solar direction, the shadow areas will start appear, and the radiance reflected to the sensor will start to decrease along with the increasing of phase angle.

It is reasonable to consider that if we take the shadow of a certain structure as a certain aggregate, its distribution on a horizontal plane follow some statistic relationships and can be described by a probability distribution function. It becomes straightforward to get the albedo, when the inherent reflectance of the elements and the distribution of the shadows are known. Moreover, because the hot spot position changes along with the solar zenith angle, it is convenient to build the relation using phase angle as a variable. As shadow effects are considered mainly to be determined by the structures of the land cover types, it is predicted that for different cover types the distribution of shadows will appear different; the seasonal change is also thought to have impact on the shadow distributions.

3. DATA PROCESSING

The space-borne POLDER-1 sensor developed by Centre National DEtudes Spatiales (CNES), was on-board the satellite ADEOS. It has a bi-dimensional CCD array detector; this unique design makes it not only available for multi directional observations but also possible for observing the hot spot reflectance. This study utilizes the POLDER-1 BRDF Database produced which distributed by Medias France [Lacaze R. et al., 2003] to estimate albedo and extract the corresponding hot spot information. The POLDER-1 BRDF Database collected 22594 sets of BRDF, during its 8 months of POLDER operation. For all datasets, the BRDF is that on the surface, that means the molecular scattering and absorption effects have been eliminated. Still no tropospheric aerosols are corrected. The BRDF datasets have been classified into 17 types of land cover based on IGBP 1-km DISCover classification, no ocean data are included. In the present study, two IGBP cover types, snow-ice and water body cover types are excluded; IGBP cover types investigated are listed in Table 1. For every dataset, there are over 80 view directions distributed in the whole viewing hemisphere, four bands of reflectance at 443 nm, 565 nm, 670 nm, and 865 nm as well as geometries of sun and view directions are provided. Detailed information of POLDER-1 BRDF database can be obtained from http://smsc.cnes.fr/POLDER/A_produits_scie.htm.

To derive surface albedo from the above-mentioned BRDF datasets, the kernel-driven model [Roujean et al., 1992] is employed. As the kernel-driven model has the merit of less inversion parameters and with reasonable data reconstructive property for most land cover type, it is widely used for the albedo estimations. The Ross-Thick volume scattering kernel [Roujean et al., 1992] and Li-Sparse geometric–optical kernel [Li et al., 1992], which has been reported to give robust and best inversion results [Pokrovsky et al., 2002], are adopted as the inversion kernels in the present study. An example of successful inversion results is shown in Fig. 1. Since uncertainties in the model inversion bring errors to the albedo estimation, data scatter due to these errors is predicted in the empirical equation derivation.

Table 1. The IGBP Cover Types Studied in this Research

1	Evergreen needleleaf forest	6	Closed shrublands,	11	Permanent wetlands
2	Evergreen broadleaf forest	7	Open shrublands	12	Croplands
3	Deciduous needleleaf forest	8	Woody savannas	13	Urban and built-up
4	Deciduous broadleaf forest	9	Savannas	14	Cropland/natural vegetation mosaic
5	Mixed forest	10	Grasslands	15	Barren or sparsely vegetated

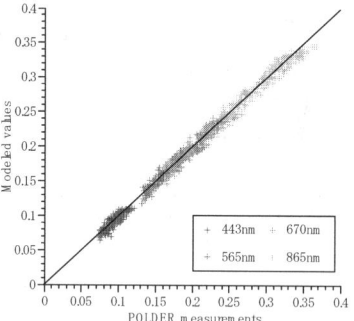

Fig. 1. An inversion example, the Land cover type of the pixel is IGBP01 (evergreen needleleaf forest), located at 33.1389°N, 112.6907°W, data collection period: 1997.6.1 ~1997.6.23, inversion accuracy: r^2(443nm)=0.761, r^2(565nm)=0.822, r^2(670nm)= 0.893, r^2(865nm)=0.967

It must be mentioned that very few scenes with reflectance of exact backscattering where the phase angle is 0 degrees can be obtained in the present datasets. Instead the viewing direction, which is most near to the exact backscattering direction, are selected by calculating and comparing the phase angle with the solar zenith angle. This makes the extracted hot spot reflectance contain some shadow information and also bring in errors when the empirical equation is derived.

For the empirical equation to be derived, in order to reduce the scatter caused by the errors mentioned above, two thresholds are set. The correlation coefficient, which expresses the inversion accuracy, is set better than 0.97; and the range of phase angle for extracting the hot spot is set to be less than 3 degrees, the great change of the reflectance in the hot spot area still brings in obvious errors. According to

Bréon's resent research, the hot spot width is mostly found at less than 2~5 degrees.

4. RESULTS AND DISCUSSIONS

For every cover type investigated, data are processed month-by-month in order to see the seasonal change effects on the hot spot-albedo relations. Fig. 2 is the scatter plots of the hot spot BRF (Bi-directional

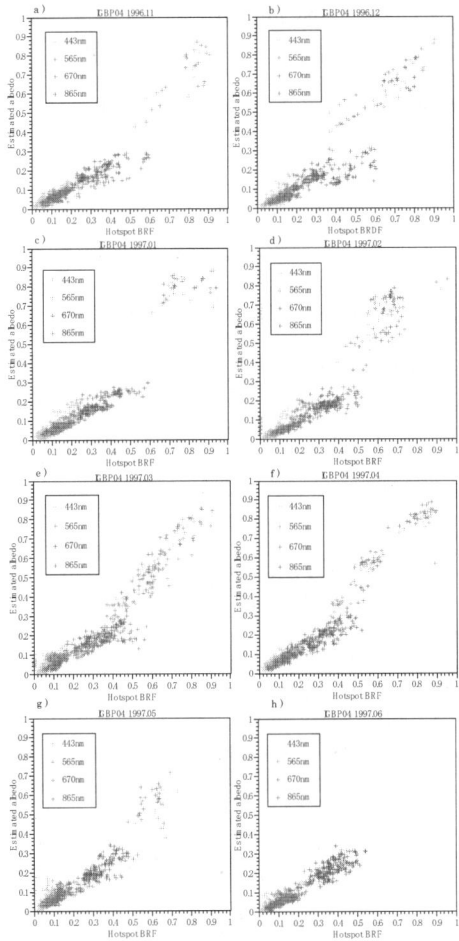

Fig. 2. The hot spot BRF vs. albedo value scattering plots, eight plots show samples in 8 months; surface type is IGBP 04 (deciduous broadleaf forest), with total 1483 samples.

Reflectance Factor) vs. estimated albedo values scattering plots. BRF is corresponding to BRDF, multiplied by a constant π, so that the unit of the directional reflectance is the same with albedo. In this figure surface investigated is deciduous broadleaf forest cover types. In each plot, data appear concentrated at the lower left part and the upper right part of the plot. Data hitting on the upper right part of each plot with albedo larger than about 0.4 are considered to be caused by the contamination of snow. By checking the locations of the samples contained in this category, most of the pixels are at high latitudes of the north hemisphere, so that in the winter season from November to May of the next year, pixels with snow contaminations are presented in the plots. Since the inversion models are not available for the surface mixed with snow, much larger scatter is observed within these data. Valid data are thought to be those on the lower left part of each plot. The inclination of these data is clearly seen, although the scatter of the data is large due to the errors mentioned in the previous section. In each plot, four spectral bands are shown; no obvious distinctions are found within the four bands. This result verifies the prediction mentioned in Bréon's previous research, that the shape of BRDF might be independent of spectral bands, because the geometric reflection dominates the BRDF shape. Moreover the eight plots show almost similar and linear relations between hot spot BRF and albedo. For deciduous forest, although, most of leaves drop off most in winter and the spatial structure changes a lot compare to that in summer, hot spot-albedo relationship keep unchanged during the eight months, which means that the relationship are not sensitive to the seasonal change of the land cover type.

Investigations on the other cover types show the same properties as in Fig. 2. Furthermore, beyond our expectations, the 15 cover types, even though they are with completely different structures, have hot spot-albedo relations that show the inclination without obvious changes. Fig. 3 is the same as Fig. 2, except for IGBP 13 (the urban and built-up). It seems that at a relatively large spatial resolution, such as 7 km of POLDER, the statistic distributions of the shadows remains relatively stable, no matter the surface cover types, therefore, the hot spot—albedo relationship are not sensitivity to the target cover types.

Fig. 4 is the sum-up of the 15 cover types. There are 215 datasets extracted from a total of 22224 datasets that fulfill the constrained conditions. The great part of the data contaminated by snow that show large scatter is filtered out. Still, there are data scatter partly due to the reason mentioned in section 3. The diagonal present the Lambertian reflector. It is clear that the magnitude of surface albedo is mainly dependent on the hot spot BRF, that is the inherent reflectance of the elements of the surface. Also, for elements with high inherent reflectance, since the multiple reflectance between elements becomes more effective, the surface albedo tends to appear brighter. Considering this reason, we fit the data with a power law function. It is shown in the following, with correlation coefficient of 0.92, and standard error of 0.022:

$$y = 0.6818 x^{1.128}$$

where x is the hot spot BRF and y is the albedo. This equation can be used for instantaneous estimation of surface albedo at global scale except snow-ice and water, applying the correspondent hot spot information. Also, if the phase function can be estimated, it is possible to estimate albedo from other viewing angles.

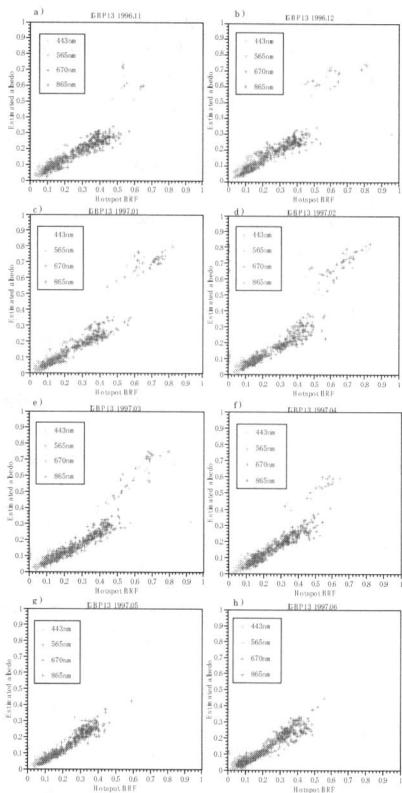

Fig. 3. The same as Fig. 2, except for the surface type of IGBP 13 (urban and built-up), with total 1198 samples.

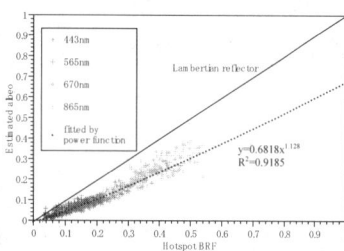

Fig. 4. The hot spot BRF vs. albedo value scattering plots with totally 215 samples extracted from the whole 15 IGBP cover types in 8 months with the constrained conditions.

6. CONCLUSIONS

In this study it was found that for relatively large spatial resolution observations, the surface BRDF is not as complex as we had expected with experience of field measurements. For 15 IGBP cover types, the hot spot-albedo relations, and the BRDF shapes, are observed to be stable within different bands, seasons and target cover types. The main property that controls the surface albedo is the inherent reflectance of the surface elements. Multiple reflectance is another factor that affects albedo. The surface structure difference is considered with very little effect on albedo. BRDF effect can be corrected using the presented power law equation with hot spot reflectance known.

7. REFERENCES

Bréon, F.-M., F. Maignan, M. Leroy, I. Grant, 2002: Analysis of the hot spot directional signatures measured from space. *Journal of Geophysical Research*, 107(D16) 10.1029/2001JD001094.

Chen, J. M., and S. G. Leblanc, 1997: A four-scale bidirectional reflectance model based on canopy architecture, *IEEE Transactions on Geoscience and Remote Sensing* 35, pp. 1316–1337.

Camacho-de Coca F., F.-M. Bréon, and F. J. Garcia-Haro, 2004: Airborne measurement of hot spot reflectance signatures, *Remote Sensing of Environments*, 90, 63-75.

Grant I. F., F.-M. Bréon, and M. M. Leroy, 2003: First observation of the hot spot from space at sub-degree angular resolution using POLDER data. *International Journal of Remote Sensing*, 24, No. 5, 1103-1110.

Hapke, B., D. Dimucci, R. Nelson, and W. Smythe, 1996: The cause of the hot spot in vegetation canopies and soils: Shadow-hiding versus coherent backscatter, *Remote Sensing of Environments*, 58, 63-68.

Hapke, B., 1993, Theory of Reflectance and Emittance Spectroscopy, *Topics in Remote Sensing III*. Cambridge University Press.

Lacaze, R., 2003: POLDER-1 BRDF Database-User Document. Medias France.

Li, X., and A. H. Strahler, 1992: Geometric–optical bi-directional reflectance modeling of the discrete crown vegetation canopy: effect of crown shape and mutual shadowing. *IEEE Transactions on Geoscience and Remote Sensing*, 30, 276–292.

Nicodemus, F., J. Richmond, J. Hsia, I. Ginsberg, and T. Limperis, 1977 Geometrical Considerations and Nomenclature for Reflectance. Washington, DC: Hemisphere.

Pokrovsky, O., and J. L. Roujean, 2003: Land surface albedo retrieval via kernel-based BRDF modeling: I. Statistical inversion method and model comparison, *Remote Sensing of Environment*, 100, 100-119.

Pinty, B., F. Roveda, M. M. Verstraete, N. Gobron, Y. Govaerts, J. V. Martonchik, D. J. Diner, and R. A. Kahn. 2000: Surface albedo retrieval from Meteosat 1.Theory. *Journal of Geophysical Research*, 105, D14, 18099-18112.

Roujean, J., M. Leroy, and P. Deschamps, 1992: A bi-directional reflectance model of the surface for correction of remote sensing data, *Journal of Geophysical Research*, Vol. 97, pp. 20455-20468.

Verstraete, M. M., B. Pinty, and R. E. Dickinson, 1990: A physical model of the bidirectional reflectance of vegetation canopies, 1. Theory. *Journal of Geophysical Research*, 95(D8): 11755-11765.

Verhoef, W., 1984: Light scattering by leaf layers with application to canopy reflectance modeling: the SAIL model, *Remote Sens. Environ.*, 16:125-141.

Walthall, C. L., J. M. Norman, J. W. Welles, G. Campbell, and B. L. Blad, 1985: Simple equation to approximate the bidirectional reflectance from vegetated canopies and bare soil surfaces, *Applied Optics*, 24, 383-387.

APPLICATION OF σ-IASI TO NAST-I

A. Carissimo, G. Grieco, and C. Serio
DIFA Dip., Università della Basilicata
85100 Potenza, Italy

V. Cuomo and G. Masiello
IMAA/CNR
Tito Scalo, Potenza, Italy

W. L. Smith
NASA Langley Research Center (Retired)
Hampton, Virginia, USA

ABSTRACT

The paper describes a new radiative transfer code for the analysis and inversion of high spectral resolution infrared observations. Its application to spectra acquired from an airplane-based Fourier Transform Spectrometer, during a flight campaign in the tropics, is presented. The campaign also deserved an in-depth validation of the radiative transfer tool.

1. INTRODUCTION

The package σ-IASI (Amato et al., 2003) is a line-by-line radiative transfer model designed for fast computation of spectral radiance and its derivatives (Jacobian matrices) with respect to a given set of geophysical parameters. It represents a compromise between the accuracy of the exact line-by-line radiative model and the fastness of the hyper-fast radiative transfer model. The ability to compute radiance derivatives gives to σ-IASI the flexibility to be used also for the inversion process of geophysical parameters. This is done through a Newton-Raphson scheme in which the Radiative Transfer Equation is step-linearized by Taylor expansion (Carissimo et al., 2005).

The tool σ-IASI has been developed to meet the spectral range of IASI (Infrared Atmospheric Sounding Interferometer). The European IASI instrument covers the spectral range between 3.6 and 15.5 micron with a spectral resolution of 0.25 cm^{-1} and it will be launched onboard of MetOp satellite on 2005.

The same spectral range is covered by the NAST-I Fourier Transform Spectrometer (NPOESS Aircraft Sounder Testbed Interferometer). NAST-I flies on board the NASA ER-2 stratospheric airplane and the Proteus, and has a spectral resolution quite similar to that of IASI. The code σ-IASI has been adapted to NAST-I and the paper shows and discusses examples of radiance inversion for temperature and water vapor. Examples from the CAMEX-3 experiment will be shown. The analysis has provided a quality check of the σ-IASI forward/inverse capability and has allowed us to assess the retrieval accuracy for temperature and water vapor (T,q) obtained by high spectral infrared resolution observations.

2. BASIC METHODOLOGY AND DATA SET

2.1 The σ-IASI Code

The forward module σ-IASI has been designed for fast computation of spectral radiance and its derivatives with respect to a given set of geophysical parameters (e.g. temperature, water vapor, ozone, carbon dioxide, nitrous oxide, surface emissivity). Although the code has been developed mostly for IASI, it is well suited for nadir viewing satellite and airborne infrared sensors with a sampling rate in the range 0.1–2 0.25 cm^{-1}. The code computes monochromatic radiance from look-up tables of monochromatic layer optical depth generated using the line-by-line model LBLRTM (Clough et al., 1995). The code σ-IASI borrows from LBLRTM the CO_2 line mixing scheme. Cross sections of heavy molecules such as chlorofluorocarbons are not parameterized in terms of look-up tables, since absorption can be computed quickly and separately: σ-IASI accounts for the presence of CFC species, namely CCl_3F (CFC-11) and CCl_2F_2 (CFC-12). The code allows one to input any user defined surface emissivity; the default is the Masuda model (Masuda et al., 1988) for sea surface emissivity. Finally, σ-IASI computes the surface emissivity Jacobian matrix and this quantity may be either a user supplied parameter or a retrieval parameter.

The main changes in the version of σ-IASI adapted for NAST-I consist in a new atmospheric layering and a new parameterization of water vapor absorption lines, which now accounts for the self-broadening effect. In order to increase the code capability of producing radiative transfer calculations for airplane observations, the number of layers has been increased up to 60; the new grid of sixty pressure levels extends from the surface up to a pressure of 0.005 hPa. The user may choose the observation altitude from the input; hence this new feature affects the input structure of the code. The new optical depth database takes into account for the water vapor self-broadening effects (details in Masiello and Serio, 2003). This last feature does not affect the input-output of the code. Moreover, the current optical depth database is generated from the last version of LBLRTM (v.8.01, June 2003), which uses the HITRAN 2K spectral line database.

2.2 The NAST-I Data Set

The NAST-I data we have analyzed refer to the CAMEX/3 (Convection and Moisture Experiment 3) Atlantic basin tropical cyclone field validation NASA ER-2 flight (during local night time, September 13-14, 1998) over Andros Island, Bahamas.

We have a total of five different sets or scan patterns of observations, corresponding to the five different aircraft tracks (Fig. 1). Each scan pattern consists of 154 scan lines for a total of 154*13=2002 observations. The five scan patterns yield 10010 observations.

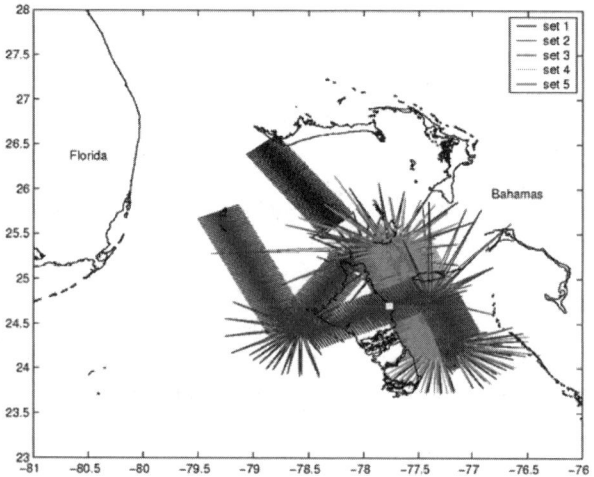

Fig. 5. Target area and the five scan patterns of observations analyzed in this study. The square box locates the position of the radiosonde station. Latitude is shown on the vertical axis, and longitude on the horizontal axis (units of degrees).

There were 4 radiosonde launched during the flight period, which will be referred to as RAOB1, 2, 3 and 4, respectively.

The radiosonde were launched from the same location at latitude 24.7° N and longitude 77.8° West (Andros, Island). NAST-I was on board ER-2 and flown over the radiosonde station to provide the data needed for the validation of the retrieved products. Once the truth profile (i.e. the radiosonde observation) is known, the retrieval can be directly compared to it and the consistency of the radiative transfer can be checked, as well.

We have found that the NAST-I clear *sky* observation, which is best co-located both spatially and temporally with the four ascents above, is the NAST-I scan #298, recorded at nadir angle. This observation was made at 00:33:30 UTC, which closely corresponds to RAOB2.

3. RESULTS

Retrieval of (T, q) has been obtained for the whole clear-sky set of NAST-I spectra by using the *inverse* physical module developed within σ-IASI (Carissimo et al., 2005). The non-linear inverse problem needs a suitable First Guess (FG) and various initialization strategies were checked and that which gave the best results consists in obtaining the FG through an Empirical Orthogonal Function (EOF) regression (e.g., see Zhou et al., 2002).

The retrieval strategy is simultaneous for the pair (T, q) and an inversion example is shown for the NAST-I observation #298 in Fig. 1 (temperature) and Fig. 2 (water vapor).

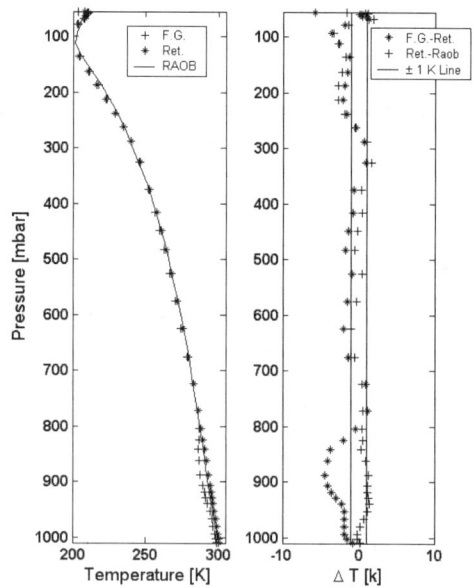

Fig. 6. Temperature inversion (left) and the difference between the retrieved profile and the radiosonde observation (right) are shown for the case of EOF-initialization; F.G. stands for First Guess and Ret. for retrieved.

It is possible to see that temperature is recovered within 1K for most of the atmospheric altitude, with a very nice performance in the lower middle part of the Atmosphere. A good retrieval performance is also achieved for water vapor

(Fig. 2). The accuracy is better than 10% in the lower to middle part of the atmosphere.

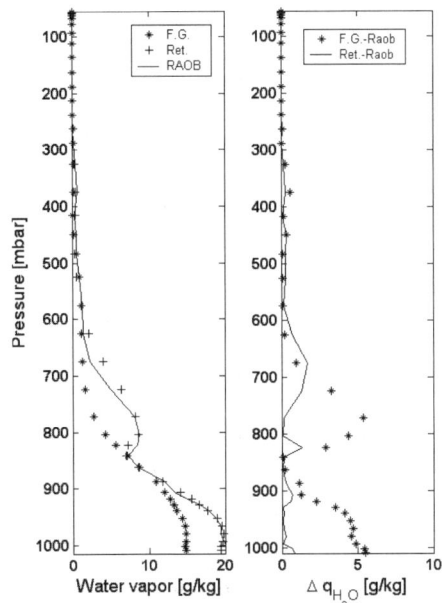

Fig. 7. As Fig. 1 but for water vapor.

It should be here stressed that the spectral ranges used for the EOF regression (FG initialization of the physical scheme) and the physical scheme itself do not perfectly coincide, in order to have for the physical inversion fresh spectral information. For the purpose of (T, q) retrieval, the EOF regression and the physical scheme use the spectral intervals listed below, respectively:

EOF Regression	Physical Scheme
670 to 800 cm^{-1}	670 to 830 cm^{-1}
1350 to 1450 cm^{-1}	1100 to 1200 cm^{-1}
2160 to 2260 cm^{-1}	1350 to 1600 cm^{-1}
	2000 to 2100 cm^{-1}
	2100 to 2260 cm^{-1}

In comparison to the EOF regression, the physical scheme uses more spectral information. In particular the atmospheric windows, 1100 to 1200 cm^{-1} and 2000 to 2100 cm^{-1}, which are rich of weak water vapor lines, are used. These two atmospheric window segments enhance the performance of the scheme in the lower part of the atmosphere.

It should be also noted that the radiosonde observations, shown for comparison in the two figures above have been reduced to the vertical pressure-grid mesh used for the retrieval products. It is possible to see from Fig. 2 that the retrieval is able to follow the inversion of water vapor that is seen in between 800–900 mbar. In other words, the inverted profile seems to be able to represent the fine structures seen in the observations. Although not shown here (see Carissimo et al., 2005 and references therein), the physical inverse scheme allows for 10–15 pieces of information or degrees of freedom for temperature and 6–10 for water vapor. From the results shown in Figs. 1 and 2, this information seems to be enough to capture details in the lower to middle part of the atmosphere.

The retrieval quality and accuracy has been also checked by analysis of the spectral residual. For the retrieval exercise shown in the two previous figures, the spectral residual is shown in Fig. 4 for the spectral range 1100 to 1210 cm^{-1}.

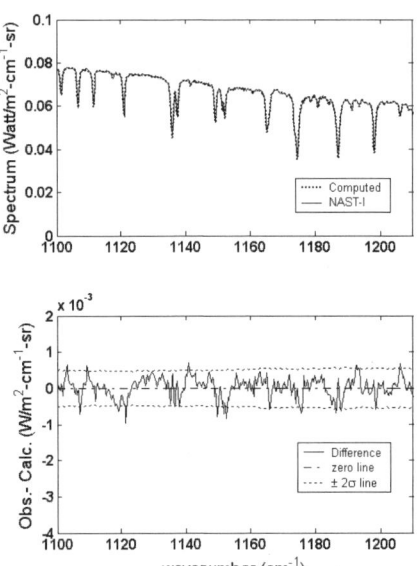

Fig. 8. Fitted and observed spectrum and spectral residual for the retrieval exercise shown in Figs. 2 and 3. The spectral range 1100 to 1200 cm^{-1} is shown.

This range is mostly sensitive to water vapor self-broadening and the absence of any systematic discrepancy in the residual testifies the effectiveness of our radiative transfer.

A very nice consistency is also seen for the case of the 6.7 μm water vapor absorption band. The spectral residual is shown in Fig. 5. Apart from the beginning of the spectral range, which is contaminated by methane for which we use climatology, the consistency between observations and computations is very good. In Fig. 4 and Fig. 5, the 2-σ tolerance interval has been obtained directly by the NAST-I radiometric noise, which, in turn, has been derived from the onboard calibration during the flight.

The performance of the inverse physical scheme has been evaluated by considering the retrieval for all the clear sky NAST-I spectra and by constructing, in the usual way, the root mean square error. In doing so, we assumed the truth to be the mean (T,q)-profile computed on the basis of the four radiosonde ascents. The results

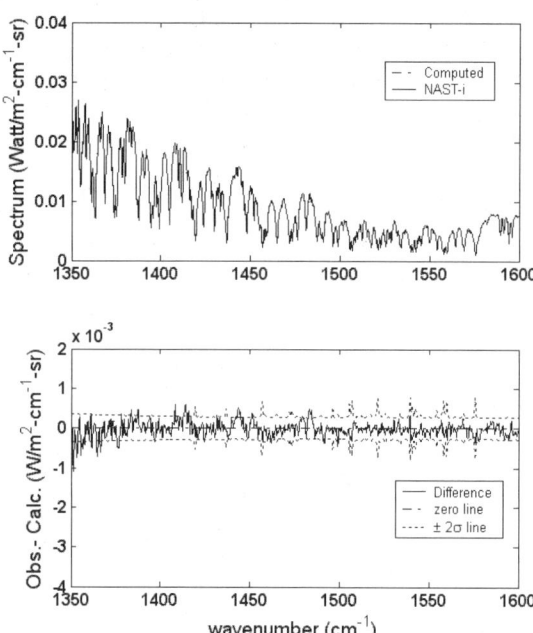

Fig. 9. Fitted and observed spectrum and spectral residual for the retrieval exercise shown in Figs. 2 and 3. The spectral range 1350 to 1600 cm^{-1} is shown.

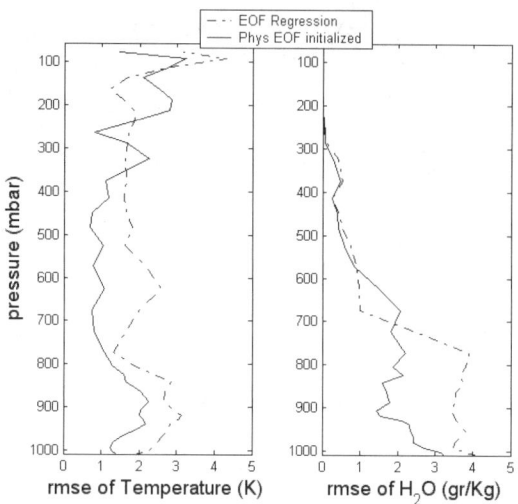

Fig. 10. Performance of the physical inverse scheme compared to the EOF regression. The EOF regression is used to initialize the physical scheme.

are shown in Fig. 6 for temperature and water vapor. The physical inversion gets improved retrieval over the EOF regression First Guess and the performance for temperature is quite close to 1 K and 10-15% for water vapor. Note that the root mean square error shown in Fig. 6 includes the atmospheric variability across the target area, so that they underestimate the goodness of the retrieval performance, which is expected to be better than that shown in Fig. 6

4. CONCLUSIONS

We have presented the last version of the σ-IASI code and discussed its application to NAST-I data. The study has demonstrated that the new σ-IASI yields radiative transfer which is highly consistent with observations. The analysis has also shown that the physical inverse scheme resolves fine structure, which may be not present in the First Guess. The performance for temperature and water vapor are coherent with the IASI expected accuracy of 1 K in 1 km layer for temperature and 10–15% accuracy in 1–2 km layer in the lower troposphere for H$_2$O.

Finally, the study has shown that high spectral resolution infrared observations, such as NAST-I data, which are of the same quality as that expected for IASI, do contain all the information to achieve the IASI accuracy goals of 1 K/km for temperature and 10% per 1–2 km layer for water vapor.

5. ACKNOWLEDGMENTS

Work supported by MURST – "PROGRAMMA OPERATIVO NAZIONALE" – misura 1.3, COS(OT).

6. REFERENCES

Amato, U., G. Masiello, C. Serio, and M. Viggiano, 2002: The σ-IASI code for the calculation of infrared atmospheric radiance and its derivatives, *Environmental Modelling and Software,* 17/7, 651-667.

Carissimo, A., I. DeFeis, and C. Serio, 2005: The physical retrieval methodology for IASI: the δ-IASI code, *Environmental Modelling and Software,* in press.

Masiello G., and C. Serio, 2003: An effective water vapor self-broadening scheme for look-up-table based radiative transfer," *SPIE* proceedings, 4882, 52-61.

Clough, S. A., M. J. Iacono, and J. L. Moncet, 1995: Line-by-line calculations of atmospheric fluxes and cooling rates *II*: Application to carbon dioxide, ozone, methane, nitrous oxide and the halocarbons, *J. Geophys. Res.,* 100, 16519-16526.

Masuda, K., T. Takashima, and Y. Takayama, 1988: Emissivity of pure and sea waters for the model sea surface in the infrared window regions, *Remote Sens. Environ.,* 24, 313-329.

Zhou, D. K., W. L. Smith, J. Li, H. B. Howell, G. W. Cantwell, A. M. Larar, R. O. Knuteson, D. C. Tobin, H. E. Revercomb, G. E. Bingham, J.-J. Tsou, and S. A. Mango, 2002: Thermodynamic product retrieval methodology for NAST-I and validation. *Appl. Opt.,* 41, No. 33, 6957-6970.

A COMPARISON OF SATELLITE DERIVED PRECIPITATION FIELDS AND METEOSAT-5 OBSERVATIONS OF MESOSCALE CONVECTIVE SYSTEMS OVER INDIAN OCEAN

T. F. Yang, I. Jobard, M. Capderou, and M. Desbois
Laboratoire de Météorologie Dynamique, Ecole Polytechnique
Palaiseau, 91128, FRANCE

ABSTRACT

METEOSAT-5 infrared images are used to infer the characteristics of mesoscale convective systems (MCSs) over Indian Ocean during INDOEX in January 1999. In order to characterize the relationship between the regional precipitation and the MCS distribution, the SSM/I data from three DMSP satellites have been used to estimate the rainfall. Rain rate over the ocean is derived from the Scattering Index algorithm [Grody, 1991] using the vertically polarized channels at 19 GHz, 22 GHz and 85 GHz. This method compares well with the Global Precipitation Climatology Project (GPCP) monthly and daily product. A monthly mean rain rate map shows the high correlation of higher rainfall with the higher frequency of MCS's distribution. The statistical properties and links of the products derived from GPCP, SSM/I and infrared channel observations are discussed.

1. INTRODUCTION

Since the use of INSAT-1B data from 1983 and the use of METEOSAT-5 images for INDOEX in 1999, the study of mesoscale convective systems (MCSs) over the Indian region has shown their important role in the regional climate. [Laing and Fritsch, 1993. Gambheer and Baht, 2001. Roca et al., 2002]. Generally the MCSs show a space scale from 25,000 to 250,000 km^2 and a time scale from several hours to more than one day [Houze, Jr., 1993]. The MCSs over the Indian Ocean have been documented using the METEOSAT-5 infrared data, taking a brightness temperature threshold of 238 K and a minimum cloud cover of 12,500 km^2 [Yang et al., 2003]. The distribution map of MCSs during the northern hemisphere winter season in this region shows three large occurrence areas: the south of Bay of Bengal; a band between the equator and 15°S (ITCZ); an area in the south-west of the Tropic of Capricorn.

In this study, we apply the Scattering Index to estimate rain rate using the SSM/I data and then compare the mean monthly estimation with GPCP monthly mean rain rate. The diurnal evolution of two 1°x1° areas has been analyzed during whole month. This study was also conducted in the frame of the preparation of the Megha-Tropiques Project. This satellite, scheduled for 2009, will have an orbit inclination of 20° at 870 km height. Any points in the band between 23°S and 23°N will be seen 3 to 6 times per day by its microwave radiometer MADRAS while it is seen only 1 or 2 times per day by a single SSM/I. Thus we use three DMSP satellites in order to approach the sampling of Megha-Tropiques. This sampling should allow examining the precipitation evolution during the MCS's life cycle.

2. DATA AND METHODS

METEOSAT-5 is a geostationary satellite which has been moved to 63°E in 1998 by EUMETSAT for the INDOEX experiment that began in January 1999. The image is taken in three channels, visible (VIS), water vapor (WV) and infrared (IR) at half-hourly intervals. The infrared channel images (wavelengths: 10.5~12.5 μm) have been converted in Laboratoire de Météorologie Dynamique to a low resolution (25 x 25 km) grid format. A total of 1,488 consecutive images are used here to extract the MCSs occurring in January and its distribution is shown in Fig. 1.

We use also here the Special Sensor Microwave Imager (SSM/I), a passive microwave instrument carried by the three DMSP Satellites: F-11, F-13, and F-14. These satellites have a 102 minutes sun-synchronous near-polar orbit at the altitude of about 860 km. SSM/I has a swath width of approximately 1400 km and its ascending equatorial crossing times is 7:25 p.m. for F-11; 5:54 p.m. for F-13, and 8:46 p.m. for F-14. There are seven channels to measure atmospheric, oceanic and land surface microwave brightness temperatures at 19.35, 37.0, and 85.5 GHz in dual polarizations (V and H); 22.235 GHz in only V polarization. Their footprint sizes vary with frequencies from 69 x 43 km to 15 x 13 km. All our SSM/I data have been projected on the METEOSAT-5 grid map into 0.25°x0.25° resolution to facilitate further analysis.

Finally the GPCP (Global Precipitation Climatology Project) provides monthly rain rate (Version 2 Combination product) obtained from a gridded analysis based on gauge measurements and multi-satellites estimates of rainfall (resolution: 2.5°x2.5°, unit: mm/day). 1-Degree daily rain rate data (1DD Combination product) based on the monthly precipitation data set was used for the diurnal evolution analysis (resolution: 1°x1°, unit: mm/day).

The principal method of a rain rate estimation is based

on the Scattering Index (SI) which is determined by subtracting the actual 85 GHz vertically polarized signal from the factor F, as derived using different combinations of the lower-frequency channels under non-scattering conditions over land and ocean in various regions of the world. Thus the *SI* at 85 GHz is defined as below

$$SI_{85v} = F - T_{85v} \quad (1)$$

where the values greater than 10 K are used to identify scattering surface. The factor F is written as

$$F = A + BT_{19v} + CT_{22v} + D(T_{22v})^2 \quad (2)$$

where the coefficients A, B, C and D are derived empirically for the water surface as 0.7152, 2.4387, –0.00504 and –174.38, respectively [Ferraro et al., 1994 and 1996]. To identify the ocean only precipitation, the following conditions should be fit: $SI>10$; $T_{22V}>38+0.88T_{19V}$. According to the results of SI, the empirical relation between rain rate (RR) over the oceanic surface and SI [Ferraro and Marks, 1995] has been applied to obtain the estimated rate,

$$RR = 0.00188 \, SI^{2.034} \quad (3)$$

The SSM/I orbit data have been projected on the METEOSAT-5 grid map. After a quality control to avoid bad scans and after exclusion of the orbits passing over the edges of the study area, we selected 201 orbits from F-11, 214 orbits from F-13 and 222 orbits from F-14, which were located between 30°E and 110°E for January 1999. Then using a resolution of 2.5°x2.5°, as in GPCP, we found that each square of the grid was seen from 40 to 60 times in a month for the ascending orbits (same for the descending orbits) by the combination of the three DMSP satellites.

While the brightness temperatures measured by SSM/I can be transformed into rain rate, we use a simple formula to calculate the mean rain rate (R_{mean}, mm/day):

$$R_{mean} = \frac{\sum_{i=1}^{n} \sum_{j=1}^{N_i} P_j}{\sum_{i=1}^{n} N_i} \times 24 \quad (4)$$

where P_j is 2.5°x2.5° spatial average rain rates per pixel for orbit j, N_i is total number of orbits (ascending and descending) during day i and n is total number of days. The result is shown in Fig. 2.

3. COMPARISON

A previous study [Yang et al., 2003] has established the distribution of MCSs over the Indian Ocean during INDOEX period. The same MCSs tracking algorithm, selecting only the oceanic clusters, has been used to build Fig. 1. It shows high MCS occurrence areas, such as the Bay of Bengal (over 60) and the north-east of Madagascar (between 5°N and 10°S, 50°E and 75°E) where a maximum occurrence of 184 is reached. The south-west of Madagascar does not show as high occurrences (around 40).

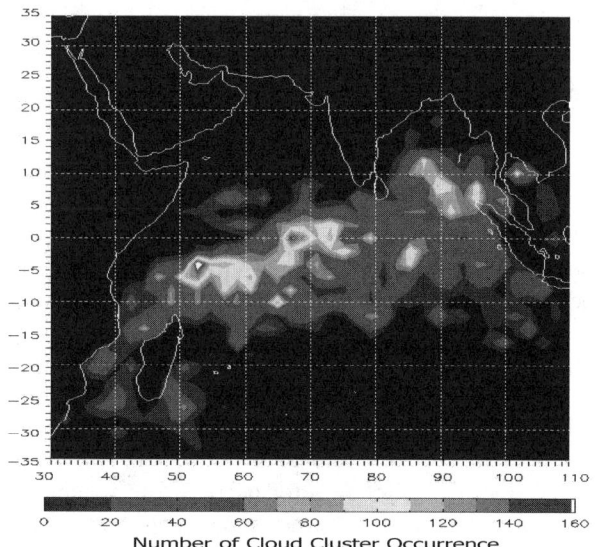

Fig. 1. Distribution of occurrences of MCSs in January 1999. The number of MCS was computed in a moving 2.5°x2.5° window. Only the purely oceanic cloud clusters have been selected.

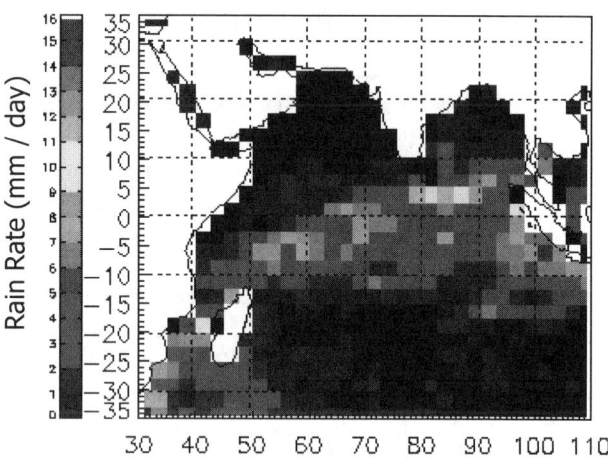

Fig. 2. Distribution of monthly mean rain rate estimation from SSM/I data.

The method of rain rate estimation using SSM/I data gives the results presented in Fig. 2. The higher rate areas (>10 mm/day) are in the south of Bay of Bengal, in the west of Madagascar and the south of Sumatra where the maximum rate of 15.6 mm/day is appeared at the area with center of 101.5°E, 6.5°S.

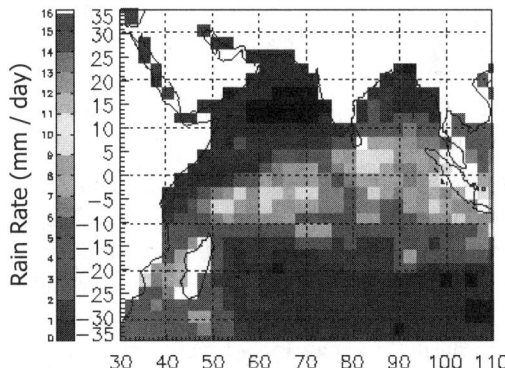

Fig. 3. GPCP monthly mean rain rate in January 1999. Three highest rain rates areas are at 108.75°E, 8.75°S with rate of 12.9 mm/day; 58.75°E, 3.75°S with rate of 12.7 mm/day and 101.25°E, 6.25°S with rate of 12.6 mm/day.

The spatial GPCP monthly rain rate distribution as shown in Fig. 3 is generally similar to the SSM/I distribution and also to the MCS distribution. It shows more high rain rate areas than the SSM/I estimation such as the region near Sumatra and the region in the north-east of Madagascar.

Comparing the two precipitation maps with the distribution of MCSs, we find that the maximum rate seen near Indonesia does not correspond to a high occurrence of MCSs. This may be explained by selecting ocean-only cloud clusters which produced an exclusion of coastal clusters, as some of the MCSs may originate on land and propagate over sea. The same situation appears in the west of Madagascar. Besides, the higher occurrence of MCSs in the open ocean corresponds better to the higher rain rate of GPCP than to the SSM/I estimation.

The comparison of combined three satellites estimation and each SSM/I satellite estimation separately with GPCP in zonal cumulated rain rates is shown in Fig. 4 which indicates well that GPCP has higher values than SSM/I estimations, about 75 mm in the equatorial region. The GPCP map shows indeed a larger extension of the rainy areas, including low rain rates areas which do not appear on the other products. The SSM/I estimations show a similar shape, but the Scattering Index fails to detect the rainfall not linked to deep convection. The difference between the estimations of the three SSM/I data increases in the south, due to less frequent sampling of this area at the border of the area studied.

Diurnal cycle of oceanic convection shows usually a maximum in the morning around 6:00 to 9:00 am local time [Roca and Ramanathan, 2000]. Over the Bay of Bengal, there is also a diurnal variation between morning and early afternoon. The maximum precipitation is associated with the higher cloud cover and deep convection during this period [Gambheer and Baht, 2001].

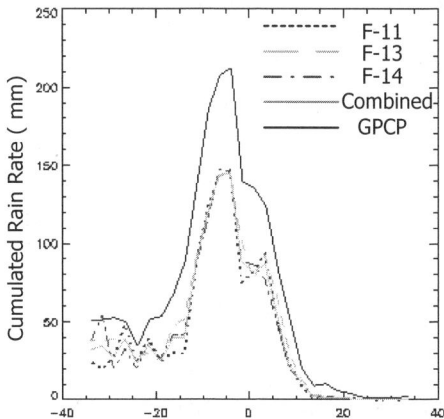

Fig. 4. Zonal cumulated rain rates of GPCP (black line), SSM/I on F-11 (blue dotted line), on F-13 (green dashed line), on F-14 (red dash-dotted line) and combined the three satellites (orange line) between 35°S~35°N.

For the three SSM/I satellites the descending orbits pass between 6:00 and 9:00 am, providing morning rain observations while ascending orbits provide evening rain observations. Fig. 5 shows the daily mean rain rates from GPCP (1DD) data and estimation for morning and evening separately from three SSM/I satellites together and individually. Two different 1°x1° areas over the ocean are selected, one near the land (Fig. 5a) and the other in the open ocean (Fig. 5b).

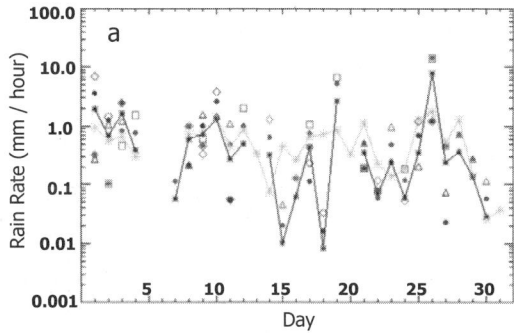

Fig. 5(a). Daily mean rain rates in 1°x1° area at 102.5°E, 5.5°S (near coast); The blue means morning time and red means evening time. The green stars are GPCP 1DD data; the black stars are the average of mean ascending and mean descending daily data. The square is for F-11 data; the triangle is for F-13 data; the diamond is for F-14 data and solid circle is the mean of morning or evening.

In case (a), there are 18 days out of 26 rainy days where morning rain is predominant. However in case (b), the morning and evening rain are equally distributed. Such phenomena suggest the effect of land-sea breezes near land area.

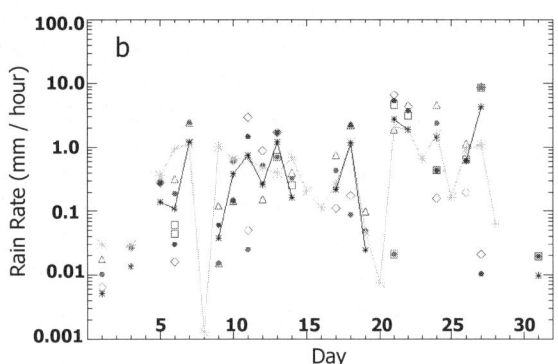

Fig. 5(b). Daily mean rain rates in 1°x1° area at 57.5°E, 3.5°S (open ocean). The same representations as in Fig. 5(a).

Besides, it is noticeable in both cases that the mean daily values obtained from three satellites together are in better agreement with GPCP values than single satellite estimation. This indicates the importance of sufficient microwave observation sampling for precipitation measurement, which is one of the advantages of the Megha-Tropiques project.

The principle of this mission is to provide measurements in the inter-tropical zone with frequent sampling for the cloud properties and precipitation; the water vapor horizontal and vertical distribution and the outgoing radiative fluxes. Satellite will be launched in 2009 with an orbit of a 20° inclination and a wide swath at 870 km altitude. The coverage is between 23°N to 23°S, repetition time from 3 to 6 times per day.

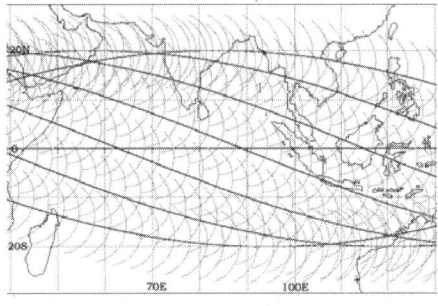

Fig. 6. The ground track of MADRAS (one of the microwave Megha-Tropiques instruments) in conical scan. Map center is at equator, 80°E.

4. CONCLUSION

The comparison of MCSs occurrence distribution over the Indian Ocean with SSM/I rain rate estimations and GPCP data shows that the higher frequency of MCS corresponds to the higher mean rainfall. Due to the selection of MCSs for only oceanic clusters, the maximum rain rate area near the continent does not reflect in the map of frequency.

Mean rain rate estimation derived from Scattering Index using SSM/I data shows relatively lower values than GPCP monthly mean rates. This may result from many areas with low rain rates on the GPCP map, corresponding to no rain areas on the SSM/I maps.

We separated the morning and evening passes to examine the daily rain rate near the equator. The case near the coast reveals more rain in the morning than in the afternoon. The area in the open ocean does not present similar tendency. Considering the need to improve the sampling during one day for a specific area in the tropical region, the passage of Megha-Tropiques satellite will provide 3 to 6 observations per day. This is useful for the study of mesoscale convective systems and the associated rainfall in different phases of their development.

5. REFERENCES

Gambheer, A. V., and G. S. Bhat, 2001: Diurnal variation of deep cloud systems over the Indian region using INSAT-1B pixel data. *Meteorol. Atmos. Phys.*, 78, 215-225.

Grody, N. C., 1991: Classification of snow cover and precipitation using the special sensor microwave imager. *J. Geophys. Res.*, 96 (D4), 7423-7435.

Ferraro, R. R., N. C. Grody, and G. F. Marks, 1994: Effects of surface conditions on rain identification using the SSM/I. *Remote Sens. Rev.*, 11, 195-209.

Ferraro R. R., and G. F. Marks, 1995: The Development of SSM/I Rain-Rate Retrieval Algorithms Using Ground-Based Radar Measurements. *J. Atmos. Oceanic Technol.*, 12, 755–770.

Ferraro, R. R., F. Weng, N. C. Grody, and A. Basist, 1996: An eight-year (1987-1994) time series of rainfall, clouds and sea ice derived from SSM/I measurements. *Bull. Amer. Meteor. Soc.*, 77, 891-905.

Houze, Jr., R. A., 1993: *Cloud dynamics.* Academic Press. 573 pp.

Laing, A. G., and J. M. Fritsch, 1993: Mesoscale convective complexes over the Indian Monsoon region. *J. Climate.*, 6, 911-919.

Roca, R., and V. Ramanathan, 2000: Scale dependence of monsoonal convective systems over the Indian Ocean. *J. Climate,* 13, 1286-1298.

Roca, R., M. Viollier, L. Picon, and M. Desbois, 2002: A multi satellite analysis of deep convection and its moist environment over the Indian Ocean during the winter monsoon. *J. Geophys. Res.*, 107, D19, 10.1029.

Yang, T. F., R. Roca, I. Jobard, and M. Desbois, 2003: Analysis of mesoscale convective system over the Indian Ocean with METEOSAT-5 during INDOEX. *Conf. of EGS-AGU-EUG*, Nice, France, 245.

SNOW PRODUCTS DERIVED FROM ADEOS-II/GLI DATA: SCIENTIFIC IMPLICATION

Teruo Aoki, Meteorological Research Institute, Tsukuba, Ibaraki, 305-0052, Japan

Masahiro Hori, Japan Aerospace Exploration Agency, Tokyo, 104-6023, Japan

Akiriho Hachikubo, Kitami Institute of Technology, Kitami 090-8507, Japan

Tomonori Tanikawa, University of Tsukuba, Tsukuba, Ibaraki, 305-8571, Japan

Hiroki Motoyoshi, Space Service, Inc. Ltd, Tsukuba, Ibaraki, 305-0052, Japan

Yasuko Iizuka, Graduate University of Advanced Studies, Tokyo 173-8515, Japan

Yukinori Nakajima and Fumihiro Takahashi, Remote Sensing Technology Center of Japan, Tokyo, 104-6021, Japan

Knut Stamnes, Wei Li, and Hans Eide, Stevens Institute of Technology, Hoboken, NJ 07030, U.S.A.

Rune Storvold, Norut Information Technology Ltd., Tromso, 9294, Norway

Jens Nieke, University of Zurich, Zurich, CH-8057, Switzerland

ABSTRACT

Global maps of snow grain size and mass concentration of snow impurities were made from April to October in 2004 with ADEOS-II/GLI data. For the calibration of the sensor and validation of the algorithms, several field campaigns were carried out in Alaska and eastern Hokkaido, Japan. Based on snow pit work, the retrieved snow grain size using the channels at 0.46μm and 0.865 μm agreed with the measured values averaged over a snow layers from surface to several-centimeter depth. However, the satellite-derived grain sizes from 1.64 μm-channel were generally smaller than those measured at the ground. The mass concentration of snow impurities retrieved from the satellite data was lower than the measured one. This is because snow impurities are assumed to be soot in the remote sensing algorithm, whereas the main composition of in situ measured impurities was generally found to be mineral dust in our sites. Our snow pit work conducted as a part of routine field experiment for GLI sciences, indicated that snow grain size and concentration of snow impurities are crucially important factors determining broadband snow albedo. An increase in snow grain size with snow surface temperature was also clearly shown both from this routine field experiment and GLI data.

1. INTRODUCTION

Since the cryosphere is the most sensitive area for global warming in the globe, the understanding of its variation is very important. Remote sensing is the most effective technique to monitor the snow/ice variation. Although it has been discussed about the snow/ice extent estimated by means of satellite remote sensing as an index of global warming, the qualitative variation of snow/ice should be detected and understood for accurate simulation of future cryosphere. Because the qualitative snow parameters such as snow grain size and impurities (insoluble solid particles with light absorption) are essentially determine the snow albedo in case of sufficiently deep snow (Fig. 1). The visible and near infrared albedos mainly depend on snow grain size and impurities. Using this we have developed the remote sensing algorithms to retrieve these snow physical parameters and have made the global maps from April to October in 2004 with ADEOS-II/GLI data. Scientific implications of these snow products are discussed with the calibration and validation (Cal/Val) results.

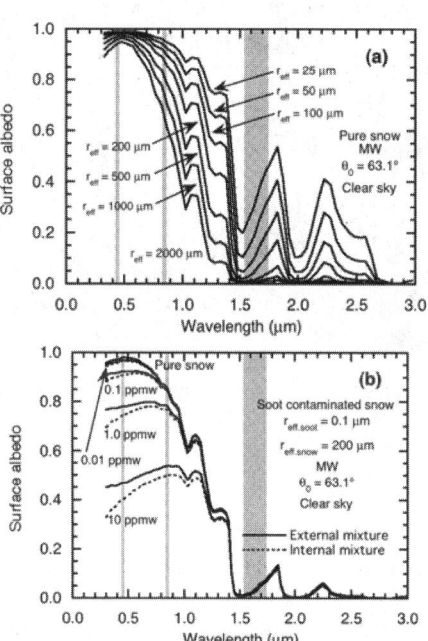

Fig. 1. Theoretically calculated spectral snow albedos depending on, (a) snow grain size, and (b) mass concentration of snow impurities (as soot). Hatched bands indicate GLI channels used for snow parameter retrievals.

2. ADEOS-II/GLI SNOW PRODUCTS

ADEOS-II/GLI snow products are two types of snow grain sizes and mass fraction of snow impurities (as soot), snow surface temperature, and snow/sea-ice cover extent.

One type of snow grain size and snow impurities are retrieved from look-up table on a space of the visible channel (0.46 µm) and the near infrared channels (0.865 µm) calculated from radiative transfer model as the functions of these snow parameters (Stamnes et al., 1999a-b, 2003a-b). Another type of snow grain size is also retrieved only from 1.64 µm-channel, where light absorption is stronger than 0.865 µm and grain size near the surface is expected to be retrieved from this channel.

GLI observes the earth completely for 4 days and the composite global maps of snow products are made from clear-sky images by making cloud discrimination. When the composite maps are made using the data in one month, almost all clouds could be eliminated. Next step is a ground discrimination in which surface types are classified into snow on sea ice, bare sea ice, open sea, ground snow, snow with vegetation, and snow-free ground (Fig. 2).

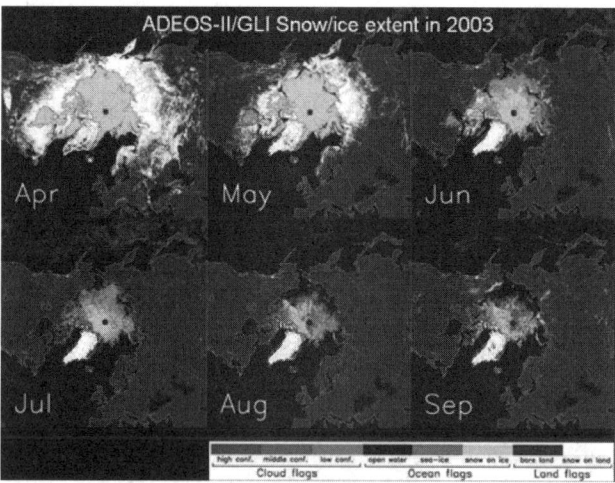

Fig. 2. Monthly snow/ice cover extent maps in the northern hemisphere in 2003.

Retrieved results of snow grain sizes and mass concentration of snow impurities are shown in Figs. 3 and 4, respectively. In general, both parameters take lower values in the high latitudinal areas and low temperature areas. In summer, snow grain size in Greenland increased up to a size range of granular snow. The preliminary investigation and an interpretation for the retrievals of snow parameters were made by Hori et al. (2001).

3. CAL/VAL ACTIVITIES AND RESULTS

For calibration of the sensor and validation of the algorithms, we carried out several field campaigns in Alaska and eastern Hokkaido, Japan. Calibration result is described in Nieke et al. (2004), where the spectral variations of sensitivities of shortwave channels of GLI sensor are investigated by cross calibration with the other sensors.

Fig. 3. Seasonal variation of snow grain size retrieved from 0.46 µm- and 0.865 µm-channels in the northern hemisphere in 2003.

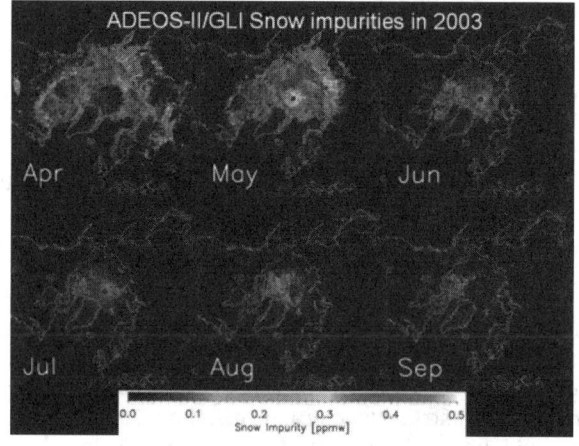

Fig. 4. Same as Fig. 3, but for mass concentration of snow impurities.

Figs. 5a-c show the validation results for snow grain size and snow impurities using GLI and MODIS data. Since these two sensors have similar channels, MODIS data were used for the validations during the periods when GLI was not operated. The retrieved snow parameters were validated with the data of in situ snow pit works and the methods of ground measurements are the same as those by Aoki et al. (2000). The total numbers of data point are not enough to conclude the characteristics of the satellite derived snow parameters. This is because of the limited chance of snow pit works on sufficiently extensive snow surface compared with satellite pixel size under clear sky condition. We thus discuss the preliminary results for the limited validation data.

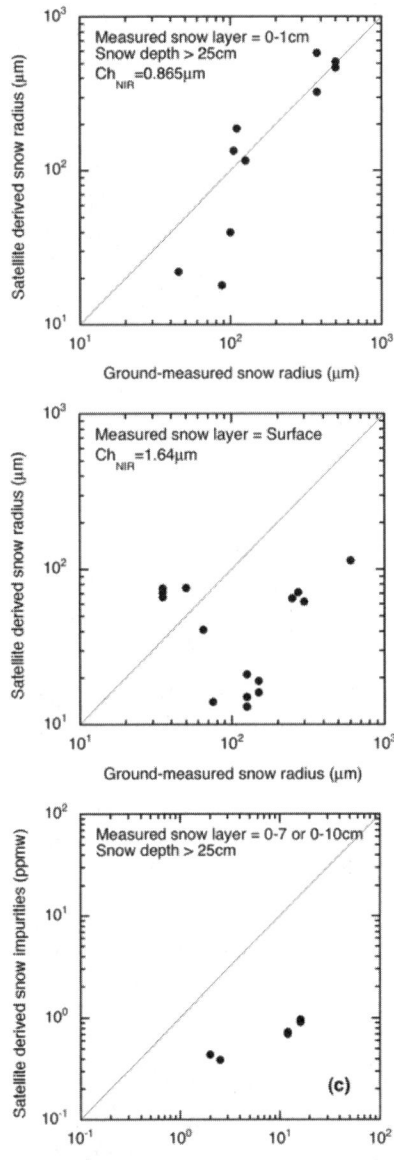

Fig. 5. Validation results for (a–b) snow grain size and (c) mass concentration of snow impurities from MODIS data in eastern Hokkaido, Japan in 2002 and 2004, and GLI data in Barrow, Alaska in 2003. Snow parameters in (a) and (c) were retrieved from the channel combination of the visible (0.46 μm) and near infrared (0.865 μm), and snow grain size in (b) only from 1.64 μm-channel.

Fig. 5a shows the satellite derived snow grain size using the channels of 0.460 μm and 0.865 μm (0.469 μm and 0.858 μm for MODIS), and the ground measured snow grain size averaged over a snow layer from surface to 1cm depth. We checked the correlation between satellite- and ground-measured grain sizes for various snow layers and confirmed that the difference between both grain sizes was a minimum for the snow layer of 0–1 cm shown in Fig. 5a, while a high correlation was kept in the snow layers from surface to several-centimeter depth.

Although the retrieved grain sizes from 1.64μm-channel are expected to agree with those measured in the layer close to surface because of stronger light absorption by ice at this wavelength as shown by Li et al. (2001) and Tanikawa et al. (2002), our satellite-derived grain sizes were generally smaller than those measured at the ground (Fig. 5b). However, the minimum difference between both grain sizes was observed at the snow surface. According to the ground based spectral measurements for granular snow in the melting season, sun crust (thin ice layer created by solar radiation under clear sky) at snow surface increased the snow reflectance by additional specular reflection. However, the reason of the disagreement in Fig. 5b requires further investigation.

Fig. 5c shows the validation results for mass concentration of snow impurities, in which satellite-derived concentrations were lower than in situ measured ones. This is because snow impurities are assumed to be soot in the algorithm from the satellite data, whereas the main composition of in situ measured impurities was generally found to be mineral dust in our sites. A part of the validation results for snow grain size and impurities with MODIS data in Japan, have been reported by Hori et al. (2003).

4. SCIENTIFIC IMPLICATIONS

Our snow pit works conducted as a part of routine field experiment for GLI sciences, indicate that snow grain size and concentration of snow impurities are crucially important factors determining broadband snow albedo needed to assess the long term impact on the radiation budget and thus climate change.

We have thus made radiation budget observations simultaneously with snow pit work during 4 winters in snow covered areas in Japan since 1999, in which snow grain size and snow impurities were measured. Using these data the effect of snow aging on broadband albedo can be investigated (Aoki et al., 2003). The dependence of albedos on elapsed time after snowfall by snow aging could be clearly classified by dry snow season and wet snow season rather than snow surface temperature. The effect of snow aging essentially attributes the increases of snow grain size and snow impurities after snowfall. The relationships between broadband albedos and these snow physical parameters were shown in Fig. 6. The measured albedos fall close to the ranges of theoretically calculated ones as functions of these snow parameters.

The mutual relationships among those snow parameters measured on the ground can be found from satellite data as well. Figs. 7 and 8 show the relationships between snow surface temperature and snow grain size obtained from in situ measured field data and GLI data, respectively. An increase in snow grain size with snow surface temperature was clearly shown both from these data. These results provide useful information required to improve the treatment of land-surface processes in regional and global climate models.

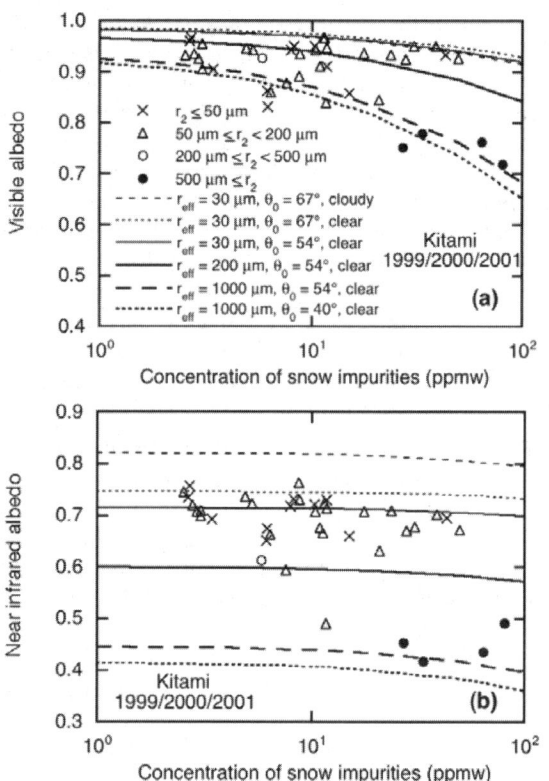

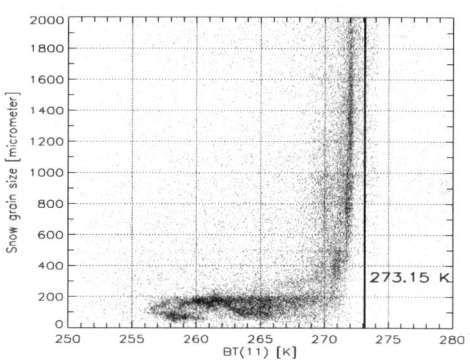

Fig. 8. Same as Fig. 7, but from MODIS data on June 18, 2000 in the northern hemisphere (after Hori et al., 2001).

Fig. 6. (a) Visible and (b) near infrared albedos as a function of mass concentration of snow impurities. Each character indicates the different ranges of optically-equivalent snow grain radius (r_2). Curves show the theoretically calculated albedos under possible conditions for clear/cloudy sky, solar zenith angle, and snow parameters (after Aoki et al., 2003).

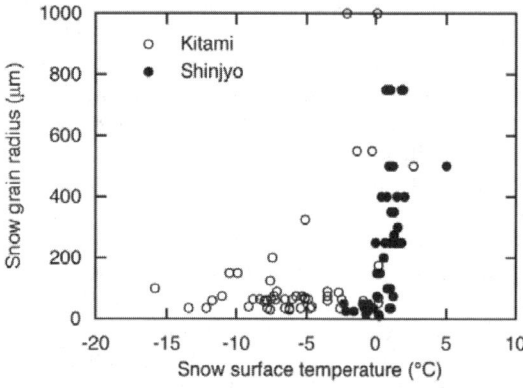

Fig. 7. Relationship between snow surface temperature and snow grain radius measured in Kitami and Shinjyo, Japan during 4 winters of 1999–2003.

5. ACKNOWLEDGMENTS

This work was conducted as part of the ADEOS-II/GLI Cal/Val experiment supported by Japan Aerospace Exploration Agency.

6. REFERENCES

Aoki, Te., Ta. Aoki, M. Fukabori, A. Hachikubo, Y. Tachibana, and F. Nishio, 2000: Effects of snow physical parameters on spectral albedo and bidirectional reflectance of snow surface, *J. Geophys. Res.*, 105, 10219-10236.

Aoki, Te., A. Hachikubo, and M. Hori, 2003: Effects of snow physical parameters on broadband albedos, *J. Geophys. Res.*, 108, 4616, doi:10.1029/2003JD003506.

Hori, M., Te. Aoki, K. Stamnes, B. Chen, and W. Li, 2001: Preliminary validation of the GLI algorithms with MODIS daytime data, *Polar Meteorol. Glaciol.*, 15, 1-20.

Hori, M., Te. Aoki, H. Ishimoto, T. Tanikawa, K. Naoki, K. Matsuoka, Y. Ogura, A. Hachikubo, R. Storvold, H. Eide, K. Stamnes, B. Chen, and W. Li, 2003: Validation of satellite derived snow physical parameters at Saroma Lagoon, Japan, *Tohoku Geophysical J.*, 36, 410-415.

Li, W., K. Stamnes, B. Chen, and X. Xiong, 2001: Snow grain size retrieved from near-infrared radiances at multiple wavelengths, *Geophys. Res. Lett.*, 28, 1699-1702.

Nieke, J., Te. Aoki, T. Tanikawa, H. Motoyoshi, and M. Hori, A satellite cross-calibration experiment, 2004: IEEE Geosci. *Remote Sens. Lett.*, 1, 215-219.

Stamnes, K., 1999a: Snow grain size/impurities (CTSK2b1), Algorithm Theoretical Basis Document, *NASDA internal document*, 27 pp.

Stamnes, K., 1999b: Cryosphere related algorithms (CTSK1), Algorithm Theoretical Basis Document, *NASDA internal document*, 28 pp.

Stamnes, K., 2003a: Snow grain size retrieved at 1.64 □m (CTSK2b1_ch28), Algorithm Theoretical Basis Document, *NASDA internal document*, 4 pp.

Stamnes, K., 2003b: Surface temperature on polar regions, Algorithm Theoretical Basis Document (CTSK2d), *NASDA internal document*, 17 pp.

Tanikawa, T., Te. Aoki, and F. Nishio, 2002: Remote sensing of snow grain-size and impurities from Airborne Mulitspectral Scanner data using a snow bidirectional reflectance distribution function model, *Ann. Glaciol.*, 34, 74-80.

AEROSOL OPTICAL THICKNESS IN FAIR-WEATHER CUMULUS IN THE AMAZON REGION ESTIMATED FROM ASTER DATA

Guoyong Wen
Goddard Earth Sciences and Technology Center
University of Maryland Baltimore Co.
1000 Hilltop Circle, Baltimore, MD 21250, USA
Robert F. Cahalan
NASA/Goddard Space Flight Center
Greenbelt, MD 20771, USA

ABSTRACT

This paper presents a method to retrieve aerosol optical thickness in clear regions of fair-weather cumulus cloud fields from high-resolution satellite images. This method is applied to ASTER images acquired in the Amazon region in Brazil.

1. INTRODUCTION

Aerosol amounts in clear regions of a cloudy atmosphere are crucial in studying the interaction between aerosols and clouds. However, it is a great challenge to retrieve aerosol amount in such regions because the structure of surrounding clouds is so complicated and cloud droplets are very effective in scattering solar radiation. Having optically thin clear regions with surrounding highly reflective clouds, the system is highly 3-dimensional (3-D). The traditional plane parallel lookup table method is evidently inapplicable.

However, aerosol optical thickness in clear regions of a cloudy atmosphere may be estimated from high-resolution satellite images under some circumstances. Among other clouds, fair weather cumulus clouds appear to be more organized than any other types of broken clouds; both vertical and horizontal dimensions of fair weather cumulus clouds are less variable. An earlier study demonstrates that the mean cloud spacing or mean distance between a clear pixel to a cloudy pixel may be used to parameterize the effects of clouds on path radiance in clear regions observed in Landsat ETM+ images at the SGP (Southern Great Plain) site of the ARM (Atmospheric Radiation Measurements) program (Cahalan *et al.*, 2001; Wen, *et al.*, 2001).

In this work we use a similar method to retrieve clear region aerosol optical thickness in a cumulus cloud image acquired by ASTER (Advanced Space Borne Thermal Emission and Reflection Radiometer) in the Amazon region in Brazil.

Corresponding author address: Dr. Guoyong Wen, NASA/Goddard Space Flight Center, Code 613.2, Greenbelt, MD 20771
E-mail: wen@climate.gsfc.nasa.gov

2. DATA ANALYSIS

The ASTER flying on Terra is a high-resolution thermal emission and reflection radiometer. The instrument covers three bands in the visible and near-infrared (VNIR) with spatial resolution of 15 m, six bands in shortwave infrared (SWIR) with spatial resolution of 30m, and five thermal infrared (TIR) bands with spatial resolution of 60m (Yamaguchi *et al.*, 1998).

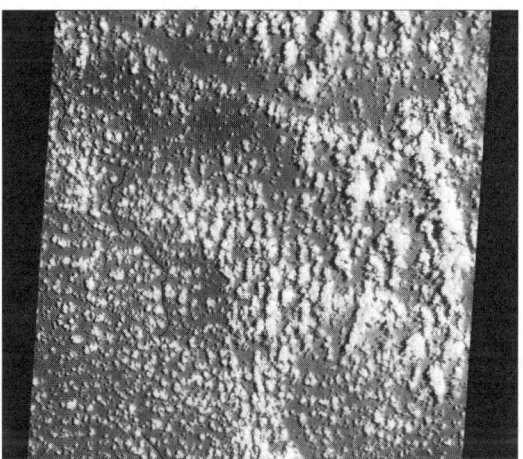

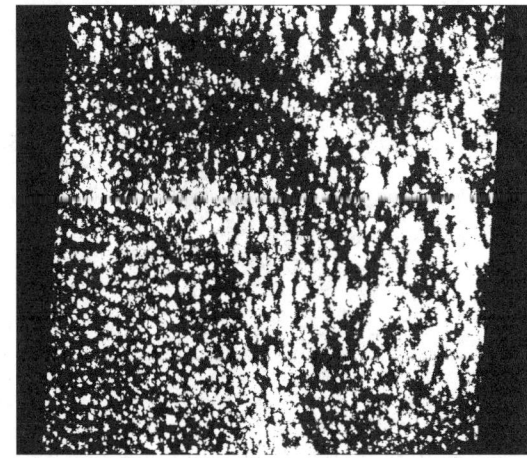

Figure 1. a) The image acquired by ASTER in Amazonian region in Brazil on October 4, 2000 with bands 5, 3, and 1 for red, green, and blue (RGB) respectively (top panel); b) The cloud mask of the above image (bottom panel).

Each ASTER image covers an area about 60 km x 60 km. Similar to MODIS (Kaufman et al., 1997) and Landsat (Wen et al., 1999) aerosol retrievals, band 2 at 0.67μm and band 5 at 2.16μm of ASTER are used in this study.

The ASTER image (Fig. 1a) used in this study was acquired on October 4, 2000 in the Amazon region of Brazil. The image is centered at (-8.70°, -59.28°) with solar zenith angle about 17°. Similar to the method in Wen et al (2001), multi spectral information is used to assign a cloud mask for the ASTER image. The cloud mask of the image is presented in Fig. 1b.

The path radiance method is applied in this study. The path radiance technique was originally introduced to retrieve aerosol optical thickness in the clear atmosphere for satellite images (Wen et al., 1999). When the path radiance method was applied to a fair weather cumulus cloudy Landsat ETM+ scene, it was found that the path radiances for visible bands were enhanced (Wen et al., 2001). The enhancement of visible path radiance is not uniformly distributed in the image. It depends on spacing in cloud street fields. For this type of field the apparent path radiance may be parameterized by an exponential function of the mean cloud-free distance for a sub-image. The true path radiance may then be determined by letting the cloud-free distance approach infinity.

For more complicated cumulus cloud fields such as those in Fig. 1 we found the relationship between cloud fraction and the apparent path radiance is better and more practical for parameterization. It has been demonstrated that the apparent path radiance depends on the cloud fraction (see Fig 4. in Wen et al., 2001). The apparent path radiance is expected to approach the true path radiance when cloud fraction approaches zero.

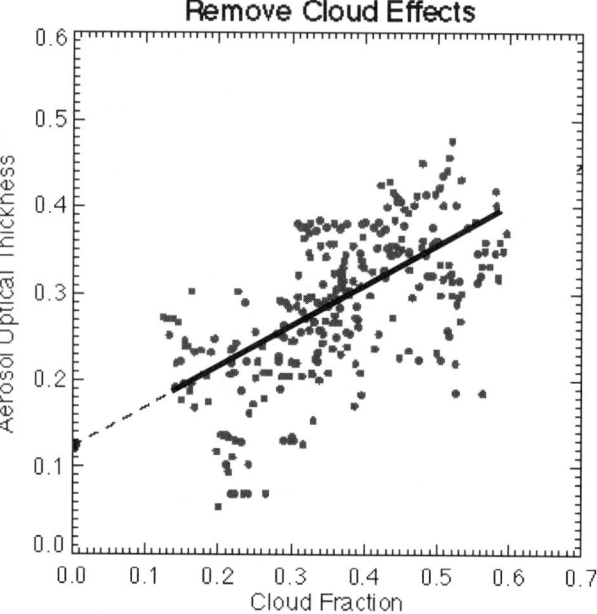

Figure 2b. The relationship between apparent aerosol optical thickness and cloud fraction. The extrapolation of the relation to zero cloud fraction yields the aerosol optical thickness in the cumulus cloud fields.

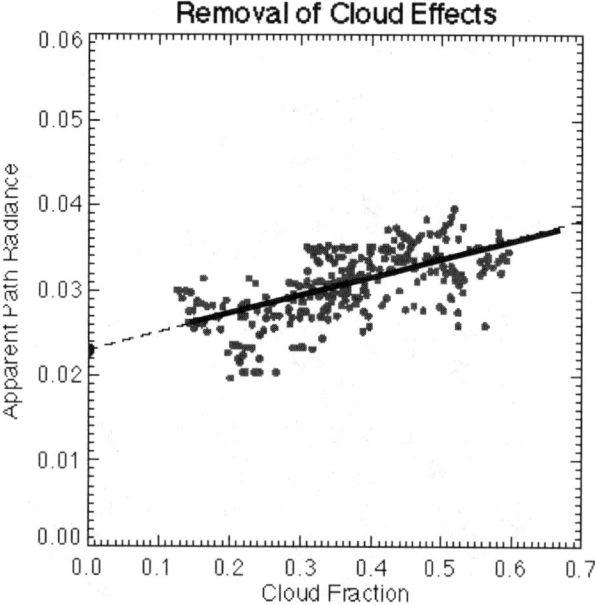

Figure 2a. The relationship between apparent path radiance and cloud fraction. The extrapolation of the relation to zero cloud fraction yields the true path radiance in the cumulus cloud fields.

Here the path radiance is calculated for a moving window of 15 km x 15 km at a step of 3km. The apparent path radiance as a function of cloud fraction is presented in Fig. 2a. The apparent path radiance is evidently a linear function of cloud fraction. Here we assume that the enhancement of the path radiance as cloud fraction increase is primarily due to 3-D effects of surrounding clouds. Hence the extrapolation of path radiance and cloud fraction relation to the imaginary point where cloud fraction approaches zero gives the expected true path radiance. Assuming a continental aerosol model, the corresponding aerosol optical thickness may be obtained for any given apparent path radiance. Similarly the extrapolation of the aerosol optical thickness and cloud fraction relation to the zero of cloud fraction yields the true aerosol optical thickness of clear regions in the cumulus cloud fields (Fig. 2b). The aerosol optical thickness is about 0.12 in the clear region of the cumulus cloud fields in Fig. 1.

3. SUMMARY AND DISCUSSION

Aerosol optical thickness in clear regions of the cumulus cloud field of the ASTER image in Amazon region is estimated to be 0.12. The apparent path radiance is found to be positively correlated with the cloud fraction. We assume that the enhancement of the path radiance as cloud fraction increases is primarily due to the 3-D effects of surrounding clouds. The physical mechanism is that the diffuse radiation from nearby clouds serves as additional radiation source to illuminate the clear area. Statistically more diffuse radiation is expected from a larger fraction cloud field than a smaller cloud fraction field, resulting in a positive correlation between path radiance and cloud fraction. Since we do not account for spatial variability of aerosol distribution, the estimated aerosol optical thickness only provides an average measure over the entire 60 km x 60 km ASTER image.

REFERENCES

Cahalan, R.F., L. Oreopoulos, G. Wen, A. Marshak, S.-C. Tsay, T. DeFelice, 2001: Cloud Chharacterization and clear-sky correction from Landsat-7, *Remote Sensing of Environment*, **78**, 83-98.

Kaufman, Y.j., A.E. Wald, L.A. Remer, B.-C. Gao, R.-R. Li, and L. Flynn, 1997: The MODIS 2.1µm channel correction with visible reflectance for use in remote sensing of aerosol, *IEEE Trans. Geosci. Remote Sensing*, vol. **35**, 1286-1298.

Wen, G., S. C. Tsay, R. Cahalan, and L. Oreopoulos, 1999: Path radiance technique for retrieving aerosol optical thickness over land. *J. Geophys. Res.*, **104**, 31321-31332.

Wen, G., R. F. Cahalan, T.-S. Tsay, and L. Oreopoulos, 2001: Impact of cumulus cloud spacing on Landsat atmospheric correction and aerosol retrieval. *J. Geophys. Res.*, **106**, 12129-12138.

Yamaguchi, Y., A. B. Kahle, H. Tsu, T. Kawakami, and M. Pniel, 1998: Overview of Advanced Spaceborne Thermal Emission and Reflection Radiometer (ASTER), *IEEE Trans. Geosci. Remote Sensing*, vol. **36**, 1062-1071.

REMOTE SENSING OF STRATOSPHERIC CHLORINE SPECIES WITH THE MIPAS/ENVISAT-EXPERIMENT

N. Glatthor, T. von Clarmann, H. Fischer, M. Höpfner, and G. P. Stiller
Forschungszentrum Karlsruhe and Universität Karlsruhe
Postfach 3460, 76021 Karlsruhe, Germany

ABSTRACT

We present observations of $ClONO_2$ and ClO, performed by the MIPAS (Michelson Interferometer for Passive Atmospheric Sounding) Fourier transform spectrometer onboard the European ENVironmental SATellite (ENVISAT). Although $ClONO_2$ and especially ClO have only weak signatures in the mid-infrared, $ClONO_2$ was successfully analysed using the Q-branch of the ν_4 band at 780.2 cm^{-1} and ClO profiles could be retrieved in regions of strong chlorine activation using the central part of the P-branch and the Q-branch ($\sim$821–846 cm^{-1}) of the 1-0 band. The radiative transfer model used is the Karlsruhe Optimized and Precise Radiative Transfer Algorithm (KOPRA). The data discussed here are southern hemispheric measurements between 8 September and 13 October 2002, covering the unusual major warming of the Antarctic vortex. They represent the first spaceborne mid-infrared observations of Antarctic ClO and of Antarctic $ClONO_2$ recovery. On 20 September, i.e. just before the onset of the major warming, MIPAS measured a pronounced $ClONO_2$ collar structure with high volume mixing ratios (vmrs) of up to 2.5 ppbv between 16 and 25 km at the vortex edge and in the upper part of the vortex (above 23 km). Correspondingly, on this day high ClO mixing ratios between 1 and 2.3 ppbv were observed in contiguous parts of the dayside lower Antarctic vortex. From 8 to 20 September, daily averages of $ClONO_2$ on the 500 K level of potential temperature exhibited high mixing ratios of up to more than 2 ppbv at the vortex edge, but low values inside the vortex. During the same period inside-vortex daytime daily averages of ClO on the 500 K level showed enhanced mixing ratios between 0.8 and 1.3 ppbv. In the course of the major warming followed a rapid ClO decrease inside the vortex until 25 September, accompanied by a $ClONO_2$ increase by 0.8 ppbv. After 25 September $ClONO_2$ started to decrease, and in October both ClO and $ClONO_2$ had decreased to outside-vortex levels.

1. INTRODUCTION

Operational processing of data obtained by the Michelson Interferometer for Passive Atmospheric Sounding (MIPAS), performed under responsibility of the European Space Agency (ESA), is restricted to the main targets temperature, H_2O, O_3, CH_4, N_2O, HNO_3 and NO_2 [*Nett et al.*, 1999]. To be able to retrieve self-consistent datasets containing considerably more trace species than enclosed in the operational dataset, an independent retrieval processor was developed at the Institut für Meteorologie und Klimaforschung (IMK) [*von Clarmann et al.*, 2003]. In this paper we present Antarctic $ClONO_2$ and ClO distributions of the period 8 September to 13 October, which was characterized by a major warming around 25 September. Such an event had never been observed before in Antarctic winters [*Allen et al.*, 2003, *Roscoe et al.*, 2005, and references therein].

Both $ClONO_2$ and ClO are key species to characterize the state of polar chlorine activation. During polar winter $ClONO_2$ and HCl, the major reservoir gases for stratospheric chlorine, are converted into molecular chlorine (Cl_2) and HNO_3 by heterogeneous reactions on polar stratospheric clouds (PSCs). In polar spring Cl_2 is photolysed and reacts with ozone to produce ClO, which destroys ozone in catalytic cycles [*Avallone and Toohey*, 2001]. In the Arctic polar vortex the recovery into the reservoir gases usually takes place via the reaction

$$ClO + NO_2 + M \longrightarrow ClONO_2 + M. \qquad (1)$$

Because of denoxification and ozone depletion generally the slower recovery path

$$Cl + CH_4 \longrightarrow HCl + CH_3 \qquad (2)$$

is predominant in the Antarctic vortex [*Douglass et al.*, 1995, *Grooß et al.*, 1997].

Stratospheric $ClONO_2$ was first detected by balloon-borne infrared solar absorption spectroscopy [*Murcray et al.*, 1979]. First spaceborne observations, also in solar occultation mode, followed in April 1985 by the Atmospheric Trace Molecule Spectroscopy (ATMOS) experiment onboard Spacelab 3 [*Zander et al.*, 1986]. Since 1992 $ClONO_2$ was observed in mid-infrared emission mode by the balloon-borne Michelson Interferometer for Passive Atmospheric Sounding (MIPAS-B) in various Arctic and midlatitude campaigns [*Oelhaf et al.*, 1994, *von Clarmann et al.*, 1997]. Spaceborne $ClONO_2$ observations were obtained over a longer period from October 1991 until May 1993 with the Cryogenic Limb Array Etalon Spectrometer (CLAES) on the Upper Atmospheric Research Satellite (UARS) [*Roche et al.*, 1993]. However, due to the measurement mode, during this period the southern polar vortex was only observed until 18 September 1992, i.e. the period of $ClONO_2$ recovery was not covered. Thus the data presented here represent the first spaceborne observations of $ClONO_2$ recovery in the Antarctic spring.

Polar ClO measurements were first performed in the microwave spectral region, starting in 1986 soon after the detection of the Antarctic ozone hole [*de Zafra et al.*, 1987]. A comprehensive global data set covering the years 1991–1997 results from the Microwave Limb Sounder (MLS) onboard the Upper Atmosphere Research Satellite (UARS) [*Santee et al.*, 2003]. Between 1992 and 1994, ClO was measured by the Millimeter-Wave Atmospheric Sounder (MAS) during three ATLAS missions on the space shuttle [*Aellig et al.*, 1996]. In February 2001, the Sub-Millimeter Radiometer (SMR), which is also capable to measure ClO, was launched on the Swedish polar-orbiting satellite Odin [*Murtagh et al.*, 2002]. ClO mesurements in the mid-infrared were performed from ground since the mid-1990s [*Kopp et al.*, 2002]. Here we present the first ClO distributions obtained by spaceborne mid-infrared emission spectroscopy.

2. THE MIPAS EXPERIMENT

MIPAS is one of the experiments onboard the ENVIronmental SATellite (ENVISAT), which was launched into a sun-synchronous polar orbit at about 800 km altitude on 1 March 2002. The satellite passes the equator at 10:00 (downward leg) and 22:00 (upward leg) local time. MIPAS is a limb-viewing Fourier transform infrared (FTIR) emission spectrometer with high spectral resolution (0.035 cm^{-1}, unapodised). The instrument covers a wide range of the mid-infrared region in five spectral bands between 685 cm^{-1} and 2410 cm^{-1} [*Fischer and Oelhaf*, 1996, *ESA-report SP–1229*, 2000], which enables simultaneous observation of more than 25 trace gases. Its field of view (FOV) is 30 km in horizontal and approximately 3 km in vertical direction. The standard observation mode, consisting of rearward scans, covers the altitude range from 6 to 68 km within 17 steps. The stepwidth is 3 km for the 13 lowermost tangent altitudes and increases to 8 km above 52 km altitude.

Operational data processing performed under ESA responsibility produces calibrated level-1B radiance spectra. Operational generation of geophysical data products is restricted to the main targets mentioned above.

3. RETRIEVAL STRATEGY

MIPAS level-1B spectra are inverted to vertical profiles of atmospheric state parameters by constrained non-linear least squares fitting [e.g. *Rodgers*, 2000, and references therein], whereby in iteration $i+1$ the vector of the unknown parameters x_{i+1} is determined as follows:

$$\mathbf{x}_{i+1} = \mathbf{x}_i + (\mathbf{K}^T \mathbf{S}_y^{-1} \mathbf{K} + \mathbf{R} + \lambda \mathbf{I})^{-1} \times$$
$$(\mathbf{K}^T \mathbf{S}_y^{-1} (\mathbf{y}_{meas} - \mathbf{y}(\mathbf{x}_i)) - \mathbf{R}(\mathbf{x}_i - \mathbf{x}_a)). \quad (3)$$

Here $\mathbf{x}_i$ is the parameter vector of iteration i, $\mathbf{y}_{meas}$ is the measurement vector and $\mathbf{S}_y$ is the error covariance matrix of the measurement. $\mathbf{K}$ is the Jacobian matrix containing the partial derivatives $\partial \mathbf{y}(\mathbf{x}_i)/\partial \mathbf{x}_i$ and $\mathbf{R}$ is a regularization matrix; $\lambda \mathbf{I}$ (scalar times unity) is the damping term proposed by *Levenberg* [1944] and *Marquardt* [1963]. $\mathbf{y}(\mathbf{x}_i)$ is the result of the radiative transfer model using the parameter vector $\mathbf{x}_i$. The vector $\mathbf{x}_a$ represents the a-priori knowledge of the unknown parameters, which is also used as first-guess in the first iteration ($\mathbf{x}_0 = \mathbf{x}_a$). A detailed description of the overall retrieval approach chosen for the IMK data processor can be found in *von Clarmann et al.* [2003].

Radiative transfer modelling was performed with the Karlsruhe Optimized and Precise Radiative Transfer Algorithm (KOPRA) [*Stiller et al.*, 2002], whereby the HIgh-resolution TRANsmission (HITRAN) database was taken as basis for spectroscopic data [*Rothman et al.*, 1998]. KOPRA is a highly precise line-by-line code, which allows modelling of various effects as non-local thermodynamic equilibrium (NLTE), CO_2 line-mixing, scattering, changes in isotopomeric abundances, finite field of view, instrumental line shape and others. Further it is able to calculate quasi-analytical derivatives of the spectral signal to atmospheric state parameters.

$ClONO_2$ and ClO retrievals were not performed using the whole spectral range of MIPAS but in microwindows, i.e. spectral regions with strong signatures of the target species and as small as possible contributions of interfering species. $ClONO_2$ was analysed between 779.5 and 781.0 cm^{-1} in the Q branch of the ν_4 band, using the cross sections provided by *Wagner and Birk* [2003]. Since the $ClONO_2$ Q branch is encircled by two strong ozone lines, O_3

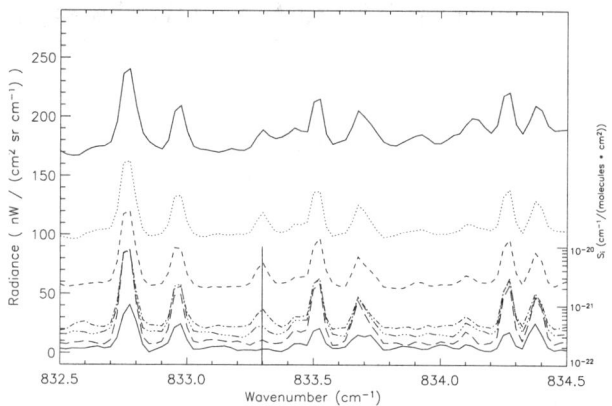

Figure 1. 77-123 coadded MIPAS spectra from Antarctic high ClO regions, 8-23 September 2002; tangent heights ∼11, 14, 17, 20, 23, 29 and 41 km (top to bottom).

Table 1. Error budget and height resolution at selected altitudes for the retrieval of a $ClONO_2$ profile inside the polar vortex on September 20, 2002. The errors are given in absolute [pptv] and relative [%] units (numbers in brackets).

Height [km]	Total error	Meas. noise	Further errors[a]	Resol. [km]
15	21(5)	15(4)	14(3)	5.4
18	45(8)	35(6)	27(4)	3.7
21	80(11)	64(9)	49(7)	3.8
24	140(6)	93(4)	98(4)	3.3
27	130(10)	99(7)	90(7)	3.4
30	110(13)	88(10)	69(8)	4.0
33	86(17)	75(15)	44(9)	5.2
36	51(15)	43(13)	27(8)	6.9

[a] Includes all error components except of measurement noise.

Table 2. Error budget and height resolution at selected altitudes for the retrieval of a ClO profile in the dayside Antarctic vortex on 20 September, 2002. The errors are given in absolute [pptv] and relative [%] units (numbers in brackets).

Height [km]	Total error	Meas. noise	Further errors[a]	Resol. [km]
16	410 (29)	270 (19)	320 (23)	4.9
18	590 (30)	400 (20)	430 (22)	4.4
20	600 (28)	450 (21)	390 (18)	4.2
22	530 (49)	470 (43)	250 (23)	5.0
25	280 (458)	280 (458)	49 (80)	4.2
30	280 (198)	280 (198)	23 (16)	9.6

[a] Includes all error components except of measurement noise.

was joint-fitted together with $ClONO_2$. The only ClO band in the mid-infrared is the weak 1-0 band centered at 844 cm^{-1}. For analysis of ClO virtually the whole P- and Q-branches between 820.975 and 846.175 cm^{-1} of this band, containig more than a dozen of the strongest ClO lines, were taken. Since the signal-to-noise ratio of the ClO signatures is rather low, single ClO lines are hardly detectable by visual inspection of single spectra. However, after coaddition of ten or more spectra from regions of high chlorine activation, which is equivalent to the signal seen by the retrieval processor in the various lines of a single spectrum, the ClO signature becomes clearly visible. In Figure 1, where up to 123 spectra from high-ClO regions have been coadded, the ClO line at 833.3 cm^{-1} (indicated by vertical line) is clearly visible in the spectra

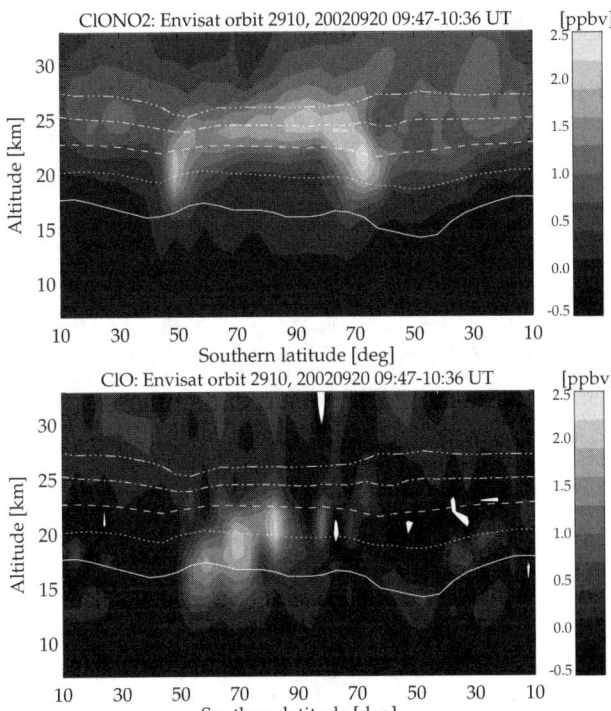

Figure 2. $ClONO_2$ (top) and ClO distribution (bottom) measured by MIPAS/ENVISAT along the southern hemispheric part of orbit 2910 of 8 September, 2002, versus latitude and altitude. The left part of the plots is the satellite's dayside downward leg and the right part the nightside upward leg. The white curves are the 400, 475, 550, 625 and 700 K levels of potential temperature (bottom to top).

from 11 to 23 km, but disappears in the upper stratosphere. Since ozone is a major interfering species in the whole region, it was also joint-fitted together with ClO.

Prior to the species retrievals, spectral shift, temperature profiles and the tangent heights were fitted. Further retrieval parameters were a microwindow-dependent radiation continuum term and a microwindow-dependent but height-independent radiometric zero level calibration correction. The continuum was forced to zero above 32 km altitude. The retrieval grid consisted of 60 levels with 1 km altitude spacing up to 44 km, expanding to 10 km at 120 km. Since in the lower atmosphere this grid is considerably finer than the tangent height spacing (~3 km), Tikhonov's first-derivative operator was applied as smoothing constraint of the retrieval [*Steck*, 2002, *von Clarmann et al.*, 2003].

4. RETRIEVAL ERROR

Tables 1 and 2 show the $ClONO_2$ and ClO retrieval errors at selected altitudes, each calculated for limb-scans from inside the Antarctic vortex. These error estimates were derived for the actual atmospheric state, i.e. for retrieved temperatures, tangent altitudes, $ClONO_2$ (ClO) and O_3 mixing ratios, whereby simulated spectra and Jacobians were taken from the final iteration. The total $ClONO_2$ retrieval error is below 0.15 ppbv at all retrieval altitudes presented. The relative error in the lower stratosphere ranges from 6 to 11%. The major error contribution results from measurement noise, but further error components, mainly uncertainties in spectroscopic data and in temperature, add up to the same order of magnitude. The total ClO retrieval error, calculated for a high-ClO scan, ranges from 0.28 to 0.6 ppbv at the retrieval altitudes

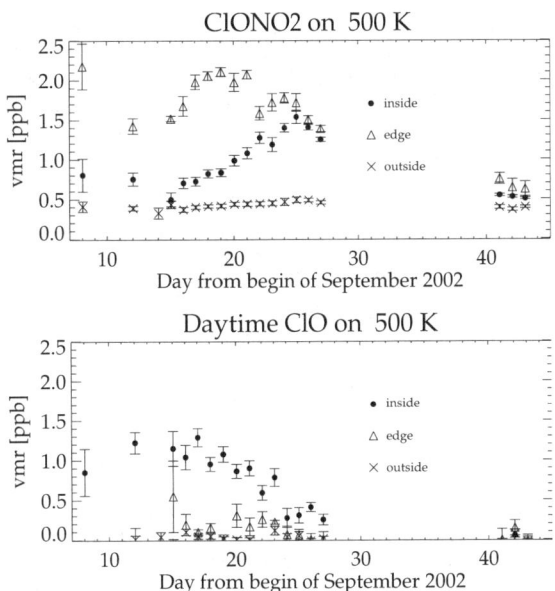

Figure 3. Daily averages of $ClONO_2$ (top) and daytime ClO (bottom) volume mixing ratios between September 8 and October 13, 2002, on the 500 K level of potential temperature, separated into measurements inside the Antarctic vortex (filled circles), in the vortex edge region (open triangles) and outside the vortex (crosses). The error bars indicate the uncertainties of the daily averages.

presented. The relative error in the region of the ClO maximum is between 28 and 49%, which shows the significance of the measurements. The height resolution, derived from the half-widths of averaging kernels of the retrievals, shows that the regions of high $ClONO_2$ (18–27 km) and ClO (16–22 km) are well resolved.

5. $ClONO_2$ AND ClO DISTRIBUTIONS

As an example for the possibility to derive $ClONO_2$ and ClO distributions from single limb-scans of MIPAS, Figure 2 shows southern hemispheric latitude-height cross sections of these species obtained on orbit 2910 from 20 September 2002, i.e. from just before the major warming. According to potential vorticity on the 475 K level, data from around 50° S (dayside) and 65° S (nightside) are from the vortex edge region (-45 to -30 K m^2 kg^{-1} s^{-1}), and data from in-between are from inside the vortex. The $ClONO_2$ distribution (top) exhibits the typical collar structure with mixing ratios of up to 2.5 ppbv above 17 km altitude (400 K) in the vortex edge region and above 23 km (550 K) inside the vortex. The asymmetry at midlatitudes above 25 km (625 K) is generally explained by daytime $ClONO_2$ photolysis. Enhanced ClO (bottom) of up to 2.3 ppbv was observed in the dayside part of the vortex (50° S – 90° S) below the region of enhanced $ClONO_2$ between 14 and 23 km (350 and 550 K). No significant amounts of ClO were found in the major part of the nighttime vortex (90° S – 70° S) because of Cl_2O_2-dimer formation as well as outside the vortex, where significant heterogeneous chlorine activation does not occur. The distribution of $ClONO_2$ shows that chlorine is deactivated in the whole altitude region from 16 to 25 km at the vortex edge and above 23 km (550 K) inside the vortex. However, the high ClO amounts below 550 K inside the vortex indicate considerable chlorine activation in this area.

Figure 3 shows daily mean mixing ratios of $ClONO_2$ (top) and of daytime ClO (bottom) on the 500 K level of potential temperature (about 21–22 km) from inside the vortex, the vortex edge and from outside the vortex for the period 8 September until 13

October, 2002, which covers the time of usual Antarctic chlorine deactivation [*Santee et al.*, 2003]. However, unfortunately MIPAS was switched off between 27 September and 11 October because of technical problems. The vortex edge on the 500 K level was defined to range from -57 to -38 K m^2 kg^{-1} s^{-1}. Except for a minimum between 12 and 15 September, the vortex edge amounts of ClONO$_2$ are above 2 ppbv until 21 September, showing advanced chlorine deactivation in this area. Thereafter they decrease to 1.4 ppbv until 27 September and to values below 0.8 ppbv in October. The ClONO$_2$ mixing ratios from inside the vortex range between 0.5 and 0.8 ppbv until 18 September, showing a slightly negative trend. During the following days they increase continuously by about 0.7 ppbv until 25 September, when they reach vortex edge levels. In the next two days they decrease by the same amount as the vortex edge values. The ClO amounts observed at the vortex edge are generally very low, also showing that chlorine deactivation had already taken place in this region. The increase in vortex edge ClO on 15 September is correlated with the minimum in ClONO$_2$, indicating temporary chlorine re-activation. Inside the vortex the ClO mixing ratios are between 0.8 and 1.3 ppbv from 8 to 17 September, which denotes considerable chlorine activation on the 500 K level. During the next 7 days the ClO amounts inside the vortex decrease rapidly to values around 0.3 ppbv. The simultaneous increase in ClONO$_2$ shows that between 17 and 25 September the main pathway for chlorine deactivation was conversion into ClONO$_2$. This kind of process usually takes place in the springtime Arctic vortex, but is untypical for the Antarctic vortex. As shown by *Höpfner et al.* [2005], the common decrease of edge and inner vortex ClONO$_2$ after 25 September is caused by subsequent conversion into HCl. This process is completed in October, since the ClONO$_2$ values from the vortex edge as well as from inside the vortex have nearly dropped to outside vortex values. More detailed discussions of the ClONO$_2$ and ClO evolution during the Antarctic major warming of 2002 are given in *Höpfner et al.* [2005] and *Glatthor et al.* [2005].

ACKNOWLEDGMENTS

Part of this work has been funded under EC contract EVG1-CT-1999-00015 (AMIL2DA), ESA contract 15530/01/NL/SF (INFLIC), BMBF contracts 07 ATF 43/44 (KODYACS), 07 ATF 53 (SACADA) and 01 SF 9953 (HGF Vernetzungsfonds). The authors like to thank ESA for giving access to MIPAS level-1 data. Meteorological analysis data have been provided by ECMWF.

REFERENCES

Aellig, C. P. et al. Latitudinal distribution of upper stratospheric ClO as derived from space borne microwave spectroscopy. *Geophys. Res. Lett.*, 23(17):2321–2324, 1996.

Allen, D. R., R. M. Bevilacqua, G. E. Nedoluha, C. E. Randall, and G. L. Manney. Unusual stratospheric transport and mixing during the 2002 Antarctic winter. *Geophys. Res. Lett.*, 30(12), 1599, doi:10.1029/2003GL017117, 2003.

Avallone, L. M. and D. W. Toohey. Tests of halogen chemistry using in situ measurements of ClO and BrO in the lower polar stratosphere. *J. Geophys. Res.*, 106(D10):10,411–10,421, 2001.

de Zafra, L. M., R. Jaramillo, A. Parrish, P. Solomon, B. Connor, and J. Barrett. High concentrations of chlorine monoxide at low altitudes in the Antarctic spring stratosphere: diurnal variation. *Nature*, 328:408–411, 1987.

Douglass, A. R, .M. R. Schoeberl, R. S. Stolarski, J. W. Waters, J. M. Russell III, A E. Roche, and S. T. Massie. Interhemispheric differences in springtime production of HCl and ClONO$_2$ in the polar vortices. *J. Geophys. Res.*, 100(D7):13,967–13,978, 1995.

European Space Agency. Envisat, MIPAS An instrument for atmospheric chemistry and climate research. ESA Publications Division, ESTEC, P. O. Box 299, 2200 AG Noordwijk, The Netherlands, SP-1229.

Fischer, H. and H. Oelhaf. Remote sensing of vertical profiles of atmospheric trace constituents with MIPAS limb-emission spectrometers. *Appl. Opt.*, 35:2787–2796, 1996.

Grooß, J.-U., R. B. Pierce, P. J. Crutzen, W. L. Grose, and J. M. Russell III. Re-formation of chlorine reservoirs in Southern Hemisphere polar spring. *J. Geophys. Res.*, 102(D11), 13,141–13,152, 1997.

Glatthor, N. et al. and T. von Clarmann and H. Fischer and U. Grabowski Spaceborne ClO observations by the Michelson Interferometer for Passive Atmospheric Sounding (MIPAS) before and during the Antarctic major warming in September/October 2002. *J. Geophys. Res.*, 109, D11307, doi:10.1029/2003JD004440, 2004.

Höpfner, M. et al. First spaceborne observations of Antarctic stratospheric ClONO$_2$ recovery: Austral spring 2002. *J. Geophys. Res.*, 109, D11308, doi:10.1029/2004JD004609, 2004.

Kopp, G., H. Berg, T. Blumenstock, H. Fischer, F. Hase, G. Hochschild, M. Höpfner, W. Kouker, T. Reddmann, R. Ruhnke, U. Raffalski, and Y. Kondo. Evolution of ozone and ozone-related species over Kiruna during the SOLVE/THESEO 2000 campaign retrieved from ground-based millimeter-wave and infrared observations. *J. Geophys. Res.*, 107, 8308, doi:10.1029/2001JD001064, 2002. [printed 108(D5), 2003].

Levenberg, K. A method for the solution of certain non–linear problems in least squares. *Quart. Appl. Math.*, 2:164–168, 1944.

Marquardt, D. W. An algorithm for least–squares estimation of nonlinear parameters. *J. Soc. Indust. Appl. Math.*, 11(2):431–441, 1963.

Murcray, D. G., A. Goldman, F. H. Murcray, F. J. Murcray, and W. J. Williams. Stratospheric distribution of ClONO$_2$. *Geophys. Res. Lett.*, 6(11), 857–859, 1979.

Murtagh, D. et al. An overview of the Odin atmospheric mission. *Can. J. Phys.*, 80:309–319, 2002.

Nett, H., B. Carli, M. Carlotti, A. Dudhia, H. Fischer, J.-M. Flaud, G. Perron, P. Raspollini, and M. Ridolfi, MIPAS Ground Processor and Data Products, *Proc. IEEE 1999 International Geoscience and Remote Sensing Symposium*, 28 June - 2 July 1999, Hamburg, Germany, 1692-1696, 1999.

Oelhaf, H., T. von Clarmann, H. Fischer, F. Friedl-Vallon, C. Frietzsche, A. Linden, C. Piesch, M. Seefeldner, and W. Völker. Stratospheric ClONO$_2$ and HNO$_3$ profiles inside the Arctic vortex from MIPAS–B limb emission spectra obtained during EASOE. *Geophys. Res. Lett.*, 21(13), 1263–1266, 1994.

Roche, A. E., J. B. Kumer, J. L. Mergenthaler, G. A. Ely, W. G. Uplinger, J. F. Potter, T. C. James, and L. W. Sterritt. The Cryogenic Limb Array Etalon Spectrometer CLAES on UARS: Experiment description and performance. *J. Geophys. Res.*, 98(D6), 10,763–10,775, 1993.

Rodgers, C. D. *Inverse Methods for Atmospheric Sounding: Theory and Practice*, volume 2 of *Series on Atmospheric, Oceanic and Planetary Physics*, F. W. Taylor, ed. World Scientific, 2000.

Roscoe, H.K., J.D. Shanklin, and S.R. Colwell. Has the Antarctic vortex split before 2002? *J. Atmos. Sci.*, 62, 582–588, 2005.

Rothman, L.S. et al. The HITRAN molecular spectroscopic database and HAWKS (HITRAN Atmospheric Workstation): 1996 Edition. *J. Quant. Spectrosc. Radiat. Transfer*, 60:665–710, 1998.

Santee, M. L., G. L. Manney, J. W. Waters, and N. J. Livesey. Variations and climatology of ClO in the polar lower stratosphere from UARS Microwave Limb Sounder measurements. *J. Geophys. Res.*, 108(D15), 4454, doi:10.1029/2002JD003335, 2003.

Steck, T. Methods for determining regularization for atmospheric retrieval problems. *Appl. Opt.*, 41(9):1788–1797, 2002.

Stiller, G. P., T. von Clarmann, B. Funke, N. Glatthor, F. Hase, M. Höpfner, and A. Linden. Sensitivity of trace gas abundances retrievals from infrared limb emission spectra to simplifying approximations in radiative transfer modeling. *J. Quant. Spectrosc. Radiat. Transfer*, 72, 249–280, 2002.

von Clarmann, T., G. Wetzel, H. Oelhaf, F. Friedl-Vallon, A. Linden, G. Maucher, M. Seefeldner, O. Trieschmann, and F. Lefèvre. ClONO$_2$ vertical profile and estimated mixing ratios of ClO and HOCl in winter Arctic stratosphere from Michelson interferometer for passive atmospheric sounding limb emission spectra. *J. Geophys. Res.*, 102(D13), 16,157-16,168, 1997.

von Clarmann, T., N. Glatthor, U. Grabowski, M. Höpfner, S. Kellmann, M. Kiefer, A. Linden, G. Mengistu Tsidu, M. Milz, T. Steck, G. P. Stiller, D. Y. Wang, H. Fischer, B. Funke, S. Gil-López, and M. López-Puertas. Retrieval of temperature and tangent altitude pointing from limb emission spectra recorded from space by the michelson interferometer for passive atmospheric sounding (MIPAS). *J. Geophys. Res.*, 108(D23), 4736, doi:10.1029/2003JD003602, 2003.

Wagner, G., and M. Birk. New spectroscopic database for chlorine nitrate. *J. Quant. Spectrosc. Radiat. Transfer*, 82, 381–407, 2003.

Zander, R., C. P. Rinsland, C. B. Farmer, L. R. Brown, and R. H. Norton. Observation of several chlorine nitrate (ClONO$_2$) bands in stratospheric infrared spectra. *Geophys. Res. Lett.*, 13(8), 757–760, 1986.

RADIATIVE CHARACTERISTICS OF CIRRUS CLOUDS AS RETRIEVED FROM AVHRR

Shuichiro Katagiri
Research Institute for Humanity and Nature
335 Takashima-cho, Kamigyo-ku, Kyoto 602-0878, Japan
katagiri@chikyu.ac.jp

Teruyuki Nakajima
Center for Climate System Research, The University of Tokyo, Japan
4-6-1 Komaba, Meguro-ku, Tokyo 153-8904, Japan
teruyuki@ccsr.u-tokyo.ac.jp

1. INTRODUCTION

Clouds have a significant impact on the climate of the earth through the radiation budget. Accordingly this radiative impact has been studied intensively, and it is thought that the climatic effect of clouds depends on its optical thickness, cloud particle size, and height; low-level and mid-level clouds have a net cooling effect but upper-level clouds such as cirrus clouds may have mostly net warming effect. To understand the radiative impact of cirrus clouds on the climate, we have to know cirrus cloud microphysical properties. However cirrus cloud measurements over the glove are still challenging studies. To detect cirrus clouds from space is very difficult owing to their semi-transparency. Cirrus clouds are often difficult to be recognized at visible wavelengths, and to be separated from the ground or sea surface or low-level clouds at infrared wavelengths. We developed a simultaneous nighttime only analysis of the optical thickness, effective radius, and cloud top temperature of cirrus clouds over a global scale. In our three-channel infrared method, three wavelengths, channel 3 (3.7 μm), channel 4 (10.8 μm), and channel 5 (12.0 μm) of Advanced High Resolution Radiometer (AVHRR) on board NOAA-9 and NOAA-11 are used, and the scheme is applied to four months data acquired from 1986 through 1994 over the latitudinal region between 60° north and 60° south.

2. ALGORITHM

It is known that the split window method is useful for detection of cirrus clouds, which uses two infrared window wavelengths, such as 10.8 and 12.0 μm. We developed the split window method to the three-channel method with 10.8, 12.0, and 3.7 μm to be more useful for retrieval of global satellite data, which were reconstructed to segmented data that contain hundred satellite pixels in each 0.5° x 0.5° latitude and longitude grid (see Fig. 1).

Figs. 2 and 3 show it is difficult to distinguish the cloud top temperature of cirrus and the surface temperature or the low-level cloud lying below cirrus clouds (shown in Fig. 5(a)) by the brightness temperature difference between 10.8 and 12.0 μm (BTD11-12). Therefore to obtain the cloud top temperature, we chose the brightness temperature difference curves between 3.7 and 10.8 μm (BTD3.7-11) that have steep inclination around the cloud top temperature indicated as A in Fig. 4. The low surface or the low cloud top temperature were obtained by using BTD3.7-11 curves illustrated in Fig. 5(b). Retrievals were carried out with the look-up table calculated by an accurate and efficient radiative transfer scheme with a combined discrete ordinate matrix operator method, named RSTAR-5b, where the effective radius, r_e, and the effective water vapor path length, w_e, are defined respectively as

$$r_e \equiv \frac{\int_0^\infty r^3 n(r) dr}{\int_0^\infty r^2 n(r) dr} \quad (1)$$

$$w_e = \int w(z)[P(z)/P_g]^{0.9}[T_g/T(z)]^{0.5} dz \quad (2)$$

A cirrus cloud model is adopted with volume equivalent spherical ice particles with the refractive index for ice and with a volume log-normal size distribution defined by

$$\frac{dV(r)}{d\ln r} = C \exp\left[-\frac{1}{2}\left(\frac{\ln r - \ln r_m}{\sigma}\right)^2\right] \quad (3)$$

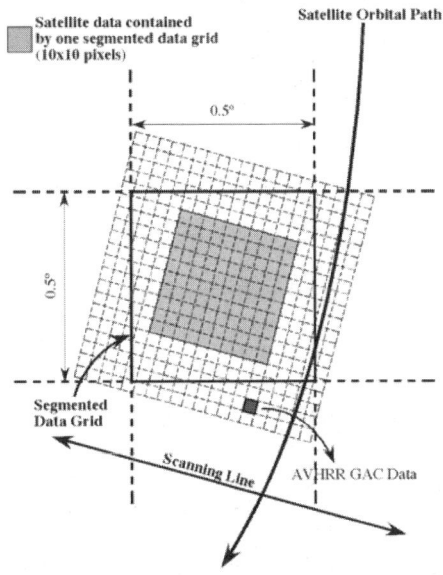

Fig. 1. 100 satellite pixels are included in each 0.5° x 0.5° latitude-longitude segment grid.

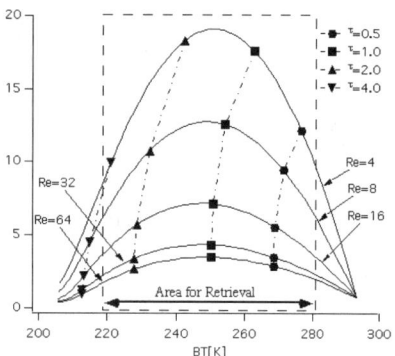

Fig. 2. Brightness temperature difference between 10.8 and 12.0 μm as a function of brightness temperature at 10.8 μm, BT11.

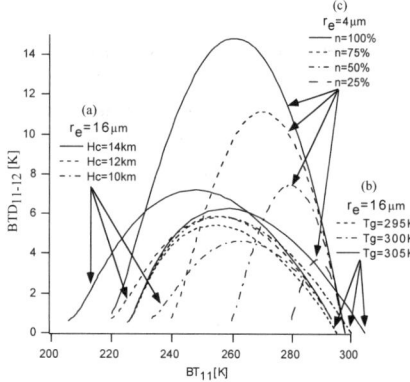

Fig. 3. BTD11-12 vs BT11 for (a) H_c=14, 12, and 10 km, with r_e=16 μm, surface temp. of T_g=295 K, and n=100 %, (b) surface temperature of T_g=295, 300, and 305 K, with r_e =16 μm, n=100 %, and H_c=12 km, and (c) n=100, 75, 50, 25 %, with r_e=4 μm, T_g=295 K, and H_c=11 km.

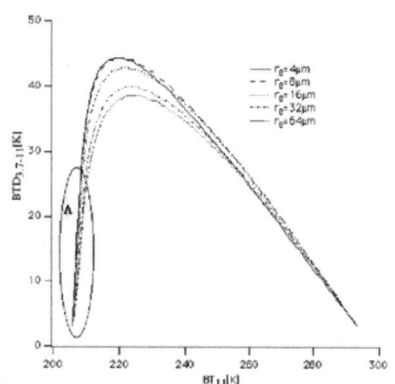

Fig. 4. BTD3.7-11 as a function of BT11.

We prepared two look-up tables consist of BTD11-12 and BTD3.7-11 with the grid system listed in Table 1. The flow is shown in Fig. 6. Pixels of cirrus are first assumed tentatively by a condition that the minimum BT11 is lower than 270K. At this time they are not settled as cirrus pixels. This ambiguous classification is used not to abandon thin cirrus clouds. Secondly we get the corresponding temperature and water vapor profiles from a reference climate data to set the initial surface temperature T_g^* and column effective water vapor w_e following Kawamoto (2001) for each 0.5° x 0.5° segment area. We also assume an initial cloud top temperature T_c^* which is low enough as compared with the actual cloud top temperature. Thirdly the actual cloud top temperature T_c and surface temperature T_g are retrieved by the iteration process with the temperature adjustment explained in Fig. 6 using the lookup table-A consists of the data of BTD3.7-11 versus BT11.

Table 1. Gridded Values in Look-up Tables

Parameter	Grid
Effective Radius	4, 8, 16, 32, 64 μm
Optical Thickness	0 ~ 2 (increase of 0.2); 2 ~ 5 (0.5); 5 ~ 20 (gradually)
Low Surface Temperature	310 ~ 230 K (decrease of 5 K)
Cloud Top Altitude	4 ~ 18 km (increase of 2 km)
Effective Water Vapor	7 kinds
Satellite Zenith Angle	0, 30, 60 degree

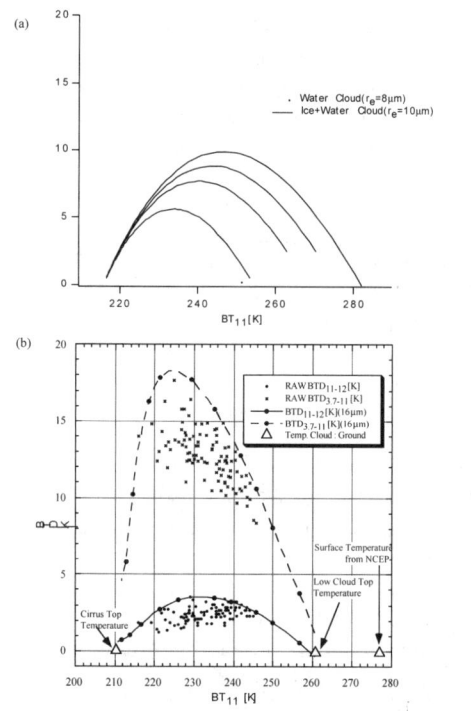

Fig. 5. (a) BTD11-12 vs BT11 curves of two layer cloud systems with a cirrus and an underlying low level water cloud. The dotted line represents the BTD curve of the low level water cloud alone. (b) An example of satellite data plot of BTD11-12 vs BT11 chart. Curves of BTD11-12 and BTD3.7-11 are also shown with the data.

Finally, the data are compared with the lookup table-B, which consists of the data of BTD11-12 versus BT11, to detect the envelope of the scatter plot. We calculate r_e and for all data scattered in the target segment box within the area for retrieval shown in Fig. 1 and get the averages of smaller ten effective radii and their optical thickness that correspond to the envelope of the scattered data in the segmented grid to avoid an imperfect cloud coverage.

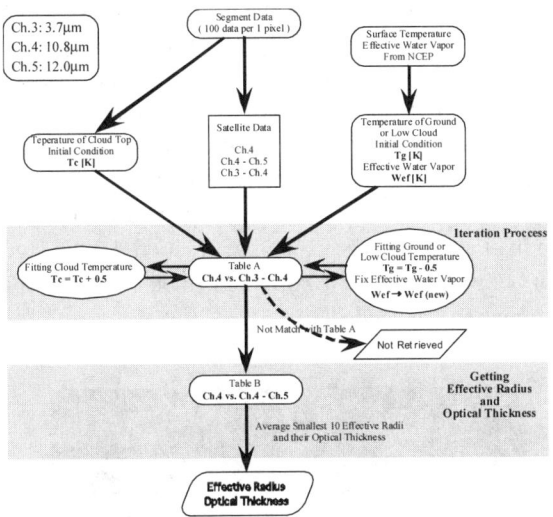

Fig. 6. The flow of the three-channel method. Table A consists of values of BTD3.7-11 versus BT11 and Table B consists of values of BTD11-12 versus BT11.

3. RESULTS

Figs. 7 and 8 show that $r_e \approx 14$ μm in the tropical region, 20 μm in the mid- and high latitudes. Very low cloud top temperature (about 215 K) appears in the West Pacific Ocean near the east region of New Guinea in January and April, and around the Philippine Islands in July.

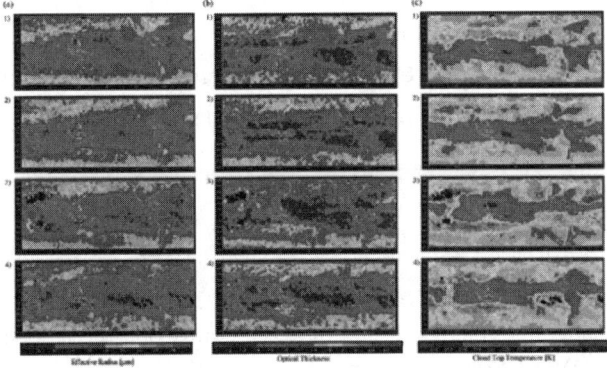

Fig. 7. Monthly mean distributions of cirrus parameters during 9 years. (a) Cirrus cloud effective radius, (b) cirrus cloud optical thickness, and (c) cirrus cloud top temperature. Panel 1), 2), 3), and 4) indicate January, April, July, and October, respectively.

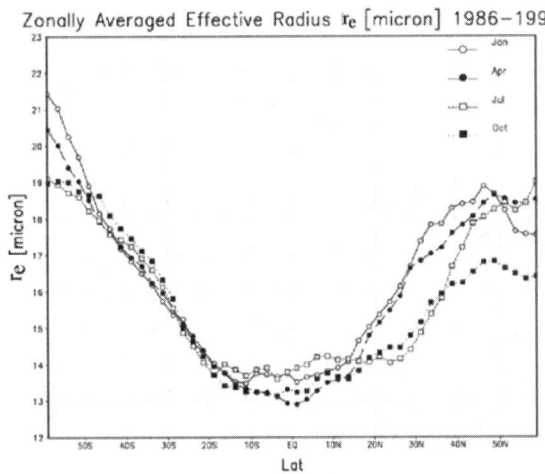

Fig. 8. Zonal mean r_e as a function of the lat. for 9 year averages of monthly mean values for January, April, July, and October.

Low cloud top temperatures appear in the other equatorial regions from the west to the center of the Pacific Ocean have low temperature (about 220 K) all through the year, and also in the regions around Central Africa. The Amazon region has a cloud top temperature as low as about 220 K except in the dry season in July. From mid- to high-latitudes, r_e increases from 14 to 19 μm. There is also a seasonal variation of r_e in mid-latitudes of the northern hemisphere similar to the tropical region with larger amplitude, whereas there is little variation in mid-latitudes of the southern hemisphere. Similar to r_e, a seasonal variation in τ_c is also seen in Fig. 8 with larger values around the high latitudinal regions of the winter hemisphere, and with smaller values in the summer hemisphere. Although the deep cumulonimbus clouds with very thick optical thickness arise throughout the year in the tropical region, cirrus clouds that spread from the top of the deep convective clouds have very small optical thickness (1.0 ~ 1.7), whereas cirrus clouds associated with the low pressures have large optical thickness. The zonal distributions of r_e in Fig. 8 display the latitudinal dependency that the effective radius of cirrus clouds in the tropical region is smaller than that of mid- and high-latitudinal regions. This contrast is thought to be caused by difference in the cirrus particle formation process in these regions.

Cirrus clouds in the tropics usually flow out from the top of deep convective cloud systems near the tropopause, and their particles grow in high altitude with low temperature and low humidity, and on the other hand, cirrus cloud particles associated with the low pressure system grow with high temperature and high humidity in mid- and high-latitudes. In Fig. 9 r_e became apparently large after the eruption of Mt. Pinatubo, and reached its maximum in July in the tropics, in October in the southern hemisphere, and in January in the northern hemisphere. Its volcanic ashes reached the altitude of 30 km across the tropopause, so that there is possibility that the volcanic ashes changed the properties of cirrus clouds. Volcanic ashes itself decrease the BTD in the infrared region causing an overestimation in the effective radius, while volcanic ashes act as ice condensation nuclei to decrease the effective radius of the cirrus clouds. It is, therefore, difficult to say if the increasing trend of r_e after the volcanic eruption is real caused by the volcanic ash effect without validation data. There is very limited in situ observation of the effective radius of cirrus clouds for validating the present satellite products. Fig. 10 which almost corresponds to Fig. 7 (c)-3) in the observation period indicates that the high cloud distributions by HIRS have similar patterns to our results.

properties are highly variable depending on time and space. The microphysical state of cirrus clouds from our analysis has a mode around r_e=14 μm, τ_c=1.6, and T_c=230 K. ISCCP results are $\tau_c = 1.0 \sim 2.5$ for and $T_c = 225 \sim 245$ K in the tropics. Compared with the aircraft observation of r_e and ISCCP analysis, these parameters have difference of Δr_e=-10 $\sim$ -15 μm, $\Delta \tau_c = -0.8 \sim +0.8$, and $\Delta T_c = -5 \sim -20$ K. Our values of r_e should be smallest, because of selecting pixels for analysis located on the largest BTD values in each 0.5° x 0.5° longitude-latitude segment. Degree of uncertainty will be getting larger according as the radius is getting larger because of its nonlinearity of the relations between effective radius and the brightness temperature differences as in Fig. 1, when we analyze all the pixels assuming the 100% sub-pixel cloud fraction. Future studies of further validation and correction method of sub-pixel cloud fraction are needed.

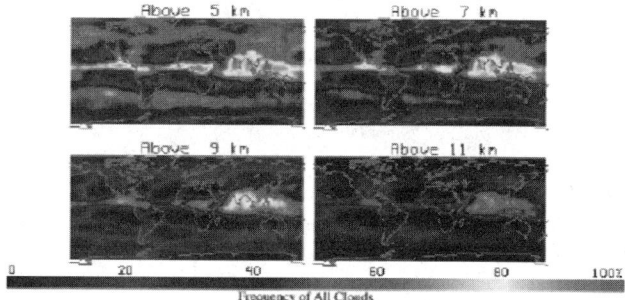

Fig. 10. All clouds frequency in July for the heights of 5, 7, 9, and 11 km detected by HIRS from '89 to '95 (Wylie and Menzel, 1999).

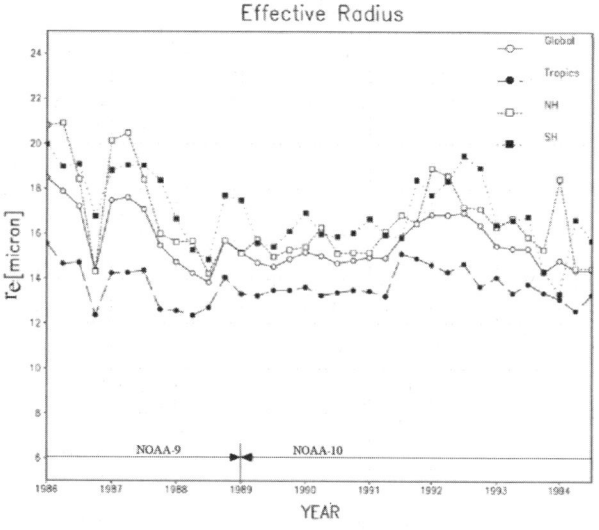

Fig. 9. Time series of r_e. Global mean, hemispherical means (60N-20N:NH, 60S-20S:SH), and tropical areal mean (20N-20S) are shown (NOAA-9: 1986–1988, NOAA-11:1989–1994).

5. ACKNOWLEDGMENTS

We are grateful to J. Tucker of NASA GSFC and Ryoichi Imasu of CCSR in ARGASS project for providing us with AVHRR GAC data used in this study.

6. REFERENCES

Kawamoto, K., T. Nakajima, and T. Y. Nakajima, 2001: A global determination of cloud microphysics with AVHRR remote sensing. *J. Climate*, 14, 2054-2068.

Wylie, D. P., W. P. Menzel, H. M. Woolf, and K. I. Strabala, 1994: Four years of global cirrus cloud statistics using HIRS. *J. Climate*, 7, 1972-1986.

4. SUMMARY

We were able to depict the main features of microphysical structure of the cirrus cloud system on global scale with long-term analysis. It is found that cirrus

DETERMINATION OF BROADBAND EMISSIVITY FOR ARID LANDS

Thomas Schmugge
New Mexico State University
Las Cruces, NM, 88003-8003, USA

Kenta Ogawa, and Shuichi Rokugawa
University of Tokyo
Bunkyo-ku, Tokyo, 113-8656, Japan

ABSTRACT

Surface emissivity in the thermal infrared region is an important parameter for studies of the radiation budget and the surface energy balance. This paper focuses on estimating the broadband emissivity using two sensors on NASA's Earth Observing System (EOS) Terra satellite, Advanced Spaceborne Thermal Emission and reflection Radiometer (ASTER) and MODerate Resolution Imaging Spectroradiometer (MODIS). We developed a regression approach to generate infrared broadband emissivity maps from ASTER or MODIS data. The regressions are to relate the broadband emissivity to the emissivities for the ASTER or MODIS channels. Both regressions were calibrated using libraries of spectral emissivities.

We applied this approach for ASTER and MODIS data acquired over North Africa and Australia. The range of the broadband emissivity was found to be between 0.86 and 0.96 for the desert area. The root mean difference between the emissivities from these two sensors is smaller than 0.015. Such an emissivity map could be used as an input for climate models and studies have shown that this contributes to improving the simulated surface and air temperature by up to 1.1 and 0.8 C respectively. The method can be applied to any arid region of the world.

Keywords: Keywords: broadband emissivity, infrared emissivity, ASTER, MODIS, energy balance, climate model

1. INTRODUCTION

Accurately characterizing the surface emissivity is necessary for correctly determining the longwave radiation leaving the surface, thus for determining surface radiation budget. This is particularly true for non-vegetated arid regions. Therefore emissivity is an important parameter for climate studies [Prabhakara and Dalu; 1976, Wilber et al.; 1999, Zhou et al.; 2003a]. For example, for a daily average surface temperature of 295 K and a clear and dry sky, the daily averaged net longwave radiation is about 135 [W m^{-2}] and net shortwave radiation is between 150 and 200 [W m^{-2}] depending on the solar geometry and the atmospheric condition. Thus an uncertainty of 10 % in broadband emissivity corresponds to a variation 15 to 20 W m^{-2} in the net longwave radiation, because the net longwave radiation is approximately proportional to the broadband emissivity times the difference in the surface and sky brightness temperatures. For the net radiation, this variation is comparable to an uncertainty of 7-9 % in albedo. The range of 10 % in broadband emissivity is observed for non-vegetated surfaces, therefore the accurate emissivity information in arid regions is important. Furthermore, a set of sensitivity tests with a climate model [Zhou et al., 2003b] indicate that a decrease of soil broadband emissivity by 0.1 could increase surface temperature by 1.1 K, air temperature by 0.8 K and net longwave radiation by 6.6 Wm^{-2} for the Sahara Desert. However, current climate models represent the land surface emissivity by a constant value or very simple parameterizations, [Bonan; 1996, 2002], because the observations of emissivity have been limited. Prabhakara and Dalu [1976] mapped the surface emissivity at 9 μm with 100 km resolution using NIMBUS-4 InfraRed Interferometer Spectrometer (IRIS), and found that emissivity varied from 0.65 to 0.90. Ogawa et al. [2003] and Ogawa and Schmugge [2004] have shown that there are large spatial variations (more than 0.1) of broadband emissivity for the desert region in the Sahara Desert. An emissivity map for the arid regions of the world could be provided by the Advanced Spaceborne Thermal Emission and reflection Radiometer (ASTER) and MODerate Resolution Imaging Spectroradiometer (MODIS) data which would be a useful improvement for our understanding of the land-atmosphere interaction.

ASTER [Yamaguchi et al., 1998] is a sensor developed by Ministry of Economy, Trade and Industry (METI) in Japan and is onboard NASA's Earth Observing System (EOS) Terra satellite launched in 1999. ASTER has five channels in the thermal infrared region (TIR) with 90 m spatial resolution. It is possible to estimate spectral emissivities for the five channels using the Temperature-Emissivity Separation (TES) algorithm [Gillespie et al., 1998]. MODIS is also onboard the Terra satellite and has ten channels in the TIR. Three of them (Channel 29, 31, and 32) are in the atmospheric window and can be used to observe the land surface with 1 km resolution. Table 1 shows the center wavelengths for the TIR channels of ASTER and MODIS. ASTER has advantages in spatial resolution and a larger number of channels in the TIR, thus, making it easier to compare with laboratory and field scale measurements and to estimate accurate broadband emissivities. MODIS has the advantages of spatial coverage and frequency of observations. We used both sensors to map the emissivity of large areas and compared the maps to access the accuracy. Jacob et al. [2004] compared the spectral emissivities and temperatures derived from ASTER with ones from MODIS and showed that they agree quite well.

In this paper, we describe the method to estimate the broadband emissivity using ASTER and MODIS data and present results for deserts in North African and Australia. We focus only on desert areas, because the largest variation of emissivity is expected to be observed there. It also helps to simplify the issue, because temporal variations of emissivity and effects of vegetation are minimum. First, we describe the regression approach, which is used for mapping the broadband emissivity from ASTER and MODIS data. This approach is used to obtain emissivity maps of the Sahara Desert and Australia from ASTER and MODIS data. Finally we compare the resulting emissivity maps.

Table 1. The center wavelengths of ASTER and MODIS in thermal infrared region

ASTER ch.	10	11	12	13	14
λ (μm)	8.29	8.63	9.08	10.66	11.29
MODIS ch.	29		31		32
λ (μm)	8.55		11.03		12.02

2. METHOD

The broadband emissivity $\varepsilon_{\lambda 1-\lambda 2}$ can be defined as,

$$\varepsilon_{\lambda 1-\lambda 2} \equiv \frac{\int_{\lambda 1}^{\lambda 2} \varepsilon_\lambda B_\lambda(T_s) d\lambda}{\int_{\lambda 1}^{\lambda 2} B_\lambda(T_s) d\lambda}, \quad (1)$$

where T_s is surface temperature, B_λ is Planck function, and ε_λ is spectral emissivity. $\lambda 1$ and $\lambda 2$ is the wavelength range for the integration. We used 8 and 13.5 μm for $\lambda 1$ and $\lambda 2$ respectively, this is because the atmospheric window region has the largest effect on the net radiation as described in Ogawa and Schmugge[2004].

First, we used a regression to map the broadband emissivity, $\varepsilon_{8-13.5}$, from the emissivities for the five ASTER channels, ε_{ch}. Both $\varepsilon_{8-13.5}$ and ε_{ch} are calculated using the emissivity spectral libraries and a regression relation is developed between them. The details of this approach are described in Ogawa and Schmugge [2004]. The broadband emissivity, $\varepsilon_{8-13.5}$ is given by,

$$\varepsilon_{8-13.5} = \sum_{ch=10}^{14} a_{ch}\varepsilon_{ch} + c_{AST} \quad (2)$$

where a_{ch} and c_{AST} are the coefficients of the regression. We calculated the coefficients using 314 emissivity spectra. The RMS error of the predicted $\varepsilon_{8-13.5}$ by the regression was 0.0041. The coefficients are given in Table 2. The broadband emissivity estimated by this regression is used to validate the regression for MODIS data described below.

Table 2. Coefficients used for the regression (2).

a10	a11	a12	a13	a14	c_{AST}
0.000	0.121	0.194	0.323	0.113	0.242

The same approach is used to estimate the broadband emissivity from MODIS data.

$$\varepsilon_{8-13.5} = a_{29}\varepsilon_{29} + a_{31}\varepsilon_{31} + a_{32}\varepsilon_{32} + c_{MOD} \quad (3)$$

where a_{ch} and c_{MOD} are the coefficients of the regression. The ε_{29}, ε_{31} and ε_{32} are the emissivity of MODIS channel 29, 31 and 32. We calculated the coefficients using the same spectra described above. The coefficients are given in Table 3. Because the correlation between ε_{31} and ε_{32} is very high (correlation coefficient is 0.85), the ε_{32} is dropped from this regression.

Table 3. Coefficients used for the regression (3). The coefficient of a32 was dropped.

a29	a31	a32	c_{MOD}
0.329	0.518	0.000	0.147

3. RESULTS

We used these two regressions with the ASTER and MODIS data acquired over the Sahara Desert, Arabian Peninsula, and Australia for mapping the broadband emissivity. The ASTER and MODIS products were from the U.S. Geological Survey's EROS Data Center[2] (EDC).

The MODIS data and the Eq. (3) were used to estimate broadband emissivity. We used MODIS products (MOD11B1[3] and MOD11C3[4] for eight-day

---

[2] http://lpdaac.usgs.gov/
[3] MODIS/Terra land surface temperature/emissivity, one day, global, 0.05 resolution, SIN projection V4

composite emissivity data). The algorithm to retrieve temperature and emissivity from MODIS data are described in Wan and Li [1997]. The resulting broadband emissivity map is shown in Fig. 1 for North Africa and the Arabian Peninsula. The range of emissivity is approximately the same as that for the ASTER data around Tunisia, i.e. between 0.86 and 0.96 for the desert area in North Africa. The areas with low emissivity (0.86-0.90) correspond to sand dunes in North Africa on the soil map of the Natural Resources Conservation Service (NRCS) of United States Department of Agriculture[5] (USDA). Such low values are not observed in the emissivity map of Australia shown in Fig. 2. The range of emissivity is between 0.91 and 0.96 for the Australian deserts. We note that the regression was applied just for the desert and open shrub area and the fixed values, 0.99 and 0.97, were given for water area and vegetated area.

For validation of the MODIS maps we used 258 scenes of the ASTER Level 2 emissivity product[6] collected around Tunisia including a part of Algeria and Libya in 2001 and 2002. The temporal emissivity variation caused by the changes of soil moisture, vegetation fraction and the atmospheric profile uncertainty could be a problem when we use data acquired at different times. However, the observed temporal variations of the emissivity weren't significant (less than 0.02 in most cases), when we considered the differences of emissivity for overlapping areas of ASTER products observed on the different days. We applied Eq. (2) to these products and generated the broadband emissivity.

We compared the emissivity derived from MODIS data with one from ASTER data in North Africa and Australia. In addition to the scenes around Tunisia, other 115 scenes of ASTER data acquired at separate location in North Africa were used for the comparison. Totally the 373 scenes were used for this comparison. For Australia, 37 scenes of ASTER data were used. Table 4 shows the comparison between the emissivity estimated from MODIS data and that determined directly with ASTER data. The RMS difference was 0.012 for North Africa and 0.014 for Australia. The bias was smaller than 0.01 for both areas. Despite the difference of resolutions between ASTER and MODIS, the results indicate that the regression provides a good estimation of the broadband emissivity.

The resulting broadband emissivity maps and the methodology for generating them will contribute to future climate modeling. An example is given in the recent paper by Zhou et al. (2003b) who found that a decrease of 0.1 in emissivity caused about a 1 C change in both ground and air temperatures.

Table 4. Broadband emissivity of desert and open shrub area derived from MODIS data and its comparison with one from ASTER data

Region	Average Emissivity $\pm \sigma*$	Range	RMSD**	Bias
North Africa	0.907±0.028	0.86-0.96	0.012	-0.005
Australia	0.936±0.019	0.91-0.96	0.014	0.004

ACKNOWLEDGEMENTS

This study was supported by the ASTER Project of NASA's EOS-Terra Program. ASTER Spectral Library was provided by the Jet Propulsion Laboratory, California Institute of Technology, Pasadena, California. Much of the work was completed while the authors (Schmugge & Ogawa) were at the USDA/ARS Hydrology and Remote Sensing Lab in Beltsville, Maryland.

REFERENCES

Belward, A. and T. Loveland, The DIS 1-km land cover data set, Global Change, The IGBP Newsletter, 27, 1996.

Bonan, G.B., A land surface model (LSM version 1.0) for ecological, hydrological, and atmospheric studies: technical description and user's guide, NCAR Technical Note NCAR/TN-417+STR., National Center for Atmospheric Research, Boulder, Colorado. 150 pp, 1996.

Bonan, G. B., K.W. Oleson, M. Vertenstein, S. Levis, X. Zeng, Y. Dai, R. E. Dickinson, and Z. Yang, The land surface climatology of the NCAR, community land model coupled to the NCAR Community Climate Model, J. Clim., 15, 3123. V3149, 2002.

Gillespie, A., S. Rokugawa, T. Matsunaga, J.S. Cothern, S. Hook, A.B. Kahle, A temperature and emissivity separation algorithm for Advanced Spaceborne Thermal Emission and Reflection radiometer (ASTER) images, IEEE Trans. on Geosci. and Remote Sens., 36, 1113-1126, 1998.

Jacob F., F. Petitcolin, T. Schmugge, E. Vermote, A. French, K. Ogawa, Comparison of land surface emissivity and radiometric temperature derived from MODIS and ASTER sensors, Remote Sens. Environ., 90, 137-152, 2004.

Ogawa K., T. Schmugge, F. Jacob and A. French, Estimation of land surface window (8-12 μm) emissivity from multi-spectral thermal infrared remote sensing - A case study in a part of Sahara Desert, Geophysical Research Letters, 30(2), doi:10.1029/2002GL016354,1067, 2003.

---

[4] MODIS/Terra land surface temperature/emissivity, 8 days composite, 0.05 resolution, SIN projection V4
[5] http://www.nrcs.usda.gov/technical/worldsoils/mapindx/order.html
[6] ASTER On-Demand Level 2 Surface Emissivity (AST_05), 90 m resolution, Version 3

Ogawa, K., Schmugge, A., Mapping of broadband emissivity (8-13.5 μm) using ASTER and MODIS data, Earth Interactions, 8, Paper 7, doi: 10.1175/1087-3562, 2004.

Prabhakara, C., and G. Dalu, Remote sensing of the surface emissivity at 9 mm over the globe, J. Geophys. Res., 81, 21, 3719-3724, 1976.

Salisbury, J.W. and D.M. D1Aria, Emissivity of terrestrial materials in the 8-14 μm atmospheric window, Remote Sens. Environ., 42, 83-106, 1992.

Wilber, A.C., D.P. Kratz, and S.K. Gupta, Surface emissivity maps for use in satellite retrievals of longwave radiation, NASA/TP-1999-209362, pp. 30, 1999.

Wan, Z. and Z. L. Li, A physics-based algorithm for retrieving land-surface emissivity and temperature from EOS/MODIS data, IEEE Trans. Geosci. Remote Sens., 35, 980-996, 1997.

Yamaguchi, Y., A.B. Kahale, H. Tsu, T. Kawakami, and M. Pniel, Overview of Advanced Spaceborne Thermal Emission and Reflection radiometer (ASTER), IEEE Trans. on Geosci. and Remote Sens., 36, 1062-1071, 1998.

Zhou, L., R.E. Dickinson, Y. Tian, M. Jin, K. Ogawa, H. Yu, T. Schmugge, A sensitivity study of climate and energy balance simulations with use of satellite derived emissivity data over Northern Africa and the Arabian Peninsula, J. Geophys. Res. Atmosphere, 108, D24, 4795, doi:10.1029/2003JD004083, 2003b.

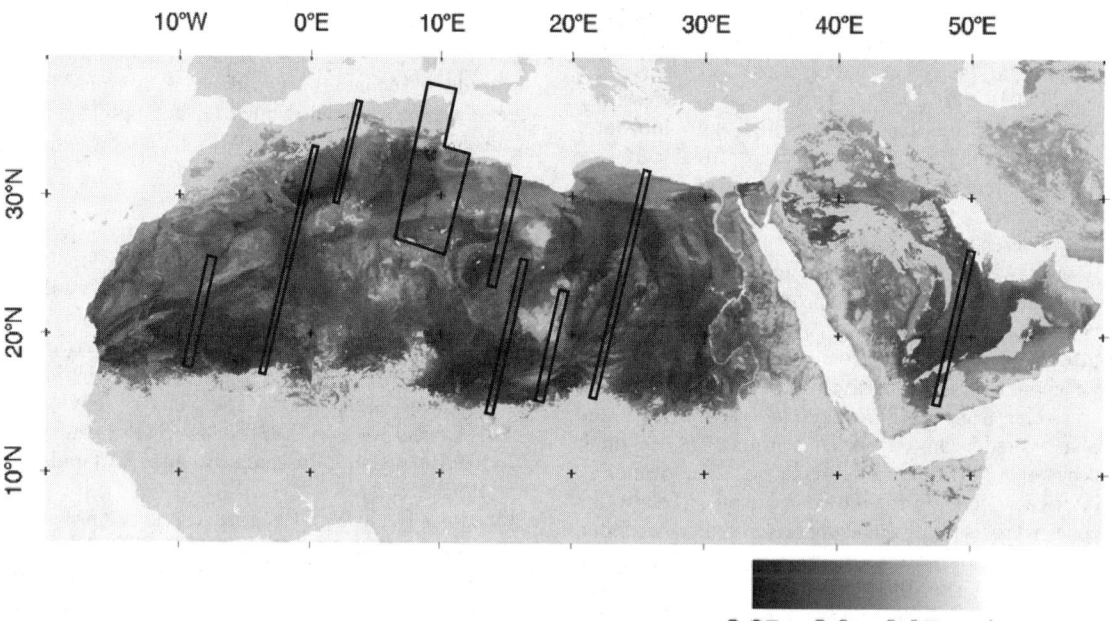

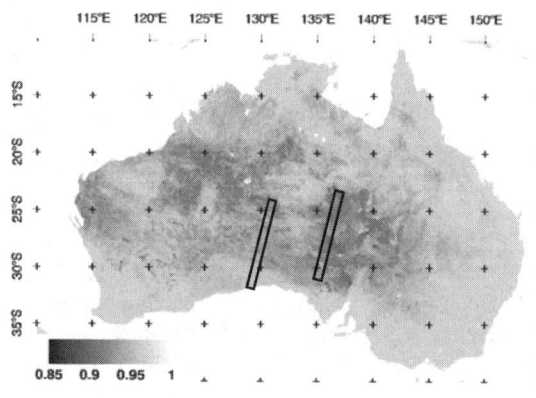

Figure 1. (Above) The broadband emissivity map of Sahara Desert derived from MODIS data and Eq (3). Water and vegetated area are masked using a land cover map by USGS (http://edcdaac.usgs.gov/glcc/glcc.html) and given fixed values. MODIS temperature/emissivity product MOD11B1 observed on November 1-5, 2001 was used. The resolution of this map is 0.05°. Boxes indicate coverage of ASTER data used in this study.

Figure 2. (Right) Broadband emissivity map derived from MODIS data. MODIS temperature/emissivity product observed on May 2003 and Eq. (3) was applied. Boxes indicate coverage of ASTER data used in this study.

BRIGHT-BAND HEIGHT STATISTICS OBSERVED BY THE TRMM PRECIPITATION RADAR

Ken'ichi Okamoto*, Hiroshi Sasaki*, Eri Deguchi**, Merhala Thurai,*** and Kayo Matsukawa*

*Osaka Prefecture University, Sakai, Osaka, 599-8531, Japan
**JSAT Corporation, Tokyo, 100-6218, Japan
***Colorado State University, Fort Collins, Colorado 80523, USA

ABSTRACT

Six year (1998–2003) TRMM PR data were analyzed to obtain averages of bright-band heights within the latitude range covered by TRMM orbit. Examined are the year-to-year variability and seasonal variation of the bight-band heights over the ocean and the land. Seasonal variations increased as the latitude increased. These values were compared with annual averages of freezing level heights by ECMWF. Comparisons showed that the PR bright-band heights typically occurred at about 400 meters below the zero degree heights. The correlation coefficients between the bright-band heights and zero degree heights were about 0.78.

1. INTRODUCTION

A bright-band is a melting layer of snow and ice that usually lies above the stratiform type of rain. On the meteorological radar, a bright-band can be seen as a nearly horizontal thin bright echo layer with a large radar reflectivity factor Z. The bright-band height gives an important indication in estimating the height of the stratiform type of rain and the 0°C isotherm height, i.e., the freezing height. The bright hand height always lies below the 0°C isotherm height because snow and ice begins to melt at the 0°C isotherm height and the melting layer grows as the melting snow and ice falls down from the 0°C isotherm height to form the bright band eventually. The rain height is one of the important parameters to retrieve the rainfall rate from the brightness temperature data obtained by the microwave radiometer.

In the design of the satellite communication link of the microwave and millimeter radio waves, the rain height is also considered to be an indispensable parameter to evaluate the effect of rain attenuation, which is a main factor of deterioration of the communication link. Therefore, the International Telecommunication Union (ITU) uses the database of the 0°C isotherm height with the grid spacing of 1.5 degrees (latitude) by 1.5 degrees (longitude), which is based on the long term meteorological data provided by the European Centre for Medium-Range Weather Forecast (ECMWF).

The ECMWF database of the 0°C isotherm height, however, is composed mainly of the data observed on the ground at the mid-latitudes and the climatic numerical models, and scarcely includes the data observed on the ground of the tropical and subtropical regions, most of which, as a matter of course, are covered with ocean.

We focus our attention on utilizing the precipitation radar (PR) 3A25 data of the Tropical Rainfall Measuring Mission (TRMM), which aims at observing the tropical and sub-tropical precipitation, and compare the bright-band height statistics calculated from the TRMM PR data with the database of the ECMWF 0°C isotherm height (Thurai et al., 2003).

2. COMPARISONS OF BRIGHT-BAND HEIGHT YEARLY VARIATION

We have analyzed the six-year (1998–2003) annual mean variations of bright-band heights over the entire globe (at the fixed latitude and the longitude from 0 to 360 degrees), which is shown in Fig. 1. From the results of the analysis, we could learn the zonal averages of the bright-band height of 1998 are higher than those of the other five years around the regions from the equator to the latitude of approximately 20 degrees south. This can be considered the effects of the 1998 El Nino event, which was the largest in the 20th century.

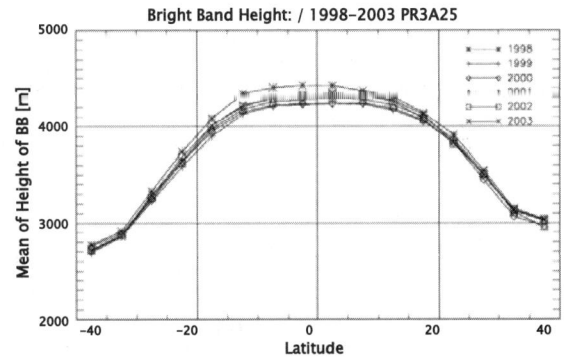

Fig. 1. Zonal average of bright-band height (1998–2003).

We have also found that the yearly variations of the bright-band heights are smaller at the higher latitude in both hemispheres. Apart from the year of El Nino 1998, the bright-band heights show almost the same figures and change in a similar way as a function of latitude in the other five years (1999–2003). However, if we look at the figures carefully, the height of bright-band is a little higher in 2002 than in the other years. This may be caused by the small El Nino, which was seen during March 2002 to March 2003. The bright-band heights are virtually constant at the low latitude (+/-10 degrees) and the figures decrease gradually as the latitudes goes higher in both hemispheres.

These downward trends of the bright-band height in each hemisphere are, however, asymmetric, and the bright-band heights are slightly higher in the northern hemisphere than in the southern hemisphere. Each of the yearly variations of the ocean and the land shows the similar trend to that of the whole globe.

3. YEAR TO YEAR VARIATION OF MONTHLY BRIGHT-BAND HEIGHT

We have analyzed six-year (1998–2003) variations of the mean bright-band heights on a monthly basis over the whole globe. The zonal bright-band height averages from January to May in 1998 are higher than those of the other five years, especially in the tropical area. This anomalous phenomenon might be associated with El Nino event, which had continued since 1997 and suddenly came to an end in May 1998. Fig. 2 shows the variation of the bright band height of every January of six years (1998–2003). It is clearly shown that the bright-band height in 2003 is the second highest, which may be the effect of the small El Nino of 2002–2003.

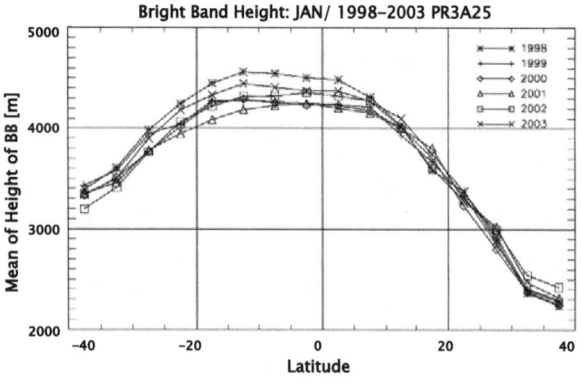

Fig. 2. **Variations of bright-band height of every January of 1998–2003.**

4. THE SEASONAL VARIATION OF BRIGHT-BAND HEIGHTS AND 0°C ISOTHERM HEIGHTS

We have focused on a seasonal aspect of the variations of the mean bright-band heights and 0°C isotherm heights over the whole globe. Fig. 3 shows seasonal variations of bright-band height (1998–2003). The zonal averages of the bright-band heights are the lowest in the winter season (December through February in the northern hemisphere, and June through August in the southern hemisphere) and are the highest in the summer season (June through August in the northern hemisphere, and December through February in the southern hemisphere). While almost no seasonal variations can be seen in the tropical area, the bright-band heights show larger seasonal variations at the higher latitudes; about 1.5 km of the bright-band height variations are recorded at the latitude of around 35 degrees north and about 1km of the bright-band height variations at the latitude of around 35 degrees south. Particularly in the northern hemisphere, the seasonal variation is larger over the land than over the ocean.

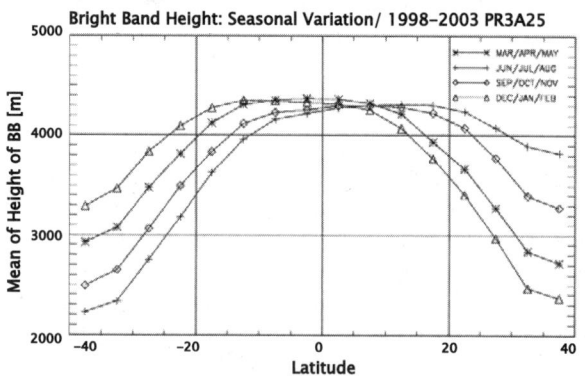

Fig. 3. Seasonal variations of bright-band height (1998–2003).

Fig. 4 shows the seasonal variation of the 0°C isotherm by the ITU database. We have looked into the seasonal variations of the 0°C isotherm heights and have found out that zonal averages of the 0°C isotherm heights are the lowest in the winter season (December through February in the northern hemisphere, and June through August in the southern hemisphere) and are the highest in the summer season (June through August in the northern hemisphere, and December through February in the southern hemisphere) just like the bright-band height variations. There is almost no seasonal variations of the 0°C isotherm heights around the tropical area, and on the other hand, the 0°C isotherm height variations are larger at the higher latitudes.

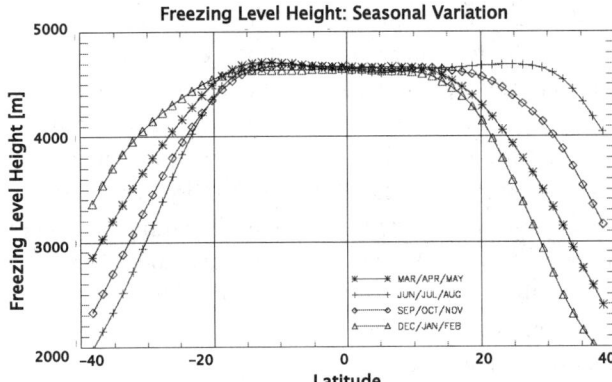

Fig. 4. Seasonal variations of 0°C isotherm by ITU database.

5. COMPARISON OF THE BRIGHT-BAND HEIGHT WITH THE 0°C ISOTHERM HEIGHTS

We have made a scatter diagram composed of a horizontal axis containing the mean bright-band heights for every 1.5 degrees of latitude and longitude and a vertical axis representing ECMWF 0°C isotherm heights (Fig. 5).

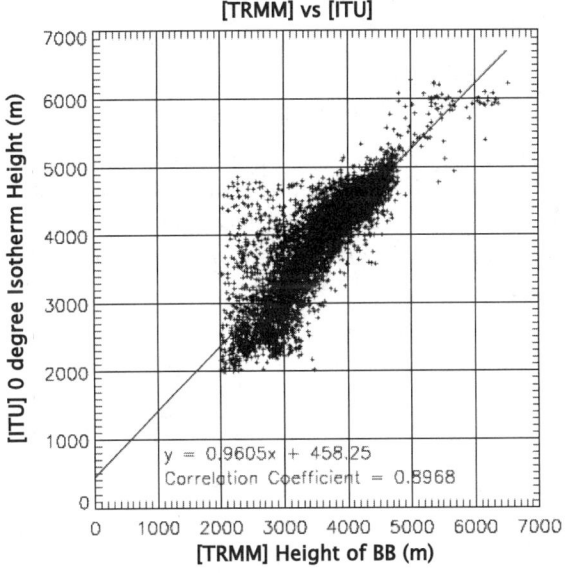

Fig. 5. [TRMM] versus [ITU].

The data do not include the bright-band heights less than 2 kilometers. The correlation coefficient is 0.90, which means there is a fairly good relationship between these heights. Fig. 6 is a histogram of the differences between the 0°C isotherm heights and the bright-band heights. FLH stands for Freezing Level Height which is the same as the 0°C isotherm height, and BBH stands for the Bright Band Height. Approximately 95% of the calculated differences between the 0°C isotherm heights and the bright-band heights range between 0 and 600 meters, and its mean value is about 304 meters.

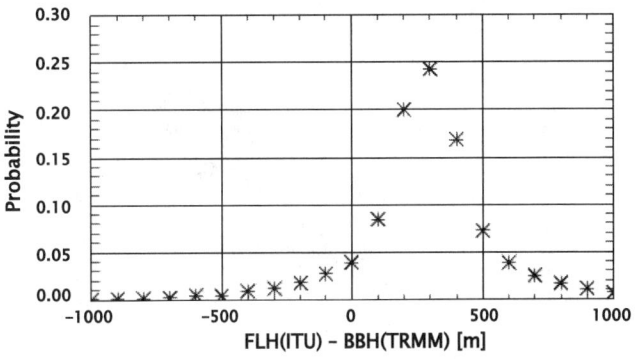

Fig. 6. Histogram of differences between 0°C isotherm height and bright-band height.

The values of differences between 0°C isotherm heights and bright-band heights are plotted as a function of latitude in Fig. 7. In the tropics and in the areas at the latitude of up to ±10 degrees, the differences are low in magnitude and their spread of variation are also low. However, the spread of the differences are noticeably larger at the higher latitudes.

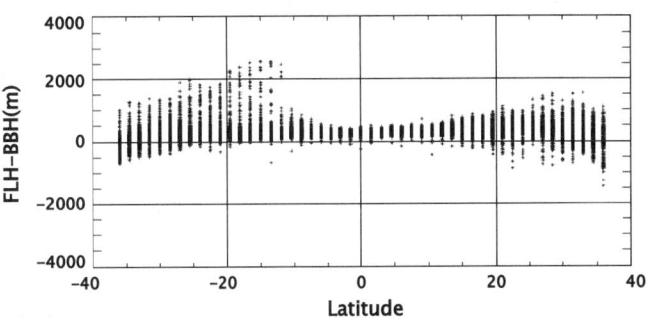

Fig. 7. Latitude dependence of the differences between 0°C isotherm height and bright-band height.

6. COMPARISON OF BRIGHT-BAND HEIGHT WITH SST

The variation of the bright band height is largely due to the variation of the SST (Sea Surface Temperature). Fig. 8 is a scatter diagram for every 5 degrees of latitude and longitude, of the monthly mean SST on the horizontal axis and monthly mean bright-band height on the vertical axis. The SST are retrieved from TMI and 5° x 5° grid data are obtained by averaging the original 0.25° x 0.25° grid data.

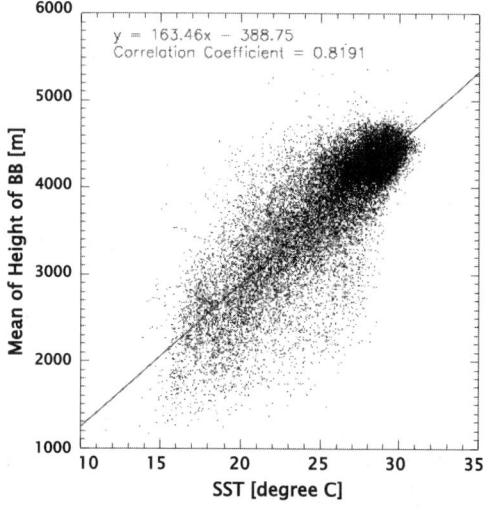

Fig. 8. TRMM/TMI derived SST versus TRMM/PR derived bright-band height (Jan. 1998–July 2001).

The correlation coefficient is 0.82 and good correlation can be seen from the scatter diagram. The bright-band height increases by about 1km every 6°C increase of the SST.

7. CONCLUSIONS

TRMM PR data provide much information on the bright band height statistics, which is useful for estimating the 0°C isotherm heights.

We have analyzed TRMM PR 3A25 six-year data (1998–2003) to obtain the monthly, yearly and seasonal variations of the bright-band heights over the ocean and the land within the latitude of +/- 40 degrees and have compared the mean values of the bright band heights of those six years with the ECMWF 0°C isotherm heights. The bright-band heights were slightly higher in the tropics during the periods of the El Nino 1998 and 2002. The mean value of the differences between the 0°C isotherm heights and the bright-band heights is 304 m, which is consistent with the previously reported value.

The TRMM satellite is continuing its mission in very good condition as of January 2005, at the time of writing.

It is a great pity that NASA and JAXA has decided to terminate the operation of TRMM in the very near future.

8. ACKNOWLEDGEMENT

The author would like to express their sincere thanks to Japan Science and Technology Corporation for providing the financial support for this research.

9. REFERENCES

Thurai, M., E. Deguchi, T. Iguchi, and K. Okamoto, 2003: Freezing height distribution in the tropics. *In International Journal of Satellite Communication Network,* 21, 533-545.

EFFECTS OF MULTIPLE SCATTERING FOR MILLIMETER-WAVELENGTH WEATHER RADARS

Satoru Kobayashi, Simone Tanelli, and Eastwood Im
California Institute of Technology, Jet Propulsion Laboratory
4800 Oak Grove Dr., MS. 300-243
Pasadena, CA 91109, USA

ABSTRACT

Effects of multiple scattering on the reflectivity measurement for millimeter-wavelength weather radars are studied, in which backscattering enhancement may play an important role. In the previous works, the backscattering enhancement has been studied for plane wave injection, the reflection of which is received at the infinite distance. In this paper, a finite beam width of a Gaussian antenna pattern along with spherical wave is taken into account. A time-independent second order theory is derived for a single layer of clouds of a uniform density. The ordinary second-order scattering (ladder term) and the second-order backscattering enhancement (cross term) are derived for both the copolarized and cross-polarized waves. As the optical thickness of the hydrometeor layer increases, the differences from the conventional plane wave theory become more significant, and essentially the reflectivity of multiple scattering depends on the ratio of mean free path in hydrometeors to radar footprint radius. This effect must be taken into account for remote sensing applications.

1. INTRODUCTION

From the early 1970's to the early 1990's, multiple scattering in randomly distributed particles was intensively studied through the analytical method of electromagnetic wave (de Wolf, 1971; Kravtsov and Saichev, 1982; Kuga and Ishimaru, 1984; Tsang and Ishimaru, 1985; Tsang and Kong, 2001). In the course of study, two main contributions of multiple scattering to reflective intensity were revealed. In general, the electromagnetic field reflected from a random medium can be represented as a sum of fields from many portions of the medium. Among this sum, only pairs of fields that have strong correlation can give contributions to a measured intensity. A first possible pair consists of a field E_A and its self-complex-conjugate field E_A^* as depicted in Fig. 1a in the case of the second order scattering. The incident field E_A is emitted from the antenna T, and scattered at points 'b' and 'a' successively in the random medium, being returned to the antenna T. This process is represented by the ladder diagram in Fig. 1b, which can be proven to be the basis of radiative transfer theory (Tsang and Kong, 2001). Another possible pair occurs in the backscattering condition depicted in Fig. 2a, in which the field E_A travels in the same path as that in Fig. 1a, while the conjugate field E_B, takes the reversal path of E_A. This process can be represented by the cross diagram in Fig. 2b. Since the path lengths of E_A and E_B are equal for the right backscattering, it always gives finite contribution to a measured intensity. As the scattering angle deviates from the right backscattering condition, the fields E_A and E_B have different path lengths, and random distributions of particles 'a' and 'b' will cause strong decorrelation, generally giving negligible contribution to the measured intensity. In short, the cross term is measured only in the vicinity of the right backscattering angle, which is the reason we refer to this phenomenon as backscattering enhancement.

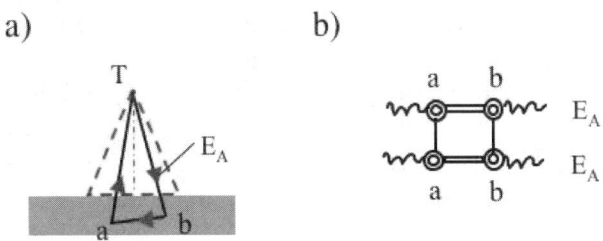

Fig. 1. Second order ladder term. An incident field E_A is emitted from the antenna T, and scattered at points 'b' and 'a' successively in the random media, being returned to the antenna T. (a): Geometry. (b): Diagram.

In all the previous theoretical works, a plane wave is injected to randomly distributed particles, and the reflected wave is collected by a receiver at infinite range. On the other hand, in remote sensing, a spherical wave with a finite beam width, usually approximated as a Gaussian antenna pattern, is injected, and the reflected wave is received by an antenna at a finite range. For the single scattering, the plane wave theory can be applied to a radar of finite beam width along with slight corrections concerning to range and gain. While for multiple scattering, we cannot adopt the plane wave theory, because a finite size footprint can be considered to give a smaller reflectivity than the plane wave theory predicts,

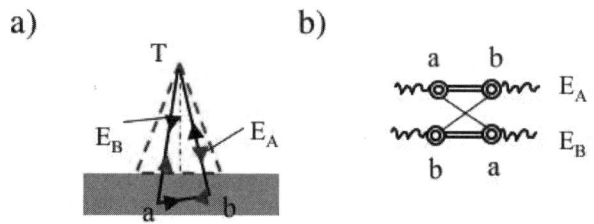

Fig. 2. Second order cross term. An incident field E_A takes the same path as in Fig. 1a, while the conjugate field E_B is emitted from the antenna T, and scattered at points 'a' and at 'b' successively (i.e., time-reversal path of E_A), being returned to the antenna T. (a): Geometry. (b): Diagram.

especially when the footprint size is much smaller than the mean free path of an illuminated body. The mean free path in a layer of hydrometeors often reaches of the order of 1000 meters for millimeter wavelength wave, while typical foot print sizes are of order of hundreds meters. In this study, a time-independent multiple scattering theory is formulated for a spherical wave along with a Gaussian antenna pattern, based on the plane wave theory of Mandt et al. (1990). Our analysis considers only a single layer of spherical water particles of a uniform diameter and a uniform number density.

2. FORMALISM

In this section, the formalism of Ishimaru and Tsang (1988) and Mandt et al. (1990) is expanded to a spherical wave with a finite beam width represented by a Gaussian function. To deal with complication introduced by the finite beam width, further simplifications of integrals are performed for the second order ladder and cross terms.

We shall consider a single layer of hydrometeors of thickness d constituted of spherical water particles of a uniform diameter D and a uniform density N_0. A 3-dB beam width θ_d and a range r_s are assumed to be $|\theta_d| \ll 1$ and $r_s \gg d$, respectively. An incident direction $\hat{k}_i$ is almost parallel to the scattering direction $\hat{k}_s$ ($\hat{k}_i \approx -\hat{k}_s$). The wavenumber in the medium is calculated self-consistently through the Fold-Twersky approximation. The 3-dB footprint radius σ_r can be defined as

$$\sigma_r^2 = r_s^2 \theta_d^2 / 2^3 \ln 2 \qquad (1.1)$$

The complete forms and derivations up to the second order terms are skipped due to the limit of allowed pages, and interested readers should refer to Kobayashi et al. (2004). Instead, some characteristic features in the formulations will be mentioned. The derived cross term has a real valued form rather than a complex valued form described in Mandt et al. (1990). It also includes an oscillation term proportional to

$$\exp[i(k_{dz}+t)\varsigma] \qquad (1.2)$$

in which ς represents the relative coordinate in the vertical direction from a random point to another point in the medium, and the deviation vector $\mathbf{k_d}$ and the variable t have been introduced based on definitions by Tsang and Ishimaru (1985):

$$\mathbf{k_d} \equiv k_{dx}\hat{x} + k_{dy}\hat{y} + k_{dz}\hat{z} = k(\hat{k}_i + \hat{k}_s) \qquad (1.3)$$
$$t \equiv k_{dx}\tan\theta\cos\varphi + k_{dy}\tan\theta\sin\varphi \qquad (1.4)$$

In Equation (1.4), θ and φ represent relative spherical coordinates between the two random points. Equation (1.2) means that decorrelation caused by random particles becomes more serious as deviation from the right backscattering angle increases. The condition of finite beam width brings the following exponential term into both the ladder and cross terms:

$$\exp[-\varsigma^2 \tan^2\theta / 4\sigma_r^2] \qquad (1.5)$$

Equation (1.5) will play an important role when the mean free path is large comparing to a radar footprint size.

3. RESULTS

When hydrometeors consist of spherical particles, the first order ladder term $I_L^{(1)}$ has only the copolarized component, i.e., $I_L^{(1)} = I_L^{(1)}(CO)$. Furthermore $I_L^{(1)}$ is almost constant in the vicinity of $\theta_s = 0$ ($\theta_s < 0.3$ degree), within which the backscattering enhancement occurs. For this reason, the intensity of a multiple scattering term will be normalized by $I_L^{(1)}$ to be converted to an effective reflectivity in the rest of paper.

The sums of only the second order terms $L_2^{co} + C_2^{co}$ in copolarization and $L_2^{cx} + C_2^{cx}$ in cross-polarization are plotted with the solid and dashed lines respectively in Fig. 3 as functions of the foot print radius normalized by the mean free path $l_0 = 1/\kappa_e$. Since the monostatic radar is our main concern, only the backscattering $\theta_s = 0$ will be considered hereinafter. Spherical water particles of a uniform diameter $D = 1$ mm with a particle number density $N_0 = 5 \times 10^3$ m^{-3} are used for calculation along with a frequency of 95 GHz, which gives the mean free path $l_0 = 77.2$ m. Fig. 3 shows that the reflectivities rapidly decrease in the region $\sigma_r / l_0 < 1$, while these are almost constant in the region $\sigma_r / l_0 > 2$.

In Fig. 4, the terms $L_2^{co} + C_2^{co}$ (solid lines) and $L_2^{cx} + C_2^{cx}$ (dashed lines) are plotted as a function of optical thickness $\tau_d = d / l_0$ for several values of normalized footprint radius σ_r / l_0. Here, only the frequency of 95 GHz and the particle diameter $D = 1$ mm are fixed, while the particle number density N_0 can be chosen arbitrarily. Since the mean free path l_0 is uniquely defined for a certain N_0 with the fixed D, the value of τ_d should be considered to be changed by varying the physical layer thickness d. Alternatively, if a layer thickness d, and a particle number density N_0 are known with information of the particle diameter $D = 1$ mm, we can determine the reduction factor from the plane wave theory (i.e., $\sigma_r / l_0 = \infty$) to a finite footprint radius σ_r, using Fig. 4. Fig. 4 also indicates the following fact: As the optical thickness of the hydrometeor layer increases, the differences from the conventional plane wave theory become more significant, and essentially the reflectivity of multiple scattering depends on the ratio of mean free path in hydrometeors to radar footprint radius. Inversely saying, the plane wave theory can be applied to a smaller value of σ_r / l_0, as the optical thickness decreases. For instance, the reflectivity for $\sigma_r / l_0 = 1$ can be approximated with the plane wave theory ($\sigma_r / l_0 = \infty$) at $\tau_d = 1$.

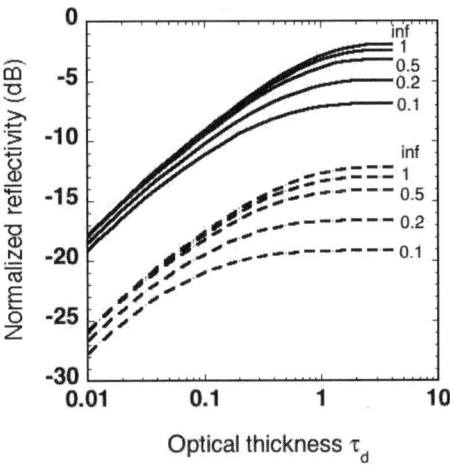

Fig. 4. The reflectivity $L_2^{co} + C_2^{co}$ in copolarization (solid lines) and the reflectivity $L_2^{cx} + C_2^{cx}$ in cross-polarization (dashed lines) as functions of the optical thickness τ_d for the backscattering $\theta_s = 0$. Hydrometeor diameter is set at $D = 1$ mm. Particle number density N_0 is arbitrary. The parameters of footprint radius σ_r / l_0 are set at ∞ (inf), 1, 0.5, 0.2, and 0.1 from top to down. The differences of finite footprint radius ($\sigma_r / l_0 = 1$-0.1) from the plane wave theory ($\sigma_r / l_0 = \infty$) reduce, as τ_d decreases.

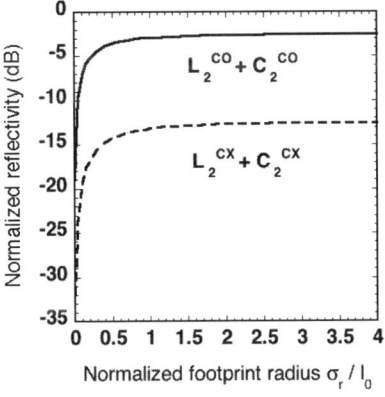

Fig. 3. The reflectivity $L_2^{co} + C_2^{co}$ in copolarization (solid line) and the reflectivity $L_2^{cx} + C_2^{cx}$ in cross-polarization (dashed line) as functions of the normalized footprint radius σ_r / l_0 for the backscattering $\theta_s = 0$. Spherical water particle of diameter $D = 1$ mm and particle number density $N_0 = 5 \times 10^3$ m^{-3} are used, which give the mean free path $l_0 = 77.2$ m.

4. ACKNOWLEDGEMENTS

The research described in this paper was carried out at the Jet Propulsion Laboratory, California Institute of Technology, under a contract with the National Aeronautics and Space Administration.

5. REFERENCES

de Wolf, D. A., 1971: Electromagnetic reflection from an extended turbulent medium: Cumulative forward scatter single-backscatter approximation. *IEEE Trans. Antenna and Propag.*, Ap-19, 254-262.

Ishimaru, A., and L. Tsang, 1988: Backscattering enhancement of random discrete scatterers of moderate sizes. *J. Opt. Soc. Am. A.*, 5, 228-236.

Kobayashi, S., S. Tanelli, and E. Im, 2004: Backscattering enhancement with a finite beam width for millimeter-wavelength weather radars. *Proc. SPIE*, 5654, *Microwave Remote Sensing. Atmos. Environment IV*, 106-113.

Kravtsov, Yu., and A. I. Saichev, 1982: Effects of double passage of waves in randomly inhomogeneous media. *Sov. Phys. Usp.*, 25, 494-508.

Kuga, Y., and A. Ishimaru, 1984: Retroflectance from a dense distribution of spherical particles. *J. Opt. Soc. Am. A*, 1, 831-835.

Mandt, C. E., L. Tsang, and A. Ishimaru, 1990: Copolarized and depolarized backscattering enhancement of random discrete scatterers of large size based on second-order ladder and cyclical theory. *J. Opt. Soc. Am. A.*, 7, 585-592.

Tsang, L., and A. Ishimaru, 1985: Theory of backscattering enhancement of random discrete isotropic scatterers based on the summation of all ladder and cyclical terms. *J. Opt. Soc. Am. A.*, 2, 1331-1338.

Tsang, L., and J. A. Kong, 2001: *Scattering of electromagnetic waves: Advanced topics*, John Wiley & Sons, Inc., New York, 413 pp.

A COMBINED TMI-SSM/I RAINFALL ESTIMATION TECHNIQUE FOR LAND SURFACES OVER WESTERN AFRICA

J. Schulz
Deutscher Wetterdienst, Satellite Application Facility on Climate Monitoring
P.O. Box 100465, D-63004 Offenbach, Germany

C. Simmer and M. Diederich
Meteorologisches Institut, Universität Bonn
Auf dem Hügel 20, D-53121 Bonn, Germany

ABSTRACT

The paper describes a Precipitation Radar calibrated combined TMI – SSM/I algorithm and its comparison to standard TRMM products. The algorithm consists of a two-stage approach to distinguish precipitation signatures from other effects. Rain free background brightness temperature maps are used to isolate the slowly varying parts of background emissivity. Surface temperature and moisture effects are reduced by means of principal component analysis, where the general signatures of temperature and moisture variations are deduced from radiative transfer simulations. The technique is applied to TRMM's TMI measurements where rain rates and rain water contents are estimated by using co-located PR measurements to effectively calibrate the microwave precipitation index. The calibration is extended to the SSM/I instrument by using TMI–PR co-locations employing SSM/I footprint characteristics. A comparison to TRMM standard products shows that the algorithm has an improved potential for the detection of light rain rates less than 2 mmh^{-1} also showing smaller rms deviations.

1. INTRODUCTION

Rainfall algorithms over land surfaces employing microwave window frequencies may only be improved if local surface conditions are accounted for and if the signal to rain rate relationship is based on data which is representative of global rainfall system variability.

The combination of the TRMM Microwave Imager and the Precipitation Radar provides an excellent tool for new retrieval developments. The difficult interpretation of microwave signals over land surfaces has favoured somewhat the implementation of empirical algorithms which make use of co-located satellite and ground based (radar) observations to derive a calibration of satellite measurements by reference data (e.g., Ferraro and Marks, 1995). For large-scale regional or even global applications this approach is hampered by the limited representativity of available calibration data sets. Thus co-located radiometric and radar measurements from TRMM provide an outstanding data source covering the tropical and subtropical latitudes over various climatological regimes.

The technique used in this paper follows the ideas of Conner and Petty (1998) and uses the methodology described in Bauer et al. (2002) to combine TMI and PR measurements by using Heidke Skill score diagrams. Section 2 provides a brief summary of the methodology and describes the calibration of the precipitation index (PI) to a rain intensity by using PR data. The calibration technique is then extended to the SSM/I instrument configuration still employing TMI-PR co-locations. Section 3 examines the application of the algorithm to a case of a meso-scale convective cluster over northwest Africa. The derived product is compared to the TRMM standard product 2A12 which delivers surface rainfall and 3D structure of hydrometeors and latent heating over the TMI swath (Kummerow et al., 2000). Additionally, the calibrated PI is compared to the calibration source (2A25) which gives an estimate of the remaining errors after the calibration. Additionally the application to TMI and SSM/I data is shown in conjunction to Meteosat 7 data over Northwest Africa. This is a pre-stage of the temporal integration of the microwave estimates using techniques by Joyce et al. (2004) or Turk et al. (2001). All data used in this study are version 5 data, so the TMI brightness temperatures are calibrated as described in Kummerow et al. (2000).

2. METHODOLOGY

The methodology has been described in detail in Bauer et al. (2002). Fig. 1 shows a flow chart depicting the main features of the algorithm. The technique used here makes use of a first order approach to remove seasonally varying surface contributions from instantaneous TMI measurements by generating maps of clear sky temporal averages of brightness temperatures using the screening technique of Ferraro et al. (1998). The resulting brightness temperatures are mapped onto a grid with a spatial resolution of 0.4° x 0.4.

To determine the integration time for the background maps running means with a different length in time were computed over Northwest Africa, which are 30 and 7 days, respectively. Over both lengths a band of high standard deviations appears along 15°N which is related to a region where it is on one hand difficult to detect rainfall (because of the semi arid surface) and on the other hand the quickest changes in the vegetation coverage can be expected. Higher standard deviations are also found over mountainous areas related to different viewing conditions of the radiometer. For the longer average the regions with increased standard deviation become larger, whereas the standard deviation of other regions doesn't change. To restrict the investigation to homogenous surfaces and to ensure as short as possible computation time we decided to take 7 day running averages and exclude grid boxes from the further analysis where the standard deviation is larger than 10 K in the 19 GHz channel.

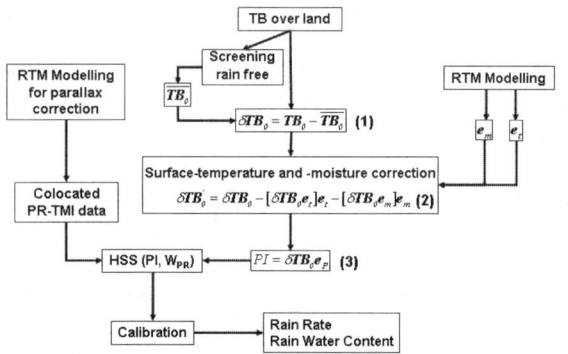

Fig. 1. TMI algorithm flow, details are described in the text. Bars over symbols denote temporal averages and [...] denote inner products. Equation (1) describes the departure of the actual brightness temperature vector and the background values. Equation (2) gives the correction of this departure for temperature and moisture related surface effects projected to the brightness temperature space. Equation (3) defines the precipitation index (PI) in units of K as a inner product of the departure defined in (1) and the precipitation related eigenvector e_P. The derivation of the PI is also applicable to SSM/I or AMSR data.

Brightness temperatures (TB) at different wavelengths are usually positively correlated in particular over land surfaces where the background signal is rather strong. Only in case of scattering by precipitating ice the higher frequency measurements are negatively correlated to those at lower frequencies. The full set of available channels carries therefore some redundant information. Thus, the dimension of the input data set (9 for the TMI) may be reduced by means of principal components.

The collection of TB vectors from the observations was therefore decomposed into independent vectors by calculation of the covariance matrix of differences between the TBs and their background values (equation (1) in Fig. 1). From this a matrix of eigenvectors can be computed. The major advantage of EOFs is that they are orthogonal to each other in vector space while brightness temperatures are not. Moreover, the associated eigenvalues provide the contribution to the total variance in decreasing order. The first EOF represents the largest contribution, the second EOF the second largest and so on. In case of TMI TBs, the eigenvalues of the first three EOFs usually sum up to more than 98% of the total variability.

The idea of surface effect correction by generating data sets where either temperature or moisture dominate the signal and which are subtracted from the actual observation signal by orthogonalisation follows the approach by Conner and Petty (1998) (equation (2) in Fig. 1). Bauer et al. (2002) used radiative transfer simulations to derive the correction eigenvectors e_t and e_m. This approach is followed here because both eigenvectors are supposed to show only the signature of either temperature or moisture. If the correction data sets are constructed from measurements as in Conner and Petty (1998) then they may contain unscreened precipitation which would suppress the precipitation signal in later analysis. The used simplified emissivity model considers only effects due to variation in skin temperature, fractional cover of vegetation and water as well as surface roughness. Thus the derived static signature of temperature and moisture may not be representative for every situation encountered during its application. Especially in situations with extreme gradients between surface temperature and low level air temperature as well as in situations with surface roughness outside the assumed range 0–0.1 cm the correction may not work properly.

For the calibration of the PI into rain water contents a co-location procedure of TMI and PR pixel values which are recorded along different scan geometries and resolutions is performed. For the TMI a reference resolution was defined which is set to the resolution of the 19 GHz channel. The effective field of view (EFOV) is enlarged by a factor of 2.5 to cover the area of which ~98% of the signal is received. The EFOVs are approximated by ellipses and the antenna gain function by Gaussian functions following the orientation of the real EFOVs along the TMI scan. PR pixels at a height of 2 km with valid PR retrievals covering more than 80% of the TMI EFOV were used to minimise sampling problems. Before the averaging a parallax correction (Bauer et al., 2002) is applied if the centre of gravity of the weighting function gives values larger than 5 km. This is important because in those cases where the signal contribution originates from higher altitudes, the comparison to radar data at the TMI footprint location may lead to errors in the calibration data set.

Heidke skill scores were employed to quantify the accuracy of rain detection from both TMI and PR and to give a calibration tool. The skill scores are computed for classes of rain intensity with a lower threshold resulting in skill score diagrams as presented in Fig. 2 for one exemplary day. To build up statistics it is necessary to sample at least a week of data to fill the diagram which is important for a stable calibration. Thus skill score diagrams are computed for each day for 10°x10° regions employing the same 7 days of data used for the construction of the background map.

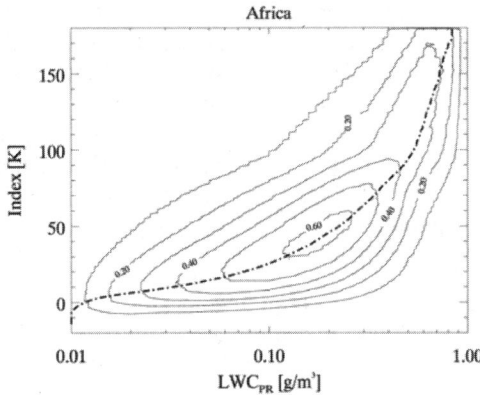

Fig. 2. Heidke Skill Score Diagram for one grid box covering the region 0°–10° E and 0°–10° N on 28 July 2000. The dash-dotted line indicates the used calibration function.

Comparing those diagrams for different regions exhibit some common features like there is almost no skill in determining rain water contents less than 0.03 gm^{-3}. The errors of the fit itself are varying with time between 0.02 gm^{-3} and 0.06 gm^{-3} where higher errors occur if singular events dominate the skill score diagram. To determine the calibration for the SSM/I the PR-TMI co-locations are still used but the PR pixel are averaged employing the viewing geometry of the SSM/I. Because of the different altitudes of the DMSP and the TRMM satellite the precipitation index for SSM/I is computed at 37 GHz ground resolution which match best with the 19 GHz resolution of the TMI.

The calibration procedure creates look-up tables for each grid box and each day that are applied to the PI index derived from the radiometers of different platforms. Because extreme values for precipitation seldom occur all TMI-PR co-locations indicating more than 20 mmh^{-1} rain are kept and used in each look-up table.

3. EXAMPLES AND UNCERTAINTY ESTIMATION

Fig. 3 (a–d) presents a typical example of the results obtained by using the polynomial fits to calibrate the PI into rainwater contents. The case represents the passage of a meso-scale convective cluster over southern parts of Northwest Africa. The measurement was taken in the decaying stage of most of the convective cells. Compared to the standard 2A12 version 5 product the calibrated PI shows a very similar spatial pattern at medium and high intensities but exhibit a negative bias in the convective cores of the meso-scale system. This systematic difference is introduced by using the PR as calibration source and is of the same size as described elsewhere for version 5 data, e.g., in Masunaga et al. (2002) or Kummerow et al. (2000). Additionally, the PI product shows more pixel-to-pixel variability and a higher rain area for low rain rates. The rainwater content derived assuming the SSM/I viewing geometry (Fig. 3d) does not differ very much from the original TMI-PI result.

However, it shows generally smaller maximum rainwater contents and a larger rain area especially at low rainwater contents which is caused by the coarser spatial resolution.

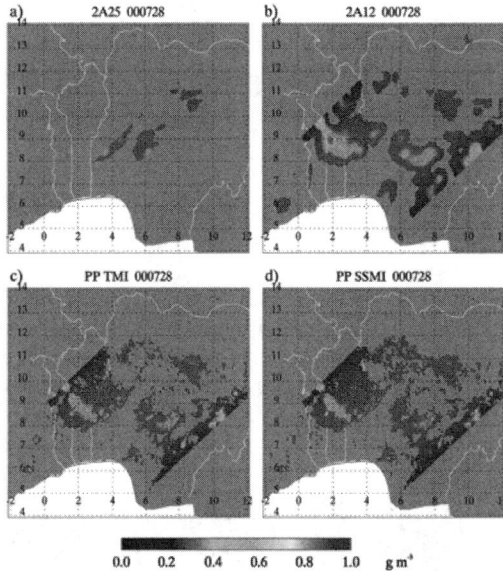

Fig. 3. Comparison of different rainwater contents (a) 2A25, (b) 2A12, (c) PP_TMI, and (d) PP_SSMI. All products are averaged to the TMI 19 GHz resolution (18 km x 30 km), except (d) which is at 37 GHz (37 km x 28 km). SSM/I resolution.

Fig. 4 (a–c) shows scatter plots of the TMI-PR derived rain water content w.r.t. the 2A25 product and the 2A12 product as well as a scatter plot of the 2A12 vs. the 2A25 product for the data of one day within one of the 10°x10° grid boxes corresponding to Fig. 3. Fig. 4d shows the binned RMS for both the 2A12 and the TMI-PR combination. The binned rms error for both considered products is in the order of 100% for rainwater contents lower than 0.1 gcm-3 and is decreasing to values between 50 and 25% for higher rainwater contents. The PI shows generally a slight improvement for the low rainwater contents and is of comparable quality for other rain water contents.

Finally, Fig. 5 gives an impression of the application of the algorithm to all available microwave radiometers for one day in July 2002. Fig. 5a presents a TMI overpass paired with a Meteosat 7 infrared brightness temperature whereas Fig. 5b shows a match between the SSM/I on DMSP-F13 and the Meteosat brightness temperature. In both images the microwave rain rates are well located w.r.t. to the coldest infrared brightness temperatures (light gray areas) as expected for convective clouds. With the aid of techniques by Joyce et al. (2004) and Turk et al. (2001) the microwave estimates can be integrated over a day to give reasonable daily rainfall estimates.

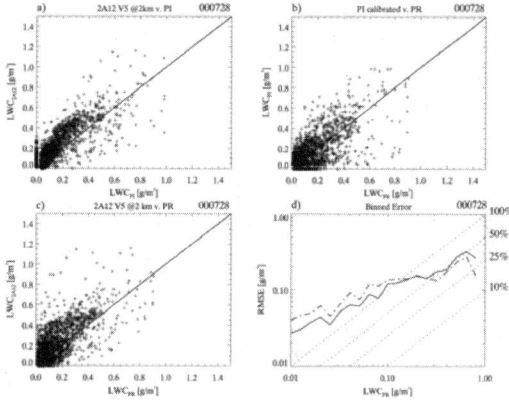

Fig. 4. Scatter plots of (a) calibrated PI estimates versus 2A12 V.5 estimates, (b) calibrated PI estimates versus averaged PR estimates, (c) 2A12 V.5 versus averaged PR estimates, and (d) remaining standard deviation between calibrated PI and averaged PR pixels for the same case as in Fig. 5.

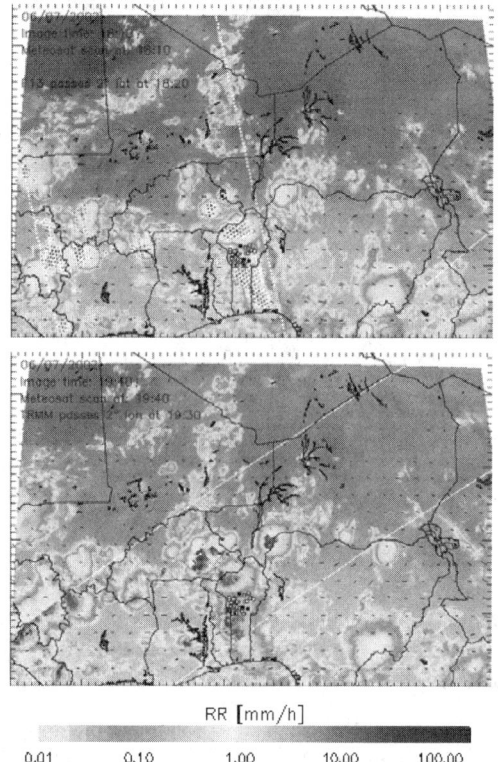

Fig. 5. Rain rate estimates from SSM/I on DMSP F13 (top panel) and TMI (lower panel) over Northwest Africa on 6 July 2002.

4. CONCLUSION

The presented rainfall retrieval algorithm shows detection skills comparable to the 2A12 version 5 product. RMS errors are slightly improved at low rain rates and are similar at high rain rates when compared to 2A12.

Although this algorithm is of empirical character it is competitive to current TRMM standard products. Because the cloud model database used for TRMM standard products is rather crude, the presented algorithm might not be competitive to future full parametric algorithms. However, the method is open for adding other radiometers like SSMIS and AMRS and can also be used within the GPM configuration.

The validation of the combined algorithm using surface based rainfall estimates is still an open task but data are rare and often of doubtful quality over Africa. The European AMMA project started in 2005 will give us more suitable data to do this type of validation.

5. ACKNOWLEDGEMENTS

The authors are grateful for free access to TRMM data through the NASA TRMM program. This research was supported by the Federal German Ministry of Education and Research (BMBF) under grant No. 07 GWK 02 and by the Ministry of Education, Science and Research (MSWF) of the federal state of Northrine-Westfalia under grant No. 514-21200200.

6. REFERENCES

Bauer, P., D. Burose, and J. Schulz, 2002: Rain detection over land surfaces using passive microwave satellite data. *Meteorologische Zeitschrift*, 11, 37-48.

Conner, M. D., and G. W. Petty, 1998: Validation and intercomparison of SSM/I rain-rate retrieval methods over the continental United States. *J. Appl. Meteor.*, 37, 679-700.

Ferraro, R. R., E. A. Smith, W. Berg, and G. J. Huffman, 1998: A screening methodology for passive microwave precipitation retrieval algorithms. *J. Atm. Sci.*, 55, 1583-1600.

Ferraro, R. R., and G. F. Marks, 1995: The development of SSM/I rain-rate retrieval algorithms using ground-based radar measurements. *J. Atm. Ocean. Technol.*, 12, 755-770.

Joyce, R. J., J. E. Janowiak, P. A. Arkin, and P. Xie, 2004: CMORPH: A method that produces global precipitation estimates from passive microwave and infrared data at high spatial and temporal resolution. *J. Hydrometeorology*, 5, 487-503.

Kummerow, C., et al., 2000: The status of the Tropical Rainfall Measuring Mission (TRMM) after two years in orbit. *J. Appl. Meteor.*, 39, 1965-1982.

Masunaga, H., T. Iguchi, R. Oki, and M. Kachi, 2002: Comparison of rainfall products derived from TRMM Microwave Imager and Precipitation Radar. *J. Appl. Meteor.*, 41, 849-862.

Turk, J., C.-S. Liou, S. Qui, R. Scofield, M. Ba, and A. Gruber, 2001: Capabilities and characteristics of rainfall estimates from geostationary and geostationary + microwave-based satellite techniques. Preprints. *Symp. of Precipitation Extremes: Prediction, Impacts, and Responses*, Albuquerque, NM, Amer. Meteor. Soc. 191-194.

PRECIPITATION RETRIEVAL FROM DUAL-VIEW SPACEBORNE PASSIVE MICROWAVE RADIOMETERS

F. Joseph Turk
Naval Research Laboratory, 7 Grace Hopper Ave., Monterey, California 93955 USA, turk@nrlmry.navy.mil

Sabatino Di Michele and Peter Bauer
European Centre for Medium-Range Weather Forecasts (ECMWF), Shinfield Park, RG2 9AX Reading, UK

Frank S. Marzano
Dipartimento di Ingegneria Elettrica, Università dell'Aquila, 67040 L'Aquila, Italy

Alberto Mugnai
Istituto di Scienze dell'Atmosfera e del Clima, Consiglio Nazionale delle Ricerche, 00044 Frascati, Italy

Laura Roberti
British Telecom, IP5 3RE Ipswich, UK

Alessandra Tassa
European Space Research Institute (ESRIN), 00044 Frascati, Italy

ABSTRACT

By design, conically-scanning passive microwave (PMW) observations from sensors onboard low Earth-orbiting (LEO) environmental satellites are engineered to provide observations using a nearly constant Earth incidence angle across the scan swath. This greatly simplifies the visual and analytical interpretation of these data over scenes of clouds and radiometrically polarized surface features such as lakes. For satellite-based retrievals of cloud and precipitation structure only view the upwelling microwave radiation from one direction, whereas the clouds themselves are three-dimensional. In this article we examine 3-dimensional (3-D) radiative transfer through mesoscale model output and simulate a PMW sensor with foreward and aft viewing capabilities. Comparisons are made with fore and aft-viewing data gathered by the Coriolis-Windsat sensor over precipitating clouds. Differences in precipitation retrieval results from the fore and aft data are compared with those from the model simulations in order to examine sources of precipitation retrieval error owing to the variable satellite viewing direction.

1. INTRODUCTION

Clouds are one of the most difficult atmospheric constituents to model owing to their detailed microphysical state, which is constantly changing in response to the cloud environment. Atmospheric cloud-resolving mesoscale models are capable of simulating the 3-D structure and dynamic time evolution of cloud structures such as mesoscale convective complexes and tropical cyclones. If the 3-D microphysical output is coupled as input to microwave radiative transfer models, the combination can be adapted to simulate the upwelling radiation along slant viewing paths, similar to a satellite perspective. This methodology has been employed for a number of recent studies (Di Michele et al., 2005), stimulated by the successful NASA Tropical Rainfall Measuring Mission (TRMM) and EOS-Aqua satellites, the latter carrying the Advanced Microwave Scanning Radiometer (AMSR-E).

The 3-D cloud structure affects the quantitative use of PMW data. Conically-scanning PMW radiometers view clouds along a slant path, thereby the instrument field-of-view (FOV) intersect a variety of rain, ice, cloud, snow and other hydrometeors. Plane parallel radiative transfer models cannot fully describe vertically extended clouds with finite sizes and horizontal inhomogeneities. Haferman et al. (1996) used a 3-D radiative transfer model to examine the effects of horizontally finite clouds upon precipitation retrieval techniques. Similar investigations were undertaken by Roberti et al. (1994) and Hong et al. (2000). Since PMW radiometers view clouds from a single direction, there is a certain degree of uncertainty in PMW-based precipitation retrievals owing to the unknown cloud geometry. However, the recently-deployed Coriolis low-Earth orbiting satellite with its WindSat scanning PMW radiometer has the capability to measure the full Stokes vector at three frequencies, as well as a unique fore-aft viewing capability to analyze upwelling radiation from two directions.

In this article, we examine mesoscale cloud model simulations using the University of Wisconsin Non-

Hydrostatic Mesoscale Modeling System (UW-NMS) and corresponding microwave radiative transfer simulations along forward and aft-viewing directions. Comparisons are made with several WindSat overpasses of tropical cyclones during autumn 2004. Using an emission-based precipitation retrieval algorithm (WindSat lacks an 85-GHz capability) (Ferraro et al., 1997) the differences between fore and aft retrievals of precipitation are perfomed. These results are in close agreement with those obtained from the model simulations, suggesting that the uncertainty in precipitation retrieval from conventional single-view PMW sensors may be estimated.

2. CLOUD/RADIATION MODELING

The UW-NMS mesoscale model carries cloud drops, rain drops, pristine crystals, aggregated crystals, snow, graupel. For the modeling described here, the simulations were run for simulation of Hurricane Bonnie which occurred in August 1998 (Di Michele et al., 2005). The horizontal grid resolution was set to 2.3-km with a 200 x 200 size, with averaging done to be commensurate with the TMI 37 GHz channel on-Earth resolution (10-km). The vertical dimension extended from 0.1 to 20 km in 39 vertical levels.

In order to simulate the fore-aft viewing observations available from WindSat, 3-D radiative transfer computations were performed using the backward Monte Carlo simulator originally developed by Roberti et al. (1994). This method follows received photons backwards through the medium in a probabilistic manner, whereby photons are either scattered or absorbed at each interaction. A simplified Henyey-Greenstein phase function is used to determine the direction of travel after each scattering event. The radiative properties of all hydrometeors are computed using Mie theory and exponential drop size distributions. The satellite zenith angle was fixed at 53° from nadir. Only vertically-polarized (V) modeling is shown for brevity. Fig. 1 depicts the cloud model geometry used to simulate upwelling equivalent blackbody brightness temperature (T_B) simulations along the 0° and 180° azimuth angle directions from Hurricane Bonnie. Fig. 2 shows the upwelling 37 and 85 GHz vertically polarized T_B results (Kelvin units) from one time step of the simulation, in this example for a azimuthal viewing angle of 270°. We next repeat the same radiative transfer calculations, but with an azimuthal angle of 90°. The differences between the 270° and 90° upwelling 37 and 85 GHz T_B are shown in Fig. 3. Note the alternating positive-negative bands that surround the eyewall structure. This occurs because the fore and aft view paths alternatively peer into lower (warm T_B) and high (cold T_B) portions of the cloud. These features have been analyzed and described by Hong et al. (2000).

The surface rainrate from the NMS model simulation is shown in Fig. 4. The rainrate retrievals were performed using the multichannel TB simulated 270° and 0° azimuth data, and differencing the images (left side of Fig. 4). The right panel of Fig. 4 shows the scatterplot of all rainrate pixels from this image (90° azimuth rainrates along the abscissa and 270° along the ordinate). Note that the standard deviation tends to increase with increasing rainrate. We will refer to this figure in the next section to compare against actual satellite observations.

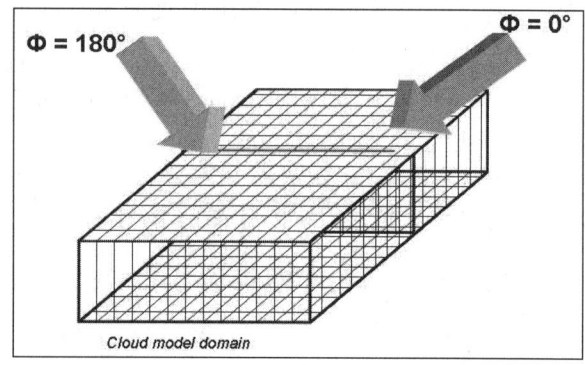

Fig. 1. Depiction of fore (0° azimuth) and aft (180° azimuth) viewing TB simulations from the UW-NMS model grid. Horizontal grid spacing is 2.3-km.

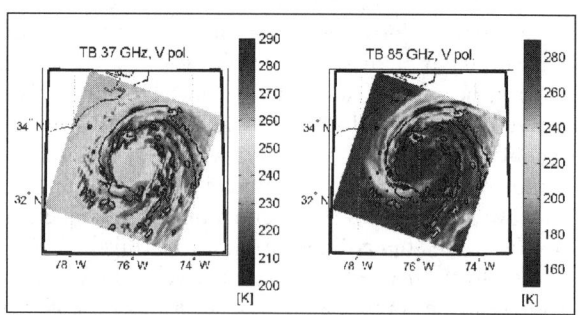

Fig. 2. Upwelling 37 and 85 GHz vertically polarized T_B results (Kelvin) from one time step of the hurricane simulation, for a azimuthal viewing angle of 270°.

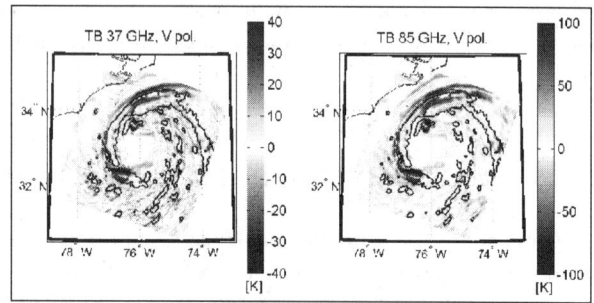

Fig. 3. Differences between the 270° and 90° azimuthal view for the upwelling 37 and 85 GHz vertically polarized T_B (Kelvin) from one time step of the hurricane simulation.

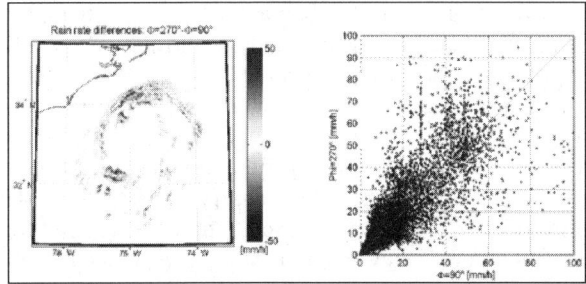

Fig. 4. (Left) Differences between the retrieved rainrate (mm hr^{-1}) from the 270° and 90° azimuthal view from one time step of the simulation. (Right) Scatterplot of all pixels from the image, 90° azimuth along the abscissa and 270° along the ordinate.

3. WINDSAT OBSERVATIONS

In January 2003, the launch of the Coriolis satellite deployed the WindSat sensor into a 833-km altitude low-Earth orbit. WindSat is a 19-channel conically scanning PMW radiometer designed by the Naval Research Laboratory (NRL) (Gaiser et al., 2004). Coriolis orbits in a sun-synchronous pattern with an approximate 6 PM local time of ascending node (LTAN), close to the crossing time of the NASA QuikScat satellite. The sensor has full Stokes vector measurements capabilities at 10.7, 18.7, and 37 GHz via direct measurement of vertical (V), horizontal (H), ±45°, left hand circular (LHC) and right hand circular (RHC) polarized radiation. The channels at 6.8 and 23.8 GHz have V/H measurement capability only. The specifications of the sensor are listed in Table 1. The fully-polarized channels were designed for PMW-based retrievals of ocean wind vectors, in support of a similar capability which will be available on the National Polar Orbiting Environmental Satellite System (NPOESS) Conical Microwave Imager and Sounder (CMIS) sensor. Unlike other PMW sensors, WindSat records data on both the forward (fore) and back (aft) side of each spin of the conically scanning feedhorn structure. Fig. 5 depicts the conical scanning mode of WindSat. The narrower aft scan data covers only a 350-km wide swath, so across a limited region of the fore scan the sensor gather dual views of the image scene separated in time by 4–5 minutes, depending upon scan position.

Fig. 6 depicts the WindSat 37H channel image from the ascending overpass over typhoon Nock-Ten centered at 0821 UTC on 22 October 2004. The right panel depicts the associated 37H GHz T$_B$ difference between the fore- and aft-viewing scans. The same alternating positive and negative TB bands as noted in the hurricane model simulation are present, wherever there are cloud edges.

To examine how the difference in the fore-aft WindSat viewing geometry depicted in Fig. 5 would impact a satellite-based precipitation retrieval algorithm, we have adapted the emission-based SSMI algorithm of Ferraro et al. (1997) to the 37, 23.8, and 18.7 GHz WindSat channels (the algorithm was designed for the 37.1, 22.235, and 19.35 GHz SSMI channels, so algorithm is not tuned for WindSat and only approximate). Using all over-ocean Windsat data between ±20 degrees latitude from 12 satellite orbits (≈ 18 hours) on 4 August 2004, the Windsat satellite data records (SDR) were processed through the emission-based algorithm separately for the fore and aft scan data. The rainrate differences are depicted in the scatterplots (fore vs. aft) of Fig. 7. As with the hurricane model simulations (right panel of Fig. 4), the standard deviation increases with increasing rain rate, approaching 50% difference between fore and aft retrievals near 25 mm hr^{-1}. Beyond this, the emission-only assumption begins to break down as larger ice particles in the radiometer FOV begin to affect the 37 GHz T$_B$. For rain rates under ≈ 5 mm hr^{-1}, the fore-aft WindSat standard deviation is larger and more variable than in the simulated fore-aft differences in Fig. 4.

There are several explanations for these model-vs.-satellite differences. One reason is the coarser scale of the WindSat data compared to the simulated TMI T$_B$ represented by the hurricane simulation radiative transfer calculations. In addition, the WindSat fore and aft pixels never exactly view the identical scene due to differences in on-Earth footprint orientation. Furthermore, scenes that are captured near swath edges have smaller relative azimuth angles (between fore and aft) and time offsets, whereas the model simulations assume a 180° azimuthal viewing difference and no time difference. Therefore, when the fore and aft WindSat data are differenced, there is a slight horizontal displacement that occurs due to natural storm advection (however its magnitude is much less than that owing to the radiometric differences due to the different viewing paths). Finally, if a predominant ocean surface wind direction is present, a certain amount of fore-aft difference over less optically thick clouds will be evident owing to surface effects. This will be especially true for the smaller rainrates under 5 mm hr^{-1}.

Table 1. WindSat Specifications. EIA = Effective Incidence Angle, LHC = Left Hand Circular, RHC = Right Hand Circular, IFOV = Instantaneous Field of View

Frequency (GHz)	Channels	EIA (deg)	IFOV (km)
6.8	V, H	53.5	40 x 60
10.7	V, H, ±45, LHC, RHC	49.9	25 x 38
18.7	V, H, ±45, LHC, RHC	55.3	16 x 27
23.8	V, H	53.0	12 x 20
37.0	V, H, ±45, LHC, RHC	53.0	8 x 13

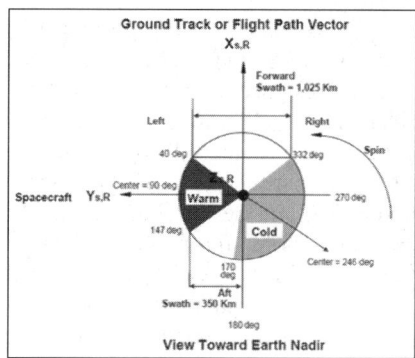

Fig. 5. WindSat scanning geometry. Figure adapted from Gaiser et al. (2004).

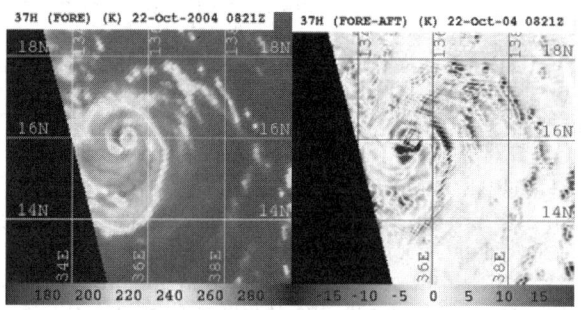

Fig. 6. (Left) 37H channel from the WindSat ascending overpass over typhoon Nock-Ten centered at 0821 UTC on 22 October 2004. Color scale is in units of T_B in Kelvin. (Right) Associated 37H GHz T_B difference between the fore- and aft-viewing scans.

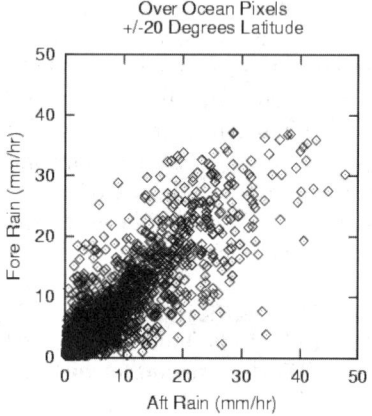

Fig. 7. Scatterplot between the WindSat-retrieved rainrate from the aft scan (abscissa) and the fore scan (ordinate) for all over-ocean Windsat data between ±20 degrees latitude from 12 satellite orbits on 4 August 2004.

4. CONCLUSIONS

In this article we used a combination of three-dimensional microwave radiative transfer simulations of a hurricane and WindSat fore-aft viewing satellite passive microwave radiometric observations in an attempt to explain and quantitatively analyze errors in satellite precipitation retrieval owing to unknown underlying cloud 3-dimensional geometry. Standard deviations of the rainrate under 25 mm hr^{-1} demonstrated considerable agreement, despite known differences between model simulations and the satellite observing system. The improved error estimates of precipitation from spaceborne observations are a key component of the planned Global Precipitation Mission (GPM) for later this decade. Dual-view measuring capabilities have been proposed for one of the proposed constellation members, the European GPM satellite (EGPM) satellite (Ingmann et al., 2003).

5. ACKNOWLEDGMENTS

This work was sponsored by the Office of Naval Research, Program Element (PE-0602435N) and the Oceanographer of the Navy through the program office at the PEO C4I&Space/PMW-180 (PE-0603207N), and the National Aeronautics and Space Administration Earth Sciences Division under grant NNG04HK11I.

6. REFERENCES

DiMichele, S., A. Tassa, A. Mugnai, F. S. Marzano, P. Bauer, and J. P. V. Poiares Baptista, 2005: Bayesian algorithm for microwave-based precipitation retrieval: Description and application to TMI measurements over ocean. *IEEE Trans. Geosci. Remote Sens.*, in press.

Ferraro, R. R., 1997: Special sensor microwave imager derived global rainfall estimates for climatological applications. *J. Geophys. Res.*, 102, D14, 16715-16735.

Gaiser, P. W., E. M. Twarog, Li Li, Karen, M. St. Germain, G. A. Poe, W. Purdy, Z. Jelenak, P. S. Chang, and L. Connor, 2004: The WindSat Spaceborne Polarimetric Microwave Radiometer: Sensor Description and Mission Overview. *Proc. of IGARSS 2004*, Anchorage, AK, USA, Sept. 20-24, CD-ROM.

Haferman, J., E. N. Anagnostou, D. Tsintikidis, W. F. Krajewski, and T. F. Smith, 1996: Physically based satellite retrieval of precipitation using 3D passive microwave radiative transfer model. *J. Atmos. Ocean. Tech.*, 13, 832-850.

Hong, Y., J. L. Haferman, W. S. Olson, and C. D. Kummerow, 2000: Microwave brightness temperatures from tilted convective systems. *J. Appl. Meteor.*, 39, 983-998.

Ingmann, P., 2003: The new Earth Explorer Opportunity missions in support of atmospheric science and application. *Proc. 2003 EUMETSAT Meteorological Satellite Users Conference*, 29 September-3 October, Weimar, Germany, 50-57.

Roberti, L., J. Haferman, and C. D. Kummerow, 1994: Microwave radiative transfer through horizontally inhomogeneous precipitating clouds. *J. Geophys. Res.*, 99, 16707-16718.

ANALYSIS OF RADIATION MEASUREMENTS AT THE COVE SEA PLATFORM

Zhonghai Jin[1], Thomas P. Charlock[2], and Ken Rutledge[1]

[1]Analytical Services and Materials, Inc., Hampton, Virginia 23666, USA
[2]NASA Langley Research Center, Hampton, Virginia 23681-2199, USA

ABSTRACT

Spectral and broadband solar radiation, along with atmospheric and oceanic optical properties, were measured at a sea platform. The measurement data are analyzed with the coupled atmosphere-ocean radiative transfer model. The results show that the ocean surface albedo (OSA) is highly variable and is sensitive to four physical parameters: solar zenith angle, wind speed, atmospheric transmission, and ocean chlorophyll concentration. An OSA look up table is created in terms of these four parameters. The result is a fast and accurate parameterization of OSA for radiative transfer and climate modeling.

1. INTRODUCTION

Several instruments were installed at the NASA's CERES Ocean Validation Experiment (COVE) site to measure a variety of radiative properties. These include broadband and narrowband downwelling and upwelling irradiances, and spectral radiances and irradiances. COVE is also a site for NASA's Aeronet to measure the atmospheric aerosol and for NOAA to measure the wind, temperature, precipitable water and ocean surface waves (Holben et al., 1998). Ocean optical properties were also measured for waters at COVE. COVE is located at 36.90°N and 75.71°W in the Atlantic Ocean, about 25 km off the coast of Virginia Beach, Virginia, of the United States. The comprehensive observations on the radiation and the physical and optical properties of atmosphere and ocean provide an excellent database for validation of radiative transfer models and remote sensing retrieval algorithms.

2. DATA ANALYSIS AND COMPARISON WITH MODEL

In this section, we will show some measurement data and compare them with those calculated by the Coupled Ocean-Atmosphere Radiative Transfer (COART) model. COART is based on the Coupled DIScrete Ordinate Radiative Transfer (CDISORT) code and evolved from Jin and Stamnes (1994). Many more new features have been implemented in the code since 1994. Because the radiative transfer equations now include the refractive index and the windblown ocean surface roughness, our solution for the coupled atmosphere-ocean system becomes consistent and rigorous. This enables COART to consider ocean layers as just additional atmospheric layers but with greatly different optical properties. COART treats absorption and scattering processes in the atmosphere and ocean explicitly. These include the scattering and absorption by atmospheric molecules, aerosols, and clouds in the atmosphere, and by water molecules, soluble (e.g., CDOM) and particulate (e.g., phytoplankton particles) materials in the ocean. Optical properties of aerosol in the atmosphere and of particulate and soluble materials in the ocean for model input here are also from in-situ measurement. COART calculates spectral or broad radiance and irradiance at any levels in the atmosphere and ocean, including the water-leaving radiance and OSA. More information on COART can be found online at http://www-cave/larc.nasa.gov/cave/.

Fig. 1 shows a model-observation comparison of the broadband shortwave fluxes from local noon to near sunset for a clear day. The morning data are not used because of shading of the platform. UTC 17:00 is the local noon when the solar elevation is highest. This figure shows that the solar zenith dependences for the downwelling and upwelling fluxes are very different. Unlike the downwelling flux, the surface upwelling flux for clear conditions does not decrease monotonically with solar elevation; the maximum upwelling flux is not at local noon but at around UTC 22:00 (17:00 local time). This is because the ocean surface albedo increases as solar elevation decreases, and this compensates for the decreased incidence at the surface due to a smaller solar elevation.

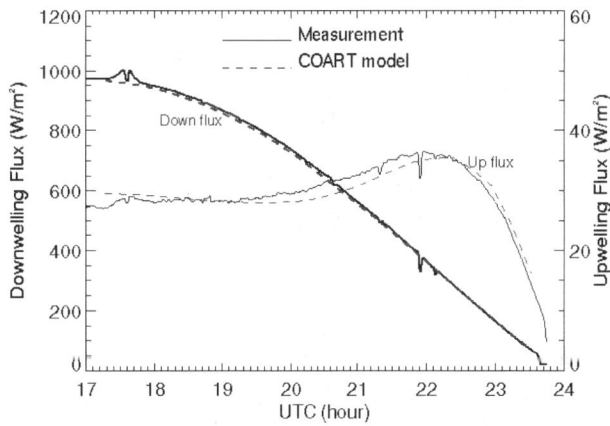

Fig. 1. Comparison of measured and modeled shortwave fluxes.

Fig. 2 shows the observed and modeled clear sky broadband OSA for two full years (2000–2001). The data are 30 minute averages; each was screened as cloud free with minute by minute measurements. The model captures the albedo variations very well.

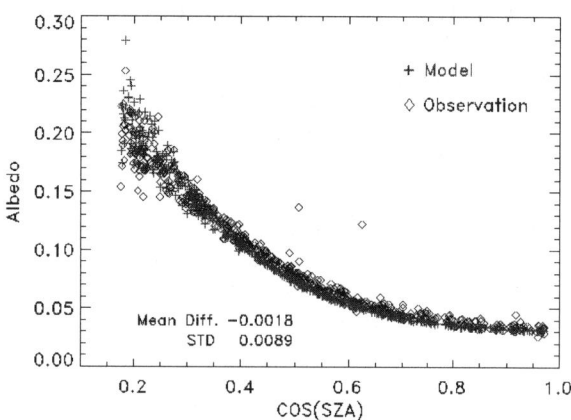

Fig. 2. Comparison of modeled and measured ocean surface albedo.

Fig. 3 compares the modeled and aircraft measured spectral albedo at COVE. In the visible spectrum, the albedo observed at altitude 3057 m is much higher than that observed at 20 m, indicating the pronounced effect of aerosol and Rayleigh scattering in the shortwave.

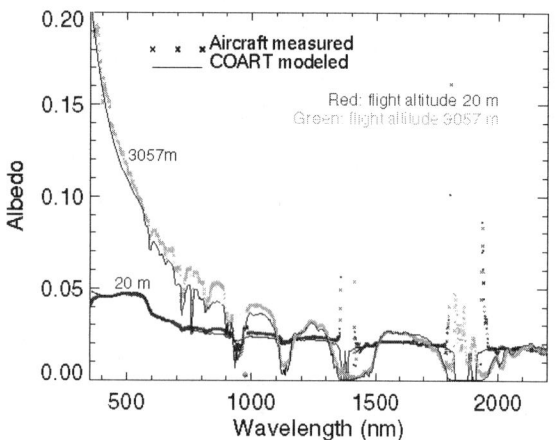

Fig. 3. Comparison of COART modeled and aircraft measured ocean surface albedo.

Fig. 4 shows a comparison of the modeled and CERES measured broadband shortwave radiances at the top of atmosphere. CERES observations were from NASA's Tera satellite. The CERES footprints selected were within 15 km of COVE and were cloud-screened. View zenith angles range from about 12° to 61°. Many of them fell into the sun-glint region, where the radiances are significantly larger due to specular reflection of the solar beam from the ocean surface. The general agreement between CERES observations and model results is fairly good. Except in sun-glint, the uncertainties in the aerosol optical properties used in the model calculations are likely the main source for the discrepancies between observed and computed radiances, because most of the TOA radiances are contributed by atmosphere instead of ocean. The spectral aerosol optical depth (AOD) used in the model is based on Cimel observations on a path between COVE and the sun. AOD from COVE to satellite may be different, especially if the view zenith angle is large, due to potential horizontal variability of aerosol. In addition, aerosol properties measured at COVE are limited to a few individual wavelengths instead of covering the whole solar spectrum CERES. The different surface coverage in size and location from different view angles also contribute to differences. More description on COART model and comprehensive validation of model using in situ, airborne, and satellite measurements are referred in Jin et al. (2005).

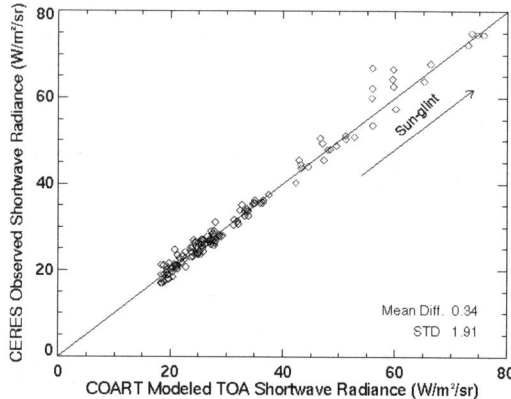

Fig. 4. Comparison of COART modeled and CERES measured TOA shortwave radiances.

3. OCEAN SURFACE ALBEDO

The OSA is a significant physical parameter controlling the solar radiative energy reflected back to atmosphere and penetrated into the ocean at the air-water interface. Hence OSA has various applications in satellite remote sensing, as well as in climate modeling. Most parameterizations for OSA have considered the solar zenith angle and wind effects (e.g., Hansen et al., 1983; Briegleb et al., 1986). However, atmospheric optical depth (of aerosols or clouds) and bulk ocean scattering (from below the sea surface) can have similar or even larger effects on ocean albedo, depending on solar zenith angle (SZA) and wavelength; but to date they have not been explicitly included in widely used parameterizations for OSA. Today, aerosol and cloud properties and ocean color data are readily available over the globe and their effects should be incorporated in the OSA parameterization.

Fig. 5 shows how the aerosol affects the OSA. The albedos were observed on two clear days from noon to late afternoon. Wind speeds during both days were small and similar, but the AODs (triangles) were very different. These measurements show that increasing AOD increases the albedo when the sun is high, but decreases albedo when the sun is low. The aerosol affects albedo primarily by altering the partition of direct and diffuse beams incident to the surface.

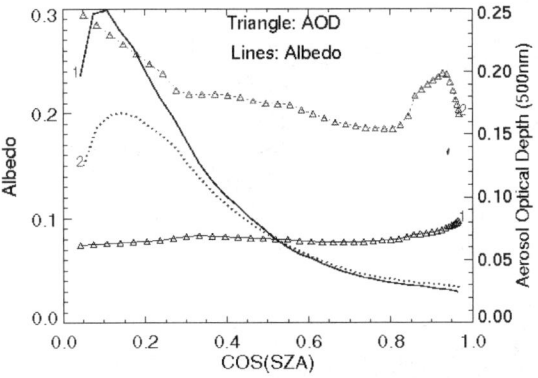

Fig. 5. Measured AOD and corresponding ocean surface albedo at COVE.

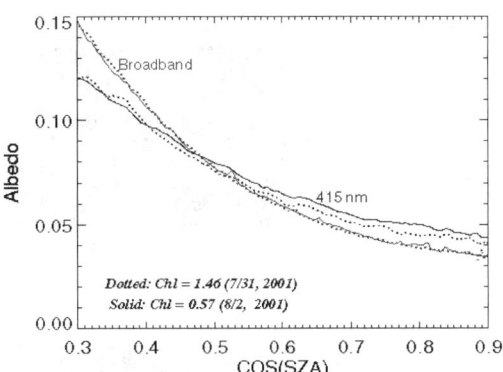

Fig. 7. Measured 415 nm albedo and broadband albedo on two clear days with different Chl in the water.

Similarly, Fig. 6 shows how the wind affects the OSA. The albedos were measured on three days having similar AODs but very different wind speeds (asterisks). These results indicate that the wind speed has small effect on albedo at high sun, but its effect increases as SZA increases. Wind affects OSA mainly by changing the slopes of surface wave facets.

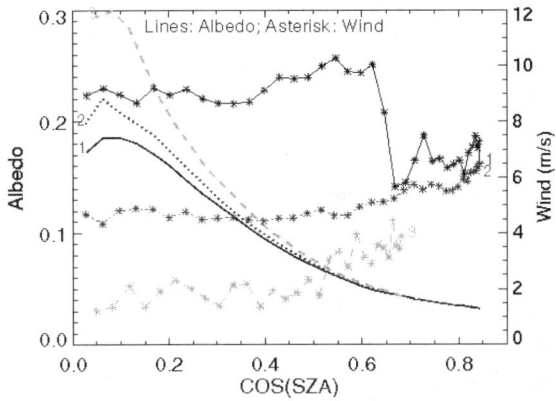

Fig. 6. Measured wind speed and corresponding ocean surface albedo at COVE.

Phytoplankton and associated products influence the ocean optics (Morel and Maritorena, 2001). Fig. 7 shows albedos for broadband and for 415 nm measured on two days; both days have small (and similar) wind speeds and AODs, but the chlorophyll concentrations (Chl in mg/m^3, the phytoplankton biomass) differ. With small SZA, the spectral albedo at 415 nm is smaller for the day with higher Chl, because high Chl causes more absorption. As SZA increases, the Chl effect on 415 nm albedo diminishes because the subsurface contribution falls, while the effect of Fresnel reflection from wave facets becomes dominant. The differences in broadband albedos between the two days are much smaller, however, because phytoplankton particles increase albedo in green and decrease it in blue.

Different applications require OSA for various atmospheric and oceanic conditions throughout the solar spectrum. While measurements are limited, our strategy is to use the validated model to create an albedo look up table (LUT) for various conditions; and span the whole spectrum having significant solar insolation at coarse but appropriate resolution.

Based on model calculations and measurements presented above, OSA is most sensitive to four parameters: SZA, wind speed, atmospheric transmission (represented by aerosol/cloud optical depth), and Chl. Other parameters, such as ozone, water vapor, and cloud height, have much smaller impacts on OSA under most conditions. Ocean sediments may also be important for OSA, but significant sediment loadings are constrained to some coastal regions. We tabulate OSA as a function of the aforementioned four parameters only.

The albedo table is available online at http://www-cave.larc.nasa.gov/cave/. For convenience, an interpolation code for the LUT is also attached. The code obtains the albedo using any combination of the four variables, for any discrete band within 0.25–4.0 μm specified by the user.

Fig. 8 displays the color contours of LUT values for broadband OSA versus the cosine of SZA and wind speed; with panels for a pristine atmosphere (AOD of 0.0), for a marine AOD of 1.0, and for cloud optical depths of 5.0 and 20.0. All panels in Fig. 8 assume Chl of 0.2 mg/m^3, about the average for the global ocean. Fig. 8 shows that the sensitivity of albedo to SZA decreases quickly as aerosol/cloud optical depth increases. The broadband OSA for a pristine sky (top-left panel) has a smaller minimum, and a larger maximum, than any other sky condition. While for a thick cloud sky (lower-right panel), the albedo is barely sensitive to SZA and slightly sensitive to optical depth, because the incident radiation is almost diffuse in this case. Note that for the moderate optical depths of 1.0 and 5.0, albedo does not vary monotonically with SZA. With no foam effect included, albedo decreases as wind increases (except for small SZA under very clear skies).

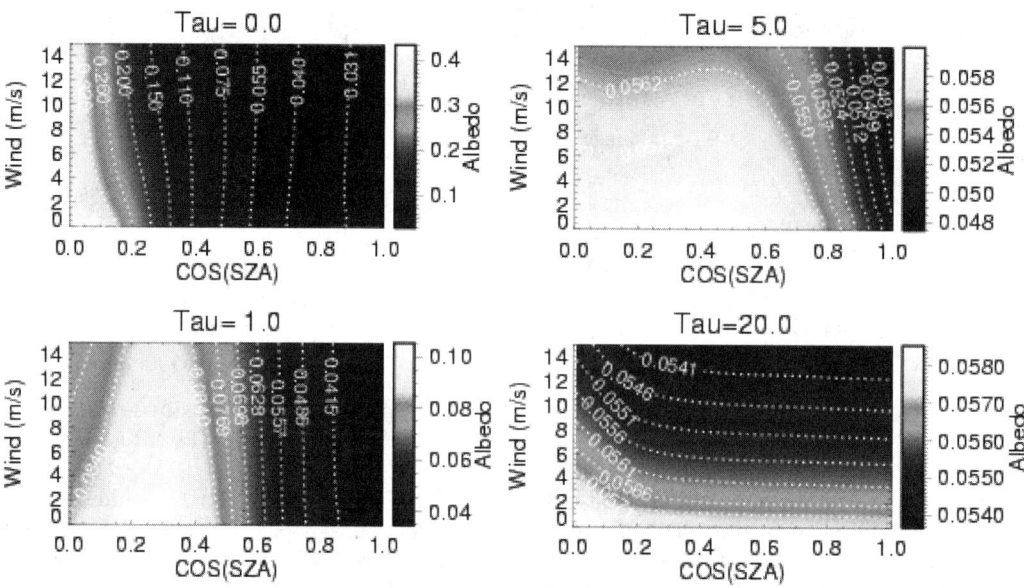

Fig. 8. Broadband ocean albedo versus wind speed and cosSZA using the LUT. Different color scale for each panel.

The wind effect on albedo is more significant for large SZA, but the SZA corresponding to the maximum wind effect depends on aerosol/cloud optical depth. These theoretical results are consistent with observations. The dependence of albedo on Chl in the LUT is also consistent with observations shown in Fig. 6.

4. CONCLUSION

- Broadband and spectral solar radiation have been carefully measured at COVE.
- Given the comprehensive measurements of atmospheric and oceanic properties as model input, the radiation measurements can be well simulated by the coupled radiative transfer model.
- Unlike the downwelling flux, the upwelling flux over ocean for clear conditions does not decrease monotonically with solar elevation, but reaches the maximum at SZA around 70°, depending on aerosol loading.
- Unlike most land surfaces, the OSA is very dynamic. It varies greatly with SZA, wind speed, atmospheric transmission, wavelength, and bulk ocean scattering.
- The clear sky broadband OSA could vary from around 0.03 to 0.4, but overcast albedo is almost constant.
- Increasing AOD will increase albedo at high sun but decrease albedo at low sun. The wind has little impact on the albedo at high sun but has a significant impact at low sun. The SZA corresponding to the maximum wind effect depends on atmospheric transmission.
- The ocean phytoplankton, indexed by the Chl, have a small effect on the broadband albedo but may change the spectral shape of ocean reflectance significantly.
- Based on measurements and model calculations, a ocean albedo LUT is created in terms of four important parameters. This provides a fast and accurate parameterization of ocean albedo for radiative transfer and climate modeling.

5. REFERENCES

Briegleb, B. P., P. Minnis, V. Ramanathan, and E. Harrison (1986), Comparison of regional clear-sky albedos inferred from satellite observations and model computations, *J. Climate Appl. Meteor.*, 25, 214-226.

Hansen, J., G. Russell, D. Rind, P. Stone, A. Lacis, S. Lebedeff, R. Ruedy, and L. Travis (1983), Efficient three-dimensional global models for climate studies: Models I and II. *M. Weather Rev.*, 111, 609-662.

Holben, B. N., and Coauthors (1998), Aeronet-A federated instrument network and data archive for aerosol characterization. *Remote Sens. Environ.*, 66, 1-16.

Jin, Z., and K. Stamnes (1994), Radiative transfer in nonuniformly refracting layered media: Atmosphere-ocean system. *Appl. Opt.*, 33, 431-442.

Jin, Z., T. P. Charlock, K. Rutledge, G. Cota, R. Kahn, J. Redemann, T. Zhang, D. Rutan, and F. Rose (2005), Radiative transfer modeling for the CLAMS experiment. In press in *J. Atmos. Sci.*

Morel A., and S. Maritorena (2001), Bio-optical properties of oceanic waters: a reappraisal. *J. Geophys. Res.*, 106, 7163-7180.

ANALYSIS OF CLOUD VARIABILITY AND SAMPLING ERRORS IN SURFACE AND SATELLITE MEASUREMENTS

Zhanqing Li*, M. C. Cribb, and Fu-Lung Chang
Department of Meteorology and ESSIC
University of Maryland, College Park, MD 20742, USA
*Email: zli@atmos.umd.edu

Alexander Trishchenko and Yi Luo
Applications Division, Canada Centre for Remote Sensing
Ottawa, Canada, K1A 0Y7

ABSTRACT

Ground-based radiation measurements are frequently used for validating the performance of a model in simulating clouds. Important questions are often raised such as: 1) how well do the measurements represent model grid mean values?, 2) how much of model-observation differences can be attributed to inherent sampling errors?, and 3) what scale does modeling need to be performed in order to capture the cloud variation? We attempt to address these questions using surface solar irradiance data retrieved from the Geostationary Operational Environmental Satellite and measured at the Atmospheric Radiation Measurement Southern Great Plains site. The satellite retrievals are used to mimic ground measurements with various spatial densities and temporal frequencies from which the sampling errors of the ground observations are quantified and characterized. In March 2000, for example, the sampling error is 16 Wm^{-2} for instantaneous irradiances averaged over an area of 10x10 km^2. It increases to 46 Wm^{-2} and 64 Wm^{-2} if the model grid size is enlarged to 200x200 km^2 and 400x400 km^2, respectively. The sampling uncertainties decrease rapidly as the time-averaging interval increases up to 24 hours, and then level off to a relatively small and stable value. Averaging over periods greater than 5 days reduces the error to a magnitude of less than 15 Wm^{-2} over all grid sizes. The sampling error also decreases as the number of ground stations increases inside a gridbox, but the most substantial reduction occurs as the number of ground sites increases from 1 to 2 or 3 for a grid size of 200x200 km^2.

1. INTRODUCTION

Radiation measurements have been widely employed for evaluating cloud parameterization schemes and model simulation results. The Atmospheric Radiation Measurement (ARM) program has an essential goal to make observations on the scale of a General Circulation Model (GCM) grid box, so as to define the physics underlying some of the important parameterizations in the GCMs used in climate change studies. While ARM has a network of radiation stations spread over a domain of 400x400 km^2, extensive radiation and cloud measurements are taken at a single location within a domain of 100x100 km^2. An important question is thus raised as to whether these measurements are adequate to represent grid mean values given the high variability of cloud and the surface. Fine-resolution cloud system models are playing an increasingly important role in revealing the fundamentals of cloud-radiation interactions and raise another question: on what scale does modeling need to occur in order to capture the physical properties that drive the system. Answers to both questions hinge on cloud variability and observation density (Li et al., 2005. Taking advantage of the high spatial and temporal resolution of Geostationary Operational Environmental Satellites (GOES) data, we mimic ground-based measurements of varying density and temporal frequency and characterize their observation uncertainties caused by cloud variability at different scales in different seasons. Such scale-dependent statistics of observation uncertainties provide critical constraints on model-observation comparisons, and are thus valuable for improving and validating cloud parameterization schemes.

2. MODELS AND INPUT DATA

The model of Li et al. (1993) was used to calculate the surface solar net irradiance (SSNI). Correction factors accounting for the effects of cloud-top altitude and aerosol loading, developed by Masuda et al. (1995), were applied to the SSNI. The top-of-the-atmosphere (TOA) albedo data are derived from high-resolution (~4-km pixel-level data) GOES data developed by NASA Langley's Cloud and Radiation Research Group. It encompasses a domain of approximately 400x400 km^2 centered on the Southern Great Plains (SGP) Central Facility (CF) site in north-central Oklahoma and produces output for about every half-hour on a daily basis. Data sets covering the periods of March, May, July, August, September, and October of the year 2000 were used in this study. The data set also provides cloud properties such as cloud-top height, pixel cloudiness and effective radius.

Input parameters to the models include the precipitable water content and aerosol optical depth at 550 nm. The former quantity was interpolated from the 2.5°x2.5° global National Centers for Environmental Prediction reanalysis (Kalnay et al., 1986) and the latter quantity was derived from multi-wavelength observations of aerosol optical depth taken at the CF (Holben et al., 2001). The optical depths at 550 nm generated from data at the CF were assumed to be representative of the whole SGP domain, given that aerosol loading is not expected to vary greatly over this generally rural area. Exceptions can occur during the spring season when localized increases in aerosol load can occur within the SGP domain due to the agricultural practice of burning fields in preparation for reseeding.

3. METHODOLOGY

Areal means of SSNI were calculated over domains of various spatial size and temporal intervals. The model of Li et al. (1993) was first applied to each individual pixel and the correction factors accounting for aerosol and cloud-top height on atmospheric absorption were added, depending on whether the pixel was identified as clear or cloudy. Areal mean net surface irradiance was calculated from all pixels within a region. Different domain sizes were selected centered on the CF. They were computed for every half-hour during the daytime for all the days when satellite data are available. Domain sizes were chosen to be representative of typical scales used in various single-column and GCM modeling schemes and range from 10x10 km^2 to 400x400 km^2. Averaging intervals chosen in this study were 1, 2, 4, 8 hours, together with daily means as well as 5 and 10-day means, and monthly means.

4. RESULTS

To ensure that the satellite-inferred SSNI can reasonably reproduce surface measurements, satellite-retrieved values are first validated against ground-based observations. Mean surface net irradiance was computed over an area of 4x4 km^2 centered on the CF and averaged over one hour. These mean fluxes were compared to ground measurements averaged over an hour (Fig. 1). Observed surface net irradiance was derived from measurements of downwelling surfaces fluxes from the Solar Infrared Radiation Station and measurements of broadband surface albedo from upward and downward pointing radiometers deployed at the CF. In general, the two sets of data agree fairly well, especially in terms of relative differences. Biases range from 2.2 Wm^{-2} to 27.0 Wm^{-2} and the root-mean-square error (RMSE) from 30 to 48 Wm^{-2} with the smallest biases occurring in March and the smallest RMSE in September. Seasonable changes in surface albedo may play an important role in the biases, while the RMSE is dictated primarily by cloud variability. To a large extent, the scatter in the plots is caused by the mismatch of satellite estimates and ground measurements in time and space. To gain further insight into the discrepancies related to data sampling, a comparison was made between one retrieval (Ret1- the mean of the satellite retrievals over a spatial domain of 400x400 km^2) and the mean of the surface observations made at all radiation stations within this domain. While the biases are generally similar to those shown in Fig. 1, the RMSEs are reduced substantially, ranging from 19 to 24 Wm^{-2}. The reduced RMSEs, albeit still significant, is attributed almost exclusively to the sampling uncertainties. This is clearly seen from comparisons between Ret1 and the mean of the retrievals over 4x4 km^2 areas surrounding each radiation station located within the domain. Since these two data sets were all retrieved from satellite data using the same algorithm, their differences attest to the sampling uncertainties. The fact that the RMSE values are similar indicates that the bulk of scattering in the satellite-surface comparisons originated from a combination of sampling and cloud variability. This finding suggests a useful tool to investigate observation sampling errors by *simulating* ground observations and comparing them with areal mean fluxes using satellite retrieval data alone.

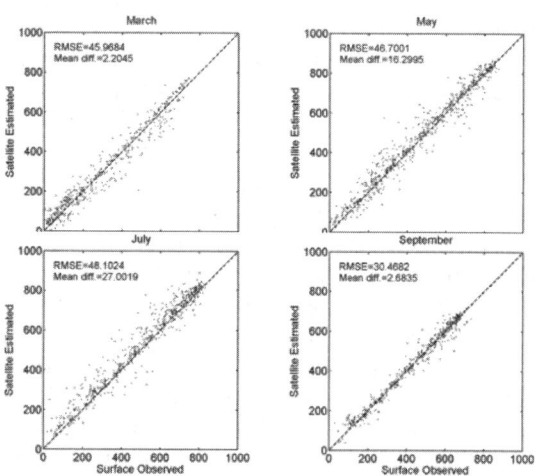

Fig. 1. Comparisons of surface net radiation observed at the CF and estimated from satellite for March, May, July, and September of 2000.

4.1 Spatial and Temporal Averaging

We first investigate sampling uncertainties incurred by using single-point data to represent a gridbox of varying size. Such single-point measurements have been widely used in validating GCMs with domain sizes of ~ 200x200 km^2. Point measurements of surface net irradiance at the CF were simulated and compared with areal means in order to evaluate their representativeness over regions of varying spatial scales. Fig. 2 shows an example of the March 2000 comparisons of satellite-estimated net SSNI averaged over a 4-km gridbox surrounding the CF (a proxy for ground observations) against those retrieved over areas of 10x10 km^2 to 400x400 km^2. As the domain size increases, the

RMSE increases dramatically; more variability is incurred by changes in the cloud conditions. For a cloud-resolving model with a 10x10 km^2 gridbox, the best accuracy one may achieve would be 16 Wm^{-2} to within the hourly ground-based observations, which increases to 46 and 64 Wm^{-2} for a GCM model of 200x200 and 400x400 km^2, respectively. A similar trend exists for the other months. The RMSE also changes with temporal averaging intervals and cloud regimes. The latter is echoed partially in the comparisons for different months. Fig. 3 shows the RMSE as a function of spatial domain size for various averaging intervals.

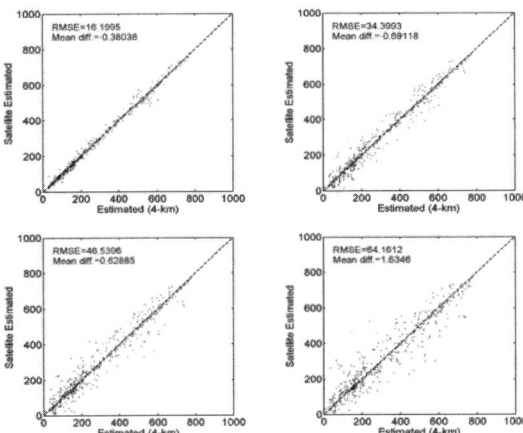

Fig. 2. Modelled surface net irradiance (averaged over 1-hr and over 10x10 (upper left), 100x100 (upper right), 200x200 (bottom left), 400x400 (bottom right) km^2) as a function of modelled surface net radiation (averaged over 1-hr and over a 4x4 km^2 domain) for March 2000.

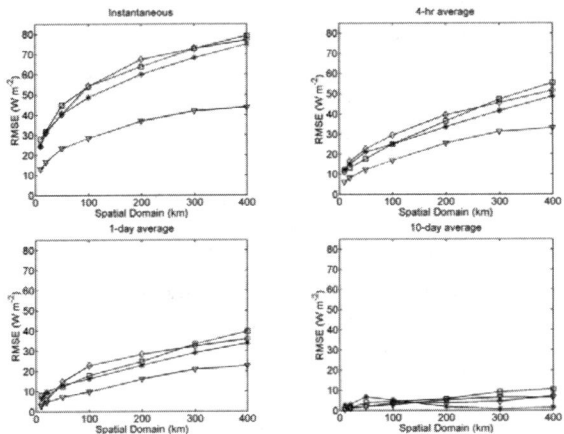

Fig. 3. Standard deviation as a function of domain size for the SSNI differences between satellite-simulated "point measurements" and areal mean values over domains of varying size and different averaging intervals: March (stars), May (squares), July (diamonds), and September (triangles).

For averaging under a day and for a 400x400 km^2 gridbox, the RMSE for September are less than those for other months by about 30 Wm^{-2}. July is the most variable month, presumably caused by the prevalence of small convective clouds. As the averaging interval increases, the difference in RMSE among different months diminishes. For a one-day averaging interval and for typical GCM gridboxes of 200x200 km^2, the daily sampling error ranges from 16 Wm^{-2} to 28 Wm^{-2} in September and July, respectively; they decrease to 10 Wm^{-2} or less if the model grid is reduced to 10x10 km^2. For the 10-day average, however, the differences among the months almost vanish.

Fig. 4 shows the difference between the monthly mean SSNI over different domain sizes and the monthly mean surface net irradiance measured at the CF for the months of March, May, July, and September. The magnitude of the inherent errors due to sampling in surface net irradiance is less than 10 Wm^{-2} for all months and domain sizes and the error diminishes to less than 3 Wm^{-2} for typical GCM grids of 200x200 km^2 or less. While this number agrees well with the general requirement of 5 Wm^{-2} for climate studies (Suttles and Ohring, 1986), it could be an unrealistic goal for regions/seasons, such as September 2000 over the larger 400x400 km^2 SGP domain. For this month, the inherent sampling error in monthly mean surface observations is 8 Wm^{-2}, substantially exceeding the required 5 Wm^{-2} unless multiple stations are deployed.

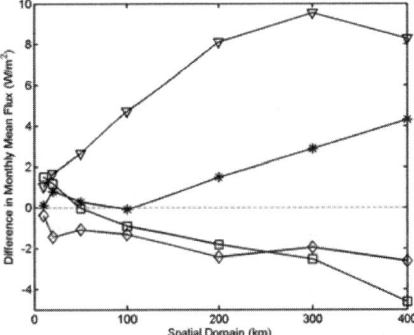

Fig. 4. Difference between monthly mean SSNI simulated for "point measurements" and areal mean values as a function of domain size varying from 10x10 km^2 to 400x400 km^2 for March (stars), May (squares), July (diamonds), and September (triangles).

4.2 Multiple Stations

Since single-point measurements do not represent well areal means over a large domain, multiple radiation stations distributed over a large domain around the CF may improve the spatial representation of the surface net radiation. General improvement is expected as more observations capture more of the variability in the SSNI, but the question remains as to how much improvement is gained as the number of stations increases, and how many stations are really needed to meet certain accuracy requirements. To address these questions, domains of different sizes centered on the CF were selected: 100x100, 200x200, 300x300, and 400x400 km^2, containing 1, 7, 12, and 21 radiation stations, respectively. A diagram of the domains used and the locations of the stations are given in Fig. 5. For each particular domain size, satellite-estimated surface net irradiance was calculated within a 4-km gridbox centered on each site as a proxy for actual surface measurements. The mean values averaged over all sites inside a particular domain are compared to the satellite-estimated areal means over the entire domain. Fig. 5 shows that the sampling error is the largest for the 100x100 km^2 gridbox in which there is only one station (the CF). The magnitudes of the RMSE for other gridbox sizes vary significantly with month but in general, there is a decrease in the RMSE as more observation sites are included in the calculations over the increasingly larger domain sizes.

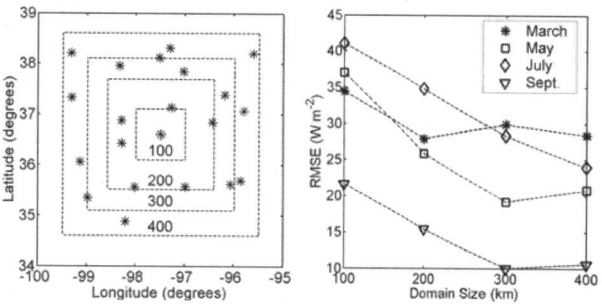

Fig. 5. Left panel: Location of radiation stations within the GOES domain with the Central Facility at the center. Dashed boxes from small to large represent grid sizes of 100x100, 200x200, 300x300, and 400x400 km^2, respectively. Right panel: Error in surface net irradiance (between satellite-estimated areal mean and observed) as a function of domain size where the number of observation sites increases with increasing domain size.

5. SUMMARY

Radiation measurements have been widely employed for evaluating cloud parameterization schemes and model simulation results. In this study, we take advantage of the high spatial and temporal resolution of the Geostationary Operational Environmental Satellite dataset of cloud properties to mimic ground-based measurements of surface net irradiance of varying density and temporal frequency and characterize their observation uncertainties caused by cloud variability at different scales in different seasons. Such scale-dependent statistics of observation uncertainties provide critical constraints on model-observation comparisons, and are thus valuable for improving and validating cloud parameterization schemes. In terms of spatial averaging, a single observation site (the Central Facility) does an increasingly poor job of representing areal means of surface net radiation as the domain size increases. Averaging the surface net radiation at more observation sites results in a decrease in error as the domain size (and number of observation sites) increases. As for temporal averaging, increasing the time interval over which means of surface net irradiance are taken leads to a general decrease in error for all domain sizes and all seasons. Instantaneous measurements incur the greatest error in surface net radiation. Some ramifications for climate and cloud modeling studies are that for fine-scale models with short integration intervals, such as cloud-resolving models, point measurements at a single site like the Central Facility do not provide the best representation of radiative quantities. For large-scale models such as GCMs, use of multiple observation sites does a reasonable job of capturing the surface net irradiance field over a large domain. When modeled radiation quantities are compared against ground observations, the inherent uncertainties due to sampling errors must be taken into consideration. Such inherent uncertainties are simulated from our satellite retrievals, which are given as functions of model domain size, averaging periods, number of observation stations, and month. If the difference between modeled and observed radiation quantities is comparable or less than the corresponding inherent uncertainty, no insight may be gained with regard to the model's performance. Such statistics are thus valuable for validating models when testing their parameterization schemes.

6. REFERENCES

Holben, B. N., et al., 2001: An emerging ground-based aerosol climatology: Aerosol optical depth from AERONET. *J. Geophys. Res.,* 106, 12067-12097.

Kalnay, E., et al., 1996: The NCEP/NCAR reanalysis 40-year project. *Bull. Amer. Meteor. Soc.,* 77, 437-471.

Li, Z., H. G. Leighton, K. Masuda, and T. Takashima, 1993: Estimation of SW flux absorbed at the surface from TOA reflected flux. *J. Climate,* 6, 317-330.

Li, Z., M. Cribb, F.-L. Chang, A. Trishchenko, and L. Yi, 2005 Natural variability and sampling errors in solar radiation measurements for model validation over the ARM/SGP region, *J. Geophy. Res.* (in press).

Masuda, K., H. G. Leighton, and Z. Li, 1995: A new parameterization for the determination of solar flux absorbed at the surface from satellite measurements. *J. Climate,* 8, 1615-1629.

Suttles, J. T., and G. Ohring, 19986: Report of the workshop on surface radiation budget for climate applications. *Technical Report WCP-115,* World Meteorological Organization.

UV-RADIATION AND CLOUDS: FIRST RESULTS FROM THE INSPECTRO PROJECT

Ronald Scheirer and Bernhard Mayer
Institut für Physik der Atmosphäre, German Aerospace Center
D-82234 Wessling, Germany
Email: ronald.scheirer@dlr.de

Sebastian Schmidt
Leibniz-Institut für Troposphärenforschung
D-04318 Leipzig, Germany

ABSTRACT

Photochemical processes are primarily driven by the actinic flux in the ultraviolet wave-range. In turn the abundance and spectral distribution of UV actinic flux depends on the atmospheric composition itself. This is particularly pronounced for clouds. In order to understand the qualitative and quantitative connection between atmospheric optical properties and tropospheric photochemical processes the EC project INSPECTRO (Influence of clouds on the spectral actinic flux in the lower troposphere) was set up. In September 2002 and May 2004 two field campaigns took place. The data set collected during these campaigns provides a powerful base for the validation of actinic flux parameterizations throughout the troposphere. Here we show how aircraft measurements of liquid water content and equivalent (or effective) radius were extrapolated to a complete 3D field, used as input for 3D radiative transfer calculations.

1. INTRODUCTION

The formation of ozone due to NO_2 photolysis is one of the most important processes in Two of the essential processes in tropospheric photochemistry (Thompson and Steward, 1991). In a simplified view this reads:

$$NO_2 + O_2 + h\nu = O_3 + NO. \qquad (1)$$

This equilibrium is controlled mainly by the ultraviolet (UV) radiation (hn, equation), more precisely the spherically integrated radiance or actinic flux (Madronich, 1987). The abundance and spectral distribution of UV actinic flux is strongly modulated by the atmospheric composition, in particular, by clouds which are highly variable in space and time. Clouds have the capability to decrease or increase the local actinic flux in comparison to the clear-sky measurements.

The EC project INSPECTRO aims at gaining a qualitative and quantitative understanding of the relationship between clouds and the actinic flux.

In September 2002 a first field campaign took place at Norwich, East Anglia, UK and a second one in May 2004 in a rural area close to Munich, Bavaria, Southern Germany. The collected data provide a comprehensive set of measured aerosols, cloud microphysics, corresponding actinic fluxes, and chemical radicals to set up a solid experimental base for the theoretical models.

2. FIELD CAMPAIGNS

The key-features of the two field campaigns are listed in Table 1. The measurement site of the first campaign was close to the coast. During the whole experimental time we found pristine air with negligible aerosol concentrations. Only periods without cirrus clouds were selected to keep the theoretic calculations both realistic and feasible. The topography was homogeneous (except for the coastline) and flat.

In contrast the area of the second campaign was large and inhomogeneous. The river Danube divides the area into two almost equally sized parts, from the northwesterly to the southeasterly corner. While the landscape is nearly flat within the southwesterly triangle it becomes hilly with more and more mountains towards northeast. Consequently, it was not expected to find homogeneous cloud conditions within the entire domain as it would be the case in a chemistry model box. Typically for this region and season, the dominating cloud type during the day is broken or scattered cumuli.

Table 1. A Brief Overview of the Experimental Phase of INSPECTRO

	First Campaign	Second Campaign
Date	September 2002	May 2004
Location	East Anglia, UK	Bavaria, Germany
Area Size	20 X 20 km^2	50 X 50 km^2
Aircraft	2 Partenavia	2 Partenavia
	1 Cessna	1 Cessna
	1 Hot Air Balloon	1 Tethered Balloon
	1 Ultralight	
Surface	4 Groundstations	5 Groundstations

3. NEEDS FOR RADIATIVE TRANSFER CALCULATIONS

In order to obtain a 3D radiation field, optical properties of relevant atmospheric components have to be derived first. These are:

1. Profiles of gaseous constituents (1D)
2. Underlying surface (0D / 2D)
3. Cloud distribution and properties (3D)

3.1 Atmospheric Profile

The background atmosphere incorporates all the radiatively relevant gaseous components. Within the given domains it is a reasonable assumption to consider the molecular composition as horizontal homogeneous (Scheirer and Macke, 2000). Thus a standard profile for the mid-latitude summer (McClatchey et al., 1972) was chosen.

3.2 Surface Reflection

In general the surface reflection in the UV is very small (around 3%); hence variations between different surface types have little influence on the radiation field. A constant lambertian surface reflection is therefore sufficient for calculations of actinic fluxes in the UV.

For the calculation of one case study (including radiance simulations) at September 13th 2002, the approach was different. Amount and direction of photons reflected from the surface were exemplarily defined by bi-directional reflectance distribution functions (BRDFs). These BRDFs depend on the land-use and the time dependent normalized differential vegetation index (NDVI). A data set from the polarization and directionality of the earth's reflectances (POLDER) instrument is available for June 1997 (Lacaze, 2003). These reflection data were parameterized following the suggestion of Rahman et al. (1993). The test area around Norwich was attributed to 12 land-use classes using the Enhanced Thematic Mapper (ETM+) on board LANDSAT7. The NDVI data were derived from AVHRR data.

3.3 Cloud Fields

To get a realistic (as far as possible) distribution of cloud optical properties during the experiment, an automated algorithm CLABAUTAIR (**cl**oud liquid w**a**ter content and effective radius retrieval **b**y an **au**tomated use of **air**craft measurements) was developed which scans aircraft measurements for characteristic patterns and extrapolates them to the whole 3D field (Scheirer and Schmidt, 2004). Fig. 1 shows a sketch of how CLABAUTAIR works. First, the main sampling directions are identified. Along these, autocorrelation functions are calculated. To extrapolate the observations, individual empty boxes are selected at random. Liquid water content and equivalent radius of each single box are calculated by a weighted (according to the autocorrelation coefficient) mean over the already filled boxes in the main directions. This step is repeated until all boxes are filled.

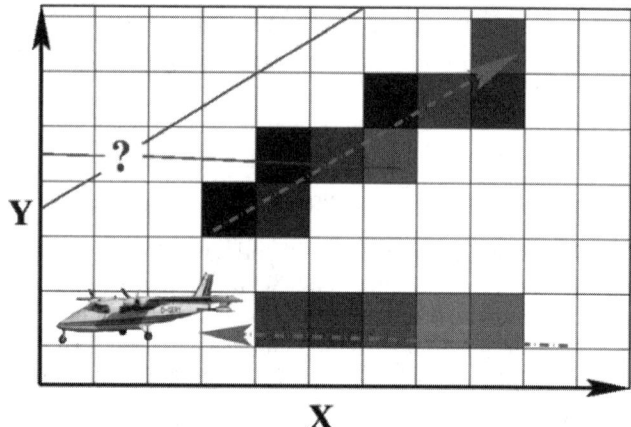

Fig. 1. Sketch of the functional principle of CLABAUTAIR.

4. RADIATIVE TRANSFER

To complete the preparation for the radiative transfer calculations the input data described in 3.1 to 3.3 were combined. For the calculation of radiation fields the MYSTIC (**M**onte Carlo code for the ph**y**sically correct **t**racing of photons **i**n **c**loudy atmosphere) solver (Mayer, 2000), operated as part of the libRadtran model (Mayer and Kylling, 2005), was used.

5. RESULTS

For the broken cloud case of September 13, 2002, aircraft measurements within the cloud layer were extremely sparse due to air traffic safety reasons. Therefore CLABAUTAIR was used to generate a horizontal slab-like cloud layer with small variations of the microphysics. A cloud mask obtained from a concurrent LANDSAT observation (cf. Fig. 3, left) was then used to define cloudy and cloud-free columns. The geometrical vertical cloud extent is horizontally constant and set to 80 m (cloud bottom is found at 900 m and cloud top at 980 m), separated into four layers with 20 m thickness each.

The total actinic flux within these layers resulting from the radiative transfer calculations is shown in Fig. 2. Note that for all four parts of this figure the grey scale is the same to retain the intercomparability among the figures. The lighter the grey—the larger the actinic flux. The largest

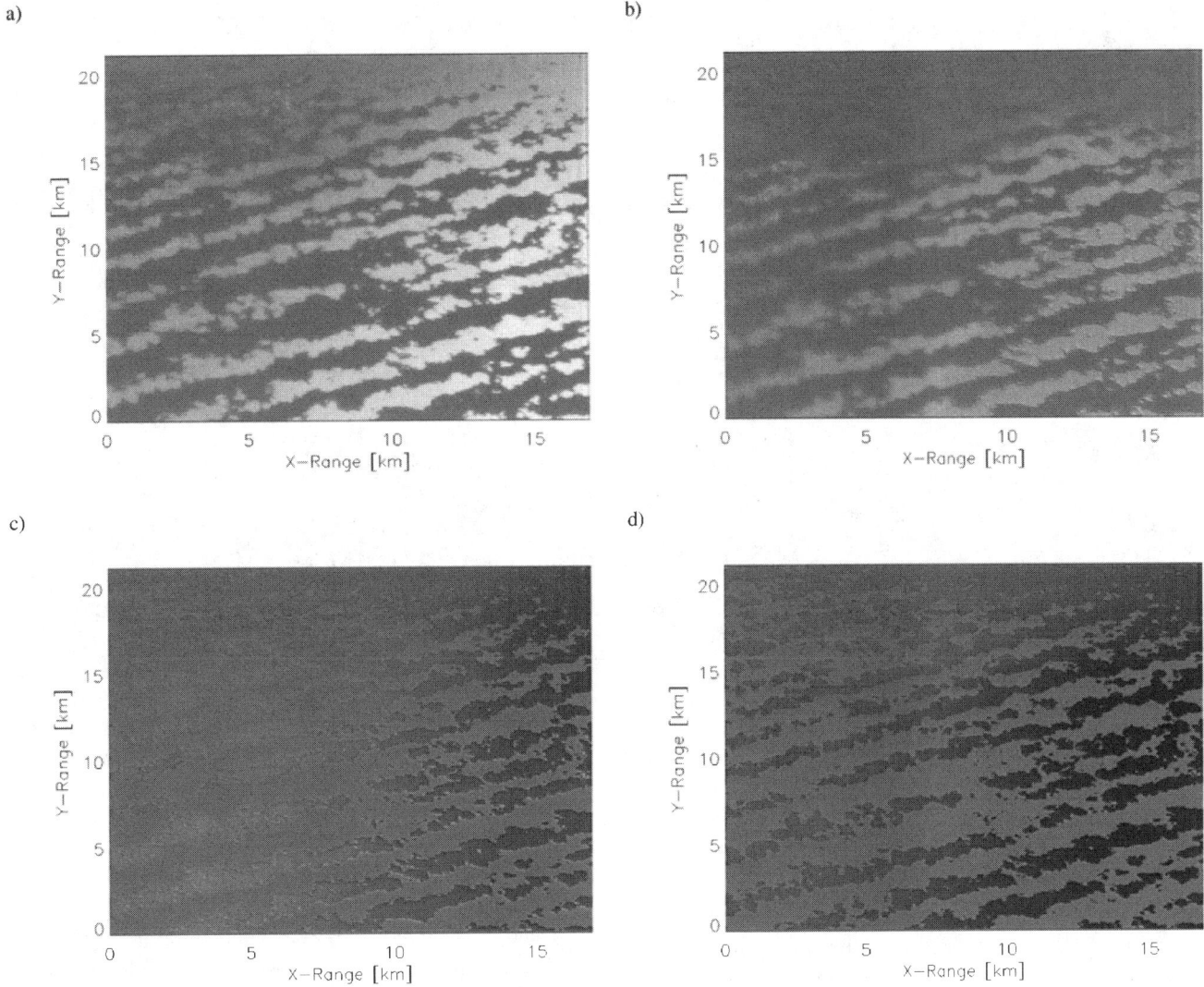

Fig. 2. Actinic fluxes within the four cloudy layers. 960 m – 980 m (a), 940 m – 960 m (b), 920 m – 940 m (c), and 900 m – 920 m (d).

values of actinic flux are found in the uppermost layer (Fig. 2a). Those appear within the clouds and the actinic flux is increasing with increasing optical thickness. In the layer between 940 m and 960 m it becomes difficult to distinguish optically thinner but cloudy regions from cloud-free parts. Illuminated cloud sides of optically thick regions provide the largest values. In the next lower layer the simple relation between optical thickness and actinic flux is lost. The fourth layer (Fig. 2d) shows a distribution with low actinic fluxes in cloudy columns and higher fluxes in cloud-free regions. This is the layer with the lowest flux on average. Note that regions with lowest values at cloud bottom correspond to those with the highest values in the top layer. At cloud top the in-cloud actinic flux is considerably enhanced compared to the value in the central layer, while at cloud bottom the situation is reversed.

The left hand side of Fig. 3 shows a LANDSAT7 scene. On the right hand side simulated top-of-the-atmosphere radiances are shown. The lack of structure in the simulated radiance is due to the lack of cloud top variations which were not available from these types of observation. It is not surprising that position and amount of clouds fit well for both figures, because the clouds used in the simulation were derived directly from this ETM+ scene, but consistence in this scene affirms that further investigations based on these simulations are likely to be very realistic.

Fig. 3. Comparison of the measured radiances by ETM+ (left) with the simulated radiances by MYSTIC (right) vs. spatial resolution and SZA.

6. ACKNOWLEDGMENTS

This work was funded by the European project INSPECTRO, contract EVK2-2001-00135.

7. REFERENCES

Lacaze, R. 2003: POLDER-1 BRDF DATABASE – User Document, Medias France, CNES BPI 2102, see http://smsc.cnes.fr/POLDER/A_produits_scie.htm.

Madronich, S., 1987: Photodissociation in the atmosphere. 1. Actinic flux and the effects of ground reflections and clouds, *J. Geophys. Res.*, 92, D8, 9740-9752.

Mayer, B., 2000: I3RC phase 2 results from the MYSTIC Monte Carlo model. Extended abstract for the I3RC (Intercomparison of 3D radiation codes) workshop, Tucson, Arizona, November 15-17, available at http://i3rc.gsfc.nasa.gov.

Mayer, B., and A. Kylling, 2005: Technical Note: The libRadtran software package for radiative transfer calculations: Description and examples of use, *Atmos. Chem. Phys. Diss.*, accepted.

McClatchey, R. A., R. W. Fenn, J. E. A. Selby, F. E. Volz, and J. S. Garing, 1972: Optical properties of the atmosphere (Third Edition), Publ. AFCRL-72-0497, Air Force Cambridge Research Lab., Hanscom, 108 pp.

Rahman, H., B. Pinty, and M. M. Vertraete, 1993: Coupled Surface-Atmosphere Reflectance (CSAR) Model 2. Semiempirical Surface Model Usable With NOAA Advanced Very High Resolution Radiometer Data, *J. Geophys. Res.*, 98, D11, 20791-20801.

Scheirer, R., and A. Macke, 2000: Influence of the Gaseous Atmosphere on Solar Fluxes of Inhomogeneous Clouds, *Phys. Chem. Earth* (B), 25(2), 73-76.

Scheirer, R., and S. Schmidt, 2004: CLABAUTAIR: a new algorithm for retrieving three-dimensional cloud structure from airborne microphysical measurements, *Atmos. Chem. Phys. Diss.*, 4, sref 1680-7375/acpd/2004-4-8609, 8609-8625.

Thompson, A. M., and R. W. Stewart, 1991: Effect of Chemical Kinetics Uncertainties on Calculated Constituents in an Tropospheric Photochemical Model, *J. Geophys. Res.*, 96, D7, 13089-13108.

REFIR MEASUREMENTS IN THE WATER VAPOUR ROTATIONAL BAND AND COMPARISON WITH A BOMEM AERI-TYPE FOURIER TRANSFORM SPECTROMETER

F. Esposito, R. Restieri, and C. Serio
DIFA/Dip., Università della Basilicata, 85100 Potenza, Italy

V. Cuomo, G. Masiello, and G. Pavese
IMAA/CNR, Tito Scalo, Potenza, Italy

G. Bianchini, L. Palchetti, and M. Pellegrini
IFAC/CNR, Firenze, Italy

T. Maestri and R. Rizzi
Fisica/Dip., Università di Bologna, Bologna, Italy

ABSTRACT

The paper presents the first field campaign of a novel Fourier Transform Spectrometer aiming at measuring in the spectral region dominated from the water vapor rotational band. The spectral radiance observed by the instrument has been compared to that recorded by a BOMEM spectrometer. A comparison with calculations is provided, as well.

1. INTRODUCTION

The Radiation Explorer in the Far InfraRed (REFIR) mission (Rizzi et al., 2002) is the result of a in-depth feasibility study, funded by European Union in 1998, of a novel space-borne instrumental package that addresses: a) the need for data in the Far InfraRed (FIR), a spectral range (100–1100 cm-1 or 9–100 µm) as yet unexplored from space, and b) the need to improve our knowledge on the distribution of the atmospheric constituents that modulate to a large extent the FIR emission to space, namely

- mid and upper tropospheric water vapour, using the strong spectral signature of water vapour in the FIR,
- mid and upper- tropospheric clouds, such as cirrus.

The REFIR Science Team blends expertise in broadband radiometry, Fourier spectroscopy, physics of detectors and optical devices, cloud and radiation remote sensing and climate modelling. The project team has identified a mature design of the REFIR spectrometer and has built-up a breadboard version of the instrument.

The instrument, which is a Fourier Transform Spectrometer, has been used in a field campaign with the objective to demonstrate feasibility, reliability and capability of the REFIR design and concept. In particular, the campaign has allowed us to check and test:

1. the radiometric calibration system
2. the data acquisition and control system
3. the interferogram sampling strategy.

The field campaign has been performed at a mountain site in South Italy. During the filed campaign REFIR has acquired spectra in the spectral range 100 to 1000 cm^{-1}. The measurements have been complemented with radio sonde ascents for the determination of temperature and water vapour profiles. In addition, during the campaign, a BOMEM Fourier Transform spectrometer has been used to observe the down welling earth's emission spectrum in the range 500 to 3000 cm^{-1}.

The paper gives an account of the field campaign and summarizes the present REFIR bread-board configuration and its main characteristics.

2. REFIR BASIC CONCEPT AND DESIGN

2.1 Concept and Optical Design

The REFIR scientific requirements are outlined in (Rizzi et al., 2002). They lead to a package of four instruments of which that of interest here is the far infrared Fourier Transform Spectrometer. This is the REFIR mission primary instrument capable of resolving the up-welling radiation from the Earth with a spectral resolution of 0.5 (apodized) cm^{-1} and a signal-to-noise ratio of S/N>100 in broadband operations (100–1100 cm^{-1}). The main parameters of the instrument are here listed:

- **Type of FTS:** Polarizing interferometer of the Martin-Puplett type modified for the double input/output ports option
- **Beam splitter type:** Wire-grid 2 µm/1 µm on 1.5 µm thick Mylar substrate
- **Spectral bandwidth:** 100–1100 cm^{-1}

- **Spectral resolution:** (max) 0.25 cm^{-1} (double-sided interferogram), 0.1 cm cm^{-1} (single-sided interferogram)
- **Optical throughput:** 0.0097 cm^{2}sr
- **Field-of-view:** 30 mrad
- **Detector type:** Room temperature DLATGS pyroelectric
- **Operational temperature:** Room temperature

The details of the optical design and REFIR concept can be found in Carli et al. (1999) and Palchetti et al. (2005).

In contrast to the beam splitter type planned for the space mission, the REFIR bread board, built up in the laboratory has been provided with an amplitude mylar-type beam-splitter. A picture of the REFIR bread board is shown in Fig. 1.

Fig. 1. Picture of REFIR bread board.

2.2 Data Acquisition and Control System

Because of the need for correction of the slow response of the pyroelectric detector, the equal-time sampling with filter compensation and numerical re-sampling (see Palchetti and Lastrucci, 2001) is the sampling technique, which has been implemented. This approach, which is based on the equal time sampling method described by Brault, 1996, enables high radiometric performance without increasing the requirements on the mechanical accuracy of the mirror drive.

We choose to use NI LabView software for all acquisition/file management/on-line analysis operations. This choice is due to the fact that by using LabView we can acquire analog channels, drive position and mirror pointing rotary stages, read counters, make preliminary analysis and save data on disk. The data acquisition unit was developed in FORTRAN language and the resulting codes were linked to LabView acquisition program.

More in details the data acquisition unit allow us to perform the following operations:

- Move and check the position of the linear stage, visualisation on a monitor display
- Acquire and save interferometer and Reference LASER signals
- Acquire, process, visualise and save House-keeping signals
- Synchronise all processes (linear stage movement, interferometer signals acquisition, House-keeping signals acquisition, pre-processing, read linear stage position, visualise samples of recorded data)
- Check all boards, in order to visualise error messages in case of errors
- Visualise interferometer and reference signals.

The software module we have developed is capable of acquiring data simultaneously from different channels, then saved in binary files. Data from house-keepings are elaborated, in order to compute Black Bodies temperatures, and visualised in real time. All information regarding each single scanning (date and hour, temperature, direction, sampling rate, etc.) are saved in a file.

2.3 Calibration

The radiometric calibration of REFIR spectra collected at Castelgrande was obtained by following the standard procedure of measuring two reference blackbodies at different temperatures (HBB, 354 K and CBB, 280 K). A reference black body (RBB) is located at first interferometer input. The three sources are thermoelectrically controlled, stabilized with a precision of ±0.1 K; and their temperatures are sensed with calibrated platinum resistance embedded in each BB. The three BB are internally coated with a Xylane-based paint, with a theoretical efficiency estimated as better then 99.9% at the operating spectral region. The two reference blackbodies HBB and CBB are viewed by rotating the input mirror, collecting reference spectra each hour. Double-sided interferograms are collected for both reference blackbodies and atmospheric measurement, and real part of their Fourier transform are taken as uncalibrated spectra. A two-point complex calibration is applied, under the hypothesis of a linear instrument response:

$$C_v = r_v(L_v - L_{0v}) \qquad (1)$$

where C_v is the complex uncalibrated spectrum, L_v is atmospheric radiance, L_{0v} is an offset from instrument emission that, in the case of REFIR, L_{0v} is due principally to the emission of RBB. The parameter r_v is the spectral responsivity of the instrument, and v represents the wavenumber. The spectral responsivity and the instrument offset are computed by writing eq. (1) for both HBB and CBB and solving for r_v and L_{0v}:

$$r_v = (C_{hv} - C_{cv})/[B_v(T_h) - B_v(T_c)]$$
$$L_{0v} = C_{hv}/r_v - B_v(T_h)$$

where $B_v(T)$ is the Planck blackbody radiance, and subscripts h and c label the quantities associated with hot and cold blackbodies. Solving eq. (1) for the source radiance and substituting spectral responsivity and offset yields the basic calibration relationship:

$$L_v = C_v/r_v - L_{0v} \qquad (2)$$

The calibrated spectrum is obtained by taking the real part of (2). As a check of calibration procedure, the emission $B_v(T_r)$ by RBB is computed, and compared to real part of the offset L_{0v}; here $B_v(T_r)$ and T_r are the Planck blackbody radiance emitted by RBB and its temperature.

The procedure has been applied to both detectors (one detector for each output port), and the final spectrum is taken as a mean of the two calibrated spectra. Another check of calibration procedure is taken by a statistical consistency test on the calibrated spectra obtained by the two channels.

3. RESULTS FROM THE FIELD CAMPAIGN

The present REFIR bread-board configuration has been tested in a field campaign which has been performed at a mountain site in South Italy: Toppo di Castel Grande at latitude 40.46° N, latitude 15.27° E, elevation 1258 m above sea level. The campaign took place in two days, 9th and 10 June 2004. During the filed campaign REFIR acquired spectra in the spectral range 100 to 1000 cm^{-1}. The measurements were complemented with radio three sonde ascents for the determination of temperature and water vapour profiles. In addition, a BOMEM Fourier Transform spectrometer has been used to observe the downwelling earth's emission spectrum in the range 500 to 3000 cm^{-1}.

Fig. 2 shows a typical REFIR spectrum over the full spectral range 270 to 1050 cm^{-1}. The sampling rate is 0.5 cm^{-1} and the acquisition time is 12 minutes (average over 120 spectra). The upward spikes which are seen in the long wave side of the spectrum are due to calibration errors and correspond to the most opaque water vapor absorption lines where the signal is nearly zero. Nevertheless, the short wave side of the water vapour rotational band, 460 to 600 cm^{-1} is nicely resolved.

Fig. 3 provides a comparison between REFIR and BOMEM for the spectral range 500 to 800 cm^{-1} for which the two overlap (note that BOMEM does not measure below 500 cm^{-1}, as evidenced in Fig. 3). The comparison says that REFIR resolves spectral structures exactly how BOMEM does, so that showing that the REFIR concept does work. This is further confirmed by Fig. 4 which is limited to short-wave side of the water vapour rotational band which corresponds to the infrared emission peak of the earth and, therefore, to the side of the spectrum where we have the highest signal for the given atmospheric state. The difference between the two spectra (Fig. 4) shows that there is no important systematic pattern, although a slight bias exists, which is likely the result of a difference in the temperature of the environment close to the two instruments.

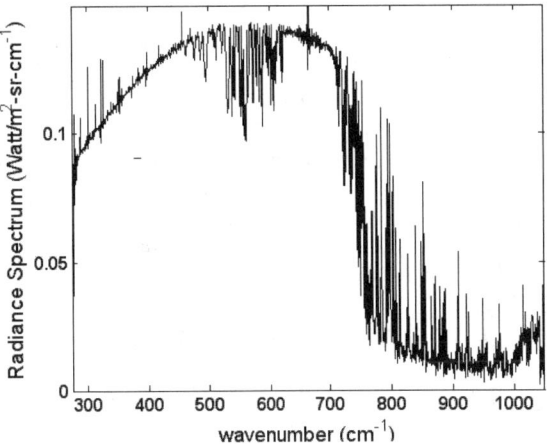

Fig. 2. Calibrated REFIR spectrum for the day of 10 June 2004, 22:14 UTC.

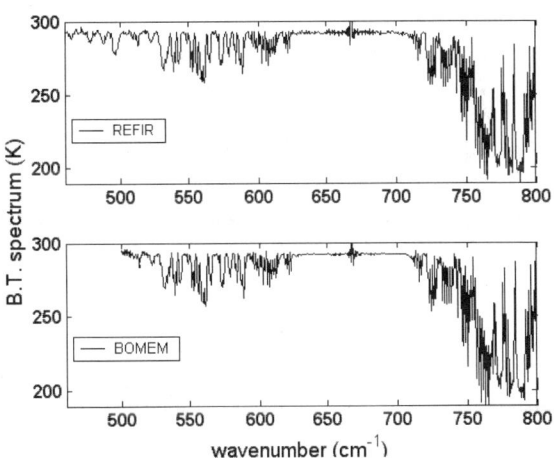

Fig. 3. Comparison between REFIR and BOMEM for the spectral range 500 to 800 cm^{-1}.

To better analyze the level of noise for the two instruments, we have also performed a comparison with an independent noise-free source, which consists in a synthetic spectrum computed by means of LBLRTM and the radiosonde observations. The results are shown in Fig. 5 and Fig. 6 for REFIR and BOMEM, respectively.

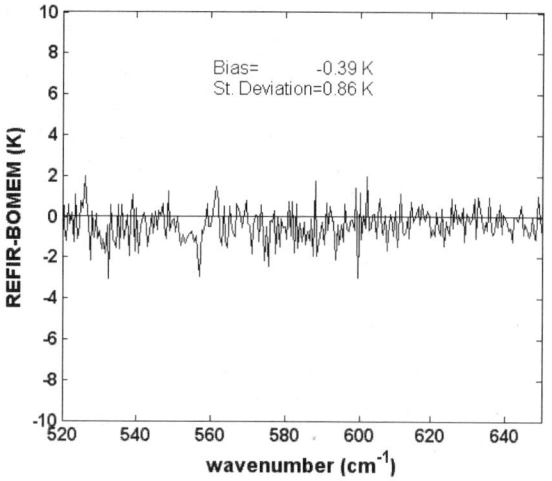

Fig. 4. REFIR-BOMEM for the range 520-650 cm^{-1}.

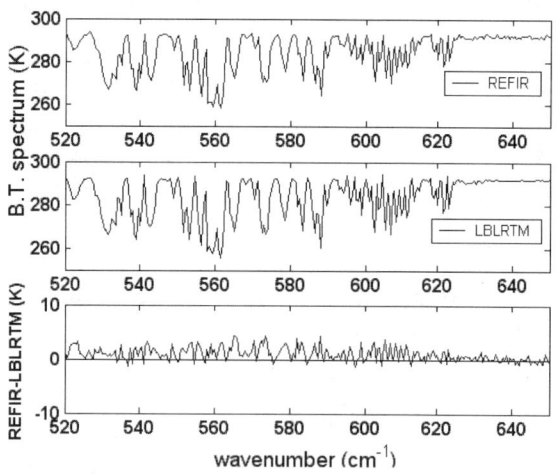

Fig. 5. REFIR vs. LBLRTM synthetic spectrum.

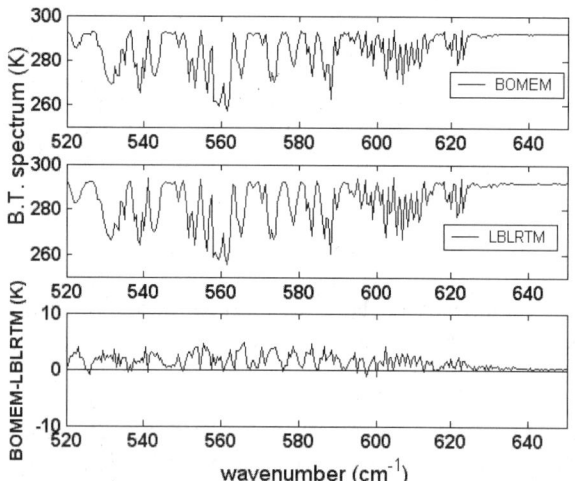

Fig. 6. BOMEM vs. LBLRTM synthetic spectrum.

It is seen that both BOMEM agrees with LBLRTM to the same extent that REFIR does. LBLRTM is slightly downward biased with respect to the calculations which could well be the result of the difficulty to characterize the local environment close to the instruments. On overall, LBLRTM agrees slightly better with REFIR (bias=0.99 K, st. deviation= 1.25 K) than with BOMEM (bias=1.39 K, st. deviation=1.26 K), although the behaviour of BOMEM against LBLRTM nearly parallels that of REFIR against LBLRTM.

4. CONCLUSIONS

A bread-board of REFIR has been developed and it has been used for the first time to acquire down-welling spectra in a field campaign.

REFIR spectra have been compared to BOMEM spectra and to synthetic spectral radiance in the BOMEM/REFIR overlapping region 500-1000 cm^{-1}.

While offering a valuable comparison between radiative transfer modelling in the far infrared and observations, the field campaign has demonstrated that the overall REFIR concept and design do work.

5. ACKNOWLEDGMENTS

Work supported by MURST – "PROGRAMMA OPERATIVO NAZIONALE" – misura 1.3, COS(OT).

6. REFERENCES

Rizzi R., L. Palchetti, B. Carli, R. Bonsignori, J. E. Harries, J. Leotin, S. Peskett, C. Serio, and A. Sutera, 2002: Feasibility study of the space-borne Radiation Explorer in the Far InfraRed (REFIR), in *Proceedings of SPIE* Vol.4485, 202-209.

Carli, B., A. Barbis, J. E. Harries, and L. Palchetti, 1999: Design of an efficient broad band far infrared FT spectrometer. *Appl. Opt.* 38, 3945–3950.

Clough, S. A., M. J. Iacono, and J. L. Moncet, 1995: Line-by-line calculations of atmospheric fluxes and cooling rates *II*: Application to carbon dioxide, ozone, methane, nitrous oxide and the halocarbons, *J. Geophys. Res.* 100, 16519-16526.

Brault, J. W., 1996: New approach to high-precision Fourier transform spectrometer design. *Appl. Opt.* 35, 2891–2896.

Palchetti, L., G. Bianchini, F. Castagnoli, B. Carli, C. Serio, F. Esposito, V. Cuomo, R. Rizzi, and T. Maestri, 2005: The breadboard of the Fourier transform spectrometer for the Radiation Explorer in the Far Infrared (REFIR) atmospheric mission, *Appl. Opt.*, in press.

Palchetti, L., and D. Lastrucci, 2001: Spectral noise due to sampling error in Fourier transform spectroscopy. *Appl. Opt.* 40, 3235-3243.

AUTOMATED RETRIEVAL OF ATMOSPHERIC AEROSOL AND TRACE GASES PROPERTIES FROM LARGE MFRSR DATASETS

M. Alexandrov[1,2], B. Carlson[2], A. Lacis[2], B. Cairns[1,2], and A. Marshak[3]

[1]Columbia University, 2880 Broadway, New York, NY 10025, USA
[2]NASA Goddard Institute for Space Studies, 2880 Broadway, New York, NY 10025, USA
[3]NASA Goddard Space Flight Center, Code 913, Greenbelt, MD 20771, USA

ABSTRACT

An upgrade of our previously developed Multi-Filter Rotating Shadowband Radiometer (MFRSR) data analysis algorithm (Alexandrov et al., 2002) is presented. The method is illustrated on MFRSR data collected from the dense local network (21 instruments) operated by US DOE Atmospheric Radiation Measurement (ARM) program and located in Oklahoma and Kansas (the Southern Great Plains (SGP) site). The new version of the retrieval algorithm features an automated cloud screening procedure based on optical thickness variability analysis. The technique is objective, computationally efficient and is able to detect short clear-sky intervals under broken cloud cover conditions. The main features of the new method are the ability to separate between the fine and coarse mode aerosol optical thicknesses (AOT) and to retrieve the fine mode size (effective radius). A bimodal gamma distribution with fixed coarse mode size is adopted as aerosol particle size model. The products of our analysis also include time series of column amounts of ozone and nitrogen dioxide. To validate the separation of the total AOT into fine and coarse parts by our method we compared its results with those of AERONET almucantar retrievals made from a colocated CIMEL sunphotometer's data. Results of analysis of a year of data from the whole SGP MFRSR network exhibit certain seasonal and geographical patterns, in particular differences in aerosol fine mode size between the measurement locations.

1. INTRODUCTION

The increased emphasis on satellite aerosol retrievals in order to constrain direct and indirect radiative forcings of atmospheric aerosols has produced a need for ground-based, ground-truth aerosol measurements. In addition, measurements from ground-based sunphotometer networks can be also used to produce a land-based aerosol climatology which is complementary to satellite retrievals that currently are being performed mostly over ocean. Ground-based MFRSR networks with their high density and wide geographical coverage can provide data indispensable for these goals. Thus, development of retrieval algorithms allowing to reveal more information contained in MFRSR measurements is important.

The MFRSR (cf. Harrison et al., 1994) makes precise simultaneous measurements of the direct solar beam extinction and horizontal diffuse flux, at six wavelengths (nominally 415, 500, 615, 670, 870, and 940 nm) at short (20 sec in our dataset) intervals throughout the day. Besides water vapor at 940 nm, the other gaseous absorbers within the MFRSR channels are NO_2 (at 415, 500, and 615 nm) and O_3 (at 500, 615, and 670 nm). Aerosols and Rayleigh scattering contribute atmospheric extinction in all MFRSR channels.

In our older paper (Alexandrov et al., 2002) we described a retrieval algorithm for MFRSR data that was designed to provide a self-consistent retrieval of aerosol and gas information from the spectral dependence of the MFRSR data. In that algorithm we adopted the simplest aerosol size distribution, namely a mono-modal distribution. While the spectral trade-offs between small particle aerosol extinction and NO_2 absorption are the largest source of uncertainty in our retrievals, the assumed shape of the aerosol size distribution appeared to play role in these trade-offs. In particular, applying a monomodal aerosol model for retrievals of a bimodal aerosol may result in an overestimation of NO_2 column amounts. To address this problem we widened the class of assumed size distribution shapes by including bimodal model with fixed (1.5 µm) effective radius of the coarse mode and retrievable fine mode size. Fine and coarse mode AOT are also retrievable, as well as column amounts of ozone and nitrogen dioxide.

Development of an automated cloud screening procedure based on optical thickness variability analysis allowed us to make the whole data processing more objective and fully automatic. In addition to the method description we present intercomparisons of our results with AERONET almucantar scan retrievals (Dubovik and King, 2000), as well as a qualitative testing of how well our retrievals capture known aerosol events (Pinatubo volcanic aerosols) and ambient properties (geographic variations in aerosol size and composition across the SGP site from northern Kansas to south-eastern Oklahoma).

2. METHOD

The automatic cloud screening algorithm (Alexandrov et al., 2004) used in our method is based on the (scale dependent) parameter (Cahalan, 1994; Cairns et al., 2000):

$$\varepsilon = 1 - \exp(<\ln \tau>) / <\tau>,$$

which characterizes the degree of horizontal inhomogeneity of an atmospheric field. Here τ denotes the optical thickness, the angular brackets indicate a moving average (over 15 data points = 5 min in our case). As a measure of atmospheric variability this parameter is complementary to the ratio of the standard deviation and the mean. To separate clear sky from thin clouds we need to modify the original optical thickness by bringing AOT and cloud optical thickness (COT) variability to the same mean, while retaining the structure and size of their fluctuations. This is done by subtracting from the optical thickness time series τ its moving average $<\tau>$ and adding back a constant τ_{const} = 0.2 which is more typical as AOT value than COT value. Thus, we obtain a renormalized

$$\tau' = \tau - <\tau> + \tau_{const},$$

and a renormalized ε' computed with τ' instead of τ. We have selected $\varepsilon' = 2 \cdot 10^{-4}$ as the operational threshold value for separating clear sky from clouds. An enveloping technique is used to include some additional clear sky points that are located between these initially selected and have similar AOT values.

The calibration technique used in the new method is similar to that used in (Alexandrov et al., 2002): we separate and individually calibrate each of the physical parameters we retrieve. In the fine mode effective radius range below 0.2 μm size retrievals are not possible because of spectral trade-offs between aerosol and NO_2, thus we take the value of 0.2 μm, as the lower limit for our retrievals. This means that the cases when fine mode particles tend to be smaller than this limit are considered separately, and a default value 0.2 μm of the effective radius is used. In the opposite case of large (greater than 0.5 μm) fine particles the spectral separation of aerosol modes in visible range becomes unreliable, thus, retrievals in these cases are made with monomodal aerosol model.

Our algorithm is applied simultaneously to a set of daily MFRSR records covering at least a month of measurements and runs level by level: first all days are cloud screened, then all 870 nm records are calibrated using compatibility between the direct and diffuse measurements, etc. This approach allows for stabilization of the daily calibration constants on each level using a robust smoothing technique.

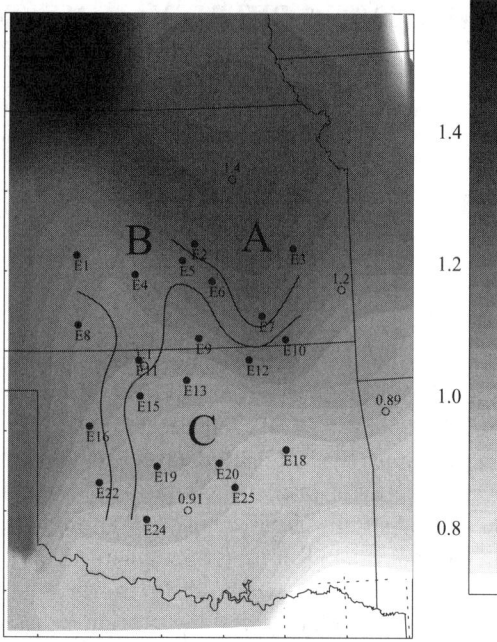

Fig. 1. Contour plot of the NO_3/SO_4 ion concentration ratios (mean values for the year 2000) obtained from National Atmospheric Deposition Program/National Trends Network (NADP/NTN) precipitation monitoring sites. The MFRSR instrument locations are shown by filled circles.

3. RESULTS

We applied our separation method to the MFRSR dataset from the local network at the U.S. Southern Great Plains (SGP) run by the DOE Atmospheric Measurement (ARM) Program (Fig. 1). The network consists of 21 instruments located at SGP's Extended Facilities (EFs) and covers the area of approximately 3 by 4 degrees in northern Oklahoma and southern Kansas with average spacing of 80 km between neighboring measurement sites.

We compared our results with the total, fine, and coarse AOTs obtained from AERONET almucantar retrievals (Dubovik and King, 2000). We used the data from the CIMEL sunphotometer labeled "Cart Site" on the AERONET web site, which is colocated with the two SGP's Central Facility MFRSRs (C1 and E13). The results of the intercomparison at 870 nm wavelength for June–September 2000 show that biases of MFRSR over AERONET of 0.009 in total, 0.015 in fine, and –0.007 in coarse AOT. Application of the constrained (zero NO_2) version of our algorithm to the same dataset showed much better agreement with AERONET: 0.009 in total, 0.0045 in fine, and 0.0044 in coarse AOT. This indicates, that the larger positive bias in fine mode AOT for unconstrained comparison relative to the smaller bias in the constrained (zero NO_2) case is a consequence of the larger particle size that we retrieve by accounting for the spectral contribution of NO$_2$ absorption. Disagreement with AERONET on fine mode particle size (AERONET's particles are smaller) appeared to have the same origin: fixing NO$_2$ amounts to zero brings our size retrievals down to AERONET values. In addition to possibility of comparison with AERONET the constrained aerosol size retrievals provide the lower bound for physically justified values of fine mode effective radius.

Applying our algorithm to the dataset from the SGP site for the 1993-1997 time period was an important test. This time period includes measurements of the well-documented decay of the stratospheric aerosol following the eruption of Mt. Pinatubo in June 1991. Our retrievals for this time period are shown in Fig. 2. It is seen, that the Pinatubo aerosols tend to be in the coarse mode. The coarse mode AOT values decrease during 1993-1995 and exhibit little seasonal variability. In contrast, the fine mode AOT values show no decreasing trend during this time period but do exhibit seasonal variability with AOT maxima in summer time. These results are in qualitative agreement with the 1991-1993 in situ measurements (Goodman et al., 1994) demonstrating that in 1993 (2 years after the eruption) the size distribution of stratospheric aerosols had become essentially bimodal.

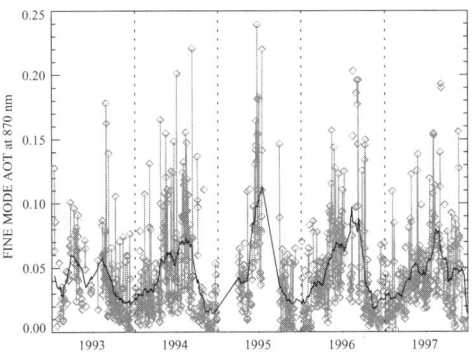

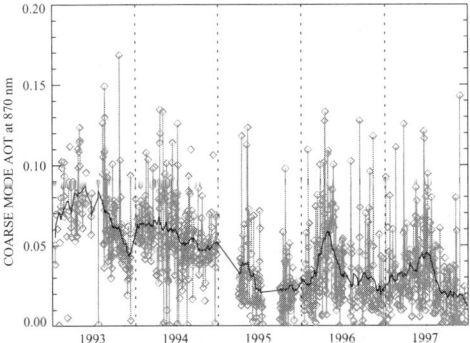

Fig. 2. Time series of daily mean fine (top) and coarse (bottom) AOT retrieved from SGP Central Facility MFRSR dataset for 1993-1997. The black curves show results of moving 3rd degree polynomial smoothing with 30 day window.

Analysis of a year of MFRSR data obtained at the SGP Extended Facilities reveals systematic geographical differences in the retrieved size of fine mode aerosols. Fig. 3 presents plots of the time series of the daily mean fine mode effective radius retrievals for the year 2000 from the Central Facility (E13) and three Extended Facilities (E2, E4, and E24). These particular sites represent three different aerosol size variability patterns characteristic of these different areas of the SGP site. We identify these types of variability as A, B, and C. The respective areas dominated by these types are shown on the SGP site maps in Fig. 1. Type A is characteristic of the North-East and South-West of the site and corresponds to larger 0.3-0.5 µm particles. Type C dominates the South-East and corresponds to very small aerosol particles with effective radius less than or equal to our detection limit 0.2 µm. Type B is intermediate between A and C and is encountered in the North-West and the central parts of the SGP where both very small and larger fine mode particles are present. The larger fine mode effective radii exhibit a seasonal maximum in Winter. We suggest that the above types of aerosol size variability may reflect a balance between different aerosol species at the SGP site, in particular, sulfates and nitrates. Aerosol sampling studies measuring size distributions of nitrate and sulfate particles have been performed in different environments indicate that nitrate particles are generally larger than sulfates. The described geographical differences in retrieved aerosol fine mode size are in agreement with the spatial variations in aerosol composition measurements, in particular NO$_3$/SO$_4$ ion concentration ratios obtained from National Atmospheric Deposition Program/National Trends Network (NADP/NTN, http://nadp.sws.uiuc.edu) precipitation monitoring sites in Oklahoma, Kansas and all of the states having a common border with these two. The result of interpolation between the mean values for the year 2000 is shown as a contour plot in Fig. 1. This plot shows a trend corresponding to domination of nitrates in A-zones (especially in the North) as opposed to domination of sulfates in the C-zone in the South-East of the SGP site. EPA AirData reports (http://www.epa.gov/air/data/) also indicate that among power plants in the area surrounding the SGP site these located in northeastern Kansas (i.e. in the area where we retrieve larger aerosol sizes) have largest fraction of NO$_x$ in their emissions.

4. CONCLUSIONS

A new improved version of the MFRSR retrieval algorithm of Alexandrov et al. (2002) is presented. This version features an automated cloud screening procedure based on optical thickness variability analysis. The new method is capable of separating fine and coarse mode AOT, as well as retrieving fine mode effective radius. The coarse mode effective radius is assumed to be fixed.

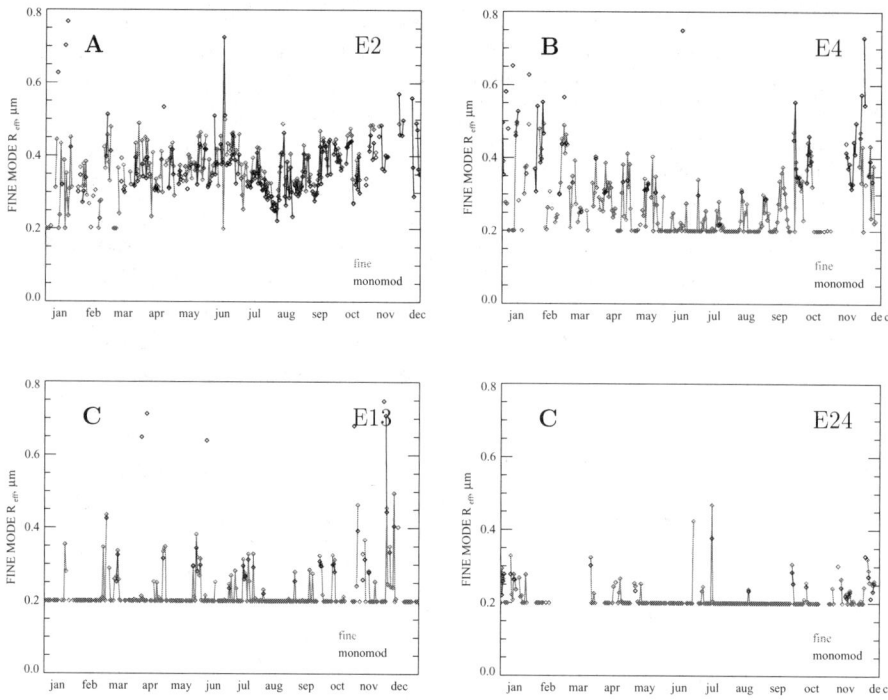

Fig. 3. Time series of daily mean fine mode effective particle radius for SGP Extended Facilities E2, E4, E13, and E24 for the year 2000. These illustrate the three characteristic types of variability denoted by A, B, and C.

The new algorithm can also perform retrievals with a monomodal size distribution (determining total AOT and effective radius) when the coarse mode is absent or when the fine mode size becomes too large to make fine and coarse mode extinction indistinguishable in the visible spectral range. In the cases when the fine mode size is smaller than 0.2 µm, the retrievals of fine and coarse AOT are performed with both fine and coarse particle sizes fixed. As in the earlier algorithm, column amounts of ozone and nitrogen dioxide are also retrieved, and the instrument calibration constants are determined from the data. The retrievals made with the new algorithm using the extensive DOE ARM program SGP site multi-instrument dataset are compared with aerosol measurements made using other instruments and analysis techniques. Our comparison with AERONET almucantar retrievals revealed that difference in retrieval approach in dealing with NO_2 absorption is the main reason for disagreement between the two datasets. The retrievals performed on the 1993–1997 dataset from the SGP Central Facility shows that our algorithm correctly interprets the multi-year decrease in AOT and increase in Angstrom parameter during 1993–1995 to the dissipation of coarse mode particles introduced in the stratosphere by the 1991 eruption of Mt. Pinatubo. The time series of fine mode effective radius retrieved from the data obtained at the SGP Extended Facilities during the year 2000 appear to reflect geographic and temporal changes in the balance between sulfate and nitrate aerosol fractions (the latter having larger size particles). All of the data processing steps required in the algorithm are fully automated. The algorithm performs the cloud screening, calibration and retrieval steps as well as saves the retrieval results in netCDF ARM-style format.

5. REFERENCES

Alexandrov, M. D., A. Marshak, B. Cairns, A. A. Lacis, and B. E. Carlson, 2004: Automated cloud screening algorithm for MFRSR data, *Geophys. Res. Lett.*, *31*, L04118, doi:10.1029/2003GL019105.

Alexandrov, M., A. Lacis, B. Carlson, and B. Cairns, 2002: Remote sensing of atmospheric aerosols and trace gases by means of Multi-Filter Rotating Shadowband Radiometer. Part I: Retrieval algorithm, *J. Atmos. Sci.*, *59*, 524-543.

Dubovik, O., and M. D. King, 2000: A flexible inversion algorithm for retrieval of aerosol optical properties from Sun and sky radiance measurements, *J. Geophys. Res.*, *105*, 20673-20696.

Goodman, J., K. G. Snetsinger, R. F. Pueschel, G. V. Ferry, and S. Verma, 1994: Evolution of Pinatubo aerosol near 19 km altitude over western North America, *Geophys. Res. Lett.*, *21*, 1129-1132.

Harrison, L., J. Michalsky, and J. Berndt, 1994: Automated multifilter shadow-band radiometer: instrument for optical depth and radiation measurement, *Appl. Opt.*, *33*, 5118-5125.

AEROSOL OPTICAL CHARACTERISTICS IN ASIA FROM MEASUREMENTS OF SKYNET SKY RADIOMETERS

K. Aoki

Faculty of Education, Toyama University, 3190 Gofuku, Toyama 930-8555, Japan

T. Nakajima

CCSR, University of Tokyo 4-6-1 Komaba, Meguro-ku, Tokyo 153-8904, Japan

T. Takamura

CEReS Chiba University, 1-33 Yayoi-cho, Inage-ku, Chiba 263-8522, Japan, & SKYNET Sky Radiometer Observation Group

ABSTRACT

Optical characteristics of aerosol were investigated using the measurements from ground-based sky radiometers. We started the long-term monitoring of aerosols with sky radiometers deployed for the SKYNET observation network. In this study we examined aerosol optical characteristics in terms of its temporal and spatial variability in Asian region and applied to validation of satellite retrievals or numerical model simulations of aerosol. The obtained aerosol optical characteristics clearly showed spatial variablities in a short period of time (e.g., Asian dust event and Siberia forest fire event at Sapporo, Nagasaki and Amami-Ohsima of Japan, and at Sri-samrong of Thailand). Especially variability was large at Sri-samrong, compared to the other three Japanese sites. Intercomparison was made between SKYNET and AERONET radiometers at Shirahama Japan, and results showed that optical thickness is in good agreement.

1. INTRODUCTION

Atmospheric aerosols play an important role in the climate change through two effects, i.e., direct and indirect effects. Understanding these aerosol effects requires knowledge of the optical characteristics and temporal and spatial variability. The global distributions of aerosol optical characteristics have been derived from satellite measurements (e.g., Higurashi and Nakajima, 2002), and have been simulated in AGCM (e.g., Takemura et al., 2000). However, it is difficult to assess the accuracy those obtained variability in time and/or space. Thus it is of need to use ground-based measurements of aerosol to validate results from satellite measurements and numerical outputs.

We have monitored aerosols optical characteristics since 1994 using sky radiometers on the SKYNET network (http://atmos.cr.chiba-u.ac.jp/aerosol/skynet/). SKYNET project is a grass roots effort to measure atmospheric radiation and related atmospheric parameter to analyze radiation impact of aerosols. Fig. 1 shows the geographical location of the Asian SKYNET sites and other observation sites.

We establish automatic data transfer and analysis system and web site (SKYNET Sky radiometer Archives web: http://skyrad.edu.toyama-u.ac.jp/~kazuma/skyrad/).

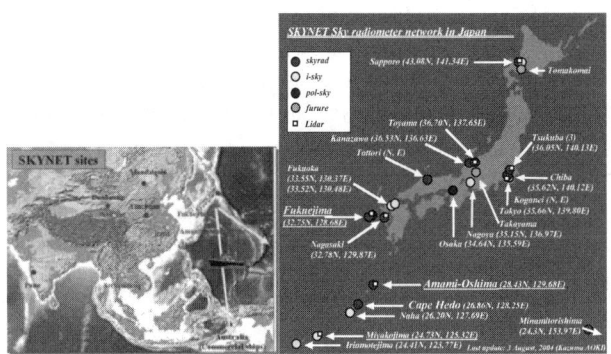

Fig. 1. Maps showing the location of the main Asian SKYNET and other related sites.

In this study, we present the seasonal and temporal variations of the relationship of Ångström parameter (aerosol optical thickness and Ångström exponent) to single scattering albedo, and the comparison of aerosol optical thickness from SKYNET and AERONET radiometer.

2. INSTRUMENTATION AND OBSERVATION

The sky radiometer (Fig. 2) is a portable instrument that takes measurements only during daytime under clear sky condition. It observes both direct solar irradiance and diffuse sky solar radiation at every 10 or 15 minutes. The aerosol optical characteristics are computed using the SKYRAD.pack developed by Nakajima et al. (1996). The analysis method is described in detail in Nakajima et al. (1996) and Aoki and Fujiyoshi (2003). The Ångström parameters (aerosol optical thickness at each wavelength ($\tau(\lambda)$) and Ångström exponent (α)) are derived using following equation:

$$\tau(\lambda) = \tau(0.5)(\lambda/0.5)^{-\alpha} \quad (1)$$

For the estimation of Ångström parameters, we use optical thickness at five wavelengths (0.4, 0.5, 0.675, 0.87, and 1.02 μm).

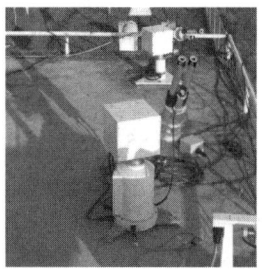

Fig. 2. Sky radiometers in Sri-samrong, Thailand.

3. RESULTS AND DISCUSSION

3.1 Aerosol Optical Characteristics

The aerosol optical characteristics are derived from direct solar irradiances and diffuse radiance measured by sky radiometer. The measurements were carried out in Sapporo (43.08N, 141.34E), Nagasaki (32.78N, 129.87E), and Amami-Oshima (28.43N, 129.68E), Japan and Sri-samrong (17.16N, 99.87E), Thailand. Aerosol optical thickness at 0.5 μm [AOD(0.5)], Ångström exponent (α), single scattering albedo at 0.5 μm [SSA(0.5)] are simultaneously retrieved for the period from March to October 2003, coincident with ADEOS-II/GLI operation period.

Fig. 3 shows the relationship of Ångström parameter (AOD(0.5) and Alpha) to SSA(0.5) from March to October 2003 at Sapporo, Nagasaki and Amami-Oshima of Japan, and Sri-samrong of Thailand during the analysis period. In Sapporo it is shown that SSA(0.5) is relatively low (i.e., 0.8 to 0.9) and Alpha seems high (around 1.5), likely due to small and absorption particles in Sapporo. These seem to be associated with forest fire events occurred in Siberia. However, all sites show low Alpha as well as low AOD(0.5) in the spring, indicating turbid atmosphere associated with Asian dust events although year 2003 is thought to be a weak event year.

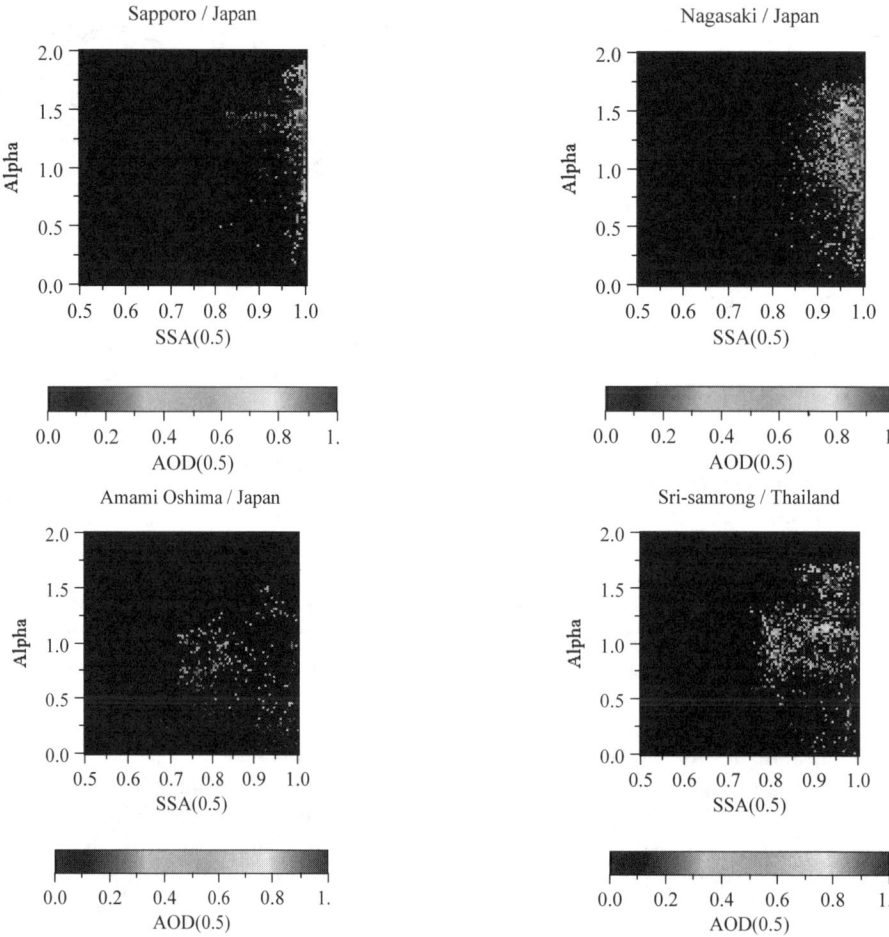

Fig. 3. Relationship of Ångström parameter (AOD(0.5) and Alpha) to SSA(0.5) for the period from March to October 2003 at Sapporo, Nagasaki and Amami Oshima of Japan, and Sri-samrong of Thailand, coincident with ADEOS-II/GLI operation period.

On the other hand, SSA(0.5), Alpha and AOD(0.5) at Sri-samrong show large variability in comparison to the other three Japanese sites. In particular AOD(0.5) appears to be high while SSA(0.5) seems low, suggesting that aerosols found in Thailand area create more turbid atmosphere with more absorption of solar radiation. It seems to be due to anthropogenic aerosols related biomass burning. Our knowledge of such aerosol characteristics is necessary to better understand the complex earth climate system.

3.2 Comparison of SKYNET Aerosol Optical Thickness with from AERONET

For the better understanding of the aerosol optical characteristics shown in global observation network, SKYNET sky radiometer results were compared with those from AERONET sunphotometer measurements—see http://aeronet.gsfc.nasa.gov/ for the detailed information on the AERONET.

Aerosol optical thickness at 0.5 μm [AOT(0.5] from 26 September to 1 November, 2003 at Shirahama (33.69 N, 135.36 E) of Japan was retrieved from a SKYNET sky radiometer compared with AOT retrieved from the AERONET radiometer (Fig.4). Fig. 4 clearly shows that AOT(0.5) is good correlation (correlation coefficient about 0.99). However, our preliminary results indicate that SKYNET AOT is comparatively low with varying dependence of bias with AOT itself. Although the current study does not provide why there exist such discrepancy, this type of intercomparison is much needed for the better understanding of error characteristics of retrieved optical properties of aerosols.

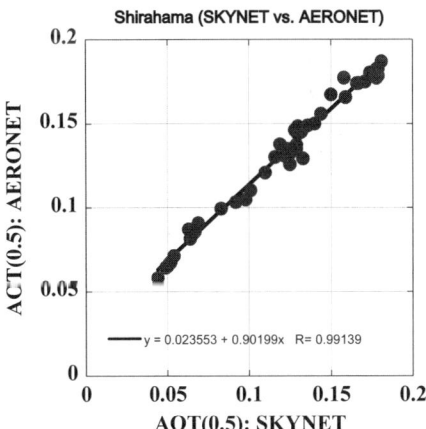

Fig. 4. Comparison of aerosol optical thickness from SKYNET and AERONET in Shirahama, Japan, during 26 September to 1 November 2003.

4. SUMMARY

We examined aerosol optical characteristics retrieved from solar radiation measurement at four SKYNET sites for the period from March to October 2003. Also compared is the aerosol optical thickness estimated from two difference instruments at Shirahama, Japan. Following conclusions are drawn:

(1) The aerosol optical characteristics show large variability in space and time in particular associated with special events such as Asian dust storm or Siberian forest fire. During the spring, associated with dust events, Alpha was low and AOD(0.5) was high.

(2) Intercomparison aerosol optical thickness at 0.5 μm from two different instruments (here SKYNET sky radiometer and AERONET sunphotometer) indicated that they are in good correlation, but SKYNET values are lower.

5. ACKNOWLEDGMENTS

We are grateful to the SKYNET community group and AERONET group for the data that they shared openly. The authors are grateful to Prof. B.J. Sohn for their constructive comments. This research has been supported by the APEX project of JST.

6. REFFERENCES

Aoki, K, and Y. Fujiyoshi, 2003: Sky radiometer measurements of aerosol optical properties over Sapporo, Japan. *J. Meteor. Soc. Japan*, 81, 493-513.

Higurashi, A., and T. Nakajima, 2002: Detection of aerosol types over the East China Sea near Japan from four-channel satellite data. *Geophys. Res. Lett.*, 29, 1836, doi:10.1029/2002GL015357.

Nakajima, T., Tonna, G., Rao, R., Boi, P., Kaufman, Y., and Holben, B., 1996. Use of sky brightness measurements from ground for remote sensing of particulate polydispersions, *Appl. Opt.* 35, 2672-2686.

Takemura, T., Okamoto, H., Maruyama, Y., Numaguti, A., Higurashi, A., and Nakajima, T., 2000. Global three-dimensional simulation of aerosol optical thickness distribution of various origins, *J. Geophys. Res*, 105, 17,853-17,873.

OPTICAL PROPERTIES OF ASIAN DUST MEASURED AT SEVERAL SITES IN JAPAN

H. Kobayashi, University of Yamanashi, Kofu, 400-8511, Japan
K. Arao, Nagasaki University, Nagasaki, 852-8521, Japan
T. Murayama, Tokyo University of Marine Science and Technology, Tokyo, 135-8533, Japan
K. Iokibe and R. Koga, Okayama University, Okayama, 700-8530, Japan
M. Yabuki and M. Shiobara, National Institute of Polar Research, Tokyo, 173-8515, Japan

ABSTRACT

Optical properties of Asian dust, such as the size distribution and complex refractive index, which are assessed in this study, are required for estimation of its effect on climate change. Size distribution was measured using a Coulter Counter after aerosol filter sampling at several sites in Japan. The imaginary part of the refractive index and the single scattering albedo of the aerosol mixture in Asian dust events were also estimated from measurements using optical particle counters, integrating nephelometer, and particle soot absorption photometer at Kofu and Tokyo. An Asian dust phenomenon was observed in Japan during 11–14 April 2003. During that period, volume-geometrical mean radii of Asian dust ranged 0.941–1.01 μm at Nagasaki, 0.841–0.994 μm at Okayama, 0.735–0.960 μm at Kofu and 0.764–0.988 μm at Tokyo. These results indicate that the size distribution of Asian dust varies spatially and temporally throughout Japan. The imaginary part of the refractive index of the bulk aerosol during the event was around 0.007 lower than the other periods. In addition single scattering albedos during the event were as high as 0.83 and 0.85 in Kofu and Tokyo, respectively. During Asian dust phenomena, therefore, the aerosol mixture absorbability becomes low.

1. INTRODUCTION

Mineral dusts, including Asian dust, have the possibility of either positive or negative radiative forcing (IPCC, 2001). Accurate understanding of radiative forcing by aerosols requires better information related to optical properties, especially the size distribution and complex refractive index.

Many methods exist for measurement of size distribution of mineral dusts, but each measurement has limitations (Sokolik et al., 2001). We demonstrate only those measurements of water-insoluble particles such as Asian dust using a Coulter Counter. Aerosol particles collected on a filter are suspended in an electrolyte solution because this method applies only to suspended water-insoluble particles, allowing measurement of aerosol particles such as mineral dust. In addition, this measurement is unaffected by particles' respective complex refractive indexes, shapes, and orientations when they pass through an aperture. Filter samplings were conducted at several sites in Japan to examine size differences and distributions among sites.

On the other hand, the complex refractive index of Asian dust, especially the imaginary part, is not well-known. We estimate the imaginary part and the single scattering albedo not of individual Asian dust particles, but of aerosol mixtures including them because of the difficulty of separating Asian dusts from other aerosols. This estimation uses measurements of optical particle counters, integrating nephelometer, and particle soot absorption photometer. We discuss the absorbability of Asian dusts based on variation of the imaginary part during Asian dust events and other periods.

2. METHODS

2.1 Sampling and Analysis of Size Distribution

Filter sampling was conducted at Nagasaki (32°45′N, 129°52′E), Okayama (34°39′N, 133°54′E), Kofu (35°39′N, 138°34′E), and Tokyo (35°41′N, 139°45′E) in Japan during spring 2003 (Fig. 1). Aerosols were collected on Nuclepore filters (Whatman plc.) with pore size 0.4 μm in diameter through a 10 mmφ tube from outside air. Blank samples were taken periodically for a qualitative check of measurement. Simultaneously, to catch the variation of aerosol particle abundance, number concentrations of aerosol particles were measured with optical particle counters every 10 min at three sites, except Okayama. At Okayama, measurements of suspended particulate matter (SPM) concentrations of the Minamigata Motor Vehicle Exhaust Gas Monitoring Station, which is located 2 km south from Okayama University, were used instead of OPC measurements.

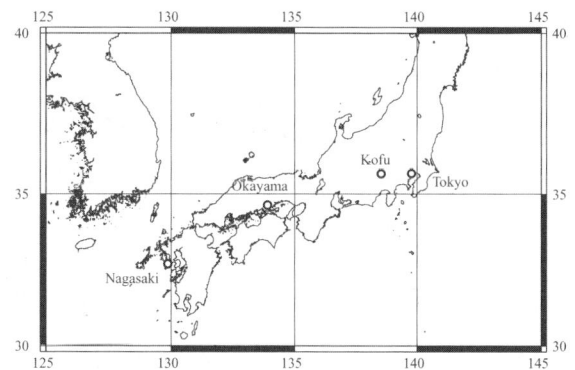

Fig. 1. Sampling sites in Japan.

Size distributions were determined using the Coulter Principle. The Nuclepore filters with the collected aerosol particles were soaked in a 90 ml electrolyte (Isoton II;

Beckman Coulter, Inc.) and washed using an ultrasonic cleaner with stirring for 10 min to extract collected aerosol particles to the electrolyte. Size distribution was measured with a Multisizer III (Beckman Coulter, inc.) equipped with 20 µm in diameter pore size aperture. The measuring volume per time was 0.05 cm^3. Measurement was repeated 6–10 times for each sample. Particles were counted in 256 size classes within 0.2–6 µm in radius. Subsequently, counts in four size classes were aggregated into one size bin to smooth the measured size distribution.

2.2 Estimation of the Imaginary Part and Single Scattering Albedo

Aerosol optical measurements were conducted using optical particle counters (OPC, KR-12A at Kofu and KC-01D at Tokyo; RION Co., Ltd.), integrating nephelometers (IN, M903; Radiance Research), and particle soot absorption photometers (PSAP; Radiance Research) at Kofu and Tokyo. Complex refractive index and single scattering albedo in 530 nm were retrieved from these simultaneous measurements (Yabuki, 2003).

3. RESULTS AND DISCUSSION

Results of size distributions and OPC measurements at four sites around 13–14 April 2003, when an Asian dust phenomenon was recorded in west Japan, are shown in Fig. 2–5. Size distributions were fitted with a lognormal distribution function. The function is

$$\frac{dV}{d\ln r} = A \exp\left(-\frac{(\ln r - \ln r_{gV})^2}{2\ln^2 \sigma_g}\right), \quad (1)$$

where where r is the particle radius, r_{gV} is the geometrical volume mean radius, and σ_g is the geometrical standard deviation.

Fig. 2 shows the Nagasaki result. All counts increased during period 2. At 1400, 13 April JST (UT+0900), the maximum of the OPC counts in ranges larger than >0.5 µm was recorded (period 3). The size distribution was also recorded as maximum at that time. The respective geometrical volume mean radii were 1.06 µm in period 2, 0.95 µm in period 3, and 1.02 µm in period 4; these geometrical standard deviations were around 2. Results for Okayama are shown in Fig. 3. The SPM concentration increased at 1700, 12 April, but it decreased soon thereafter. After 1000, 13 April, the SPM concentration increased again and the maximum peak height of the size distribution was recorded. The respective geometrical volume mean radii were 0.82 µm in period 3 and 0.90 µm in period 4; these geometrical standard deviations were around 1.94 and 2.22. Fig. 4 shows results at Kofu. After 2300 JST, 13 April, the increase of the OPC counts in the ranges of >0.15 µm and >0.25 µm continued, but the larger ranges counts did not vary. Afterward, the peak height of the size distribution became the maximum (period 5). The geometrical volume mean radius was 0.76 µm. In the following period, the OPC counts in the ranges of >0.5 µm and larger began to decrease, but the smaller range counts were unchanged (period 6). The geometrical volume mean radius shifted smaller of 0.72 µm. Finally, the result for Tokyo is shown in Fig. 5. The OPC counts in the ranges of >0.15 µm and >0.25 µm decreased to 2330 JST, 13 April, subsequently, these counts began to increase again. In contrast, the OPC counts in the ranges of >0.5 µm and larger increased steeply at 2030 JST, 13 April, before 3 h of the beginning to increase of the smaller range counts. Then, the larger range counts recorded the maximum value (period 2). The peak of the size distribution in this period was also the highest value. The geometrical volume mean radius was 0.92 µm. These results indicate that the size distribution of Asian dust varies spatially and temporally over Japan.

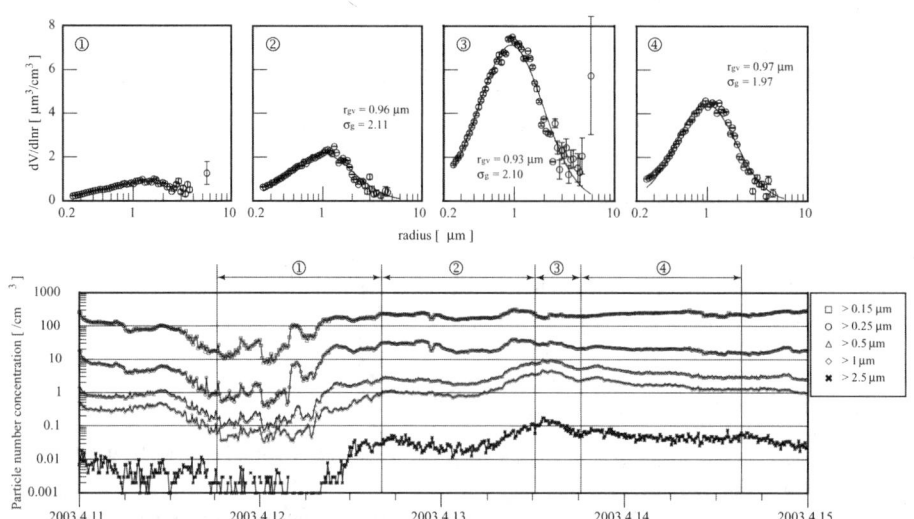

Fig. 2. Size distributions of water-insoluble particle and aerosol number concentration in Nagasaki.

SURFACE MEASUREMENTS AND FIELD EXPERIMENTS

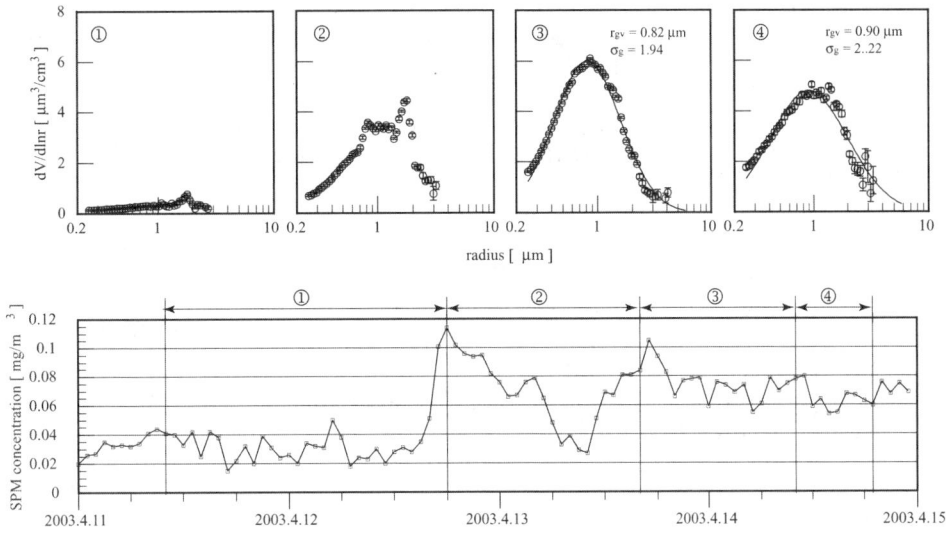

Fig. 3. Size distributions of water-insoluble particle and SPM concentration in Okayama.

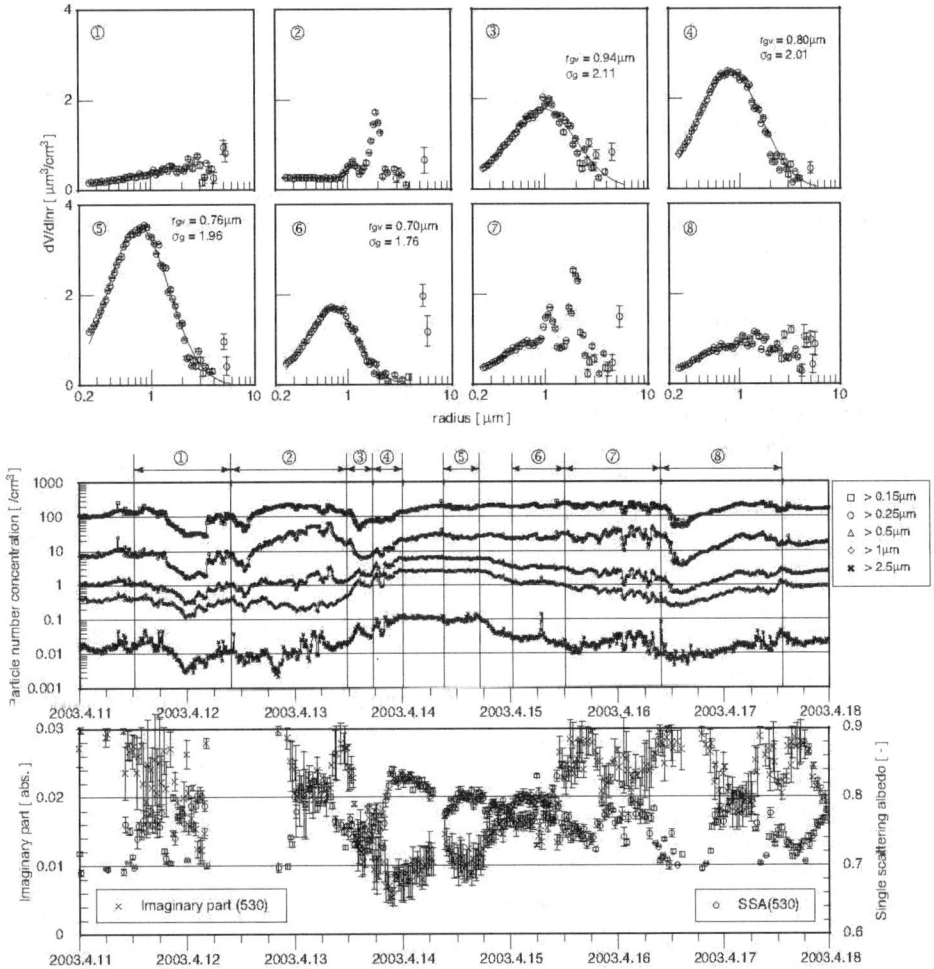

Fig. 4. Size distributions of water-insoluble particle, aerosol number concentration, imaginary part of complex refractive index and single scattering albedo in Kofu.

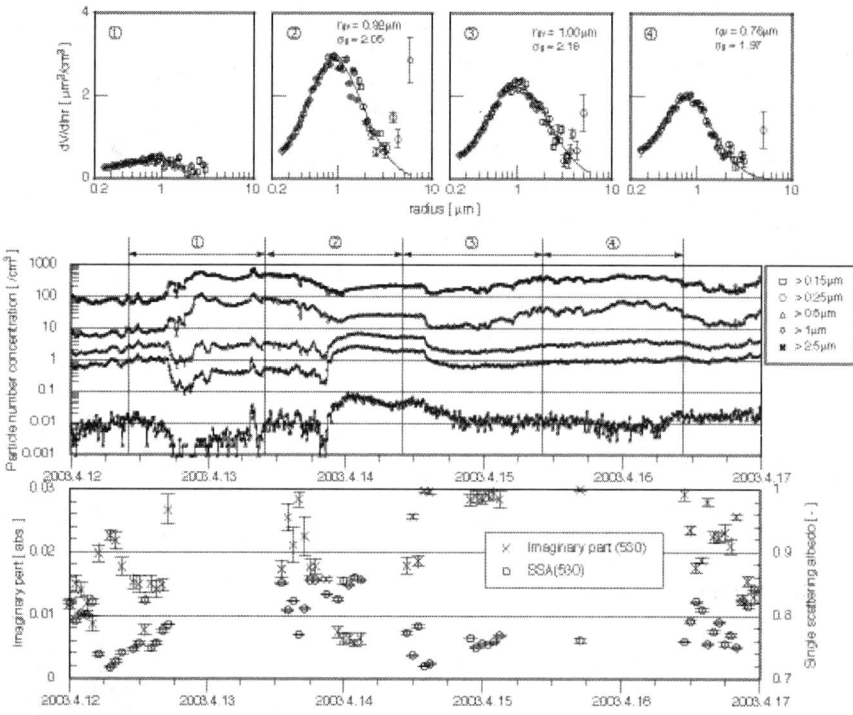

Fig. 5. Size distributions of water-insoluble particle, aerosol number concentration, imaginary part of complex refractive index and single scattering albedo in Tokyo.

The retrieved imaginary part and single scattering albedo are also shown in Figs. 4 and 5. The imaginary part during Asian dust event, especially at the initial stage, was lower than in the other periods at both sites. Absolute values were around 0.007. On the other hand, the single scattering albedo also showed a similar tendency, but of opposite swell. The respective values during the event were around 0.82 and 0.86 at Kofu and Tokyo, which are higher than in other periods. Therefore, the absorbability of the aerosol mixture during Asian dust phenomenon becomes low. It is possible that the Asian dust absorbability is weak.

4. CONCLUSION

Size distribution of water-insoluble particle such as Asian dust was measured using a Coulter Counter. The imaginary part of the refractive index and the single scattering albedo of the aerosol mixture in Asian dust events were also estimated from measurements of OPC, IN, and PSAP. Results indicate that the size distribution of Asian dust varied spatially and temporally over Japan and that the absorbability of the aerosol mixture during Asian dust phenomena is low.

5. ACKNOWLEDGMENTS

This study was supported by a Grant-in-Aid for Scientific Research on Priority Areas under Grant No. 14048228 from Ministry of Education, Culture, Sports, Science and Technology, in Japan.

6. REFERENCES

Intergovernmental Panel on Climate Change (IPCC), Ed., 2001: *Climate Change 2001: The Scientific Basis.* Cambridge Univ. Press, 896 pp.

Sokolik, I. N., D. M. Winker, G. Bergametti, D. A. Gillette, G. Carmichael, Y. J. Kaufman, L. Gomes, L. Schuetz, and J. E. Penner, 2001: Introduction to special section: Outstanding problems in quantifying the radiative impacts of mineral dust. *J. Geophys. Res.*, 106, 18015-18028.

Yabuki, M., M. Shiobara, H. Kobayashi, K. Hara, K. Osada, H. Kuze, and N. Takeuchi, 2003: Optical properties of aerosols in the marine boundary layer during a cruise from Tokyo, Japan to Fremantle, Australia. *Journal of the Meteorological Society of Japan,* 81, 151-162.

AUTOMATIC IMAGE RECORDING NETWORK OF SAND STORMS AND DUSTY AIRS IN NORTHEAST ASIA

Kisei Kinoshita, Naoko Iino, Satoshi Hamada, and Hiroyuki Kikukawa, Kagoshima University, 890-0065, Japan
Tsatsaral Batmunkh and Jugder Dulam, Institute of Meteorology and Hydrology, Ulaanbaatar 210646, Mongolia
Wang Ning and Zhang Gang, Environmental Science Department, Northeast Normal University, Changchun, China
Andrew Tupper, Darwin Volcanic Ash Advisory Centre, Bureau of Meteorology, Australia

ABSTRACT

Digital cameras to detect dusty airs by long-time interval recordings are operating in Changchun, China since the middle of March 2003, and in Ulaanbaatar, Mongolia since the middle of March 2004, by using photo and video cameras. Digital photo camera recordings were also done at two stations, Bulgan and Dalanzadgad in southern Gobi, Mongolia in the spring of 2004. Web-camera recording in Kagoshima, Japan started in December 2000. Heavy dust was observed on 10 March 2004 in Changchun, in conformity with the satellite data, while moderate or light dust were observed in other days and stations in the records of the spring 2004.

1. INTRODUCTION

Recently, great efforts have been made to develop ground observation and lidar networks in China, Korea and Japan, covering the source regions by the Aeolian Dust Experiment on Climate impact (ADEC) collaboration (Mikami et al., 2005). ADEC now covers 80E in Northwest China to 140E in Japan, and Shapotou (37.5N, 104.6E) for the source observation, Hohhot (40.9N, 111.4E) and Sapporo (43.1N, 141.3E) for the lidar network (Sugimoto et al., 2005), and other southern stations. It may be important to extend the observation points further north for the full understanding of the transport of the Asian dust, as some heavy dust episodes take northern routes from northwest China and Mongolia through northeast China, the Siberian Maritime Provinces and the Sea of Japan toward northern Pacific Ocean, while most of the others take southeast routes to Korea and Japan toward the Pacific, as seen by satellite images (Kagoshima Kosa Analysis Group, 2001; Iino et al., 2003). The northern routes were often responsible for the trans-Pacific episodes (e.g., Uno et al., 2001).

In order to observe the dust phenomena from the ground in Northern Asia, digital photo and video cameras have been set in Changchun (43.9N, 125.2E), Northeast China since the middle of March 2003, and also in Ulaanbaatar (47.9N, 106.9E), Mongolia since the middle of March 2004. Digital photo cameras were also set at Bulgan (44.1N, 103.7E) and Dalanzadgad (43.5N, 104.4E) in south Gobi, Mongolia in the spring of 2004. On the other hand, the web-camera recording in Kagoshima (31.6N, 130.5E), southwest Japan, started in the end of 2000 to monitor volcanic clouds, and is also useful for the study of dusty air. There are dust reports from 118 meteorological stations in Japan by visual observations. A heavy dust episode affecting Changchun and Kagoshima in November 2002 was reported in Kinoshita et al. (2003), together with the results of ground observations and satellite imagery of dust events in 2003. In this paper, we discuss the ground observation results in three countries concerning the dust events in 2004, supplemented with the satellite imagery. Preliminary reports were given in Kinoshita et al. (2005a and 2005b).

2. LONG-TERM DIGITAL CAMERA RECORDINGS

The locations of camera observation stations are shown in Fig. 1. Digital photo and video cameras have been set at Northeast Normal University in Changchun, China since 18 March 2003, and also at the Institute of Meteorology and Hydrology, Ulaanbaatar, Mongolia since 16 March 2004. At both stations, the cameras have been powered by AC power sources through uninterrupted power supplies (UPS), as shown in Fig. 2. Digital photo cameras were also set at two meteorological stations, Bulgan and Dalanzadgad in south Gobi, Mongolia on 18 March 2004. The cameras at four stations are listed in Table 1, together with the directions and floors in buildings.

Fig. 1. Location map of camera observation stations (+).

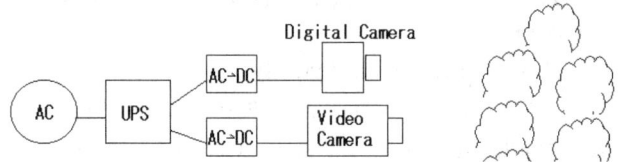

Fig. 2. Camera system for long-term recording.

Table 1. List of Cameras in Four Stations
Focusing length with 35 mm camera equivalent is f_e.

Station	Camera	Type	f_e(mm)	Direction	Floor
Changchun	Photo	Casio QV-R4	37.5–112.5	N	5
Changchun	Video	Sony DCR-TRV40E	48–480	N	5
Ulaanbaatar	Photo	Ricoh Caplio G4wide	28–85	W	3
Ulaanbaatar	Video	Sony DCR-TRV900	41–492	W	3
Bulgan	Photo	Casio QV-R4	37.5–112.5	S	1
Dalanzadgad	Photo	Sharp MD-PS1	43, 86	W	1

The digital photo cameras in all of the above stations were set to take pictures with one-hour interval, and the video cameras were set to take 0.5 sec. records in ten-minute interval. They are expected to run automatically for a few to several months without changing media, as long as the power supply continues without surges and no mechanical trouble. There are many important points to adjust for the initial set-up at the restart (Kinoshita et al., 2004). The white-balance of the color was set to the sunlight mode for all the cameras in order to see the air turbidity from the color information of the sky. The zooming of the cameras was set as wide as possible, within the allowance of the window frames in the respective scenes. At the station in Ulaanbaatar, the photo camera has been tilted by 90 degrees to take wide vertical view, and a semi-fish eye converter lens (0.45 x) has been attached to the video camera.

The following data were obtained in the spring of 2004.

- Changchun: photos, 10 March – 21 August; video records, 10 March – 21 August.
- Ulaanbaatar: photos, 16 March – 20 June; video records, 16 March – 2 June.
- Bulgan: photos, 18 March – 12 April.
- Dalanzadgad: photos, 27 March – 17 June (lacking, March 31, April 1–7 and 14–19, May 6, 12–13 and 20–31, June 8–10).

The performances of automatic recordings were almost without trouble in Changchun and Ulaanbaatar. At Bulgan, isolated in Gobi desert, the camera was powered by a battery-pack outside so as to avoid the long-time shortage of the AC power supply. Only 26 days record was obtained at Bulgan, short by more than 100 days as planned, possibly because of an insufficient battery charge. At the Dalanzadgad station, restarts of the camera were necessary for many times, possibly because of mechanical troubles of the camera.

All of the photo records were edited by html to see a few days at a glance from dawn to dusk. The video records were converted into mpeg files for each day separately. These full data were copied in CDs for research purposes. Three photos per day at local standard time 9, 12, and 15 hours are shown in the homepages linked from http://arist.edu.kagoshima-u.ac.jp/adust/kosa-e/2004kosa/. The Aerosol Vapour Index (AVI) images of NOAA/AVHRR data during the period between 8 February and 31 May 2004 received at Kagoshima University are also displayed there. These are enhanced images of the difference of the brightness temperature of thermal infrared band at 12 and 11 μm given by AVHRR-5 and 4 (Kagoshima Kosa Analysis Group, 2001; Iino et al., 2003). NOAA dust images of Mongolia are displayed by Information and Computer Center (ICC), Mongolia (http://env.pmis.gov.mn).

At Kagoshima University in Japan, a web-camera monitoring Sakurajima volcano has been running since December 2000, taking photographic records of the eastern sky with five-minute interval all through the year at the top of a seven-floored building and archived at http://volceye.edu.kagoshima-u.ac.jp/webcam/archive/.

3. HEAVY DUST EVENT ON 10 MARCH 2004 IN CHANGCHUN

Interval recordings by digital photo and video cameras in Changchun in 2004 started in the evening on March 9. On the next day, very dusty air was observed almost all the day. This is the most prominent event of dusty scenes among the camera records at five stations in the spring of 2004. The photo and video images were significantly different from ones in other days and stations. The event is clearly seen in a NOAA AVI image in Fig. 3a, where heavy dust area without cloud is almost white, extending from Shandong Peninsula to the north in Northeast China, behind very low pressure system shown in Fig. 3b. Heavy dust area is also seen in MODIS images of the brightness temperature differences between the 11 and 12 μm, and also 8.6 and 12 μm channels.

SURFACE MEASUREMENTS AND FIELD EXPERIMENTS

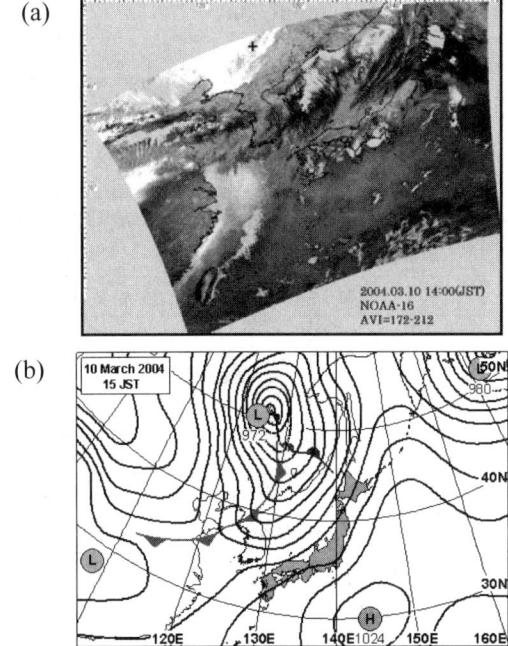

Fig. 3. (a) NOAA AVI image on 10 March at 1300 CST. The cross indicates the location of Changchun. (b) Surface weather map at 1400 CST provided by Japan Weather Association.

In order to identify dusty days from camera images, we compare typical photos of dusty, clear and cloudy skies in the daytime in Fig. 4a, b and c respectively. The RGB (Red, Green and Blue) values in 8 bit along a common vertical line near the center of each image, as indicated in Fig. 4a, are shown in Fig. 5. For dusty skies, the relationship among the colour components is R > G > B, which is reverse to the clear sky case. For cloudy skies, the differences become small, especially between R and G, but the relation is as clear sky case. Scatter diagrams of the values B/G and R/G normalized by the Green component in the upper part of the section in Fig. 4b are shown in Fig. 6, where three clusters are clearly separated corresponding to the air turbidity. Similar result was obtained for the 2003 data (Kinoshita et al., 2003). We may also study the decrease of the visibility in dusty air in the photographic data, from the decrease of the contrast between the buildings far away and the background sky (Kinoshita et al., 2005b).

Fig. 4. The photo scenes at the Changchun station in March: (a) Dusty scene on 10 March, 14 CST, (b) Fine scene on 13 March, 11 CST, and (c) Cloudy scene on 16 March, 12 CST. Vertical lines in (a) and (b) indicate the sections for the for the RGB analyses in Figs. 5 and 6.

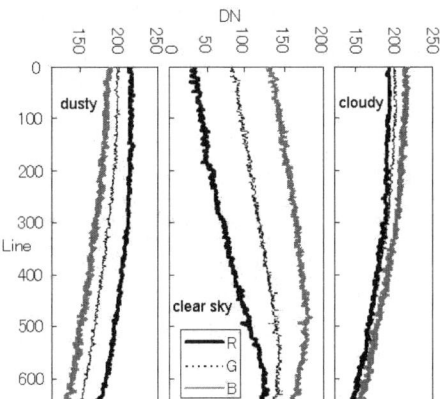

Fig. 5. RGB profiles at sky regions in Fig. 4a, b and c along a common line shown in Fig. 4a.

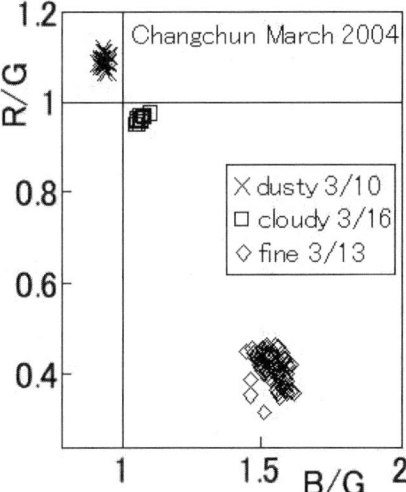

Fig. 6. The scatter diagram of R/G vs. B/G for three scenes in Fig. 4 at the upper line indicated in Fig. 4b.

4. OTHER DUST EVENTS IN THE SPRING OF 2004

We discuss dusty events in the camera records in our stations, the NOAA AVI images, and meteorological information in Mongolia. On 24 and 26 February, dusty scenes in AVI images were seen in northern China. They correspond to synoptic situation in Mongolia summarized in Table 2, with cold front over Mongolia on 23 and 27. The days of dust observation in Japan at more than 4 points of meteorological stations are the followings: Feb: 26, Mar: 11, 12, 14, 15, Apr: 3, 17, 21, 22, May: 7, 11. The last column of Table 2 refers to these days. The cold front over Mongolia on 8–9 March corresponds to the prominent dusty events in Changchun discussed in sec. 3, and dust and sand storm in south Gobi on 9 March in NOAA image processed by ICC (Dulam and Baasandai, 2005).

On 27 March, a dense dust area was observed around the Chinese-Mongolian border in a NOAA image processed by ICC.

Table 2. Dusty Days in Mongolia in Spring 2004 (Dulam and Baasandai, 2005)

2004/ Month	Day	Synoptic situation	No. of stations with dust storms	Gusty wind m/s	Commentary
February	23-24	Cold front over Mongolia on 23	19	12-28	Arrived at Japan on 26 Feb.
February	27	Cold front over Mongolia on 27	14	14-22	
March	8-9	Cold front over Mongolia on 8-9	38	12-28	Arrived at Japan on 11 March.
March	27-28	Cold front over Mongolia on 27-28	14	10-28	
April	13-14	Cold front over Mongolia on 14	36	12-24	Arrived at Japan on 17 April.
April	16-17	Zonal Cold front over Mongolia on 17	25	12-34	
April	20	Cold front over Mongolia on 20	22	12-19	Arrived at Japan on 22 April
April	27-28	Cold front over Mongolia on 27-28	23	10-24	
May	1-2	Cyclone and cold front	39	9-24	
May	4	Cold front over the east of Mongolia on 4	16	12-20	Arrived at Japan on 7 May.
May	6-8	Cyclone sector on 6 and cold front on 7-8	21	10-20	Arrived at Japan on 11 May.
May	11	Cyclone and cold front on 11	12	12-20	
May	18	Cold front over Mongolia on 18	16	12-34	

The camera data in the afternoon at Bulgan and Dalanzadgad stations indicate that there was dusty area southward to them. In the NOAA-AVI images, dusty air in Northeast China was often seen on 28–30. In the camera data in Changchun, it was dusty in almost always on 28 and 29, while it was occasionally dusty on 27 and 30. (Light dusts were also seen on 15-16, 22-23 March there.) In Ulaanbaatar, light dusts were seen in the camera data occasionally in March on 21, 22, 26–28 and 31.

In April and May, somewhat dusty scenes were observed at some of the stations. In general, sandstorms and dusts in the photo records in the spring of 2004 were not so heavy, with the exception of 10 March in Changchun. One reason may be the snow-cover and wetness in Gobi. The snow line was around central Gobi in the middle of March, and snowfall to cover the ground was observed at Bulgan on 31 March.

5. CONCLUSIONS

Interval recording method by photo and video cameras, in conjunction with satellite imagery and other related data, is confirmed to be useful. The results may be utilized for the studies of not only dusty air but also many other aspects of weather changes. RGB analysis applied to interval records is effective to study the turbidity of the air. There are correlations of dusty weather in different countries.

6. REFERENCES

Dulam, J., and E. Baasandai, 2005: Dust storm observations in Mongolia in spring 2004, *Proc. CEReS Int. Symp. Remote Sensing*, Chiba, Japan, Feb. 2005, pp. 100-107.

Iino, N., K. Kinoshita, R. Iwasaki, T. Masumizu, and T. Yano, 2003: NOAA and GMS observations of Asian dust events during 2000-2002, *Proc. SPIE*, Vol. 4895, pp. 18-27.

Kagoshima Kosa Analysis Group, 2001: Satellite Imagery of Asian Dust Events, Kagoshima University, 159 pp.

Kinoshita, K., N. Wang, G. Zhang, S. Hamada, S. Tsuchida, A. Tupper, and N. Iino, 2003: Long-time observation of Asian dust in Changchun and Kagoshima, 2nd Int. Workshop on Sandstorms, Nagoya (to be published in Water, Air and Soil Pollutions).

Kinoshita, K., C. Kanagaki, A. Minaka, S. Tsuchida, T. Matsui, A. Tupper, H. Yakiwara, and N. Iino, 2004: Ground and Satellite Monitoring of Volcanic Aerosols in Visible and Infrared Bands, *Proc. CEReS Int. Symp. Remote Sensing*, Chiba, Japan, Dec. 2003, pp. 187-196.

Kinoshita, K., S. Hamada, N. Iino, H. Kikukawa, J. Dulam, T. Batmunkh, N. Wang, and G. Zhang, 2005a: Interval Camera Recordings of 2004 Asian Dusts in Mongolia, Northeast China and Southwest Japan, 4th ADEC Workshop, Jan. 2005, Nagasaki, Japan, pp. 349-352.

Kinoshita, K., H. Kikukawa, N. Iino, N. Wang, G. Zhang, J. Dulam, T. Batmunkh, and S. Hamada, 2005b: Properties of long-time digital camera records in Changchun and Ulaanbaatar, *Proc. CEReS Int. Symp. Remote Sensing*, Chiba, Japan, Feb. 2005, pp. 136-141.

Mikami, M., et al., 2005. An introduction to Japanese-Sino Joint Project ADEC (Aeolian Dust Experiment on Climate impact), 4th ADEC Workshop, Jan. 2005, Nagasaki, Japan, pp. 1-2.

Sugimoto, N, A. Shimizu, I. Matsui, et al., 2005: Observations of aerosols and clouds with Mie-lidar network, 4th ADEC Workshop, Jan. 2005, Nagasaki, Japan, pp. 33-36.

GROUND-BASED MEASUREMENTS OF UV DOSES AND OZONE AMOUNTS USING FILTER INSTRUMENTS

H. Eide
Light and Life Laboratory
Department of Physics and Engineering Physics
Stevens Institute of Technology
Hoboken, NJ 07030, USA

A. Dahlback
University of Oslo, Norway

K. Stamnes
Stevens Institute of Technology, USA

B.-A. Høyskar
Norwegian Institute for Air Research (NILU), Norway

R. O. Olsen, F. J. Schmidlin, and S. C. Tsay
NASA, USA

Jakob Stamnes
University of Bergen, Norway

ABSTRACT

Recent technology advances have made ground-based measurements of UV doses and ozone column amounts with inexpensive multi-channel filter instruments not only possible, but also an attractive alternative to other more labor-intensive and weather dependent methods. Filter instruments can operate unattended for long periods of time, and it is possible to obtain accurate ozone column amounts even on cloudy days. We present results from extensive comparisons of the performance of several ground-based instruments. These include the NILU-UV and GUV filter instruments, as well as the Dobson and Brewer instruments. We also compare data obtained with the ground-based instruments with data from the EP-TOMS instrument. Our results show that ozone column amounts obtained with current filter-type instruments are just as good as those obtained with the Dobson instrument, and certainly out-performs the Dobson instrument on cloudy days. We also find that the results compare well with EPTOMS, even for high zenith angles. Here we briefly review the results of the intercomparisons and discuss data quality issues of importance to operating filter UV instruments. Our assessment is that this type of instruments can supplement, or even replace, the more labour intensive and older ground-based UV instruments for long-term data records, provided that reasonable care is taken when operating the instruments.

1. INTRODUCTION

We present results from intercomparisons between various ground-based instruments for determination of total ozone column amounts. These multi-channel filter-instruments are an attractive alternative to more expensive and resource consuming alternatives since they are capable of operating unattended and can provide high precision data even on cloudy days. Multi-channel filter radiometers have no moving parts and their calibration is performed using a spectroradiometer and standard procedures. A calibration against another filter instrument can also be done. The instruments are primarily designed to measure irradiances in a few channels in the UV and visible part of the solar spectrum. A method for derivation of biologically weighted UV doses, cloud effects, and total ozone column amounts (Dahlback et al., 1996, Høyskar et al., 2003), allows one to use these instruments to provide value-added products that are of great interest to the scientific community. We have used data from filter instruments at stations covering latitudes from 38° N to 79° N and compared them with data from standard ozone instruments like the Brewer, Dobson, and the TOMS instrument aboard the Earth Probe satellite. The data used in the comparisons are from five different sites: the University of Oslo, Norway, the NASA Goddard Space Flight Center facilities at Wallops Island, Virginia, Stevens Institute of Technology, New Jersey, Longyearbyen on Spitsbergen, and the University of Alaska, Fairbanks (during the TOMS3F campaign). A few of the instruments have been in continuous operation since 1995 and the stability if these instruments are discussed. The results show that ozone column amounts derived from this type of instrument are accurate and that the influence of clouds on ozone column retrievals is small.

2. RESULTS

A comparison of daily total ozone measurements from NILU-UV #13 and AD direct sun measurements from Dobson #56 in Oslo for 2000–2003 is shown in Fig. 1. The distance between the instruments is approximately 20 km. The Dobson measurements are performed within 1 hour from local solar noon and the NILU-UV measurements are based on a 1-hour average around local solar noon (normally 60 single measurements). The mean difference (NILU-UV – Dobson)/Dobson for the 3 year period is 0.4% ± 1.9% (1 standard deviation). A similar comparison for the Oslo GUV #9222 and the Dobson was done for the same time period. The mean difference here is 1.7% ± 1.4%. No pronounced seasonal variations in the ratios are observed for the NILU-UV or the GUV.

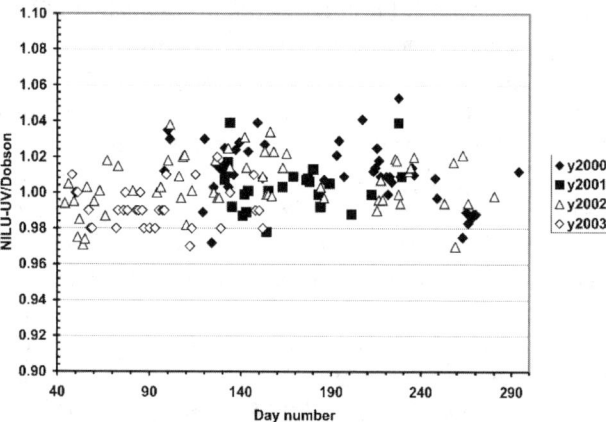

Fig. 1. Comparison of daily ozone measurements from NILU-UV #13 and Dobson #56 in Oslo 2000–2003. The mean difference is 0.4% ± 1.9%.

GUV instruments have been in continuous operation in Oslo (60° N) and Ny-Ålesund, Spitzbergen (79° N) since 1995. Comparisons of ozone column amounts derived from these GUV's with satellite measurements from the Earth Probe TOMS instrument for 1996–1999 are shown in Figs. 2 and 3. The data contain cloudy as well as clear skies. In Oslo for day 80–260 (which is the season when the noon SZA is less than 60°) the relative difference is 0.0% ± 2.5%. For the entire year the difference is -0.4% ± 3.9%, but this includes the two-month period when the SZA at noon is larger than 80° and the column amounts derived from both instruments depend on the ozone profile. The relative difference for Ny-Ålesund for the period March 20 to September 20 when the SZA is less than 80° is 0.1% ± 3.3%. The results of the comparisons of NILU-UV and GUV with Dobson and TOMS are shown in Table 1.

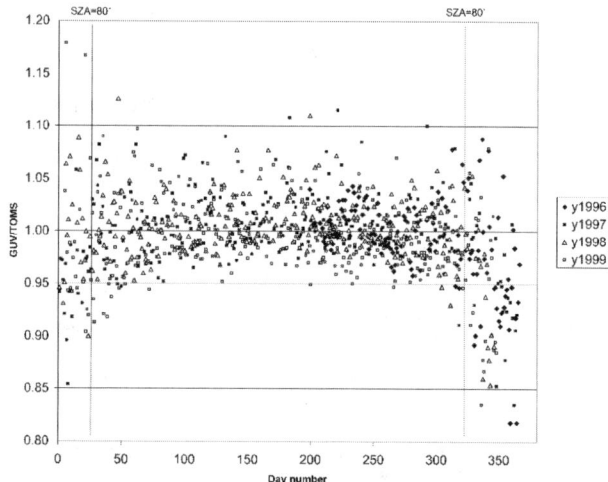

Fig. 2. Comparison of daily GUV and TOMS ozone measurements in Oslo 1996–1999. Cloudy days are included.

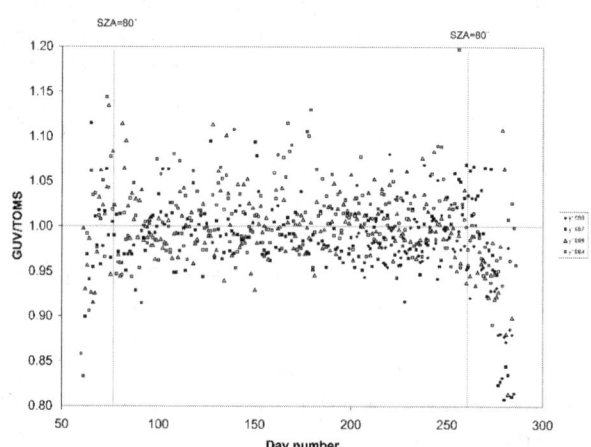

Fig. 3. Comparison of daily GUV and TOMS ozone in Ny-Ålesund 1996–1999. Cloudy days are included.

To summarize; an extensive intercomparison of data from the NILU-UV and GUV filter instruments, with data from the Dobson, Brewer, and EP-TOMS instruments, shows that the percentage differences for total ozone amount is generally less than 2% with standard deviations of about 2%. The ozone retrievals are found to be little affected by clouds for cloud transmission greater than about 40%.

3. FILTER INSTRUMENT OPERATION

Special design features in and around the diffuser and detectors of the filter UV instruments are used to improve the cosine response of the instruments. The photo-detectors report photon counts to the data-logger.

Table 1. Comparison of NILU-UV and GUV with Dobson and TOMS (Relative Differences in Percent ± 1σ). [1]Dobson AD Direct sun 2000–2003. [2]Days with SZA < 80°, Cloudy Days Included, 1996–1999. [3]Days with SZA < 60° Cloudy Days are Included, 1996–1999

Site	NILU-UV vs Dobson[1]	GUV vs Dobson[1]	GUV vs TOMS[2]	GUV vs TOMS[3]
Oslo	0.4 ± 1.9	1.7 ± 1.4	0.2 ± 3.2	0.0 ± 2.5
Ny-Åselund			0.1 ± 3.3	0.4 ± 3.3

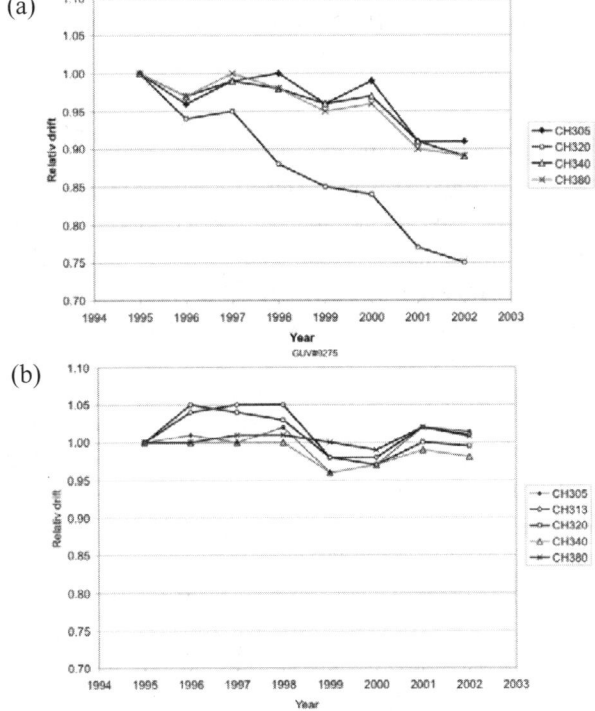

Fig. 4. Relative drift for the four UV channels in GUV #9222 in Oslo, Norway (a) and for the five UV channels in GUV #9275 at Ny-Ålesund, Spitzbergen, Norway (b). The drift is relative to a frequently calibrated traveling reference GUV instrument.

Since photo-detectors (and possibly the filters) are affected by the operating temperature, the photo-detectors are heated to ensure a stable operating temperature. The detector operating temperature is typically set to 40°C, but it could be set to some other temperature as long as the instrument calibration is performed at the same temperature. Dark-currents are also recorded at calibration time.

Ozone column amounts derived from the GUV and NILU-UV instruments are nearly insensitive to clouds if the cloud optical depth is below a certain value (or the cloud transmission CLT is above a certain value). We have found that if the CLT > 30–40% the influence of clouds on the measured ozone column amount is negligible. Please see Dahlback et al., 2003 for details. Ozone column amounts derived from the GUV and NILU-UV instruments are based on comparison of a measured irradiance ratio with the same ratio computed with a radiative transfer model. Our results show that the measured ozone column amount is independent on the profile for SZA < 65°. For large SZA the derived ozone column amount depends strongly on the choice of ozone profile. Using the right ozone profile we have found that ozone measurements typically are good for SZA up to at least 80°.

Filter-instruments usually have several channels centered at selected wavelength bands. The filters and detectors are subject to influence by environmental conditions that can change their characteristics. For this reason filter instruments are designed to minimize these influences. Unfortunately, some instruments experience degradation over time due to filter drift. There are several ways to detect filter drift. One obvious method is to compare calibration factors between two consecutive calibrations. If there was no drift in any of the components from one calibration to the next these factors will be the same. Comparisons of ozone column amounts derived from different channel combinations provide useful information about the stability of the instrument. Also other derived products can be used to detect filter drift. As an example we can look at two GUV instruments, GUV #9222 and GUV #9275. These instruments have been used in the Norwegian UV monitoring network since 1995 and they are annually (normally in June) compared to a traveling reference GUV instrument. This way any drift over time in the individual channels is detected. The results for 1995 to 2002 are shown in Fig. 1.

While #9275 is a quite stable instrument #9222 shows a pronounced degradation with time particularly in the 320 nm channel. However, we may account for this drift in the data analysis. The ozone column amount is derived from a channel ratio and different channels can be used. Normally the 320/305 ratio is used, but 340/305 and 380/305 also give good results (however with slightly larger cloud effect than the 320/305 ratio). The effect of the large drift in the

320 nm channel for #9222 can be avoided by using the 340/305 or the 380/305 ratio in the periods when the sensitivity in channel 320 changes rapidly.

The accuracy of an instrument that is in need of drift correction is largely determined by our ability to characterize the drift and the effect the drift has on the instrument sensitivity. Tests performed with recently calibrated instruments show that drift can be completely corrected for, and thus the instrument capable of producing its nominal 2% accuracy for the derived ozone amounts and dose rates. In other cases the instrument sensitivity is degraded and drift correction can only partially relieve the problem.

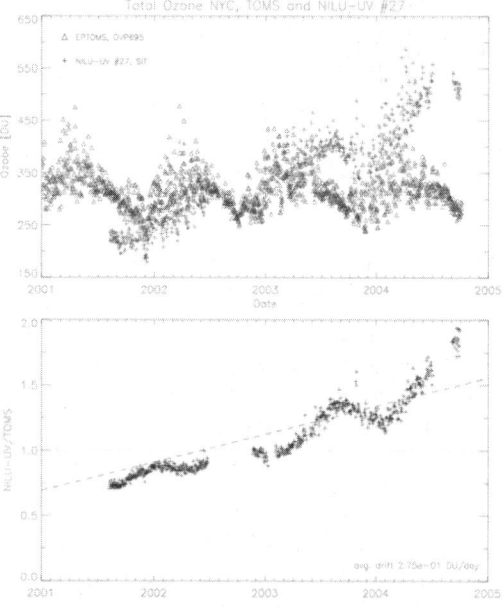

Fig. 5. Lacking other sources for comparison of total ozone amounts, one can compare ground-based data sources with EP-TOMS data and over time possibly detect drift in instrument calibration. This figure shows a comparison of ozone amounts derived from NILU-UV #27 at Stevens Institute of Technology in the New York City area. The data in this series was "normalized" using data obtained in mid March 2003 from a reference NILU-UV instrument (#21). Drift in #27 is clearly present throughout all of the period, leading to a progressively worsening sensitivity to ozone variations as compared to TOMS. The reason for the drift is unknown and currently under investigation.

4. SUMMARY

We have shown that total ozone column amounts derived from irradiance measurements with NILUUV and GUV instruments are of high quality for clear as well as cloudy skies. The percentage differences between the filter-instruments and the Dobson's are less than 2% and with standard deviations less than 2%. The influence of clouds on ozone measurements with the filter-instruments is found to be small for CLT > 30–40%. We have found that the ozone measurements are insensitive to ozone profile for SZA < 60°. The stability of filter-instruments seems to be variable. Some are very stable over years, while other starts to drift without any clear reason. Like for any instrument designed to measure ozone column amounts frequent calibrations or inter-comparisons are required to detect any drift. This can be done with a traveling reference instrument or standard lamps. By correcting the data for the measured drift the quality of the data is assured. Our assessment is that these instruments are accurate and stable enough to contribute to the long-term UV data record.

5. REFERENCES

A. Dahlback, H. Eide, B. Høiskar, R. Olsen, F. Schmidlin, S.-C. Tsay, and K. Stamnes, 2003: Comparison of data for ozone amounts and uv doses obtained from simultaneous measurements with various standard uv instruments, *Proc. SPIE*, Volume: 5156.

Dahlback, A. 1996: Measurements of biologically effective uv doses, total ozone abundances, and cloud effects with multichannel, moderate bandwidth filter instruments, *Appl. Opt.* 35, pp. 6514–6521.

Høyskar, B. A. K., R. Haugen, T. Danielsen, A. Kylling, K. Edvardsen, A. Dahlback, B. Johnsen, M. Blumthaler, and J. Schreder, 2003: Multichannel moderate-bandwidth filter instrument for measurement of the ozone-column amount, cloud transmittance, and ultraviolet dose rates, *Appl. Opt.*, 42, pp. 3472–3479.

NEW RADIATION AND ENERGY BALANCE OF THE WORLD AND ITS VARIABILITY

Atsumu Ohmura

Institute for Atmospheric and Climate Science, Swiss Federal Institute of Technology (E.T.H.)
CH-8057 Zurich, Switzerland

ABSTRACT

A new global energy (heat) balance is presented, based on a composite database of surface measurements, satellite-based observations and GCM computations. The present picture differs greatly from its predecessors in larger atmospheric absorption of solar radiation, smaller global radiation at the earth's surface and larger terrestrial (longwave) atmospheric radiation. The causes of the errors in earlier works are explained. Furthermore, the long-term variability of radiation and its effect on global energy balance are discussed.

1. INTRODUCTION

Energy balance on the surface plays a key role in determining the thermal conditions of the climate system. Radiation in particular plays an important role within the entire energy balance, because 1) the energy cycle begins with radiation; 2) radiative components are large; and 3) the energy flow ends with radiation. Furthermore, most climate changes are initiated by changes in radiation. The Milankovitch hypothesis is an example of solar radiation causing a substantial climate change. The enhanced greenhouse effect is another example being caused however by terrestrial radiation. Throughout the 20th century, a number of works were presented for the quantitative evaluation of the flux components. In its early stages the energy balance was primarily a computational science. It was mainly during the International Geophysical Year (IGY, 1957–58) that the global network of instrumental observations was established. Today we have observational data of about half a century, which can withstand scientific scrutiny, revising earlier misconceptions and thus opening a new chapter. The author presents in this article a new global energy balance, using observations wherever possible, but also integrating satellite measurements and model computations as needed. Finally, the observed long-term variability in energy balance will be discussed.

2. SOLAR RADIATION

The earlier radiation studies no doubt underestimated the atmospheric absorption in the solar spectral range. In the following discussion, the absorption will be expressed as a relative quantity with respect to the TOA (Top Of Atmosphere) solar irradiance. The most frequently published absorption values cluster around 17%. This bias was partly caused by overestimating the planetary albedo, an example of which is 0.40 by Budyko (1956). This trend was subsequently corrected. The bias was also caused by overestimating surface solar radiation (global radiation). The larger global radiation (about 200 Wm^{-2}) is still widely upheld in the literature, for example Kiehl and Trenberth (1997) which is reproduced in the 3rd IPCC (Intergovernmental Panel on Climate Change) report (Houghton et al., 2001). These errors are large and remained to date unquestioned mainly because many researchers have not seriously looked at experimental results from the laboratory and the field. The unnaturally high transparency of the atmosphere was conceived at a very early stage. Abbot and Fowle (1908) set out with 18% atmospheric absorptance based on the Fowle's equation which is now known to underestimate water vapour absorption. The underestimation of absorption (15 to 17%) was subsequently presented in a number of key textbooks, such as Sellers (1965), Flohn (1968) and Fortak (1971). The transparency of the atmosphere for solar radiation was often exaggerated, as this feature conveniently stressed the importance of the greenhouse effect (Houghton et al., 1990). One of the earliest attempts to correct this tendency was Budyko's (24%) world atlas of heat balance (1963), although the larger absorption in this work went largely unnoticed by majority of scientists. London and Sasamori (1971) also envisaged larger absorption (22%). These earlier works relied heavily on computation as observations did not provide the necessary degree of accuracy. Later, Budyko et al. (1988) and Ohmura and Gilgen (1993) obtained much larger total absorption of 26% and 28%, respectively. These works used shortwave net radiation at TOA and the earth's surface, both of which were observed.

The simulations by GCMs also underestimated atmospheric absorption, hence overestimated surface global radiation. One of the earliest GCM intercomparisons was made among NCAR, GFDL and GISS, resulting in a good agreement of global radiation within 10 Wm^{-2} around the mean of 200 Wm^{-2} (Gutowski et al., 1988). This amount is almost equivalent to the annual mean global radiation for the Hawaiian Islands. It is difficult to imagine that sunny Hawaii could possibly represent the global mean. Obviously earlier GCMs were tuned to mimic the belief of the time. The first comprehensive radiation codes intercomparison (Fouquart et al., 1991) also indicated that the median value of the absorption calculated by the 26 participating models was significantly smaller than the absorption computed by the line-by-line model. The underestimation by the GCMs was already noticeable by examining the absorption only by water vapour. It was, however, not mentioned that the reference computation by two line-by-line models was producing about 20% absorption by water vapour alone in the cloudless atmosphere. This would have been a good opportunity to recognise the missing absorption in water vapour rather than in clouds (Cess et al., 1995).

Two independent methods were used to find realistic atmospheric absorption. Simple but accurate model estimations were made based on MODTRAN. The absorption of solar radiation was simulated for various precipitable water vapour (from 0 to 60 mm) with the water vapour alone mode. This small experiment shows that about 75 Wm^{-2} solar radiation is absorbed with a mean precipitable water vapour of 25 mm, corresponding to about 20% of the TOA solar irradiance. The second experiment presented in Table 1 is also based on MODTRAN with plausible concentrations of radiatively active gases and material. The absorption by individual components is evaluated with the adding and subtracting method and the absorption is expressed for the global mean. The result is the absorption of 96 Wm^{-2}, or 28%. If a certain amount of dark aerosol and strong cloud-radiative effect are added, the total atmospheric absorption of solar radiation can exceed 100 Wm^{-2}.

Table. 1. Atmospheric Absorption of Solar Radiation Simulated with MODTRAN for the Globe (in Wm^{-2}) (Cloud Radiative Effect is Due to Computations made Based on ISCCP D2)

[H$_2$O]=25 mm water equivalent	70.1
[O$_3$]=300 DU	13.1
[CO$_2$]=370 ppm	3.8
[CH$_4$]=1.7 ppm	1.1
Aerosol : Optical depth=0.1	
Singlescattering albedo=0.95	5.1
Cloud (ISCCP)	2.8
Total absorption	96

The second method to estimate the absorption of solar radiation by the entire atmosphere is to calculate the divergence (more appropriately "the difference") of shortwave net radiation at the TOA and the earth's surface. The global mean TOA shortwave net radiation is accurately known and lies between 236 (Rossow and Zhang, 1995) and 240 Wm^{-2} (Barkstrom et al., 1990; B. Barkstrom, personal communications). The geographical distributions of global radiation and albedo at the earth's surface have recently been re-evaluated. The re-evaluation was based on long-term observations at about 900 sites for land surfaces and the computations by ECHAM4 with AMIP (Atmospheric Model Inter-comparison Project) SST (Sea surface Temperature) and sea-ice distribution for sea surfaces. The ECHAM4 computations for sea surfaces are regionally adjusted to observations done on islands. This analysis gave global radiation of 170 Wm^{-2}, albedo of 0.14 and shortwave net radiation of 145 W^{-2}. This new result is very close to the earlier work by Ohmura and Gilgen (1993) with global radiation of 169 Wm^{-2}, albedo, 0.15 and shortwave net radiation 144 Wm^{-2}. Since the uncertainty is about ± 3 Wm^{-2}, these two results can be regarded as identical. The divergence of shortwave net radiation at TOA and the earth's surface, hence the total absorption by the entire atmosphere is 95 Wm^{-2}. These two quite different approaches gave virtually identical results. The main reason for the smaller absorption of earlier works is the underestimation of the absorption by water vapour and aerosol. There are recently a number of works supporting the larger absorption of solar radiation in the atmosphere (Wild et al., 1997; Hatzianastassiou and Vardavas, 2001). However many GCMs and regional models are still equipped with radiation codes that significant underestimate solar absorption. Many models participating in the Atmospheric Model Intercomparison Project II (AMIPII) have not improved radiation codes. The mean underestimation by the 20 participating GCMs is 9 Wm^{-2} (Wild, 2005). Many these models are also participating in the IPCC AR4. Obviously, the systematic error of this magnitude which is much larger than the effect of doubled CO_2, is not taken seriously by many modellers.

3. TERRESTRIAL RADIATION

Accurate observations of terrestrial (longwave) radiation became possible only after the introduction of precision pyrgeometers. This type of an instrument required detailed characterisations, before being put in operation (Albrecht and Cox, 1977; Philipona et al., 1995). These developments helped detect a systematic underestimation inherent to pyrradiometers that have been in use since 1950s (Ohmura and Gilgen, 1993). Pyrradiometers underestimate terrestrial atmospheric radiation during the day, as the solar heating of the up-facing sensor induces convective heat loss. This principle applies also to net radiometers which are widely used today. For the last 15 years a network with accurate observation of terrestrial radiation has been operating under the Baseline Surface Radiation Network (BSRN) of the World Climate Research Programme (WCRP). This network is providing accurate and stable norm values for 36 sites around the world. The most important finding from this network is the detection of the systematic underestimation of the terrestrial atmospheric radiation in earlier climatology as shown in Table 2. In fact, some climatological calculations came close to realistic values in 1960s. Presently the most reliable global distribution of the terrestrial atmospheric radiation is obtained by using ERA15, but adding corrections for the regions with the BSRN observations. Using this method, the best estimate for the global mean terrestrial atmospheric radiation is 345 Wm^{-2}. Unlike global radiation, the observational records of terrestrial atmospheric radiation are not long enough to detect temporal changes.

A number of GCMs still underestimate this component. In case of ECHAM the code improvement experienced a quantum jump from ECHAM3 to ECHAM4 by re-coding the H_2O continuum (Giorgetta and Wild, 1995). This is an example of a successful code-improvement. Rather than blindly tuning the code, they identified the causal problem of the underestimation and systematically corrected the errors. A number of GCMs with significant underestimations are participating in the IPCC AR4. Some of these models exhibit errors larger than 10 Wm^{-2}. These models are incapable of simulating the future climate under increased greenhouse gases. Therefore, for a meaningful model intercomparison in the future, only the best models should be used.

Table 2. Global Mean Terrestrial (Longwave) Atmospheric Radiation Proposed by Various Authors and Models (in Wm^{-2})

Dines (1917)	194	Satellite	
Albrecht (1949)	328	ISCCP (D2)	343
Budyko (1956)	333		
London (1957)	336	GCMs	
Budyko (1963)	344	ECHAM3	335
Sellers (1965)*	273	NCAR	311
Flohn (1968)*	328	GFDL	326
Fortak (1971)*	272	GISS	335
London & Sasamori (1971)	342	CCC	315
Budyko et al. (1988)	345	LMD	342
Ohmura & Gilgen (1993)	350	ECHAM4	345
Salby (1996)*	327		
Ohmura (1997)	345		
Kiehl & Trenberth (1997)	324		
Pavlakis et al. (2004)	342		

*) Text books

Sources in this table are not listed in references unless they are mentioned in the text.

4. NET RADIATION AND NON-RADIATIVE COMPONENTS

Summing up the above discussion, and noting that the mean surface temperature is 14°C, the most likely global mean net radiation is 105 ± 5 Wm^{-2}. Since the subsurface net heat flux which is mainly used for heating the ocean is still a fraction of 1 Wm^{-2}, 105 Wm^{-2} can safely be assumed to be spent for latent heat of evaporation and sensible heat flux. These are the components for which we have no possibility of evaluating global means based on measurements. Long-term observations of turbulent fluxes are rare and virtually non-existent over the ocean. This is the type of problem for which model simulations are suited. Comparing the latent and sensible heat fluxes, the former has several advantages. Firstly, the latent heat flux is observed at many more sites than sensible heat flux. Secondly, latent heat flux, when integrated for the whole globe should be the same as the global mean precipitation which is observed. However, several well executed GCM simulations show a wide range of global mean latent heat flux. Latent heat fluxes computed in GCMs, however, correlate very well with net radiation which is a heat source for evaporation. Since the net radiation is better known than latent heat flux, and annual global precipitation is close to 1000 mm y^{-1}, mean latent heat flux is the order of 85 Wm^{-2}. The remaining 20 W m^{-2} is attributed to sensible heat flux. When global means calculated with various GCMs are compared, ERA15 comes closest to the observed values. Curiously the re-analyses with more years such as ERA40 and NCEPRA fell behind ERA15. This might indicate that the longer re-analyses are not necessarily better for computing regional fluxes. The input data of earlier days are probably not suited for accurate computation of fluxes. The newly evaluated global energy balance is presented in Fig. 1.

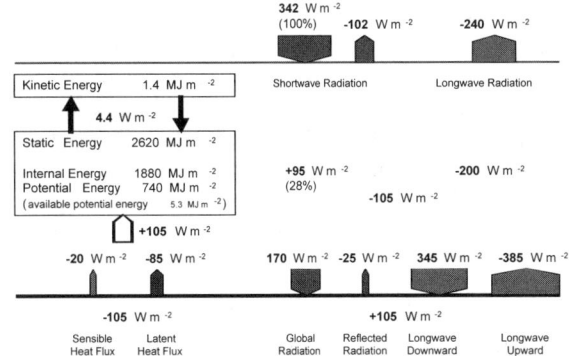

Fig. 1. New global energy balance of the atmosphere and earth's surface.

5. LONG-TERM CHANGES OF FLUXES

The irradiances were believed to be fairly constant. There is an authoritative and convincing argument by Budyko (1971) that global radiation is a stable meteorological element which observed over several years can be used as a climate norm. It was the geographical variations which were considered important in early days and not the temporal changes. The possibility of a long-term variation of energy fluxes was first identified with respect to global radiation (Ohmura and Lang, 1989). In fact most well kept time series of global radiation showed a systematic change, lasting for a longer period and overlapping short-term, year-to-year fluctuations. This type of change has a longer trend than several years, which is characteristic for the volcanic effect. For example, the stations with more than 50 years records in Europe and Japan show a peak in 1940s and 50s and a declining trend thereafter. This declining trend subsequently called global dimming has not lasted to present in many regions. In Europe, Japan, North America, and many parts of China the trend reversed in late 1980s and is presently on way to brightening (Wild et al., 2005). In less extensive areas in Africa, India and the western Pacific global dimming is continuing. To detect recent trend-reversals, the time series of global and tropical cloud amounts prepared by ISCCP (International Satellite Cloud Climatology Project) and interpreted by W. B. Rossow (NASA) was most instrumental. The magnitude of the changes in global radiation is 5 to 10 W m^{-2} over a period of 20 to 30 years. This variation was explained with the changes in cloud amount by Ohmura and Lang (1989). Liepert et al. (1994) related the variations of global radiation in Germany to changes in aerosol. Recently, however, it was found that the clear-sky total transmissivity of the atmosphere is highly correlated not only with the clear sky global radiation but also with the mean global radiation (Wild et al., 2005). What we see here is a concerted change among aerosol, clouds and global radiation. The effect of the variable solar radiation on actual evapotranspiration was discussed by Ohmura and Lang (1989).The effect of global radiation upon the variation in pan evaporation was recently pointed out by Roderick and Farquhar (2002). Processes between radiation and other terrestrial exchanges remain to be clarified. The magnitude of the variations discovered by observations tell, however, that the global energy balance as discussed above and illustrated in Fig. 1 should not be taken as constant but rather as variable. It is also possible that another component such as the steady flow of heat into the sub-surface may have to be introduced, if the on-going warming intensifies (Ohmura, 2004; Hansen et al., 2005).

6. REFERENCES

Abbot, C.G., and F. W. Fowle, 1908: Radiation and terrestrial temperature. *Ann. Astrophys. Obs. Smithsonian Inst.*, 2, 125-189.

Albrecht, B., and S. K. Cox, 1977: Procedures for improving pyrgeometer performance. *J. Appl. Meteorol.*, 16, 188-197.

Barkstrom, B., E. F. Harrison, and R. B. Lee, III, 1990: Earth Radiation Budget Experiment. *EOS*, 71, 297-305.

Budyko, M. I., 1956: *Heat Balance of the Earth's Surface.* (German translation by Pelzl, E.), Fachliche Mitteilungen 1/100, East German Air Force, Porz-Wahn, 282 pp.

Budyko, M. I. (Ed.), 1963: *Atlas Teplovogo Balansa Zemnogo Shara.* Akademiya Nauk SSSR, Moscow. 69 pp.

Budyko, M. I., 1971: *Climate and Life. International Geophys. Ser.*, 18, Academic Press, NY and London, 508 pp.

Budyko, M. I., G. S. Golitsyn, and Y. A. Izrael, 1988: *Global Climatic Catastrophes.* Springer Verlag, Berlin, 99 pp.

Cess, R. D., and co-authors, 1995: Absorption of solar radiation by clouds: observations versus models. *Science*, 267, 496-499.

Flohn, H., 1968: *Vom Regenmacher zum Wettersatelliten.* Kindlers Universitäts Bibliothek, Kindler Verlag, Munich, 254 pp.

Fortak, H., 1971. *Meteorologie.* Carl Habel Verlag, Berlin and Darmstadt, 287 pp.

Fouquart, Y., and B. Bonnel: 1991: Intercomparing shortwave radiation codes for climate studies. *J. Geophys. Res.*, 96, D5, 8955-8968.

Giorgetta, M., and M. Wild, 1995: The Water Vapour Continuum and its Representation in ECHAM4. Report 162, Max Planck Institute, Hamburg, 38 pp.

Gutowski, W. J., D. S. Gutzler, D. Portman, and W. Wang, 1988: Surface Energy Balance of Three General Circulation Models: Current Climate and Response to Increasing Atmospheric CO_2. TRO42, DOE/ER/60422-H1, U. S. Department of Energy, 119 pp.

Hansen, J., L. Nazarenko, R. Ruedy, M. Sato, J. Willis, A. Del Genio, D. Koch, A. Lacis, K. Lo, S. Menon, T. Novakov, J. Perlwitz, G. Russell, G. A. Schmidt, and N. Tausnev, 2005: Earth's energy imbalance: Confirmation and implications. *Sciencexpress*, 1/10.1126/science.1110252.

Hatzianastassiou, N., and I. M. Vardavas, 2001: Shortwave radiation budget of the Southern Hemisphere using ISCCP C2 and NCEP/NCAR climatological data. *J. Climate*, 14, 4319-4329.

Houghton, J. T., G. J. Jenkins, and J. J. Ephraums, (Eds.), 1990: *Climate Change, The IPCC Scientific Assessment.* Cambridge Univ. Press, Cambridge, 364 pp.

Houghton, J. T., Y. Ding, D. J. Griggs, M. Noguer, P. J. van der Linden, X. Dai, K. Maskell, and C. A. Johson, (Eds.), 2001: *Climate Change 2001: The Scientific Basis.* Cambridge Univ. Press, Cambridge, 881 pp.

Kiehl, J. T., and K E. Trenberth, 1997: Earth's annual global mean energy budget. *Bull. Am. Met. Soc.*, 78, 197-208.

Liepert, B., P. Fabian, and H. Grassl, 1994; Solar radiation in Germany-Observed trends and an assessment of their causes, Part I: Regional approach. *Beitr. Phys. Atmosph*, 15-29.

London, J., and T. Sasamori, 1971; Radiative energy budget of the atmosphere. In Matthews, W. H., W. W. Kellogg, and G. D. Robinson (Eds.): *Man's Impact on the Climate*, M.I.T.Press, Cambridge, 141-155.

Ohmura, A., and Lang, H., 1989: Secular variation of global radiation in Europe. In Lenoble, J. and Geleyn, J.-F. (Eds.): *IRS'88: Current Problems in Atmospheric Radiation*, A. Deepak Publ., Hampton, VA, 298-301.

Ohmura, A. and Gilgen, H., 1993: Re-evaluation of the global energy balance. *Geophys. Monogr.*, 75, 93- 110.

Ohmura, A., 2004: Cryosphere during the Twentieth Century. *Geophys. Monogr.*, 150, 239-257.

Philipona, R., C. Fröhlich, and C. Betz, 1995: Characterization of pyrgeometers and the accuracy of atmospheric long-wave radiation measurements. *Appl. Optics*, 34, 1598-1605.

Roderick, M. L., and G. Farquhar, 2002: The cause of decreased pan evaporation over the past 50 years. *Science*, 298, 1410-1411.

Rossow, W. B., and Y. Zhang, 1995: Calculation of surface and top of atmosphere radiative fluxes from physical quantities based on ISCCP data sets. Part II: Validation and first results. *J. Geophys. Res.*, 100, D1, 1167-1197.

Sellers, W. D., 1965: *Physical Climatology.* Univ. Chicago Press, Chicago and London, 272 pp.

Wild, M., Ohmura, A., and Cubasch, U., 1997: GCM-simulated surface energy fluxes in climate change experiments. *J. Climate,* 10, 3093-3110.

Wild, M., 2005: Solar radiation budgets in atmospheric model intercomparisons from a surface perspective. *Geophys. Res. Lett.*, 32, doi:10.1029/2005GL022421.

Wild, M., Gilgen, H., Roesch, A., and Ohmura, A., 2005: From dimming to brightening: Decadal changes in solar radiation at the Earth's surface, *Science*, 308, 847-850.

CONTINUOUS MEASUREMENT OF AEROSOL CHARACTERISTICS BY ADEC SKY-RADIOMETER NETWORK

Akihiro Uchiyama and Akihiro Yamazaki
JMA Meteorological Research Institute
1-1 Nagamine, Tsukuba, Ibaraki 305-0052, JAPAN

Hiroki Togawa
JMA Niigata District Observatory
4-4-1 Saiwainishi, Niigata, 950-0908, JAPAN

Jun'ichi Asano
JMA Meteorological Satellite Center
3-235 Nakakiyoto, Kiyose, Tokyo 204-0012, JAPAN

ABSTRACT

Aeolian dust drifting from arid and semi-arid region in the central region of continents is known to cause dust storms. Aeolian Dust Experiment on Climate impact (ADEC) started in April 2000 to investigate the supply of dust to the atmosphere and to estimate the radiative forcing of dust. In this project, a sky-radiometer network was developed in order to investigate the characteristic of aeolian dust on the way of transportation from the source region to the Japan area. Furthermore, we planed and implemented Intensive Field Observation in March and April, 2002 and in April, 2003 and obtained some results. After the first ADEC Intensive Observation period (IOP) in 2002, the measurement by the sky-radiometer has been continued. We have accumulated the continuous observation data more than 2 years by the ADEC sky-radiometer network.

1. INTRODUCTION

As described in the report of Intergovernmental Panel on Climate Change (IPCC), there are many studies on the radiative effect of aerosol on climate but its level of scientific understanding is very low (IPCC, 2001). The effect of mineral dust on radiative forcing is also one of the most uncertain factors. Aeolian dust drifting from arid and semi-arid region in the central region of Asian continent is known as the cause of dust storms "Kosa" event. This event causes not only serious damage to agriculture, economics and human health in these regions but also a climatic impact on the global scale.

Aeolian Dust Experiment on Climate impact (ADEC) started in April 2000 to investigate the supply of dust to the atmosphere and estimate the radiative forcing of dust using the aerosol transportation model. In this project, a sky-radiometer network is developed in order to investigate the characteristics of aeolian dust on the way of transportation from the source region to the Japan area. The first intensive field observation was performed during the period from April 8 to April 21, 2002, (ADEC IOP1). In this period, aeolian dust came flying to Japan area and dust events were often observed (Kurosaki and Mikami, 2003).

The observations of aeolian dust by the sky-radiometer have already been reported on some papers (e.g., Aoki and Fujiyoshi 2003, Kim et al., 2004). The previous studies analyzed the data measured at one or two observation sites. The observation of ADEC sky-radiometers network was the first network observation of aeolian dust in the eastern Asia region. In this paper, we briefly describe ADEC sky-radiometer network and results of IOP-1 and continuous measurements.

2. SKY-RADIOMETER NETWORK

The sky-radiometer (PREDE Co., Ltd. POM-01, POM-02) is a radiometer to measure not only the solar direct irradiance but also the radiance from the sky. The wavelength position of the sky-radiometer used in the ADEC project is 340, 380, 400, 500, 675, 870, 940 (H_2O), 1020, 1225, 1600, 2200 nm. The interference filters were used to resolve the spectral positions. The band width of channels less than 1225 nm is 10 nm and that for channel greater than 1600 nm is 20 nm. The accuracy of the band center is +/−2 nm. In this study, we use the channels of the wavelength less than 1020 nm, except 940 nm, which is the water vapor channel.

The sky-radiometers were setup at the five observation sites in China and the four observation sites in Japan. The observation sites in China are Qira (80.73E, 37.02N), Aksu (80.83E, 40.62N), Shapotou (105.00E, 37.46N), Qingdao (120.33E, 36.07N) and Beijing (116.37E, 39.97N). The observation sites of Japan are Naha (127.69E, 26.20N), Fukuoka (130.37E, 33.55N), Nagoya (136.97E, 35.15N) and Tsukuba (140.13E, 36.05N).

Qira and Aksu are in the source region of aeolian dust. Shapotou is in the source region and also on the way of transportation from the western part of China. Qingdao is in the coast of the continent. Instruments were set up at each site in March 2002, except Beijing, which was set up in September 2003.

In Qira, Aksu, and Shapotou sites, only sky-radiometers were set up. In others sites, the pyranometers and pyrgeometer were set up as well in order to measure the downward irradiances. The location of the sites is shown in Fig. 1. As seen from the figure, the observation sites are placed to catch the aeolian dust transported from the western part of China.

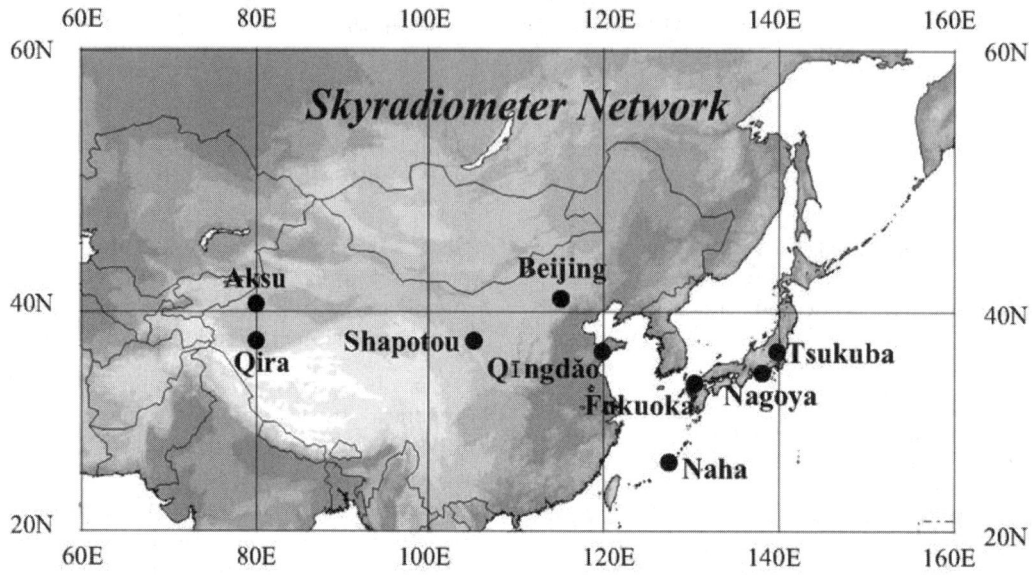

Fig. 1. ADEC Sky-radiometer network site map.

The Qira, Shapotou, Naha, Fukuoka, and Nagoya sites are connected through telephone lines and data are transferred to Tsukuba site everyday. The data transferred to Tsukuba site is processed and analyzed every day. The processed data could be seen on the ADEC web site and was utilized for monitoring the present status of aerosol. After IOP1, data transfer and data processing have continued, and the analyzed data has been opened through the ADEC web site.

3. DATA ANALYSIS

In this study, we used the latest version of the Skyrad package, which is software to analyze the sky-radiometer data and firstly developed by Nakajima et al. (1996). The retrieval software to analyze sky-radiance were developed by Nakajima et al. (1996) and Dubovik and King (2000). The latest version of this software can retrieve the complex refractive index as well as the optical thickness and particle size distribution (Nakajima et al., 2003 and Dubovik and King, 2000). However, in this study, we used the constant complex refractive index; m=1.50−0.005i.

The analysis of sky-radiometer needs the total amount of ozone and the surface pressure to take account of ozone absorption and Rayleigh scattering of air molecules. The total amount of ozone is taken from TOMS data. The surface pressure is taken from the nearest SYNOP data from observation site. The surface albedo was assumed to be 0.1 and Lambert surface for all channels. The sensitivity test for the change of surface albedo showed that the sensitivity for change of albedo was small.

The radiance from the sky is measured by the horizontal scan (equi-zenith angle scan). The data between 3 and 30 degrees from the sun are used. The retrieved data are optical thickness at each channel and volume size distribution from 0.0121 to 16.5 μm in radius. As shown in Figs. 2 and 3, Ångström exponent is used as an index of size distribution; lower Ångström exponent means that the relative amount of large particles in the size distribution is high and the larger Ångström exponent means that the relative amount of small particles in the size distribution is high. The Ångström exponent α is defined as exponent when we approximated the wavelength dependence of optical thickness as power of wavelength (λ),

$$\tau(\lambda) = \tau_{500}(\lambda/\lambda_{500})^{-\alpha}.$$

where τ_{500} is the optical thickness at wavelength 500 nm. α and τ_{500} are determined by least squares method from all channel optical thickness data.

4. RESULTS OF ADEC IOP-1

Several dust events occurred during the period of ADEC IOP1 in April 2002 and they were observed by ADEC sky-radiometer network. In this paper, we describe only summary of IOP-1 (Uchiyama et al., 2005).

(1) The data with Ångström exponent between 0.0 and 0.5 corresponds to the dust event day data. The major differences of size spectra between dust event day and non dust event day are the large increase of particle volume spectra greater than 0.5 μm in radius. (2) The contribution of the particles with radius greater than 0.5 μm, which correspond to coarse dust particles, to the total optical thickness frequently exceeds more than ~70% in the source region, and exceeds in Qingdao and sites in Japan on the dust event day. (3) The contribution of coarse dust particles to the total volume is more than ~80% on the dust event day. (4) The retrieved volume spectrum in the source region, Aksu and Qira, is not dependent on the optical thickness. This means that the floating aerosols mainly consist of dust particles in the source region. (5) The total volume observed in Japan sites is one third of that in the source region. (6) When the size distribution for the coarse mode is approximated by log-normal size distribution, effective radius r_{eff} is 2.1 to 2.3 μm in China sites and 1.6 to 1.8 μm in Japan sites. r_g is about 0.7 μm at Aksu, Qira, and Shapotou and about 0.5 μm at other sites ($\ln r_g$ is the center of log-normal size distribution). The width of size distribution (σ_g) is scattered between 0.67 and 0.87. These difference of size distribution among the observation sites are caused by the modification of air mass including aeolian dust; coarser dust particles are partially removed during the transportation, and the air mass including aeolian dust as a main component are partially contaminated by the aerosol into the atmosphere from anthropogenic activities. The observation network of ADEC caught this change clearly.

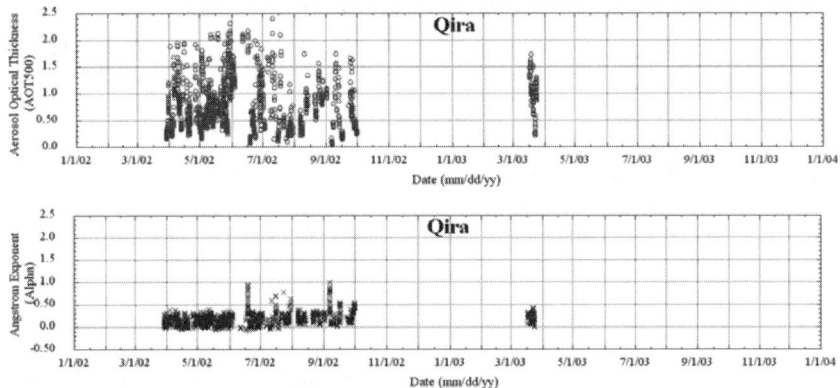

Fig. 2. Time series of optical thickness at 500 nm (upper part) and Ångström exponent (lower part) at Qira in 2002 and 2003.

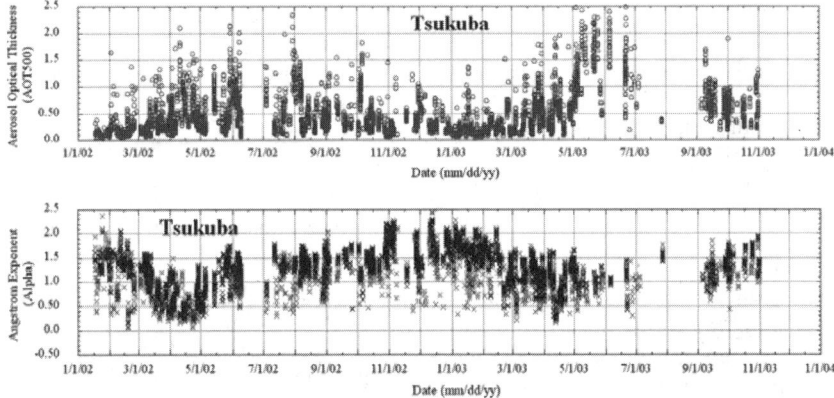

Fig. 3. Time series of optical thickness at 500 nm (upper part) and Ångström exponent (lower part) at Tsukuba in 2002 and 2003.

5. RESULTS OF CONTINOUS OBSERVATION

After the first ADEC Intensive Observation period (IOP) in 2002, the measurement by the sky-radiometer has been continued except the calibration, maintenance, and malfunction periods as much as we can. We have accumulated the continuous observation data by the ADEC sky-radiometer network for 3 years.

In this paper, all data cannot be shown. Time series of optical thickness at 500nm (τ_{500}) and Ångström exponent (α) at Qira and Tsukuba are shown as examples.

The following results can be seen from the time series of optical thickness at 500nm and Ångström exponent. The optical thickness has seasonal variation. In the source region (Qira, Aksu, and Shapotou), optical thickness is larger in spring and optical thickness is smaller in summer than in spring. The sites in Japan (Fukuoka, Nagoya, Tsukuba) have also seasonal variation; optically thick in summer, and optically thin in winter. Ångström exponent in Qira is almost constant (0.0 to 0.5) all over the season. This means that in the source region the major part of aerosol consists of dust all over the season. At the other sites, Ångström exponent in spring is lower than that in the other season. The lower Ångström exponent in spring is observed even in Naha site, which is located in the most southern region and surrounded by the ocean. The lower Ångström exponent events were more frequently observed in the spring 2002 than in the other year. This corresponds to more frequent occurrence of dust event in spring 2002 than in the other years. The sky-radiometer network detected the phenomena other than the dust event; for example Siberia forest fire in May 2003.

6. SUMMARY

The sky-radiometer network was developed in order to investigate the aeolian dust from the source region to the Japan area, and intensive observations were implemented in April 2002. We could not trace a specified dust storm due to the coarse network. However, the ADEC observation sites are located in the source region of aeolian dust and on the way of transportation, and therefore the average feature expresses the change of characteristics of aerosol. The change of size spectrum for aeolian dust from the western part of China to the Japan area by anthropogenic activities could be seen.

We have accumulated sky-radiometer data since March 2002 after ADEC first IOP. From these data, we can clarify the dust properties by comparing the spring dust season and the others. Furthermore, since there is a possibility that the sky-radiometer network can detect phenomena other than dust event, we can clarify the aerosol properties in the East Asian area. At some observation sites, pyranometers and pyrgeometer are installed as well as the sky-radiometer. Analyzing the solar irradiances and the retrieved data from sky-radiometer, we will be able to estimate the radiative effect of aerosol in East Asian area.

7. ACKNOWLEDGEMENTS

This work was supported in the project of Aeolian Dust Experiment on Climate impact (ADEC) funded by the Ministry of Education, Culture, Sports, Science and Technology of Japan. The network observation was made under the cooperation of many Chinese and Japanese scientists and engineers. The authors would like to express special thanks to Prof. Shi Guangyu, Institute of Atmospheric Physics, Chinese Academy of Science (CAS). The observation in China would be never performed without his cooperation and assistance.

8. REFERENCES

Aoki K, and Y. Fujiyoshi, 2003: Sky radiometer measurements of aerosol optical properties over Sapporo, Japan, *J. Meteor. Soc.*, Japan, 81, 493-513.

Dubovik O., and M. D. King, 2000: A flexible inversion algorithm for retrieval aerosol optical properties from Sun and sky radiance measurements, *J. Geophys. Res.*, 105, No. D16, 20, 673-20696.

IPCC 2001, Climate Change 2001: The Scientific Basis, Contribution of Working Group I to the Third Assessment Report of the Intergovernmental Panel on climate change (IPCC) (edited by J. T. Houghton, Y. Ding, D. J. Griggs, M. Noguer, P. J. Van der Linden, X. Dai, K. Maskell, and C. A. Johnson), *Cambridge University Press.*

Kim D.-H., B.-J. Sohn, T. Nakajima, T. Takamura, T. Takemura, B.-C. Choi, and S.-C. Yoon, Aerosol optical properties over east Asia determined from ground-based sky radiation measurements, *J. Geophys. Res.*, 109, D02209, doi:10.1029/2003JD003387,2004.

Kurosaki Y., and M. Mikami, 2003: Recent frequent dust events and their relation to surface wind in East Asia, *Geophys. Res. Lett.*, 30, No. 14, 1736, doi:10.1029/2003GL017261.

Nakajima, T., G. Tonna, R. Rao, Y. Kaufman, and B. Holben, 1996: Use of sky-brightness measurements from ground for remote sensing of particulate polydispersions, *Appl. Opt.*, 35, 2672-2686.

Nakajima T., M. Sekiguchi, T. Takemura, I. Uno, A. Higurashi, D. Kim, B. J. Sohn, S.-N. Oh, T. Y. Nakajima, S. Ohta, I. Okada, T. Takamura, and K. Kawamoto, Significance of direct and indirect radiative forcing of aerosols in the East China Sea region, *J. Geophys. Res.*, 108, No. D23, 8658,doi:1029/2002JD003261,2003.

Uchiyama A., A. Yamazaki, H. Togawa, and J. Asano, 2005: Characteristics of Aeolian dust observed by sky-radiometer in the Intensive Observation Period 1 (IOP1), accepted in *J. Meteorol. Soc. Japan.*

AEROSOL OPTICAL DEPTH MEASUREMENT WITH MFRSR, UV-MFRSR, AND SUN PHOTOMETER AT GWANGJU, KOREA

Jeong Eun Kim and Young Joon Kim
Advanced Environmental Monitoring Research Center (ADEMRC)
Department of Environmental Science and Engineering
Gwangju Institute of Science and Technology (GIST)
1 Oryong-dong, Buk-gu, Gwangju 500-712, Republic of Korea

ABSTRACT

Aerosol optical depth (AOD) and Ångström exponent were determined with two multifilter rotating shadowband radiometers; visible-MFRSR, UV-MFRSR, and a CIMEL sun photometer at the Advanced Environment Monitoring Research Center (ADEMRC), Gwangju Institute of Science and Technology (GIST), Korea (35°13`N, 126°50`E) in February 2004. MFRSR measures global and diffuse radiation with 1 broadband and 6 narrowband channels in the visible and near IR ranges. UV-MFRSR is an UV version of MFRSR, which has 7 narrowband UV channels. CIMEL sun photometer which has been employed at Gwangju since February 2004 as a part of NASA AERONET network gives the AOD information in 440, 670, 870, and 1020 nm. During a study period of 4 ~ 27 February 2004 mean spectral AOD varied from 0.15±0.16 at 1020 nm to 0.73±0.30 at 304 nm. For the same observation period sun photometer retrievals of AOD exceeded MFRSR AOD values by 0.025 with a correlation coefficient of 0.985 except an Asian dust case. Average Ångström exponent derived from MFRSR and sun photometer data was 1.20±0.28 and 1.24±0.25, respectively. However, mean Ångström exponent in UV range was 2.16±0.87. When Asian dust event occurred on 24 February 2004, AOD from sun photometer data varied from 1.10 at 440 nm to 1.11 at 1020 nm with an Ångström exponent of 0.01. MFRSR-derived AODs and Ångström exponent were determined to be 0.80 (870 nm) ~ 0.88 (415 nm) and 0.18, respectively. The difference in AODs and Ångström exponent were due to inconsistency of measurement time on this date. UV-MFRSR based AOD was 0.98 (367 nm) ~ 1.27 (310 nm) with an Ångström exponent of 0.88 on an Asian dust day. Aerosol size distributions, asymmetry parameter, and single scattering albedo were determined from the sun photometer data using the AERONET inversion algorithm (Dubovik et al., 2000). Aerosol bimodal size distribution retrieved from the sun photometer data showed a fine mode peak at around 0.1μm and a coarse mode peak at around 3 μm with the single scattering albedo of 0.90 ~ 0.98 at 440 nm.

1. INTRODUCTION

Atmospheric aerosol is known to impact the earth's climate by directly scattering and absorbing solar radiation and by indirectly modifying cloud lifetime and cloud albedo. However, to quantify these effects is still uncertain compared to other factors affecting the earth's climate in predicting climate change (IPCC, 2001). Ground-based radiation measurements provide column aerosol radiative properties such as aerosol optical depth, Ångström exponent, single scattering albedo, phase function, etc. Temporal and spatial radiation data for long time can evaluate the numerical models of aerosol optical behaviors to predict the effects of aerosol on climate change.

Ground-based aerosol optical depth measurements provide global aerosol climatology and validate satellite aerosol retrieval. The well-known methodologies to acquire aerosol optical depth at the ground level are sun photometers and shadowband radiometers. Aerosol Robotic Network (AERONET) operated by NASA employed Cimel sun photometer for aerosol optical depth retrieval, and U.S. Department of Agriculture (USDA) UV-B Monitoring Program is using shadowband radiometer (MFRSR by Yankee Environmental System). However, instrument comparisons of two types at a single site for the same period are not common. Therefore, a comparison of two different types of instruments, sun photometer and shadowband radiometer can improve the quality of data product.

2. METHODOLOGY

2.1 Research Site and Instrumentation

The measurement site, Gwangju Institute of Science and Technology (GIST) at Gwangju (35.13°N, 126.50°E) is located in the southern part of Republic of Korea and is surrounded by urban area, agricultural area, and an industrial area. However, the effect by local anthropogenic sources such agricultural biomass burning and vehicle emission is not significant at this site. Long range transported air pollutants are main reason to increase aerosol optical depth (AOD) at Gwangju site. This site began sun photometer operation from February 2004 as a permanent AERONET site. A visible-

MFRSR and UV-MFRSR are also operated since June 1998 and February 2002, respectively, at the same location.

CIMEL sun photometer gives the information about aerosol optical properties such as AOD in 440, 670, 870, and 1020 nm, Ångström exponent, single scattering albedo, etc. AOD is derived from direct sun radiance measurements. A visible-MFRSR and an UV-MFRSR measure the global and diffuse spectral irradiances to retrieve direct normal irradiance and to determine aerosol optical properties, such as the aerosol optical depth and Ångström exponent. The UV-MFRSR measures the spectral UV irradiances centered at wavelengths 304, 311, 317, 323, 331, and 367 nm while visible-MFRSR measures the spectral irradiances centered at 415, 500, 615, 673, and 870 nm. The measurement outputs from UV-MFRSR and visible-MFRSR are sampled every 20 seconds and 15 seconds, respectively, and were averaged over one-minute intervals and saved into two separate YESDAS (Yankee Environmental Systems) data loggers.

2.2 Aerosol Optical Depth and Ångström Formula

Total atmospheric optical depth was determined by analyzing direct normal solar irradiance from the total and diffuse solar irradiances measured from the UV-MFRSR. Beer-Lambert law provides the total atmospheric optical depth. Aerosol optical depth is determined by subtracting the contributions by Rayleigh scattering, total column ozone, and water vapor. Water vapor absorption is not considered since the wavelengths used in the this study are free from absorption by water vapor.

$$I = I_0 \exp(-\tau_{total} m), \quad (1)$$

$$\tau_{total} = \tau_{ozone} + \tau_{watervapor} + \tau_{Rayleigh} + \tau_{aerosol}, \quad (2)$$

where I_o is the solar irradiance at the top of the atmosphere, I is the direct normal solar irradiance, τ is the total optical depth, and m is the air mass. Ozone optical depth is retrieved from the product of ozone absorption cross section at 226K by Molina and Molina (1986) and total column ozone retrieved from UV-MFRSR data (Gao et al., 2001).

The Ångström formula determines the spectral dependence of aerosol optical depth.

$$\tau(\lambda) = \beta \lambda^{-\alpha} \quad (3)$$

where $\tau(\lambda)$ is the aerosol optical depth (AOD) at each wavelength, λ, β is a turbidity coefficient, and α is the Ångström exponent[4]. In this study the Ångström exponents were calculated from spectral AODs from the linear regression fit in the UV and visible-near IR ranges for UV-MFRSR and visible-MFRSR, and visible-near IR range for sun photometer.

3. RESULTS AND DISCUSSION

3.1 Aerosol Optical Depths

Variations of aerosol optical depth (AOD) from visible-MFRSR, UV-MFRSR, and sun photometer are plotted in Fig. 1. One min data were averaged to daily value. The AOD results showed a distinct anticorrelation with wavelength. Daily mean values of AOD varied from 0.39 to 1.20 at 304 nm and from 0.04 to 0.56 at 870 nm. The average values of AOD during the observation period were 0.74±0.27, 0.74±0.30, 0.60±0.27, 0.57±0.27, 0.54±0.27, 0.46±0.26, 0.39±0.23, 0.33±0.18, 0.29±0.18, 0.27±0.16, 0.24±0.15, 0.18±0.13 at 304, 310, 317, 323, 331, 367, 415, 500, 615, 673, and 870 nm. Ångström exponents at visible range (α_{vis}) and at UV range (α_{uv}) were 1.10±0.32 and 2.81±1.05, respectively.

During the study period Asian Dust event was observed on 26 February. On this date AOD increased from 0.36 to 0.85 and 0.16 to 0.56 compared to non Asian Dust period average at 415 nm and 870 nm, respectively, while α_{vis} decreased from 1.12 to 0.59. AOD at higher wavelength (2.5 times) decreased more distinctly than at shorter wavelength (1.4 times). Comparing to the minimum AOD case, AOD increased by 4~5 times when Asian Dust event occurred on 26 February. Data at UV range were missing on 26 February.

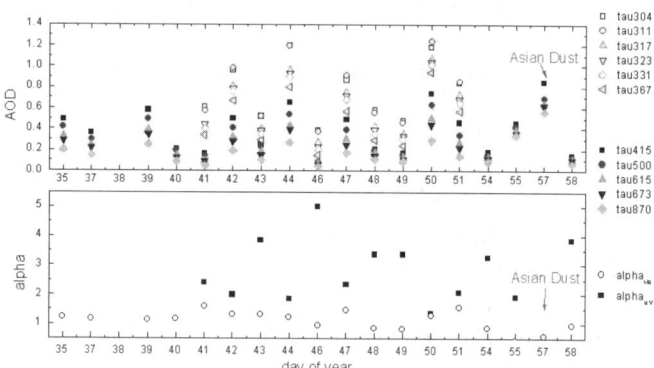

Fig. 1. Variation of aerosol optical depths (top) and Ångström exponents (bottom) retrieved from MFRSR and UV-MFRSR in February 2004 at Gwangju, Korea.

3.2 Sun Photometer

Fifteen min AOD data were averaged into daily mean values from sun photometer direct sun radiance and were plotted in Fig. 2. The spectral AOD retrieved from sun photometer also decreased with wavelength. The AOD varied from 0.13 to 0.57 and 0.05 to 0.25 at 440 nm and 1020 nm, respectively. Daily average values during the period were 0.31±0.15, 0.24±0.14, 0.18±0.09, 0.15±0.07, 0.13±0.06 at 440, 670, 870, and 1020 nm, respectively. Ångström exponent was retrieved from the spectral AODs at 440, 670, and 870 nm. The average value was 1.09±0.27. When Asian Dust occurred only one set of AOD was retrieved and the values were 0.44, 0.31, 0.27, and 0.25 at 440, 670, 870, and 1020 nm. Ångström exponent was 0.74 which was much lower than the average value for non Asian Dust period (1.12).

AOD at 550 nm was estimated from AOD at 500 nm and Ångström exponent for visible-MFRSR and at 440 nm and Ångström exponent for sun photometer in order to compare the AOD results. The comparison result of AOD at 670 nm and alpha between visible-MFRSR and sun photometer was shown in Fig. 3. The results of collocated visible-MFRSR and sun photometer show good coincidence with 5% and 3% difference in AOD and alpha retrieval, respectively. The magnitude of difference is plotted in Fig. 4 for AODs at 670 nm and 870 nm and alpha. The regression analysis gives us a linear relationship of AOD and Ångström exponent between visible-MFRSR and sun photometer as

$$\tau_{sun} = 0.9021 \times \tau_{MFR} + 0.0013 \quad (4)$$
$$\alpha_{sun} = 0.9626 \times \alpha_{MFR} + 0.0895 \quad (5)$$

The correlation coefficients are 0.90 and 0.91 for AOD and Ångström exponent, respectively. Namely, visible-MFRSR retrieval is slightly higher than sun photometer results. The instrument types and retrieval processes might attribute the difference. MFRSR is thermostat type and sun photometer is not. MFRSR determines I_0 from measured irradiance and Langley method, while sun photometer applies reference values of I_0 from calibration procedure.

Single scattering albedo (total particulates) and asymmetry parameter (total particulates) were also retrieved from sky radiance of sun photometer. These parameters were retrieved only for five cases. Fig. 5 shows the values of single scattering albedo and asymmetry parameter calculated from sun photometer at Gwangju, Korea. Single scattering albedo and asymmetry parameter at shorter levels is usually higher than at longer wavelength. For five days observed, single scattering albedo varied from 0.95 to 0.08 except on 16 February when the value was exceptionally low. Low single scattering albedo means higher absorption by particles. Usually single scattering albedo is believed to be near 1.0 for sulfate and marine aerosols and ranges from 0.5 to 0.7 for dust and soot aerosols (Lacis and Mishcenko, 1995; Lenoble 1993). Thus lower single scattering albedo indicates the higher portion of dust and soot aerosols. However, spectral AODs were close to mean and Ångström exponent was higher than usual values which showed low possibility of dust effect. The study site is located outskirts of urban city and near agricultural area. Usually the atmosphere in February over the study site is affected by agricultural activities, particularly field burning to prepare rice cultivation. Carbonaceous aerosols, especially elemental carbon, emitted from field burning might absorbed the radiation and lower the single scattering albedo. We should be very cautious to conclude the reason for the low single scattering albedo since the evidence is not enough. Therefore, continuous monitoring of single scattering albedo and agricultural activities are necessary. Asymmetry parameter usually relates to particle size distribution. The values ranged from 0.6 to 0.73 for 5 cases.

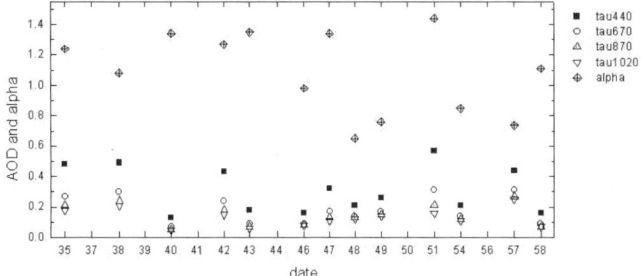

Fig. 2. Variation of aerosol optical depths and Ångström exponents retrieved from sun photometer direct sun radiance in February 2004 at Gwangju, Korea.

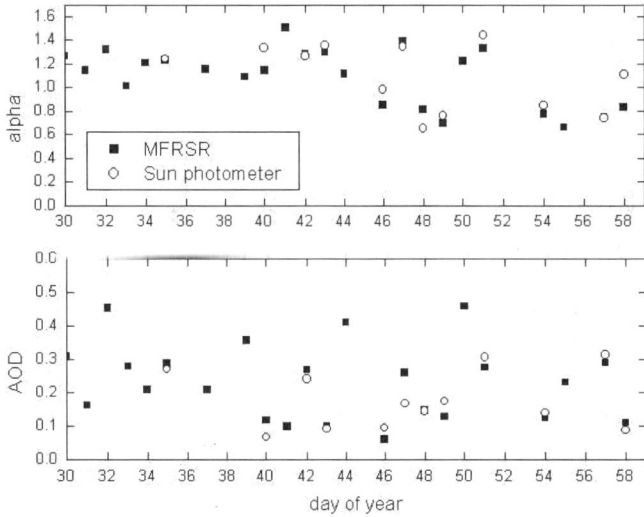

Fig. 3. Comparison of AOD at 670 nm alpha from visible-MFRSR and sun photometer retrievals.

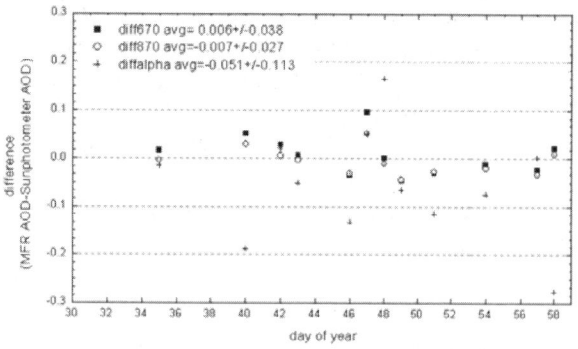

Fig. 4. Magnitude of difference between daily mean of MFR AOD/alpha and sun photometer AOD/alpha.

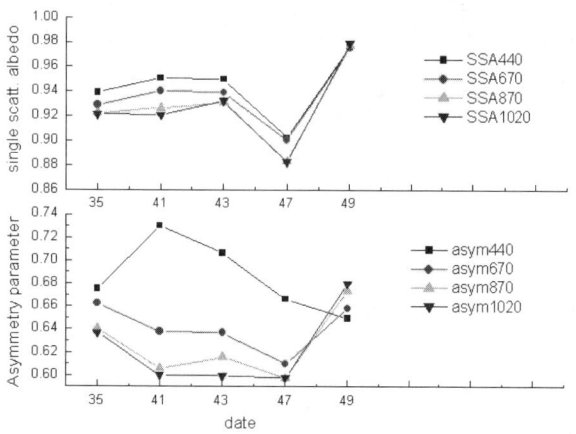

Fig. 5. Single scattering albedo and asymmetry parameter retrieved from sun photometer in February 2004 at Gwangju, Korea.

4. CONCLUSIONS

The aerosol radiative properties from three different radiation instruments were retrieved and intercompared for some parameters. The correlation coefficients for aerosol optical depth at 550 nm and Ångström exponent were 0.77 and 0.91, respectively. The difference may be attributed to the retrieval process and instrument type. Asian Dust event on 26 February increased 4~5 times of AOD compared to background condition of atmosphere and lowered the Ångström exponent to 0.59 which is half of background value of Ångström exponent. Due to cloud single scattering albedo and size distribution couldn't be retrieved. Only five cases of single scattering albedo, and asymmetry parameter were retrieved. For some cases the inferred results could be confirmed by the other parameters.

Newly installed sun photometer was proved to be produced similar values of aerosol optical depth and Ångström exponent with existing visible-MFRSR and provides more aerosol optical properties. Continuous measurements of sun photometer at the study site will contribute to clarify the local aerosol optical properties more clearly.

5. ACKNOWLEDGEMENT

This work was supported in part by the Korea Science and Engineering Foundation (KOSEF) through the Advanced Environmental Monitoring Research Center (ADEMRC) and Ministry of Education & Human Resource Development through the Brain Korea 21 (BK21) at Gwangju Institute of Science and Technology.

6. REFERENCES

Ångström, A., 1961: Techniques of determining the turbidity of the atmosphere, *Tellus,* 13, 214-223.

Gao, W. J. Slusser, J. Gibson, G. Scott, D. Bigelow, J. Kerr, and B. McArthur, 2001: Direct-sun column ozone retrieval by the ultraviolet multifilter rotating shadow-band radiometer and comparison with those from Brewer and Dobson spectrophotometers, *App. Opt.*, 40(19), 3149-3155.

Intergovernmental Panel on Climate Change, *Climate Change 2001: The Scientific Basis,* Cambridge Univ. Press, New York (2001).

Lacis, A. A., and M. I. Mishcenko, 1995: Climate forcing, cloud sensitivity and climate response: A radiative modeling perspective on atmospheric aerosols," in *Aerosol Forcing of Climate*, edited by R. J. Charison and J. Heintzenberg, John Wiley, New York.

Lenoble, J., 1993: *Atmospheric Radiative Transfer*, 532 pp., A. Deepak Publishing, Hampton, Va.

RELATIONSHIP BETWEEN AEROSOL PROPERTIES AND AIR POLLUTANTS

S. Mukai, I. Sano, and M. Yasumoto
Kinki University, Faculty of Science & Technology
3-4-1 Kowakae, Higashi-Osaka, 577-8502, Japan

ABSTRACT

For understanding urban aerosols, sun/sky photometry has been undertaken over Higashi-Osaka with multi-spectral photometers CE-318 (Cimel Electronique) for an AERONET (Aerosol Robotic Network) site since 2002. A new instrument SPM-613D (Kimoto Electric) began taking measurements of suspended particles matter (SPM) as TSP (total suspended particulates), PM10, PM2.5 and OBC (optical black carbon) on March 15, 2004 at the same AERONET site.

The relationship is examined between aerosol properties obtained from radiometry and the SPM measurements. It is found that there is a strong correlation between the SPM concentrations and aerosol properties, which indicates that aerosol characteristics can be estimated from SPM data, and vice versa.

1. INTRODUCTION

Higashi-Osaka is famous for its heavy air pollution. The aerosol properties here are especially complicated due to mixing of anthropogenic and natural compounds. Higashi-Osaka is located between Osaka Bay and the Ikoma Mountains, and one of the industrial cities comprising the so-called Keihanshin Industrial Zone. Multi-spectral photometer CE-318-2 (Cimel Electronique) was set up for an AERONET site at Higashi-Osaka in 2002 to make measurements for understanding the characteristic features of the urban aerosols. This instrument has four observing channels for photometry whose central wavelengths are 0.44, 0.67, 0.87, and 1.02 μm, and polarimetric facilities at 0.87 μm. The radiometer was calibrated using a standard AERONET procedure (Dubovik et al., 2000a).

It is difficult to relate SPM data directly to aerosol properties, but suspended particles matter (SPM) data approximately represents the mass concentration of atmospheric particles at the surface. In other words, air pollutants could bear some relations to the emission and transportation of aerosols. In order to elucidate the correlation between aerosol properties and concentrations of in-situ atmospheric particles at the surface (Mukai et al., in press a), a new instrument SPM-613D (Kimoto Electric) has been undertaken the measurements of SPM concentrations as TSP (total suspended particulates), PM10 ($<\approx 10$ μm), PM2.5 ($<\approx 2.5$ μm), and OBC (optical black carbon) since March 15, 2004 at the AERONET/Higashi-Osaka site (Goloub et al., 2000).

2. SPM SAMPLING

It has been shown that anthropogenic aerosol particles dominate the air over urban cities due to local emissions such as from diesel vehicles and, chemical industries (Wang et al., 2003). It is also well known that small atmospheric particles play an important role in human health and climate change (Balasubramanian et al., 2003).

The new SPM-613D instrument makes it possible to determine the relationship between aerosol properties and the particulate mass, since it can separate the contributions of fine particles (PM2.5), coarse particles (PM10/TSP), and OBC. The SPM sampler works continuously, even on cloudy/rainy days, while radiometry is only available during daytime on a clear day. A replaceable Teflon tape filter is used to collect particles entering into the tip of SPM-613D, and ICP-MASS analysis is employed for extraction. The mass concentrations of PM2.5 and PM10/TSP are measured using the beta gauge method. TSP is only measured for the yellow sand season. The OBC is measured by optical density.

Fig. 1 shows the measurements of TSP/PM10 (colored black) and PM2.5 (denoted by light color) in the upper panel, and OBC in the lower panel. Roughly speaking, the SPM measurements over Higashi-Osaka are divided into two seasons, spring and others. Each season has each character, for example in spring from March to mid-June the coarse mode particles, which are calculated from the difference betweenTSP/PM10 and PM2.5, are dominant. These data indicate that several dust events occurred at Higashi-Osaka during spring of 2004. It is of interest to mention that the dust events appear with approximate periodicity of 4 days, and are attributable to soil dust transported from China in the westerly wind. During the period shown, the air pressure system typically changed every 4 days. And after June, the fine mode particles (PM2.5) are dominant in usual (see from June to December in Fig. 1). The U.S. Environmental Protection Agency defines the value of PM2.5 concentration as the air quality index (AQI), e.g., nearly zero in a very clean atmosphere. The AQI at Higashi-Osaka site is occasionally larger than 65.5 μgm^{-3}), which means not good for health (Sarigiannis et al., 2004).

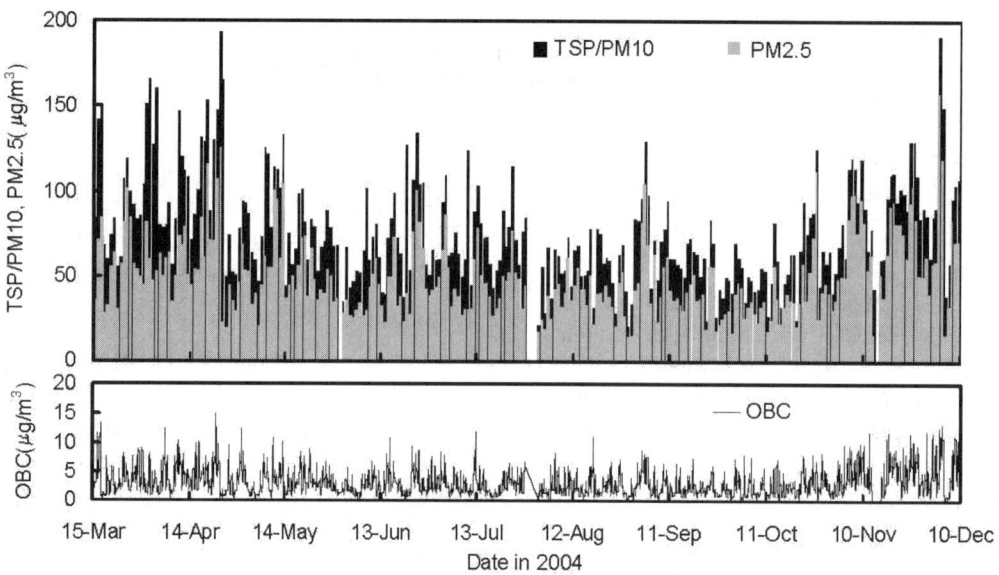

Fig. 1. Measurements with SPM-613D from 15 March to 8 December 2004 over Higashi-Osaka, where TSP/PM10 (solid curve) and PM2.5 (shaded part) were observed.

3. AEROSOL RADIOMETRY

To characterize urban aerosols, sun/sky photometry and polarimetry with PSR-1000 (Opto Research) have been performed over Higashi-Osaka since 1996. Multispectral photometer CE-318-2 (Cimel Electronique) was subsequently set up for an AERONET site in 2002. Another AERONET instrument (CE-318-1) has been operating at Shirahama since 2000 as a part of NASA/AERONET (Hollben et al., 1998). The radiometers at both Osaka and Shirahama were calibrated using a standard AERONET procedure. The Shirahama site is in the middle of Japan far from large cities and faces the Pacific Ocean, and hence the aerosols are normally oceanic in nature though occasionally contaminated with aerosol events such as continental dusts transported from China. Therefore Shirahama data are representative of typical background aerosols in Japan (Sano et al., 2003). In contrast aerosol loading in the Higashi-Osaka atmosphere involving considerable anthropogenic aerosols is larger than at Shirahama on ordinary days (Mukai et al., in press b).

The column aerosol optical thickness (AOT: τ_λ) is an important aerosol parameter that is measured by direct sun photometry of the sun. The Ångström exponent (α) is calculated from the spectrum tendency of AOT, and α' is derived from the second derivative of AOT (Eck et al., 1999).

$$\alpha = -\ln(\tau_{\lambda 1}/\tau_{\lambda 2}) / \ln(\lambda_1/\lambda_2), \quad (1)$$

$$\alpha' = d\alpha / d\ln\lambda$$
$$\alpha' = -2\{\ln(\tau_{\lambda 1}/\tau_{\lambda 2}) / \ln(\tau_{\lambda 1}/\tau_{\lambda 2}) - \ln(\tau_{\lambda 2}/\tau_{\lambda 3}) / \ln(\tau_{\lambda 2}/\tau_{\lambda 3})\} / \ln(\tau_1/\tau_3). \quad (2)$$

where wavelengths λ_1, λ_2, and λ_3 correspond to the central wavelengths of three observing channels. The values of α are closely related to the aerosol size. The aerosol index (AI), a good indicator of the dominance of anthropogenic particles, is defined as

$$AI = AOT \cdot \alpha \quad (3)$$

Other aerosol properties, such as the size distribution, complex refractive index, and single scattering albedo, are retrieved based on the AERONET standard inversion method (Dubovik et al., 2000b).

Fig. 2 shows the AOT (0.87 μm), α and the α' for all days for which measurements are available from 15 March to 8 December, 2004, over Higashi-Osaka. The dashed line at 0.2 in the AOT (0.87 μm)-panel, and each solid line at $\alpha = 1.0$ and $\alpha' = 0$ in other each panel are referred to below. The dashed line denotes the reference value for the discrimination of aerosol events, which are defined as an unusually heavy aerosol loading, namely AOT (0.87 μm) > ~0.2. It is found that the aerosol events occasionally happen at Higashi-Osaka in 2004. Now focusing on the aerosol events at Higashi-Osaka, the value of α is small (i.e., $\alpha < 1.0$) in spring, on the contrary that is large ($\alpha > 1.0$) in other season. In addition, the sign of α', which is an index to distinguish carbonaceous aerosols from the desert dust, is changing from spring events to other events, i.e., $\alpha' < 0$ for spring and $\alpha' > 0$ after June. These results coincide with the SPM measurements in Fig. 1. It is also found from Fig. 1 that the ratio of OBC to PM2.5 is high in the case of aerosol events after June. Therefore, spring aerosol events are due to the yellow sand dust and usual aerosol events at Higashi-Osaka are caused by fine particles from anthropogenic emissions.

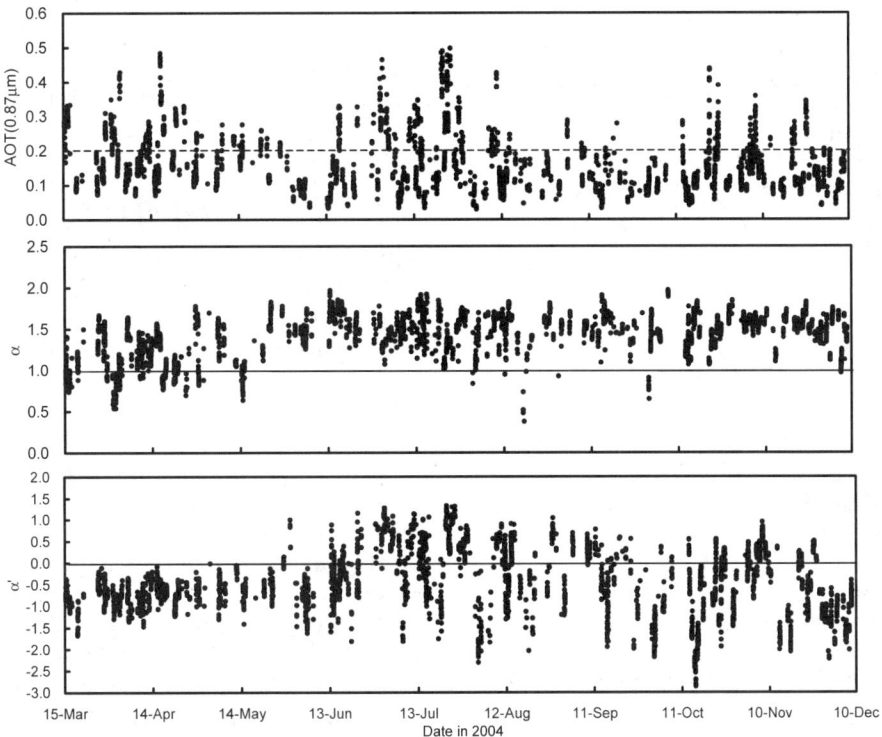

Fig. 2. Radiometric data of aerosols with AERONET from 15 March to 8 December 2004 over Higashi-Osaka. AOT (0.87 μm), α and α' are shown in the upper, middle and lower panels respectively. The dashed lines indicating in AOT (0.87 μm) value of 0.2, and the solid lines for α=1.0 and α'=0.0 are discussed in the text.

In order to clarify the two kind of aerosol events at Higashi-Osaka, Fig. 3 is prepared to present AOT (left panel) and α' (right panel) against α for typical three type of aerosols; type-A: AOT>0.2 after mid-June, type-B: AOT<0.1, and type-C: AOT>0.2 during 15 March to mid June. The averaged values of (AOT, α, α') for each type are presented in Table 1 and their positions are denoted by the open squares in Fig. 3. It is possible to say that type-A represents the anthropogenic aerosol events with (AOT>0.2, α>1.0, α'>0), type-B corresponds to the background clear atmosphere of (AOT<0.1, α>1.0, α'<0), and type-C shows the coarse dust events with (AOT>0.2, α<1.0, α'<0).

Table 1. Averaged Values of (AOT, α α') for each Type

type	AOT	α	α'
A	0.30	1.43	0.47
B	0.09	1.43	-1.25
C	0.30	0.91	-0.71

Fig. 3 shows the relationship between AOT (0.87 αm), α and α'. Since small values of α are representative of coarse dust, yellow sand events produce high values of AOT and low values of α which corresponds to the type-C in Fig. 3. In the contrast, some other aerosol events have no distinct features of dust events. It can be seen that fine particles dominate at Higashi-Osaka. In other word, aerosol events at Higashi-Osaka seem to be caused by anthropogenic small haze particles (type-A) except yellow sand season. For reference clear atmosphere, i.e., AOT<0.1, is shown by type-B.

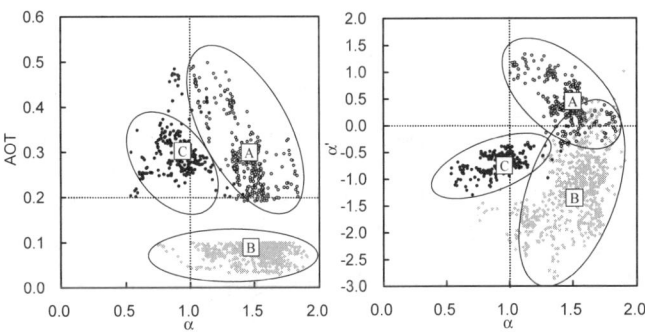

Fig. 3. Scattergrams of AOT (in the left panel) and α' (in the right panel) against α. The characters A, B, and C are discussed in the text.

The complicated features of aerosols at Higashi-Osaka are demonstrated in the second derivatives (α') in Fig. 3. During dust events, α' is negative, and for almost all cases α' takes negative values. It is of interest to mention that the positive values of α' appear with large α and large AOT occurred in the case of type-A.

4. CORRELATION BETWEEN AEROSOLS AND SPM

The SPM-613D measurements indicate the dominance of fine anthropogenic (PM2.5) and coarse (PMc) particles over Higashi-Osaka during the aerosol events. A scatter gram of PM2.5 against aerosol index (AI) shows clear correlation in Fig. 4.

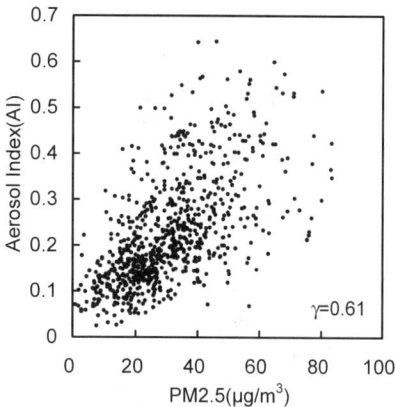

Fig. 4. A scatter gram of PM2.5 against aerosol index (AI).

Chemical analysis of the Teflon tape filter used in the SPM-613D combined with the obtained OBC data reveal the compound elements of particulate matter. These results all indicate that the high-level anthropogenic particles in the air over the Osaka site. It is well known the air quality of Higashi-Osaka site is very poor, and our results indicate that the local air quality is influenced not only by local emissions from sources such as diesel vehicles and chemical industries, but also by large scale climatic condition such as dust particles coming from continental desert areas.

5. CONCLUSION

We have examined the relationship between the aerosol properties derived from radiometry and the particulate mass simultaneously obtained with the new SPM-613D instrument. Strong correlations were found between the PM2.5 concentrations and the aerosol index AI values. SPM measurements allow the dominate particle size to be determined (i.e., PM2.5, TSP/PM10, or BOC). Combining radiometric aerosol information and the surface-level particulate mass is useful when studying air quality and aerosol properties. For example, the linear correlation between SPM and AI allows the aerosol properties to be estimated from SPM data in areas without an AERONET site, on cloudy/rainy days, or at night (Sarigiannis et al., 2004). Alternatively, satellite-derived aerosol information is useful for indicating air quality in a global scale.

Chemical analysis of the SPM-613D measurements will help determine the aerosol composition.

6. ACKNOWLEDGMENTS

This work was supported in part by Grants-in-aid for Scientific Research on Priority Areas from MEXT (no. 14048224), Scientific Research from JSPS (no. 13573017), and JAXA (no. JX-PSPC-132120).

7. REFERENCES

Balasubramanian, R., W.-B. Qian, S. Decesan, M. C. Facchini, and S. Fuzzi, 2003: Comprehensive characterization of PM2.5 aerosols in Singapore, *J. Geophys. Res.*, 108, 4523, doi:10,1029/2002JD002517.

Dubovik, O., A. Smirnov, B. N. Holben, M. D. King, Y. J. Kaufman, T. F. Eck, and I. Slutsker, 2000: Accuracy assessments of aerosol optical properties retrieved from AERONET sun and sky-radiaometric measurements, *J. Geophys. Res.*, 105, 9791-0806.

Dubovik, O., and M. D. King, 2000: A flexible inversion algorithm for retrieval of aerosols optical properties from sun and sky radiance measurements, *J. Geophys. Res.*, 105, 20673-20696.

Eck, T. F., B. N. Holben, J. S. Reid, O. Dubovik, A. Smirnov, N. T. O'neill, I. Slutsker, and S. Kinne, 1999: Wavelength dependence of the optical depth of biomass burning, urban, and desert dust aerosols, modal, *J. Geophys. Res.*, 104, no. D24, pp. 31,333-31,349.

Goloub, P., J. L.Deuze, M. Herman, A. Marchand, D. Tanre, I. Chiapello, B. Roger, and R. P. Singh, 2000: Aerosol Remote Sensing Over Land From The Spaceborne Polarimeter POLDER, *Proc. IRS 2000*, 115-116.

Holben, B. N., T. F. Eck, I. Slutsker, D. Tanré, J. P. Buis, A. Setzer, E. Vermote, J. A. Reagan, Y. Kaufman, T. Nakajima, F. Lavenu, I. Jankowiak, and A. Smirnov, 1998: AERONET - A federated instrument network and data archive for aerosol characterization, *Rem. Sens. Environ.*, 66, 1–16.

Mukai, S., I. Sano, Y. Okada, and B. N. Holben (in press a), Comparison of aerosol properties with air pollutants, *Adv. Space Rev.*

Mukai, S., I. Sano, and B. N. Holben (in press b), Aerosol properties over Japan by sun/sky photometry, *Air, Water and Soil Pollution*.

Sano, I., S. Mukai, Y. Okada, B. N. Holben, S. Ohta, and T. Takamura, 2003: Optical properties of aerosols during APEX and ACE-Asia experiments, *J. Geophys. Res.*, 108, 8649, doi:10.1029/2002JD003263.

Sarigiannis, D., N. Sifakis, N. Soulakennis, M. Tombrou, and K. Schafer, 2004: Satellite-derived determination of PM10 concentration and of the associated risk on public health, *Proc. SPIE*, 408-416.

Wang, J., and S. A. Christopher, 2003: Intercomparison between satellite-derived aerosol optical thickness and PM2.5 mass: Implications for air quality studies, *Geophys. Res. Letter*, 30, 2095, doi:10.1029/2003GL 018174.

RETRIEVAL OF AEROSOL MICROPHYSICAL PROPERTIES FROM MFRSR OBSERVATIONS

E. I. Kassianov, J. C. Barnard, and T. P. Ackerman
Pacific Northwest National Laboratory
Richland, WA, 99354, USA

ABSTRACT

Multi-filter Rotating Shadowband Radiometers (MFRSRs) are widely deployed over the world. These radiometers measure the total, direct, and diffuse components of shortwave, narrowband irradiance at 6 wavelengths. For 5 of these wavelengths, aerosol optical depths and single scattering albedos can be retrieved. We describe here a simple retrieval technique that can significantly extend the capability of the MFRSR to study atmospheric aerosols and can provide a means for simultaneous retrieval of the mean particle radius, total number of particles (for an assumed size distribution), and the imaginary refractive index. The analysis of our initial numerical results shows that accurate retrievals of aerosol characteristics can be achieved. In addition, we successfully applied the suggested technique to derive temporal variations of aerosol microphysical properties from ground-based MFRSR measurements performed in an urban region (Mexico City).

1. BACKGROUND

If we assume that aerosol particles are homogeneous spheres, then the Lorenz-Mie theory can be applied to relate the aerosol microphysical properties to aerosol optical properties. The equation for the height-dependent extinction coefficient is

$$\sigma_\lambda(z) = \int_{a_{min}}^{a_{max}} \varphi(z,a) K_{e,\lambda}(a) da \qquad (1)$$

where $\varphi(z,a)$ denotes the aerosol size distribution as a function of height and a is the radius. The weighting function has the form

$$K_{e,\lambda}(a) = \pi a^2 Q_e(n,m; a/\lambda) \qquad (2)$$

where Q_e is the extinction efficiency, n and m are the real and imaginary parts of the refractive index, respectively; and λ is the wavelength.

The optical depth of entire aerosol column can be calculated from the vertical profile

$$\tau_\lambda = \int_0^{z_\infty} \sigma_\lambda(z) dz \qquad (3)$$

where z_∞ is the top of the atmosphere. From (1) and (3), one can obtain

$$\tau_\lambda = \int_{a_{min}}^{a_{max}} f(a) K_{e,\lambda}(a) da \qquad (4)$$

where $f(a)$ represents the *columnar* aerosol size distribution (with units, say, of $\#/cm^3$)

$$f(a) = \int_0^{z_\infty} \varphi(z,a) dz \qquad (5)$$

Equation (4) forms the basis for retrievals of $f(a)$ from a set of τ_λ. To derive $f(a)$, one must use observations for a large number of wavelengths (ten or more values τ_λ). In principle, two different wavelengths can be used for $f(a)$-retrievals, and these wavelengths can be chosen to avoid ozone and water vapor contamination. However, when using only two wavelengths one has to assume a shape for the size distribution; then, given τ_λ at the two wavelengths, one can derive two parameters that describe the distribution. For example, the Junge power law has been used to model the aerosol size distribution.

2. APPROACH

We propose here another approach for $f(a)$-retrievals by using two wavelengths. We again must assume the shape of the size distribution, however, the shape of the assumed distribution is different from the Junge power law and is thought to be a more realistic representation of aerosol size distribution shapes. We assume that the shape of a normalized aerosol size distribution

$$\varphi^*(a) = \frac{1}{z_\infty} \int_0^{z_\infty} \varphi(z,a) dz \qquad (6)$$

can be described by a combination of three lognormal distributions. The parameters of such distributions are specified to represent different types of tropospheric aerosols.

From (5) and (6) it follows that $f(a) = \varphi^*(a) z_\infty$. Note, z_∞ can be considered as a scaling parameter and its value is unknown. Since the specification of this value does not affect our $f(a)$-retrieval, we set $z_\infty = 1$ km. Below we outline a simple way of adjusting the assumed distribution $f(a)$ to the spectral observations of τ_λ.

2.1. Bulk Parameters

We adjust the assumed distribution so that the modeled optical thickness $\tau_{mod,\lambda}$ values match the spectral observations of $\tau_{obs,\lambda}$. Specifically, taking two observed spectral values of $\tau_{obs,\lambda}$ denoted as ($\lambda = 1$ and 2) we estimate two bulk parameters of the assumed size distribution: the *total number of particles N* and the *mean particle radius R*. These two parameters are the 0th moment and 1st moment of the size distribution, respectively:

$$N = \int_{a_{min}}^{a_{max}} f(a)\,da, \quad R = \frac{1}{N}\int_{a_{min}}^{a_{max}} a\,f(a)\,da \qquad (7)$$

Referring to the modeled optical thickness as $\tau_{mod,\lambda}(N, R)$ (here we explicitly note the dependence of $\tau_{mod,\lambda}$ on N and R), the matching procedure varies N and R in $f(a)$, until $\tau_{mod,\lambda}$ and $\tau_{obs,\lambda}$ are equal for each of the two wavelengths.

This matching procedure amounts to a simple shifting of the size distribution in two orthogonal directions (the shape of $f(a)$ is fixed). The first transformation is a *vertical* shift (along the y-axis) of $f(a)$. This vertical shift simply changes N, the total number of particles, while keeping the mean particle radius R, fixed. The second transformation is a *horizontal* shift (along the x-axis) of $f(a)$. This horizontal shift simply changes R, the mean particle radius, while keeping the total number of particles N, fixed.

Fig. 1 shows an example of the complex relationship between τ_λ and two parameters (N and R).

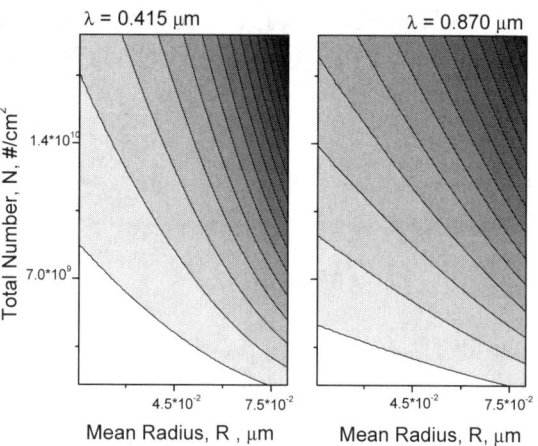

Fig. 1. Two-dimensional diagrams of model optical depth τ_λ in terms of the total number N of particles (in the column) and their mean radius R for two wavelengths (0.415 μm and 0.870 μm). These diagrams are obtained from Mie calculations for urban aerosol model (Hobbs et al., 1993) and a given refractive index ($n = 1.5$ and $m = 0$). Dark and white colors represent high and low values of τ_λ, respectively.

Once the shape of the size distribution is decided, finding values of N and R is a simple matter. We assume that both the real and imaginary parts of the refractive index are known, and given the two inputs—the observed optical depths $\tau_{obs,\lambda}$ at two wavelengths—we have two equations and two unknowns (N and R). These can be solved for N^* and R^*, the values of N and R, such that $\tau_{mod,\lambda}$ is equal to $\tau_{obs,\lambda}$ for both wavelengths.

2.2. Imaginary Refractive Index

In practice, the above description is lacking because we do not know in advance either the shape of the aerosol size distribution or the refractive index. However, one can make a reasonable assumption about the aerosol type (e.g., urban, or desert dust) and select an appropriate aerosol model (shape) for a particular case (e.g., Hobbs et al., 1993). Such selection should be performed by analysis of possible aerosol sources (e.g., their location, strength) and local meteorology (e.g., wind direction). Different kinds of the tropospheric aerosols have different values of the real refractive index n; but, for a given aerosol type, these n values do not vary much (e.g., Dubovik et al., 2002). Therefore, one can set a plausible value of the real refractive index (e.g., urban aerosol).

The imaginary refractive index, m, varies through a large range, up to several orders of magnitude. Therefore, this parameter cannot be assumed and must be determined from observations. In order to provide a third observational constraint, we include observations of the spectral *diffuse* irradiance (D). To reduce the effect of ozone and water vapor contamination, we select wavelengths where ozone and water vapor absorb very weakly or not at all (e.g., 0.415 μm and 0.870 μm), and to reduce the effect of surface albedo on the diffuse irradiance, we select a wavelength where the surface albedo is small (e.g., 0.415 μm, for surfaces free of snow and ice). Our retrieval now requires three assumptions: (i) the shape of aerosol size distribution, (ii) the real part of the refractive index, and (iii) the surface albedo at wavelength 0.415 μm. Given these assumptions, we now have a closed problem of *three inputs* ($\tau_{obs,1}$, $\tau_{obs,2}$ and D_1), and *three unknowns* (N, R, m).

2.3. Sensitivity Tests

The technique consists of three steps that compose an iterative scheme. The first step obtains the aerosol size distribution from the spectral measurements of the direct irradiance (for a given complex refractive index). The second step determines the effective value of the imaginary refractive index from the diffuse irradiance (for the aerosol size distribution determined during the first step). The third step determines whether to stop the iterations or not. For a given iteration step, the value of the imaginary index is compared with its previous value. If the relative difference exceeds the given threshold, then we repeat the first and second steps. If not, the iteration is considered to be converged.

To estimate the sensitivity of the suggested retrieval to the uncertainties in the real refractive index, the surface albedo, and the shape of the size distribution, we perform a few numerical experiments (Kassianov et al., 2004). Seven different models of aerosol size distribution are used (Hobbs et al., 1993). The analysis of our initial results shows that successful retrievals of aerosol properties can be achieved for the atmospheric aerosols composed mainly of small and large (accumulation) particles. If the relative contribution of coarse or giant particles (1 μm and larger) is significant (e.g., a dust storm), then the aerosol optical depth becomes almost spectrally independent, and the technique fails to retrieve the aerosol properties. Fortunately, these situations are rare.

2. MFRSR DATA RETRIEVAL

Here we illustrate the performance of our technique by applying it to the MFRSR observations taken during April 27, 2003, in Mexico City, as part of the Mexico City Metropolitan Area (MCMA) field campaign. The environment in this region is urban, so we assume that the urban aerosol model is appropriate for our retrieval. Also, we assume n is equal to 1.4 (Dubovik et al., 2002), and surface albedo is 0.07 (0.415 μm).

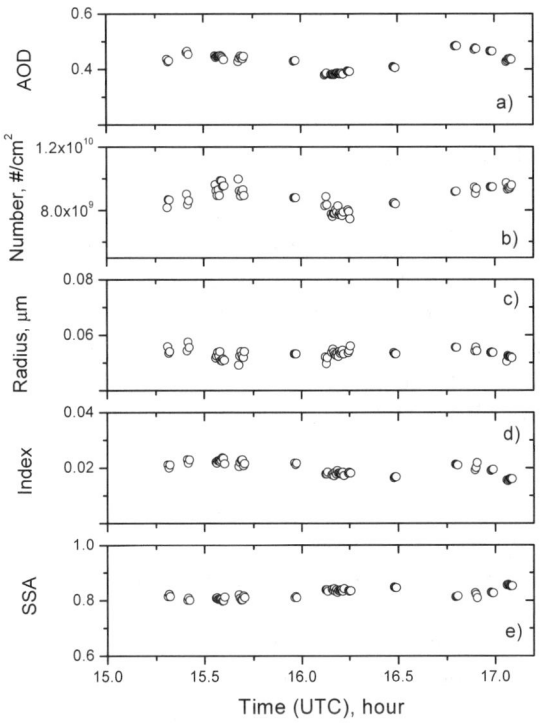

Fig. 2. Time series of (a) aerosol optical depth, (b) the total number of particles, (c) the mean particle radius, (d) the imaginary part of refractive index, and (e) the single scattering albedo derived in Mexico City on April 27, 2003.

3. AERONET DATA RETRIEVAL

To further illustrate our retrievals, we compare MFRSR and AERONET retrievals of aerosol properties in Mexico City on April 27, 2003. In contrast to the AERONET-derived aerosol optical depth, other AERONET products (e.g., the refractive index and size distribution) are obtained with coarse temporal resolution (varies from 0.3 hour to 1 hour): additional time is required for sky-scanning (solar almucantar during low sun and in the principal plane during high sun).

Since MFRSR-retrievals are performed with high temporal resolution (20 seconds), we average them, and then the averaged values are compared with available AERONET data (Fig. 3). For given temporal window (from 1520 UTC to 1720 UTC), there are available AERONET products obtained at 1636 UTC only.

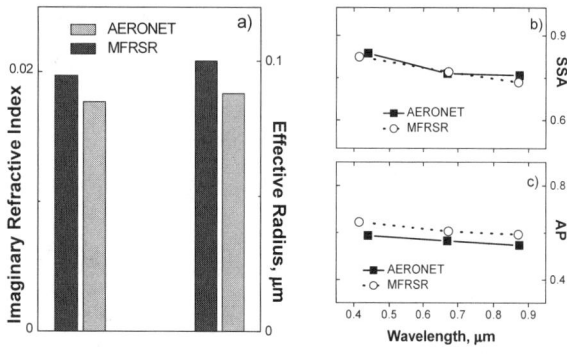

Fig. 3. Aerosol properties derived from the MFRSR (black) and AERONET (gray) data: (a) the imaginary refractive index (two left columns) for 0.415 μm (MFRSR) and 0.441 μm (AERONET), and the effective radius (two right columns); (b) the single scattering albedo and (c) the asymmetry parameter.

4. DISCUSSION

The aerosol retrievals discussed here were performed for hemispherically cloud free periods, which is the optimal condition. Since this technique relies in part on the diffuse solar radiation scattered by aerosol, any diffuse scattering by clouds is a source of error. Thus, the usefulness of the approach is directly related to the amount of time that the sky is hemispherically cloud-free. An analysis of about nine years of data from the Southern Great Plains (SGP) site of the ARM Program, shows that significant periods of cloudless sky occur on approximately 30% of all days at the SGP (Oklahoma). For this analysis, we use the algorithm of Long and Ackerman (2000). Applying this algorithm to data from the ARM site in Barrow Alaska produces a roughly comparable result.

5. SUMMARY

We propose a simple retrieval technique that allows for simultaneous retrieval of the mean particle radius, total number of particles (for an assumed size distribution), and the imaginary refractive index from MFRSR measurements.

The technique has been successfully applied to derive temporal variations of aerosol microphysical properties derived from ground-based MFRSR measurements performed during one particular day in Mexico City. The MFRSR-derived aerosol properties are in good agreement with independent AERONET retrievals made also in Mexico City (April 27, 2003). This favorable comparison suggests that a relatively inexpensive instrument, widely used over the world, can be applied to derive columnar aerosol optical and microphysical properties.

6. ACKNOWLEDGMENTS

This research was supported by the US Department of Energy (DOE) under the auspices of the Atmospheric Science Program of the Office of Biological and Environmental Research, under contract DE-AC06-76RLO 1830 at the Pacific Northwest National Laboratory (PNNL). PNNL is operated for the US DOE by Battelle Memorial Institute.

7. REFERENCES

Dubovik, O., et al., 2000: Accuracy assessments of aerosol optical properties retrieved from Aerosol Robotic Network (AERONET) Sun and sky radiance measurements. *J. Geophys. Res.*, 105, 9791-9806.

Hobbs, P., et al., 1993: *Aerosol-Cloud-Climate Interactions*, 233 pp., Academic Press, Inc.

Kassianov, E., J. Barnard, and T. Ackerman, 2005: Retrieval of Aerosol Microphysical Properties by Using Surface MFRSR Data: Modeling and Observations. *J. Geophys. Res.* (in press).

Long, C. N., and T. P. Ackerman, 2000: Identification of clear skies from broadband pyranometer measurements and calculation of downwelling shortwave cloud effects, *J. Geophys. Res.*, *105*, 15609-15626.

MEASUREMENTS OF AEROSOL OPTICAL PROPERTIES AT PALLAS GAW STATION IN NORTHERN FINLAND

Veijo Aaltonen, Heikki Lihavainen, Juha Hatakka, and Yrjö Viisanen
Finnish Meteorological Institute
Sahaajankatu 20 E, FI-00880, Helsinki, Finland
Email: Veijo.Aaltonen@fmi.fi

ABSTRACT

Measurements of optical properties of atmospheric aerosols have been carried out in 2000–2004 at Pallas-Sodankylä GAW station located in northern Finland. Measured parameters were σ_{sp}, σ_{bsp}, σ_{ap}, ω_0, $å$ and b, of which the aerosol scattering coefficient had highest seasonal variation. The diurnal variation of the parameters observed were very weak.

1. INTRODUCTION

The Earth's radiation budget, and thus the predicted climate change, is affected through radiative forcing by the changes in the concentration and composition of aerosol particles. It can be as large as, but opposite in sign to, the effect of increased greenhouse gas concentrations. However, owing to their high spatial and temporal variability (seasonal dust storms or biomass burning), their globally and annually averaged climatic forcing is associated with a large uncertainty. Radiative transfer models suffer from uncertainties arising from the lack of accurate input parameters. The lack of data on the aerosols is one of the main problems. There is thus growing need of aerosol measurements. When estimating the direct aerosol radiative forcing, both the amount of aerosol and their optical properties are important.

2. METHODS

Measurements of light scattering coefficient with integrating nephelometer (TSI 3563) in Pallas have been going on continuously since February 2000. The instrument measures scattering coefficient σ_{sp} and backscattering coefficient σ_{bsp} at three wavelengths, 450, 550, and 700 nm. Measurements of black carbon (BC) concentration with Aethalometer (Magee Scientific) in Pallas have been going on since June 1996. Light absorption coefficient, σ_{ap}, was calculated from BC concentration using equation by Bodhaine (1995):

$$\sigma_{ap} = -A \ln (I_2 / I_1) / C Q (t_2 - t_1), \quad (1)$$

where C is a constant representing the enhancement of aerosol absorption for the aerosol embedded in the filter matrix over that in the atmosphere, I_1 and I_2 are the intensity ratios of the sample beam to the reference beam at times t_1 and t_2, Q is the flow rate and A is the filter area. Used values for C were 1.5, 1.9, and 2.4 for wavelenghts of 450, 550, and 700 nm, respectively. The values of σ_{ap} were calculated assuming that σ_{ap} is inversely proportional to λ for BC. Single scattering albedo ϖ_0 was calculated from σ_{sp} and σ_{ap}.

Ångström exponent has been calculated from scattering coefficients at 700 and 550 nm using equation

$$å = -\log \sigma_{sp(700)} - \log \sigma_{sp(550)} / \log 700 - \log 550. \quad (2)$$

This can be also expressed by $\sigma_{sp} = K\lambda^{-å}$, where λ is wavelength and K is a constant. Practically, large $å$ value implies smaller particles and vice versa.

3. RESULTS AND DISCUSSION

Annual variation of σ_{sp} and σ_{bsp} is very clear. σ_{sp} and σ_{bsp} start to rise early in spring, reaching in summer their highest values, which are almost twice as high as their average values (6.32 Mm^{-1} for total scattering at 550 nm). This indicates that largest amount of accumulation mode particles is present in summer in the particulate phase. After July the values decrease very rapidly, and stay quite constant in autumn. Values of σ_{ap} are high in early spring, after which they start to decrease rapidly, and stay relatively constant in autumn (yearly averaged value of 1.06 Mm^{-1} at 550 nm). High values in late winter and early spring may be related to higher BC concentrations which originate from heating of the houses or the start of agricultural activity. As a result, ϖ_0 is almost constant year-round (Fig. 1).

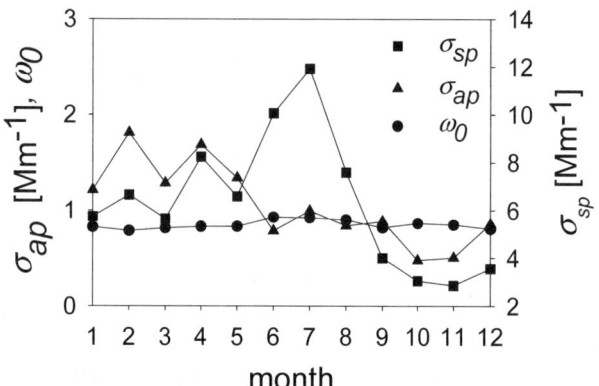

Fig. 1. Seasonal variation of σ_{sp}, σ_{ap}, and σ_0 at 550 nm.

Ångström exponent is quite constant in spring, and starts to increase in late summer. After September it starts to decrease again. The values of hemispheric backscattering fraction b stay constant throughout the year getting only slightly higher values in autumn (Fig. 2).

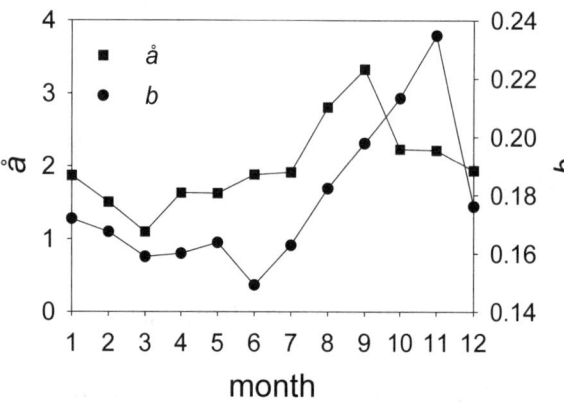

Fig. 2. Seasonal variation of å and b at 550 nm.

The year to year variation of both the scattering and backscattering coefficient displayed a clear seasonal cycle with an autumn minimum and 5–6 times higher summer maximum. Both σ_{sp} and σ_{bsp} started to rise early in spring, reaching their highest values in summer. After July the values decreased very rapidly and stayed relative constant during the autumn. The autumn minimum in scattering may be related to the fact that both cloudiness and precipitation have maximum in autumn, and therefore air is expected to be cleaner. The scattering coefficient reached its maximum usually in July and minimum in November. The backscattering coefficient behaved in the same way, though with lower absolute values. The absorption coefficient has higher values in spring than in other seasons, but does not have remarkable variation in yearly level. High values of σ_{ap} in 2004 might be due to the instabilities in the flow measurement inside the instrument. The single scattering albedo is very constant year-round (Fig. 3).

The Ångström exponent had a clear year to year variation with a minimum in spring and about two times higher maximum in late summer. Based on the absolute values of å, we may conclude that scattering was dominated by submicron aerosols in Pallas, and especially so during the late summer and autumn. The hemispheric backscattering fraction b varied by roughly a factor two between different months but did not show a consistent seasonal pattern (Fig. 4).

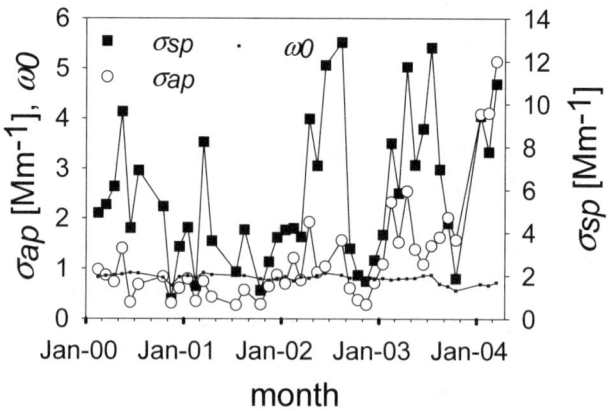

Fig. 3. Year to year variation of σ_{sp}, σ_{ap}, and ω_0 at 550 nm.

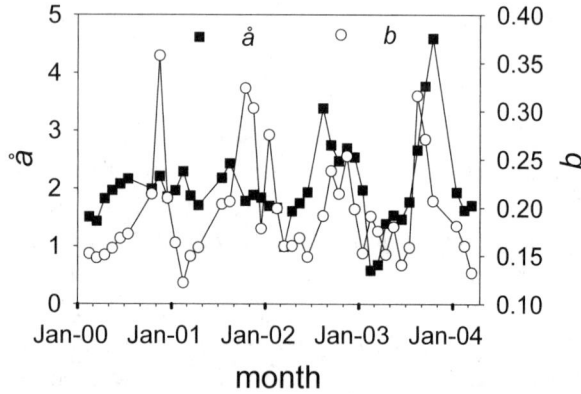

Fig. 4. Year to year variation of å and b at 550 nm.

The diurnal behaviour of σ_{sp} and σ_{bsp} varies very weakly, being slightly higher at daytime (averaged value of 6.3 Mm^{-1} for total scattering at 550 nm). Daily variation of σ_{ap} (averaged value 1.1 Mm^{-1} at 550 nm) follows the same pattern as scattering coefficient. ϖ_0 is almost constant throughout the day, with averaged value of 0.85 at 550 nm (Fig. 5).

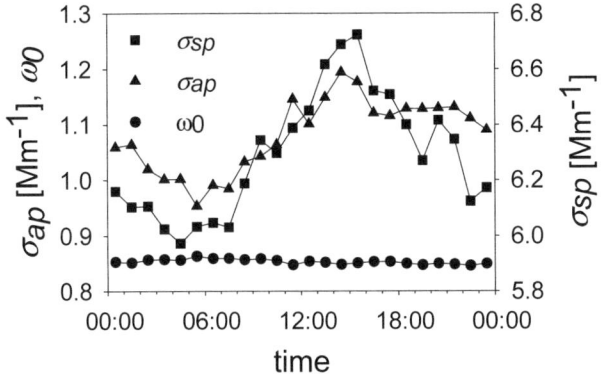

Fig. 5. Diurnal variation of σ_{sp}, σ_{ap}, and ω_0 at 550 nm.

Ångström exponent and hemispheric backscattering fraction stay relatively constant throughout the day (Fig. 6).

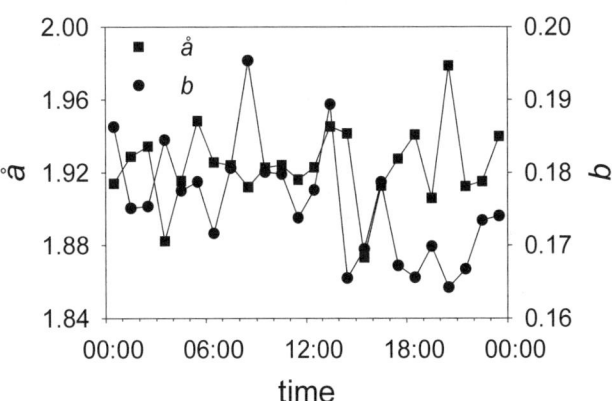

Fig. 6. Diurnal variation of $å$ and b at 550 nm.

4. REFERENCE

Bodhaine, B., 1995: Aerosol absorption measurements at Barrow, Mauna Loa and the South Pole. *J. Geophys. Res.* 100, 8967-8975.

UNCERTAINTY ANALYSIS OF DATA FROM THE POLAR ATMOSPHERIC EMITTED RADIANCE INTERFEROMETER (PAERI) DURING THE SOUTH POLE ATMOSPHERIC RADIATION AND CLOUD LIDAR EXPERIMENT (SPARCLE)

Michael S. Town

Department of Atmospheric Sciences – University of Washington, Seattle, Washington, 98195, USA

Von P. Walden

Department of Geography – University of Idaho, Moscow, Idaho, 83844, USA

ABSTRACT

An uncertainty analysis for the Polar Atmospheric Emitted Radiance Interferometer (Polar AERI, or PAERI) is presented for its deployment during the South Pole Atmospheric Radiation and Cloud Lidar Experiment (SPARCLE). The uncertainty analysis involves a Taylor series expansion of the AERI calibration equation, and treatment of several biases and systematic uncertainties in the SPARCLE data set. Most of the random error in the PAERI data set is due to detector noise and lies in spectral regions of water vapor emission. An independent calibration check was developed for the PAERI that has the potential to save power and weight in future instrumentation.

1. INTRODUCTION

The Atmospheric Emitted Radiance Interferometer (AERI) family of instruments was developed by the Space Science and Engineering Center (SSEC) of University of Wisconsin to assist the Atmospheric Radiance Measurement (ARM) Program in its pursuit of characterizing and monitoring the Earth's radiation budget (Knuteson et al., 2004a, 2004b). Knuteson et al. (2004b) thoroughly characterized the general performance of the AERIs under laboratory conditions. Their results found that the AERIs meet the requirements set by the Department of Energy (U.S. Department of Energy, 1990).

This work contains an uncertainty analysis for the *Polar* Atmospheric Emitted Radiance Interferometer (PAERI) for its 2001 deployment in the field during the South Pole Atmospheric Radiation and Cloud Lidar Experiment (SPARCLE).

2. POLAR ATMOSPHERIC EMITTED RADIANCE INTERFEROMETER

The PAERI was developed in collaboration with the SSEC for deployment at the South Pole. It is based around a two-input, two-output interferometer (Bomem's MR100), with a sandwich MCT/InSb detector and two calibration blackbodies developed by the SSEC. Performance details and sample spectra of the AERIs are given in Knuteson et al. (2004a,b). Significant details of the PAERI performance are its spectral bandwidth (450-3000 cm^{-1}), spectral resolution (0.48 cm^{-1}), and the 1σ uncertainty in the blackbody emissivity ($\pm$0.00027) and temperature ($\pm$0.032K). The PAERI differs from the standard AERI in that it uses an extended range detector (like the Arctic AERIs) and a scene mirror that is capable of viewing 180 degrees from zenith to nadir.

The PAERI was operated nearly continuously for 10 months during SPARCLE. During that time it observed the sky and snow surface at many angles. The surface temperature at the South Pole ranged from –30°C to –70°C during SPARCLE and downwelling broadband irradiances ranged from 240 Wm^{-2} to 80 Wm^{-2} (Town et al., 2005). Wintertime (Apr.–Oct.) measurements were plagued significantly by blowing snow events which obstructed the scene viewing mirror and calibration blackbodies. Intense wind also occasionally caused the blackbody and interferometer reference temperatures to drift.

The calibration sequence of the PAERI is such that every few scenes are bracketed by measurements of ambient and hot blackbodies (ABB and HBB, respectively). The ABB was allowed to drift at ambient temperatures for the first three months of SPARCLE after which it was set to a few K above ambient to prevent the formation of frost in the ABB. The HBB was set between 30 and 40K above the ambient surface temperature. A sample calibration sequence is: ABB HBB SKY HBB ABB SKY ABB HBB. This sequence is designed to bracket the observed spectra between known calibration points frequently; preempting any potential calibration drift between adjacent observations.

3. CALIBRATION

The calibration equation employed by the PAERI is introduced by Revercomb et al. (1988) and is given below as eq. (1). It a spectral equation and is applied to the PAERI observations at each frequency after the raw interferogram has been "Fourier transformed" into spectral space.

Here L_s is the calibrated scene radiance, C_s the uncalibrated scene radiance, C_c the uncalibrated ABB radiance, C_w the uncalibrated HBB radiance, and L_w and L_c

are the theoretically predicted HBB radiances and ABB radiances based on measured blackbody temperatures and emissivities. C_s, C_w and C_c are complex quantities due to asymmetries in the observed, uncalibrated radiances. The complex ratio of the difference spectra is computed to minimize the effects of asymmetry in the interferogram on the phase of the final spectra (Revercomb et al., 1988).

In the Bomem MR100, the scene radiance is interfered with radiation from a reference blackbody (known as the 2nd port) inside the MR100. The raw signals C_s, C_w and C_c all individually depend on the difference in brightness temperature (T_B) between the 2nd port and the scene. If the T_B of sky scene and the 2nd port T_B are similar for a particular frequency then C_s will be small for that frequency.

The calibration equation assumes that the instrument responsivity varies linearly between blackbody views. This is not always true for parts of the SPARCLE data set. High winds associated with blowing snow events sometimes caused the 2nd port temperature to drift dramatically.

Figs. 1 and 2 show the Taylor series expansion of eq. (1). We believe that all *random* sources of error are accounted for by the Taylor series expansion. Blowing snow and the accumulation of frost and snow on the scene mirror affect C_s, C_w, and C_c in a manner similar to the effect of detector noise; they lower the signal-to-noise ratio through absorption and reemission of infrared light.

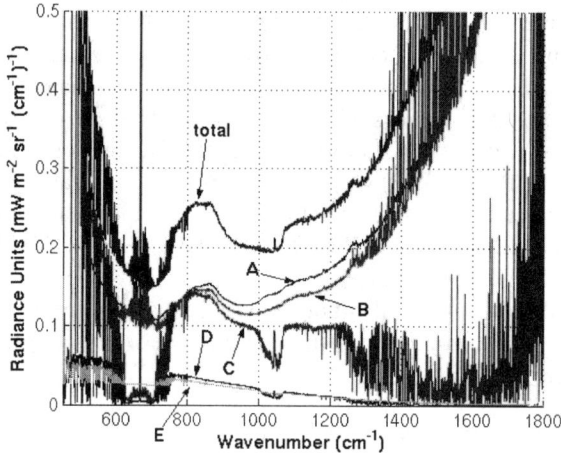

Fig. 1. Uncertainty in PAERI calibration due to random error for 450–1800 cm^{-1} (MCT detector bandwidth). The top curve represents the total random uncertainty as the square-root of the sum of variances from the terms of the Taylor series expansion of eq. (1). The rest of the curves represent error from: A = Detector noise in the sky scene, B = detector noise in the ABB scene, C = detector noise in the HBB scene, D = uncertainty in the HBB T and emissivity, E = uncertainty in the ABB T and emissivity.

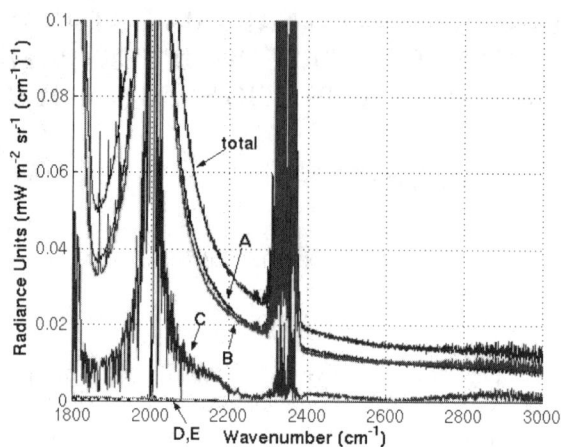

Fig. 2. Same as Fig. 1 but for 1800–3000 cm^{-1} (InSb detector bandwidth).

$$L_s = \Re\left\{\frac{C_s - C_c}{C_w - C_c}\right\}(L_w - L_c) + L_c \qquad (1)$$

4. QUALITY CONTROL

The quality control (QC) of this data set was divided into two categories: correctable and uncorrectable errors. Correctable errors include MCT detector nonlinearity and an effective laser wavenumber. Uncorrectable errors discovered in the data set include repeated direct and indirect solar shock to the detectors, a secondary field-of-view (SFOV) in the MR100 optics, times of excessive detector noise, and other miscellaneous sources of random error that we were able to eliminate based on the time of day or logbook history. Penny Rowe discovered an error in the finite field-of-view (FOV) correction, a part of the calibration procedure. The spectral effects of this error are minor and are discussed in Rowe (2004).

4.1 Correctable Errors

Most mid-range MCT detectors have some sort of nonlinear response to incident radiation. In fact, each MCT detector tends to have a slightly different nonlinear response due to manufacturing differences. There are many ways to correct for a nonlinear detector response. The AERIs employ a post-calibration correction based on prior laboratory characterization of the particular MCT detector.

The MCT detector deployed in the PAERI during SPARCLE had a nonlinearity effect as large as 0.1 mWm^{-2} str^{-1} (cm^{-1})$^{-1}$ (radiance units, RU) in the 450-1800 cm^{-1} range. The nonlinearity was characterized prior, during and after the field season and found to be constant.

The next source of correctable error in our data set was from the MR100's internal HeNe laser. Each MR100 uses the zero points in the self-interfered HeNe laser to precisely dictate the sampling resolution of the observed interferogram. Any error in the knowledge of the MR100 laser wavenumber results in subtle errors in the final spectra. The laser wavenumber of the PAERI shifts slightly every time the optical path of the PAERI is realigned. Due to the finite FOV of the PAERI, zero points dictated by the HeNe laser are not exactly the mean zero points for all light in the PAERI FOV. Thus, an effective laser wavenumber is necessary. Simply using the manufacturer's specified laser wavenumber can cause an error in spectral radiances of up to 0.5 RU.

Effective laser wavenumbers were determined by Penny Rowe (personal communication, 2004) for this data set by minimizing residuals between observed clear-sky spectra and line-by-line radiative transfer model (LBLRTM) simulations of downwelling radiance.

4.2 Uncorrectable Errors

A major source of uncorrectable error in the SPARCLE data set was due to direct, or nearly direct, viewing of the Sun. This occurred accidentally several times during the late summer and fall seasons. The solar shock caused erratic spectra temporarily but also caused a semi-permanent bias of approximately –1 RU in all spectra after the solar shock which is evident in the time series of daily mean uncertainty for 2001 (not shown). Times when the PAERI would potentially view the Sun were eliminated from the data set but the –1 RU bias was not removed. This problem only exists in the data from January–March 2001, after which the scene mirror was set to avoid zenith angles close to the Sun.

Another source of uncorrectable error was the SFOV. For each scene and blackbody view, the PAERI was also seeing off-axis light that amounted to approximately 0.6% of the total signal. The geometry of the SFOV was such that it saw different amounts of the scene mirror, scene mirror holder and inside of the support structure at different angles. Therefore, the effects of the SFOV were not removed by the calibration procedure. Current estimates put the effects of the SFOV at +0.2K for a blackbody source at 260K and –0.4K for a source at 100K.

Other sources of error were twice daily instrument checks that perturbed observations and a daily reboot of the data acquisition computer. These data were eliminated based on the time of each check or reboot. Times of excessive detector noise were also flagged using thresholds set for the instrument responsivity and an independent calibration check (ICC) discussed below. Other miscellaneous issues with the data (primarily closure due to blowing snow) were eliminated based on logbook history.

4.3 Independent Calibration Check (ICC)

A method for checking the calibration of the PAERI independently of the standard calibration sequence was developed for the purpose of providing QC of the data set. The ICC is based on the raw signal from the interference of scene radiation with radiation from the 2nd port. The magnitude of the raw signal is related nonlinearly to the temperature difference between the 2nd port and the scene. If the data collection system (interferometer, blackbodies, and data collection computer) is stable throughout a field season then one should be able to predict the uncalibrated signal with only the knowledge of the T_B of the scene and the 2nd port. Alternatively, one should also be able to determine the T_B of the scene with only knowledge of the T_B of the 2nd port and the raw instrument signal.

Fig. 3 shows the relationship between the difference between T_B of the 2nd port and the scene and the complex uncalibrated instrument signal at 600 cm^{-1}. The data are from four months of blackbody data taken during the 2001 SPARCLE field season. Each data point, or 'scene', in Fig. 3 represents a time when the PAERI was observing a blackbody. The temperatures and emissivities of the blackbodies are known precisely, allowing us to develop nonlinear relationships like Fig. 4 for each frequency observed by the PAERI.

This method of checking the calibration of the PAERI has two major caveats. The first is that presently the nonlinear curve is not accurate at high values of ΔT. This is due to the lack of data with which to constrain the nonlinear fit at high values of ΔT. ΔT could be constrained at its extremes through laboratory observations of extremely cold sources (i.e., LN$_2$ and CO$_2$-ice), but only if the temperature of the cold source can be determined accurately. The second caveat is if the detector response drifts significantly with time, then one will not be able to create a consistently valid relationship like the one shown in Fig. 3. Fig. 3 implies that the latter caveat is probably not an issue for the AERIs because the curve fits are good for this time period.

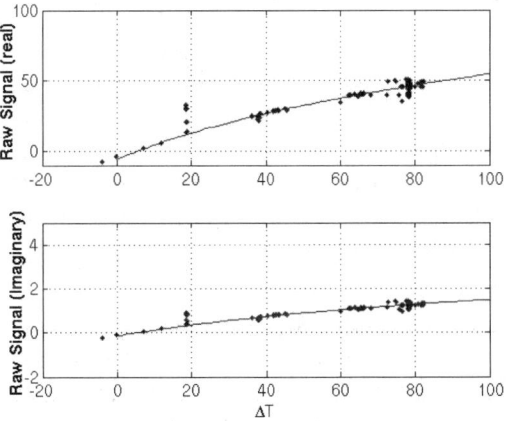

Fig. 3. Real and imaginary uncalibrated signals at 600 cm^{-1} plotted as a function of the difference between the 2nd port T_B and the scene T_B. This figure was generated with over four months of blackbody data from SPARCLE. The curve is of the form a*log(b*(ΔT + c)) + d. 'c' is an arbitrary constant to avoid the singularity at $\Delta T = 0$.

If a comprehensive set of nonlinear curves are developed for each wavenumber, then this technique could be useful in developing a new field instrument that does not require blackbody looks bracketed around every few scenes. In fact, it is believed that the AERIs should be able to observe for hours to days on end without a calibration look. This ICC therefore allows for instrumentation that would consume less power since the blackbodies need only be heated for infrequent calibration checks. Under special circumstances, the blackbodies may not even be required in the day-to-day operation of the instrument, resulting in a much lighter instrument.

4.4 Summary of Statistics

The amount of data eliminated for each of the uncorrectable errors but the SFOV are listed in Table 1. The bulk of the data eliminated for 'miscellaneous' reasons are due to times of blowing snow events when the PAERI was not able to observe.

Table 1. Percentage of Data Removed from the SPARCLE Data Set by QC Category. The Total does not Equal the sum of the Elements Because some Criteria Overlapped in Time

QC Class	Fraction of spectra eliminated (%)
Direct Solar	3
Daily Reboot	3
Daily Check	5
ICC	4
Miscellaneous	16
Total	24

Table 2 lists errors in fluxes calculated from the data shown in Figs. 1 and 2. It turns out that most of the errors in fluxes calculated in the MCT detector bandwidth occur in water vapor emission bands. This is because of the instrument response at the edges of the detector's bandwidth rather than any effect that water vapor may have on the measurement process.

Table 2. Errors in Fluxes Calculated Through Spectral and Angular Integration of Data Presented in Figs. 1 and 2. The MCT and InSb Totals are Square-root Sums of Squares of Other Elements of this Table

Band	Bandwidth (cm^{-1})	Error (W m^{-2})
MCT	450–1800	0.7
InSb	1800–3000	0.3
H_2O	450–550 and 1375–1800	0.698
CO_2	550–800	0.088
Window Region	800–950 and 1100–1200	0.006
O_3	950–1100	0.005
N_2O/CH_4	1200–1375	0.008

5. CONCLUSIONS

A Taylor series expansion of the AERI calibration equation is presented here. Detector noise is responsible for approximately 95% of the error in the PAERI spectra. The other 5% is due to uncertainty in blackbody temperature and emissivity. Uncertainty in PAERI spectra is predominantly in regions of water vapor emission. This is due to poor signal-to-noise in those regions of the spectrum.

The random errors and biases in the SPARCLE data set that cannot be corrected are minor and can often be eliminated from the data set through quality control procedures without compromising the data set. The most significant factor in data loss during SPARCLE was the instrument's inability to operate during times of blowing snow.

An independent calibration check (ICC) was developed for the PAERI and applied to the SPARCLE data set. Despite its present weaknesses at accurately calibrating low radiances, the ICC may provide an opportunity for the development of lighter, more energy-efficient interferometers in the future.

6. REFERENCES

Knuteson, R. O., H. E., Revercomb, F. A., Best, N. C., Ciganovich, R. G., Dedecker, T. P., Dirkx, S. C., Ellington, W. F., Feltz, R. K., Garcia, H. B., Howell, W. L., Smith, J. F., Short, and D. C., Tobin, 2004a. Atmospheric Emitted Radiance Interferometer (AERI) Part I: Instrument design. *J. Atmos. Oceanic Technol.*, 21(12), 1763-1776.

Knuteson, R. O., H. E., Revercomb, F. A., Best, N. C., Ciganovich, R. G., Dedecker, T. P., Dirkx, S. C., Ellington, W. F., Feltz, R. K., Garcia, H. B., Howell, W. L., Smith, J. F., Short, and D. C., Tobin, 2004b. Atmospheric Emitted Radiance Interferometer (AERI) Part II: Instrument performance. *J. Atmos. Oceanic Technol.*, 21(12), 1777-1789.

Revercomb, H. E., H., Buijs, H. B., Howell, D. D., LaPorte, W. L., Smith, and L. A., Sromovsky, 1988. Radiometric calibration of IR Fourier transform spectrometers: Solution to a problem with the High-Resolution Interferometer Sounder. *Appl. Opt.* 27, 3210-3218.

Rowe, P. M., 2004. *Measurements of the Foreign-Broadened Contiuum of Water Vapor in the 6.3 micron band at -30° Celsius*. Ph.D. thesis, University of Washington, Seattle, Washington.

Town, M. S., V. P. Walden, S. G. Warren, 2005. Spectral and broadband longwave downwelling radiative fluxes, cloud radiative forcing and fractional cloud cover over the South Pole, *J. Climate* (in press).

AIRBORNE MEASUREMENTS OF UPWELLING AND DOWNWELLING SPECTRAL ACTINIC FLUX DENSITIES DURING THE INSPECTRO CAMPAIGN

E. Jäkel, M. Wendisch, and S. Schmidt
Leibniz-Institute for Tropospheric Research (IfT), Leipzig, D-04318, Germany
(jaekel@tropos.de; wendisch@tropos.de; schmidt@tropos.de)

T. Trautmann
Deutsches Zentrum für Luft- und Raumfahrt e.V. (DLR), Oberpfaffenhofen, D-82234, Germany
(thomas.trautmann@dlr.de)

A. Kniffka
Institute for Meteorology, University of Leipzig, Leipzig, D-04103, Germany
(kniffka@server1.rz.uni-leipzig.de)

ABSTRACT

An airborne system for fast measurements of spectral actinic flux densities in the UV-A and VIS spectral range is introduced. The AFDM (Actinic Flux Density Meter) is designed to measure up- and down-welling spectral actinic flux densities separately with a time resolution of less than one second. Thus the AFDM is capable to resolve fast changing conditions of inhomogeneous clouds or surface reflection. The optical inlets mounted at the top and the bottom of the aircraft are actively stabilized in a horizontal position with respect to the Earth-fixed coordinate system during the flight. Field data collected during the INSPECTRO (INfluence of clouds on the SPECtral actinic flux in the TROposphere) campaign in 2002 are presented and the horizontal variability of measured spectra is evaluated. Furthermore profiles measured in overcast situations are discussed to characterize the cloud impact on the actinic radiation field in conjunction with model calculation.

1. INTRODUCTION

Most photochemical processes are driven by solar radiation, mainly in the ultraviolet (UV) and visible (VIS) spectral ranges. This radiation is influenced by absorption and scattering processes by gases, aerosol particles and cloud droplets. Through photodissociation, different reactive products are formed, such as the hydroxyl radical OH, which initiates most of the oxidative removal of trace gases.

Airborne measurements of actinic flux densities, which quantify the radiation energy available for photochemical processes help to understand the atmospheric photo-chemistry. Actinic flux densities can be used to derive photolysis frequencies J, which quantify the first-order rate coefficient of any photochemical dissociation process. The J-values are derived for each photochemical reaction from integrating the product of the spectral absorption cross section, quantum yield (both for the respective species), and the actinic flux density over wavelength.

Ground-based spectrometer measurements have been described by several authors (Müller *et al.*, 1995; Eckstein *et al.*, 2003; Edwards and Monks, 2003). Unlike chemical actinometers and filter radiometers, spectrometers permit the calculation of J-values for any chemical species whose absorption cross section and quantum yield significantly contribute within the respective spectral range.

To evaluate modeled actinic flux densities, particularly in the presence of clouds, airborne measurements are required. Such measurements were performed by the Scanning Actinic Flux Spectroradiometer (SAFS) (Shetter and Müller, 1999; Shetter *et al.*, 2002), which guarantees a reliable stray light rejection and therefore accurate measurements in the UV spectral range. The disadvantages of the SAFS are the poor time resolution (~10 s) and low wavelength stability. For measurements in rapidly changing conditions such as in cloudy environments the coarse time resolution may lead to serious problems in interpretation of the data. Therefore developing a new, fast measuring system with a time resolution of less than 1 second and high temporal wavelength stability were our main motivations.

2. INSTRUMENTATION

The AFDM, installed on a small research aircraft (Fig. 1), measures spectral down- and upwelling actinic flux densities ($F_a^\downarrow$, $F_a^\uparrow$). The optical inlets mounted at the top and the bottom of the aircraft are actively stabilized in a horizontal position with respect to the Earth-fixed coordinate system during the flight with an accuracy of ±0.2° for aircraft pitch- and roll angles of less than ±6°. More details are given by Wendisch *et al.* (2001). The horizontal stabilization system assures (i) a clear separation between the measurements of the upper and lower hemisphere, and (ii) the possibility to correct the non-ideal angular response of the optical inlets.

Fig. 1. Partenavia P68B with radiation inlets.

Each of the inlets (manufactured by Metcon, Königstein, Germany) is connected via optical fibers with a multi-channel spectrometer (MCS). The MCS modules (manufactured by Zeiss GmbH, Jena, Germany) consist of a fixed grating for wavelength splitting and a photo diode array (manufactured by Hamamatsu, Japan) with 512 pixels for detection within a spectral range between 280 and 700 nm (spectral resolution of about 0.8 nm). Each spectrometer is mounted in a ceramic body; therefore thermal and mechanical stability is assured. The radiation passes through an entrance slit (70 µm) and is dispersed by a flat field diffraction grating (248 lines/nm), which is blazed at 450 nm. The two spectrometers work nearly simultaneously using a multiplexer. Two separate integration times can be adjusted. Typical scan times are 250–600 ms for the shorter integration and 500–1200 ms for the longer integration, depending on atmospheric conditions.

The AFDM has proven to work accurately within 305–700 nm wavelength. Because of the small number of counting statistics masked by possible stray light and optical-thermal offsets, the sensitivity of the actinic flux density drops below 305 nm. An overall uncertainty of ±8.0% in the UV and ±4.9% in the VIS spectral range was calculated with Gaussian error propagation which includes uncertainties of the calibration and transfer lamps, the determination of the receiving plane, the wavelength calibration, and the remaining angular response error.

Stray light that is inevitable in a single monochromator, as well as dark current were treated by subtracting an averaged signal of all counts from 280–290 nm from each spectrum with the assumption of a wavelength-independent stray light/dark current signal. Laboratory measurements of the dark current showed only a slight wavelength dependence, which can be neglected. Tests with spectral cut-off filters determined that the amount of stray light in the total signal was less than 1% above 430 nm wavelength. Plots of the stray light signals below 430 nm revealed no distinct wavelength dependence.

The capability of the AFDM to determine photolysis frequencies is restricted to species [for example, nitrogen dioxide (NO_2), hydrogen peroxide (H_2O_2), formaldehyde (CH_2O), and nitrate (NO_3)], whose spectral products of quantum yield, absorption cross section, and actinic flux densities are significant mostly above 305 nm. A comparison of photolysis frequencies calculated for the whole spectra and for a so-called cut spectral range above 305 nm showed differences of less than 3% for these species.

3. MEASUREMENT EXAMPLES

3.1 Overview

The AFDM was tested in a measurement campaign in September 2002. The experiment was held in England in the framework of the European project INSPECTRO. Airborne and ground-based radiation and microphysical (aerosol and droplets) measurements were performed. 11 measurement flights were carried out under clear sky (to determine the surface albedo) and cloudy conditions. The main emphasis was to measure actinic flux densities in the presence of clouds above, within, and below isolated cumulus clouds, broken stratocumulus, and compact stratus cloud layers. The following sections show some examples of actinic flux densities measurements for various atmospheric conditions. Furthermore, photolysis frequencies of NO_2 derived from these measurements are presented by applying the literature data on spectral quantum yield and absorption cross section of DeMore et al. (1997).

3.2 Horizontal Variability

The horizontal variability of $F_a^\downarrow$ and $F_a^\uparrow$ measured above land and sea in clear sky conditions is presented in Fig. 2a–d. 10 km flight legs in a constant flight altitude were investigated. The averaged down- and upwelling actinic flux densities are plotted separately; the error bars indicate the standard deviation of the averaged spectra. The change of the solar zenith angle (SZA) is in the range of ±1°. Therefore the influence of the SZA variability during the flight track can be neglected. Obviously $F_a^\downarrow$ is nearly independent of surface reflection properties (Fig. 2a,b). In contrast $F_a^\uparrow$ is very sensitive to changes in the surface albedo (Fig. 2c,d), where the spectral shape of the averaged spectra is similar to the spectral course of the surface albedo. The percentage of spectral relative variability rv (ratio of standard deviation and average) shows a nearly wavelength-independent course for the downwelling component, which is lower than the wavelength-dependent rv of $F_a^\uparrow$. One would expect a lower rv of $F_a^\uparrow$ above sea than above land because of the gray surface albedo of the sea and the more wavelength-dependent albedo of land surfaces. Fig. 2c and 2d do not show this behavior. There are two possible reasons for this: (i) irregular whitecaps caused by wind were observed on the sea, which could increase the surface reflection and lead to higher variabilities than a calm sea. (ii) The flight track above sea was flown near the coastline. Hence the downward-looking optical inlet was also influenced by the land surface.

albedo of a compact cloud layer (Webb et al., 2004). Whereas above the stratocumulus layer rv increases with increasing wavelength.

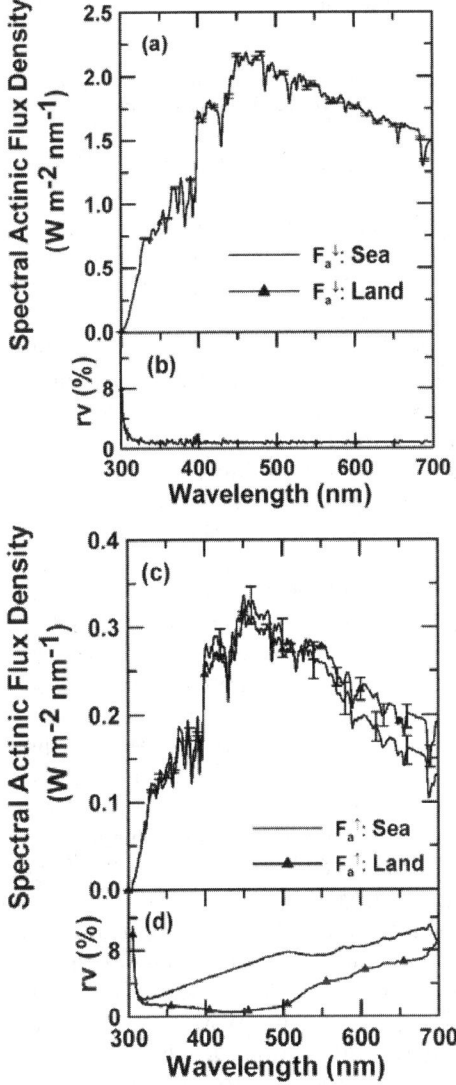

Fig. 2. (a), (c) Averaged down- and upwelling actinic flux densities measured above land and sea in a flight altitude of 2 km at a mean SZA of 56°; (b), (d) Relative variability (rv) = ratio of standard deviation and average (after Jäkel et al., 2005).

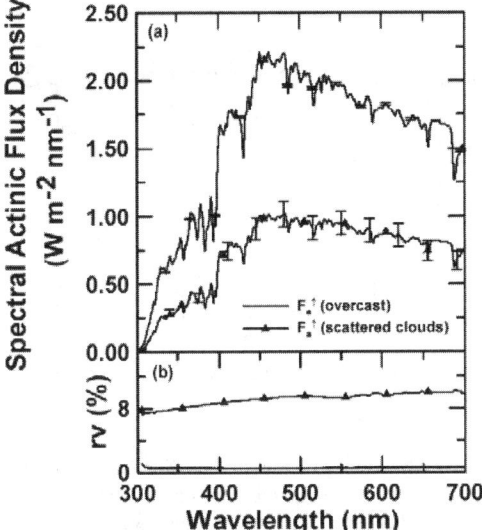

Fig. 3. (a) Averaged upwelling spectral actinic flux densities above a stratus and a broken cloud layer; (b) rv-plot (after Jäkel et al., 2005).

3.3 Vertical Variability

A profile of the downwelling actinic flux density measured at a descending flight leg through the compact stratus cloud layer is presented. The liquid water content (*LWC*) of the cloud measured with a Fast-Forward Scattering Spectrometer Probe, as well as the $F_a^\downarrow$ at a representative wavelength (400 nm) is plotted in Fig. 4a. The *LWC* marks a cloud extension of ~ 400m. The error bars indicate the measurement uncertainty. In general $F_a^\downarrow$ strongly decreases within the cloud. At the cloud top the increasing downward diffuse component leads to a peak in this part of the cloud that is due to multiple-scattering. This enhancement depends on SZA and wavelength. At lower SZA and higher wavelengths the increase is more distinct than for higher SZAs and lower wavelengths. Therefore only a slight enhancement is noticed in the $J(NO_2)$ plot, as displayed in Fig. 4b, where the error bars give only the measurement uncertainty, not the uncertainties of the molecule data (absorption cross section and quantum yield).

Fig. 3 shows the averaged spectra of $F_a^\uparrow$ in the presence of different cloud layers (overcast and broken). The SZA of both flight periods was nearly the same (50°–53°). Both averaged spectra were calculated for flight legs of 10 km length. As expected the variability of $F_a^\uparrow$ above the broken cloud layer is higher than above the compact stratus layer. Comparing the rv of both cases a factor of ~15 is noticed. The spectral rv of the averaged stratus spectrum is nearly wavelength-independent due to the gray cloud top

In addition one-dimensional (1-D) model simulations based on DISORT (Stamnes et al., 1988) were carried out by the University of Leipzig. Crucial input data for the simulation as droplet and aerosol properties stem from aircraft and ground-based measurements. Fig. 4 contains the simulated 400 nm profile of $F_a^\downarrow$. In general the model simulation agrees quite well with the measured profile. Larger deviations were found in altitudes below the cloud layer. Also the NO_2 photolysis frequencies, which represent an integral of spectral actinic flux densities, agree within the range of measurement uncertainties. For the given conditions the enhancement of $F_a^\downarrow$ in the upper part of the cloud is not visible in the results of the 1-D model.

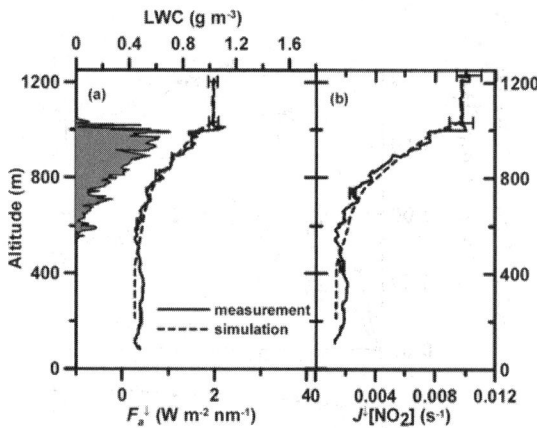

Fig. 4. (a) Vertical profile of the *LWC* (gray area), the measured and simulated downwelling actinic flux density at 400 nm; (b) Vertical profile of NO_2 photolysis frequency derived from measurement and simulation.

4. SUMMARY

A new airborne system for fast measurements of upwelling and downwelling actinic flux densitites has been introduced. The system is operated together with a similar irradiance measurement unit on a horizontally stabilized platform. Therefore the radiation portions of both hemispheres can be sampled separately, which is important for studies of the influence of cloud and surface albedo inhomogeneities. The advantages of the AFDM compared with scanning actinic flux densitiy measurement systems are the increased time resolution of less than 1 s, the high wavelength stability, and the extended wavelength range, which allows calculations of photolysis frequencies for a wider range of species. The AFDM does not cover the UV-B spectral range (below 305 nm). Therefore $J[O(1D)]$ cannot be derived directly from AFDM measurements.

Measurements of up- and down-welling actinic flux densities in clear sky and cloudy conditions have shown the applicability of the AFDM in the field. The highest variability of the reflected actinic flux density was observed above a scattered cloud layer, the lowest above a compact stratus cloud layer. Measured and simulated $F_a^\downarrow$ ($J[NO_2]$) were compared for a vertical cloud penetration. The agreement was within the measurement uncertainties.

5. ACKNOWLEDGMENT

The research of E. Jäkel and A. Kniffka was funded by the Deutsche Forschungsgemeinschaft (contracts WE 1900/6-1 and -2 and TR315/3-1 and -2). The INSPECTRO project was funded by contract EVK2-CT-2001-00130 from the European Commission. We are grateful to *enviscope GmbH* (Frankfurt/Main, Germany) for its support with the aircraft measurements.

6. REFERENCES

DeMore, W., S. Sander, D. Golden, R. Hampson, M. C. Howard, A. Ravishankara, C. Kolb, and M. Molina, 1997: Chemical kinetics and photochemical data for use in stratospheric modeling, evaluation number 12. *JPL Publ. 97-4*, California Institute of Technology, Pasadena, Calif.

Eckstein, E., D. Perner, C. Brühl, and T. Trautmann, 2003: A new 4π-spectroradiometer: instrument design and application to clear sky conditions. *Atmos. Chem. Phys.*, **3**, 1965-1979.

Edwards, G. D., and P. S. Monks, 2003: Performance of a single-monochromator diode array spectroradiometer for the determination of actinic flux and atmospheric photolysis frequencies. *J. Geophys. Res.*, 108, D16 doi:10.1029/2002JD002844.

Jäkel, E., M. Wendisch, T. Trautmann, and A. Kniffka, 2005: Airborne system for fast measurements of upwelling and downwelling spectral actinic flux densities. *Appl. Opt.* 44 (3), 434-444.

Müller M., A. Kraus and A. Hofzumahaus, 1995: O3→O(1D) photolysis frequencies determined from spectroradiometer measurements of solar actinic UV radiation: comparison with chemical actinometer measurements. *Geophys. Res. Lett.*, 22, 679-682.

Shetter R., and M. Müller, 1999: Photolysis frequencies using actinic flux spectroradiometry during PEM-Tropics mission: Instrumentation description and some results. *J. Geophys. Res.*, 104, D5, 5647-5661.

Shetter R. E., L. Cinquini, B. L. Lefer, S. R. Hall, and S. Madronich, 2002: Comparison of airborne measured and calculated spectral photolysis frequencies during the PEM Tropics B mission. *J. Geophys. Res.* 107, doi:10.1029/2001JD001320.

Stamnes, K., S. C. Tsay, W. Wiscombe, and K. Jayaweera, 1988: Numerically stable algorithm for discrete-ordinate-method radiative transfer in multiple scattering and emitting layered media. *Appl. Opt.*, 27, 2502-2509.

Webb, A. R., A. Kylling, M. Wendisch, and E. Jäkel, 2004: Airborne measurements of ground and cloud spectral albedos under low aerosol loads. *J. Geophys. Res.*, 109, doi:10.1029/2004JD004768.

Wendisch, M., D. Müller, D. Schell, and J. Heintzenberg, 2001: An airborne spectral albedometer with active horizontal stabilization. *J. Atmos. Oceanic Technol.*, 18, 1856-1866.

THE FRENCH NETWORK FOR SPECTRAL MEASUREMENT OF SOLAR UV IRRADIANCE

J. Lenoble, C. Brogniez, M. Houët, and M. Legrand,
LOA (UMR 8518), USTL
Villeneuve d'Ascq, 59655, France

J. Lenoble, A. de La Casinière, and T. Cabot
IRSA, UJF, Grenoble, 38000, France

F. Guirado
CEMBREU
Villard Saint Pancrace, Briançon, 05100, France

ABSTRACT

The paper presents the two French stations equipped for the measurement of spectral solar ultraviolet (UV) radiation; one is located in the Northern France, in a flat urban region, and the other in a high altitude valley in the Southern Alps. The instruments are briefly described, and some results are shown.

1. INTRODUCTION

The high energy photons of UV solar radiation have important biological and chemical effects. The depletion of stratospheric ozone has raised concern on a possible increase of UV in the short wavelength range (280–330 nm) at the Earth's surface. Various other parameters, cloudiness, aerosols, tropospheric ozone, and ground reflection, have also an impact on UV radiation that can be analyzed only by combining spectral measurements and modelling.

The scientific objectives of a network for measuring solar UV radiation are: i) analysis of its natural variability and of the various modulating factors, ii) detection of possible long term trends related to human activity, iii) validation of UV climatology provided by satellite observations, and iv) providing of data to various user communities.

The network for Detection of Stratospheric Changes (NDSC) records UV data from several stations around the world, including the two French stations. The European Union has organised intercomparisons between the instruments operating in different European countries, and has supported the establishment of a UV database at the Finish Meteorological Institute (FMI) in Helsinki, Finland.

Section 2 presents the two French sites with their instruments, and the radiation codes used for analysing the data. In section 3, a selection of results is shown; they concern the retrieval of total ozone column, and of aerosol optical depth, the analysis of the impact of a snow covered ground, and a model for retrieving daily erythemal doses from a few simple parameters. Conclusion and plans for future developments are in section 4.

2. PRESENTATION OF THE SITES AND INSTRUMENTS

At Villeneuve d'Ascq, in the plain region of Northern France, near the industrial city of Lille, the spectroradiometer SPUV01 is installed since 1997, and operates since 1999. At Briançon-Villard Saint Pancrace (1310m asl), in the French Southern Alps, two spectroradiometers, SPUV02 and IRSA are operational since 1999. The apparent redundancy of two similar instruments at the same place has two justifications: avoiding gaps of data when an instrument needs repair, and allowing cross validations. The characteristics of SPUV and IRSA spectroradiometers are summarized in Table 1. The three instruments are calibrated with standard lamps traceable to the National Institute of Standards and Technology (NIST) or to the National Physical Laboratory (NPL). They have been successfully intercompared with other European instruments during a few campaigns (Bais et al., 2001). All spectroradiometers are equipped with a movable shadow disc, and global and diffuse irradiances are measured alternatingly; their difference provides the direct solar irradiance.

In both stations a broadband radiometer is operating regularly; it allows to follow the rapid oscillations due to clouds. In Villeneuve d'Ascq, the YES instrument provides the erythemal UV, whereas in Briançon a Scintec radiometer gives integrated UV-A (315–400nm) and erythemal UV.

Table 1. Spectroradiometer Characteristics

Characteristics	SPUV type	IRSA type
Monochromator	JobinYvon HD10	Bentham DM150
Spectral range	280-450 nm	290-400 nm
Step	0.5 nm	0.5 nm
Resolution FWHM	0.75 nm	0.80 nm
Sampling frequency	15 min	30 min
Cosine error	<5% at SZA75°	<5% at SZA75°
Detector	photomutiplier	photomutiplier
Threshold	10^{-6} $Wm^{-2}nm^{-1}$	10^{-5} $Wm^{-2}nm^{-1}$
Temperature	28°C	25°C

Modelling of UV irradiance is done, for cloudless sky, with a successive order of scattering (SOS) code for the Briançon station, and with DISORT code at Villeneuve d'Ascq. Both codes have successfully taken part in a model intercomparison (Van Weele et al., 2000). The main uncertainty in modelling is due to the uncertainty on input parameters, mainly on aerosol optical depth. This parameter is measured, at Briançon site, by a manual sunphotometer CIMEL. Villeneuve d'Ascq station is part of the Aerosol Robotic Network (AERONET), using an automatic CIMEL instrument.

A first check of data quality is obtained, for cloudless days, by comparing the measured irradiance with the model irradiance. The broadband irradiance computed by integration of the spectral data is also compared to the broadband radiometer data, and this provides a second quality control.

3. SELECTION OF RESULTS

We will present in this section some selected examples of results deduced from the UV spectral irradiance measurements performed at both stations; this choice is somewhat arbitrary, and more results can be found in the published litterature.

3.1 Total Ozone Column (TOC) Retrieval from Global Irradiance

The ratio of global irradiance around 305 nm (in the ozone absorption band), to global irradiance around 340 nm (no ozone absorption) is compared to the same ratio computed in cloudless conditions, for different TOC, in look up tables (LUT). The second parameter in the LUT is the solar zenith angle (SZA); climatological aerosol values are used. The method supposes that the cloud reduction factor is not wavelength dependent, and does not vary during the spectral scan, and this is the major cause of uncertainty.

Fig. 1 presents the TOC observed at Briançon station during the year 2000 (Masserot et al., 2002) by the two spectroradiometers, compared with TOMS ozone values. The agreement is generally within about 4%, or 12 DU, which is consistent with the estimated uncertainty of both methods.

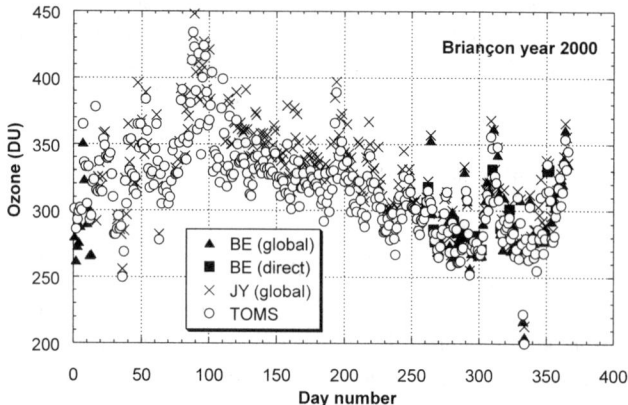

Fig. 1

3.2 Total Spectral Aerosol Optical Depth (AOD) Retrieval from Direct Irradiance

The ratio of direct solar irradiance at the Earth's surface to the extraterrestrial solar irradiance provides the total atmospheric optical depth. Subtracting the molecular optical depth for the station altitude, and the ozone optical depth (known from TOC) in the short wavelength range, leads to the AOD. The major uncertainty is due to the calibration of the instrument, which reflects on the irradiance ratio. The relative uncertainty can become large for small AOD; it increases for smaller SZAs, because of the decrease of the atmospheric pathlength.

Fig. 2 (Houët, 2003) shows the time variation of AOD at Villeneuve d'Ascq on 21 February, 2003, for 3 wavelengths, compared with AOD measured by the CIMEL instrument. The agreement is very good for this case of rather high turbidity.

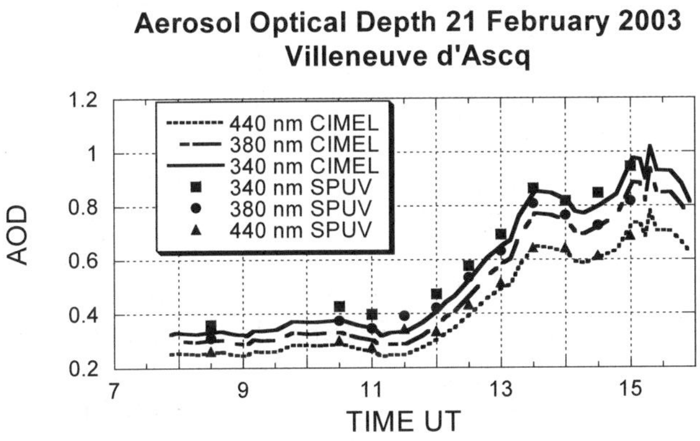

At Briançon station we have observed this enhancement, for cloudless days during winter, by ratioing the observed global irradiance to the irradiance computed with the SOS radiation code, for a black surface, all other conditions being the same. This enhancement can not easily be modelled, because it does not depend only on the surface reflectance near the station, but a rather large area extending as far as 50 km can contribute to the amplification. The topography and the type of surface (rocks, trees, roads,.etc) make the refectance of such a large area very inhomogeneous, and an effective albedo has to be defined, from a detailed 3-D model.

Fig. 3 compares the observed enhancement for erythemal UV at Briançon for cloudless days in winter 2002, with the enhancement computed with a 3-D code at two wavelengths in the erythemal range (Lenoble et al., 2004). In the model, the snow albedo has been fixed at 0.8 above the tree line, and at 0.3 between the snow line and the tree line.

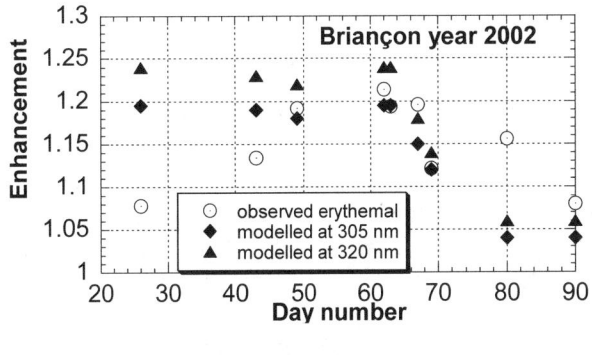

Fig. 2

Fig. 3

3.3 Impact of a Snow Covered Ground on UV Global Irradiance

In UV most ground surfaces have a very low reflectance (a few %), with the noticeable exception of snow covered ground. Clean fresh snow may have a reflectance reaching 90%. Multiple reflections between the surface and the backscattering atmosphere can strongly enhance the diffuse, and therefore the global, irradiance at the Earth's surface, when it is covered with snow. This effect is important in high latitude regions, and also in mountainous areas during winter.

3.4 Reconstruction of Daily Erythemal Doses from Simple Parameters

A multilinear correlation (de La Casinière et al., 2002) is established between the atmospheric transmissivity T for different biologically active radiation daily doses and 3 simple parameters (daily sunshine fraction, minimum daily SZA, and TOC). T is defined as the ratio of the daily dose at the Earth's surface to its extraterrestrial value. Using the coefficients obtained for year 2000, permits to retrieve the doses for 2001 with an error between 3 and 9% for monthly mean values. Fig. 4 presents the measured and reconstructed erythemal doses at Briançon in year 2001.

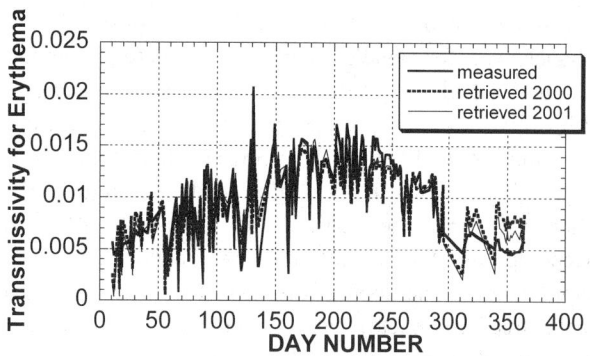

Fig. 4

4. CONCLUSION AND FUTURE PLANS

Two stations measure regularly, since 1999, the spectral solar UV irradiance, in the North of France and in the Southern French Alps. Their data are stored in two international databases, and the analysis has lead to several publications. In the future, the irradiance measurements will be completed by measurements of the spectral actinic flux that commands the photochemical reactions.

5. ACKNOWLEDGMENTS

This work has been supported by the French Ministry of Environment, and by three European Union contracts: Scientific UV Data Management (SUVDAMA), European Database for Ultraviolet Climatology and Evaluation (EDUCE), Characteristics of UV Radiation field in the Alps (CUVRA).

6. REFERENCES

Bais, A. F., B. G. Gardiner, H. Slaper, M. Blumthaler, G. Bernhard, R. McKenzie, A. R. Webb, G. Seckmeyer, B. Kjeldstad, T. Koskela, P. Kirsch, J. Gröbner, J. B. Kerr, S. Kazadzis, K. Leszczynski, D. Wardle, C. Brogniez, W. Josefsson, D. Gillotay, H. Reinen, P. Weihs, T. Svenoe, P. Eriksen, F. Kuik, and A. Redondas, 2001: The SUSPEN intercomparison of ultravioletspectroradiometers, *J. Geophys. Res.*, 106, 12509-12525.

de La Casinière, A., M. L. Touré, D. Masserot, T. Cabot, and J. L. Pinedo Vega, 2002: Daily doses of biologically active UV radiation retrieved from commonly available parameters, *Photochemistry and Photobiology*, 76, 171-175.

Houët, M., 2003: Spectroradiométrie du rayonnement solaire UV au sol: Améliorations apportées à l'instrumentation et au traitement des mesures. Analyse pour l'évaluation du contenu atmosphérique en ozone et aérosol. *Thèse Université des Sciences et Technologies de Lille*, 4 Décembre 2003.

Lenoble, J., A. Kylling, and I. Smolskaia, 2004: Impact of snow cover and topography on ultraviolet irradiance at the Alpine station of Briançon, *J. Geophys. Res.*, 109, D16209, doi:10.1029/2004JD004523.

Masserot, D., J. Lenoble, C. Brogniez, M. Houet, N. Krotkov, and R. McPeters, 2002: Retrieval of ozone column from global irradiance measurements and comparison with TOMS data. A year of data in the Alps, *Geophysical Research Letters*, 29, 10.1029/2002GL014823.

Van Weele, M., T. J. Martin, M. Blumthaler, C. Brogniez, P. N. den Outer, O. Engelsen, J. Lenoble, B. Mayer, G. Pfister, A. Ruggaber, B. Walravens, P. Weihs, B. G. Gardiner, D. Gillotay, D. Haferl, A. Kylling, G. Seckmeyer, and W. M. F. Wauben, 2000: From model intercomparison towards benchmark UV spectra for six real atmospheric cases. *J. Geophys. Res.*, 105, 4915-4926.

UV FILTER RADIOMETERS: CALIBRATION, FIELD MEASUREMENTS AND APPLICATIONS

A. Los
Kipp & Zonen
Röntgenweg 1, 2624 BD Delft
The Netherlands

ABSTRACT

Atmospheric ultraviolet radiation measurements are difficult to perform due do the drastic decrease of UV-B irradiance caused by strong stratospheric ozone absorption. Since a couple of decades precise UV measurements are made with spectrophotometers, most successfully with double monochromators. As these instruments are expensive and labour intensive, UV filter radiometers are frequently used as alternative or additional devices, offering the possibility to maintain a UV radiation monitoring network at reasonable costs. The most widely used type of instrument for this purpose is the broad band filter radiometer. Unfortunately, measurements made with such instruments suffer from non-ideal physical instrument properties, making UV measurements inaccurate if instrument discrepancies are not properly handled.

From July 2003 until the end of 2004, various broad band UV filter radiometers were evaluated at the European Reference Centre for Ultraviolet Radiation Measurements (ECUV). The intension of the investigation is to establish a reference group composed of broad band UV filter radiometers from institutions which may benefit from the uniform and well maintained UV irradiance scale realised at the ECUV. We present results for Erythema weighted UV irradiances obtained from these collocated measurements with the broad band UV radiometer reference group and the reference UV spectrophotometer. Besides the spectral characteristics of broad band UV radiometers, their angular response and stability are also considered in order to make a comprehensive uncertainty estimate. Measurements from 2003 and 2004 show that the selected group of broad band radiometers agree within 10% to the reference spectrophotometer.

1. INTRODUCTION

One of the most widely used instruments to measure atmospheric UV radiation is the broad band UV filter radiometer. Since the introduction of the Robertson-Berger radiometer in 1976 (Berger, 1976) a number of commercial broad band UV radiometers have been developed. Similar to the Robertson-Berger instrument the broad band radiometers in this study are used as sun-burning UV irradiance (CIE-1987) measurement devices only. Although the broad band UV radiometer is an accepted monitoring instrument it must be used carefully in order to make accurate and valuable measurements. While careful instrument calibration and dedicated measurement correction are fundamental requirements to make accurate UV measurements the environmental factors may also affect the measurement accuracy. A former study, presented at the IRS in 2000 (Lehmann et al., 2000), shows how significant these environmental effects can be when using broad band UV radiometers in the field.

There are a number of critical instrument properties that have to be considered during calibrations and measurements: (i) the spectral response function of broad band UV radiometers deviates considerably from the theoretical response curve, (ii) the cosine response function of the entrance optics is not perfect, (iii) the long term stability of broad band UV radiometers can be poor, and (iv) the optical components can be sensitive to temperature and humidity variations. The measurements performed at the ECUV with a number of different broad band UV radiometers are used to investigate the measurement quality and long term stability of this Erythema weighting instrument type. In this study we present results obtained with two YES Inc. and three Kipp & Zonen instruments. The measurement period used for the investigation started in July 2003 and ended in December 2004. The Brewer MKIII spectrophotometer of the ECUV was used as the reference instrument, i.e., all broad band UV radiometer calibrations are traceable to this spectrophotometer.

2. INSTRUMENTATION AND MEASUREMENTS

Fig. 1 shows the relative spectral response functions, Suvs, of the two YES Inc. and the three Kipp & Zonen instruments on a logarithmic scale (y-axis) as a function of wavelength in nm (x-axis). As one can see, deviations between the response functions of the individual instruments are large and substantially different from the theoretical curve, given as a composition of three linear functions (CIE-1987). The quality of the spectral response function measurements has been verified recently in an intercomparison of several laboratory facilities (Schreder et al., 2004). Fig. 2 shows the measurement platform of the ECUV located at the Joint Research Centre (JRC), Italy, with the UV spectrophotometer (Brewer, MKIII) at the right hand side and some broad band UV radiometers in the back. Suvs is shown on a logarithmic scale (y-axis) as a function of wavelength in nm (x-axis).

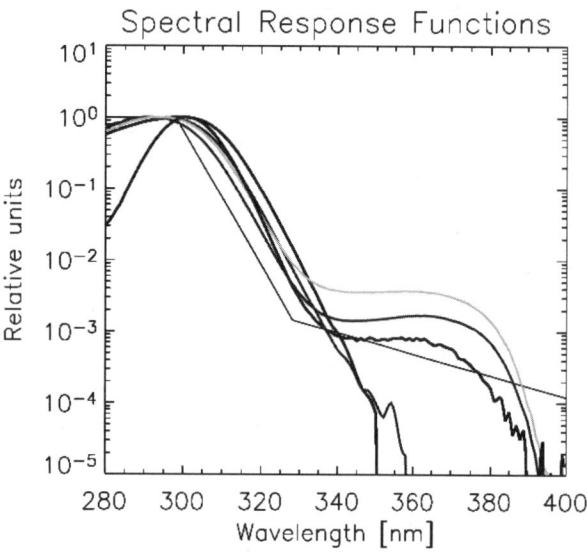

Fig. 1. Relative spectral response functions, Suvs, of the five broad band Erythemal radiometers used in this study. The curve, composed of three linear functions, is the theoretical Erythema weighting (CIE-1987) function.

Fig. 2. Measurement platform of the ECUV located at the Joint Research Centre (Italy) with the UV spectrophotometer (Brewer, MKIII) at the right hand side and some broad band UV radiometers in the back.

2.1 Calibration

The purpose of the first step in the calibration process is to allocate a sensitivity to the broad band UV radiometer in Volts per W/m2, referred to as the "radiometric calibration factor", ρ_{uvs} The index "uvs" stands for "UV Sensor" which represents one of the five broad band UV radiometer used in this study. To determine ρ_{uvs} the broad band UV radiometer has to measure atmospheric UV radiation under (ideally) cloud-free measurement conditions side-by-side to the reference spectrophotometer. Only synchronised spectral and broad band measurements (time lag less than 5 sec.) are used to determine ρ_{uvs} according to

$$\rho_{uvs} = \frac{U_{uvs}}{\int E_{Bre}(\lambda) S_{uvs}(\lambda) d\lambda}. \qquad (1)$$

U_{uvs} denotes the radiometer readings (in Volts) and $E_{Bre}(\lambda)$ represents the spectrophotometric measurements. The calibration formula given in Equation 1 yields a classic sensitivity, i.e., the ratio between the reading of the radiometer (enumerator) and the UV radiative flux which is physically detected by the radiometer (denominator). Note, that the denominator does not account for the deviation between physical and theoretical response functions. The radiometric calibration factor ρ_{uvs} therefore yields the UV irradiance, I_{uvs}, as it is physically detected by the broad band UV radiometer, according to

$$I_{uvs} = \frac{U_{uvs}}{\rho_{uvs}}. \qquad (2)$$

As broad band UV radiometers usually have spectral response functions which do not match the theoretical function (see Fig. 1) the UV irradiance obtained according to Equation 2 is physically unsatisfactory. Therefore, a measurement correction is required to minimise the so-called "spectral mismatch error", i.e., the error between the UV irradiance obtained according to Equation 2 and the true, Erythema weighted (CIE-1987) irradiance.

2.2 Measurement Correction Methods

Without any measurement correction, the broad band UV radiometer can provide results that deviate by a factor of 2 or more from the true values. The magnitude of the deviation depends mainly on the extent of the spectral mismatch and the measurement condition. In addition to the spectral mismatch error, other factors as mentioned in Section 1 potentially reduce the measurement quality of broad band UV radiometers.

The variable measurement conditions taken into account in this study include the solar zenith angle, Θ_0, and the total Ozone column density, $[O_3]$. Other atmospheric factors affecting UV irradiances, such as extinction of UV radiation due to aerosols, are not explicitly accounted for as they are assumed to be small compared to the effects that varying solar zenith angles and the Ozone column densities have on the spectral distribution of the UV radiation. In this second step of the calibration process we use two correction methods which both can significantly reduce the spectral mismatch induced measurement error of broad band UV filter radiometers.

2.2.1 Model-Based Correction Method

The model-based correction method improves the broad band UV radiometer measurements for atmospheric conditions typically encountered during field applications. The variable model parameters used to determine the various correction factors are the solar zenith angle, Θ_0, and the Ozone column density, $[O_3]$. Other fixed model parameters were chosen in order to represent the measurement conditions at the ECUV as close as possible.

The modelled spectra are used to determine the conversion factors, $\gamma(O_3, \Theta_0)$, defined as

$$\gamma(O_3,\Theta_0) = \frac{\int E_{TUV}(O_3,\Theta_0,\lambda) S_{uvs}(\lambda) d\lambda}{\int E_{TUV}(O_3,\Theta_0,\lambda) S_{cie}(\lambda) d\lambda}. \qquad (3)$$

The solar zenith angles, Θ_0, are varied between 0° and 85° (using steps of 5°) and the Ozone column densities, $[O_3]$ are varied between $200 DU$ and $500 DU$ (using steps of $10 DU$), yielding 18·31=558 conversion factors. Hence, the effects of other atmospheric compounds on the UV radiation, e.g., extinction due to aerosols, are not explicitly included in the model-based correction method. If broad band UV irradiances under exceptional conditions have to be measured with broad band UV radiometers, it is recommended to calculate new conversion factors using model parameters that are representative for the exceptional condition (e.g., snow covered land surface at a location which is mostly snow free).

2.2.2 Observation-Based Correction Method

The observation-based correction method uses spectrophotometric measurements to infer the final calibration factors, which include the correction of the spectral mismatch error. In Bodhaine et al. (1998) a detailed description of the observation-based correction method can be found. The observation-corrected Erythema weighted irradiance, Iuvs, is determined according to

$$I_{uvs} = \frac{U_{uvs}}{\delta_{uvs}(O_3,\Theta_0)}, \qquad (4)$$

where $\delta_{uvs}(O_3,\Theta_0)$ denotes the spectral mismatch corrected calibration factor of the broad band UV radiometer. As the observation-based correction method uses measured spectra to determine the final calibration factors, the atmospheric parameters that affect spectral UV irradiances will also affect the individual calibrations. Therefore, the observation-based correction method is - strictly spoken - only valid for the location at which the calibration has taken place. However, a large number of measurements under many measurement conditions (covering a large number of solar zenith angles and Ozone column densities) should improve the statistics of the corrections sufficiently to provide an universal corrected calibration factor table as a function of Θ_0 and $[O_3]$.

3. RESULTS AND CONCLUSIONS

From July 2003 until the end of 2004 a total of 12864 useful broad band UV radiometer measurements could be collected with the five radiometers at the ECUV. To get the best results out of the measurements the YES instruments had to be corrected according to the observation-based method while for the Kipp & Zonen instruments the model-based correction method had to be used. It is likely that the observation-based correction method removes certain instrument-specific measurement errors more efficiently than the model-based correction method. This, however, does not legitimate to draw conclusions about the quality of the instruments. As will be shown later, both correction methods can reduce measurement errors equally well, despite the significant differences in optical properties that exist among all five instruments. For the assessment of the measurement accuracy all corrected broad band UV radiometer measurements were subtracted from the integrals of the spectrally measured Erythema weighted irradiances. With the differences, which represent the best possible estimate of the Erythema weighted irradiance measured by broad band UV radiometers, a probability distribution function (PDF) is determined. Fig. 3 shows the individual PDFs, which are normalised for better comparability in the figure only. The statistical moments of the broad band UV radiometer measurements are determined with the enveloping function, i.e. the sum of all PDFs. The mean value, i.e., the mean difference between the spectrally derived and the broad band radiometer based irradiances, is as small as 5.6 10-4 W/m2. This difference corresponds to a UV Index of only 0.02. However, the standard deviation at a 2σ-level is as large a 10%, meaning that although dedicated measurement corrections were applied large deviations are still possible.

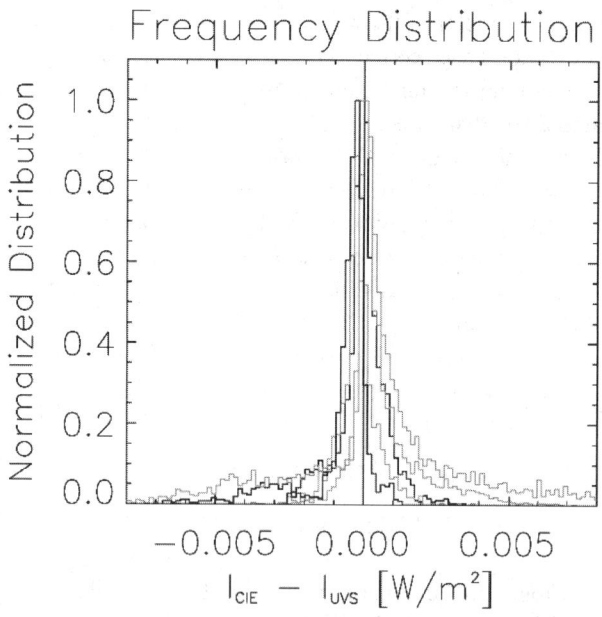

Fig. 3. Probability distribution functions of the difference between spectral and corrected broad band radiometer measurement of the Erythema weighted irradiances for all five broad band UV radiometers.

3.1 Conclusions

To measure Erythema weighted UV irradiances with broad band UV filter radiometers, careful calibrations and measurement corrections must be applied. Under ideal measurement conditions the differences between spectrally derived and properly corrected broad band radiometer measurements of the Erythema weighted irradiance can be arbitrarily low. However, it is far more difficult to determine the accurate UV irradiance under variable atmospheric measurement conditions. Nevertheless, it can be concluded that the broad band UV radiometer is a suitable instrument for UV monitoring, especially under fair weather conditions, if regularly calibrated and corrected properly.

4. ACKNOWLEDGMENTS

Dr. J. Gröbner is acknowledged for providing spectral data and for making the measurement platform available for UV filter radiometer measurements at the JRC (Italy).

5. REFERENCES

Berger, D. S., 1976: The sunburning ultravioletmeter: Design and performance. *Photochem. Photobiol*, 24:587 – 593.

Bodhaine, B. A., E. G. Dutton, R. L. McKenzie, and P. V. Johnston, 1998: Calibration broadband UV instruments: Ozone and solar zenith angle dependence. *J. Atmos. Ocean. Technol.*, 15:916 – 926.

CIE-1987, McKinley, A. F., and B. L. Diffey, 1987: A reference action spectrum for ultraviolet induced erythema in human skin. *CIE J.*, 6:17 – 22.

Lehmann, A., et al., 2000: Comparison between models and measurements of direct and diffuse UV erythemal irradiance under clear sky conditions – analysis of the environmental impacts on the measurements. In *IRS 2000: Current Problems in Atmospheric Radiation*. A. Deepak Publishing, Hampton, Virginia, 547 – 550.

Schreder, J., J. Gröbner, A. Los, and M. Blumthaler, 2004: Intercomparison of monochromatic source facilities for the determination of the relative spectral response of erythemal broadband filter radiometers. *Optics Letters*, 29(13).

COMPARISON BETWEEN OBSERVED SPECTRAL ALBEDOS AND THEORETICAL ONES FOR ARTIFICIAL SNOWPACK

T. Tanikawa and M. Aniya
Graduate School of Life and Environmental Sciences, University of Tsukuba, Tsukuba, Ibaraki, 305-8572, Japan

T. Aoki
Meteorological Research Institute, Tsukuba, Ibaraki, 305-0052, Japan

A. Hachikubo
Kitami Institute of Technology, Kitami, Hokkaido, 090-8507, Japan

M. Hori
Japan Aerospace Exploration Agency, Harumi, Tokyo, 104-6023, Japan

O. Abe
National Research Institute for Earth Science and Disaster Prevention, Shinjyo, Yamagata, 996-0091, Japan

ABSTRACT

Optical properties of artificial snowpacks composed of spherical and non-spherical particles were investigated by the measurements of spectral albedo in a cold laboratory. The measured spectral albedos in the visible and near-infrared wavelengths (0.35–2.5 μm) were compared with theoretically calculated ones, in which Monte Carlo radiative transfer model was employed for multiple scattering combined with the Mie theory and ray-tracing technique for single scattering. This model also takes into account the influence of the setup of the cold laboratory. The measured spectral albedo of the spherical snow particles agreed with the theoretical one for the snow grain size measured from the snow pit work. For the non-spherical particles which were dendrites, the snow albedo were found to be more influenced by the branch width than the branch length of the snow crystals from the comparison of the measured spectral albedo with the theoretical one using the shape of circular cylinder. The relationship between the spherical and non-spherical particles showed that the spectral albedos of the non-spherical particles could be obtained using equal-Volume/Area sphere.

1. INTRODUCTION

Most of the radiative transfer models are based on the assumption of the spherical shape of snow grains, which is quite unrealistic. The snow shape changes from new snow to other types (e.g., compacted snow or granular snow). In general, the snow crystals have a complex structure and consist of irregularly shaped ice grains in contact with other grains. In order to obtain the optical properties of the various snow types, we measured spectral albedo of artificial snowpacks composed of spherical and non-spherical particles in a cold laboratory. The measured spectral albedos in spectral region (λ = 0.35–2.5 μm) were compared with theoretically calculated ones, in which Monte Carlo radiative transfer model was employed for multiple scattering combined with the Mie theory and ray-tracing technique for single scattering by snow particles.

2. INSTRUMENT SETUP

Spectral snow albedo measurements and snow pit work were carried out in the cold laboratory in National Research Institute for Earth Science and Disaster Prevention, Japan (Fig. 1). Two types of snow were prepared for this experiment: the spherical particles (case-S) and the dendrites (case-D) (Fig. 2). Snow depths of case-S and case-D were 31 cm and 57 cm, respectively. Two kinds of dimensions of grain size were measured: one-half the length of major axis of crystals or dendrites (r_1), and one-half the branch width of dendrites or one-half the dimension of the narrower portion of broken crystals (r_2). For case-S, the snow grain size r_1 and r_2 in the layer from surface to 5 cm were less than 100 μm and 10–30 μm, respectively and those below 5 cm depth were less than 100 μm and 20–40 μm, respectively. For case-D, the first layer (0–31.5 cm) consisted of the snow grains of r_1 < 500 μm and r_2 = 10–30 μm, and the second layer (31.5–57 cm) was similar to that for case-S. Spectral albedo was observed using a grating spectrometer, FieldSpec FR, made by ASD Inc. (USA). In this measurement, the effect of the shadow of the White Reference Standard (WRS) over the snow surface on the observed albedo cannot be ignored. These measurements have some differences from the field measurements: the snow area is finite, and some photon escapes from the snow-wall; the area of light source is finite and the emitted light is not parallel (Fig. 1). These effects on observed albedo are discussed in Section 4.1.

3. MONTE CARLO RADIATIVE TRANSFER MODEL

The radiative transfer model based on the Monte Carlo method (MC-RTM) was employed, which is based on the Mie theory for case-S, and ray-tracing technique for case-D (Macke and Mishchenko, 1996) for single scattering. A photon from any lamp was determined by the directivity of the light intensity. Photon entering into the snow was propagated under the Beer's law. When a photon hits snow particles, single scattering occurs. The type of "scattering" or "absorption" is selected by the single scattering albedo. If scattering, the new direction is determined by the phase function of snow grains. The distance from one scattering point to the next scattering point is calculated again. These procedures are repeated until the photon is either absorbed or escapes from the snowpack. The snow albedo is defined as the ratio of the numbers of down-welling and up-welling photons that reach the WRS.

4. RESULTS AND DISCUSSIONS

4.1 Effects of Instrument Setup on Snow Albedo

4.1.1 Finite Snowfield and Snow Wall

Firstly the effects of finite snowfield on the snow albedo are examined by MC-RTM and Doubling and Adding radiative transfer model (DA-RTM) for case-S. For MC-RTM, we assume that snow wall is completely covered with the snow container (hereafter MC-RTM1). The snow albedos calculated by MC-RTM1 are lower than those by DA-RTM at all wavelengths (Fig. 3). The effect of the finite snowfield causes the snow albedo reduction because not only the snow but also the snow container absorbed some photon.

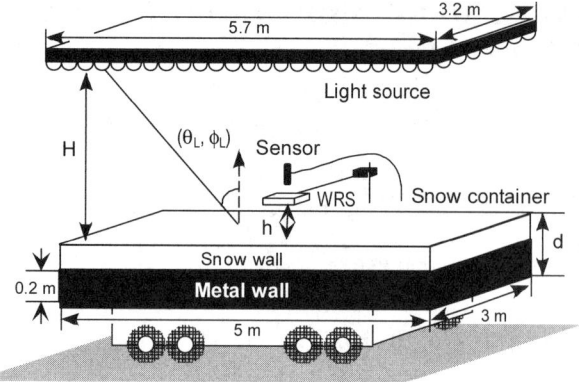

Fig. 1. Schematic illustration of the instrument setup of a snow container and light source in the cold laboratory. The snow was produced by an artificial snowfall system, which was set beneath the ceiling of the cold laboratory. The snow lies on snow container of 3 × 5 m in size with metal wall of 20 cm in height. Meaning of parameters: d; snow depth: h; height of WRS above snow surface: (θ_L, ϕ_L); emitted and azimuth angles of light, which are equal to incident angle to snow container. WRS (6 × 6 cm) is placed at $h = 10$ cm above the snow surface and the position 1 m away from side of the snow container.

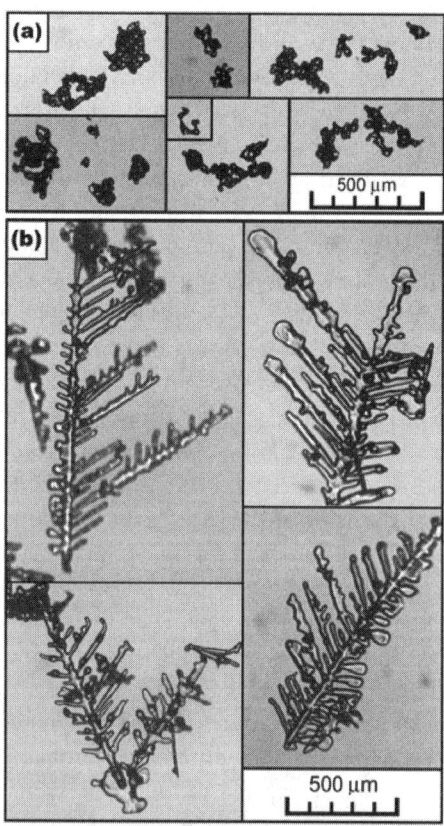

Fig. 2. Micrographs of artificial snow crystals; (a) spherical particles (case-S), and (b) dendrites (case-D).

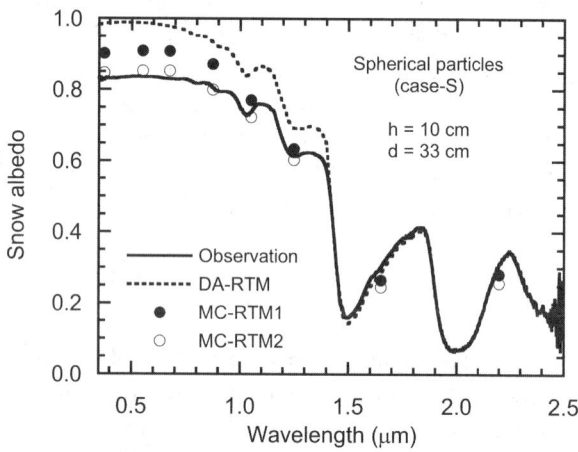

Fig. 3. Comparison of theoretically calculated albedos using the DA-RTM and the MC-RTM for $r_{eff} = 20$ μm (case-S). For reference, the observed albedos are shown in this figure. MC-RTM1 indicates the calculated albedo under the assumption that snow wall is completely covered with a metal wall and MC-RTM2 under the assumption of real instrument conditions.

Secondly the effects of snow wall on the snow albedo are examined. The spectral albedo calculated by MC-RTM under the real observation condition is also shown in Fig. 3 (hereafter MC-RTM2). The spectral albedo calculated by the MC-RTM2 was lower than those by the MC-RTM1 at all wavelengths. This is because a part of the lights penetrating into the snow escapes from the snow wall at the sides of the snow container. The effect of the snow wall of the snow container also acts to reduce the albedo at all wavelengths.

4.1.2 Finite Light Source and the Height of the WRS

The effect of the finite light source and the height of the WRS on the snow albedo were examined for case-S. The spectral albedos are simulated as a function of height of the WRS h (Fig. 4). The spectral albedos are highest for all the wavelengths when the WRS height $h_M = 12$–15 cm. The spectral albedos increase with $h < h_M$. This is mainly because of the effect of WRS's shadow on the snow surface, which is large (small) for lower (higher) WRS. For $h > h_M$, the snow albedo decreases gradually. This can be explained by the relationships of the WRS with the finite snowfield and the light source. When the WRS height increases, the WRS gets close to the light source and away from the snow surface. This results in the effects: (1) the shadow effect of the WRS relatively decreases; (2) the downward flux measured with WRS increases because the WRS approaches the lamps; and (3) the measured upward flux decreases because the field of view of the snow surface from the under surface of the WRS becomes narrower. Therefore, the spectral albedo depends on the height of the WRS, in relation to the area of light source and snow surface at all wavelengths.

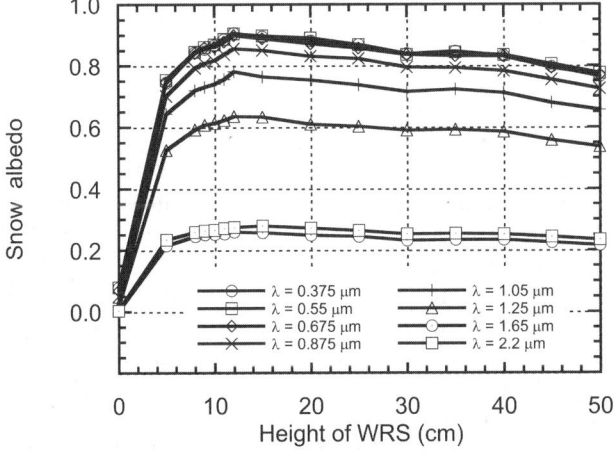

Fig. 4. Theoretically calculated albedos using the MC-RTM as a function of the WRS's height h for the spherical particle with $r_{eff} = 20$ μm.

4.2 Comparison Between the Theoretically Calculated Albedo and Observed One

4.2.1 Spherical Particles (Case-S)

Observed spectral albedo and the theoretically calculated ones for three kinds of snow grains ($r_{eff} = 10$, 20, and 50 μm) are shown in Fig. 5. The snow grain radius for which the theoretical albedo agrees with the observed albedo is estimated to be just less than $r_{eff} = 20$ μm. This value was consistent with the measured snow grain size r_2 (10–30 μm) from snow pit work. It was, thus, found that the optically-equivalent snow grain size of the spherical snow particles is r_2 as shown by Aoki et al. (2000).

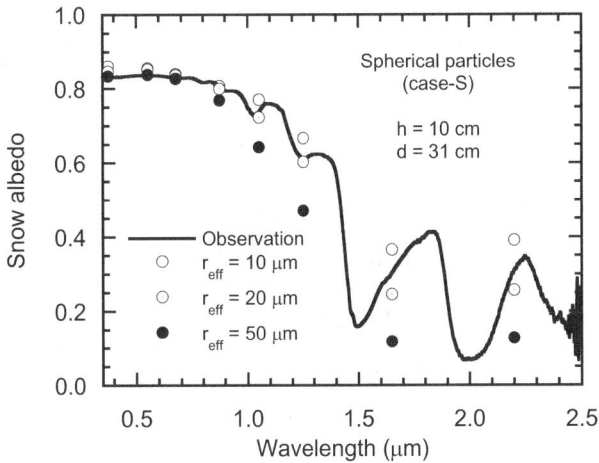

Fig. 5. Comparison between measured spectral albedos for the spherical particles (case-S) and theoretically calculated ones using the MC-RTM for the spherical particles with $r_{eff} = 10$, 20, and 50 μm.

4.2.2 Dendrites (Case-D)

To simulate the characteristics of long and narrow branches of dendrite (Fig. 2b), we assumed the randomly oriented circular cylindrical ice particles as non-spherical particles for spectral albedo calculations. The size of the non-spherical ice particles was determined from the micrographs to estimate the lengths of circular cylinder by means of the image-processing software, Image Hyper Pro (Inter Quest Inc., Japan), where the averaged values of R and L are 14.2 μm, and 189.3 μm, respectively. Based on the result, the spectral albedos of five kinds of size and shape for circular cylinder were simulated (Fig. 6). From the comparison between the theoretically calculated albedos and the measured ones, best-fitted cases are (ii) and (iii), where $R = 14$ μm and L is 190 μm and 380 μm. This value of $R = 14$ μm is the same as the measured value from the image processing and the values of L are the same or longer than the measured branch length. For dendrites, the snow albedo in the near-infrared region essentially depends on the radius R and the length L needs to be sufficiently long. If we assume the optical equivalent snow grain size is $r_2 = 14$ μm, it falls in the measured range of r_2 (10–30 μm) for snow pit work.

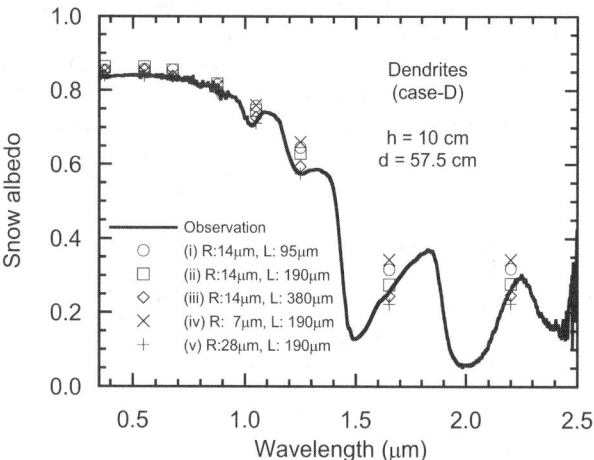

Fig. 6. Comparison between measured spectral albedos for the dendrites (case-D) and theoretically calculated ones using the MC-RTM for five kinds of cylindrical ice particles. The spectral albedos were calculated with a two-snow-layer model (first layer: surface—31.5 cm, and second layer: 31.5 – 57 cm).

4.2.3 Optical Property Relationship Between Case-S and Case-D

If the snow albedo composed of non-spherical ice particles can be simulated by spherical model of ice particles, such a method could be very useful, because the non-spherical particles have generally several dimensions to define the shapes, while the spherical particles has only one dimension. We estimate a radius of equal volume-area ratio (V/A) sphere by assuming our snow grains to be circular cylinder (Grenfell and Warren, 1999). The radii of the equal-V/A sphere (r_{VA}) are approximately 20 μm for (i)–(iii), and 10 and 37 μm for (iv) and (v), respectively. As shown in the previous section, the best-fitted case of the theoretically calculated albedos with the measured ones for case-D were (ii) and (iii), where $r_{VA} \sim 20$ μm. This value is close to the optically equivalent snow grain size (14 μm) and falls within the range of the branch width r_2 (10–30 μm) measured in the snow pit works.

The observed albedo for case-S is almost the same as that of case-D (Figs. 5 and 6). This is a coincidence. The observed albedo for case-S agreed with the theoretically calculated albedo for r_{eff} just less than 20 μm. This value

agrees well with $r_{VA} \sim 20$ μm for case-D. This means that the spectral albedo for the snow of dendrites can be represented by the radius of the equal-V/A sphere for circular cylinder size, which supports the results of Grenfell and Warren (1999).

5. CONCLUSION

The optical properties of the artificial snowpacks with the spherical (case-S) and the non-spherical particles (case-D) were investigated by observing the spectral albedo in the cold laboratory. The theoretically calculated albedos using the Monte Carlo radiative transfer model, in which the influence of the instrument setup of the cold laboratory is taken into account, were compared with the observed ones. The theoretically calculated albedo for case-S for the same snow grain size of *in situ* measured from the snow pit work agreed with the observed one, suggesting that the optically equivalent snow grain size is the one-half dimension of the narrower portion of broken crystals. For case-D, the most relevant dimension for the optical properties was found to be the shortest dimension of crystals rather than branch length of dendrites by the comparison between the theoretically calculated albedo and the observed one.

The measured spectral albedo for case-S was similar to that for case-D. Estimated snow grain size from the comparison with the theoretically calculated albedos for spherical particles was just less than 20 μm. For non-spherical particles, which were dendrites, the radius of the equal-V/A sphere by assuming the snow particles to be circular cylinder was approximately 20 μm. These values agree with the measured grain size for r_2 (10–30 μm) for both cases. The optical properties relationship between the spherical and the non-spherical snow particles show that the spectral albedos of the non-spherical particles can be represented using equal-V/A sphere.

6. REFERENCES

Aoki, Te, Ta. Aoki, M. Fukabori, A. Hachikubo, Y. Tachibana, and F. Nishio, 2000: Effects of snow physical parameters on spectral albedo and bi-directional reflectance of snow surface. *Journal of Geophysical Research,* 105, 10219–10236.

Grenfell, T. C, and S. G. Warren, 1999: Relationship of a non-spherical ice particle by a collection of independent sphere for scattering and absorption of radiation. *Journal of Geophysical Research,* 104, 31697–31709.

Macke, A., and M. I. Mishchenko. 1996: Applicability of regular particles shapes in light scatterign calculations for atmospheric ice particles. *Applied Optics*, 35, 4291–4296.

CORRECTION OF THE DIFFUSE INFLUENCE IN SPECTRODIRECTIONAL MEASUREMENTS

J. T. Schopfer[a], S. Dangel[a], J. W. Kaiser[a], M. Kneubühler[a], J. Nieke[a], G. Schaepman-Strub[b], M. E. Schaepman[c], and K. I. Itten[a], jschopfer@geo.unizh.ch

[a]Remote Sensing Laboratories (RSL), Dept. of Geography, University of Zurich, Winterthurerstr. 190, CH-8057, Zurich
[b]Nature Conservation and Plant Ecology, Wageningen University and Research Centre
Bornsesteeg 69, NL-6708 Wageningen
[c]Laboratory of Geo-Information Science and Remote Sensing, Wageningen University
Droevendaalsesteeg 3, NL-6708 Wageningen

ABSTRACT

Spectro-directional surface measurements can either be performed in the field or within a laboratory setup. Laboratory measurements have the advantage of constant illumination and neglectable atmospheric disturbances. On the other hand, artificial light sources are usually less parallel and less homogeneous than the clear sky solar illumination. To account for these differences and for determining for which targets a replacement of field by laboratory experiments is indeed feasible, a quantitative comparison is a prerequisite. Currently, there exist no systematic comparisons of field and laboratory measurements using the same targets. In this study we concentrate on the difference in spectro-directional field and laboratory data of the same target due to diffuse illumination and applied a correction term proposed by (Martonchik, 1994). Spectro-directional data were obtained with a GER3700 spectroradiometer. Additionally, a MFR sun photometer directly observed the total incoming diffuse irradiance. In the laboratory, a 1000W brightness-stabilized quartz tungsten halogen lamp was used. For the first direct comparison of field and laboratory measurements, we used an inert and highly anisotropic target with high angular anisotropy. Analysis showed that the diffuse illumination in the field is leading to a higher total reflectance and less pronounced angular anisotropy.

1. INTRODUCTION

The goniometer system of the Remote Sensing Laboratory (RSL) can be used for spectro-directional field measurements (Field Goniometer System FIGOS) and spectro-directional laboratory measurements (Laboratory Goniometer System LAGOS) (Dangel et al., 2003). However, there are obvious differences between the two cases, which have to be considered:

- In field experiments the target is left in its natural environment and is exposed to the natural direct and diffuse illumination. Diffuse illumination is depending on the illumination zenith angle and the atmospheric conditions. It is present in the field also under clear sky conditions, but is usually neglected in the laboratory.
- The direct illumination by the sun can be treated as being parallel (within 0.5°) and homogeneous over the area and height profile of the target, while laboratory illumination is usually non-parallel, non-homogeneous and not constant as a function of the target height.
- The illuminated area in the laboratory is limited; adjacency and multiple scattering effects can be very different from field experiments.
- The spectrum of artificial light sources differs from that of the sun, which is additionally attenuated by the atmosphere. This is usually neglected since reflectance measurements are normalized using a reference target.
- The polarization of the natural and artificial light sources can be different.
- Living plants may behave differently under field and laboratory conditions.

Taking these differences into account, the advantage of laboratory measurements lies in the independence of weather conditions, time of day or seasonal conditions. The illumination intensity and angles can be held constant over time and freely chosen. Currently, there exist no systematic comparisons of field and laboratory measurements using the same artificial targets and therefore it is not known for which targets a replacement of field by laboratory measurements is indeed feasible. This study has been performed focusing on the effects of the diffuse illumination as the main difference between spectro-directional field and laboratory measurements.

The directional surface reflectance properties are by definition characterized by the bidirectional reflectance distribution function (BRDF), or equivalently, the bidirectional reflectance factor (BRF) and depend on the surface properties only (Martonchik et al., 2000). However, spectro-directional field experiments with goniometer systems are only able to observe approximations of the bidirectional reflectance factor. The directly observed quantity in field experiments is called hemispherical conical reflectance factor (HCRF), corresponding to hemispherical illumination, which depends on the atmospheric conditions, and conical observation. Laboratory experiments suffer from imperfect illumination resulting in a rather biconical than bidirectional reflectance factor. In this preliminary study the conicality on the illumination and observation side has been neglected. This is acceptable for the observation side since the field of view (FOV) of the sensor is quite small (3°). Current studies at RSL pay attention to the conicality of both the illumination source and the sensor. Additionally, the changing size and position of the sensor's footprint as a function of the observation angle have to be considered, especially if the target is not very large or exhibits different BRDF's at different parts.

In order to make measurement results of field and laboratory spectro-directional experiments directly comparable, we need to retrieve the BRDF for both cases. For the field case we followed the well known procedures proposed by Martonchik (1994) and Lyapustin and Privette (1999), which correct the measurements only for the diffuse illumination and not for any other imperfections. For these methods, the diffuse radiation has to be measured over the complete hemisphere at the same angular resolution as the reflected radiation of the target. Since we are not yet able to measure the incoming diffuse radiation at angular resolution, we used a simplified approach measuring the diffuse irradiance with a MFR sun photometer. For the laboratory case, the approximated BRF is used since the standard retrieval schemes do not apply because they rely on the separation of direct and diffuse illumination.

2. METHODOLOGY

2.1 Comparison requirements

For comparison purposes of spectro-directional field and laboratory measurements it is necessary to hold as many parameters as possible constant. So, the target, the measurement instruments, the experiment setup, the illumination and observation geometries, directions and areas remain the same. As mentioned, a basic difference of the two measurement cases is that in the laboratory we obtain BRF data and in the field HDRF data, using the approximations discussed above. Field data is influenced by atmospheric conditions, especially by the diffuse irradiance, which has to be corrected. For spectral analysis we compare the averaged nadir reflectances from 400 to 2500 nm. Directional analysis is mainly done in the solar principle plane at a wavelength of 496 nm.

- **A) Target:** For the first direct comparison of spectro-directional field and laboratory measurements we used an artificial, inert target, borrowed from JRC (Joint Research Center). The target size is 25 cm x 25 cm and it consists of a matrix of cubes, carved out of a thick plate of sanded duralumin. The spectro-directional properties show a high angular anisotropy due to the cast shadows of the cubes as a function of the illumination angles. Furthermore, its BRDF is not rotationally symmetric (only 90° symmetry), it depends on the illumination and view azimuth angles. In order to reduce adjacency effects due to the limited size of the target, a black aluminum plate (size 1.2 m x 1.2 m) was used as background in both the laboratory and field case.

- **B) Instruments and experiment setup:** The field and laboratory experiments were performed using the same measurement setup: a GER3700 sensor, mounted on the goniometer system, measuring the spectro-directional reflectances over the whole hemisphere at an azimuthal angular resolution of 30° and a zenithal angular resolution of 15°. For a detailed description of RSL's goniometer system please refer to (Sandmeier et al., 1999). In the field case, additionally, the total and diffuse illumination is permanently measured with a MFR-7 sun photometer (Yankee Environmental Systems Inc.) at 6 wavelengths (415, 500, 615, 673, 870, and 940 nm). The direct illumination is then obtained computing the difference between the total and diffuse illumination. In the laboratory case, a 1000W brightness-stabilized quartz tungsten halogen lamp was used as illumination source (Dangel, 2003). The lamp is mounted on an adjustable tripod, which allows the use of the same illumination directions of the target as in the field case.

- **C) Illuminated area:** The illumination distance (distance from the light source to the centre of the target) in the laboratory was held constant at 1.54 m for all illumination angles. For the smallest used illumination angle (28.5°) the illumination ellipse shows a size of about a = 32.25 cm (short half axis) and about b = 37 cm (semi-major axis). However, for larger illumination zenith angles the semi-major axis is changing, which leads to an increase of the inhomogeneity and non-parallelism over the illuminated area. These effects were neglected in this study since the illumination distance remains the same and those effects particularly appear in the forward direction and at a great distance from the central part of the beam.

- **D) Observed area:** Similar effects of a changing instantaneous ground field of view (IGFOV) also occur on the observation side. In order to reduce adjacency effects we concentrated on observation angles from +45° to –45° in the analysis of the data.

2.2 Correction for Diffuse Irradiance

In this study we followed a procedure (Martonchik, 1994), where the incidence irradiance is split up into a direct and diffuse component $E_{dir}^{inc}(\mu_0)$ and $E_{diff}^{inc}(\mu_0)$. The diffuse influence then is accounted for in a correction term which is subtracted from the reflected field radiances $L(\mu,\mu_0,\varphi,\varphi_0)$. The resulting BRF_Δ then is

$$BRF_\Delta = \frac{L(\mu,\mu_0,\varphi,\varphi_0) - \Delta(\mu,\mu_0,\varphi,\varphi_0)}{\pi^{-1}[E_{dir}^{inc}(\mu_0) + E_{diff}^{inc}(\mu_0)]} \quad (1)$$

where

μ, μ_0 is the cosine of the view and illumination zenith angle and

φ, φ_0 is the view and illumination azimuth angle.

$E_{dir}^{inc}(\mu 0)$ and $E_{diff}^{inc}(\mu 0)$ are measured by the MFR and the diffuse influence is described by

$$\Delta(\mu,\mu 0,\phi,\phi 0) = \pi^{-1}\int_0^1\int_0^{2\pi}R(\mu,\mu',\phi,\phi')L_{diff}^{inc}(\mu',\mu 0,\phi',\phi 0)d\Omega \quad (2)$$

$$-\pi^{-1}R(\mu,\mu 0,\phi,\phi 0)\int_0^1\int_0^{2\pi}L_{diff}^{inc}(\mu',\mu 0,\phi',\phi 0)d\Omega,$$

Where

R is the BRF of the target,

L_{diff}^{inc} is the diffuse incident radiance [Wm^{-2}sr^{-1}]

$\mu'd\mu'd\phi'$ is the projected solid angle.

In our case we assume that L_{diff}^{inc} is constant over the angles (since the MFR only observes the total incoming diffuse irradiance), and therefore the integral

$\int_0^1\int_0^{2\pi}L_{diff}^{inc}(\mu',\mu 0,\phi',\phi 0)d\Omega$ becomes the constant factor

$E_{diff}^{inc}(\mu 0)$:

$$\Delta \cong \pi^{-1}E_{diff}^{inc}(\mu 0)(\pi^{-1}\int_0^1\int_0^{2\pi}R(\mu,\mu',\phi,\phi')d\Omega - R(\mu,\mu 0,\phi,\phi 0), \quad (3)$$

The Δ term in equation (3) is the product of the diffuse irradiance and the target anisotropy. The anisotropy is determined using the difference of the target BRF and the BRF integrated for a specific illumination angle.

3. DATA

The field data has been acquired in July 2002 at Oberpfaffenhofen in Gilching (D). With FIGOS, a total of 6 hemispheres (Hem) of the artificial JRC target were measured at different illumination angles. The MFR sun photometer was recording irradiance data permanently from 11:48h until 18:30h. For LAGOS, also 6 hemispheres (Labhem) under the same illumination angles have been measured in the goniometer laboratory at RSL. Fig. 1 shows an overview of the spectro-directional dataset:

Hemisphere	zenith	azimuth
Hem/Labhem_b	37.8°	5.4°
Hem/Labhem_c	33.3°	17.6°
Hem/Labhem_d	28.7°	42.2°
Hem/Labhem_f	28.5°	11.5°
Hem/Labhem_i	40°	29.5°
Hem/Labhem_j	59.4°	32.6°

Fig. 1. Spectro directional dataset.

3.1 Quality Assessment

To fulfil the comparison requirements, only spectro-directional reflectance data from +45° to –45° zenith angle for both LAGOS and FIGOS are considered for analysis. Due to shadowing of either the sensor (field) or the lamp (laboratory), no measurement near the hotspot is possible. In the laboratory, even measurements in the principal plane at zenith angles larger than the actual illumination zenith angle are affected by shadowing of the tripod in the backscattering region and have to be omitted.

In order to compare field and laboratory measurements with respect to the changing influence of the diffuse illumination, spectro-directional data at different times of the day were obtained. In the following, we consider the diffuse irradiance at 496 nm. Fig. 2 shows the ratio of the diffuse irradiance to the direct irradiance, along with the measuring times of the hemispheres c, d, f, i, and j.

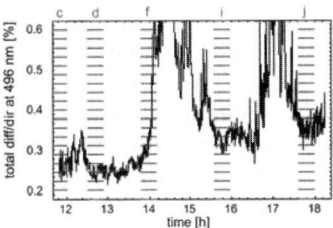

Fig. 2. Diffuse quantity at measurement times.

The diffuse influence is increasing with the illumination zenith angle, but also depending on the sky cover (overpassing clouds at 14h–15h and 17h). The hemispheres f and j underlie a strong diffuse influence and therefore a strong discrepancy to the corresponding laboratory measurements is expected.

4. RESULTS

Generally, the nadir reflectance over the whole spectrum (400 nm to 2500 nm) is decreasing with an increasing illumination zenith angle. Maximal reflectance in the dataset is measured for zn=28.7° (Hem/Labhem_d) and minimal reflectance for zn=59.4° (Hem/Labhem_j), resulting from the increasing cast shadow of the target cubes at larger illumination zenith angles. Nadir reflectances of FIGOS (Hem) show higher values than of LAGOS (Labhem), but the differences depend on the illumination zenith angle (Fig. 3).

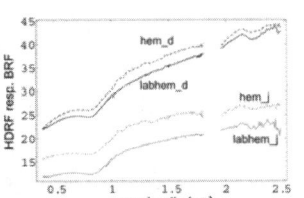 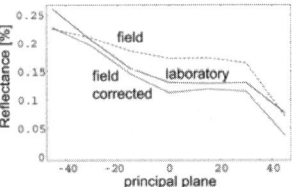

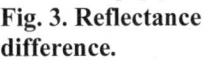

Fig. 3. Reflectance difference.

Fig. 4. Hem_j corrected (BRF_Δ)

A larger illumination zenith results in a longer path of the solar radiation through the atmosphere and therefore in more diffuse light which is illuminating the strong cast shadows of the cubes of the JRC target. Imagine yourself looking at the target from the nadir position: the shadowed area will grow with increasing illumination zenith. And therefore, more dark (shadowed) area is available to be illuminated by the diffuse irradiance in the field, but not in the laboratory.

A comparison of the mean reflectances of the corrected BRF_Δ data to the original field and laboratory data reveals, that for large illumination zenith angles the correction is better than for small illumination zenith angles. However, the significance of the mean reflectance is minor, since only zenith angles from +45° to −45° are considered. The correction quality is therefore discussed in the solar principal plane. The BRF_Δ of hem_j, which is strongly influenced by diffuse irradiance, exhibits a very good correction as shown in Fig. 4. However, for hem_f (also strongly influenced by diffuse irradiance) the correction term fails. An explanation might be that the diffuse irradiance here is caused by moving clouds, instead of a large illumination zenith angle as for hem_j. This might lead to an inhomogeneous diffuse irradiance, which is not accounted for with our approximation for equation (3) for the incident diffuse radiance.

5. CONCLUSIONS

In this study a direct comparison of spectro-directional field and laboratory measurements using an artificial target has been performed for the first time. We concentrated on the difference due to the diffuse illumination and applied a correction method following the well known approach by (Martonchik, 1994). For the comparison, an inert (no variation over time) and highly anisotropic (large Δ since stronger directional effects due to diffuse light), artificial target was chosen. The conclusions of the obtained results are depicted as follows:

The spectral analysis shows a typically about 2% higher reflectance in field measurements than in the laboratory. This difference increases with increasing illumination zenith angle and occurs due to illumination of shadowed areas in the field case.

An assessment of the correction method seems difficult, since it is not sensitive enough for field measurements underlying only little diffuse influence. For field measurements with large illumination zenith angles good results were obtained. Obviously the angular distribution of the diffuse irradiance may differ depending on its origin, either caused by a long solar radiation path or by a changing atmosphere (sky cover).

For future investigations concerning the influence of diffuse irradiance in spectro-directional field measurements a large dataset with varying atmospheric conditions is necessary. Better correction can be obtained by measuring the incoming diffuse radiation at the same angular resolution and time as the spectro-directional reflectance. Therefore a goniometer system with two spectro-radiometers, one looking upwards and one looking downwards, is proposed (Abdou et al., 2000).

For this comparison study approximations concerning illumination and observation geometries and areas have been made for the laboratory case. Further research is currently done at RSL to account for the non-parallelism of the illumination and inhomogeneity of the illuminated area.

6. ACKNOWLEDGEMENTS

The authors wish to thank Brian Hosgood and Michel Verstraete of JRC for lending their panel for the comparison measurements.

7. REFERENCES

Abdou W. A., Helmlinger M. C., Donel J. E., Bruegge C. J., Pilorz S. H., Martonchik J. V., and Gaitley B. J., 2000: Ground measurements of surface BRF and HDRF using PARABOLA III. In *Journal of Geophysical Research,* Vol. 106, 11967-11976.

Dangel S., J. Schopfer, M. Kneubühler, K. I. Itten, 2004: Towards a direct comparison of field- and laboratory goniometer measurements. To be submitted to *IEEE Transactions on Geoscience and Remote Sensing*.

Dangel S., M. Kneubühler, R. Kohler, M. Schaepman, J. Schopfer, G. Schaepman-Strub, K. I. Itten, 2003: Combined field and laboratory goniometer system – FIGOS and LAGOS. In *IGARSS 2003, International Geoscience and Remote Sensing Symposium*.

Govaerts Y., M. Verstraete, and B. Hosgood, 1997: Evaluation of a 3D radiative transfer model against goniometer measurements on an artificial target. In *Journal of Remote Sensing,* Vol. 1, 131-136.

Lyapustin A. I., J. L. Privette, 1999: A new method of retrieving surface bidirectional reflectance from ground measurements: Atmospheric sensitivity study. In *Journal of Geophysical Research,* Vol. 104, 6257-6268.

Martonchik J., C. Bruegge, and A. Strahler, 2000: A review of reflectance nomenclature used in remote sensing. In *Remote Sensing Reviews,* Vol. 19, 9-20.

Martonchik, J. V., 1994: Retrieval of surface directional reflectance properties using ground level multiangle measurements. In *Remote Sensing of Environment,* Vol. 50, 303-316.

Sandmeier S. R., K. I. Itten, 1999: A field goniometer system (FIGOS) for acquisition of hyperspectral BRDF data. In *IEEE Transactions on Geoscience and Remote Sensing,* Vol. 37 (2), 978-986.

Sandmeier S. R., 2000: Acquisition of bidirectional reflectance factor data with field goniometers. In *Remote Sensing of Environment,* Vol. 73, 257-269.

THE RADIATE AND LIGHT CHARACTERISTICS OF THE CLIMATE CHANGE DURING THE SECOND HALF OF 20th CENTURY IN MOSCOW

G. M. Abakumova, E. V. Gorbarenko, E. I. Nezval', and O. A. Shilovtseva
Meteorological Observatory of Geographical Department of Moscow State University Vorobévy Gory
Moscow, 119992, Russia

ABSTRACT

In the second half of the 20th Century in the Meteorological Observatory of Moscow State University the analysis of long-term actinometric measurements was done. This analysis included the examination of multiannual variability of the aerosol optical thickness at $\lambda=550$ nm (AOT), total (N_T) and low (N_L) cloud amount, sunshine duration, net radiation and its components, surface albedo, natural illuminance (E), ultraviolet (UV, $\lambda \leq 380$ nm) and photosynthetically active (PAR, 380–710 nm) solar radiation.

Besides, there the presence of "clear" period in Moscow was revealed, characterized by low values of AOT at 1994–2001. Compared with the norm the increase of integral (300–5000 nm) and UV radiation was estimated in clear sky condition during this period.

On the example of one month (March) the strong connection between monthly means of surface albedo and air temperature was showed.

1. INTRODUCTION

The studies in long-term trends of atmospheric transparency, radiation fluxes, cloudiness and temperature have the great importance for the investigation of the climate change.

Meteorological Observatory of Moscow State University named after M.V.Lomonosov (MO MSU) was founded in 1954 by the famous soviet climatologist Boris P.Alisov and actinometrist Mikhail S. Averkiev. As a result of their efforts nowadays MO MSU is a unique station in Russia, where the complex of long-term measurements of integral (since 1955) and spectral solar radiation (ultraviolet radiation since 1968, photo synthetically active radiation since 1980), natural illuminance (since 1964) is held. In this work the analysis of trends based on these long-term measurements are presented.

2. INSTRUMENTS AND METHODS

The thermoelectric instruments constructed by Yu.D. Yanishevsky are used as the receivers of integral solar radiation: direct radiation—actinometer A-80; diffuse, global and reflected radiation—pyranometer M-80M; net radiation—net radiometer (balance meter) M-10M (Revised Instruction Manual, 1986). During 1960th the recorders of UV radiation and natural illuminance were constructed in MO MSU. They are the unique devices. The recorder of UV radiation utilized a special selenium barrier-layer photocell with higher response in UV region of the spectrum and UV glass filter (Chubarova and Nezval, 2000). Because of using a filtered selenium photocell in the recorder of natural illuminance the spectral responsivity curve of it becomes similar to the average human eye standard observer curve (Shilovtseva, 2001). Being a diffuser an integrating sphere applied in both instruments improves the cosine function.

The measurements of PAR are carried out with the help of modified Yanishevsky pyranometer, so called "colored" or white-red pyranometer (Makhotkina, 1983).

All devices are mounted on the special platforms on the roof of the MO MSU. They are inspected every hour and are cleaned from dust, moisture, dew and frost by an observer. The registration of the all fluxes of solar radiation is realized every minute by a special soft-hardware system, constructed in MO MSU.

Direct solar radiation data has been used for the calculation of atmospheric transparency characteristics. Cloud amount and type have been determined visually.

For the estimation of the long-term changes in time series they were approximated to a linear or polynomial equation. Statistical significance of linear trend has been estimated by the use of Student's parameter **t**. Only the linear trends having the significance level more than 95% have been taken into account further in our analysis.

3. RESULTS

3.1 The Multiannual Variability of Yearly Aerosol Optical Thickness

Aerosol optical thickness is determined using the measurements of direct solar radiation according to the method worked out by T.A. Tarasova and Ye.V. Gorbarenko (Gorbarenko, 1997). During the 1955–2003 the tendency of AOT decrease was observed: –28% with statistical significant level P=0.90. The highest values of AOT were observed in 1982–1983, and 1992 and were caused by volcanoes El Chichon and Pinatubo eruptions (Fig.1). The period of 1994–2001 is characterized by the highest atmospheric transparency due to the absence of significant volcanic eruptions, which influence on the global air pollution, and due to the reduction of anthropogenic air pollution following the industry recession on the European part of Russia. Some increase of AOT in 1972 and 2002 was caused by forest and peat bog fires in the Moscow region.

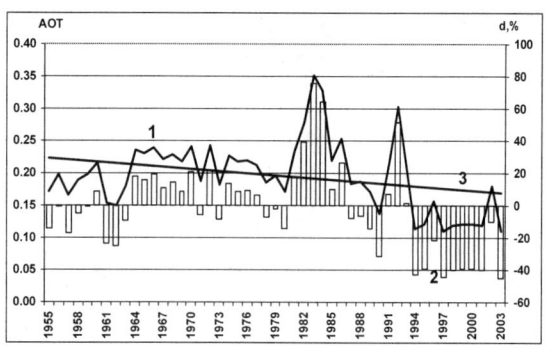

Fig. 1. Annual mean aerosol optical thickness (1), its anomalies d (2) and linear AOT trend (3).

3.2 Solar Radiation in Clear Sky Atmosphere

Direct solar radiation S, incident on the normal surface, under the cloudiness condition is the primary characteristic of atmospheric transparency. There is a statistically insignificant increasing tendency (5%, P=0.85) of the multiannual variations of noon values of S in % from the norm calculated for the period 1955–1999. The most important decrease of S from the norm was caused by the El Chichon and Pinatubo eruptions (–14% in 1984 and –12.5% in 1992–1993). On the contrary, the most remarkable increase of S from the norm was observed during the period of 1994–2001: 8–14. The influence of smoke from the forest and peat bog fires also caused the decrease of noon S, during the warm period (May–September), by nearly 10% and was comparable with Pinatubo eruption effect.

The ratio of direct solar radiation coming the horizontal surface (S') to diffuse radiation (D) is the sensitive characteristic of atmospheric transparency too. There is the statistical significant linear trend of S'/D (May-September, 1955–2003, solar elevation h=40°): 26%, P>0,95. The range of S'/D is quite wide: from 1.9 (1983) – 2.2 (1992) after volcanic eruptions to 6 (2000) (Fig. 2). When the smoke from the forest and peat bog fires was observed the S'/D varied from 2.4 in 1972 to 2.8 in 2002.

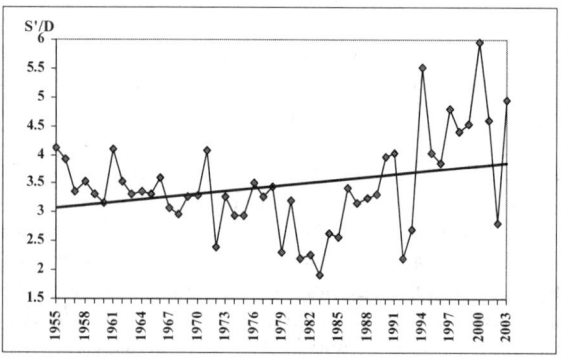

Fig. 2. Annual mean ratio S'/D.

The analysis of the long-term variations of direct (S'), diffuse (D), global (Q) and the ratio D/Q for two spectral ranges of solar radiation—integral and ultraviolet (UV)— under clear sky conditions (May–September, h=40°) has shown that the significant linear tendency is observed only for D_{UV} (–5%, P=0.95) and $(D/Q)_{UV}$ (–7%, P=0.95). The character of change of S'_{UV}, Q_{UV} and $(D/Q)_{UV}$ is synchronous with that of S', Q, and D/Q: the correlation coefficients between these parameters are 0.88, 0.73, and 0.87 correspondingly.

The deviation of solar radiation in 1994–2001 from the norm under clear sky condition is shown in Table 1.

Table 1. Deviation of Solar Radiation (dS',dD, dQ) in 1994–2001 from the Norm Under Clear Sky Condition, %

h,°	10	20	30	40	50
	Integral Radiation				
DS'	14.9	11.9	10.2	9.1	8.5
dD	-10.2	-16.1	-19.2	-21.8	-22.6
dQ	3.2	2.8	2.4	2.1	1.8
d(S'/D)	28.4	32.2	33.7	36.1	37.1
	Ultraviolet Radiation				
dS'_{UV}	-	18.0	18.6	18.4	18.1
dD_{UV}	-	2.7	0.4	-1.1	-2.2
dQ_{UV}	-	5.8	5.7	5.5	5.4

3.3 Solar Radiation in Cloudy Sky Atmosphere

Before analyzing the time series of solar radiation sums the same explorations of the sunshine duration and cloud amount were done.

The statistically significant linear increase tendency for total N_T (+8%) and low N_L (+17%) cloud amount with P=0.999 is observed for the period 1958–2003 (Fig.3). There is no statistically significant linear trend for sunshine duration (SD). But during the last decade of 20th Century the decrease of cloud amount and increase of sunshine duration is remarkable.

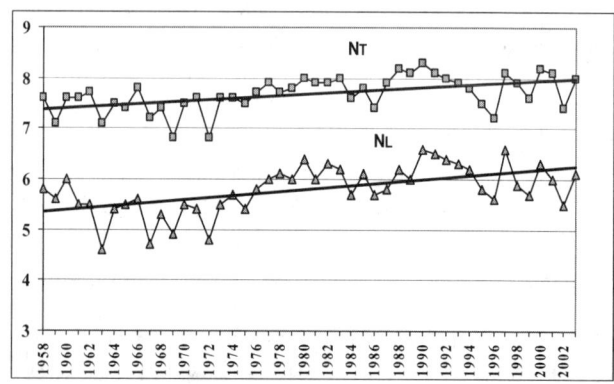

Fig. 3. Annual mean total (N_T) and low (N_L) cloud amount.

During 1958–2003 the weak tendency to decrease was characteristic for direct (–1%), diffuse (–5%, P=0.94) and global (–3%) integral radiation. Accordingly to the cloud amount variance during the last decade (1991–2000) the increase tendency for S' and Q is evident (Fig. 4).

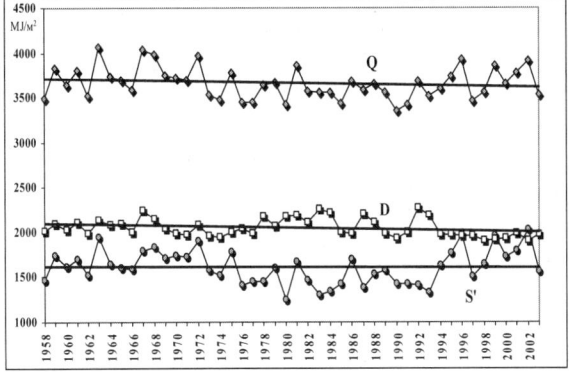

Fig. 4. Direct (S'), diffuse (D) and total (Q) solar radiation.

The most statistically significant decrease tendency was observed for reflected solar radiation R_k (–16%, P=0.95). Also there was observed a weak increase trend for absorbed radiation B_k (+2%) and net radiation B (+7%, P=0.90) (Fig. 5). The most significant increase of B was characteristic for the winter period (39%), when in Moscow the increase tendency of air temperature was mostly evident.

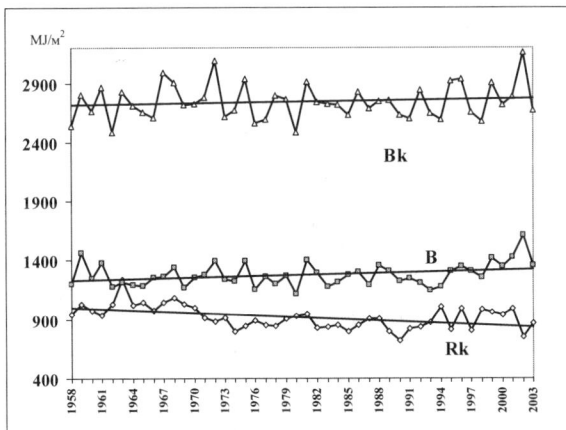

Fig. 5. Absorbed (B_k) and reflected (R_k) radiation, net radiation (B).

According to the R_k change the surface albedo A_k has the statistically significant negative linear trend (–15%). Such remarkable decrease of reflected surface properties was caused by the increase of surface pollution and obstruction height due to urban development.

The air temperature is connected with mean monthly albedo most closely in March, October and November, when formation or destruction of snow cover strongly depends on air temperature variations. For example, the good inverse agreement between monthly mean values of albedo A_k and air temperature in March was observed in MOMSU (Fig.6). During the studied period A_k decreased on 16%, and T, in opposite, increased on 3.7° (correlation coefficient is r=–0.61). The range of A_k variation is sufficient: from 68% in 1963 to 19% in 2002, the years when the lowest (–9.2°C) and highest (+2.4°C) monthly air temperature was observed during the last 46 years.

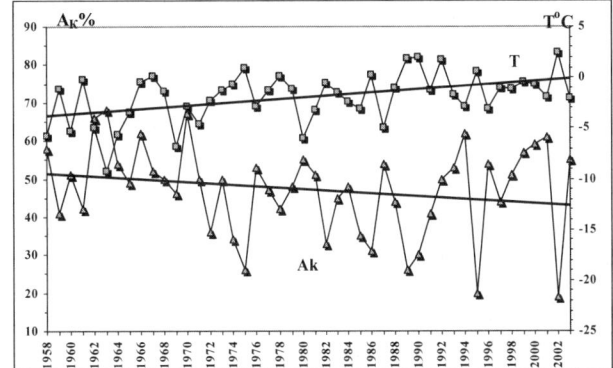

Fig. 6. Monthly mean air temperature (T) and surface albedo (A_k, %) in March.

The mean and extreme yearly sums of various solar fluxes are presented in the Table 2.

Table 2. The Mean and Extreme Yearly Sums of Solar Radiation (MJ/m^2), Natural Illuminance (Mlx•h), Sunshine Duration (SD, hour) and Cloud Amount

	Mean value	Maximum		Minimum	
		value	year	value	year
N_T	7.7	8.3	1990	6.8	1969/72
N_L	5.8	6.6	1990	4.6	1963
SD	1742	2126	2002	1478	1980
S'	1611	2020	2002	1244	1980
D	2048	2267	1992	1896	2002
Q	3659	4065	1963	3346	1990
R_K	917	1239	1963	717	1990
B_K	2742	3164	2002	2484	1962
B	1279	1610	2002	1117	1980
Q_{UV}	153	170	1968	141	1980
Q_{PAR}	1737	1937	1996	1534	1990
E_Q	112	129	1996	99	1974

The study of multiannual variability of solar radiation in various spectral intervals showed that there had been no statistically significant linear tendency in changes of ultraviolet radiation during 1968–2003, photosynthetically active radiation during 1980–2001 and natural illuminance during 1964–2003. The multiannual variability of these components are described quite well by polynomial equations of third order (Fig. 7).

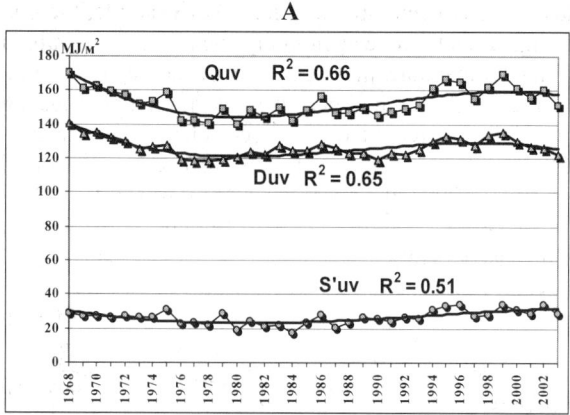
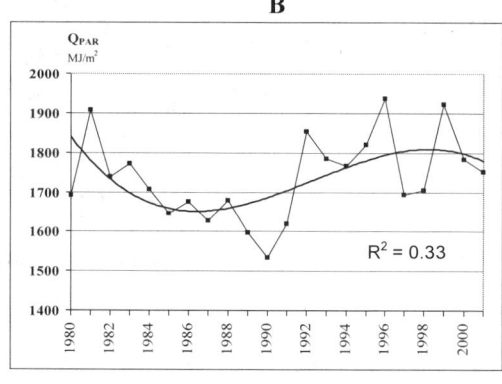
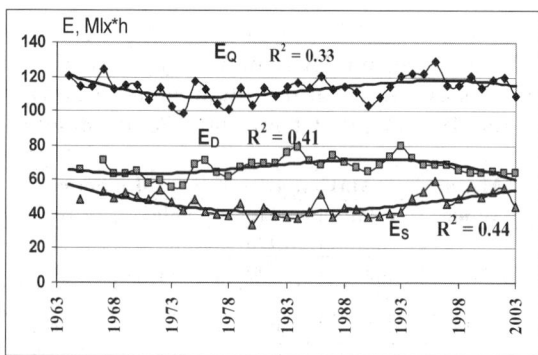

Fig. 7. The multiannual variability of ultraviolet (A), photosynthetically active (B) solar radiation and natural illuminance (C).

Table 3. Deviation of Monthly Sums of Global and Diffuse Integral Radiation (dQ, dD), Photosynthetically Active Radiation (dQ_{PAR}) and Natural Illuminance (dE_Q, dE_D) in 1994–2001 from the Norm, %

Month	1958-2003		1964-2003		1980-2003
	dQ	dD	dE_Q	dE_D	dQ_{PAR}
I	-10	-8	3	3	0
II	-3	-3	0	1	-2
III	6	-6	8	-1	6
IV	6	-1	8	3	10
V	-2	-5	3	-4	-1
VI	0	-7	3	-5	3
VII	1	-5	5	-5	3
VIII	1	-8	1	-2	4
IX	3	-4	5	-3	8
X	-1	-7	5	1	1
XI	-3	0	5	2	2
XII	0	-3	5	4	5

4. CONCLUSIONS

The analysis of time series of atmospheric transparency and solar radiation in different ranges of spectrum in clear-sky atmosphere has shown that the period of 1994–2001 was the most transparent period in Moscow. In the case of cloudy sky atmosphere—when cloud amount in some months of 1994–2001 exceeded monthly norm - the influence of clouds on radiation has exceeded the influence of atmospheric transparency on it.

This study was supported by the Russian Foundation for Basic Research (Project RFBF 05-05-64696).

5. REFERENCES

Chubarova, N., and Ye. Nezval', 2000: Thirty year variability of UV irradiance in Moscow. *Journal of Geophysical Res.*, V. 105, ND10, p. 12,529-12,539.

Gorbarenko E., 1997: Spatial and temporal variations of atmospheric aerosol optical thickness on the territory of the former Soviet Union. In *IRS1996: Current Problems in Atmospheric Radiation*, A. Deepak Publishing, Hampton, Virginia, 774-777.

Makhotkina Ye. L., 1983: The colored pyranometers. *Transactions of the Main Geophysical Observatory*, Issue 456, pp. 71-77 (in rus.).

Revised Instruction Manual on Radiation Instruments and Measurements, 1986: *WCRP Publications,* Series, N7, WMO/TD- N 149, October 1980,. 140 p.

Shilovtseva, O., 2001: The experience of the visual solar radiation measurements in the Moscow State University Meteorological Observatory. In *IRS2000: Current Problems in Atmospheric Radiation*, A. Deepak Publishing, Hampton, Virginia, 1117-1120.

The comparison of the mean values of cloud amount derived for "clear period" with the norm has shown that the most significant differences in N_L and N_T were characteristic for January, February and May. The increase of total cloud amount in 1994–2001 is 14, 10, and 16% and of N_L—6, 5 and 7% accordingly. The remarkable decrease of N_L without change of N_T was observed in March and November (5 and 13% accordingly). The change of solar radiation and E is presented in Table 3. UV radiation increases during the warm period (May-September) of 1994–2001 for 4% in comparison with the period of 1968–2003.

WATER VAPOR ESTIMATED WITH GPS ON TIBETAN PLATEAU AND ITS INFLUENCE ON RADIATION BUDGET

Zhian Sun
Bureau of Meteorology Research Centre
Melbourne, VIC. 3001, Australia

Jingmiao Liu and Hong Liang
Chinese Academy of Meteorological Sciences
Beijing, 100081, China

ABSTRACT

Precipitable water (PW) determined by Global Positioning System (GPS), radiosonde and operational NWP system analysis at three stations (Naqu, Gaize and Deqin) on the Tibetan Plateau are compared. The results show that the PW determined by radiosonde and NWP analysis are systematically smaller than those determined by the GPS measurements. These differences have a large effect on the surface radiation budget. The modeled radiative flux difference at the surface due to the difference in PW is about 20 Wm^{-2} in the shortwave and 30 Wm^{-2} in the longwave.

1. INTRODUCTION

Water vapor is an important element in numerical weather prediction (NWP) and climate studies. It plays a crucial role in the global energy and hydrological cycles. Water vapor is a dominant greenhouse gas in the atmosphere and is a strong absorber of both solar and infrared radiation. The distribution of water vapor is highly variable in both space and time. Lack of precise and continuous water vapor measurements is one of the major sources of error in short-range weather prediction (Kou et al., 1996). In recent years, a new observational technique, based on the Global Positioning System (GPS) that is sensitive to the spatial and temporal distribution of the water vapor content in the atmosphere has made it possible to retrieve precise and continuous estimates of water vapor with spatial density governed by the number of receivers deployed. Many GPS based estimates of water vapor have been compared with the estimates based on radiosondes, water vapor radiometers (WVR) and NWP analysis (Kopken, 2001) and the results indicate that the GPS estimates of water vapor are in generally good agreement with the measurements of radiosondes and WVR.

In this study, we present the comparison of PW over the Tibetan Plateau determined by GPS, radiosonde, and analyses from the Global Assimilation and Prediction System (GASP) used at the Australian Bureau of Meteorology and analyses from the European Centre for Medium-Range Weather Forecasts (ECMWF) operational NWP system. In addition to the PW comparison, we further investigate the effect of PW determined by different methods on the surface radiation budget and compare the results with observations. Since the air mass in this region is only about half of that at sea level and the aerosol load in the atmosphere is lower, the downward solar irradiance at the surface may be largely determined by water vapor in the atmosphere. Thus this comparison may provide an indirect validation for the water vapor measurements as an accurate water vapor amount used in the calculations should lead to results comparable to the observations.

2. GPS AND RADIATION OBSERVATIONS

A long-term observational project for the measurement of water vapor on the Tibetan Plateau has been conducted as a Japanese-China collaboration since 1999. The GPS receivers were installed at Gaize (32.3 N, 84.06 E, 4420 m) in the western part of the plateau, Naqu (30.48 N, 92.06 E, 4518 m) in the eastern part of the plateau and Deqin (28.65 N, 99.17 E, 3594.9 m) at the south-east edge of the plateau. The GPS receiving stations consist of Dorne-Margolin antenna-plus-chokering assemblies with a radome. The GPS receivers recorded the data from 7–8 satellites in view every 30 seconds. The operational radiosonde measurements are carried out twice a day at Naqu. An Automatic Weather Station (AWS) was installed at Gaize and started operation in October 1997. Total downward and upward solar radiation and infrared thermal radiation are measured at 5 seconds intervals by upward-looking and downward-looking high precision pyranometers (MS-802, EKO instruments trading CO., LTD., Japan) and a precision pyrgeometer (MS-202) mounted on a 1.5-meter-high horizontal platform. The data sets were averaged to a one-hour time step to reduce random error. All radiation sensors are calibrated and repaired every year. The data collected in year 2001 are used in this study.

3. COMPARISON OF PW BETWEEN GPS, RADIOSONDE AND MODEL ANALYSIS

We first compare PW measurements determined by the GPS and radiosonde operated at the national meteorological observational station at Naqu. The results at the radiosonde launch times (00:00 and 12:00 UTC) for year 2001 are shown in Fig. 1a. The GPS data for a period between 25 July and 24 August are missing. The first sign one can see is that the water vapor amount at Naqu is very low. Even in the rainy season (JJA) PW is only about 20 mm, less than half of that in the tropical regions (Liou et al., 2001). The majority GPS based PW are higher than those from radiosonde measurements. The rms difference between the GPS and radiosonde is 1.75 mm. Given the fact that PW itself is relatively small, this rms is in fact proportionally larger than those at other regions reported in literature. In the winter season, for example, the relative difference in PW between the GPS and radiosonde can be 100%.

respectively. These results are very similar to those shown in Fig. 1a, with the PW from the analysis being smaller than the GPS estimates. This is expected as the model analysis is largely dependent on the available radiosonde observations. If the radiosonde has a dry bias then the same bias in the analysis is almost inevitable.

Gaize and Deqin do not have radiosonde observations. Thus we only compare the GPS PW with the analysis and the results for Gaize are shown in Fig. 2. It is seen that the PW determined by the two NWP model analysis is again systematically smaller than the GPS estimate. The difference is even bigger than that found at Naqu. The rms difference is 2.2 mm for GASP and 2.4 mm for ECMWF. The GPS PW data at Deqin were obtained from 26 April 2002 to 22 August 2003. A scatter plot for this period is shown in Fig.3. PW from the two-model analysis are significantly less than the GPS estimates. The rms difference is 7.7 mm for GASP analysis and 7.3 mm for ECMWF analysis.

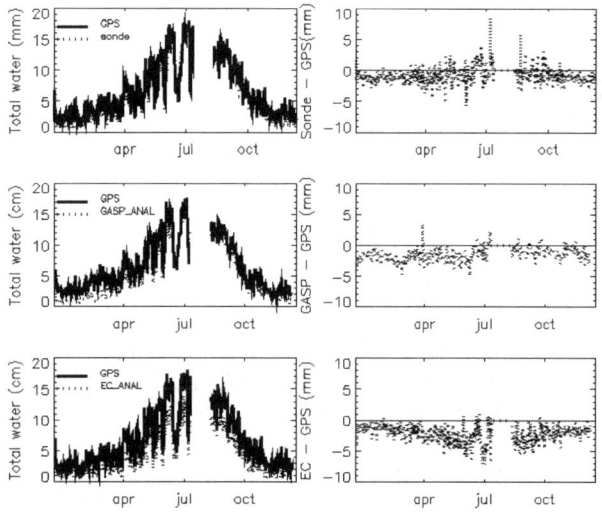

Fig. 1. Comparison of PW in 2001 determined by GPS at Naqu with radiosonde (top); GASP analysis (middle) and ECMWF analysis (bottom). The right column shows the corresponding difference.

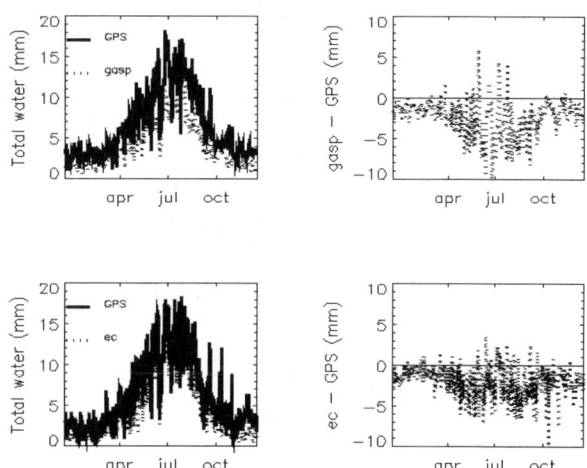

Fig. 2. Comparison of PW determined by the GPS and analysis of the GASP and ECMWF operational models at Gaize.

The type of radiosonde used at Naqu is 701A home made product (model 701A) designed in the early 70s with a frequency of 400 Hertz. The accuracy of this instrument is highly questionable because of the age of its design; no cross comparison has been done before. It is most likely that the radiosonde is subject to a dry bias.

The comparison of GPS based PW with those determined by the GASP analysis at 11.00/23.00 UTC and ECMWF analysis at 00/12 UTC are shown in Fig. 1b and c,

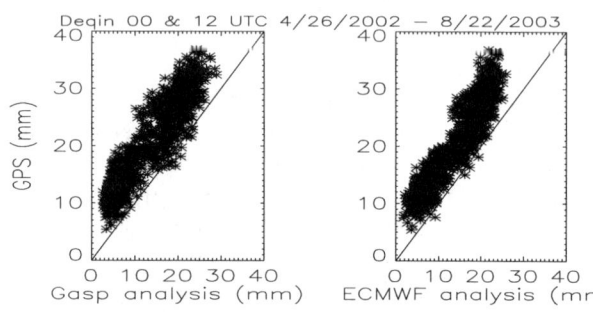

Fig. 3. Same as Fig. 2 but for Deqin.

The lack of observations in these regions is responsible for the large discrepancy. Note that the above results are similar to the findings reported by Wu et al. (2003) for the mountainous area of Sumatra Island. This similarity is probably related to the cold temperature at high elevation for which the radiosondes exhibit a dry bias.

4. IMPACT ON RADIATION BUDGET AT SURFACE

The previous comparisons have shown a relatively large difference between the GPS based PW and those determined using the radiosonde and NWP model analysis, but it is not clear which results are more reliable as no independent benchmark is available. Here we perform radiative transfer calculations to examine the impact of these water vapor differences on the surface radiation budget and compare the results with radiation measurements at the surface. As mentioned earlier, the radiation measurements are generally more accurate and the atmosphere over the Tibetan Plateau is relatively clean with much less air pollution and aerosol loading in the atmosphere compared with other regions. Therefore the water vapor should play a dominant role on the surface radiation budget under cloud free conditions. It is expected that the modeled radiation using more accurate water vapor should be closer to the observations.

The calculations were first performed for Gaize as the radiation observations from the AWS are available. The radiation model used in the calculations is the Bureau version of Edwards and Slingo code (Sun and Rikus, 1999). The GASP analysis profiles at 10:30 local time were used and the ozone profile was scaled to match the total column amount from TOMS. The surface albedo from the AWS measurements is used in the calculations. The analysis water vapor profile is then scaled by the GPS PW to force the analysis PW to match the GPS values. The effect of cloud is not included in the calculations as cloud information is not available. The rural aerosol optical properties from GADS (Hess et al., 1998) were included in the calculation as background. Fig. 4 shows a plot of the time series of the total downward solar irradiance at the surface at 10:30 local time for the year 2001 (upper panels). The observations show a large day-to-day variation due to the effect of clouds. The upper limit of the observational envelope can be regarded as irradiance under clear sky. The modeled results generally follow this envelope but not close enough. This implies that the model atmosphere is not optically thick enough, possibly due to uncertainties in water vapor and aerosols. The results from using the scaled PW are closer to the observations, indicating that the GASP moisture analysis in this region may be too dry. The upper right panel of Fig. 4 shows the irradiance difference between the two-modeled results. It is seen that with the scaling of water vapor amount the downward solar irradiance at the surface can be reduced by about 20 Wm^{-2} (about 3% of total value), which is large and will in turn affect the surface energy balance.

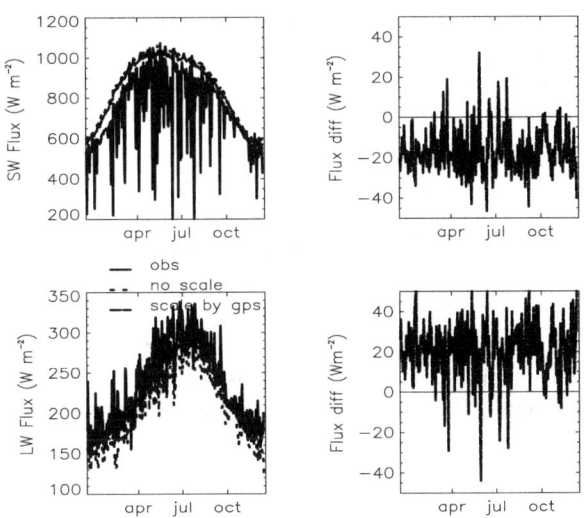

Fig. 4. Comparison of modelled and observed radiation in 2001 at Gaize. The upper panels represent the downward solar radiation at the surface and the lower panels denote the downward longwave radiation at the surface. The observations at 10:30 local time are plotted in the solid curve and the corresponding model results are plotted as dotted and dashed curves, respectively.

The lower panel in the Fig. 4 shows the comparison for the downward longwave irradiance at the surface. Again the large observed values are due to the presence of clouds. Of the two modelled results, the values determined with the water vapor scaled by the GPS PW are clearly closer to the observations. The differences in the longwave irradiance due to the PW differences in the calculations range up to 40 Wm^{-2}, as shown in the lower right panel of the Fig. 4. The same calculations are also performed for Naqu and Deqin and the results are similar to those shown in Fig. 4.

Chen et al. (2003) have recently performed a sensitivity study using the NCAR regional climate model (Giorgi et al., 1993) to examine the effect of the doubling CO_2 on the surface warming on the Tibetan Plateau region. Their results show that doubling CO_2 in this region can cause a change in the surface absorbed solar radiation of about −10 and 10 Wm^{-2} and a change in the downward longwave radiation of about 5–15 Wm^{-2}, depending on the season and surface elevation. The reduction of solar radiation and enhancement of longwave radiation due to the uncertainty of water vapor found in this study are both larger than the effects due to the doubling of CO_2. It should be emphasized that the results of Chen *et al.* (2003) were obtained using a climate model. Whereas the results from this study are from a radiation code alone. If run an offline radiation code as for a case of doubling CO_2 the radiative forcing at the surface is smaller than the values from climate model. On the other hand, the results shown in Fig. 4 were obtained at the relatively high solar elevation angle (local solar time 10:30). The results for the daily average will be smaller. Therefore, it may not be appropriate to directly compare these two findings. A similar study to Chen *et al.* (2003) should be performed for water vapor using a regional climate model to appropriately assess the impact of water vapor uncertainty in this region. Nevertheless, the problem of the uncertain water vapor measurements in this region is considered to be significant and caution is needed when discussing model results associated with water vapor in this region.

5. CONCLUSION

The precipitable water over the Tibetan Plateau determined by GPS, radiosonde and NWP operational model analysis are compared in this study. It is found that the PW from the radiosonde measurements is systematically smaller than that of the GPS estimates. Similarly, the PW determined by the GASP and ECMWF operational model analysis are also smaller than the GPS values. These systematic differences are likely due to the old type of radiosonde used in the Tibetan Plateau, which may be subject to a dry-bias at cold temperatures.

The effects of the uncertainty in water vapor on the surface radiation budget in this region were investigated using the radiation model with GASP model analysis profiles. The results show that the downward solar radiation at the surface can be reduced by about 20 Wm^{-2} and the downward longwave radiation at the surface can be increased by up to 40 Wm^{-2} if the GASP water vapor profiles are corrected to have PW matching the GPS measurements. These changes are much larger than those due to the effect of doubling CO_2 in this region.

Many studies have shown that the water vapor amount in this region is an important factor influencing the Asian summer monsoon. The large uncertainty in water vapor found in this study is a clear concern. It is necessary to examine the accuracy of the radiosonde used in this region and if possible to install more GPS receivers to improve the moisture analysis. It would also be useful to investigate the effect of the water vapor uncertainty on NWP forecasts and climate simulations and this will be the subject of our study in the future.

6. ACKNOWLEDGMENTS

The authors would like to acknowledge Dr. Bourke for useful discussion and encouragement, Dr. Rikus and Mr. Tomasz for reviewing the manuscripts.

This study was supported by the Chinese National High Technology Research and Development Plan under contract No. 2002AA135360.

7. REFERENCES

Chen, B., W. C. Chao, and X. Liu, 2003: Enhanced climate warming in the Tibetan Plateau due to doubling CO_2: a model study, *Climate Dynamics*, 20, 401-413.

Giorgi, F., M. R. Marinucci, and G. T. Bates, 1993: Development of a Second-Generation Regional Climate Model (RegCM2). Part I: Boundary-Layer and Radiative Transfer Processes, *Mon. Wea. Rev*, 121, 2794-2813.

Hess, M., P. Koepke, and I. Schult, 1998: Optical properties of aerosols and clouds: The software package OPAC, *Bull. Amer. Met. Soc.*, 79, 831-844.

Kopken, C., 2001: Validation of integrated water vapor from numerical models using ground-based GPS, SSM/I, and water vapor radiometer measurements, *J. Appl. Meteorol.*, 40, 1105-1117.

Kou, Y.-H., X. Zou, and Y. R. Guo, 1996: Variational assimilation of precipitable water using non-hydrostatic mesoscale adjoint model. Part I: Moisture retrieval and sensitivity experiments, *Mon. Wea. Rev.*, 124, 122-147.

Liou, Y.-A., and Y.-T. Teng, 2001: Comparison of precipitable water observations in the near tropics by GPS, Microwave radiometer, and radiosondes, *J. Appl. Meteorol.*, 40, 5-15, 2001.

Sun Z., and L. Rikus, 1999: Improved application of ESFT to inhomogeneous atmosphere, *J. Geophys. Res.*, 104, 6291-6303.

Wu, P., J.-I. Hamada, S. Mori, Y. I. Tauhid, and M. D. Yamanaka, 2003: Diurnal variation of precipitable water over a mountainous area of Sumatra Island, *J. Appl. Meteorol.*, 42, 1107-1115.

VALIDATION EXPERIMENT FOR SATELLITE REMOTE SENSING AND NUMERICAL MODELS OF 'YAMASE' CLOUDS

Masaya Kojima[1], Shoji Asano[1], Hironobu Iwabuchi[2], and Shin-ichi Otake[1]

[1]Center for Atmospheric and Oceanic Studies, Tohoku University
Sendai, 980-8578
JAPAN

[2]Frontier Research System for Global Change
Yokohama, 236-0001
JAPAN

1. INTRODUCTION

'Yamase' clouds are typical maritime atmospheric boundary-layer (ABL) clouds that occur over the sea to the east off the Sanriku coast (the eastern part of the Northeastern District of the Main Island of Japan) in summer season under easterly cool wind, i.e., the so-called *Yamase* (Kawamura, 1995), blown out from Okhotsk high (see Fig. 1). The *Yamase* brings cloudy and cool weather with low insolation to the Sanriku district, and, sometimes, results in severe crop damages. The state-of-the-art performance of atmospheric general circulation models and weather prediction models is not good enough to simulate properly such low-level stratiform clouds mainly because of the coarse spatial-resolution. We are studying the formation mechanism of the *Yamase* clouds by using a non-hydrostatic model (Nagasawa et al., 2004; Iwasaki et al., 2002). To validate the model performance, we have also carried out satellite remote sensing and surface observations on board a ship for the *Yamase* clouds. Here we present the physical parameters of the *Yamase* clouds from the shipboard observation and satellite remote sensing. We compare liquid water path (*LWP*) and shortwave radiation flux at the surface (*SWFLUX*) estimated from satellite remote sensing and numerical simulation for low-level stratiform clouds in the *Yamase* events in the summer of 2003.

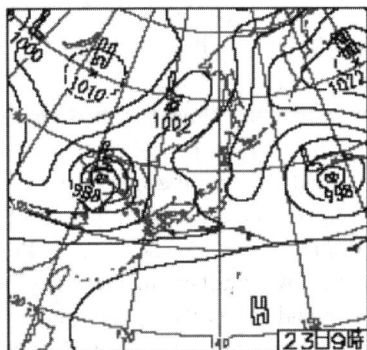

Fig. 1. Surface pressure pattern for a *Yamase* event on 23 June 2003.

2. OBSERVATION ONBOARD A SHIP

The shipboard experiment of the *Yamase* clouds was carried out during the period from 19 to 28 June 2003. In addition to the routine marine weather observations and intensive GPS-sonde launchings, we conducted clouds observations using various radiometric instruments such as a laser ceilometer (ImpulsePhysik, LD-25), a dual-frequency microwave radiometer (Radiometric Co., WVR-1100), and the total-band and near-infrared-band pyranometers (EKO Co., MS-801). The measured physical parameters can be used to validate the products from satellite remote sensing and numerical simulations. Among them, *LWP* and *SWFLUX* are particularly useful parameters for model validation. Fig. 2 shows the course variations of temperature and humidity profiles in the lower troposphere, observed by GPS sondes during the *Yamase* event. The figure shows that over the ship site, the lower part of ABL became cool and humid with the inflow of *Yamase* wind in the evening of 22 June, and very low *Yamase* clouds (might be fogs) appeared with the cloud-base heights of a few tens meters in the night through the next morning. This indicates the beginning of the *Yamase* event, which continued to the morning of 25 June 2003, when the ABL was well mixed, and became warmer and drier. The mean cloud-base height averaged over the duration was about 270 m, and the corresponding mean *LWP* was about 60 gm^{-2} for the *Yamase* clouds. Thus, for the first time we could observe the time variation of the marine ABL with *Yamase* clouds from their formation to dissipation.

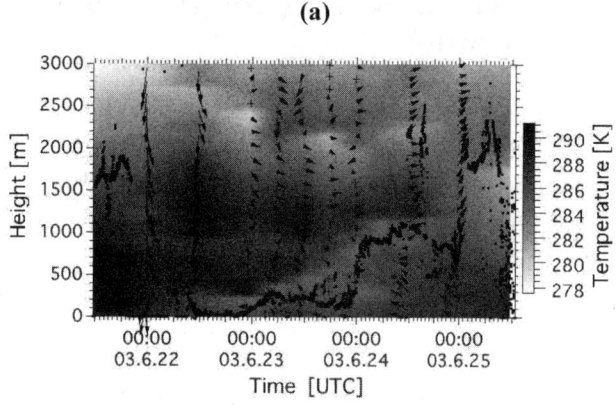

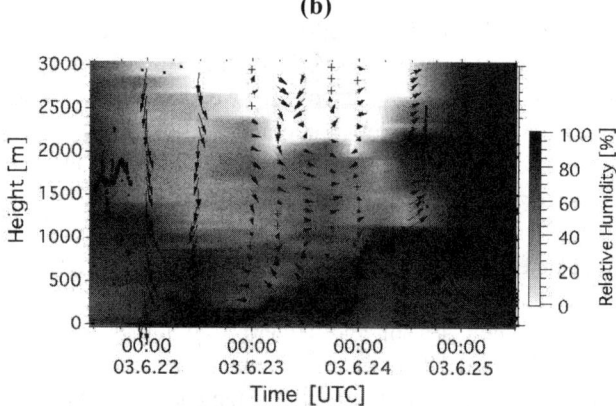

Fig. 2. Time variation in UTC of the temperature *(a)* and relative humidity *(b)* profiles, interpolated from those measured by the GPS-sondes launched from the ship, in the marine lower atmosphere from 21 June to 25 June 2003. In the figure, the cloud-base heights (*black dots*) measure by the ceilometer and the wind profiles (*arrows*) measured by the GPS-sondes are also superimposed.

3. SATELLITE REMOTE SENSING

To study spatial distribution of loud physical parameters of *Yamase* clouds in the Western North Pacific region, we used AVHRR/3 (Advanced Very High Resolution Radiometer version 3) data from NOAA-17 satellite. The cloud optical thickness (τ) and effective particle radius (R_{eff}) were retrieved from the daytime AVHRR/3 data of the visible channel (ch.) 1 centered at wavelength 0.63 μm and the near-infrared channel 3A at 1.61 μm, by modifying the two-channel algorithm of Nakajima and Nakajima (1995) to use ch. 3A instead of their ch. 3 at 3.7 μm. The infrared-channel data (ch.4 at 10.8 μm, and ch.5 at 12.0 μm) were used to estimate the cloud-top temperature and to screen out pixels contaminated by higher clouds. The *LWP* can be estimated as products of the retrieved τ and R_{eff} through the approximate relation, $LWP = 2\rho\tau_c r_{eff}/3$, where ρ is the density of liquid water. The cloud physical parameters were retrieved in every grid of each 15x15 km^2 width. We analyzed satellite data for the periods from 19 June to 31 July 2003, and from 19 June to 31 July 2004. In stead of direct validation of the satellite remote sensing by comparison with in-situ measurement, we compared the satellite-derived *LWP* with those measured by the shipboard microwave radiometer (MWR) for the collocated cloud scenes, as shown in Fig. 3. Although the satellite-derived *LWP* are, in general, larger than the MWR-measured *LWP*, both are agreed within their measurement uncertainties. This suggests that the results of satellite remote sensing are reasonable.

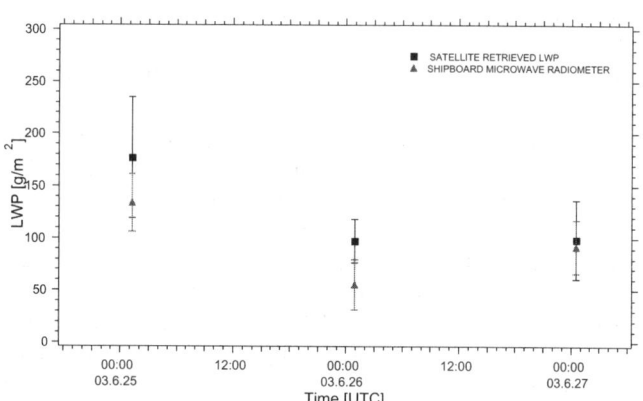

Fig. 3. Comparison of LWP estimated from satellite and ship for the same cloud scenes.

Fig. 4 shows the horizontal distribution of the retrieved τ and R_{eff} for the *Yamase* clouds on 25 June 2003. The easterly surface winds are also depicted by the arrows superimposed on the panels. The figures shows optically thin and rather uniform, stratiform low-level clouds widely distributed over region. However, in some locations, the cloud distribution exhibits band-like features and cellular structures, particularly, in the leeward places. For the clouds in the area enclosed by the points (38°N, 142°E), (38°N, 146°E), (42°N, 142°E), and (42°N, 146°E), we found that both R_{eff} and τ were rather small at the generation stage of the *Yamase* clouds, then they grew gradually with time. The correlation between the retrieved R_{eff} and τ was generally in positive correlation. In the decaying stage of the Yamase clouds, however, the correlation partly broke down.

SURFACE MEASUREMENTS AND FIELD EXPERIMENTS

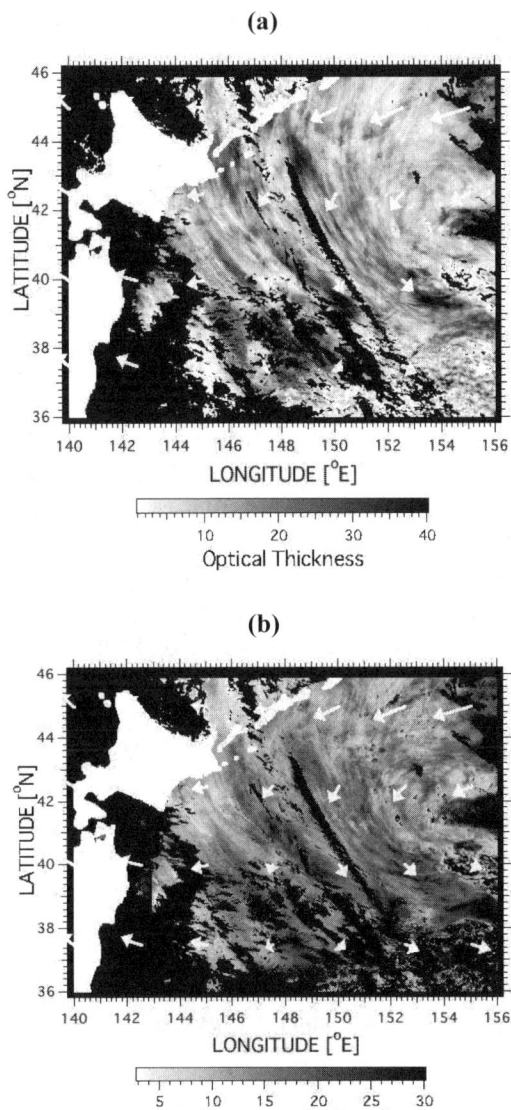

Fig. 4. Optical thickness *(a)* and effective radius *(b)* retrieved from the AVHRR data of NOAA-17 for the low-level clouds on 25 June 2003. The black areas indicate the areas of no data and/or covered by higher clouds. The white arrows indicate wind velocity at 1000 hPa from the NCEP/NCAR reanalysis data.

4. COMPARISON OF YAMASE CLOUDS FROM SATELLITE REMOTE SENSING AND NUMERICAL SIMULATION

The satellite-retrieved cloud distribution and physical parameters can be used to validate the performance of numerical models. In Fig. 5, we compare the horizontal distribution of the satellite-derived *LWP* with those simulated by a non-hydrostatic model (Ujiie et al., 2004), for the *Yamase* clouds on 18 July 2003. The simulation was calculated with a horizontal resolution of 5 km. The comparison shows fairly good agreement in gross distribution pattern of the horizontal distribution of *Yamase* clouds. However, there are significant discrepancies between the quantitative *LWP* values from the satellite remote sensing and model simulations. The model generally overestimated *LWP*, compared with the satellite. This means that the cloud formation parameterization isn't inadequate. Therefore the numerical model should be further improved in the cloud formation parameterization.

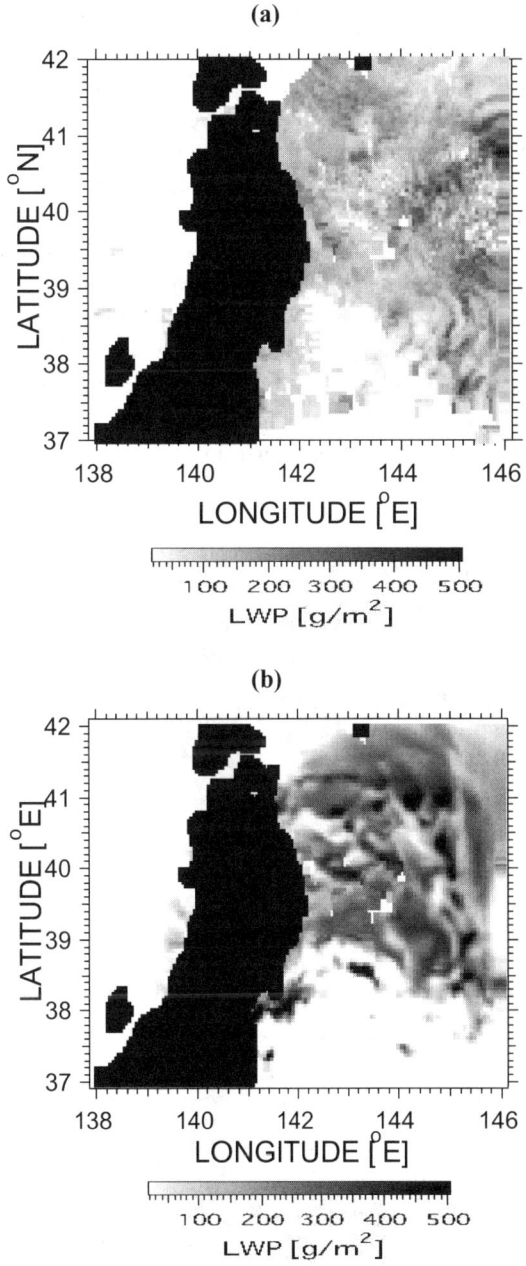

Fig. 5. Horizontal distributions of *LWP* produced by *(a)* satellite remote sensing and *(b)* numerical simulation.

5. CONCLUSION

In this study, for the first time onboard the ship, we could observe the time variation of marine ABL in the *Yamase* event from its beginning to ending. The observational data of the ABL time variation and the associated cloud fields will be useful to validate the performance of numerical simulations of *Yamase* event. We have also retrieved such cloud physical parameters as cloud optical thickness, effective particle radius, and liquid-water-path from satellite remote sensing using the NOAA-17/AVHRR/3 data for low-level water clouds in the Western North Pacific region. The satellite remote sensing was validated by comparing the satellite-derived *LWP* with those measured by the shipboard MWR for the collocated cloud scenes. It is shown that most clouds were fairly thin with optical thicknesses between 4 and 20 and with rather uniform particle radii between 8 and 16 µm. We compared the satellite-derived *LWP* distribution of low-level clouds with those simulated by the non-hydrostatic model. Although the model could roughly reproduce the horizontal distribution of low-level clouds, the model generally overestimated the *LWP* values. Therefore the numerical model needs to be further improved, especially, in the parameterization of cloud formation.

6. ACKNOWLEDGEMENTS

The shipboard observation was carried out within the *Yamase* Intensive Experiment under the cooperation with the Sendai District Meteorological Observatory and the Hakodate Marine Observatory. We are grateful to Dr. F. Sakaida for his kind assistance in processing the NOAA/AVHRR data. We thank Mr. M Ujiie and Atmospheric Dynamics Group of Tohoku University for providing the model simulation results. The study was partly supported by the Research Revolution 2002 Project.

6. REFERENCES

Iwasaki, T., S. Asano, H. Okamoto, and R. Nagasawa, 2002: A cloud study system using a nonhydrostatic multi-nested regional climate model. *Proc. EarthCARE Workshop* (Harumi, Tokyo, 17–18 July 2002), 171-174.

Kawamura, H. (*Ed.*), 1995: YAMASE, *Meteorol. Res. Note* (Japan Met. Soc.), No. 183, 179 pp (*in Japanese*).

Nagasawa, R., T. Iwasaki, S. Asano, K. Saito, and H. Okamoto, 2004: A numerical study of low-level cloud formation in 'Yamase' with a nonhydrostatic multi-nested regional climate model, *J. Meteor. Soc. Jpn* (submitted).

Nakajima, Y. T., and T. Nakajima, 1995: Wide-Area Determination of Cloud Microphysical Properties from NOAA AVHRR Measurements for FIRE and ASTEX Regions. *J. Atmos. Sci.*, 23, 4043-4059.

Ujiie, M., and T. Iwasaki, 2004: Development of a shallow cloud parameterization scheme: Part 2. Implementation into a meso-scale model. In *IRS '04*.

COMPARISON OF LONGWAVE EFFECTIVE CLOUD FRACTION WITH ARM CLOUDINESS MEASUREMENTS

E. E. Takara and R. G. Ellingson

The Florida State University, Department of Meteorology, Tallahassee, FL 32306-4520, USA

ABSTRACT

The longwave effective cloud fraction is derived from surface longwave measurements by pyrgeometer, interferometer, and cloud base height measurements. Results from two sets of radiance thresholds, one for low clouds and another more sensitive to high clouds, are presented and compared to the opaque total sky cover from a sky imager. With thresholds set for the low clouds there was some agreement with the sky imager for lower clouds, but agreement decreased as cloud base height increased.

When the threshold was set for high clouds there was increased high cloud detection, at the cost of overestimating cloud amounts for middle and low clouds. As with any method for cloud detection from surface longwave measurements, there are two problems. First, clouds are harder to detect as temperature and water vapor amount increases because of masking by increased gaseous emission. Second, multiple cloud layers make it difficult to determine proper threshold values for cloud detection.

1. INTRODUCTION

The Atmospheric Radiation Measurement (ARM) program is one of the climate research efforts of the United States (US) Department of Energy. As the name implies, the ARM program seeks to measure quantities relevant to atmospheric radiation, archiving measurements over time scales suitable for climate studies. ARM takes surface based measurements at sites in the US Southern Great Plains (SGP), the North Slope of Alaska (NSA), and the Tropical Western Pacific (TWP). ARM also conducts field campaigns, often in co-operation with other agencies.

The Cloudiness Intercomparison field campaign took place from February through May 2003 at the SGP Central Facility (36° 37'N 97° 30'W). Its purpose was to compare several different instruments and methods of measuring cloud amount. While it may seem to be a simple quantity, cloud amount is somewhat elusive. Different types of instruments placed next to each other can give different cloud amounts because they use different parts of the spectrum, have different fields of view, sampling rates, etc. Here the longwave effective cloud fraction (N_e) is described and compared to measurements from the Total Sky Imager (TSI), (Long, 2003).

Another consideration is that cloud amount depends on the physical scale under consideration. The cloud amount appropriate for comparison to a single pyrgeometer is not likely to be useful for a grid square with 100 km sides.

In terms of N_e, the average longwave surface flux F, over an area that is large compared to individual clouds, is

$$F = (1 - N_e)F_{clear} + N_e F_{overcast}. \quad (1)$$

F_{clear} is the clear sky flux; the flux that would occur if the broken cloud field was removed. $F_{overcast}$ is the flux that would occur if the broken cloud field became completely overcast. N_e is the fractional sky coverage of flat black plates.

2. INSTRUMENTS AND MEASUREMENTS

Re-arranging (1) to solve for N_e and substituting F_{pyr} for F yields:

$$N_e = \frac{F_{pyr} - F_{clear}}{F_{overcast} - F_{clear}}. \quad (2)$$

In Han and Ellingson (1999) longwave parameterizations were found for cumulus cloud fields. In that study, F_{pyr}, F_{clear}, and $F_{overcast}$ were obtained through measurements, fixing N_e. That process is used here for all skies.

Several data sets from the ARM SGP central facility were used to derive N_e. The longwave (0–3000 cm^{-1}) downward flux, F_{pyr}, was measured directly by pyrgeometer. The radiances and surface emitting temperature from the Atmospheric Emitted Radiance Interferometer (AERI), (Felz et al., 1998), and cloud base from Active Remote Sensing of Cloud Layers (ARSCL), (Clothiaux et al., 2000), were used to derive the F_{clear} and $F_{overcast}$ throughout the day. The resulting N_e are compared to the TSI opaque total sky cover, the fraction of the TSI hemispherical view taken up by opaque clouds.

This method, like others based on longwave surface measurements (e.g., Durr and Philipona, 2004) has the advantage of being continuous; the same algorithm is valid in daytime and nighttime. The disadvantage of surface longwave methods is that clouds become difficult to detect during the summer or in the Tropics because of larger gaseous emission due to greater water vapor amounts and higher temperatures. This increased emission masks high/cold clouds. Active measurements at other wavelengths such as radar and lidar have the advantage of being able to "illuminate" high clouds. Shortwave methods (e.g., Long and Ackerman, 2000) are pseudo-active, with the Sun for illumination.

3. COMPUTING F_{clear} AND $F_{overcast}$

The AERI takes calibrated radiance measurements of the downwelling radiance from 520 to 3020 cm^{-1} at approximately 0.5 cm^{-1} resolution. Adding the 5200 measured radiances, I_{AERI}, from 520 to 3020 cm^{-1} gives the total measured AERI radiance I^{tot}_{AERI}:

$$I^{tot}_{AERI} = \sum I_{AERI} . \quad (3)$$

Assuming that the downwelling radiance outside the 833–1250 cm^{-1} window can be approximated by the Planck function (B_ν), pseudo-window radiance (I_w) can be defined as:

$$I_w = I^{tot}_{AERI} - I_{opq} . \quad (4)$$

$$I_{opq} = \int_{520\,cm^{-1}}^{833\,cm^{-1}} B_\nu(T_s,\nu)d\nu + \int_{1250\,cm^{-1}}^{3020\,cm^{-1}} B_\nu(T_s,\nu)d\nu . \quad (5)$$

The combination of a low standard deviation in the AERI 990cm^{-1} radiance and a high I_w indicates the presence of clouds. If an overcast threshold radiance is used, the observed I_w is considered an overcast radiance, $I_w^{overcast}$, when $I_w \geq I_{thold}$ (overcast). A low standard deviation and low I_w indicates clear skies. When using a clear threshold, I_w is considered to be a clear radiance, I_w^{clear}, when $I_w \leq I_{thold}$ (clear). The values of I_{thold} (overcast/clear) change according to the surface temperature and water vapor amounts.

The clear and overcast fluxes can be computed from the I_w according to:

$$I^{tot,clear}_{AERI} = I_w^{clear} + I_{opq} . \quad (6)$$

$$I^{tot,overcast}_{AERI} = I_w^{overcast} + I_{opq} . \quad (7)$$

$$F_{clear} = LI^{tot,clear}_{AERI} + F_{0-520} + F_{offset} . \quad (8)$$

$$F_{overcast} = LI^{tot,,overcast}_{AERI} + F_{0-520} + F_{offset} . \quad (9)$$

$$F_{0-520} = \pi \int_{0\,cm^{-1}}^{520\,cm^{-1}} B_\nu(T_{surf},\nu)d\nu . \quad (10)$$

$$F_{offset} = \delta_{pyr-AERI} . \quad (11)$$

Where L is a conversion factor ranging from π at the surface to 2.45 for clear skies, F_{0-520} is the flux from the 0–520 cm^{-1} opaque spectrum not measured by the AERI, and $\delta_{pyr-AERI}$ is the instrument offset between the AERI and pyrgeometer. By iteration, clear and overcast radiances can be found and used to construct the clear and overcast flux envelopes throughout the day. Using Eq. 2, this leads to N_e.

4. COMPARING N_e TO THE TSI

Two sets of threshold radiances were used to derive N_e. In the first, clear and overcast threshold radiances I_{thold} (clear) and I_{thold} (overcast) were specified based on monthly values.

In order to increase sensitivity to clouds, the second used only the previous clear threshold I_{thold} (clear).

4.1 N_e for Clear and Overcast Threshold Radiances

Fig. 1 is a binned scatter plot comparing the derived TSI opaque total sky cover on the x-axis to the longwave N_e on the y-axis, for ARSCL clear sky and lowest cloud base heights (Z_b) less than 4 km. The x and y points are in eleven bins, nine of width 0.1 starting at 0.05, and two of width 0.05 starting at 0 and 0.95 respectively. There will be a single diagonal line of black boxes from (0,0) to (1,1) if the N_e and TSI agree within the limits of the bins.

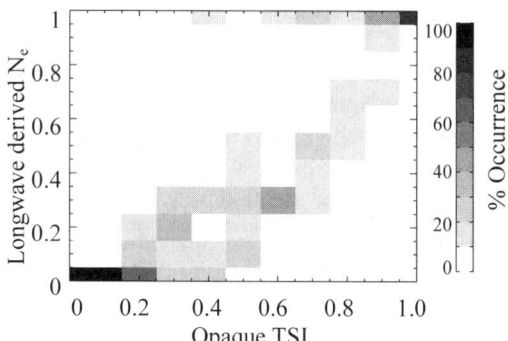

Fig. 1. Binned longwave N_e vs. opaque TSI for ARSCL clear and clouds with $Z_b < 4$ km, clear and overcast thresholds.

The N_e were derived using the first set of thresholds, using both clear and cloudy threshold radiances. This underestimates the cloud amount in most cases because some actual cloud radiances are below the cloudy threshold. The binned plot shows significant agreement at opaque TSI ≥0.95. So, the longwave N_e can identify overcast low clouds. The agreement at TSI ≤0.05 is somewhat spurious because an overly high overcast threshold will overestimate the clear sky amount.

Fig. 2 compares the clear and overcast threshold N_e with TSI for medium clouds: 4 km ≤ Z_b < 8 km. Again, the N_e consistently underestimates the cloud amount. There is some agreement at opaque TSI ≥0.95, but much less than for Zb < 4 km. This indicates that there are problems with longwave N_e as the clouds get higher.

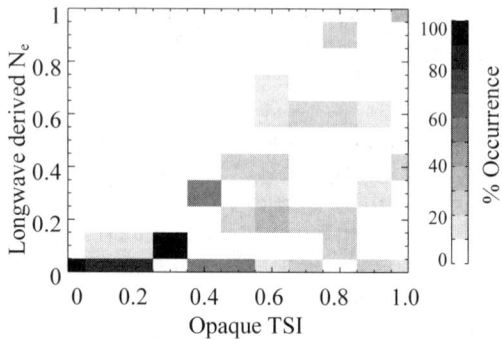

Fig. 2. Binned longwave N_e vs. opaque TSI for ARSCL clouds with 4 km ≤ Z_b < 8 km, clear and overcast thresholds.

Fig. 3 shows the clear and overcast threshold N_e- TSI comparison for high clouds $Z_b \geq 8$ km. This illustrates the problem with detecting higher clouds using longwave instruments. For the most part the longwave N_e shows clear skies. In fact, the number of clear sky cases is overestimated by a factor of 2.

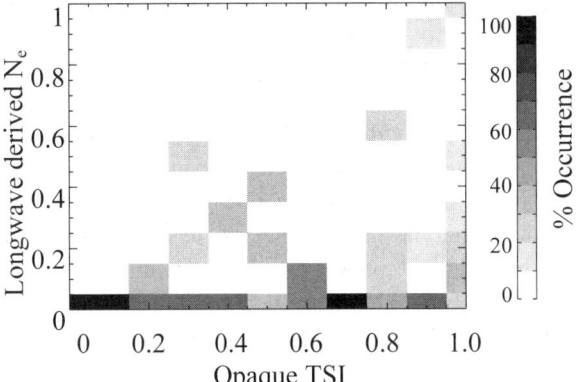

Fig. 3. Binned longwave N_e vs. opaque TSI for ARSCL clouds with $Z_b \geq 8$ km, clear and overcast thresholds.

Fig. 4 is a combination of Figs. 1–3, showing the comparison of clear and overcast threshold N_e for all sky conditions. The longwave N_e derived using clear and overcast thresholds underestimates the cloud amount for all Z_b.

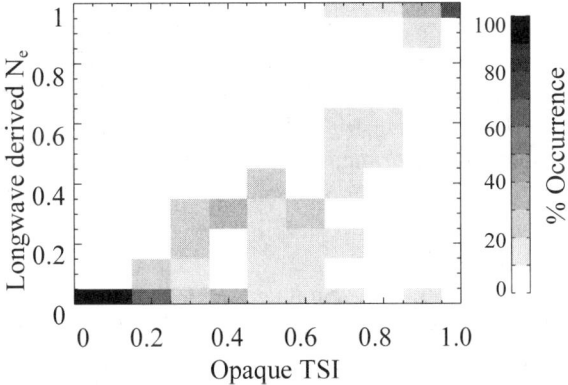

Fig. 4. Binned longwave N_e vs. opaque TSI for all skies, clear and overcast thresholds.

4.2 N_e for Clear Threshold Radiance

Fig. 5 shows the threshold clear and overcast N_e- TSI comparison for clear sky and $Z_b < 4$ km with N_e derived using only the previous clear sky threshold. This should be more sensitive to less emitting (colder or thinner) clouds.

The increased sensitivity has the disadvantage of making large clear sky radiances appear to be cloud radiances. This can be seen along the top of the figure, where there is a significant number of bins at $N_e=1$ when the opaque TSI values are less than 0.8. As before, the binned plot shows significant agreement at opaque TSI ≥ 0.95.

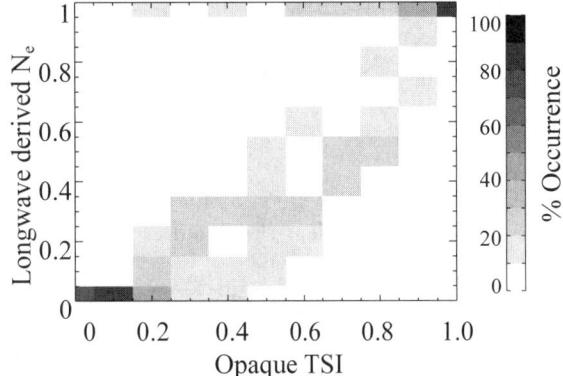

Fig. 5. Binned longwave N_e vs. opaque TSI for ARSCL clear and clouds with $Z_b < 4$ km, clear

Fig. 6 shows the comparison for medium clouds: 4 km $\leq Z_b < 8$ km. The bins in Fig. 6 are much more scattered than in Fig. 2; the similarity is the bins along the bottom, possibly due to problems with higher clouds. As with Fig. 5, the bins along the top indicate too much sensitivity, leading to overestimates of cloud amount.

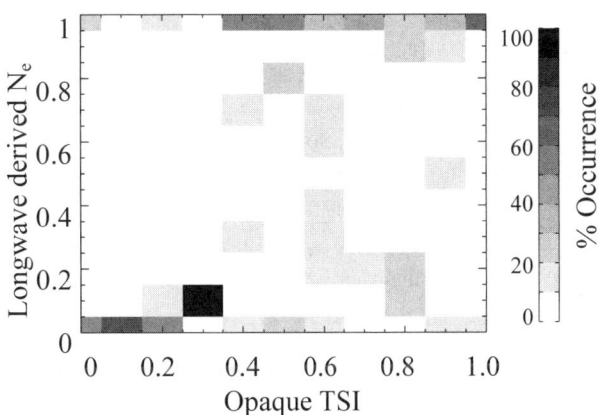

Fig. 6. Binned longwave N_e vs. opaque TSI for ARSCL clouds with 4 km $\leq Z_b < 8$ km, clear threshold.

Fig. 7 shows the N_e- TSI comparison for high clouds; $Z_b \geq 8$ km. It shows some success in detecting higher clouds—the upper right quadrant is more populated than in Fig. 3. Fig. 8 shows the comparison for all sky conditions.

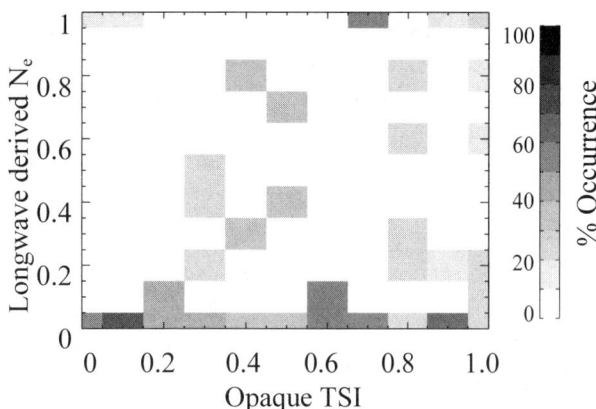

Fig. 7. Binned longwave N_e vs. opaque TSI for ARSCL clouds with $Z_b \geq 8$ km, clear threshold.

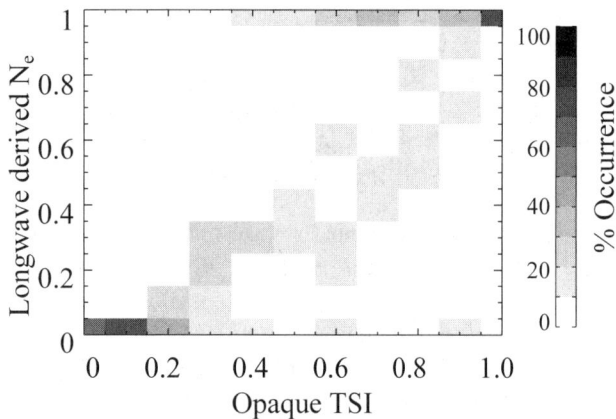

Fig. 8. Binned longwave N_e vs. opaque TSI for all skies, clear threshold.

Using only a clear sky threshold shows some improvement in detecting clouds with low emission, but the figures indicate that sensitivity to low-emitting clouds comes at the cost of overestimating the cloud amount. This may be addressed by more sophisticated thresholds for clear and overcast radiance.

5. SUMMARY AND CONCLUSIONS

The results from two sets of thresholds for computing effective cloud fraction (N_e) were compared with the total sky cover for opaque clouds measured by the TSI. The first set used threshold values to determine clear sky and cloud radiances. The second used only the previous clear sky threshold. The N_e found using clear and cloud thresholds were almost always lower than the TSI opaque cloud amount. This N_e showed agreement with the TSI for clouds below 4 km. The agreement between N_e and TSI decreased for middle and high clouds. The N_e found using only the clear sky threshold was more sensitive to low-emitting clouds, noticeable for the high clouds ($Z_b \geq 8$ km). The increase in sensitivity caused an overestimate of cloud amount, for middle (4 km $\leq Z_b <$ 8 km) and low clouds ($Z_b < 4$ km).

As the Cloudiness Intercomparison progressed through spring and approached summer, gaseous emission increased as the average temperature and water vapor amount increased, masking the cloud emission. Making the threshold limits more flexible will improve these results. Another related problem as the atmosphere became more active in the spring and summer was multiple cloud layers. Multiple cloud layers will require a more sophisticated algorithm which can properly maintain multiple overcast flux envelopes. Until the threshold limits can be made more flexible, this will be an awkward problem.

6. ACKNOWLEDGMENTS

This paper was sponsored in part by the U.S. Department of Energy's Atmospheric Radiation Measurements (ARM) program under grant DEFG0202ER63338.

7. REFERENCES

Durr, B., and R. Philipona, 2004: Automatic cloud amount detection by surface longwave downward radiation measurements. *J. Geophys. Res.*, 109, No. D5, D05201.

Clothiaux, E. E., T. P. Ackerman, G. G. Mace, K. P. Moran, R. T. Marchand, M. Miller, and B. E. Martner, 2000: Objective determination of cloud heights and radar reflectivities using a combination of active remote sensors at the ARM CART sites. *J. of Appl. Meteor.*, 39, 645-665.

Felz, W. F., W. L. Smith, R. O. Knuteson, H. E. Revercomb, H. M. Woolf, and H. B. Howell, 1998: Meteorological applications of temperature and water vapor retrievals from the ground-based Atmospheric Emitted Radiance Interferometer (AERI). *J. Appl. Meteor.*, 37, 857-875.

Han, D., and R. G. Ellingson 1999: Cumulus cloud parameterizations for longwave radiation calculations. *J. Atmos. Sci.*, 56, 837-851.

Long, C. N., and T. P. Ackerman, 2000: Identification of clear skies from broadband pyranometer measurements and calculation of downwelling shortwave cloud effects. *J. Geophys. Res.*, 105, No. D12, 15609-15626.

ALL-SKY AEROSOL DIRECT FORCING TO SW AND LW AT TOA AND SURFACE USING CERES TERRA AND THE MATCH ASSIMILATION

T. P. Charlock
NASA Langley Research Center, Hampton, Virginia 23681-2199, USA

F. G. Rose and D. A. Rutan
Analytical Services and Materials, Inc., Hampton, Virginia, USA

D. W. Fillmore and W. D. Collins
National Center for Atmospheric Research, Boulder, Colorado, USA

ABSTRACT

A method for computing direct aerosol forcing has been applied to an extensive Terra data set, spanning every second CERES footprint at surface and top-of-atmosphere (TOA). Calculations for the Surface and Atmosphere Radiation Budget (SARB) use cloud properties from MODIS and aerosol properties from MODIS and the MATCH assimilation. Fluxes are routinely compared with completely independent broadband radiometric measurements at approximately 50 ground sites; here reported for March 2000 to December 2001. For 25 March 2000, daily mean all-sky aerosol forcing (SW plus LW) has been estimated as -0.5 Wm^{-2} at TOA (i.e., aerosols cool the planet), $+5.3$ Wm^{-2} for the atmosphere (heating), and -5.9 Wm^{-2} for the surface net (cooling); the comparable TOA forcing for a theoretically clear ocean is -3.1 Wm^{-2}. The record is publicly available as Terra CERES CRS Edition 2A. Production of a revised set of fluxes and forcings is underway.

1. INTRODUCTION

Direct aerosol forcing is considered as the radiative flux including the effects scattering, absorption, and emission by aerosols, minus the flux without aerosols. The magnitude of direct aerosol forcing—and especially its anthropogenic component—is uncertain (IPCC, 2001). Like trace gases, aerosols affect the planetary radiation balance at TOA. Absorption of SW by aerosols a yields a larger forcing at the surface; this can also potentially spin down the hydrological cycle (i.e., Liepert et al., 2004) and have impacts quite different from those of increased CO_2. A climate model simulates the response to a given forcing. In part because we do not know the correct aerosol forcing (natural and anthropogenic) of recent decades, it has not been possible to rigorously validate any model simulation of global mean tropospheric temperature spanning the same interval.

The CERES observations of TOA fluxes (Wielicki, et al., 1997) includes a program to also compute the fluxes at TOA, within the atmosphere and at the surface, and also to validate the results with independent ground based measurements (Charlock and Alberta, 1996). To permit the user to infer cloud forcing and direct aerosol forcing with the computed SARB, CERES includes surface and TOA fluxes that have been computed for cloud-free (clear) and aerosol-free (pristine) footprints; this accounts for aerosol effects (SW and LW) to both clear and cloudy skies.

2. COMPUTATION OF FLUXES

The major inputs to the flux calculation are the instantaneous scene identification; cloud properties from MODIS (Minnis et al., 2002); TOA radiation from the CERES instrument in large (~20 km) footprints; 6-hourly gridded fields of temperature, humidity, wind (GEOS4); and ozone from NCEP (Yang et al., 2000). Aerosol information is taken from MODIS and from the NCAR Model for Atmospheric Transport and Chemistry (MATCH), an assimilation that here also employs aerosol retrievals from MODIS (Fillmore et al., 2005). The archive includes flux profiles calculated by algorithms that partially constrain to CERES TOA observations; and the "untuned" fluxes first calculated by the original inputs. Forcings based on the untuned record are used here.

We use a fast, plane parallel correlated-k radiative transfer code (Fu and Liou, 1993, Fu et al., 1998, 1999) which has been highly modified. A 2 stream calculation is used for SW. LW employs a 2/4 stream version, wherein the source function is evaluated with the quick 2-stream approach, while radiances are effectively computed at 4 streams. Constituents for the thermal infrared include H_2O, CO_2, O_3, CH_4 and N_2O. A special treatment of the CERES 8.0–12.0 µm window includes CFCs (Kratz and Rose, 1998) and uses the Clough CKD 2.4 version of the H_2O continuum. In collaboration with Dr. Qiang Fu, the code was modified to include 10 separate bands between 0.2–0.7 µm. In cooperation with Dr. Seiji Kato, we have included the HITRAN2000 data base for the determination of correlated k's in the SW (Kato et al., 1999). We make a first order accounting for inhomogeneous cloud optical thickness (using the gamma weighted two stream approximation of Kato et al., 2004) in the SW, fitting a 13-element histogram of cloud optical thickness in each footprint. An external mixture of aerosols, clouds, and gases is assumed. All-sky aerosol forcing is determined by running with clouds (if present), gases, and aerosols, and subtracting the flux from a run with no aerosols. A theoretical clear-sky aerosol forcing is computed for all footprints as the difference of the cloud-free flux with aerosols minus the cloud-free flux with no aerosols. Aerosol forcing includes the effects scattering (SW and LW), absorption (SW and LW) and emission (LW) by aerosols.

Land surface albedo is explicitly retrieved for clear footprints using a quick table look-up to the Langley Fu-Liou code that relates observed CERES TOA albedo, surface albedo, solar zenith angle (SZA), precipitable water, and aerosol optical thickness (AOT). The spectral shape of the surface albedo is assumed as per the International Geophysical Biospherical Project (IGBP) land type (see http://www-surf.larc.nasa.gov/surf). When cloudy, the land surface albedo is taken from a gridded record of clear-sky retrievals during the same month; and adjusted to account for an effective diffuse SZA beneath clouds. Ocean spectral albedo is obtained using a look up table (LUT) based on discrete ordinate calculations with a sophisticated coupled ocean atmosphere radiative transfer code (Jin et al, 2004). Inputs for ocean spectral albedo include SZA, wind speed, chlorophyll concentration (which has a minor effect on broadband flux), and SW optical depth of clouds and aerosols. There is an empirical correction for surface foam based on wind speed.

AOT is taken from MODIS (MOD04 described by Kaufman et al., 1997) when available. Over the ocean, MOD04 is used for 7 wavelengths; the AOT is interpolated to the remainder of the spectrum using the selected aerosol type, as specified below. Over the land, MOD04 provides AOT at 3 wavelengths, and the MOD04 Angstrom exponent is used to guide the extension over the spectrum. If the MOD04 instantaneous AOT is not available (i.e., footprint is overcast), we temporally interpolate from a file of the MODIS Daily Gridded Aerosol. When cloudiness in the footprint exceeds 50%, or when there is no MODIS AOT, we use AOT from the NCAR MATCH. When AOT is taken from MATCH, we assume it for one wavelength only (0.63 μm). MATCH AOT is apportioned to 7 types (small dust, large dust, soot, soluble organic, insoluble organic, sulfate, and sea salt) on a daily basis over the globe for all sky conditions. The Terra CERES CRS Edition 2A results described here assume a global climatological scale height for each of the 7 aerosol types. [A subsequent Edition 2B uses explicit height profiles for the 7 types that vary for each gridbox, each day.]

Aerosol type is always taken from MATCH; this guides the selection of the asymmetry factory (g) and the single scattering albedo (SSA). Asymmetry factors and SSA are assumed from the Tegen and Lacis (1996) and OPACS-GADS (Hess et al., 1998) models.

Reading Dubovik et al. (2002) on AERONET, we infer that the dust optical properties we have selected may be too strongly absorbing. [The subsequent Edition 2B results, which are not plotted in Figs. 1–3 below, use updated optical properties for dust, from A. Lacis of NASA GISS.]

3. VALIDATION OF FLUXES AT SGP

How credible are the computed surface fluxes? The web site http://www-cave/larc.nasa.gov/cave / (search for "CERES CAVE") is a gateway to a point and click version of the radiative transfer code used here; time series of subset results at selected sites; and compares the surface fluxes retrieved by CERES with independent measurements at over 50 sites scattered around the globe (Rutan et al, 2001). Fig. 1 shows the bias (computed minus observed) for a cluster of 22 Atmospheric Radiation Measurement (ARM) sites in the Southern Great Plains (SGP). The absolute magnitudes of biases in Fig. 1 for clear sky (those footprints screened as cloud free by MODIS) are small and generally less that the corresponding magnitudes for clear-sky aerosol forcing in Fig. 2. This gives confidence in the clear-sky forcing to SW insolation, which has a peak during summer at SGP.

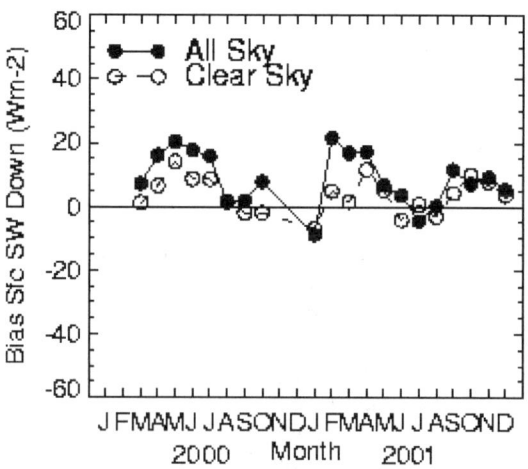

Fig. 1. Bias (computed minus observed) in surface insolation (monthly mean of daylight SGP overpasses).

The insolation at all "CAVE" sites (not just SGP) for 2001 gave results similar to Figs. 1 and 2. For all sites, the mean all-sky (clear-sky) observed insolation for daytime Terra overpass was 482.6 Wm^{-2} (713.3 Wm^{-2}), bias was 7.5 Wm^{-2} (−5.6 Wm^{-2}), and aerosol forcing −10.7 Wm^{-2} (−15.3 Wm^{-2}).

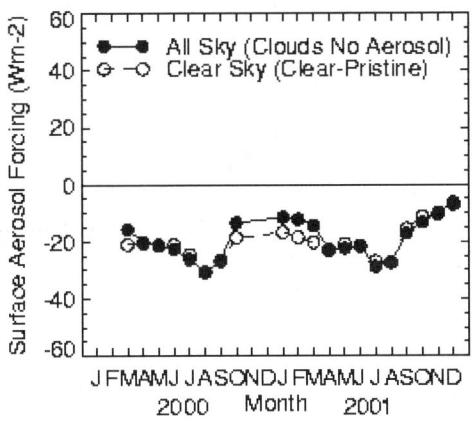

Fig. 2. Aerosol forcing to surface insolation (monthly mean of daylight SGP overpasses).

Returning to only the SGP sites, we consider surface downward LW flux (DLF), which has a smaller range of variation than does surface insolation. The bias for clear-sky DLF in Fig. 3 is somewhat disappointing. The DLF bias due is due, in turn, to a bias in the surface air temperature inputs from GEOS4. However, the all-sky DLF bias is no larger than the clear-sky DLF bias, attesting to the high fidelity of the cloud property inputs from Minnis et al. (2002).

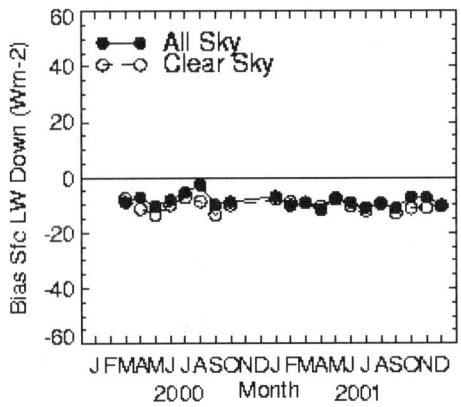

Fig. 3. Bias (computed minus observed) in surface DLF (monthly mean of day and night SGP overpasses).

4. GLOBAL FORCING FOR 1 DAY

The aerosol forcing for the approximately two million CERES footprints on 25 March 2000 in Table 1 have been weighted, to compensate for the differing sizes of the footprints and the increased frequency of sampling at higher latitudes. Table 1 estimates 24-hour mean forcing as a simple average of Terra overpasses, which occur at ~1030 L and ~2230 L over most of the globe. Net flux in Table 1 is taken as downwelling minus upwelling; the atmosphere forcing is the forcing to TOA net minus the forcing to surface net. To facilitate a diagnosis of the effects of clouds on aerosol forcing, Table 1 uses the designation "as if clear" for a theoretically clear forcing, which is computed whether or not the footprint is clear or cloudy. This is different than the "clear" qualification in Figs. 1, 2, 3, wherein clear denotes the subset of footprints that have been screened as cloud free by MODIS. Clouds substantially reduce TOA aerosol forcings but have proportionately less effect on the larger aerosol forcing to the atmosphere, which ultimately affects the hydrological cycle. The substantial atmosphere forcing to all-sky global SW plus LW (5.3 Wm^{-2} in Table 1) has been significantly reduced from the SW only value (6.6 Wm^{-2}) by atmosphere cooling due to LW (–1.3 Wm^{-2}).

Table 1. Global Aerosol Forcing on 25 March 2000 as Mean of Terra Overpasses (~1030 L and ~2230 L)

SW Aerosol Forcing (Wm^{-2})

level	Globe	Globe as if clear	Ocean	Ocean as if clear
TOA Net	–1.1	–2.6	–2.2	–3.6
Atmosphere	6.6	6.9	3.2	3.3
Surface Net	–7.7	–9.4	–5.4	–6.9

LW Aerosol Forcing (Wm^{-2})

level	Globe	Globe as if clear	Ocean	Ocean as if clear
TOA Net	0.6	0.8	0.3	0.5
Atmosphere	–1.3	–1.9	–0.7	–1.4
Surface Net	1.8	2.7	1.0	1.8

SW+LW Aerosol Forcing (Wm^{-2})

level	Globe	Globe as if clear	Ocean	Ocean as if clear
TOA Net	–0.5	–1.8	–1.9	–3.1
Atmosphere	5.3	5.0	2.5	1.9
Surface Net	–5.9	–6.7	–4.4	–5.1

5. DISCUSSION OF FORCINGS

A comparison of the biases in surface insolation (Fig. 1) and values of respective aerosol forcings (Fig. 2) suggests that the CERES SARB clear-sky forcing to insolation may be accurate over SGP to within a factor of two. More detailed study is needed to rigorously judge the corresponding forcings to all-sky insolation. This is certainly the case for aerosol forcing to DLF, which is small relative to the bias in Fig. 3.

What is the quality of the retrieved forcing displaced from SGP, as over the whole globe in Table 1? Note that all results mentioned to this point use "Edition 2A" scale heights for aerosols; and the older Tegen and Lacis properties for dust (when present). The more sophisticated treatment of both in the coming Edition 2B yields substantial changes for SW forcing: All-sky global SW TOA forcing is -1.1 Wm^{-2} (-2.5 Wm^{-2}) in Edition 2A (2B); and corresponding SW surface forcings are -7.7 Wm^{-2} (-6.6 Wm^{-2}) in Edition 2A (2B). These large changes (Edition 2A to 2B) point to the importance of dust optical properties over the globe. In contrast, relatively little dust is found over SGP (Figs. 1–2). Uncertainties in the optical properties of dust are likely to challenge the estimation of direct aerosol forcing to the global atmosphere and hydrological cycle for some time.

6. ACKNOWLEDGEMENTS

The provision of an updated set of spectral optical properties for desert dust by Dr. Andrew Lacis (NASA GISS in New York) is much appreciated. We are grateful for production support by the CERES Data Management Team, and especially the SARB component led by L. H. Coleman.

7. REFERENCES

Charlock, T. P., and T. L. Alberta, 1996: The CERES/ARM/GEWEX Experiment (CAGEX) for the retrieval of radiative fluxes with satellite data. *Bull. Amer. Meteor. Soc.*, 77, 2673-2683.

Dubovik, O., B. Holben, T. Eck, A. Smirnov, Y. Kaufman, M. King, D. Tanre, and I. Slutsker, 2002: Variability of absorption and optical properties of key aerosol types observed in worldwide locations. *J. Atmos. Sci.*, 59, 590-608.

Fillmore, D. W., W. D. Collins, and A. J. Conley, 2005: Aerosol direct radiative forcing—estimates from a global aerosol climatology constrained by MODIS assimilation, in preparation.

Fu, Q., W. B. Sun, and P. Yang, 1999: Modeling of scattering and absorption by nonspherical cirrus ice particles at thermal infrared wavelengths. *J. Atmos. Sci.*, 56, 2937-2947.

Fu, Q., and K.-N. Liou, 1993: Parameterization of the radiative properties of cirrus clouds. *J. Atmos. Sci.*, 50, 2008-2025.

Hess, M., P. Koepke, and I. Schult, 1998: Optical Properties of Aerosols and Clouds: The software package OPAC. *Bull. Amer. Meteor. Soc.*, 79, 831-844.

Intergovernmental Panel on Climate Change (IPCC), 2001: *Climate Change 2001: The Scientific Basis*, edited by J. T. Houghton et al., Cambridge Univ. Press, New York.

Jin, Z., T. P. Charlock, W. L. Smith, Jr., and K. Rutledge, 2004: A look-up table for ocean surface albedo. *Geophys. Res. Lett.*, 31, L22301.

Kato, S., T. P. Ackerman, E. G. Dutton, N. Laulainen, and N. Larson, 1999: A comparison of modeled and measured surface shortwave irradiance for a molecular atmosphere, *J. Quant. Spectrosc. Radiat. Transfer.*, 61, 493-502.

Kato, S., F. G. Rose, and T. P. Charlock, 2004: Computation of domain-averaged irradiance using satellite-derived cloud properties. In press for *Journal of Atmospheric and Oceanic Technology*.

Kaufman, Y. J., D. Tanre, H. R. Gordon, T. Nakajima, J. Lenoble, R. Frouin, H. Grassl, B. M. Herman, M. D. King, and P. M. Teillet, 1997: Passive remote sensing of tropospheric aerosol and atmospheric correction for the aerosol effect. *J. Geophys. Res.*, 102, 16,815-16,830.

Kratz, D. P., and F. G. Rose, 1999: Accounting for molecular absorption within the spectral range of the CERES window channel. *J. Quant. Spectrosc. Radiat. Transfer*, 48, 83-95.

Liepert, B. G., J. Feichter, U. Lohmann, and E. Roeckner, 2004: Can aerosols spin down the water cycle in a warmer and moister world? *Geophys. Res. Lett.*, 31, L06207, doi:10.1029/2003GL019060.

Minnis, P., D. F. Young, B. A. Wielicki, D. P. Kratz, P. W. Heck, S. Sun-Mack, Q. Z. Trepte, Y. Chen, S. L. Gibson, and R. R. Brown: 2002: Seasonal and Diurnal Variations of Cloud Properties Derived from VIRS and MODIS Data. Extended abstract for 11th Conference on Atmospheric Radiation (AMS), 3–7 June 2002 in Ogden, Utah.

Rutan, D. A., F. G. Rose, N. Smith, and T. P. Charlock, 2001: Validation Data Set for CERES Surface and Atmospheric Radiation Budget (SARB). *GEWEX News,* 11, No. 1 (February), pp. 11–12. Available at http://www.gewex.com under "Newsletter".

Tegen, I., and A. A. Lacis, 1996: Modeling of particle size distribution and its influence on the radiative properties of mineral dust aerosol. *J. Geophys. Res.*, 101, 19,237-19,244.

Wielicki, B. A., B. R. Barkstrom, E. F. Harrison, R. B. Lee, G. L. Smith, and J. E. Cooper, 1996: Clouds and the Earth's Radiant Energy System (CERES): An Earth Observing System Experiment. *Bull. Amer. Meteor. Soc.,* 77, 853-868.

Yang, S.-K., S. Zhou, and A. J. Miller, 2000: SMOBA: A 3-dimensional daily ozone analysis using SBUV/2 and TOVS measurements. http://www.cpc.ncep.gov/products/stratosphere/SMOBA/smoba_doc.html.

LONG-TERM TREND OF SURFACE SHORTWAVE RADIATION OVER CHINA

T. Hayasaka and K. Kawamoto
Research Institute for Humanity and Nature, Kyoto 602-0878, Japan

J.-Q. Xu
Frontier Research System for Global Change, Yokohama 236-0001, Japan

G.-Y. Shi
Institute of Atmospheric Physics, Chinese Academy of Sciences, Beijing 100029, China

ABSTRACT

Downward surface shortwave (SW) radiative flux in China was evaluated by using pyranometer data, products derived from satellite cloud data and those calculated by parameterization with ground-based meteorological data. These radiation data are in general consistent with each other for monthly average values although the satellite derived data have positive bias for the big city areas whereas negative bias for the desert area in the west part of China. One of the reasons for these biases is ascribed to inappropriate assumption of aerosols in the satellite derived data. It is shown from the pyranometer data analysis that the surface SW radiation tends to decrease for the period 1971–2000 although more comprehensive analyses of the related meteorological data are still needed.

1. INTRODUCTION

Downward shortwave (SW) radiative flux on the surface is one of the most important factors, which affect on the Earth's climate. Variation of the surface SW radiative flux is related to cloud amount, cloud optical thickness, water vapor amount, aerosol optical thickness, etc. It is pointed out that aerosols have been increasing with an increase in energy consumption due to the recent rapid growth of economy in China. These aerosols may give rise to both the direct and indirect effects in this region. On the other hand, the surface SW radiation is affected by natural, as well as the anthropogenic, factors.

In the present study, we evaluate the surface SW radiative flux in China for the past few decades using pyranometer measurement data, products derived from satellite cloud data and meteorological data, and those derived from parameterization with operational meteorological data such as water vapor and sunshine duration. Long-term and seasonal variations of surface SW radiation are also discussed.

2. RADIATION DATA SET

2.1 Pyranometer Data

Chinese Meteorological Administration (CMA) has been measuring downward SW radiative flux at more than 122 stations, of which more than 60 stations continue the measurements from late 1950's or early 1960's to 2000. All of them are originally daily accumulated data. In this study monthly average data based on daily accumulated data were used for the analysis.

Calibration before early 1990s was carried out every 5 year by using the Eppley blackbody cavity radiometer PMO-6 which is calibrated in Japan Meteorological Agency. CMA participates in the intercomparison with the international calibration standard of World Radiation Center in Davos, Switzerland. There still exist some errors even if the calibration is successfully carried out, for example, error factors belonging to the instrument are cosine response, azimuth response, temperature response, spectral selectivity, stability, and non-linearity. The error factors in operation also exist such as contaminations due to dust, snow, dew, water-droplet on the glass dome cover, and incorrect sensor leveling.

There are several unknown error factors in the past data for long period so that another procedure is necessary for evaluating the data quality. The quality of the pyranometer data in this study was evaluated in principle according to Younes et al. (2004). This method is a physical check to avoid unrealistic values. 3% of all data were judged to be incorrect by this check.

There remain some uncertainties if the quality check was carried out for the past data. In this study we checked even those data suspected to be erroneous, and compared them with satellite derived data and parameterization data. Pyranometer data consistent with satellite derived data were then used for the long-term and seasonal variation analysis.

2.2 Satellite Data

There are several products of surface SW radiation data set derived from satellite cloud data. We used ISCCP FD monthly averaged data provided by NASA GISS and NASA/GEWEX SRB data by NASA Langley Research Center.

ISCCP FD data were calculated by applying the radiative transfer model of NASA GISS GCM to the ISCCP cloud data and ancillary data sets (Zhang et al., 2004). The spatial resolution is 280 km and temporal resolution is originally 3 hours. Although the total period of the FD data is 18 years, we used monthly averaged data for 11 years from 1984 to 1994 to be consistent with GEWEX SRB data and parameterization data set.

GEWEX SRB data have a higher spatial resolution than ISCCP FD data. The algorithm of SRB is originally based on Pinker and Laszlo (1992). Since the surface radiation is not measured by satellite, it is derived from measured radiance data and meteorological input data by using radiative transfer algorithm. The quality check SW algorithm known as the Langley Parameterized Shortwave Algorithm (LPSA) provides shortwave surface radiative flux for daily average and 1° x 1° in spatial resolution (Gupta et al., 2001) although monthly average data were used in this study. Cloud properties derived from ISCCP DX data set (Rossow and Schiffer, 1999) are used in LPSA instead of direct measurements of radiance from space.

2. 3 Parameterization Data

Monthly average data of surface SW radiation are also obtained from operational meteorological data by using parameterization method. The surface SW radiation is basically estimated by using the following formula for clear sky condition and cloudy sky condition (Xu et al., 2005).

For clear sky condition, downward SW flux is estimated by using meteorological data,

$$\frac{S_{df}}{S_{0d}} = (C_1 + 0.7 \times 10^{-m_d F_1})(1 - i_3)(1 + j_1)$$

$$C_1 = 0.21 - 0.2\beta_{DUST}, \quad \text{for } \beta_{DUST} < 0.3$$
$$= 0.15, \quad \text{for } \beta_{DUST} \geq 0.3$$

$$F_1 = 0.056 + 0.16(\beta_{DUST})^{0.5}$$
$$i_3 = 0.014(m_d + 7 + 2\log_{10} w)\log_{10} w$$
$$j_1 = [0.066 + 0.34(\beta_{DUST})^{0.5}](ref - 0.15),$$

where S_{df}: average downward SW flux on the Earth's surface, S_{0d}: SW flux at the top of atmosphere, β_{DUST}: turbidity factor, m_d: daily mean optical airmass, w: precipitable water, ref: surface albedo.

For cloudy sky condition, downward SW flux is estimated also by using sunshine duration,

$$\frac{S_d}{S_{0d}} = a + b\frac{N}{N_0}, \quad \text{for } 0 < \frac{N}{N_0} \leq 1$$

$$= c, \quad \text{for } \frac{N}{N_0} = 0$$

$$a = 0.179 + 0.32\left(1 - \frac{p_s}{1000}\right),$$
$$b = 0.55,$$
$$c = 0.114 + 0.32\left(1 - \frac{p_s}{1000}\right),$$

where S_d: average downward SW flux on the Earth's surface, S_{0d}: SW flux at the top of atmosphere, N: sunshine duration, N_0: maximum sunshine duration, p_s: surface pressure.

3. DATA COMPARISON AND ANALYSIS

The pyranometer measurement data were compared with parameterization from meteorological data, GEWEX SRB data, and ISCCP FD data. All these data were compared month by month for the period from 1984 to 1994. A simple linear regression analysis was carried out for the comparison of SRB data and parameterization data with pyranometer data by using the following formula,

$$Q = a_0 + aPYR,$$

with $Q = SRB$ and $Q = XU$, where SRB, XU, and PYR are SRB data, parameterization data, and pyranometer data, respectively.

In total 65 meteorological observatories have long-term records of pyranometer measurement for more than 30 years. The comparisons without parameterization data were carried out for all these locations. Among them, the results of 18 locations are summarized in Table 1 because the parameterization method could be applied to the meteorological data at 18 observatories. R2 in the table shows the coefficient of determination. The period for the comparison is 1984–1994.

It is found in Table 1 that satellite derived data, e.g., GEWEX SRB and ISCCP FD, tend to have large positive biases against pyranometer data in large cities such as Beijing, Guangzhou, Hefei, Jinan, Shanghai, and Xian. Large negative biases are found for SRB data in desert areas such as Dunhuang and Golmud. These tendencies are found in other data sets not shown in this table. On the other hand, parameterization data seem to be more consistent with pyranometer data than satellite data resulting from the values of bias and the coefficients a, a0, and R2.

Table 1. Summary of the Surface SW Radiation Comparison Among Pyranometer (PYR), Satellite (SRB, FD) and Parameterized (XU) Data

Name	Lat.(deg.)	Long.(deg.)	SRB-PYR (W/m2)	FD-PYR (W/m2)	XU-PYR (W/m2)	a(SRB)	a0(SRB)	R2(SRB)	a(XU)	a0(XU)	R2(XU)
Altay	47.73	88.08	-15.29	-11.38	-1.87	0.929	-2.996	0.969	1.046	-9.830	0.979
Beijing	39.93	116.28	23.48	16.90	7.50	1.045	16.459	0.965	1.036	1.821	0.978
Changsha	28.22	112.92	10.64	22.96	7.68	0.856	36.857	0.948	0.880	23.195	0.952
Dunhuang	40.15	94.68	-28.59	-5.05	-0.19	0.848	2.488	0.948	0.960	7.953	0.957
Golmud	36.42	94.90	-35.14	-6.13	1.73	0.868	-5.713	0.960	1.026	-4.213	0.976
Guangzhou	23.13	113.32	30.22	40.25	13.23	0.978	32.987	0.834	0.932	21.954	0.850
Hailaer	49.22	119.75	-2.26	2.42	-2.98	1.010	-3.770	0.966	1.015	-5.317	0.966
Harbin	45.75	126.77	9.86	9.15	-2.53	0.908	23.621	0.879	0.919	9.641	0.851
Hefei	31.87	117.23	24.46	26.60	13.68	0.945	31.898	0.872	0.952	20.240	0.871
Hotan	37.13	79.93	-2.57	16.45	7.66	1.053	-12.168	0.934	1.021	3.874	0.955
Jinan	36.68	116.98	34.53	29.40	20.42	1.072	23.920	0.937	1.032	15.797	0.934
Lanzhou	36.05	103.88	8.40	33.25	15.60	0.724	53.143	0.943	1.024	11.725	0.977
Lhasa	29.67	91.13	-13.44	12.82	22.31	0.491	96.328	0.529	0.720	84.206	0.639
Nanchang	28.60	115.92	15.16	22.81	15.36	0.870	32.643	0.926	0.970	19.365	0.956
Shanghai	31.40	121.48	21.96	18.36	7.25	0.990	26.695	0.920	1.011	8.870	0.926
Shenyang	41.73	123.45	21.39	11.13	5.93	1.015	19.122	0.932	1.021	2.793	0.928
Turpan	42.93	89.20	3.44	10.92	0.09	0.861	27.282	0.951	0.988	2.174	0.977
Xi'an	34.30	108.93	29.50	32.74	0.27	0.894	43.456	0.910	0.955	6.239	0.882

These discrepancies are ascribed to the aerosols assumed in the calculation of SW radiation from satellite cloud data. In SRB data, aerosols are estimated by using clear sky pixels compared with an assumed surface albedo while the aerosols in FD data are climatological profile value of monthly average (Zhang et al., 2004). Therefore the discrepancies in the desert areas are smaller for FD data than SRB data. On the other hand, absorbing aerosols in the sub-cloud layer is important for SW radiation on the surface. However, these aerosols cannot be observed from space and incorrect assumption might be used for both SRB and FD data sets. We carried out simple calculations of the effect of absorbing aerosols below the cloud layer, and it is shown that about 20~30 W/m2 bias is easily ascribed to these aerosols. Other factors for the discrepancy are effect of cloud inhomogeneity (Iwabuchi and Hayasaka, 2002). However, this effect seems to be small if the cloud properties are retrieved from various viewing angle radiance data.

Seasonal variations and long-term trends of surface SW radiation were studied by using pyranometer measurement data after checking the quality of the data by comparison with SRB, FD and parameterization data sets.

It is found that an amplitude of seasonal variation of SW radiation is large and stable in the north and west regions, while it is small but complicated in south and southeast regions. SW radiation almost all over China has a tendency to decrease for 1971–2000 as shown by Fig. 1. Although only the results of the pyranometer data are shown in Fig. 1, parameterization data for 18 locations also have a similar trend during the same period. The average of the SW radiation linear trend for the pyranometer data is –3.48 W/m2/decade and that for the parameterization data is –2.69W/m2/decade, respectively. The average trend for all payranometer data shown by Fig. 1 is –2.56 W/m2/decade. Therefore, it is plausible that the decreasing trend is real if the uncertainties are taken into account.

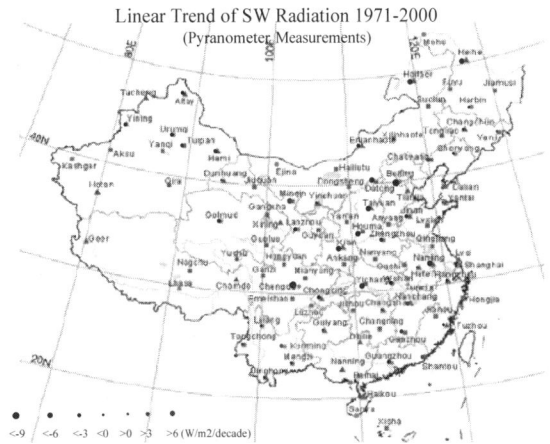

Fig. 1. Linear trend in surface SW radiation over China, 1971–2000.

4. DISCUSSION

As presented by the previous section, there is a decreasing trend of surface SW radiation during the period 1971–2000. Major meteorological factors related to the surface SW radiations are cloud amount, cloud optical depth, water vapor and aerosols.

It is pointed out by Kaiser (1998) that a decreasing trend in total cloud amount was observed over much of China for the period of 1951–1994. The decreasing rate is 1–3% sky cover per decade with a statistical significance. The recent decreasing trend of total cloud amount in this region is also found with seasonal variation from ISCCP D2 data as shown by Fig. 2 although the period is different from Kaiser's analysis. According to Ding et al. (2004), this decreasing trend was found for high, middle and low clouds. On the other hand, it is shown in Fig. 2 that an irregular variability of cloud optical thickness. The cloud optical thickness appears to increase particularly after 1996. The change of 0.5~1.0 in optical thickness gives rise to 10~30 W/m2 daily average SW radiative flux at the ground surface in this region. Therefore the effect of decreasing trend of cloud amount on the surface SW radiation might be compensated by changes in cloud optical thickness.

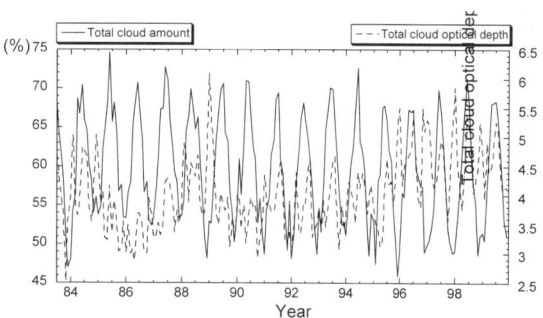

Fig. 2. Variations of total cloud amount (solid line) and cloud optical thickness (broken line) averaged over China derived.

In addition to the changes in cloud properties, increasing aerosol loading is observed in most of China. Luo et al. (2001) estimated optical thickness at 0.75 μm from 1960 to 1990 by applying the method of Qiu (1998) to the pyrheliometer data. They found the increase in aerosol optical thickness particularly for the period 1960-1980. The aerosol direct effect, i.e., scattering and absorption of solar radiation by aerosol particles, is also one of the reasons for decrease in the surface SW radiation. Although surface water vapor pressure tends to increase over China for 1954–1994, its effect on the SW radiation seems to be small (Kaiser, 2000).

5. SUMMARY

Downward shortwave radiative flux on the surface in China were evaluated by using pyranometer data, products derived from satellite cloud data and calculated by parameterization with ground-based meteorological data such as sunshine duration and water vapor. These radiation data are in general consistent with each other for monthly average values although the satellite derived data have positive bias for the big city areas whereas negative bias for the desert area in the west part of China. One of the reasons for these biases is ascribed to inappropriate assumption of aerosols in the satellite derived data. As for long-term monitoring of radiation, the surface SW radiation tends to decrease for the period 1971–2000 although a synthetic analysis should be discussed because the satellite data are limited after 1980s and long-term pyranometer measurements have some difficulties to evaluate those qualities.

6. ACKNOWLEDGEMENTS

We would like to thank Drs. P. W. Stackhouse Jr. and S. Gupta for providing surface radiation data set calculated from ISCCP data.

7. REFERENCES

Ding, S., G. Shi, and C. Zhao, 2004: Analyzing global trends of different cloud types and their potential impacts on climate by using the ISCCP D2 dataset. *Chinese Sci. Bull.*, 49, 1301-1306.

Gupta, S. K., N. A. Ritchey, A. C. Wilber, C. H. Whitlock, G. G. Gibson, and P. W. Stackhouse Jr., 1999: A climatology of surface radiation budget derived from satellite data. *J. Climate*, 12, 2691-2710.

Iwabuchi, H., and T. Hayasaka, 2002: Effects of cloud horizontal inhomogeneity on the optical thickness retrieved from moderate-resolution satellite data. *J. Atmos. Sci.*, 59, 2227-2242.

Kaiser, D. P., 1998: Analysis of total cloud amount over China, 1951–1994. *Geophys. Res. Lett.*, 25, 3599-3602.

Kaiser, D. P., 2000: Decreasing cloudiness over China: An updated analysis examining additional variables. *Geophys. Res. Lett.*, 27, 2193-2196.

Luo, Y., D. Lu, X. Zhou, and W. Li, 2001: Characteristics of the spatial distribution and yearly variation of aerosol optical depth over China in last 30 years. *J. Geophys. Res.*, 106, 14501-14513.

Pinker, R. T., and I. Laszlo, 1992: Modeling surface solar irradiance for satellite applications on a global scale. *J. Appl. Meteor.*, 31, 194-211.

Qiu, J., 1998: A method to determine atmospheric aerosol optical depth using total direct solar radiation. *J. Atmos. Sci.*, 55, 734-758.

Rossow, W. B., and R. Schiffer, 1999: Advances in understanding clouds from ISCCP. *Bull. Amer. Meteor. Soc.*, 80, 2261-2287.

Xu, J., T. Hayasaka, K. Kawamoto, and S. Haginoya, 2005: An estimation of downward surface radiation over China. *J. Meteor. Soc. Japan* (in press).

Younes, S., R. Claywell, and T. Muneer, 2004: Quality control of solar radiation data: Present status and proposed new approaches. *Energy*, 30, 1553-1549.

Zhang, Y.-C., W. B. Rossow, A. A. Lacis, V. Oinas, and M. I. Mishenko, 2004: Calculation of radiative fluxes from the surface to top of atmosphere based on ISCCP and other global data sets: Refinements of the radiative transfer model and the input data. *J. Geophys. Res.*, 109, D19105, doi:10. 1029/2003JD004457.

AEROSOL OPTICAL/RADIATIVE FORCING PROPERTIES OVER EAST ASIA DETERMINED FROM SKYNET RADIATION MEASUREMENTS

B. J. Sohn and Do-Hyeong Kim
School of Earth and Environmental Sciences, Seoul National University, Seoul, Korea

Teruyuki Nakajima
Center for Climate System Research, University of Tokyo, Tokyo, Japan

Tamio Takamura
Center for Environmental Remote Sensing, Chiba University, Chiba, Japan

ABSTRACT

From the analysis of SKYNET measurments, it was found that aerosols in East Asia have smaller single scattering albedos (i.e., 0.89 for Asian dusts in Dunhuang, 0.9 for urban type aerosols in Yinchuan, and 0.88 for biomass burning aerosols in Sri-Samrong), compared to the single scattering albedo for the same type of aerosols found in other areas, suggesting that aerosols over East Asia absorb comparatively more solar radiation.

1. INTRODUCTION

Despite the importance of the Asian continent as an aerosol source region there have been few measurements, especially in the desert area of north China. Moreover, due to the large spatiotemporal variations of aerosols, long-term monitoring near the source regions or their downwind areas is needed. Focusing on long-term monitoring of aerosols and an assessment of aerosol impact on the climate system over East Asia, a ground-based aerosol/radiation observation network, named the Skyradiometer Network (SKYNET), has been operating since 1997. Instruments used for observing surface solar radiation and aerosol characteristics include skyradiometer, pyranometer, and pyrheliometer.

In this study, we investigate the magnitude and varying degrees of aerosol influence on the solar radiation budget at the surface and TOA over East Asia using surface solar radiation measurements taken at SKYNET sites. Effort will also be made to interpret the relationship of these measurements to retrieved aerosol optical parameters. To this end, we developed a method of estimating ARF using retrieved aerosol optical parameters from skyradiometer measurements. In doing so, adjusted imaginary parts of refractive indices (and then single scattering albedos) estimated from the diffuse/direct method, originally introduced by King and Herman (1979) and extended by Nakajima et al. (1996), were used as inputs into the exact radiative transfer (RT) model instead of using direct retrievals from skyradiometer measurements.

2. OBSERVATION AND ANALYSIS

Direct and diffuse solar radiation were measured using a sky radiometer (POM-01L, Prede Co. Ltd.) in daytime at seven wavelengths of 315, 400, 500, 675, 870, 940, and 1020 nm. Aerosol optical thickness, single-scattering albedo at five wavelengths (400, 500, 675, 870, and 1020 nm), Ångström exponent, and volume size distribution [$dV/d\ln r$ (cm^{3}cm^{-2})] were retrieved using an inversion software (i.e., SKYRAD.pack version 3) developed by Nakajima et al. (1996). A detailed explanation about quality control, retrieval methods, and sensitivity analysis can be found in Kim et al. (2004).

For the radiative forcing calculation, surface solar radiation measurements in the wavelength band between 0.3 and 4.0 μm were used in conjunction with aerosol optical properties obtained from the skyradiometer measurements. Direct solar fluxes were measured by the pyrheliometer, which measures the intensity of a radiant beam at normal incidence coming only from the solar disk with about 5 degrees of the full opening view angle. A shaded pyranometer was used to measure diffuse solar radiation, as both the pyranometer and the shading disk were mounted on an automated solar tracker to ensure that the pyranometer was continuously shaded. Then global downwelling solar fluxes were determined from direct solar radiation measured by pyrheliometer multiplied by the cosine of the solar zenith angle plus diffuse radiation by the shaded pyranometer. Although global downward solar fluxes were also measured by the pyranometer, the pyranometer tends to be more uncertain when the solar zenith angle is high because the detector responds differently to the solar incident angle (Satheesh et al., 1999). Thus, the global fluxes determined from both pyrheliometer and shaded pyranometer were preferable to those measured by the pyranometer.

Measurements were taken at seven sites, given in Fig.1. Among them, four observation sites, i.e., Mandalgovi, Dunhuang, Yinchuan, and Sri-Samrong, form the SKYNET, while the Anmyon, Gosan sites of Korea, and the Amami-Oshima site of Japan are for the special radiation measurements made during the spring. Results are shown only for the SKYNET sites.

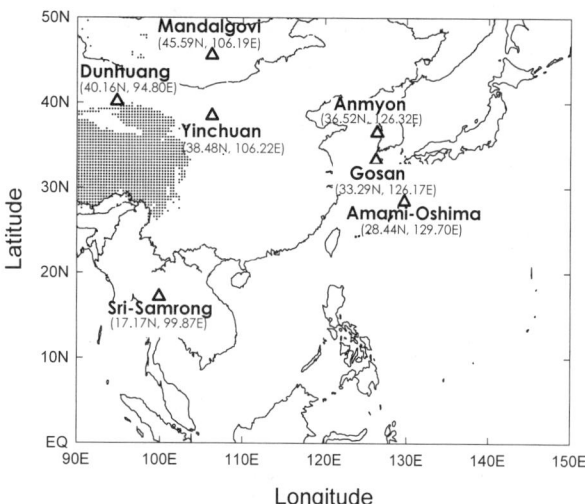

Fig. 1. Geographical locations of observation sites.

3. METHODOLOGY FOR RADIATIVE FORCING CALCULATION

For the radiative forcing calculation, we have developed an algorithm, adjusting aerosol parameters to bring in calculated values of diffuse and direct solar radiation close to observed values. Retrieval procedures used for the calculation of aerosol radiative forcing from ground-based solar radiation measurements are provided in Fig. 2. As illustrated at the top of Fig. 2, aerosol optical parameters and atmospheric profiles are used as inputs into a RT model for simulating surface solar fluxes. For the flux calculation, we use the real part of refractive index retrieved from sky radiance measurements. Thus values vary with space and time, ranging from 1.45 to 1.70. During the retrieval, the columnar water vapor amount and the imaginary part of the refractive index (hereafter absorption index) are obtained by minimizing the difference between the calculated and measured fluxes, as suggested in King and Herman (1979), and Nakajima et al. (1996). In doing so, the absolute accuracies of the pyrheliometer and pyranometer are used as the convergence criteria (ε) in Fig. 2. The adjusted absorption index and columnar water vapor amount are then used for the forcing calculation.

4. AEROSOL RADIATIVE FORCING OVER EAST ASIA

For an ARF calculation, simultaneous measurements of sky radiation and diffuse/direct solar radiation flux at the surface are required. Because of limited match-ups of sky radiation vs. solar flux data due to the inhomogeneous data coverage, the ARF coverage is also limited. Therefore, ARF is given with the number of match-ups used for the forcing calculation in each month—see the left panels of Fig. 3. The middle panels of Fig. 3 display the 24-hour averaged surface ARFs (Wm^{-2}) given with AOT at 0.5 µm at Dunhuang, Yinchuan, and Sri-Samrong. The match-ups available for Mandalgovi were not enough to represent aerosol forcing characteristics in that area, so that results obtained from measurements for Mandalgovi are not presented.

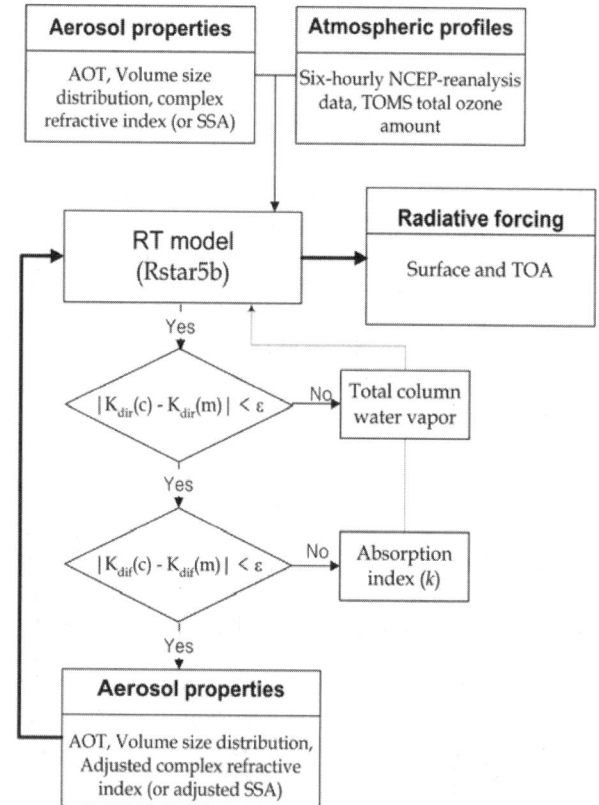

Fig. 2. A schematic diagram for the procedures used in aerosol radiative forcing calculation from skyradiometer and surface solar flux measurements. K_{dir} and K_{dif} represent direct and diffuse fluxes, c and m in parentheses represent calculated and measured values, respectively.

The relatively small number of available data for each month was largely due to the cloud contamination. However, it was also contributed by the employed assumption that the SSA from sky radiation measurents should be consistent with the adjusted SSA, allowing only a small adjustment to satisfy the measured surface radiation fluxes. The retrieved SSA from the flux method was compared against the one from sky radiation measurements only, and then the retrieval was discarded if the difference was greater than 0.05. In Fig. 3, calculated ARFs (represented by cross hairs) are given as a function of optical thickness at 0.5 µm. Also provided are theoretically calculated SSAs (represented by curves) at 0.5 µm, which range from 1.0 (top curve) to 0.6 (bottom curve) with a 0.05 interval. For the simpler demonstration using the reduced number of parameters, theoretical SSAs at each site were calculated using mean values of aerosol volume size distribution and surface albedo, but with varied optical thickness. Because of this, some points appear to have unrealistic SSAs greater than unity. Also given in the middle panels of Fig. 3 are the mean SSA and g. In addition, mean β at the TOA, atmosphere, and surface are provided in the right panels of Fig. 3.

As expected Dunhuang is influenced mainly by dust particles all year around. In Yinchuan, the general aerosol characteristics are related to the accumulation processes of urban aerosols throughout the year, and dusts during the springtime. Sri-Samrong showed patterns for biomass burning aerosols during mid-October to mid-February. The results obtained in this study over those three sites will be interpreted with the aid of typical aerosol types found at those observation sites. Details of aerosol characterization and their seasonal variations for those sites are found in Kim et al. (2004).

Because Dunhuang is largely affected by frequent Asian dust episodes during the spring to early summer period, ARFs for Dunhuang found in this study are largely due to Asian dust aerosols. The mean SSA and g are 0.89 and 0.71, respectively. The SSA obtained in this study at 0.5 μm is slightly smaller than 0.92~0.93 at 0.44 μm found over the Saharan desert or Saudi Arabia (Dubovik et al., 2002a), implying that Asian dusts are more absorbing aerosol. On the other hand, the obtained g (≈0.71) is close to the values found in the Saharan desert (≈0.72) or Saudi Arabia desert (≈0.68) (Dubovik et al., 2002).

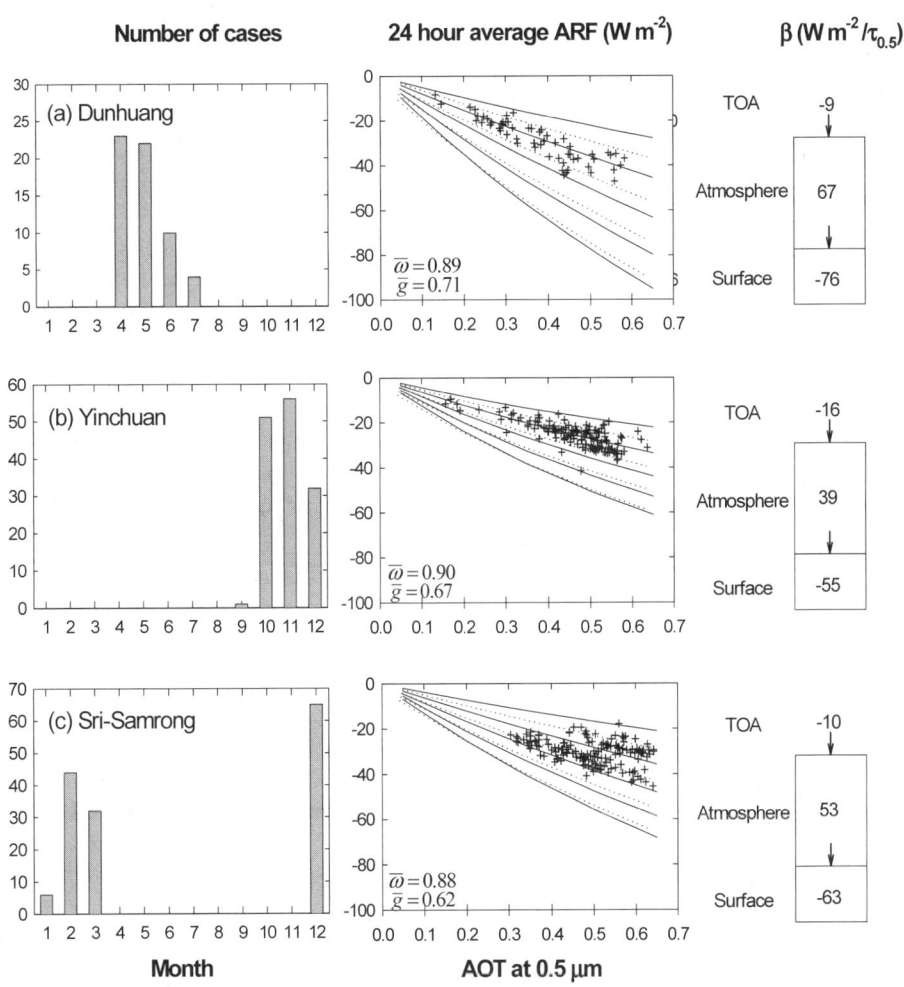

Fig. 3. Number of cases used in each month (left), 24-hour averaged surface radiative forcing (Wm^{-2}) with AOT at 0.5 μm (middle), and forcing efficiency (β) given in Wm^{-2} (right) at three stations. Curves represent the theoretical radiative forcing calculated with SSAs of 1.0 (top) to 0.6 (bottom), with a 0.05 interval. Cross hairs represent observed forcing. $\overline{\omega}$ and $\overline{g}$ represent mean single scattering albedo and asymmetry factor at 0.5 μm, respectively.

The resultant forcing efficiency in the atmosphere is about 67 Wm^{-2}, suggesting that Asian dusts over the Dunhuang area absorb a significant amount of solar radiation. Combining with the relatively smaller forcing efficiency at the TOA (≈ -9 Wm^{-2}), the overall surface forcing efficiency is estimated to be -76 Wm^{-2}, indicating that the surface is deprived of a substantial amount of solar energy in the presence of Asian dusts. The smaller TOA forcing efficiency seems to be associated with high surface reflectivity over the Dunhuang area, since the negative TOA forcing can be reduced by the aerosol absorption of surface-reflected upwelling radiation, in particular under high surface albedo conditions, as noted in Sahara dust studies. More discussion on the influences of Asian dusts on the ARF is found in Kim et al. (2004).

Yinchuan represents the urban type aerosol because the match-up data only cover the September-December period. The obtained SSA of around 0.9 is similar to the SSA (0.87 to 0.90) found over the Indian Ocean during the INDOEX period (Satheesh and Ramanathan, 2000). However, 0.9 of SSA in Yinchuan is much smaller than 0.98 at 0.44 μm for urban type aerosols observed at the NASA Goddard Space Flight Center (GSFC) (Dubovik et al., 2002). This variability of SSA for urban type aerosols may be due to differences in fuel types and emission conditions. Automobile traffic should be the main local source of the pollution around the GSFC. On the other hand, the highly absorbing optical properties in the INDOEX are mainly due to the presence of soot type aerosols from inefficient fossil fuel combustion for heating and cooking, and biomass burning (Satheesh et al., 1999). Considering that aerosols from inefficient fossil fuel combustion are also abundant over China (Chameides et al., 1999), we suspect that the small SSA value in Yinchuan is also due to the contributions of soot type aerosols to SSA.

5. ACKNOWLEDGEMENTS

This research has been supported by ADD/KAIST of Korea.

6. REFERENCES

Chameides, W. L., H. Yu, S. C. Liu, M. Bergin, X. Zhou, L. Mearns, G. Wang, C. S. Kiang, R. D. Saylor, C. Luo, Y. Huang, A. Steiner, and F. Giorgi, 1999: Case study of the effects of atmospheric aerosols and regional haze on agriculture: An opportunity to enhance crop yields in China through emission controls, *Proceedings of the National Academy of Sciences*, *96*, 13,626-13,633.

Dubovik, O., A. Smirnov, B. N. Holben, M. D. King, Y. J. Kaufman, T. F. Eck, and I. Slutsker, 2000: Accuracy assessment of aerosol optical properties retrieved from Aerosol Robostic Network (AERONET) sun and sky radiance measurements, *J. Geophys. Res.*, *105*, 9791-9806.

Dubovik, O., B. N. Holben, T. F. Eck, A. Smirnov, Y. J. Kaufman, M. D. King, D. Tanre, and I. Slutsker, 2002: Variability of absorption and optical properties of key aerosol types observed in worldwide locations, *J. Atmos. Sci.*, *59*, 590-608.

Eck, T. F., B. N. Holben, I. Slutsker, and A. Setzer, 1998, Measurements of irradiance attenuation and estimation of aerosol single scattering albedo for biomass burning aerosols in Amazonia, *J. Geophys. Res.*, *103*, 31,865-31,878.

Kim, D. H., B. J. Sohn, T. Nakajima, T. Takamura, T. Takemura, B. C. Choi, and S. C. Yoon, 2004, Aerosol optical properties over East Asia determined from ground-based sky radiation measurements, *J. Geophys. Res. 109*, D02209, doi:10.1029/2003 JD003387.

King, M. D., and B. M. Herman, 1979: Determination of the ground albedo and the index of absorption of atmospheric particulates by remote sensing. Part I: Theory, *J. Atmos. Sci.*, *36*, 163-173.

Nakajima, T., T. Hayasaka, A. Higurashi, G. Hashida, N. Moharram-Nejad, Y. Najafi, and H. Valavi, 1996: Aerosol optical properties in the Iranian region obtained by ground-based solar radiation measurements in the summer of 1991, *J. Appl. Meteor.*, *35*, 1265-1278.

Ramanathan, V., and Co-authors, 2001: Indian Ocean Experiment: An integrated analysis of the climate forcing and effects of the great Indo-Asian haze, *J. Geophys. Res.*, *106*, 28,371-28,398.

Satheesh, S. K., V. Ramanathan, Xu Li-Jones, J. M. Lobert, I. A. Podgorny, J. M. Prospero, B. N. Holben, and N. G. Loeb, 1999: A model for the natural and anthropogenic aerosols over the tropical Indian Ocean derived from Indian Ocean Experiment data, *J. Geophys. Res.*, *104*, 27,421-27,440.

Satheesh, S. K., and V. Ramanathan, 2000: Large differences in the tropical aerosol forcing at the top of the atmosphere and Earth's surface, *Nature, 405*, 60-63.

Tanre, D., Y. J. Kaufman, B. N. Holben, B. Chatenet, A. Karnieli, F. Lavenu, L. Blarel, O. Dubovik, L. A. Remer, and A. Sminov, 2001: Climatology of dust aerosol size distribution and optical properties derived from remotely sensed data in the solar spectrum, *J. Geophys. Res.*, *106*, 18,205-18,217.

AVHRR OBSERVATIONS OF THE AEROSOL INDIRECT EFFECT FOR SUMMERTIME STRATIFORM CLOUDS IN THE NORTHEASTERN ATLANTIC

Mark A. Matheson, James A. Coakley, Jr., and William R. Tahnk
Oregon State University
Corvallis, OR, 97333, USA

1. INTRODUCTION

Human activity has increased the concentration of aerosols in the atmosphere. Aerosols interact directly with solar radiation (aerosol direct radiative forcing) as well as modify the optical properties of clouds causing changes to the Earth's radiative energy budget (aerosol indirect radiative forcing). Indirect forcing is the largest source of uncertainty in assessing the anthropogenic forcing of climate.

An increase in the number of aerosols is likely to lead to an increase in the number of droplets activated at the time of cloud formation. The simplest assumption is that the amount of liquid water at the time of formation is not influenced by the number of condensation nuclei so that the same amount of liquid water is distributed over more cloud droplets, making them smaller. This change in droplet size will cause the cloud to be more reflective and thus have a cooling effect. This process is known as the "Twomey effect" (Twomey, 1974).

The Twomey effect reduces the mean cloud droplet size and increases the droplet number concentration. These changes in turn affect the ability of droplets to coalesce to precipitable size. Drizzle suppression may increase the lifetime of clouds affected by increases in aerosol concentration. An increase in cloud lifetime could lead to an increase in cloud cover, which would also be a cooling effect (Albrecht, 1989).

The goal of the current research is to use satellite radiances to empirically measure the changes in cloud radiative and microphysical properties related to changes in aerosol loading.

Other studies (Sekiguchi et al., 2003 and references therein) have also used satellite retrievals of clouds and associated aerosols to investigate the aerosol indirect effect. In this study, associated observations of aerosols and clouds are limited to small geographic regions and to the same satellite overpass. These conditions of simultaneity and collocation avoid the possibility of aerosols occurring on one day in one location being compared with cloud properties on another day or in another location as was possible in some of the earlier studies. The current study provides independent confirmation of some of the findings of earlier results at higher spatial resolution and also investigates how relationships between cloud properties and fractional cloud cover may bias the interpretation of cloud-aerosol interaction.

2. DATA

Global Area Coverage (GAC) radiances, with a nominal resolution of 4 km, measured by the Advanced Very High Resolution Radiometer (AVHRR) for the months of May through August of 1995 are analyzed. All daytime satellite overpasses from the four months are used.

The analysis area chosen for this study is in the Northeastern Atlantic bounded by 35°–55° N latitude and 20° W–0° longitude. Radiances from over land are not used. The study region contains both coastal and open ocean regions. In addition, the meteorological conditions during the four months include periods of both sustained onshore and offshore flow. These geographic and meteorological conditions allow for the analysis of a wide range of aerosol burdens and their effects on clouds.

The radiative transfer equations used in the processing of satellite radiances are strictly valid only for homogenous parallel layers of clouds. For this reason, the analysis is limited to single-layer, low-level clouds, which are also the clouds most likely to be affected by cloud-aerosol interaction. 1° × 1° latitude-longitude regions that contain data that are not from a single-layer, low-level cloud system or from cloud-free ocean are excluded from the analysis.

0.55-μm aerosol optical depth is retrieved in imager pixels identified as cloud free (Coakley et al., 2002). In pixels identified as containing clouds, the following cloud properties are retrieved: droplet effective radius, 0.64-μm cloud optical depth, cloud-top altitude, and fractional cloud cover (Coakley et al., 2005).

3. INFLUENCE OF AEROSOLS ON CLOUDS

Satellite data for individual cloud and aerosol properties are averaged over all satellite passes. The air in the Northeastern Atlantic is more polluted near the coast and cleaner over the ocean (Fig. 1). Similarly, the droplet effective radii of the clouds are small near the coast and large over the open ocean (Fig. 2). In the southern part of the study region, aerosol optical depth decreases and droplet effective radius increases with distance from land, trends which are qualitatively consistent with the Twomey effect.

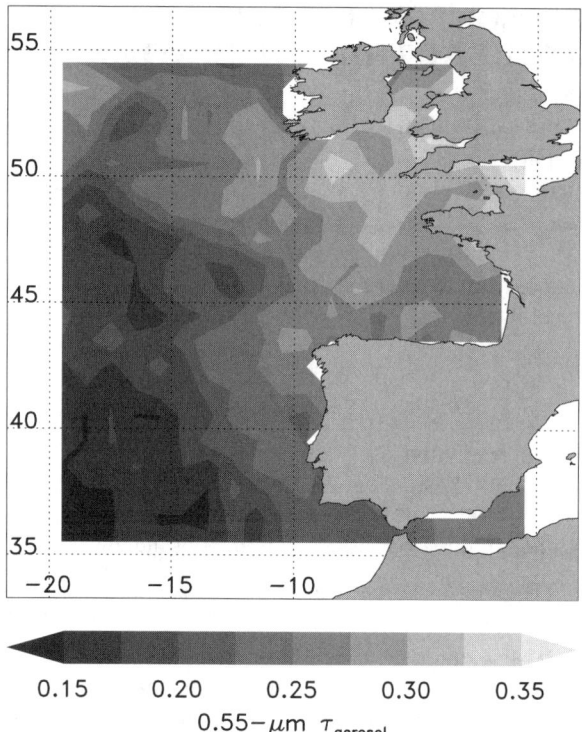

Fig. 1. Summertime climatology of 0.55-μm aerosol optical depth.

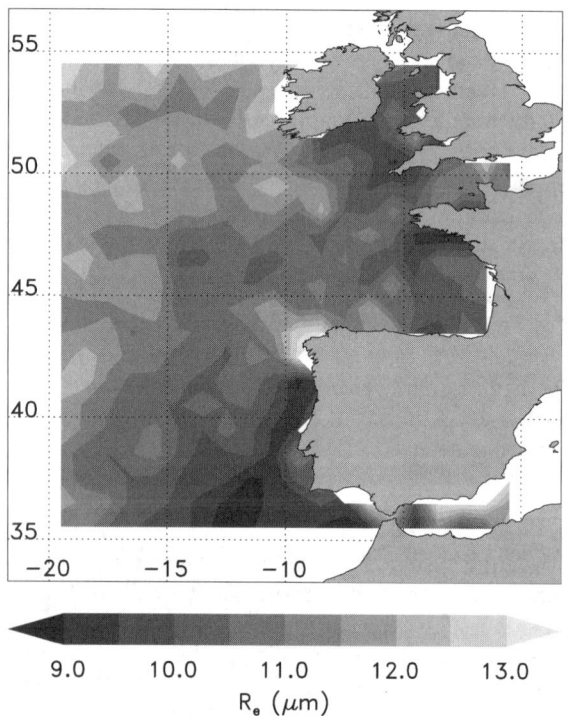

Fig. 2. Summertime climatology of droplet effective radius (μm) for low-level, single-layered cloud systems.

The mean aerosol optical depths in 1° × 1° latitude-longitude regions are associated with mean cloud properties in the same region on the same day. It is presumed that aerosol properties in the clear portions of the region are well correlated with the aerosol properties in the cloudy portions of the same region on the same day.

All data are divided into 5° × 5° latitude-longitude regions and displayed simultaneously on the plots in their appropriate geographic location (Figs. 3–6). Plots for regions that contain land use data only from the pixels identified as being over ocean. Limiting the data to small regions ensures that cloud and aerosol properties are restricted to similar geographic and meteorological settings.

Error bars represent the standard error, given by the standard deviation of the daily means contributing to that bin divided by the square root of the number of days contributing to that bin. A linear fit, inversely weighted by the standard error, is plotted and the slope is reported for each 5° region.

Fig. 3 shows the relationship between droplet effective radius and aerosol optical depth. In almost all 5° regions there is a decrease in droplet size as aerosol burden increases. The statistical significance of these slopes in some coastal regions, however, is small.

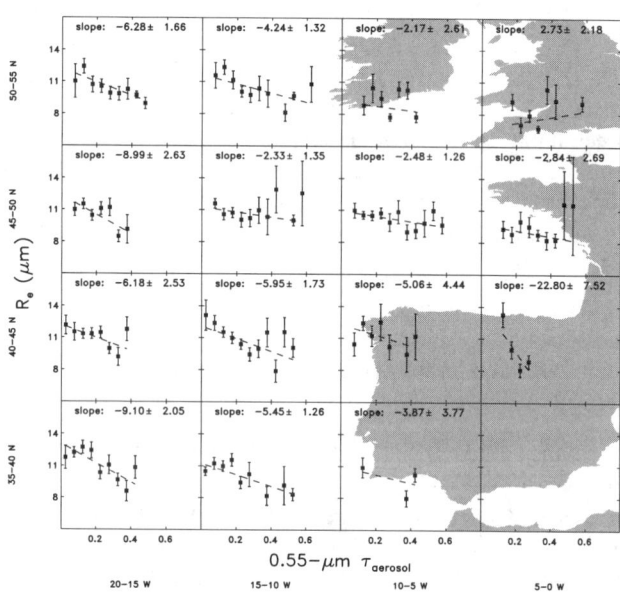

Fig. 3. Mean 0.55-μm aerosol optical depth and associated mean cloud droplet effective radius (μm) in the same 1° × 1° latitude-longitude region. Each of the sub-panels contains data from that 5° × 5° latitude-longitude region.

RADIATIVE BUDGET AND FORCING

Fig. 4 shows the relationship between cloud optical depth and aerosol optical depth. Regions in the west show an increase in cloud optical thickness as aerosol burden increases. However, coastal regions show no trend or possibly even negative trends. An increase in cloud optical depth and a decrease in droplet effective radius implies an increase in droplet number concentration as aerosol optical depth increases, consistent with the Twomey effect.

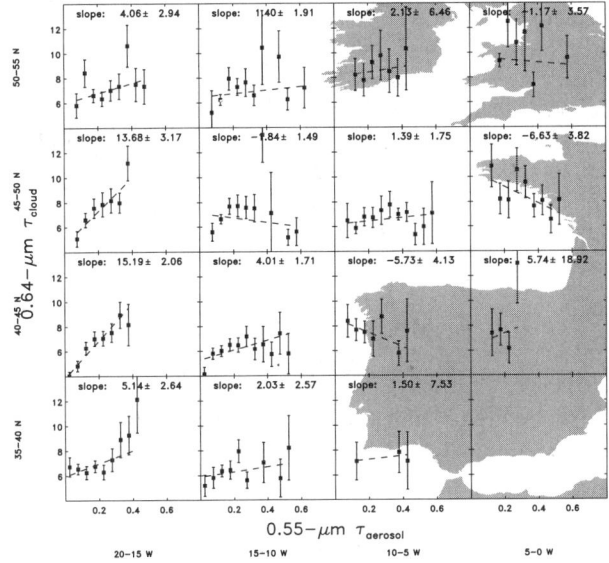

Fig. 4. Same as Fig. 3, but for cloud optical depth.

Clouds in some regions are gaining liquid water as aerosol burden increases (Fig. 5). The gain in liquid water is consistent with drizzle suppression and the cloud lifetime effect. In some regions, however, clouds have less liquid water when there is high aerosol burden when compared to cases where there is low aerosol burden.

If an increase in aerosol burden leads to increased atmospheric heating, it may hinder cloud formation, a phenomenon referred to as the aerosol semi-direct effect. It is possible that the increase in aerosol burden near the coast may be causing broken clouds to evaporate more quickly, thus leading to a decrease in cloud optical depth and cloud liquid water as aerosol burden increases.

An alternate explanation for the trends of decreased cloud optical depth and cloud liquid water as aerosol burden increases posits that higher aerosol burdens are associated with incursions of dryer air. Perhaps, high aerosol situations are associated with air originating over the continent, which is likely to be dryer. Clouds that form in this air are starved for water and quickly evaporate. Clouds that formed in less polluted conditions may have formed in moister, oceanic air with higher humidity and are not as likely to dry out (Ackerman et al., 2004).

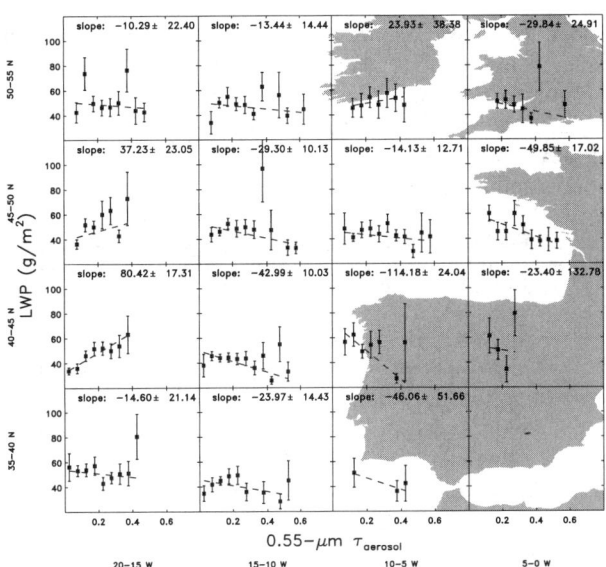

Fig. 5. Same as Fig. 3, but for liquid water path (g/m^2).

4. INTRINSIC CLOUD PROPERTIES

As cloud cover increases within a geographic region, the cloud properties retrieved by satellite change in predictable ways: cloud optical depth, cloud droplet effective radius, and liquid water path all get larger as cloud cover increases (not shown). Here, the fraction of pixels containing clouds within the 1° × 1° regions is used as a measure of cloud cover.

In some regions, aerosol optical depth appears to increase as the cloud field becomes more extensive (Fig. 6). This trend indicates that in clear areas near clouds, the aerosol burden appears to increase. The apparent increase in aerosol burden could be caused by cloud contamination. A single AVHRR GAC pixel is representative of a region that is approximately 3 × 5 km at nadir. It is possible that pixels identified as being cloud-free actually contain sub-pixel resolution clouds. These unidentified clouds will cause the retrieved aerosol optical depths to be erroneously high. Presumably, sub-pixel scale clouds will be more common in the presence of detectable clouds, explaining the trend toward larger optical depth as the number of cloudy pixels increases within a 1° region. Better algorithms for detecting sub-pixel scale resolution clouds and higher spatial resolution would both help alleviate this source of error.

The apparent increase in aerosol burden with increasing cloud cover could also be caused by the swelling of aerosol particles in the vicinity of clouds. Clouds form in environments with high relative humidity and aerosol particles swell as relative humidity increases. If the increase in aerosol optical depth is due entirely to the swelling of aerosol particles, then there may not be an increase in CCN number as the aerosol optical depth is increasing, so the change in cloud properties due to increased aerosol optical depth is not, in fact, the aerosol indirect effect as it is currently understood. There are currently no satellite data products that accurately report aerosol size. Resolution of this issue may depend on continued *in situ* observations and future satellite instruments such as the CALIPSO lidar.

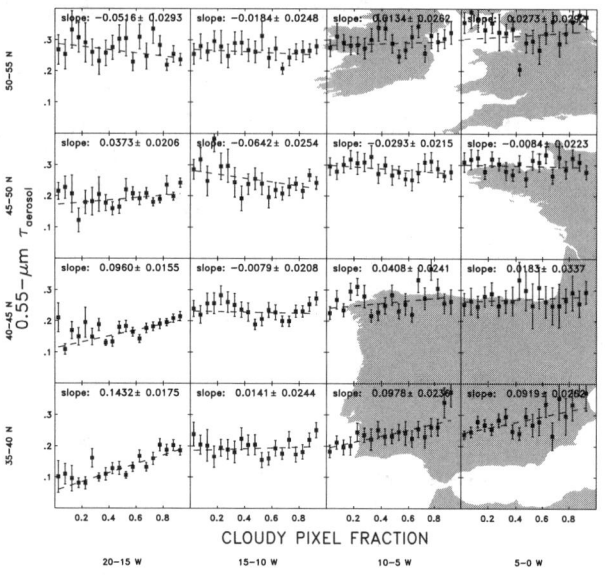

Fig. 6. Fraction of pixels in a 1° × 1° latitude-longitude region identified as containing clouds and the mean 0.55-μm aerosol optical depth in that same square. Each of the sub-panels contains data from that 5° × 5° square.

5. RADIATIVE FORCING ESTIMATES

The influence of increased aerosol burden on the region's radiation balance is investigated by using the measured cloud and aerosol properties as inputs to a broadband radiative transfer model. The broadband model accounts for Rayleigh scattering, gaseous absorption, aerosols, and clouds (Coakley et al., 2002). The aerosol optical depth is arbitrarily changed from 0.15 to 0.25. The values of the cloud optical depth are calculated from the trends shown on Fig. 4, but accumulated for the entire 20° region. For an increase in aerosol optical depth of 0.1, direct radiative forcing for cloud-free oceans is calculated to be -3.9 W/m^2. Indirect radiative forcing for overcast oceans is 72% larger at -6.7 W/m^2.

The above calculations are repeated using the observed change in droplet effective radius combined with an assumption of fixed liquid water to calculate a new cloud optical depth. With fixed liquid water, indirect aerosol radiative forcing is -7.1 W/m^2. Many climate model simulations show increased cloud water content with increased aerosols (e.g., Feichter et al. 2004) whereas the current study indicates a decrease in liquid water path (averaged for the entire 20° region). Because many climate models use calculated liquid water to derive cloud optical depth, these models may be overestimating aerosol indirect radiative forcing.

A profiling instrument, such as the lidar to be flown on the CALIPSO satellite would aid the current study as follows. A lidar could determine if observed clouds and aerosols are at the same altitude, and thus likely to be interacting. A space-borne lidar would also be able to better detect tenuous upper-level cirrus clouds that may be influencing the cloud and aerosol property retrievals. In addition a lidar in combination with a multi-spectral imager may be able to distinguish between sub-pixel scale cloud contamination and aerosol swelling near clouds in order to explain the increase in observed aerosol optical depth as cloud fraction increases (Fig. 6).

6. ACKNOWLEDGMENTS

This work was supported in part by NOAA's Office of Global Programs and NASA's CALIPSO project through the NASA Langley Research Center.

7. REFERENCES

Ackerman, A. S., M. P. Kirkpatrick, D. E. Stevens, and O. B. Toon, 2004: The impact of humidity above stratiform clouds on indirect aerosol climate forcing. *Nature*, 432, 1014-1017.

Albrecht, B. A., 1989: Aerosols, cloud microphysics, and fractional cloudiness. *Science*, 245, 1227-1230.

Coakley, J. A., Jr., W. R. Tahnk, P. K. Quinn, C. Devaux, and D. Tanre, 2002: Aerosol optical depths and direct radiative forcing for INDOEX derived from AVHRR: Theory. *J. Geophys. Res.*, 107, doi:10.1029/2000JD00182.

Coakley, J. A., Jr., M. A. Friedman, and W. R. Tahnk, 2005: Retrieval of cloud properties for partly cloudy imager pixels. *J. of Atmos. and Ocean. Tech.*, in press.

Feichter, J., E. Roeckner, U. Lohmann, and B. Liepert, 2004: Nonlinear aspects of the climate response to greenhouse gas and aerosol forcing. *J. Climate*, 17, 2384-2398.

Sekiguchi, M., T. Nakajima, K. Suzuki, K. Kawamoto, A. Higurashi, D. Rosenfeld, I. Sano, and S. Mukai, 2003: A study of the direct and indirect effects of aerosols using global satellite data sets of aerosol and cloud parameters. *J. Geophys. Res.*, 108, doi:10.1029/2002JD003359.

Twomey, S., 1974: Pollution and the planetary albedo. *Atmos. Environ.*, 8, 1251-1256.

ESTIMATES OF THE DIRECT AND INDIRECT RADIATIVE FORCING OF CLIMATE BY ANTHROPOGENIC AEROSOLS

J. A. Coakley, Jr.
Oregon State University, Corvallis, OR 97331-5503

N. G. Loeb and T. P. Charlock
NASA Langley Research Center, Hampton, VA 23681-2199

A. S. Ackerman
NASA Ames Research Center, Moffett Field, CA 94035

Y. J. Kaufman
NASA Goddard Space Flight Center, Greenbelt, MD 20771

U. Lohmann
ETH Zürich CH-0893, Switzerland

ABSTRACT

Strategies for using multispectral satellite imagery data to determine both the aerosol direct and indirect radiative forcing for the combination of natural and anthropogenic aerosols are examined. Empirical assessments of both the direct and indirect forcing are hampered by an inability to distinguish between cloud-free and cloud-contaminated pixels in cloudy regions. Estimates of the indirect radiative forcing derived from correlations between aerosol and cloud properties are further confounded by spatial-scale dependent relationships among the cloud properties. In addition, the radiative heating of the atmosphere by aerosols has to be accounted for in any assessment of radiative forcing.

1. INTRODUCTION

As discussed in the most recent Intergovernmental Panel on Climate Change assessment (IPCC, 2001), the forcing of the climate by anthropogenic aerosols is highly uncertain. The effect of the aerosols is expected to counter somewhat that due to the buildup of the greenhouse gases, but whereas the optical properties for the greenhouse gases are well known, those for aerosols are not. Aerosol optical properties depend on composition, size, and shape which are highly variable, even from particle to particle within a haze layer from a single source. Likewise, whereas the greenhouse gases have long residence times in the atmosphere (10 to 10^2 years) and are uniformly distributed over the Earth, aerosols are subject to rainout and have short residence times. Aerosols are thus highly variable in both space and time, making their characterization seemingly impossible. The large uncertainties estimated for the aerosol forcing in IPCC (2001) reflect this difficulty, and they potentially pose major challenges for assessments of climate change.

Aerosol particles scatter and absorb incident sunlight. They scatter, absorb, and emit infrared radiation, but owing to their sizes, such interactions are generally, but not always, relatively small compared to their interactions with sunlight. These interactions constitute the aerosol direct radiative forcing. Some aerosol particles act as cloud condensation nuclei (CCN) and ice nuclei around which cloud droplets and ice crystals form. Such particles affect the numbers and sizes of cloud droplets and ice crystals, and consequently, the interaction of clouds with the radiation field. This modification of clouds leads to the aerosol indirect radiative forcing. While the scattering of sunlight by aerosols leads to an increase in the planetary albedo, and thus cooling, absorption by the aerosols heats the atmosphere and reduces the sunlight absorbed by the surface. Both processes could significantly alter the hydrologic cycle (e.g., Menon et al., 2002, among others).

For water clouds, as the number of haze particles increases, the number of CCN is expected to increase leading to clouds with more droplets, but the droplets are smaller. Twomey (1974) pointed out that if the amount of water in the clouds remained constant while droplet numbers increased and droplet radii decreased, then polluted clouds would reflect more sunlight than unpolluted clouds. This enhancement in reflectivity is commonly referred to as the "Twomey Effect." Others pointed out that the formation of precipitation would be suppressed in clouds with smaller droplets (Albrecht, 1989). Polluted clouds are therefore expected to retain more water in condensed form and to last longer than their unpolluted counterparts. As such, the aerosol indirect radiative forcing is also expected to counteract the forcing due to the buildup of the greenhouse gases, but its magnitude is completely uncertain and could be as large as the greenhouse gas forcing.

IPCC (2001) estimates of aerosol radiative forcing depended solely on climate model simulations. Here the capabilities of observational approaches to determining the total aerosol radiative forcing, the combination of the forcing for the natural and anthropogenic components, are surveyed to assess the possibility of using observations to constrain the model estimates. A major difficulty of the current observing techniques appears to be an inability to distinguish between thin, broken cloud, and haze. Errors in observational estimates of the direct aerosol radiative forcing may be comparable to the forcing now estimated for the human contribution, which is roughly one fourth of the sum of the natural and anthropogenic aerosol forcing. Strategies to observationally determine the indirect radiative forcing of aerosols have resorted to correlations of collocated aerosol and cloud properties within confined geographic regions. While such correlations might provide some insight on the aerosol indirect effect, owing to errors in the retrieved aerosol and cloud properties and to systematic changes in the cloud properties that appear to be functions of spatial scale, such correlations may easily be misinterpreted.

2. DIRECT RADIATIVE FORCING

Given the variability that aerosol particles exhibit in composition, shape, and size, modeling their effect on the incident sunlight and emitted infrared radiation would seem impossible. Nevertheless, dozens of case studies involving *in situ* observations collected with aircraft-borne instruments have demonstrated that the extinction of sunlight by the particles determined from measurements of chemical composition, numbers, and sizes can be characterized to within about 20% (e.g., Schmid et al., 2003). Extinction is undoubtedly easier to characterize than the amount of sunlight absorbed and the amount scattered; nonetheless, such agreement offers hope. Similarly, estimates of aerosol optical depth derived from reflected sunlight measured with MODIS on Aqua and Terra fall within the anticipated accuracies as demonstrated through direct comparison with surface-based observations (Remer et al., 2004). Uncertainties in the 0.55-μm optical depth appear to be constrained to the pre-launch estimates: $\Delta \tau = \pm 0.03 \pm 0.05\tau$ for oceans and $\Delta \tau = \pm 0.05 \pm 0.15\tau$ for land, with satellite estimates tending to underestimate the optical depths particularly when the optical depths are large. For maritime aerosols over oceans the forcing in cloud-free regions is approximately -35 Wm^{-2} for unit optical depth and for continental aerosols over land the cloud-free forcing is approximately -25 Wm^{-2} for unit optical depth. For typical global average optical depths of 0.1–0.2, the uncertainties in the retrieved optical depths suggest errors in the forcing may be of order 1 Wm^{-2} for cloud-free regions when considering all aerosols, not simply anthropogenic aerosols.

A problem that remains with such observations, however, is the degree of cloud contamination that enters into the remotely sensed aerosol products. In retrieving aerosol optical depth from satellite imagery data, only pixels identified as being cloud-free are used. This practice leads to an unintended bias. The pixels used for the aerosol retrievals are chosen because a scene identification scheme has identified them as being utterly free of cloud. The assumption is that aerosol concentrations are the same between the clouds in partially covered pixels. As shown in Fig. 1, however, aerosol optical depth in a region of ~25 km scale appears to increase as the fractional cloud cover within the region increases (Loeb and Manalo-Smith, 2004). The observations shown in the figure are from the CERES Single Satellite Footprint (SSF) product from the Terra satellite and cover the Earth, 60°S–60°N for DJF 2000–2001. Aerosol optical depths are shown for two retrieval schemes: the multi-channel scheme used in the MODIS MOD04 product, and a single-channel, NOAA-Like retrieval. Approximately half the difference in optical depth is due to differences in the retrieval schemes and half the difference is due to differences in cloud screening. Others have likewise observed a positive correlation between aerosol optical depth and cloud cover fraction on regional scales (Sekiguchi et al. 2003; Matheson et al. 2004).

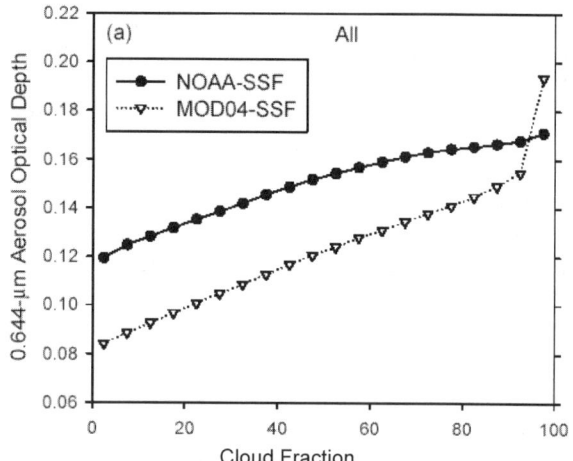

Fig. 1. **Aerosol optical depth and cloud cover fraction for CERES SSF derived using the multi-channel MOD04 aerosol retrievals and a NOAA-Like single-channel aerosol retrieval. The observations are for DJF 2000–2001 (Loeb and Manalo-Smith, 2004).**

Of course, the increase in aerosol optical depth with increasing cloud cover might arise for the following reasons: 1) aerosol particles grow and thus optical depth increases with increasing relative humidity and regional cloud cover also generally increases with increasing relative humidity; 2) haze in the vicinity of clouds may be subjected to the extra illumination from the cloud sides thereby creating the appearance of increased aerosol optical depth as cloud cover increases, 3) photochemical reactions within clouds are expected to increase the size of aerosol particles left when droplets evaporate, 4) as the fractional cloud cover within a region increases, pixels identified as being cloud-free are more likely to be contaminated by clouds, and 5) increasing aerosol burdens could lead to increasing cloud cover (Ackerman et al., 2003; Sekiguchi et al., 2003).

The dependence of the retrieved aerosol optical depth on the regional cloud cover suggests that differences in cloud screening techniques will lead to differences in aerosol optical depths, and consequently to different estimates of the aerosol direct radiative forcing. Given the apparent sensitivity of aerosol optical depth to cloud cover, differences in cloud screening could easily lead to differences in optical depth (~0.02) and thus give rise to order 0.7 Wm^{-2} differences in estimates of the cloud-free radiative forcing. Clearly, the behavior of aerosols in cloudy environments requires investigation.

Lidar observations combined with high resolution multispectral imagery, as will become available for the CALIPSO mission, should help to characterize aerosol properties in the vicinity of cloud systems (Winker et al. 2003). In addition, estimates of the direct forcing should be improved by integrating ground-based observations (e.g., Liepert, 2002) with satellite-based assessments. CERES has a pilot project combining the Model for Atmospheric Transport and Chemistry (Collins et al., 2001) with Baseline Surface Radiation Network radiometer data to from long-term, integrated satellite and surface observations of radiative fluxes and aerosol properties that, at least at regional and seasonal scales, should help to constrain the top-of-the atmosphere and surface aerosol direct radiative forcing.

3. INDIRECT RADIATIVE FORCING

The most common strategy used to determine the aerosol indirect radiative forcing is to correlate satellite-derived aerosol properties with properties retrieved for low-level clouds in geographic regions containing both aerosols and clouds. Ideally, the comparisons should be made between collocated aerosol and cloud properties (e.g., Kaufman and Fraser, 1997, among others), but time and space averages have been used in some studies (e.g., Sekiguchi et al., 2003, among others). Clearly, with time and space averaging the interpretation of the correlations becomes dubious. Clouds on one day or one region might be compared with aerosols on another day in another region. In the correlation studies, clouds which show a decrease in droplet radii and an increase in column droplet number concentrations with increasing aerosol burden are taken to be affected by the nearby aerosols. In most of the studies, such clouds exhibit an increase in visible optical depth that is consistent with either constant cloud liquid water (Kaufman and Fraser, 1997, among others) or an increase in cloud liquid water (Sekiguchi et al., 2003, among others) thus supporting to some extent the results of GCM modeling studies (Lohmann and Lesins, 2003).

Fig. 2 shows cloud droplet radius, visible optical depth, and column cloud liquid water amount derived for low-level marine stratus in the North Atlantic observed in the summer of 1995. The results were obtained using NOAA-14, 4-km Advanced Very High Resolution Radiometer (AVHRR) data (Matheson et al., 2004). The results in the figure are composites of aerosol and cloud properties collocated within $1° \times 1°$ latitude-longitude regions. Composites are accumulated for $5° \times 5°$ latitude-longitude regions. For maritime regions, where the atmosphere is moist, the clouds show the trends in droplet radius, visible optical depth, and column water amount found in the other studies. The results in Figs. 1 and 2 suggest that since aerosol optical depth and cloud fraction are positively correlated, cloud optical depth should also be positively correlated with cloud fraction, which indeed, it is. The positive correlation of cloud cover and cloud optical depth, on the other hand, could result from the thickening of clouds in regions that become heavily cloud covered. What parts of the relationships shown in Fig. 2 are due to the indirect effect of the aerosols and what parts are due to an apparent link between cloud horizontal scales and other cloud properties will be difficult to ascertain.

Fig. 2 also shows that near the European continent, where the maritime atmosphere is dried through intrusions of continental air, the cloud liquid water is diminished as aerosol burdens increase. The change in the optical depth, and consequently, cloud reflectance is correspondingly reduced. These results demonstrate the dangers of the simplistic use of correlations to determine the aerosol indirect radiative forcing. Not only are the clouds responding to the particles, but they are also responding to the environment in which the haze is imbedded. Regions that appear to have repeatable meteorological conditions despite the effects of haze might serve as natural laboratories for studying the effects of aerosols on clouds. Koren et al. (2004) claim that the Amazon basin may provide an example of such a laboratory.

The loss of liquid water for polluted clouds in the coastal region shown in Fig. 2 has also been found in studies of ship tracks off the west coast of the U.S. In the case of the ship tracks, polluted and unpolluted clouds separated by less than 10 km can be compared, thereby ensuring that both the polluted and unpolluted clouds are subject to the same thermodynamic environment. The loss of liquid water in the polluted clouds counters the results of most climate model simulations of the aerosol indirect radiative forcing. Similar losses of cloud liquid water in polluted clouds, however, have been simulated in a Large Eddy Simulation cloud model when the free troposphere above the polluted cloud is dry (Ackerman et al., 2004), as is often the case for marine stratus off California and for the region in Fig. 2 near the European continent.

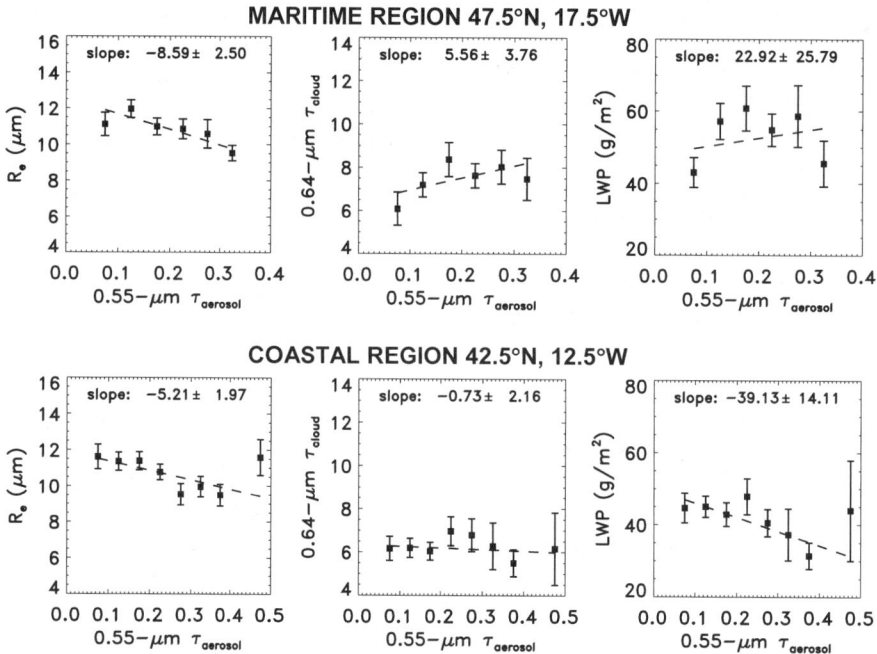

Fig. 2. Aerosol optical depth, cloud droplet effective radius, cloud optical depth, and cloud liquid water amount composited for 5° × 5° latitude-longitude regions of the North Atlantic. The latitudes and longitudes at the center of the regions are indicated. The upper panels give results for a region some distance from the coast of Europe; the bottom panels give results for a region adjacent to the Iberian peninsula.

4. REFERENCES

Ackerman, A. S., et al., 2003: Enhancement of cloud cover and suppression of nocturnal drizzle in stratocumulus polluted by haze. *Geophys. Res. Lett.*, 30, doi:10.10292002GL016634.

Ackerman, A. S., et al., 2004: The impact of humidity above stratiform clouds on indirect aerosol climate forcing. *Nature*, 432, 1014-1017.

Albrecht, B. A., 1989: Aerosols, cloud microphysics and fractional cloudiness. *Science*, 245, 1227-1230.

Collins, W. D., et al., 2001: Simulating aerosols using a chemical transport model with assilimiation of satellite aerosol retrievals: Methodology for INDOEX. *J. Geophys. Res.*, 106, 7313-7336.

Intergovernmental Panel on Climate Change (IPCC), 2001: *Climate Change 2001: The Scientific Basis*, edited by J. T. Houghton et al., Cambridge Univ. Press, New York.

Kaufman, Y. J., and R. S. Fraser, 1997: Control of the effect of smoke particles on clouds and climate by water vapor. *Science*, 277, 1636-1639.

Koren, I., et al., 2004: Measurement of the effect of Amazon smoke on inhibition of cloud formation. *Science*, 303, 1342-1345.

Liepert, B. G., 2002: Observed reductions of surface solar radiation at sites in the United States and worldwide from 1961 to 1990. *Geophys. Res. Lett.*, 29, 1421, 10.129/2002GL014910.

Loeb N. G., and N. Manalo-Smith, 2004: Top-of-atmosphere direct radiative effect of aerosols over global oceans from merged CERES and MODIS observations. *J. Climate* (submitted).

Lohmann, U., and G. Lesins, 2003: Stronger constraints on the anthropogenic indirect aerosol effect. *Science*, 298, 1012-1015.

Menon, S., et al., 2002: Climate effects of black carbon aerosols in China and India. *Science*, 297, 2250-2253.

Matheson, M. A., et al., 2004: AVHRR observations of the aerosol indirect effect for summertime stratiform clouds in the Northeastern Atlantic. *IRS 2004* (submitted).

Remer, L. A., et al., 2004: The MODIS aerosol algorithm, products and validation. *J. Atmos. Sci.*, (in press).

Schmid, B., et al., 2003: Column closure studies of lower tropospheric aerosol and water vapor during ACE-Asia using airborne sun photometer and airborne in situ and ship-based lidar measurements. *J. Geophys. Res.*, 108(D23), 8656, doi:10.1029/2002JD003361.

Sekiguchi, M., et al., 2003: A study of the direct and indirect effects of aerosols using global satellite data sets of aerosol and cloud parameters. *J. Geophys. Res.*, 108(D22), 4699, DOI:10.1029/2002JD003359.

Twomey, S., 1974: Pollution and the planetary albedo," *Atmos. Environ.*, 8, 1251-1256.

Winkler, D. M., et al., 2003: The CALIPSO mission: Spaceborne lidar for observation of aerosols and clouds. *Proc. SPIE Int. Soc. Opt. Eng.*, 4839, 1-11.

SATELLITE OBSERVATIONS OF CLOUD RADIATIVE FORCING FOR THE AFRICAN TROPICAL CONVECTIVE REGION

J. M. Futyan, J. E. Russell, and J. E. Harries

Space and Atmospheric Physics Group
Blackett Laboratory, Imperial College, London, SW7 2BZ, UK

ABSTRACT

Previous results have highlighted the difficulty in understanding the delicate balance between longwave and shortwave cloud effects in convective regions using monthly mean gridded data. Here, a method is presented to separate the contribution of individual cloud types occurring during a month to the mean cloud forcing. A significant contribution to the net cloud forcing in the convective region is found to result from low non-convective clouds, which cannot be separated using standard monthly mean products.

1. INTRODUCTION

Clouds strongly influence the Earth's radiation balance, acting both to increase the reflection of solar radiation back to space, causing surface cooling, and to reduce the emission of thermal radiation to space, giving a warming effect. In the annual global mean, clouds have been shown to have a net cooling of $\sim 20 Wm^{-2}$ (*Ramanthan et al.,* 1989), locally and on shorter timescales their effects can be much larger. The impact of cloud on the energy balance is quantified by the longwave (LW) and shortwave (SW) cloud forcing (CRF) (equation 1).

$$LWCRF = OLR_{clear} - OLR_{cloudy}$$
$$SWCRF = F^{ref}_{clear} - F^{ref}_{cloudy} \quad (1)$$

Here OLR is outgoing longwave radiation flux, and F^{ref} is the reflected solar flux. The subscripts clear and cloudy refer to clear sky (cloud free) and cloudy (all-sky) conditions. The net forcing is the sum of the LW and SW components.

Monthly average LW forcings reach their maximum values of 50–100 Wm^{-2} in the convectively disturbed regions of the tropics; however this heating effect is nearly cancelled by a correspondingly large SW forcing. This observation of near cancellation between large counteracting cloud effects in tropical convective regions led to debate as to the cause of this behaviour (*Kiehl* 1994; *Hartmann et al.,* 2001). *Hartmann et al.* (2001) suggest that this behaviour may indicate feedbacks in the climate system influencing the ensemble of cloud types associated with convection.

Near cancellation between LW- and SWCRF has been assumed to be a generic property of all tropical convective regions, over both land and ocean (e.g., *Kiehl,* 1994). However the majority of regional scale studies focus on the western tropical Pacific warm pool region, and those which exist for other regions do not necessarily find a similar degree of cancellation (e.g., *Rajeevan and Srinivasan,* 2000). Recently, *Futyan et al.* (2004) used data from the Clouds and the Earth's Radiant Energy System (CERES) instrument (*Wielicki et al.,* 1995) to show that, on average, a similar degree of cancellation is observed for the convective region over tropical Africa as for the Pacific warm pool. However, larger departures from cancellation, resulting in negative net forcing, occur in the Atlantic ITCZ region. *Futyan et al.* (2004) also found greater variability within the African and Atlantic regions, which they argue is in part due to the inclusion of the radiative effects of low clouds present on some days during the month.

Here we present a method to estimate the contribution to the monthly mean cloud forcing for individual cloud types using new high temporal resolution data from the Meteosat-8 satellite. The method is demonstrated for June 2004, and the resulting separation of the effects of high, mid and low level clouds used to show that low level clouds do indeed make a significant contribution to the net cloud forcing within the African and Atlantic convective regions.

2. DATA

The Geostationary Earth Radiation Budget (GERB) radiometer (*Harries et al., 2004, submitted manuscript*) makes accurate broadband measurements of both the LW and SW components of the top of atmosphere energy budget every 15 minutes. The data are enhanced through synergy with SEVIRI (the Spinning Enhanced Visible and Infra-Red Imager (*Schmetz et al.,* 2002)), which provides information for unfiltering and scene identification in the GERB data processing.

As validation of GERB data is ongoing, here we demonstrate the method and its possible applications using GERB-like fluxes. These data are produced at the Royal Meteorological Institute of Belgium (RMIB) as part of the GERB processing via a narrow-to-broad band regression on the SEVIRI channels (*Clerbaux and Dewitte,* 1999), and represent an estimate of the GERB measurement. A comparison of the monthly mean cloud forcing with that observed by CERES (section 3 below) is used to confirm the suitability of these data for this preliminary study.

Before the cloud forcing can be calculated, estimates of the monthly mean clear and all sky flux are required. For GERB, the excellent temporal sampling means that the all sky mean flux can be calculated as a simple average of the observed data. However, for the clear sky flux, forming an accurately sampled mean is not a straightforward task due to the limited sampling in some regions. Here we follow the approach of *Futyan and Russell (submitted manuscript, 2004)* to estimate the monthly mean clear sky flux.

To estimate the contribution to the cloud forcing for individual cloud types, we also require a cloud type classification. Here, we use the CLA cloud classification produced at EUMETSAT (the European organisation for the exploitation of Meteorological satellites) from SEVIRI data (*Lutz, 1999*). This classification defines 10 cloud classes based on meteorological cloud types, rather than via rigid cloud height and optical depth bins. Here the simplified classification of low, mid and high-level clouds is used. These data are currently available at three-hourly intervals so the monthly averages calculated in section 4 are for the average over these eight time-steps.

3. COMPARISON OF MONTHLY MEAN CLOUD FORCING WITH CERES

Fig. 1 shows the monthly mean LW, SW, and net cloud forcing for June 2004, calculated using GERB-like data and for CERES ERBE-like data (ES-4, Edition1, FM1 only). CERES ERBE-like data are available on a 2.5 degree grid, while the GERB-like data has a resolution of approximately 0.5 degrees. Good general agreement is found, both in the magnitude and pattern of cloud forcing. However, the GERB-like data slightly underestimate both the LW and SW forcings in the convective region compared to CERES, producing a slightly less negative net forcing. These differences may relate to differences in sampling and clear sky identification methods, but are likely to be primarily due to uncertainties in the narrow to broadband conversion. However, the good general agreement suggests that the GERB-like data are sufficient to investigate the partitioning of cloud forcing by cloud type, although absolute magnitudes of the forcings inferred in subsequent sections should be treated with some caution.

4. CLOUD FORCING BY CLOUD TYPE

Cloud forcing is usually calculated as the difference between the monthly average all-sky flux and the monthly average clear-sky flux for a particular region, as in section 3 above. However, if an estimate of the clear sky flux can be made for each footprint, then the cloud forcing can be estimated on an instantaneous basis as the difference between the observed flux in that footprint F_{obs}^i, and the clear sky flux estimate F_{clear}^i.

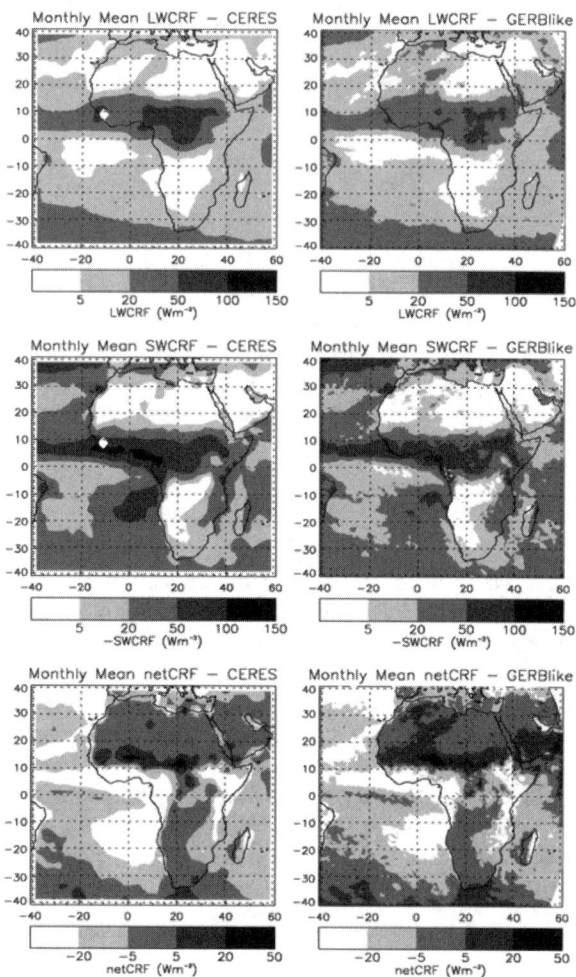

Fig. 1. LW, SW, and net CRF for June 2004 for CERES ERBE-like data at 2.5 degree resolution (left) and for GERB-like data at ~0.5 degree resolution (right).

Using co-registered cloud classification information from SEVIRI, each footprint can be assigned to one of the pre-defined cloud classes. The forcing for that cloud class is equal to the observed forcing while the contribution by other cloud types is zero for that footprint at that time.

The instantaneous high cloud forcing (*highCRF*) is therefore defined by equation 2, with similar definitions for other cloud classes.

$$highCRF = F_{clear}^i - F_{obs}^i \quad \text{if flag is high cloud}$$
$$highCRF = 0 \quad \text{otherwise} \quad (2)$$

By averaging these instantaneous forcings over time, the monthly mean forcing due to each class of cloud can be calculated. Individually, these quantities provide an estimate of the relative importance of different cloud types to the radiation budget of a region, while their sum is a measure of the overall cloud forcing, consistent with the standard definition.

This method requires an estimate of the clear sky flux at all footprints at all times. As no clear estimate can be made on an instantaneous basis for cloudy regions, here we use the monthly timestep mean clear sky flux. This is used directly in the LW, while in the SW, the average albedo is used with the instantaneous incident solar flux to estimate the reflected SW component.

Fig. 2 shows maps of the monthly mean LW- and SWCRF for high, mid and low level clouds for June 2004. As expected, high and mid level clouds dominate the LWCRF, while all types of cloud contribute significantly to the SWCRF. High and mid level clouds dominate in the convective belt and at around $40°$ while low clouds provide most of the forcing over the sub-tropical oceans. The region shown in cross-hatching in Fig. 2 is that with (overall) LWCRF greater then $30 Wm^{-2}$. This definition was used by *Futyan et al.* (2004) to select the convective region and excludes regions dominated by low cloud throughout the month. Fig. 2 suggests that this definition is reasonable as most of the area where the effects of high and mid level clouds dominate is included, although some regions with significant mid or high level cloud forcing are excluded. However, low clouds also contribute significantly to the SWCRF within the convective region selected, primarily over the Atlantic and south coast of West Africa. This corresponds to the region with strongly negative overall net cloud forcing in Fig. 1.

Fig. 3 shows the LW, SW, and net cloud forcings averaged over this convective region for CERES and GERBlike data. For the GERB-like data the breakdown of these forcings according to cloud type is also shown. Separate plots are shown for the African land region (land footprints within 20W–50E, 40S–20N) and the Atlantic region (ocean footprints within 40W–20E, 10S–20N). The overall forcings agree well for the CERES and GERB-like data, particularly over the ocean region. The breakdown of the effects of the individual cloud types highlights the additional information available using this method. Over land, high clouds have a small positive net forcing, but contributions from mid, low and mixed level clouds lead to a negative overall net forcing. Over ocean, the net forcing is negative even when only high clouds are considered, but contributions from low and mixed mid and low level clouds result in a substantially more negative value overall.

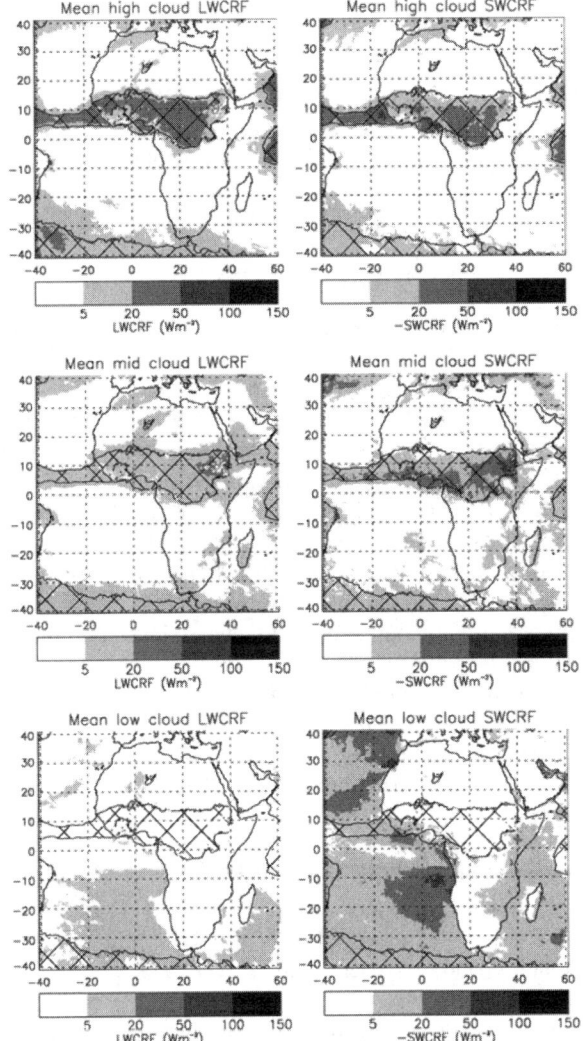

Fig. 2. Maps of LW- (left) and SW- (right) CRF due to high (upper panel), mid (middle panel) and low (lower panel) level clouds for June 2004. The region marked with cross-hatching is the convective region selected using a LW forcing limit of 30 Wm^{-2}

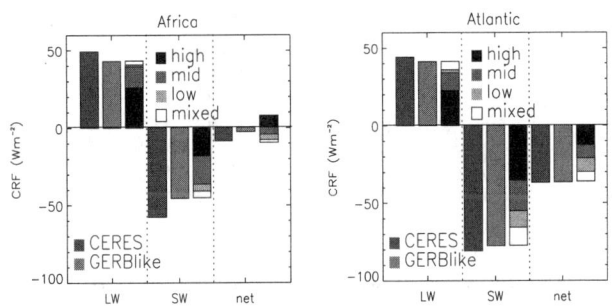

Fig. 3. Bar charts of the area average LW, SW, and net cloud forcing for the African land convective region (left) and for the Atlantic convective region (right), showing the overall forcings for CERES and GERBlike data, and the breakdown according to cloud type for the GERBlike data. Convective regions are defined to contain all gridboxes with LWCRF greater than 30 Wm^{-2}

The cloud forcing ratio $R = -SWCRF/LWCRF$ allows the balance between the LW- and SWCRF to be assessed in a manner independent of the cloud fraction. Fig. 4 shows the impact on the area average forcing ratio of the definition chosen for the convective region. If a straightforward geographical box is used to define the convective region (definition 1), the effects of regions dominated by low clouds throughout the month, such as the stratocumulus deck off the west coast of Africa, are included, resulting in high values of R. The use of a threshold on the LWCRF at 30 Wm^{-2} (definition 2) excludes these regions, and so results in a lower value of R, particularly over the Atlantic region. However, the effects of low clouds which occur on some days or at some times during the month cannot be excluded using only monthly mean radiation budget data.

By taking advantage of the high temporal resolution of GERB (or GERB-like) data, the effects of these clouds can be explicitly excluded. In definition 3, the region is selected using the 30 Wm^{-2} threshold on the overall LWCRF, but only the effects of high and mid level cloud are included. This results in a further reduction in the area average cloud forcing ratio, confirming that low clouds do contribute significantly to the behaviour within this region as suggested by *Futyan et al.* (2004). Finally, as the effects of low clouds can now be explicitly excluded by only averaging the high and mid cloud forcings there should no longer be any need to impose a LW forcing limit to select the convective region. In definition 4 the spatial constraint imposed by the LW forcing limit is removed. The R value found is essentially the same as that for definition 3 over land, but decreases slightly over ocean, presumably as more thin high cloud, which has a net warming effect, is included around the edge of the convective region.

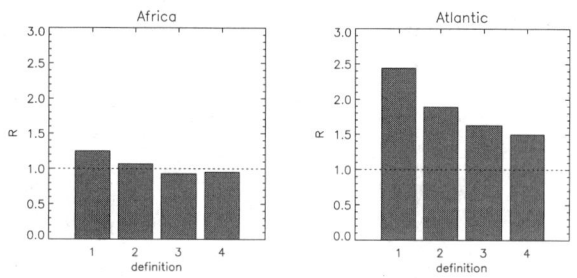

Fig. 4. Bar charts of the area average cloud forcing ratios for the African land convective region (left) and for the Atlantic convective region (right) for various definitions of the 'convective region', as given in the text.

5. SUMMARY

The high temporal resolution of GERB data, and the synergy with SEVIRI, allow the contribution of individual cloud types to the monthly mean cloud forcing within the GERB field of view to be determined. A method to achieve this is proposed and its application demonstrated using GERB-like data. The suitability of these data is verified through comparison of the overall monthly mean cloud forcing with data from CERES. Overall good agreement is found, although differences, presumably due to inaccuracies in the narrow-to-broad band conversion are seen in some regions. The method is applied to the convective regions over tropical Africa and the Atlantic, and is shown to provide a more complete separation of the effects of convective and low non-convective cloud than can be achieved using standard monthly mean gridded products. Once validated GERB data are available, this method will provide accurate quantitative assessment of the importance of different cloud types to the regional radiation balance.

6. ACKNOWLEDGEMENTS

With thanks to Nicolas Clerbaux at RMIB for providing the GERBlike data. The CERES data were obtained from the NASA Langley Research Center Atmospheric Sciences Data Center.

7. REFERENCES

Clerbaux, N., and S. Dewitte (1999), RGP-SP: Spectral Modelling, *RMIB technical report MSG-RMIB-GE-TN-0005*, RMIB, available from http://gerb.oma.be/.

Futyan, J. M., J. E. Russell, and J. E. Harries (2004), Cloud Radiative Forcing in Pacific, African and Atlantic Tropical Convective Regions, *J. Climate*, 17, 3192–3202.

Hartmann, D. L., L. A. Moy, and Q. Fu (2001), Tropical Convection and the Energy Balance at the Top of the Atmosphere, *J. Climate*, 14, 4495–4511.

Kiehl, J. T. (1994), On the Observed Near Cancellation between Longwave and Shortwave Cloud Forcing in Tropical Regions, *J. Climate*, 7, 559–567.

Lutz, H. J. (1999), Cloud Processing for METEOSAT Second Generation, *Technical Memorandum 4*, EUMETSAT, available from http://www.eumetsat.de/.

Rajeevan, M., and J. Srinivasan (2000), Net Cloud Radiative Forcing at the Top of the Atmosphere in the Asian Monsoon Region, *J. Climate*, 13, 650–657.

Ramanthan, V., R. D. Cess, E. F. Harrison, P. Minnis, B. R. Barkstrom, E. Ahmad, and D. Hartmann (1989), Cloud-Radiative Forcing and Climate: Results from the Earth Radiation Budget Experiment, *Science*, 243, 57–62.

Schmetz, J., P. Pili, S. Tjemkes, D. Just, J. Kerkmann, S. Rota, and A. Ratier (2002), An Introduction to Meteosat Second Generation (MSG), *Bul. Amer. Meteorol. Soc.*, 83(7), 977–992.

Wielicki, B. A., R. D. Cess, M. D. King, D. A. Randall, and E. F. Harrison (1995), Mission to Planet Earth: Role of Clouds and Radiation in Climate, *Bull. Amer. Meteorol. Soc.*, 76(11), 2125–2153.

THE VISIBLE AND NEAR INFRARED COMPONENTS OF THE SHORTWAVE RADIATION BUDGET

I. Laszlo
NOAA/NESDIS/Office of Research and Applications
Camp Springs, Maryland, 20746, USA

R. T. Pinker
Department of Meteorology, University of Maryland
College Park, Maryland, 20742, USA

ABSTRACT

The University of Maryland Global Energy and Water Cycle Experiment, Shortwave Radiation Budget algorithm (V2.1) was used with the International Satellite Cloud Climatology (ISCCP) D1 satellite data to estimate monthly values of shortwave, visible and near infrared top of atmosphere (TOA) and surface radiative fluxes for the period 1984–2000. Based on these data, the global and zonal averages of absorption and cloud radiative effects are examined.

1. INTRODUCTION

Over the past two decades, a number of studies employed various satellite data to infer the total or partial radiation budget. For many applications information on the total shortwave radiative energy is sufficient. However, recent models of the hydrologic cycle, the interactions and feedbacks between the atmosphere and the terrestrial biosphere, and estimation of oceanic and terrestrial net primary productivity require spectrally resolved radiative forcings. A more detailed evaluation of climate models is also possible with satellite-derived radiative fluxes that are spectrally resolved.

In this paper we describe a spectral radiation budget data set and present global and zonal analyses derived from such information.

2. THE RADIATIVE FLUXES

The data set includes top of atmosphere (TOA) and surface (SRF) clear-sky and all-sky fluxes in three broad spectral intervals: visible (VIS: 0.4–0.7 µm), near infrared (NIR: 0.7–4.0 µm) and shortwave (SW: 0.2–4.0 µm). Ultraviolet fluxes can be obtained by subtracting the sum of VIS and NIR fluxes from the SW. The fluxes are derived from the D1 product of the International Satellite Cloud Climatology Project (ISCCP) (Rossow and Schiffer, 1999) for the period July 1983 – September 2001 using Version 2.1 of the University of Maryland (UMD), Global Energy and Water Cycle Experiment (GEWEX) Shortwave Radiation Budget (SRB) algorithm (hereafter referred to as the UMD/SRB algorithm). Details on the UMD/SRB algorithm can be found in Pinker and Laszlo, 1992; Laszlo and Pinker, 1997. The D1 parameters used as input to the algorithm are the mean visible clear and cloudy radiances, the clear-sky composite radiance, cloud fraction, snow fraction, column amount of ozone and precipitable water, solar and satellite zenith and relative azimuth angles at three hourly intervals.

Radiation budget data are provided at the same time intervals for each day at each of the 6596 equal-area grid cells representing an area of about 280 by 280 km. Daily and monthly averages for each grid cell are also included. The data cover the period July 1983 – September 2001.

3. RESULTS

In the current study, we present results from the analysis of monthly means. These were derived from the three-hourly estimates of the radiative fluxes. In order to represent each month with equal weight, data from January 1984 – December 2000 (17 years) only were used.

3.1 Global Annual Mean

Annual means were obtained by calculating global averages of the 6596 equal-area monthly values, and subsequently averaging the 204 monthly mean global fluxes. At high latitudes, observations for some cells are missing. To reduce their effect on the global average, monthly average of TOA downward flux was calculated both analytically and numerically from available cells, and all fluxes were multiplied by the ratio of numerical to analytical TOA flux. To facilitate comparisons with other studies, we assumed a solar constant of 1367 Wm^{-2} for the analytical TOA downward flux; while in the UMD/SRB algorithm the solar constant is 1372.6 Wm^{-2}. This led to a global annual mean TOA SW irradiance of 342 Wm^{-2}. Global annual mean radiation budget parameters at the top of atmosphere, at the surface and of the atmosphere in the three spectral bands (VIS; NIR; SW) are summarized in

Table 1. The parameters listed in this table include the all-sky downward flux (DN), net flux (NF), albedo (A) and cloud radiative forcing (CRF). TOA irradiance in the near infrared spectrum is about 35% larger than that in the visible. Yet, because of the stronger absorption in the NIR, equal amount of radiation reaches the surface in these two spectral intervals. The global average of the SW radiation absorbed at the surface is 167 Wm^{-2}. About 48% of this energy is absorbed in the visible spectrum. Since the surface albedo is smaller in the visible than in the NIR, the energy absorbed in the visible is by 3 Wm^{-2} more than that absorbed in the NIR. In respect to the atmosphere-surface system, the visible albedo at TOA is larger than the NIR albedo, and therefore, the energy absorbed in the visible is 35% less than that absorbed in the NIR. Global annual SW cloud radiative forcing at the surface is 2 Wm^{-2} larger than that at the TOA. Most of this enhancement comes from the NIR spectral region, where the atmospheric forcing is 4 Wm^{-2}; a positive value indicating heating of the atmosphere by clouds. This heating is reduced somewhat in the visible, resulting in a SW atmospheric cloud forcing of about 3 Wm^{-2}.

Table 1. Global Annual Mean All-sky SW, VIS and NIR Downward Flux (DF), Net Flux (NF), Albedo (A), and Cloud Radiative Forcing (CRF) at the Top of Atmosphere (TOA), at the Surface and of the Atmosphere. Fluxes and Cloud Radiative Forcing are in Wm^{-2}

	DF	NF	A	CRF
TOA				
SW	342	241	0.295	−46
VIS	134	88	0.338	−19
NIR	181	136	0.252	−25
Surface				
SW	189	167	0.118	−48
VIS	89	80	0.101	−18
NIR	89	77	0.137	−29
Atmosphere				
SW		74		3
VIS		8		−1
NIR		59		4

Global annual mean SW radiation budget parameters have been estimated in a number of studies. The SW TOA albedo of ~30% in the current data set is in agreement with the value obtained from the Earth Radiation Budget Experiment (ERBE) (Barkstrom et al., 1989) and from the French-Russian-German Scanner for Radiation Budget (ScaRaB) project (Kandel et al., 1998). The SW surface absorption in the current data set is only 2 Wm^{-2} larger than the one obtained by Rossow and Zhang (1995) who used the ISCCP C1 data, and very close to the 168 Wm^{-2} reported by Kiehl and Trenberth (1997) who used model calculations constrained by satellite observations at TOA.

It is, however, 25 Wm^{-2} larger than the surface absorption estimated by Ohmura and Gilgen (1993) who used ground observations of radiative fluxes and estimates of surface albedo. It is also 21 Wm^{-2} larger that the value obtained by Major (1998) who combined surface data with global satellite observation of cloudiness. The SW atmosphere absorbed flux of 74 Wm^{-2} in the current data set is 7 Wm^{-2} larger than that obtained by Kiehl and Trenberth (1997), and 9 Wm^{-2} larger than that received by Rossow and Zhang (1995), but it is 9 Wm^{-2} smaller than the one estimated by Li, et al. (1997). The global annual mean value of the shortwave TOA cloud radiative forcing (CRF) agrees well with those derived from ERBE and ScaRaB (−48 Wm^{-2}). It is however, smaller than those estimated by Rossow and Zhang (−54 Wm^{-2}) and Kiehl and Trenberth (−50 Wm^{-2}). A more detailed comparison of the global SW absorption is presented in Laszlo and Pinker (2002). As yet, global scale estimates of spectral radiative fluxes are not available for use in this comparison.

Table 2. Net SW, VIS and NIR Flux for Land, Ocean, and Coast

	all	land	ocean	coast
TOA				
SW	241	224	247	237
VIS	88	83	91	87
NIR	136	125	140	133
Surface				
SW	167	142	177	160
VIS	80	70	84	77
NIR	77	64	82	73
Atmosphere				
SW	74	82	71	77
VIS	8	13	7	9
NIR	59	61	58	60

Table 2 presents the net flux separately for land, ocean and coastal areas. These parameters were obtained by calculating the annual averages separately for land, ocean and coastal ISCCP grid cells. For comparison, Table 2 includes the global values representing all ISCCP cells. The annual surface-absorbed SW radiation averaged for ocean cells is 177 Wm^{-2}. Solar radiation absorbed at the land surface is 35 Wm^{-2} smaller than over oceans. The reasons for this: 1) the average SW land surface albedo is about three times that of the ocean; 2) the amount of radiation available for absorption at the surface is less because the average optical depth is about three times larger over land than over ocean, even though the cloud cover averaged for land cells is smaller than that averaged for ocean cells. As expected, there is little difference (2 Wm^{-2}) between the VIS and NIR flux absorbed at the ocean surface. For land cells the differences are larger (6 Wm^{-2}) mostly because of the different VIS and NIR surface reflectances. These VIS/NIR absorbed fluxes account for 49%/45% and 47%/46% of the SW surface-absorbed flux for land and ocean, separately. The atmosphere over land areas absorbs 11 Wm^{-2} more than the atmosphere over the oceans. The difference between the VIS and NIR atmosphere-absorbed flux is somewhat larger over land than over ocean. This is due to the higher optical depths of clouds and aerosols over land than over oceans. The VIS/NIR atmosphere-absorbed fluxes represent 16%/46% and 10%/82% of the SW fluxes over land and ocean, respectively. Absorbed flux values for coastal cells lie between the values for land and ocean cells.

It is of interest to note that the SW surface net flux of 142 Wm^{-2} for land is the same as reported in Ohmura and Gilgen (1993) for the global average. Ohmura and Gilgen (1993) used a relationship based on radiation data measured at a large number of ground sites and conventional meteorological data to estimate the SW radiation budget on a global scale. These sites are concentrated over land, and as such, might not represent accurately the radiation budget on a global scale.

3.2 Zonal Averages

Annual mean SW absorption by the atmosphere-surface system varies significantly with latitude (Fig. 1). It is about 60 Wm^{-2} at the polar latitudes and it reaches 320 Wm^{-2} in the tropics. Most of the SW absorption occurs at the surface. It is the surface absorption that defines the latitudinal variation of the column absorption, since the latitude dependence of the atmospheric absorption, relative to that of the surface absorption, is small. SW atmospheric absorption is dominated by absorption in the NIR. The VIS atmospheric absorption is less than 10 Wm^{-2} at all latitudes with a minimum in the tropics and an increase towards the poles. This pattern is due primarily to the larger ozone amount and increased atmospheric pathlength at high latitudes. Atmospheric absorption in the near infrared shows an opposite change with latitude; it is larger at low latitudes. This can be attributed to the zonal distribution of the water vapor, the major absorber in the atmosphere in this spectral interval. Zonal variations of visible and near infrared absorption at the surface are similar in magnitude. They are almost equal at mid-latitudes, while NIR absorption is somewhat smaller in the tropics. The latter is due in part to the reduction of the NIR downward flux by stronger atmospheric absorption in this region, and in part to the relatively high NIR surface albedos of vegetation and deserts.

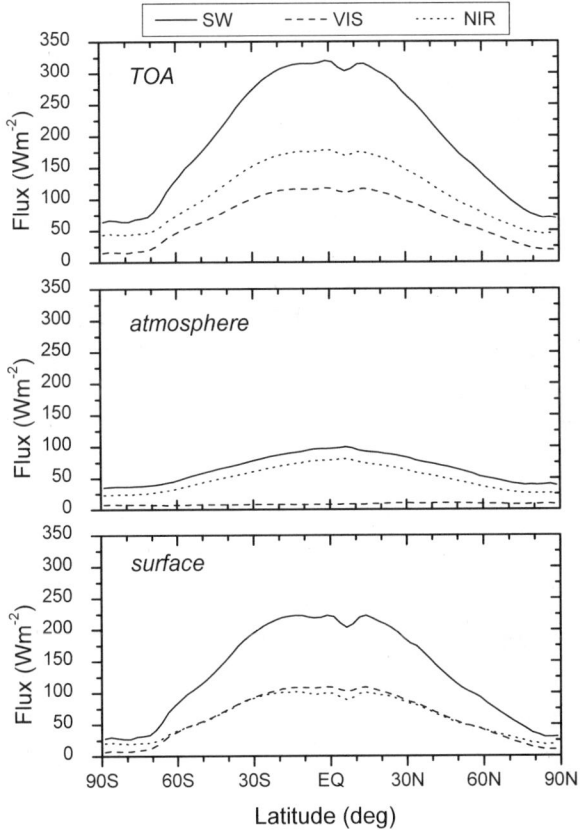

Fig. 1. Zonal mean of annual mean SW, VIS, and NIR absorption by the atmosphere-surface system, by the atmosphere and by the surface.

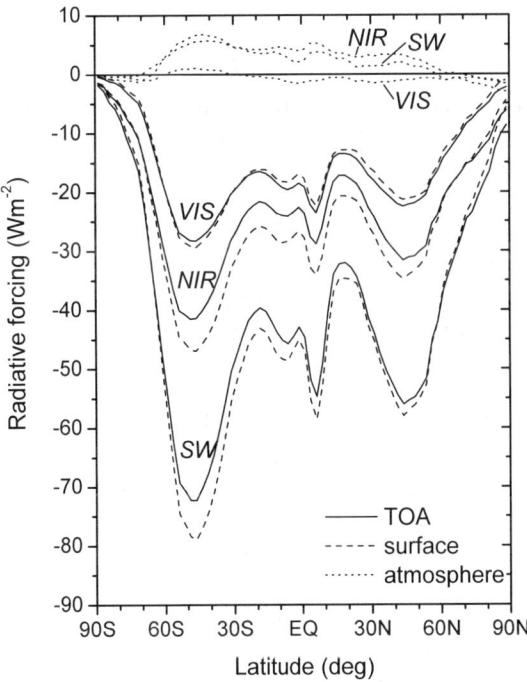

Fig. 2. Zonal mean of annual mean SW, VIS, and NIR cloud radiative forcing at TOA, of the atmosphere and at the surface.

The SW, VIS and NIR cloud radiative forcing also exhibits a strong latitudinal dependence (Fig. 2). The largest forcing (more negative) values occur in the northern and southern hemisphere storm tracks (around 45°N and 45°S). Except for high latitudes, SW and NIR surface cloud forcings are stronger than the TOA forcing. VIS cloud forcing is about one third of the SW forcing, and except for latitude zones around 45°S, surface VIS forcing is smaller than TOA forcing. Globally, SW and NIR atmosphere forcings are positive. Over land, SW and VIS atmosphere forcings are negative (not shown).

4. SUMMARY

We used the ISCCP D1 data as input to Version 2.1 of the UMD/SRB algorithm to estimate monthly values of the shortwave, visible and near infrared TOA and surface net fluxes for the period of January 1984 – December 2000. Global annual mean SW radiation budget parameters from this data set compare well with similar parameters from other satellite-based studies. However, the SW surface/atmosphere absorption is ~20 Wm^{-2} larger/smaller than that from studies that use ground measurements. The results indicate that 56% percent of the total 241 Wm^{-2} absorbed by the system is absorbed in the near infrared, and only 37% of it is absorbed in the visible. Annual global mean atmospheric and surface absorption for land and ocean areas differ by 11 and 35 Wm^{-2}, respectively.

The visible, near infrared and shortwave radiation budget data derived here can be used, for example, in hydrological models for quantifying radiative forgings. The data also provide additional constraints for general circulation models. The partitioning of the shortwave radiation budget into visible and near infrared components is based on radiative transfer modeling (not on direct measurements). The radiative transfer modeling and the inversion procedure assume plane parallel clouds and account for the current state of knowledge on cloud-radiation interactions applicable to global scale computations. As our understanding of this interaction improves the values of the spectral components cited in this paper may change.

5. ACKNOWLEDGMENTS

This work was supported by grant NAG59634 from the National Aeronautics and Space Administration (NASA) EOD/IDS. The ISCCP D1 data were obtained from the NASA Langley Research Center Atmospheric Sciences Data Center.

The views, opinions, and findings contained in this report are those of the authors and should not be construed as an official National Oceanic and Atmospheric Administration (NOAA) or U.S. Government position, policy, or decision.

6. REFERENCES

Barkstrom, B., E. Harrison, G. Smith, R. Green, J. Kibler, R. Cess, and the ERBE Science Team, 1989: Earth Radiation Budget Experiment (ERBE) archival and April 1985 results. *Bull. Amer. Meteor. Soc.*, 70, 1254-1262.

Kandel, R., M. Viollier, P. Raberanto, J. Ph. Duvel, L. A. Pakhomov, V. A. Golovko, A. P. Trishchenko, J. Mueller, E. Raschke, R. Stuhlmann, and the International ScaEaB Scientific Working Group (ISSWG), 1998: The ScaRaB earth radiation budget dataset. *Bull. Amer. Meteor. Soc.*, 79, 765-783.

Kiehl, J. T., and K. E. Trenberth, 1997: Earth's annual global mean energy budget. *Bull. Amer. Meteor. Soc.*, 78, 197-208.

Laszlo, I., and R. T. Pinker, 1997: Impact of changes to water vapor parameterization on surface short-wave radiative fluxes. In *IRS '96: Current Problems in Atmospheric Radiation*, A. Deepak Publishing, Hampton, Virginia, USA, 526-529.

Laszlo, I., and R. T. Pinker, 2002: Shortwave radiation budget of the Earth: Absorption and cloud radiative effects, *Időjárás*, 106, 189-205.

Li, Z., L. Moreau, and A. Arking, 1997: On solar energy disposition: A perspective from observation and modeling. *Bull. Amer. Meteor. Soc.*, 78, 53-70.

Major, G., 1998: On surface-absorbed solar radiation. *Bull. Amer. Meteor. Soc.*, 79, 92-93.

Ohmura, A., and H. Gilgen, 1993: Re-evaluation of the global energy balance. *Interactions between Global Climate Subsystem: The Legacy of Hann, Geophys. Monogr.* No. 75, Amer. Geophys. Union, 93-110.

Pinker, R. T., and I. Laszlo, 1992: Modeling surface solar irradiance for satellite applications on a global scale, *J. Appl. Meteor.*, 31, 192-211.

Rossow, W. B., and R. A. Schiffer, 1999: Advances in understanding clouds from ISCCP. *Bull. Amer. Meteor. Soc.*, 80, 2261-2287.

Rossow, W. B., and Y.-C. Zhang, 1995: Calculation of surface and top of atmosphere radiative fluxes from physical quantities based on ISCCP data sets 2. Validation and first results. *J. Geophys. Res.*, 100, 1167-1197.

THE NASA/GEWEX SURFACE RADIATION BUDGET DATASET: RESULTS AND ANALYSIS

S. J. Cox, S. K. Gupta, J. C. Mikovitz, M. Chiacchio, and T. Zhang
AS&M, Inc., Hampton, VA 23666, USA

P. W. Stackhouse, Jr.
NASA Langley Research Center, Hampton, VA 23681, USA

ABSTRACT

Release 2 of the NASA/GEWEX Surface Radiation Budget project (SRB) is described. Global average and time series results are presented. Validation shows good agreement with BSRN observations.

1. INTRODUCTION

The NASA/GEWEX SRB project (Stackhouse et al., 2004) provides estimates of shortwave and longwave surface fluxes on a 1° x 1° grid for the period July 1983 to October 1995. A variety of temporal resolutions is supported: 3-hourly, daily, monthly, and monthly/3-hourly. For a global product such as this, measured radiances from satellite-borne sensors provide the primary source of input data. Radiances and cloud properties are taken from the International Satellite Cloud Climatology Project (ISCCP; Rossow and Schiffer, 1999) DX (pixel-level) dataset. Radiation algorithms are used to calculate surface fluxes from the satellite data and meteorological inputs.

For both the shortwave and longwave datasets, SRB makes use of two sets of algorithms: primary and quality-check (QC). The primary SW algorithm is derived from Pinker and Laszlo (1992). The QC shortwave algorithm is described in Gupta et al., 2001 and provides only daily and monthly temporal resolution. For longwave, the primary algorithm is an adaptation of Fu et al. (1997), and the QC is taken from Gupta et al. (1992).

The meteorological input uses temperature and humidity profiles from the Goddard Earth Observing System-1 (GEOS-1) reanalysis product, and column ozone from the Total Ozone Mapping Spectrometer (TOMS) archive.

The current data set is described as Release 2 and is available to the community from the NASA Langley Atmospheric Sciences Data Center at the website: eosweb.larc.nasa.gov/PRODOCS/srb/table_srb.html

Efforts are well underway to extend the dataset through the 2001 and beyond time frame, and to improve and further validate the algorithm and input datasets.

2. RADIATION DATA

Table 1 compares the global 12 year mean surface fluxes from SRB with previous products. The most notable difference between SRB and Ohmura and Gilgen (1993)

Table 1. Representative Globally Averaged Fluxes for NASA/GEWEX SRB and Earlier Products

Parameter	Ohmura & Gilgen (1993) GEBA Surf. Obs.		Kiehl and Trenberth (1997) ERBE/CCM3		Zhang & Rossow (latest) 5 yr Mean ('85-'89)		NASA/GEWEX SRB Rel. 2* (NASA LaRC) 12 yr Mean (July '83 – June '95)			
							SW, LW		SW, LW QC	
	Flux	% F_0	Flux	% F_0	Flux	% F_0	Flux	% F_0	Flux	% F_0
SW Down	169.0	49.4	198	57.9	189.4	55.4	186.2	54.5	184.2	53.9
SW Net	142.0	41.6	168	49.2	164.7	48.2	164.6	48.0	160.9	47.1
LW Down	345	100.9	324	94.8	344.6	100.8	342.6	100.6	345.2	101.0
LW Net	-40.0	-11.7	-66	-19.3	-50.9	-14.9	-50.8	-14.9	-47.2	-13.8
Total Net	102.0	29.8	102	29.8	113.8	33.3	113.8	34.2	113.7	33.3
SW CRF	--	--	--	--	-59.0	-17.3	-56.9	-16.6	-58.5	-17.1
LW CRF	--	--	46	13.5	31.1	9.1	36.5	10.4	35.6	10.4
Total CRF	--	--	--	--	-27.9	-8.2	-20.4	-6.2	-22.9	-6.7

appear in the shortwave with SRB 17 Wm^{-2} higher. Zhang and Rossow (personal communication) is also based on ISCCP data and show general agreement with SRB. Kiehl and Trenberth (1997) made calculations based on ERBE results and climate model radiation algorithms and gave a SW surface flux higher higher still, indicating there is still a substantial uncertainty in the community for this fundamental parameter.

The SRB data set is able to capture relevant temporal features. Fig. 1 shows time series anomalies for global and tropical averages of SW and LW surface fluxes. The reduction in SW radiation reaching the surface after the Pinatubo eruption in 1991 is a prominent feature, especially in the tropics. The recovery from the eruption is likely to be overshot, due to satellite orbital drift factors. Efforts to correct are ongoing. Spatial trends are featured in Fig. 2. There is a pronounced negative trend in July in the upper Midwest of North America, due to an increase in cloud amount seen by ISCCP.

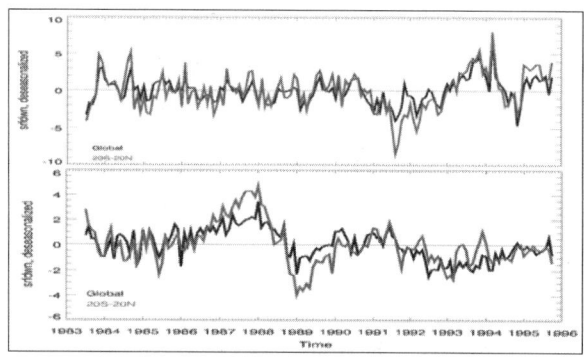

Fig. 1. Monthly anomalies of surface shortwave (above) and longwave (below) radiation fluxes.

3. VALIDATION WITH SURFACE MEASUREMENTS

The Release 2 data overlaps the start of the period covered by the Baseline Surface Radiation Network (BSRN) project by a few years. Validation results against BSRN are presented in Fig. 3. RMS error is highest at the 3-hrly resolution, and lowest at monthly resolution. Differences in cloudiness seen by the satellites at their single observation time over a 1° grid box and instruments at a fixed point likely account for the 3-hrly RMS values. Longwave RMS values are lower, partly due to the inherently lower intraday variability of longwave radiation. Both longwave and shortwave show a slight negative bias, though within BSRN instrument uncertainty. Future releases of SRB will more fully overlap the BSRN period, including many more observation sites, providing outstanding validation opportunities.

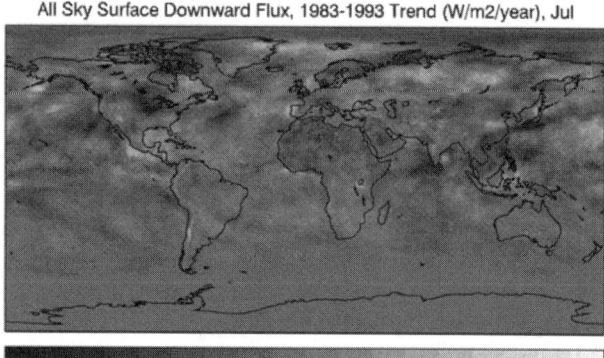

Fig. 2. Trends in monthly average surface shortwave flux over the 1983–1993 time period for January and July.

Further validation results are presented by Zhang et al. (2004).

The regional representativeness of a single point or grid box is a subject of recurring interest and has been investigated here. Fig. 4 shows an example. The correlation of anomalies in the surface shortwave time series between a grid box containing Bermuda and the surrounding grid boxes is shown. As would be expected, correlation is highest in the immediate vicinity of the subject box. In the case of Bermuda, the correlation is highest along a southwest-to-northeast axis. There are negative correlations as close as the southeastern United States. This is an apparent manifestation of recurring circulation patterns involving the position and strength of the Bermuda high and the subtropical jet. A strong Bermuda high will lead to clear conditions and high solar fluxes in Bermuda, and will be correlated to troughs, high cloud amount, and low solar fluxes in the SE US.

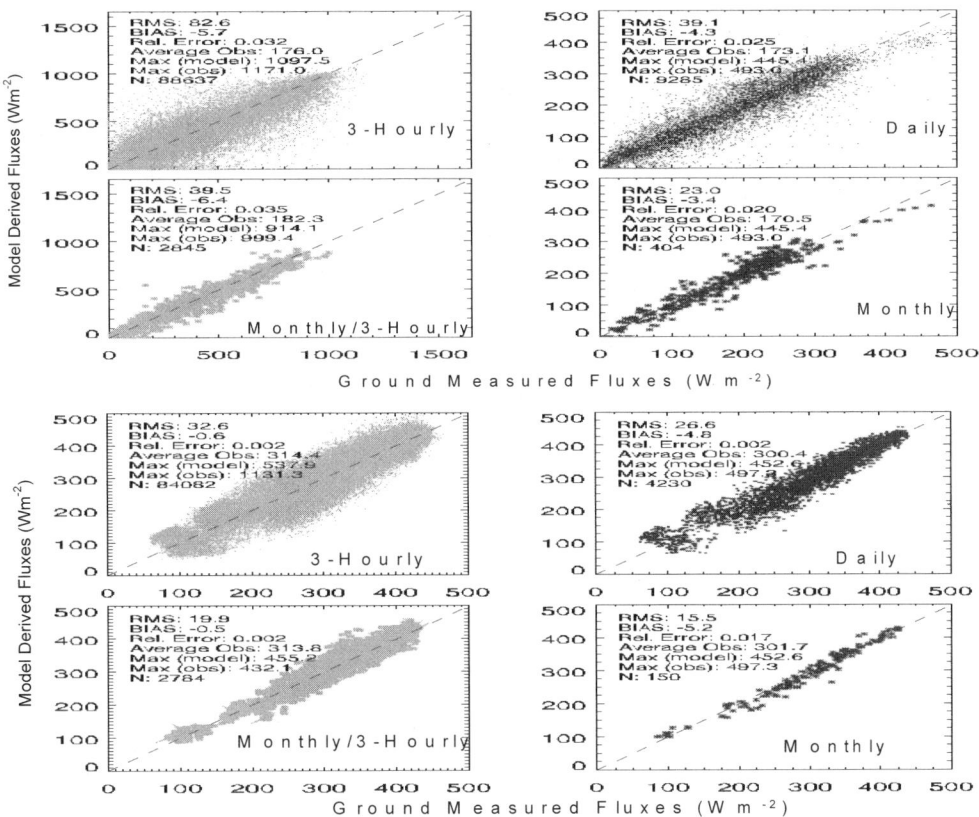

Fig. 3. Validation of SRB 1°x 1° grid box fluxes against BSRN measured station fluxes for shortwave (above) and longwave (below).

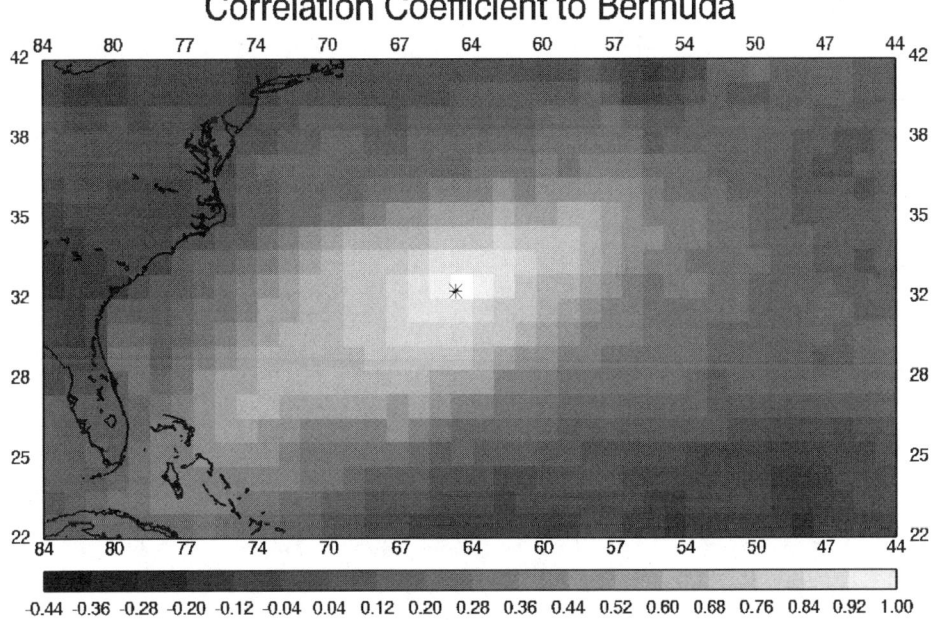

Fig. 4. Correlation of SW surface flux anomalies to Bermuda anomalies.

4. REFERENCES

Fu, Q., K. N. Liou, and A. Grossman, 1997: Multiple scattering parameterization in thermal infrared radiative transfer. *J. Atmos. Sci.,* 54, 2799-2814.

Gupta, S. K., W. L. Darnell, and A. C. Wilber, 1992: A parameterization for longwave surface radiation from satellite data: Recent improvements. *J. Appl. Meteor.,* 31, 1361-1367.

Gupta, S. K., D. P. Kratz, P. W. Stackhouse, Jr., and A. C. Wilber, 2001: The Langley Parameterized Shortwave Algorithm for surface radiation budget studies (Vers 1.0). *NASA/TP-2—1-211272,* 31 pp. Available online at http://techreports.larc.nasa.gov/ ltrs/ltrs.html.

Kiehl, J. T. and K. E. Trenberth, 1997: Earth's annual global mean energy budget. *Bull. Amer. Meteor. Soc.,* 78, 197-208.

Ohmura, A., and H. Gilgen, 1993: Re-evaluation of the global energy balance. Interactions between global climate subsystems: The legacy of Hann. *Geophys. Monogr.,* No. 75, International Union of Geodesy and Geophysics, 93-110.

Pinker, R., and I. Laszlo, 1992: Modeling surface solar irradiance for satellite applications on a global scale. *J. Appl. Meteor.,* 31, 194-211.

Rossow, W. B., and R. Schiffer, 1999: Advances in understanding clouds from ISCCP. *Bull. Amer. Meteor. Soc.,* 80, 2261-2287.

Stackhouse, P. W., Jr., S. K. Gupta, S. J. Cox, J. C. Mikovitz, T. Zhang, and M. Chiacchio, 2004: 12-Year surface radiation budget data set. *GEWEX News,* 14, 10-12.

Zhang, T., P. W. Stackhouse, Jr., S. K. Gupta, S. J. Cox, and J. C. Mikovitz, 2004: Signals of climate variations in the WCRP/GEWEX SRB datasets and their connections with other climate indices. *IRS2004: Current Problems in Atmospheric Radiation.*

CM-SAF SURFACE RADIATION BUDGET: FIRST RESULTS FROM THE INITIAL OPERATIONS PHASE

R. Hollmann, R. W. Müller, and A. Gratzki

Deutscher Wetterdienst, Kaiserleistr. 35, 63067 Offenbach, Germany

Rainer.hollmann@dwd.de

ABSTRACT

As part of the ground segment of EUMETSAT, the Satellite Application Facility (SAF) on Climate Monitoring (CM-SAF) will derive operationally consistent cloud and radiation parameters in high spatial resolution for an area that first covers Europe and part of the North Atlantic. The cloud and surface radiation products will be based on data from the polar orbiting satellites NOAA and METOP for the northern latitudes, and on data from METEOSAT second generation (MSG) for the mid latitudes. Later the area will be extended to the full MSG disk covering as well all of Africa. The cloud and radiation (top of the atmosphere and surface) products will be provided in different time and spatial resolutions, ranging from daily mean to monthly means. The spatial resolution is 15 by 15 km^2. Starting 2004 the CM-SAF has reached the Initial Operations Phase (IOP) which will be followed by the Full Operations Phase (FOP) starting in 2007. This paper focuses on the surface radiation budget products and will give an overview and presents first validation results.

1. INTRODUCTION

The surface radiation budget and its components are key parameters for climate monitoring. A non-balanced radiation budget is the cause for atmospheric motions in different scales. The components of the radiation balance are influenced by the surface properties, the atmosphere and by energy transports between the ocean and the atmosphere. Surface radiation budget parameters can be used to validate regional and global climate models, agro-meteorological models and NWP models.

The provision of basic monitoring products will form the basis for value added services of NMSs towards key user communities such as government agencies, power supply industries, agriculture, or water management. These value added services include the provision of climate atlases, monthly monitoring reports, regional expertise and information services and methods for dedicated applications. Especially for solar energy use, information concerning the distribution of the shortwave radiation components is very important. In the field of hydrometeorology surface radiation data are needed for computation of hydrological components, especially evaporation. In agro-meteorology, the solar irradiance has a strong influence on the growing of plants. Improved agro-climatological charts and precise agro-meteorological advice will be very helpful for individual farmers for planning purposes.

This article describes the products of the surface radiation budget of the Satellite Application Facility on Climate Monitoring (CM-SAF). Satellite Application Facilities (SAFs) are specialised development and processing centres within the EUMETSAT Applications Ground Segment. They complement the production of standard meteorological products derived from satellite data at EUMETSATs Central Facilities.

The next section gives an overview of the algorithms and needed input data sets, the section thereafter presents first pre-processing results and initial comparisons with other data sets. The paper will end with conclusion and future work.

2. ALGORITHMS AND INPUT DATA

The CM-SAF is using existent well validated algorithms to obtain the surface radiation budget (SRB). Four different components are derived independently but consistently, the surface solar incoming radiation (SIS), surface downwelling longwave radiation (SDL), surface outgoing longwave radiation (SOL) and surface albedo (SAL).

SIS is defined as the shortwave solar radiation flux reaching a horizontal unit earth surface in the 0.3–4 µm wavelength band, SDL as the thermal radiance from the atmosphere and the clouds reaching the ground surface in the 4–100 µm wavelength band and SOL is the thermal emission of the Earth's surface in the 4–100 µm wavelength band. All fluxes are expressed in Wm^{-2}. Fluxes will be calculated as the energy integrated over a certain period of time divided by this time length. For short wave fluxes, it must be noticed that a daily value in Wm^{-2} will represent the daily energy divided by 24 hours and not by the daylight duration. SAL is the ratio of the solar flux reflected from the Earth's surface to the surface incoming solar flux. It is expressed in percent.

From the individual components the net fluxes for longwave and shortwave as well as the surface radiation budget are derived. The net shortwave and longwave fluxes are defined as the difference between incoming and outgoing fluxes, and the total surface radiation budget is the sum of net shortwave and net longwave flux. The surface radiation algorithms use CM-SAF cloud information as cloud mask, cloud top pressure and cloud type as input into the retrieval algorithms.

3. SURFACE RADIATION FLUXES

For the longwave surface fluxes the Gupta algorithm (Gupta et al. (1989, 1992)) was adapted. It is a parameterised model which requires information on the temperature profile of the lowest layers of the atmosphere, the water vapour profile and cloud base height information. All temperature and humidity information used in the surface flux retrieval as well as the surface albedo calculations are taken from Numerical Weather Prediction (NWP) models. In the operational CM-SAF processing the DWD NWP model GME is used. The cloud base height is determined from the satellite retrieved cloud top pressure, assuming a fixed cloud thickness of 50 hPa.

To obtain the surface outgoing longwave flux the Stefan-Boltzmann equation with a surface dependent emissivity factor is used. The surface temperature is taken from NWP analysis data. In the future this might be replaced by satellite retrieved surface temperatures.

For the calculation of the surface incoming shortwave flux (SIS) an algorithm similar to the one developed by Pinker (e.g., Pinker and Laszlo, 1992) is used. The basic idea for the algorithm is that the relationship between the broadband (0.2–4.0 µm) atmospheric transmittance and the reflectance at the top of the atmosphere (TOA) can be described by state of the art radiative transfer models. With a radiative transfer model (e.g., LibRadtran (www.libRadtran.org), SHDOM (Evans, 1998)), the broadband atmospheric transmittance is calculated once in relationship to the broadband TOA albedo for a variety of atmospheric and surface states and stored in look-up tables. The actual computation of the surface irradiance involves two steps. First the broadband TOA albedo and the surface broadband albedo are determined from the satellite measurement. Then the atmospheric transmittance is determined from the TOA albedo together with information on the atmospheric and surface state from the pre-computed tables. To make this interpolation more time efficient and also take advantage of the inherent symmetries a modified Lambert-Beer (MLB) approach after Mueller et al. (2004) is used. Therefore it is only necessary to compute the SIS at two different sun zenith angles of 0° and 60°. For these angles a MLB relation is fitted and their values are stored and used in the interpolation to the measured sun zenith angle. For the aerosol, model computed single scattering albedo, asymmetry factor and aerosol optical thickness are used. Both are taken from OPAC/GADS climatology (Hess et al., 1998, Koepke et al., 1997) or alternative from global aerosol model climatologies (e.g., Kinne, 2004, pers. communications).

In summary, input parameters into the SIS-algorithm are the broadband top of the atmosphere (TOA) albedo, solar zenith angle, surface albedo, cloud properties, aerosol and ozone data and the total water vapour content of the atmosphere. The broadband TOA albedo for the area covered by the MSG satellite, the surface albedo and the cloud properties are derived by other groups within the CM-SAF (see other CM-SAF articles in this report). All other input data are either from climatological data sets (aerosol, ozone) or NWP-data (water vapour). The broadband TOA albedo for AVHRR is calculated using the narrow-to-broadband conversion from Hucek and Jacobowitz (1995) and the angular distribution models (ADM) from Suttles et al. (1988) for the conversion from radiances to fluxes.

Daily averages of longwave fluxes are derived by a linear average of all available data. The daily mean value of SIS is derived following the method by Möser (described in Diekmann et al., 1988) by taking into account the diurnal variation of the solar incoming clear-sky flux.

4. SURFACE RADIATION PRODUCT CHARACTERISTICS AND EXAMPLES

All surface radiation products will be provided as monthly means and, for MSG based products, also as monthly mean diurnal cycles. Additionally daily averages of shortwave fluxes and a weekly composite of the surface albedo will be available. Initially products will be based on locally received data from the polar orbiting NOAA satellites. The locally received data (receiving station Offenbach, Germany) do not cover the CM-SAF baseline area (30°N–80°N, 60°W–60°E) completely. Additionally a low number of daily NOAA overpasses affects the quality of the daily averages. Version 2 surface radiation products will additionally use data from the Meteosat Second Generation (MSG) satellite, which provides full disk images every 15 minutes and therefore allows a much better representation of the diurnal cycle and a better coverage of the CM-SAF baseline area. Version 3 products finally will combine NOAA and MSG data, using MSG data for mid- and low latitudes and NOAA data for higher latitudes outside the field of view of MSG. The spatial resolution of the surface radiation products is 15 x 15 km^2.

The pre-operational processing of NOAA-AVHRR data started in February 2004 and is changing into an operational system in 2005. Fig. 1 shows the monthly mean of land surface radiation fluxes for July 2004. The different spatial coverage for the shortwave and longwave fluxes is due to the fact that a minimum number of images per day is required for the calculation of daily averages, which is less frequently achieved for daytime overpasses.

5. INITIAL VALIDATION RESULTS FOR INSTANTANEOUS SURFACE FLUXES

Two reference months, June 2003 and October 2002, and the pre-operational month March 2004 have been selected for a first validation study using CM-SAF cloud data from the Advanced Very High Resolution Radiometer (AVHRR) onboard the polar-orbiting NOAA satellites. For this study, data from single satellite images (instantaneous data) were compared with surface measurements. The validation is done as follows: for each instantaneous overpass an area of 15 by 15 pixels corresponding to about

15 x 15 km² symmetrically around the surface site has been extracted from the satellite derived data and their statistics was computed. The surface measurements are available in different temporal resolutions from hourly mean values down to one minute averages, thus the decision was taken to average all surface data within one hour and compare this to the satellite spatial mean as described above. All measurements were taken and the monthly mean values have been calculated and compared. The results of the comparison are shown in Table 1 for SDL and in Table 2 for the shortwave fluxes.

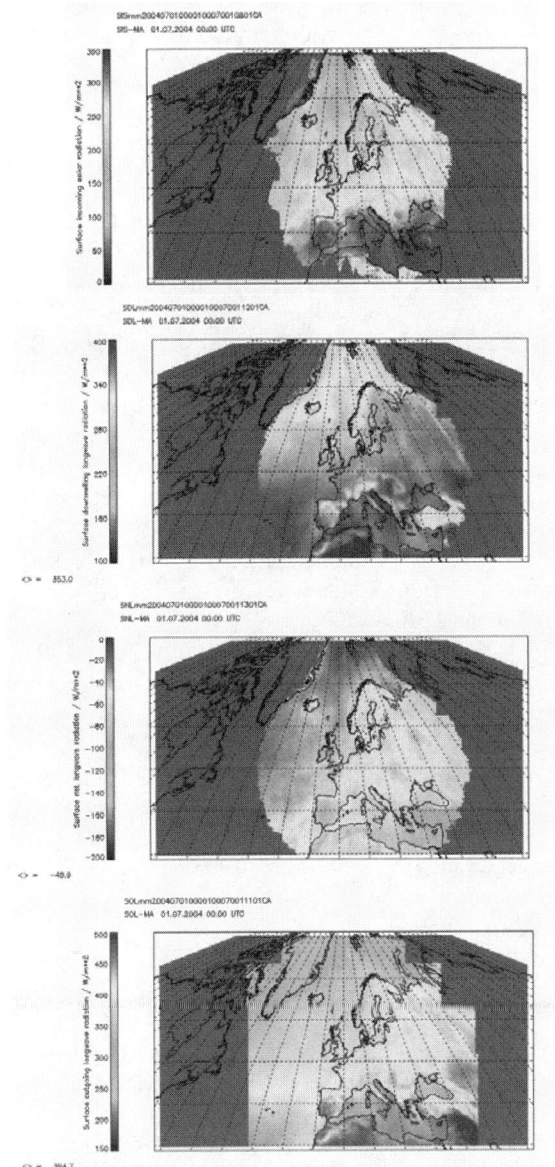

Fig. 1. Monthly mean of surface radiation products in W/m² for July 2004 for the CM-SAF area based on NOAA-AVHRR data: Surface incoming solar radiation downwelling longwave radiation, outgoing longwave radiation and net longwave radiation (from top to bottom).

Table 3. Validation Results for SDL for Different Surface Sites and Months. Given is the Monthly Mean Difference in W/m² of Satellite Estimate Minus Measurement Value

Station	Oct. 2002	Jun. 2003	Mar. 2004
Carpentras	–14.2	–8.1	–17.2
Cabauw	–14.0	–11.6	–9.5
Lindenberg	–3.7	8.2	–2.8
Payerne	–15.4	–14.1	–27.5
Average	**–11.9**	**–5.5**	**14.3**

The validation of the instantaneous satellite derived vs. hourly averaged surface measurements of the longwave components shows a good agreement which is close to (SDL) or within (SOL) the target accuracy of 10 W/m² for monthly means with some exceptions (Table 1). The water vapour amount as well as the surface temperature is provided by the NWP analysis which might be an error source for specific regions. The reasons for the biases have to be analysed in more detail, which is in progress.

Table 2. Results of Validation for SIS for Different Surface Sites and Months. Given is the Mean Difference in W/m² (and in Percentage of the Monthly Mean) of Satellite Estimate Minus Measurements

Station	Jun. 2003		Mar. 2004	
Carpentras	5.8	(1.0 %)	0.3	(0.1 %)
Cabauw	14.8	(2.9 %)	18.8	(8.2 %)
Lindenberg	10.8	(2.4 %)	–6.6	(2.9 %)
Payerne	17.3	(3.2 %)	21.9	(8.8 %)
Average	**12.2**	**(2.4 %)**	**8.6**	**(3.6 %)**

General sources of error for the longwave fluxes include the emissivity map which has no temporal (i.e., seasonal) variation. Also elevation effects (resulting from a different spatial resolution of the NWP and satellite data) are not taken into account, which may explain some of the higher biases for Payerne with its elevation of 600 m MSL.

The SIS results are encouraging but some months and stations (e.g., Cabauw and Payerne stations in March 2004) show larger biases (Table 2).

At the moment no winter month has been processed, where very bright snow covered surface will be a critical point for the SIS derivation as well as for good cloud detection. Of general interest for the quality of SIS will be the selection of appropriate aerosol information.

At all, the validation of the first results of the processing indicates that the target accuracy can be achieved. It is important to note that the derived biases can not be transferred one-to-one to monthly or daily means, because of the averaging process. In this context the percentage biases are the relevant quantities. They are small demonstrating the usability of the SIS products for climate monitoring purposes.

6. CONCLUSION AND OUTLOOK

Currently, global trends cannot be projected to regional scales (e.g., Dorn et al., 2003) which in reverse emphasises the importance of high spatial and temporal resolution data products to have an appropriate regional climate monitoring and analysis. Better understandings of the processes involved in regional climate change are important for the understanding of the climate system. Therefore the monitoring of radiative forcing of the atmosphere in high spatial and temporal resolution is of great importance for the analysis and understanding of climate dynamics and its changes. The CM-SAF with its products is aiming at this to provide the user with such a high quality data-set.

In February 2004, the CM-SAF started the pre-operational processing of NOAA-AVHRR data to derive climate data sets for the target area in Europe. This paper describes the algorithms which are used to derive surface radiation budget products on high spatial scale of 15 x 15 km². First results of the ongoing validation with selected sites have been shown.

For all components of the surface radiation budget the validation results are encouraging and indicate that the targeted accuracies on monthly mean scales can be reached. Individual flux cases with higher differences to surface measurements need further critical investigation. The validation demonstrates the potential of the radiation algorithms to retrieve radiation products usable for climate monitoring and analysis of regional climate change. Several algorithm improvements are currently in progress, aiming to improve the data accuracy. An operational continuous validation and product monitoring will provide a homogenous and consistent product with high quality.

Generally, one has to take into account that with the overpasses of AVHRR providing a poor temporal coverage during the day is an important source of errors. With the use of all new features of the Meteosat Second Generation satellite in 2005 it is expected that the statistics will improve and it is expected that the biases especially of SIS will be reduced.

7. ACKNOWLEDGEMENTS

The surface radiation data have been kindly provided by the Swiss Meteorological Service (MeteoSwiss), Alfred Wegener Institute (Germany), Deutscher Wetterdienst, KNMI, Météo-France. We thank Arne Kylling (NILU) and Bernhard Mayer (DLR) for providing the LibRadtran RTM package. LibRadtran is available from http://www.libradtran.org. We thank Frank Evans for providing SHDOM. References

8. REFERENCES

Diekmann, F.-J., 1988: An operational estimate of global solar irradiance at ground level from METEOSAT data: results from 1985 to 1987. *Meteor. Rundsch.*, 41, 65-79.

Dorn, W., K. Dethloff, A. Rinke, and E. Roeckner, 2003: Competition of NAO regime changes and increasing greenhouse gases and aerosols with respect to Arctic climate projections. *Climate Dyn.*, 21, 447-458.

Evans, K. F., 1998: The spherical harmonics discrete ordinate method for three-dimensional atmospheric radiative transfer. *J. Atmos. Sci.*, 55, 429-446.

Gupta S., 1989: A parameterization for longwave surface radiation from sun-synchronous satellite data. *J. Climate*, 2, 305-320.

Gupta S. K., W. L. Darnell, and A. C. Wilber, 1992: A parameterization for longwave surface radiation from satellite data: recent improvements. *J. Appl. Meteor.*, 31, 1361-1367.

Hess, M., P. Koepke, and I. Schult, 1998: Optical properties of aerosols and clouds: the software package OPAC. *Bull. Amer. Meteor. Soc.*, 79, 831-844.

Hucek, R., and H. Jacobowitz, 1995: Impact of scene dependence on AVHRR albedo models. *J. Atmos. Oceanic. Technol.*, 12, 697-711.

Koepke, P., M. Hess, I. Schult, and E. Shettle, 1997: Global aerosol data set. MPI Meteorologie Hamburg Tech. Rep. 243, 44 pp.

Mueller, R. W., and Coauthors, 2004: Rethinking satellite-based solar irradiance modelling: The SOLIS clear sky module. *Remote Sens. Environ.*, 91, 160-174.

Ohmura, A., and Coauthors, 1998: Baseline surface radiation network (BSRN/WCRP): new precision radiometry for climate research. *Bull. Amer. Meteor. Soc.*, 79, 2115-2136.

Pinker, R. T., and I. Laszlo, 1992: Modelling surface solar irradiance for satellite applications on a global scale. *J. Appl. Meteor.*, 31, 194–211.

GLOBAL SPHERICAL MODEL OF EARTH RADIATION IN PLANET SCALE

T. A. Sushkevich, S. A. Strelkov, E. V. Vladimirova, A. N. Volkovich, V. V. Kozoderov,
A. K. Kulikov, and S. V. Maksakova

Keldysh Institute of Applied Mathematics, Russian Academy of Sciences
4, Miusskaya Ploschadj, Moscow, 125047 Russia, Email: tamaras@keldysh.ru

ABSTRACT

Adequate understanding of the radiation processes is necessary for providing scientific and technological progress and for preventing possible negative consequences of the climate changes or significant deviations of the spectral-radiation balance of the planet. Unfortunately, no possible climate and biophysical changes can be predicted at present for certain. This is so, in particular, because of low accuracy of the description of radiation in the climate models as well as of the atmospheric and oceanic circulation.

This paper continues our long-term research on the development of methods and algorithms for numerical solution of the problems of radiation transfer in scattering, absorbing, and emitting spherical systems with a complex structure. This work has been stimulated, to a great extent, by a significant change in the information technologies due to implementation of high-efficiency multiprocessor computers with parallel structures.

1. INTRODUCTION

Radiation transfer in the Earth's atmosphere is investigated on the scales of the entire planet. A specified method is proposed of numerical solution of a general boundary condition problem for the theory of radiation transfer in a spherical layer with a mosaic reflecting underlying surface to simulate Earth's radiation field. Models of influence functions have been formulated for the boundary problem of transfer theory with spherical geometry. The global radiation field of the planet as the atmosphere—mosaic Earth's surface (land,ocean) is calculated by means of a functional—the "superposition integral" which describe link between the "scenarios" of surface and the image. The influence function is the kernel of this functional which is called the optical transfer operator (Sushkevich et al., 1990).

Of great importance the radiation transfer problem in the atmosphere has got recently due to a necessity of a comprehensive analysis of physical, chemical, meteorological, biological processes responsible for the Earth's radiation field formation. Radiative processes do play a central role in the heat and energy exchange in the atmosphere and consequently in global and local climate formation of the planet. A certain misbalance of radiative processes in the atmosphere—surface system under anthropogenic and natural influences may result in a destruction of self-restoring potential of the biosphere and cause the relevant catastrophic consequences. Uncertainty is still large nowadays of forecasts concerning possible climate and biophysical changes due to, in particular, low accuracy of the radiation description in models of climate, circulation of the atmosphere and ocean, etc., is still large nowadays.

Mechanisms of variability (dynamic processes: circulation, convection, turbulent fluxes; radiation processes; photochemical processes) of geophysical, meteorological, climate conditions of the Earth are affected by radiation field; these mechanisms possess complex non-linear links that make difficult forecasting the listed effects, their estimates and significance. Electromagnetic field of the Earth registered by various instruments while remote sensing of natural targets carries information about the current condition of the environment. Practically all this information about the Earth's surface and atmosphere is contained in the spectral distribution of electromagnetic waves which are emitted or reflected by the Earth into space and are measured by space vehicles (SV). The Earth's surface and atmosphere are an entire "atmosphere—underlying surface" system both components of which are permanently vary in time and space.

The described in (Sushkevich, 1966; Sushkevich et al., 1997; Sushkevich, 1999; Sushkevich, 1999a) approach based on an analysis of the relevant equations for characteristics in curvature coordinates and on different means to take the anisotropy effect into account along with speeding up of iterative convergence in the sub-areas of calculation enables to proceed to numerical solution of the 3D-inhomogeneous spherical problem that simulates conditions closely related to natural ones on the planet Earth. This approach gets urgent for realization due to the necessity to solve problems of photo-radiation chemistry of the atmosphere (troposphere and ozone-sphere in Sun-uprising, twilight and occultation conditions, for polar regions) as well as to obtain information products regarding tomography research for the Earth's atmosphere including refractometric techniques and applied to the space systems operating in observation conditions with horizontal routes. Other applications are concerned polar regions, elaboration of models of spectral & radiation budget of the Earth, phase radiance of the Earth for the space navigation instruments (return of the SV on the Earth, navigation of the SV along the Earth), specified projects realization dealing with additional sources of energy onboard the SV by using solar radiation reflected by the Earth. Monitoring of environmental disasters and their consequences is also among these applications.

New perspective opportunities of mathematical modeling of the atmospheric radiation for the planet Earth are connected with qualitative alterations of information technologies resulted from updated high-productive computer systems as well as with software for the wide areas of applications while utilizing super-computers with parallel calculations.

2. METHODOLOGICAL CONCEPTION

Transfer problem is considered for optical (solar and long-wave) radiation in the atmosphere—Earth system (AES) while using a spherical shell approach with an incident external parallel flux. The optical transfer operator (OTO) is constructed within the linear-system approach (Sushkevich and Vladimirova, 2003; Sushkevich and Vladimirova, 2003a) to take into account the effect of spatial inhomogeneity (mosaic) of underlying surface (land surface, upper layer of cloudiness or hydrometeors) into account. The influence function (IF) of the general boundary-value problem (GBP) of radiation transfer theory that serves as the OTO kernel is considered as a universal characteristic invariant relative to particular structures of inhomogeneity for the relevant reflecting and emitting surfaces.

The total intensity of monochromatic (the wave length λ is fixed) or quasi-monochromatic (under fixed λ and the width $\Delta\lambda$) stationary radiation $\Phi_\lambda(\mathbf{r},\mathbf{s})$ (λ is omitted below) in any point $A(\mathbf{r})$ with its radius-vector $\mathbf{r}$ in any direction $\mathbf{s}$ is found as the solution of the general radiation transfer boundary-value problem

$$K\Phi = F^{in}, \quad \Phi\big|_t = F^t, \quad \Phi\big|_b = \varepsilon R\Phi + F^b \quad (1)$$

within the phase area Γ with the relevant linear operators: transfer operator

$$D \equiv (\mathbf{s}, \text{grad}) + \sigma_{tot}(\mathbf{r}), \quad (2)$$

for the 3-D spherical geometry problem

$$(s, \nabla\Phi) = \cos\vartheta \frac{\partial\Phi}{\partial r} + \frac{\sin\vartheta\cos\varphi}{r}\frac{\partial\Phi}{\partial\psi} - \frac{\sin\vartheta}{r}\frac{\partial\Phi}{\partial\vartheta} +$$
$$+ \frac{\sin\vartheta\sin\varphi}{r\sin\psi}\frac{\partial\Phi}{\partial\eta} - \frac{\sin\vartheta\sin\varphi\,\text{ctg}\,\psi}{r}\frac{\partial\Phi}{\partial\varphi}; \quad (3)$$

the collision integral or the source function is given by

$$B(\mathbf{r},\mathbf{s}) \equiv S\Phi = \sigma_{sc}(\mathbf{r})\int_\Omega \gamma(\mathbf{r},\mathbf{s},\mathbf{s}')\Phi(\mathbf{r},\mathbf{s}')d\mathbf{s}', \quad (4)$$

$$d\mathbf{s}' = \sin\vartheta'd\vartheta'd\varphi';$$

integral & differential operator is written as $K \equiv D - S$; the reflectivity operator is given in general case by

$$[R\Phi](\mathbf{r}_b,\mathbf{s}) = \int_{\Omega^-} q(\mathbf{r}_b,\mathbf{s},\mathbf{s}^-)\Phi(\mathbf{r}_b,\mathbf{s}^-)d\mathbf{s}^-, \quad (5)$$

$$\mathbf{s}\in\Omega^+.$$

The function $F^{in}(\mathbf{r},\mathbf{s})$ represents here the radiation sources density inside the area G; $F^b(\mathbf{r}_b,\mathbf{s}^+)$ and $F^t(\mathbf{r}_t,\mathbf{s}^-)$ are radiation sources on boundaries of the spherical shell determined for the rays $\mathbf{s}$ directed inside the area G.

The operator R describes the law of radiation reflection on the underlying surface that is situated on the lower layer of the boundary G_b; the parameter $0 \leq \varepsilon \leq 1$ fixes the related act of interaction between radiation and the underlying layer. If $R \equiv 0$ (or $\varepsilon = 0$) the first boundary-value problem (FBP) of radiation transfer is dealt with

$$K\Phi_a = F^{in}, \quad \Phi_a\big|_t = F^t, \quad \Phi_a\big|_b = F^b \quad (6)$$

for any spherical layer with transparent non-reflecting absolutely "black" boundaries or with "vacuum" boundary conditions.

Three types of radiation problems can be outlined which require the Earth's surface effect be taken into account. The first type is given by the problems of energy and Earth's radiation budget when radiation of the Sun serves as the source. These problems are solved mainly in the approach of a planar model of the Earth shell with an implicit account of a homogeneous underlying surface (lambertian or non-orthotropic). The second type is represented by remote sensing problems of the atmosphere and cloudiness when the Earth's surface is considered as a clutter. The third type is concerned remote sensing problems for the Earth's surface when it is necessary to remove the effect of the atmosphere (to conduct atmospheric correction) or at least to take its effect into account.

The following components are always present in any active or passive remote sensing system of the Earth surface: a category of "scenarios" or "scene", i.e. a radiance distribution of observed targets or landscape; an atmospheric channel of imagery transformation; a measuring instrument for electromagnetic waves; a hardware complex of data processing and imagery recognition. The effect of the atmosphere is within the three listed components (besides the instrument): atmospheric & optical mechanisms do influence on the "scenarios" formation, on the relevant image transfer via natural media and are taken into account in radiation correction procedures while "scene" analysis.

The atmospheric channel is considered as an element of the optical system radiation transfer and the optical transfer operator is formulated using mathematical techniques of the linear-system approach and the superposition integral.

The general boundary problem (1) with the operators (2)–(5) is linear on the sources and its solution can be

RADIATIVE BUDGET AND FORCING

looked for in the form of the superposition $\Phi = \Phi_a + \Phi_q$. The background radiation Φ_a is determined as a solution of the FBP problem (6). The problem of path radiance determination Φ_q due to the reflecting effect of the underlying surface is just what is defined as GBP

$$K\Phi_q = 0, \quad \Phi_q\big|_t = 0, \quad \Phi_q\big|_b = \varepsilon R\Phi_q + \varepsilon E, \quad (7)$$

where the underlying surface illumination (radiance, flux) created by the background radiation $E = R\Phi_a$ serves as a source of the incident solar radiation (insolation).

3. RESULTS

Theoretical constructions and algorithms of the optical transfer operator calculation are based on theory of generalized solutions, theory of integral transformations for the generalized functions and on series of general theory of regular disturbances (an asymptotic method). The approach elaborated on these rigorous mathematical bases is called as the influence function method.

Solution of first boundary-value problem

$$K\Phi = 0, \quad \Phi\big|_t = 0, \quad \Phi\big|_b = f(\mathbf{s}^h; r_\perp, \mathbf{s}), \quad (8)$$

where $r_\perp = (\psi, \eta) \in \Omega$, $dr_\perp = \sin\psi \, d\psi \, d\eta$, being distribution, can be written in the form of linear functional – superposition integral

$$\Phi(\mathbf{s}^h; r, r_\perp, \mathbf{s}) \equiv (\Theta, f) \equiv$$
$$\equiv \frac{1}{2\pi} \int_{\Omega^+} d\mathbf{s}_h^+ \frac{1}{4\pi} \int_\Omega \Theta(\mathbf{s}_h^+; r, r_\perp - r'_\perp, \mathbf{s}) \times$$
$$\times f(\mathbf{s}^h; r'_\perp, \mathbf{s}_h^+) \sin\psi' \, d\psi' \, d\eta'.$$

Its kernel is influence function $\Theta(\mathbf{s}_h^+; r, r_\perp, \mathbf{s})$ — solution of FBP (Model 1)

$$K\Theta = 0, \quad \Theta\big|_t = 0, \quad \Theta\big|_b = f_\delta$$

with parameter $\mathbf{s}_h^+ \in \Omega^+$ and source

$$f_\delta(\mathbf{s}_h^+; r_\perp, \mathbf{s}) = \delta(r_\perp)\delta(\mathbf{s} - \mathbf{s}_h^+).$$

If source $f(r_\perp)$ – isotropic and horizontally-nonhomogeneous then the solution of FBP (8) is linear functional – convolution

$$\Phi(r, r_\perp, s) = F_c(f) \equiv (\Theta_c, f) \equiv$$
$$\equiv \frac{1}{4\pi} \int_\Omega \Theta_c(r, r_\perp - r'_\perp, s) f(r'_\perp) \sin\psi' \, d\psi' \, d\eta'$$

with kernel – influence function

$$\Theta_c(r, r_\perp, \mathbf{s}) = \frac{1}{2\pi} \int_{\Omega^+} \Theta(\mathbf{s}_h^+; r, r_\perp, \mathbf{s}) d\mathbf{s}_h^+,$$

which satisfies FBP (8) with axial symmetry (Model 2)

$$K\Theta_c = 0, \quad \Theta_c\big|_t = 0, \quad \Theta_c\big|_b = \delta(r_\perp).$$

For anisotropic and horizontally-uniformity source

$$\Phi(\mathbf{s}^h; r, \mathbf{s}) = F_r(f) \equiv (\Theta_r, f) \equiv$$
$$\equiv \frac{1}{2\pi} \int_{\Omega^+} \Theta_r(\mathbf{s}_h^+; r, \mathbf{s}) f(\mathbf{s}^h; \mathbf{s}_h^+) d\mathbf{s}_h^+$$

with linear functional kernel

$$\Theta_r(\mathbf{s}_h^+; r, \mathbf{s}) = \frac{1}{4\pi} \int_\Omega \Theta(\mathbf{s}_h^+; r, r_\perp, \mathbf{s}) \sin\psi \, d\psi \, d\eta.$$

The influence function Θ_r is the solution of one-dimensional spherical FBP with azimuth dependence (Model 3)

$$K_r\Theta_r = 0, \quad \Theta_r\big|_t = 0, \quad \Theta_r\big|_b = \delta(\mathbf{s} - \mathbf{s}_h^+).$$

At isotropic and horizontally homogeneous source the solution of FBP (8)

$$\Phi(r, \mathbf{s}) = fW(r, \mathbf{s}), \quad f = \text{const}$$

is calculated by influence function

$$W(r, \mathbf{s}) =$$
$$= \frac{1}{2\pi} \int_{\Omega^+} d\mathbf{s}_h^+ \frac{1}{4\pi} \int_\Omega \Theta(\mathbf{s}_h^+; r, r_\perp, \mathbf{s}) \sin\psi \, d\psi \, d\eta =$$
$$= \frac{1}{4\pi} \int_\Omega \Theta_c(r, r_\perp, \mathbf{s}) \sin\psi \, d\psi \, d\eta =$$
$$= \frac{1}{2\pi} \int_{\Omega^+} \Theta_r(\mathbf{s}_h^+; r, \mathbf{s}) d\mathbf{s}_h^+,$$

which is called also transport function taking account of multiple scattering and must be determined as the solution of one-dimensional spherical FBP (Model 4)

$$K_r W = 0, \quad W\big|_t = 0, \quad W\big|_b = 1.$$

On the basis of regular perturbation theory by means of series

$$\Phi_q(\mathbf{s}^h; \mathbf{r}, \mathbf{s}) = \sum_{k=1}^\infty \varepsilon^k \Phi_k.$$

GPB (7) is reduced to recursive system of FBP (8):

$$K\Phi_k = 0, \quad \Phi_k\big|_t = 0, \quad \Phi_k\big|_b = E_k \quad (9)$$

with sources $E_k = R\Phi_{k-1}$ for $k \geq 2$, $E_1 = E$.

An operation is introduced that describes a single act of radiation interaction with underlying surface by defined influence functions:

$$[Gf](\mathbf{s}^h;\mathbf{r}_b,\mathbf{s}) \equiv R(\Theta,f) =$$
$$= \int_{\Omega^-} q(\mathbf{r}_b,\mathbf{s},\mathbf{s}^-)(\Theta,f)d\mathbf{s}^-.$$

Solutions of system FBP (9) are found as linear functionals

$$\Phi_1 = (\Theta,E), \quad \Phi_k = (\Theta,R\Phi_{k-1}) = (\Theta,G^{k-1}E).$$

Asymptotically exact solution of GBP (7) is obtained as linear functional—optical transfer operator

$$\Phi_q = (\Theta,Y),$$

where optical image "scenarios" or underlying surface brightness

$$Y \equiv \sum_{k=0}^{\infty} G^k E = \sum_{k=0}^{\infty} R\Phi_k \quad (10)$$

are given by the sum of Neumann series in accordance with multiple reflectivity of radiation from the underlying surface taking multiple scattering within the medium into account.

The "scenarios" (10) would satisfy to the known Fredholm equation of second kind

$$Y = R(\Theta,Y) + E,$$

which is usually called an equation for the "near-ground image". Total radiation or "space image" of the AES are described by the linear functional

$$\Phi = \Phi_a + (\Theta,Y). \quad (11)$$

This is the linear functional (11) that gives an adequate to initial GBP (1) mathematical model of radiation transfer within the AES for various structures of the E source and types of underlying surface not depending on the AES dimension (1-, 2- or 3-D). It is sufficient to calculate a finite Neumann series only for the "scenarios" (10) instead of the series calculation on the multiple reflection in the total phase volume of the GBP (1) solution.

All the variety of existing approaches can be derived to the following basic three approaches as our analysis has shown of the problem of taking into account the relevant effects and remote sensing of the Earth's surface. An implicit technique of taking the surface reflectivity into account was the first approach that emerged earlier. The second approach deals with functionals and conjugated equations. The third approach is given by an explicit technique of the indicated IF method. The IF terminology unifies all types of singularity and diffusivity of the source and all types of surfaces.

4. CONCLUSION

As a result of the outlined approaches, the initial GBP (1) has been reduced to the linear functional (11) with the formulation of the linear-system approximation to solve remote sensing problems and to take into account the contribution of reflected and emitted spherical Earth's surface in 1-, 2- and 3-D geometry according to **r** and with 2-, 3-, 4-, and 5-metric phase spaces (**r**, **s**). Any manifestation of non-linear effects due to multiple re-reflection of radiation from the Earth surface has been rigorously determined in the process of the "scenarios" formation; these effects are described by the influence functions which are represented by the linear transfer characteristics of any isolated shell of the atmosphere. It is necessary to note that the IF calculations in the multi-dimension case of the spherical problem can effectively be realized by Monte-Carlo techniques as well as in a small-angular approach.

5. ACKNOWLEDGMENTS

The work has been supported by the Russian Foundation for Basic Research (project 03-01-00132).

6. REFERENCES

Sushkevich, T. A., 1966: An axis-symmetrical problem of radiation propagation in spherical system of coordinates. Report No. 0-572-66, IAM USSR AS.

Sushkevich, T. A., S. A. Strelkov, and A. A. Ioltukhovsky, 1990: Characteristic Method to the Atmospheric Optics Problems. (in Russian) Moscow, Nauka. 296 p. (Book review in *Transport theory and statistical physics*, Vol. 22, 4, 587-591 (1993)).

Sushkevich, T. A., E. V. Vladimirova, E. I. Ignatijeva, A. K. Kulikov, S. V. Maksakova, and S. A. Strelkov, 1997: A spherical model of radiation transfer in the Earth atmosphere. 3. Approach. Method of solution. Reprint No. 85, KIAM RAS, 32 p.

Sushkevich, T. A., 1999: About atmospheric correction problem of satellite information. In *Studies of the Earth from Space*, No. 6, 49-66.

Sushkevich, T. A., 1999a: About modeling of solar radiation transfer in spherical atmosphere of the Earth and in clouds. In *Optics of the Atmosphere and Ocean*, Vol. 12, No. 3, 251-257.

Sushkevich, T. A., and E. V. Vladimirova, 2003: About a model of taking reflectivity on boundaries into account in radiation transfer problems in a spherical shell. In *Siberian Journal of Computational Mathematics*, Vol. 6, No. 1, 73-88.

Sushkevich, T. A., and E. V. Vladimirova, 2003a: Optical transfer operator for the spherical "atmosphere – Earth" system. In *Optics of the Atmosphere and Ocean*, Vol. 16, No. 4, 305-310.

UV RADIATION IN PAST AND FUTURE MODELLING WITH ALL ATMOSPHERIC PARAMETERS INCLUDING CLOUDINESS

J. Reuder
Geophysical Institute, University of Bergen, Allegaten 70, N-5007 Bergen, Norway

P. Koepke and J. Schween
Meteorologisches Institut der Universität München, Theresienstrasse 37, 80333 München, Germany

ABSTRACT

A reconstruction of spectral UV irradiance for the time period 1968–2000 with a temporal resolution of one hour has been performed for various sites in Central Europe. Algorithms for the derivation of the required input parameters total ozone, aerosol, surface albedo and clouds have been developed to determine corresponding input data sets for the radiation transfer calculations from routine observations. The resulting spectral UV irradiances have been weighted by different action spectra and with respect to various biologically relevant time intervals. The results for Central Europe show an increase of 2 % to 5 % per decade for UV-B radiation due to the stratospheric ozone reduction.

Future UV levels have been modelled for $2xCO_2$ climatic conditions. Total ozone content for this scenario has been derived from results of the coupled climate chemistry model ECHAM4/CHEM. Cloud effects on UV irradiances have been described by cloud modification factors, derived from cloud cover by a parameterization using the liquid and ice water content determined by the regional climate model MM5/MCCM. Surface albedo values have been converted from snow cover data of this model. The results show that UV is expected to decrease in the future due to the recovering ozone layer, but a further UV enhancement is predicted for summertime due to a decreasing cloud amount in Central Europe.

1. INTRODUCTION

Model calculations are the only way to estimate future UV radiation levels. But also for the past, reconstruction of UV radiation by means of radiation transfer models is necessary, since UV irradiance is measured only at few stations and only for a limited number of years. Long term data sets are required for the investigation of biological and medical effects of UV radiation. Both, UV scenarios for future climate conditions and UV reconstruction for the past decades will be presented for Central Europe.

2. METHOD

2.1 Radiation Transfer Model STAR

All UV calculations for past and future conditions have been performed with STAR (**S**ystem for the **T**ransfer of **A**tmospheric **R**adiation), a 1-dimensional radiation transfer model developed at the Meteorological Institute, University of Munich (Ruggaber et al., 1994, Schwander et al., 2002). It is based on the matrix-operator method (Nakajima and Tanaka, 1988) and considers absorption and scattering of all UV relevant atmospheric components as well as surface albedo. The quality of the model is proved by various intercomparisons with other multiple scattering models and measurements (Koepke et al., 1998, de Backer et al., 2001, van Weele et al., 2000). The version STARneuro (Schwander et al., 2002) additionally enables the consideration of the effect of inhomogeneous and broken cloudiness on spectral UV radiation. Cloud cover and cloud type are input information for an algorithm based on a neural network.

2.2 Total Ozone Content

Besides solar elevation and clouds, the total ozone content (TOC) is the crucial atmospheric parameter for UV-B radiation reaching the Earth's surface. Ground based Dobson and Brewer measurements of TOC from around 20 stations all over Europe have been used for the presented UV reconstruction starting 1968. An interpolation algorithm has been developed and applied, which additionally considers an orographic correction due to the effects of tropospheric ozone near the earth surface on the TOC at different altitudes above sea level (Reuder and Koepke, 2005).

The model calculation of future UV levels is based on TOC scenarios derived from coupled climate chemistry simulations by ECHAM/CHEM (Reuder et al., 2001).

2.3 Aerosols

Measurements of the average aerosol optical depth (AOD) in Europe show an annual course with higher values in summer and a minimum during winter (Weller et al., 1998). Accordingly the AOD has been considered with its typical annual behaviour (Reuder and Koepke, 2005). To describe the aerosol properties, spectral behaviour of scattering coefficient and single scattering albedo, the type 'average continental' from OPAC (Hess et al., 1998) is used.

Aerosol conditions, especially in densely populated areas, show high spatial and temporal variability with distinct effects on UV radiation (e.g., Reuder and Schwander, 1999). However, the aerosol description given above is used unchanged for past and future UV calculations, since average UV conditions are analysed here.

2.4 Surface Albedo

Surface albedo for UV calculations has been derived from information on snow cover. A neural network algorithm (Schwander et al., 1999) is used to transform actual snow height and the number of days since last snow fall into spectral albedo values. For UV reconstructions the necessary information on snow cover comes from daily observations. For the purpose of modelling future UV the model output of regional climate simulations by MM5/MCCM has been taken into account.

2.2 Clouds

The average cloud effect on the irradiance is usually described by a cloud modification factor CMF that represents the ratio between the radiation in the presence of clouds and the radiation under cloud free but otherwise identical atmospheric conditions (Borkowski et al., 1977, Koepke et al., 2002, Schwander et al., 2002).

CMFs for UV reconstruction in the past have been determined from hourly observations of cloud type and cloud cover performed by the German Meteorological Service (see Reuder and Koepke, 2005). The CMF information for future conditions has been derived from the humidity profiles from regional climate simulations using MM5/MCCM.

3. RESULTS

3.1 Reconstruction of UV Irradiance in the Past

The reconstruction of the different weighted UV irradiances is presented on the example of two Central European stations, Hoher Peissenberg (47.8°N, 11.0°E, 980 m a.s.l.) and Würzburg. (49.8°N, 10.0°E, 170 m a.s.l.). Fig. 1 shows the daily average exposures of UV-A (315 nm–400 nm, H_{UVA}), UV-B (280 nm–315 nm, H_{UVB}) and erythemally weighted UV irradiance (H_{ERY}) on a monthly basis. The vertical bars denote the reconstructed data including cloud influence, the symbols above correspond to identical atmospheric input parameters, however for cloud-free conditions. The different colouring represents 3 time intervals 1968–80, 1981–90 and 1991–2000.

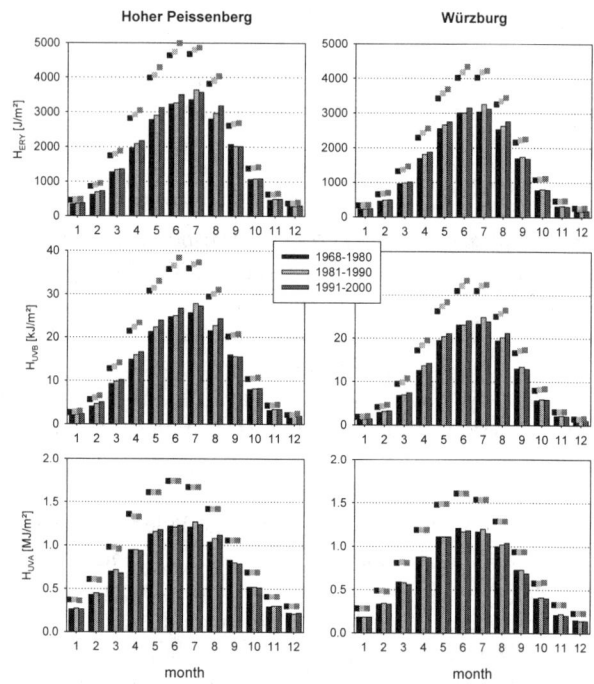

Fig. 1. Reconstructed UV exposure for the time period 1968–2000. Average daily exposure values of integral weighted UV radiation are shown for Hoher Peissenberg (left) and Würzburg (right). The bars represent radiation levels with consideration of cloudiness for the averaging intervals 1968–1980, 1981–1990 and 1991–2000. The small square marks above indicate the corresponding radiation levels under assumption of a cloud free atmosphere.

Due to its higher altitude and the geographical location of 2 degrees to the South, Hoher Peissenberg holds higher average UV levels than Würzburg. Under cloud-free summertime conditions the differences reach values of about 15 % for H_{UVB} and 10 % for H_{ERY}. In the UVA spectral range the effect is slightly smaller. Basic reason for the spectral dependence of the effect is given by the reduction of the strongly wavelength dependent Rayleigh scattering. In addition H_{UVB} and H_{ERY} are influenced by systematic differences in total ozone content due to geographical position and the effects of tropospheric ozone. The average TOC in the period 1968–2000 has been 329.8 DU for Hoher Peissenberg and 332.6 DU for Würzburg.

For cloud-free conditions it can be seen that H_{UVB} and H_{ERY} show a distinct increase due to the reduction of the stratospheric ozone layer during the past decades. The most pronounced increase in radiation is observed during spring and summer. H_{UVA}, nearly unaffected by TOC, remains constant for the three decades during most time of the year. Only during winter and early spring a slight decrease in radiation can be noticed. This is associated with a reduction of average surface albedo due to an observed decrease in snow cover during the reconstruction period.

Considering cloud influence H_{UVA} shows a weak positive tendency during all summer months for Hoher Peissenberg which has to be attributed to a reduced cloud effect in Southern Bavaria. For other seasons no, or a weak negative trend in UVA can be found here and in Würzburg. However at Würzburg only for August an increase in UVA can be observed. In case of H_{UVB} and H_{ERY} the general increase caused by ozone reduction remains visible also for the reconstruction under consideration of clouds. Only for the months of July and September the UV reduction due to an increase of clouds overcompensates for the radiation increase by ozone reduction.

Table 1. Trends in UV radiation in % per decade derived from reconstructed spectral UV irradiance from 1968 to 2000

	Würzburg		Hoher Peissenberg	
	cloud free	including clouds	cloud free	including clouds
H_{UVA}	0.0	−0.8	−0.3	+0.6
H_{UVB}	+3.1	+2.3	+2.9	+3.7
H_{ERY}	+2.9	+1.9	+2.6	+3.4

The derived linear trends from 1968 to 2000 have been calculated and are presented in Table 1. For cloudless conditions the magnitude of the trend is about +3 %/decade for H_{UVB} and H_{ERY} with no significant difference in the trends between Würzburg and Hoher Peissenberg. Consideration of cloud effects results in regional differences. The UV-B trends for Würzburg are reduced by cloud effects of about 1 %/decade. This indicates an increase in cloud cover and cloud optical depth during the time interval 1968–2000 for Northern Bavaria. In contrast the trends of Hoher Peissenberg are increased in the same order of magnitude, due to reduced cloud cover and average cloud optical depth over Southern Bavaria. This cloud effect can be seen clearly in the trend of the H_{UVA} annual exposure. As expected no trend is found for cloud-free conditions. Including clouds into the reconstruction, the UV-A trend for the site Würzburg gets negative (−0.8 %/decade), while Hoher Peissenberg holds a slight positive tendency of +0.6 %/decade.

3.2 Future UV Scenarios

Fig. 2a presents average daily erythemally weighted UV exposure (H_{ERY}) for Southern Germany on a monthly basis. Different symbols denote different climatic scenarios, $1 \times CO_2$, corresponding to conditions around 1990 and a doubled CO_2 scenario ($2 \times CO_2$), each for a cloud free and an atmosphere including cloud effects.

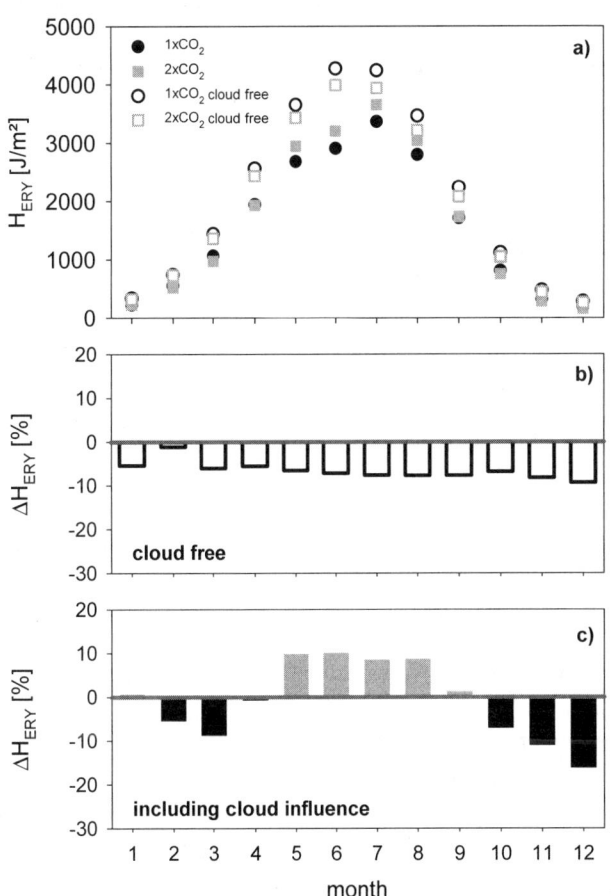

Fig. 2. Annual course of daily averages of erythemally weighted UV exposure H_{ERY} for different scenarios (a) together with estimated percentage changes from $1 \times CO_2$ to $2 \times CO_2$ conditions for a cloud free atmosphere (b) and including cloud effects (c).

Under cloud free conditions (b) an average reduction in H_{ERY} can be expected all over the year within the next decades due to predicted slow recovery of the stratospheric ozone layer. This effect is in the order of 5% to 10%. Taking cloud effects into account, the situation changes distinctly. Especially during summertime a further enhancement in average UV exposure in the order of 10% can be expected. This can be clearly associated to a reduction in average cloudiness due to a forecasted shift in typical summertime weather patterns over Central Europe.

4. ACKNOWLEDGEMENTS

The authors are very grateful to S. Trepte and P. Winkler from the Meteorological Observatory Hoher Peissenberg for provision of ozone and cloud observations. Synoptic observations for Würzburg and snow cover reports all over Bavaria have been provided by the German Meteorological Service. Data on total ozone content of other European stations have been taken from the World Ozone and Ultraviolet Data Center (WOUDC) at Toronto. The presented investigations have been performed under project D1 in the framework of the Bavarian Joint Research Programme on Elevated UV Radiation (BayFORUV) funded by Bayerisches Staatsministerium für Landesentwicklung und Umweltfragen (StMLU).

5. REFERENCES

De Backer, H., P. Koepke, A. Bais, X. de Cabo, T. Frei, D. Gillotay, C. Haite, A. Heikkilä, A. Katzanzidis, T. Koskela, RE. Kyrö, B. Lapeta, J. Lorente, K. Masson, B. Mayer, H. Plets, A. Redondas, A. Renaud, G. Schauberger, A. Schmalwieser, H. Schwander, and K. Vanicek, 2001: Comparison of measured and modelled UV indices for the assessment of health risks, *Meteorol. Appl.*, 8, 3, 267-277.

Borkowski, J. L., A.-T. Chai, T. Mo, and A. E. O. Green, 1977: Cloud effects on middle ultraviolet global radiation, *Acta Geophys. Pol.*, 25, 287-301.

Hess, M., P. Koepke, and I. Schult, 1998: Optical Properties of Aerosols and Clouds: The Software Package OPAC, *Bull. Amer. Meteor. Soc.*, 79, 5, 831-844.

Koepke, P., A. Bais, D. Balis; M. Buchwitz, H. De Backer, X. De Cabo, P. Eckert, P. Eriksen, D. Gillotay, A. Heikkilä, T. Koskela, B. Lapeta, Z. Litynska, J. Lorente, B. Mayer, A. Renaud, A. Ruggaber, G. Schauberger, G. Seckmeyer, P. Seifert, A. Schmalwieser, H. Schwander, K. Vanicek, and M. Weber, 1998: Comparison of Models Used for UV Index Calculations, *Photochem. Photobiol.*, 67, 6, 657-662.

Koepke, P., J. Reuder, and H. Schwander, 2002: Solar UV radiation and its variability due to the atmospheric components, *Recent Research Developments in Photochemistry and Photobiology*, ed. By S. G. Pandalai, Volume 6, 11-34.

Nakajima, T., and M. Tanaka, 1988: Algorithms for radiative intensity calculations in moderately thick atmospheres using a truncation approximation, *J. Quant. Spectrosc. Radiat. Transfer*, 40, 1, 51-69.

Reuder, J., M. Dameris, and P. Koepke, 2001: Future UV radiation modelled from ozone scenarios, *J. Photochem. Photobiol., B.*, 61, 3, 94-105.

Reuder, J., and P. Koepke (2005), Reconstruction of UV radiation over Southern Germany for the past decades, *Met. Z.*, in print.

Reuder, J., and H. Schwander, 1999: Aerosol effects on UV radiation in nonurban regions, *J. Geophys. Res.*, 104, D4, 4065-4077.

Ruggaber, A., R. Dlugi, and T. Nakajima, 1994: Modelling radiation quantities and photolysis frequencies in the troposphere *J. Atmos. Chem.*, 18, 171-210.

Schwander, H., P. Koepke, and A. Ruggaber, 1997: Uncertainties in modeled UV-irradiances due to limited accuracy and availability of input data, *J. Geophys. Res.*, 102, D8, 9419-9429.

Schwander, H., P. Koepke, A. Kaifel, and G. Seckmeyer, 2002: Modification of spectral UV irradiance by clouds, *J. Geophys. Res.*, 107, 10.1029/2001JD001297, AAC 7-1 to AAC 7-12.

Schwander, H., B. Mayer, A. Ruggaber, A. Albold, G. Seckmeyer, and P. Koepke, 1999: Method to determine snow albedo values in the UV for radiative transfer modelling, *Appl. Opt.*, 38, 18, 3869-3875.

van Weele, M., T. J. Martin, M. Blumthaler, C. Brogniez, P.N.d. Outer, O. Engelsen, J. Lenoble, G. Pfister, A. Ruggaber, B. Walravens, P. Weihs, H. Dieter, B. G. Gardiner, D. Gillotay, A. Kylling, B. Mayer, G. Seckmeyer, and W. Wauben, 2000: From model intercomparison towards benchmark UV spectra for six real atmospheric cases, *J. Geophys. Res.*, 105, D4, 4915-4925.

Weller, M., E. Schulz, U. Leiterer, T. Naebert, A. Herber, and L. W. Thomason, 1998: Ten years of aerosol optical depth observation at the Lindenberg Meteorological Observatory, *Contr. Atm. Phys.*, 71, 4, 387-400.

COMPARATIVE STUDY OF UVB AND TOTAL RADIATION ATTENUATION OBSERVED IN A SOUTH AMERICAN REGION

M. P. Corrêa, J. C. Ceballos, and M. J. Bottino
Satellite and Environmental Systems Division, Centro de Previsão de Tempo e Estudos Climáticos
Instituto Nacional de Pesquisas Espaciais, Rod. Dutra, km 40, Cachoeira Paulista, SP, 12.630-000, Brazil

G. Coronel
Laboratorio de Investigación de la Atmósfera y Problemas Ambientales, Universidad Nacional de Asunción
Campus Universitario, San Lorenzo, Paraguay

ABSTRACT

Some cases showing the relationship between UVB and total solar radiation under cloudless, cloudy and forest fires conditions are presented. The analysis is focused on data collected since 1996 in Asunción City (Paraguay—25.33S, 57.52W; altitude 130m) during wet (with low aerosol loading and cloudy days) and dry seasons (with biomass burning aerosol predominance in clear-sky days). For the wet season—February/April—a clustering method applied to GOES-8 imagery allowed the identification of different types of clouds. Aerosol content during the dry season—August/September—was assessed using TOMS Aerosol Index. Results suggest a functional relationship between UVB and total irradiances under clear-sky and overcast local conditions. Under fair weather cumulus presence the comparison requires further analysis.

1. INTRODUCTION

Paraguay is located in the south-central area of South America. This region has rainy summers and relatively dry winters. Wet seasons are characterized by deep convection and severe thunderstorms, while clear-sky days are typical for dry seasons. During the "dry months", anthropogenic biomass burning is a usual agriculture activity. However, the most part of burning foci is not only located in Paraguay region, but mainly in the large territorial extension of its neighbor, Brazil. Smoke clouds produced in these burnings are carried over Paraguay, causing different problems as visibility reduction or increasing the number of respiratory disease cases among the population.

These seasonal characteristics and their consequences provide important research subjects discussed in several works. First of all, smoke clouds attenuate solar radiation, mainly in the UV and PAR spectral regions (Yamasoe et al., 2003). Smoke clouds can attenuate UV radiation as strongly as deep-convective clouds (Correa and Coronel, 2002). Aerosol particles in the smoke clouds can also exert indirect effects, for instance in cloud formation and development (Silva Dias et al., 2002). A better knowledge about influences of these particles on radiative fluxes can be relevant for a better understanding of cloud dynamics and other meteorological phenomena.

It is also important to improve knowledge concerning the influence exerted by clouds on the UV fluxes. Deep convection or stratified conditions can act as a shield for UV radiation. On the other hand, fair weather cumulus clouds can provide a situation with temporary increase of these fluxes.

Finally, this work shows the analysis of theoretical assessments and their relationship with measurements carried through in UV and total spectrum. Studies on the correlation between measurements carried out in different parts of the spectrum can provide an important tool for UV evaluation in the regions where UV measurements are not available because of much higher price of UVR sensors. The next section presents the instruments, radiative algorithms and the methodology. The results are analyzed in the following section and the conclusions and suggested topics of future studies are shown in the last section.

2. INSTRUMENTATION AND METHODS

2.1 Measurement Instruments

UVB and total radiation measurements were performed at Asunción (Paraguay), 25.33S; 57.52W, during 1996 and 2002. UVB radiation data were obtained using a UVB-1 pyranometer of Yankee Environmental Systems (htpp://www.yesinc.com). This device measures global (direct + diffuse) irradiance between 280 and 320 nm on a horizontal plane. Total solar irradiance was measured using a Li-Cor LI-200SZ pyranometer (http://www.licor.com/). This instrument has performance compatible with a precision pyranometer measuring solar radiation in the 280–3000 nm interval (Ceballos et al., 2004). Data were obtained at 15 minute intervals.

2.2 Radiative Transfer Models Used for Clear-sky Calculations

UVB irradiance was assessed using the Ultraviolet Global Atmospheric Model—UVGAME (Corrêa et al., 2003). UVGAME is an UV multiple scattering spectral model for radiative transfer based on the discrete-ordinate method. It is built in FORTRAN for running under Microsoft Windows as well as UNIX environment. Total solar radiation was assessed using a simplified model for clear-sky conditions (Ceballos et al., 2004). Daily cycles of UVB (UVB_o) and total (G_o) irradiances do not consider aerosol presence.

2.3 Method for Identification of Different Types of Clouds and Clear-sky Pixels in GOES-8 Imagery

Pixels in full resolution GOES 8 images were considered as 4-dimensional vectors with components given by reflectance, brightness temperature and their respective textures. Classification procedure labels a pixel in one of 30 classes (reference vectors, or centroids), choosing the minimal Euclidean distance. The reference vectors had been found by Bottino and Ceballos (2003) for a training set of images, using an iterative clustering procedure ("dynamic groups") which lead to 30 different classes of pixel environment (defined by the same number of centroids). Hierarquical clustering of the centroids showed the existence of 5–8 main types of groups; nephanalysis of classified images of the training set made evident the correspondence of centroids with five main types of scenes (surface, cumulus, cirrus, stratus and deep convective multi-layered cloud). In this manner each centroid became labeled with a scene type, simultaneously providing information about corresponding reflectance, temperature and texture. Classification by minimal Euclidean distance was applied to GOES imagery provided by CPTEC/INPE, for the period analyzed.

3. DISCUSSION

In this work we show two sets of analyses. The first one is a preliminary study related to UV attenuation caused by smoke from biomass burning. These results illustrate strong UVB attenuation caused by fire episodes that occur during dry seasons. The second analysis explores relationships between UVB and total irradiance measurements with cloud cover characteristics.

Fig. 1 shows Erythemal UVB attenuation caused by smoke from biomass burning in cloudless days.

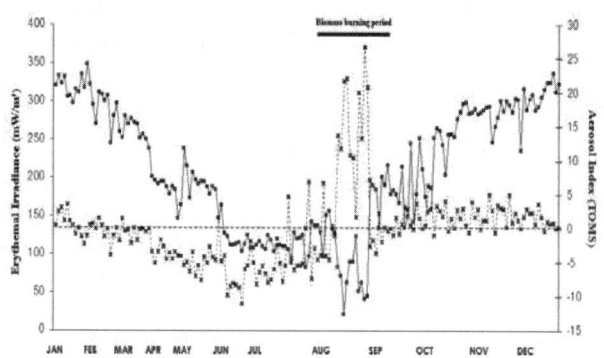

Fig. 1. UVB irradiance attenuation caused by smoke (represented by TOMS aerosol index) from biomass burning in cloudless days (1999).

In this figure, Aerosol Index [AI] provided by NASA indicates the amount of aerosol in the atmosphere during 1999. In particular situations, smoke clouds [SC] can attenuate UVR as strongly as water clouds. In South America (mainly in Brazil and Bolivia) these episodes can occur due to agricultural activities in dry seasons (August/September). These SC are frequently transported over South America reaching Paraguay. Particularly in 1999, a steady and very thick smoke cloud provoked almost 90% of UV attenuation compared with calculations considering aerosol free conditions. This is about the same attenuation observed in stratified optically thick clouds (see Fig. 2a–d). Additional studies related to this influence are being carried out in São Paulo metropolitan area to verifying the influence of intense urban pollution.

Fig. 2 illustrates two daily cycles of measurements and calculations of ultraviolet [UV] (left) and total solar [G] (right) irradiances for the "wet" period. Fig. 3 shows the relationship between UV/UVo and G/Go quotients; and Fig. 4 shows cloud classification in GOES-8 imagery for morning and afternoon time. They have evident clear-sky predominance in the morning with an increasing afternoon cloudiness over Asunción region. TOMS ozone data was used in simulations. March 1, 2005 shows 252 Dobson Units over Asunción. However, there is no available ozone data for March 6, 2001. An interpolation considering all available data during 3 days after and 3 days before shows 256 Dobson units for this data. Therefore, ozone data can not lead a noticeable influence.

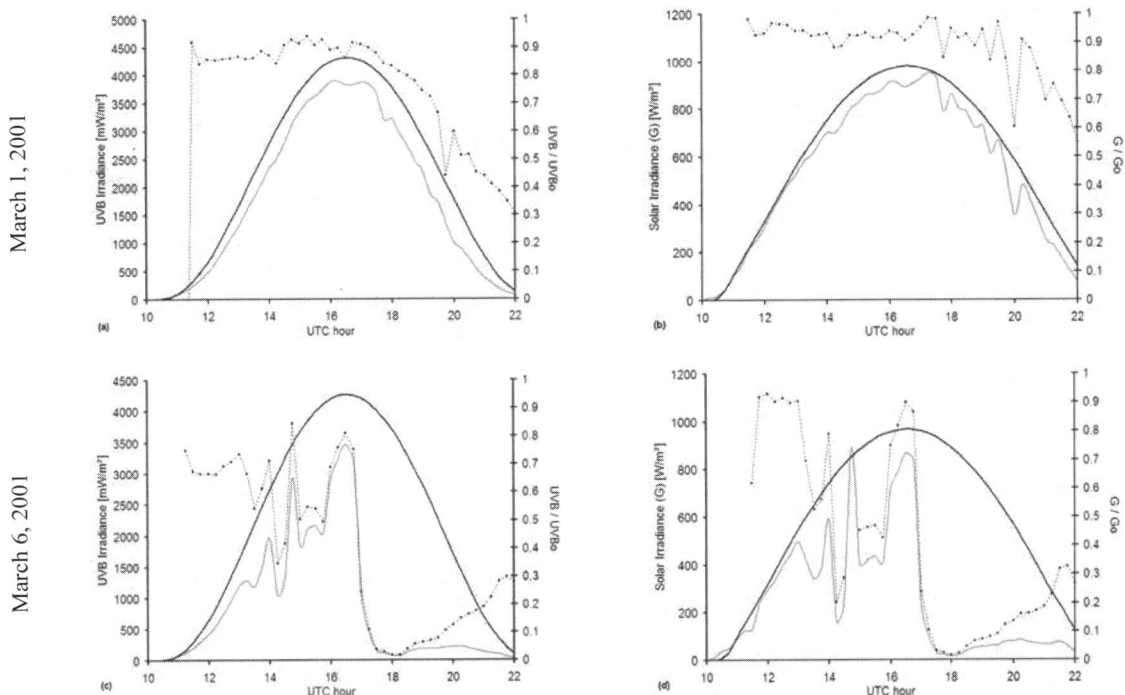

Fig. 2. UVB (a,c) and total (b,d) irradiances. Solid lines: black—simulations (UV_o and G_o) for aerosol free clear-sky conditions; gray—measurements (UV and G); dashed lines: attenuation (simulations/measurements—UV/UV_o and G/G_o).

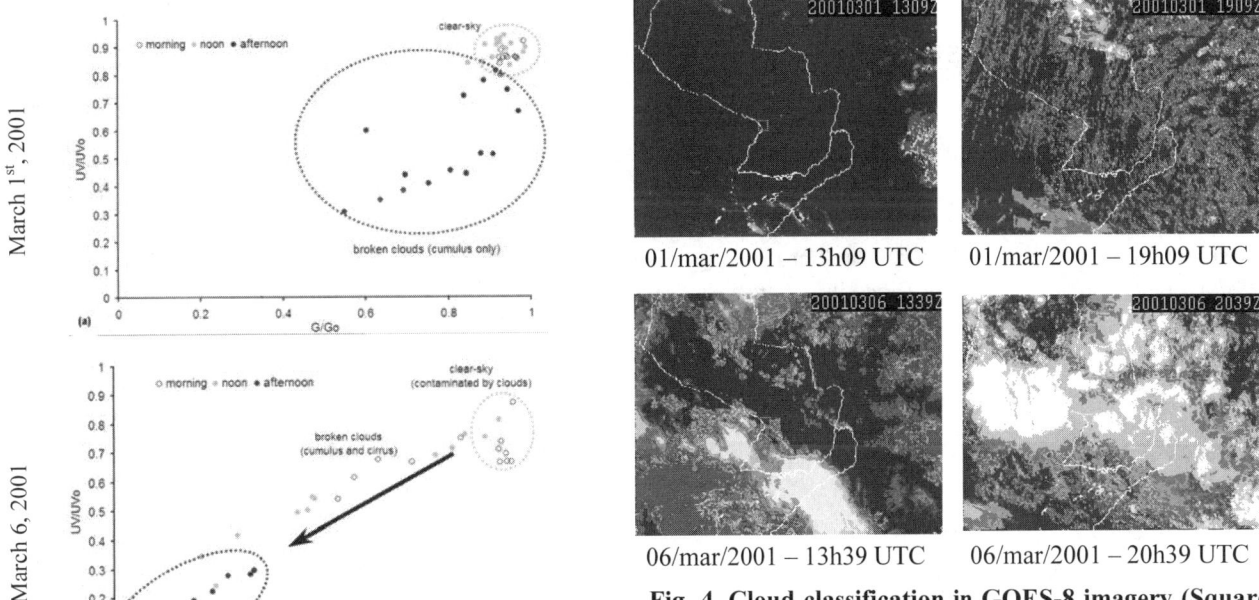

Fig. 3. Relationship between UVI/UVI_o and G/G_o, a) March 1, 2001; b) March 6, 2001.

Fig. 4. Cloud classification in GOES-8 imagery (Square box identifies Asunción city).

Quotients u = G/Go and v = UV/UVo are related to attenuation induced by additional atmospheric elements (aerosol or cloud). As expected, points (u,v) tend to be concentrated at the upper right corner of Fig. 4 for clear-sky conditions. Fluctuations can be explained by cloud contamination or aerosol presence (not detected by classification). Pairs (u,v) appear fairly aligned between (0,0) and (1,1) during March 6, 2001 (from morning to afternoon). The type of cloud is clearly associated to in situ coverage; i.e., cumulus and/or cirrus for partial coverage and deep convection for full coverage. The same type of u and v variation is observed in Fig. 2 (March 6) during the daily cycle. These results suggest that under deep convection UVB and total radiation has equivalent attenuations and its intensity depends on cloud optical depth. On the other hand, Fig. 3 makes evident that a broken cloud field (case of fair weather cumuli in March 1, 2001) introduces a complex geometry of interactions between radiation and clouds. Points (u,v) suggest that UV fluxes could be more affected than total radiation, with different but proportional intensity. Further analysis using separate measurements of ultraviolet, visible and infrared irradiances is recommended, in order to better separate effects of lateral scattering from those of cloud and water vapor transmittance. This kind of analysis will also allow verifying the influence of aerosol and ozone.

4. FINAL CONSIDERATIONS

The discussion involving both UV and total radiation suggests a functional relationship between them, under clear-sky and overcast local conditions. Relationship in fair weather cumulus conditions requires further analysis. Cloud classification in satellite imagery can provide useful information about cloud cover and its characteristics. Additional comparative analysis of attenuation produced by smoke and water cloud yield valuable information for UV monitoring in populated areas. Aerosol optical properties will be studied using surface measurements (AERONET). This work is being carried to provide information about the relationship between particle properties and radiative fluxes attenuation. These measurements, including UV and total radiation, are being carried in São Paulo city, the largest and most populous city of South America.

Contact: mpcorrea@cptec.inpe.br

5. ACKNOWLEDGMENTS

The participation of the main author in IRS'04 was granted by Brazilian Research Foundation FUNCATE and Korea Meteorological Society. The authors also wish to thank Dr. Carlos Frederico de Angelis.

6. REFERENCES

Bottino, M. J., and J. C. Ceballos, 2003: Classification of scenes in mustispectral GOES-8 IMAGERY. In: *Proceedings of XI Remote Sensing Brazilian Symposium*, Belo Horizonte.

Ceballos, J. C., M. J. Bottino, and J. M. Souza, 2004: A simplified physical model for assessing solar radiation over Brazil using GOES-E imagery. *J. of Geophys. Res.*, 109, D02211, doi:10.1029/2003JD003531.

Corrêa, M. P., P. Dubuisson, and A. Plana-Fattori, 2003: An overview about the ultraviolet index and the skin cancer cases in Brazil. *Photochemistry and Photobiology*, 78(1), 49-54.

Corrêa, M. P., and G. Coronel, 2000: Variabilidade das medidas de irradiâncias UV-B e eritêmica em períodos de queimadas. In: *Proceedings of XII Brazilian Congress of Meteorology*, 1771-1777.

Silva Dias, M. A., et al., 2002: Clouds and rain processes in a biosphere-atmosphere interaction context in the Amazon region. *J. of Geophys., Res.*, 107, 23, 1-46.

Yamasoe, M., A. Plana-Fattori, M. P. Corrêa, P. Dubuisson, B. Holben, and P. Artaxo, 2003: Measurement of global and direct photosynthetic active radiation – the effect of smoke. In: *EGS-AGU-EUG Joint Assembly*, Nice, France, 06-11 April 2003 - ID-NR: EAE03-A-12276.

THE INFLUENCE OF FOREST AND PEATBOG FIRES ON THE OPTICAL AND RADIATIVE REGIMES OF THE ATMOSPHERE AND RADIATIVE FORCING OVER CENTRAL RUSSIA

N. Ye. Chubarova, G. M. Abakumova, E. V. Gorbarenko, E. I. Nezval, and O. A. Shilovtseva
Moscow State University Geographical Department, Meteorological Observatory, Vorobyovy Gory
Moscow, 119992, Russia, chubarova@imp.kiae.ru

A. N. Rublev
Russian Research Center "KURCHATOV INSTITUTE"
Moscow 123182, Russia

ABSTRACT

We analyzed radiation and optical properties of the atmosphere during severe fire events over Central Russian plain in 2002. The fire "cloud" was characterized by high aerosol optical thickness (AOT) up to 3 at 500 nm with mean AOT500 of 1.02, high values of single scattering albedo (SSA=0.95) according to retrievals from AERONET CIMEL sun and sky photometer as well as by high concentration of optically effective gas species. Distinctive spectral features of solar irradiance attenuation include larger drops in the UV, when compared with typical clear sky conditions (−24% for total and −38% for UV irradiance) with decreases in both direct and diffuse components of UV. This is in agreement with model results if we take into account additional absorption by gases (mainly, O_3 and NO_2) and reduce aerosol single scattering albedo in the UV (0.89–0.9 for <320 nm). Radiation forcing efficiency is about −39 W/m² at TOA and −80 to −100 W/m² at ground. The radiation budget of the mean fire "cloud" averaged over August 2002 is distributed among reflection, transmission and absorption as 26%, 58%, and 29%. The absorption due to smoke aerosol is about 5% and the absorption by NO_2 comprises about 2%. Both reflection and absorption increase by approximately 4%, compared with typical conditions without fires.

1. INTRODUCTION

Summer-fall 2002 was characterized by significant forest and peatbog fires in Central Russia, which changed the optical and radiation regimes of the atmosphere. During this period a complex program of aerosol, gaseous and solar radiation measurements including total 300–5,000 nm (IR), visible 400–700 nm, erythemally-weighted UV irradiance (Qer), UV irradiance 300–380 nm (UV380), and natural illuminance 380–780 nm (E, in lux) has been in operation at the Meteorological Observatory of Moscow State University (MO MSU), and at its site in a Moscow suburb (50km to the west) at the MSU Zvenigorod biostation. The information about the measurement program can be found at MO MSU home page: http://momsu.newmail.ru/english. The radiative parameters of the atmosphere observed in 2002 were compared with those measured during intensive fires in 1972. Using assumed radiative characteristics of smoke aerosols and gas species as input parameters for a model, we simulated their impacts on the attenuation of solar irradiance in different spectral ranges and compared them with measurements. We show the radiation budget and radiative forcing of the fire "cloud", and compare it with typical conditions as per a standard continental aerosol model.

2. RESULTS

2.1 Meteorological and Optical Conditions.

The intensities of fires over the Moscow region is often characterized by Nesterov's flammability indices. Fire indices are calculated according to Nesterov's equation: $G=\sum(T^*d)$, which takes into account the day-to-day accumulation of the joint effect of maximum temperature at midday (T) and dew point depression ($d=T-Td$) in the absence of precipitation higher than 3mm. Fire risks are classified with K fire class values from 0-5, where K=3, K=4, K=5 relate to dangerous, high dangerous and extremely dangerous fire conditions. Fig. 1 shows the frequency distribution of these classes in typical situation (2001) as well as in fire conditions. One can see higher frequency of the classes K=4 and K=5 during periods with fires (especially in 2002) compared with typical conditions without fires.

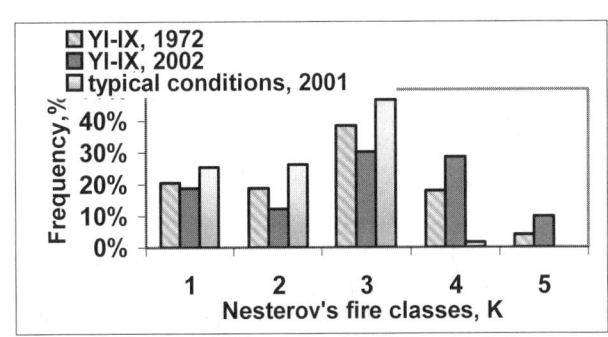

Fig. 1. Frequency distribution of Nesterov's fire classes, K, in typical (2001) and in fire conditions (1972 and 2002). June–September periods.

The main forest and peatbog fires during summer 2002 were located to the east of Moscow city. By late August they were distributed in other Moscow regions. Fig. 2 shows a high resolution satellite image of the Moscow region during the fires on 30.07.02 with 12-hour backward trajectories at different heights for cloudless conditions and high aerosol optical thickness (AOT500=1.6). Most fires were 50-100 km to the east of Moscow, this produced strong heterogeneity over eastern regions but well diffuse plumes over Moscow (see Fig. 2). The high correlation (r=0.91) between simultaneous AOT500 records measured at Zvenigorod biostation (by Hazemeters) and in Moscow (by Cimel) has indicated a quite homogeneous distribution of the fire "cloud" to the west of the main fire locations.

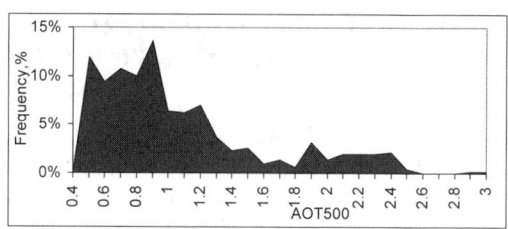

Fig. 3. Frequency distribution of aerosol optical thickness at 500nm in Moscow during fires in 2002 from AERONET Cimel data.

Table 1. Statistics of Some Optical Parameters: AOT at 500 nm and at 380 nm, Angstrom Parameters in Two Spectral Intervals (n=513), Single Scattering Albedo and Factor Asymmetry at 440 nm (only Clear Sky Cases, n=36), and Ground Concentration of Different Gaseous Species ($\mu g/m^3$, n=36) in Fire Conditions. Moscow. 2002

	Mean	Sigma
AOT500	1.02	0.53
AOT380	1.43	0.66
Angstrom parameter (440-870 nm)	1.65	0.15
Angstrom parameter (340-380 nm)	0.99	0.28
SSA at 440nm	0.95	0.01
Factor asymmetry at 440nm	0.68	0.02
NO_2	99.7	49.9
O_3	62.6	63.8
SO_2	9.5	16.7
HCHO	21.4	16.8

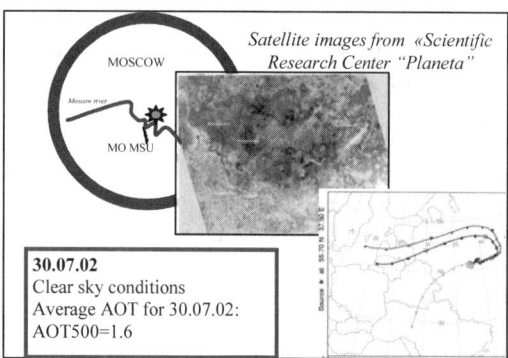

Fig. 2. The satellite image of the fires over eastern part of Moscow region and 12-hour NOAA backward trajectories at approximately 500 m(red), 1000 m(blue) and 1500 m (green) heights.

The AOT500 distribution in typical conditions without fires during warm period is characterized by the near absence of AOT500>0.4 (less than 1% of cases in May-June 2002, see Uliumdzhieva et al., 2005). To identify conditions with the presence of fire "cloud" we thus use the simple filter AOT500>0.4. The distribution of AOT500 is shown in Fig. 3. Table 1 presents the statistics of optical parameters, which can represent the characteristics of a regional fire "cloud" model. High concentrations of nitrogen dioxide (NO_2), formaldehyde (HCHO) and sulphuric dioxide (SO_2) were also observed during the fires. It is necessary to mention that the maximum values of gas concentration (except SO_2) were 1.5–3 times higher than instantaneous maximum allowable concentrations.

The correlations of AOT500 with the NO_2 and HCHO concentrations as well as with daily maxima of surface ozone, all exceed 0.5. But the correlation of AOT500 with SO_2 is much less (0.37), indicating the existence of other anthropogenic sources for SO_2. Evidently, the correlation between the concentration of optically active gases and aerosol produced additional attenuation of solar irradiance during the fires.

2.2 The Changes in Solar Irradiance at Ground.

The attenuation of solar irradiance at surface during the fires has significant spectral features. Fig. 4 presents solar elevation dependences of diffuse and direct components of UV380, total irradiance (IR) and illuminance (E) normalized to the values measured in typical clear sky conditions. We can mark the growth of the diffuse component of total irradiance and illuminance, which has an effective wavelength at 550 nm, up to 140–160% and the decrease of diffuse UV up to –26% at small solar elevations compared with typical clear sky conditions. At the same time there is a distinct drop in the direct component for all spectral regions and especially for UV (up to 80%). The mean changes in diffuse irradiance are 116%, 100%, and -10% respectively for total, illuminance, and UV irradiance, while the losses of their direct components are about 52%, 70%, and 80%. There are also pronounced spectral changes in the attenuation of global (Q) irradiance: average losses are about 23%, 31%, 36%, and 42% respectively to QIR, Qvisible, QUV380 and Qer. The mean relative change in attenuation of global illuminance (28%) is similar to that obtained for global visible irradiance (31%) due to the proximity of their effective wavelengths.

RADIATIVE BUDGET AND FORCING

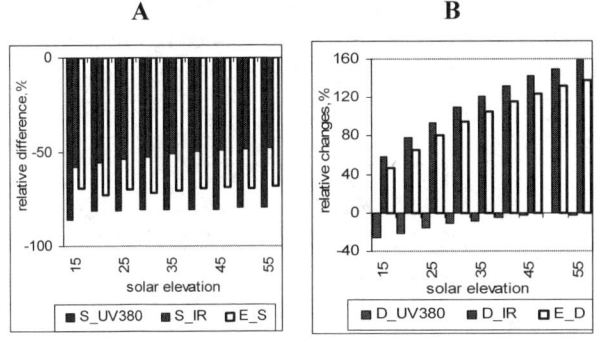

Fig. 4a,b. The changes in diffuse D (a) and direct S (b) components of ultraviolet 300-380nm (UV380), total (IR) irradiance and illuminance (E) in cloudless fire smoke conditions compared with typical situation.

Fig. 5 presents QIR and QUV380 losses during the fires in 2002 and in 1972. There were larger attenuations of both fluxes in 2002 due to more severe fires (see higher K values observed in 2002 in Fig. 1). We can also mark stronger decrease in UV irradiance in both fire seasons.

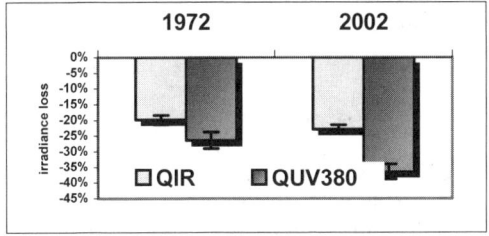

Fig. 5. Attenuation of QUV380 and QIR during fire smoke events in 1972 and 2002; cloudless conditions.

In order to explain these results, we calculated global irradiance in different spectral regions (Qer, QUV380 and Qvisible) and compared them with the measurements. In addition to aerosol parameters we considered the concentrations of gas species which are optically effective in UV and visible spectral region. It was shown that for erythemally weighted irradiance the rather high observed concentrations of SO_2, and HCHO do not play vital role (less than 1-2%). Surface ozone can attenuate by up to 2–3%. Extremely high concentration of NO_2 can attenuate up to 10–15% in longwave UV and up to 5–6% in visible region of spectrum. The use of observed gas species concentrations in the model improved the agreement with experimental data in all spectral regions with standard deviation falling from 0.16 to 0.14 for Qer, from 0.11 to 0.08 for QUV380 and from 0.037 to 0.032 for Qvisible. In shortwave UV region, however, we still have some overestimation by the model, especially at high aerosol optical thickness, possibly due to the application of Cimel SSA retrievals at 440 nm, beyond the UV range. According to our estimates the systematic discrepancies at high aerosol optical thickness would be eliminated only if SSA=0.92 is taken for QUV380 and 0.89–0.9—for Qer. These values are noticeably less than SSA retrievals at 440 nm (see Table 1). This suggests the existence of an additional absorber, like soot, with strong effects in UV region. The larger attenuation of UV is thus explained by both the effects of multiple scattering (especially at longer paths), and higher aerosol and gas species absorption in this spectral range.

2.3 Radiation Forcing and Radiation Budget.

To evaluate the changes in shortwave radiation budget we will use "radiative forcing" to account for the effects of aerosol and gas species. In this case radiative forcing (R) and radiative forcing efficiency are calculated as:

$$R = Q_{net(gas,aerosol)} - Q_{net(no\ gas,no\ aerosol)} \quad (1)$$

$$Reff = R/dAOT500 \quad (2)$$

Therefore, R_{eff} characterizes relative changes in solar irradiance without and with gas and aerosol content for $dAOT500=1$.

To analyze the role of different atmospheric components on radiation budget we chose 5 days with fire "cloud" situations and recalculated the optical properties of aerosol, specially accounting for gaseous NO_2 absorption (Chubarova and Dubovik, 2004). Shortwave flux calculations used Monte-Carlo model developed by Rublev, which has been carefully validated against radiation measurements (Chubarova et al., 1999). We should note that during fire periods surface albedo A is low. For example, in August monthly mean A comprised 0.15 in 2002 and 0.17 in 1972 compared with A of 0.2 in typical conditions. The changes in reflectance of the surface during fires (possibly due to soot sedimentation at ground and/or the simple burning of grasses, etc) will also affect the estimates of radiative forcing. Taking the computed value for the molecular atmosphere $Q_{net(no\ gas,no\ aerosol)}$ we calculated a model R and a measured R radiative forcing at the ground using $Q_{net(gas,aerosol)}$ respectively from model and measurements, both in fire conditions. Table 2 shows mean, minimum, and maximum radiative forcing at the ground calculated using radiative forcing efficiency from measurements ($Reff\_meas$) with AOT variations as well as pure model R values (to avoid $Reff$ non-linear AOT dependence). One can see significant variation of fire "cloud" radiation forcing at ground reaching $-210 \div -240$ W/m². Note, that in typical conditions (mean AOT500 = 0.16) R at ground is much smaller (-25 W/m²). We also note that the calculated effect of NO_2 on R values at ground is about $-6 \div -10$ W/m²; for $Reff$ it is close to -5 W/m².

Radiative forcing at the top of the atmosphere (TOA) can reach -91 W/m² at maximum AOT. Due to comparatively small effective size of smoke aerosol, AOT in the longwave (thermal infrared) region is not very high, therefore, aerosol longwave radiative forcing is small. Hence, we can speak about significant cooling of the planet at high smoke aerosol loading compared with typical conditions when R at TOA is $-10 \div -20$ W/m^2.

The absorption in the atmosphere is calculated according to the following equation:

$$Q_a = S_0 \cos\theta - F^{\uparrow}_{\lambda} - \int F^{\downarrow}_{\lambda}(1-A_{\lambda})d\lambda \qquad (3)$$

where $S_0 = 1367$ W/m^2, θ - zenith angle, $F^{\uparrow}_{\lambda}$ and $F^{\downarrow}_{\lambda}$ – upward and downward fluxes, A_{λ} - surface albedo.

Table 2. Statistics of Shortwave Radiative Forcing in W/m^2 at the TOA and at the Ground. Cosz≈0.6. Fire Smoke Conditions (July–September 2002); A = 0.17

	Ground model/ measurements	TOA MODEL
Rmean(AOT500=1.02)	–100 / –82	–39
Rmax(AOT500=0.4)	–49 / –32	–12
Rmin(AOT500=2.96)	–210 / –239	–91
Average Reffl	–90±10/ –81±12	–39

Fig. 6 illustrates the sensitivity of absorption in the atmosphere due to different factors: aerosol, NO$_2$ and the sum of absorption by other gaseous species (H$_2$O, etc.). The calculations were done for mean, maximum and minimum AOT in fire conditions. On average, the solar absorption due to both aerosol and NO$_2$ in the observed fire "cloud" is about 28% but can reach almost 35% at the highest AOT. The mean shortwave absorption by NO$_2$ is about 2% and can reach 4%.

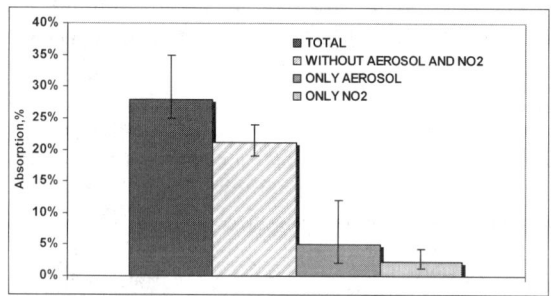

Fig. 6. The solar absorption in the atmosphere in fire smoke conditions. Monte-Carlo simulations, cosZ≈0.6. Mean fire conditions are characterized by AOT500=1.02, NO$_2$=3.8 matm.cm.

On the whole, there is a distinct redistribution of radiation budget in the atmosphere in the presence of fire "cloud". Fig. 7 illustrates the changes in the components of radiation budget in conditions with the total absence of aerosol, with continental aerosol at AOT500=0.16 and in the presence of mean fire "cloud" with AOT500=1.02 and NO$_2$=3.8 matm.cm. The calculations were fulfilled for one month (August 2002) and reflect the climatic changes in the radiation budget.

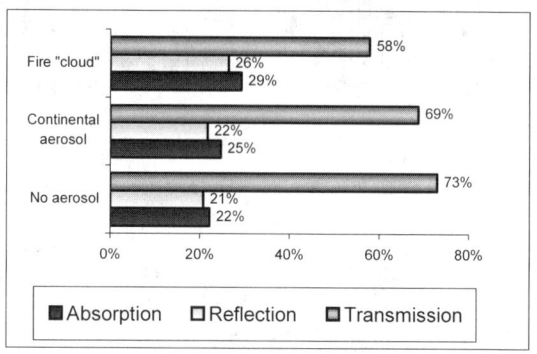

Fig. 7. The redistribution of radiation budget in different conditions. August 2002. (See the details in the text.)

While typical aerosol decreases the transmission only by 4%, the presence of fire "cloud" result in its 15% decrease relative to the transmission in the molecular atmosphere. The ratio of absorbed (by the atmosphere) to reflected (at TOA) changes little from typical to fire conditions. But during fires there are increases in both absorption and reflection (≈4%).

3. ACKNOWLEDGMENTS

The authors would like to thank Mosecomonitoring Agency and Institute of Atmospheric Physics for providing data on concentration of some gas species; Dr. Wei Min Hao (USDA Forest Service) for providing the Hasemeters, the Scientific Research Center "Planeta" for providing satellite images. Dr. Rublev was partly supported by Russian Science foundation for Basic Research, grant 04-05-64579.

4. REFERENCES

Chubarova, N., and O. Dubovik, 2004: The sensitivity of aerosol properties retrievals from AERONET measurements to NO2 concentration over industrial region on the example of Moscow, *Optica Pura y Aplicada*, No. 3, 3315-3319. http://www.sedoptica.es/indice.html.

Chubarova N., N. Rublev, A. Trotsenko, and V. Trembach, 1999: "The comparisons between modeled and measured surface shortwave irradiances in clear sky conditions," *Izvestiya, Atmospheric and Oceanic Physics*, 2, 201-216.

Uliumdzhieva, N., N. Chubarova, and A. Smirnov 2005: "Aerosol characteristics of the atmosphere over Moscow from Cimel sun photometer data," "*Meteorology and Hydrology,*" No. 1, 48-57.

CORRESPONDENCE OF THE LOW CLOUD MICROPHYSICS TO THE AEROSOL AMOUNT OVER CHINA

K. Kawamoto and T. Hayasaka
Research Institute for Humanity and Nature
Kyoto, Kyoto, 602-0878, Japan

I. Uno
Research Institute for Applied Mechanics, Kyushu University
Kasuga, Fukuoka, 816-8580, Japan

ABSTRACT

To examine the anthropogenic aerosol effect on low cloud microphysics, a comparison between the aerosol amount M_a and the low cloud effective particle radius r_e over China is made. We use a numerical simulation for the aerosol amount and a satellite retrieval for the cloud microphysics. The comparison of both annual-mean values reveals that r_e is negatively correlated with M_a. This behavior can be understood from the Twomey effect. If we have more aerosols, the radius of each cloud droplet becomes smaller through interaction with aerosols.

1. INTRODUCTION

Cloud-radiation issues attract much attention in climate change studies. In particular, the aerosol indirect effect, that is, influence on climate through cloud modification is thought to be one of the most uncertain processes (IPCC, 2001). In order to examine the effect of anthropogenic aerosols on clouds, Chameides et al. (2002) made comparisons of model-simulated aerosol loading with International Satellite Cloud Climatology Project (ISCCP)-derived cloud optical depth and area extent (cloud amount) over China; they found that cloud optical depth is positively correlated with aerosol loading; on the other hand, the relationship between aerosol loading and cloud amount was not so clear. They could not, however, close the aerosol-to-cloud microphysical loop, because cloud droplet size is not retrieved by ISCCP. Filling this gap, we make comparisons between aerosol amount and water cloud particle size. Low-lying water clouds are the focus because most of the aerosol particles reside in the lower part of the atmosphere. In this work, we use satellite retrievals for cloud properties and a numerical simulation for aerosol amount.

Since the 1980s, China has increased its industrial production due to a substantial political changeover, and this has resulted in the emission of considerable amounts of gases (e.g., SO_2 NO_x, and so on) (Streets and Waldhoff, 2000). Kawamoto et al. (2004) compared the anthropogenic SO_2 emission inventory with the low cloud effective particle size over China and found a negative correlation between these two quantities. We assume they are correlated because once SO_2 is converted to sulfate aerosols, some of the sulfate aerosols then serve as cloud condensation nuclei (CCN). This was the motivation for our study on the effect of human activities on cloud properties using the SO_2 emission inventory as a proxy. Here we try to investigate the relationship between aerosols and low cloud effective particle radius, since aerosols are more closely connected to clouds than SO_2 emission.

2. DATA

2.1 Aerosol Amount

We used the following numerical model named the Chemical Weather Forecast System (CFORS) to simulate the aerosol loading. CFORS is designed as a multi-tracer, online, system built within the Regional Atmospheric Modeling System (RAMS) (Pielke et al., 1992). A unique feature of CFORS is that multiple tracers are run online in RAMS, so that all the meteorological information (originally generated from European Center for Medium Range Weather Forecasting data) from RAMS is directly used by the tracer model. As a result, CFORS produces high time resolution, 3-dimensional fields of tracer distributions and major meteorological parameters. The chemical species that were simulated in this work are SO_4, NO_3, NH_4, OC, and elemental carbon. The aerosol loading M_a ($\mu g/m^2$) is defined as the vertical integration of all these five species.

2.2 Low-cloud Microphysics

We retrieve water cloud properties from AVHRR (Advanced Very High Resolution Radiometer) data on board NOAA satellites. Detailed description of the algorithm should be referred to Kawamoto et al. (2001). We use three channels, the visible (0.64 micron), near-infrared (3.73 micron) and infrared (11 micron) to retrieve, respectively, the optical depth at visible wavelength τ_{low}, effective particle radius re and cloud top temperature T_c. Monthly-averaged (January, April, July and October) datasets for each property are produced at 0.5-degree spatial resolution. In the processing, pixels whose viewing zenith angles are larger than 25 degrees are discarded. According to Iwabuchi and Hayasaka (2002), this constraint can substantially reduce retrieval errors associated with three-dimensional cloud geometry.

3. RESULTS

Fig. 1 illustrates the annual-mean aerosol amount with numerical model over China. We notice that there are mainly two remarkable aerosol sources. One is the highly-populated eastern coastal area (Beijing and Shanghai), and the other is inland industrial area (Chengdu and Chongqing). Fig. 2 indicates the annual-mean r_e from satellite retrievals. Generally, r_e over ocean is larger than that over land primarily because fewer CCN are available over ocean (Kawamoto et al., 2001).

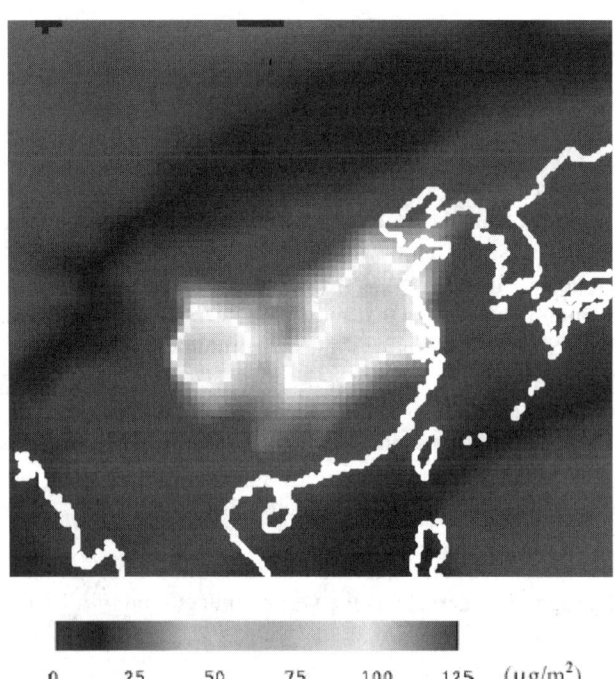

Fig. 1. The annual-mean aerosol loading (μg/m2).

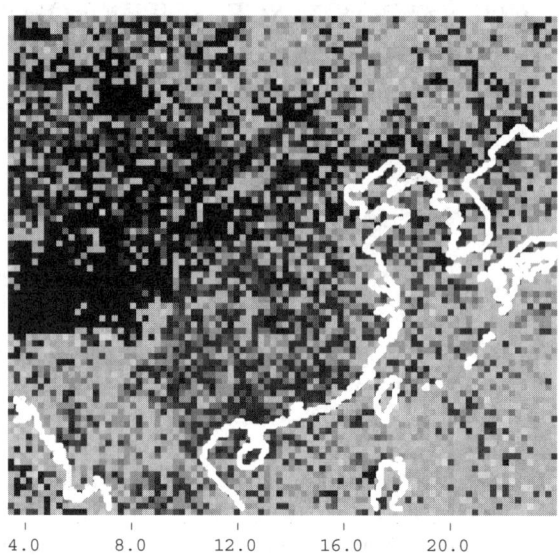

Fig. 2. The annual-mean low cloud effective particle radius *(μm)*.

The following procedure was adopted to examine the relationship between M_a and r_e. Values of M_a were divided into bins, and r_e values on the same geographical grid were collected for the corresponding aerosol amount bin. Then averages of r_e for M_a bins were calculated. Fig. 3 shows the result of such comparison. We find that r_e gets smaller as M_a becomes greater. This behavior can be explained by the Twomey effect (1977). Where many soluble aerosol particles are available as CCN, cloud droplets cannot grow as large, because more growing droplets then compete for the same supply of water vapor. This finding is analogous to SO_2 - r_e relationship studied by Kawamoto et al. (2004).

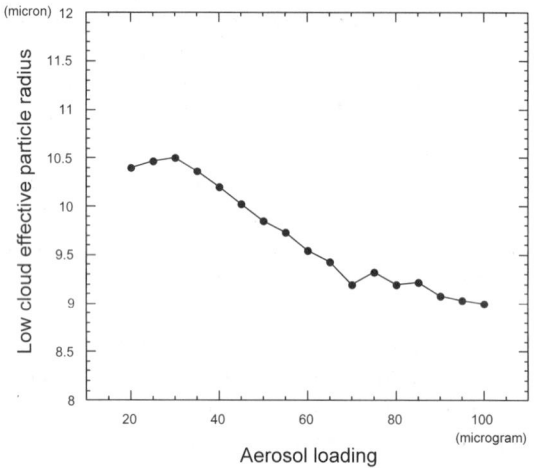

Fig. 3. The relationship between aerosol loading and low cloud effective particle radius.

4. CONCLUDING REMARKS

In order to examine the aerosol indirect effect over China, we compared model-simulated aerosol loading and satellite-derived cloud particle size. The results support the Twomey effect (Twomey, 1977): cloud particle sizes are smaller where aerosol amounts are larger.

Together with the finding of relationship between SO2 emission and cloud particle size by Kawamoto et al. (2004), this study also provides the observational evidence of cloud modification (the aerosol indirect effect of the first kind). Furthermore, this completes the aerosol-to-cloud microphysical linkage reported partially by Chameides et al. (2002), who did not have access to information on cloud particle size.

For future tasks, analyses at finer time resolution such as seasonally would be needed in addition to the annual-mean comparison. Relationships of aerosol loading to other cloud properties (i.e., lifetime and areal extent) are also relevant to climate studies. Finally, we caution that cloud droplet size is affected by updraft velocity and entrainment rate, as well as by the population of available CCN. The effects of updraft velocity and entrainment rate are not analyzed here.

5. ACKNOWLEDGEMENTS

One of the authors (K. K.) is partly supported by Special Coordination Funds for Promoting Science and Technology of Japan.

6. REFERENCES

Chameides, W. L., C. Luo, R. Saylor, D. Streets, Y. Huang, M. Bergin, and F. Giorgi, 2002: Correlation between model-calculated anthropogenic aerosols and satellite-derived cloud optical depths: Indication of indirect effect? *J. Geophys. Res.*, 107, NO. D10, 10.1029/2000JD000208, 2002

IPCC, 2001: Climate Change 2001: The scientific basis. J. T. Houghton et al., ed., *Cambridge University Press*, Cambridge, 884 pp.

Kawamoto, K., T. Nakajima, and T. Y. Nakajima, 2001: A Global Determination of Cloud Microphysics with AVHRR Remote Sensing, *J. Climate*, 14, 2054-2068.

Kawamoto, K., T. Hayasaka, T. Nakajima, D. Streets, and J.-H. Woo, 2004: Examining the aerosol indirect effect over China using an SO_2 emission inventory, *Atmos. Res.*, 72, 353-363.

Pielke, R. A., et al., 1992: A comprehensive meteorological modeling system – RAMS, *Meteorol. Atmos. Phys.*, 49, 69-91.

Twomey, S., 1977: The influence of pollution on the shortwave albedo of clouds. *J. Atmos. Sci.*, 34, 1149-1152.

STUDY OF CLOUD MICROPHYSICAL STRUCTURE WITH CLOUD PROFILING RADAR AND LIDAR: *MIRAI* CRUISE

H. Okamoto and T. Nishizawa
Center for Atmospheric and Oceanic Studies, Graduate School of Science, Tohoku University
Sendai, 980-8578, Japan

H. Kumagai
National Institute of Information and Communications Technology, Koganei, Tokyo 184-8795, Japan

N. Sugimoto
National Institute for Environmental Studies, Tsukuba, Ibaraki 305-8505, Japan

T. Takemura
Research Institute for Applied Mechanics, Kyushu University, Kasuga, Fukuoka 816-8580, Japan

T. Nakajima
Center for Climate System Research, The University of Tokyo, 4-6-1 Komaba, Meguro-ku, Tokyo 153-8904, Japan

ABSTRACT

We observed the vertical distribution of clouds over the Pacific Ocean near Japan and also over tropics using lidar and a 95-GHz radar on the Research Vessel *Mirai*. Cloud analyses derived from observations using cloud mask schemes showed that the radar represented clouds better than the lidar did.

The vertical structure of clouds observed with the radar/lidar system was compared to clouds in the aerosol transport model SPRINTARS, which is based on the CCSR-NIES Atmospheric General Circulation Model. The cloud amount, radar reflectivity factor, and lidar backscattering coefficient were simulated by the model and compared to observations using height-time cross-sections. The overall pattern was well reproduced, although the model underestimated (overestimated) mean cloud amount below 2 km (above 8 km). Agreement was best between 2 and 7 km. Cloud microphysics in the model could be validated through comparison of derived model radar and lidar signals with observations.

1. INTRODUCTION

Despite considerable effort using satellite-borne passive sensors, however, uncertainties remain in the assessment of climate impacts due to clouds. The accuracy of satellite measurements is reduced by the presence of multi-layered structure and by vertical non-homogeneity in cloud microphysics. Most retrieval algorithms for cloud radiative properties assume a single homogeneous layer. Retrieved properties can include errors in clouds that have many layers.

Active instruments such as cloud radar and lidar are powerful tools that can provide detailed vertical profiles of macroscale and microphysical properties. However, few studies, except for some ground-based radar and lidar studies, have covered wide areas. In this study, we used shipborne radar and lidar data from over the Pacific Ocean near Japan. The radar and lidar were installed on the Research Vessel *Mirai*, which is operated by the Japan Marine Science and Technology Center (JAMSTEC). The MR01/K02 cruise, conducted in May 2001, was the first cruise with both cloud radar and lidar on board. Shipborne experiments subsequent to MR01/K02 also recorded cloud radar and lidar data, including a cruise in the tropics from September to December 2001. Methods have been developed that combine cloud radar with other sensors. The key issue when using lidar data with visible or near infrared wavelengths is attenuation correction. Okamoto et al. (2003) determined the vertical distribution of the effective radius and ice water content using a forward-type algorithm that uses look-up tables for 95-GHz radar and lidar at 532 nm. We applied the method to the data taken in *Mirai* cruise.

Radar and lidar observations have been planned from two satellites that are scheduled to be launched in 2005 as part of the NASA Earth System Science Pathfinder (ESSP) program. CloudSAT will carry a 95-GHz radar (Stephens, 2000). Cloud-Aerosol Lidar and Infrared Pathfinder Satellite Observations (CALIPSO; Winker, 2002) will include lidar at 0.532 μm and 1.064 μm. Cloud-profiling radar (CPR) and lidar satellite measurements will follow the CloudSAT and CALIPSO experiments. It is therefore vital to understand the information latent in combining 95-GHz radar and lidar systems.

2. VERTICAL STRUCTURE OF CLOUDS

The received power is affected by particle size between the cloud and the radar or lidar such that the radar is sensitive to large particles and the lidar is sensitive to small particles. Furthermore, extinction due to clouds at lidar wavelengths is greater than extinction for radar. These differences affect how each sensor can be used and interpreted. How one sensor outperforms the other depends on macroscale properties of clouds and cloud microphysics. Four different cloud mask schemes were developed to assess how the radar and the lidar retrievals performed when clouds were present. We test the performance of radar and lidar on the basis of the four cloud mask schemes and found that the radar represented clouds better than the lidar did. Radar detected 95% of clouds between 2 and 11 km; lidar detected <50%. Lidar was able to detect some clouds below 2 km and above 11 km that the radar could not, suggesting that the use of lidar data can improve observations. Further details of the analyses are found in Okamoto et al., 2005 (in preparation).

Fig.1 shows the height-time cross-section of cloud occurrence during the second cruise from September to December 10, 2001 over tropics.

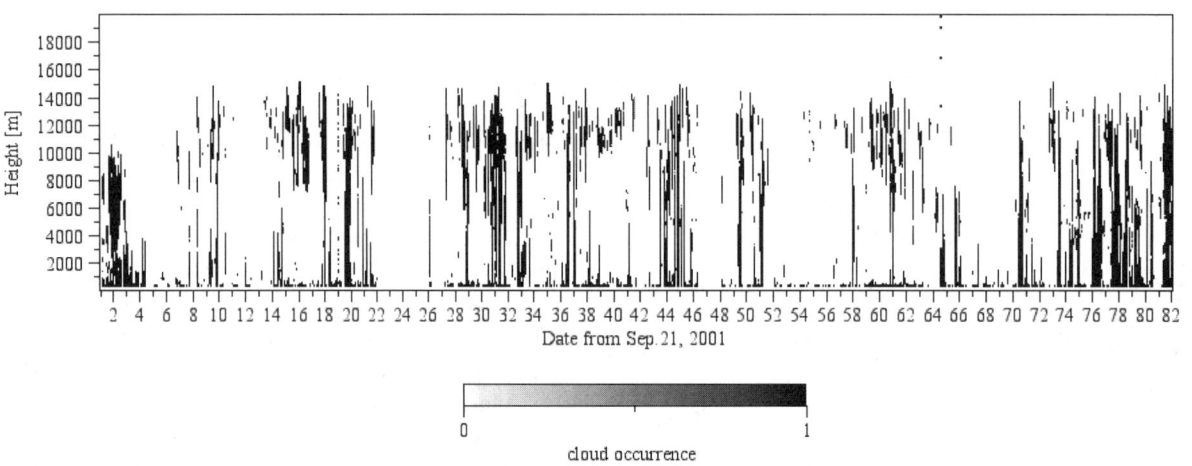

Fig. 1. The height-time cross-section of cloud occurrence during the second Mirai cruise from September 21 to December 10, 2001 in the tropics.

Then we derived the average vertical distribution of clouds from these data sets (Fig.2). There are some differences between the cloud properties in mid-latitude (May), and those in the tropics (September, October and November). That is, two local maxima in mid-latitude are found at 7 and 10 km in mid-latitude, while, the location is much higher in tropics and is at around 12 km. In the tropics, clouds have somehow homogeneous structure between 2 and 8 km, contrary to the mid-latitude where there is a local minimum at 2 km.

We also derived the average number of layers. The average number of layers turns to be 1.5. The mean cloud cover (83%) with the average number (1.5) yielded an average number of layers of 1.8 when clouds existed. The single layer assumption is usually made in retrieval algorithms for instruments, such as AVHRR, MODIS, or GLI. This suggested some cautions that the cloud properties derived from these algorithms used in the passive sensors should be carefully tested against estimates from active sensors.

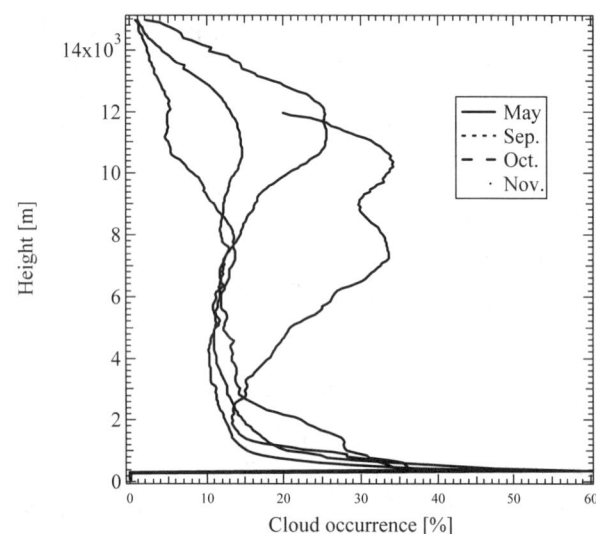

Fig. 2. Height-time cross section of cloud amount derived from the radar and lidar observations in mid-latitude (May) and in tropics (Sep., Oct. and Nov.).

3. VALIDATION OF GENERAL CIRCULATION MODEL

We compared the results of a General Circulation Model to radar and lidar data along the *Mirai* cruise track. We used SPRINTARS, a model based on the CCSR-NIES GCM, and it can simulate the cloud fields as well as concentration of four types of aerosols such as sulfates, carbonaceous aerosols, sea salt, and dust, and this model has been used to assess the effects of aerosols on climate (Takemura et al., 2005). The model in this study had T106 truncation, corresponding to a horizontal resolution of around 100 km, and 20 vertical levels. Temperature, pressure, and relative humidity estimated in SPRINTARS were nudged with 6-hour-interval NCEP/NCAR reanalysis data. In mid-latitude, observations and the model showed fairly good agreement from 2 to 7 km. However model underestimated low clouds below 2 km and significantly overestimated cloud amounts (fig. is not shown). We also performed similar comparisons for the tropics data. Fig.3 showed the vertical distribution of the mean cloud amount as observed and modeled in the tropics. The data for September, October, and November are used to derive mean cloud amount in the region. The similar trends were found, i.e., the model underestimated/overestimated low/high cloud amount. This may be related to the deficiencies of cloud prediction in the model. For the high clouds, the model seemed to fail to predict the proper lifetime of the clouds, due to the failure in the estimates of proper terminal velocity of ice particles in the altitude.

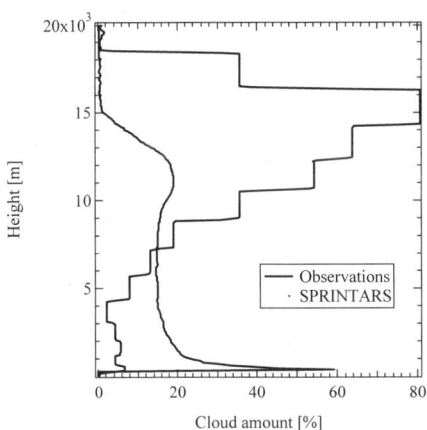

Fig. 3. Comparison between averaged vertical profiles of mean cloud amount deduced from radar or lidar data and from SPRINTARS in the tropics. The periods are from Sept., to Nov. in 2001.

Besides the cloud amount, comparisons of the radar reflectivity factors and the lidar backscattering coefficient facilitate the validation of model cloud microphysical properties such as liquid/ice water content (hereafter LWC/IWC) and effective radius (R_{eff}). The combination of radar and lidar data is a powerful tool to help retrieve ice microphysics. Applicability is limited for ground-based measurements to cases in which there are no water clouds below the ice clouds of interest. It is shown that the probability that the radar and the lidar can record the same clouds at 10 km was 38% at 8 km and 12% at 10 km in midlatitude. That is, the remaining clouds may not be analyzed by the radar/lidar method.

We did not use a retrieval algorithm in this section. Instead, radar and the lidar signals were simulated from SPRINTARS output and directly compared to observations as an alternative approach to validate cloud microphysics in the model. When simulated signals differ from observations, other discrepancies can arise, such as cloud position, cloud amount, particle size, and LWC/IWC. When cloud position is accurately predicted, the combined use of radar and lidar signals may help reveal problems in predicting/assuming cloud microphysics and mean IWC in the model. When simulated radar and lidar signals are both larger than the observations, IWC is predicted to be larger than actual values, and particle size is larger or the same.

From the comparisons, we found that simulated particle sizes in water clouds of 10 μm were smaller than observed. LWC may have been higher in the model at lower levels. At upper levels, the assumed ice particle size in the model (40 μm) was smaller than the observed value.

4. CLOUD MICROPHYSICS FROM RETRIEVAL ALGORITHM

In this section, we applied the retrieval algorithm to the radar and lidar data in order to obtain remotely sensed ice microphysics. The details of the algorithm and the application to the ground-based data are found in Okamoto et al. (2003). The following parameters were obtained in the observations; radar reflectivity factor, Doppler velocity and LDR (linear depolarization ratio) from cloud profiling radar, and backscattering coefficients at two wavelengths, i.e., 532 nm and 1064 nm, and depolarization ratio from the lidar. Therefore the combinational use of the two active sensors provided a unique opportunity to study, 1) vertical distribution of cloud microphysics such as effective radius and cloud water content, 2) Relationship between temperature, radius and fall velocity.

Fig. 4 showed the results of the microphysics in May 21, 2001 over Pacific Ocean near Japan. Particle size tended to be larger as the altitude decreased. This may suggest the relation between the temperature and particle size. The particle effective radius ranges from 50 μm to 200 μm. While, IWC ranges from 0.001 to $0.1 g/m^3$.

The average relation between the size and velocity is estimated from the whole observational data (Fig. 5). As expected from the terminal velocity, the size increased as velocity increased for effective radius> 100 μm. However, there is almost no size dependence for small particles <100 μm. and the steep slope for the particles with larger one.

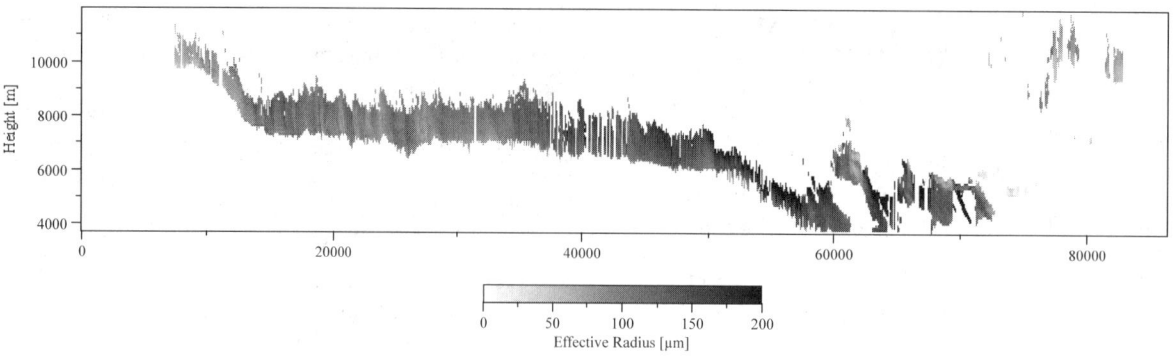

Fig. 4. The height-time cross sections of effective radius of ice particles over Pacific Ocean near Japan in May 21, 2001.

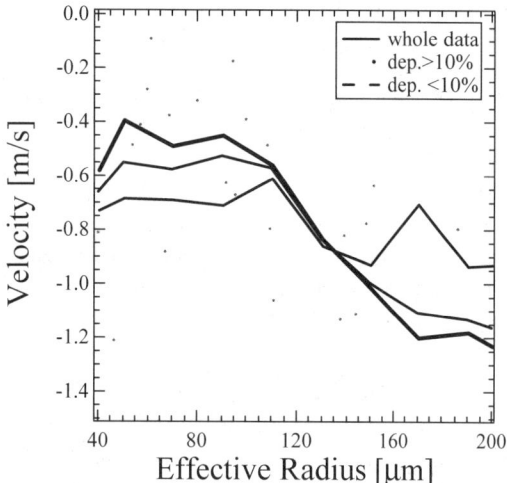

Fig. 5. Relation between the fall velocity and particle size for the whole data in mid-latitude.

When we choose the particles that show the lidar depolarization ratio >10%, the fall velocities for the particles are smaller than those for the whole. While, the particles with smaller depolarization ratio (<10%) shows the larger velocity compared with the average one from the whole data. This suggests that the terminal velocity of non-spherical particles is smaller than that of sphere with the same mass as the non-spherical particles due to the larger air resistance of the non-spherical particle.

5. SUMMARY

We developed the cloud mask schemes for the analyses of cloud macroscale properties of clouds. Some differences between macroscale properties in mid-latitude and those in tropics were found. The mean cloud amount derived from radar and lidar observations was used to test the SPRINTARS model, which is based on the NIES-CCSR GCM with nudging by NCEP-NCAR reanalysis data. The model patterns resembled the observations in mid-latitude and also in tropics. The model overestimated clouds above 8 km by 80% and underestimated them below 2 km. The observations and the model showed fairly good agreement between 2 and 7 km. The tendency to overestimate cloud cover increased as the altitude increased. Similar trends were found in tropics and the performance of the model in the region turned to be worse.

The comparisons of simulated radar/lidar signals by the model with observations were performed. In addition to the radar/lidar retrieval methods, we demonstrated that these two approaches might be used to infer cloud microphysics and may lead to solve the problems mentioned above in the model.

6. REFERENCES

Okamoto, H., S. Iwasaki, M. Yasui, H. Horie, H. Kuroiwa, and H. Kumagai, 2003: An algorithm for retrieval of cloud microphysics using 95-GHz cloud radar and lidar, *J. Geophys. Res.*, 108(D7), 4226, doi:10.1029/2001JD001225.

Stephens, G. L., and the CloudSat Science Team, 2002: The CloudSat Mission and the A-train. A new dimension of space-based observations of clouds and precipitation. *Bull. Amer. Meteor. Soc.*, 83, 1771-1790.

Takemura, T., T. Nozawa, S. Emori, T. Y. Nakajima, and T. Nakajima, 2005: Simulation of climate response to aerosol direct and indirect effects with aerosol transport-radiation model, *J. Geophys. Res.*, 110, D02202, doi:10.1029/2004JD005029.

Winker, D. M., J. Pelon, and M. P. McCormick, 2002: The CALIPSO Mission: Spaceborne lidar for observation of aerosols and clouds, *Proc. SPIE*, 4893, 1-11.

USING SATELLITE OBSERVATIONS AND REANALYSES TO EVALUATE CLIMATE AND WEATHER FORECAST MODELS

Richard P. Allan

Environmental Systems Science Centre, University of Reading
Reading, Berkshire, RG6 6AL, UK; Email: rpa@mail.nerc-essc.ac.uk

ABSTRACT

Satellite observations of water vapour and radiative fluxes are used in combination with reanalyses data to evaluate the Met Office weather and climate prediction models. Using reanalysis vertical motion data, it is established that much of the climate model error in radiative fluxes relates to errors in dynamical fields. Radiative feedback processes are to a large extent independent of these positional errors in the large-scale circulation. It is therefore of great value to analyse the relationships between water vapour, clouds and radiation in terms of dynamic regime or by averaging over large-scale circulation systems. Using the latter approach, it is found that the monthly and interannual relationships between water vapour and clear-sky radiation are consistent between models and satellite data. However, the variation in cloud radiative effect appears much larger in the observations than the model. To further understanding of radiative feedback processes involving water vapour and cloud, in is necessary to examine sub-monthly time-scales. Comparisons between the Met Office weather forecast model and Geostationary Earth Radiation Budget (GERB) data are currently in progress and illustrate the potential of this strategy.

1. INTRODUCTION

Satellite radiance measurements provide valuable information on the distribution and variability of clouds, water vapour and the Earth's radiative energy budget. Using this information to initialise, evaluate and improve global climate and weather forecast models is vital in reducing the uncertainty in predictions. The following paper shows some examples in which data from satellite instruments and reanalyses are used to evaluate model simulations and understand radiative feedback processes.

2. SAMPLING DIFFERENCES

In evaluating models using satellite data, it is vital to first ensure that similar physical quantities are being compared. For example, outgoing longwave radiation (OLR) estimated from satellite instruments is readily comparable with the corresponding model diagnostic. It is also important that the spatial and temporal sampling is consistent. For example, the spatial and temporal sampling of clear-sky OLR (OLRc) in models and satellite data are often inconsistent (e.g., Allan et al., 2003). This is because satellite estimates of OLRc can only be made for cloud-free regions while models can compute OLRc diagnostically for all grid points at all times.

Further difficulties arise when using regional measurements of radiative fluxes or humidity to evaluate model processes. This is because locally, model errors are highly sensitive to small positional errors in the large-scale circulation. This is illustrated using model simulations from the Hadley Centre climate model, HadAM3 (see Allan et al. 2003 for details), with OLR data from the Clouds and the Earth's Radiant Energy System (CERES; e.g., Wielicki et al., 2002) and 500 hPa vertical motion (ω) products from the ECMWF 40-year reanalysis (ERA-40; Uppala et al., 2005). Fig. 1 shows the positive spatial correlation between model OLR error and model ω error over the tropical Pacific for April 1998.

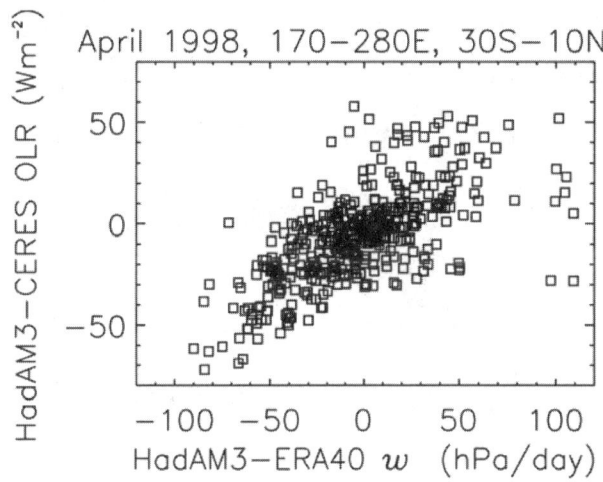

Fig. 1. Errors in model OLR (HadAM3 minus CERES) as a function of model vertical motion error at 500 hPa (HadAM3 minus ERA40) over the tropical Pacific (170–280°E, 30°S–10°N) for April 1998.

While errors in atmospheric circulation are important for local predictions of atmospheric conditions, these positional errors are not generally crucial to the radiative feedback processes operating in models. It is therefore of great value to remove this effect. To address this issue the adopted approach has generally been either to (i) analyse spatial averages over entire circulation systems (e.g., Soden et al., 2002; Allan and Slingo, 2002) or (ii) to examine dynamical regimes using vertical motion fields from reanalyses (e.g., Bony et al., 2004; Ringer and Allan, 2004). In the following two sections we adopt approach (i) to diagnose large-scale radiative feedback processes involving water vapour and cloud.

3. WATER VAPOUR FEEDBACK

The radiative properties of water vapour dictate that absorption of radiation increases with the logarithm of water vapour concentration over much of the longwave spectrum. Additionally, the Clausius-Clapeyron equations describe a quasi-exponential increase in the water vapour holding capacity of the atmosphere as temperatures rise. Combined, these theoretical constraints predict a strongly positive water vapour feedback providing that the water vapour concentration remains roughly at a constant fraction of the saturation specific humidity (unchanging relative humidity, RH).

The largest uncertainty in water vapour feedback is arguably related to how tightly the relationship between humidity and temperature is controlled by thermodynamics. Recent evidence from models and satellite data suggest that water vapour variations are strongly constrained by constant RH (e.g., Wentz and Schabel, 2000; Soden et al., 2002; Allan et al., 2003). However, it remains vital to verify the invariant nature of global mean RH in models.

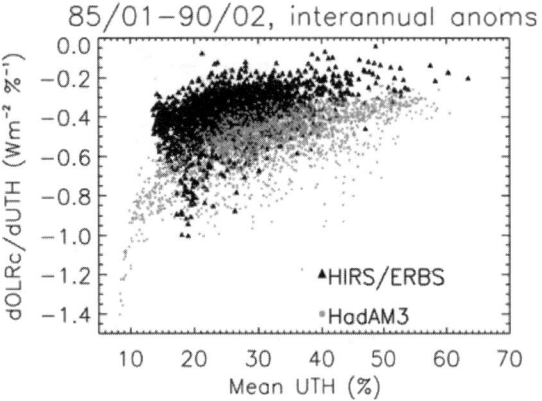

Fig. 2. Sensitivity of clear-sky OLR to interannual changes in upper tropospheric humidity (UTH) as a function of mean UTH for a model (HadAM3) and combination of satellite data (ERBS and HIRS).

It is also important that models correctly simulate relationships between water vapour, cloud and radiation. For example, Fig. 2 shows simulations and observations of the sensitivity of OLRc to upper tropospheric RH (UTH) as a function of mean UTH. The satellite measurements were made by the Earth Radiation Budget Satellite (ERBS) scanner (e.g., see Wielicki et al., 2002) and the High Resolution Infrared Sounder (HIRS) instrument (e.g., see Allan et al., 2003). The sensitivity was calculated at each grid-point for the region 30°S–30°N using interannual monthly anomalies relative to a seasonal climatology. Both model and satellite data show a more strongly negative sensitivity of OLRc to changes in UTH for low humidity regions. The HadAM3 model appears to produce larger magnitude sensitivity for most grid-points. Sampling inconsistency between the ERBS and HIRS measurements may explain the tendency for a larger spread in dOLRc/dUTH for a given UTH. It is unclear why the observed UTH does not display values below 13%. Measurement error relating to calibration and orbit drift are also likely to influence the comparisons, despite careful calibration of the HIRS record. Further work is required to extend these analyses to evaluate regional radiative feedbacks involving water vapour and other components of the hydrological cycle including cloud and precipitation.

4. CHANGES IN LOW-LATITUDE RADIATION BUDGET AND CLOUDINESS

Crucial to the evaluation of radiative feedback simulated by models is the monitoring of the Earth's radiation budget at the top of the atmosphere. Fig. 3 displays monthly variations in components of the low latitude radiative energy balance for the HadAM3 model (black) and satellite data (grey lines and symbols). The observations comprise data from ERBS (1985–1990), ScaRaB (1994/5) and CERES (1998) which are described in Wielicki et al. (2002) and references therein. Consistent with the previous section dealing with water vapour feedback, OLRc simulated by the HadAM3 model appears broadly consistent with the variation observed using a variety of satellite instruments (Fig. 3a). However, CERES OLRc is up to 2 Wm^{-2} larger than model values during summer 1998; this primarily results from differences over land regions and may relate to errors in model land surface temperature.

Also shown in Fig. 3a is the model all-sky OLR (dashed line). This variability is almost identical to the clear-sky variability indicating that, for the low-latitude mean, changes in the longwave radiation budget relating to variations in cloudiness are small. Comparing the all-sky OLR (Fig. 3b) and the all-sky reflected shortwave radiation (RSW) however (Fig. 3c) indicates that the decadal changes observed by the satellite data are much larger than simulated by the models (Wielicki et al., 2002).

The satellite data suggests a reduction in cloud radiative effect from the 1980s to the 1990s, which is not reproduced by the model. Although it is possible the data is in error, the agreement between a number of independent satellite instruments (Wielicki et al., 2002) and the differing responses of clear-sky and cloudy-sky OLR on the same scanning instruments suggests that the changes are real. The ERBS non-scanner data employed is similar to the data presented in Wielicki et al. (2002) but has been corrected for errors relating to diurnal aliasing in the seasonal cycle and orbital degradation (T. Wong, personal comm.).

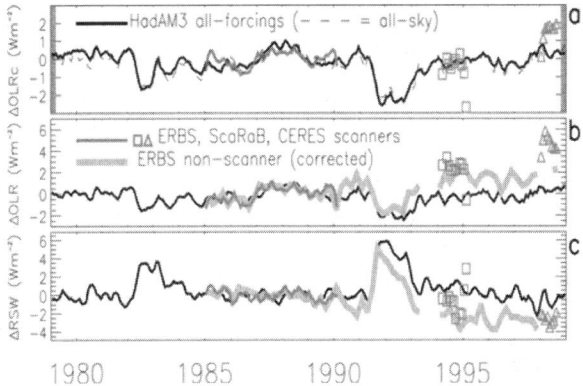

Fig. 3. Time series of 40°S–40°N mean (a) clear-sky OLR, (b) OLR and (c) RSW anomalies with respect to the 1985–1989 mean seasonal climatology. Bold line denotes HadAM3 while grey lines and symbols represent satellite observations (see also Allan and Slingo, 2002; Wielicki et al., 2002).

The reason for the differing changes in low-latitude cloudiness between models and data remain unclear. While monthly or longer averages are useful for monitoring large-scale changes in clouds, water vapour and the Earth's radiation budget, it is difficult to relate these variations to the physical processes generally operating on much shorter time-scales. It is therefore important to assess the performance of models on time-scales ranging from the model time-step up to hours and days. During these time-scales, model parameterizations are directly contributing to the evolving processes important for weather forecasts and climate prediction. The next section presents preliminary work comparing the Met Office forecast model with new geostationary satellite data on such time-scales.

5. EVALUATION OF AN NWP MODEL USING GEOSTATIONARY EARTH RADIATION BUDGET (GERB) DATA

New radiative flux data from the GERB instrument on the Meteosat-8 satellite is being exploited in comparisons with the Met Office forecast model. The GERB instrument measures broadband radiances, which are converted to fluxes using angular dependence models. These measurements contribute to data products every 15 minutes at a sub-satellite resolution of about 40 km, covering the African/Atlantic hemisphere. For further details of the instruments and preliminary validation results, see Harries et al. (2005).

GERB observations of OLR and RSW are being routinely accessed from the Royal Meteorological Institute of Belgium (RMIB) and compared with model simulations based on the forecast analysis at 00, 06, 12 and 18 hours UTC. These comparisons are displayed at http://www.nerc-essc.ac.uk/~rpa/GERB/gerb.html and the data stored for further analysis. The methodology, first results and validation of GERB clear-sky fluxes over the ocean are described in Allan et al. (2005).

Fig. 4 shows an example comparison of GERB and model albedo (RSW normalised by the calculated incoming solar radiation) for 12 UTC, 27 October 2004. The bright regions indicate thick cloud or bright surfaces, such as the Sahara desert. Dark regions of low albedo correspond with clear-sky ocean regions. Pole-ward from the tropics, the distribution of cloud, relating to large-scale mid-latitude weather systems, appears well simulated by the model. Cloud data is not assimilated as part of the model initialisation (or analysis). However, the large-scale atmospheric dynamics are well constrained by the model assimilation system which utilises observed variables such as water vapour, temperature and wind-fields to produce a reasonable distribution of humidity. The model cloud parametrizations convert this information into realistic cloud fields.

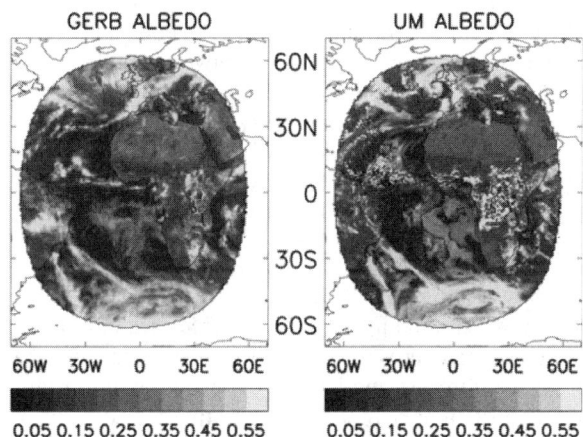

Fig. 4. Shortwave albedo measured by GERB satellite data (left) and simulated by the Met Office forecast model (right) for 27 October 2004, 1200 UTC.

Over lower latitudes, errors in surface albedo are apparent over the Sahara, while errors in cloud-related albedo are present over Brazil (the model underestimates cloud cover) over tropical Africa (the amount and spatial variation in cloud cover is overestimated by the model) and over the southeast Atlantic marine stratocumulus region (the model albedo is too large). There is also a tendency for simulated convection to lack the organization shown in the GERB observations. Further model evaluation using GERB data is currently underway. It is planned to use the high temporal resolution of the Meteosat-8 data in evaluation of the model by conducting numerical forecast experiments that explore the sensitivity of the model simulations to changes in surface properties and new physical parameterizations. It is also planned to use the GERB data, in combination with narrow-band radiance data from the SEVIRI instrument, also onboard Meteosat-8, in studies of water vapour and cloud radiative processes (e.g., Futyan et al., 2004) and aerosol (e.g., Haywood et al., 2005).

6. CONCLUSIONS

Satellite radiance measurements provide valuable information on clouds, water vapour, aerosol and surface properties that are vital in evaluating and improving weather and climate prediction models. It is important to ensure that the observed quantity and its spatial and temporal sampling are consistent with corresponding model diagnostics in such comparisons between models and data. It is also informative to separate model errors relating to positional errors in the large-scale atmospheric circulation from the errors relating to physical processes (e.g., Bony et al., 2004), although errors in physical processes (e.g., water vapor transport or convection) can also lead to positional errors.

The large-scale variation in water vapour and clear-sky radiation appears consistent between models and satellite data, suggesting that water vapour feedback processes are well represented by such models. However, this is not the case for the radiative effect of cloud, which exhibits greater variability in the satellite data compared to model simulations. The reasons for this discrepancy are not yet clear (Wielicki et al., 2002).

A strategy to more closely relate radiative feedbacks involving water vapour and other variables such as cloud, precipitation and aerosols is described. Analyses from the Met Office forecast model are compared with instantaneous geostationary broadband radiative flux data from the GERB instrument. These data are preliminary and currently under validation. However, initial comparisons suggest this approach may elucidate radiative forcing and feedback processes on the time-scales at which the model parameterizations are directly contributing to the evolving forecasts, thereby enabling model improvement.

7. ACKNOWLEDGMENTS

This work was funded through the NERC/Met Office Connect-B grant NER/D/S/2002/00412. The input and support of Tony Slingo is gratefully acknowledged. Thanks also to the SINERGEE project collaborators at the Met Office and the GERB International Science Team, in particular colleagues at Imperial College and also the Royal Meteorological Institute of Belgium, who process the GERB data. The ERBS and CERES data were retrieved from the NASA Langley DAAC, the ERA-40 data was obtained from the ECMWF, the ScaRaB data was taken from the Centre National d'Etudes Spatiales, Toulouse and the HIRS data was provided by Darren Jackson.

8. REFERENCES

Allan, R. P., and A. Slingo, 2002: Can current climate forcings explain the spatial and temporal signatures of decadal OLR variations? *Geophys. Res. Lett.,* 29(7), 1141, doi:10.1029/2001GL014620.

Allan, R. P., M. A. Ringer, and A. Slingo, 2003: Evaluation of moisture in the Hadley Centre Climate Model using simulations of HIRS water vapour channel radiances, *Q. J. Roy. Met. Soc.,* 129, 3371-3389.

Allan, R. P., A. Slingo, S. F. Milton, and I. Culverwell, 2005: Exploitation of geostationary Earth radiation budget data using simulations from a numerical weather prediction model: Methodology and data validation, *J. Geophys. Res., in press.*

Bony, S., J.-L. Dufresne, H. Le Treut, J. J. Morcrette, and C. A. Senior, 2004: On dynamical and thermodynamical components of cloud changes, *Clim. Dyn.,* 22, 71-86

Futyan, J. M., J. E. Russell, and J. E. Harries, 2004: Cloud Radiative Forcing in Pacific, African, and Atlantic Tropical Convective Regions, *J. Climate,* 17, 3192–3202.

Harries, J. E., and co-authors, 2005: The Geostationary Earth Radiation Budget (GERB) Experiment, *Bull. Amer. Meteorol. Soc, in press.*

Haywood, J. M., R. P. Allan, I. Culverwell, A. Slingo, S. F. Milton, J. M. Edwards, and N. Clerbaux, 2005: Can desert dust explain the outgoing longwave radiation anomaly over the Sahara during July 2003? *J. Geophys. Res.,* 110, D05105, doi:10.1029/ 2004JD005232.

Ringer, M. A., and R. P. Allan, 2004: Evaluating climate model simulations of tropical cloud, *Tellus,* 56A, 308-327.

Soden, B. J., R. T. Wetherald, G. L. Stenchikov, and A. Robock, 2002: Global cooling after the eruption of Mount Pinatubo: a test of climate feedback by water vapor, *Science,* 296, 727-730.

Uppala, S. M., and co-authors, 2005: The ERA-40 Re-analysis, submitted to *Q. J. Roy. Meteorol. Soc.*

Wentz, F. J., and M. Schabel, 2000: Precise climate monitoring using complementary satellite data sets, *Nature,* 403, 414-416.

Wielicki, B. A., T. M. Wong, R. P. Allan, A. Slingo, J. T. Kiehl, B. J. Soden, C. T. Gordon, A. J. Miller, S.-K. Yang, D. A. Randall, F. Robertson, J. Susskind, and H. Jacobowitz, 2002: Evidence for large decadal variability in the tropical mean radiative energy budget, *Science,* 295, 841-844.

MODES OF TROPICAL WATER CYCLE VARIABILITY

John J. Bates
NOAA/NESDIS National Climatic Data Center
Asheville, North Carolina

1. INTRODUCTION

In order to separate the natural and anthropogenic effects of climate change, it is important to understand and quantify feedback mechanisms. Any process that changes the sensitivity of the climate response to an imposed anthropogenic forcing is called a feedback mechanism. Feedback mechanisms can either increase (a positive feedback) or decrease (a negative feedback) the magnitude of the climate response to forcing agents. For example, given the magnitude of a man-made greenhouse gas such as CO_2, the relationship between the magnitude of this anthropogenic climate forcing and the magnitude of the climate change response in a general circulation model, perhaps a global warming of 2°C, defines the climate sensitivity.

Large uncertainties occur in the feedback processes associated with clouds and water vapor, particularly in the Tropics. This is because in the Tropics there is a very large range in outgoing longwave radiation (OLR) between cold, high clouds near the Tropical tropopause and the clear, warm and dry atmospheres in the subtropical deserts. Recent papers have documented large variability in outgoing longwave radiation on both short and long time scales. Wielicki *et al.* (2002) find both large temporal (white in frequency space) variability in broadband OLR and a significant decadal trend. The white frequency variability is approximately ±2 Wm^{-2} for the tropical strip 20N–20S. The trends they find include a drop of about 2 Wm^{-2} from the late 1970s to the mid 1980s and a rise of about 4 Wm^{-2} from the late 1980s to the late 1990s. Work by Chen *et al.* (2002) examine possible mechanisms for the OLR trends documented by Wielicki and find this trend in OLR is consistent with changes in cloudiness and upper tropospheric humidity that suggest a decadal-time-scale strengthening of the tropical Hadley-Walker circulation. A closer examination of the Wielicki *et al.* (2002) time series of tropical OLR, however, also reveals high variability on subseasonal to seasonal time scales. Causes for this scale of variability were not fully examined by Wielicki et al. nor by Chen *et al.* (2002).

My examination of multiple satellite-derived indices and re-analysis over the past two decades suggests that there are important interactions between the different temporal scales due to the unique dynamical wave modes in the tropics. In the tropics, a dynamical wave duct opens in the northern winter and spring seasons such that mid-latitude Rossby waves can propagate deep into the subtropics and, sometimes, between hemispheres. The amount of this wave activity strongly affects the water and energy budget of the tropics and is a function of El Niño-Southern Oscillation (ENSO) state and mid-latitude transient activity. This discovery indicates that one must examine both changes in the transient tropical circulation modes, not just changes in the mean Hadley-Walker circulation, when seeking mechanisms to explain the large observed changes in the tropical water and energy cycle.

2. INTRASEASONAL TO INTERANNUAL VARIABILITY

Tropical (30°N–30°S) average time series of sea surface temperature, SST, [Reynolds, 1994], tropospheric temperature from the TOVS Microwave Sounding Unit (MSU channel 2) [Christy et al., 1998; Christy et al., 1995], upper tropospheric humidity (UTH), and CLear-sky OLR from ECMWF 15-year Re-Analysis (CLERA) [Slingo et al., 1998] and Clouds and the Earth's Radiant Energy System (CERES) [Wielicki et al., 1998] are shown in Fig. 1. For this study of short frequency variability, the OLR data sets have been detrended and normalized. Also included in these plots are SST indices for the tropical Pacific regions sensitive to ENSO, the Niño 4 region in the western equatorial Pacific and the Niño 3 region in the central equatorial Pacific.

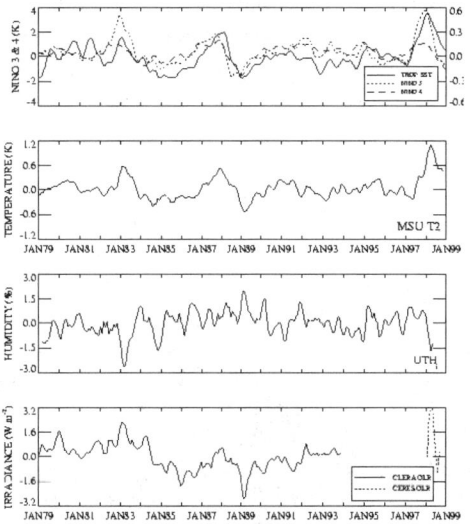

Fig. 1. Tropical (30N–30S) indices of a) sea surface temperature, b) lower tropospheric temperature (MSU2R), c) upper tropospheric humidity, and d) outgoing longwave radiation (OLR).

Although the correlations between SST and UTH and between SST and clear-sky OLR are not significant, there does appear to be some association between the different time series. The two most extreme negative values of UTH (most extreme positive value of clear-sky OLR) occur during the mature phase of the extreme ESNO events of 1982–83 and 1997–98. The most extreme positive value of UTH (most extreme negative value of clear-sky OLR) occurs during the extreme cold event of 1989. These UTH extremes are of much shorter duration than the SST and tropospheric temperature anomalies. This suggests that anomalies in tropical UTH and clear-sky OLR occur on both seasonal and interannual time scales in contrast to SST and tropospheric temperature which show only interannual variability.

In an attempt to identify which regions of the tropics contribute most significantly to the tropical wide time series of Fig. 1c, a one-point correlation map between the tropical-wide UTH time series and the UTH time series was computed at each grid point.

This map (Fig. 2; note due to truncation of the graphics, the magnitude of the 1998 ENSO is truncated) shows that the largest contribution comes from the eastern Pacific (Bates et al., 2001; Bates et al., 1996. Several studies have indicated that transient eddy activity is high in this area [Kiladis and Weickmann, 1997] and that the transient eddies are associated with large plumes of moisture [Iskenderian, 1995]. This transient eddy activity has a strong seasonal component [Kiladis and Weickmann, 1997] and also shows a strong interaction with ENSO events [Matthews and Kiladis, 1999]. Thus, variations in eddy activity are a possible candidate to explain some extremes of the tropical interannual anomaly time series of UTH and clear-sky OLR.

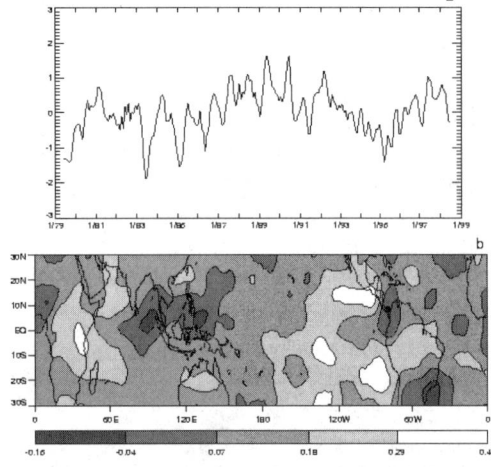

Fig. 2. Interannual anomaly of UTH (a) and one-point correlation map (b).

3. PROPAGATION OF MID-LATITUDE ROSSBY WAVES INTO THE TROPICS AND THE ROLE OF THE WESTERLY DUCT

To examine the changes in transient eddy activity in this region during the large warm event of 1982–83 and the large cold event of 1989–90, we employed the dynamical analysis of 200 hPa winds as outlined in Kiladis [1998]. The horizontal E vector is a pseudo-vector constructed by calculating time-mean covariances between the perturbation zonal, u', and meridional, v', wind components:

$$\vec{E} = \left(\overline{v'^2 - u'^2}, \overline{-u'v'} \right) \quad (1)$$

The term $\overline{v'^2 - u'^2}$ is a measure of the mean anisotropy of Rossby waves. For example, if v'^2 is consistently larger than u'^2, the Rossby waves are preferentially elongated in the meridional direction and the E-vector points eastward. The term $\overline{-u'v'}$ is the negative of the time-mean northward flux of westerly momentum associated with perturbations. Together the two components approximate the preferred direction of the group velocity of the Rossby waves using suitable approximations, including the assumption of quasi-geostrophy. For this work, we use the NCEP re-analysis data and bandpass filter for 6–30 day transients.

Another useful diagnostic for representing the mean background state in which the transients are embedded is the stationary Rossby wavenumber:

$$K_s = \left(\frac{\beta_*}{\overline{U}} \right)^{1/2} \quad (2)$$

where $\beta_* = \beta - \dfrac{\partial^2 \overline{U}}{\partial y^2}$ \quad (3)

is the meridional gradient of absolute vorticity associated with the basic flow, $\overline{U}$ is the monthly mean 200 hPa zonal wind, and $\beta = \dfrac{\partial f}{\partial y}$ is the meridional gradient of planetary vorticity. K_s is the total wavenumber at which a barotropic Rossby wave is stationary at a particular location in a given background zonal flow. Low values of the stationary Rossby wave number (below about 10) indicate regions of strong eddy activity and high values (above 15) indicate regions of weak eddy activity.

E-vectors and the stationary Rossby wavenumber were computed for all months and plots of the minima UTH (April 1983) and maxima UTH (February 1989) months are shown in Fig. 3.

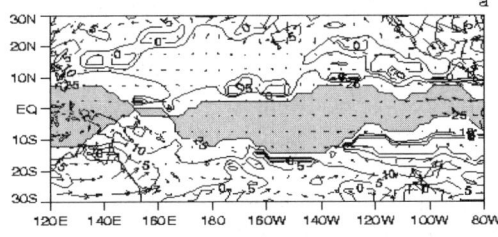

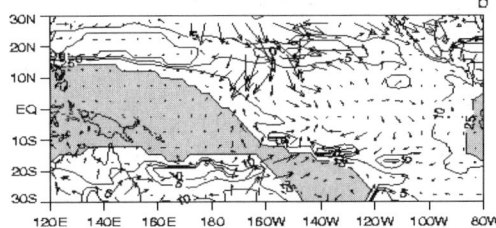

Fig. 3. E-vectors and stationary Rossby wave numbers for February of 1983 (a) and January of 1989 (b).

In the northern hemisphere, mid-latitude interactions with the tropics are greatest in boreal winter and spring when transient Rossby wave activity with periods between 5 and 30 days is at a peak. During strong cold events, such as 1989 (Fig. 3b), tropical convection occurs only over the far western Pacific Ocean since the western Pacific warm pool shrinks and moves to the west during cold events. In the tropical upper troposphere near the equator, this creates strong westerly winds in the outflow from this convection over the central and eastern Pacific. This allows the opening of a westerly duct in the eastern Pacific and supports propagation of Rossby waves deep into the subtropics [Webster and Holton, 1982]. This is evidenced in Fig. 3b where values of stationary Rossby wave numbers less than 10 are found in the regions between the dateline and the west coast of South America. Large values of E-vectors pointing toward the equator near Hawaii are indicative of equatorially-propagating Rossby waves from the mid-latitudes into the subtropics. As shown by Kiladis [1998], the Rossby waves then propagate to the east and are associated with large plumes of moisture extending from near Hawaii to the U.S. west coast sometimes dubbed the 'pineapple express'.

Conversely, during the large warm event of 1982 (Fig. 3a), deep convection extends far to the east in the equatorial Pacific Ocean. Upper tropospheric westerlies over the equator weaken dramatically or even reverse in the central and eastern Pacific. This is confirmed by calculations of the stationary Rossby wave number for these events. Large values of the stationary Rossby wave number are found over the eastern tropical Pacific, effectively shutting down the westerly duct. Virtually no E-vectors pointing toward the equator are found in this month,

indicating an almost complete absence of Rossby waves in the subtropical North Pacific.

The statistical characteristics of the relationship between the UTH extremes and Rossby wave activity were examined by computing the seasonal mean Rossby stationary wave number. I computed the mean of all months within the season for only those months within the season with tropical average UTH anomalies greater than 0.7% and UTH anomalies less than –0.7%. For the boreal spring season (Fig. 4), low values of the stationary Rossby wave number extend further west versus the mean along the equator when UTH anomalies exceed 0.7% (Fig. 4b) and are found further east when UTH anomalies are less than – 0.7% (Fig. 4c). The situation is similar for the boreal winter season. Thus, the westerly duct is larger when UTH anomalies exceed 0.7% and is much smaller when UTH anomalies are less than –0.7%.

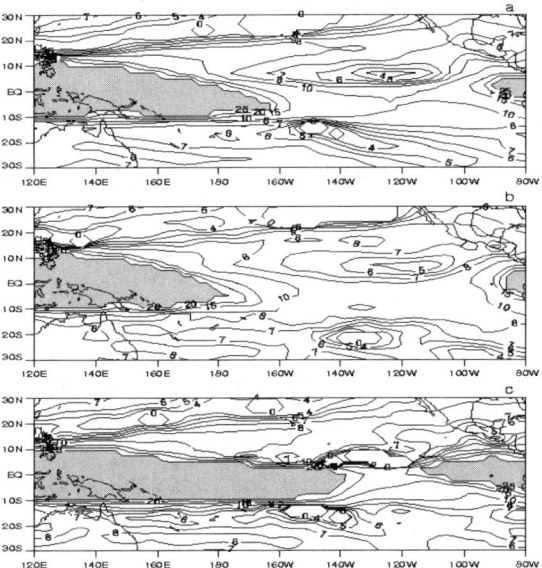

Fig. 4. Stationary Rossby wave number corresponding to normal UTH (a), high extremes of UTH (b), and low extremes of UTH (c).

4. EFFECTS ON THE TROPICAL WATER BUDGET

The budget equation for water vapor is

$$\frac{Dq}{Dt} = s(q) + D \qquad (4)$$

where s(q) is the rate of generation or destruction of water vapor and D is the molecular and eddy diffusion through the boundaries. Expanding (4) in sigma coordinates we get

$$\frac{Dq}{Dt} = \left(\frac{\partial q}{\partial t} + \vec{v} \cdot \nabla q\right)_\sigma + \dot{\sigma}\frac{\partial q}{\partial \sigma} \qquad (5)$$

where the total derivative of specific humidity has been decomposed into the local time rate of change in specific humidity, horizontal and vertical advection terms. Introducing mass continuity and writing in flux form, the water vapor balance equation can be written

$$\frac{Dq}{Dt} = \frac{\partial q}{\partial t} + \nabla \cdot (\vec{v}q) + \frac{\partial(\dot{\sigma}q)}{\partial \sigma} + q\left[\frac{\partial \ln p_s}{\partial t} + \vec{v}\cdot\nabla \ln p_s\right] \quad (6)$$

The local time tendency, horizontal and vertical flux components of (6) are analogous to the water vapor equation in pressure coordinates; however, an additional surface pressure tendency term and a horizontal advection of surface pressure term account for variations in surface pressure. These terms are negligible over the ocean.

Water vapor mass flux was computed for an arbitrary sized box defined by longitude coordinates λ_1 and λ_2, latitude coordinates N_1 and N_2, and vertical coordinates σ_1 and σ_2. The zonal water vapor mass flux through a meridional plane was defined

$$\int_{\phi_1}^{\phi_2} \frac{p_s(\lambda,\phi)}{g}\left(\int_{\sigma_1}^{\sigma_2}(uq)_\lambda d\sigma\right) r\, d\phi \quad (7)$$

where P_s is surface pressure, g is acceleration due to gravity, and r is the earth's radius. Likewise for the meridional water vapor flux through a zonal plane

$$\int_{\lambda_1}^{\lambda_2} \frac{P_s(\phi,\lambda)}{g}\left(\int_{\sigma_1}^{\sigma_2}(\overline{vq})_\phi d\sigma\right) r\cos\phi\, d\lambda \quad (8)$$

and the vertical flux through a surface defined on sigma level (sigma)

$$\int_{\phi_1}^{\phi_2}\left(\int_{\lambda_1}^{\lambda_2}\frac{\sigma p_s(\lambda,\phi)}{g}(\dot{\sigma}q)_\sigma d\lambda\right) r^2 \cos\phi\, d\phi \quad (9)$$

This study focuses on the horizontal region defined by 160E–150W, 15N–25N and the vertical region defined between the sigma levels of 0.5 and 0.2; a region with climatological high OLR indicating significant energy loss to space. It is generally accepted that the vertical flow in this region is dominated by subsidence, which will dry the upper troposphere and allow significant amount of energy to escape to space. To balance this drying process, water vapor needs to be transported either horizontally or vertically into this region. To investigate this balance, we to computed the monthly water vapor mass flux for each side of the box over a 17-yr period (a longer time series will be available soon) using the computed monthly mean total moisture flux data. Summation of the six terms results in a net divergence/convergence of water vapor in the region.

A detailed examination of the monthly moisture flux in this region indicates 39% of the months over the 17-yr period have a mean downward vertical motion field but a mean upward vertical moisture flux in the upper troposphere. Fig. 5 gives time series of the region-mean vertical velocity and the total, stationary and transient vertical moisture flux for the 0.5–0.2 sigma layer. Whereas the climatology of the vertical velocity is downward, the vertical moisture flux is upward with all the mean flux contribution coming from the transient vertical flux. Therefore, transient disturbances of less than 30 days contribute almost entirely to the climatological mean upward water vapor flux found in this subtropical region. In fact, 79% of the months with downward motion and upward moisture flux also have downward stationary moisture flux. Therefore, it is entirely the transient moisture flux which contributes to upward moisture flux in these cases. This transient upward flux is strong enough to contribute significantly to the total vertical moisture flux such that upward flux occurs. Further verification of this result was performed by computing each term of water vapor balance equation at each grid point in the box. Integration of the water vapor tendency, horizontal and vertical divergence terms achieved the same result as the flux calculation.

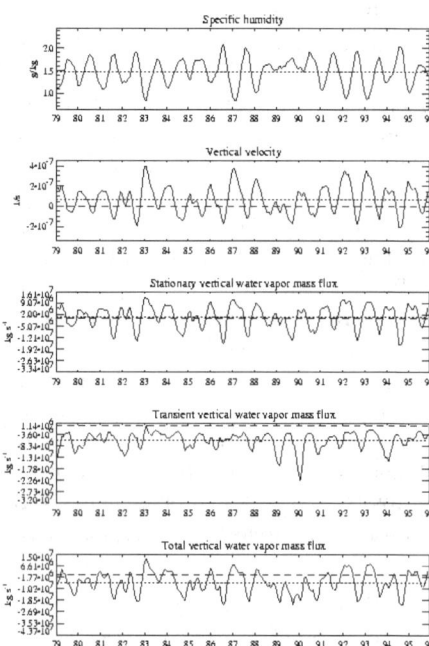

Fig. 5. Components of the vertical water vapor mass flux for the subtropical North Pacific.

5. CONCLUSIONS

Atmospheric dynamical states, resulting from the interaction of the tropics with the mid-latitudes during boreal winter and spring, are responsible for most of the observed extremes in the tropical UTH time series. These states are referred to as the westerly duct in the eastern Pacific Ocean. The two extreme states of the westerly duct, and their influence on UTH, are illustrated schematically in Fig. 6.

During extremes of high UTH (Fig. 6a), strong westerlies flowing out from deep convection in the western equatorial Pacific create a long fetch of westerlies over the equatorial eastern Pacific. This opens the westerly duct and allows Rossby waves (strong eddy activity) to propagate into the subtropics and re-hydrate the subtropical upper troposphere. Conversely, when deep convection extends into the central and eastern Pacific (Fig. 6b), westerly winds over the tropical eastern Pacific are weak and Rossby waves are blocked from propagating into the subtropics (no eddies).

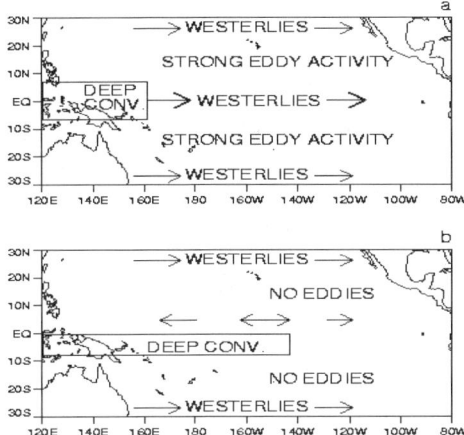

Fig. 6. Schematic of tropical basic states for extremes of high UTH (a) and low UTH (b).

Hypotheses for mechanisms that control the water and energy budget of the tropics often refer only to the effects of the Hadley and Walker cells in causing the observed variability. This is equivalent to ignoring the effects of transient eddies. Analysis of the water and energy budgets, however, demonstrates that transient eddies contribute significantly to the water and energy budgets and, thus, can not be ignored. Instead, the contribution of transient eddies must be included, but these contributions may be understood within the lower frequency regimes of seasonal to interannual variability.

6. REFERENCES

Bates, J. J., D. L. Jackson, F.-M. Breon, and Z. D. Bergen, Variability of upper tropospheric humidity 1979-1998, *Journal of Geophysical Research*, 106 (D23), 23,271-23,282, 2001.

Bates, J. J., X. Wu, and D. L. Jackson, Interannual variability of upper-tropospheric water vapor band brightness temperature, *Journal of Climate*, 9 (2), 427-438, 1996.

Chen, J., B. E. Carlson, and A. D. Del Genio, Evidence fo strengthening of the Tropical general circulation in the 1990s. *Science*, 295, 838-841, 2002.

Christy, J. R., R. W. Spencer, and E. S. Lobl, Analysis of the merging procedure for the MSU daily temperature time series, *Journal of Climate*, 11, 2016-2041, 1998.

Christy, J. R., R. W. Spencer, and R. T. McNider, Reducing Noise in the MSU Daily Lower-Tropospheric Global Temperature Dataset, *Journal of Climate.*, 8 (4), 888-902, 1995.

Iskenderian, H., A 10-year climatology of Northern Hemisphere tropical cloud plumes and their composite flow patterns, *Journal of Climate*, 8, 1630-1637, 1995.

Kiladis, G. N., Observations of Rossby Waves Linked to Convection over the Eastern Tropical Pacific, *Journal of the Atmospheric Sciences*, 55 (3), 321, 1998.

Kiladis, G. N., and K. M. Weickmann, Horizontal Structure and Seasonality of Large-Scale Circulations Associated with Submonthly Tropical Convection, *Monthly Weather Review*, 125 (9), 1997-2008, 1997.

Matthews, A. J., and G. N. Kiladis, Interactions between ENSO, Transient Circulation, and Tropical Convection over the Pacific, *Journal of Climate*, 12 (10), 3062-3074, 1999.

Reynolds, R. W. S., T. M., Improved sea surface temperature analysis using optimum interpolation, *Journal of Climate*, 7, 929-937, 1994.

Slingo, A., J. A. Pamment, and M. J. Webb, A 15-year simulation of the clear-sky greenhouse effect using the ECMWF reanalyses: fluxes and comparisons with ERBE, *Journal of Climate*, 11 (4), 690-706, 1998.

Webster, P. J., and J. R. Holton, Cross-equatorial response to middle-latitude forcing in a zonally varying basic state, *Journal of the Atmospheric Sciences*, 39 (4), 722-733, 1982.

Wielicki, B. A., B. R. Barkstrom, B. A. Baum, T. P. Charlock, R. N. Green, D. P. Kratz, R .B. Lee Iii, P. Minnis, G. L. Smith, T. Wong, D. F. Young, R. D. Cess, J. A. Coakley Jr., D. A. H. Crommelynck, L. Donner, R. Kandel, M. D. King, A. J. Miller, V. Ramanathan, D. A. Randall, L. L. Stowe, and R. M. Welch, Clouds and the Earth's Radiant Energy System (CERES): Algorithm Overview, *IEEE Transactions on Geoscience and Remote Sensing*, 36 (4), 1127-1136, 1998.

Wielicki, B. A., and co-authors, Evidence for large decadal variability in the Tropical mean radiative energy budget, *Science*, 295, 841-844.

REGIONAL CLIMATE RESPONSE INDUCED BY AEROSOL RADIATIVE FORCING OVER EURASIA DURING BOREAL SPRING

Maeng-Ki Kim and Woo-Seop Lee
Dept. of Atmospheric Science, Kongju National University, Gongju, 314-701, Korea (mkkim@kongju.ac.kr)

K. M. Lau
Laboratory for Atmospheres, NASA Goddard Space Flight Center, Greenbelt, Maryland 21046, USA

Kyu-Myong Kim
Science System Applications, Inc, Lanham, Maryland, USA

Y. C. Sud
Climate and Radiation Branch, Laboratory for Atmospheres, NASA GSFC, Greenbelt, Maryland, USA

Greg K. Walker
SAIC/General Sciences Operation, Beltsville, Maryland, USA

Mian Chin
Atmospheric Chemistry and Dynamics Branch, Laboratory for Atmospheres, NASA GSFC, Greenbelt, Maryland, USA

ABSTRACT

Global and regional climate impacts of present day aerosol-loading are investigated using NASA finite volume General Circulation Model (fvGCM) simulations spanning across Spring season. For aerosol loading, three-dimensional distribution of five species of tropospheric aerosols viz, sulfate, black carbon, organic carbon, soil dust, and sea salt obtained from the Goddard Ozone Chemistry Aerosol Radiation and Transport model (GOCART) simulations were prescribed to the fvGCM. This study reveals that atmospheric heating anomalies caused by the absorbing aerosols reorganize the stationary part of sea-level-pressure, vertical temperature profiles, and associated geopotential heights over broad areas adjacent to, and remote from regions of heavy aerosol loading. Thus aerosols impact not only local climates, but also continental and global climate through atmospheric transport and teleconnection patterns.

1. INTRODUCTION

Recent studies have shown that aerosols play an important role in determining the magnitude of global warming, because of their ability to alter the energy balance of the earth-atmosphere system through the interaction with solar and terrestrial radiation (Jacobson, 2002; Menon et al., 2002; Takemura et al., 2002). The effect of aerosols is not only limited to cooling by scattering, but also heating by absorption of solar radiation. The sign of the temperature change induced by aerosol forcing can vary depending on the aerosols' radiative properties and their distribution over the dark ocean and reflective land (Kaufman et al., 2002). Moreover, regional climate can be easily influenced by circulation change that may be induced by aerosol radiative forcing. Thus, the effects of aerosols on regional climate change need to be consistently integrated into local heating and cooling, as well as the effects of dynamic circulation induced by three-dimensional aerosol radiative forcings.

2. MODEL EXPERIMENTS

In this study, climate impacts of aerosol radiative forcing are investigated using the NASA finite volume General Circulation Model (fvGCM) with Relaxed Arakawa Schubert Scheme (McRAS). The model has a horizontal resolution of 2 by 2.5 degrees and 55 vertical levels. The McRAS prognostic cloud scheme has been developed with the aim of improving moist processes, microphysics of clouds, and cloud-radiation interactions in GCMs (Sud and Walker, 1999). Radiative transfer model is described Chou et al. (1999). Aerosol optical thicknesses of five aerosols, i.e., black carbon, organic carbon, dust, sulfate, and sea salt, are prescribed as three-dimensional monthly mean values for two years (2000–2001) from the GOCART model (Chin et al., 2002). Anthropogenic sources of sulfate and carbonaceous aerosols are also considered (Chin et al., 2002). The extinction coefficient, single scatter albedo, and asymmetric factor of each aerosol type are obtained as function of relative humidity and wave length of 11 broaden bands, based on Mie theory in radiation package.

In the control experiment, no aerosol forcing is included (NA), sea surface temperature (SST) is prescribed from observed weekly mean data during the simulation period (10 years) from September 1987 to December 1996. In the anomaly experiment, all the prescribed boundary conditions are the same as in control run except for including direct effect of aerosol loading on radiative forcings of the Earth-atmosphere system using all aerosols types (AA) described above. To understand the role of absorbing aerosols, we carry out three additional model simulations, which are otherwise identical to NA and AA; one without black carbon (experiment NB) and one without dust (ND).

3. GLOBAL SCALE RESPONSE

Fig. 1 shows the aerosol radiative forcing induced by all aerosols in AA, at TOA, within the atmosphere, and at the surface. The pattern of aerosol radiative forcing in the atmosphere is similar to that at the surface with the opposite sign, implying that if there is no aerosol in the atmosphere, the surface would have absorbed the corresponding amount of atmospheric absorption. Large positive SLP (sea level pressure) anomaly found over East Asian monsoon region with the center around Korean peninsula while negative anomaly is seen over Tibetan Plateau, which is mainly due to surface warming induced by high surface albedo-aerosol interaction (Fig. 2). The aforementioned anomalies form an east-west dipole pattern, yielding geostrophic southwesterlies, which leads to warm advection toward the central and northern part of Asian continent. This pattern also implies a strengthening of the western Pacific subtropical high, which governs the onset and evolution of the East Asian summer monsoon. Large SLP change is also seen in the Africa-European region with the positive anomaly in the northern part and negative anomaly in the southern part of Mediterranean.

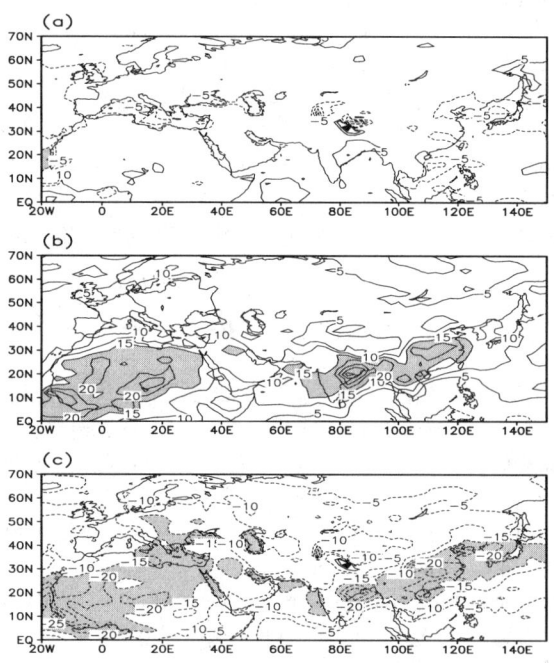

Fig. 1. Aerosol radiative forcing (a) at the top of the atmosphere, (b) in the atmosphere, and (c) at the surface in Wm^{-2}. Regions larger than 10 or smaller than –10 Wm^{-2} are shaded.

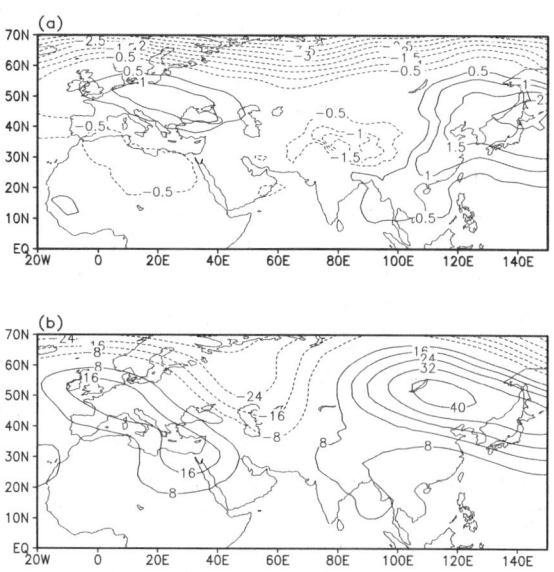

Fig. 2. (a) Sea level pressure and (b) geopotential height change at 500 hPa by all aerosols. Contour intervals in (a) and (b) are 0.5 hPa and 10 m, respectively.

4. REGIONAL IMPACTS OF BC AND DUSTS

Fig. 3 shows the surface air-temperature anomalies for the three cases. With all-aerosol forcing (Fig. 3a), the most pronounced and robust (exceeding 1% confidence level) surface cooling is found in a broad region over the Caspian Sea, which however, is not a region of high aerosol loading. Hence the strong cooling here is not due to direct radiative effect of aerosols, but rather a response in the large-scale circulation (Fig. 2a). In contrast, the negative anomalies found over the aerosol source region of the Bay of Bengal

(BC), East Asia (BC and sulfate), and western Africa (dust) are consistent with the direct radiative forcing by aerosols. The strongest surface warming is found over an extensive region northeastern Asia, again in an area with low aerosol loading, and in any case, not consistent with the surface cooling expected from aerosol forcing. From Fig. 3b, it is obvious that the pronounced cooling over eastern Europe and the Caspian Sea region is associated with an atmospheric teleconnection forced by dust aerosol heating over northern and western Africa as shown in Fig. 2. This wavetrain produces warming over southern Europe, and northern Siberia. As shown in Fig. 3c, BC produces surface cooling over eastern India and East Asia, and in doing so induces a meridional circulation over the East Asian sector with strong subsidence in northern latitudes (40–60° N). The most pronounced effect is to enhance the surface warming over central and eastern Siberia.

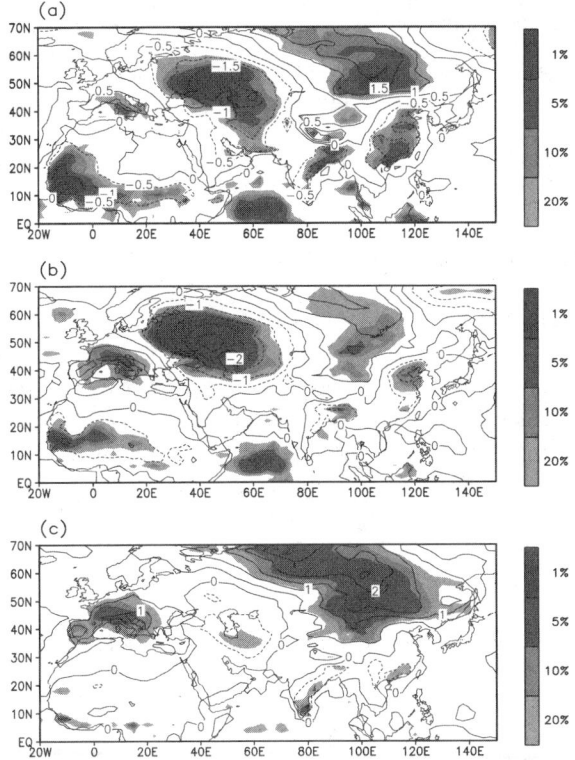

Fig. 3. Surface air temperature anomaly induced by (a) the effects of all aerosols (AA–NA), (b) dust aerosol with the background of four aerosols (AA–ND), and (c) BC aerosol with the background of four aerosols (AA–NB). Contour intervals are 0.5 °C. Significance levels are shown as half-tone scale on the right.

5. CONCLUSIONS

Temperature anomalies induced by aerosol radiative forcing show wave train patterns spanning from the northern Africa to the north Eastern Pacific via the central Asia continent. Even though the surface cooling is not linearly proportion to the atmospheric heating by absorbing aerosols, surface cooling appears in most of the regions with large aerosol loading, i.e., China, India, and Africa. But, much strong surface temperature change appear in two broaden regions with cooling around Caspian Sea and warming in Baikal Lake of Eurasian continent where does not have large aerosol loading, as shown in both observation (Fig. 4) and model (Fig. 3a), indicating that the regional climate can be influenced by not only radiative forcing itself but also circulation change. The additional experiments confirm the cause of significant circulation change to be the two absorbing aerosols, dust and BC. It implies that strong atmospheric heating by absorbing aerosols can redistribute mass over the broaden region around aerosol loading layer. As a result, wave train structure in temperature is a product of the harmony of the contribution induced by spatial distributions and optical properties of each aerosol, especially in dust and BC. Therefore, in the regional and even in broaden region, temperature change can cause much larger cooling than directly induced amount by radiative forcing. On the contrary large surface warming can be induced inconsistent with the role of aerosol radiative forcing.

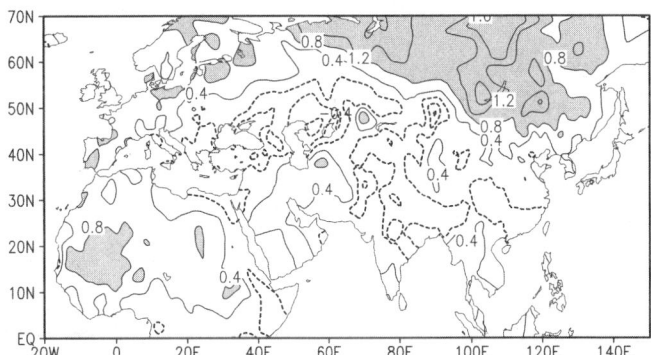

Fig. 4. Change in observed land surface temperature for 40 years. The change is calculated by the difference between 1980–1998 and 1961–1979. Positive values over 0.8 °C per 20 years are shaded.

6. ACKNOWLEDGEMENTS

This work is supported by the NASA Modeling and Analysis Program. Most of the work was carried out while the first author is visiting the Laboratory for Atmospheres, on a joint GSFC and Goddard Earth Sciences and Technology Center (GEST) fellowship. M. K. Kim was partially supported by the SRC program of Korea Science and Engineering Foundation.

7. REFERENCES

Chin, M., P. Ginoux, S. Kinne, O. Torres, B. N. Holben, B. N. Duncan, R. V. Martin, J. A. Logan, A. Higurashl, and T. Nakagima, 2002; Tropospheric aerosol optical thickness from the GOCART model and comparisons with satellite and sun photometer measurements. *J. Atmos. Sci.,* 59, 461-483.

Chou, M.-D., and M. Suarez, 1999: A solar radiation parameterization for atmospheric studies. NASA/TM-1999-104606, Vol. 15, 40p

Jacobson, M. Z., 2002: Control of fossil-fuel particulate black carbon and organic matter, possibly the most effective method of showing global warming. *J. Geophys. Res.,* 2002, 107, D19, doi:10.1029/2001JD001376.

Kaufman, Y. J., D. Tanre, and O. Boucher, 2002: A satellite view of aerosols in the climate system. *Nature,* 419, 215-223.

Menon, S., J. Hansen, L. Nazarenko, and Y. Luo, 2002: Climate effects of black carbon aerosols in China and India. *Science,* 297, 2250-2253.

Sud, Y. C., and G. K. Walker, 1999: Mcrophysics of clouds with the relaxed Arakawa-Schubert scheme (McRAS). Part I: Design and evaluation with GATE phase III data. *J. Atmos. Sci.,* 56, 3196-3220.

Takemura, T., T. Nakajima, O. Bubovik, B. Holben, and S. Kinne, 2002: Single scattering albedo and radiative forcing of various aerosol species with a global three dimensional model. *J. of Climate,* 15, 333-352.

EFFECTIVE RADIUS OF CLOUD DROPLETS BY GROUND-BASED REMOTE SENSING: IMPLICATION TO AEROSOL INDIRECT EFFECT

Byung-Gon Kim

Department of Atmospheric and Environmental Sciences, Kangnung National University
Gangnung, 210-702, Korea

ABSTRACT

Aerosol indirect radiative forcing of climate change is considered the most uncertain forcing of climate change over the industrial period, despite numerous studies demonstrating such modification of cloud properties and several studies quantifying resulting changes in shortwave radiative fluxes. We have previously used ground-based remote sensing of cloud optical depth (τ_c) by narrowband radiometry and liquid water path (LWP) by microwave radiometry to demonstrate substantial (factor of 2) day-to-day variation in cloud drop effective radius (r_e) at the ARM Southern Great Plains site (Kim et al., 2003). Here we extend the previous one-year study to 3-years (1999–2001) with special emphases on the relationship of cloud drop effective radius with light-scattering coefficient (σ_{sp}), and the corresponding association of radiative forcing. The results showed the significant and more improved correlation of r_e with σ_{sp}, and the resultant enhancement of calculated net shortwave irradiance at the top of atmosphere in conjunction with a decrease in cloud drop radius.

1. INTRODUCTION

Aerosols affect global climate indirectly by altering cloud microphysical and radiative properties, so-called aerosol indirect radiative forcing, which is considered the most uncertain forcing of climate change over the industrial period (IPCC, 2001), despite numerous studies demonstrating such modification of cloud properties and several studies quantifying resulting changes in shortwave radiative fluxes.

Studies relating to the enhancement of cloud droplet concentrations to the increase of cloud albedo have typically limited to in-situ and remotely sensed characterization of cloud microphysics during intensive field campaigns (Radke et al., 1989; Albrecht et al., 1995). It is desirable to use ground-based remote sensing to understand aerosol indirect effect, because field campaigns are limited in spatial and temporal dimensions. Therefore, ground-based remote sensors have recently been used to examine the relation of aerosols with cloud microphysics at a few selected sites Feingold et al., 2003; Kim et al., 2003; Garrett et al., 2004) while satellite remote sensors have been employed to provide systematic aerosol effects on clouds in a regional and global view (Nakajima et al., 2001; Breon et al., 2002). As a result, cloud drop effective radius (r_e) was weakly associated with variation in aerosol loading as characterized by light-scattering coefficient at SGP (Kim et al., 2003; hereafter KSMM) and North Slope Alaska (Garrett et al., 2004).

Here we extend KSMM by investigating the aerosol effect on r_e in a relatively homogeneous stratus cloud covering the surface measurement site at SGP for 1999 ~ 2001. Accordingly we estimate the corresponding radiative forcing which is associated with the variations in r_e, mostly attributable to aerosol loading, using a radiative transfer model.

2. METHODOLOGY

Basically, most of methods and data are similar to KSMM except for three-year analysis from 1999 to 2001. The primary instruments used in this study at the SGP site (97.48°W, 36.61°N) are summarized in Table 1.

Cloud boundaries are retrieved every 10 seconds from the active remotely-sensed cloud locations value-added product (Clothiaux et al., 2000). Liquid water path (LWP) is determined by microwave radiometer (MWR), which measures time series of column-integrated liquid water based on the microwave emissions of liquid water molecules at mainly 31.4 GHz.

Table 1. Summary of primary instrumentation and value-added products[a]

Instrument	Measured Quantities	Comments	Temporal resolution
MFRSR (Multi-Filter Rotating Shadowband Radiometer)	Cloud optical depth (τ_c)	Measures direct and total-horizontal irradiances at 415 nm.	20 s
MWR (Microwave Radiometer)	Liquid water path (LWP)	Uses microwave brightness temperature, Accuracy 30 g m^{-2}	20 s
Nephelometer	Scattering coefficient (σ_{sp})	At 450, 550, 700 nm for the size of aerodynamic diameter less than 1μm	1 min
Active Remotely-Sensed Cloud Locations	Cloud boundaries	Best estimates from MMCR, Ceilometer and Lidar	10 s

[a] Value-added products refer to data sets resulting from assimilation and analysis of data from multiple instruments.

Cloud optical depth has been determined by a Multi-Filter Rotating Shadowband Radiometer (MFRSR) at 415 nm, which is limited to complete overcast conditions. Min and Harrison (1996a) and Min et al. (2001) have developed a family of inversion methods to infer cloud optical properties from MFRSR and MWR. As is standard, the cloud radiative properties are parameterized in terms of a cloud averaged drop effective radius r_e, and total liquid water path LWP, based on Mie theory. Using total horizontal transmittance at 415 nm, together with LWP, one can simultaneously retrieve cloud optical depth and r_e through the use of a nonlinear least squares minimization in conjunction with an adjoint method of radiative transfer (Min and Harrison, 1996b).

The measurement uncertainty of r_e was discussed in KSMM. All τ_c, LWP, and r_e values are reproduced as 5-minute averages, since the 5-minute averaging period permits the narrow field of view (4.5°) measurement of LWP to better correspond to the wide field of view (120°) measurement of τ_c by the MFRSR (Min et al., 2001) and reduce the bias induced by the different measurement volume. The most favorable cloud type for testing relations between radiation and microphysical properties is a widespread low-level non-precipitating, liquid water cloud layer without interference from higher-level ice clouds. The additional criterion to KSMM is that in order to consider more reliable relationship of surface aerosol measurement to cloud drop size at the cloud level, the cloud top height should be lower than 2 km above the ground. Since surface aerosol measurement are likely to be most appropriate for particularly a measure of accumulation aerosol concentration under well-mixed conditions (Feingold et al., 2005), the days are additionally excluded in the present analysis through the examination of the vertical soundings of meteorology when a mixed layer did not develop even during the daytime, possibly resulting in vertical decoupling of the cloud layer and the surface.

Instead of CCN, accumulation mode aerosols with their size of 0.1 to 1.0 μm diameter are measured by integrating nephelometers at SGP (Sheridan et al., 2001), which measure σ_{sp} as a function of relative humidity (RH) at three visible wavelengths (nominally 450, 550, and 700 nm). We use measurements at the 550 nm wavelength and low RH ≤ 40%, representative of the light scattering coefficient of the dry aerosol. Measurements are available as 1-minute averages for five 6-minute intervals per hour, which is finally interpolated to measurement time (5-minute average) of cloud drop effective radius.

3. RESULTS

There are 16 analysis days, which has suitable overcast episodes meeting the established criteria lasting at least 3 hours or more. To relate the cloud drop effective radius to aerosol loading, aerosol light scattering coefficient was used as a surrogate, which can be made at the surface simultaneously with the remote sensing of the cloud properties. The relationships between r_e and σ_{sp} are shown in Fig. 1, which indicates general decrease in r_e with the increasing σ_{sp}, as expected for the Twomey mechanism and quite consistent with KSMM. Note that different symbols indicate the 16 selected days. It is certain that a relation between r_e and σ_{sp} becomes clearly discernible only when examining the entire data set. The sensitivity of r_e to aerosol can be described by the indirect effect parameter IE.

$$IE = -d\log r_e / d\log \sigma_{sp}$$

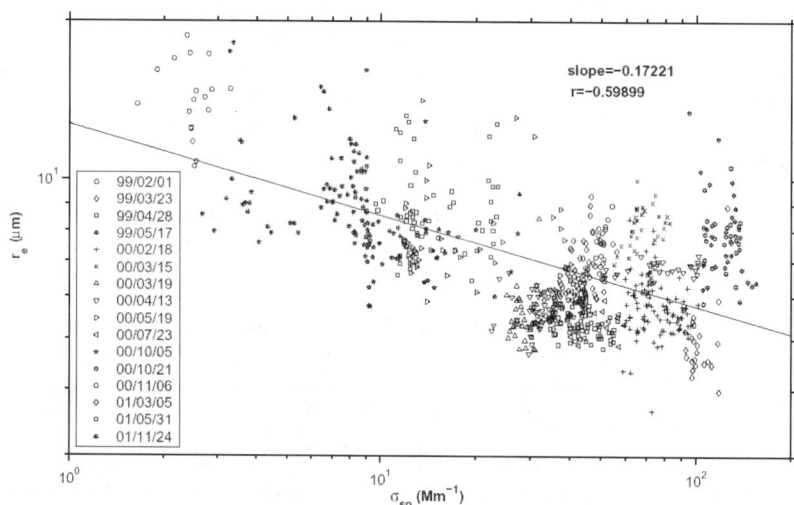

Fig. 1. Scatterplot (decimal logarithmic axes) of 5-minute average cloud drop effective radius (r_e) versus light scattering coefficient (σ_{sp}) for sub-micrometer aerosol at 550 nm. Data for individual days are distinguished by symbol. Data for σ_{sp} are interpolated to measurement time of r_e and gaps in σ_{sp} are filled by interpolation.

The *IE* value emphasizes relative rather than absolute sensitivities, which is useful in remote sensing applications where retrieved quantities may exhibit bias but accurately reproduce trends (Feingold et al., 2003; Garrett et al., 2004). For the data set as a whole we obtain *IE*=0.17 ($r^2 = 0.36$), slightly greater and better correlated than 0.13 ($r^2 = 0.24$) in KSMM, which could suggest data sets pertaining to much longer sampling periods improve the statistics. This is also supportive for the previous works by ground-based approaches; 0.02 ~ 0.16 between r_e and aerosol light extinction coefficient in Feingold et al. (2003), 0.13 ~ 0.19 between r_e and σ_{sp} in Garrett et al. (2004). These values obtained from ground-based remote sensing are somewhat greater than those (0.04 ~ 0.08) from global satellite data (Breon et al., 2002).

The change in broadband shortwave radiation budget (0.25 ~ 4 μm) associated with changes in r_e was evaluated using the SBDART (Santa Barbara DISORT Atmospheric Radiative Transfer; Ricchiazzi, et al., 1998) that is based on the algorithm for discrete-ordinate-method radiative transfer in multiple scattering and emitting layered media (Stamnes et al., 1988). The three days (2/18, 10/21, and 10/26 of 2000) were selected according to the distinct segregation of r_e.

The increase in spherical albedo was closely associated with decreasing drop radius for a given LWP (e.g. see Fig. 11 in KSMM). The average value (10.2 μm) of r_e on 10/26 is assumed as a reference value for the estimation of radiative forcing. The calculated total radiant flux at the surface closely agreed within 5% with observations obtained using the broadband radiometer.

The difference in upward flux (Fig. 2) means the difference between upward shortwave fluxes for the given days (2/18 and 10/21) of the observed LWP and r_e and those for the condition of the same LWP and the reference value of r_e, which represented the remarkable diurnal trend with the highest at noon. The net upwelling solar irradiance at the top of the atmosphere corresponding to a decrease in r_e from the reference value to the observed r_e, increased up to 70~80 Wm^{-2} and 30~40 Wm^{-2}, respectively with somewhat scattered distribution. In spite of the subjunctive calculation, the magnitude of this effect, which is attributable to changes in r_e for realistic values of cloud LWP and other parameters, must be regarded as substantial in any quantitative consideration of the local shortwave radiation budget.

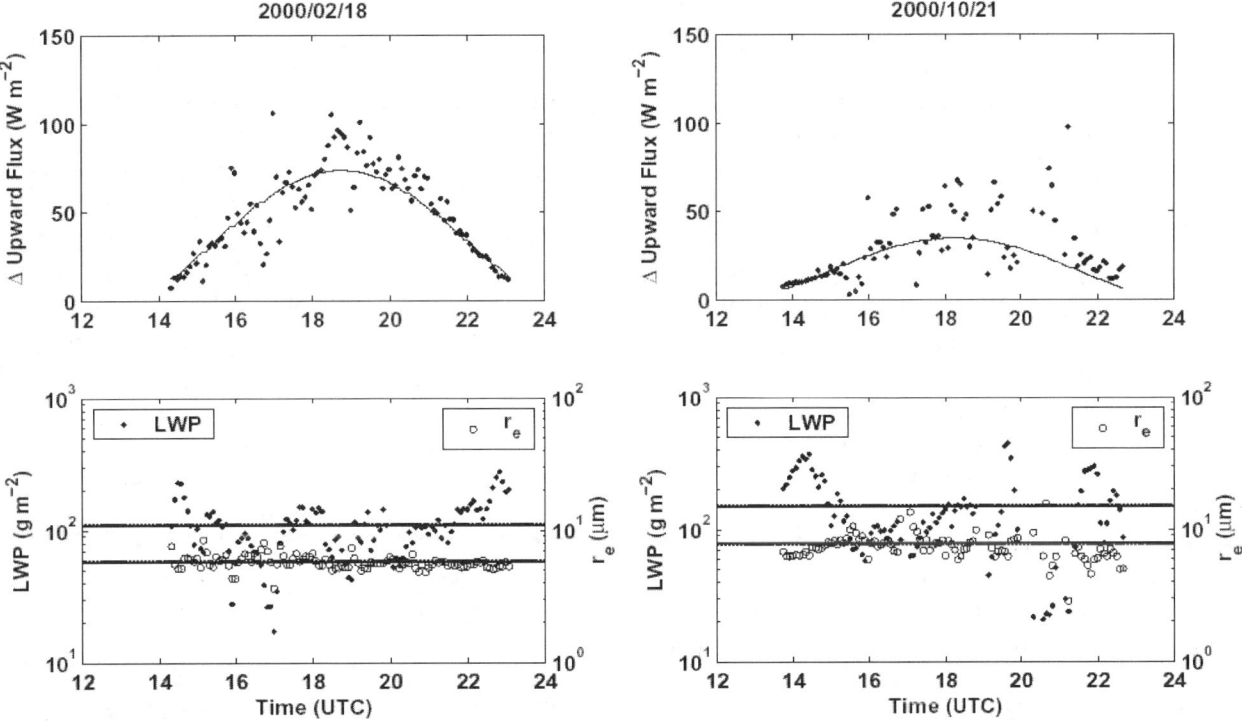

Fig. 2. The difference between resultant upward solar irradiance (top panel) for the observed LWP and r_e and for the same LWP and the reference value of r_e on 2/18 and 10/21, 2000, together with the solid line representing the difference of mean values. Temporal variations of the observed LWP and r_e are displayed in the bottom panel with the solid lines indicating mean values of LWP and r_e.

4. CONCLUSION

Ground-based remote sensing of cloud optical depth and LWP has been used to examine the relation of cloud droplet effective radius (r_e) to aerosol and the associated Twomey forcing, using SGP ARM data for 1999–2001 and a radiative transfer model (SBDART).

Cloud droplet effective radius was found to exhibit a negative correlation with σ_{sp} at the surface with much better association than that of KSMM, although r_e still appeared to be influenced by various kinds of factors such as meteorological influence which should be more examined in the near future. Notably, a decrease in r_e results in the increase in calculated net upward irradiance at the top of the atmosphere in the use of the radiative transfer model. This study demonstrates the feasibility to evaluate the aerosol indirect effect using ground-based remote sensing and still limitations to improve.

5. ACKNOWLEDGEMENTS

Byung-Gon Kim acknowledges partial support under the Atmospheric and Oceanic Sciences Program of Princeton University. This work was also supported in part by US DOE's ARM program. Byung-Gon Kim would like to thank Stephen Schwartz, Mark Miller and Qilong Min for their fruitful comments and suggestions.

6. REFERENCES

Albrecht, B. A., C. S. Bretherton, D. Johnson, W. H. Schubert, and A. S. Frisch, 1995: The Atlantic Stratocumulus Transition Experiment – ASTEX. *Bull. Amer. Meteor. Soc.,* 76, 889-904.

Breon, F.-M., D. Tanre, and S. Generoso, 2002: Aerosol effect on cloud droplet size monitored from satellite. *Science,* 295, 834-838.

Clothiaux, E. E., T. P. Ackerman, G. G. Mace, K. P. Moran, R. T. Marchand, M. A. Miller, and B. E. Martner, 2000: Objective determination of cloud heights and radar reflectivities using a combination of active remote sensors at the ARM CART sites. *J. Appl. Met.,* 39, 645-665.

Feingold, G., W. Eberhard, D. E. Lane, and M. Previdi, 2003: First measurements of the Twomey effect using ground-based remote sensors. *Geophys. Res. Lett.,* 1287, doi:10.1029/2002GL016633.

Feingold, G., R. Furrer, P. Pilewskie, L. A. Remer, Q. Min, and H. Jonsson, 2005: Aerosol indirect effect studies at Southern Great Plains during the May 2003 intensive operation period, submitted to *J. Geophys. Res.*

Garrett T. J., C. Zhao, X. Dong, G. G. Mace, and P. V. Hobbs, 2004: Effects of varying aerosol regimes on low-level Arctic stratus. *Geophys. Res. Lett.,* 31, doi:10.1029/2004GL019928.

International Panel on Climate Change, 2001: *Climate change 2001: the scientific basis.* Cambridge University Press.

Kim, B.-G., S. E. Schwartz, M. A. Miller, and Q. Min, 2003: Effective radius of cloud droplets by ground-based remote sensing: Relationship to aerosol. *J. Geophys. Res.,* 108, doi:10.1029/2003JD003721.

Min, Q., and L. C. Harrison, 1996a: Cloud properties derived from surface MFRSR measurements and comparison with GOES results at the ARM SGP site, *Geophys. Res. Lett.,* 23, 1641-1644..

Min, Q., and L. C. Harrison, 1996b: An adjoint formulation of the radiative transfer method. *J. Geophys. Res.,* 101, 1635-1640.

Min, Q., L. C., Harison, and E. Clothiaux, 2001: Joint statistics of photon path length and cloud optical depth: case studies. *J. Geophys. Res.,* 106, 7375-7386.

Nakajima, T., A. Higurashi, K. Kawamoto, and J. E. Penner, 2001: A possible correlation between satellite-derived cloud and aerosol microphysical parameters. *Geophys. Res. Lett.,* 28, 1171-1174.

Radke, L. F., J. A. Coakley, Jr., and M. D. King, 1989: Direct and remote sensing observations of the effects of ships on clouds. *Science,* 246, 1146-1149.

Ricchiazzi, P., S. Yang, C. Gautier, and D. Sowie, 1998: SBDART: A research and teaching software tool for plane-parallel radiative transfer in the Earth's atmosphere. *Bull. Amer. Meteorol. Soc.,* 79, 2101-2114.

Sheridan, P. J., D. J. Delene, and J. A. Ogren, 2001: Four year of continuous surface aerosol measurements from the Department of Energy's Atmospheric Radiation Measurement program Southern Great Plains cloud and Radiation testbed site, *J. Geophys. Res.,* 106, 20735-20747.

Stamnes, K., S. Tsay, W. Wiscombe, and K. Jayaweera, 1988: Numerically stable algorithm for discrete-ordinate-method radiative transfer in multiple scattering and emitting layered media. *Appl. Opt.,* 27, 2502–2509.

SIMULATION OF CLIMATE CHANGE BY AEROSOL DIRECT AND INDIRECT EFFECTS WITH AEROSOL TRANSPORT-RADIATION MODEL

Toshihiko Takemura
Research Institute for Applied Mechanics, Kyushu University
Kasuga, Fukuoka 816-8580, Japan

Teruyuki Nakajima
Center for Climate System Research, University of Tokyo
Meguro-ku, Tokyo 153-8904, Japan

Toru Nozawa
National Institute for Environmental Studies
Tsukuba, Ibaraki 305-8506, Japan

ABSTRACT

The radiative forcing and changes in cloud and precipitation by the aerosol direct and indirect effects are simulated with a global aerosol transport-radiation model, SPRINTARS. A microphysical parameterization on the cloud-aerosol interaction is adopted in the model, and then the cloud droplet effective radius and precipitation rate are diagnosed, which relate to the aerosol indirect effect. The global mean radiative forcing of the direct and indirect effects at the tropopause by anthropogenic aerosols are calculated to be -0.1 Wm^{-2} and -0.9 Wm^{-2}, respectively. The simulated results indicate that a decrease in the cloud droplet effective radius by anthropogenic aerosols occurs globally, while changes in the cloud water and precipitation are strongly affected by a variation of the dynamical hydrological cycle induced by the radiation budget change due to the aerosol direct and first indirect effects rather than the second indirect effect. However, the cloud water can increase and the precipitation can simultaneously decrease in regions where a large amount of anthropogenic aerosols and cloud water exist.

1. INTRODUCTION

Intergovernmental Panel on Climate Change (IPCC) (2001) estimated that the radiative forcing due to anthropogenic aerosols is -0.5 Wm^{-2} with an uncertainty factor of 2 for the direct effect and from 0 to -2.0 Wm^{-2} without a plausible value for the first indirect effect. The second indirect effect could not be estimated because of a much lower confidence. These results indicate that there is a larger uncertainty in the evaluation of the aerosol radiative forcing, especially of the indirect effect, than of the greenhouse gases which exist more homogeneously than aerosol particles both spatially and temporally. The IPCC [2001] mainly compiled past modeling studies on estimating the aerosol radiative forcing, while recently the indirect radiative forcing was estimated from the assimilation between satellite retrievals and modeling (e.g., Nakajima et al., 2001; Lohmann and Lesins, 2002). The information of cloud properties from satellites is useful for reducing the uncertainty of the aerosol indirect radiative forcing.

In this study, the direct and indirect radiative forcing by anthropogenic aerosols are calculated by the latest version of a global aerosol transport-radiation model, Spectral Radiation-Transport model for Aerosol Species (SPRINTARS), which is fully validated with a lot of measured data on aerosol parameters (Takemura et al., 2000, 2002, 2005). Changes in the meteorological field due to the aerosol effects are also simulated by SPRINTARS completely coupled with an atmospheric general circulation model (AGCM) including the mixed layer ocean.

2. MODEL DESCRIPTION

SPRINTARS is driven by the AGCM jointly owned by the Center for Climate System Research (CCSR)/University of Tokyo, National Institute for Environmental Studies (NIES), and Frontier Research Center for Global Change (FRCGC). The model predicts mass mixing ratios of the main tropospheric aerosols, that is, carbonaceous (black carbon (BC) and organic carbon (OC)), sulfate, soil dust, and sea salt, and the precursor gases of sulfate, that is, sulfur dioxide (SO_2) and dimethylsulfide (DMS). The aerosol transport processes include emission, advection, diffusion, sulfur chemistry, wet deposition, and gravitational settling. The radiation scheme in CCSR/NIES/FRCGC AGCM (Nakajima et al., 2000) is extended for the aerosol direct effect related to scattering and absorption by aerosol particles and for the indirect effect related to a change in cloud droplet size. The detailed description of SPRINTARS is given by Takemura et al. (2000, 2002, 2005).

Time series of the emission fluxes for anthropogenic BC, OC, and SO_2 from 1850 to 2000 are provided by our research group (T. Nozawa et al., personal communication, 2004). The present study considers both the first and second indirect effects only for water stratus cloud. A parameterization based on the Köhler theory is introduced into SPRINTARS, which diagnoses the cloud droplet number concentration depending not only on the aerosol particle number concentration but also on the size distributions and chemical properties of each aerosol species, updraft velocity, and saturation condition of the water vapor (Takemura et al., 2005). The horizontal resolution of the triangular truncation is set at T42 (approximately 2.8° by 2.8° in longitude and latitude) and the vertical resolution at 20 layers. Two equilibrium experiments are carried out, present-day and preindustrial aerosol emissions, with the mixed layer ocean model and the fixed greenhouse gas concentrations. Each experiment is integrated for 50 years and analyzed for the last 30 years. The radiative forcing is, however, estimated with the prescribed tropospheric temperature, sea surface temperature, and sea ice experiments according to the general method of calculating the radiative forcing with AGCMs.

3. RESULTS

Fig. 1 shows the annual mean distributions of the aerosol optical thickness at 0.55 μm in the present-day simulation. Sulfate and carbonaceous aerosols concentrate

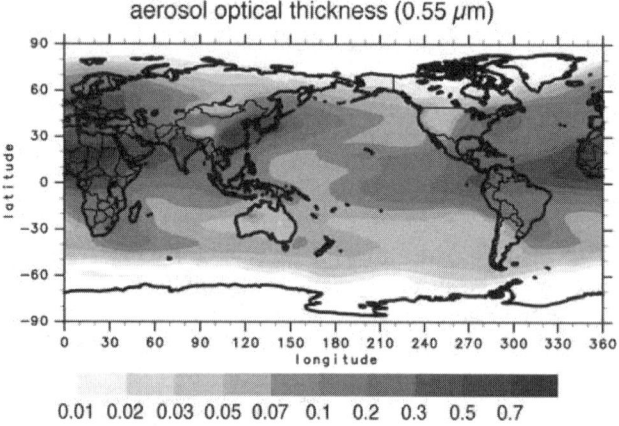

Fig. 1. Annual mean distributions of the simulated total aerosol optical thickness at 0.55 μm.

over populous regions in the midlatitudes of the Northern Hemisphere because of consumption of fossil and bio fuels and agricultural activities. Carbonaceous aerosols also exist over biomass burning regions in central and southern Africa and South America. A large amount of soil dust aerosols are emitted from Saharan, Arabian, and Asian regions. The simulated aerosol optical properties of not only optical thickness but also Ångström exponent that is an index of particle size and single scattering albedo are in reasonable agreement with ground-based, satellite, and aircraft measurements (Takemura et al., 2000, 2002, 2003).

Fig. 2 is an example of comparison with Aerosol Robotic Network (AERONET) at principal and characteristic sites for the aerosol optical thickness. They are in reasonable agreement, but the mean bias is approximately 30%. It may one of the reasons for uncertainty of the estimation of the radiative forcing.

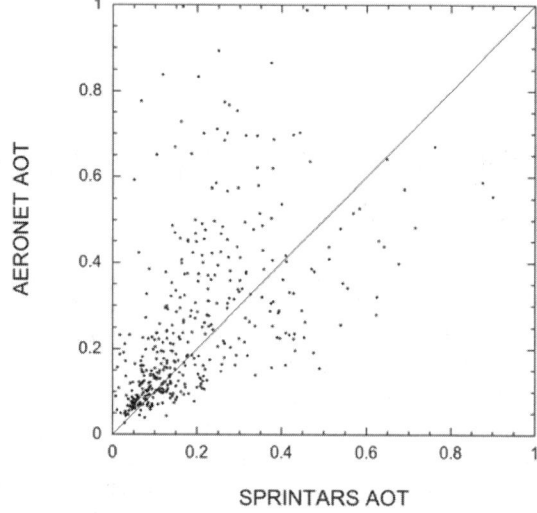

Fig. 2. Comparison of the monthly mean aerosol optical thickness at 0.55 μm between SPRINTARS and AERONET at 36 locations.

The direct and indirect radiative forcing at the tropopause by anthropogenic aerosols are shown in Fig. 3, and the global mean forcing of the direct effect by anthropogenic aerosols is given in Table 1. The direct forcing exceeds -1 Wm^{-2} in East and South Asia, Europe, and North America because of industrial and domestic OC and sulfate aerosols though the anthropogenic BC produces a positive forcing. The negative forcing is also large over most of southern Africa and South America mainly because of biomass burning aerosols which is a mixture of BC, OC, and sulfate. On the other hand, the outflow of carbonaceous aerosols from biomass burning and industrialized regions makes the direct radiative forcing positive over the Saharan and Arabian deserts with their high surface albedo. Biomass burning aerosols also cause the positive forcing off central and southern Africa and South America because the multiple scattering of the solar radiation by the lower cloud layer than the aerosol layer enhances its absorption by biomass burning aerosols. If it is under clear-sky conditions or the cloud layer is relatively higher than the aerosol layer,

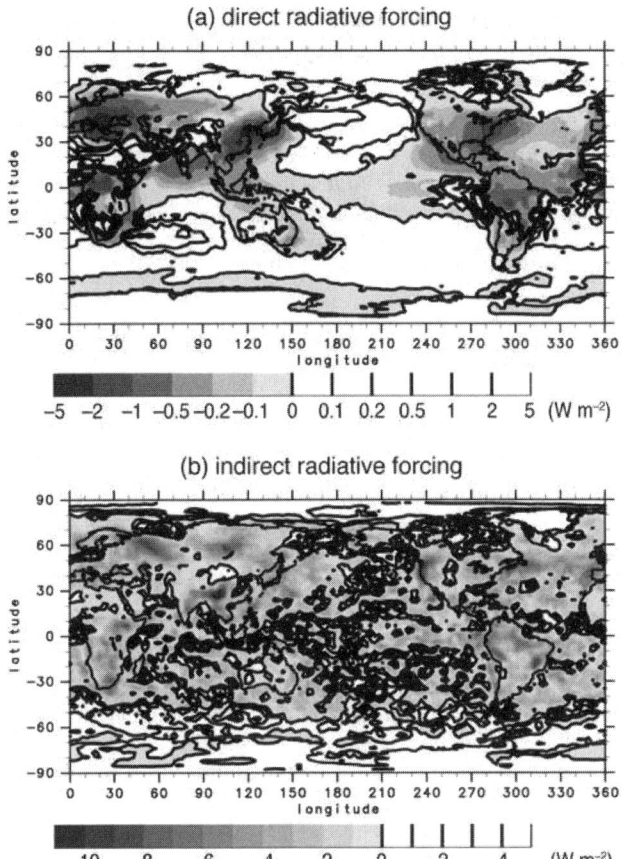

Fig. 3. Simulated aerosol (a) direct and (b) indirect radiative forcings at the tropopause from preindustrial to present day.

Table 1. Annual Global Mean Direct Radiative Forcing from Preindustrial to Present Day for All-sky and Clear-sky Conditions at the Tropopause and the Surface in Wm^{-2}

	Tropopause		Surface	
	All-sky	Clear-sky	All-sky	Clear-sky
BC	+0.42	+0.26	–0.76	–0.94
OC	–0.27	–0.51	–0.36	–0.57
Sulfate	–0.21	–0.52	–0.16	–0.41
Total	–0.06	–0.77	–1.28	–1.92

the direct radiative forcing by biomass burning aerosols is negative. The global mean anthropogenic forcing at the tropopause by the aerosol direct effect is calculated to be –0.06 Wm^{-2} and –0.77 Wm^{-2} under all-sky and clear-sky conditions, respectively. The indirect radiative forcing by anthropogenic aerosols in this study is calculated as a comparison of a difference in the cloud radiative forcing between two experiments, preindustrial and present runs. The indirect radiative forcing is generally negative except in the polar region and areas less affected by anthropogenic aerosols. The large negative forcing is seen over the northwestern Pacific, Southeast Asia, Eurasia, North America, South America, and southern Africa. The global mean anthropogenic forcing by the aerosol first plus second indirect effects is calculated to be –0.94 Wm^{-2}. If the other formula is used for calculating the precipitation rate, which doesn't depend on the aerosol number concentration, only the first indirect effect can be estimated to be –0.52 Wm^{-2}. Most past studies cited by the IPCC (2001) estimated the indirect radiative forcing over –1 Wm^{-2} for the global mean from the preindustrial to present day. If it is over –1 Wm^{-2}, however, an unrealistic temperature decrease by the aerosol effects may be simulated (Takemura et al., 2005).

Fig. 4 illustrates the changes in the cloud droplet effective radius at cloud top above 273K, liquid water path, and precipitation for the period from preindustrial to present day. The cloud droplet effective radius, which is related to the first aerosol indirect effect, decreases almost all over the globe. Large changes from 2 to 5 μm are simulated in East and South Asia, Europe, and North America due to their industrial and domestic activities, and in central and southern Africa and South America due to the expansion of anthropogenic biomass burning. The global mean change in the cloud droplet effective radius is calculated to be –1.1 μm. An increase in cloud water results in an increase in precipitation over a large part of the globe, and vice versa. Especially, there is a strong constraint in tropical regions though the cloud microphysical treatment is not included in the convective scheme of the model. It is indicated that an increase (decrease) in liquid water in the southern (northern) tropics is due to convergence (divergence) of water vapor at the lower troposphere (Takemura et al., 2005). Therefore it is suggested that changes in liquid water and precipitation in the tropics is a feedback effect due to tropospheric cooling by the aerosol direct and first indirect effects. The difference in the horizontal water vapor flux is much smaller at the middle troposphere than the lower troposphere (not shown). It is also suggested that a change in water vapor around cloud base is essential for this feedback mechanism. In East and Southeast Asia and the Atlantic in the tropics and mid-latitude of the Northern Hemisphere, on the other hand, an increase in the cloud water and a decrease in precipitation simultaneously occur. This suggests that the aerosol second indirect effect on the cloud microphysical process is stronger than the effect of the dynamic hydrological cycle in the regions where a large amount of anthropogenic aerosols and cloud water exist.

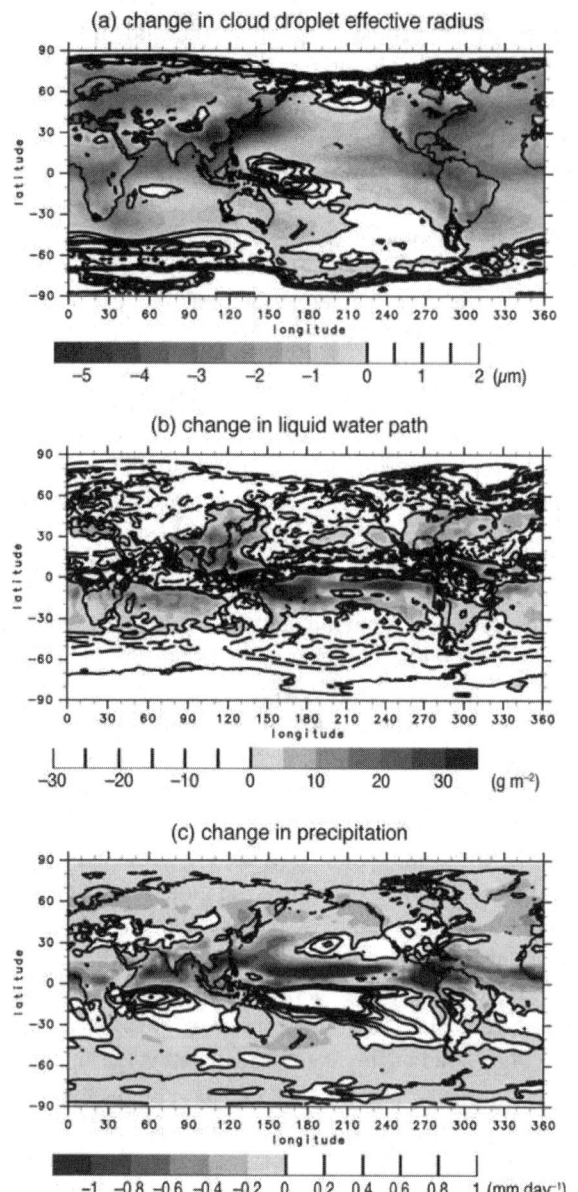

Fig. 4. Change in the simulated (a) cloud droplet effective radius at the cloud top above the temperature of 273K, (b) liquid water path, and (c) precipitation from preindustrial to present day.

4. CONCLUSIONS

A global aerosol transport-radiation model, SPRINTARS, coupled to a general circulation model simulated climate response to the direct and indirect effects of anthropogenic aerosols. The diagnosing scheme based on the Köhler equation calculated the cloud droplet effective radius, cloud water, and precipitation. This study indicated that the global mean direct and indirect radiative forcing are −0.1 and −0.9 Wm^{-2}, respectively. The cloud droplet effective radius reduces about 10% on the global mean because of the first indirect effect, while changes in the liquid water path and precipitation mainly depend on the change in the hydrological cycle due to cooling by the aerosol direct and first indirect effects.

We have to continue modeling studies on the aerosol indirect effect in order to reduce the uncertainty of estimating the climate change. One factor to consider is the role of aerosol particles as ice nuclei. The other factor is to simulate the continuative growth process both of aerosol particles and cloud droplets with the bin model. Their vertical profiles obtained from active sensors can also make models for the assessment of the aerosol indirect effect more reliable.

5. ACKNOWLEDGMENTS

We thank the contributors to the development of the CCSR/NIES/FRCGC GCM and the AERONET. This study is partly supported by the APEX project of the Core Research for Evolutional Science and Technology of the Japan Science and Technology Agency. The simulation in this study was done on NEC SX-6 of the NIES.

6. REFERENCES

Intergovernmental Panel on Climate Change (IPCC), 2001: *Climate Change 2001: The Scientific Basis*. J. T. Houghton et al., Eds., 896 pp., Cambridge Univ. Press, New York.

Lohmann, U., and G. Lesins, 2002: Stronger constraints on the anthropogenic indirect aerosol effect. *Science*, 298, 1012–1015.

Nakajima, T., M. Tsukamoto, Y. Tsushima, A. Numaguti, and T. Kimura, 2000: Modeling of the radiative process in an atmospheric general circulation model. *Appl. Opt.*, 39, 4869–4878.

Nakajima, T., A. Higurashi, K. Kawamoto, and J. E. Penner, 2001: A possible correlation between satellite-derived cloud and aerosol microphysical parameter. *Geophys. Res., Lett.*, 28, 1171–1174.

Takemura, T., H. Okamoto, Y. Maruyama, A. Numaguti, A. Higurashi, and T. Nakajima, 2000: Global three-dimensional simulation of aerosol optical thickness distribution of various origins. *J. Geophys. Res.*, 105, 17853–17873.

Takemura, T., T. Nakajima, O. Dubovik, B. N. Holben, and S. Kinne, 2002: Single scattering albedo and radiative forcing of various aerosol species with a global three-dimensional model. *J. Clim.*, 15, 333–352.

Takemura, T., T. Nakajima, A. Higurashi, S. Ohta, and N. Sugimoto, 2003: Aerosol distributions and radiative forcing over the Asian Pacific region simulated by Spectral Radiation-Transport Model for Aerosol Species (SPRINTARS). *J. Geophys. Res.*, 108(D23), 8659, doi:10.1029/2002JD003210.

Takemura, T., T. Nozawa, S. Emori, T. Y. Nakajima, and T. Nakajima, 2005: Simulation of climate response to aerosol direct and indirect effects with aerosol transport-radiation model. *J. Geophys. Res.*, 110, D02202, doi:10.1029/2004JD005029.

SIGNALS OF CLIMATE VARIATIONS IN THE WCRP/GEWEX SRB DATASETS AND THEIR CONNECTIONS WITH OTHER CLIMATE INDICES

Taiping Zhang[1], Paul W. Stackhouse, Jr.[2], Shashi K. Gupta[1], Steve J. Cox[1], and Colleen Mikovitz[1]

[1]AS & M, Inc., One Enterprise Parkway, Suite 300, Hampton, VA 23666-5845
[2]NASA Langley Research Center, 21 Langley Boulevard, Hampton, VA 23681-2199

ABSTRACT

The datasets, Release 2, from the surface radiation budget (SRB) project of the Global Energy and Water Cycle Experiment (GEWEX) under the World Climate Research Program (WCRP), along with those of the Global Precipitation Climatology Project (GPCP) under WCRP for the same period, are the subjects of this study. Here we employ the empirical orthogonal functions (EOFs) to study the spatial and temporal variations of the global radiation fields, and then we use canonical correlation analysis (CCA) in the EOF space to quantitatively study the connections between the different SRB variables and other climate indices. The results show that the WCRP/GEWEX SRB data and the GPCP data exhibit distinct signals that are closely correlated with the El Nino/Southern Oscillation (ENSO) and North Atlantic Oscillation (NAO).

1. INTRODUCTION

Longwave and shortwave radiation processes play critical roles in the energetics and dynamics of the global climate system, and long-term observations and records of such processes are indispensable in studying and gaining insight into the climatology on global scale. Designed for this purpose, the SRB project of GEWEX under WCRP has up till now produced in its new version, Release 2, datasets that cover the period from July 1983 to October 1995.

These datasets, along with those of the GPCP under WCRP for the same period, are the subjects of this study. As an integral part of the global climate system, these statistical fields of radiation and precipitation are presumed to have also recorded footprints of climatic variations on various regional scales and global scale as well as patterns on different parts of the world that are telepathically connected. Analysis by means of EOFs has been proved to be efficient to identify such variations and patterns. It is found that the rhythms of the ENSO (e.g., Torrence and Webster, 2004) and the North Atlantic Oscillation (e.g., Hurrell *et al.*, 2003) both make strong presence in these datasets.

But how is one statistical field correlated with another? What can we do with the correlations of different statistical fields? The answers lie in CCA.

We will first describe the datasets we have, and then briefly the mathematical methods we use in this study. Finally, we show selected results in figures and describe them with conclusions.

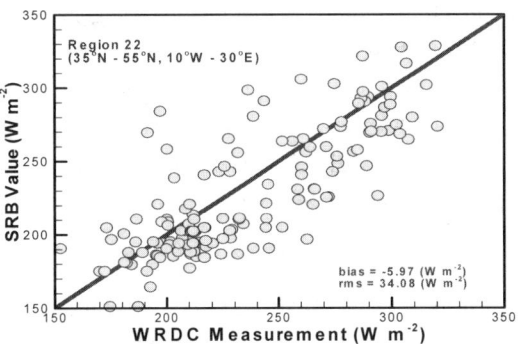

Fig. 1. SRB *vs.* WRDC Data: Monthly means of all-sky surface downward shortwave flux for July, 1987 in Region 22 (35°N – 55°N, 10°W – 30°E). Note that July is usually the month that has the largest RMS. In other words, RMSs for other months of the year are generally smaller.

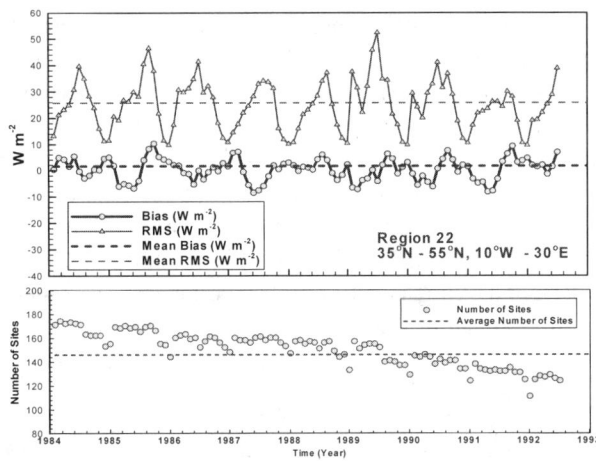

Fig. 2. Bias and root-mean-square (RMS) between the SRB and WRDC data sets over Region 22 (35°N – 55°N, 10°W – 30°E) from January, 1984 to December, 1992 and number of sites. Note that July is usually the month that has the largest RMS.

2. THE DATA

The SRB variables include, among others, all-sky surface shortwave/longwave downward/upward fluxes, cloud fraction, and cloud optical depth over 44,016 roughly equal-area grids which are 1°x1° at the Equator (Gupta et al., 1999; Stackhouse et al., 2002). Comparison of the satellite-derived data and their ground-observed counterparts, as presented in Figs. 1 and 2, shows generally very good agreement. Please note that the selected Region 22 (35°N – 55°N, 10°W – 30°E), which covers a significant part of Europe, is in the Northern Hemisphere, and July is around the warmest period of the year, and as a consequence, the RMS reaches about its possible maximum. Other months in the same region generally show smaller RMS values.

The GPCP datasets have a 2.5°x2.5° resolution. The time coverage thereof is much longer than that of the SRB datasets, but we will take only the time window that corresponds to the SRB time coverage.

3. THE MATHEMATICAL METHOD

The EOF has been used to identify spatial patterns that change simultaneously (Bess et al., 1992; Peng and Fyfe, 1996; Fraedrich et al., 1997), and CCA, combined with EOF, can be used to find coupled patterns which can further be used for prediction on different time scales (Barnett and Preisendorfer, 1987; Bretherton et al, 1992; Torrence and Webster, 2004).

The EOFs of a temporal-spatial dataset can be obtained by solving a positive definite matrix derived from the dataset. More specifically, let $\mathbf{X}$, a matrix whose elements are x_{ij}'s, represent such a dataset, where i, the spatial index, runs from 1 to p, and j, the temporal index, runs from 1 to n. Then, $\mathbf{X}$ can be decomposed as $\mathbf{X}=\mathbf{EC}$ where $\mathbf{E}$ is a $p\mathrm{x}p$ matrix whose columns, which are orthonormal to each other, are often called EOFs; $\mathbf{C}$ is a $p\mathrm{x}n$ matrix whose rows are often called EOF coefficients. The multiplication of an EOF and its corresponding EOF coefficient represents a certain amount of variance of $\mathbf{X}$.

This is an eigensystem problem, because $\mathbf{S}=\mathbf{XX}^T$ is a positive definite matrix and therefore can be written as $\mathbf{S}=\mathbf{ECC}^T\mathbf{E}^T=\mathbf{E}\Lambda\mathbf{E}^T$, where Λ is a diagonal matrix whose elements are eigenvalues of $\mathbf{S}$ and, $\mathbf{E}$ is a matrix consisted of the corresponding eigenvectors. $\mathbf{E}$ is normalized, i.e., $\mathbf{EE}^T=\mathbf{E}^T\mathbf{E}=\mathbf{I}$, where $\mathbf{I}$ is the identity matrix. So $\mathbf{C}=\mathbf{E}^T\mathbf{X}$.

The CCA, on the other hand, concerns the correlation of two spatial–temporal fields. Let $\mathbf{X}$ be a $p\mathrm{x}n$ matrix as before, and $\mathbf{Y}$ a $q\mathrm{x}n$ matrix where q is the spatial dimension. Now the goal is to find a pair of vectors, $\mathbf{c}$ and $\mathbf{d}$, such that the correlation coefficient mathematically defined as $\rho = \mathrm{Cov}(\mathbf{u},\mathbf{v})/\sqrt{\mathrm{Var}(\mathbf{u})\cdot\mathrm{Var}(\mathbf{v})}$ is maximized, where $\mathbf{u}=\mathbf{c}^T\mathbf{X}$ and $\mathbf{v}=\mathbf{d}^T\mathbf{X}$.

The CCA, which involves solving a complex eigensystem, can possibly be performed directly on the data in its original time and space, but it is much more straightforward to do it in EOF space, especially when we already have the EOFs, and we chose the latter in this study. The mathematical details, which are not given here due to limited space and for the sake of brevity, can be found in such as von Storch and Zwiers (1999).

The concept of EOF is useful under many conditions. However, there are situations where EOFs are not the optimum choice to depict temporal-spatial patterns (Gamage and Blumen, 1993).

4. RESULTS AND CONCLUSIONS

In order to evaluate the quality of the SRB data, the shortwave downward flux from the set was compared with ground-based observations conducted at hundreds of sites across the world. The world was first divided into 22 regions, and the comparison was made on a region-by-region and month-by-month basis, and bias and RMS were computed for each region each month. The comparisons show generally very good agreement. Shown in Figs. 1 and 2 are the scatter plot for Region 22 (35°N – 55°N, 10°W – 30°E), July, 1987 and bias and RMS for all months in the region.

The extension of the data sets in both time (148 months) and space (global) enables the data to contain appreciable signals of regional as well as global climatic variabilities. The current study shows that all the SRB variables show strong signals of ENSO over the western Pacific, and the signals are identifiable on a global scale by their close correlation with the ENSO Index. Although the energy balance of the Earth's surface depends on other forms of energy fluxes in addition to the longwave and shortwave radiation, the CCA of the net longwave (LWNT) and net shortwave (SWNT) show considerable correlations, and their canonical variates based on the deseasonalized data are strongly associated with the ENSO Index. As shown in Figs. 3 and 4, their correlation coefficients are, respectively, as high as 0.5440 and 0.5725, implicating considerable climatological significance. The first pair of canonical vectors (not shown here), which explain, respectively, 14.25% of the total variance of LWNT and

17.96% of that of SWNT display a correlation coefficient 0.9562. CCA pairs of other SRB variables show similar correlations.

The NAO Index is often presented for the mean winter, i.e., December-March average, the SRB data over the Atlantic sector for EOF analysis are also similarly rendered. The first EOF coefficients of shortwave downward flux (SWDW), longwave downward flux (LWDW), cloud fraction, precipitation as well as the sea level pressure (Source: http://dss.ucar.edu/datasets/) show rhythms synchronous with the NAO Index. The variances explained vary but all are climatologically significant. For instance, the first EOF of SWDW explains 24.11% of its total variance, and the first EOF of precipitation explains 30.74% of its total variance. And Figs. 5 and 6 show their relations with the NAO Index with correlation coefficient, respectively, 0.7227 and 0.6173. Also shown in these plots is the first EOF of sea-level pressure, a significant indicator of NAO. Due to these interdependent relations, the advances in prediction of NAO will also contribute to the prediction of surface solar insolation and precipitation for renewable energy and agricultural purposes in the eastern part of North America and throughout Europe.

The wide spatial and temporal coverage of the SRB data provides possibilities for seasonal predictions by means of statistics. The current study attempted to do so by using intra- as well as inter-variable correlations between one season and another. In the case of intra-variable prediction, the predictor and predictand are the same variable, except that the predictor is from the season prior to the season to be predicted. The predictor in intra-variable prediction is called persistence (Barnett and Preisendorfer, 1987). The inter-variable prediction, on the other hand, refers to the case when predictor and predictand are different variables. CCA is apparently an efficient tool for such statistical predictions. Fig. 7 shows a sample result from this study. In this case, we attempt to establish the correlation between the December-January-February (DJF) SWDW flux and the same variable in the next season, i.e., March–April–May.

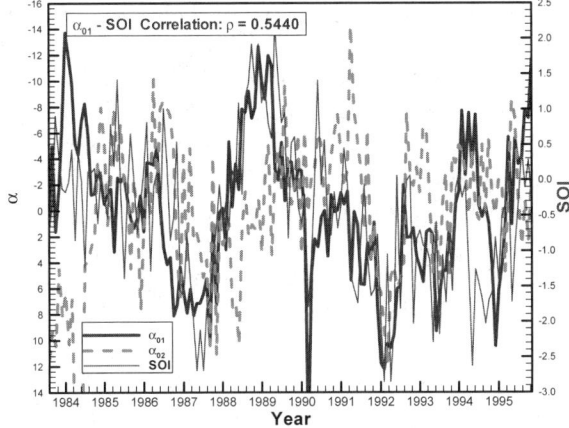

Fig. 3. CCA of monthly means of deseasonalized LWNT and SWNT: Canonical variates of LWNT from the first two pairs. (Data Coverage: July 1983 – October 1995).

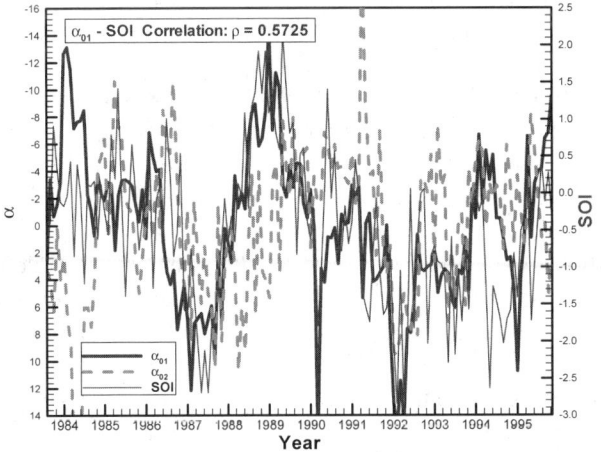

Fig. 4. CCA of monthly means of deseasonalized LWNT and SWNT: Canonical variates of SWNT from the first two pairs. (Data Coverage: July 1983 – October 1995).

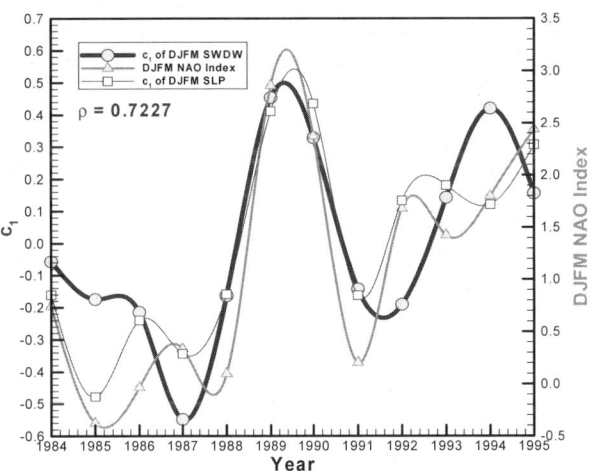

Fig. 5. The 1st EOF coefficient of the deseasonalized DJFM SWDW flux over the Atlantic sector (20°N – 70°N, 90°W – 40°E) in comparison with the DJFM NAO Index as well as the 1st EOF coefficient with the sea-level pressure.

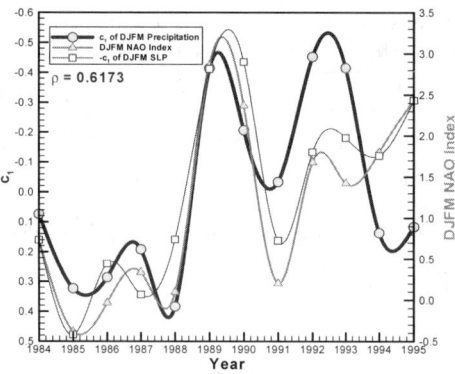

Fig. 6. The 1st EOF coefficient of the deseasonalized DJFM precipitation over the Atlantic sector (20°N – 70°N, 90°W – 40°E) in comparison with the DJFM NAO Index as well as the 1st EOF coefficient of the sea-level pressure.

Shown in Fig. 7 are the first three EOF coefficients derived from the original MAM SWDW record (the *thick* curves) in comparison with the "predicted" corresponding ones (the *thin* curves) which are computed using the regression relation based on the CCA. The good agreement implies possibly close correlation between DJF SWDW and MAM SWDW. We also carried out CCA for other pairs of variables, such as DJF precipitation and MAM precipitation, DJF SWDW and MAM precipitation, etc.

The quality of such statistical prediction depends on how well the time series of the two seasons involved correlate, and if the time coverage of the datasets is long enough to establish the proper correlation. For practical applications, further work is required to carefully evaluate the statistical significance and forecast skill scores of the scheme with data over longer period of time.

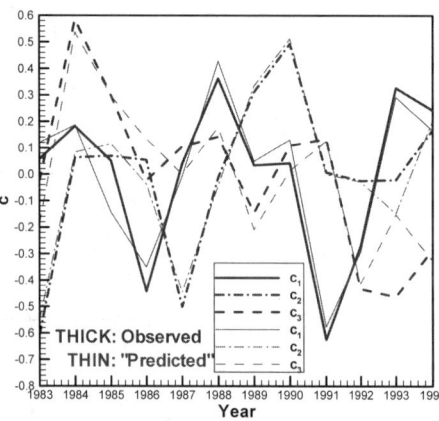

Fig. 7. Global March-April-May (MAM) Shortwave Downward Flux in EOF Space as "Predicted" Using Persistence from the Three Previous Months (DJF). (Data Coverage: July 1983 – October 1995).

REFERENCES

Barnett, T. P. and R. Preisendorfer, 1987. Origins and Levels of Monthly and Seasonal Forecast Skill for United States Surface Air Temperatures Determined by Canonical Correlation Analysis. *Monthly Weather Review*, 115, 1825-1850.

Bess, T. D., et al., 1992. Annual and Interannual Variations of Earth-Emitted Radiation Based on a 1-Year Data Set. *Journal of Geophysical Research*, 97(D12), 12,825-12,835.

Bretherton, C. S., C. Smith, and J. M. Wallace, 1992. An Intercomparison of Methods for Finding Coupled Patterns in Climate Data. *Journal of Climate*, 5, 541-560.

Fraedrich, K., J. L. McBride, W. M. Frank, and R. Wang, 1997. Extended EOF Analysis of Tropical Disturbances: TOGA COARE. *Journal of the Atmospheric Sciences*, 54, 2363-2372.

Gamage, N., and W. Blumen, 1993. Comparative Analysis of Low-Level Cold Fronts: Wavelet, Fourier, and Empirical Orthogonal Function Decompositions. *Monthly Weather Review*, 121, 2867-2878.

Gupta, S. K., et al., 1999. A Climatology of Surface Radiation Budget Derived from Satellite Data. *Journal of Climate*, 12, 2691-2710.

Hurrell, J. W., Y. Kushnir, G. Otterson, and M. Visbeck, 2003. An Overview of the North Atlantic Oscillation. In: The North Atlantic Oscillation: Climatic Significance and Environmental Impact. *Geophysical Monograph*, 134. American Geophysical Union.

Peng, S. and J. Fyfe, 1996. The Coupled Patterns between Sea Level Pressure and Sea Surface Temperature in the Midlatitude North Atlantic. *Journal of Climate*, 9, 1824-1839.

Stackhouse Jr., P. W., et al., 2002. New Results from the NASA/GEWEX Surface Radiation Budget Project: Evaluating El Nino Effects at Different Scales. *The 11th Conference on Atmospheric Radiation*, June 3–7, 2002, Ogden, Utah.

Torrence, C. and P. J. Webster, 1998. The annual cycle of persistence in the El Nino/Southern Oscillation. Quarterly *Journal of the Royal Meteorological Society*, 124, 1985-2004.

von Storch, H., and F. W. Zwiers, 1999. *Statistical Analysis in Climate Research*. Cambridge University Press.

AUTHOR INDEX
SUBJECT INDEX

AUTHOR INDEX

-A-

Aaltonen, V., 347
Abakumova, G. M., 375, 439
Abe, O., 367
Aben, I., 27
Acarreta, J. R., 215
Ackerman, A. S., 407
Ackerman, T. P., 343
Adachi, A., 235
Alados-Arboledas, L., 75
Alexandrov, M., 111, 307
Allan, R. P., 451
Andreae, M. O., 123
Aniya, M., 367
Anwender, D., 71
Aoki, K., 311
Aoki, T., 95, 167, 255, 367
Arao, K., 315
Asano, J., 331
Asano, S., 59, 195, 383

-B-

Badosa, J., 75
Barnard, J. C., 343
Bates, J. J., 455
Batmunkh, T., 319
Bauer, P., 287
Bianchini, G., 191, 303
Bonnel, B., 55
Borde, R., 231
Bottino, M. J., 435
Boudak, V. P., 47
Brazile, J., 151
Brogniez, C., 359

-C-

Cabot, T., 359
Cachorro, V., 75
Cahalan, R. F., 259
Cairns, B., 111, 307
Calbá, J., 75
Campmany, E., 75
Capderou, M., 251
Carissimo, A., 247
Carli, B., 191
Carlson, B., 111, 307

Castagnoli, F., 191
Ceballos, J. C., 63, 435
Chaboureau, J. P., 115
Chan, P. K., 239
Chang, F.-L., 155, 295
Charlock, T. P., 291, 391, 407
Cheremisin, A. A., 131
Chiacchio, M., 419
Chin, M., 461
Chiou, L., 187
Cho, H. K., 223
Choi, G. H., 223
Choi, S. H., 223
Choi, W., 207
Chou, M. D., 239
Chou, S. H., 239
Chu, D. A., 159
Chubarova, N. Y., 439
Coakley, J. A., 403, 407
Collins, W. D., 391
Coronel, G., 435
Corrêa, M. P., 63, 67, 435
Cortesi, U., 191
Cox, S. J., 419, 473
Cribb, M. C., 295
Cui, Y., 243
Cuomo, V., 247, 303

-D-

Dahlback, A., 323
Dangel, S., 371
de Cabo, X., 75
de La Casinière, A., 359
de la Morena, B., 75
Deguchi, E., 275
Desbois, M., 251
Diaz, A. M., 75
Diederich, M., 283
Di Michele, S., 287
Dubovik, O., 127
Dubuisson, P., 55, 115, 127
Duforet, L., 55
Dulam, J., 319

-E-

Edwards, D., 223
Eide, H., 167, 171, 175, 255, 323

Eisinger, M., 183
Ellingson, R. G., 35, 43, 387
Esposito, F., 303

-F-

Fetzer, E., 211
Fillmore, D. W., 391
Fischer, H., 263
Fomin, B. A., 63, 67
Fournier, N., 183
Fujieda, T., 95
Fukabori, M., 95
Futyan, J. M., 411

-G-

Gang, Z., 319
García, S. G., 51
Giraud, V., 115
Glatthor, N., 263
González, J., 75
Gorbarenko, E. V., 375, 439
Govaerts, Y., 231
Govaerts, Y. M., 147
Granger, S., 211
Gratzki, A., 423
Grieco, G., 247
Grossmann, K., 163
Guirado, F., 359
Gupta, S. K., 419, 473
Gusev, O., 163

-H-

Hachikubo, A., 255, 367
Hamada, S., 319
Hamann, U., 83
Harries, J. E., 411
Hatakka, J., 347
Hayasaka, T., 395, 443
Hearty, T., 211
Helas, G., 123
Herrmann, H., 123
Hollmann, R., 423
Holmlund, K., 231
Höpfner, M., 9, 263
Hori, M., 167, 255, 367
Horvath, H., 75

AUTHOR INDEX

Houët, M., 359
Høyskar, B.-A., 323
Huang, H. L., 187
Hungershöfer, K., 123

-I-

Ignatijeva, E. I., 79
Iino, N., 319
Iinuma, Y., 123
Iizuka, Y., 255
Im, E., 279
Imasu, R., 203
Iokibe, K., 315
Ionov, D. V., 179
Ishida, H., 59
Itten, K. I., 151, 371
Iwabuchi, H., 383

-J-

Jäkel, E., 51, 355
Jeong, M. J., 159
Jin, Z., 291
Jobard, I., 251

-K-

Kaiser, J. W., 151, 371
Kassianov, E. I., 31, 343
Katagiri, S., 267
Kaufman, Y. J., 407
Kaufmann, M., 163
Kawamoto, K., 395, 443
Kikukawa, H., 319
Kim, B. G., 465
Kim, D. H., 399
Kim, J., 223
Kim, J. E., 335
Kim, K. M., 461
Kim, M. K., 461
Kim, Y., 207
Kim, Y. J., 335
Kim, Y. M., 227
Kim, Y.-S., 199
Kinoshita, K., 319
Knap, W. H., 215
Kneubühler, M., 371
Kniffka, A., 51, 355
Knuteson, R., 143
Kobayashi, H., 315
Kobayashi, S., 279

Kobayashi, T., 235
Koepke, P., 23, 71, 431
Koga, R., 315
Kojima, M., 383
König, M., 231
Kostsov, V. S., 163
Kozoderov, V. V., 427
Krol, M. C., 13
Kulikov, A. K., 79, 427
Kumagai, H., 447

-L-

Labajo, A., 75
Lacis, A., 111, 307
Lambert, J. C., 179
Lambrigtsen, B., 211
Landgraf, J., 27
Larar, A. M., 19, 143, 187
Laszlo, I., 219, 415
Lattanzio, A., 147
Lau, K. M., 461
Lee, H. C., 223
Lee, K., 199
Lee, K.-M., 207
Lee, S. H., 223
Lee, S.-Y., 211
Lee, W. S., 461
Legrand, M., 103, 359
Lenoble, J., 359
Lenoble, J., 359
Li, J., 187
Li, W., 167, 171, 175, 255
Li, Z., 1, 155, 159, 295
Liang, H., 379
Lihavainen, H., 347
Lim, H. S., 223
Lin, P. H., 239
Liu, J., 379
Liu, X., 19, 143, 187
Loeb, N. G., 407
Lohmann, U., 407
Lorente, J., 75
Los, A., 363
Luo, Y., 295
Lutz, H. J., 231

-M-

Ma, Q., 87, 91
Ma, Y., 35

Maestri, T., 303
Maksakova, S. V., 79, 427
Mallet, M., 127
Mango, S. A., 143, 187
Manning, E. M., 211
Marshak, A., 111, 307
Martinez-Lozano, A., 75
Marzano, F. S., 287
Masiello, G., 247, 303
Massie, S. T., 207
Matheson, M. A., 403
Matrosov, S. Y., 135
Matsui, I., 99, 139
Matsukawa, K., 275
Mayer, B., 5, 83, 299
Mech, M., 71
Melnikova, I. N., 107
Mikovitz, C., 473
Mikovitz, J. C., 419
Motoyoshi, H., 255
Mugnai, A., 287
Mukai, S., 339
Müller, R. W., 423
Murayama, T., 315
Musat, I. C., 43

-N-

Nakajima, H., 207
Nakajima, T., 17, 267, 311, 399, 447, 469
Nakajima, Y., 255
Nezval, E. I., 375, 439
Nieke, J., 151, 255, 371
Ning, W., 319
Nishizawa, T., 99, 447
Nozawa, T., 469

-O-

Ogawa, K., 271
Ohmura, A., 327
Okamoto, H., 99, 119, 447
Okamoto, K., 275
Olsen, E., 211
Olsen, R. O., 323
Oppenrieder, A., 71
Ota, Y., 203
Otake, S., 383

-P-

Pagano, T., 211
Palchetti, L., 191, 303
Pancrati, O., 103
Park, J. H., 207
Parmar, R. S., 123
Parol, F., 55
Pavese, G., 75, 303
Pellegrini, M., 191, 303
Pinker, R. T., 415
Pinty, J. P., 115
Pujadas, M., 75

-R-

Raasch, S., 83
Rakitin, A. V., 163
Restieri, R., 303
Reuder, J., 71, 431
Revercomb, H., 143
Rizzi, R., 303
Roberti, L., 287
Roger, J. C., 127
Rokugawa, S., 271
Rose, F. G., 391
Rublev, A. N., 439
Ruggaber, A., 71
Russell, J. E., 411
Rutan, D. A., 391
Rutledge, K., 291

-S-

Sano, I., 339
Sasaki, H., 275
Sasano, Y., 207
Sato, K., 119
Schaepman, M. E., 371
Schaepman-Strub, G., 371
Scheirer, R., 51, 299
Schläpfer, D., 151
Schmetz, J., 231
Schmid, O., 123
Schmidlin, F. J., 323
Schmidt, S., 51, 299, 355
Schmugge, T., 271
Schopfer, J. T., 371
Schreier, M., 71
Schulz, J., 283
Schwander, H., 71
Schween, J., 71, 431
Schween, J. H., 23
Seckmeyer, G., 83
Sekiguchi, M., 17
Serio, C., 247, 303
Shalamiansky, A. M., 179
Shi, G. Y., 395
Shilovtseva, O. A., 375, 439
Shimizu, A., 99, 139
Shiobara, M., 315
Silva, A. M., 75
Simmer, C., 283
Smith, W. L., 19, 143, 187, 247
Sohn, B. J., 399
Sola, Y., 75
Spada, F., 13
Spurr, R., 167, 171, 175
Stackhouse, P. W., 419, 473
Stammes, P., 183, 215
Stamnes, J., 171, 175, 323
Stamnes, K., 167, 171, 175, 255, 323
Stiller, G. P., 263
Storvold, R., 255
Strelkov, S. A., 79, 427
Sud, Y. C., 461
Sugimoto, N., 99, 139, 447
Sun, Z., 379
Sushkevich, T. A., 79, 427

-T-

Tahnk, W. R., 403
Takahashi, F., 255
Takamura, T., 243, 311, 399
Takara, E. E., 387
Takemura, T., 99, 447, 469
Tanelli, S., 279
Tanikawa, T., 255, 367
Tassa, A., 287
Thurai, M., 275
Tilstra, L. G., 215
Timofeyev, Yu. M., 163, 179
Tipping, R. H., 87, 91
Togawa, H., 331
Town, M. S., 351
Trambusti, M., 191
Trautmann, T., 51, 123, 355
Trentmann, J., 123
Trishchenko, A., 295
Tsay, S.-C., 159, 323
Tupper, A., 319
Turk, F. J., 287

-U-

Uchiyama, A., 331
Uno, I., 443

-V-

van Deelen, R., 27
van Gent, J. I., 39
Vassilyev, Y. V., 131
Vesperini, M., 55
Viisanen, Y., 347
Vladimirova, E. V., 79, 427
Volkovich, A. N., 427
von Clarmann, T., 263

-W-

Walden, V. P., 351
Walker, G. K., 461
Wang, K. H., 239
Watanabe, T., 95
Wen, G., 259
Wendisch, M., 51, 355
Wiedensohler, A., 123

-X-

Xu, J. Q., 395

-Y-

Yabuki, M., 315
Yamazaki, A., 331
Yang, T. F., 251
Yasumoto, M., 339
Yokota, T., 207
Yoo, J. M., 227
Yoshida, Y., 195
Yuan, T., 1

-Z-

Zeromskiene, K., 123
Zhang, T., 419, 473
Zhao, T.X.-P., 219
Zhou, D. K., 19, 143, 187
Zinner, T., 5

SUBJECT INDEX

This index was derived from Key Words provided by individual authors. While an effort has been made to avoid duplication, cross referencing and elimination of alternative terminology have not been attempted.
*Page number refers to first page of a paper if specific page numbers within the paper were not given.

0°C isotherm height, 275
2 stream, 391
2/4 stream, 391
3D
    geometrical structure of the cloud fields, 35
    radiative transfer calculation, 83
    radiative transfer, 5, 6, 51
95 GHz, 119, 120, 121, 122, 447, 448, 449, 450
σ-IASI, 247, 248, 250

A

Absorption
    atmosphere, 417
    coefficient, 347, 348
    in the atmosphere due to different factors: aerosol, NO_2, 442
    surface, 417
Actinic flux density, 51, 52, 53, 54
ADEOS-II, 255
Advanced Very High Resolution Radiometer (AVHRR), 403, 409
Aeolian Dust Experiment on Climate impact (ADEC), 331, 332, 333, 334
AERONET, 221, 222, 308, 309, 339, 340, 341, 342
Aerosol, 99, 100, 101, 102, 131, 139, 140, 141, 223, 224, 225, 226, 231, 232, 291, 399, 400, 401, 402, 432, 465, 466, 467, 468
    amount, 443, 444, 445
    anthropogenic, 127, 129, 130
    -cloud interaction, 407, 408
    direct effect, 469, 470
    direct radiative forcing, 407, 408, 409
    indirect effect (AIE), 1, 2, 469, 470, 471, 472
    indirect forcing (AIF), 1, 3
    indirect radiative forcing, 403, 405, 406, 407, 409, 410
    layer, 131
    loading, 461, 462, 463
    microphysical properties, 343*
    optical depth (AOD), 2, 3, 43, 46, 335, 336, 337, 403, 404, 405, 406, 408
    optical properties, 123, 129, 347
    optical thickness (AOT), 111, 112, 113, 114, 159, 160, 161, 259, 260, 261, 307, 312, 313, 314, 340, 342, 375, 376, 378
    particle swelling, 405
    radiative forcing, 400, 461, 462, 463
    refractive index, 343, 345
    retrieval, 219, 220, 221, 222
    semi-direct effect, 405
    single-scattering albedo, 343, 345
    size distribution, 307
    stratification, 134
    type(s), 159, 160
Aerosols, 435, 436, 437, 438
    and clouds, 403*
    nitrate, 309, 310
    sulfate, 309, 310
    volcanic, 309
AFDM, 355, 356, 358
Air pollution, 152
Airborne
    imaging spectrometer, 151
    measurements, 51
Albedo, 243
    near infrared (NIR), 416
    shortwave (SW), 416
    surface, 147, 148, 167, 168, 170, 416
    top of atmosphere (TOA), 417
    visible (VIS), 416
Algorithm, 99, 100, 102
Altitude effect, 75
Ångström
    exponent (AE), 159, 160, 161, 312, 332, 333, 334, 335, 336, 337, 338, 340, 347, 348, 349
    parameter, 312, 313
Anisotropic sea surface, 55, 57
Anomalous absorption, 108, 109
Anthropogenic, 391
APEX, 151
Approximated equations for the mean radiance, 32, 33, 34
ARM observation, 35
Asian dust, 315
ASTER, 259, 260, 261, 271, 272, 273
Asymmetry factor (g), 392
Atmosphere-smoke system, 79
Atmospheric
    composition, 191
    correction, 167, 168, 171, 172, 175, 176, 177, 178, 199
    instability, 231, 232, 233
    radiance, 191
    Radiation Measurement (ARM), 295, 387, 390
    radiation modeling, 419
    state, 187
    turbidity, 75
AVHRR, 219, 220, 221, 222, 267, 403, 409

B

Backscattering, 119
    enhancement, 279, 280
Barley, 23
Baseline Surface Radiation Network (BSRN), 409
Bidirectional reflectance distribution function, 5
Biologically effective irradiance, 75
Biomass burning aerosol, 123
BOMEM Fourier Transform spectrometer, 303, 305
BRDF, 243
bright-band, 275
Broadband emissivity, 271, 272, 273
Broad-band radiation scheme, 17
BSRN, 328, 409
Bulk density, 120

C

Cabannes
    line, 28
    scattering, 28
CALIPSO, 406, 409
Canonical correlation analysis, 473
CCA, 473, 474, 475, 476
CERES, 411
    SSF, 408
China, 395, 397, 398, 443, 444, 445, 446
CIMEL sun photometer, 335, 336
Cirrus, 155, 156, 157, 158, 267, 268, 269, 270
Clay, 105
Climate
    data set, 426
    forcing, 403, 406
    impact, 461, 462
    model, 271, 273
    monitoring, 423
Climatology, 423
ClO/ClONO2 (Stratospheric chlorine species), 263, 264, 265, 266
Cloud, 139, 140, 141, 187, 235, 465, 466, 467, 468
    amount, 375, 376, 377, 378, 448
    contamination of aerosol retrievals, 405
    droplet number concentration, 405, 407, 409
    droplet radius, 403, 404, 405, 406, 409
    effective radius, 237
    fraction, 408, 410
    free line of sight, 24
    inhomogeneity, 59, 60, 61, 62
    liquid water amount, 407, 409, 410
    liquid water path, 404, 405, 406
    occurrence, 448
    optical depth, 403, 404, 405, 406, 410
    optical properties, 115, 117
    optical thickness, 24
    parameters, 135, 136, 137, 138
    properties, 155, 156, 195, 196, 197, 198
    radar, 119, 120, 121, 122, 447, 448, 449, 450
    radiation modeling, 287, 288, 289, 290
    radiative forcing, 411, 416, 417
    reflectivity, 407
    screening, 308
    top pressure, 183, 184, 185
    type, 412
    -aerosol interaction, 403
    -top altitude, 403
Clouds, 187, 432, 435, 436, 437, 438
CM-SAF, 423
CO, 223, 224, 225, 226
Collision
    -induced absorption, 91, 92
    -induced fundamental bands, 91
Controlled combustion experiments, 123
Correlated-k, 391
    distribution method, 67
COST model, 199
CRISTA-1, 163, 165
Cryosphere, 255

D

Depolarization, 99, 100, 450
Diffuse
    illumination, 371, 372, 373, 374
    influence, 372, 373, 374
Direct
    and diffuse irradiances, 343, 345
    forcing, 391
Discrete dipole approximation (DDA), 119, 120, 121, 122
DOE ARM Program, 307, 308, 309, 310
Doppler velocity, 121
DOS (Dark Object Subtraction), 199
Doubling-adding method, 27
Drizzle, 235
    suppression, 403, 405
Droplet effective radius (DER), 2, 3

E

Earth
    radiation, 427
    Radiation Budget (ERB), 191, 194, 419
East Asia, 399, 400, 401
ECHAM, 328
EFEU, 123
Effective radius, 465, 466, 467, 468
El Nino, 239, 240, 241, 242
El Nino/Southern Oscillation (ENSO), 473, 474

SUBJECT INDEX

Elastic scattering, 27
Empirical orthogonal function, 473
Energy balance, 271, 327
Envisat, 215, 263, 264, 265, 266
EOF, 473, 474, 475, 476
EOS Aqua, 211, 214
EUMETSAT, 149
Evaporation, 329
Evaporative cooling, 239, 240, 241, 242
Expectation value, 24
External mixture, 109
Extinction-to-backscattering ratio, 99

F

Fair weather cumulus, 259
Fall velocity, 450
Far infrared, 191
Far-infrared region, 91
Fast k-distributions model, 67
Filling-in, 27
Filter
    instruments, 323, 324, 325
    ozonometer, 179
Fog, 227, 228, 229
Forecast, 476
Forest
    and peatbog fires in Central Russia, 439
    fire, 223, 224, 225, 226
Forward model, 19, 20, 21, 22
Fourier Transform Spectrometer, 191, 247
Fractional cloud cover, 403, 405
Fraunhofer lines, 27
FTIR spectrometers, 9
FTS, 203, 204, 205
Future UV scenarios, 431, 433

G

Gas
    profile, 163, 164, 165, 166
    species concentrations, 441
GCM, 327
GCOS, 150
General Circulation Model, 17, 447
Geochemical nature, 103
GEOS4, 391
Geostationary
    Earth Radiation Budget (GERB), 411
    Operational Environmental Satellite (GOES), 295, 298
    satellite, 147
GEWEX, 473
GEWEX/SRB, 415

GLI, 255
Global
    aerosol products, 159, 160, 162
    climate change, 403*
    dimming, 329
    Positioning System (GPS), 379, 380, 381, 382
GMS-5, 147, 148
GOME, 27
    and TOMS instruments, 179, 180, 181, 182
Gravito-photophoretic, 131
Ground-based
    MFRSR measurements, 343, 345
    multi-sensor, 135, 136, 137, 138
    network, 179, 182
    UV measurements, 323

H

Heat balance, 327
Hydrological cycle, 391
High-resolution Fourier transform spectrometer, 95
HITRAN database, 95, 96, 98
Hot spot, 243
Hyper spectra, 19
Hyperspectral, 167, 168, 170, 171, 172, 173, 175

I

Ice
    cloud, 119
    clouds, 447
    water content (IWC), 122
IGBP, 243
ILAS-II, 207, 208, 209
Illuminance, 440
Imaginary part, 315
IMG spectrum, 204, 205, 206
Indirect effect, 403, 405, 406
Inelastic scattering, 27
Influence function, 79, 80, 82, 427, 428, 429, 430
Infrared
    emissivity, 271*
    radiance, 203, 204
    spectrometer, 211, 214
Inhomogeneous
    cloud, 391
    cloud cover, 53
INSPECTRO, 51, 52, 54, 299, 302, 355, 356, 358
Intensive Observation period (IOP), 331, 332, 333, 334
Intergovernmental Panel on Climate Change (IPCC), 407
Internal mixture, 109
International Satellite Cloud Climatology Project (ISCCP), 328, 415, 416, 418
Interval recording, 319, 320, 322

IR, 227, 228, 229
    spectral radiance, 143
ISCCP, 328

K

KOMPSAT (KOrea Multi-Purpose SATellite), 199
KOPRA, 9

L

La Nina, 239, 240, 241, 242
Lanczos algorithm calculations, 87
Land surface albedo, 392
Leaf angle distribution, 24
Lidar, 99, 100, 101, 102, 447, 448, 449, 450
Limb observation, 39, 40, 41
Line strengths and half-widths, 95
Linear depolarization ratio (LDR), 120
Line-by-line method, 63, 65
Liquid water path (LWP), 383, 384, 385, 386
Long-term, 395, 396, 397, 398
    UV data records, 326
Longwave, 63, 391
    effective cloud fraction, 387
    radiation, 419
Low cloud microphysics, 443
LW (longwave), 391

M

Marine strato cumulus, 232
Mark's boundary conditions, 50
Marshak's boundary conditions, 48
Matrix operator, 23
Mesoscale
    convective system (MCS), 251, 252, 253, 254
    model, 115, 116, 117
Mesosphere, 131
METEOSAT, 147, 148
    -5, 251, 252
    -8, 231, 232, 233
MFRSR, 111, 112, 113, 114, 307, 308, 309, 310, 335, 336, 337
Microwave-radiometer, 195, 197, 198
Mie theory, 367, 368
MIE-LIDAR, 139, 140, 141
Millimeter water vapor continuum, 87
Mineral
    composition, 105, 106
    dust, 103, 104, 105, 106
MIPAS, 263, 264, 265, 266
Mixed-phase cloud, 195, 196, 197, 198
MLT, 163

Model, 451, 452, 453, 454
    for Atmospheric Transport and Chemistry (MATCH), 391, 409
    inversion, 244
MODIS, 155, 156, 157, 158, 219, 220, 221, 222, 223, 225, 226, 227, 228, 271, 272, 273, 391
    MOD04, 408
Monte Carlo
    method, 87, 88
    radiative transfer model, 13, 367, 368, 370
    model, 441
MOPITT, 223, 224, 225, 226
MSC (Multi-Spectral Camera), 199
Multilayer broken clouds, 31
Multiple
    Raman scattering, 27
    scattering, 27, 279, 280

N

NAO, 473, 474
NAST-I, 143, 144, 145, 247, 248, 249, 250
Nesterov's fire classes, 440
Net
    flux, 416
    radiation, 375, 377
    surface heating, 239, 240, 241, 242
Nitrate aerosols, 309, 310
Nitric Acid Trihydrate (NAT), 207, 208, 209, 210
Non parasitic leaf spots, 23
Nonsphericity, 120
Non-stellar diffuse light, 43
North Atlantic Oscillation, 473

O

Ocean
    albedo, 72
    color, 171, 172, 173, 175, 176
    spectral albedo, 392
    surface albedo, 291, 292, 293, 294
OMI, 27
OPAC-GADS, 392
Optical
    and microphysical properties of aerosols, 129
    thickness, 107, 108, 109, 110, 332, 333, 334
    transfer operator, 79, 80, 82
Orientation, 120
Oxygen A-band, 183, 184
Ozone
    amounts, 326
    chemistry, 191
    column density, 364, 365

P

Pacific warm pool, 411
Particle
  movement, 131
  size, 449
Photolysis frequencies, 73
Photophoretic, 131
Photosynthetically active radiation, 375, 377, 378
Planetary radiation budget, 391
Plant recovery, 24
PM10, 339, 340, 342
PM2.5, 339, 340, 341, 342
Polar
  Atmospheric Emitted Radiance Interferometer (Polar AERI, or PAERI), 351, 352, 353, 354
  Stratospheric Cloud (PSC), 9, 207, 208, 209, 210
Polarization, 55, 56, 57, 58
POLDER, 107, 110, 243
PR, 275, 283, 284, 285, 286
Precipitable water (PW), 2, 3, 379, 380, 381, 382
Precipitation, 1, 3, 251, 252, 253, 254, 283, 284, 285, 286, 287, 288, 289, 290
Probability of clear line of sight (PCLOS), 35
Promt, 83, 84, 85, 86
PSC properties (Composition and Size Distribution), 207, 208, 209, 210
Pyranometer, 395, 396, 397, 398

Q

Quartz, 105, 106

R

Radar, 195, 196, 197, 198, 235, 283, 284
  reflectivity factor, 120
Radiance, 291, 292
  composite, 415
  visible (VIS), 415
Radiation, 327
  and optical properties of the atmosphere, 439
  budget, 379, 381, 382, 439
  Forcing and Radiation Budget, 441
  on tilted surfaces, 72
  transfer, 427, 428, 429, 430
  transfer equation (RTE), 47, 49, 50
  transfer model STAR, 71, 431
Radiative
  fluxes, 63
  forcing, 127
  transfer, 55, 65, 291, 294
  transfer code, 115, 117, 127, 128, 130
  Transfer Model, 19
Radiometer, 103, 235

Rain rate, 251, 252, 253, 254
Rayleigh scattering, 27, 83, 84, 85
Ray-tracing technique, 367, 368
Reanalyses, 451, 452
Redistribution of radiation budget, 442
REFIR, 303, 304, 305, 306
Reflectance, 215, 216, 227, 228, 229
Refractive index, 124, 208
Regional fire "cloud" model, 440
Remote
  measurement, 191, 194
  sensing, 19, 20, 103, 167, 171, 178, 403*
Repair mechanism, 24
Retrieval, 187
  of trace gas abundances, 9
Retrievals of inhomogeneous clouds, 5
Ring effect, 27
Rossby wave, 455, 456, 457, 458
Rotational Raman scattering, 27

S

Sahel, 106
Sampling uncertainties, 295, 296
Sand storm 1, 321
Satellite, 227, 228, 395, 396, 397, 398, 451, 452, 453, 454
  climatology, 426
  data validation, 179*
Scale invariance, 111, 112, 113
Scaling function, 59, 60, 61
Scattering coefficient, 347, 348, 349
SCIAMACHY, 27, 39, 183, 185, 215, 216
Ship tracks, 410
S-HIS, 143, 144, 145
Shortwave, 63, 391
  radiation, 419, 420
  radiation flux at the surface (SWFLUX), 383
Single scattering albedo (SSA), 107, 108, 109, 110, 312, 315, 347, 348, 392
Size distribution, 315
SKYNET, 311, 312, 313, 314
Small angle approximation (SAA), 47, 48, 49, 50
Snow
  and ice, 167, 168
  grain size, 255
  impurities, 255
Snowpack, 367, 368, 370
Solar
  heating, 239, 240, 241, 242
  irradiance, 440
  radiation, 31, 32, 327, 399, 400, 401, 402, 435, 436, 437
  radiation at the surface, 423
  UV radiation, 359
  zenith angle (SZA), 392

South America, 435, 438
South Pole Atmospheric Radiation and Cloud Lidar Experiment (SPARCLE), 351, 352, 353, 354
Southern Great Plains (SGP), 295, 296, 297
Spectral
    actinic flux densities, 355, 357, 358
    albedo, 367, 368, 369, 370
    signature, 103, 105, 106
    solar UV irradiance, 361
Spectral Radiation Budget
    global, 415
    near infrared (NIR), 416
    shortwave (SW), 416
    surface, 415, 416
    top of atmosphere (TOA), 415, 416
    visible (VIS), 416
    zonal, 417
Spectro-directional field measurements, 371, 372, 374
Spherical harmonics (SH) method, 47, 48, 49, 50
SPM, 339, 340, 342
SPRINTARS, 469, 470, 471
SRB, 473
SSM/I, 251, 252, 253, 254, 283, 285, 286
STAR, 23, 431
Stochastic radiative transfer, 31
Stratosphere, 131
Sulfate aerosols, 309, 310
Sun flecks, 24
Sunshine duration, 376, 377
Surface albedo, 72, 147, 148, 167, 168, 170, 377, 432
    and Atmosphere Radiation Budget (SARB), 391
    downward LW flux (DLF), 393
    emissivity, 203, 204, 205
    insolation, 392
    Radiation Budget, 419, 423
    radiation measurement, 419
    Reflectance, 199
    shortwave radiation, 395
    solar net irradiance (SSNI), 295, 296, 297, 298
SW (shortwave), 391

T

Terra CERES CRS, 391
Theoretical retrieval precision, 152
Thermal infrared, 103, 104
Three-dimensional radiative transfer, 59
Tibetan Plateau, 379, 381, 382
TMI, 283, 284, 285, 286
Top-of-atmosphere (TOA), 391
Total ozone, 179, 180, 182, 431
Transmittance-Ratio Technique, 207, 208, 209
TRMM, 275, 283, 286
Tropical
    Africa, 411
    Convection, 411*
Tropics, 122, 447, 448, 449, 450
TSP, 339, 340, 342
Turbid water / coastal water, 174, 175, 176, 177, 178
Twomey effect, 403, 404, 405, 406, 407

U

Upper
    level wind divergence, 233
    troposphere/lower stratosphere (UTLS), 191, 194
User-friendly software, 63, 64, 65
UTH, 455, 456, 457, 458
UV
    actinic flux, 299
    doses, 323, 326
    filter radiometer, 363, 364, 365, 366
    index, 75
    irradiance, 440
    radiation, 363, 364, 365
    radiation 1, 378
    reconstruction, 431, 432
    solar radiation, 75
    spectral range, 355, 356
UV-A spectral range, 355
UV-B spectral range, 358
UV-impact, 24, 26
UV-vis retrievals, 13
UVB radiation, 435, 436, 437

V

VIS spectral range, 355, 356
Visible, 227
Volcanic aerosols, 309

W

Water
    -insoluble particle, 315
    vapor, 379, 381, 382
    vapour, 191, 193, 194
Weather radar, 279
Web-camera recording, 319
Westerly duct, 455, 456, 457, 458
Whole sky imager, 43
Windsat, 287, 288, 289, 290

Y

Yamase
    clouds, 383, 384, 385, 386
    events, 383, 386